Transfusion Medicine and Hemostasis

Transfusion Medicine and Hemostasis

Clinical and Laboratory Aspects

Third Edition

Edited by

Beth H. Shaz

Chief Medical and Scientific Officer and Executive Vice President, New York Blood Center, New York, NY, United States

Christopher D. Hillyer

President and Chief Executive Officer, New York Blood Center, New York, NY, United States
Professor, Department of Medicine, Weill Cornell Medical College, New York, NY, United States

Morayma Reyes Gil

Director, Hematology and Coagulation Laboratories, Montefiore Medical Center, Bronx, NY, United States
Associate Professor, Department of Pathology, Albert Einstein College of Medicine, New York, NY, United States

ELSEVIER

Elsevier
Radarweg 29, PO Box 211, 1000 AE Amsterdam, Netherlands
The Boulevard, Langford Lane, Kidlington, Oxford OX5 1GB, United Kingdom
50 Hampshire Street, 5th Floor, Cambridge, MA 02139, United States

Library of Congress Cataloging-in-Publication Data

A catalog record for this book is available from the Library of Congress

British Library Cataloguing-in-Publication Data

A catalogue record for this book is available from the British Library

ISBN: 978-0-12-813726-0

For information on all Elsevier publications visit our website at
https://www.elsevier.com/books-and-journals

Publisher: Stacy Masucci
Acquisition Editor: Tari K. Broderick
Editorial Project Manager: Tracy I. Tufaga
Production Project Manager: Stalin Viswanathan
Cover Designer: Matthew Limbert

Typeset by TNQ Technologies

Contents

Donor Testing

Transfusion Medicine

Testing

Antigens

Blood Products

List of Contributors

SUCHITRA S. ACHARYA, MD
Program Head, Bleeding Disorders and Thrombosis Program; Cohen Children's Medical Center of New York; Director, Long Island Jewish Hemophilia Treatment Center Northwell Health; *and* Professor of Pediatrics, Hofstra North Shore-Long Island Jewish School of Medicine, Hempstead, NY, United States

JUDITH AESCHLIMANN, MSc
New York Blood Center, New York, NY, United States

AHMAD AL-HUNITI, MD
University of Iowa Hospital and Clinics, Carver College of Medicine, Iowa City, IA, United States

ANA G. ANTUN, MD, MSc
Department of Hematology and Medical Oncology, Emory University School of Medicine, Atlanta, GA, United States

SUZANNE A. ARINSBURG, DO
Assistant Professor, Department of Pathology, Icahn School of Medicine at Mount Sinai, New York, NY, United States

SCOTT T. AVECILLA, MD, PhD
Medical & Scientific Director, Cell Therapy Laboratory Services, Department of Laboratory Medicine, Memorial Sloan Kettering Cancer Center, New York, NY, United States

BURAK BAHAR, MD
Department of Laboratory Medicine, Yale University School of Medicine, New Haven, CT, United States

DEBRA JO BAILEY, SBB
American Red Cross Biomedical Services Southeastern Michigan Region, Detroit, MI, United States

GLAIVY BATSULI, MD
Assistant Professor, Department of Pediatrics, Emory University, Division of Hematology/Oncology/BMT, Aflac Cancer and Blood Disorders Center, Children's Healthcare of Atlanta, Atlanta, GA, United States

HENNY H. BILLETT, MD
Division of Hematology, Department of Oncology, Montefiore Medical Center and the Albert Einstein College of Medicine, Bronx, NY, United States

EVAN M. BLOCH, MD, MS
Johns Hopkins University School of Medicine, Baltimore, MD, United States

MICHAEL A. BRIONES, DO
Aflac Cancer and Blood Disorders Center, Children's Healthcare of Atlanta/Emory University School of Medicine, Atlanta, GA, United States

JAMES B. BUSSEL, MD
Weill Cornell Medical College, New York, NY, United States

MARCUS A. CARDEN, MD
Clinical Fellow Hematology, Emory University School of Medicine, Atlanta, GA, United States

WAYNE L. CHANDLER, MD
Department of Laboratories, Seattle Children's Hospital, University of Washington, Seattle, WA, United States

DONG CHEN, MD, PhD
Division of Hematopathology, Mayo Clinic, Rochester, MN, United States

MARIE CSETE, MD, PhD
Chief Executive & Science Officer, Cell & Tissue Biology, Huntington Medical Research Institutes, Pasadena, CA, United States; Professor, Anesthesiology, USC Keck School of Medicine, Los Angeles, CA, United States; *and* Visiting Associate Medical Engineering, California Institute of Technology, Pasadena, CA, United States

ADAM CUKER, MD, MS
Department of Medicine and Department of Pathology & Laboratory Medicine, Perelman School of Medicine, University of Pennsylvania, Philadelphia, PA, United States

MELISSA M. CUSHING, MD
Weill Cornell Medicine and New York-Presbyterian Hospital, Department of Pathology and Laboratory Medicine, New York, NY, United States

JENNIFER DAVILA, MD
Department of Pediatrics, Albert Einstein College of Medicine, Bronx, NY, United States

ROBERT A. DESIMONE, MD
Weill Cornell Medicine and New York-Presbyterian Hospital, Department of Pathology and Laboratory Medicine, New York, NY, United States

ROGER Y. DODD, PhD
American Red Cross, Rockville, MD, United States

NANCY M. DUNBAR, MD
Dartmouth-Hitchcock Medical Center, Lebanon, NH, United States

AMY L. DUNN, MD
Nationwide Children's Hospital, Columbus, OH, United States; *and* Ohio State University, Columbus, OH, United Sates

PATRICK A. ERDMAN, DO
Memorial Sloan-Kettering Cancer Center, New York, NY, United States

IVO M.B. FRANCISCHETTI, MD, PhD
Department of Pathology, Montefiore Medical Center, Albert Einstein College of Medicine, Bronx, NY, United States

RICHARD O. FRANCIS, MD, PhD
Assistant Professor, Department of Pathology and Cell Biology, Columbia University Medical Center-New York Presbyterian Hospital, New York, NY, United States; *and* Medical Director, Special Hematology and Coagulation Laboratory, Columbia University Medical Center-New York Presbyterian Hospital, New York, NY, United States

YELENA Z. GINZBURG, MD
Associate Professor, Hematology Oncology, Icahn School of Medicine at Mount Sinai, New York, NY, United States

ELIZABETH A. GODBEY, MD
Department of Pathology, Virginia Commonwealth University, Richmond, VA, United States

RUCHIKA GOEL, MD, MPH
Assistant Professor, Assistant Medical Director of Transfusion Medicine and Cellular Therapy, Pathology and Pediarics (Pediatric Hematology Oncology), New York Presbyterian Hospital, Weill Cornell Medical College, New York, NY, United States

AMIT GOKHALE, MD
Department of Laboratory Medicine, Yale University School of Medicine, New Haven, CT, United States

YITZ GOLDSTEIN, MD
Albert Einstein College of Medicine, New York, NY, United States

JED B. GORLIN, MD, MBA
Physician Services, Innovative Blood Resources, St. Paul, MN, United States

CHERYL A. GOSS, MD
Memorial Sloan Kettering Cancer Center, New York, NY, United States

JANIS R. HAMILTON, MS, MT(ASCP)SBB
American Red Cross, Detroit, MI, United States

JEANNE E. HENDRICKSON, MD
Department of Laboratory Medicine, Yale University School of Medicine, New Haven, CT, United States; *and* Department of Pediatrics, Yale University School of Medicine, New Haven, CT, United States

RONG HE, MD
Division of Hematopathology, Mayo Clinic, Rochester, MN, United States

CHRISTOPHER D. HILLYER, MD
President and Chief Executive Officer, New York Blood Center, New York, NY, United States; *and* Professor, Department of Medicine, Weill Cornell Medical College, New York, NY, United States

ELDAD A. HOD, MD
Associate Professor, Pathology and Cell Biology, Columbia University Medical Center, New York, NY, United States

YEN-MICHAEL S. HSU, MD, PhD
Assistant Professor, Weill Cornell Medical College, New York, NY, United States

HEATHER A. HUME, MD, FRCPC
Clinical Professor, University of Montreal and Department of Pediatrics, Centre Universitaire Ste-Justine, Montreal, QC, Canada

JACK JACOB, DO
Albert Einstein College of Medicine, New York, NY, United States

SHILPA JAIN, MD, MPH
Division of Pediatric Hematology-Oncology, Women and Children's Hospital of Buffalo and Hemophilia Center of Western New York, Buffalo, NY, United States

FLORENCIA G. JALIKIS, MD
Anatomic and Clinical Pathology, University of Washington, Seattle, WA, United States

ALEXANDRA JIMENEZ, MD
Medical Director, Transfusion and Laboratory Services, New York Blood Center, New York, NY, United States

SHAWN M. JOBE, MD, PhD
Associate Investigator, Blood Research Institute, Blood Center of Wisconsin, Milwaukee, WI, United States; *and* Associate Professor, Pediatrics and Cell Biology, Neurobiology, and Anatomy, Medical College of Wisconsin, Milwaukee, WI, United States

CASSANDRA D. JOSEPHSON, MD
Center for Transfusion Medicine and Cellular Therapies, Department of Laboratory Medicine and Pathology, Emory University School of Medicine, Atlanta, GA, United States; *and* Aflac Cancer and Blood Disorders Center, Children's Healthcare of Atlanta/Emory University School of Medicine, Atlanta, GA, United States

MATTHEW S. KARAFIN, MD, MS
Associate Medical Director, Medical Sciences Institute, Blood Center of Wisconsin, Milwaukee, WI, United States; *and* Assistant Professor, Pathology, Medical College of Wisconsin, Milwaukee, WI, United States

LOUIS M. KATZ, MD
Chief Medical Officer, SMTQR, America's Blood Centers, Washington DC, DC, United States; *and* Adjunct Clinical Professor, Division of Infectious Diseases, Roy and Lucille Carver College of Medicine, Iowa City, IA, United States

CHRISTINE L. KEMPTON, MD, MSc
Department of Hematology and Medical Oncology, Emory University School of Medicine, Atlanta, GA, United States

DEBRA A. KESSLER, RN, MS
Vice President, Medical Programs and Services, New York Blood Center, New York, NY, United States

MARGARITA KUSHNIR, MD
Division of Hematology, Department of Oncology, Montefiore Medical Center and the Albert Einstein College of Medicine, Bronx, NY, United States

MICHELE P. LAMBERT, MD, MSTR
Assistant Professor, Pediatrics, Perelman School of Medicine at the University of Pennsylvania, Philadelphia, PA, United States; *and* Attending Physician, Division of Hematology, The Children's Hospital of Philadelphia, Philadelphia, PA, United States

MARISSA LI, MD
Medical Director, Medical Affairs, United Blood Services, Scottsdale, AZ, United States; *and* Assistant Medical Director, LifeStream Blood Center, Scottsdale, AZ, United States

DEEPA MANWANI, MD
The Children's Hospital at Montefiore, Albert Einstein College of Medicine, Bronx, NY, United States

IRINA MARAMICA, MD, PhD, MBA
Quest Diagnostics Nichols Institute, San Juan Capistrano, CA, United States

SUSANNE MARSCHNER, PhD
Terumo BCT, Lakewood, CO, United States

EMELINE MASSON FRENET, PhD
Cord Blood Distribution Manager, National Cord Blood Program, New York Blood Center, New York, NY, United States

CATHERINE E. MCGUINN, MD
Assistant Professor, Pediatrics, Weill Cornell Medicine, New York, NY, United States

SHANNON L. MEEKS, MD
Associate Professor, Department of Pediatrics, Emory University, Division of Hematology/Oncology/BMT, Aflac Cancer and Blood Disorders Center, Children's Healthcare of Atlanta, Atlanta, GA, United States

CONNIE H. MILLER, PhD
Division of Blood Disorders, National Center on Birth Defects and Developmental Disabilities, Centers for Disease Control and Prevention, Atlanta, GA, United States

CATERINA P. MINNITI, MD
Department of Medicine, Division of Hematology, Albert Einstein College of Medicine, Bronx, NY, United States

WILLIAM B. MITCHELL, MD
Division of Pediatric Hematology/Oncology, Icahn School of Medicine at Mount Sinai, New York, NY, United States

GRACE F. MONIS, MD, PhD
Assistant Professor, Pathology, University of California Davis Medical Center, Sacramento, CA, United States

THERESA NESTER, MD
Associate Professor, Laboratory Medicine, University of Washington Medical Center, Seattle, WA, United States; *and* Medical director of Integrated Transfusion Services, Bloodworks Northwest, Seattle, WA, United States

PHILIP NORRIS, MD
Blood Systems Research Institute and the University of California, San Francisco, CA, United States

ANGELA NOVOTNY, MT(ASCP)SBBCM
Innovative Blood Resources, Memorial Blood Center Division, St. Paul, MN, United States

MONIKA PARODER-BELENITSKY, MD, PhD
Department of Pathology, Montefiore Medical Center, Albert Einstein College of Medicine, Bronx, NY, United States

SHIBANI PATI, MD, PhD
University of California and Blood Systems Research Institute, San Francisco, CA, United States

KIM PECK, MA, MT(ASCP)SBB
Senior Director, Technical Services, Community Blood Center, Kansas City, MO, United States

HUY P. PHAM, MD, MPH
Department of Pathology, Keck School of Medicine of the University of Southern California, Los Angeles, CA, United States

ALLYSON PISHKO, MD
Department of Medicine, Perelman School of Medicine, University of Pennsylvania, Philadelphia, PA, United States

EVA D. QUINLEY, MS, MT(ASCP) SBB, CQA (ASQ)
Regional Director, Administration, LifeSource/ITxM, Chicago, IL, United States

SABRINA RACINE-BRZOSTEK, MD, PhD
Department of Pathology, Montefiore Medical Center, Bronx, NY, United States

LYNSI RAHORST, MHPE, MT(ASCP) SBB
Community Blood Center of Greater Kansas City, Kansas City, MO, United States

HANNA RENNERT, PhD
Weill Cornell Medicine and NewYork-Presbyterian Hospital, Department of Pathology and Laboratory Medicine, New York, NY, United States

JOSEPH S.A. RESTIVO, DO
Fellow, Blood Banking & Transfusion Medicine, BloodCenter of Wisconsin, Milwaukee, WI, United States

MORAYMA REYES GIL, MD, PhD
Department of Pathology, Montefiore Medical Center, Albert Einstein College of Medicine, Bronx, NY, United States

LIZ ROSENBAUM-MARINARO, MD
Medical Director, Medical Affairs, United Blood Services, Albuquerque, NM, United States

MIKHAIL ROSHAL, MD, PhD
Hematopathology Service, Memorial Sloan Kettering Cancer Center, New York, NY, United States

SARA RUTTER, MD
Department of Pathology, Yale University School of Medicine, New Haven, CT, United States; *and* Department of Laboratory Medicine, Yale University School of Medicine, New Haven, CT, United States

BRUCE S. SACHAIS, MD, PhD
New York Blood Center, New York, NY, United States

SURBHI SAINI, MBBS
Penn State Children's Hospital, Hershey, PA, United States; *and* Penn State University, College of Medicine, Hershey, PA, United States

SUSMITA N. SARANGI, MBBS, MD
Novant Prince William Medical Center, Manassas, VA, United States

WILLIAM J. SAVAGE, MD, PhD
Harvard Medical School, Waban, MA, United States

ANDROMACHI SCARADAVOU, MD
Medical Director, National Cord Blood Program, New York Blood Center, New York, NY, United States; *and* Associate Attending, Bone Marrow Transplant Service, Department of Pediatrics, Memorial Sloan-Kettering Cancer Center, New York, NY, United States

JOSEPH SCHWARTZ, MD, MPH
Department of Pathology and Cell Biology, New York-Presbyterian Hospital – Columbia University Irving Medical Center, New York, NY, United States

JANSEN N. SEHEULT, MB, BCH, BAO
University of Pittsburgh Medical Center, Pittsburgh, PA, United States

ERIC SENALDI, MD
New York Blood Center, New York, NY, United States

SALIMA SHAIKH, MD
Assistant Medical Director, Blood Centers of the Pacific, San Francisco, CA, United States

ANJALI SHARATHKUMAR, MD
University of Iowa Hospitals and Clinics, Carver College of Medicine, Iowa City, IA, United States

BETH H. SHAZ, MD
Chief Medical and Scientific Officer and Executive Vice President, New York Blood Center, New York, NY, United States

PATRICIA A. SHI, MD
Clinical Services, New York Blood Center, New York, NY, United States; *and* Division of Hematology, Department of Medicine, Albert Einstein College of Medicine, New York, NY, United States

SIERRA C. SIMMONS, MD, MPH
Department of Pathology, University of Alabama at Birmingham, Birmingham, AL, United States

ELIZABETH M. STALEY, MD, PhD
Department of Pathology and Immunology, Washington University School of Medicine in St. Louis, St. Louis, MO, United States

EMILY K. STORCH, MD
Food and Drug Administration, Silver Spring, MD, United States

DONNA STRAUSS, MS
Executive Director, Core Operations, Laboratories and Hospital Services, New York Blood Center, New York, NY, United States

YVETTE C. TANHEHCO, PhD, MD, MS
Department of Pathology and Cell Biology, Columbia University Medical Center, New York, NY, United States

AARON A.R. TOBIAN, MD, PhD
Johns Hopkins University, Baltimore, MD, United States

CHRISTOPHER A. TORMEY, MD
Department of Laboratory Medicine, Yale University School of Medicine, New Haven, CT, United States; *and* Pathology & Laboratory Medicine Service, VA Connecticut Healthcare System, West Haven, CT, United States

MARY TOWNSEND, MD
Senior Medical Director, Corporate Medical Affairs, Blood Systems, Scottsdale, AZ, United States

DUC Q. TRAN, MD, MSc
Department of Hematology and Medical Oncology, Emory University School of Medicine, Atlanta, GA, United States

NANCY L. VAN BUREN, MD
Associate Medical Director, Innovative Blood Resources, St. Paul, MN, United States

SUNITHA VEGE, MS
New York Blood Center, New York, NY, United States

RANDALL VELLIQUETTE, MS
New York Blood Center, New York, NY, United States

FRANCESCA VINCHI, PhD
New York Blood Center, New York, NY, United States

RONA S. WEINBERG, PhD
Director, Cellular Therapy Laboratory, New York Blood Center, New York, NY, United States

STUART P. WEISBERG, MD, PhD
Department of Pathology and Cell Biology, Vagelos College of Physicians and Surgeons, New York, NY, United States

CONNIE M. WESTHOFF, PhD, SBB
Executive Scientific Director, Immunohematology and Genomics, New York Blood Center, New York, NY, United States

MICHAEL WHITE, MD
Fellow Physician, Pediatric Hematology and Oncology, Emory University, Atlanta, GA, United States

ANNE M. WINKLER, MD
Director of Medical Affairs, Instrumentation Laboratory, Bedford, MA, United States

LUCIA R. WOLGAST, MD
Assistant Professor, Department of Pathology, Albert Einstein College of Medicine, Director, Clinical Laboratories– Moses Division, Associate Director, Hematology Laboratories, Montefiore Medical Center, Bronx, NY, United States

KALINDA WOODS, MD
Department of Obstetrics and Gynecology, Emory University School of Medicine, Atlanta, GA, United States

LINA Y. DIMBERG, PhD
Terumo BCT, Lakewood, CO, United States

MARK H. YAZER, MD
University of Pittsburgh Medical Center, Pittsburgh, PA, United States; *and* The Institute for Transfusion Medicine, Pittsburgh, PA, United States

CAROLYN T. YOUNG, MD
Vice President, Chief Medical Officer, Rhode Island Blood Center, Providence, RI, United States; *and* Clinical Assistant Professor, Pathology and Laboratory Medicine, Brown Alpert Medical School, Providence, RI, United States

PATRICIA E. ZERRA, MD
Center for Transfusion Medicine and Cellular Therapies, Department of Laboratory Medicine and Pathology, Emory University School of Medicine, Atlanta, GA, United States; *and* Aflac Cancer and Blood Disorders Center, Children's Healthcare of Atlanta/Emory University School of Medicine, Atlanta, GA, United States

KAREN L. ZIMOWSKI, MD
Pediatric Hematology/Oncology Fellow, Aflac Cancer and Blood Disorders Center, Children's Healthcare of Atlanta, Emory University School of Medicine, Atlanta, GA, United States

About the Editors

Beth H. Shaz, MD, is Chief Medical and Scientific Officer, Executive Vice President at New York Blood Center, and Adjunct Assistant Professor, Department of Pathology and Cell Biology, Vagelos College of Physicians and Surgeons, Columbia University. Beth is responsible for all medical and scientific activities throughout the NYBC enterprise, which includes Rhode Island Blood Center, Innovative Blood Resources (Memorial Blood Center and Nebraska Blood Center), Community Blood Center of Greater Kansas City, and Blood Bank of Delmarva. Medical activities comprise hemophilia services, clinical apheresis services, perioperative autologous transfusion services, cellular therapy, medical education, medical consultation, transfusion services, bone marrow donor recruitment, and donor management. Scientific activities include basic science at the Lindsley F. Kimball Research Institute and Comprehensive Cell Solutions (CCS), which contains translational and clinical research. NYBC's scientific and medical activities focus on transfusion medicine, cell therapy, regenerative medicine, infectious disease, hematology, and personalized medicine. Beth is an editor of 10 books in transfusion medicine, author of over 130 articles pertaining to transfusion medicine. She is an associate editor of *TRANSFUSION* and on the editorial board of *BLOOD*. Previously, she was an Associate Professor at Emory University School of Medicine and director of the transfusion service at Grady Memorial Hospital. Also, she was an instructor at Harvard Medical School and associate director of the transfusion service at Beth Israel Deaconess Medical Center in Boston. Beth received her MD with research distinction from the University of Michigan and BS in chemical engineering with distinction from Cornell University. She completed a general surgery internship at Georgetown University, an anatomic and clinical pathology residency at Beth Israel Deaconess Medical Center, and a transfusion medicine fellowship at Harvard Medical School.

Christopher D. Hillyer, MD, is President and CEO of New York Blood Center and Professor, Department of Medicine, Weill Cornell Medical College, New York, NY. Previously, he was the tenured, endowed Distinguished Service Professor, Department of Pathology and Laboratory Medicine, Emory University School of Medicine and served as director of the Emory Center for Transfusion and Cellular Therapies with responsibility for all aspects of clinical and academic transfusion medicine at Emory's seven principle hospitals. Dr. Hillyer is an editor of 12 textbooks in transfusion medicine, author of over 160 articles pertaining to transfusion, human immunodeficiency virus, and herpes viruses, most notably cytomegalovirus. Nationally recognized as an expert in hematology and blood transfusion, Dr. Hillyer is also a past president, board of directors of AABB, and a former trustee of the National Blood Foundation. Dr. Hillyer has been awarded many million dollars in research funding from the National Institutes of Health, the Centers for Disease Control and Prevention, and other agencies.

He was an associate editor of *TRANSFUSION* and served on several other editorial boards. Dr. Hillyer was formally recognized for his work in Africa as part of the AABB/Emory cooperative agreement from the President's Emergency Plan for AIDS Relief (PEPFAR) and is a recipient of two Tiffany Awards from the American Red Cross where he also served as a medical director and a member of their national Medical Advisory Board. He also received the 2014 Emily Cooley Award from AABB for his "significant commitment and contributions to the field of transfusion medicine through extensive teaching, mentoring and professional leadership, and the countless clinical, scientific, and innovative resource materials he has created to educate others." He is a cofounder of Transfusion and Transplantation Technologies, Inc (3Ti) and holds over 20 patents or patents pending. Dr. Hillyer is board certified in transfusion medicine, hematology, medical oncology, and internal medicine. He received his BS from Trinity College, and his MD from the University of Rochester School of Medicine, with postgraduate training and fellowships in hematology–oncology, transfusion medicine, and bone marrow transplantation at Tufts-New England Medical Center.

Morayma Reyes Gil, MD, PhD, is an associate professor in the Department of Pathology at Montefiore Medical Center (MMC) and Albert Einstein College of Medicine. As the director of hematology and special coagulation laboratories, Dr. Reyes Gil oversees all hematology and coagulation laboratories at all MMC hospitals, collaborates in clinical research and clinical trial studies, and continues translational research in vascular biology. She started her career in academic pathology at the University of Washington as an assistant professor in the Department of Pathology and Laboratory Medicine. Dr. Reyes Gil has been a recipient of several awards and research grants from private research foundations and from the National Institutes of Health. Dr. Reyes Gil is board certified in clinical pathology and has authored over 50 peer-reviewed articles in the field of vascular biology. Dr. Gil is a graduate of the University of Puerto Rico. She received her MD and PhD from the University of Minnesota. She then completed a residency in clinical pathology and specialized in coagulation and benign hematology at the University of Washington.

Preface

The editors are thrilled at the success of the first and second editions and present the third edition with great excitement. With the third edition we have updated and expanded some areas, including quality and regulatory, blood management, cellular therapy, regenerative medicine, coagulation testing, and molecular testing. The vision remains to provide clinical and laboratory information as regards these two fields that have high degrees of overlap into a concise easy-to-use book that would serve both as a comprehensive textbook and as a reference guide. We believe that *Transfusion Medicine and Hemostasis: Clinical and Laboratory Aspects, Third Edition* has the depth of information to be helpful to all physicians and allied health professionals who order and administer blood components, cellular therapies, and specialized factors for hemostatic abnormalities; who order coagulation testing; and those who consult and care for these often very ill patients. We expect the breadth of the book would be ideal for pathology, medicine, surgery, and anesthesia residents; transfusion, hematology, and anesthesia fellows; and certified and specialized practitioners; as well as medical technologist in transfusion, cellular therapy, hematology, and coagulation. We have chosen to employ a standardized format throughout the book which allows each chapter to be focused on a well-defined subject, four to eight pages, which is easy to read, and the information is both precise and concise. Although extensive reference lists are valuable in larger texts, we chose to have key, recent publications as Recommended Reading, most often from the past few years. A general reference list to larger textbooks and standards in the fields is included and indexes that cross reference diagnostic, clinical, and therapeutic commonalities. We thank you for your support and welcome observations, criticisms, and suggestions so that we can continue to offer the most outstanding book.

We, the editors, would also like to acknowledge the outstanding expertise and guidance of Tracy Tufaga and other team members at Elsevier who have played an instrumental role in the creation of this textbook.

B.H. Shaz
C.D. Hillyer
M. Reyes Gil

Preface

The editors are thrilled at the success of the first and second editions and present the third edition with great excitement. With the third edition we have updated and expanded some areas, including quality and regulatory, blood management, cellular therapy, regenerative medicine, coagulation testing, and molecular testing. The vision remains to provide clinical and laboratory information as regards these two fields that have high degrees of overlap into a concise easy-to-use book that would serve both as a comprehensive textbook and as a reference guide. We believe that *Transfusion Medicine and Hemostasis: Clinical and Laboratory Aspects, Third Edition* has the depth of information to be helpful to all physicians and allied health professionals who order and administer blood components, cellular therapies, and specialized factors for hemostatic abnormalities; who order coagulation testing; and those who consult and care for these often very ill patients. We expect the breadth of the book would be ideal for pathology, medicine, surgery, and anesthesia residents, transfusion medicine, hematology, and anesthesia fellows; and certified and specialized practitioners; as well as medical technologists in transfusion, cellular therapy, hematology, and coagulation. We have chosen to employ a standardized format throughout the book, which allows each chapter to be focused on a well-defined subject, four to eight pages, which is easy to read and the information is both precise and concise. Although extensive reference lists are valuable in larger texts, we chose to have key recent publications as Recommended Reading, most often from the last few years. A general reference list of larger textbooks and standards in the fields is included and indexes that cross-reference [illegible] abnormalities. We thank you for your support and welcome observations, criticisms, and suggestions so that we can continue to offer the most outstanding book.

We the editors would also like to acknowledge the outstanding expertise and guidance of Tracy Tufaga and other team members at Elsevier who have played an instrumental role in the creation of this textbook.

B.H. Shaz
C.D. Hillyer
M. Reyes Gil

PART I

Blood Banking and Transfusion Medicine

CHAPTER 1

Introduction to Blood Banking and Transfusion Medicine

Christopher D. Hillyer, MD and Beth H. Shaz, MD

A safe and reliable blood supply is critical to the function of complex healthcare systems worldwide. Blood transfusion is one of the most common therapeutic medical practices. The field of transfusion medicine (blood banking and transfusion services) has expanded to therapeutic apheresis, regenerative medicine, cellular therapy, tissue banking, and coagulation.

Blood Transfusion History: In 1667, Jean Denis published the transfusing of lamb blood (because of its presumed soothing qualities) to an agitated man (resulting in hemolytic transfusion reaction). In 1818, James Blundell first successfully transfused human blood to a patient with postpartum hemorrhage.

Blood Groups: Karl Landsteiner, in 1900, demonstrated the presence of the ABO blood group system. In the 1920s, ABO testing became routine. The Rh blood group system was discovered during 1939–40 by Landsteiner, Weiner, Levine, and Stetson, explaining many unexpected transfusion reactions cause. In 1945, Coombs, Mourant, and Race described antihuman globulin sera use to detect IgG antibodies in compatibility testing, thus providing the Coombs test.

Blood Storage: Direct transfusion (donor artery anastamosed to recipient vein) was performed in 1908, and direct transfusion using a three-way stopcock was used until World War II. While sodium citrate as an anticoagulant use was considered in 1914 and used (with glucose) some during World War I to set up blood depots, blood could be typically stored for a few days. In 1943, acid citrate dextrose solution allowed storage for weeks, facilitating blood "banking." Additionally, acidification of anticoagulant-preservative solution allowed autoclaving and reduced bacterial contamination.

Blood Derivatives: In 1940, Cohn developed cold ethanol fractionation process, allowing plasma to be divided into albumin, gamma globulin, and fibrinogen, among other proteins (called Cohn fractionation). In 1961, Pool and Shannon recognized that the precipitate (cryoprecipitate) that formed when plasma was thawed in the cold contained factor VIII, revolutionizing hemophilia A treatment. In 1985, dry-heated, lyophilized factor VIII and IX concentrates became available. Genetically engineered (recombinant) factor VIII became available in 1993 and factor IX in 1998. Most recently, factor products are engineered without any human components.

In 1967, Rh immune globulin was introduced commercially, near eliminating Rh hemolytic disease of the fetus and newborn.

Transfusion Medicine and Hemostasis. https://doi.org/10.1016/B978-0-12-813726-0.00001-5

Blood Component Therapy: Introduction of plastic bags to replace glass bottles for collection and storage in 1950 allowed development of component therapy with use of refrigerated centrifuges to separate components by density and precollection attached satellite bags to store the prepared components. This enabled optimal storage of each component and treatment of patients only with the component they needed.

Apheresis: In the 1950s, Cohn designed a centrifuge to separate cellular components from plasma. Donor apheresis allowed collection of therapeutic doses of platelets, granulocytes, RBCs, and plasma from a single donor. Automation of therapeutic apheresis devices has expanded its use, vital to the treatment of many diseases (e.g., thrombotic thrombocytopenic purpura, sickle cell disease).

Adverse Effects of Transfusion: Blood safety is continuously improved. In the 1960s, blood banks became increasingly aware that paid donors were associated with higher rates of hepatitis transmission, and by 1970, transition to an all-volunteer US blood supply began. In 1971, commercial testing for hepatitis B surface antigen began, further reducing posttransfusion hepatitis rate. In 1985, HIV antibody test was introduced. By 1990, testing for hepatitis C became routine and then HIV antigen testing was introduced. By 2000, nucleic acid testing for HIV and HCV in the developed world further reduced residual risk to ~1:2,000,000 screened units. Subsequently, transmission of other emerging infection has become apparent, such as West Nile virus, *Trypanosoma cruzi* (agent of Chagas' disease), and *Babesia* sp. Additional tests, donor screening, and pathogen reduction technologies have been implemented to sustain blood safety. Mitigation strategies for noninfectious transfusion complications, such as leukoreduction for febrile nonhemolytic transfusion reactions, donor screening and testing for transfusion-related acute lung injury, and irradiation for transfusion-associated graft versus host disease, have been implemented.

Donor Safety: Processes and studies have continued to improve donor safety, such as changing height and weight requirements for young donors to decrease vasovagal reactions. More recently, methods to mitigate iron deficiency in young and frequent donors are being researched and implemented.

Decade of "Right-Sizing" Utilization: Before 2008, blood utilization was increasing annually in most developed nations. By 2009, US blood transfusion approached 50 RBC units/100,000 inhabitants, while Canada was ~30/100,000, Denmark ~60/100,000 inhabitants, and Kenya <2/100,000. In 2009, significant decrease in blood utilization occurred in the United States and worldwide as patient blood management programs were implemented. US blood unitization decreased to ~35/100,000 by 2016.

Blood Industry: The *blood industry* is the business relating to the pipeline (Fig. 1.1), while the *discipline* is the medical field relating to the many processes in the pipeline. Blood industry includes manufacturers (also called suppliers) of information systems, reagents, appliances, and devices used by blood establishments, blood banks, transfusion services, and nongovernmental organizations (AABB, College of American Pathologists (CAP), National Marrow Donor Program (NMDP), and Foundation

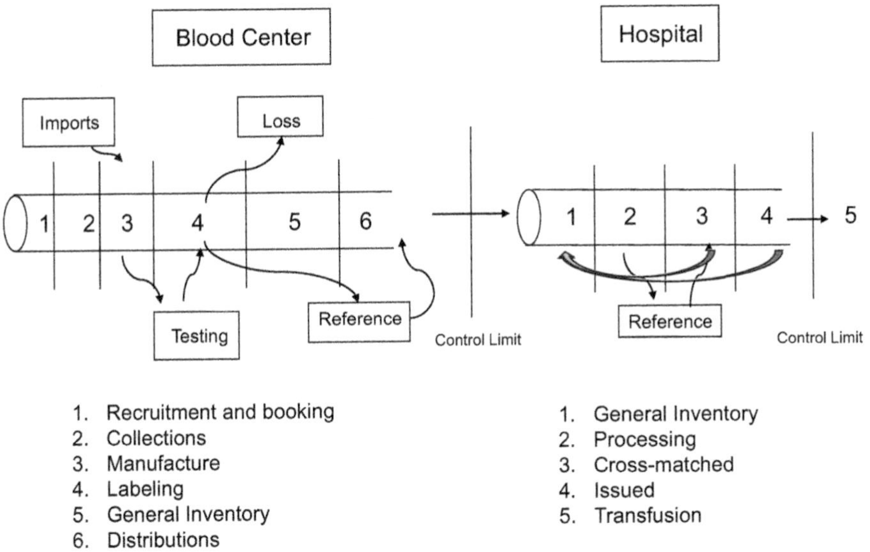

FIGURE 1.1 The blood pipeline.

for the Accreditation of Cellular Therapy (FACT)). Advanced Medical Technology Association (AdvaMed), a trade association made up of medical technology companies, which has taken a leadership role in defining appropriate corporate–customer relationships for compliance with federal trade, financial and tax regulations, understanding of the challenges facing the industry as regard to government and third-party insurer reimbursement for blood and blood components.

Discipline: Blood banking and transfusion medicine has expanded to related laboratory disciplines, services, and therapeutics, including perioperative transfusion (perioperative autologous donation and cell salvage), therapeutic apheresis and phlebotomy, coagulation or specialized laboratories, hospital tissue banking, regenerative medicine, and cellular therapy—which itself includes collection, processing, storage, and distribution of hematopoietic progenitor cell products, pancreatic islet cells, and related minimally and highly manipulated cells. While credentialed as a single entity by the American Board of Pathology, blood banking and transfusion service terms have different meanings (transfusion medicine encompasses both terms).

Blood Banking: Blood banking refers to collection, processing, storage, and distribution of blood and blood components at a blood collection facility, defined by the FDA during registration or licensure as a community blood bank, although a small percentage of units are collected in the hospital setting, defined by the FDA as a hospital blood bank.

Transfusion Service: Transfusion service encompasses pretransfusion and compatibility testing; postmanufacture processing, including irradiation, washing, and volume reduction; and administration of appropriate products to appropriate patients at the appropriate time. The transfusion service is mostly in hospitals (few centralized transfusion services exist in the United States serving multiple hospitals) usually under the FDA designation at registration as a hospital transfusion service. The transfusion service is typically responsible for clinical consultation regarding complex transfusion and coagulation issues, choice of specialized products, including coagulation factor concentrates, intravenous gamma globulin, and albumin, development of guidelines, and review of blood component therapy.

National Structure: Worldwide, the predominant healthcare structure is a national medical system with a national blood service. One example is the UK National Health Service Blood and Transplant, which provides blood (collection of blood and cellular therapies), transplantation, diagnostic, and therapeutic services as well as research and clinical trials. National blood and transfusion services typically formulate national transfusion guidelines and establish hemovigilance programs.

United States' Structure: The United States has a network of ~65 blood centers, some of which operate in multiple states (American Red Cross Biomedical Services, Blood Systems, New York Blood Center). Blood centers are members of several organizations, including America's Blood Centers (ABC), Blood Centers of America (BCA), and AABB.

AABB creates standards and performs accreditation in transfusion medicine and cellular therapies. AABB, in collaboration with other organizations, published guidelines for platelet, plasma, and RBC transfusions and developed a hemovigilance system for monitoring of donor and recipient adverse reactions and quality control incidents.

Further Reading

Ellingson, K. D., Sapiano, M. R. P., Haass, K. A., Savinkina, A. A., Baker, M. L., Chung, K. W., et al. (2017). Continued decline in blood collection and transfusion in the United States-2015. *Transfusion, 57*(Suppl. 2), 1588–1598.

Sapiano, M. R. P., Savinkina, A. A., Ellingson, K. D., Haass, K. A., Baker, M. L., Henry, R. A., et al. (2017). Supplemental findings from the national blood collection and utilization surveys, 2013 and 2015. *Transfusion, 57*(Suppl. 2), 1599–1624.

Shaz, B. H., & Hillyer, C. D. (2010). Transfusion medicine as a profession: Evolution over the past 50 years. *Transfusion, 50*, 2536–2541.

CHAPTER 2

Quality Principles in Transfusion Medicine

Eva D. Quinley, MS, SBB, CQA (ASQ)

Quality, as defined in the field of blood banking and transfusion medicine, has evolved from a focus on retrospective auditing and review of sampling and testing. This evolution began with a focus on compliance and meeting minimal regulations, standards, and guidelines of accrediting and regulating agencies. More recently, it moved to a focus on compliance leveraged with proactive and direct involvement in, and construction of, quality systems including change management systems and others that have a positive effect on operating efficiencies throughout an organization. Through implementation of quality management systems and voluntary accreditation, blood banks and transfusion services have developed processes designed to ensure safe and effective products and services for their customers. A well-designed and integrated quality program provides guidance, support, and oversight to all aspects of an organization. Continual improvement in quality is an important aspect of creating and maintaining efficient and effective operations.

The meaning of the term "*quality*" continues to evolve over time. Initially, quality centered on *quality control*, sampling, and testing, but with increasing attention from governmental agencies, including the US Food and Drug Administration (FDA), and focus from organizations such as AABB in the early 1990s, quality became a discipline of its own. For a time, "quality" became synonymous with "compliance," that is, adherence to documented regulations and standards that defined and influenced the activities and priorities of quality departments. Then, accrediting organizations such as AABB promulgated quality system elements and standards for facilities to follow. Some facilities sought to become certified by ISO (International Organization for Standardization), thus adhering to a different and perhaps more stringent set of criteria in their compliance. Defining quality as just compliance was somewhat shortsighted, however, as it limited quality's contribution to the overall business and focused on external agencies and organizations to ensure internal quality. Today quality has become internally defined and driven and has taken on a notion of a culture that is understood and upheld by every employee, while retaining full compliance with accrediting, registering, and licensing agencies. The quality emphasis now extends to efficiencies in operations, with quality staff providing input into business decisions where appropriate. This has not changed the requirement for independence of the quality unit but provides for the appropriate involvement of quality early on in process design and critical business practices. These additional roles have increased the likelihood of positive results for organizations. Fig. 2.1 depicts the evolution of quality as a hierarchy.

Basic Principles: Quality is defined in many ways. The Webster's Dictionary defines quality as a distinguishing attribute, implying that it could be good or bad or just mediocre. In transfusion medicine, quality of products and services must be as high as possible, and such a definition requires that the quality of the *processes* by which those

Transfusion Medicine and Hemostasis. https://doi.org/10.1016/B978-0-12-813726-0.00002-7

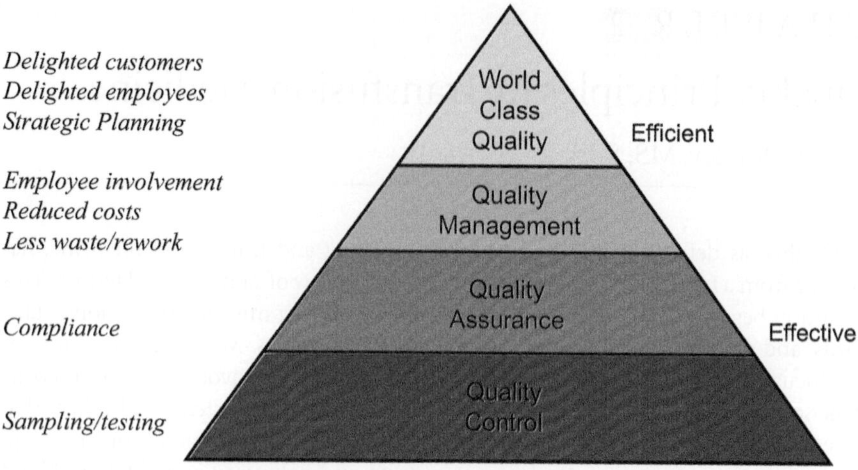

FIGURE 2.1 Quality hierarchy.

products and services are manufactured and delivered must be as high as possible. The safety of donors and patients is dependent on such actions and decisions, which are made every day. A quality definition that incorporates the evolving nature of the field of quality is sometimes used by this author: "Quality is doing the right things right while utilizing the least resources possible." In other words, quality is doing things correctly each and every time in the most efficient way possible. This definition allows quality to include effectiveness and efficiency.

Quality control (QC) is a relatively narrow term, including activities such as sampling and testing, that provides information about the quality of a product and can provide assurance that something or someone is functioning at a given time as it is supposed to function. QC is performed on reagents, equipment, and products; QC also includes annual competency assessments of employees. QC activities may include reviews of documentation for accuracy and completeness, visual inspections, and measurements of product attributes. It is important to note that QC activities may be performed by staff in operations, by quality staff, or by both. The key fact here is that review of the work must be done by someone other than the person who did the work.

Quality assurance is a broader term, including all activities to ensure that processes are developed and implemented, as they are designed and assures that they achieve the anticipated results. A quality assurance program is far reaching and includes activities such as QC testing, development of standard operating procedures, deviation management, validation, training, and internal auditing. Traditionally, these activities are under the realm of the quality department and are associated with the implementation of robust quality systems. The goal of a good quality assurance program is to ensure consistency in high-quality output and to decrease errors. It is important to highlight that the quality department is the champion of quality in an organization, *and* that quality is everyone's responsibility, individually and collectively. This allows a quality culture to emerge, which then provides enhanced "quality" in all areas with enhancement in operational efficiencies, thus leading to cost reductions and improvements.

TABLE 2.1 AABB Quality System Essentials

Organization	Documents and records
Resources	Deviations
Equipment	Assessments: internal and external
Supplier and customer issues	Process improvement through corrective and preventive action
Process control	Facilities and safety

Quality Management Systems: Blood banks and transfusion services should have an organized *quality management system*. A quality management system is a series of processes that are linked together and controlled centrally to increase assurance of product and manufacturing process quality. As part of the quality management system, blood banks and transfusion services should have a *quality policy* describing their overall intentions and direction with respect to quality. They should also have a *quality manual*, which details the various processes in the quality management system and how the organization achieves success in implementing quality throughout its operations. The number of elements in a quality system and the way they are grouped together varies depending on the organization.

With mergers and acquisitions, it is important that all of the organizations adhere to the same quality principles as soon as possible. Thus, the integration of quality becomes a high priority in these instances.

The American Association of Blood Banks: The AABB has organized its quality management system around 10 quality system essentials (QSEs) as found within the AABB *Standards for Blood Banks and Transfusion Services,* 31st edition (Table 2.1). AABB assesses compliance with its *Standards* and accredits blood banks and transfusion services. This accreditation is voluntary on the part of the blood bank or transfusion service. Nonconformances or deficiencies in meeting the *Standards*, which are found during an AABB assessment, are issued, and correction is required for continued accreditation. A brief overview of each of the QSEs follows.

Organization: This QSE is intended to ensure that blood banks and transfusion services have an organizational structure in place, which is well defined and assures that quality management is implemented and working throughout the organization, including in the administrative, medical, technical, and quality areas. It is leadership's responsibility to ensure that the organization is compliant and safe, not only for donors and patients but also for employees.

Regular reviews of the quality management system by leadership are an important element of this QSE. Reports on activities within the quality management system provide evidence as to whether the quality management system is effective. Such reports may include quality indicators that apply to both quality and operational areas. For example, a discarded product is both a quality and operational issue, as is a customer complaint; monitoring these occurrences over a period of time may lead to opportunities for quality improvements and increased operational efficiencies. Reviewing reports and discussing issues regularly can lead to valuable insights and help identify areas where process improvement is needed. Regular management reviews are encouraged.

The quality organization must have authority and responsibility for the quality management system, and there must be a process for individuals to communicate quality concerns anonymously. Most facilities accomplish this through an anonymous "hotline" where callers can report quality issues without fear of retribution.

Resources: This QSE relates to staff, requiring well-written job descriptions with job qualifications that are clearly defined. It also addresses the need for staff to have orientation training, which includes information about the company and how it operates, and training in job-related/specific tasks as well as in quality and safety. Additionally, staff should be evaluated regularly and provided opportunity for continuing education. This QSE is essential to successful quality operations.

Equipment: Equipment that is used in the blood banks and transfusion services must be identified and qualified for its intended use. If applicable, equipment should be calibrated against known standards on a routine basis according to written procedures. Equipment should be on a documented schedule for maintenance, including cleaning. If repaired, the equipment should be requalified before use. Equipment that is in disrepair or fails QC should be removed from service and clearly marked to prevent its unintentional use. There should be an investigation of the cause of any equipment failures.

Any modifications to computer systems, including software, hardware, and databases should be well planned with inclusion of an assessment of risk and validation (see Chapter 3). Modifications or repairs should be carefully evaluated for unrecognized adverse impact to operations before implementation and again after the modification/repair is in place to make sure that there are no untoward issues.

Storage devices must be maintained and monitored to ensure that they meet criteria for storage of supplies/reagents and products. Such storage devices should have alarms to alert users when temperatures approach unacceptable levels. They should be included in scheduled cleaning as well, and records of cleaning should be maintained.

Devices used to warm blood must have a sensor to detect malfunctions to prevent hemolysis or damage to blood and blood components during the warming process. These devices must be well maintained for patient safety.

The blood bank and transfusion service shall have processes that support the implementation and modification of software, hardware, and databases and shall maintain records of activities such as validation and monitoring of data integrity. There should also be an alternate system for backup, which is routinely tested. Computer systems must be managed in a compliant manner and access to such systems should be only to those who have authorization.

Supplier and Customer Issues: To have quality products and services, blood banks and transfusion services must ensure that their suppliers can provide materials that meet clearly defined quality criteria. This process is called *supplier qualification*. It defines expectations of both the blood banks and transfusion services and their respective suppliers, through written agreements, and provides a process that offers a high degree of assurance. Supplier qualification may include onsite audits of the supplier. Another key aspect is the requirement for regular reviews to ensure continued compliance with any supplier agreements.

There should also be a documented process whereby incoming materials are inspected before use, to make sure that they are indeed acceptable for use. With storage of supplies, a first in, first out (FIFO) approach is needed, along with adherence to storage requirements as specified by the manufacturer of the supplies. Finally, blood banks and transfusion services must have a methodology to know when and where supplies are used, i.e., traceability to the process or service in which the supplies were used and traceability from the donor to the patient.

Process Control: Control of processes should be engineered into those processes from development through implementation and is critical to ensuring quality from the outset. These process controls ensure that the process itself is implemented to produce quality results. Change control is essential, and processes should be well validated to provide a high degree of assurance that any changes will produce desired results. Communication of change must be clear, timely, and delivered to all impacted by the change. Once processes are implemented, controls must be in place to ensure they remain in a validated state. Accommodation for planned deviations must be incorporated; however, because in any normal operation, there may be a justifiable need to deviate from a procedure for a period of time. Such deviations must be appropriately reviewed and approved by quality staff.

Quality control and inspection are also essential to this QSE. These activities, when performed at regular intervals, ensure that criteria are met and that the processes are "in control," that is, operating as they should be in accordance with well-written procedures and producing expected results. Additionally, records that allow traceability (details such as who, when, where) and trackability (documented logical sequence(s) of events that can be followed) are necessary to provide evidence of the work that was done and how, when, etc. it was done.

It is important that all materials and equipment are used in accordance with manufacturers' instructions. It is also important that measures are taken to prevent contamination of blood or blood products during collection, processing, and transfusion. The blood bank or transfusion service should have methods to detect bacterial contamination, and if present, a mechanism to identify the organisms.

Labeling: Blood banks and transfusion services must have strict control over labeling. The label should be unique and allow complete traceability back to the source. This is true for both donor identification and product identification. Labels must be in compliance with standards and regulations and validated for their intended use to ensure they adhere well under a variety of conditions. In the transfusion service, labeling includes patient identification and identification of samples for testing. All labels used must be clear, accurate, and complete.

Collection: Donor education and consent is required. Donors must understand the importance of providing accurate information to the blood center and must be thoroughly informed about the donation process and the testing that will be done on their blood. Donors must be notified of certain results of the donor screening process in accordance with FDA regulations as found in 21 CFR 630.6.

Care of the donor is of primary importance both during and after a donation. Facilities that collect blood must have procedures to describe both of these activities, and staff must be trained to detect signs and symptoms of donor reactions. Donors must be qualified according to written procedures, which clearly define acceptance criteria based on the type of donation procedure. Blood should be collected using techniques to minimize risk of bacterial contamination, and samples should be collected for testing in a manner to maintain sterility of the unit. Once collected the blood should be maintained at appropriate temperatures for the type of product to be made. Products collected by apheresis should also be maintained at appropriate temperatures for the type of product.

Component Production: The methods used to prepare various blood components should maintain the quality of the product and should be compliant with FDA requirements. Such methods should be documented in written procedures. Prepared components must meet certain QC criteria to be considered acceptable for transfusion. Processes such as irradiation and leukoreduction must be performed using effective methods, and the product label must reflect the modified condition.

Testing of Donor Blood: Blood that is intended for transfusion must be tested by FDA-approved test methodologies. Test results must be available and appropriate before labeling units. Tests include ABO, Rh, and infectious disease testing. In emergent situations, blood may be released before completion of testing, but this should be done rarely and only with documented medical need and physician approval. Platelets are also tested for bacterial contamination, which is especially important due to their storage at room temperature.

Transfusion Service–Related Issues: Appropriate physician order is a requirement when blood and blood products are needed. Patient samples must be labeled completely, accurately, and legibly. Pretransfusion testing includes ABO, Rh, and tests for unexpected antibodies to red cell antigens. There should also be comparison with historical records if available. Discrepancies in results should be investigated and appropriate action taken before transfusion. Blood and blood components that are compatible with the patient's type and antibody status should be selected for transfusion. Crossmatches may be performed by computer, following written procedures, which ensure that blood selected is appropriate for the intended recipient. Blood should be issued following written procedures and, should blood be returned, it must be inspected and found suitable before it can be reissued.

When blood and blood components are transfused, the recipient or an appropriate designee should consent to receipt of the transfusion. Verification of identification of the intended recipient is critical, and the recipient should be observed for reaction during and for a time after the transfusion.

Documents and Records: Only current versions of documents should be used, and there should be control over document distribution. Procedures should be reviewed at least every other year and revised when necessary. An archived copy of documents no longer in use should be maintained. Documents should at all times be accessible to individuals doing the work.

Records created within blood banks and transfusion services must be clearly and uniquely identified and reviewed by appropriate individuals. As records provide evidence of the work that was done, it is important these are accurate and complete. Effective record review is essential to ensuring this. Changes to records should be strictly controlled whether those records are paper or electronic. Records must be stored in a manner that protects their integrity and should be retained in accordance with a written retention policy.

Procedures should be in place to ensure that all information, including data, is maintained in a manner to protect its integrity and to maintain confidentiality where required.

Deviations, Nonconformances, and Adverse Events: It is expected that nonconformances will occur, and it should be determined what is done with those nonconformances when they occur. Blood banks and transfusion services must have procedures to detect, document, investigate, correct, and monitor nonconformances. Certain nonconformances must be reported to outside agencies. 21 CFR 606.171(b) requires reporting of certain events to the FDA. Many states also have their own reporting requirements. It is critical to ensure that nonconforming products are segregated and cannot be distributed.

Adverse events that occur during the donation process or during transfusion of blood and blood components are types of nonconformances that must be documented and investigated. Transfusion services should report adverse events associated with donor blood to the blood collection agency from which the product was obtained. This includes evaluation of immediate and delayed reactions, including potential disease transmission. Fatalities associated with transfusion must be reported promptly to the FDA. If repeat donors are found to be positive or at risk for certain infectious diseases, the blood collection agency must perform a lookback in accordance with FDA regulations.

Assessments: Internal and External: Blood banks and transfusion services need to assess their quality systems through monitoring quality indicators and, as applicable, through proficiency testing programs. Data from this monitoring should be assessed for trends and for opportunities to improve processes. In addition, internal assessments should include audits of processes and systems. There should be procedures describing the monitoring system, which include a process for responding to any issues found. External assessments can be performed by a number of organizations, including AABB, the state, FDA, CMS, etc.

Through the use of data, the blood banks and transfusion services should find opportunities for improvement. Corrective and preventive actions that are implemented should be evaluated for effectiveness, and blood bank and transfusion service management should regularly review data to determine the status of the quality management system and operational processes.

Facilities and Safety: The physical space in which work is done must be conducive to performing the work in a safe and efficient manner and in compliance with all applicable laws and regulations while minimizing health or safety risks. This includes ensuring the work area is clean and well maintained, and that access is limited only to individuals who have a justified reason to be in the area. There should be procedures for safe storage of materials that could pose a risk to health. The facility should have a well-defined plan of action, which will ensure the safety of donors, employees, and the continuance of services to patients as much as is reasonably possible, in the event of an emergency.

Other Quality Management Systems

The Joint Commission: The Joint Commission (TJC) develops standards (performance measures) and accredits or certifies health-care organizations and programs in the United States, including transfusion services. TJC, through its Blood Management Performance Measures Technical Advisory Panel, identified and published blood-related performance measures as a tool for health-care organizations to evaluate the transfusion consent process, blood utilization, and blood administration documentation. This tool is also helpful to identify processes related to elective surgery that may decrease the need for blood and improve patient safety. TJC requires reporting of sentinel events, which are unexpected occurrences involving death or serious physical or psychological injury, or the risk thereof. For example, erroneous transfusion of incompatible blood to a patient would be considered a sentinel event. TJC conducts surveys (inspections) of transfusion services at least every 3 years. TJC, AABB, and CAP (see next section) have deemed status for CLIA inspections (CMS) (see Chapter 3).

College of American Pathologists: College of American Pathologists (CAP) also accredits transfusion services and hospital-based blood banks. Inspections are conducted using a well-defined checklist addressing a variety of transfusion medicine issues and occur every 2 years with a self-assessment in the alternate year. Accredited facilities must also participate in proficiency testing. As with AABB accreditation, CAP accreditation is voluntary, but if a facility does not have CAP accreditation, it is subject to state inspections. Facilities desiring accreditation by both CAP and AABB may request a joint inspection.

Clinical and Laboratory Standards Institute: Clinical and Laboratory Standards Institute (CLSI), formerly known as the National Committee on Clinical and Laboratory Standards (NCCLS), provides guidelines for a quality management system based on AABB *Standards*. The CLSI QMS includes pertinent laboratory regulations, standards, and accreditation requirements.

ISO 15189: This ISO Standard specifies the quality management system requirements applicable to medical laboratories. The standard provides advice related to the use of the laboratory service, the collection of patient samples, the interpretation of test results, acceptable turnaround times, how testing is to be provided in a medical emergency and the lab's role in the education and training of health-care staff. The standard is based on ISO/IEC 17025 and ISO 9001, but it is a unique document that takes into consideration the specific requirements of the medical environment and the importance of the medical laboratory to patient care. Several countries outside the United States base their laboratory QMS on this standard.

Process Improvement: A true quality organization continually improves, and the ultimate goal of any quality program should be to improve. Arthur Ash stated: "*Success is a journey, not a destination.*" To paraphrase that quote "*True quality is a journey, not a destination.*" There is always room for improvement. Many blood banks and transfusion services have implemented programs such as Six Sigma or Lean Manufacturing to find opportunities to improve quality, both in terms of effectiveness and efficiency.

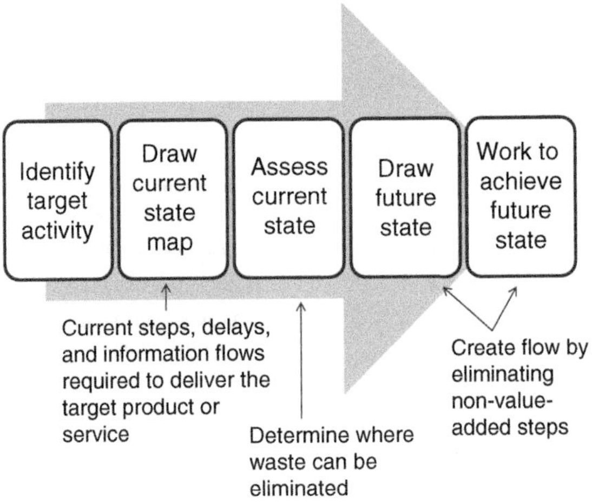

FIGURE 2.2 **Basic steps in value stream mapping.**

In some organizations, these are combined into a Lean Six Sigma program. These programs involve reduction in variation of processes and removal of non–value-added steps. *Value stream mapping* has become an important tool in these efforts. Value stream mapping is a lean manufacturing technique used to analyze and design the flow of materials and information required to bring a product or service to a consumer. It can be applied to nearly any activity. The steps in value stream mapping are outlined in Fig. 2.2. Applying this technique to activities within transfusion medicine and acting on the results can produce significant quality improvements.

New Role of Quality: The role of quality in transfusion medicine has evolved. Quality no longer is just a "police force" finding wrongs, but now is actively involved in finding opportunities to improve. Quality is still involved in traditional quality activities such as quality control and standard operating procedure review but its scope is expanding into the business and safety arenas. Through monitoring various performance measures related to the National Patient Safety Goals of TJC, quality works to ensure that transfusion medicine activities are conducted so as to ensure the safety of patients.

Serving as champions of process improvement efforts and working side by side with operations are two illustrations of how quality now plays an important role in the business of transfusion medicine. More and more quality is truly providing guidance and support along with oversight. This new role leads to better overall quality in all aspects of transfusion medicine.

Further Reading

Anyaegbu, C. C. (2011). Quality indicators in transfusion medicine: The building blocks. In *ISBT science series* (6) (pp. 35–45).

Berte, L. M. (2007). Laboratory quality management: A roadmap. *Clin Lab Med, 27,* 771–790.

Blaylock, R., & Lehman, C. M. (2011). Managing transfusion service quality. *Arch Pathol Lab Med, 135*, 1415–1424.

Dolcemascolo, D. (2006). *Improving the extended value stream: Lean for the entire supply chain*. New York: Productivity Press.

Foss, M. L., Stubbs, J. R., & Jones, G. (2011). Integrating quality, education, lean and performance management into a culture of continuous improvement. *Transfusion, 51*, 1598–1603.

Rhamy, J. (2010). Synergies between blood center and hospital quality systems. *Transfusion, 50*, 2793–2797.

World Health Organization. *Aide-memoire. Quality systems for blood safety*. 2002. Available from: http://apps.who.int/medicinedocs/documents/s17255e/s17255e.pdf.

CHAPTER 3

Regulatory Issues in Transfusion Medicine

Eva D. Quinley, MS, SBB, CQA (ASQ) and Joseph Schwartz, MD, MPH

Transfusion medicine is a highly regulated industry, and regulatory oversight has and continues to be a major factor in most aspects of operations. Although in the United States, the Food and Drug Administration (FDA) is the primary regulator, a number of other state and federal agencies participate in regulating both blood banks and transfusion services. FDA registration and/or licensure is a requirement for those blood banks and transfusion services that manufacture blood and blood products or wish to participate in interstate commerce. Blood banks and transfusion services are equivalent to manufacturers of both drugs and biologics and must implement strict compliance with current good manufacturing practices (cGMPs) as found in both Parts 200 and 600 of the Title 21 of the Code of Federal Regulations (CFR). The FDA inspects these organizations to ensure compliance with the law and can issue sanctions if the level of noncompliance warrants such action (Fig. 3.1).

Increased regulatory pressures have prompted blood banks and transfusion services to make dramatic changes in the way in which they operate. To understand the evolution of regulation in transfusion medicine, it is helpful to understand the history of the regulation of biologics and drugs. Table 3.1 provides a brief outline of this history, including those applicable to transfusion medicine.

In the midnineties, the FDA began to inspect blood banks and, in some cases, transfusion services with intensified scrutiny. Furthermore, in 1993, the FDA issued a consent decree to the American Red Cross, which was followed by the levying of additional consent decrees in subsequent years to other blood collection organizations. In another move to emphasize its intent, the FDA, in 1995, published Quality Assurance Guidelines for Blood Establishments. These Guidelines outlined the FDA's expectations for quality assurance in blood banks and transfusion services. Thus, compliance became a major driver in the operations of blood banks and transfusion services. This focus continues today.

Regulation Versus Accreditation: *Regulation* and *accreditation* are not the same thing. Regulation involves laws (rules) that must be followed, while accreditation is a seal of approval from an independent accrediting body, such as AABB or The Joint Commission, certifying that an organization or individual has met specific standards. Accreditation is voluntary; regulation is law. Both accreditation and regulation are drivers of safety in transfusion medicine. It is interesting to note that in transfusion medicine, accreditation is frequently so essential that many accreditation requirements have the same effect as regulations, but accreditation is based on standards, which are not legally binding.

The Food and Drug Administration: The FDA's objective is to ensure a safe and effective blood supply and to provide regulatory oversight of the blood supply in the United States. This agency regulates blood establishments under the authority of Title

Transfusion Medicine and Hemostasis. https://doi.org/10.1016/B978-0-12-813726-0.00003-9

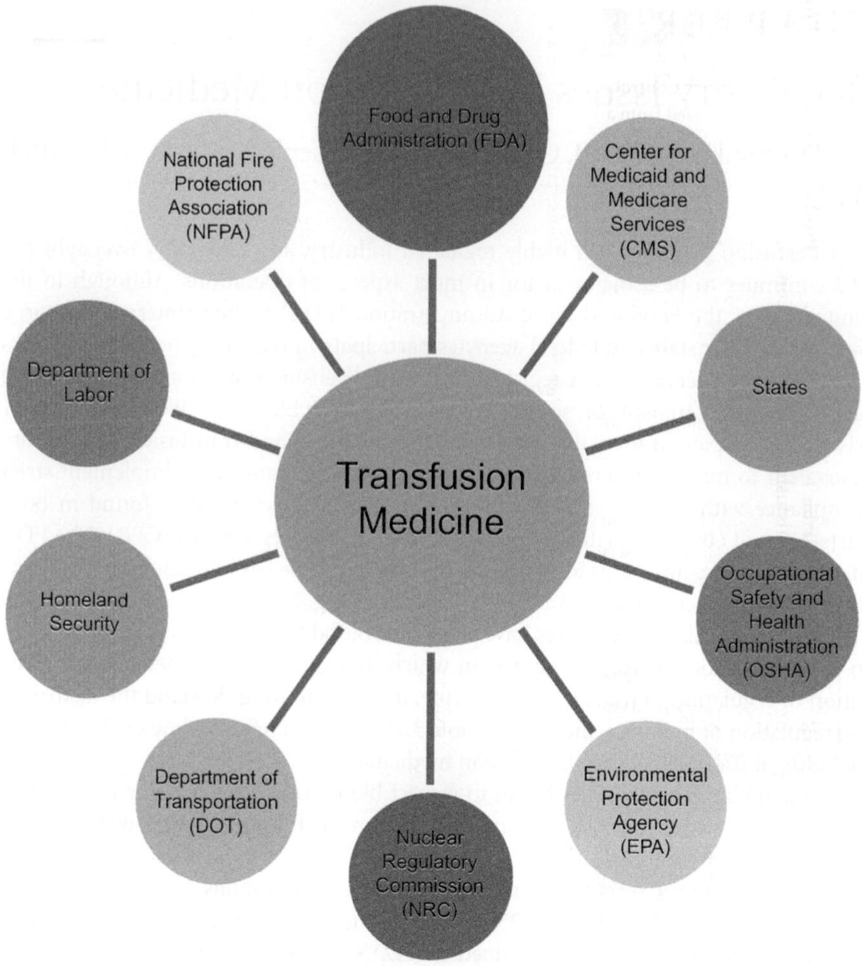

FIGURE 3.1 Agencies regulating transfusion medicine.

21 of the CFR, Parts 211 through 226 and Parts 600 through 680. These sections include good manufacturing practices for drugs and biologics. The descriptor *blood establishment* includes blood banks and hospital transfusion services as well as other facilities, such as plasma collectors and laboratories that perform testing on donor blood. Blood establishments that manufacture medical devices such as blood establishment computer software (BECS) are also regulated under Title 21 Part 820, but these regulations are not discussed in this chapter.

The FDA has three major functions:

1. Development and promulgation of regulations
2. Registration and licensure of blood establishments
3. Inspection and compliance actions

TABLE 3.1 History of United States Regulation of Transfusion Medicine

1902	Biologics Control Act passed as a result of 13 children contracting tetanus from a vaccine harvested from a tetanus-infected horse
1906	The Jungle—expose of unsafe practices in the meat-packing industry—Pure Food and Drug Act
1912	Public Health Service Act—blood defined as a biologic product
1937	Elixir sulfanilamide—use of diethylene glycol (analogue of antifreeze) in a drug leads to mass poisoning and deaths
1938	Food, Drug, and Cosmetic Act passed, requiring products must be tested and found safe
1944	Public Health Service Act requires licensure for biologic products
1956	AABB takes steps toward voluntary regulation of blood banks
1963	21 CFR 210 and 211 published
1972	Regulation of blood banks transferred to FDA
1975	FDA classifies blood as a drug and a biologic
1985	First FDA approved test kit for HIV
1988	Blood bank executes voluntary agreement with FDA relating to its computer system
1990	Increased FDA scrutiny of blood banks
1993	FDA issues a consent decree against the American Red Cross, the country's largest blood provider
1994	FDA requires 510 (k) clearance for blood establishment software (BECS)
1995	FDA issues its Quality Assurance Guideline for Blood Establishments
2015	After almost 23 years, the American Red Cross consent decree is lifted
2016	FDA mandates testing for the Zika virus under an IND, the first time a noncleared test was mandated

The FDA's Center for Biologics Evaluation and Research (CBER) works closely with other parts of the Public Health Service (PHS) to establish blood regulations and to identify and respond to potential threats to blood safety or supply. Regulations promulgated by the FDA are binding both on the agency and the blood establishment industry. The FDA also issues guidance documents that, while not laws, do represent the FDA's current thinking on the handling of a particular subject.

FDA oversight reflects and supports five overlapping layers of safety in the manufacture of blood and blood products: donor screening; testing of the blood for transfusion transmissible agents; maintenance of donor deferral lists; quarantine to prevent the distribution of unsuitable blood and blood products; and investigation of breaches in safety with required and appropriate corrective action.

Registration and Licensure: Section 510 of the Federal Food, Drug, and Cosmetic Act (the Act) requires drug and device manufacturers, including manufacturers of biological products, to register with the FDA. Title 21 CFR Part 607 lists and describes the registration requirements for manufacturers of human blood and blood products. Transfusion services are commonly registered, as opposed to being licensed, while most blood banks in which blood is collected are both licensed and registered. This has changed somewhat as transfusion services have adopted methodologies to test platelets on day 4 and 5 or to extend platelet life to 6 or 7 days. These transfusion services must now register with the FDA as a manufacturer. Unless a product crosses states lines, licensure is not required. Some transfusion services are exempt from registration if they do not engage in activities that the FDA has defined as requiring registration. Activities that are exempt include routine compatibility testing, product pooling, product thawing, and transfusion. Activities generally requiring registration include, but are not limited to, collection, processing (including product manipulations), testing, storage, and distribution.

Section 351 of the PHS Act requires specific licensure for some biological products, specifically those involved in interstate commerce. Blood banks that participate in interstate commerce must be licensed with the FDA, and licensure applies to both facilities and products. To obtain a license, a blood bank must submit a *Biologics License Application* (BLA). This may require the submission of data for agency review. The FDA will also conduct an inspection of the requesting facility before issuing a license. Once a license is granted, the FDA will routinely inspect the facility to ensure continued compliance with cGMPs.

Once approved, changes to a BLA require FDA notification and, in some cases, the changes may require FDA approval before an organization can implement them. The FDA considers changes in three broad categories. Those that have a substantial potential to adversely affect the final product are known as prior approval supplement changes; the FDA may take up to 12 months to act on such changes. Changes with a moderate potential to adversely impact the final product can be submitted in the second category, CBE 30. This is shorthand to describe a change to be effective not less than 30 days from submission to the FDA. Minor changes, such as a modification to standard operating procedures (SOPs) that do not affect a major manufacturing step, can be implemented immediately and must be described in the organization's annual report to the FDA.

Inspections and Compliance Actions: The Food, Drug, and Cosmetic Act provides authority for the FDA to inspect blood banks and transfusion services. Through a memorandum of understanding issued in 1980, hospital transfusion services that are exempt (as above) are covered by and may be inspected by the Center for Medicaid and Medicare Services (CMS), but the FDA reserves the right to also inspect these facilities. Under the oversight of the Office for Regulatory Affairs, FDA investigators, also known as consumer safety officers, inspect blood banks and transfusion services. The inspections are typically unannounced, but they are typically performed at reasonable times of day and in a reasonable manner. The FDA presents Form 482, Notice of Inspection, on arrival at a facility. This form provides information about the inspection, including the legal authority by which the inspection is authorized. If the investigator sees evidence of noncompliance with FDA regulations, a written statement of the noncompliance, referred to as an observation, is provided on Form 483 and reviewed with facility leadership at the conclusion of the inspection. Although a written response it not required by law, most blood banks and transfusion services will respond to the Form 483 by writing a corrective action plan. Once an inspection is completed, the FDA issues an Establishment Inspection Report, which is a narrative that provides a detail about a particular inspection. Routine inspections usually occur every two years, although the FDA can choose to inspect more frequently. The FDA may also conduct inspections to follow up on an issue from a prior inspection or as the result of a complaint.

Food and Drug Administration Sanctions: If the FDA finds that an organization is not compliant with the regulations, sanctions can be levied against the organization, part of the organization or even individuals within the organization. These sanctions fall into two categories: advisory/administrative and judicial. Advisory/administrative sanctions include Form 483 observations, recalls, warning letters, and intent to revoke or the actual revocation of an organization's license. Judicial sanctions include product seizure, injunctions, consent decrees, fines, and criminal prosecution. These more serious sanctions are issued only when there is continued noncompliance and/or disregard for the regulations by an organization's management.

Current Good Manufacturing Practices: cGMPs applicable to blood banks and transfusion services are found in 21 CFR, Parts 211 and 606. These represent minimal standards expected by the FDA to ensure that blood and blood products and services are produced in a manner that protects the safety, quality, identity, purity, and potency (SQuIPP) of the product or service. A third set of regulations, formerly known as the medical device GMPs but now known as the Quality System Regulations, are applicable to blood organizations only if they manufacture a medical device, such as a BECS. These regulations are found in 21 CFR 820 but as indicated earlier are not discussed in this chapter. The FDA inspects blood banks and transfusion services for compliance with cGMPs (Table 3.2).

GMPs are applicable to those blood banks and transfusion services that perform manufacturing. In 21 CFR 600.3(u), the FDA defines *manufacturing* as: all steps in propagation or manufacture and preparation of products and includes but is not limited to filling, testing, labeling, packaging, and storage by the manufacturer. cGMPs are minimal requirements, and strict compliance is expected. The following sections briefly highlight cGMPs areas found in both the 200 and 600 series of 21 CFR that are applicable to blood banks and transfusion services. The reader is referred to the CFR for specific regulations.

Organization and Personnel: Both blood banks and transfusion services rely on many human resource factors to meet regulatory requirements. Individuals who work in blood banks and transfusion services must be adequate in number and must have education, training, or experience to perform their assigned functions. They should receive not only job skills training but also training in cGMPs, and, if in a supervisory role, in those managerial skill sets as well. 21 CFR 21.22 addresses the requirement for a *quality control unit* that has the responsibility and authority to reject unsuitable

TABLE 3.2 Areas Covered by Current Good Manufacturing Practices

Documentation	Training
Validation	Deviation management
Standard operating procedures	Labeling
Facilities	Auditing
Equipment	Process control
Personnel	

components of manufacturing or unsuitable products that are manufactured. It also addresses the need to describe in writing the responsibilities of the quality control unit and the procedures that it follows. The Guideline for Quality Assurance in Blood Establishments provides the FDA's thinking on the roles and responsibilities of the quality assurance or quality control unit. Lastly, consultants used by blood banks and transfusion services should have education, training, or experience adequate to advise on the subject for which they are retained. Access to the manufacturing areas should be limited only to those who have justified reason to be there.

Facilities: Work environment is also a key consideration. cGMPs require that a facility has adequate space to prevent mix-ups, both in the manufacture of blood and blood products, their quarantine until all testing is complete, their storage until release, and in the quarantine and release of incoming materials. Adequate space for private and accurate examination of donors to ensure that they are eligible is another requirement. Inadequate space to provide confidential donor screening is a common FDA observation.

Facilities must be maintained and cleaned according to written procedures. Maintenance and cleaning actions must be documented, and records must be retained. The regulations also require adequate lighting and handwashing/toilet facilities. Disposal of trash and unsuitable blood and blood components must occur in a safe and sanitary way.

Equipment: Equipment used in blood banking and transfusion services impacts the quality of products and services. Equipment used in the manufacturing process must be maintained and cleaned according to written procedures and must be qualified for its intended use. In other words, the equipment is tested to prove that it can do what it is supposed to do as described by the manufacturer. If the equipment is used for measurement, it must also be calibrated against a known standard to ensure it is working appropriately. Records must be maintained for all of these activities.

With computers or related systems, 21 CFR 211 addresses requirements to maintain control over such systems, including their input and output. Control includes requiring verification of the input and output on a frequency that is sufficient to ensure reliability of the data and a mechanism to back up data.

Production and Process Controls: A fundamental quality and compliance concept is the need for written procedures for those activities that constitute manufacturing in transfusion medicine, including those related to manufacturing blood and blood products and those related to investigation of deviations. 21 CFR 606.100 provides a list of required procedures (Table 3.3). Documented training on these procedures is necessary for personnel who perform the tasks within the procedures.

The FDA allows organizations to develop their own procedures or to use procedures from another organization as long as those procedures are compliant with cGMPs. The most common FDA observation is failure to follow SOPs.

Validation: A process must produce consistent results over time. The FDA requires that processes must be validated for their intended use. *Validation* provides a high

TABLE 3.3 Examples of Procedures Required in Transfusion Medicine

Recruitment of donors	Deviation management, including adverse events
Collection, including donor screening and selection	Reporting
Processing, including testing	Quality systems (validation, document management)
Compatibility testing, including sample collection	Facilities management
Labeling	Equipment management
Storage	Cleaning
Distribution	

degree of assurance that a process is working and consistently producing a desired result. Sometimes, validation is confused with *qualification*. Validation challenges the process to determine, with a high degree of assurance, that the process is working and consistently producing a desired result. Qualification is usually a one-time activity. With validation, the key concept is repeatability, the ability to determine that consistent, reliable results can be achieved over and over again. This includes worst-case scenarios, or testing at the limits, to aggressively challenge the process and each of its steps. For example, if a transport box is intended to contain and convey a blood product and maintain a specific temperature, validation testing would be done at the extremes of expected temperatures (hot and cold). Once a process has been validated, controls must be in place to ensure that the process is maintained in a validated state. While validation provides a high degree of assurance, the desired results are only repeatable when procedures are followed to the letter. Otherwise, the results are suspect. The human factor in transfusion medicine is a huge variable in maintaining a validated state. Well-written SOPs, well-trained staff who understand the importance of their work, and good supervision are the best ways to control this variable.

Laboratory Control: Testing of blood products is a key requirement addressed in 21 CFR 606, Subpart H. This includes ensuring that samples are identified appropriately. This section of the regulations also addresses compatibility testing, requiring written procedures and test methods to determine incompatibility between the recipient and the donor as well as a method for handling the need to expedite transfusion in life-threatening emergencies.

Labeling: Labeling supports the identity of the product, also a fundamental requirement. cGMPs require that labeling be performed in an area that is physically or spatially separated from operations. This requirement points to the importance of error-free labeling. cGMPs require strict control over the labeling process, from the receipt of the labels to their application. Labels must be inspected on receipt and found to be accurate and complete before use. Labels should be stored in a way that prevents mix-ups, and access to the labels should be limited to those who are trained to perform labeling operations. There should also be a mechanism to control and reconcile labels used against labels received. This reduces the likelihood of a label mix-up. A misplaced label could have serious consequences if not discovered.

A blood bag label must meet specific requirements. The label must contain certain information that supports the concept of traceability back to a donor. 21 CFR 640 provides additional labeling requirements for blood and blood products. A circular of information that provides instructions for use of the blood product is required and also is considered part of the labeling activity. AABB has produced a standard circular of information (Circular of Information for the Use of Human Blood and Blood Components; http://www.aabb.org/resources/bct/Documents/coi0809r.pdf), which is revised regularly and commonly used in the industry.

Records and Reports: Documentation must be timely, accurate, and complete. Records should be created concurrently with work performance and should be legible and recorded in indelible ink to stand the test of time. Records should allow for complete traceability (details such as who, what, when, and where) and trackability (follow a logical sequence of steps). 21 CFR 606, Subpart I, provides a list of the types of records required by cGMPs for blood bank and transfusion services. In addition to production records, cGMPs require that records are maintained of adverse reactions and the investigation of those reactions. Reports should accurately reflect information, and queries from electronic systems should be tested to ensure that all required data are present.

Deviations: A deviation is a departure from procedures or other requirements. cGMPs require that records are maintained related to deviations, their investigation, and corrective actions. Blood banks and transfusion services are expected to have systems in place to detect, document, investigate (including root cause analysis), correct, and monitor all deviations, whether or not the deviations have a negative impact on the product. The FDA understands that there will be deviations, but it is important to the FDA that organizations learn from their deviations so that they will not recur. The ultimate goal is the prevention of other similar deviations. The reporting of certain types of deviations is addressed in 21 CFR 606.171. Licensed manufacturers, unlicensed registered blood establishments, and transfusion services must report certain deviations and relevant information about the event within 45 calendar days of the date of discovery. Table 3.4 lists the criteria for reporting a deviation.

Laboratory Laws and Regulations: CMS regulates all medical laboratories in the United States in accordance with Section 353 of the PHS Act, as amended by Clinical Laboratory Improvement Amendments (CLIA). CLIA regulations are found in 42 CFR 493. These regulations provide the requirements for certification and reimbursement for Medicare/Medicaid payments. They provide minimal standards for facilities, equipment, and personnel and also require participation in a proficiency testing program. Inspections by CMS or one of its "deemed status" partners occur every two years. The level of personnel required is based on the level of complexity of the testing performed in an organization. Tests are classified as either waived, moderate complexity or high complexity. An example of a moderate complexity test is compatibility testing using manual reagents. Infectious disease testing is classified as high complexity. CMS can remove certification or impose fines for failure to comply with the regulations.

TABLE 3.4 Criteria for Reporting of Deviations

1. Either
   (i) represents a deviation from current good manufacturing practice, applicable regulations, applicable standards, or established specifications that may affect the safety, purity, or potency of that product; or
   (ii) represents an unexpected or unforeseeable event that may affect the safety, purity, or potency of that product.
2. Occurs in your facility or another facility under contract with you.
3. Involves distributed blood or blood components.

Other Regulatory Agencies: Many states have regulations of their own, which apply to blood banks and transfusion services. These states usually perform their own inspections and have sanctions for failure to comply. Other agencies such as the Department of Transportation (DOT), the Occupational Safety and Health Administration (OSHA), the Nuclear Regulatory Commission (NRC), and the Environmental Protection Agency (EPA) also regulate transfusion medicine. It is beyond the scope of this chapter to discuss each of these in detail; the reader is referred to the relevant agency's website for more information.

Further Reading

Berte, L. M. (1995). A quality system for transfusion medicine … and beyond. *Med Lab Obs*, *27*, 61–64.

Code of Federal Regulations Title 21 parts 600, 640, 210, 211. (2009). Washington, DC: United States Government.

Food and Drug Administration. (1995). *Quality assurance guideline for blood establishments*. Rockville, MD: CBER Office of Communication, Training, and Manufacturers Assistance.

Shoos-Lipton, K., & Otter, J. (2010). AABB and FDA: A shared history of patient safety. *Transfusion*, *50*, 1643–1646.

CHAPTER 4

Role of the Physician in the Blood Center

Jed B. Gorlin, MD, MBA

Physicians play a central role in blood centers with direct and indirect responsibilities. The physician's role actively participates in the blood center activities, including collection, manufacturing, storage, and distribution of biologics (cellular therapies and blood components), and, in some cases, infectious disease testing, reference laboratories, therapeutics (apheresis, cell salvage), and research (Fig. 4.1). Blood centers serve a humanitarian need, and mission statements of blood centers often include performing at the highest level regarding donor and patient care. However, blood centers are corporate entities and have financial responsibilities. As physicians, the ultimate responsibility is to advocate for the best donor and patient outcomes and not blood center revenues (Box 4.2).

Blood Center Table of Organization: Physicians in blood centers are often given the title of Chief Medical Officer (CMO) or Medical Director with a reporting line to the Chief Executive Officer (CEO). They may have business units or areas where they are directly responsible; however, their responsibility often extends beyond documented lines of authority as they are responsible for medical aspects of the organization. Furthermore, the CMO may also be CEO, Clinical Laboratory Improvement Amendments (CLIA) director, and/or FDA responsible head.

Specific Roles: Blood collection is performed under the direction of the medical director. The physician must be qualified by education, training, and/or experience and licensed. The physician is responsible for donor selection and safety, including donor eligibility For example, New York state has specific requirements for medical director in that they are responsible for direction and operation of the blood bank, compliance with all applicable regulations and requirements, establishment and implementation of written standard operating procedures, ensuring competency of and delegating responsibilities to blood bank staff, investigation of cases of possible transfusion-transmitted diseases, and reporting adverse events as required by regulatory agencies and accrediting organization. The medical director per American Association of Blood Banks has responsibility and authority for all medical and technical policies, processes, and procedures, and for the consultative and support services that relate to the care and safety of donors and/or transfusion recipients. As such, physician's roles may include determining appropriate equipment, adjusting donation intervals, methods to decrease donor or patient adverse events, and implementing new donor tests or testing algorithms.

Quality Systems and Laboratories: The blood center physician is directly involved in the quality systems function of the institution, and again he or she is well suited to understanding the impact of procedural changes and evaluating product acceptability following deviations from procedures or errors. This last responsibility is often

Transfusion Medicine and Hemostasis. https://doi.org/10.1016/B978-0-12-813726-0.00004-0

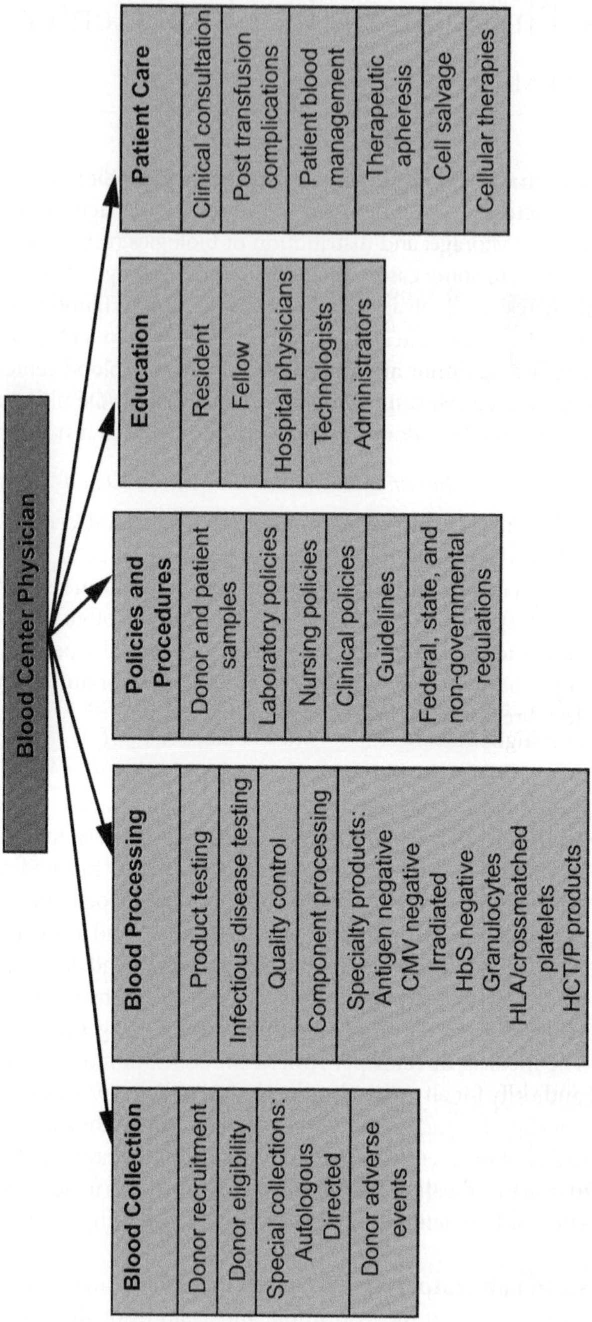

FIGURE 4.1 Blood center physician oversight.

BOX 4.1 Material Review Board Process

- Regulatory/quality leadership: typically, personnel within the regulatory or quality department of a blood center will be responsible for the MRB process overall and focus on regulatory, FDA-related, and procedural aspects of deviations and incidents that occur.
- Medical office leadership: typically the CMO, the medical office element of the MRB will focus on medical, SQuIPP, and specific donor or patient issues as it affects donor, unit, and patient safety.
- Operations leadership: typically the COO, the operational element of the MRB will focus on the assessment of impact to inventory.
- Administrative leadership: typically the CEO, the administrative element of the MRB is available if consensus cannot be reached.

BOX 4.2 Model Blood Center Mission Statement

To provide the highest level of transfusion care and therapy to patients via optimal collection, process and distribution of safe, effective, and available cellular therapy and blood products, with respect, appreciation and care for the donor, and excellent customer service.

accomplished by a material review board (MRB) process (Box 4.1). The physician is also directly involved with inspections and accreditation, internal audits and problem management, and the plan(s) for resolution of any noted deficiencies.

Risk Management: Physicians are often called on to help mitigate risk to the blood center by ensure adherence to regulatory and community standard requirements. Having physician input on development of donor education materials may assist in diminishing adverse donor reactions. Finally, physicians should review the consent statement to ensure it highlights both the common risks (fainting and falls) and less common but more severe outcomes (nerve injury).

Clinical and Laboratory Services: These may include the immunohematology reference (IRL), RBC genomics, histocompatibility (HLA), platelet laboratories, clinical apheresis services, hematopoietic progenitor cell/tissues (HCT/P) collection, cellular therapies laboratory, and perioperative transfusion services. The IRL and HLA laboratory are regulated under current good manufacturing practices and good laboratory practices and require CLIA certification. HCT/P collection and processing are regulated under cGTP (current good tissue practices) and the FDA. Therapeutic apheresis is considered the practice of medicine and is not under the regulatory authority of the FDA. The medical director usually serves also as CLIA laboratory director, holds states' licensure, and may require other certifications.

Clinical Apheresis: Physicians determine procedural aspects of clinical services (therapeutic apheresis, HCT/P, and other cellular therapies apheresis collections) and provide consultation to ordering physicians (the ordering physician is responsible for patient care). Clinical apheresis is performed by registered nurses or specially qualified medical technologists, and the procedures may occur outside the blood center.

Immunohematology Reference, Platelet, Histocompatibility, and Hematopoietic Progenitor Cell/Tissues Laboratories: In the IRL, the physician may contribute both medical and technical input, including interpretation and decision-making regarding immunohematology and RBC genomics testing, blood typing (both phenotyping and genotyping), and testing for unexpected antibodies. The physicians may provide clinical consultation to hospital transfusion medicine and transfusing clinicians regarding complicated donor–recipient matching, platelet transfusion refractoriness, sickle cell disease, and posttransfusion complications.

Research: Blood centers offer rich environments for research and development (the Blood Research Institute of the BloodCenter of Wisconsin, the Jerome H. Holland Laboratory of the American Red Cross, the Blood Systems Research Institute of Blood Systems, and the Lindsley F. Kimball Research Institute of New York Blood Center). The medical director may need to act as an advocate for continued funding of research efforts. Major NIH-funded research programs could not function without significant blood center and blood center physician/scientist involvement, such as the Recipient Epidemiology and Donor Evaluation Study (REDS-III) consortium (https://reds-iii.rti.org/), which includes domestic and international efforts dedicated to blood donor and recipient safety and blood availability. The physician may act as an advisor or interpreter for local researchers' requests for blood components and samples for research, compliance with donor consents for research, and the blood centers role in community health.

Further Reading

McCullough, J. (2006). The role of physicians in blood centers. *Transfusion, 46*, 854–861.

McCullough, J., Benson, K., Roseff, S., et al. (2008). Career activities of physicians taking the subspecialty board examination in blood banking/transfusion medicine. *Transfusion, 48*, 762–767.

Shaz, B. H., & Hillyer, C. D. (2010). Transfusion medicine as a profession: Evolution over the past 50 years. *Transfusion, 50*, 2536–2541.

CHAPTER 5

Blood Donation

Mary Townsend, MD and Salima Shaikh, MD

In the United States, ~14 million units of whole blood (WB) are collected each year from ~7 million donors. The blood is processed into blood components (RBCs, platelets, plasma, and cryoprecipitate) and used in the treatment of surgical, obstetric, medical, and other patients. The blood donation process includes donor recruitment, eligibility, screening, collection, and postdonation care and is (1) overseen by a licensed, qualified physician, (2) regulated by the FDA, AABB (and state), and (3) designed to provide protection for both the blood donor and the transfusion recipient by ensuring appropriate donor eligibility, donation procedures, and testing.

Currently, there are no FDA-approved, or clinically equivalent, substitutes for blood. One blood substitute, a hemoglobin-based oxygen carrier, is available for "compassionate use" only, for certain indicated populations such as Jehovah's witnesses and others who refuse blood transfusions, and sickle cell and other patients who need rare antigen-negative blood that is not available. Indeed, without blood donations of RBCs, platelets, and plasma, modern transfusion therapy, which allows complex surgery, damage control resuscitation for trauma, transplantation of major organs, care of the complicated postpartum patient, and chemotherapy treatment for cancer patients, would not be possible. Thus, the blood donor and blood donation is critical to the function of advanced healthcare systems.

Blood Donors: The eligible donor pool was estimated to be 41% of all US individuals, with lower eligibility rates for African-Americans compared with whites, and females compared with males. Blood donation is considered an act of selflessness or altruism that has little tangible benefit to the donor. Thus, blood donors are afforded special protection under federal and state regulations, and via cautious and confidential practices of blood collection facilities.

Changes in population demographics and blood center outreach programs have led to increased participation in community blood donation programs by all racial and ethnic groups, but minorities have significantly lower participation. Understanding motivation and barriers to blood donation, both in general and based on cultural differences of various groups, is essential for an adequate blood supply. Motivators for blood donation include altruism, humanitarianism, awareness of need, sense of social obligation, social pressure, need to replace blood used, and increased self-esteem. Barriers include fear, inconvenience, not having time, perceived (or real) medical disqualification, not knowing there is a need for blood, distrust, and apathy. Repeat blood donors are advantageous, as they have fewer adverse reactions and lower risk of transmitting disease. Of special concern are the shifting demographics by which older, repeat donors are reaching the age at which they are moving from donors to recipients. Efforts to recruit new donors, especially young donors, are hampered by changing societal norms regarding volunteerism and societal duty.

Transfusion Medicine and Hemostasis. https://doi.org/10.1016/B978-0-12-813726-0.00005-2

Donation Process: Blood donation is heavily regulated by AABB and FDA (and some states), and consists of sequential steps before, during and after phlebotomy to ensure both donor and recipient safety. These steps include recruitment materials, educational materials, donor health history questionnaire (DHQ), physical examination, informed consent, phlebotomy, postdonation care, and management of donor reactions. AABB website contains FDA-approved DHQ, education materials, medication deferral list, user brochure, and other donor eligibility materials (http://www.aabb.org/tm/questionnaires/Pages/dhqaabb.aspx).

Recruitment Materials: Donor recruitment materials (1) highlight reasons to give blood using phrases such as "give the gift of life" or "save a life," (2) include donation criteria such as weight, age, and common reasons for deferral, and (3) may offer donation incentives, such as free health screening or small gifts.

Educational Materials: The AABB task force created blood donation education materials, which provide an overview of the donation process and donor eligibility, focusing on high-risk behaviors, signs/symptoms of AIDS, infectious disease transmission, and blood donation risks. Potential donors are informed of the importance of providing accurate information and withdrawing themselves from the donation process if appropriate to maximize recipient safety. Educational materials may also include fact sheets for emerging and potential transfusion-transmitted infectious agents (http://www.aabb.org/tm/eid/Pages/default.aspx). In addition, AABB recommends that blood collectors provide donors with information on iron stores and blood donation, including facts on normal iron intake, replacement of iron lost during blood donation via healthy diet and/or iron supplements, and symptoms of iron deficiency.

Properly designed educational materials may mitigate donor adverse reactions by providing information regarding the process of blood donation (to decrease anxiety) and methods to decrease vasovagal reactions.

Registration: Donors must provide their name, address (to contact donor for future donations or to provide test result or other information), and date of birth. The FDA Final Rule in the Code of Federal Regulations (CFR) states that the blood collector "must obtain proof of identity of the donor and a postal address where the donor may be contacted for 8 weeks after donation" (21 CFR § 630.10 2015). At the time of presentation, centers search their donor deferral database to prevent an individual who is ineligible from donating. Blood centers are required to be able to completely trace all donations from donor to hospital, which traces product to recipient.

Health Screening: Determination of donor eligibility lies at the core of donor and blood safety. This is a multistep process that uses mini physical exam (vital signs, weight) and questionnaire to assess well-being, medications, vaccinations, medical conditions, and risk behaviors.

Mini Physical Exam: This screens for conditions that could adversely impact donor (hemoglobin/hematocrit, blood pressure, pulse) and recipient (skin lesions or stigmata of intravenous drug use) (Table 5.1).

TABLE 5.1 Donor Criteria for Donation

Reference Standard 5.4.1A—Requirements for Allogeneic Donor Qualification	
Criteria	**Acceptable Value(s)**
Age	≥17 years or conforming to applicable state law (many states allow 16-year-old donors with parental permission or consent)
Weight	≥110 lbs; if weight <110 lb, then a maximum of 10.5 mL/kg, including samples, can be collected. Anticoagulant in the bag must be adjusted if less volume is drawn
Donation interval	≥56 days after whole blood donation
	≥112 days after 2-unit RBC donation
	≥28 days after infrequent plasmapheresis
	≥2 days after plasma-, platelet-, or leukapheresis
Hemoglobin; hematocrit	≥12.5 g/dL; 38% (not by earlobe puncture) for females
	≥12.0 g/dL; 36% (not by earlobe puncture) for females with FDA approval
	≥13.0 g/dL; 39% (not by earlobe puncture) for males
Blood pressure (required by FDA to be within normal limits, no AABB requirement; standard values given)	90–180 mmHg systolic 50–100 mmHg diastolic
Heart rate required by FDA to be within stated range; no AABB requirement)	50–100 bpm <50 bpm if otherwise healthy athlete (responsible blood center physician must be contacted on first occurrence)
Temperature	≤37.5°C (99.5°C) measured orally
Antecubital fossa	Free of lesions, "track marks," scars (signs of IV drug use)

Modified from *AABB Standards for Blood Banks and Transfusion Services* (31st ed.).

Donor History Questionnaire: The DHQ contains questions asked of donors before starting each donation process. The Uniform Donor History Questionnaire (uDHQ), developed by an interorganizational task force, is recommended by AABB and endorsed by the FDA in May 2016 in a guidance https://www.fda.gov/downloads/BiologicsBloodVaccines/GuidanceComplianceRegulatoryInformation/Guidances/Blood/UCM273685.pdf. While it is not required that the uDHQ is used, center-specific DHQs, which can be in verbal, written, or electronic format, must comply with FDA regulations and AABB Standards and may be more (but not less) restrictive. Questions cover topics intended to protect both donor and recipient health (Table 5.2).

The FDA Final Rule (21 CFR § 630.10 201), implemented in May 2016, changed some longstanding donor eligibility criteria. Changes include infectious disease criteria (elimination of the donor deferral for hepatitis after the 11th birthday), donor vital sign criteria (minimum and maximum values for blood pressure and pulse, and responsible physician oversight of approval for donors with values outside these limits), requirement for blood centers to have procedures to control platelet bacterial contamination risk, and record keeping for deferred donors. Eligibility criteria are subject to change, as emerging infectious diseases are recognized and labeled as threats to blood safety. Some

TABLE 5.2 Donor Deferrals for Infectious Disease Risk and Donor Safety

Deferral Period	Criteria	Rationale
Indefinite	Stigmata of Intravenous Drug Abuse (IVDA)	Infectious disease
	Present or past clinical or lab evidence of HIV, HBV, HCV, HTLV	
	History of babesiosis or Chagas	
	Use of needle to inject nonprescribed drugs	
	Receipt of money, drugs, or payment for sex	
	Receipt of a xenotransplant	
	Blood relative of someone with CJD	CJD
	Receipt of a dura mater graft	vCJD
	Time in Europe >5 years cumulative	
	Time in United Kingdom >3 months cumulative	
	Member of US military, civilian employee, or dependent with time in specific military bases	
	Receipt of a transfusion in United Kingdom or France	
3 years	Diagnosis of malaria	Malaria
	Lived 5 consecutive years in a malarial country (see AABB malarial flowchart)	
12 months	Receipt of a blood transfusion	Infectious disease
	Receipt of organ, tissue, bone marrow, or graft of human origin	
	Exposure to someone else's blood	
	Sexual contact with HIV, HBV, symptomatic HCV	
	Sexual contact with a prostitute	
	Sexual contact with someone who used a needle to inject nonprescribed drugs	
	Male with sexual contact with another male	
	Sexual contact of a female with a male who had sex with another male	
	Ear or body piercing not performed in a regulated state using sterile needle(s) and nonreusable ink	
	Diagnosis or treatment for gonorrhea or syphilis	
	Incarceration for more than 72 consecutive hours	
	Travel to a malarial endemic area (see AABB malarial flowchart)	Malaria
16 weeks	Donated apheresis double red cell	Donor safety
8 weeks	Donation of whole blood, platelets, plasma (variable deferrals apply based on combination of component)	
6 weeks	Pregnant now or in the past 6 weeks	
48 hour	Taken aspirin or an aspirin-containing medication	Platelet function
Variable	Had problems with heart or lung, cancer, blood or bleeding disorder	Donor safety, other

TABLE 5.3 Medication Deferrals

Indication/Type of Medication	Name of Medication	Deferral Period
Antiplatelet agent	Feldene	2 days
	Effient, Brilinta	7 days
	Pavix, Ticlid, Zontivity	14 days
Anticoagulant	Xarelto, Fragmin, Lovenox, Pradaxa, Eliquis, Savaysa	2 days
	Coumadin, Warfilone, Jantoven, Heparin, Arixtra	7 days
Acne treatment	Accutane, Amnesteem, Absorica, Caravis, Myorisan, Sotret, Zenatane	1 month
Hair loss remedy	Propecia	1 month
Prostate symptoms	Proscar	1 month
	Avodart, Jalyn	6 months
Basal cell skin cancer	Erivedge	24 months
Relapsing multiple sclerosis	Aubagio	24 months
Psoriasis	Soriatane	36 month
	Tegison	Permanent
Hepatitis exposure	Hepatitis B immune globulin	12 months
Experimental medication or unlicensed vaccine		12 months
Growth hormone from human pituitary glands		Permanent
Insulin from cows (bovine or beef insulin) manufactured in the United Kingdom		Permanent

deferral periods are set by AABB Standards and/or FDA Guidance while others are at medical director discretion.

General well-being: Questions to determine general well-being include those to confirm the donor is feeling healthy and well, not pregnant or recently pregnant, and free of cancer, heart or lung conditions, or bleeding disorders.

Medications: Medication deferral list is mostly comprised of medication that can potentially harm recipients, such as teratogens, or reduce product potency, such as anticoagulants and antiplatelet drugs (Table 5.3). Deferral periods after discontinuing specific drugs vary depending on pharmacokinetics. Medications other than those shown on Table 5.3 are at the medical director discretion.

Vaccinations and immunizations: See Table 5.4.

Abbreviated Donor History Questionnaire: Introduction of an abbreviated questionnaire resulted from longstanding requests by frequent donors for which the repetitiveness and length of questionnaire was a disincentive to donation, and from studies by behavioral scientists showing that recall was improved for more recent behaviors. Studies by Kamel et al. demonstrated increased donor satisfactions using such an approach. In 2011, the FDA issued a draft guidance (updated in 2016) accepting the AABB abbreviated DHQ (aDHQ). Use of the aDHQ is for frequent donors who have donated at least two times using the full-length form, the last donation being within 6 months.

TABLE 5.4 Donor Deferral Period for Vaccinations

Vaccine	Deferral Period
Receipt of toxoids, or synthetic or killed viral, bacterial, or rickettsial vaccines if donor is symptom-free and afebrile (anthrax, cholera, diphtheria, hepatitis A, hepatitis B, influenza, Lyme disease, paratyphoid, pertussis, plague, pneumococcal polysaccharide, polio [Salk/injection], rabies, Rocky Mountain spotted fever, tetanus, typhoid [by injection])	None
Receipt of recombinant vaccine	Refer to AABB *Standards for Blood Banks and Transfusion Services*, 30th edition
Receipt of live attenuated viral and bacterial vaccines (measles [rubeola], mumps, polio [Sabin/oral], typhoid [oral], yellow fever)	2 weeks
Receipt of live attenuated viral and bacterial vaccines (German measles [rubella], chicken pox [varicella zoster])	4 weeks
Smallpox	Refer to most recent FDA Guidance
Receipt of other vaccines, including unlicensed vaccines	12 months unless otherwise indicated by medical director

From *AABB Standards for blood banks and transfusion services* (30th ed.).

Donor Informed Consent: Donor informed consent is an essential part of the donation process. While AABB Standards spell out minimal requirements for consent, getting informed consent is a legal process subject to requirements specified by the state and FDA. AABB Standards states that informed consent must be performed on the day of donation, before collection and must contain at a minimum: procedure elements, procedure risks, tests performed, and requirements to report donor information including test results to state or local health departments as required by state law. The donor should be given a chance to ask questions and have them answered as well as the ability to withdraw from the donation process at any time. This information may be presented in the educational materials. Some states require that minor donors (i.e., those under the age of 18) have a parent or guardian give written consent to donate.

Technical Aspects

Identification: Identification of the donor, and all donated blood product(s), from collection to final disposition, is critical to the safety and protection of both donor and recipient. Blood collection bags and blood samples used for testing, and all components that are prepared from the initial donation must be labeled in accordance with ISBT 128 requirements and must be trackable back to the source donor through the donor record.

Venipuncture Site Preparation: The antecubital area is the best site for locating a suitable vein from which to collect blood. This area must be free of skin lesions. An inspection of both antecubital fossa is required to identify any skin disease, scarring, or evidence of intravenous drug use. The use of a tourniquet or a blood pressure cuff inflated to 40–60 mmHg, together with having the donor open and close the hand, can enhance the prominence of the veins. The blood pressure cuff or tourniquet should be removed before preparing the skin for phlebotomy. Adequate skin cleansing and diversion pouches are required to protect against bacterial contamination.

Phlebotomy: Blood must be collected by a trained phlebotomist using aseptic methods, and a sterile, closed system. The blood unit should be collected from a suitable vein in a single venipuncture after the pressure device (blood pressure cuff or tourniquet) has been deflated and reinflated. WB is collected into a closed blood collection set consisting of a sterile collection bag containing anticoagulant (typically 70 mL of anticoagulant in a container approved to collect 500 ± 50 mL of WB), integrally attached tubing, and a large bore needle.

The total amount collected from the donor, including segments and specimen tubes, should not exceed 10.5 mL/kg of donor weight. During collection, blood should be thoroughly mixed with anticoagulant to prevent clotting, either manually or automatically.

Postdonation Care: A pressure dressing is applied to venipuncture site for hemostasis. The donor should proceed to the refreshment area to hydrate and rest for 15 minutes. In addition, the donor should receive postdonation instructions, which include care of venipuncture site, hydration recommendations, appropriate postdonation activity level, and actions to take in the event of a reaction. Contact information for the donor center should be provided to report any adverse reactions, to modify his or her donor history information, or to request his or her blood not be used for transfusion.

Further Reading

Eder, A., Dy, B., Kennedy, J., Notari, I. V. E., Strupp, A., Wissel, M., et al. (2008). The American Red Cross donor hemovigilance program: Complications of blood donation reported in 2006. *Transfusion, 48*, 1809–1819.

France, C., France, J. L., Himawanl, K., et al. (2012). Assessment of donor fear enhances prediction of presyncopal symptoms among volunteer blood donors. *Transfusion, 52*, 375–380.

Fridey, J. L., Townsend, M. J., Kessler, D. A., & Gregory, K. R. (2007). A question of clarity: Redesigning the American Association of Blood Banks blood donor history questionnaire – a chronology and model for donor screening. *Transfus Med Rev, 21*, 181–204.

Kamel, H. T., Bassett, M. B., Custer, B., et al. (2006). Safety and donor acceptance of an abbreviated donor history questionnaire. *Transfusion, 46*, 1745–1753.

Newman, B. H., Pichette, S., Pichette, D., & Dzaka, E. (2003). Adverse effects in blood donors after whole blood donation: A study of 1000 blood donors interviewed 3 weeks after whole blood donation. *Transfusion, 43*, 598–603.

Newman, B. H., Satz, S. L., Janowicz, N. M., & Siegfried, B. A. (2006). Donor reactions in high-school donors: The effects of sex, weight, and collection volume. *Transfusion, 46*, 284–288.

Riley, W., Schwei, M., & McCullough, J. (2007). The United States' potential blood donor pool: Estimating the prevalence of donor-exclusion factors on the pool of potential donors. *Transfusion, 47*, 1180–1188.

Steele, W. R., Schreiber, G. B., Guiltinan, A., et al. (2008). Retrovirus Epidemiology Donor Study. Role of altruistic behavior, empathetic concern, and social responsibility motivation in blood donation behavior. *Transfusion, 48*, 43–54.

Wehrli, G., & Sazama, K. (2010). Universal donor education and consent: What we know and where we should go. *Transfusion*, *50*, 2499–2502.

Zou, S., Eder, A. F., & Musavi, F. (2007). Implementation of the Uniform Donor History Questionnaire across the American Red Cross Blood Services: Increased deferral among repeat presenters but no measurable impact on blood safety. *Transfusion*, *47*, 1990–1998.

CHAPTER 6

Apheresis Blood Component Collections

Mary Townsend, MD and Marissa Li, MD

In the United States, 14% of RBC, 12% of plasma, and 91% of platelet products collected in 2015 were collected by apheresis procedures (for granulocyte products, see Chapter 7).

Manufacture of blood components by apheresis provides many advantages over whole blood (WB) collection. Apheresis collections are already leukoreduced, available for component therapy without further processing, and product type, volume, dose, and number controlled. Apheresis collection is an effective and efficient method to maximize blood component collection from a single donor by collecting products that are most needed. Using apheresis technologies allows blood centers to collect multiple components during a single donation (both different components and multiple units of the same component) with an added benefit of cost reduction of individual donor testing. Finally, use of a smaller caliber needle may enhance donor comfort.

Disadvantages of apheresis collection include increased expense of kits (which may be offset by overall increase in efficiency), additional phlebotomist training, inability to collect large number of products quickly due to procedure time, and some adverse events not observed in WB collection, i.e., citrate-related reaction and infiltration.

Most regulations and standards are the same for donors undergoing apheresis as WB collection (Chapter 5); however, there are some unique qualifications. Maximum allowable rolling 12-month loss is 1400 mL of RBCs and, based on donor weight, 12,000 mL (for donors weighting ≤175 lb) and 14,400 mL (for donors weighing >175 lb) of plasma.

RBC Apheresis Collections: RBC apheresis allows for collection of two RBC products (double RBC procedure). In contrast to WB, minimum hemoglobin and hematocrit for double RBC collections are higher, at 13.3 g/dL or 40%, respectively. Deferral period after double RBC collection is 112 days. Gender-specific minimum height and weight collection requirements based on FDA guidance and apheresis device manufacturer recommendations apply.

RBC apheresis donations are well tolerated, with lower vasovagal rates than WB donations (0.13% apheresis vs. 5.3% WB). Reasons cited for lower vasovagal rates in RBC apheresis compared with WB donation include saline replacement, reinfusion of plasma, and longer procedure times (30–40 minutes for apheresis vs. 8–10 minutes for WB collection).

Plateletpheresis Collections: Apheresis platelet products are more common than WB-derived platelet (WBDP) products in the United States. An apheresis platelet usually contains the equivalent of 4–6 units of WBDP units, with the benefit of elimination of pooling, reduction of donor exposures, and reduction in risk of bacterial contamination. Apheresis technology allows for the collection of one to three platelet

Transfusion Medicine and Hemostasis. https://doi.org/10.1016/B978-0-12-813726-0.00006-4

products during a single donation, with or without additional concurrent red cells and/or plasma.

Plateletpheresis donors must meet the same criteria as WB donors with exception of frequency: plateletpheresis donors can donate 72 hours after a WB donation and as frequently as two times per week (with intervals between donations of two or more days). The maximum number of platelet donations is 24 per rolling 12 months with restrictions for annual cumulative total RBC and plasma losses. If RBC loss exceeds 100 mL for any given procedure, 56 days must lapse before the next donation.

Additionally, apheresis platelets should not be collected from donors taking certain antiplatelet medications. Specific deferral criteria are defined by the medical director, but at minimum preclude donation within 48 hours after ingestion of aspirin with longer deferral periods for other antiplatelet medications: Feldene (piroxicam)—48 hours; Effient (prasugrel) and Brilinta (ticagrelor)—7 days; and Plavix (clopidogrel), Ticlid (ticlopidine), and Zontivity (vorapaxar)—14 days.

Qualification of a plateletpheresis donor includes a precollection platelet count to ensure that platelet count is ≥150,000/μL. If procedure precollection platelet count is unavailable, a default platelet count may be used to program the apheresis device. Repeat donors can be qualified using platelet count from their last donation. Donors with a platelet count <150,000/μL must be deferred.

The maximum total plasma volume removed in any given platelet collection is 500 mL for donors weighing <175 lb or 600 mL for donors weighing ≥175 lb. A platelet additive solution may be substituted for 65% of the plasma product volume, thereby reducing the amount of plasma in the final collect product, enabling collection of more concurrent platelet or plasma products.

Detailed records must be kept for each plateletpheresis procedure performed, including adverse events and platelet count. All apheresis laboratory values and collection records should be reviewed periodically by a medical director to detect trends such as declining platelet count or hemoglobin.

Plasmapheresis Collections: Plasmapheresis products are generally collected either as *transfusable plasma* or *source plasma* for further manufacturing into human-based plasma derivatives, such as intravenous immunoglobulin, albumin, or clotting factor concentrates. Source plasma is usually collected by large commercial plasmapheresis collection centers from paid plasma donors. Source plasma is subject to specific and distinct requirements per the FDA, and additional quality standards and certification programs are provided by Plasma Protein Therapeutics Association (PPTA).

Plasmapheresis programs can either be *infrequent plasmapheresis* (plasma donation no more frequently than once every four weeks) or *frequent plasmapheresis* (donation is more often than once every four weeks) programs. Infrequent plasmapheresis donors have the same donor eligibility requirements as WB donors.

For *frequent plasmapheresis*, there are additional requirements to ensure donor safety, including the following:

- Donors must be observed closely during the procedure. Emergency medical care must always be available (e.g., physician or physician substitute must be on site at time of collection).

- Physical exam performed and documented by physician or physician substitute at first donation and annually thereafter.
- Sample for total serum protein before each and every collection, and value must be normal (6.0–9.0 g/dL).
- Serum protein electrophoresis or quantitative immunodiffusion at initial and 4-month intervals thereafter.
- At least 48 hours should elapse between successive procedures. Donors should not undergo more than two procedures within 7-day period for a maximum of 104 donations within a year.

Further Reading

Burgstaler, E. A. (2006). Blood component collection by apheresis. *J Clin Apher, 21*, 142–151.

Food and Drug Administration. (January 30, 2001). *Guidance for industry: Recommendations for collecting red blood cells by automated apheresis methods.* Rockville, MD: CBER Office of Communication, Training, and Manufacturers Assistance.

Food and Drug Administration. (December 17, 2007). *Guidance for industry and FDA review staff: Collection of platelets by automated methods.* Rockville, MD: CBER Office of Communication, Training, and Manufacturers Assistance.

Smith, J. W., & Gilcher, R. O. (2006). The future of automated red blood cell collection. *Transfus Apher Sci, 34*, 219–226.

CHAPTER 7

Recipient-Specific Blood Donation

Mary Townsend, MD and Liz Rosenbaum-Marinaro, MD

Recipient-specific blood components are typically donated in three circumstances: exceptional medical need, directed donation, and autologous donations. Aside from specific donor qualifications, evaluating need and appropriate use of these specific components requires balanced medical oversight and optimal communication between ordering physician, transfusion service medical director, and collection facility medical director.

Exceptional medical need exists when blood or appropriate components are collected for a specific patient who requires a rare product. Directed donors donate to a specific patient, who is often a family member. Autologous donors donate for themselves, usually for a specific procedure that will require blood (typically RBCs).

Exceptional Medical Need: There are patient-specific situations for which the blood center medical director may accept a blood donor and override typical criteria, such as allowing for shorter interdonation time interval. For RBC donation, this can be as frequent as every 3 days, provided the predonation hemoglobin and hematocrit are ≥12.5 g/dL and 38%, respectively, for females and ≥13.0 g/dL and 39%, respectively, for males. Donors who give multiple units to a specific patient/recipient are sometimes called designated donors. The donor must fulfill all other allogeneic blood donation criteria. Circumstances for which blood products are collected for this reason include the following:

- *RBC products for a patient with multiple alloantibodies or alloantibodies to high-frequency RBC antigens, when rare RBC products are needed.*
- *Platelet products for a neonate with NAIT who requires platelet antigen-negative products.* NAIT is associated with alloantibodies to platelet-specific antigens. In the white population, this antibody is most commonly toward HPA-1a, which has an antigen frequency of 98% in this population. Collection of a plateletpheresis product from a donor with this rare alloantigen negative type may be an exceptional medical need. In addition, use of maternal (washed) platelets, which are alloantigen-negative, classifies as exceptional medical need due to waiver of the postpartum restriction of blood donation.
- *IgA-deficient plasma or platelet products for a patient with anaphylactic transfusion reactions secondary to antibodies to IgA.* Anaphylactic reactions occur in 1 in 20,000 to 50,000 transfused blood components. Reactions related to antibodies to IgA occur less frequently but are the most common cause of transfusion-associated anaphylaxis in the United States. The management of patients with antibodies to IgA requires IgA-deficient products, such as washed RBCs and platelets, and platelets and plasma products from IgA-deficient donors. Of note, anaphylactic reactions can also occur in patients who are deficient in haptoglobin and have developed antihaptoglobin antibodies. These patients require washed products or products from haptoglobin-deficient donors.

Transfusion Medicine and Hemostasis. https://doi.org/10.1016/B978-0-12-813726-0.00007-6

- *Granulocytes for infected, immune-suppressed recipient with neutropenia.* Indications for granulocytes are limited and include neonatal sepsis, neutrophil function defects, and neutropenic patients with fungal infections. Granulocyte donors are typically stimulated with G-CSF and/or steroids and may be recruited to give more than one donation (see Chapter 37).

Directed Donation: Directed donors give their blood to be used for a friend, family member, or another specific individual in need of blood products. Directed donors must meet all criteria for allogeneic blood donation and therefore if these products are not used by the designated recipient, they can be released into the general inventory if policies and procedures are in place.

Units donated by directed donors are not safer (and potentially less safe) than units from nonremunerated volunteer donors because they are frequently first-time donors and may be unwilling to divulge high-risk behaviors during medical history screening. In addition, directed donors should be aware that if their ABO blood type is not compatible with the recipient, their donations may not be appropriate for the intended recipient's use. Directed donor units are more expensive due to costs associated with procuring, labeling, and ensuring that they are reserved for a specific recipient. The 2015 National Blood Collection and Utilization Survey (NBCUS) continues to demonstrate a significant decrease in directed units transfused, 70,000 compared with 126,000 units in 2006 ($P < .05$).

Autologous Donation: Autologous donors donate their own blood for their personal use, usually before scheduled elective surgeries. Use of autologous blood peaked in the late 1980s and early 1990s when risk of transfusion-transmitted diseases, especially HIV, was high, making autologous donation a safer option. Collection of autologous units has declined since; per the 2015 NBCUS, 25,000 autologous RBC products were collected and 20,000 transfused. As with directed units, autologous units are more costly to produce due to specific procurement and labeling requirement and the need to reserve them for a specific patient. Further, hospital reimbursement for autologous units may be problematic.

To prevent their collection and subsequent discard, autologous units should only be collected in advance of planned procedures that have a high likelihood of RBC use. These include major orthopedic procedures, vascular surgery, and cardiothoracic surgery or if the patient has an extremely rare blood type, multiple alloantibodies, or alloantibodies to high-frequency RBC antigens that would make finding compatible units from the general population difficult. Procedures for which autologous collection is contraindicated include uncomplicated obstetric delivery (including cesarean section), cholecystectomy, and hysterectomy.

As well as mitigating the patient's risk of contracting transfusion-transmitted diseases, potential benefits of autologous blood include minimizing exposure to foreign RBCs that may stimulate alloimmunization.

To collect autologous blood, the donor/patient's physician must write an order/prescription that includes date of anticipated transfusion (usually date of the surgery) and number and type of blood products needed (typically RBC products are needed, but some complicated surgeries may also require other products). Eligibility criteria are less stringent because the primary purpose of autologous screening is to identify any individual who might be harmed by donation (Table 7.1). The blood center medical

TABLE 7.1 Comparison of Criteria for Allogeneic and Autologous Blood Donors

Criterion	Allogeneic Donors	Autologous Donors
Age	Conform to applicable state law or ≥16 years old	As defined by medical director
Whole blood (WB) volume collected	≤10 mL/kg, including samples	≤10 mL/kg, including samples; collection of small volume units require adjustment of the anticoagulant volume to obtain the appropriate ratio
Hemoglobin/hematocrit	Female ≥12.5 g/dL 38% Male ≥13.0 g/dL 39%	11.0 g/dL 33%
Intervals between donations of RBCs	56 days after each WB donation	72 hours after each WB donation[a]
Blood pressure (BP)	Not required per AABB, but FDA requires BP within normal limits typically defined as Systolic: ≥90 mmHg; ≤180 mmHg Diastolic: ≥50 mmHg ≤100 mmHg	As defined by medical director
Donor history	Donor health history questionnaire (DHQ) questions	Abbreviated, as defined by medical director: Heart and lung diseases; blood disorders, bleeding conditions; pregnancy
Medication and vaccination deferrals	Current medication deferral list[b] Current AABB Standards	As defined by medical director
Temperature	≤37.5°C (99.5°F) orally	As defined by medical director; defer for conditions that present risk of bacteremia
Pregnancy	Defer if currently pregnant or pregnant within last 6 weeks	As defined by medical director
Transfusion/transplant history	Receipt of dura mater, human pituitary growth hormone, blood, human tissue, human-derived factors, xenotransplant	As defined by medical director
Venipuncture site	Free from skin lesions and punctures/scars indicative of self-injectable drug (narcotic, etc.) use	As defined by medical director
Platelet count (for platelet donors)	≥150,000/μL before procedure, if performed more frequently than once every 4 weeks; not required before first procedure, if procedure is not performed more often than every 4 weeks	As defined by medical director
Contraindications	As per DHQ and blood center criteria	As defined by medical director, examples include among others: unstable angina; recent Myocardial Infarction (MI) or Cerebrovascular accident (CVA); symptomatic heart or lung disease; untreated aortic stenosis

[a]Last donation no less than 72 hours before date and time of need.
[b]http://www.aabb.org/tm/questionnaires/Documents/dhq/v2/DHQ%20Medication%20Deferral%20List%20v2.0.pdf.
Modified from Eder, A. (2007). Allogeneic and autologous blood donor selection. In J. R. Roback, M. R. Combs, B. J. Grossman, & C. D. Hillyer (Eds.), *Technical manual* (16th ed.). Bethesda, MD: AABB Press.

director may modify or set criteria for autologous donors based on discussions with the donor/recipient's ordering physician including a risk/benefit analysis. In some cases, health risk to the donor of donating blood may outweigh benefit of having his or her own blood available for transfusion.

Further Reading

Food and Drug Administration. (2015). *Federal Register 21 CFR Parts 606, 610, 630, et al. Requirements for blood and blood components intended for transfusion or for further manufacturing use; final rule*. Rockville, MD: CBER Office of Communication, Training and Manufacturers Assistance.

CHAPTER 8

Adverse Donor Reactions

Jed B. Gorlin, MD, MBA

About 2%–5% of whole blood donors experience some form of reaction although most are mild vasovagal reactions and small hematomas. Reactions are more likely to occur in young (≤20 years old), low weight/blood volume, first-time, and female donors. Vasovagal reactions are most common, which, when moderate to severe, can lead to loss of consciousness (LOC), seizure-like activity, and severe injury secondary to a fall. More rarely, adverse reactions such as nerve damage and injury from falling can lead to permanent disabilities. Donors who experience an adverse event are less likely to return.

Complications of Whole Blood Donation: One-third of donor reactions and one-half of syncope-related injuries occur in adolescents and young adults (16–25 years). While interventions to decrease these reactions have been implemented, such as establishing minimum predicted donor blood volumes for donation, they have resulted in ~20% decrease in adverse events.

Hematoma: About 30% of donors have an arm complication: 23% report bruise (contusion/hematoma) and 10% pain. These complications can result in pain, swelling, tenderness, redness, and warmth. They can be treated with warm compresses and mild analgesics and generally resolve completely within 7–14 days.

Nerve Injury: Nerve injury incidence by active follow up is 0.9%, but 0.16% by voluntary postdonation reports. Complaints include sensory changes in the forearm, wrist, hand, or shoulder and radiating pain. 15% of affected donors will also report decreased arm or hand strength. Chronic nerve damage from the phlebotomy needle is an unusual event, yet interaction of the phlebotomy needle with branches of superficial nerve branches likely occurs frequently from a study of cadavers. Nerve injury is usually transient: 70% disappear within a month, >99% resolve within one year. Rare cases of complex regional pain syndrome (reflex sympathetic dystrophy) have been reported.

Vasovagal: Vasovagal reaction occurs in 2%–30% donations, depending on donor group with wide variation among reporting centers. Predisposing factors include first-time donors, donors with low weight, and history of previous donor adverse reaction. In a recent study of high school donors, anticipatory fear and length of draw time were strongest predictors of observed reaction rates. For donors with >10-minute draw time, observed reaction rates were 31% among donors self-reporting fear versus 10% for those denying fear. For draw times less than 6 minutes, the reported rates were 16% versus 5%, respectively.

Proposed mechanisms for higher rates in young and first-time donors include changes in central thalamic pathways, vascular baroreceptor sensitivity, and age-dependent

Transfusion Medicine and Hemostasis. https://doi.org/10.1016/B978-0-12-813726-0.00008-8

responses to physical and emotional stress. Anxiety related to blood donation, blood or needles in general contribute to vasovagal reactions, as they can precipitate the reaction even before donation, and clusters of fainting ("epidemic fainting") can occur when prospective donors witness an untoward event at a blood drive. Asking donors about their fears does not increase the rate of reaction and indeed allows staff to focus distraction techniques on fearful donors at highest risk of reactions.

Symptoms include chills or cold extremities, feeling of warmth, light-headedness, nausea, pallor, weakness, hyperventilation, and declaration of nervousness. Signs include hypotension, rapid or slow pulse, sweating, and twitching. Vasovagal reactions can progress to LOC and seizure-like activity (tonic–clonic movements, tetany, with loss of bladder or bowel control). Vasovagal events are classified as prefaint, no LOC (uncomplicated or minor), LOC (uncomplicated), LOC (complicated), and injury.

Interventions include the following: asking about fear and offering distraction, calmness and reassurance, elevation of the donor's legs above the heart, ensuring donor is lying down or in safe place in case of fainting, cold compresses to the neck or forehead, and monitoring blood pressure, pulse, and respirations (vital signs) periodically until recovery.

Prolonged Recovery: Prolonged recovery is defined as presyncopal symptoms, with or without LOC, that do not resolve within 30 minutes. If the donor has not recovered or has unstable vital signs during the typical 5–30 minutes expected for complete recovery, most blood collection centers not located within a hospital (in the United States) will call 911 for emergency medical services (EMS) response. Staff members should also notify both their facility medical director either before or after the 911 call and local host site, depending on the severity of symptoms and/or resulting injury that occurs.

Arterial Puncture: Arterial puncture occurs at a rate of 1/34,000 donations and is more common among inexperienced phlebotomists than those with experience. Most common indicator is rapid filling of collection bag, total phlebotomy time of less than 4 minutes and bright red color of blood. This event can occur with the initial puncture, or after needle adjustment. Proper treatment requires immediate needle removal and prolonged application of pressure at injury and arm elevation. Rarely, pseudoaneurysms develop at the site of arterial punctures.

Iron Loss: Whole blood or red blood cell (RBC) donation results in iron loss (200–250 mg of iron per 500 mL of blood or 250 mL of RBCs). Repeated donation correlates with iron store depletion. Symptoms of iron deficiency include pica, typically craving of ice (pagophagia), feeling weak or tired, difficulty concentrating, dyspnea, light-headedness, and possibly restless leg syndrome. In the RISE (REDS-II Donor Iron Status Evaluation) study, 15% of frequent whole blood donors had absent iron stores (AIS, defined as ferritin <12 ng/mL) and 42% had iron deficiency anemia (IDE, defined as log of ratio of soluble transferrin receptor to ferritin was ≥2.07): 0% and 2.5% of first-time/male donors had AIS and IDE, respectively; 6% and 25% of first-time/reactivated females had AIS and IDE, respectively; 16% and 49% of frequent males had AIS and IDE, respectively; and 27% and 66% of frequent females had AIS and IDE, respectively. Young donors have a higher baseline rate of low iron stores and are therefore at particular risk of early depletion. Subsequent studies have documented that iron replacement dramatically shortens time to both time to return to baseline hemoglobin and recovery of iron stores. Low dose of iron (18–19 mg), typically found in multivitamins with iron, was equivalent to higher doses and had minimal side effects. Blood centers are introducing programs to educate donors, measure ferritin, or provide iron replacement to those with documented or at risk for low iron stores, such as frequent donors or younger females.

Complications of Apheresis Donations: Apheresis collections are less prone to vasovagal reactions. Volume removed is replaced with 0.9% NaCl, resulting in less frequent (<50/10,000 collections) reactions. In addition, apheresis RBC donors are required to have higher blood volumes than whole blood donors. Donor reactions specific to apheresis include hypocalcemia (from citrate), allergic reactions, and machine malfunction leading to hemolysis, air emboli, and leakages.

Citrate-Induced Hypocalcemia: Citrated anticoagulants are used in apheresis donor collections. Citrate binds calcium, inhibiting clotting in the apheresis circuit. Transient hypocalcemia associated with apheresis collections is usually well tolerated. Decreases in ionized calcium may be manifested by perioral or peripheral paresthesias. Less frequently, donors may express an "unusual taste" or experience transient nausea or light-headedness. Other signs and symptoms include cyanosis, spasms, chills/shivering, confusion, tetany, pallor, change in pulse, chest pain, shock, dyspnea, tachycardia, and tremors. Severe hypocalcemia can result in LOC, convulsions, or cardiac arrhythmia. Hypocalcemia occurs in up to 1000/10,000 of collections but is severe in less than 3/10,000 of collections. Treatment of hypocalcemia includes slowing collection rate/reducing citrate infusion rate, administering calcium tablets and, for severe cases, terminating the collection immediately, notifying emergency care personnel and the medical director.

Machine Malfunction: Machine malfunction may lead to hemolysis, air embolism, or leakage, which occurs infrequently, but effects can be severe. Newer apheresis devices have improved safety features to prevent excessive citrate or air infusion to protect the donor and mitigate these complications.

Allergic Reactions: Donors may be hypersensitive to sterilizers, especially ethylene oxide, although sterilization by irradiation has largely replaced chemical sterilization. Some series have noted a disproportionate number of apheresis donors with allergic symptoms to be on calcium channel blockers, such as lisinopril. Symptoms include hives, difficulty in breathing, wheezing, hypotension or hypertension, anxiousness, tachycardia or bradycardia, facial swelling or flushing, and burning eyes. Symptoms tend to manifest early in the procedure during the reinfusion cycles and generally dissipate on stopping the procedure.

Granulocytapheresis: Use of various starch compounds increases separation of granulocytes from RBCS, enhancing granulocyte yield. Starch side effects include allergic reactions, edema, and weight gain. Granulocytapheresis donors receiving corticosteroids should be screened for hypertension, diabetes mellitus, and peptic ulcer disease, as this medication may aggravate these conditions. Donors should be advised of risk of headache, elation, sleeplessness, or nausea. Generally, use of a single dose of corticosteroids is not associated with serious short- or long-term side effects. More common side effects of G-CSF are bone/joint ache or pain, muscular ache and/or headache. Pain is usually mild to moderate and treated with NSAIDs. No long-term side effects of G-CSF in normal donors have been reported.

Approach to the Donor and Donation Process: For all significant adverse reactions, staff should terminate donation. It may be preferable both for donor comfort and prevention of anxiety in other donors to shield the reacting donor from observation by other donors. It is important to reassure the donor, contact the physician, and perhaps call for EMS, depending on the length and severity of the reaction. Interventions to prevent injury from falling should also be implemented when possible.

Prevention

Predonation Education: Efforts should be in place to educate the potential donor on the "what to expect" of blood donation, especially among first-time donors. Addressing fear prospectively has been shown to decrease donor reactions.

Drive Setup and Environment: Well-planned, adequately staffed, and organized layout, especially at high schools, decreases wait times, as prolonged waiting has been shown to exacerbate fear among younger first-time donors.

Selection Criteria: Whole blood donor eligibility tables for young donors (<20 years old) using sex, height, and weight equation insure minimum estimated blood volume of at least 3.5 L. One study observed 20% drop in symptomatic rates after adopting this approach.

Distraction: Distraction techniques have been shown to aid in relaxing some donors based on their methods of coping.

Water Ingestion: Administration of 500 mL (16.9 fl oz) of water 30 minutes before whole blood donation, usually given to the donor at the time of their arrival at the blood drive, has been supported by a few studies. One study involving around 9000 high school students (17–19 years old) over a 2-year period reported a decrease in vasovagal reactions by 21% in both male and female donors. The administration of popular high-energy sports drinks (250 mL) before and after whole blood donation has been shown to be of benefit based on its sodium and glucose replacement value and expansion of intravascular space. The mechanism is related to increase gastric distension, which increases sympathetic tone and overall total peripheral resistance.

Applied Muscle Tension: Applied muscle tension (AMT) in combination with water hydration has been shown to diminish presyncopal and syncopal reactions. AMT involves repetitive contraction of major muscle groups of the arms and legs thereby promoting venous return and cardiac output, which affects cerebral blood flow.

TABLE 8.1 Donor Hemovigilance Reaction Types and Categories

Reaction Type	Reaction Category
Vasovagal	Prefaint, no loss of consciousness (LOC) (uncomplicated or minor) LOC, any duration (uncomplicated) LOC, any duration (complicated) Injury
Local injury related to needle	Nerve irritation hematoma/bruise arterial puncture
Apheresis	Citrate hematoma/bruise air embolus
Allergic	Local systemic anaphylaxis
Other	Other

Salty Snacks: Beneficial effect of increased dietary sodium to improve orthostatic tolerance in patients with symptomatic orthostatic hypotension due to autonomic failure is well documented.

Salty snacks before or immediately after donation is a low-cost addition to reaction mitigation.

Biovigilance and Reporting: Death after blood donation and transfusion has to be reported to the FDA. During 2015, the FDA investigated a total of 41 fatality reports occurring in transfusion recipients. About 13 million whole blood and voluntary apheresis donations and more than 32 million paid apheresis plasma donations, only a single fatality was considered possibly related to the donation and that was linked to a source plasma donation.

AABB established donor hemovigilance with definitions in line with international organizations (Table 8.1) (http://www.aabb.org/research/hemovigilance/Pages/donor-hemovigilance.aspx).

Further Reading

AABB Association Bulletin #08-04. (August 28, 2008). *Strategies to reduce adverse reactions and injuries in younger donors.*

AABB Association Bulletin #12-03. (March 16, 2017). *Updated strategies to limit or prevent iron deficiency in blood donors.*

Cable, R. G., Glynn, S. A., Kiss, J. E., et al. (2011). Iron deficiency in blood donors: Analysis of enrolment data from the REDS-II donor iron status evaluation (RISE) study. *Transfusion, 51*, 511–522.

Diekamp, U., Gneissl, J., Rabe, A., et al. (2014). Donor hemovigilance with blood donation. *Transfus Med Hemother, 42*, 181–192.

Eder, A. F. (2012). Improving safety for young blood donors. *Transfus Med Rev, 26*, 14–26.

France, C. R., France, J. L., Frame-Brown, T. A., et al. (2016). Fear of blood draw and total draw time combine to predict vasovagal reactions among whole blood donors. *Transfusion, 56*, 179–185.

Horowitz, S. H. (2000). Venipuncture-induced causalgia: Anatomic relations of upper extremity superficial veins and nerves, and clinical considerations. *Transfusion, 40*, 1036–1040.

Newman, B. (2013). Arm complications after manual whole blood donation and their impact. *Transfus Med Rev, 27*, 44–49.

CHAPTER 9

Component Preparation and Manufacturing

Donna Strauss, MS

Blood components, including red blood cell (RBC), plasma, platelet, and granulocyte products, are prepared either from whole blood (WB) or by automated apheresis donation. Component preparation and manufacturing allow each component to be manufactured and stored under optimal conditions. Component therapy has largely replaced WB due to ability to choose components that target specific patient's needs and enable each component to be stored optimally. Additionally, components can be modified (e.g., irradiated) or selected (e.g., hemoglobin S negative) to meet specific patient requirements. Components and modifications are discussed in more detail in their separate chapters (Chapters 33–48).

Whole Blood: WB is the most common starting product for component preparation and manufacturing. WB is generally not found in most transfusion services today because component therapy is more appropriate to target patient's specific indications for transfusion (e.g., RBC products for symptomatic anemia, plasma products for multiple coagulation factor deficiencies, and platelet products for thrombocytopenia). Additionally, in WB, platelet survival is lost due to refrigerator storage, and coagulation factor activity, especially for labile factors (Factors V and VIII), deteriorates over time. Usually 500 mL of WB is collected into bag with 70 mL of anticoagulant-preservative solution (see below), creating a product with final hematocrit of ~38%. WB is stored at 1–6°C. Depending on the manufacturer's system and a country's regulations, different hold times are allowed before preparation of components from WB. If platelet products are to be manufactured, WB product must be stored at room temperature until the platelets are removed.

Fresh WB is occasionally being used by the US military via "walking blood donors" (donors who are ABO/D typed, infectious disease tested and available to donate in times of need). This product is transfused as soon as possible and has a shelf life of 24 hours when stored at room temperature, allowing maintenance of platelet function, while still minimizing the risk of bacterial overgrowth. Primary use of this product is to infuse viable platelets when other platelet products are not available.

This practice has led to recent use of refrigerated stored WB for civilian trauma and other massive transfusion clinical indications. This product may be leukoreduced using platelet-sparing filter. Additionally, low-titer group O products are being used emergently in trauma patients for initial resuscitation until laboratory-guided (such as Thromboelastography) component therapy can be implemented.

Component Manufacturing: When WB is manufactured into components, it is collected into a primary bag containing an anticoagulant-preservative solution. Primary bag has up to three attached satellite bags for RBC, platelet, and plasma or

Transfusion Medicine and Hemostasis. https://doi.org/10.1016/B978-0-12-813726-0.00009-X

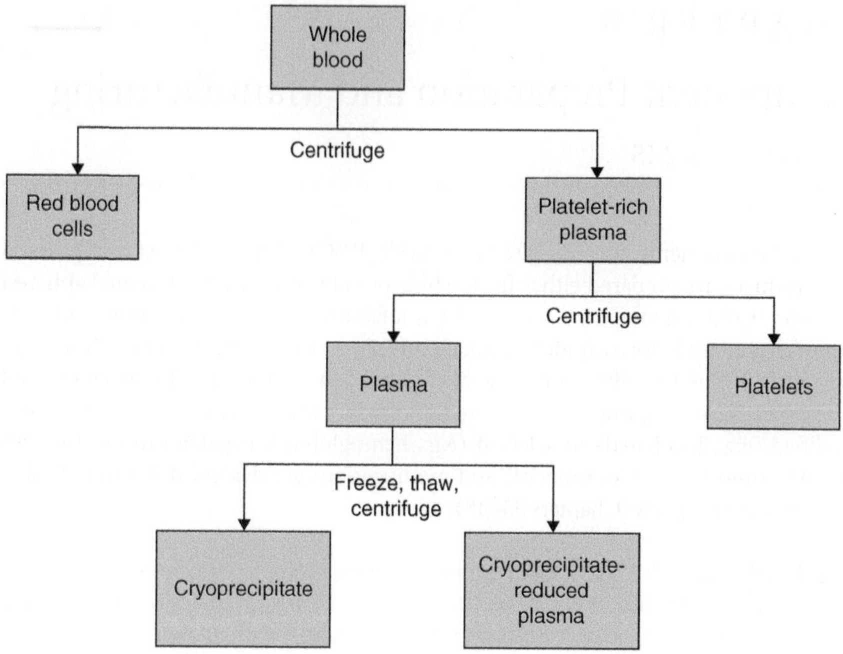

FIGURE 9.1 Manufacturing of blood components from whole blood.

cryoprecipitate component manufacturing. Because of the different specific gravities of RBCs (1.08–1.09), plasma (1.03–1.04), and platelets (1.023), differential centrifugation of WB unit is used to prepare components. Optimal component separation requires specific centrifugation variables, such as rotor size, speed, and duration of spin (Fig. 9.1). Automated laboratory-based component manufacturing systems are available but not in widespread use in the United States, as they are better suited for buffy coat platelet production.

Anticoagulant-Preservative Solutions: WB is collected into containers with anticoagulant-preservative solutions. Anticoagulant-preservative solutions include acid–citrate–dextrose (ACD), citrate–phosphate–dextrose (CPD), citrate–phosphate–dextrose–dextrose (CP2D), and citrate–phosphate–dextrose–adenine (CPDA-1). Citrate (sodium citrate and citric acid) acts as an anticoagulant, and phosphate (monobasic sodium phosphate and trisodium phosphate), adenine, and dextrose are substrates for cellular metabolism.

RBC Components: RBC products are used primarily for the treatment of symptomatic anemia or hemorrhage to increase tissue oxygenation. Anticoagulant-preservative solutions allow RBC components to be stored for extended periods of time at 1–6°C without a significant detrimental effect on RBC quality. Shelf life of given product is determined by solution used and based on criterion that <1% hemolysis in product at the end of storage and 75% of RBCs transfused remain viable 24 hours posttransfusion. RBCs stored in CPD or CP2D have shelf life of 21 days and CPDA-1 of 35 days.

RBCs stored in CPD, CP2D, or CPDA-1 have hematocrit of ~80% with final volume of 225–350 mL and ≤110 mL of plasma. Majority of RBC products are stored in additive solutions.

Additive Solutions: Additive solutions, such as AS-1, AS-3, and AS-5, can be added to primary anticoagulant-preservative solution to increase the shelf life of RBCs to 42 days. All of these solutions contain dextrose, adenine, and sodium chloride; AS-1 and AS-5 contain mannitol, and AS-3 contains monobasic sodium phosphate, sodium citrate, and citric acid. Additive solution products have final volume of 350–400 mL and hematocrit of 55%–65%, containing 100–110 mL of additive solution and 10–40 mL of plasma. In general, additive solutions must be added to the product within 24 hours of collection, depending on manufacturer's specifications. Outside of the United States, saline–adenine–glucose–mannitol (SAGM), which is hypertonic, or phosphate–adenine–glucose–guanosine–saline–mannitol (PAGGSM), which is isotonic, is used.

RBC Modification: RBC products can be further modified for specific patients' needs. Multiple modifications can be performed on a single product. RBCs can be frozen, volume reduced, washed, rejuvenated, leukoreduced, irradiated, pathogen inactivated, and aliquoted. Additionally, RBCs can be labeled CMV, HbS, or antigen negative.

Plasma Components: Plasma components are primarily used to treat multiple plasma factor deficiencies or prevent development of a coagulopathy. Plasma contains coagulation factors, albumin, and fibrinogen found in WB. Plasma products that are used in the transfusion service include fresh frozen plasma (FFP), plasma frozen within 24 hours of phlebotomy (FP24), plasma frozen within 24 hours after phlebotomy held at room temperature up to 24 hours after phlebotomy (PF24RT24), and thawed plasma (TP). FFP must be frozen within 6–8 hours of collection, while FP24 and PF24RT24 are frozen within 24 hours of collection. FP24 and PF24RT24 have lower levels of Factors V and VIII than FFP, but this reduction is not clinically significance. Plasma products are stored at −18°C or colder for up to 1 year (with FDA approval, FFP can be stored at −65°C or colder for up to 7 years). Before transfusion, plasma must be thawed at 30–37°C, which requires ~20 minutes; it is then stored at 1–6°C if not transfused immediately. Once thawed, plasma must be transfused within 24 hours; otherwise, it becomes classified as TP, which may be stored at 1–6°C for up to 5 days. During these 5 days of storage, Factor VIII activity levels decrease by ~30%. All plasma products can be used interchangeably.

Liquid plasma (LP) is plasma that has never been frozen, is stored at 1–6°C, and has a shelf life of 21 days. LP has decreased and variable clotting factor activity levels and may be used for massive transfusion in patients with life-threatening hemorrhage. LP has viable lymphocytes.

Plasma products may be further manufactured into cryoprecipitate and cryoprecipitate-reduced plasma. Additionally, large pools of plasma are manufactured into solvent/detergent-treated plasma, which has minimal (if any) risk of transmitting lipid-enveloped viruses (HIV, HCV, and HBV). Other PI plasma products available outside of the United States, which are not pooled, include methylene blue (MB), riboflavin, and ultraviolet light. PI through amotosalen and ultraviolet light is FDA approved.

Plasma (stored at −18°C or below) and LP (stored at 1–6°C) are manufactured from WB no later than 5 days after the WB expiration date and can be converted into an unlicensed product termed "recovered plasma" that can shipped for further fractionation into albumin, immune globulin (Rh immune globulin and intravenous immunoglobulin), and coagulation factor concentrates (Factors II, V, VII, VIII, III, IX, X, and XI), Protein C, Protein S, alpha2-antiplasmin, fibrinogen, ADAMTS13, antithrombin III, and von Willebrand factor (VWF).

Cryoprecipitate: Primary indication for cryoprecipitate (FDA term is cryoprecipitated antihaemophilic (AHF)) is for fibrinogen replacement. Cryoprecipitate is prepared by slowly thawing FFP to 1–6°C, thereby leading to formation of precipitate, which is collected and refrozen to make cryoprecipitate. Each unit of cryoprecipitate, which is ~15 mL, must contain fibrinogen (>150 mg) and Factor VIII (>80 IU) at these specified levels. Additionally, each unit contains VWF (80–120 IU), Factor XIII (40–60 IU), and fibronectin. Cryoprecipitate is usually administered to adults in doses of 5–10 units, which are pooled before transfusion (known as "cryopool"). Some blood centers manufacture prepooled cryoprecipitate, where five units are pooled together before storage. This product is easier for transfusion service to use than pooling before issue (post-thaw). Cryoprecipitate units can be stored for up to 1 year at −18°C and must be transfused within 4 hours of pooling using open system and within 6 hours of thawing using closed system.

Cryoprecipitate-Reduced Plasma: Cryoprecipitate-reduced plasma is used exclusively for plasma exchange or transfusion in patients with thrombotic thrombocytopenic purpura. Cryoprecipitate-reduced plasma is the residual component fluid of FFP from which cryoprecipitate has been removed and therefore contains decreased amounts of VWF, Factor VII, Factor XIII, and fibrinogen. Cryoprecipitate-reduced plasma is stored at −18°C with an expiration date of 1 year and once thawed at 30–37°C is stored at 1–6°C for up to 5 days.

Platelet Components: Platelet products are used for prophylaxis or treatment of bleeding secondary to thrombocytopenia or dysfunctional platelets. WB-derived platelets have volume of ~40–70 mL and must contain $>5.5\times10^{10}$ platelets. Usual adult dose of WB-derived platelets is pool of 4–6 concentrates. WB-derived platelets can be pooled before issue and subsequently tested for bacterial contamination; prepooled, leukoreduced, and bacterial contamination-tested product is also approved and must contain $>2.2\times10^{11}$ platelets in 180 mL of plasma. Apheresis platelets have volume of ~300 mL and must contain $>3.0\times10^{11}$ platelets in 75% of products tested. Most apheresis systems use ACD-A as anticoagulant-preservative solution, are leukoreduced (termed *process* leukoreduction) and have very low RBC contamination (which is most relevant for transfusion of D-positive products to D-negative females of child-bearing potential).

Apheresis platelets can also be collected into platelet additive solution (PAS). In the United States, two PAS are FDA approved, and other PAS are approved outside of the United States. Current PAS replace 65% of plasma and are composed of a combination of citrate, phosphate, acetate, magnesium, potassium, gluconate, and/or glucose.

Platelets stored in PAS versus plasma are associated with fewer acute transfusion reactions, particularly allergic reactions. Newer PAS under study replace 95% of plasma.

Platelets are stored at room temperature (20–24°C) with gentle agitation for a maximum of 5 days (7 days with extra bacteria mitigation). Once the system has been opened, the product must be transfused within 4 hours.

Buffy Coat Platelets: Outside of the United States, platelets can be prepared from WB using a buffy coat method, which differs from the US method (platelet-rich plasma [PRP] method). Instead of removing PRP after softly centrifuging WB followed by separation of platelets from plasma using hard spin (PRP method), WB first goes through hard spin, after which RBCs and plasma are removed above and below buffy coat. Buffy coat is then softly spun, after which residual white blood cells (WBCs) are removed, leaving platelet concentrate. Four to six buffy coat platelet concentrates are then pooled and resuspended in one of donor's plasma or PAS to create single-pooled platelet product. No clinical difference between these products are known; however, buffy coat method has several advantages, including TRALI mitigation through PAS or male donor plasma, decrease in residual WBC, and automation of platelet preparation from WB.

Platelet Modification: Platelet products can be further modified for specific patients' needs. Platelets can be volume reduced, washed, leukoreduced, irradiated, pathogen inactivated, and aliquoted. Additionally, platelets can be labeled CMV or antigen negative.

Granulocytes: Granulocytes are administered for treatment of patients with neutropenia and life-threatening infection. Granulocytes are usually collected using apheresis devices, but they may also be prepared from buffy coat of centrifuged WB for neonatal transfusions. Apheresis-derived granulocytes have volume of ~200 mL and contain RBCs (20–50 mL), platelets ($\sim 3 \times 10^{11}$), and plasma in addition to granulocytes ($>1 \times 10^{10}$). Modalities used to increase granulocyte dose are to stimulate the donor with granulocyte-colony stimulating factor (G-CSF) and/or corticosteroids, and/or to use hydroxyethyl starch to improve sedimentation and collection during apheresis. The use of all three modalities should increase product yield to $>1 \times 10^{11}$ granulocytes. Granulocytes should be transfused as soon as possible but may be stored for up to 24 hours after collection at 20–24°C without agitation. All granulocyte products should be irradiated to prevent transfusion-associated graft versus host disease, but they cannot be leukoreduced and, therefore, CMV-seronegative products may be indicated for at-risk patients.

Component Labeling: Blood components must be labeled in accordance with FDA regulations and AABB Standards. Labeling must conform with most recent version of the United States Industry Consensus Standard for the Uniform Labeling of Blood and Blood Components using ISBT 128.

- Original label and added portions must be attached to container, including ABO/Rh type, donation identification number, product code, and facility identification.

- All modifications must be specified.
- If new label is applied, process must be accurate.
- Labeling process must include a second check to ensure accuracy.

Further Reading

Circular of information for the use of human blood and blood components. https://www.aabb.org/tm/coi/Documents/coi1113.pdf.

Levin, E., Culibrk, B., Gyöngyössy-Issa, M. I., et al. (2008). Implementation of buffy coat platelet component production: comparison to platelet-rich plasma platelet production. *Transfusion*, *48*(11), 2331–2337.

CHAPTER 10

Serologic Testing of Donor Products

Kim Peck, MA, MT(ASCP)SBB

Each blood component must have its ABO and D antigen status tested as a primary mechanism of preventing ABO-incompatible blood transfusions and D antigen sensitization. Additionally, components must be tested for the presence of unexpected, clinically significant antibodies to prevent potential recipient hemolysis. Red blood cell (RBC) products often undergo further RBC antigen characterization to supply blood products for patients with alloantibodies or patients with chronic anemia (e.g., sickle cell disease [SCD]) who require phenotype-matched RBC components. Most blood centers use serologic methods to perform these tests, but the use of molecular testing is growing. Serologic methods include tube testing, gel method (also known as column agglutination), or solid phase. These tests, and the various methods, are discussed in more detail in the chapter on pretransfusion testing (Chapter 21).

ABO Group Typing: ABO blood group phenotypes include A, B, AB, and O. Reciprocal antibodies are consistently present in the majority of individuals' sera without previous RBC exposure (e.g., anti-B antibodies in blood group A patients), and these antibodies may result in severe intravascular hemolysis after transfusion of ABO-incompatible blood components. Prevention of ABO-incompatible transfusion is the primary objective of pretransfusion testing.

ABO is determined by testing donor RBCs with anti-A and anti-B (known as "forward" or "front" type) and donor plasma with group A_1 and group B RBCs (known as "reverse" or "back" type). Discrepancies between front and back type must be resolved before labeling a blood component (Chapter 25). RBC genotyping can be used to resolve typing discrepancies.

D Antigen Phenotype: The D antigen (also termed Rh or Rh(D)) is the most immunogenic RBC antigen and therefore the D antigen is second to the ABO blood group in importance in transfusion medicine. D-negative products must be appropriately labeled, such that a recipient of a D-negative product does not form anti-D. Currently, donor centers use typing reagents to detect the D antigen, and these reagents must be sensitive enough to detect the presence of weak D. Weak D testing can either be performed by including the antihuman globulin phase, which takes additional time, reagents, and controls, or through the use of certain automated techniques, which are sensitive enough for "weak D" detection such that a separate "weak D" test is not required. D-negative donors in Germany and Austria were *RHD* tested by molecular methods, and about 1 in 1000 have the *RHD* gene and express the D antigen but at levels that are not detected by routine serologic tests (e.g., D_{el}). In addition, transfusion of these products with low levels of D antigen can result in sensitization of the D-negative recipient. Therefore, these donors were removed from the D-negative donor pool (Chapter 26).

Transfusion Medicine and Hemostasis. https://doi.org/10.1016/B978-0-12-813726-0.00010-6

Antibody Screening for the Presence of Unexpected RBC Antibodies: Alloantibodies other than anti-A and anti-B (which are expected RBC antibodies), are found in <1% of the US donor population. Most clinically significant alloantibodies to RBC antigens require RBC exposure as a result of prior pregnancy or transfusion; rarely RBC antibodies are naturally occurring, such that no previous RBC exposure is necessary. Most blood centers test all donated blood for unexpected antibodies. While the AABB Standards state that blood from donors with histories of transfusion or pregnancy must be antibody tested, it is often not practical to separate donor specimens by these criteria. In addition, automated platform use eases testing by incorporating ABO and D testing, and antibody screening on a single instrument.

In contrast to the patient setting, antibody testing in a donor may use donor plasma or serum tested against either pooled or individual reagent RBCs of known phenotypes (because of lower sensitivity of using pooled RBCs, patient pretransfusion testing must be performed on at least two individual screening RBCs) (Chapters 21 and 22). Blood components that contain RBC antibodies can be used for transfusion, given transfusion service has appropriate policies in place—either transfusing to antigen-negative individuals or washing the product.

Phenotyping RBC Products: In the United States, the current practice is to only match for the ABO and D antigens, but there are two situations when multiple antigen-negative RBC products are requested: (1) when recipients have corresponding alloantibodies and (2) for special, chronically transfused patient populations, such as patients with SCD, thalassemia, and autoantibodies. Alloimmunization occurs in approximately 2%–6% of patients who receive RBC transfusions, but the rate of alloimmunization may be as high as 36% in patients with SCD. Recipients with multiple alloantibodies impair the ability to provide antigen-negative, compatible RBCs, because of the difficulty finding antigen-negative units. The number of products screened multiplies depending on RBC antigen prevalence and number of negative antigens necessary. Mass-scale genotyping has allowed for the identification of more antigen-negative units and thus enabling phenotype/genotype matching to more patients, thereby improving patient care.

The reason to phenotype-match products is to prevent alloantibody formation and the subsequent negative consequences of hemolytic transfusion reactions. However, prophylactic matching makes transfusion more difficult. Therefore, currently, phenotype matching is only applied to specific patient populations.

Donor centers currently screen and stock RBC products to keep a pool of frequently needed antigen-negative products. Usually, donor centers screen products from repeat donors and family members of patients who have formed alloantibodies to high-prevalence antigens for rare blood types. High-throughput instruments are available to phenotype for C, E, and K. Batch serologic screening is technologist time-intensive to perform, typing reagents are expensive, and appropriate controls are required. As a result, donor centers have algorithms to help determine the likelihood of an antigen-negative product based on limited phenotyping, and donor race and ethnicity.

The American Rare Donor Program (ARDP) is a list of more than 45,000 individuals compiled from AABB and the American Red Cross (ARC) who are active blood donors with a blood type that occurs in <1:10,000 people. ARDP supplies these rare

RBCs worldwide, but mostly within the United States. In addition, the International Society of Blood Transfusion (ISBT) maintains a rare blood donor program, which is compiled and maintained by the International Blood Group Reference Laboratory (IBGRL).

The use of large-scale genotyping methods has enabled increased identification of these RBC products and corresponding donors. There are multiple mass-scale genotyping platforms. The above concept was demonstrated when 2355 donors were genotyped to predict for the presence of K, k, Jk^a, Jk^b, Fy^a, Fy^b, M, N, S, s, Lu^a, Lu^b, Di^a, Di^b, Co^a, Co^b, Do^a, Do^b, Jo^a, Hy, LW^a, LW^b, Sc^1, Sc^2, and HgbS and identify 21 rare donors (Co(a−b+), Jo(a−), S−s−, and K+k−).

Antibody Titer Anti-A, Anti-B: Blood products, especially platelet products from group O donors, may be tested for presence of high-titer anti-A (more commonly) and/or anti-B to prevent hemolytic transfusion reactions secondary to ABO incompatibility. Currently in the United States, this is inconsistently practiced with a variety of methods and cutoff values to determine which products are high-titer and therefore at higher risk of causing a hemolytic reaction.

International: Outside the United States, more extensive RBC phenotype matching and other serological testing may occur. For example, in the United Kingdom (UK), K-negative RBC products are recommended for the transfusion of females with child-bearing potential (i.e., <55 years) and patients with positive DAT. In addition, the UK screen blood components for the presence of high-titer anti-A and/or anti-B using a 1/100 dilution with an AB group RBC on an automated platform. Components that are negative are labeled appropriately. Components with high-titer anti-A and/or anti-B are reserved for ABO group-specific transfusions.

Further Reading

Anstee, D. J. (2009). Red cell genotyping and the future of pretransfusion testing. *Blood, 114*, 248–256.

Westhoff, C. M. (2013). The concept of "confirmatory testing" of donors for ABO and RhD. *Transfusion, 53*, 2837–2839.

CHAPTER 11

Overview of Infectious Disease Testing

Alexandra Jimenez, MD and Evan M. Bloch, MD, MS

Safe and available blood products are critical requirements for the optimal functioning of any advanced medical system. Ensuring blood product safety relies on active surveillance and timely recognition of emerging transfusion-transmissible infections, recruitment and selection of low-risk donors, robust donor screening using risk based deferral and laboratory testing for infectious disease markers, and application of good manufacturing practices, rigorous quality, accreditation, and inspection systems. Collectively, these measures have contributed to a high level of blood transfusion safety (Table 11.1). This chapter focuses on laboratory-based infectious disease testing of whole blood and apheresis-derived allogeneic donations collected from nonremunerated volunteer donors in the United States. Outside the United States, infectious testing practices are similar, although individual requirements and regulatory bodies may differ. Infectious testing in low- to middle-income countries is more variable: while the same principles apply, limited financial resources, infrastructure, and technical expertise impose limitations both on the scope and quality of available testing.

Background: A number of infectious diseases can be transfusion transmitted, and the agents responsible for these infectious diseases have four elements in common (Table 11.2).

Recognition of transfusion-associated infectious risk began in the 1940s after cases of transfusion-transmitted syphilis; this led to routine donor testing for syphilis. Later, viral hepatitis (hepatitis B and C) and HIV would become the major infectious risks to the blood supply. Advances in testing methodologies and strategies, immunology, and molecular biology have led to improved understanding of the biology of these infectious agents; they have also spurred development and subsequent implementation of highly sensitive and specific tests to mitigate transfusion-transmission risk.

Approach to Testing: In the United States, blood products and prerelease tests that are used in component manufacturing are classified as "biologics" by the FDA. As such, infectious disease tests used in the blood industry undergo extensive clinical trials before FDA licensure. A list of current FDA-licensed tests can be found at www.FDA.gov/cber/products/testkits.htm.

Generally, the following scheme is used (Table 11.3):

1. Each blood donation is tested: testing is performed on a sample drawn at time of donation, and the donated units are placed in quarantine until test results are completed, deemed negative, and reviewed. Exception is serologic testing for *Trypanosoma cruzi*, agent of Chagas disease, which requires testing only once at the time of first donation.

Transfusion Medicine and Hemostasis. https://doi.org/10.1016/B978-0-12-813726-0.00011-8

TABLE 11.1 Data on Infectious Disease Window Period and Residual Risk

Test	Window Period (Days)	Estimated Residual Risk of Transfusion Transmission and/or Number of Reported Cases in the United States
HIV MP-NAT	7–10	1 case per 2 million (transfused products)
HIV Ab	21	
HCV MP-NAT	7	1 case per 2 million
HCV Ab	51–58	
HBsAg	30–38	1 case per 1.7 million
HBcAb		
HBV MP-NAT	18.5–29.2	
CMV Ab		Rare; concurrently addressed by leukoreduction
HTLV Ab	80	1 case per 3 million
WNV MP-NAT	6.9	Rare after screening; about 1 case per year
ZIKV ID-NAT		4 cases reported in Brazil; none in the United States
Trypanosoma cruzi Ab (Chagas disease)		10 reported cases in the United States and Canada before testing
Treponema pallidum Ab (syphilis)		Last reported transfusion-associated case in the United States in 1966
Bacterial contamination of platelets (sepsis)		<1:100,000 apheresis platelets depending on product and testing

CMV, cytomegalovirus; *HBcAb*, antibody to hepatitis B core antigen; *HBsAg*, hepatitis B surface antigen; *HBV*, hepatitis B virus; *HCV*, hepatitis C virus; *HIV*, human immunodeficiency virus; *HTLV*, human T-cell lymphotropic virus; *ID-NAT*, individual donor–nucleic acid testing; *MP-NAT*, minipool–nucleic acid testing; *WNV*, West Nile virus; *ZIKV*, Zika virus.

TABLE 11.2 Criteria for Transfusion-Transmission of Infectious Agents

There must be an asymptomatic, infectious phase in the blood donor.
Agent viability must be maintained during storage.
There must be a seronegative recipient population.
The agent must be capable of inducing disease after transfusion

2. With exception of most viral nucleic acid testing (NAT), each donor sample is tested individually. NAT is an expensive yet sensitive technology that reduces the window period (period when the patient is first infected and viremic, but current testing methodologies are unable to detect the virus). To optimize the cost-effectiveness of screening, NAT is performed, mostly, using a pooled aliquot (i.e., minipool) from 6 to 16 donor samples. Minipool size is informed both by the incidence of the target infection in the donor population and assay sensitivity.
3. Nonreactive tests are considered to be negative for infectious marker, and related blood products may be released from quarantine and issued.
4. Samples with initially reactive screening tests are typically retested in duplicate. If both repeat test results are nonreactive, sample is classified as negative and

TABLE 11.3 Screening Test Results

Analyte	Screening Result		Result Interpretation	Suitable for Transfusion	Sample to Confirmation	Product Label	Product Disposition	
	Initial	Repeat[a]					RBC/PLT	Plasma
HBc Ab	NR	None	NR	Yes	N/A	ABO/Rh	Inventory	Inventory
HBs Ag	R	N + N	IR	Yes	N/A	ABO/Rh	Inventory	Inventory
	R	R + R or N + R	RR	No	N/A	Biohazard	Discard	Further manufacture
HIV-1/2 plus O Ab	NR	None	NR	Yes	No	ABO/Rh	Inventory	Inventory
HCV Ab	R	N + N	IR	Yes	No	ABO/Rh	Inventory	Inventory
HTLV-I/II Ab	R	R + R or N + R	RR	No	Yes	Biohazard	Discard	Discard
Trypanosoma cruzi Ab								
HIV-1 RNA	MP-NAT: NEG	N/A	NR	Yes	No	ABO/Rh	Inventory	Inventory
HCV RNA	MP-NAT: POS	Resolution POS	R	No	No	Biohazard	Discard	Discard
HBV DNA	ID-NAT: NEG	N/A	NR	Yes	No	ABO/Rh	Inventory	Inventory
WNV RNA[b]	ID-NAT: POS	N/A	R	No	No	Biohazard	Discard	Discard
ZIKV RNA								
Syphilis Ab	NEG	None	NEG	Yes	No	ABO/Rh	Inventory	Inventory
	POS/IND[c]	R/POS/IND	POS/IND	No	Yes	Biohazard	Discard	Discard
Atypical RBC antibody	NEG	None	NEG	Yes	N/A	ABO/Rh	Inventory	Inventory
	POS	None	POS	RBC—yes if washed PLT/plasma—no	Yes	RBC—ABO/Rh	RBC to frozen blood PLT to discard	Discard

ID-NAT, individual donor–nucleic acid testing; *MP-NAT*, minipool–nucleic acid testing; *N/A*, not applicable; *NEG*, negative; *NR*, nonreactive; *POS*, positive; *R*, reactive.

[a]Initial reactive (IR) samples are retested in duplicate. (Except initial reactive ID-NAT samples using a licensed test are not retested.)

[b]NAT is performed in minipools of 16 samples for HIV, HCV, and HBV and for WNV (unless ID-NAT is triggered). Positive pools are resolved to individual donations. Currently, individual donations are being tested for ZIKV.

[c]Indeterminate syphilis results are treated as positive.

corresponding products may be released from quarantine. If one or both repeat tests are reactive (termed repeatedly reactive), products cannot be released for transfusion and are destroyed.

5. Samples that are repeatedly reactive undergo confirmatory with, or without, supplementary testing. Confirmatory testing is typically performed using assays with different sensitivity and specificity characteristics than the initial screening tests. Some screening tests, such as hepatitis B core antigen, do not have a confirmatory test. Results of confirmatory testing improve accuracy and predictive value of results while supplementary tests aid significantly in donor notification and counseling.

Determination of Need and Requirement for Testing: While a number of infectious agents can be transfusion transmitted, testing is performed for majority, but not all, of these pathogens. Many factors are considered in determining the requirement for testing, including pathogen prevalence in donor population, ability to be transfusion transmitted, clinical severity of infection, and testing technology, reagent, methodology, and test kit availability.

In the United States, most tests are mandated in the FDA Code of Federal Regulations (CFR) or under FDA Guidance documents. Currently, Babesia tests are available under investigational use and have been adopted by some centers in high endemic areas.

Selection of Testing Methodology: Test methodologies used as screening tests for donors have higher sensitivity, and correspondingly reduced specificity, as compared with diagnostic tests. Donors are lower risk than general or patient populations. Unlike clinical testing, donor screening is performed on individuals who feel well, often reflecting early stage of infection or low infectious burden (e.g., viremia, bacteremia, or parasitemia) thus requiring added analytic sensitivity.

Biology of Infection and Testing Strategies: Biology of any given infection will impact clinical risk and inform optimal screening approach and testing strategy. Early infection, target virus/infectious particle may only be detected through NAT; this is followed by the appearance of detectable antigen and finally evidence of antibody formation. Thus, testing strategies may target antibody, antigen, and DNA or RNA as is illustrated for HIV in Fig. 12.1. Before detection, the window period exists for any viral infection and includes both the tissue and eclipse phase, in which there is early infection in tissues but no viremia, and preseroconversion window period, in which viral particles are present in blood stream but are below limits of detection by current screening test methodologies.

- Serological testing is used to detect antibody (Ab), antigen (Ag), or both Ag and Ab combination testing (Ag/Ab).
- NAT detects RNA or DNA using either polymerase chain reaction or transcription-mediated amplification approach. NAT has been instrumental to infectious risk mitigation, by decreasing the window period.

Product and Donor Management: Results of infectious disease testing are used to release product, retrieve previously released products, determine donor management, and trigger lookback (i.e., recipient notification) (Tables 11.3–11.5).

TABLE 11.4 Donor Management

Screening Test Result	Confirmatory[a]	Surveillance	Deferral Type	Reentry Available
HBc RR	N/A	Yes—1st hit	2 hits—indefinite	Yes
HBsAg RR	Confirmed	No	Permanent	No
	Not confirmed	No	Temporary: 8 weeks	Automatic after 8 weeks
			Permanent if HBc is RR now or on subsequent donations	No
HBV NAT reactive	N/A	No	Indefinite	Yes
	Negative	No	Indefinite	Yes
HCV RR	Indeterminate	No	Indefinite	Yes
	Positive	No	Permanent	No
HCV NAT reactive	N/A	No	Indefinite	Yes
	Negative or indeterminate and HIV-2 NR	No	Indefinite	Yes
HIV-1/2 RR	Negative/HIV-2 RR HIV-2 negative or indeterminate	No	Indefinite	Yes
	Positive	No	Permanent	No
HIV-1 NAT POS	N/A	No	Indefinite	Yes
HTLV 1/2 RR	Negative	Yes—1st hit	2 hits—permanent	No
	Positive	No	Permanent	No
Syphilis RR or Indeterminate	Syphilis G-positive	No	Indefinite	Yes—defer donor for at least 12 months
	Syphilis G-negative	No	None	N/A
Chagas (*Trypanosoma cruzi*)	Negative,	No	Indefinite	Yes
	Indeterminate, positive	No	Permanent	No
WNV RNA POS ZIKV RNA POS	N/A	No	Temporary:120 days	Automatic after 120 days

[a]For HBsAg RR, HCV RR, and HIV RR, if relevant NAT is reactive, no further confirmation is required. If NAT is nonreactive, additional confirmation testing is performed. *N/A*, not applicable.

TABLE 11.5 In Date Product Retrieval and Lookback

Positive Test	In Date Product Retrieval[a]		Lookback[a]
	Blood Components	Recovered Plasma	Blood Components
HIV-1/2 Ab[b]	All products	Unpooled only	10 years or 1 year before last NR
Triplex (HIV/HCV/HBV) before discrimination	All products	Unpooled only	N/A
HIV NAT[c]	See Triplex	See Triplex	1 year
HBV NAT	See Triplex	See Triplex	None required by FDA
HBsAg	All products	Unpooled only	None required by FDA
HTLV	All products except plasma[d]	None	None required by FDA
HBcAb	All products	None	None
HCV Ab	All products	Unpooled only	10 years or 1 year before last NR
HCV NAT[c]	All products	Unpooled only	1 year
West Nile virus	Products in prior 120 days	Products in prior 120 days	None
Zika virus	Products in prior 120 days	Products in prior 120 days	120 days
Trypanosoma cruzi (Chagas)	All products	None	All products

[a]In date product retrieval does not involve recipient notification. Lookback may involve recipient notification.
[b]HIV-2 RR; may consider unlicensed HIV-2 Western Blot results if performed.
[c]Current donation is NR for antibody.
[d]Pooled and unpooled plasma; Pooled plasma refers to source or recovered plasma distributed for further manufacture that has been already been added to a multiunit pool at the derivatives manufacturer. By extension, an unpooled unit is still held individually and could be discarded without affecting other units.

Further Reading

Perkins, H. A., & Busch, M. P. (2010). Transfusion-associated infections: 50 years of relentless challenges and remarkable progress. *Transfusion*, *50*, 2080–2099.

Zou, S., Stramer, S. L., & Dodd, R. Y. (2012). Donor testing and risk: Current prevalence, incidence, and residual risk of transfusion-transmissible agents in US allogeneic donations. *Transfus Med Rev*, *26*, 119–128.

CHAPTER 12

Human Immunodeficiency Virus Screening

Debra A. Kessler, RN, MS and Alexandra Jimenez, MD

Screening donated blood for human immunodeficiency virus (HIV) is critical to maintaining a safe blood supply. HIV transmission by intravenous administration of infected blood products proved to be highly efficient and accordingly, shortly after HIV-1 discovery in 1983, donor deferral measures were initiated and, in 1985, the first test was licensed for donor screening. Donor deferral and testing has decreased blood transfusion transmission from ~1:100 transfused products in 1982 in the San Francisco area, to current estimates of <1:2 million products nationwide.

While HIV is not the most common infection found in donated blood, it is, perhaps, the most serious and is likely to be distressing to a donor. Therefore, fairly extensive and supportive counseling must be available. Additionally, false reactive screening results must be communicated to donors as well as the possibility of reentry testing to resume donating blood.

Description: HIV is a lentivirus, which is a subgroup of the retrovirus family, and the causative agent of acquired immune deficiency syndrome (AIDS). Retroviruses are RNA viruses with the presence of viral particle–associated reverse transcriptase and a unique replication cycle. Virus particles attach to the cell membrane (in the case of HIV, CD^{4+} lymphocytes), subsequently enter into the host cell, then the reverse transcriptase enzyme copy viral RNA into cDNA (complementary double-stranded DNA), where cDNA is then integrated into host cell's genome. Subsequent transcription, processing, and translation of viral genes are mediated by host cell enzymes. Particles then bud from the plasma membrane and infect other cells. In addition, virus can spread by fusion of infected and uninfected cells or by replication of integrated viral DNA during mitosis or meiosis.

Infection: HIV infection can be transmitted through sexual contact, in utero or childbirth, breastfeeding, and parenteral exposure to blood. The CDC reported that in the United States in 2015, the highest incidence of new diagnosis was in African-Americans, males, age 20–29 years old and the most common risk identified was male-to-male sex. Rate per 100,000 of new infections among women is dropped from 7.3% in 2010 to 5.4% in 2015. In comparison, among men, rate dropped from 27.3% in 2010 to 24.4% in 2015 (https://www.cdc.gov/hiv/library/reports/hiv-surveillance.html).

60% of acute HIV infections result in nonspecific flulike illness with incubation period of 2–4 weeks. Acute infection resolves in weeks to few months resulting in asymptomatic period that may last years. Eventually, the virus can no longer be controlled, as helper CD^{4+} (T4) lymphocytes are destroyed. This loss of CD^{4+} lymphocytes results in development of opportunistic infections and direct viral effects on multiple organs; together resulting in death on average after 3 years once AIDS is diagnosed,

Transfusion Medicine and Hemostasis. https://doi.org/10.1016/B978-0-12-813726-0.00012-X

if not treated. Post infection survival has improved with the advent of potent antiretroviral therapies. However, these medications do not eradicate HIV and have multiple side effects. Moreover, resistant viral strains have been known to develop, which adds to the difficulty of treatment.

HIV Types: HIV types 1 and 2 (HIV-1 and HIV-2) infections both cause AIDS. HIV-1 family is divided into main (M), outlier (O), and non-M, non-O (N) groups. Group M has 11 distinct subtypes or clades (A–K). In the United States, clade B is almost exclusively prevalent. The greatest genetic diversity exists in Central Africa. Group O is the most common in Cameroon and surrounding West African countries (where it represents 1%–2% of HIV infections). Group O infection in the United States is very rare and is usually found in African immigrants or their partners. Previous generations of HIV antibody and nucleic acid assays did not reliably detect group O, but current assays have added sensitivity to group O. HIV-2 is also rare in the United States with no reported cases of HIV-2 transfusion transmission in the United States.

Determination of Need and Requirement for Testing: FDA and AABB *Standards* require donors be screened for HIV using antibody and nucleic acid tests (NATs). The CFR and AABB *Standards* state that donors must be notified of any medically significant abnormality detected through testing. Additionally, regulatory and standard requirements involve tracing donations given before the positive unit (Lookback) and that the collection center notifies consignees within 30 days of a confirmed positive test result, that recipients of donor's previous donations must be notified and recommended to be tested in case donor was in window period when transfused unit was collected.

Anti-HIV Testing: Anti-HIV testing was, and is, accomplished by enzyme-linked immunosorbent assay (ELISA or EIA), and repeatedly reactive tests are confirmed by immunofluorescence assays (IFAs) or Western blot (WB). EIA repeatedly reactive blood units are discarded regardless of outcome of confirmatory testing. In 1992, the FDA required that donor blood be tested for HIV-2 and HIV-1; individual HIV-2 or combination tests for HIV-1/2 EIAs were added to the testing scheme.

Immediately after infection is the "window period," which is the time during which blood product may be infectious but infection cannot be detected via screening tests. Current HIV antibody tests have window period of ~22 days, while NATs narrow to ~11 days. After infection, "blips" or small bursts of viremia have been identified during window period. Whether donor can transmit infection during this period is unclear. This is followed by rapid "ramp up" or increasing level of nucleic acid, antigen, and finally antibody levels (Fig. 12.1).

Nucleic Acid Testing: HIV NAT is performed on pools of 6–16 donor samples depending on technology (minipool NAT). NAT used on single donor samples is termed individual donor NAT (ID NAT). For initial screening, minipool NAT is used, rather than ID NAT, as it significantly reduces expense without significantly decreasing sensitivity and results in fewer false positives. Any pool that is positive is "resolved" by ID NAT on individual samples from the pool to determine which donor sample(s) caused the positive pool result.

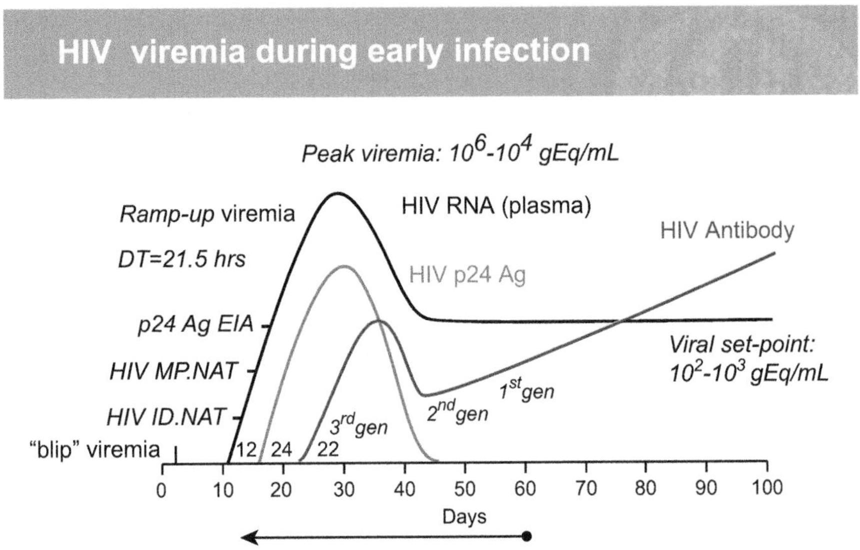

FIGURE 12.1 HIV viremia during early infection. (Courtesy Busch, M. P. (2007). Evolving approaches to estimate risks of transfusion-transmitted viral infections: Incidence-window period model after ten years. *Dev Biol, 127*, 87–112.)

HIV NAT can be done by two methods: (1) PCR or (2) transcription-mediated amplification. In both technologies, HIV-1 RNA is copied by reverse transcriptase to produce transcription complexes and amplified more than one billionfold.

Automation: Currently, much of donor testing is automated, including HIV testing. HIV NAT is currently performed in an assay that is available for HIV/HCV/HBV multiplex testing. This extends the resolution of reactive pools to individual samples that then must be tested to determine which marker is positive for each marker using individual multiplex NAT assay that discriminates between HIV, HCV and HBV or the individual marker NAT assays.

Blood Unit Management: HIV units that are nonreactive on serology (antibody) and NATs are considered to be negative, and related donated unit components may be released from quarantine and issued as appropriate. Units with either reactive HIV serology or NATs are discarded, and donor is indefinitely deferred from future donation. Serology screening tests have relatively high false reactive rate. If a donor tests reactive on HIV antibody and NAT, screening collection is considered to contain HIV and additional confirmation testing is not required. If antibody screening test is repeatedly reactive, but NAT is nonreactive, confirmation testing is performed. The confirmatory tests (WB or IFA) allow determination of infectious state for donor counseling.

Donor Management: Donors who are screening reactive for serology and NAT are permanently deferred as donors who are positive on WB or IFA confirmation testing. Donors whose donations are negative or indeterminate on confirmation testing and negative for NAT or are reactive for NAT and negative for EIA testing, are indefinitely deferred and become eligible for reentry testing after 8 weeks. Reentry testing requires new sample be drawn and test negative for EIA and ID NAT.

Donors who are being treated for HIV infection can often test negative on NAT and confirmed positive for serology because the viral load is being suppressed to undetectable levels. This can be seen in autologous HIV-infected donors.

Further Reading

Anderson, S. A., Yang, H., Gallagher, L. M., et al. (2009). Quantitative estimate of the risks and benefits of possible alternative blood donor deferral strategies for men who have had sex with men. *Transfusion, 49*, 1102–1114.

Evatt, B. (2006). Infectious disease in the blood supply and the public health response. *Semin Hematol, 43*, S4–S9.

O'Brien, S. F., Yi, Q. L., Fan, W., Scalia, V., Goldman, M., & Fearon, M. A. (2017). Residual risk of HIV, HCV and HBV in Canada. *Transfus Apher Sci, 56*, 389–391.

Zou, S., Stramer, S. L., & Dodd, R. Y. (2012). Donor testing and risk: Current prevalence, incidence, and residual risk of transfusion-transmissible agents in US allogeneic donations. *Transfus Med Rev, 26*, 119–128.

CHAPTER 13

Hepatitis B Virus Screening

Debra A. Kessler, RN, MS and Alexandra Jimenez, MD

Hepatitis B virus (HBV) is a DNA virus from the family Hepadnaviridae. It is transmitted parenterally, sexually, and perinatally. Every blood donation is tested for hepatitis B surface antigen (HBsAg), hepatitis B core antibody (anti-HBc), and HBV nucleic acid testing (NAT). Donors may be permanently deferred if confirmed positive or temporarily deferred based on various testing scenarios.

Posttransfusion hepatitis due to HBV was a common consequence from 1940 to 1970, infecting as many as 25% of transfusion recipients. In 1970, the initial mitigation step was to eliminate high-risk donors (prisoners and paid donors) and move to an all-volunteer donor blood supply. This was followed by the introduction of HBsAg testing, decreasing transfusion transmission risk to 12%–13%. In 1982, with the addition of alanine transaminase (ALT) testing, a liver function test that is elevated in some cases of HBV infection, transmission decreased to 10%. Incidence continues to decrease because of increased HBsAg testing sensitivity, introduction of anti-HBc testing in 1987, and the recent addition of HBV NAT.

In the United States, the estimated risk of TT-HBV (transfusion-transmitted HBV) is 1:592,000 to 1:754,000. However, suspected cases of TT-HBV are infrequently confirmed: Of the 2790 cases of TT-HBV reported to CDC in 2014, only one was confirmed. Therefore, calculated residual risk may be overestimated secondary to overestimates of HBV incidence, overestimates of window period, overreporting, incomplete investigations of cases, and widespread use of the HBV vaccine.

Determination of Need and Requirement for Testing: The FDA and AABB Standards require testing allogeneic donors for evidence of infection with HBV NAT, HBsAg, and anti-HBc.

HBsAg: HBsAg, an HBV viral coat antigen, is produced in large quantities in infected-cell cytoplasm and continues to be produced in patients with chronic, active HBV infection (Fig. 13.1). This has been used as the primary screening test. HBsAg can be identified in an infected donor's serum or plasma by enzyme immunoassays (EIAs) using animal antibodies (anti-HBs) as solid phase capture reagent and conjugated anti-HBs as probe. Chemiluminescent labels have replaced enzyme conjugates and chromogenic detection methods in automated testing strategies.

Sensitivities and specificities of current testing methods for HBsAg are high, but prevalence of detectable HBsAg in blood donors is low, and thus positive predictive value is also low.

Confirmation of HBsAg and interpretation of results:

- Donor with repeatedly reactive (RR) testing result for HBsAg may be confirmed by concurrent reactive HBV NAT. These donors are permanently deferred. However, if NAT is nonreactive, neutralization test is performed.

Transfusion Medicine and Hemostasis. https://doi.org/10.1016/B978-0-12-813726-0.00013-1

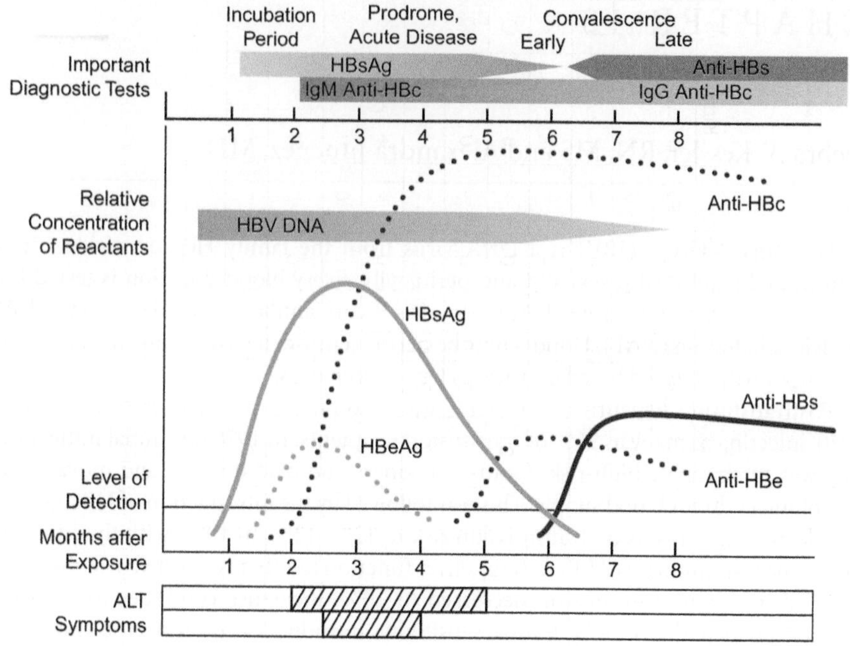

FIGURE 13.1 Emergence of markers for hepatitis B virus (HBV). *ALT*, alanine transaminase; *Anti-HBc*, hepatitis B core antibody; *HBsAg*, hepatitis B surface antigen; *anti-HBs*, hepatitis B surface antibody; *HBeAg*, hepatitis B envelope antigen; *anti-HBe*, hepatitis B envelope antibody. (From Hollinger, F. B. (2008). Hepatitis B virus infection and transfusion medicine: Science and the occult. *Transfusion 48*, 1001–1016.)

- With neutralization testing, HBsAg RR donor is confirmed as positive when anti-HBsAg is added to the serum, and HBsAg signal decreases by ≥50% from the control signal (antibody not added), such that the antigen is neutralized. Confirmed donors are permanently deferred.
  - If donor serum is not neutralized but the donor is RR and anti-HBc–reactive on current or any other donation, the donor is permanently deferred.
  - If the donor's HBsAg reactivity is unconfirmed and anti-HBc is nonreactive, then the donor may be retested after 8 weeks and reinstated if HBsAg and anti-HBc are nonreactive. This retest may occur as per routine testing of a new donation.

Anti-HBc: Anti-HBc testing was implemented in the mid-1980s as a surrogate marker to reduce the transmission of other hepatidities, especially non-A, non-B hepatitis (now called hepatitis C). Anti-HBc is present in infected individuals, both in chronic carrier state and at the end of an acute resolving infection. Tests used for donor screening detect IgM and IgG. Because typically only IgM is tested for in the clinical setting, donors believe our results are wrong. Two available test methods are (1) solid phase inhibition immunoassay, with recombinant HBc as the capture reagent and labeled, partially purified anti-HBc antibody as the probe (presence of antibody in donor serum inhibits signal) and (2) direct antiglobulin assay for anti-HBc, which has better specificity than inhibition immunoassay method. No confirmatory test method for anti-HBc exists, but using two different test methods increases probability that any individual RR is either positive or negative. In the United States, the different methodologies are not typically used together, and instead, most blood centers in the United States allow a donor to continue to donate after a single RR anti-HBc test but be indefinitely deferred after a second RR anti-HBc test.

HBV NAT: HBV is constantly evolving, and sometimes tests do not pick up mutated strains. TT-HBV occurs from donors who are negative for HBsAg yet are infected with HBV (i.e., during the window period or owing to mutant virus strains). Although HBV NAT was licensed by the FDA in 2005, recommendation for its use was not made because the newly licensed test format (testing in a pool of 24 samples) did not have additional sensitivity beyond that of the HBsAg test. Since then, two new tests have been licensed, which have increased sensitivity, resulting in a decreased window period (18.5-29.2 days MP-NAT vs 30-38 days HBsAg EIA). The new tests use 6 and 16 sample pools, respectively. These new tests are in multiplex format allowing for HIV, HCV, and HBV NAT, simultaneously.

HBV NAT detects breakthrough infections that were previously undetected by serologic methods and can occur in previously vaccinated individuals who are exposed to HBV. In one study of 3.7 million donations, nine HBV NAT–only positives were found, of which six were in vaccinated individuals. Breakthrough infections are characterized by HBV NAT positivity, presence of HBV-neutralizing anti-HBs (developed as a result of HBV vaccination), low viral load, and absence of symptoms. HBsAg and anti-HBc may not develop or may be delayed. Infectivity of units obtained from HBV-vaccinated donors with breakthrough HBV infections is unknown. Nevertheless, this will increase as more of the population, especially younger donors, gets vaccinated. HBV vaccination is recommended as part of childhood vaccination program, and if not received at that age, most schools require that it be administered by age 12 years as a pre-requisite to attending classes, and most healthcare agencies require it for employment.

Donors who solely test positive for HBV NAT are indefinitely deferred with possibility of reentry.

Donor Management: *Permanent deferral:* donors who are RR on HBsAg EIA screening and HBV NAT; donors who confirm positive for HBsAg on neutralization testing; and donors who are RR on HBsAg EIA screening, not neutralized on testing, but are RR for anti-HBc.

Indefinite deferral: donors who consistently test negative for HBV NAT and HBsAg but are RR for anti-HBc on more than one occasion and donors who are solely NAT-positive.

Reentry: To reenter a donor who has tested NAT-positive, you must collect a nondonation sample at least 6 months after the positive sample. Individual donor (ID) NAT must be performed, as the FDA requires that testing be done using a test with sensitivity <2 IU/mL at 95% detection rate. Additionally, HBsAg and anti-HBc licensed assays must be performed. If the donor is negative on all tests, they may be reentered. If they test positive by ID-NAT, they must be permanently deferred.

Donors, who test negative for NAT and HBsAg but are RR for anti-HBc on more than one occasion, are eligible for reentry testing after a minimum of 8 weeks after the last RR anti-HBc test. A nondonation sample must be used and tested using FDA-licensed tests for HBsAg, anti-HBc, and ID-NAT. If the donor is negative on all tests, they may be reentered.

Further Reading

Hollinger, F. B. (2008). Hepatitis B virus infection and transfusion medicine: Science and the occult. *Transfusion*, *48*, 1001–1026.

Janssen, M. P., van Hulst, M., Custer, B., et al. (June 22, 2017). An assessment of differences in costs and health benefits of serology and NAT screening of donations for blood transfusion in different Western countries. *Vox Sang*. https://doi.org/10.1111/vox.12543.

Stramer, S. L., Notari, E. P., Krysztof, D. E., & Dodd, R. Y. (2013). Hepatitis B virus testing by mini pool nucleic acid testing: Does it improve blood safety? *Transfusion*, *53*, 2449–2458.

Stramer, S. L., Wend, U., Candotti, D., et al. (2011). Nucleic acid testing to detect HBV infection in blood donors. *N Engl J Med*, *364*, 236–247.

Zou, S., Stramer, S. L., & Dodd, R. Y. (2012). Donor testing and risk: Current prevalence, incidence, and residual risk of transfusion-transmissible agents in US allogeneic donations. *Transfus Med Rev*, *26*(2), 119–128.

CHAPTER 14

Hepatitis C Virus Screening

Debra A. Kessler, RN, MS and Alexandra Jimenez, MD

Hepatitis C virus (HCV) is primarily transmitted through blood exposure. It is a single-stranded RNA flavivirus, distinguished by a low rate of recognized acute infection and a high rate of chronic infection that results in substantial morbidity and mortality over long periods of observation. Because most infections are asymptomatic, testing blood donors is critical.

Historically, any viral hepatitis that was not caused by hepatitis A or B viruses was referred to as non-A, non-B hepatitis, which in 1989 was discovered to be HCV. In 1987, anti-HBc was implemented and served as a surrogate marker for screening out donors with these nonspecific hepatidities. In addition, HIV high-risk behavior questions, especially those related to blood exposure and needle use, eliminated many donors at risk of HCV infection. In 1990, anti-HCV testing was implemented, followed by a much improved test in 1992. In 1999, nucleic acid testing (NAT) became available; subsequently residual risk of transfusion-transmitted HCV (TT-HCV) is estimated to be 1:1,800,000 units transfused.

In 2015, the CDC reported that in the United States, the number of acute cases of HCV has been increasing annually from 2010 through 2015, with an overall incidence rate of 0.8 cases per 100,000 (https://www.cdc.gov/hepatitis/statistics/2015surveillance/index.htm). The most common risk factor identified was intravenous drug use and occupational blood exposure from needlesticks.

Treatment of HCV infection is critical to improving long-term health outcomes particularly because of the major advancements that have been made in treatments. Previously, treatment consisted of pegylated interferon combined with oral doses of ribavirin. Since 2011, with the development of direct-acting antivirals targeting nonstructural proteins of HCV, use of a combination regimen of these agents has increased viral response rates to >90%.

Determination of Need and Requirement for Testing: FDA guidance and AABB *Standards* require testing allogeneic donors for evidence of infection with HCV using anti-HCV and HCV RNA testing. If either of the tests are reactive, the unit must be discarded. In addition, donors of these units must be notified, educated about the meaning of the test result, and placed on the collection center's donor deferral registry. Lastly, an investigation of previous units donated by a confirmed positive donor must be subjected to lookback per FDA. Lookback requires that the collection center notify consignees within 45 days of a confirmed positive test result and that recipients of the donor's previous donations must be notified and recommended to be tested in case the donor was in the window period when the transfused unit was collected.

Anti-HCV: The primary screening tests for HCV antibodies are a third-generation enzyme immunoassay (EIA) and a chemiluminescent immunoassay (ChLIA). The currently available tests have recombinant HCV antigens representing four viral sequences,

Transfusion Medicine and Hemostasis. https://doi.org/10.1016/B978-0-12-813726-0.00014-3

with 98% sensitivity approximately 10 weeks (window period) after infection with HCV. If initial screening for anti-HCV is reactive, the test is repeated in duplicate. If two out of three of the screening tests are reactive, the sample is considered repeatedly reactive (RR).

HCV NAT: NAT is performed concurrently to antibody screening test. Two HCV NAT assays are FDA licensed and commercially available: (1) PCR test performed on RNA by chemical methods and (2) transcription-mediated amplification (TMA) test on RNA in a solid phase (probe-capture) method. Both test methods can identify HCV RNA in an average of 11 days after infection. Both are available in a multiplex format (HIV, HCV, and HBV). Multiplex testing occurs in pools of 6 or 16. If a pool is positive, individual samples are tested with multiplex assay some of these assays can discriminate between the three infections otherwise an individual NAT for each marker must be performed. If a donor is HCV NAT–positive and EIA or ChLIA–RR, the likelihood that he or she has acute or chronic HCV infection approaches 100%. This donor will be permanently deferred, and no additional testing is needed. If the donor tests negative for HCV NAT, but RR for EIA or ChLIA, then supplemental testing may be performed.

Supplemental Testing: For many years, confirmatory testing for an RR HCV screening test was performed using the only FDA-licensed antibody supplemental/confirmatory test, recombinant immunoblot assay (RIBA), against HCV antigens. In October 2012, RIBA became commercially unavailable in the United States. Today testing uses an alternate EIA for supplemental testing.

Irrespective of supplemental testing results, if a donor is RR by HCV antibody screening or NAT, his or her donation cannot be used for transfusion.

Donor Management: Donors who are: (1) NAT nonreactive, RR on EIA screening test, and alternate ELISA supplemental test reactive or non reactive; (or historically RIBA indeterminate, negative or not performed), or (2) reactive for HCV NAT but nonreactive for screening EIA may be tested for reentry. The reentry sample may be drawn at a minimum of 6 months after reactive collection. Reentry testing may not be performed on a donation. The sample is tested with individual donor multiplex NAT that discriminates between HIV, HCV and HBV or specific individual HCV NAT and two different EIA tests. If all are nonreactive, the donor may be reentered to the pool of eligible donors.

Further Reading

Custer, B., Kessler, D., Vahidnia, F., et al. (2015). Risk factors for retrovirus and hepatitis virus infections in accepted blood donors. *Transfusion*, *55*(5), 1098–1107.

Kleinman, S. H., Stramer, S. L., Brodsky, N. P., et al. (2006). Integration of nucleic acid amplification test results into hepatitis C virus supplemental serologic testing algorithms: Implications for donor counseling and revision of existing algorithms. *Transfusion*, *46*, 695–702.

Stramer, S. L., Glynn, S. A., Kleinman, S. H., et al. (2004). Detection of HIV-1 and HCV infections among antibody-negative blood donors by nucleic acid-amplification testing. *N Engl J Med*, *351*, 760–768.

CHAPTER 15

West Nile Virus Screening

Debra A. Kessler, RN, MS and Alexandra Jimenez, MD

West Nile virus (WNV), a single-stranded RNA flavivirus, is primarily spread by the mosquito *Aedes albopictus*. The only screening test used for WNV is nucleic acid testing (NAT), although IgG and IgM antibody testing may be used for donor counseling. Many donors have low viral copy numbers of circulating WNV, which result in false-negative testing by minipool (MP) NAT but can be identified by individual donor (ID) NAT. Thus, during local WNV activity, testing is switched from MP-NAT to ID-NAT.

WNV transmission involves birds and mosquitoes, and thus humans are incidental hosts. In 1999, WNV was first reported in the United States in New York and subsequently spread westward throughout the continental United States where it caused (and still causes) significant seasonal epidemics. While ~80% of human WNV infections are asymptomatic, symptomatic infections result in fever, muscle ache and headache, nausea, and vomiting. About 1 in 150 infected individuals will have severe disease resulting in meningitis and/or encephalitis (convulsion, coma, paralysis) and, less frequently, death.

Transfusion transmission of WNV was confirmed in 23 patients after RBC, platelet, and plasma transfusions in 2002. Of the 16 donors of these 23 units, 9 had viral symptoms before or after donation and 5 were asymptomatic (2 were lost to follow-up).

In October 2002, the FDA issued guidance on WNV for deferral of donors with suspected or acute WNV infection and for retrieval and quarantine of any blood products from donors with postdonation illness that could be from WNV infection (deferral of donors based on donor questionnaire was retracted in 2005). By summer of 2003, it was recognized that a screening test for WNV was needed and an MP-NAT was implemented. From 2003 to 2005, >1000 viremic donors were documented and seven cases of probable or confirmed transfusion transmission occurred. It became increasingly recognized that many donors had low viral copy numbers of circulating WNV, which result in false-negative testing by MP-NAT but could be identified by ID-NAT. Thus, the testing algorithm was changed from MP testing only to establishing a trigger (and detrigger) for ID-NAT; that is, MP-NAT was used until viremic donations were identified in the geographic area of the donor's residence, then ID-NAT was used in that geographic area to increase sensitivity. In 2006, at least one case of transfusion-transmitted WNV was reported, resulting in further changes in the testing algorithm, most notably triggering ID-NAT on only one positive donor in the geographic area. This was followed by FDA guidance supporting this scenario. Area of residence could be determined by zip codes, county, or other comparable well-specified regions. These steps have mitigated WNV transfusion transmission.

Transfusion Medicine and Hemostasis. https://doi.org/10.1016/B978-0-12-813726-0.00015-5

Determination of Need and Requirement for Testing: FDA guidance and AABB *Standards* require screening allogeneic blood donors for WNV RNA.

NAT Screening: Transfusion-associated WNV infection during the 2002 US epidemic prompted rapid development of two NAT assays to screen for WNV viremia, first implemented under an IND (investigational new drug) in 2003 (now FDA-approved). Year-round MP-NAT is performed, and when an appropriate epidemiologic trigger is reached, ID-NAT is initiated.

WNV testing has been estimated to cost ~1.5 million US dollars per quality-adjusted life year gained. Together, AABB WNV Biovigilance Network (http://www.aabb.org/programs/biovigilance/Pages/wnv.aspx), which collects positive WNV donor activity, and local department of health, which tracks WNV infections in humans, mosquitos, birds, and other animals, allow for efficient data communication between donor centers. This communication provides blood collectors with the information needed for the implementation of ID-NAT based on geographic criteria.

In the final guidance issued in November 2009, the FDA allows individual centers to define a threshold for switching from MP-NAT to ID-NAT screening during high WNV activity and for reverting to MP-NAT screening when the high WNV activity subsides. In 2013, AABB published recommendations for conversion from MP-NAT to ID-NAT based on identification of a presumptive viremic donation (PVD) (http://www.aabb.org/programs/publications/bulletins/Pages/ab13-02.aspx). PVD is defined as a donation from a donor whose blood tests positive for WNV by NAT.

The criteria recommended by AABB is the use of one of the following:

a. One PVD with evidence of other WNV activity in a collection region.
b. Two PVDs in a 7-day rolling period if there is no other WNV activity reported in a collection region.
c. One PVD if the collection facility decides not to include considerations of other WNV activity in the region in its triggering decision.

AABB defines "other WNV activity" as the identification of PVDs at another blood center in the collection region, if human WNV cases have been reported to county/state health departments or CDC and/or if there are reports of WNV activity in mosquitoes or animals. Once one of the trigger criteria has been identified, ID-NAT should occur within 48 hours. To revert back to MP-NAT, AABB recommends that there must be a minimum of 14 days without a PVD and an absence of other indicators of regional WNV activity. This can be assessed at 7-day intervals corresponding with updates to public health surveillance sites.

Confirmatory Test: WNV infectivity of the donor may be confirmed by repeat NAT reactivity on a follow-up sample or IgM and IgG antibody reactivity on either the original (index) sample or follow-up sample. Although this information plays no role in donor deferral, it can be useful in donor counseling.

Donor Management: WNV NAT-reactive donations cannot be used for transfusion, and all prior in-date components collected from a positive donor within 120 days before the positive donation should be discarded. The donor is then deferred for

120 days after the WNV-positive donation. Studies have shown that the virus is cleared by this time, IgG is present, and generally IgM has cleared.

Further Reading

Busch, M. P., Caglioti, S., Robertson, E. F., et al. (2005). Screening the blood supply for West Nile virus RNA by nucleic acid amplification testing. *N Engl J Med, 353*, 460–467.

Francis, R., Strauss, D., Dunn, W., et al. (2012). West Nile virus infection in blood donors in the New York City area during the 2010 seasonal epidemic. *Transfusion, 52*, 2664–2670.

Groves, J. A., Shaft, H., Normura, J. H., et al. (2017). A probable case of West Nile virus transfusion transmission. *Transfusion, 57*, 850–856.

Stramer, S. L., Fang, C. T., Foster, G. A., et al. (2005). West Nile virus among blood donors in the United States, 2003 and 2004. *N Engl J Med, 353*, 451–459.

CHAPTER 16

Zika Virus Screening

Carolyn T. Young, MD

In 2016, the association between maternal Zika viral infection and microcephaly in babies was recognized. Zika virus is spread through mosquitoes, sexual transmission, and potentially through transfusion. Owing to its associated morbidity, FDA recommended screening the blood supply by nucleic acid testing (NAT) or performing pathogen inactivation. To date, no Zika virus infection has been reported through transfusions in the United States; however, three infections have been reported through platelet transfusions worldwide without clinical sequelae in the recipients.

Infection: Zika virus, discovered in 1947 and named after the Zika forest in Uganda, is a primarily *Aedes aegypti* mosquito–borne ribonucleic acid (RNA) flavivirus. The first human Zika infection occurred in 1952. The first epidemic outside of Africa and Asia occurred in 2007 in the Yap Islands. A Zika epidemic occurred in Brazil in 2015–16. In February 2016, the World Health Organization declared the suspected link between Zika virus disease and microcephaly—a Public Health Emergency of International Concern. Data from the French Polynesian Zika outbreak that occurred in 2013–14 strongly supported the hypothesis that Zika infection in the first trimester of pregnancy was associated with an increased risk of microcephaly. Furthermore, 3% of samples from asymptomatic blood donors contained detectable Zika virus RNA during the outbreak, which caused concern for transmission by blood transfusions.

In 2017, data from the US Zika Pregnancy Registry reported that the rate of Zika virus–associated birth defects was 11%–15% if the infection occurred within the first trimester. Laboratory evidence of possible Zika virus infection included a positive Zika virus RNA NAT or detection of recent Zika virus infection or recent unspecified flavivirus infection by serologic tests, i.e., positive or equivocal Zika virus IgM and Zika virus plaque reduction neutralization test (PRNT) titer ≥10 and Dengue virus PRNT <10. Zika disease is nationally reportable.

Although pregnant women were asymptomatic in 62% infected, birth defects were reported in similar proportions whether or not the mother acknowledged symptoms of Zika virus disease. Birth defects associated with Zika are defined by the Registry as microcephaly, calcium deposits in the brain indicating possible brain damage, excess fluid in the brain cavities and surrounding the brain, absent or poorly formed brain structures, abnormal eye development, confirmed hearing loss, and other problems from damage to the brain that affects nerves, muscles, and bones, such as club foot or inflexible joints. The most recent update on data from 2017 stated that there were 2191 completed pregnancies, i.e., live births, miscarriages, stillbirths, and terminations, with or without Zika-associated birth defects. Of those, 106 live-born infants and 9 pregnancy losses had Zika-associated birth defects, which constitute ~5% of the completed pregnancies.

Transfusion Medicine and Hemostasis. https://doi.org/10.1016/B978-0-12-813726-0.00016-7

Zika virus infections are asymptomatic in about 80% of individuals. Mild symptoms may consist of fever, rash, headache, myalgia, conjunctivitis, and arthralgias. Guillain–Barre may occur.

Determination of Need and Requirement for Testing: The FDA issued final guidance recommending either NAT or pathogen reduction of blood products.

NAT Screening: Two commercial platforms for blood donor NAT are available in the United States; one is still under an investigational new drug (IND) protocol, whereas the other was FDA-approved in 2017. The assays use either real time reverse transcription polymerase chain reaction (PCR) or transcription-mediated amplification. Individual or combinations of minipool and individual donor NAT were used in INDs in April 2016. The approved test is for individual samples. Minipools of three donor samples were used for PCR testing when Zika virus testing was performed November 2013–February 2014 in French Polynesia. Research is continuing on the use of pooled compared with individual samples. One test has a clinical specificity greater than 99%. The other test has analytical sensitivity of 95% and detection of 10–30 copies/mL. Specificities are similar to current blood screening NAT at greater than or equal to 99.90%. They detect genetic variants of Zika virus, including African and Asian strains. Tests use two-region amplification and detection.

Confirmatory Test: Serologic assays such as immunofluorescent antibody and PRNT have been used for confirmation of results. However, donor and product management are based on initial result of NAT.

Blood Unit Management: All blood donations are tested for Zika by NAT although it is not required if pathogen reduction is performed. Logistically, it is not feasible to separate apheresis platelets which will require testing from those which will not. If a donation has a positive result, all blood products from the associated donation are discarded. All in-date blood and blood components collected from the donor in the 120 days before the donation that is Zika NAT–reactive is quarantined, recalled, and discarded. If blood components were transfused, the transfusion service is advised to inform the recipient's physician of record regarding the possible need for monitoring and counseling the recipient for a possible Zika virus infection.

Donor Management: Donors who have reactive results for Zika NAT or recent history of Zika virus infection are deferred for 120 days after a positive test or symptom resolution, whichever is longer. Separate reentry testing is not required after the deferral. Testing is performed on every donation year round.

Further Reading

Barjas-Castro, M. L., Angerami, R. N., Cunha, M. S., Suzuki, A., Noqueira, J. S., & Roco, I. M. (2016). Probable transfusion-transmission of Zika virus in Brazil. *Transfusion*, *56*(7), 1684–1688.

Galel, S. A., Williamson, P. C., Busch, M. P., Stanek, D., Bakkour, S., Stone, M., et al. (2017). First Zika-positive donations in the continental United States. *Transfusion*, *57*, 762–769.

Jimenez, A., Shaz, B. H., & Bloch, E. M. (2017). Zika virus and the blood supply: What do we know? *Transfus Med Rev*, *31*(1), 1–10.

Jimenez, A., Shaz, B. H., Kessler, D., & Bloch, E. M. (2017). How do we manage blood donors and recipients after a positive Zika screening result? *Transfusion*, *57*(9), 2077–2083.

Johansson, M. A., Mier-y-Teran-Romero, L., Reefhuis, J., Gilboa, S. M., & Hills, S. L. (2016). Zika and the risk of microcephaly. *N Engl J Med*, *375*(1), 1–4.

Katz, L. M., & Rossman, S. N. (2017). Zika and the blood supply. *Arch Pathol Lab Med*, *141*, 85–92.

Motta, I. J. F., Spencer, B. R., Cordeiro da Silva, S. G., Arruda, M. B., Dobbin, J. A., Gonzaga, Y. B., et al. (2016). Evidence for transmission of Zika virus by platelet transfusion. *N Engl J Med*, *375*, 1101–1103.

Musso, D., Nhan, T., Robin, E., Roche, C., Bierlaire, D., Zisou, K., et al. (2014). Potential for Zika virus transmission through blood transfusion demonstrated during an outbreak in French Polynesia, November 2013 to February 2014. *Euro Surveill*, *19*(4) pii: 20761.

Williamson, P. C., Linnen, J. M., Kessler, D. A., Shaz, B. H., Kamel, H., Vassallo, R. R., et al. (2017). First cases of Zika virus-infected US blood donors outside states with areas of active transmission. *Transfusion*, *57*, 770–778.

CHAPTER 17

Babesia Screening

Carolyn T. Young, MD

Babesia microti is the leading red blood cell (RBC) transfusion-transmitted pathogen reported to the FDA. Babesiosis most commonly occurs after a tick bite but can be transfusion-transmitted and vertically transmitted. Owing to the increasing incidence of babesiosis and transfusion-transmitted babesiosis (TTB), testing of blood donors was implemented. Currently, there is no licensed test for blood donor screening for *B.microti* which can be performed at a blood center, consequently, blood centers performing testing for *B. microti* are doing so under investigational new drug (IND) protocols. Testing has prevented TTB.

Description: *B. microti* has resulted in 0-2 deaths annually reported to the FDA since FY 2012. Babesiosis has been nationally reportable since 2011. Incidence of babesiosis and TTB has increased. Hundreds of cases of TTB have been reported (1%–2% of babesia cases in New York State are TTB cases). TTB has occurred after administration of RBC-containing blood products, primarily RBCs since whole blood-derived platelets contain minimal red cells. Some blood centers in high-risk, endemic areas began screening whole blood and RBC donors for *B. microti.*

Infection: Babesiosis is caused by intraerythrocytic protozoan parasites, usually *B. microti*, via deer tick, *Ixodes scapularis*. *B. microti* is endemic to the northeast and upper midwest of the United States, and its geography is expanding. The first case of babesiosis in humans was identified in 1957 in Croatia, in an asplenic farmer who had grazed his cattle in tick-infested pastures. He presented with fever, anemia, and hemoglobinuria and expired owing to renal failure. The agent was either *Babesia bovis* or *Babesia divergens*, both pathogens of cattle. Other species causing babesiosis in humans include *Babesia duncani* (WA1) reported in the north Pacific coast, *B. divergens*–like (MO1) organisms from Missouri, *Babesia odocoilei* (EU1) a species seen in Europe, *Babesia venatorum* in China, TW1 in Taiwan, and KO1 in Korea. Originally a disease identified in Europe, babesiosis has since been reported globally.

Patients with babesiosis may experience no symptoms, mild to moderate flulike symptoms, or severe symptoms leading to death; symptoms typically occur within 6 weeks. Severe disease is more likely in people who are immunosuppressed including patients with malignancies, asplenia, and HIV infection. Asymptomatic infections contribute to the variance between seroprevalence and the number of reported cases. On Block Island, RI, 19% (13 of 67) of infected adults and 40% (4 of 10) of the infected children were asymptomatic. *B. microti* immunofluorescence antibody (IFA) titers were ≥1:64. When symptomatic, babesiosis may resolve within months but can last over a year. Treatment is usually a course of antibiotics using atovaquone and azithromycin or clindamycin and quinine; severe parasitemia may require RBC exchange.

Transfusion Medicine and Hemostasis. https://doi.org/10.1016/B978-0-12-813726-0.00017-9

The greatest risk of tick infection occurs during warmer months when the ticks are most active. However, TTB occurs year-around, and clinical laboratory testing shows positive results occur year-around.

Determination of Need and Requirement for Testing: Testing for babesia is not mandated by the FDA or AABB. Testing for *B. microti* in blood donors started as seroprevalence surveillance studies using IFA without affecting blood management in real time and demonstrated that a history of tick bites was not a reliable indicator of serologic status. While 0.6% CT blood donors were seropositive, only 0.4% of those reporting bites had *B. microti* antibodies. The first IND for *B. microti* screening was submitted in 2009 after TTBs in infants. There were two additional INDs testing for *B.microti*, one using IFA and PCR, the other using a peptide-based enzyme immunoassay (EIA) before the Blood Products Advisory Committee (BPAC) meeting in 2015 agreed 11 to 3, that scientific data and the FDA analysis support the concept of nationwide, year-round testing of blood donations for Babesia risk by an antibody-based test. BPAC agreed unanimously that NAT-based testing should be performed on blood donations in certain high-risk states. Eight members voted in favor of all nine known endemic states—CT, MA, RI, NY, NJ, MN, WI, NH, and ME—having NAT-based testing. Six members voted that MD, DC, VA, VT, PA, DE, and FL also be included. Alternative options included adding PA to the known state list. There is no licensed test for blood donor screening for *B. microti* which can be performed by a blood center, although the FDA approved an antibody and a NAT-based test in March 2018, all testing performed by blood centers are under INDs. Since testing is licensed, TTB is a relevant transfusion transmitted infection by definition, and the FDA will provide guidance.

NAT Screening: Babesia Testing Investigational Containment (BTIC) study used a NAT-based assay and an IFA assay for donor screening in 2010. The polymerase chain reaction (PCR) used primers and probes targeting the *B. microti* 18S ribosomal ribonucleic acid (RNA) gene. Target DNA extraction and concentration from whole blood were performed with a magnetic bead–based isolation and purification system. Sensitivity of PCR detection was 66 piroplasms per mL. Results were positive if *B. microti* DNA was detected in one or more PCR amplifications from a replicate extraction of the sample of donor blood. The IFA used a cutoff of ≥1:128. Infants, pediatric SCD and thalassemia patients, were the target patient populations to receive these babesia tested units. Using the same assays, Blood Donation Screening for Evidence of Babesia Microti by Real-time PCR and IFA Assays (BNATIFA) was approved in 2012. BTIC joined BNATIFA in 2013. The reactive rates for the studies were 0.4%–0.6%; most reactive results were due to IFA. PCR comprised 20% of the reactive results, and 13% of the PCR positive samples were IFA-negative in BNATIFA. In follow-up samples, the median time to PCR-negative status was 4.7 months; 86% of the donors with PCR-positive results resolved to negative results within 1 year. The median time for donors remaining seropositive by IFA regardless of PCR status was about 17 months. Blood donor screening for *B. microti* under IND has shown positive results year-round. No reported case of TTB occurred from any *B. microti* tested negative unit under these INDs. However, TTB cases have been reported from *B. microti* untested blood at 1 in 18,074.

Two high-throughput NAT-only screening assays began in 2017 under IND. Both assays are designed to detect the presence of four clinically relevant species of Babesia, i.e., *B. microti, B. duncani, B. divergens,* and *B. venatorum*. They are qualitative systems that detect Babesia DNA and RNA in whole blood samples from human blood donors.

Antibody Screening: A peptide-based EIA detecting both IgM and IgG antibodies to *B. microti* was implemented in 2013 under IND. Seroprevalence with EIA was about 1% in a high-risk area but decreased to <0.5% over time. Evidence suggests that screening donors has decreased incidence in TTB in areas where EIA has been implemented.

Confirmatory Testing: Under IND, additional assays may be performed to confirm positive screening results.

All donor management is based on the initial screening results, which are the tests of record.

Blood Unit Management: All in-date blood and blood components from donors with positive Babesia testing results are discarded. Consignee notification is performed on all cellular products donated within 12 months of a positive test result. Recipient testing may be indicated.

Donor Management: Donors who give RBCs are tested at each donation. Currently, donors with positive testing results are deferred indefinitely. Donors are notified of positive testing results performed as an investigational test and advised to see their physician. Donor follow-up testing may be performed to determine duration of positive tests.

Donors who present to donate with a history of babesiosis or provide postdonation information of a history of babesiosis are deferred indefinitely. There is no FDA-approved donor reentry testing algorithm; however, BPAC members supported a deferral period of at least 2 years and recommended that a reentry algorithm include antibody testing and NAT.

Further Reading

AABB. *Association bulletin #14-05 Babesiosis.* Available from: http://www.aabb.org/programs/publications/bulletins/documents/ab14-05.pdf.

Bloch, E. M., Levin, A. E., Williamson, P. C., Cyrus, S., Shaz, B. H., Kessler, D., et al. (2016). A prospective evaluation of chronic *Babesia microti* infection in seroreactive blood donors. *Transfusion, 56*(7), 1875–1882.

Food and Drug Administration. (May 13, 2015). *Blood Products Advisory Committee Meeting Summary Minutes.* Available from: http://wayback.archive-it.org/7993/20170723130835;https://www.fda.gov/AdvisoryCommittees/CommitteesMeetingMaterials/BloodVaccinesandOtherBiologics/BloodProductsAdvisoryCommittee/ucm457554.htm.

Herwaldt, B., Linden, J. B., Bosserman, E., Young, C., Olkowska, D., & Wilson, M. (2011). Transfusion-associated babesiosis in the United States: A description of cases. *Ann Intern Med, 155*(8), 509–519.

Leiby, D. A., Chung, A. P., Cable, R. G., Trouen-Trend, J., McCullough, J., Homer, M. J., et al. (2002). Relationship between tick bites and the seroprevalence of *Babesia microti* and *Anaplasma phagocytophilia* (previously *Ehrlichia* sp.) in blood donors. *Transfusion, 42*(12), 1585–1591.

Levin, A. E., Williamson, P. C., Bloch, E. M., Clifford, J., Cyrus, S., Shaz, B. H., et al. (2016). Serologic screening of United States blood donors for *Babesia microti* using an investigational enzymeimmunoassay. *Transfusion, 56*(7), 1866–1874.

Moritz, E. D., Winton, C. S., Tonnetti, L., Townsend, R. L., Berardi, V. P., Hewins, M., et al. (2016). Screening for *Babesia microti* in the U.S. blood supply. *N Engl J Med, 375*, 2236–2245.

Young, C., Chawla, A., Berardi, V., Padbury, J., Skowron, G., Krause, P., et al. (2012). Preventing transfusion-transmitted babesiosis: Preliminary experience of the first laboratory-based blood donor screening program. *Transfusion, 52*, 1523–1529.

CHAPTER 18

Syphilis, Human T-Cell Lymphotropic Virus, and Chagas Screening

Debra A. Kessler, RN, MS and Alexandra Jimenez, MD

Testing blood donors for syphilis, human T-cell lymphotropic virus (HTLV), and *Trypanosoma cruzi* infection is mandated by the FDA and AABB Standards. Syphilis was the first transfusion-transmitted disease for which the blood supply was screened in the United States. HTLV is a retrovirus that can be transmitted by transfusion, and donor testing has mitigated the risk since put into place in 1988. Transfusion-transmitted Chagas disease, which is caused by *T. cruzi*, has rarely been reported.

Syphilis: Syphilis was the most commonly recognized transfusion-transmitted disease pre–World War II. For more than 50 years, donors have been screened for evidence of infection by *Treponema pallidum*, the organism responsible for causing syphilis. The last reported case of transfusion-associated syphilis was in 1966. Disappearance of transfusion-associated syphilis resulted from decline in incidence, antibiotic use, increased control of the disease spread, cessation of direct donor-to-patient transfusion, storage of blood components (*T. pallidum*, which is anaerobic, is not infectious after ~72 hours of refrigeration), and donation testing.

Background: Syphilis is transmitted primarily through sexual or vertical transmission; the spirochete passes through intact mucous membranes or compromised skin. The primary stage of syphilis is a chancre, which appears 10–90 days (average 21 days) after infection. The organism then disseminates to the blood. Four to ten weeks after the primary stage of infection, untreated individuals enter the secondary stage, developing a rash, fever, sore throat, malaise, weight loss, hair loss, and headache. At this stage they will become seropositive. Then, untreated individuals enter the latent stage, lasting years. The late stage (neurosyphilis and other damage to internal organs) presents in ~15% of untreated individuals and can appear 10–20 years after infection. Both treponemal and nontreponemal antibodies are present during this stage.

Determination of Need and Requirement for Testing

Serologic Screening Tests: Antibody detection is either via nontreponemal or treponemal tests. Nontreponemal tests identify active or recent infections and become negative after disease treatment; treponemal tests identify current and distant infection. Until 2018, most blood centers used treponemal tests because they were automated and had better performance characteristics. The frequency of reactive treponemal screening tests of US donors is ~0.05%. In January 2018, an automated nontreponemal test received approval for donor screening. Given that the vast majority of donors with reactive treponemal tests are nonreactive on nontreponemal testing, this should save many donors from deferral.

Transfusion Medicine and Hemostasis. https://doi.org/10.1016/B978-0-12-813726-0.00018-0

Nontreponemal Antibody Tests (Rapid Plasma Reagin Test or Venereal Disease Research Laboratory Test): Nontreponemal tests detect antibodies that react against cardiolipin phospholipid produced in response to infected host tissue. Nontreponemal tests have a false-positive rate of 1%–2% owing to pregnancy, other infections (such as HIV, mononucleosis, tuberculosis, rickettsial infection, and other spirochetal and bacterial endocarditis), and disorders of immunoglobulin production (such as rheumatoid arthritis and ulcerative colitis cirrhosis). Rapid plasma reagin (RPR) test uses cardiolipin-coated carbon particles, which agglutinate with antibodies.

Treponemal Antibody Tests: Treponemal tests detect antibodies that specifically target *T. pallidiun*. Microhemagglutination for *T. pallidum* (MHA-TP) system has been widely used for donor screening because of its automation. MHA-TP uses sensitized sheep erythrocytes coated with *T. pallidum*, which agglutinate with anti–treponemal IgM and IgG antibodies.

Confirmatory/Supplemental Tests: Fluorescent treponemal antibody absorption (FTA-ABS) test and syphilis-G enzyme immunoassay (Syphilis-G) are automated treponemal antibody tests specific to *T. pallidum*. Using FTA-ABS test, antibodies bind *T. palladium* antigen, and resulting antigen–antibody complex reacts with fluorescein-tagged anti–human globulin, which is visualized by green fluorescence under UV light. Syphilis-G uses biotinylated anti–human IgG labeled with streptavidin–peroxidase to detect antibody. False positivity occurs most commonly in individuals with autoimmune disease.

Donor Management: Donor units that test reactive for screening serologic tests may not be used for allogeneic transfusion. Donors who screen reactive, but are confirmatory negative, are not deferred. If confirmatory test is positive, donor is deferred indefinitely and is eligible for reentry after 12 months if evidence of completion of a known effective treatment for syphilis is obtained from a physician or public health clinic.

Human T-Cell Lymphotropic Virus: Human T-cell leukemia viruses (HTLV-I and HTLV-II) were the first human retroviruses discovered and can be transmitted by transfusion of cellular blood products (whole blood, RBCs, and platelets). Probability of transmission from RBCs diminishes during storage. Before blood screening, seroconversion rates of 44%–63% had been reported in recipients of HTLV-I–infected cellular components in HTLV-I endemic areas (e.g., Japan, the Caribbean). Lower rates (~20%) had been reported in recipients of contaminated cellular components in the United States.

Half of US donors seropositive for HTLV-I/II are infected with HTLV-II. HTLV-II–infected donors often report either history of injecting drug use or sexual contact with injecting drug user. With testing, residual risk for transfusion transmission is 1:3 million units transfused.

Background: Major routes of transmission are intravenous drug use, unprotected sexual intercourse, from mother to child via breastfeeding, and blood transfusion containing infected lymphocytes. Most infections are asymptomatic, but there is a

2%–4% risk of developing disease up to 40 years after infection with HTLV-1 (lesser risk with HTLV-2). Risk of adult T-cell leukemia/lymphoma (ATLL) in individuals infected at birth is 4% in their lifetime; risk is lower in those infected during adulthood. Clinical course of ATLL is aggressive, with a median survival of <12 months in acute and lymphoma forms. HTLV is also associated with HTLV-1–associated myelopathy (HAM), which is a slowly progressive myelopathy, characterized by spastic paraparesis of lower extremities, hyperreflexia, sphincter dysfunction, and urinary incontinence. Risk of HAM is ~2% in HTLV-1–infected individuals. Other diseases associated with HTLV-1 infection include lymphocytic pneumonitis, uveitis, polymyositis, arthritis, bronchitis, dermatitis, and other infectious syndromes. HTLV-2 has not definitively been associated with any disease entity; however, it has been linked with mild neurologic disorders.

HTLV-1 predominately infects CD^{4+} lymphocytes, whereas HTLV-2 preferentially infects CD^{8+} lymphocytes and, to lesser extent, CD^{4+} lymphocytes, B lymphocytes, and macrophages. Therefore, leukoreduction is theoretically able to prevent HTLV transfusion transmission. Seroprevalence for HTLV-I/II in blood donors in the United States is <0.01%. Questions have been raised about the need for HTLV testing in the age of 100% leukoreduction.

Determination of Need and Requirement for Testing

Screening Tests: FDA has approved enzyme-linked immunosorbent assays (ELISA), which detect both HTLV-I and HTLV-II antibodies. These tests typically use viral lysates as the capture reagent, and adherent donor antibodies are identified with antiglobulin conjugate.

Confirmatory/Supplemental Tests: One approach to supplemental testing had been to confirm a repeatedly reactive (RR) anti-HTLV ELISA test by repeating test using an alternate manufacturer's ELISA assay. As of December 2014, a Western blot assay (MP Diagnostics HTLV Blot 2.4) was FDA-approved for supplemental HTLV testing on RR screening samples. This assay uses combination of envelope recombinant proteins and viral lysate (envelope rgp46-I, rgp46-II, and GD21, and gag p24 and p19 antigens) to confirm presence of antibodies against HTLV-I/II and differentiate between HTLV-I and HTLV-II infection (gag p24 and p19 antigens). Determining if the infection is HTLV-1 or HTLV-2 is helpful for counseling purposes.

Donor Management: Donors who are RR for anti–HTLV-I/II screening test and are positive by supplemental testing are permanently deferred. Donors who are RR for anti–HTLV-I/II screening test but negative or indeterminate on supplemental testing on more than one occasion are indefinitely deferred.

Counseling about HTLV antibody–positive donors is difficult owing to the lack of clarity over developing disease. One study, with 138 HTLV-I–infected, 358 HTLV-II–infected, and 759 uninfected control US blood donors, after adjusting for known and potential confounders, found that HTLV-II infection was associated with increased mortality (hazard ratio 2.8), although no statistically significant increase was seen in HTLV-I–infected individuals.

Chagas Disease: With increased immigration from Central and South America, seroprevalence of *T. cruzi* antibodies in the United States has steadily risen in the past three decades. It is estimated that ~300,000 people and 1: 25,000 donors may be infected. Seroprevalence is higher in Miami (1:9000) and Los Angeles (1:2000). Since 1989, 10 cases of transfusion-transmitted Chagas disease have been reported in the United States and Canada. Two of these cases were identified by performing lookback on positive donors.

Description: *T. cruzi* is a protozoan parasite, endemic to the Americas, predominantly Central and South America, although infected persons can be found worldwide. *T. cruzi* is transmitted by infected triatomine bug. In the acute phase, patients may demonstrate a chagoma with local lymphadenopathy, Romaña sign (swelling around the eye), or painless unilateral edema of palpebrae and periocular tissues, malaise, fever, anorexia, and edema of face and lower extremities in addition to generalized lymphadenopathy and hepatosplenomegaly. On many occasions, the acute phase will resolve spontaneously without the patient having realized he or she is infected. The chronic phase may present years to decades after initial infection in up to 30% of patients. Patients may develop arrhythmias, cardiomyopathy leading to heart failure, or thromboembolism. Additionally, they may develop megaesophagus and megacolon, which may be complicated by airway obstruction/aspiration pneumonitis, colonic obstruction, volvulus, septicemia, and/or death.

Determination of Need and Requirement for Testing

Screening Test: FDA-approved ELISA has been available to screen donors for *T. cruzi* antibodies since 2007. An ELISA by a second manufacturer was FDA-approved in 2010. In 2015, the FDA determined *T. cruzi* to be a relevant transfusion-transmitted infection and subsequently recommended that blood establishments test for *T. cruzi*.

Currently the FDA has issued a guidance recommending one-time testing of each US donor for *T. cruzi*. To date there have been only a few cases of positive donors who appear to have been infected in the United States.

Blood products from donors testing RR by ELISA should be quarantined and removed from supply. Additionally, in-date products from the donor should be quarantined. Donor should be notified and indefinitely deferred regardless of confirmation testing. The Chagas Disease Biovigilance Network records screening and confirmatory results from donor testing: http://www.aabb.org/programs/biovigilance/Pages/chagas.aspx.

Confirmatory Tests: Before 2011, confirmatory testing was not required and, if performed, could either be through a second ELISA test, radioimmunoprecipitation assay (RIPA), or immune fluorescence assay (IFA). Now that an FDA-licensed confirmatory/supplemental test (a multistep enzyme strip immunoassay) for *T. cruzi* is available, donations found to be RR by a licensed screening test must be tested further for confirmation.

Donors whose samples test positive or indeterminate on licensed supplemental test must be deferred permanently.

Donor Management: In 2017, the FDA issued final guidance recommending a reentry algorithm where donors who were indefinitely deferred owing to a history of

Chagas disease or who were RR for *T. cruzi* on a licensed screening test but either not tested or negatively tested on an investigational or licensed supplemental test would be eligible for reentry testing. A follow-up sample provided at least 6 months after the RR donation must be negative on two alternate licensed screening tests and a licensed supplemental test for the RR donor to be reentered.

Further Reading

Bern, C., & Montgomery, S. (2009). An estimate of the burden of Chagas disease in the United States. *CID*, *49*, e52–e54.

Kane, M. A., Bloch, E. M., Bruhn, R., Kaidarova, Z., & Murphy, E. L. (2015). Demographic determinants of syphilis seroprevalence among US blood donors, 2011–2012. *BMC Infect Dis*, *15*(1), 63.

Orland, J. R., Wang, B., Wright, D. J., HOST Investigators, et al. (2004). Increased mortality associated with HTLV-II infection in blood donors: A prospective cohort study. *Retrovirology*, *24*, 4.

Sobata, R., Matsumoto, C., Uchida, S., et al. (2015). Estimation of the infectious viral load required for transfusion-transmitted human T-lymphotropic virus type 1 infection (TT-HTLV-1) and of the effectiveness of leukocyte reduction in preventing TT-HTLV-1. *Vox Sang*, *109*, 122–128.

Steele, W. R., Hewitt, E. H., Kaldun, A. M., et al. (2014). Donors deferred for self-reported Chagas disease history: Does it reduce risk? *Transfusion*, *54*, 2092–2097.

Stramer, S. L., Foster, G. A., & Dodd, R. Y. (2006). Effectiveness of human T-lymphotropic virus (HTLV) recipient tracing (lookback) and the current HTLV-I and -II confirmatory algorithm 1999 to 2004. *Transfusion*, *46*, 703–707.

Wilson, L. S., Ramsey, J. M., Koplowicz, Y. B., et al. (2008). Cost-effectiveness of implementation methods for ELISA serology testing of *Trypanosoma cruzi* in California blood banks. *Am J Trop Med Hyg*, *79*, 53–68.

CHAPTER 19

Bacterial Screening

Jed B. Gorlin, MD, MBA

Bacterial contamination of blood components may result in septic transfusion reactions (Chapter 67), which can be fatal. Platelet products, particularly those on day 4 or 5 of storage, are most commonly associated with these reactions because they are stored at room temperature in nutrient media (plasma) enabling bacteria growth. Technologies have been developed to mitigate risk of bacterial contamination of platelet products, including methods to avoid or decrease bacterial contamination, inactivate bacteria, and detect contamination. Each method offers advantages and limitations. In addition, new strategies are being developed.

Risk of septic reaction was first addressed by AABB with Standard 5.1.5.1, requiring blood collection agencies implement methods to limit and detect bacteria in platelet products. This resulted in improved collection methods, with skin cleansing and diversion pouches, and use of bacterial detection methods, including culture. These steps decrease septic reaction rate from reported range of 10–400 to 7–25 in 1,000,000 and fatal septic reaction rate from reported range of 2–63 to 2–12 in 1,000,000 apheresis platelet products transfused. Thus, implementation of bacterial screening has apparently decreased the overall risk by approximately 2/3, but the FDA has issued several iterations of a draft guidance intended to further reduce this risk. This chapter will address methods used to limit and detect bacterial contamination in platelet products (Table 19.1).

Methods to Minimize Bacterial Contamination

Donor Screening: Donor eligibility questions are used to exclude donors with fever, with symptomatic infections, or receiving antibiotics for infection. Donor screening does not eliminate donors with asymptomatic infections, which may include gram-negative bacteria, resulting in severe septic transfusion reactions.

Skin Preparation: Skin is the source of contamination for ~80% of bacterially contaminated units. Skin decontamination with iodine decreases bacterial contamination. Chlorhexidine has been documented to be more effective in reducing bacterial contamination in studies measuring residual colonies, although large-scale studies showing fewer septic reactions are lacking. Consequently, many centers offer both, with many using iodine–povidone as default and chlorhexidine after donor reports of skin reaction.

Diversion Pouch: Bacteria within hair follicles and scar tissue (typically observed in repeat donors) may not be completely eliminated by skin preparation. Sample diversion pouch collects skin plug and first ~50 mL of blood, which further reduces product contamination; *Staphylococcus* species bacterial contamination decreased from 0.14% to 0.03% of units. Blood samples collected from diversion pouch are used for infectious disease testing.

Transfusion Medicine and Hemostasis. https://doi.org/10.1016/B978-0-12-813726-0.00019-2

TABLE 19.1 Methods to Reduce Bacterial Contamination

Bacterial contamination avoidance	Donor eligibility
	Collection platform
	Skin preparation
Bacterial contamination reduction	Diversion pouch or, on automated collections, return of initial 50-100 ml of collection volume to donor
Bacterial contamination inhibition	Cold, cryopreserved, or lyophilized storage[a]
	Storage solutions
Bacterial contamination inactivation	Pathogen inactivation
Bacterial contamination detection	Culture
	Nucleic acid testing[a]
	Point of release immunoassays

[a]Currently not available.

While cold storage of platelets is not currently licensed, several centers are preparing and transfusing cold stored platelets under IND.

Collection Platform: More than 90% of platelets transfused in the United States are collected using apheresis technology. Retrospective data report significantly higher rate of contamination and septic transfusion reactions on components collected on the Amicus (Fresenius Kabi) as opposed to Trima (Terumo BCT) apheresis platform. Subsequently, changes in Amicus platform collections have been made intending to mitigate this increased risk.

Methods to Inhibit or Inactivate Bacteria

Cold Storage: Cold storage of platelets would decrease bacteria growth. Cold storage techniques are actively being investigated (with storage for up to 14 days, as 3 days of storage has already been FDA-approved), particularly for use in trauma patients. Cold storage causes irreversible activation, and platelets are rapidly cleared from circulation. Thus they may be suitable for actively bleeding patients but not for prophylactic transfusions.

Storage Solutions: Platelet storage additive solution (PAS) allows decreases of 65% in plasma volume in platelet products. Bacteria grow up to fourfold faster in PAS than those in plasma, which may be advantageous owing to enhanced detection by culture earlier in storage. Conversely, higher bacterial load may result in higher clinical sepsis rate.

Pathogen Inactivation: Pathogen inactivation technologies target any nucleic acid–containing organism including a broad variety of bacteria as well as viruses and parasites (Chapter 48).

Although both technologies yield multiple log reduction of various bacterial strains and no cases of septic reactions following administration of treated products, they both result in measurable decreases in platelet count increment and the resulting platelet function as measured by need for additional platelet transfusions.

Methods to Detect Contaminated Product: Current AABB *Standards* and College of American Pathology (CAP) require the use of enhanced methods to detect

TABLE 19.2 Limits of Detection for Assays

Culture	10^1–2 CFU/mL
eBDS	10^2–3 CFU/mL
Platelet Pan Genera	10^3–5 CFU/mL
BacTx	10^3–4 CFU/mL
Gram stain	10^6 CFU/mL

bacteria in platelet components; these methods have to be FDA-approved or validated to provide equivalent sensitivity.

Culture-Based Methods: One culture based system (BacT/ALERT, bioMérieux, Durham, NC) is FDA-approved for *quality control* testing in leukoreduced platelet products (eBDS, another culture method, is no longer manufactured). Culture detects an inoculum of 10^{1-2} CFU/mL of bacteria (Table 19.2). Typically 8 mL of the product is inoculated into aerobic bottle at least 24 hours after collection. However, culture at 24 hours may miss slow-growing bacteria, such as *Propionibacterium acnes* or coagulase-negative staphylococcal species, which are usually of low pathogenicity. Additional anaerobic bottle can be used, but many centers do not use because of increased sample collection volume, cost, and higher false positivity rate. Recently, the National Health Service in the United Kingdom documented that increasing culture volume to 8 mL in aerobic and anaerobic cultures for each final apheresis component and delaying culture until 36–48 hours after collection further diminished the observed clinical sepsis rate about 90% to ~1:1,000,000.

If culture is negative for at least 12 hours, per manufacturer's instruction, product can be released for transfusion; however, culture bottle is retained until unit expiration date. Positivity is triggered by increase in CO_2. If culture turns positive, then the blood center notifies receiving transfusion service of potentially contaminated product, which can notify the patient's physician, if transfused, or discard product.

True positive bacterial cultures require organism identification. Blood center may wish to notify donor implicated in contaminated unit with non–skin flora organism. Owing to diseases associated with asymptomatic bacteremia, medical referral of such donors may be warranted in accordance with public health recommendations. For example, *Streptococcus bovis* and *Streptococcus* G infections have been reported in donors with occult malignancy, especially colon carcinoma.

Immunoassay: There are two FDA-approved point of release assays.

Platelet Pan Genera Detection (PGD) test (Verax Biomedical, Worcester, MA) is a qualitative rapid lateral-flow immunoassay FDA-approved as an additional safety measure. It is used as standalone quality control test for bacterial detection in pools of up to six whole blood–derived platelet concentrates and as an adjunct device for bacterial detection for leukoreduced apheresis platelet products. In the latter case, it must be used with another approved quality control bacterial detection method (i.e., culture method presented earlier).

PGD uses 0.5 mL of product, takes 30–40 minutes, and has positive/negative color change readout. It detects presence of lipoteichoic acid and lipopolysaccharide antigens found on aerobic and anaerobic gram-positive and gram-negative bacteria, respectively, making it possible to detect bacterial species most frequently implicated in contaminated platelet samples. Its detection limit is 10^{3-5} CFU/mL of bacteria, which is lower than the sensitivity rate of culture, but it is performed at later storage time when bacteria number has expanded (after bacteria have entered their exponential growth phase) and product must be tested within 24 hours of transfusion. It has ~0.5% false-positive rate. PGD testing can be used to extend storage time of apheresis platelets up to 7 days.

Another immunoassay, BacTx Assay System (Immunetics, Boston, MA), is FDA-approved for bacterial testing of leukocyte-reduced whole blood–derived platelet units. This system detects peptidoglycan from cell wall of both gram-positive and gram-negative bacteria and can be performed in 45 minutes using 0.5 mL product sample. Peptidoglycan-binding protein initiates series of enzymatic reactions that convert substrate to a visible color. In the presence of bacteria, color will develop at a rate proportional to bacterial contamination level. Automated reader is used to monitor color intensity changes, identifying any contaminated platelet units. This system has a detection limit of 10^{3-4} CFU/mL for many clinically relevant strains.

Future Considerations: FDA draft guidance and Blood Product Advisory Committee meeting outcomes suggest that bacterial mitigation strategies will include pathogen inactivation, large volume delayed culture, and primary culture with additional safety measure testing (e.g., PGD testing).

Lyophilized and cryopreserved platelets are currently undergoing early phase clinical studies. Other bacterial detection methods are currently under development, notably fluorescent cytometric techniques and nucleic acid (PCR) methods.

Further Reading

Brecher, M. E., Jacobs, M. R., Katz, L. M., et al. (2013). Survey of methods used to detect bacterial contamination of platelet products in the United States in 2011. *Transfusion, 53*(4), 911–918.

Capocelli, K. E., & Dumont, L. J. (2014). Novel platelet storage conditions: Additive solutions, gas, and cold. *Curr Opin Hematol, 21*(6), 491–496.

Dumont, L. J., Wood, T. A., Housman, M., et al. (2011). Bacterial growth kinetics in ACD-A apheresis platelets: Comparison of plasma and PAS III storage. *Transfusion, 51*(5), 1079–1085.

Eder, A., & Goldman, M. (2011). How do I investigate septic transfusion reactions and blood donors with culture-positive platelet donations? *Transfusion, 51*, 1662–1668.

Eder, A., Dy, B. A., DeMerse, B., et al. (2017). Apheresis technology correlates with bacterial contamination of platelets and reported septic transfusion reactions. *Transfusion, 57*(12), 2969–2976.

FDA. Bacterial risk control strategies for blood collection establishments and transfusion services to enhance the safety and availability of platelets for transfusion: Draft guidance for industry. (March 2016).

Jacobs, M. R., Smith, D., & Heaton, W. A. (2011). Detection of bacterial contamination in prestorage culture-negative apheresis platelets on day of issue with the Pan Genera Detection test. *Transfusion, 51*, 2573–2582.

Kreuger, A. L., Middelburg, R. A., Kerkhoffs, J. H., et al. (2017). Storage medium of platelet transfusions and the risk of transfusion-transmitted bacterial infections. *Transfusion, 57*, 657–660.

McDonald, C., Allen, J., & Brailsford, S. (2017). Bacterial screening of platelet components by National Health Service Blood and Transplant, an effective risk reduction measure. *Transfusion, 57*, 1122–1131.

CHAPTER 20

Role of the Transfusion Service Physician

Bruce S. Sachais, MD, PhD and Eric Senaldi, MD

According to the AABB *Standards* (Standard 1.0), the transfusion service (TS) must have a clearly defined structure and documentation of individuals responsible for key functions, including executive management. TS must have a medical director who is responsible for medical and technical policies, processes, and procedures. Hospital TS is typically within the departments of pathology and laboratory medicine (or clinical pathology).

Specific Roles: TS physician has multiple roles, responsibilities, and functions, including overseeing specific laboratories, laboratory and administrative policies and procedures, direct and indirect patient care, education, and research (Fig. 20.1).

Transfusion Service: TS physician not only directs the TS, where pretransfusion testing and product modifications and dispensing occur, but also may direct other laboratories and/or clinical services, including human leukocyte antigen (HLA), cellular therapy, coagulation laboratories, and therapeutic apheresis, phlebotomy, infusion, collection, perioperative, patient blood management, and tissue banking services. Each of these requires policies, procedures, and medical, administrative, and technical oversight.

Transfusion Service Management: TS physician functions include strategic planning, project management, determination of scope and complexity of services offered, information technology oversight, clinical and technical consultation, utilization review, formation and promulgation of clinical guidelines, regulatory compliance, quality systems and assurance, quality control, human resources, and customer service.

Patient Care: TS physician can provide both direct and consultative patient care. Direct patient care, where services can be reimbursed through either Evaluation and Management (E&M) or Current Procedural Terminology (CPT) codes, is performed through therapeutic apheresis, therapeutic phlebotomy, and hematopoietic progenitor cell (HPC) collection, as well as interpretation of complex serologic and patient matching issues and evaluation of transfusion complications. Consultative services are often requested regarding provision of appropriate or specialized products, transfusion reactions, blood management, hemostasis and thrombosis-related test result interpretation, and apheresis. TS physicians improve patient care through policies and procedures that enhance quality, and through hospital committee participation (e.g., transfusion committee, patient safety committee).

Transfusion Committee: The Joint Commission (TJC) Standards (PI.1.10) require hospitals to collect and monitor their performance regarding blood product use, which

Transfusion Medicine and Hemostasis. https://doi.org/10.1016/B978-0-12-813726-0.00020-9

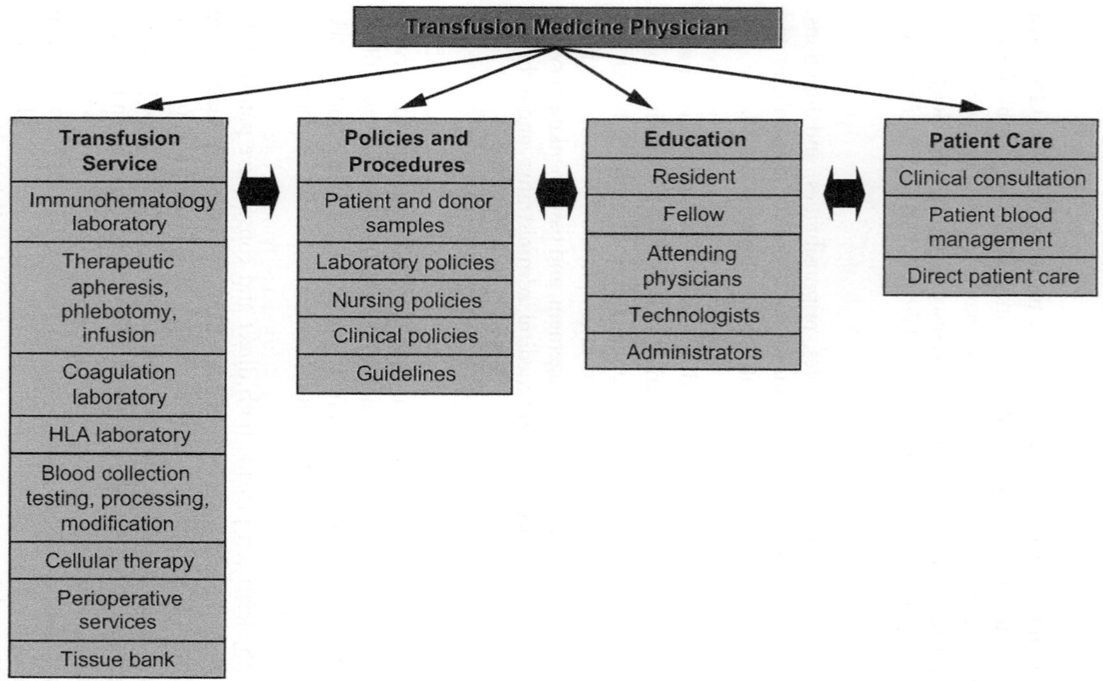

FIGURE 20.1 Transfusion service physician oversight. *HLA*, human leukocyte antigen.

can be overseen by hospital transfusion committee or its equivalent. Data must be analyzed and tracked over time to identify levels of performance, patterns, trends, and variation, compared with external sources, and results must identify opportunities to improve. Areas of improvement must be prioritized, actions taken to improve, changes evaluated to ensure that they result in improvement, and actions taken when improvement is not achieved or sustained. These requirements can be met by reviewing ordering, distribution, handling, dispensing, administration, and effects of blood products.

Membership: The members of the transfusion committee include representatives from departments that use blood products, such as hematology, surgery, anesthesia and obstetrics, and from TS, clinical laboratory, and nursing and hospital administration.

Goals: Primary goal is to improve patient outcome. The committee should approve blood supplier and monitor blood usage, adverse events, errors, accidents, and TS quality plan. The committee should ensure that TS has the resources it needs to carry out its responsibilities. The committee also oversees transfusion practice guidelines, criteria for blood product use, and blood administration procedures and blood management program.

Transfusion Guidelines: Transfusion guidelines encompass transfusion of products (i.e., RBCs, plasma, platelets, and cryoprecipitate), product modification (e.g., irradiation), and transfusion in special clinical scenarios (i.e., sickle cell disease, massive transfusion). Their creation requires communication between ordering physician and TS physician, as well as logistics in providing products to the patient. A model process is as follows:

1. An audit is performed to understand current and established transfusion practices.
2. Draft guidelines are created by multidisciplinary team based on evidence in the literature.
3. Education of ordering physicians and other members of the healthcare team ensues and is required to determine understandability and likelihood of implementation and compliance.
4. Guidelines are revised, approved, adopted, and implemented.
5. Periodically, repeat auditing is required to guarantee that guidelines are being followed and continue to be appropriate.

Many hospitals now use computerized physician order entry (CPOE) for blood product ordering. CPOE can be used for prospective audits to determine appropriate blood use. Appropriate indications for transfusion (based on hospital policy) can be listed with requirement to choose applicable indication. Educational "pop up" windows can be incorporated to provide additional information and guidance to clinicians.

Maximum Surgical Blood Order: Maximum surgical blood order schedule (MSBOS) establishment for common surgical procedures enables determination of amount of blood that will be available for particular operation. One approach to MSBOS development is to design the orders to meet ≥80% of patient operative blood requirements.

Informed Consent: TS physician is required by the College of American Pathologists (CAP) and AABB *Standards* to participate in development of policies, processes, and procedures regarding transfusion recipient informed consent. At a minimum, informed consent must communicate risks, benefits, and alternatives of transfusion, right to refuse transfusion, and ability to ask questions.

Inventory Management: Although the US blood supply is usually adequate, the margin of supply to demand can at times be low. TS physician is significantly involved in inventory management, especially when the supply is low (such as during winter holidays and summer), and is responsible for a communication plan to make stakeholders aware of significant blood shortages. TS physician is also responsible for triaging units during times of significant shortage and communicating recalls and lookbacks to ordering physicians. TS physicians also oversee product inventory management for specific patients or patient groups, including patients with multiple alloantibodies who may require large numbers of RBC products for transfusion.

Education: TS physician educates many groups of individuals. Formal education of residents and fellows is overseen through the ACGME (Accreditation Council for Graduate Medical Education), which requires learning opportunities, assessment and outcomes in six core competencies: patient care, medical knowledge, system-based practice, practice-based learning, professionalism, and interpersonal and communication skills. TS physician educates technologists, ordering physicians and hospital administrators. Education is ongoing and via multiple methods, including presentations, patient cases, journal club, quality assurance projects, and one-on-one conversations.

Research: TS physician may participate in research activities, ranging from NIH-funded to quality improvement projects. For example, Recipient Epidemiology and Donor Evaluation Study (REDS-III) consortium (https://reds-iii.rti.org/), which includes domestic and international efforts dedicated to donor safety and blood availability issues, transfusion practices, education, and training, is funded by the NHLBI. Transfusion therapy clinical trials improve patient care directly by increasing understanding of their risks and benefits. Basic science research is needed to better understand beneficial and adverse effects of transfusion and improve transfusion management. Collaboration between basic scientists and clinicians allows for the "bench to bedside and back" development of optimal patient care. Lastly, quality improvement projects affect patient care by highlighting areas of improvement, such as turn-around time, patient identification, and blood utilization review.

Transfusion Medicine as a Career: Blood banking and transfusion medicine (BB/TM) physicians work in a variety of settings, including blood centers, hospital TS, industry, government, and large commercial laboratories. BB/TM physicians continue to participate in expanding areas outside the hospital TS and donor program, such as HLA laboratory, collection of human cells, tissues, and cellular and tissue-based products (HCT/P) collection, cellular therapies (including regenerative medicine) laboratory, therapeutic apheresis services, tissue banking, and coagulation laboratory. A survey demonstrated that approximately 70% of BB/TM fellowship graduates from

1995 to 2004 had published articles within the previous 3 years, and 12% had published more than 10 publications within that timeframe. BB/TM physicians should be advocates for best care of patients and donors, teachers and mentors, and devoted to the continual improvement of the BB/TM field.

Further Reading

Fung, M. K., Crookston, K. P., et al. (2007). A proposal for curriculum content in transfusion medicine and blood banking education in pathology residency programs. *Transfusion, 47*(10), 1930–1936.

McCullough, J. (2006). The role of physicians in blood centers. *Transfusion, 46*, 854–861.

McCullough, J., Benson, K., Roseff, S., et al. (2008). Career activities of physicians taking the subspecialty board examination in blood banking/transfusion medicine. *Transfusion, 48*, 762–767.

Shaz, B. H., & Hillyer, C. D. (2010). Transfusion medicine as a profession: Evolution over the past 50 years. *Transfusion, 50*, 2536–2541.

Szczepiorkowski, Z. M., & AuBuchon, J. P. (2006). The role of physicians in hospital transfusion services. *Transfusion, 46*, 862–867.

CHAPTER 21

Pretransfusion Testing

Suzanne A. Arinsburg, DO

Pretransfusion testing is performed to select compatible blood components for transfusion, thereby preventing harm to recipient and maximizing posttransfusion survival.

AABB *Standards* require the following:

1. Positive patient identification and their corresponding blood specimen;
2. ABO group and D typing of patient's specimen;
3. Testing of patient's serum/plasma for unexpected, clinically significant RBC antibodies, which are defined as non-ABO antibodies known to cause hemolytic transfusion reactions and/or hemolytic disease of the fetus and newborn (HDFN);
4. Comparison of current findings with previous results;
5. Confirmation of ABO group of RBC components (i.e., whole blood and RBCs);
6. Confirmation of D-type of D-negative RBC components;
7. Selection of ABO- and D-appropriate components for patient;
8. Serologic or computer crossmatching (of components containing >2 mL of RBCs); and
9. Labeling of component for issue, with patient's identifying information.

Specific tests performed and frequency and methodology of such testing differ based on patient and component to be issued and are done in accordance with the establishment's standard operating procedures (SOPs).

Patient Identification: Accurate patient identification at time of specimen collection for type and screen and before blood component administration is essential for safe transfusion practice. Mistransfusion is defined as transfusion of blood component (1) with incorrect blood group, (2) intended for another patient, and (3) that was not prescribed. Causes include wrong blood in tube (mislabeling specimen), patient misidentification, and testing errors.

Pretransfusion Orders: Requests must contain two independent patient identifiers to uniquely identify each patient. Other required information on request includes component requested, amount, any special requirements (e.g., irradiation), ordering physician name, and patient age and gender. Patient diagnosis and transfusion and pregnancy history are helpful in guiding the testing and/or component selection.

Type and Screen Request: Type and screen order laboratory to determine the patient's ABO group, RhD-type, and alloantibody screen. If antibody screen is positive, alloantibody identification is then performed. Patient sample is stored in transfusion service laboratory for crossmatching if transfusion is needed.

Transfusion Medicine and Hemostasis. https://doi.org/10.1016/B978-0-12-813726-0.00021-0

Request for Transfusion: If RBCs are requested, appropriate units will be selected and crossmatched. If other components are requested, appropriate units are selected and released.

Patient Specimen: If patient specimen has not been submitted to the laboratory before transfusion request, it must be submitted with transfusion request.

Pretransfusion testing can be performed on either serum or plasma.

Positive patient identification must occur before specimen collection and labeling. If possible, patient should be involved in identity verification process. Specimen must be labeled after collection but before leaving patient's bedside, with two independent patient identifiers and collection date. In addition, there must be a mechanism to identify phlebotomist (e.g., initial on the specimen label). When pretransfusion specimen is received in the laboratory, an appropriately trained individual must confirm that all identifying information on request is identical to information on specimen itself. In case of any discrepancy or doubt, the specimen is rejected and new specimen must be obtained for testing. Transfusion service should accept only those specimens that are completely, accurately, and legibly labeled with indelible ink. As hemolyzed or lipemic specimens may create problems in evaluating test results, it is important to draw a new specimen whenever possible. Each laboratory must have procedures and policies that define acceptable criteria for accepting specimens for pretransfusion testing and describe how to document and handle unsatisfactory specimens.

Specimen Storage and Retention: Patient specimens and samples of donor RBC components must be stored at refrigerated temperature for at least 7 days after transfusion. This allows for repeat or additional testing if the patient develops transfusion reaction.

Specimen Age: Crossmatching is required for RBC components and other components containing >2 mL of RBCs (i.e., granulocytes). If patient has been transfused or pregnant within past 3 months or transfusion and pregnancy history is unknown, then specimen used for crossmatch must have been collected within 3 days from the time of transfusion event. Each institution should have policy that defines the length of time specimens may be used for patients not transfused or pregnant within the last 3 months. Plasma, platelet, and cryoprecipitate components do not require crossmatching, and therefore, new specimens are not required for these components if results of the patient's ABO and D typing are on record.

Pretransfusion Testing

Serologic Testing: In pretransfusion serologic tests, in vitro RBC antigen–antibody reactions are demonstrated, most often, with final endpoint of agglutination. Antibodies against RBC antigens are either IgM or IgG. IgM antibodies can cause direct agglutination of RBCs having corresponding antigens. IgG antibodies usually do not produce direct agglutination after attaching to their corresponding antigens on RBC surface. Antiglobulin test is used to detect RBCs coated with IgG antibodies. This test uses secondary antibody directed against human globulins (anti–human globulin [AHG;

also known as antiglobulin]) that attaches and agglutinates sensitized RBCs. AHG can be anti-IgG only or polyspecific containing anti-IgG and anti-C3d and may contain anti-C3b and other Ig and complement antibodies (Fig. 21.1). AHG may be used in direct antiglobulin test (DAT; also known as direct Coombs test) to detect in vivo coating of RBCs or in indirect antiglobulin test (IAT; also known as indirect Coombs test) to demonstrate in vitro reactions between RBCs and antibodies. IAT is used in antibody detection (antibody screen), antibody identification, crossmatching, and RBC phenotyping. Negative AHG test (both DAT and IAT) must be followed by control system of IgG-sensitized RBCs, termed "check cells" (also known as Coombs control), to confirm that the result is not a false-negative. If check cells do not agglutinate, then test must be repeated.

Some RBC antibodies (e.g., anti-Jk^a and anti-Jk^b) are capable of in vitro hemolysis of corresponding antigen-positive RBCs; therefore, reactions should also be evaluated for presence of hemolysis.

Method

Immediate Spin Testing: Either reagent RBCs with known RBC antigen expression are mixed with patient plasma or reagent antisera with known antibody specificity is mixed with patient RBCs. After mixing at room temperature, sample is centrifuged and agglutination or hemolysis is detected.

Indirect Antiglobulin Testing: For IAT, RBCs and plasma or antisera are mixed as mentioned earlier but then incubated at 37°C to enhance binding of IgG. Sample is then washed to remove unbound antibody followed by addition of AHG, centrifugation, and detection of agglutination or hemolysis.

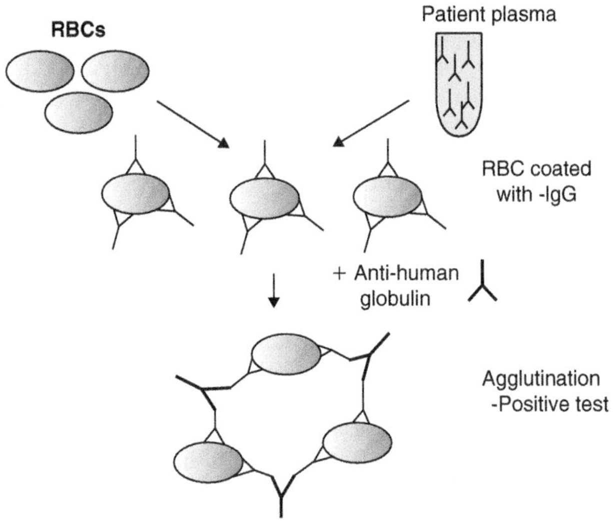

FIGURE 21.1 Anti–human globulin (AHG) test. (Modified from Hillyer, C. D., Silberstein, L. E., Ness, P.M., et al. (Eds). (2007). *Blood banking and transfusion medicine: Basic principles & practice* (2nd ed.). Philadelphia, PA: Churchill Livingstone, Elsevier.)

Reactions are interpreted based on degree of agglutination or presence of hemolysis. Alternative methods that do not require agglutination have also been developed, including solid phase and flow cytometry. Many of these are automated.

Detection Techniques: Variety of techniques can be used for detection of RBC antibodies; in the United States, tube and gel methods are most commonly used followed by solid phase methods. Regardless of method, test sensitivity is usually >97%–99% when performed under FDA-approved conditions. Each laboratory should determine how testing will be performed, including what methods to use as primary and alternative methods.

Tube Test: Tube method can be performed with no enhancement reagents (i.e., only with saline), although, more commonly, enhancement media are used to increase test sensitivity. Enhancement media include albumin and low–ionic strength solution (LISS), which decrease zeta potential (repulsive electric potential between RBCs that prevents their aggregation) to bring the RBCs closer together. Polyethylene glycol (PEG), which removes water and concentrates antibodies in sample, is also be used. Each enhancement media has its limitations; for example, LISS and albumin enhance cold autoantibodies and PEG enhances warm autoantibodies. Different methods have varying sensitivity and specificity. Tube test is flexible and relatively inexpensive, but reactions are unstable, grading of reactions is subjective, test requires increased technologist time, and testing is not amenable to automation.

Gel Test: Gel method (also known as column agglutination) has similar sensitivity to PEG tube test, requires less technologist time, produces reactions that are stable for up to 24 hours, and allows for more standardized grading of reactions; however, it is more expensive than tube test, as special equipment must be purchased to accommodate gel cards. In this method, RBCs interact with antibodies at top of the column or microtube. RBCs are then forced into column medium (gel matrix) by centrifugation. Agglutinated RBCs, which are too large to go through, remain at top part of gel column, while nonagglutinated RBCs move through gel to the bottom of the column. Automated platforms are available.

Solid Phase Test: Solid phase method has benefits similar to gel method, but solid phase method requires that technologists are carefully trained to interpret reactions because unlike tube and gel methods, solid phase reactions are not based on agglutination. RBCs or their membranes, of known phenotype, are immobilized to microplate wells. After plasma is added, incubation followed by washing is performed. Bound antibody is then detected by indicator RBCs that adhere diffusely over the well if reaction is positive or collect at the bottom as pellet if negative. Automated platforms are available.

Automation: Both gel and solid phase methods have been automated and require smaller sample volumes than tube test method. Other advantages of automation include freeing staff to perform other duties, error reduction by reducing number of steps that require human intervention, stable reaction endpoint that can be reviewed as necessary, use of interface between instrument and the laboratory information system to download test requests and upload test results, and better turnaround times in batch modes.

Molecular Methods: Molecular testing is not routinely performed as part of pretransfusion testing. However, DNA-based method for blood typing is used in certain clinical situations in which serologic tests may not be sufficient. These include fetal genotyping in management of HDFN, prediction of antigen status when no serologic reagents are available, resolving serologic typing discrepancies, and determination of blood groups of patients who have been recently transfused, have positive DAT, autoantibodies, or multiple or high-frequency alloantibodies.

Recipient Testing: Pretransfusion testing for allogeneic transfusion must include ABO group and D type of patient RBCs and detection and identification of unexpected RBC antibodies in patient plasma or serum. Pretransfusion testing for autologous transfusion only requires ABO group and D type of patient RBCs. When plasma, platelets, or cryoprecipitate is ordered, historical test results may be used, although it is preferable that test results from at least two different specimens are available and there is no result discrepancy.

ABO Blood Type: ABO antibodies are the most clinically significant of RBC antibodies. Antibodies against antithetical isohemagglutinin ABO antigens are consistently present in sera of individuals without any previous RBC exposure because they are produced in response to environmental stimulants, such as bacteria. These antibodies are mostly IgM that may result in severe intravascular hemolysis after transfusion of ABO-incompatible blood. ABO-incompatible transfusion prevention is the primary objective of pretransfusion testing. ABO group is determined serologically using immediate spin testing. Patient RBCs are tested with anti-A and anti-B (known as the front/forward type) and patient plasma with A_1 and B RBCs (known as the back/reverse type) (Table 21.1). If discrepancy is detected either between forward and back typing or between current and historical type and transfusion is necessary before resolution, only group O RBCs should be issued.

D Type: After ABO group system, D is the most clinically important RBC antigen. Immunization to D antigen only occurs in D-negative individuals after antigen exposure via pregnancy or transfusion. D antigen has the greatest immunogenicity of RBC antigens; ~20% of hospitalized D-negative individuals who receive D-positive RBC component will form anti-D. Therefore, patient RBCs must be tested with anti-D reagents during pretransfusion testing. Newer monoclonal reagents allow anti-D testing to be

TABLE 21.1 ABO Typing

Reaction With RBCs Tested With:		Reaction With Plasma Tested With:		Interpretation
Anti-A	Anti-B	A1 cells	B cells	ABO group
0	0	+	+	O
+	0	0	+	A
0	+	+	0	B
+	+	0	0	AB

performed using immediate spin testing. If D typing cannot be interpreted, especially in females of childbearing potential, then the patient should receive D-negative RBCs. Testing patient RBCs for weak D (where the D test is taken to AHG phase) is not necessary, except in neonates. Without this additional step, weak D individuals are treated as D-negative and there is no harm in giving these individuals D-negative RBCs. On the other hand, if a neonate born to a D-negative mother is not tested for weak D after being tested as D-negative, the mother will not be considered an Rh immune globulin (RhIg) candidate and possibly be immunized if the neonate is actually D-positive.

Antibody Screen: Antibody screen is performed to identify unexpected clinically significant antibodies against foreign (nonself) RBC antigens (RBC alloantibodies). Antibody is considered clinically significant if it has been associated with HDFN and/or hemolytic transfusion reaction. Because these alloantibodies are primarily IgG, AHG must be used for maximal sensitivity. Reagent RBCs used for antibody detection must express D, C, E, c, e, M, N, S, s, P_1, Le^a, Le^b, K, k, Fy^a, Fy^b, Jk^a, and Jk^b-antigens. If antibody screen is positive, further tests must be performed to identify antibody (Chapter 22).

Comparison With Previous Records: Results from current tests must be compared with previous records (if there has been prior testing), and any discrepancies must be resolved before using test results for blood component selection. Records also are reviewed for past history of clinically significant RBC antibodies, transfusion reactions, or required product modifications. Even if current antibody detection test is negative, AHG crossmatch is required for patients with history of clinically significant antibodies.

Donor Unit Testing: Additional tests must be performed on RBC units before transfusion. Whole blood and RBC components must be tested for confirmation of ABO group, while only units labeled as D-negative must have confirmation of D type. Confirmatory testing for weak D is not required (collection centers must perform weak D testing for component labeling). Any discrepancies must be resolved before issue of the blood for transfusion. Plasma, platelet, and cryoprecipitate components do not require confirmation of ABO/D typing.

Blood Component Selection

ABO Compatibility: Blood components should be ABO-*identical* or ABO-*compatible* with recipient (Table 21.2). If the component contains ≥2 mL of RBCs (e.g., RBC and granulocyte components), then donor RBCs must be compatible with recipient plasma (Table 21.3). Because plasma containing components (e.g., plasma, platelets, and whole blood) have ABO antibodies, donor plasma should be compatible with recipient RBCs (Table 21.4). Whole blood must be ABO-identical with the recipient. Low titer (typically defined immediate spin anti-A/anti-B titer <1:254) whole blood and plasma products and low titer (typically defined immediate spin anti-A/anti-B titer <1:100) platelet stored in plasma are used to decrease risk of hemolytic reaction due to ABO incompatibility.

D Compatibility: Only RBC-containing components (RBCs, granulocytes, whole blood, and platelets) need to be matched for D antigen. D-negative components

TABLE 21.2 ABO Selection of Blood Components

Component	ABO Selection of the Component
Whole blood	Identical to patient
RBCs	Compatible with patient plasma
Granulocytes	Compatible with patient plasma
Plasma	Compatible with patient RBCs
Platelets	All ABO groups acceptable; preferable to give components compatible with patient RBCs
Cryoprecipitate	All ABO groups acceptable

TABLE 21.3 ABO Group Selection for RBC Transfusion

Patient ABO Group	RBC-Containing Component ABO Group			
	1st Choice	2nd Choice	3rd Choice	4th Choice
AB	AB	A	B	O
A	A	O		
B	B	O		
O	O			

TABLE 21.4 ABO Group Selection of Plasma Component Transfusion

Patient ABO Group	Plasma-Containing Component ABO Group			
	1st Choice	2nd Choice	3rd Choice	4th Choice
O	O	A	B	AB
A	A	AB		
B	B	AB		
AB	AB			

should be reserved for D-negative recipients to prevent anti-D formation. Occasionally D-negative components are not available for D-negative patients. In this situation, it is important to weigh alloimmunization risk against transfusion urgency. Anti-D formation is most detrimental to female with childbearing potential, but prevention of antibody production may also be important in patients who may require ongoing RBC support (such as patients undergoing transplantation or with chronic anemia). Anti-D formation is highest with RBC components (80% in healthy volunteers and ~20% in hospitalized patients) and is very low in patients with malignancy undergoing chemotherapy and receiving apheresis platelet components. In D-negative patients receiving D-positive platelet components, anti-D formation risk is <2%. In patients receiving platelet products or other low-volume RBC-containing products, RhIg can be given to prevent anti-D formation after transfusion of D-positive RBCs. It has been reported that RhIg can prevent anti-D formation after D-positive RBC transfusion in D-negative patient, but there is risk of hemolysis with large doses of RhIg in presence of large volumes of D-positive RBCs and red cell exchange with D-negative RBCs should be considered. In emergency situations, there may not be adequate time to acquire D-negative component for D-negative patient.

Other Blood Groups: RBC antigens, other than ABO and D, are not routinely considered in blood product selection of nonalloimmunized patients in the United States. However, many institutions elect to transfuse phenotypically matched RBCs for Rh (i.e., C and E) and K antigens in frequently transfused populations such as patients with sickle cell disease. Other countries match for Rh and/or K routinely, or in select populations.

When a patient is found to have clinically significant unexpected antibody(ies), currently or historically, corresponding antigen(s)-negative units must be selected for crossmatching.

Crossmatch: RBC components (including whole blood) and components containing >2 mL of RBCs (i.e., granulocytes) must be crossmatched with the patient specimen before issue except in emergency situations. Crossmatching may be performed serologically or electronically to demonstrate compatibility between antigens on donor RBCs and recipient plasma antibodies.

Immediate Spin Crossmatch: Immediate spin crossmatch is a serologic crossmatch designed to detect ABO incompatibility between donor and recipient. If patient has no history of clinically significant antibodies and current antibody screen is negative, then immediate spin crossmatch is adequate. Advantages of an immediate spin–only crossmatch are decreased workload, turnaround time, and reagent costs.

AHG Crossmatch: AHG crossmatch is used to detect incompatibility between patient plasma and both ABO and non-ABO antigens on donor RBCs. For patients with alloantibodies, current or historical, AHG crossmatch must be performed.

Electronic Crossmatch: Electronic crossmatching can be used with validated computer system in patients who have no previous or current clinically significant alloantibodies and whose ABO group has been determined at least twice (once on a current sample). Advantages of electronic crossmatch are decreased workload, decreased volume requirement of patient samples, improved blood inventory, and decreased turnaround time for component issue.

Labeling and Release of Blood Components: AABB *Standards* require the following to occur at the time of issue:

1. Tag or label with patient's two independent identifiers, donor unit number, and compatibility testing results (if performed) must be securely attached to the container.
2. Final check of records (patient name and identification number, patient's ABO/D type, donor component number, donor's ABO/D type, interpretation of crossmatch [if performed], date and time of issue, and identification of special

transfusion requirements [such as irradiation, antigen-negative, and cytomegalovirus (CMV)-reduced risk]).

3. Confirmation that identifying information on request, records, and components are in agreement. Prior to issue, component must be acceptable for use and should be checked for abnormal color, leakage, and outdate.

Final identification of patient and component must be performed by transfusionist and another individual, who must independently identify patient and donor unit and certify that all identifying information on forms, tags, and labels are in agreement. Record of transfusion should be made part of the patient's medical record.

Special Clinical Situations

Neonates: Initial pretransfusion specimen is required to determine ABO/D type. ABO typing only includes forward type, as anti-A and/or anti-B antibodies are not formed during the first 4–6 months. Antibodies detected soon after birth are from placental transfer. Patient or maternal plasma can be used for unexpected RBC antibody detection and crossmatch. If clinically significant antibodies are not detected, then repeat testing and crossmatching are not required in the first 4 months of life during a single hospitalization. If clinically significant unexpected RBC antibodies are detected, RBC units selected for transfusion must be corresponding antigen–negative or AHG crossmatch–compatible until antibodies are no longer detectable in the neonate's plasma. If non–group O infant is to receive non–group O RBCs that are not compatible with maternal ABO group, then the neonate's plasma must be tested for presence of ABO antibodies (reverse type) using method that includes AHG phase. If anti-A or anti-B is detected, RBCs lacking the corresponding antigen are transfused. Neonates that test D-negative on initial testing must be tested for weak D if they are born to D-negative mothers. This identifies neonates that are actually D-positive, so their mothers can receive RhIg prophylaxis.

Urgent Requirement for Blood: When blood is urgently needed before pretransfusion testing is complete, the patient's physician must weigh risk of transfusing uncrossmatched or partially matched blood against risk of delaying transfusion. When blood is released before completion of pretransfusion testing, records must contain signed statement from requesting physician indicating that clinical situation was sufficiently urgent to require release of blood. The statement does not need to be obtained before issuing lifesaving transfusions. When urgent release is requested, RBC units issued should be either of the following:

1. Group O, if the patient's ABO group is unknown. It is preferable to give D-negative blood if the recipient's D type is unknown, especially if the patient is a female of childbearing potential (typically defined as <50 years old).
2. ABO- and D-compatible, if ABO/D typing on a current sample has been completed.

Tag or label attached to these units must clearly indicate that compatibility testing was not completed at the time of issue. Compatibility testing must be completed as soon as possible after releasing the unit, using patient sample collected as early as possible in transfusion sequence. If incompatibility is detected, the patient's physician and medical director of transfusion service should be notified immediately.

Quality Assurance: Clinical Laboratory Improvement Amendments of 1988 (CLIA'88) regulates ABO group and D typing, antibody detection, and crossmatching. Proficiency testing must be performed at least twice a year. A program of quality control must be established to ensure that reagents, equipment, and methods function as expected. Results need to be reviewed, and when appropriate, corrective action taken. Reagents and other materials must be stored and used in accordance with the manufacturer's written instructions.

Further Reading

Chapman, J. F., Elliott, C., Knowles, S. M., et al. (2004). Working party of the British Committee for standards in haematology blood transfusion task force. Guidelines for compatibility procedures in blood transfusion laboratories. *Transfus Med, 14*, 59–73.

Shulman, I. A., Maffe, L. M., & Downes, K. A. (2005). North American pretransfusion testing practices, 2001–2004: results from the College of American Pathologists Interlaboratory comparison program Survey Data, 2001–2004. *Arch Pathol Lab Med, 129*, 984–989.

CHAPTER 22
Antibody Identification

Suzanne A. Arinsburg, DO

Antibodies to RBC antigens except anti-A and anti-B are called unexpected antibodies. Unexpected RBC antibodies may be alloantibodies reacting against foreign antigens or autoantibodies reacting against self-antigens. Depending on the patient population, 1%–35% of patients have unexpected alloantibodies.

Once an alloantibody is detected on antibody screen, antibody specificity must be determined. Some unexpected alloantibodies may be clinically significant with potential to cause shortened survival of transfused RBCs, acute or delayed hemolytic transfusion reactions (HTRs), and hemolytic disease of the fetus or newborn (HDFN). Patients with clinically significant alloantibodies should receive RBCs that are negative for corresponding antigen.

Clinical Significance: RBC antibodies are termed clinically significant if they can cause hemolysis of corresponding antigen-positive RBCs resulting in HTR or HDFN.

Clinical significance of a specific unexpected RBC alloantibody is unknown but is presumed based on published reports. Alloantibodies may simply bind to transfused RBCs detected serologically, may lead to shortened RBC survival, or cause mild to severe acute or delayed HTR. Detection of alloantibodies and determination of their specificity (identification) is, therefore, important (Tables 22.1 and 22.2). RBC autoantibodies are generally reactive against reagent and autologous RBCs. Transfused RBCs are generally expected to survive as well as patient's own RBCs. Autoantibodies can cause difficulties in detection and identification of underlying clinically significant alloantibodies.

Unexpected Antibody Identification: Unexpected RBC antibodies can be detected initially as ABO typing discrepancy, positive antibody screen, incompatible crossmatch, or through eluate. Once detected, unexpected antibody(ies) must be identified to determine their clinical significance and requirement for antigen-negative RBCs.

Preanalytical Considerations

Age: Infants who are <6 months old usually do not produce antibodies, and identified antibodies are generally maternal due to placental transfer. Maternal plasma is used for crossmatching.

Gender: Women are more often sensitized to RBC antigens due to sensitization from paternally inherited antigens on fetal RBCs. Pregnancy history may be important in antibody identification.

Transfusion History: Prior transfusion of RBC-containing component is a possible sensitizing event. Timing of most recent transfusion is critical, as transfused RBCs can circulate for up to 3 months potentially complicating serologic testing results and interpretation.

Transfusion Medicine and Hemostasis. https://doi.org/10.1016/B978-0-12-813726-0.00022-2

TABLE 22.1 Summary of Major Blood Group Antigens

Antigens	Systems	IgM	IgG	Transfusion Reactions	Hemolytic Disease of the Fetus or Newborn (HDFN)	Prevalence (%)	
						White	African-American
A	ABO	X	X	Mild–severe	None–moderate	40	27
B	ABO	X	X	Mild–severe	None–moderate	11	20
D	Rh	X	X	Mild–severe	Mild–severe	85	92
C	Rh		X	Mild–severe	Mild	68	27
E	Rh	X	X	Mild–moderate	Mild	29	22
c	Rh		X	Mild–severe	Mild–severe	80	96
e	Rh		X	Mild–moderate	Rare	98	98
K	Kell	X	X	Mild–severe	Mild–severe	9	2
k	Kell		X	Mild–moderate	Mild–severe	99.8	>99
Kp^a	Kell		X	Mild–moderate	Mild–moderate	2	<1
Kp^b	Kell		X	None–moderate	Mild–moderate	>99	>99
Ja^a	Kell		X	None–moderate	Mild–moderate	<1	20
Jp^b	Kell		X	Mild–moderate	Mild–moderate	<99	99
Fy^a	Duffy		X	Mild–severe	Mild–severe	66	10
Fy^b	Duffy		X	Mild–severe	Mild	83	23
Jk^a	Kidd		X	None–severe	Mild–moderate	77	92
Jk^b	Kidd		X	None–severe	None–mild	74	49
M	MNS	X	X	None	None–mild	78	74
N	MNS	X		None	None	70	75
S	MNS		X	None–moderate	None–severe	52	31
s	MNS		X	None–mild	None–severe	89	94
U	MNS		X	Mild–severe	Mild–severe	100	>99
Le^a	Lewis	X		Few	None	22	23
Le^b	Lewis	X		None	None	72	55
Lu^a	Lutheran	X	X	None	None–mild	8	5
Lu^b	Lutheran	X	X	Mild–moderate	Mild	>99	>99
Do^a	Dombrock		X	Rare	+DAT/No HDFN	67	55
Do^b	Dombrock		X	Rare	None	82	89
Co^a	Colton		X	None–moderate	Mild–severe	>99.9	>99
Co^b	Colton		X	None–moderate	Mild	10	10
P_1	P	X		Rare	None	79	94

Modified from Hillyer, C. D., Stauss, R.G., & Luban, N. L. C. (Eds.). (2004). *Handbook of pediatric transfusion medicine*. San Diego, CA: Elsevier Academic Press.

Ethnicity: Certain antigens may be of high or low frequency in specific populations. Knowledge of the patient's ethnic origin may provide clues to antibody specificity.

Medical History: Certain diseases have been associated with RBC antibodies. Cold agglutinin syndrome, Raynaud phenomenon, and infections with *Mycoplasma pneumoniae* are often associated with anti-I autoantibodies. Infectious mononucleosis is

TABLE 22.2 Clinical Significance of Antibodies to the Major Blood Group Antigens

Usually Clinically Significant	Sometimes Clinically Significant	Clinically Insignificant If Not Reactive at 37°C	Generally Clinically Insignificant
A and B	At^a	A_1	Bg
Diego	Colton	H	Chido/Rogers
Duffy	Cromer	Le^a	Cost
H in O_h	Dombrock	Lutheran	JMH
Kell	Gerbich	M, N	Knops
Kidd	Indian	P_1	Le^b
P, PP_1P^k	Jr^a	Sd^a	Xg^a
Rh	Lan		
S, s, U	LW		
Vel	Scianna		
	Yt		

Reproduced from Hillyer, C. D., Stauss, R.G., & Luban, N. L. C. (Eds.). (2004). *Handbook of pediatric transfusion medicine*. San Diego, CA: Elsevier Academic Press.

associated with anti-i autoantibodies. Patients with paroxysmal cold hemoglobinuria may demonstrate autoantibodies with anti-P specificity. Systemic lupus erythematosus, multiple myeloma, chronic lymphocytic leukemia, and lymphoma are associated with warm autoantibodies. Patients who have received solid organ or hematopoietic stem cell transplants may demonstrate passive antibodies that originate from donor passenger lymphocytes.

Drugs: Drugs are known to cause antibody identification problems. Other sources of drug-related problems during antibody testing include recent administration of intravenous immune globulin and Rh immune globulin.

Specimen Requirements: Serum or plasma can be used for antibody testing; however, EDTA-anticoagulated specimens are preferred due to potential for in vitro uptake of complement components by RBCs in clotted specimens.

Reagents

Antibody Identification Panel: Antibody identification requires testing the patient's plasma against panel of selected RBC samples (typically 8–14 reagent RBCs) with known antigen expression of major blood groups (Rh, Kell, Kidd, Duffy, and MNS) (Fig. 22.1). Reagent RBCs obtained from commercial vendors are supplied with accompanying panel with antigen expression profile of each reagent RBC. Panel RBCs are generally group O, thereby allowing plasma of any ABO group to be tested. Reagent RBCs are selected so that if one takes all the examples of RBCs into account, a distinctive pattern of positive and negative reactions exists for each of the many antigens (including D, C, E, c, e, M, N, S, s, P1, Le^a, Le^b, K, k, Fy^a, Fy^b, Jk^a, and Jk^b). Selected RBCs in a panel should allow for identification of single specificities of common alloantibodies with exclusion of most others.

In patients with known alloantibody history, selected cell panel of corresponding antigen-negative RBCs is a better approach to identify any newly formed antibodies. It is not necessary to reconfirm the previously identified antibody(ies). If the patient's phenotype is known, a selected panel may be tested to demonstrate the presence or absence of possible alloantibodies minimizing amount of testing required.

Antiglobulin Reagent: Indirect antiglobulin test (IAT) using antihuman globulin (AHG) use is the most common method of antibody identification.

Testing: For initial panel testing, it is common to use same method (i.e., tube, gel, or solid phase) used in antibody screen test.

Immediate Spin (Room Temperature) Phase: Some serologists may choose to include a room temperature reading after centrifugation of the specimen. This allows for detection of cold-reacting antibodies such as anti-M, -N, -P1, -I, -Le^a, or -Le^b. Many institutions omit this reading, as most cold reacting antibodies are clinically insignificant, but some clinically significant antibodies may be missed (e.g., anti-Vel, IgM only anti-Fy^a).

37°C Phase: Test reading after 37°C incubation before addition of AHG may detect some antibodies (e.g., potent anti-D, -E, or -K) that can cause direct agglutination of RBCs. Other antibodies (e.g., anti-Jk^a) may be detected by lysis of antigen-positive RBCs during 37°C incubation if serum is tested. This phase is often omitted, as these antibodies are generally also detected after addition of AHG.

AHG Phase: Limiting the reading to only after addition of AHG would limit positive reactions caused by clinically insignificant cold-reactive antibodies, omits unnecessary testing, and decreases turnaround time. Automated techniques (gel and solid phase) only detect antibodies reacting in this phase.

Interpretation of Results: Results are interpreted as positive or negative according to presence or absence of reactivity (i.e., agglutination, hemolysis, or detection of antigen–antibody binding). Usually strength of reaction is graded 1–4+, with 4+ being the strongest agglutination reaction.

How to Read a Panel?: A widely used first approach to the interpretation of panel results is to tentatively exclude specificities based on nonreactivity of the patient's plasma with RBCs that express the antigen. This is referred to as a "cross-out" or "rule-out" method. Pattern of reactivity for each specificity that was not excluded is compared with pattern of reactivity obtained with the patient's plasma. If there is a pattern that matches the patient's plasma pattern exactly, that is most likely the antibody specificity in the plasma. If there are remaining specificities that were not excluded, additional testing is needed to eliminate remaining possibilities and to confirm the suspected specificity. This process requires testing plasma with selected RBCs that express certain specific antigens and lack others. To ensure that an observed pattern of reactions is not the result of chance alone, conclusive antibody identification requires plasma to be tested against a sufficient number of reagent RBCs that lack or express the antigen that corresponds with the apparent specificity of the antibody. A standard approach

	Rh								MNS				Lutheran		P_1	Lewis		Kell		Duffy		Kidd			LISS		Check
CELL	D	C	E	c	e	f	v	C^w	M	N	S	s	Lu^a	Lu^b	P_1	Le^a	Le^b	K	k	Fy^a	Fy^b	Jk^a	Jk^b	IS	37C	AHG	cells
1. r′r-2	0	+	0	+	+	+	0	0	+	+	0	+	0	+	0	0	+	0	+	+	0	+	+	∅	∅	2+	
2. $R1^wR1$-1	+	+	0	0	+	0	0	+	+	+	+	0	+	+	+	0	+	0	+	0	+	+	0	∅	∅	2+	
3. R1R1-6	+	+	0	0	+	0	0	0	0	+	0	+	0	+	+	+	0	+	+	0	+	+	0	∅	∅	2+	
4. R2R2-8	+	0	+	+	0	0	0	0	+	+	+	+	0	+	+	0	+	+	0	0	+	0	+	∅	∅	∅	✓
5. r′r-3	0	0	+	+	+	+	0	0	+	+	+	0	0	+	+	+	0	0	+	0	+	0	+	∅	∅	∅	✓
6. rr-32	0	0	0	+	+	+	+	0	+	0	+	0	0	+	+	0	0	0	+	+	+	+	0	∅	∅	∅	✓
7. rr-10	0	0	0	+	+	+	0	0	+	+	+	+	0	+	0	0	+	+	+	0	+	+	+	∅	∅	∅	✓
8. rr-12	0	0	0	+	+	+	0	0	0	+	0	+	0	+	+	0	+	0	+	+	0	0	+	∅	∅	∅	✓
9. R_o-4	+	0	0	+	+	+	0	0	+	0	0	+	0	+	0	+	0	0	+	0	0	0	+	∅	∅	∅	✓
Cord cell	/	/	/	/	/	/	/	/	/	/	/	/	/	/	/	0	0	/	/	/	/	/	/				
Patient																								∅	∅	∅	✓

FIGURE 22.1 An example of an antibody identification panel in a patient with anti-C.

has been to require that three antigen-positive RBCs react and three antigen-negative RBCs fail to react for each specificity identified. When that approach is not possible, a more liberal approach allows the minimum requirement for a probability (*P*) value of ≤0.05 to be met by having two reactive and three nonreactive RBCs or by having one reactive and seven nonreactive RBCs (or the reciprocal of either combination). Use of two reactive and two nonreactive RBCs is also an acceptable approach for antibody confirmation. Some antibodies demonstrate dosage reacting more strongly with RBCs having double dose of the antigen (homozygous for the allele) than RBCs having single dose of antigen (heterozygous for the allele). This is most commonly seen in Kidd, Duffy, Rh, and MNS systems. Ideally, antibodies should be ruled out using reactions with RBCs that carry a double dose of antigen, but this may not be possible for low-frequency antigens with rare homologous expression (such as K).

Autologous Control (Autocontrol): The patient's plasma is reacted with autologous RBCs as part of panel testing to detect antibodies coating RBCs in the patient's circulation. Use of enhancement reagents for autocontrol can cause positivity seen only as an in vitro phenomenon. If autocontrol is positive in the IAT, a direct antiglobulin test (DAT) should be performed. If DAT is negative, antibodies to an enhancement medium constituent or autoantibodies that react only in the enhancement medium should be considered.

Positive Direct Antiglobulin Test in a Recently Transfused Patient: In a patient with an alloantibody who has been recently transfused with antigen-positive RBCs, circulating donor RBCs may be coated with the alloantibody resulting in a positive DAT or positive autocontrol. This usually happens in patients in which the alloantibody was undetectable in pretransfusion testing. If DAT is positive for IgG, an elution can be performed that dissociates the antibody from the sensitized RBCs and allows identification of the antibody when the eluate is tested against reagent RBCs in the antibody identification panel (Chapter 23).

Phenotyping: After alloantibody identification, the patient's RBCs should be phenotyped to determine if the patient is negative for antigenic specificity of antibody identified. This helps to confirm the specificity of identified antibody. Phenotyping of the patient's RBCs may be complicated by transfusion or pregnancy in the past 3 months, as fetal or transfused RBCs may be circulating. If the specificity of antibody(ies) in the patient's plasma is clear, extensive efforts to type the patient's RBCs are unnecessary. Compatible AHG crossmatch using antigen-negative donor units provides additional confirmation of the antibody's specificity.

Ideally, a pretransfusion specimen is used for phenotyping but, if not possible, the patient's RBCs may be separated from donor cells before phenotyping. Centrifugation can be used to separate autologous more recently formed RBCs with lower specific gravity than older transfused cells. Hypotonic saline can be used in patients with sickle cell disease to separate autologous hemoglobin SS or SC from transfused hemoglobin AA cells.

Cold and warm autoantibodies can cause false-positive reactions. Use of warm saline washes or dithiothreitol treatment (DTT) of RBCs can help to separate cold autoantibodies from RBCs. For RBCs coated with warm autoantibodies, gentle heat elution, chloroquine treatment, or treatment with acid glycine/EDTA can be used, or typing can be performed with direct agglutinating (IgM monoclonal) reagents.

Molecular genotyping offers an alternative to serologic typing and is especially useful in situations where the patient has been recently transfused or where the patient's RBCs are heavily coated with IgG. There are rare situations, however, where the genotype of a person may not predict the RBC phenotype (e.g., inactivating mutations, rare new alleles, or different antigen expression on different tissues in transplant patients) (Chapter 24).

Frequency of Testing: It is not necessary to repeat identification of known antibody(ies). AABB *Standards* states that in patients with previously identified antibodies, methods of testing shall be those that identify *additional* clinically significant antibodies. The patient's plasma should be tested with RBCs that are negative for antigen(s) that corresponds to the patient's known antibody(ies).

Techniques Used in Complex Antibody Identification Cases

Enhancement Media: Enhancement media (i.e., low ionic strength salne (LISS) and polyethylene glycol (PEG)) are discussed in Chapter 21.

Enzyme or Chemically Treated RBCs: Enzymes or chemicals that destroy or alter RBC antigens can be useful in antibody identification either to confirm the presence of antibody or to detect additional antibodies. Panels of enzyme-treated RBCs are typically used in complex antibody identification cases where multiple antibodies or antibodies to high-frequency antigens may exist. Ficin, papain, trypsin, and bromelin are commonly used enzymes that destroy or alter the Duffy and MNS blood group antigens and Xg^a, JMH, Ch, Rg, S, Yt^a, Mg, Mi^a/Vw, Cl^a, Je^a, Ny^a, JMH, some Ge, and In^b, and enhance the Rh, Kidd, Lewis, and ABO blood group antigens, P_1 and I. DTT destroys or alters Kell, Lutheran, Dombrock, and Cromer blood group antigens, Yt^a, JMH, Kn^a, McC^a, Yk^a, LW^a, and LW^b. ZZAP is a combination of DTT and proteolytic enzymes that alters all antigens listed above. When enzyme-treated RBCs are used for antibody identification, blood group antigens, which are destroyed or altered with enzymes, cannot be ruled out.

For example, when a patient has antibodies to JMH (high-frequency antigen) and D. Initial panel without enzymes would show broad reactivity while ficin panel would only show reactivity with D-positive RBCs.

Prewarm Technique: Cold-reactive antibodies may react in AHG phase of testing if they have broad thermal amplitude and react at 37°C. Reactivity can often be eliminated by using saline-suspended RBCs and performing 30- to 60-minute incubation followed by IAT. If reactivity remains, prewarm technique may be used in which the patient's serum and reagent RBCs are warmed to 37°C separately and then mixed. Although clinically significant antibodies that are weakly reactive can become negative in prewarm testing, risk is considered minimal. Small percentage of patients with cold agglutinins will still have reactivity. In these patients, cold adsorption or adsorption with rabbit erythrocyte stroma (REST, which removes anti-I and anti-IH and some clinically significant alloantibodies, such as anti-B, -D, and -E) may be used.

Cold Antibody Screen: The cold antibody screen is used to confirm and identify the presence of cold autoantibodies. The patient's plasma is combined with appropriate reagent RBCs (such as A1, B, O adult, O cord [I negative], and autologous RBCs) and then read for agglutination at varying temperatures (immediate spin, room temperature, 18 and 4°C). This technique can be useful in resolving antibody identification problems.

Neutralization: Neutralization of antibodies can be useful in confirming an antibody's identity or identifying other antibodies. Anti-Sd^a is neutralized by urine (Sd^a substance); anti-Ch and anti-Rg (Chido and Rodgers substance) and anti-Le^a and anti-Le^b (Lewis substances) are neutralized by plasma; and P_1 is neutralized by P_1 substance found in hydatid cyst fluid and pigeon egg whites. When using neutralized plasma for antibody identification, blood group antigens that were neutralized cannot be ruled out. For example, if a patient has anti-P_1 (high-frequency antigen) and anti-D, neutralization with P1 substance can help identify the anti-D and rule out other alloantibodies.

Adsorption: In patients with panreactive warm autoantibodies, special techniques are used to adsorb out autoantibodies from plasma to determine whether underlying alloantibodies are present. These tests are not routinely performed in laboratories and may need to be performed in a reference laboratory.

Autologous Adsorption: If the patient has not been transfused within last 3 months and has an adequate hematocrit, then autologous adsorption can be performed. In this procedure, whole blood sample is first separated into RBC and plasma fractions, and then heat or other elution techniques are used to remove autoantibodies from RBCs as well as enhance antibody adsorption. Treated RBCs, now stripped of autoantibodies, are mixed with plasma fraction to "adsorb out" any additional autoantibodies present in plasma. When performed correctly, adsorbed plasma should be free of autoantibodies, leaving only alloantibodies if present.

Allogeneic Adsorption: If the patient has been recently transfused (within the last three months) or has low hematocrit, then allogeneic adsorption must be performed. In allogeneic adsorption, allogeneic RBCs (which may or may not be enzyme-treated because enzyme destroys some antigens) are used to adsorb out autoantibodies. Because allogeneic RBCs that are antigen-positive will absorb out that antibody, at least three phenotype-appropriate RBCs must be used for allogeneic adsorption to rule out most alloantibodies. For example, plasma adsorbed with D-positive RBC cannot be used to rule out the presence of anti-D. Adsorbed plasma can then be used for crossmatch.

Immunohematology Reference Laboratory: When antibody problems cannot be resolved, or when rare antigen-negative blood is needed, Immunohematology Reference Laboratory (IRL) can provide consultation and assistance. IRLs have access to genotyping, enzyme techniques, adsorption–elution techniques, rare RBCs, which are absent of high-frequency antigens, rare sera with antibodies to antigens, as well as other methods to identify antibodies and help determine appropriate RBC product selection.

Determining Clinical Significance of Identified Antibody(ies): Antibodies that react at 37°C, by IAT, or both, are potentially clinically significant. Non-A and -B antibodies that react at room temperature and below are usually not clinically significant;

however, there are exceptions. Reported experience with examples of antibodies with same specificity can be used in assessing their clinical significance. Consultation with the IRL may also be helpful. Certain laboratory tests have been used to predict antibody'(ies) clinical significance including monocyte monolayer assay, which quantifies phagocytosis and/or adherence of antibody-coated RBCs, antibody-dependent cellular cytotoxicity, which measures lysis of antibody-coated RBCs, and chemiluminescence assay, which measures respiratory release of oxygen radicals after phagocytosis of antibody-coated RBCs. Other tests such as thermal amplitude studies or radiolabeled, antigen-positive RBC survival studies may also be useful.

Selection of Blood for Patients With Unexpected Antibody(ies): After clinically significant antibody has been identified, antigen-negative RBC units must be selected for all future transfusions, even if antibodies are no longer detectable. Additionally, AHG crossmatch must be performed. When clinically significant antibody cannot be conclusively excluded, RBC units that lack corresponding antigen should be selected for transfusion. For autoantibodies, see Chapter 51.

Patients Requiring Rare Blood: Rare blood includes RBC units that are negative for high-prevalence antigens and units that are negative for a combination of common antigens. For patients requiring such blood, IRL can provide consultation and assistance through its access to the American Rare Donor Program (ARDP) or AABB UnCommon Good Program. Family members of patients, especially siblings, with antibodies to high-prevalence antigens are another potential source of antigen-negative blood. For infants with HDFN resulting from multiple antibodies or antibody to high-prevalence antigen, mother (if ABO compatible) is potential donor. If clinical situation allows, autologous RBC transfusions should be considered for patients with rare phenotypes who are expected to need rare blood. For some patients with multiple antibodies who are not able to donate autologous units, it may be necessary to determine whether any of the antibodies is less likely to cause RBC destruction and, in a critical situation, to give blood that is incompatible for that particular antigen.

Further Reading

Daniels, G., Poole, J., deSilva, M., et al. (2002). The clinical significance of blood group antibodies. *Transfus Med, 12*, 287–295.

Shulman, I. A., Downes, K. A., Sazama, K., & Maffei, L. M. (2001). Pretransfusion compatibility testing for red blood cell administration. *Curr Opin Hematol, 8*, 397–404.

Winters, J. L., Richa, E. M., Bryant, S. C., et al. (2010). Polyethylene glycol antiglobulin tube versus gel microcolumn: Influence on the incidence of delayed hemolytic transfusion reactions and delayed serologic transfusion reactions. *Transfusion, 50*, 1444–1452.

CHAPTER 23

Direct Antiglobulin Test

Angela Novotny, MT(ASCP)SBB[CM]

Direct antiglobulin test (DAT [e.g., Coombs']) detects in vivo sensitization of RBCs with immunoglobulins and/or complement. DAT is useful in investigation of acute and delayed hemolytic transfusion reaction (AHTR and DHTR), hemolytic disease of the fetus and newborn (HDFN), autoimmune hemolytic anemia (AIHA), and drug-induced immune hemolysis (DIHA) (Table 23.1). Positive DAT can occur with or without hemolysis. There are many causes of positive DAT; interpretation of a positive or negative result should include the patient's history, clinical data, and results of other laboratory tests. Laboratory test results demonstrating evidence of increased RBC destruction include anemia, increased reticulocytes, increased LDH, decreased haptoglobin, hemoglobinemia, and hemoglobinuria. DAT can help differentiate immune from nonimmune causes of hemolysis. DAT may be positive in healthy individuals. Positive DATs are reported in 1:1000 up to 1:14,000 healthy blood donors, and 1:6 to 1:100 of hospitalized patients without signs of hemolysis or hemolytic anemia.

Indications

Antibody Identification: DAT is performed as part of the evaluation of unexpected RBC antibodies involved in AIHA (HDFN, AHTR, DHTR, and delayed serologic

TABLE 23.1 Causes of a Positive Direct Antiglobulin Test
Autoantibodies (warm autoimmune hemolytic anemia [WAIHA], cold agglutinin disease)
Alloantibodies to recently transfused antigen-positive RBCs (acute or delayed hemolytic or serologic transfusion reaction)
Passively transfused alloantibodies against the patient's RBCs resulting from plasma-containing components (plasma, platelets) or a plasma derivative (intravenous immunoglobulin or Rh immune globulin)
Maternal alloantibodies coating the fetal RBCs (hemolytic disease of the fetus and newborn)
Drug-dependent antibodies reactive with drug-treated RBCs (penicillin, cefotetan)
Drug-dependent antibodies reacting with untreated RBCs in the presence of drug (ceftriaxone, piperacillin)
Drug-independent antibodies, indistinguishable from WAIHA (fludarabine, cladribine, methyldopa, levodopa, procainamide)
Nonimmunologic protein adsorption (cephalothin, diglycoaldehyde, cisplatin, oxaliplatin, clavulanate, tazobactam)
Antibodies derived from passenger lymphocytes as a result of either solid organ or hematopoietic stem cell (HSC) transplantation
Elevated levels of IgG or complement in patients with sickle cell disease, β-thalassemia, renal disease, multiple myeloma, autoimmune disorders, AIDS, etc.

Transfusion Medicine and Hemostasis. https://doi.org/10.1016/B978-0-12-813726-0.00023-4

transfusion reaction [DSTR]). DAT and eluate may aid in antibody identification. Newly formed antibody can be bound to the circulating transfused cells and only detected in the eluate.

Autoimmune Hemolytic Anemia: A DAT is performed to determine whether hemolysis has an immune basis due to IgM or IgG. This differentiation is important because treatment may vary. Positive DAT with IgG (with or without complement) and panagglutinin in the eluate sample is consistent with diagnosis of warm autoimmune hemolytic anemia (WAIHA). DAT positive with complement only is seen in patients with cold agglutinin disease (Chapter 51).

Drug-Induced Immune Hemolytic Anemia: DAT can be used to evaluate the presence of DIHA. Eluates in these cases are commonly nonreactive (Chapter 51) unless the cells are incubated with the drug itself.

Hemolytic Disease of the Fetus and Newborn: DAT is performed to evaluate for HDFN. DAT may also be positive when ABO incompatibility is present between mother and baby (Chapter 49). If antibody is to a low-frequency antibody, consider testing maternal or neonatal sample against paternal RBCs. A false-negative DAT may be present when high-titer anti-D is present.

Hemolytic Transfusion Reactions: DAT is performed on posttransfusion sample to evaluate a possible AHTR or DHTR (Chapters 63 and 64). Positive DAT result should be compared with result from pretransfusion sample. If DAT is positive for IgG, then an eluate should be performed. If patient is non–group O, eluate should include group A and group B cells. DAT may be negative in HTR if all transfused RBCs have been hemolyzed.

Method: DAT uses antihuman globulin (AHG) reagent, anti-IgG, and anti-C3 (Fig. 23.1). DAT can initially be performed with polyspecific AHG reagent, which contains both anti-IgG and anti-C3d. If positive with the polyspecific reagent, sample can be retested with monospecific anti-IgG and anti-C3d/C3dg reagents to further characterize reactivity. IgG has four different subtypes (IgG1, IgG2, IgG3, and IgG4). Twenty-nine different isoallotypes (genetic variations within IgG subclasses) have been detected. Not all monoclonal AHG reagents are able to detect all IgG subclasses (e.g., if an antibody is only IgG4, reagents lacking anti-IgG4 will be negative). When DAT is positive for anti-IgG and anti-C3, RBCs should be tested with inert control reagent (e.g., 6% albumin or saline). Reactivity with control reagent invalidates test result.

Evaluation of a Positive Direct Antiglobulin Test: Positive DAT evaluation depends on clinical context of patient, such as underlying diagnosis, medication history, pregnancy, transfusion history, and presence of hemolytic anemia. Clinical significance of positive DAT needs to be assessed in conjunction with the clinical and laboratory information. Further evaluation may be indicated in patients who have evidence of hemolytic anemia, with transfusion within the last 3 months, receiving medications associated with positive DAT and hemolysis, who have received a solid organ

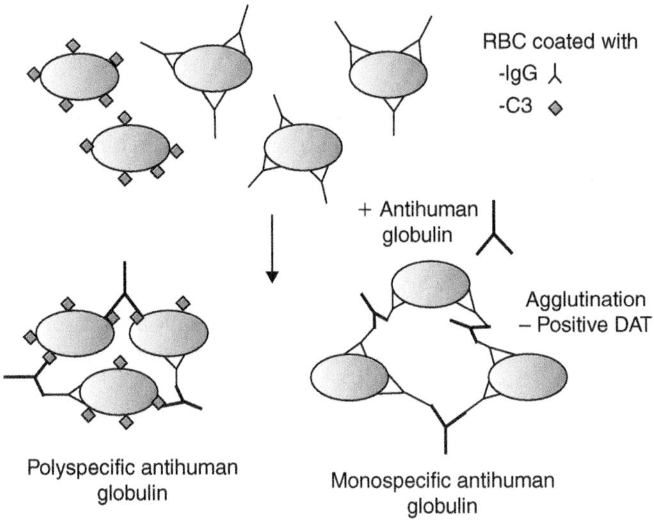

FIGURE 23.1 Direct antiglobulin test. (Modified from Hillyer, C. D., Silberstein, L. E., Ness, P. M., et al. (Eds.). (2007). *Blood banking and transfusion medicine: Basic principles and practice* (2nd ed.). San Diego, CA: Elsevier Academic Press.)

or HSC transplantation, and receiving IVIG or Rhlg. Additional tests include testing patient's plasma for presence of auto- or alloantibodies, and eluate testing.

Eluate: Elution of antibodies from sensitized RBCs can be accomplished using a variety of techniques, such as heat, freeze-thaw, cold acid, digitonin acid, and dicholoromethane glycine-HCl/EDTA. The eluate, which contains the removed antibodies, is used for antibody identification. Eluate preparations concentrate the antibody, which may aid in antibody identification.

Panagglutinin: When the eluate reacts with all RBCs tested then an autoantibody is most likely the reason for positive DAT. Antibody to high-incidence antibody should be considered in cases where panagglutinin is seen with a negative autocontrol.

Nonreactive Eluate: Nonreactive eluate may occur if the sample is tested against RBCs that do not express the appropriate antigen(s). This may be the case with antibodies against low-frequency or ABO antibodies when patients receive ABO-out of group components; DIHA commonly results in nonreactive eluates. Approximately, 80% of hospitalized patients with positive DAT will have nonreactive eluate.

False-Negative Direct Antiglobulin Test: If antibody causing immune hemolysis is not IgG or IgM (e.g., IgA or IgM not accompanied by complement fixation) it may not be identified. IgG bound to RBCs may also be at a concentration too low for reagent to cause agglutination. Approximately, 3%–11% of patients with clinical signs of AIHA have negative DAT. In case of DAT-negative hemolytic anemia, more sensitive techniques can be used. These "enhanced DAT," can include use of anti-IgA, monomeric anti-IgM for IgM antibodies not accompanied by complement fixation and/or cold 4°C or low ionic strength wash. Additionally, incorrect washing or resuspension, or delayed testing may result in false-negative DATs.

False-Positive Direct Antiglobulin Test: RBCs from refrigerated clotted specimens may have positive DAT due to postcollection complement binding. Repeat testing using EDTA specimen. Cord blood specimens contaminated with Wharton's jelly may cause false-positive DAT results; increase washing of cord cells, or use of a heel/venous specimen may be required.

Further Reading

Howie, H. L., Delaney, M., Want, X., et al. (2016). Serological bind spots for variants of human IgG3 and IgG4 by a commonly used anti-immunoglobulin reagent. *Transfusion, 56*, 2953–2962.

Kamesaki, T., Toyotsuji, T., & Kajii, E. (2013). Characterization of direct antiglobulin test-negative autoimmune hemolytic anemia: A study of 154 cases. *Am J Hematol, 88*, 93–96.

Klein, H. G., & Anstee, D. J. (Eds.). (2014). *Mollison's blood transfusion in clinical medicine* (12th ed.) John Wiley & Sons, Ltd.

Petz, L. D., & Garraty, G. (2004). In *Immune hemolytic anemias* (2nd ed.). Philadelphia: Churchill Livingstone.

Pierce, A., & Nester, T. (2011). Pathology consultation on drug-induced hemolytic anemia. *Am J Clin Pathol, 136*, 7–12.

Segel, G. B., & Lichtman, M. A. (2013). Direct antiglobulin ("Coombs") test-negative autoimmune hemolytic anemia: A review. *Blood Cells Mol Dis*. 2013.12.003.

CHAPTER 24
Molecular DNA–Based Blood Group Typing

Connie M. Westhoff, PhD, SBB

DNA-based testing for blood group antigens has become commonplace in number of clinical situations. These include typing for minor antigens in multiple transfused immunized patients to determine risk for production of additional blood group antibodies, patients with positive direct antiglobulin test (DAT) and serum autoantibody, patients facing chronic transfusion therapy, and for locating antigen-negative blood when no serologic reagent is available, as well as in prenatal medicine to assess risk for hemolytic disease of the newborn (HDFN) and to guide Rh immune globulin (RhIg) therapy for pregnant women.

Determination of blood group antigens by DNA methods (genotyping) is an indirect method for predicting individual's blood group phenotype, in contrast to direct testing by serologic methods using specific antibody (phenotyping). DNA typing results are often reported as *predicted* type to distinguish results from testing done by serologic methods.

Most blood group antigens result from single-nucleotide gene polymorphisms (SNPs) inherited in a straightforward Mendelian manner, making assay design and interpretation fairly straightforward. However, ABO and Rh blood groups are more complex. There are >200 different alleles encoding glycosyltransferases responsible for ABO type, and single point mutation in A or B allele can result in inactive transferase, i.e., group O phenotype. Next-generation sequencing (NGS) technology holds promise for routine ABO typing by DNA methods. For Rh system, testing for common antigens D, C/c, and E/e is fairly straightforward in most individuals, but antigen expression is more complex in diverse ethnic groups. There are >200 RHD alleles encoding weak D or partial D phenotypes, and >100 RHCE alleles encoding weak, altered, or novel hybrid Rh proteins. RH genotyping, particularly in minorities, requires sampling of multiple regions of the gene(s) and algorithms for interpretation.

The most commonly used methods for determination of RBC human erythrocyte antigens (HEAs) and human platelet antigens (HPAs) are semiautomated polymerase chain reaction (PCR) using florescent probes with automated readout. Automated methods increase number of target alleles in PCR, allowing determination of numerous antigens in single assay. Most platforms currently available are based on fluorescent bead technology.

Real-Time Polymerase Chain Reaction: Real-time PCR allows automated detection of amplification products and does not require handling of PCR reaction products postamplification. The method is also quantitative, which allows determination of gene copy number. The most common design uses sequence-specific fluorescent probe (TaqMan) that binds to target SNP of interest. The probe has reporter dye (fluorophore) attached to 5′ end and quencher at 3′ end that prevents reporter dye from fluorescing. As target locus is amplified, DNA polymerase encounters bound

Transfusion Medicine and Hemostasis. https://doi.org/10.1016/B978-0-12-813726-0.00024-6

TaqMan probe and degrades it, allowing reporter to fluoresce when freed from proximal 3′ quencher. Fluorescence amount is directly proportional to release of reporter and PCR product amount.

Bead Technology: This method use multiplexing (amplification of multiple target loci in one assay) with automated detection and interpretation. Many use 96-well format multiplex design that allows for typing multiple samples for many different antigens. Allele-specific capture probes are affixed to beads of many fluorescent colors (>100). Multiple beads are used, with each color targeting different SNPs. Amplified DNA fragments are allowed to anneal to allele-specific capture probes and are then elongated using fluorescent-labeled nucleotides. Beads and associated signals are analyzed by flow cytometer or fluorescence microscopy.

Testing platforms that use bead, or another, DNA array technology, can readily sample multiple genes, and/or regions of genes, and apply automated multifaceted algorithms for accurate interpretation of alleles.

Discrepancies Between Serology (Phenotype) and DNA (Genotype): When differences between serologic and DNA testing occur, it is important to investigate. This can indicate the presence of novel allele or genetic variant, particularly when testing individuals from diverse ethnic groups. Primary cause of discrepancies between serologic phenotype and DNA genotype when testing donors by large-scale DNA typing has been traced to manual recording errors. Other common causes of discrepancies include the presence of variant alleles encoding weak antigen expression. One example is FYX allele, which encodes amino acid change causing a Fy(b+^{w}) weak phenotype. RBCs type as Fy(b−) with most serologic reagents. Prevalence of *FYX* encoding Fy(b+^{w}) in Caucasians is nearly 2%.

DNA testing interrogates single or few SNPs associated with antigen expression and cannot sample every nucleotide in the gene. Consequently, discrepancies will occur when typing patients who have rare or novel silencing mutations that cause loss of antigen expression, resulting in false-positive-predicted antigen type. Silencing mutations can be familial, or common, in particular ethnic group. For example, silencing mutations responsible for S-s-U- phenotypes are common in African black ethnic groups. Fy(a−b−) phenotype found in African blacks is caused by mutation in promoter region of *FYB*, which disrupts binding site for erythroid transcription factor GATA-1 and results in loss of Duffy expression on RBCs. For accuracy, GATA-1 mutation must be included when typing for Duffy in African blacks. Expression of the protein on endothelium is not altered and Fy(a−b−) individuals with GATA-1 mutations are not at risk for anti-Fy^{b}. Silencing mutations associated with loss of Kidd antigen expression occur more often in Asians, while nucleotide changes encoding amino acid changes that weaken Kidd expression are seen in blacks.

For resolution of discrepancies and to identify new alleles, specialty referral laboratories use methods similar to those used for high-resolution HLA typing, i.e., gene-specific amplification of coding exons followed by sequencing, or alternatively, gene-specific cDNA amplification and sequencing.

Exon-Specific Gene Sequencing: Primers specific for the exon encoding allelic polymorphism, or for each exon of the gene when performing full gene sequencing, are used for amplification. Exon-specific products can be visualized by gel electrophoresis, separated from excess primers, and sequenced by standard Sanger sequencing methods. Computer software programs are used to compare nucleotide sequence obtained to known reference sequence to detect nucleotide changes. Translation programs are used to determine if any nucleotide changes discovered encode amino acid changes in the protein, or alternatively, are synonymous and predicted to be silent.

Gene Sequencing by cDNA Analysis: When gene sequencing is required, usually when investigating the presence of new or novel alleles, sequencing of cDNA synthesized from mRNA is the preferred approach. This requires isolation of mRNA from RBCs. It is important to note that commercial kits available for RNA isolation from blood samples isolate RNA from WBCs, discarding RBC lysate. At least, 0.5 mL of RBCs is needed, as most of the residual mRNA is present in reticulocytes, which represent the minority of cells in nonanemic patient samples. This approach offers advantage that noncoding introns are removed (spliced) and coding region sequence can be analyzed directly. cDNA is synthesized from mRNA using reverse transcriptase and a 3′ gene-specific primer, followed by PCR amplification with 5′ and 3′ gene-specific primers, purification of products, and sequencing.

Applications of DNA-Based Molecular Testing

Transfusion Service Applications

Type Patients Who Have Been Recently Transfused: In patients receiving chronic or massive transfusion, presence of donor RBCs often makes RBC typing by serologic agglutination inaccurate. DNA typing overcomes this limitation and avoids time-consuming and cumbersome cell separation methods to isolate and type the patient's reticulocytes. DNA assays for blood groups avoid interference from donor-derived DNA by targeting and amplifying a region of the gene common to all alleles. This allows reliable blood group determination with DNA prepared from blood samples collected after transfusion. In transfusion-dependent patients who produce alloantibodies, extended antigen profile is important to determine additional blood group antigens to which the patient can become sensitized.

Type RBCs Coated With Immunoglobulin (Positive DAT): In patients with RBCs coated with immunoglobulin (IgG), with or without autoimmune hemolytic anemia, presence of bound IgG often makes RBC typing by serologic methods invalid. IgG removal techniques are not always effective and can destroy or weaken antigen of interest. For patients with serum autoantibody, DNA testing allows determination of extended antigen profile to select antigen-negative RBCs for transfusion. This avoids use of "least incompatible" blood for transfusion and allows transfusion of units "antigen-matched for clinically significant blood group antigens" to prevent delayed transfusion reactions and circumvent additional alloimmunization. Importantly, this approach can improve patient care and testing turnaround time by eliminating need for repeat adsorptions to remove autoantibody to rule out new underlying RBC alloantibodies.

Determination of D Status: Altered expression of D antigen occurs in 2% of Caucasians, <1% of Asians, and ~4% of black and Hispanic groups. Routine serologic

D-typing reagents cannot distinguish RBCs with weak D or partial D, and distinction between these is of clinical importance because latter are at risk for anti-D. *RHD* genotyping strategies that sample multiple regions of *RHD* can discriminate weak D and partial D phenotypes. Females of child-bearing age with partial D would potentially benefit from receiving D-negative RBCs for transfusion to avoid future pregnancy complications and be considered for RhIG prophylaxis. Many transfusion services err on side of caution and transfuse patients with D-negative RBCs if serologic D-typing reaction strength is weaker than expected (less than 3+ to 4+). This results in unnecessary use of limited D-negative blood supply.

Alloantibody Versus Autoantibody: For patients presenting with RBCs that type positive for antigen with apparent antibody of the same specificity in serum or plasma, DNA-based investigation is helpful for transfusion management. If the sample is predicted to be antigen positive by SNP testing, it should be further investigated by high-resolution gene sequencing. Samples may have novel amino acid change in the protein carrying the blood group antigen. These result in new epitopes and altered (or partial) expression of conventional antigen.

Sickle Cell Disease and Chronic Transfusion Therapy: Alloimmunization is serious complication of chronic transfusion, particularly in patients with sickle cell disease (SCD) and β-thalassemia requiring long-term transfusion support. Antibody production results in DHTR. Alloimmunization results in significant delay in transfusion, increases costs, and is associated with risk for production of additional antibodies and often results in chronic positive DAT with apparent warm autoantibody, i.e., "panagglutinins," which complicate further workups and transfusion therapy.

Many programs attempt to prevent or reduce the risk and incidence of alloantibody production by transfusing RBCs that are antigen-matched for D, C, E, and K (and few include $Fy^{a/b}$, $Jk^{a/b}$, and S). The most frequent antibodies encountered in patients with SCD receiving units antigen-matched for D, C, E, and K, have complex Rh specificities. These antibodies include anti-D, -C, and -e in patients whose RBCs type serologically as D+, C+, and e+. RH genotyping has revealed that these patients have RHD and/or RHCE alleles with amino acid changes that encode altered or partial antigens. For example, ~23% of SCD patients with C+ RBCs do not have a conventional RhCe protein. In these patients, C antigen is expressed from hybrid *RHD* gene that has lost expression of D, but encodes C epitope; more than a third of these patients make anti-C or -Ce. Anti-D in D+ patients with SCD is associated with RHD alleles encoding partial D antigens detected by RHD genotyping. Anti-e in e+ patients with altered Rhce proteins are more problematic to manage. Transfusion with e− blood will expose them to E antigen, and most are E− and at risk of anti-E. In summary, RH genotyping expands and extends matching for Rh in this patient population and is important for transfusion management.

Transfusion Support for Hematopoietic Stem Cell Transplantation: Although most antibodies to RBCs including ABO are not a barrier to engraftment, RBC alloantibodies can cause complications. Increasing use of nonmyeloablative conditioning results in mixed chimerism in which both recipient- and donor-derived lymphocytes and recipient plasma cells persist. Non-ABO blood group antibodies directed to minor antigens can cause delayed erythropoiesis or hemolysis of incompatible donor RBCs, and new RBC antibodies (whether donor- or patient-derived) can complicate transfusion support and recovery. Genotyping of recipient peripheral blood sample and buccal cells can be done in an attempt to determine origin of new antibodies post-transplant, i.e., donor-derived or recipient-derived, to inform selection of units for transfusion. Genotyping donor and recipient for extended blood group antigens before transplantation proactively informs transfusion management in patients with existing alloantibodies.

Prenatal Practice

Testing for RhD in Pregnant Women: Serologic typing for RhD cannot distinguish women whose RBCs lack some epitopes of D (partial D) and are at risk for anti-D. RBCs with partial D type as D+, some in direct tests and others by indirect tests, but these women would potentially benefit from receiving RhIG prophylaxis if they deliver D+ fetus. The situation is confusing for patient care, as serologic testing cannot distinguish weak reactivity due to missing epitopes (partial D) from weak reactivity due to quantitative reduced antigen (weak D). RHD genotyping can distinguish weak D from partial D is recommended to guide RhIG prophylaxis and blood transfusion recommendations.

Hemolytic Disease of the Fetus and Newborn: Clinically significant alloantibody identification in a pregnant woman relies on demonstration of IgG antibody by serologic testing, but management of the pregnancy is aided by paternal sample testing, and if indicated, fetal sample testing to assess HDFN risk.

Paternal Testing: Paternal RBCs should be tested for corresponding antigen. If RBCs are negative, fetus is not at risk. If the father is positive, zygosity testing can determine if the father is homozygous or heterozygous for the gene expressing the antigen, particularly when there is no allelic antigen, or no antisera to detect the allelic product.

DNA zygosity testing of paternal samples is most often done when testing for possible HDFN due to anti-D or anti-K. For maternal anti-K, testing paternal RBCs for expression of allelic k antigen should be performed, but not all providers have reagents available and genetic counselors often request DNA testing. For maternal anti-D, *RHD* zygosity testing by DNA methods is the only method to determine paternal gene copy number. Number of different genetic events cause D-negative phenotype and more than one assay are often done to accurately determine RHD zygosity, especially in minority ethnic groups.

If the father is *RHD* homozygous, all children will be D+, and fetal monitoring will be required. If the father is heterozygous, the fetus has a 50% chance of being at risk. D type of the fetus should be determined to prevent invasive and unnecessary testing, and so the mother need not be aggressively monitored or receive immune-modulating agents.

Fetal Testing: To determine the fetal antigen status, fetal DNA can be isolated from cells obtained by amniocentesis. Alternatively, fetal antigen status can be performed by noninvasive fetal testing from maternal plasma. The discovery that cell-free, fetal-derived DNA is present in maternal plasma by approximately 5 weeks of gestation allows maternal plasma to be used as a source of fetal DNA to determine fetal antigen status. This is particularly successful for D typing because D-negative phenotype in the majority of samples is due to the absence of RHD gene. Testing for presence or absence of a gene is less demanding than testing for single gene polymorphism. This approach is being used in Europe to test maternal plasma for the presence of fetal RHD gene, to eliminate unnecessary administration of antepartum RhIg to ~40% of D-negative women who are carrying D-negative fetus.

Neonatal Alloimmune Thrombocytopenia: Neonatal alloimmune thrombocytopenia (NAIT) diagnosis is based on demonstrating HPA-specific antibody in maternal serum and identifying incompatibility between the parents by HPA platelet genotyping. Twenty-eight HPAs have been characterized, but incompatibility in HPA-1 accounts for ~80% of cases. Platelet genotyping is used to confirm HPA status of the mother and to type paternal sample. If the father is homozygous for target HPA, all children will be positive and fetal monitoring will be required. If the father is heterozygous, the fetus has 50% chance of being at risk. Fetal HPA type should be determined to prevent invasive and unnecessary testing. To determine fetal HPA antigen status, fetal DNA can be isolated from cells obtained by amniocentesis. Noninvasive fetal testing from maternal plasma has been reported and may become more readily available in the future.

Other Clinical Applications

McLeod Syndrome: McLeod phenotype, characterized by weak expression of RBC Kell system antigens and absence of Kx antigen, is encoded by X-linked gene, *XK*. This X-linked syndrome manifests only in males and is associated with late onset clinical or subclinical myopathy, neurodegeneration, and central nervous system manifestations. The syndrome may be underdiagnosed, and the physical characteristics, which often develop only after fourth decade of life, include muscular and neurological problems. More than 30 different *XK* gene mutations associated with McLeod phenotype have been found. Different *XK* mutations appear to have different clinical effects and may account for the variability in prognosis. Sequencing of *XK* to determine specific type of mutation in individuals with McLeod phenotypes has clinical prognostic value.

ABO-Typing Discrepancies in Patients: DNA testing is useful to resolve patient typing discrepancies, confirm subgroup status, and determine original blood type of patients who were massively transfused or original blood type of transplant recipients by testing a buccal sample. Ability to determine individual's antigen status eliminates group O RBCs and AB plasma use. ABO genotyping can aid in differentiation of subgroup alleles, particularly to confirm A_2 subgroup in kidney donors who may have been transfused, or whose RBCs gave discordant reactivity in serologic testing with anti-A_1 reagents.

Donor Center Applications

Typing for Antigens for Which There Are No Commercial Reagents: DNA-based typing has become standard to identify antigen-negative units for which there are no serologic reagents. One of the most often used is to type for Dombrock (Doa/b). Antibodies to these antigens are clinically significant, but patient serum antibodies are often weak, have poor avidity, disappear over time, and are almost always present with other blood group specificities that interfere with screening donor units. Because Dombrock antibodies are difficult to detect, matching of patients with donors for

Dombrock antigens should be considered in patients when survival of transfused RBCs is compromised and complex mixtures of antibodies are present, even when no specific serum antibody to Dombrock antigens is demonstrable by serologic testing.

Confirming D Type of Donors: Donor centers must perform test for weak D to avoid labeling product as D-negative that might result in anti-D in response to transfused RBCs. It is well known that some donor RBCs with very weak D expression are not typed as D+ with current serologic reagents and are labeled as D− for transfusion. Prevalence of weak D RBCs not detected by serologic reagents is ~0.1%, and although clinical significance has not been established, donor RBCs with weak D expression have been associated with rare alloimmunization. RHD genotyping could improve donor testing by confirming D− phenotypes.

High-Throughput Screening for Donor Inventory and Rare Antigens: Ability to screen for multiple minor antigens in single-assay format has been significant aid to donor centers to provide antigen-negative products. Although DNA methods are not yet FDA licensed to label donor units, testing is a valuable screening tool requiring only negative tests to be confirmed with licensed reagent. High-throughput screening has made a major contribution by identifying uncommon and rare donors.

Resolving Donor ABO Discrepancies to Retain Donors: Donor samples with ABO-typing results that are discordant with results from a previous donation must be investigated and may be FDA reportable. Donors with depressed antigen or antibody expression and RBC reactions that do not match the plasma reactions cannot be labeled for transfusion, and the products are discarded. Determination of ABO by DNA-based methods can resolve these, and although testing is not FDA approved for labeling blood components, DNA methods offer the potential to be a confirmatory test. Confirmatory testing by DNA would avoid unit loss and retain group O donors with depressed antibody titers.

Future Perspectives: NGS, also known as massively parallel sequencing technology, is transforming the genomics field and plays a central role in health care and the movement toward personalized medicine. A number of researchers have now shown that NGS can provide adequate coverage of the genes that determine blood group–extended antigen types. This approach's strength lies in its ability to potentially detect the relevant polymorphisms including null alleles, novel mutations, and complex gene rearrangements in blood group genes simultaneously. As the cost of NGS continues to drop, patient genome sequencing as part of clinical care will become more commonplace. Secondary analysis of NGS data for patients needing transfusion therapy would represent cost-effective and practical use of this existing information. Knowing which antigens the patient lacks and is at risk for sensitization would streamline pretransfusion testing and enable informed decision-making for blood product selection.

Further Reading

Chou, S. T., & Westhoff, C. M. (2017). Application of genomics for transfusion therapy in sickle cell anemia. *Blood Cells Mol Dis*, *67*, 148–154.

Haspel, R. L., & Westhoff, C. M. (2015). How do I manage Rh typing in obstetric patients? *Transfusion*, *55*, 470–474.

Hillyer, C. D., Shaz, B. H., Winkler, A. M., & Reid, M. (2008). Integrating molecular technologies for red blood cell typing and compatibility testing into blood centers and transfusion services. *Trans Med Rev*, *22*, 117–132.

Lane, W. J., Westhoff, C. M., Uy, J. M., et al. (2016). Comprehensive red blood cell and platelet antigen prediction from whole genome sequencing: Proof of principle. *Transfusion*, *56*, 743–754.

Sandler, S. G., Flegel, W. A., Westhoff, C. M., et al. (2015). It's time to phase in RHD genotyping for patients with a serologic weak D phenotype. *Transfusion*, *55*, 680–689.

CHAPTER 25

ABO and H Blood Group System

Lynsi Rahorst, MHPE, MT(ASCP) SBB and
Connie M. Westhoff, PhD, SBB

ABO blood group system consists of A, B, and H (ABH) antigens. Group O individuals express the precursor H antigen but lack A and B antigens. Individuals form antibodies (anti-A and anti-B) to the antigens they lack. These antibodies are termed "naturally occurring," as they are present in sera of individuals without previous red blood cell (RBC) exposure. These IgM antibodies can activate complement and can thus cause severe intravascular hemolysis after transfusion of ABO-incompatible blood components, making them the most clinically significant antibodies in transfusion practice. Prevention of ABO-incompatible RBC transfusion is the primary objective of pretransfusion testing.

Antigens: ABH antigens are carbohydrate structures that are synthesized in a stepwise fashion by glycosyltransferase enzymes that sequentially add specific monosaccharides to glycoproteins and glycolipids. H antigen defines the O blood group and is precursor for A and B antigens. Group O RBCs have large amounts of H antigen and no A or B antigens. Some precursor H antigen remains on A and B RBCs, depending on the transferase efficiency, with least on A_1B RBCs. H antigen concentration by ABO blood group varies ($O > A_2 > B > A_2B > A_1 > A_1B$).

Each RBC carries more than 2 million ABH antigen sites. ABH antigens are also found on other tissues, including endothelial and epithelial cells of lung, gut, and urinary and reproductive tracts (and are therefore termed histo-blood group antigens). Hence, they are important in solid organ transplantation, where ABO incompatibility may require minimization of recipient anti-A and/or anti-B levels to prevent acute rejection. Prevalence of ABO blood groups differs in various populations (Table 25.1).

Antigen Expression: ABO antigens are not fully developed at birth. Adult levels of ABO antigens with complex branching oligosaccharide structures appear on the RBCs by 2–4 years.

A and B Subgroups: RBCs from some group A or B individuals that react moderately, weakly, or not at all with standard anti-A or anti-B sera are termed subgroups; B subgroups are encountered less frequently than A subgroups. Approximately, 80% of group A individuals are A_1, while ~20% are A_2; subgroups A_3, A_{el}, A_x, etc. are less frequently encountered. Difference between A_1 and A_2 is both quantitative (~five times fewer A antigens on A_2 than A_1 RBCs) and qualitative (structural differences). Because of the structural difference, A_2 individuals can form anti-A1 (1%–8% of A_2 individuals and ~30% of A_2B individuals have anti-A1). Anti-A1 does not usually result in hemolysis of group A_1 RBCs; however, hemolytic anti-A1 has been reported, and its presence can lead to organ rejection in some circumstances. It is prudent to transfuse

Transfusion Medicine and Hemostasis. https://doi.org/10.1016/B978-0-12-813726-0.00025-8

TABLE 25.1 ABO Blood Group Prevalence

	Prevalence (%)		
ABO Group	Caucasian	African-American	Asian
O	45	49	43
A	40	27	27
B	11	20	25
AB	4	4	5

these individuals with RBCs that have been shown to be crossmatch compatible (e.g., A_2 or O). *Dolichos biflorus* lectin is used to distinguish A_1 from A_2, as the lectin is diluted to agglutinate A_1 but not A_2 or weaker subgroup RBCs. RBC reactivity with the lectin is not always straightforward, and when it is of clinical importance to determine A_1 versus A_2 status, for example in kidney transplantation, ABO genotyping (see below) is recommended.

Bombay and Para-Bombay Phenotypes: Bombay individuals lack ABH antigens on RBCs, tissues, and secretions and make anti-A, anti-B, and potent anti-H. These individuals must be transfused with RBCs from other Bombay individuals, which are very rare. Para-Bombay individuals lack H antigen on their RBCs but have H antigen (and, depending on their *ABO* gene, A and B antigens) in their secretions. The RBCs of these individuals may express A and B antigens very weakly due to adsorption of soluble A and B antigens onto RBCs. Para-Bombay also describes individuals who have minimal H on their RBCs, regardless of secretor status.

Antibodies: Anti-A and anti-B are found in the individuals' plasma who lack corresponding antigen (group O have anti-A and anti-B; group A have anti-B; group B have anti-A; group AB have neither). These antibodies are produced in response to environmental stimuli, such as plant and bacterial moieties (e.g., *Escherichia coli* sugars) and are therefore termed "naturally occurring." Antibody production begins after birth and is usually detectable by 4–6 months of age, reaches peak at age 5–10 years, and then declines with increasing age. Immunodeficient patients may not produce detectable levels of anti-A and/or anti-B. These antibodies are primarily IgM with some IgG; IgM is responsible for agglutination at room temperature and for activating complement, which, in combination with numerous ABO antigens on RBCs, is responsible for the severe transfusion reactions that may result from ABO-incompatible transfusion. Hemolytic disease of the fetus and newborn (HDFN) caused by ABO antibodies is usually mild because only IgG crosses the placenta, fetal ABO antigens are not fully developed, and ABO tissue antigens provide additional targets for the antibody. HDFN is most often seen in non–group O infants of group O mothers, because group O individuals can have increased amounts of IgG anti-A, anti-B, and anti-A,B.

Antibody Titers: Anti-A and anti-B titer results are used in a variety of clinical situations, including ABO-incompatible solid organ or hematopoietic stem cell (HSC) transplantation. Titers are also evaluated in blood components, such as ABO-mismatched platelets, group O whole blood, and group A plasma used in trauma settings. There is not currently a uniform method for titering these antibodies nor is there a uniform critical value. Antibodies may be tested in both room temperature phase to detect IgM antibody levels and antihuman globulin phase after incubation at 37°C to detect the IgG levels. Further differentiation between IgG and IgM antibody titers can be determined by treating plasma with dithiothreitol to inactivate IgM, followed by testing with antihuman IgG.

Plasma-Rich Blood Components: Rare examples of fatal acute hemolysis have been reported in patients transfused with ABO-incompatible components containing high titers of anti-A or anti-B. Typically, these are group O components, with significant amounts of plasma, such as apheresis platelets. Because of inadequate platelet inventories, it is not always possible to transfuse ABO-matched platelets. Anti-A and/or anti-B titers are performed on group O platelet components by some institutions; a critical titer cutoff is used to identify units that are at higher risk for producing acute hemolysis and those are transfused to group O recipients only. Alternatively, volume-reducing or volume-storing platelets in platelet-additive solution are other strategies to mitigate risk when transfusing ABO-mismatched platelets. ABO antibody titering is also relevant in transfusion of group O whole blood or group A plasma to recipients of unknown blood type in a trauma setting, or to relieve shortages of AB plasma.

Solid Organ Transplantation: As a limited number of donor organs are available, ABO-incompatible solid organ transplantation is sometimes performed. There are several reports of comparable outcomes of ABO-incompatible kidney and liver transplants to ABO-compatible transplants. Successful ABO-incompatible solid organ transplants involve monitoring recipient ABO antibody titers and using plasma exchange and immunosuppression to reduce titers. Each institution typically develops an ABO titer protocol, including both IgG and IgM phases of testing, and determines the critical titer at which ABO-incompatible transplantation can proceed. Besides organ rejection due to recipient anti-A and/or anti-B, another risk associated with incompatible solid organ transplantation is passenger lymphocyte syndrome, which occurs when lymphocytes in the solid organ produce antibodies against the recipient (i.e., group O organ into a group A patient). For kidney transplantation, Group A_2 or weaker A subgroup organs transplanted into group O or B recipients have been shown to be equivalent to group O organ donors transplanted into non–O recipients.

Incompatible Hematopoietic Stem Cell Transplantation: Antibody titers can be used to determine need for HPC product modification, or need to lower antibody levels in patients (Chapter 53). Delayed hemolysis can occur when donor lymphocytes produce antibodies against patient RBCs, usually within 7–14 days after transplantation. Hemolysis can be severe and fatal, especially in group A patients with a group O donor, but this is minimized in patients receiving methotrexate or similar medication.

ABO-Typing Discrepancies: Any discrepancy in the results of ABO typing, which occurs when the RBC reactions (i.e., front type) do not match the plasma reactions (i.e., back type) should be resolved before issuing blood components to patients (Table 25.2). If transfusion is necessary before resolution, or results cannot be resolved, group O

TABLE 25.2 ABO Typing Discrepancies and Approaches to Resolve

Resolving ABO Discrepancies				
	Discrepancy	Description	Explanation	Resolution
Unexpected RBC reactions	Weakened expression of A and B antigens	Acquired hematologic diseases and other conditions.		• Increase incubation of RBCs and antisera at room temperature.
	Loss of antigens	Somatic chromosomal deletion of ABO locus.		• Incubate at 4°C with controls (group O and autologous RBC). • Treat cells with proteolytic enzymes. • ABO genotyping. • Adsorption of anti-A or anti-B followed by elution (with appropriate A, B, and O RBC controls).
	Unexpected reactions with anti-A/anti-B	B(A) Phenotype	Group B individuals with high levels of galactosyltransferase resulting in detectable levels of the A-sugar (GalNAc) on RBCs. RBCs are agglutinated by anti-A that contains murine monoclonal antibody MHO4.	• Use alternate anti-A reagent (one that does not contain MHO4).
		Acquired B Phenotype	Phenomenon in group A patients when microbial deacetylating enzymes modify the A-sugar (GalNAc) to resemble the B-sugar (galactose). Associated with bacteremia secondary to intestinal obstruction or gastric/intestinal malignancy.	• Observed with monoclonal anti-B containing the ES-4 clone, and was shown to be associated with acid pH. The manufacturer has since modified the reagent.

	Polyagglutination	RBCs agglutinate with all human sera, which can result from genetic inheritance or infection. Polyagglutinable RBCs can be characterized by lectin typing.	*T activation:* Bacteria or virus produce neuraminidase that cleaves N-acetylneuraminic acid and exposes hidden antigen (cryptantigen) on RBCs. All normal sera contain anti-T. Resolves on elimination of causal organism. *T_n polyagglutination:* HPC mutations cause defective synthesis of oligosaccharides which exposes cryptantigens on RBCs. Treat RBCs with proteolytic enzymes to eliminate reactivity.
	Mixed-field agglutination	Occurs when there are two or more cell populations in the patient's circulation, i.e., after transfusion of non-ABO identical RBCs or hematopoietic stem cell transplant from a non-ABO identical donor, or chimera (intrauterine exchange of erythropoietic tissue by fraternal twin or mosaicism from dispermy). Some subgroups (A_3 subgroup, for example) also demonstrate mixed-field agglutination.	• Obtain patient transfusion/transplant history.
	Spontaneous agglutination of RBCs	Cold reactive autoagglutinins coating RBCs.	• Incubate RBCs at 37°C, followed by washing with warm saline. • DTT treat RBCs to remove IgM antibody (DTT also weakens/destroys some blood group antigens).

Continued

TABLE 25.2 ABO Typing Discrepancies and Approaches to Resolve—cont'd.

Resolving ABO Discrepancies				
	Discrepancy	Description	Explanation	Resolution
Unexpected serum reactions	Extra reactivity in serum/plasma	Group A patient's serum reacts with A_1 cells	A subgroup with anti-A1.	• Test serum with A_2 cells (should not react). • Test cells with anti-A1 lectin, *Dolichos biflorus* (A1-negative).
		Serum reacts with all reagent cells (A_1, A_2, B, and O) and auto control	Cold autoantibody.	• Prewarm to 37°C cells and serum before testing. • Remove autoantibody by adsorption.
			Rouleaux: high protein in serum causes "stack-of-coins" aggregation of red cells.	• Saline replacement.
		Serum reacts with all reagent cells (A_1, A_2, B, and O), auto control negative	Cold-reacting alloantibody (i.e., anti-M, anti-P1).	• Identify the alloantibody; prewarm cells and serum before testing, incubate tests at 37°C. • Test serum with A_1 and B cells that lack the corresponding antigen.
	Missing reactivity in serum/plasma	Serum does not react with expected reagent cells (i.e., group O patient's serum does not react with A_1 or B cells)	Neonate.	• Serum testing not performed on patients <4 months old.
			Immunosuppression, elderly patient.	• Extend incubation time at room temperature. • Increase serum/plasma to RBC ratio.

DTT, dithiothreitol.

RBCs and group AB (or low-titer group A) plasma can be issued. For donors, units cannot be labeled or released if there is a discrepancy. Discrepancies may result from problems with the RBCs or plasma, testing problems, or technical errors. Negative results may be obtained when positive results are expected, or positive results may be seen when negative results are expected. To resolve a discrepancy, sample mix up should first be ruled out and testing repeated after washing patient RBCs (and reagent cells if applicable). It is important to obtain a patient history, including previous transfusions or HSC transplantation.

***ABO* and *FUT* Genes:** *H(FUT1)* and *Se(FUT2)* genes encode fucosyltransferase enzymes that form H antigen on RBCs and in secretions, respectively. *H(FUT1)* encodes a fucosyltransferase enzyme preferentially expressed in erythroid cells, which adds fucose in α (1,2) linkage on type 2 glycoproteins to form H antigen on RBCs. *Se(FUT2)* encodes a fucosyltransferase, which is preferentially expressed in epithelial cells and adds fucose in α (1,2) linkage to type 1 glycoprotein chains (Fig. 25.1) to form ABH antigens in secretions and fluids (secretor phenotype). Approximately, 20% of individuals lack ABH in secretions due to a defective fucosyltransferase *Se(FUT2)* indicated as *sese* (nonsecretor phenotype). The rare Bombay phenotype (group O_h) results from homozygosity for inactive *H(FUT1)* and *Se(FUT2)*. Para-Bombay individuals are homozygous for a null allele at *H(FUT1)* but have at least one functional *Se(FUT2)*.

ABO gene encodes glycosyltransferases that add either *N*-acetylgalactosamine (GalNAc) or galactose (Gal) to H antigen, resulting in A and B antigens, respectively (Fig. 25.2). ABO alleles are codominantly expressed, and the transferase enzymes responsible for A and B antigens differ by 4 out of 354 amino acids (p.Arg176Gly, Gly235Ser, Leu266Met, and Gly268Ala). Mutations in transferase genes that cause decrease in enzyme efficiency result in reduced number of antigens and altered branching structure responsible for A and B subgroup phenotypes. At the gene level, the most common A_2 allele differs from A_1 allele in that it encodes a single amino acid change (p.Pro156Leu) and extraneous residues at terminal end of transferase. Group O is an autosomal recessive trait, resulting from inheriting two nonfunctional *ABO* genes. The most common O allele has a nucleotide deletion, designated c.261delG, which results in a truncated product with no enzyme activity. A large number (>200) of different A and B subgroup alleles and inactive O alleles have been reported.

ABO Genotyping: DNA-based testing is useful to resolve typing discrepancies, confirm subgroup status, and determine blood type of massively transfused patients or HSC transplant recipients (buccal swab), as examples. Ability to accurately determine an individual's ABO type eliminates use of group O RBCs and AB plasma for transfusion due to ABO typing discrepancy. Determination of ABO by DNA-based methods is not approved for labeling blood components; however, these methods offer future potential to avoid loss of donor's blood components due to serologic ABO discrepancy with inability to appropriately label them. ABO genotyping can aid in differentiation of subgroup alleles, particularly to confirm A_2 subgroup in kidney donors who may have been transfused or whose RBCs give discordant reactivity in serologic testing with anti-A1 lectin reagents.

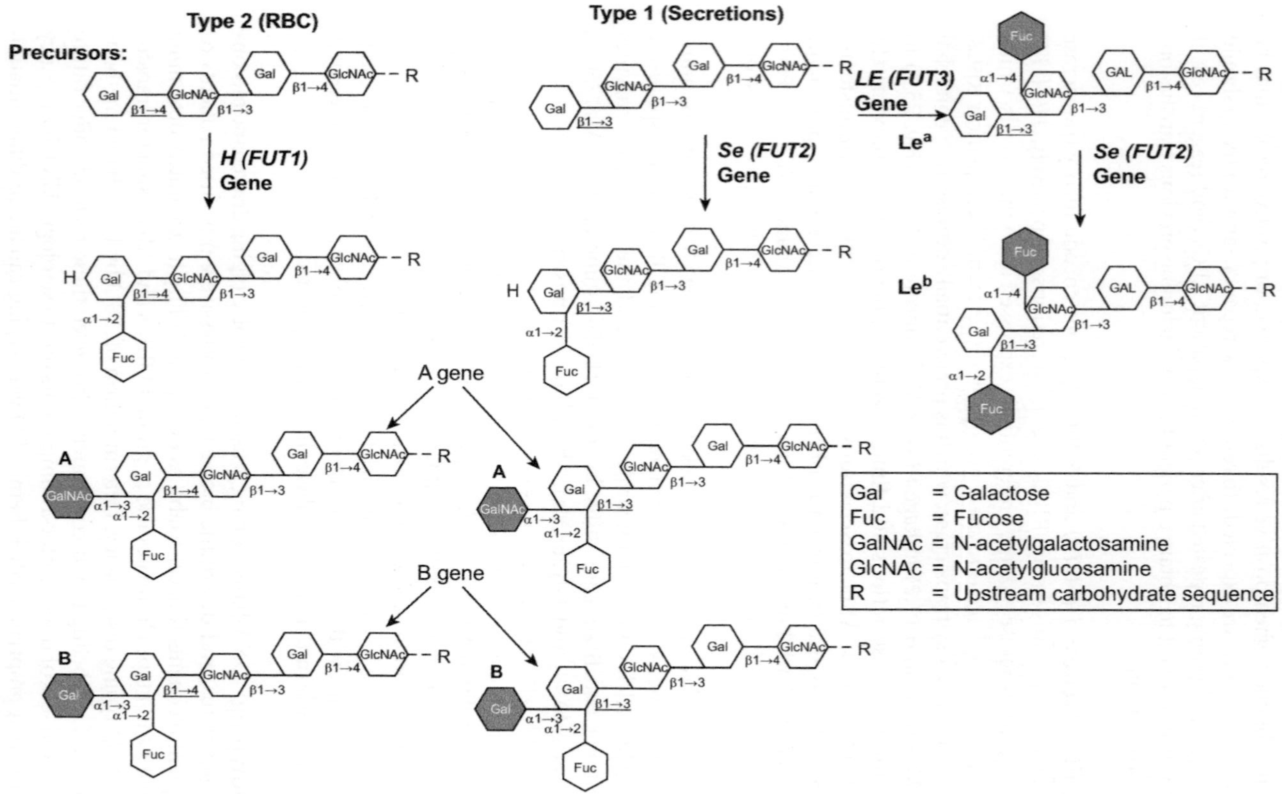

FIGURE 25.1 Synthesis of A, B, H, and Lewis antigens. Oligosaccharide precursor core type 1 and type 2 structures differ only in the linkage between the terminal galactose (Gal) and the *N*-acetylglucosamine (GlcNAc), shown underscored. Terminal carbohydrates that define the antigens are shown in black. (Modified from Hillyer, C. D., Silberstein, L. E., Ness, P. M., et al. (Eds.). (2007). *Blood banking and transfusion medicine: Basic principles and practice* (2nd ed.). San Diego, CA: Elsevier Academic Press.)

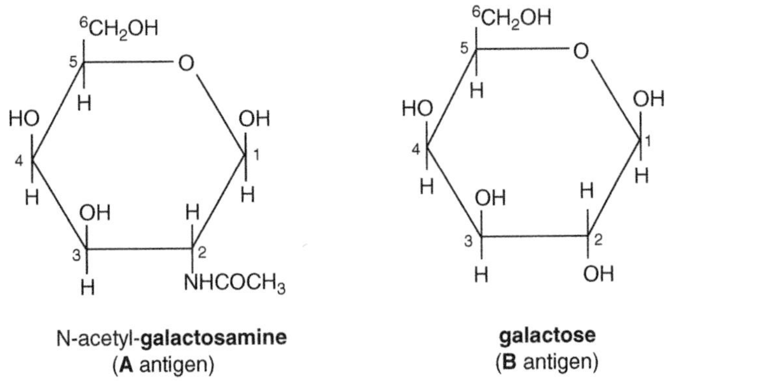

FIGURE 25.2 **Terminal carbohydrates that define the A and B antigens.** The terminal galactose residues differ only in that the A antigen has substituted the aminoacetyl group on carbon number 2.

Further Reading

Agaronov, M., DiBattista, A., Christenson, E., Miller-Murphy, R., Strauss, D., & Shaz, B. H. (2016). Perception of low-titer group A plasma and potential barriers to using this product: A blood center's experience serving community and academic hospitals. *Transfus Apher Sci*, *55*, 141–145.

Berséus, O., Boman, K., Nessen, S. C., & Westerberg, L. A. (2013). Risks of hemolysis due to anti-A and anti-B caused by the transfusion of blood or blood components containing ABO-incompatible plasma. *Transfusion*, *53*(Suppl. 1), 114S–123S (Review).

Josephson, C. D., Mullis, N. C., Van Demark, C., & Hillyer, C. D. (2004). Significant numbers of apheresis-derived group O platelet units have 'high-titer' anti-A/A, B: Implications for transfusion policy. *Transfusion*, *44*, 805–808.

Karafin, M. S., Blagg, L., Tobian, A. A., King, K. E., Ness, P. M., & Savage, W. J. (2012). ABO antibody titers are not predictive of hemolytic reactions due to plasma-incompatible platelet transfusions. *Transfusion*, *52*, 2087–2093.

O'Donghaile, D., Kelley, W., Klein, H. G., & Flegel, W. A. (2012). Recommendations for transfusion in ABO-incompatible hematopoietic stem cell transplantation. *Transfusion*, *52*, 456–458.

Redfield, R. R., Parsons, R. F., Rodriguez, E., et al. (2011). Underutilization of A_2 ABO incompatible kidney transplantation. *Clin Transplant*, *26*, 489–494.

Rowley, S. D. (2001). Hematopoietic stem cell transplantation between red cell incompatible donor-recipient pairs. *Bone Marrow Transplant*, *28*, 315–321.

Storry, J. R., & Olsson, M. L. (2009). The ABO blood group system revisited: A review and update. *Immunohematology*, *25*, 48–59.

McVey, J., Baker, D., Parti, R., Berg, R., Gudino, M., & Teschner, W. (2015). Anti-A and anti-B titers in donor plasma, plasma pools, and immunoglobulin final products. *Transfusion*, *55*(Suppl. 2), S98–S104.

CHAPTER 26

Rh and RhAG Blood Group Systems

Sunitha Vege, MS and Connie M. Westhoff, PhD, SBB

The Rh blood group system (C, c, E, e, D, and more than 50 other antigens) is second only to ABO in clinical importance because the Rh antigens, especially D, are highly immunogenic and the antibodies can result in delayed hemolytic transfusion reactions (HTRs) and hemolytic disease of the fetus and newborn (HDFN). The RH locus consists of two homologous genes, *RHD* and *RHCE,* located in close proximity on chromosome 1 and encoding proteins consisting of 417 amino acids. The D-negative phenotype is associated with deleted or mutated/inactive *RHD* gene. The system is complex, especially in African black and Hispanic ethnic groups both serologically and genetically, as point mutation(s) and genetic exchange between the two genes generate new epitopes on the Rh proteins and novel antigens. While antisera is commercially available to detect the common D, C/c, and E/e antigens, serologic reagents are not available to identify the many other antigens; however, detection is possible by DNA methods, i.e., RH genotyping. The Rh system is associated with production of antibodies with multiple and complex specificities that can make it difficult to find compatible red blood cell (RBC) products. DNA testing to determine the patient's RH genotype aids in component selection and transfusion management particularly for patients with sickle cell disease (SCD).

Common Antigens: The Rh system has more than 50 antigens, but those of routine concern are D, carried on RhD protein and encoded by the gene designated *RHD*, and C, c, E, and e antigens, carried on the RhCE protein encoded by the gene designated *RHCE* (Table 26.1). Antigens are codominantly expressed. *RHD* and *RHCE* genes, each consisting of 10 coding regions (exons), are 97% homologous encoding proteins differing by between 32 or 35 amino acids. The two genes are inherited as an Rh haplotype. Prevalence of Rh haplotypes differs by ethnic group (Table 26.1). A third gene, *RHAG,* is 47% identical in the coding region to *RHD* and *RHCE* and encodes the ancestral protein RhAG. Rh antigens are carried on hydrophobic 12-pass transmembrane proteins (Fig. 26.1). RhAG is important for trafficking the RhD and RhCE proteins to the membrane, and lack of RhAG results in the absence of Rh antigen expression (Rh-null phenotype) or marked reduction of Rh antigen expression (Rh-mod phenotype). Rh and RhAG proteins form the Rh core complex in the membrane, and they interact with CD47, glycophorin B (carry SsU antigens), LW, and AE1 (also known as Band 3). This complex is linked to the membrane skeleton via Rh/RhAG–ankyrin interaction and CD47–protein 4.2 association. RhAG is involved in ammonia/ammonium transport and is important for cation balance in RBCs. Function of the Rh blood group proteins, RhD and RhCE is not known, but RBCs lacking all Rh antigens have structural abnormalities (i.e., stomatocytes). RhAG is a highly conserved protein but carries several antigens; two of high prevalence (RHAG1, RHAG3) and one of low prevalence (RHAG2).

Transfusion Medicine and Hemostasis. https://doi.org/10.1016/B978-0-12-813726-0.00026-X

TABLE 26.1 Nomenclature and Prevalence of Rh Haplotypes

Haplotype	Shorthand for Haplotype	Prevalence (%) White	Black	Asian
DCe	R_1	42	17	70
DcE	R_2	14	11	21
Dce	R_0	4	44	3
DCE	R_z	<0.01	<0.01	1
ce	r	37	26	3
Ce	r′	2	2	2
cE	r″	1	<0.01	<0.01
CE	r^y	<0.01	<0.01	<0.01

Modified from Hillyer, C. D., Strauss, R. G., & Luban, N. L. C. (Eds.). (2004). *Handbook of pediatric transfusion medicine.* San Diego, CA: Elsevier Academic Press.

D Antigen: Presence or absence of the D antigen confers the Rh-positive or Rh-negative status commonly used in lay and scientific parlance. In the United States, ~85% of the white population is D-positive (Rh-positive), and 15% is D-negative (Rh-negative), while only 5%–8% of blacks and less than 1% of Asians are D-negative. Most D-negative individuals, especially of European descent have a deletion of the *RHD* gene. D-negative phenotype can also occur as a result of various mutations in *RHD*, including premature stop codons, insertions, deletions, or *RHD/RHCE* hybrid alleles. The inactive *RHD* pseudogene, which results from a 17-bp duplication and premature stop codon, is frequently found in blacks. D-negative is rare in Asia, and D typing is not routinely done in some Asian countries.

Weak D: The term "serologic weak D" is applied to RBCs that carry low levels of D antigen (formerly called D^u, which is an obsolete term, and its use is discouraged), which are not detected, or give a less than robust reactions by direct agglutination with anti-D. Weak D can be detected after incubation with anti-D and use of antiglobulin (AHG) reagent. However, if a patient has a positive DAT, weak D test will not be valid, as the test may be a false positive due to immunoglobulin (IgG) coating the cells. Because manufacturers' reagents contain different clones and formulations, reactivity of weak D RBCs may be variable and depend on the specific reagent and the method used for detection. Alternatively, DNA testing can be used to accurately determine the D status.

More than 95 different RHD alleles account for the molecular basis of weak D expression, with weak D types 1, 2, and 3 being the most common in Caucasians. Weak D alleles most often have a single amino acid change. Serologic weak D can also result from decreased D antigen expression when *RH*C* is in *trans* to *RHD* (phenotype Dce/Ce). A very weak form of D, termed D_{el}, is only detected by adsorption and elution of anti-D and is more prevalent in Asians.

RBCs with weak D are less immunogenic than normal D-positive RBCs, but anti-D stimulation can occur. Components from individuals with weak D are labeled as

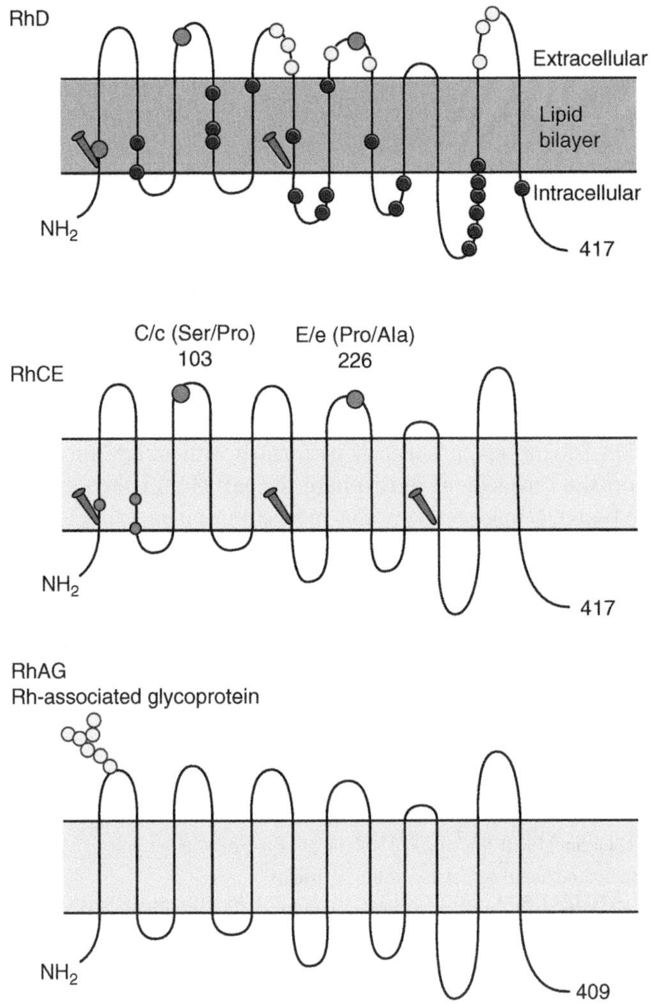

FIGURE 26.1 Predicted 12-transmembrane span model of the RhD, RhCE, and RhAG proteins in the red blood cell membrane. The amino acid differences between RhD and RhCE are shown as *symbols*. The eight extracellular differences between RhD and RhCE are indicated as *open circles*. The nail structures represent the location of possible palmitoylation sites. Positions 103 and 226 in RhCE, which are critical for C/c and E/e expression, respectively, are indicated. The *N*-glycan on the first extracellular loop of the Rh-associated glycoprotein is indicated by the branched structure. (From Hillyer, C. D., Silberstein, L. E., Ness, P. M., et al. (Eds.). (2007). *Blood banking and transfusion medicine: Basic principles and practice* (2nd ed.). San Diego, CA: Elsevier Academic Press.)

D-positive, but not all weak D RBCs are detected by donor testing. It is important to investigate if a D-negative patient given an apparent D-negative product makes anti-D.

Patient samples do not need to be tested for weak D, with the exception of newborns, as the mother is a candidate for Rh immune globulin (RhIG) administration if the baby is weak D-positive. If testing for weak D is performed and found to be positive, or RBCs give less than a robust reaction with anti-D, further consideration should be given to assess the presence and risk of anti-D, especially in females of child-bearing potential. If the positive test is due to a weak D phenotype, the patient can generally receive D-positive RBCs without risk of immunization, but if the weak D reactivity is due to a partial D phenotype (below), then the patient is much more likely to become immunized against D. Serology reagents cannot differentiate a serologic weak D from a partial D phenotype, as both can present with weaker than expected reactivity with anti-D. *RHD* genotyping is recommended to determine the D status (below).

Partial D: Individuals with partial D (previously termed D mosaic) lack epitope(s) of the RhD protein, and when exposed to D antigen through transfusion or pregnancy can produce allo anti-D. There are numerous different partial D phenotypes. Routine D-typing reagents cannot distinguish partial D phenotypes. RBCs expressing partial D can give strong, variable, or weak reactions with anti-D, depending on the specific type of partial D. Patient typing is usually performed with an IgM monoclonal anti-D reagent that does not detect, in the direct phase, the partial DVI phenotype, which is the most common partial D in Caucasians. The most common partial D, DIIIa, in African-Americans usually goes undetected in serologic tests as the RBCs give strong reactions with anti-D. Many partial D phenotypes (also known as D categories) have arisen as a result of nucleotide exchange between the two genes *RHCE* and *RHD*, and less often from single-nucleotide changes in *RHD*.

Females of child-bearing potential who have a partial D phenotype benefit from receiving D-negative RBCs for transfusion and potentially benefit from RhIG prophylaxis. In practice, most are frequently typed as positive and are recognized only after they form anti-D. Importantly, DNA-based testing can distinguish these and should be considered in females of child-bearing potential, especially if they present with variable or weak RBC typing with anti-D reagents.

C/c and E/e Antigens: C and c differ by six nucleotide substitutions causing four amino acid changes. Only the Ser103Pro position strictly correlates with C/c antigenicity and expression of the C or c antigen, respectively. E and e differ by one nucleotide change, resulting in one amino acid difference, c.676C>G, p.Pro226Ala.

G Antigen: G antigen is expressed on both RhD and RhC proteins and results from the three amino acids shared between these two proteins. Anti-G presents as anti-D plus anti-C. In pregnant women, it is important to correctly distinguish anti-G (resulting from sensitization to D-negative, C-positive RBCs) from separable anti-D and anti-C, as these women are candidates for RhIG to prevent anti-D.

V and VS Antigens: Low-frequency V and VS antigens are expressed on the RBCs of ~30% of African-Americans, and result from an amino acid substitution Leu245Val in Rhce protein. When an additional substitution, Gly336Cys, is present with Leu245Val, expression of V antigen is lost but VS expression remains. V/VS antigens are not present on routine screening cells, so antibodies are primarily detected as an incompatible crossmatch or found as part of panel testing and antibody identification in samples with additional alloantibodies.

Partial Rhce Antigens: Individuals of African descent often have *RHCE*ce* genes that encode altered c or altered e, and/or a hybrid *RHD* that doesn't encode D antigen but encodes a partial C antigen present in 1/3 of C-positive patients with SCD. Individuals with these partial antigens can make allo anti-C or e-like specificities, such as anti-hr^S, -hr^B, -RH18, and -RH34. Many different RH alleles are associated with these partial antigens, resulting in variation in the fine specificity of the antibody, which makes it difficult to find compatible RBCs. Sometimes only rare D — — (dash, dash, i.e., lacking expression of CcEe) RBCs appear to be compatible. Often the altered *RHCE* are linked to altered *RHD* encoding partial D, so these individuals can also make anti-D. RH genotyping can be used to find genotype-matched compatible donors.

Rh-Null: Rh-null RBCs lack expression of Rh antigens, are stomatocytic and spherocytic, and affected individuals have variable degrees of anemia. *Regulator type* is caused by mutations in *RHAG* gene, so they have no or very reduced Rh or RhAG proteins, and *amorph type* is caused by mutations in *RHCE* gene on a deleted *RHD* background, so they have no Rh proteins and reduced RhAG proteins. RBCs that lack or have severely reduced RhAG have a cation defect.

D-Typing Discrepancies: In general, D reagents used in the United States for typing donors at blood centers on automated platforms differ in specificity and sensitivity from those used when typing patients in hospital transfusion service. The goal is to detect weak D phenotypes as being D-positive in donors. In contrast, the goal at transfusion service, especially for females of child-bearing potential, is to avoid detection of the more common partial D phenotypes to treat these conservatively as D-negative for transfusion. Different typing reagents, which are monoclonal or monoclonal blends of anti-D, may type weak D or partial D phenotypes differently. Therefore, individuals may have discrepancy in their D type (e.g., an individual may type as D-negative in one center and D-positive with an alternate reagent in another center). Discrepancies should be resolved to rule out sample mixup, and in the case of women of child-bearing potential, DNA testing for *RHD* should be considered.

***RHD* Genotyping to Manage RhIG in Pregnancy:** *RHD* genotyping should be considered to determine candidacy for RhIG for pregnant women whose RBCs type weaker than expected or have variable serologic D typing. Individuals determined by genotyping to be weak D types 1, 2, and 3 can be safely managed as D-positive and are not candidates for RhIG. This avoids unnecessary RhIG administration and avoids unnecessary use of D-negative donor units.

RH Genotyping: DNA testing for *RH* is very helpful for patients with SCD needing long-term chronic transfusion support to aid antibody identification and transfusion decision-making. African black ethnic groups have an increased incidence of altered or variant alleles that encode partial Rh antigens. RH comprehensive genotyping is available in some larger centers in the United States for research, use to guide transfusion, and the use continues to grow in transfusion medicine practice (Table 26.2).

TABLE 26.2 Applications of RH Genotyping for Transfusion Practice

For patients with sickle cell disease when C, E, and K antigen-matching to detect partial antigens and evaluate risk for alloantibodies to D, C, and e. (Patients who are C+ with partial C antigen are better served with C-negative transfusion protocol)
In the prenatal setting to distinguish weak D and partial D to determine candidates for Rh immune globulin and to inform transfusion management
To determine the Rh type of a fetus at risk for hemolytic disease of the fetus and newborn due to maternal anti-D or other Rh antibodies
Patients who have been recently transfused due to mixed population of red blood cells (RBCs)
IgG-coating RBCs which makes phenotyping suboptimal
Patients with warm autoimmune hemolytic anemia to provide extended-matched unit for transfusion
When RBCs type as antigen-positive and serum antibody with the corresponding specificity present, to distinguish presence of altered (partial) antigen to determine if the antibody is allo or auto

Expression: RhD, RhCE, and RhAG proteins are restricted to erythroid and myeloid cells. Expression of RhD/RhCE occurs during late erythropoiesis, with RhAG expressed earlier. Homologues of RhAG, termed RhBG and RhCG, are present in the kidney, skin, liver, and many other tissues.

Antibodies: Most Rh antibodies are IgG (some have IgM component) and are clinically significant, causing both HTRs and HDFN. Approximately, 20% of D-negative individuals who are exposed to D-positive RBC components in hospital setting form anti-D, which is in contrast to healthy male volunteers deliberately immunized, where 80% make anti-D after receiving as little as 0.5 mL of D-positive RBCs. In the maternal setting, ABO-incompatibility between mother and fetus has a partial protective effect on mother forming anti-D, as incompatible and D-positive cells are rapidly removed by the antithetical isohemagglutinin. Anti-D and anti-c can cause severe HDFN, while anti-C, anti-E, and anti-e usually cause no or mild HDFN. Anti-D made by women with partial DVI has been associated with severe or fatal HDFN, while anti-D made by women with other partial D phenotypes is associated with mild or no HDFN.

Autoantibodies: Many autoantibodies in patients with warm autoimmune hemolytic anemia have specificities that appear to be directed toward Rh antigens, which are usually demonstrated by lack of reactivity with Rh-null RBCs. However, Rh-null RBCs also lack or have decreased expression of other proteins in the Rh complex, including absence of CD47, low levels of LW, glycophorin B (SsU antigens), and RhAG. Transfusion of Rh antigen-matched RBCs has not been shown to survive better than antigen-positive RBCs in patients with warm autoimmune hemolytic anemia. If transfusion cannot be avoided, consideration should be given to transfusion with donor units extended antigen-matched for Rh and K, and if possible, also matched for additional clinically significant major blood groups. Avoiding exposure to antigens that the patient lacks typically mitigates alloantibody production and expedites laboratory investigation, which often requires multiple adsorptions to rule out underlying alloantibodies.

Further Reading

Bruce, L. J., Guizouarn, H., Burton, N. M., et al. (2009). The monovalent cation leak in overhydrated stomatocytic red blood cells results from amino acid substitutions in the Rh–associated glycoprotein. *Blood, 113*, 1350–1357.

Chou, S. T., Jackson, T., Vege, S., et al. (2013). High prevalence of red blood cell alloimmunization in sickle cell disease despite transfusion from Rh-matched minority donors. *Blood, 122*, 1062–1071.

Flegel, W. A. (2006). How I manage donors and patients with a weak D phenotype. *Curr Opin Hematol, 13*, 476–483.

Haspel, R. L., & Westhoff, C. M. (2015). How do I manage Rh typing in obstetric patients? *Transfusion, 55*, 470–474.

Sandler, S. G., Flegel, W. A., Westhoff, C. M., et al. (2015). It's time to phase-in *RHD* genotyping for patients with a serological weak D phenotype. *Transfusion, 55*, 680–689.

Yazer, M., & Triulzi, D. J. (2007). Detection of anti-D in D-recipients transfused with D+ red blood cells. *Transfusion, 47*, 2197–2201.

CHAPTER 27

Kell, Kx and Kidd Blood Group Systems

Janis R. Hamilton, MS, MT(ASCP)SBB and Connie M. Westhoff, SBB, PhD

Kell and Kx Blood Group Systems: The Kell blood group system contains more than 34 antigens, although only one (K1) is of importance in routine practice. K1 is also commonly referred to as K, or incorrectly as "Kell," as Kell is the name of the entire blood group system. Antigens are encoded by the *KEL* gene located on chromosome 7. They are found in the RBC membrane on a type 2 glycoprotein (CD238), are part of the zinc endopeptidase family and function as endothelium-3 convertases. Rare Ko (null) RBCs lack all Kell glycoprotein and all Kell system antigens. Individuals of this phenotype, however, have no obvious pathology. The Kell protein has a large extracellular C-terminal domain of 665 amino acids and a short N-terminal cytoplasmic domain (Fig. 27.1). Kell antigens are destroyed by reducing agents such as dithiothreitol and 2-aminoethylisothiouronium bromide, which reduce the multiple intrachain disulfide bonds in the extracellular domain of the Kell protein; this characteristic is useful in antibody identification.

A related blood group system Kx, contains one antigen, Kx (XK1). This protein is encoded by the X-linked gene *XK*. Mutations in *XK* lead to severely reduced expression of Kell antigens, known as the McLeod phenotype, and neuromuscular abnormalities (described below). Kx is a 10-pass membrane-spanning protein that is linked to the Kell protein by a single disulfide bond and is essential for the expression of Kell system antigens. The Kx antigen is not usually exposed on the RBC, as that anti-Kx does not react with RBCs that carry normally expressed Kell antigens.

Antigens and Their Molecular Basis: The majority of Kell antigens are caused by nucleotide changes in *KEL* that cause single amino acid substitutions in the Kell protein. There are seven sets of high- and low-incidence antithetical antigens: K and **k**; Kp^a, $\mathbf{Kp^b}$, and Kp^c; Js^a and $\mathbf{Js^b}$; **KEL11** and KEL17; **KEL14** and KEL24; KEL25 and **KEL28**; and KEL31 and **KEL38** (high-incidence antigens are in bold typeface). The polymorphisms associated with the most frequently considered Kell antigens are identified in Fig. 27.1. There are also many other nonallelic low- and high-incidence antigens that are beyond the scope of this chapter. The prevalence of the commonly encountered Kell antigens differs by ethnic group: K+ is more common in Caucasian individuals and less often seen in those of African ancestry; Kp(a+b−) phenotype is almost always found in whites; and Js(a+b−) is almost exclusively found in individuals of African ethnicity (Table 27.1). These prevalence data have relevance when searching for antigen-negative donor units and assessing antibody production in patients from different ethnic groups. Loss of Kell antigens, K_0, results from various *KEL* mutations including nucleotide deletion, defective splicing, and premature stop codons. Mutations causing weak expression of Kell antigens are "Kell-mod" phenotypes.

Transfusion Medicine and Hemostasis. https://doi.org/10.1016/B978-0-12-813726-0.00027-1

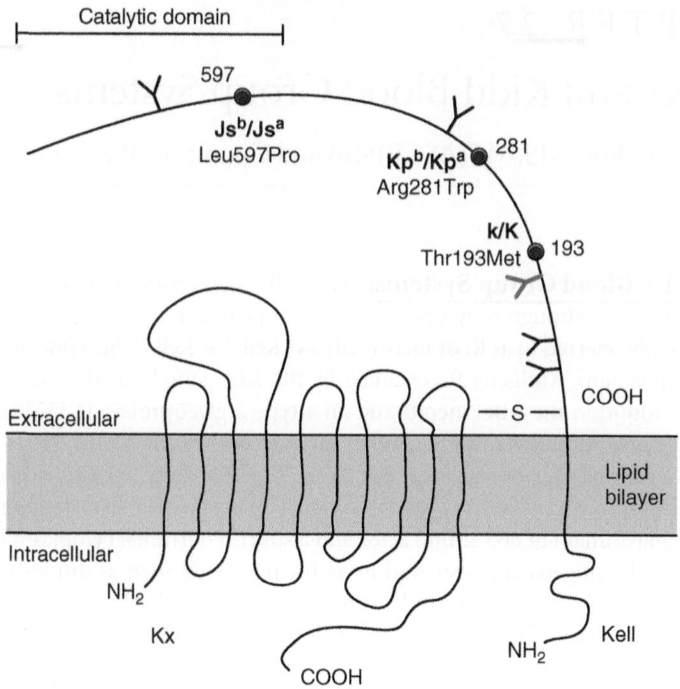

FIGURE 27.1 Kell and Kx proteins. Kell is a single-pass protein, but Kx is predicted to span the red blood cell membrane 10 times. Kell and Kx are linked by a disulfide bond, shown as —S—. The amino acids that are responsible for the more common Kell antigens are shown. The *N*-glycosylation sites are shown as Y. The gray Y represents the *N*-glycosylation site that is not present on the K (K1) protein. (From Hillyer, C. D., Silberstein, L.E., Ness, P.M., et al. (Eds.). (2007). *Blood banking and transfusion medicine: Basic principles and practice* (2nd ed.). San Diego, CA: Elsevier Academic Press.)

McLeod Phenotype: The McLeod phenotype arises through deletions and mutations in *XK*, resulting in depressed expression of the Kell system antigens. Acanthocytic RBC morphology is sometimes present. McLeod syndrome, an X-linked condition affecting males, is a multisystem degenerative disorder. The syndrome encompasses a variety of muscular, neurological, and psychiatric defects, including skeletal muscle wasting, seizures, and cardiomyopathy. Symptoms can develop as late as the fourth decade of life. McLeod syndrome, thought to be underdiagnosed, has been reported in approximately 60 males worldwide. The phenotype can also be present in individuals with chronic granulomatous disease when there is a large chromosome X deletion that includes both the *XK* and the *CYBB* loci. These RBCs lack both Km and Kx antigens.

Expression: The Kell glycoprotein is present on myeloid progenitors, testicular and lymphoid tissues, and skeletal muscle. It is expressed very early during erythropoiesis. XK protein is found in muscle, heart, brain, and hematopoietic tissue.

Antibodies: K antigen is strongly immunogenic and anti-K is frequently formed in K-negative individuals who are exposed to K-positive RBCs. Rare cases of naturally

TABLE 27.1 Kell Blood Group System Phenotypes and Prevalence

	Prevalence (%)	
Phenotype	**White**	**Blacks**
K–k+	91	98
K+k+	8.8	2
K+k–	0.2	Rare
Kp(a+b–)	Rare	0
Kp(a–b+)	97.7	100
Kp(a+b+)	2.3	Rare
Kp(a–b–c+)	0.32 Japanese	0
Js(a+b–)	0	1
Js(a–b+)	100	80
Js(a+b+)	Rare	19

Modified from Hillyer, C. D., Strauss, R. G., & Luban, N. L. C. (Eds.). (2004). *Handbook of pediatric transfusion medicine*. San Diego, CA: Elsevier Academic Press.

occurring anti-K have been reported, but most anti-K alloantibodies are identified in individuals who have received RBC transfusion or been pregnant. Antibodies to all other Kell system antigens also result from RBC exposure. Autoantibodies may occasionally have Kell specificity accompanied by transient suppression of the Kell system antigens. Anti-K and other antibodies to Kell system antigens (including anti-Kp^a, anti-Kp^b, anti-Js^a, and anti-Js^b) can result in hemolytic transfusion reactions (HTRs), both immediate and delayed, and hemolytic disease of the fetus and newborn (HDFN). Maternal antibody titer and amniotic fluid bilirubin levels are poor predictors of the severity of HDFN. Because Kell antigens are expressed at very early stages of erythroid maturation, Kell system antibodies can cause in utero suppression of erythropoiesis. Therefore, these antibodies may result in severe anemia without concomitant hemolysis and elevated bilirubin.

The antibody made by K_0 individuals is called anti-Ku, which reacts with all cells except other K_0 null phenotypes. Individuals with the McLeod phenotype having no Kx antigen, make anti-Kx and another specificity called anti-Km. It is difficult to find compatible donors for individuals of the K_0 phenotype, males with McLeod syndrome and those immunized to high-prevalence antigens other than k, Kp^b, and Js^b. In such cases autologous donation and storage is recommended.

Kidd Blood Group System: The Kidd blood group system has three antigens: Jk^a, Jk^b, and Jk3; however, Jk^a and Jk^b are the most important antigens in routine practice. Only individuals of the rare null phenotype, Jk(a–b–) lack the high-prevalence Jk3 antigen. The Kidd glycoprotein is encoded by the *JK (SLC14A1)* gene found on chromosome 18. It consists of 389 amino acids organized into five extracellular loops, spanning the RBC membrane 10 times, with cytoplasmic N-terminal and C-terminal regions. The Kidd protein functions as a urea transporter (Fig. 27.2). On RBCs, it facilitates rapid urea transport across RBC membrane ensuring structural stability, as the

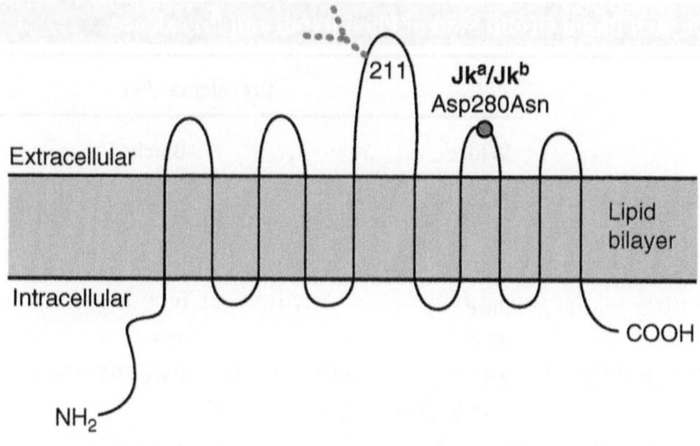

FIGURE 27.2 Predicted 10-transmembrane domain structure of the Kidd/urea transporter. The polymorphism responsible for the Kidd antigens and the site for the *N*-glycan are indicated. (From Hillyer, C. D., Silberstein, L.E., Ness, P.M., et al. (Eds.). (2007). *Blood banking and transfusion medicine: Basic principles and practice* (2nd ed.). San Diego, CA: Elsevier Academic Press.)

TABLE 27.2 Kidd Blood Group System Phenotypes and Prevalence

	Prevalence (%)	
Phenotype	**White**	**Black**
Jk(a+b–)	26	52
Jk(a–b+)	23	8
Jk(a+b+)	50	40
Jk(a–b–)	Rare	Rare

Modified from Hillyer, C. D., Strauss, R. G., & Luban, N. L. C. (Eds.). (2004). *Handbook of pediatric transfusion medicine*. San Diego, CA: Elsevier Academic Press.

cells pass through the kidney's renal medulla. RBCs from Jk-null individuals are resistant to in vitro lysis by 2M urea. The 2M urea reagent is sometimes used to lyse RBCs when counting platelets and can be used for large-scale testing of donors for the rare Jk(a–b–) phenotype. The Kidd urea transporter is also expressed in the kidney and is important in regulating urea as part of urine concentration and water conservation. Jk-null individuals have a reduced capacity to concentrate urine; however, they display no other known abnormalities.

Antigens and Their Molecular Basis: Jk^a and Jk^b antigens arise from a single amino acid polymorphism in the *JK* gene: c. 838A>G, p. Asn280Asp. This change gives rise to three common phenotypes: Jk(a+b–), Jk(a–b+), and Jk(a+b+) (Table 27.2). The Jk(a–b–) or null phenotype is rare but has an increased occurrence in Polynesian and Asian individuals due to a *JK* mutation that causes skipping of exon 6 and those of Finnish background due to a single amino acid substitution: c.871T>C, p.Ser291Pro. Other examples of the Jk(a–b–) phenotype arise from variations in the gene involving missense and splice-site mutations, exon deletions, and premature stop codons. Variant or weakened expression of both the Jk^a and Jk^b antigens caused by single amino acid substitutions has also been identified.

Expression: Kidd antigens are detected on RBCs at 11-week gestation and are well developed at birth. These antigens are expressed on endothelial cells of vasa recta in the medulla of the human kidney. The JK protein has been found on multiple other tissues including colon, brain, heart, lung, and liver although not on white blood cells or platelets.

Antibodies: Antibodies to the Kidd blood group system antigens are usually found in sera containing other antibodies. Importantly, Kidd antibodies can go undetected in routine testing because the antibody titer often decreases below the limit of detection or may react only with RBCs with a double dose of the antigen (Jk(a+b−) or Jk(a−b+)). For these reasons, Kidd antibodies are responsible for about one-third of all delayed HTRs, which may be severe. The antibodies are mainly IgG but can be partially IgM. Kidd antibodies rarely cause HDFN, and when they do, it is generally not severe. Anti-Jk3 can be produced by Jk(a–b–) individuals. Autoantibodies with JK specificity and alloantibodies in JK-antigen-positive individuals expressing a variant Jk^a or Jk^b antigen can be found. Case reports suggest a possible role for Kidd system antibodies in renal graft survival because of the presence of JK antigens on renal cells.

Further Reading

Jung, H. H., Danek, A., & Frey, B. M. (2007). McLeod syndrome: A neurohaematological disorder. *Vox Sang*, 3, 112–121.

Rourk, A., & Squires, J. E. (2012). Implications of Kidd blood group system in renal transplantation. *Immunohematology*, 3, 91–94.

Watkins, C. E., Litchfield, J., Song, E., et al. (2011). Chronic granulomatous disease, the McLeod phenotype and the contiguous gene deletion syndrome - a review. *Clin Mol Allergy*, 9, 13.

CHAPTER 28

MNS and Duffy Blood Group Systems

Judith Aeschlimann, MSc and Connie M. Westhoff, PhD, SBB

MNS Blood Group System: MNS blood group system contains 49 recognized antigens; the major antigens are M, N, S, s, and U. M and N antigens are located on glycophorin A (GPA), while S, s, and U antigens are carried on glycophorin B (GPB). Both are single-pass membrane sialoglycoproteins (Fig. 28.1). Genes encoding GPA and GPB are adjacent on the chromosome, and inheritance of MN and Ss antigens are linked (Table 28.1).

Antigens and Genetic Basis: M and N antigens are antithetical and sensitive to enzyme treatment including ficin, papain, trypsin, and pronase; these properties can aid in antibody identification. The null phenotype lacks M, N, and high-prevalence antigen(s) and is designated En(a−).

The antithetical S and s antigens are sensitive to α-chymotrypsin and pronase, with variable sensitivity to ficin and papain. U is a high-prevalence antigen on GYB. RBCs

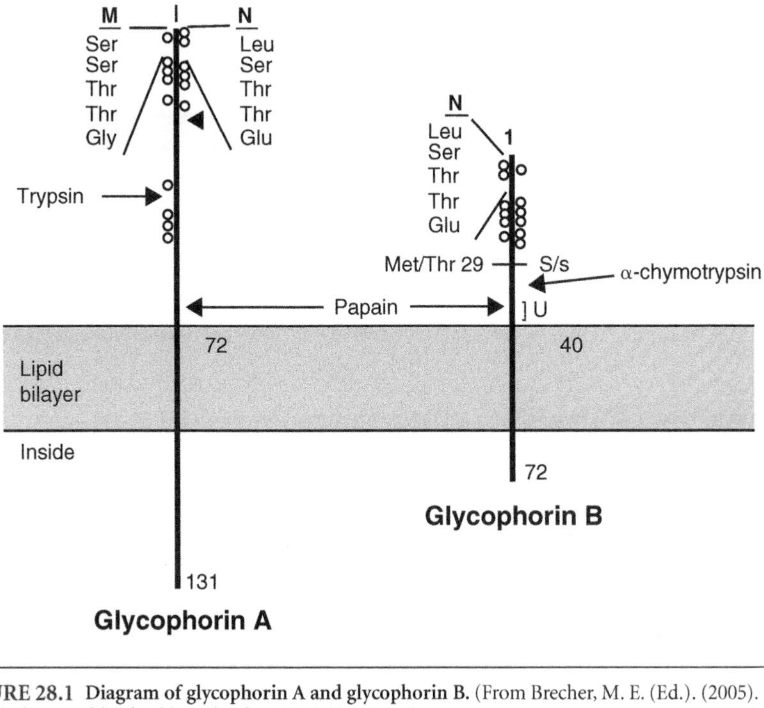

FIGURE 28.1 Diagram of glycophorin A and glycophorin B. (From Brecher, M. E. (Ed.). (2005). *Technical manual* (15th ed.). Bethesda, MD: AABB Press.)

Transfusion Medicine and Hemostasis. https://doi.org/10.1016/B978-0-12-813726-0.00028-3

with U− or U^{VAR} (U weak) phenotypes are S− and s− and are found in blacks with approximate frequency of 1.5%. The glycophorin (GYP) null phenotype, designated M^kM^k, lacks both MN and Ss antigens.

GPA is a receptor for bacteria, viruses, and *Plasmodium falciparum*, a chaperone for Band 3 transport to the RBC membrane, a major component contributing to the negatively charged RBC glycocalyx, and may also function as a complement regulator. GPB is a member of the Band 3/Rh-macrocomplex in the RBC membrane. The rare null phenotypes are not associated with any apparent health defects.

GPA and GPB are encoded by *GYPA* and *GYPB* located in close proximity on chromosome 4. The M/N-encoding alleles are designated *GYPA*M* or *GYPA*N* and differ by three nucleotide changes in exon 2 (c.59C>T; c.71G>A; c.72T>G), which encode two amino acid changes (p.Ser20Leu and p.Gly24Glu). The S/s alleles are *GYPB*S* and *GYPB*s* and differ by one change in exon 4 (c.143C>T; p.Thr48Met). The U− phenotype is caused by deletion of the coding region of *GYPB*, while the U^{VAR} (U weak) phenotype is associated with partial or complete skipping of exon 5. Most of the low-frequency antigens result from either amino acid substitutions or rearrangements between the two homologous genes giving rise to hybrid alleles.

Expression: Cord RBCs express MNSs antigens. M and N antigens are also expressed on renal endothelium.

Antibodies: Anti-M is primarily IgM but may have an IgG component that can appear to be IgM due to ability to cause direct agglutination of RBCs in absence of antihuman globulin reagent because of large number of GPA molecules on RBCs. Anti-M reactivity at 37°C has occasionally been associated with hemolytic transfusion reaction (HTR) and hemolytic disease of the fetus and newborn (HDFN), and

TABLE 28.1 MNS Blood Group System Phenotypes and Prevalence

	Prevalence (%)	
Phenotype	**White**	**Black**
M+ N− S+ s−	6	2
M+ N− S+ s+	14	7
M+ N− S− s+	10	16
M+ N+ S+ s−	4	2
M+ N+ S+ s+	22	13
M+ N+ S− s+	23	33
M− N+ S+ s−	1	2
M− N+ S+ s+	6	5
M− N+ S− s+	15	19
M+ N− S− s−	0	0.4
M+ N+ S− s−	0	0.4
M− N+ S− s−	0	0.7

Modified from Hillyer, C. D., Strauss, R.G., & Luban, N. L. C. (Eds.). (2004). *Handbook of pediatric transfusion medicine*. San Diego, CA: Elsevier Academic Press.

therefore patients with anti-M reactive at 37°C should be transfused with RBCs that are crossmatch-compatible and should be assessed for IgG component and HDFN risk. Anti-N is primarily IgM and has not been associated with HTR or HDFN. Anti-M and anti-N can be naturally occurring alloantibodies produced without previous RBC antigen exposure. They often show a dosage effect (reacting with RBCs homozygous for the antigen more strongly than cells heterozygous having a single dose). In contrast, anti-S, anti-s, and anti-U are IgG antibodies that are formed in response to RBC stimulation and are associated with HTR and HDFN. Anti-En^a, a designation used for immune antibodies reacting with common determinants on GPA, is usually IgG and has caused HTR and HDFN. Autoantibodies with apparent M, N, S, s, or U specificity are observed rarely, but autoanti-N has been found in dialysis patients when equipment was sterilized with formaldehyde.

Duffy Blood Group System: The protein carrying Duffy (Fy) antigens is a multipass transmembrane glycoprotein with an extracellular glycosylated amino terminal region (Fig. 28.2). The antigens show a dosage effect meaning there are twice as many Fy^a antigens on RBCs from an individual who is homozygous for *FY*A* than on RBCs from an individual who is heterozygous *FY*A/B* or hemizygous (has one silenced allele). Duffy antigen frequency varies between ethnic groups, as the null phenotype in blacks reflects selection pressure for resistance to malaria (Table 28.2).

Antigens and Genetic Basis: The most prevalent phenotypes in whites are Fy(a+b−), Fy(a−b+), and Fy(a+b+) (Table 28.2). Fy(a−b−) phenotype is prevalent in

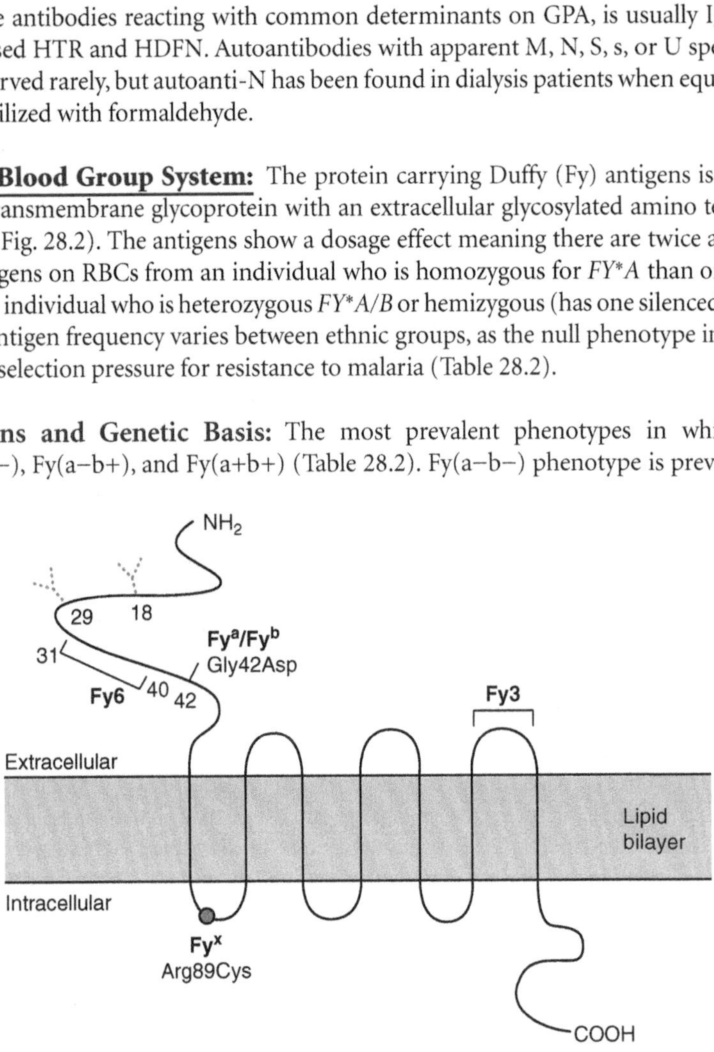

FIGURE 28.2 The predicted seven-transmembrane domain structure of the Duffy protein. The amino acid change responsible for Fy^a/Fy^b polymorphism, the mutation responsible for Fy^x, and the glycosylation sites and the regions where Fy3 (and Fy6) map are indicated. (From Hillyer, C. D., Silberstein, L. E., Ness, P.M., et al. (Eds.). (2007). *Blood banking and transfusion medicine: Basic principles and practice* (2nd ed.). San Diego, CA: Elsevier Academic Press.)

TABLE 28.2 Duffy Blood Group System Phenotypes and Prevalence

	Prevalence (%)	
Phenotype	**White**	**Black**
Fy (a+b−)	17	9
Fy (a−b+)	34	22
Fy (a+b+)	49	1
Fy (a−b−)	Rare	68
Fy(b+^{w})	1.4	0

Modified from Hillyer, C. D., Strauss, R.G., & Luban, N. L. C. (Eds.). (2004). *Handbook of pediatric transfusion medicine*. San Diego, CA: Elsevier Academic Press.

blacks and is caused by a mutation in the promoter region of *FY*B*, which disrupts a binding site for the erythroid transcription factor GATA-1 and results in loss of Duffy expression on RBCs. Expression of the protein on endothelium is not altered and Fy(a−b−) individuals with GATA-1 mutations are not at risk for anti-Fy^{b}. Fy3 is an antigen common to all Duffy proteins. Fy^{a} and Fy^{b}, but not Fy3, antigens are sensitive to proteolytic enzyme treatment.

Duffy, previously known as DARC but recently renamed ACKR1, binds excess proinflammatory chemokines released in the circulation. Chemokines are chemotactic cytokines that attract white blood cells to sites of inflammation. As a promiscuous receptor, ACKR1 binds chemokines from both the C-X-C (includes IL-8) and C-C (includes CCL5 and MCP-1) classes. The biological relevance, if any, is not clear. Several studies have linked Fy(a−b−) phenotype to benign neutropenia. The protein is also a receptor for malaria parasites *Plasmodium vivax* and *Plasmodium knowlesi*; thus, individuals who have Fy(a−b−) RBCs are resistant to infection by these malarial organisms.

ACKR1 is located on chromosome 1 and consists of two coding exons. The blood group genes are designated *FY*A* or *FY*01* and *FY*B* or *FY*02* and differ by a single nucleotide change in exon 2 (c.125A>G) encoding amino acids p.Asp42Gly. Fy(a+) phenotype is associated with 42Gly and Fy(b+) with 42Asp. An altered FYB allele found primarily in whites is characterized by weak Fy^{b} expression, termed Fy^{X} phenotype, and is caused by a c.265C>T (p.Arg89Cys) change in *FY*B*. This Fy^{b} weak type is often not detected by serology and is the most frequent cause of discrepancies between serology and molecular typing. *FYB* associated with the GATA-1 mutation in individuals with Fy(a−b−) phenotype is designated *FY02N.01* (N for null), and the change is located in the erythroid promoter 67 nucleotides upstream of the start codon (c.-67C>T).

Expression: Duffy antigens are detected as early as 6–7 weeks of gestation and are well developed at birth. Duffy antigens are expressed on endothelial cells of postcapillary venules of the kidney, spleen, heart, lung, muscle, duodenum, pancreas and placenta, and Purkinje cell neurons.

Antibodies: Duffy antibodies are almost always IgG and result from RBC stimulation. Anti-Fy^{a} may cause mild to severe HTR and HDFN. Fy^{b} is a poor immunogen, and when anti-Fy^b is present, it rarely causes HDFN but is considered a clinically significant antibody for transfusion.

Further Reading

Afenyi-Annan, A., Kail, M., Combs, M. R., et al. (2008). Lack of Duffy antigen expression is associated with organ damage in patients with sickle cell disease. *Transfusion, 48*, 917–924.

Duchene, J., Novitzky-Basso, I., Thiriot, A., et al. (2017). Atypical chemokine receptor 1 on nucleated erythroid cells regulates hematopoiesis. *Nat Immunol, 7*, 753–761.

Meny, G. M. (2010). The Duffy blood group system: A review. *Immunohematology, 26*, 51–56.

Palacajornsuk, P. (2006). Review: Molecular basis of MNS blood group variants. *Immunohematology, 22*, 171–182.

CHAPTER 29

Lewis, I, P1PK, FORS, and GLOB Blood Group Systems

Randall Velliquette, MS and Connie M. Westhoff, PhD, SBB

Lewis Blood Group System: Le^a and Le^b antigens are synthesized by two independent fucosyltransferases, and Le^a is a precursor molecule for synthesis of Le^b (Fig. 29.1).

Antigens and Genetic Basis: Lewis antigens (Le^a and Le^b) are not intrinsic to RBC membrane, but are synthesized by intestinal epithelial cells, circulate in plasma either free or bound to lipoproteins, and are then passively adsorbed onto the RBC membrane. Le^a and Le^b are synthesized in a stepwise fashion by fucosyltransferases, which add fucose moieties onto type I glycoprotein chains; these enzymes are encoded by *LE(FUT3),* which adds fucose to GlcNAC (*N*-acetylglucosamine) and *SE(FUT2),* which adds fucose on the Gal (galactose) moiety. The enzyme encoded by *LE(FUT3)* is responsible for synthesis of Le^a resulting in Le(a+b−) phenotype. The enzyme encoded by *SE(FUT2)* converts Le^a to Le^b by the addition of another fucose molecule, resulting in Le(a−b+) phenotype. Individuals who have mutations in *LE(FUT3)* do not make Le^a and therefore cannot synthesize Le^b, regardless of *SE(FUT2)* enzyme activity, resulting in Le(a−b−) phenotype. Individuals who have mutations in *SE(FUT2)* have reduced activity of fucosyltransferase and reduced conversion of Le^a to Le^b, which can result in weak expression of both Le^a and Le^b, resulting in uncommon Le(a+b+) phenotype principally found in Taiwanese and Asian individuals (Table 29.1).

Expression: Lewis antigens are adsorbed onto RBCs, platelets, and lymphocytes from plasma glycolipids. Lewis antigens are also widely distributed on human tissues, are present in soluble form in body fluids, and serve as receptors for some pathogenic bacteria. The antigens are not expressed on cord RBCs, and Le^a appears before Le^b and usually within the first few months of life. Antigen expression does not reach adult levels until 6 years of age. Antigen levels on RBCs are often diminished during pregnancy and various diseases; diminished antigens on RBCs are likely to be secondary to changes in endothelial secretion or changes in plasma lipoprotein content.

Antibodies: Lewis antibodies are predominantly found in persons with Le(a−b−) RBCs, and are often identified in plasma of pregnant women with depressed antigen expression. Antibodies are primarily IgM and reactive at temperatures below 37°C; however, mixtures of IgM/IgG or pure IgG, which are reactive at 37°C, do exist. Most Lewis antibodies are naturally occurring (present without previous exposure to antigen-positive RBCs) and are not usually clinically significant (Table 29.2). Rare cases of antibodies causing hemolytic transfusion reaction (HTR) have been reported and are more commonly due to antibodies against Le^a than Le^b. Importantly, soluble Lewis

Transfusion Medicine and Hemostasis. https://doi.org/10.1016/B978-0-12-813726-0.00029-5

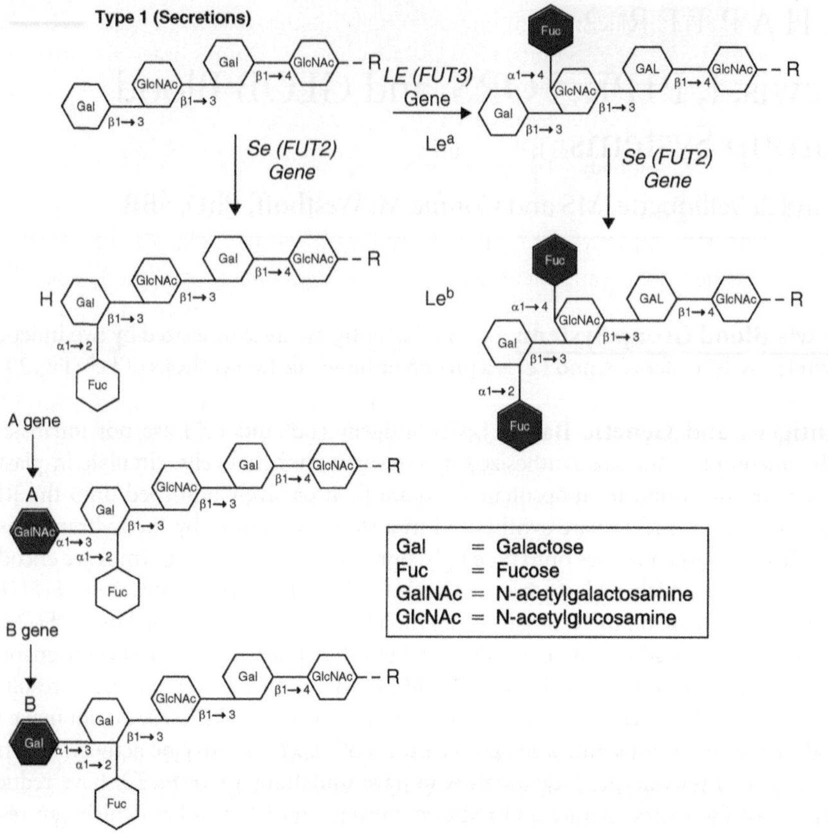

FIGURE 29.1 Synthesis of Lewis antigens. Antigens are synthesized on type 1 precursor substrates by FUT2 and FUT3 encoded fucosyltransferases. Terminal carbohydrates that define the antigens are shown in blue. (Modified from Hillyer, C. D., Silberstein, L. E., Ness, P.M., et al. (Eds.). (2007). *Blood banking and transfusion medicine: Basic principles and practice* (2nd ed.). San Diego, CA: Elsevier Academic Press.)

TABLE 29.1 Lewis Blood Group Phenotypes and Prevalence

	Prevalence (%)		
Phenotype	**Caucasian**	**African-American**	**Asian**
Le(a−b+)	72	55	72
Le(a+b−)	22	23	22
Le(a−b−)	6	22	6
Le(a+b+)	Rare	Rare	3[a]

[a]Prevalence of 10%–40% in Chinese in Taiwan, Japanese, Polynesians, and Australian Aborigines.

Modified from Hillyer, C. D., Strauss, R. G., & Luban, N. L. C. (Eds.). (2004). *Handbook of pediatric transfusion medicine.* San Diego, CA: Elsevier Academic Press.

TABLE 29.2 Clinical Significance of Lewis, I, P1PK, GLOB, and FORS System Antibodies

System	Antibody	Transfusion Reactions	Hemolytic Disease of the Fetus and Newborn
Lewis	Le^a	Usually no: several cases of severe hemolytic transfusion reaction (HTR) reported, some delayed reactions	No: one mild case reported
	Le^b	Usually no: several cases of severe HTR reported, some delayed reactions	No: one mild case reported
I	I/i	No for autoantibodies: Increased destruction of I + RBCs reported in adult i phenotype persons with alloanti-I	No
P1PK	P1	No to moderate: rare delayed	No
	P^k	No to severe	No to severe
	NOR	Unknown: No data	Unknown: No data
GLOB	P	No to severe (rare)	Usually No: Mild cases reported in P^k mothers with anti-P
	PX2	Unknown: No data	Unknown: No data
FORS	FORS1	Unknown: No data	Unknown: No data
GLOB collection	LKE	Rare: 1 reported case	No

antigens present in donor plasma neutralize the antibody, and Lewis antigens can also elute from RBCs into the plasma. Antibodies do not cause hemolytic disease of the fetus and newborn (HDFN) because Lewis antigens are not expressed on fetal RBCs and most anti-Le^a and -Le^b are IgM and do not cross the placenta. Patients with Lewis antibodies may be transfused with RBCs that are crossmatch-compatible at 37°C.

I Blood Group System: I and i antigens are located on the same carbohydrate chains that carry RBC ABH antigens. Biosynthesis of the Ii oligosaccharides begins by sequential action of β1,3-*N*-acetylglucosaminyltransferase (β3GnT5) and β1,4-galactosyltransferase (β4GalT1) to form linear i antigen. The i antigen is transformed into I by branching enzyme β1,6-*N*-acetylglucosaminyltransferase (IGnT) (Fig. 29.2).

Antigens and Genetic Basis: I and i antigens differ in their branching structure. The i antigen is found predominantly on fetal and infant RBCs, where a disaccharide unit (*N*-acetyllactosamine) is linked in linear chain. During the first 6 years of life, increased expression of acetylglucosamine transferase, encoded by *GCNT2 (IGnT)*, results in increased branching of carbohydrate structure leading to expression of I antigen. In some adults, i antigen is not converted to branched chain I antigen secondary to gene mutations leading to lack of transferase activity and rare i ("adult little i") phenotype. Congenital cataracts are associated with a lack or marked reduction of I antigen on RBCs and lens.

The gene responsible for I antigen is *GCNT2* (or *IGnT*), and the reference allele is designated *GCNT2*01*. *GCNT2* is organized into three exons, but here are three versions

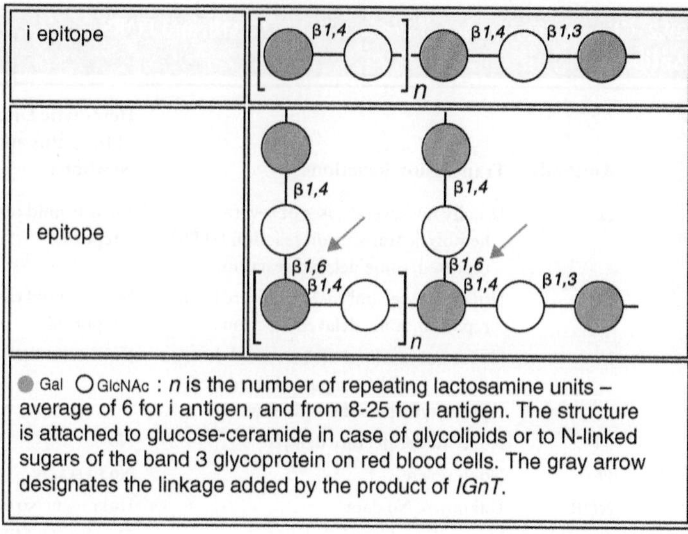

FIGURE 29.2 The I and i antigens. (From the National Library of Medicine (NLM) web pages.)

of exon 1: 1A, 1B, or 1C. As a result, three alternative mRNA transcripts may be synthesized: GCNT2A, GCNT2B, or GCNT2C. GCNT2C-transcript is present in erythroid cells and encodes 6-β-*N*-acetylglucosaminyltransferase; branching enzyme for I expression. The $I+^{W}$ phenotypes are due to homozygous or compound heterozygous single nucleotide changes in *GCNT2*. The i_{adult} (or I null phenotype) is due to homozygosity or compound heterozygosity for missense or nonsense mutations or deletion of the *GCNT2* coding region.

Expression: RBCs from newborns predominately express i, while RBCs from adults have predominantly I antigen. Adult RBCs vary in amount of I antigen expressed. Ii antigens are found on the surface of most cells and on soluble glycoproteins in saliva, plasma, and other fluids.

Antibodies

Autoantibodies: Anti-I are mainly IgM autoantibodies or cold agglutinins that react optimally at 4°C and are present in plasma of all adults at low titers. Transient polyclonal autoanti-I may arise from a *Mycoplasma pneumoniae* infection. Anti-i is associated with infectious mononucleosis and may be found in plasma of these patients and occasionally causes hemolysis. When these autoantibodies react at room temperature, they can interfere with pretransfusion testing, which can usually be circumvented by prewarm testing, i.e., performing tests at 37°C. Usually, these autoantibodies are benign; rarely, when they are high titer (>1:1000) and reactive at >30°C, they can result in a hemolytic anemia termed cold agglutinin disease.

Anti-IH is an autoantibody most commonly seen in group A_1 (less common in AB or B) individuals. Anti-IH agglutinates RBCs carrying both H and I determinants (e.g., adult group O or A_2 RBCs, which express I and the greatest amounts of H antigen).

Autoanti-IH is usually not clinically significant, but HTR reactions have been reported, usually if the antibody is reactive at 37°C.

Alloantibodies: Adults with rare inherited i phenotype make anti-I, which is usually high-titer IgM and reactive at low temperatures. Examples of hemolytic alloanti-I, which demonstrated reactivity at 37°C, have been reported (Table 29.2). Alloanti-i has not been recognized.

P1PK, Globoside (GLOB), and FORS Blood Group Systems: P1PK system is comprised of P1, P^k, and NOR antigens, which are defined by terminal galactosyl (Gal) sugars added to precursor glycosphingolipids (lactosylceramide -CDH) by action of transferase enzyme 4-α-galactosyltransferase. GLOB system contains P (GLOB1) and PX2 (GLOB2) antigens, which are terminal sugars on CDH synthesized by 3-β-*N*- acetylgalactosaminyltransferase. FORS1 antigen, i.e., Forssman glycolipid, is synthesized by 3-*α*-*N*-acetylgalactosaminyltransferase1 by addition of GalNAc to globoside (P antigen).

Antigens and Genetic Basis: In P1PK system, P1 and P^k are high-prevalence antigens, although P1 is polymorphic (P2 or P1−), and NOR is of low-prevalence. *A4GALT* controls P1PK system and is comprised of four exons and encodes a galactosyltransferase, which synthesizes P1 and P^k antigens (Fig. 29.3). Note P^k is precursor for P antigen (GLOB system below). P1PK-encoding allele is designated *A4GALT*P1.01*. P_1 versus P_2 phenotype is determined by changes in exon 2a of *A4GALT* resulting in transcriptional downregulation. P_2 (i.e., P1−) phenotype is found in 21% of European whites, 80% of Cambodians and Vietnamese, and 6% of African blacks. Uncommon p phenotype (P1PP^k−, previously Tj(a−)) is the null of the system and associated with inactivating mutations in *A4GALT*. NOR antigen, found in only two families to date, is associated with single-nucleotide change in *A4GALT* with altered enzyme reactivity. NOR+ cells are agglutinated by most human sera, i.e., polyagglutinable.

In the GLOB system, P (GLOB1) and PX2 (GLOB2) are high-prevalence antigens encoded by *B3GALNT1*, which is comprised of five exons and encodes 3-β-*N*-acetylgalactosaminyltransferase, which adds GalNAc residues to P^k for P antigen formation, and converts GalNAc to paragloboside for PX2 antigen formation (Fig. 29.3). P antigen is the most abundant neutral glycosphingolipid in RBC membrane. The reference allele encoding P is designated *GLOB*01*. P−(null phenotype) arises from various mutations in exon 5 of *B3GALNT1*. The null phenotype encoded by these alleles can be either P1+ or P1−; i.e., P_1^k or P_2^k phenotypes, respectively. PX2 antigen was added to GLOB system in 2016 when it was shown to be the product of *B3GALNT1*. PX2 antigen is expressed on virtually all RBCs across all populations (Fig. 29.3.).

Expression: P1 expression on RBCs is highly variable, is weaker in children, and does not reach adult levels until approximately 7 years of age. P1 is also expressed on lymphocytes, granulocytes, and monocytes and in a wide variety of organisms. P1 is widely distributed in nature and is a receptor for microorganisms, including strains of *Escherichia coli* and *Streptococcus suis*.

P^k antigen is expressed on all RBCs, albeit weakly. P antigen is expressed on all RBCs except P− or PP1P^k−(null or p phenotype) RBCs. P antigen is receptor for B19

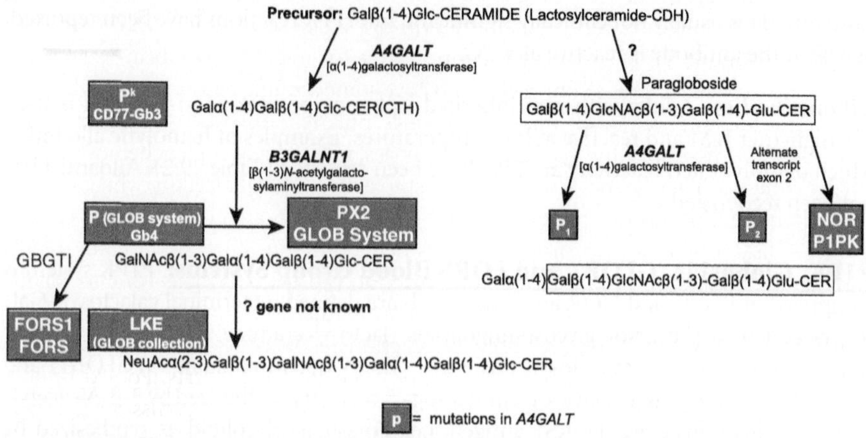

FIGURE 29.3 The structure of the P^k and P1 antigens, encoded by the gene product of *A4GALT*, and the P antigen, which is synthesized from P^k, and PX2 antigen by the gene product of *B3GALNT1*.

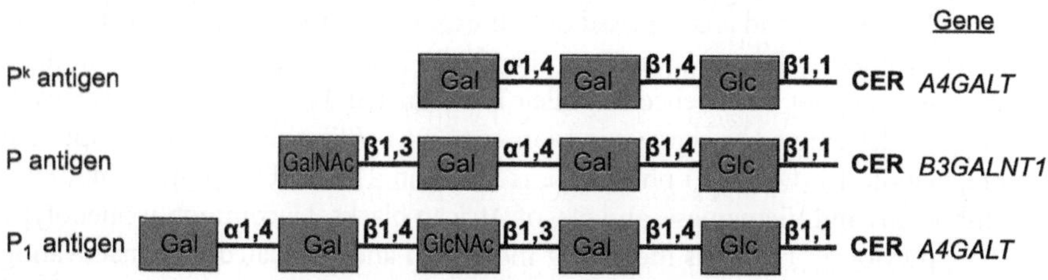

FIGURE 29.4 Biosynthesis of the P1PK system and GLOB system antigens.

parvovirus, which causes erythema infectiosum (fifth disease) and occasionally more severe disease, such as RBC aplasia. Individuals with rare null p phenotype are therefore resistant to parvovirus B19 infection.

Antibodies

Anti-P1: P1-negative individuals frequently form anti-P1, which is cold-reactive IgM antibody that is usually "naturally occurring." Anti-P1 has not caused HDFN. It has, rarely, been reported to cause HTR, and in those cases antibody demonstrates reactivity at 37°C (Table 29.2).

Autoanti-P: Autoanti-P is biphasic IgG autohemolysin typically associated with paroxysmal cold hemoglobinuria (PCH). PCH is a rare autoimmune hemolytic anemia that typically occurs in young children after viral infection. These children have biphasic IgG antibody, termed Donath–Landsteiner antibody (usually of anti-P specificity). The term biphasic antibody is used because antibody binds to RBCs at low temperature and RBC lysis occurs when temperature is raised. Donath–Landsteiner test is used to diagnose PCH.

Alloanti-P: Alloanti-P is the specificity found in persons with the P^k phenotype. Anti-P can be IgM or IgG and if reactive at 37°C can cause severe HTR (Table 29.2). Mild cases of HDFN have been reported in P^k mothers with anti-P. Cytotoxic IgM and IgG3 antibodies directed against P and/or P^k antigens are associated with higher than normal rate of spontaneous abortion in women with rare p [Tj(a−)], P1k, and P2k phenotypes.

Anti-PX2: Anti-PX2 is likely naturally occurring IgG but may be a mixture of IgM and IgG and is reactive at 37°C. Clinical significance of anti-PX2 is unknown.

Anti-PP1P^k: Alloanti-PP1P^k is usually present in rare individuals with p phenotype. Anti-PP1P^k can cause HTR and similar to anti-P, is associated with a high rate of spontaneous abortion (Table 29.2).

FORS Blood Group System: FORS became a blood group system in 2011 and is comprised of the FORS1 antigen (Forssman antigen).

Antigen and Genetic Basis: FORS1 is a low-prevalence antigen and not normally expressed on human RBCs but is widespread on RBCs of most animals. FORS1 is a product of *GBGT1*, comprised of 7 exons, that encodes globoside 3-*α*-*N*-acetylgalactosaminyltransferase1 and catalyzes the formation of Forssman glycolipids by addition of GalNAc to P antigen (Fig. 29.3). Synthesis of Forssman glycoproteins has been lost in humans, but homozygosity or heterozygosity for single nucleotide change (c.887G>A) in exon 7 of *GBGT1* activates transferase leading to synthesis of FORS1 antigen.

Antibody: Anti-FORS1 may be IgM or IgG and reactive at room temperature or 4°C. Clinical significance of anti-FORS1 is unknown.

GLOB Collection: GLOB collection is comprised of one antigen, LKE, expressed on virtually all RBCs in all populations. LKE antigen is determined by terminal sugars synthesized on precursor lactosylceramide by sequential action of transferases encoded by multiple genes. However, the gene encoding the transferases required for the final step in LKE antigen synthesis is unknown (Fig. 29.3). Three LKE phenotypes exist: $LKE+^S$, $LKE+^W$, and LKE−. LKE-negative RBCs have elevated P^k antigen. Anti-LKE is IgM cold-reactive antibody, and posttransfusion hemolysis was reported in one example (Table 29.2).

Further Reading

Kaczmarek, R., Buczkowska, A., Mikołajewicz, K., et al. (2014). P1PK, GLOB, and FORS blood group systems and GLOB collection: Biochemical and clinical aspects. Do we understand it all yet? *Transfus Med Rev, 28*, 126–136.

Lund, N., Olsson, M. L., Ramkumar, S., et al. (2009). The human P(k) histo-blood group antigen provides protection against HIB-1 infection. *Blood, 113*, 4980–4991.

Thuresson, B., Westman, J. S., Olsson, M. L., et al. (2011). Identification of a novel *A4GALT* exon reveals the genetic basis of the P_1/P_2 histo-blood groups. *Blood, 117*, 678–687.

CHAPTER 30

Other Blood Group Systems, Collections, and Series

Debra Jo Bailey, SBB and Connie M. Westhoff, SBB, PhD

Blood group systems not described in other chapters are discussed here, as well as antigens that have not yet been assigned to a blood group system. The International Society of Blood Transfusion (ISBT) classifies RBC antigens and also denotes them as low-frequency or high-frequency. *Low-frequency antigens* have a prevalence of less than 1% while *high-frequency antigens* have a prevalence of greater than 90%. Per ISBT, blood group system "*consists of one or more antigens controlled at a single gene locus, or by two or more very closely linked homologous genes with little or no observable recombination between them.*"

There are 36 blood group systems comprising more than 316 different RBC antigens. Some of the blood group systems of interest in transfusion medicine discussed in this chapter are shown diagrammatically in Fig. 30.1. A number of antigens (currently 38) cannot yet be assigned to a system due to lack of information. These antigens are grouped into a collection or series. A collection "*consists of serologically, biochemically, or genetically related antigens,*" while antigens not able to be classified into a *system* or *collection* are grouped according to prevalence in a series: low-prevalence antigens are the *700 series* and high-prevalence antigens the *901 series.*

Blood group systems in this chapter are divided into three categories:

1. Potentially clinical significant
2. Sometimes clinically significant
3. Not considered clinically significant

Importantly, there can always be exceptions to these classifications.

Potentially Clinically Significant Blood Group Systems

Dombrock System: Dombrock system currently consists of 10 antigens that include common polymorphic Do^a and Do^b, and several high-prevalence antigens including Gy^a, Hy, and Jo^a. Null phenotype, Gy(a−), is rare. Hy− and Jo(a−) are found in black individuals and are associated with weak expression of Do^a and Do^b. Dombrock antigens are carried on adenosine diphosphate ribosyltransferase (Dombrock glycoprotein), which is linked to RBC membrane by glycosylphosphatidylinositol (GPI). Antibodies to Do^a and Do^b are usually found in sera containing other RBC alloantibodies and are often weakly reactive, making them difficult to identify. In addition, these antibodies often drop over time to undetectable levels. Anti-Do^a and anti-Do^b have caused acute and delayed hemolytic transfusion reactions (HTRs); hemolytic disease of the fetus and newborn (HDFN) has not been reported. Antibodies to Gy^a, Hy, and Jo^a cause moderate delayed HTR and no HDFN.

Transfusion Medicine and Hemostasis. https://doi.org/10.1016/B978-0-12-813726-0.00030-1

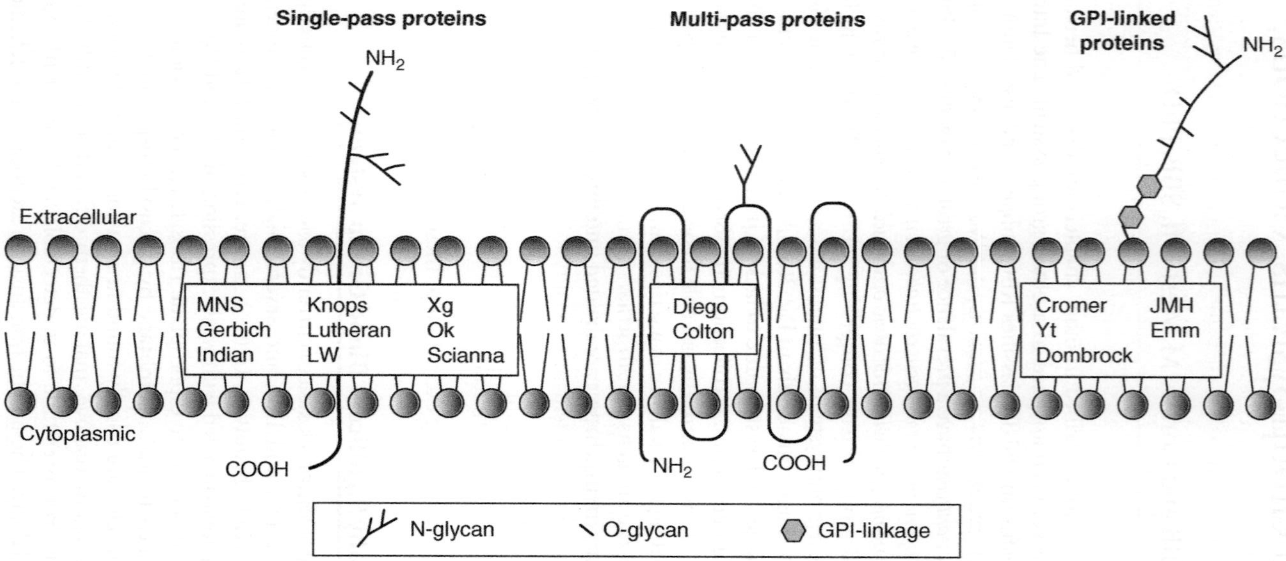

FIGURE 30.1 **Diagram of the red blood cell (RBC) membrane illustrates the type of membrane components that carry the blood group antigens. This figure does not show components carrying Chido/Rodgers or Lewis antigens because they are not integral membrane components or synthesized by the RBC.** (From Hillyer, C. D., Silberstein, L. E., Ness, P. M., et al. (Eds.). (2007). *Blood banking and transfusion medicine: Basic principles and practice* (2nd ed.). San Diego, CA: Elsevier Academic Press.)

Diego System: Diego antigens are carried on Band 3 (AE-1) that functions as the major chloride–bicarbonate exchanger in the RBC membrane important for CO_2 transport and a major structural RBC membrane protein. Absence of Band 3 is thought not to be compatible with life. Diego antigens include two pairs of antithetical antigens (Di^a/Di^b and Wr^a/Wr^b) and ~18 other low-prevalence antigens. Di^b is the high-prevalence antigen, and Di^a is rare in all but some North and South American groups where it can occur in 11%–54% of individuals. Uniquely, expression of high-prevalence Wr^b antigen is dependent on both Band 3 and glycophorin A (MN). Anti-Di^a and anti-Di^b have caused HTRs and HDFN. Anti-Wr^a is relatively common antibody reported in 1%–8% of normal donors and 30% of patients with autoimmune hemolytic anemia. Antibody is IgG and/or IgM and can occur with or without RBC exposure. Anti-Wra has caused HTRs and HDFN. Anti-Wr^b is rare but potentially causes HTR but has not caused HDFN.

Colton System: Colton antigens reside on RBC water channel protein aquaporin-1 (AQP-1). Co^a and Co4 are high-prevalence antigens, and Co^b is found in 8% of Caucasians and is uncommon in other ethnic groups. Co(a−b−) phenotype, which is also called Co:−3 or Co null is very rare. Antibodies have been implicated in HTR and mild to severe HDFN.

Gerbich System: Gerbich antigens are carried on glycophorin C and glycophorin D, which are products of a single gene (*GYPC*) using alternative initiation sites. System includes high-prevalence antigens Ge2, Ge3, and Ge4 and number of other high- and low-prevalence antigens. There are three Gerbich system phenotypes: the Yus type (Ge:−2, 3, 4; individuals with this phenotype can form anti-Ge2), the Gerbich type (Ge:−2, −3, 4; individuals with this phenotype can form anti-Ge2 or -Ge3) and the Leach type (Ge:−2, −3, −4; individuals with this phenotype are true nulls for glycophorin C and D and can form anti-Ge2, -Ge3, or -Ge4). Antibodies variably cause HTR, and anti-Ge3 has caused late onset HDFN. Antibodies to low-prevalence antigens rarely result in HDFN.

Augustine System: High-prevalence antigens AUG2 (a.k.a., At^a) and AUG1 belong to Augustine System as of 2015. Antigens are carried on RBC equilibrative nucleoside transporter 1 (ENT1) protein, which is responsible for adenosine transport across the plasma membrane. Anti-AUG2 is made by rare individuals of African ancestry whose RBCs lack the AUG2 (At^a) antigen. One individual of European ancestry whose RBCs typed AUG:−2 but plasma antibody reacted with all RBCs tested including those that were AUG2—was found to be deficient in ENT1. Total lack of ENT1 defines AUG-null or AUG:−1,−2 phenotype and antibody produced by these individuals is anti-AUG1. Anti-AUG2 has been implicated in HTRs and increased RBC clearance. Both anti-AUG1 and anti-AUG2 have caused mild or no HDFN.

Vel System: High-prevalence Vel antigen was assigned to its own blood group system in 2015. A protein called small integral membrane protein 1 (SMIM1) is lacking in individuals with Vel− phenotype. SMIM1 function is unknown. It is well documented that anti-Vel can be IgG and/or IgM and activate complement and can cause mild to severe HTRs. HDFN due to anti-Vel is not common.

Blood Group Systems Sometimes Clinically Significant

Cromer System: Cromer antigens are carried on decay accelerating factor (DAF), a complement regulatory protein that is anchored to RBC membrane by GPI and serves to regulate complement activation. DAF is missing from RBCs of patients with paroxysmal nocturnal hemoglobinuria (PNH) because they are deficient in all GPI-linked proteins. Cromer antigens are present on leukocytes, platelets, and placental trophoblasts and, in a soluble form, in plasma and urine. There are 16 high- and 3 low-prevalence antigens in the system. Individuals lacking all Cromer antigens have the Inab (Cr null) phenotype and are lacking DAF protein but have normal GPI expression in contrast to individuals with PNH. Reports of Cromer antibodies indicate variable potential clinical significance as determined by monocyte monolayer assays (MMAs), ^{51}Cr survival studies, or clinical data after transfusion of antigen-positive RBCs. They do not cause HDFN as placenta is a rich source of DAF, offering an alternative target to the RBCs for binding antibodies.

Lutheran System: Lutheran antigens are carried on glycoproteins that belong to a family of adhesion molecules known as B-CAM. The system consists of 24 antigens, but the major antigens are Lu^a (low prevalence) and Lu^b (high prevalence). The rare Lu_{null} phenotype is a result of homozygous inheritance of inactive Lutheran gene (*LU*), while severe reduction in expression of Lutheran antigens, often detectable only by adsorption/elution, is associated with heterozygous mutation in erythroid transcription factor gene *EKLF*. This "inhibitor phenotype" is termed In(Lu), and mutation in *EKLF* prevents normal expression of Lutheran and other blood group antigens (P1, AnWj, In^a, and In^b). Mutation in erythroid transcription factor GATA-1 also results in X-linked depression of Lutheran antigen expression. Lutheran antigens are poorly developed at birth. Anti-Lu^a has not been associated with HTRs and has rarely been associated with mild HDFN. Anti-Lu^b has been reported to cause mild HTRs and mild HDFN. Anti-Lu3 is found in immunized individuals with the Lu(a−b−), i.e., Lu_{null} phenotype, and may cause delayed HTRs or HDFN.

Indian System: Indian antigens, In^a (low prevalence) and In^b INFI, INJA, and INRA (high prevalence), are located on CD44, a widely distributed cell adhesion molecule. Antibodies are not generally clinically significant, but there is one report of anti-In^b associated with HTR. The high-prevalence antigen AnWj is currently assigned to 901 series of antigens and not Indian System; however, it also appears to be carried on or associated in some way with CD44. Antibodies to AnWj have caused serious HTRs, and In(Lu) RBCs (see above) are recommended for transfusion.

Scianna System: Seven Scianna antigens are carried on ERMAP (erythrocyte membrane-associated protein). Sc1 (high prevalence) and Sc2 (low prevalence) are antithetical, while Sc3, STAR, SCER, and SCAN are of high prevalence and Rd is of low prevalence. Very rare Scianna-null individuals make anti-Sc3. Antibodies to all Scianna antigens are rare. Antibodies to Scianna high-prevalence antigens have not been reported to cause HTRs or HDFN, while anti-Sc2 has resulted in mild HDFN.

Ok System: Ok System is composed of three high-prevalence antigens, Ok^a, OKGV, and OKVM. Antigens in this system are carried on CD147, also known as basigin, which has been described to be a receptor for *Plasmodium falciparum* invasion. Ok(a−) is a rare phenotype, and all individuals with this phenotype reported to date are Japanese. OKGV− and OKVM− phenotypes appear to be equally rare being first described in an individual of Iranian and Hispanic backgrounds, respectively. RBC survival studies indicated anti-Ok^a could cause reduced cell survival. Clinical significance of anti-OKGV and anti-OKVM is not known, as there has been only one report of each specificity. No HDFN has been reported.

Yt System: Yt (Cartwright) antigens, Yt^a and Yt^b, are located on RBC GPI-linked enzyme acetylcholinesterase. Anti-Yt^a has been reported to result in accelerated destruction of Yt^a-positive RBCs but has not been associated with HDFN. Anti-Yt^b has not been associated with HTRs or HDFN.

LAN System: Lan is a high-prevalence antigen carried on ABCB6 (ATP-binding cassette, subfamily B member 6), a multipass protein with wide tissue distribution involved in heme synthesis. The first Lan− phenotype was recognized in 1961, and in 2012, two groups using different approaches showed that inactivating mutations in *ABCB6* defined this null phenotype. Lan− phenotype occurs in about 1: 20,000 individuals including Caucasians, blacks, and Japanese. An eye developmental defect is associated with changes in ABCB6, but Lan− individuals are healthy and the protein does not appear to be necessary for normal erythropoiesis. Anti-Lan has caused mild to severe HTRs and mild HDFN.

JR System: Jr^a is a high-prevalence antigen on ABCG2 glycoprotein (ATP-binding cassette, subfamily G, member 2), also known as breast cancer resistance protein (BCRP), involved in transport of urate and possibly other compounds. The first examples were reported in 1970, and in 2012, it was shown that inactivating mutations in *ABCG2* define the Jr(a−) or null phenotype. Jr(a−) phenotype is found primarily in Japanese and other Asians, a few northern Europeans, Arabs, and one Mexican. Anti-Jr^a is associated with HTRs and usually mild HDFN, but one fatal case of HDFN has been reported.

Gill System: Single high-prevalence GIL1 antigen is located on aquaporin 3 (AQP3), a member of the aquaporin family of water and glycerol channels. GIL1 null phenotype has been reported in five probands, including American, French, and German. The antibody has caused HTR and positive DAT in the fetus, but no cases of clinical grade HDFN have been reported.

Blood Group Systems Not Considered Clinically Significant

Knops System: Knops antigens are located on C3b/C4b receptor, CR1, which is the primary complement receptor on RBCs. Most of the eight officially recognized Knops antigens are high prevalence (Kn^a, McC^a, Sl1, Yk^a, KCAM); however, there is a marked difference in prevalence of Sl1 and KCAM and low-prevalence McC^b antigens between blacks and Caucasians thought to be due to selection for resistance to malaria. Antibodies are IgG, show variable weak reactivity but may continue to react even at high dilutions. These antibodies are not clinically significant.

Chido/Rodgers System: Chido (Ch) and Rodgers (Rg) antigens are high-prevalence antigens present on isotypes (C4A carries Ch and C4B carries Rg) of the fourth component of complement (C4), which serves as a platform for interaction of antigen–antibody complex and complement proteins. These antigens are absorbed onto the RBC from plasma. Antibodies are not considered clinically significant and can be neutralized by plasma from antigen-positive individuals.

John Milton Hagen System: There are six antigens of high prevalence carried on GPI-linked semaphorin glycoprotein, CD108, or Sema7A. Antibodies are not considered clinically significant, as they are often found in elderly patients with acquired antigen loss and associated with weakly positive direct antiglobulin test. One example of HTR was seen in a patient who had an alloantibody due to inheritance of variant form of CD108.

Landsteiner–Wiener System: LW (Landsteiner–Wiener) antigens are carried by intercellular adhesion molecule, ICAM-4. The LW^a, LW^{ab} (high prevalence) and LW^b (low prevalence) antigens are more strongly expressed on D-positive than D-negative RBCs and therefore, anti-LW^a may be confused with anti-D. RBCs with an Rh_{null} phenotype are LW(a–b–). Antibodies are not associated with HTRs or HDFN.

Raph System: Single antigen MER2 (RAPH1) is carried on CD151, tetraspanin. Anti-MER2 was reported in three Israeli Jews on renal dialysis that were subsequently found to carry nucleotide deletions causing an RAPH-null phenotype, suggesting that CD151 deficiency is associated with renal failure. Anti-MER2 in individuals with single amino acid changes in gene causing a MER2– phenotype was not associated with disease, and these antibodies have not caused HTRs or HDFN.

Xg System: Xg^a and CD99 antigens of XG system are carried on different sialoglycoproteins but are encoded by pair of homologous genes. XG gene on X chromosome encodes for Xg^a and CD99 gene on X and Y chromosomes encode for CD99. CD99 is known to be an adhesion molecule on many tissue cells including RBCs. CD99 expression on RBCs is affected by presence or absence of Xg^a. Anti-Xg^a is uncommon, may either be RBC stimulated or naturally occurring, and has not been implicated in HTRs or HDFN. Rare antibodies to CD99 have been detected, and little is known about their significance.

CD59 System: CD59 blood group system was officially recognized in 2014. It consists of one high-prevalence antigen, CD59.1. One example of anti-CD59 has been described in an individual with CD59 deficiency carrying null allele. In this case report, transfusion with incompatible CD59+ donor RBCs was reported to be well tolerated. CD59 is responsible for binding complement components C8 and C9 and preventing formation of membrane attack complex to protect cells from complement damage. Individuals who lack CD59 are typically ill with neuropathy, strokes, and episodes of Coombs-negative hemolysis.

Blood Group Collections

Cost Collection: Cost collection consists of Cs^a and Cs^b antigens. The antibodies are not clinically significant.

Er Collection: Er collection consists of Er[a] (high prevalence) and Er[b] (low prevalence). Rare individuals with Er(a−b−) RBCs make anti-ER3, but no HTR is reported and no clinical HDFN, although DAT was positive.

High-Prevalence RBC Antigens (901 Series)

Sd[a] Antigen: Sd[a] is carried on carbohydrates on RBCs, and expression is highly variable and may diminish during pregnancy. Anti-Sd[a] is not considered clinically significant, but examples of HTRs with RBCs with strong Sd[a] expression were reported.

Emm Antigen: Emm antigen was reported in 1987 and is carried on a GPI-linked protein on RBC, but specific protein identity is not known. Six Emm− probands reported are ethnically diverse. Clinical significance for transfusion and pregnancy is not known, as all but one of the probands was male and not transfused.

MAM Antigen: MAM antigen was reported in 1993, with rare MAM− probands. Anti-MAM was associated with HDFN and neonatal thrombocytopenia, and monocyte monolayer assay suggests it is clinically significant for transfusion.

PEL Antigen: PEL antibody is associated with reduced RBC survival but no HDFN.

ABTI Antigen: ABTI antigen appears to have a phenotypic relationship to Vel antigen despite molecular evidence that it does not reside on Vel carrier protein SMIM1. Anti-ABTI has been reported in three members of one Israeli Arab family causing no HDFN. There is no data regarding its clinical significance for transfusion.

Low-Prevalence RBC Antigens (700 Series): These antigens occur in less than 1 in 500 individuals and include Batty (By), Biles (Bi), Box (Bx[a]), Christiansen (Chr[a]), HJK, HOFM, JFV, JONES, Jensen (Je[a]), Katagiri (Kg), Livesay (Li[a]), Milne, Peters (Pt[a]), Rasmussen (RASM), Reid (Re[a]), REIT, Torkildsen (To[a]). Antibodies to these low-prevalence antigens do not cause transfusion problems because donor units are not likely to carry antigen, but many have caused HDFN.

Human Leukocyte Antigens Residually Expressed on RBCs: Bg (Bennett-Goodspeed) antigens correspond to human leukocyte antigens (HLAs): Bg[a] corresponds to HLA-B7, Bg[b] corresponds to HLA-B17, and Bg[c] corresponds to HLA-A28. These antigens are expressed variably on RBCs.

Further Reading

Brunker, P. A. R., & Flegel, W. A. (2011). Scianna: The lucky 13th blood group system. *Immunohematology*, *27*, 41–57.

Daniels, G. (2016). The Augustine blood group system, 48 years in the making. *Immunohematology*, *32*, 100–103.

Halverson, G. R., & Peyrard, T. (2010). A review of the Colton blood group system. *Immunohematology*, *26*, 22–26.

Johnson, N. C. (2011). XG: The forgotten blood group system. *Immunohematology*, *27*, 68–71.

Lomas-Francis, C., & Reid, M. E. (2010). The Dombrock blood group system: A review. *Immunohematology*, *26*, 71–78.

Moulds, J. M. (2010). The Knops blood group system: A review. *Immunohematology*, *26*, 2–7.

Peyrard, T. (2013). The Lan blood group system: A review. *Immunohematology*, *29*, 131–135.

Rumsey, D. M., & Mallory, D. A. (2013). GIL: A blood group system review. *Immunohematology*, *29*, 141–144.

Smart, E. A., & Storry, J. R. (2010). The OK blood group system: A review. *Immunohematology*, *26*, 124–126.

Storry, J. R., & Peyrard, T. (2017). The Vel blood group system: A review. *Immunohematology*, *33*, 56–59.

Storry, J. R., Reid, M. E., & Yazer, M. H. (2010). The Cromer blood group system: A review. *Immunohematology*, *26*, 109–118.

Walker, P. S., & Reid, M. E. (2010). The Gerbich blood group system: A review. *Immunohematology*, *26*, 60–65.

Weinstock, C., Anliker, M., & von Zabern, I. (2015). CD59: A long-known complement inhibitor has advanced to a blood group system. *Immunohematology*, *31*, 145–151.

CHAPTER 31

Human Platelet Antigens

Scott T. Avecilla, MD, PhD

Platelet-specific antigens associated with the formation of alloantibodies in exposed patients are the basis for human platelet antigen (HPA) categorization. Over the years, these antigenic determinants have been mapped to a relatively short list of platelet-expressed adhesion/aggregation molecules, namely GPIa, GPIb, and GPIIb/IIIa.

Antigens: Table 31.1 lists current nomenclature for the various HPA groups with their associated other names, the major protein within which the antigenic epitope resides, and relative frequencies in the white population. HLA Class I and A/B/H group antigens are also found on the platelet surface. For the latest allelic frequency information, the Immuno Polymorphism Database section on HPA can be queried at http://www.ebi.ac.uk/ipd/hpa/.

Mode of Inheritance: Of the 35 HPAs, 12 have been categorized into six biallelic groups (HPA-1, 2, 3, 4, 5, 15). Designated "a" antigen of the pair is typically the high-frequency antigen, whereas "b" antigen is the low-frequency antigen. The remaining 23 HPAs do not appear to correspond to a pair of alleles and are all designated as "b" antigens due to their low population frequency. It is hypothesized that due to the very low frequency of the "b" antigen that the probability of having a "bb" homozygote would be exceedingly rare and therefore no antibodies against a hypothetical corresponding "a" antigen have been clinically documented.

Molecular Basis of Type: The majority of HPAs reside on platelet-specific adhesion/aggregation molecules, which include GPIIb, GPIIIa, GPIa, GPIbα, GPIbβ, GPIV, GPV, and CD109. Originally, typing was performed with patient-derived antisera; however, an important limitation is that patients often have additional antibodies against Class I HLA antigens and therefore the sera was not monospecific and contaminated with anti-HLA antibodies. Once the molecular determinants for HPA type were discovered to be due to amino acid substitutions at specific points, the application of molecular methods to perform typing using DNA-based techniques became the "gold standard."

Disease Associations: Alloantibody formation against HPA antigens is associated with several forms of thrombocytopenia, specifically neonatal alloimmune thrombocytopenia (NAIT), posttransfusion purpura (PTP), and platelet transfusion refractoriness (see Chapters 94, 69, and 57, respectively).

Neonatal Alloimmune Thrombocytopenia: NAIT is when fetal/neonatal thrombocytopenia occurs as a consequence of maternal transplacental transmission of HPA alloantibodies that react with fetal/neonate platelets. A mother is exposed to an HPA for which she is negative and subsequently produces an alloantibody that binds

Transfusion Medicine and Hemostasis. https://doi.org/10.1016/B978-0-12-813726-0.00031-3

TABLE 31.1 Human Platelet Antigen Table

System	Antigen	Other Names	Glycoprotein	Antigen Frequency (%)
HPA-1	HPA-1a	Zw^a, Pl^{A1}	GPIIIa (CD61)	97.9
	HPA-1b	Zw^b, Pl^{A2}		28.8
HPA-2	HPA-2a	Ko^b	GPIbα (CD42b)	>99.9
	HPA-2b	Ko^a, Sib^a		13.2
HPA-3	HPA-3a	Bak^a, Lek^a	GPIIb (CD41)	80.95
	HPA-3b	Bak^b		69.8
HPA-4	HPA-4a	Yuk^b, Pen^a	GPIIIa (CD61)	>99.9
	HPA-4b	Yuk^a, Pen^b		<0.1
HPA-5	HPA-5a	Br^b, Zav^b	GPIa (CD49b)	99.0
	HPA-5b	Br^a, Zav^a, Hc^a		19.7
	HPA-6bw	Ca^a, Tu^a	GPIIIa (CD61)	0.7
	HPA-7bw	Mo^a	GPIIIa (CD61)	0.2
	HPA-8bw	Sr^a	GPIIIa (CD61)	<0.01
	HPA-9bw	Max^a	GPIIb (CD41)	0.6
	HPA-10bw	La^a	GPIIIa (CD61)	<1.6
	HPA-11bw	Gro^a	GPIIIa (CD61)	<0.25
	HPA-12bw	ly^a	GPIbβ (CD42c)	0.4
	HPA-13bw	Sit^a	GPIa (CD49b)	0.25
	HPA-14bw	Oe^a	GPIIIa (CD61)	<0.17
HPA-15	HPA-15a	Gov^b	CD109 (CD109)	74
	HPA-15b	Gov^a		81
	HPA-16bw	Duv^a	GPIIIa (CD61)	<1
	HPA-17bw	Va^a	GPIIb/IIIa (CD61)	<0.4
	HPA-18bw	Cab^a	GPIa (CD49b)	Case report
	HPA-19bw	Sta	GPIIIa (CD61)	Case report
	HPA-20bw	Kno	GPIIb (CD41)	Case report
	HPA-21bw	Nos	GPIIIa (CD61)	Case report
	HPA-22bw	Sey	GPIIb (CD41)	Case report
	HPA-23bw	Hug	GPIIIa (CD61)	Case report
	HPA-24bw	$Cab2^{a+}$	GPIIb (CD41)	Case report (<1)
	HPA-25bw	Swi^a	GPIa (CD49b)	Case report
	HPA-26bw	Sec^a	GPIIIa (CD61)	Case report
	HPA-27bw	$Cab3^{a+}$	GPIIb (CD41)	Case report (<1)
	HPA-28bw	War	GPIIb (CD41)	Case report
	HPA-29bw	Kha^b	GPIIIa (CD61)	Case report
		Pl^T	GPV	>99.9
		Vis	GPIV	
		Pe^a	GPIbα (CD42b)	
		Dy^a	38 kDa GP	
		Mou^a	Unknown	26
		Lap^a	GPIIb (CD41)	

Modified from Metcalfe, P, Watkins, N. A, Ouwehand, W. H., Kaplan, C., Newman, P., Kekomaki, R., et al. (2003). Nomenclature of human platelet antigens. *Vox Sang, 85*(3), 240–245; Allen et al., 2005; https://www.ebi.ac.uk/ipd/hpa/table1.html.

to and causes the rapid clearance/destruction of fetal/neonate platelets that express the antigen. Clinical consequences of NAIT range from thrombocytopenia without overt bleeding to in utero fetal demise secondary to intracranial hemorrhage or other severe bleeding complications. Alloantibodies against HPA-1a have been implicated in the majority of cases of NAIT (75%) with HPA-5b being the second most frequent target.

Posttransfusion Purpura: PTP is a thrombocytopenic condition, which occurs 2–14 days after an uneventful platelet-containing transfusion (RBC or platelet transfusion). Generation of alloantibodies against a specific HPA (usually HPA-1a) not only affects transfused platelets but a bystander effect is observed where the patient's native platelets are also cleared/destroyed, resulting in profound thrombocytopenia.

Immune Platelet Refractoriness: Immune refractoriness to allogeneic platelet transfusions is most frequently associated with HLA Class I alloimmunization; however, there are cases due to alloantibodies against HPAs. HPA-1a and HPA-5b have been implicated in this immune refractoriness. In patients congenitally lacking GPIb (Bernard Soulier) and GPIIb/IIIa (Glanzmann thrombasthenia), a wide range of alloimmune HPA antibodies can result from allogeneic platelet exposure.

Laboratory Diagnosis: A variety of laboratory techniques exist to determine HPA phenotype and genotype of the patient and potential platelet donors. Additional methods to detect and determine the specificity of antiplatelet antibodies exist. Knowledge of the patient's HPA phenotype and the alloantibody specificity can be useful when providing platelets to patients with platelet alloantibodies.

Serology: Serology was the original method by which HPAs were identified using sera from patients with NAIT, PTP, and platelet refractoriness. Anti-HPA sera were mixed with patient's platelets, and if the sera caused an agglutination reaction, the patient was deemed to be positive for that HPA. Conversely, an initial method to detect the presence of an alloantibody by using the alloimmunized patient's sera mixed with HPA-typed platelets could be used. The major limitation of this method is that the sera were often polyspecific and therefore results could be ambiguous.

Molecular Techniques for Human Platelet Antigen Typing: Molecular methods have been developed, which use the genetic basis for HPA type to overcome the limitations inherent with serology. Single-nucleotide polymorphisms (SNPs) resulting in amino acid substitutions on particular platelet-specific glycoproteins have been reported. By using the published genetic information, several assays have been developed to identify the presence or absence of said SNPs as way to genotype and predict the phenotype. Polymerase chain reaction (PCR) can be used to differentially amplify specific SNPs to determine type; alternatively, the relevant DNA regions of interest can be amplified and directly sequenced to determine type. Abovementioned methods are simply examples because there are many variations of PCR, DNA array, and other technologies to detect specific genes.

Flow Cytometry for HPA Antibody Detection: Flow cytometry has been developed as a way to identify the presence of platelet antibodies. Patient serum is first incubated with washed normal donor platelets to allow the binding of platelet-specific antibodies. After the binding reaction, the platelets are washed to remove excess serum. Next, a fluorescently labeled secondary antibody is added, which binds to the platelet-bound antibodies and "tags" the platelets in a detectible way. Finally, the platelets are examined by the flow cytometer to measure if there is a relative increase in platelet-associated fluorescence, which would indicate the presence and quantity of antiplatelet antibody present in the patient's serum. Similar to serology; however, flow cytometry cannot distinguish between antibodies bound to platelet-specific antigens or to other antigens such as Class I HLA. This limitation can be overcome with the use of monoclonal antibodies directed at specific HPAs to create a competitive binding assay. The monoclonal antibody would be used to first "block" the target HPA on the platelets after which the patient's sera is added to the binding reaction, and a loss of fluorescence signal would indicate that the detected antibody in the patient's sample is against an HPA. Competitive binding flow cytometry unfortunately does not lend itself to being scaled up to examine multiple HPAs simultaneously.

Solid Phase Techniques for HPA Antibody Detection: Solid phase techniques have been used to detect the presence and specificity of platelet antibodies in addition to streamlining platelet crossmatching. Two common methods in wide use are the mixed passive hemagglutination assay (MPHA) and the monoclonal antibody-specific immobilization of platelet antigen (MAIPA) assay. MPHA relies on the immobilization of platelets onto the bottom of a round bottom microtiter plate, which is then incubated with the patient's serum. After the binding reaction, the well is washed of excess serum, and indicator RBCs are added to the well. These indicator RBCs have on their surface antihuman antibodies, which can bind to the Fc portion of platelet-bound antibodies. After the addition of the indicator RBCs, the microtiter plate is centrifuged, and if the indicator RBCs form a button on the bottom of the well that indicates a negative reaction, however, if the indicator RBCs are evenly coating the bottom of the well, that indicates the presence of platelet-bound antibodies. The same limitation found in the flow cytometry assay of not being able to distinguish the target of the platelet antibody is present with this assay. Because of the microtiter format, this assay can be run in parallel to substantially increase the test throughput. This parallel run feature is used for platelet crossmatching, in that the patient's serum is reacted with immobilized platelets of potential donor units to search for nonreactive, compatible units. Another modification to the MPHA is the use of phenotyped platelets for immobilization, and similar to RBC antibody panels, these panels of platelets can be used to infer the identity of platelet antibodies in some cases because interference from anti-HLA can still occur. The next stage in development of solid phase methods was made possible with the advent of monoclonal antibody reagents directed against specific HPAs. One widely used example is the MAIPA assay. MAIPA technique relies on mouse monoclonal antibodies, which bind to specific HPAs in a way that does not interfere with the patient's HPA–alloantibody binding. Both mouse monoclonal antibody and patient serum are incubated with normal platelets to allow binding simultaneously. Platelets are then washed, and their plasma membrane disrupted with a detergent to solubilize platelet antigens. The soluble antigens are then transferred to a microtiter plate, which has been precoated with an antimouse antibody to capture the mouse antibodies, which are bound to the desired HPA. After an incubation step, the well is washed, and an antihuman antibody coupled with a chromogen is added. If there is the presence of an anti-HPA antibody in the patient's serum that has the same target as the mouse monoclonal antibody, then a chromogenic reaction will occur. This assay has the major advantage in that each well of the microtiter place can correspond to a single HPA, and there is no interference from other antibodies such as those directed against Class I HLA.

Further Reading

Allen, D. L., Lucas, G. F., Ouwehand, W. H., & Murphy, M. F. (2005). Platelet and neutrophil antigens. In M. F. Murphy, & D. H. Pamphilon (Eds.), *Practical Transfusion Medicine* (2nd ed.). Malden: Blackwell Publishing.

Metcalfe, P., Watkins, N. A., Ouwehand, W. H., Kaplan, C., Newman, P., Kekomaki, R., et al. (2003). Nomenclature of human platelet antigens. *Vox Sang*, *85*(3), 240–245.

Robinson, J., Mistry, K., McWilliam, H., Lopez, R., & Marsh, S. G. (2010). IPD–the immuno polymorphism database. *Nucleic Acids Res*, *38*(Database issue), D863–D869.

CHAPTER 32

Human Leukocyte Antigens

Emeline Masson Frenet, PhD and Andromachi Scaradavou, MD

HLA (human leukocyte antigen) is the human major histocompatibility complex (MHC), a multigene family involved in the defense of humans (and all vertebrae) against pathogens. The HLA molecule's role is to present peptides to T cells. Depending on the peptide, antigen presentation can lead to activation of T cells and initiation of an adaptive immune response. HLA molecules interact with NK cells too, inhibiting cytotoxicity when HLA molecules (mainly HLA-C) link the NK inhibitor receptor. They also have a primordial role in lymphocyte maturation in the thymus.

HLA Genes: The HLA complex is located within the 6p21.3 region of the short arm of chromosome 6 (Fig. 32.1) and contains >240 genes of diverse functions. Many of the genes encode immune system proteins. The MHC gene family is divided into three subgroups or classes, named I, II, and III. The classical loci, routinely studied in human medicine, are HLA-A, HLA-B, and HLA-C for class I, and HLA-DRB1, HLA-DQB1, and HLA-DPB1 for class II. HLA genes are closely linked to one another and are inherited en bloc as a genetic unit. The series of HLA alleles on a single chromosome 6 is called *haplotype*. Combination of maternal and paternal haplotypes inherited creates the individual's HLA genotype.

HLA Antigens: HLA antigens are glycoproteins and belong to the immunoglobulin superfamily, meaning that they form domains (Fig. 32.1). The genes' organization in a sequence of exons and introns reflects that particular protein structure.

HLA class I molecules are composed of a polymorphic α-chain combined with a monomorphic β-globulin chain and are expressed on almost all nucleated cells. They present endogenously processed peptides (derived from cytoplasmic protein degradation from self, as well viral proteins, for example) to $CD8^+$ T cells.

HLA class II molecules are composed of two polymorphic chains (α and β) and are expressed constitutively only on professional antigen-presenting cells (such as macrophages, dendritic cells, and B cells, as well as activated T cells and thymic epithelial cells). Their expression can be induced on other cells, in case of stress, for example. They present peptides derived from endosomal and lysosomal protein degradation (from extracellular content, such as bacteria or parasites) to $CD4^+$ T cells.

The combination of peptide and peptide-binding groove of the HLA molecule forms the epitope that is recognized by the T-cell receptor. Antigen recognition is *restricted to the MHC*, as neither peptide nor MHC alone can stimulate T-cell responses, which require formation of an MHC–peptide complex.

Polymorphism: The MHC region is the most polymorphic of the human genome. Polymorphism is mainly located in the peptide-binding groove, where it affects antigen presentation and confers selective advantage to the population. More than 17,000 HLA alleles have been described (12,893 for class I and 4802 for class II as of March 2018).

Transfusion Medicine and Hemostasis. https://doi.org/10.1016/B978-0-12-813726-0.00032-5

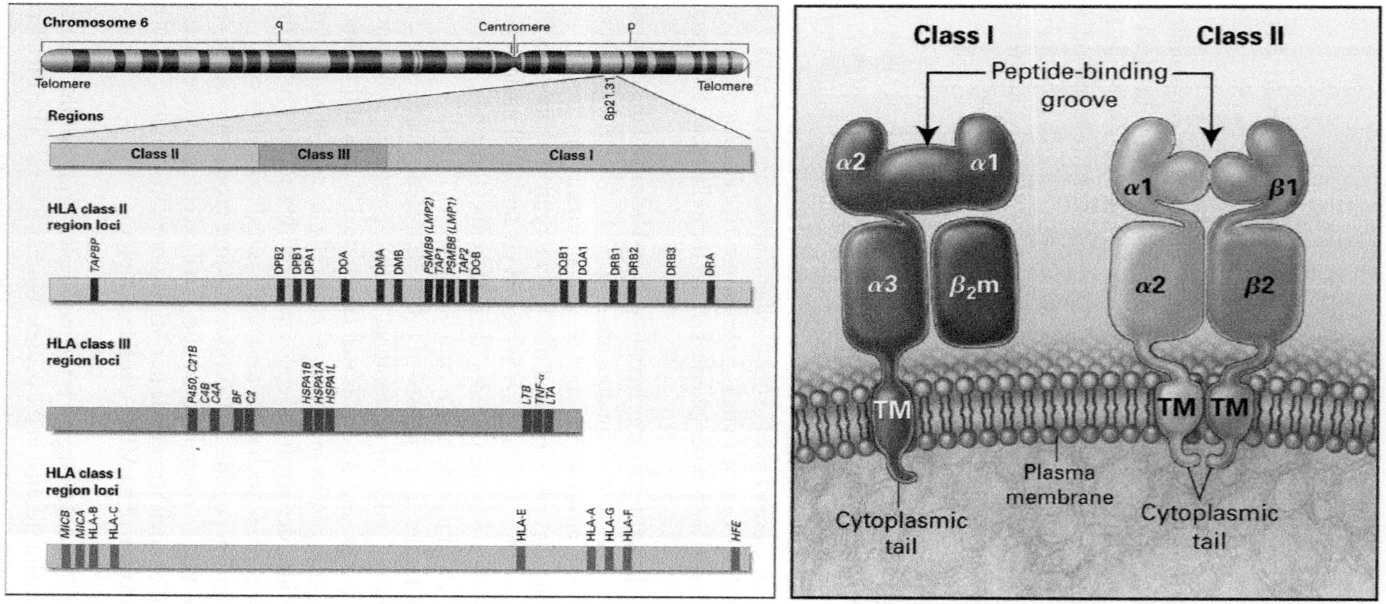

FIGURE 32.1 Genetic organization of the human leukocyte antigen (HLA) complex and structure of HLA class I and II molecules. (Klein, J., & Sato, A. (2000). The HLA system. *N Engl J Med, 343*(10), 702–709.)

Frequencies of individual HLA alleles vary greatly within a population and between populations.

Common alleles are those that appear with gene frequencies >0.001 in any reference population. Well-documented alleles are those having been described more than five times in unrelated individuals. The current common/well-documented HLA allele list is available on ImMunoGeneTics database (https://www.ebi.ac.uk/ipd/imgt/hla/).

Alleles are not randomly combined in haplotypes, as certain HLA alleles are found associated with one another more frequently than would be predicted by chance alone. This *linkage disequilibrium* is particularly found for loci that are close together (HLA-B and HLA-C, HLA-DRB1 and HLA-DQB1), defining frequent and rare allele associations that affect donor searches.

HLA Typing: Historically, HLA antigens were defined by serological methods using a complement-dependent microcytotoxicity assay and panels of alloantisera-containing specific HLA antibodies.

DNA Typing Methods: Most HLA typing methods in use are based on amplification of specific HLA gene portions from genomic DNA using polymerase chain reaction (PCR). DNA-based typing methods vary in regard to the level of discrimination (resolution) they provide in defining the nucleotide sequence of an HLA gene.

- *Sequence-specific oligonucleotide probe (SSOP) hybridization* is based on amplification of the most relevant portions of exons 2 and 3 (coding for the peptide-binding domain in class I molecules) or exon 2 (coding for the peptide-binding domain in class II molecules), followed by hybridization (on nylon membranes, plates, or flow cytometry beads) using sequences specific for a certain allele or group of alleles.
- *Sequence-specific primers (SSP)* use multiple PCR reactions, each specific for an allele or group of alleles. Presence or absence of amplification is detected by electrophoresis. The number of reactions needed depends on locus polymorphism and degree of resolution.
- *Sequencing-based typing (SBT)* involves sequencing HLA genes and comparing the sequence with published libraries (IMGT/HLA database). Degree of resolution will depend on the length of the sequence obtained. It is the only method with ability to detect and characterize new alleles.
- *Next generation sequencing (NGS)* methods have recently been applied to HLA typing, increasing throughput and allele identification capabilities, and allowing for easier suitable donor identification.

HLA Nomenclature: The WHO Nomenclature Committee for Factors of the HLA System is responsible for naming HLA genes and allele sequences (http://www.hla.alleles.org). Each HLA allele name has a unique number corresponding to up to four sets of digits (two or three), separated by colons (Fig. 32.2). The first two digits describe the type (allele family), usually corresponding to a serological antigen. The Next two or three digits are assigned in the order those sequences were determined. Subsequent digits may be used in some typings (Fig. 32.2).

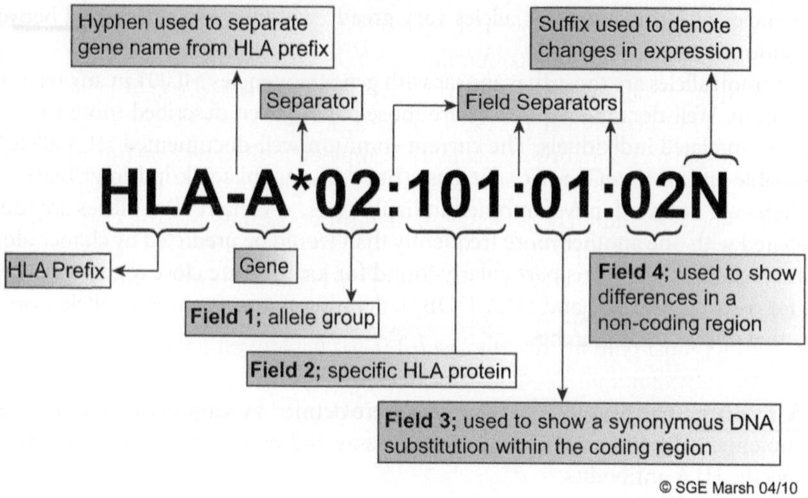

FIGURE 32.2 **Human leukocyte antigen (HLA) nomenclature.** (From Nunes, E., Heslop, H., Fernandez-Vina, M., et al. (2011). Definitions of histocompatibility typing terms. *Blood, 118*, e180–e183.)

Low Resolution: DNA-based typing result at the level of digits composing the first field in DNA-based nomenclature, or serologically defined equivalent (e.g., HLA-A2).

High Resolution: Result consistent with a set of alleles that encode the same protein sequence for the peptide-binding site, excluding alleles that are not expressed as cell surface proteins (e.g., HLA-A*02:01).

Allelic Resolution: Result consistent with a single allele (e.g., HLA-A*02:01:01:01).

Reporting of Ambiguous typing results: Resolution level between low and high resolution involves a letter code (up to five letters so far) after the first set of digits, translating into a list of possible alleles. For example, DRB1*01:AD means DRB1*01:01 or DRB1*01:04 but excludes alleles DRB1*01:02 and DRB1*01:03.

HLA alleles that have identical nucleotide sequences across exons encoding peptide-binding domains will be designated by upper case "G," which follows first three fields of allele designation (e.g., HLA-A*02:01:01G). HLA alleles having nucleotide sequences that encode the same protein sequence for peptide-binding domains will be designated by an upper case "P," which follows two fields of allele designation (e.g., HLA-A*02:01P).

HLA Antibodies: Exposure to foreign HLA antigens through pregnancy, transfusion, or transplantation may elicit HLA-specific antibody production. Additionally, HLA antibodies have been detected in the serum of nonalloimmunized healthy males, thought to be induced by cross-reactive bacterial antigens or peptides from food or allergens. Previously used lymphocytotoxicity methods have been replaced by solid phase assays (mainly flow cytometry) that have high sensitivity and specificity.

HLA and Hematopoietic Stem Cell Transplantation: Progress in unrelated hematopoietic stem cell transplantation (HSCT) has depended greatly on advances in the techniques for HLA allele definition. When there is no HLA-identical sibling (having inherited the same HLA haplotypes from the parents), unrelated donors will be evaluated. High-accuracy typing has reduced discrepancies and permits identification of the most compatible unrelated donor.

Unrelated Adult Volunteer Donors: Current practice is to perform high-resolution or allele-level typing of donor and recipient in unrelated HSCT from adult volunteer donors for HLA-A, HLA-B, HLA-C, and HLA-DRB1 (8 alleles), and sometimes also DQB1 (10 alleles). Fully HLA-matched adult donors at allele level are highly preferable, as mismatched transplants have a higher incidence of graft rejection, transplant-related mortality, and acute GvHD.

Owing to the extensive polymorphism of HLA and its different frequencies in various racial groups, probability of finding fully matched unrelated adult donors varies significantly depending on the patient's racial/ethnic background. Although chances are up to 80% for Caucasian patients, they are lower (20%–40%) for ethnic minority patients because these are underrepresented in adult volunteer registries.

A Single allele mismatch (9/10 matched donor) can be tolerated, but alternative hematopoietic cell sources have to be considered if no compatible-enough unrelated adult donor can be identified. Haploidentical family members, matched for only one HLA haplotype, are being increasingly used, using T-cell depletion/CD34 cell selection of grafts to reduce GvHD risk.

Unrelated Cord Blood Grafts: Unrelated cord blood grafts are currently matched at four to six antigens, namely HLA-A, HLA-B at the antigen level, and HLA-DRB1 at the allele level. When available, higher allele match for HLA-A, HLA-B, and HLA-C should be preferred, as it affects transplant-related mortality. Cord blood unit selection is a balancing act between best HLA match and adequate cell dose (total nucleated cells and CD^{34+} cells). When a suitable cell dose (>2.5–3 × 10e7 TNC/kg) cannot be reached with a single unit, two units can be used.

Donor Registries: Registries and cord blood banks collect and store HLA information of volunteers to identify matched unrelated donors for patients requiring HSCT. Currently, there are >30 million adult donors and 730,000 cord blood units available for patients worldwide. Registries apply matching algorithms to identify potential suitable donors. They also provide tools to assess match probabilities, using their vast resources in terms of HLA information on large numbers of individuals from different ethnic groups. As haplotype frequencies are better discriminators between populations than single allele frequencies, they are currently used to help select donors that should be contacted for additional typing.

HLA Antibodies: Patients should be screened for anti-HLA antibodies at the time of donor search, as presence of donor specific antibodies (DSAs) have been associated with primary graft failure, particularly in the setting of a high degree of mismatches (cord blood and haploidentical donors). DSA strength is evaluated by mean fluorescence intensity (MFI) using flow cytometry assays; however, threshold above which DSA become clinically significant is unclear.

HLA and Solid Organ Transplantation

HLA Matching: In the United States, the national Organ Procurement and Transplant network (OPTN) currently requires typing for HLA-A, HLA-B, and HLA-DR at the antigen level, although many laboratories also type for HLA-C, HLA-DQ, and HLA-DP.

Better HLA match leads to better graft function, longer survival (for both graft and patient), and reduced risk of sensitization. However, owing to the very low number of available organs, and to shorten wait-list time, less matched organs are often selected.

HLA Antibodies: Anti-HLA sensitization poses the most significant barrier to transplantation. Presence of pretransplant DSA is associated with hyperacute rejection of the graft, while de novo sensitization after transplant can lead to the recently characterized acute humoral rejection. It is also hypothesized that DSAs participate in the very complex process of chronic rejection. One-third of patients on the US solid organ waiting lists are sensitized to HLA when tested with sensitive techniques. Sensitized patients wait longer for a transplant, experience significantly more rejection episodes, and have overall shorter graft survival rates.

Current antibody screening methods detect and assess antibodies for three major characteristics relevant to transplant outcomes: specificity, isotype, and strength. Antibody-mediated rejection (AMR) risk clearly increases with increased antibody MFI. Use of more sensitive antibody screening assays has allowed "virtual crossmatch" to be performed before transplant, to identify unacceptable mismatches before organ allocation (predicting the results of the actual crossmatch between the patient's sera and the donor's lymphocytes). The definition of unacceptable mismatches (in terms of strength, locus, and isotype) varies with the organ, the patient's clinical situation, the conditioning regimen, and the possibility of desensitization. After transplantation, patients are regularly tested for DSA emergence or resurgence, as these may contribute to acute and chronic rejection.

HLA and Transfusion: Anti-HLA immunization (through transfusion, but also pregnancy and transplantation) is responsible for some of the complications seen in patients receiving blood and blood products. The use of more sensitive and specific techniques to detect anti-HLA antibodies has improved the investigation of transfusion reactions but also facilitated implementation of anti-HLA antibody negative product selection. These issues will be addressed in more details in another chapter.

HLA and Diseases: Presence of a specific HLA allele has been associated with an increased risk of developing some diseases (auto-immune, like ankylosing spondylitis with B27, as well as infectious and malignant), although it is often not clear what role the HLA genes play in the development of the disease.

HLA and Drug Hypersensitivity: Several forms of immune mediated drug reactions are associated with expression of specific HLA alleles. The most frequent are B*57:01 and hypersensitivity to Abacavir, Carbamazepine skin reactions and B75 (B*15:02, B*15:11, B*15:18) and A*31:01, and Allopurinol-induced cutaneous manifestations with B*58:01. Pre-treatment HLA allele screening (usually by a specific PCR-SSP assay) and avoidance of the drug in case of a positive result has been successfully applied to prevent these reactions.

Further Reading

Althaf, M. M., El Kossi, M., Jin, J. K., et al. (2017). Human leukocyte antigen typing and crossmatch: a comprehensive review. *World J Transplant*, *7*(6), 339–348.

Amore, A. (2015). Antibody-mediated rejection. *Curr Opin Organ Transplant*, *20*(5), 536–542.

Brown, C. J., & Navarrete, C. V. (2011). Clinical relevance of the HLA system in blood transfusion. *Vox Sang*, *101*, 93–105.

Kindwall-Keller, T. L., & Ballen, K. K. (2017). Alternative donor graft sources for adults with hematologic malignancies: a donor for all patients in 2017!. *Oncologist*, *22*(9), 1125–1134.

Moncharmont, P. (2018). Platelet component transfusion and alloimmunization: Where do we stand? *Transfus Clin Biol* S1246-7820(18) 30021–1.

Nunes, E., Heslop, H., Fernandez-Vina, M., et al. (2011). Definitions of histocompatibility typing terms. *Blood*, *118*(23), e180–e183.

Redwood, A. J., Pavlos, R. K., White, K. D., et al. (2018). HLAs: Key regulators of T-cell-mediated drug hypersensitivity. *HLA*, *91*(1), 3–16.

Websites: http://hla.alleles.org, http://www.ebi.ac.uk/imgt/hla (IMGT/HLA Database), https://www.wmda.info/ (HSCT donor search).

Zachary, A. A., & Leffell, M. S. (2016). HLA mismatching strategies for solid organ transplantation – a balancing act. *Front Immunol*, *7*(7), 575.

CHAPTER 33

Red Blood Cells Products

Eldad A. Hod, MD

The majority of RBC products are made from 450 to 500 mL of whole blood donated into an anticoagulant–preservative solution; about 14% of RBC products are collected via automated RBC apheresis. Whole blood is centrifuged to pack the RBCs, the platelet-rich plasma is expressed, and an additive solution (AS) is added. This product can undergo a number of modifications, including leukoreduction, freezing, rejuvenation, washing, irradiation, and/or volume reduction. Additionally, there has been recent interest in using whole blood, which contains viable platelets, in trauma patients with life-threatening hemorrhagic shock.

In the United States, approximately 10.7 million units of whole blood and 1.7 million apheresis units are manufactured into the ~11 million RBC products that are transfused each year. Worldwide, it is estimated that ~100 million units of RBC products are transfused. General characteristics of RBC products include ~130–240 mL of RBCs, ~50–80 g of hemoglobin (Hgb), and ~150–250 mg of iron. The total volume, ~250–350 mL, and hematocrit, 55%–80%, vary depending on the anticoagulant–preservative solution used; higher volume and lower hematocrit are found in AS-containing products. Small amounts of plasma, platelets, and leukocytes remain in RBC products, unless the leukocytes and platelets (for most filters) have been removed by leukoreduction, which is typical in about 80% of the RBC products used in the United States. Canada and many European countries have a universally leukoreduced blood supply.

RBC Storage Lesion: The RBC storage lesion is the term that collectively refers to the biochemical and physical changes occurring to the RBCs and supernatant during storage. In general, biochemical changes include progressive increases in free Hgb, lactate, and potassium concentrations paralleled by gradual decreases in RBC adenosine triphosphate (ATP) and pH during storage. In addition, proinflammatory cytokines accumulate during storage of nonleukoreduced units but not in leukoreduced units. Within the first week of storage, 2,3-diphosphoglycerate (2,3-DPG) declines rapidly; however, it regenerates within 24 hours after transfusion. The decreased 2,3-DPG results in a shift in the oxygen dissociation curve to the left, which leads to less oxygen release than normal RBCs at the same partial pressure of O_2. Recent metabolomic studies reveal a number of other metabolites that change during storage. In addition to biochemical changes, RBCs change from a deformable biconcave disk, to reversibly deformed echinocytes, to irreversibly deformed spherechinocytes with increased membrane stiffness. These morphologic changes may also result in decreased oxygen transport owing to the inability of these RBC to flow through the microcirculation. On transfusion, storage-damaged RBCs are cleared from the circulation. Macrophage processing of these RBCs releases large amounts of iron into the circulation, consequences of which are currently being studied. However, large randomized trials comparing the standard of care with fresher

Transfusion Medicine and Hemostasis. https://doi.org/10.1016/B978-0-12-813726-0.00033-7

RBC products have not observed differences in clinical outcome, leading to the recommendation of AABB current practice guidelines that patients, including neonates, receive RBC units selected at any point within their licensed dating period (standard issue), rather than limiting patients to transfusion of only fresh products. Controversy still exists regarding the safety of particularly old RBC products (i.e., those stored for 35–42 days), as randomized trials have not examined the safety of these products.

Transfusion Guidelines

Indications: RBCs are transfused to mitigate the signs and symptoms of anemia, reflecting a significant deficiency in oxygen-carrying capacity and/or tissue hypoxia due to an inadequate circulating RBC mass (Table 33.1).

RBC Exchange: RBCs are used during RBC exchange (erythrocytapheresis) either therapeutically or prophylactically. RBC exchange can be required in hemolytic disease of the fetus and newborn for the prevention of kernicterus and in patients with sickle cell disease for the treatment of severe complications or prophylactically (Chapter 52).

Transfusion Trigger: Because there are no precise indicators of tissue hypoxia, nor objective, measurable indicators of symptomatic anemia, it has become common practice to administer RBC products based on laboratory parameters including Hgb. These numbers constitute "transfusion triggers," at which transfusion is generally considered appropriate and above which it is not. While the use of transfusion triggers is helpful in considering a patient's general condition, there is no universal transfusion trigger, and therefore, the clinical assessment of each patient is imperative so that unnecessary transfusion can be avoided in patients who have adapted well to their current level of anemia and so that transfusion is not withheld when needed.

TABLE 33.1 Indications and Contraindications for Transfusion of Red Blood Cells

Indications[a]
To mitigate tissue hypoxia due to decreased oxygen-carrying capacity associated with an inadequate RBC mass
As a source of replacement RBCs during RBC exchange (erythrocytapheresis)
Hgb < 7.0 g/dL in hemodynamically stable or burn patients
Hgb < 8.0 g/dL with orthopedic surgery or cardiac surgery and those with preexisting cardiovascular disease
Hgb < 9 g/dL in oncologic patients with septic shock[b]
Contraindications
To correct anemia due to iron deficiency
As a source of nutritional supplementation
For volume expansion or to increase oncotic pressure
To improve wound healing, recovery, or a sense of well-being

[a]These recommendations do not apply to acute coronary syndrome, severe thrombocytopenia, and chronic transfusion–dependent anemia.
[b]Requires confirmation in future trials.

Guidelines: As a general concept only, transfusion of RBC products in adults is rarely indicated when the Hgb concentration is >10 g/dL and is almost always considered to be indicated when the Hgb concentration is <6 g/dL.

Restrictive Versus Liberal Transfusion Strategies: There is some evidence that restrictive transfusion (vs. liberal) policies result in decreased patient complications after allogeneic RBC transfusion. In a number of studies, a restrictive transfusion strategy (i.e., not transfusing until the Hgb reaches 7–8 g/dL) does not increase morbidity and mortality and results in fewer transfusions and, thus, transfusion reactions, than a liberal transfusion strategy (i.e., transfusing to keep the Hgb above 9–10 g/dL). Current AABB clinical practice guidelines recommend a Hgb trigger of 7 g/dL for hospitalized adult patients who are hemodynamically stable, including critically ill patients. A restrictive RBC transfusion threshold of 8 g/dL is recommended for patients undergoing orthopedic surgery and cardiac surgery and those with preexisting cardiovascular disease, although a threshold of 7 g/dL is likely comparable, but there is a lack of evidence for these patient categories. Furthermore, owing to a lack of quality evidence, these recommendations do not apply to patients with acute coronary syndrome, severe thrombocytopenia (patients treated for hematological or oncological reasons who are at risk of bleeding), and chronic transfusion–dependent anemia.

Special RBC Products and Circumstances: Details on each of the special blood product processes and modifications are discussed elsewhere in this book (please see Chapters 42–48 for product modification and Chapters 49–59 for specific patient populations).

Administration: RBCs are administered only with a physician's order. A process for obtaining informed consent for RBC (and other blood product) administration should be in place. This process should allow individuals who will not accept RBC transfusion to document their refusal.

RBC products are transfused through tubing that contains a microaggregate filter capable of removing particulate matter >170 μm in size. In general, it is not necessary to prewarm the products, although some patients with cold agglutinin disease may require the use of blood warmers and patients who are transfused at rates >100 mL/minutes are at increased risk of cardiac events and hypothermia unless the blood is warmed.

ABO/D Compatibility: RBC products must be ABO-compatible with the recipient's plasma to avoid potentially immediate and life-threatening immune hemolysis. Ideally, crossmatched ABO-type–specific or ABO-type–compatible products (if type-specific products are not available) are issued (Table 21.3). In emergent circumstances, group O products can be issued without crossmatch or knowledge of the recipient's ABO type.

To minimize alloimmunization to the D antigen and thus limit the risk of future hemolytic disease of the fetus and newborn or hemolytic transfusion reactions, it is expected that D-compatible products will be administered. In emergent situations, where the D typing is unknown, D-negative products should be given to all recipients, but it is especially important in D-negative women of childbearing potential. In situations where the D-negative inventory is inadequate to meet the transfusion needs, transfusion of D-positive products to D-negative individuals should be directed by transfusion service policy (typically men and women older than 50 years to receive D-positive units) and medical director decision after weighing the risk of alloimmunization against the risk of withholding transfusion.

Quality Assurance: The requirements for RBC product preparation are dictated by the AABB *Standards* (Standard 5.7.5) and the CFR (Code of Federal Regulations). In the United States, current regulations focus primarily on the volume of blood collected, ensuring that at least 75% of erythrocytes are viable 24 hours after transfusion. As per the Council of Europe guidelines, the level of free Hgb should not exceed 0.8% of the RBC mass, while per the FDA guidelines, the level of free Hgb should not exceed 1.0% in 95% of units tested. It is the responsibility of the blood component laboratory to ensure that quality control is performed for all components (e.g., RBC product volume/hematocrit, white blood cell count for leukoreduced units). Failure to meet the product specifications should lead to investigation and correction as part of the overall quality assurance program.

Dose: In adult recipients, RBC products are typically given as one to two products per transfusion episode with a push to release only one unit at a time for optimal patient blood management. In an average-sized adult without ongoing bleeding or hemolysis, each RBC product can be expected to increase Hgb by 1 g/dL or hematocrit by ~3%. In children, dosing is often based on recipient weight, and a dose of 10–15 mL/kg is typical.

Adverse Events: Acute hemolytic transfusion reactions can result from inadvertent ABO-incompatible RBC product transfusion and may result in hypotension, disseminated intravascular coagulation (DIC), renal failure, and death. Alloimmunization to minor RBC antigens can also occur and may result in delayed hemolytic transfusion reactions on subsequent transfusion with antigen positive cells. Transmission of infectious diseases is possible. Further details about adverse events associated with blood transfusion can be found in later sections of this book.

Alternatives and Future Considerations: Alternatives to standard allogeneic RBCs include preoperative autologous donation, intraoperative and postoperative cell salvage, and acute normovolemic hemodilution. The administration of erythropoietin and other medications to optimize the patient's preoperative Hgb level can reduce the need for allogeneic RBC transfusion in some clinical settings. Discontinuation of anticoagulation and antiplatelet medications can decrease blood loss and thus need for transfusion. Pharmacologic agents and surgical techniques to minimize blood loss can also be helpful in minimizing RBC transfusion (Chapter 59).

Hgb-based oxygen carriers (HBOCs) are Hgb solutions in a number of formulations under investigation. Currently none of these are generally available for use in the United States; however, it is possible to obtain formulations on a compassionate use basis for care of patients in which RBC transfusions are deemed unacceptable. Pathogen-inactivated RBC products are also under investigation, although no process is currently in use for RBCs. In addition, enzymatic removal of the terminal sugars of the A and B antigen molecules to produce chemically modified, universal group O RBC products and manufacturing RBCs from induced pluripotent stem cells are under investigation.

International Considerations: In countries where the risk of transfusion-transmitted infectious diseases is high, strategies have evolved that are quite divergent from those in more developed countries. For example, in highly endemic areas for malaria, RBC transfusions are not typically initiated until the Hgb ≤ 3 g/dL owing to blood inventory limitations and the high prevalence of HIV in some donor populations.

In some countries, the volume of whole blood collection varies (e.g., 200–400 mL whole blood unit donation in Japan). Many countries outside of the United States store RBC products in an AS termed SAG-M, which contains saline (sodium chloride; for isotonicity), adenine (for ATP generation), glucose (for metabolism), and mannitol (for decreasing RBC lysis).

Further Reading

Bergamin, F. S., Almeida, J. P., Landoni, G., et al. (2017). Liberal versus restrictive transfusion strategy in critically ill oncologic patients: The transfusion requirements in critically ill oncologic patients randomized controlled trial. *Crit Care Med*, *45*, 766–773.

Carson, J. L., Guyatt, G., Heddle, et al. (2016). Clinical practice guidelines from the AABB: Red blood cell transfusion thresholds and storage. *JAMA*, *316*, 2025–2035.

Cooper, D. J., McQuilten, Z. K., Nichol, A., et al. (2017). Age of red cells for transfusion and outcomes in critically ill adults. *N Engl J Med*, *377*, 1858–1867.

Heddle, N. M., Cook, R. J., Arnold, D. M., et al. (2016). Effect of short-term vs. Long-Term blood storage on mortality after transfusion. *N Engl J Med*, *375*, 1937–1945.

Hess, J. R. (2012). Scientific problems in the regulation of red blood cell products. *Transfusion*, *52*, 1827–1835.

Lacroix, J., Hébert, P. C., Fergusson, D. A., et al. (2015). Age of transfused blood in critically ill adults. *N Engl J Med*, *372*, 1410–1418.

Mazer, C. D., Whitlock, R. P., Fergusson, D. A., et al. (2017). Restrictive or liberal red-cell transfusion for cardiac surgery. *N Engl J Med*, *377*, 2133–2144.

Palmieri, T. L., Holmes, J. H., 4th, Arnoldo, B., et al. (2017). Transfusion requirement in burn care evaluation (TRIBE): A multicenter randomized prospective trial of blood transfusion in major burn injury. *Ann Surg*, *266*, 595–602.

Simon, G. I., Craswell, A., Thom, O., & Fung, Y. L. (2017). Outcomes of restrictive versus liberal transfusion strategies in older adults from nine randomised controlled trials: A systematic review and meta-analysis. *Lancet Haematol*, *4*, e465–e474.

CHAPTER 34

Plasma Products

Joseph S.A. Restivo, DO and Matthew S. Karafin, MD, MS

Product Names: Plasma products in the United States include fresh frozen plasma (FFP), plasma frozen within 24 hours after phlebotomy (PF24 or FP24), plasma frozen within 24 hours after phlebotomy held at room temperature up to 24 hours after phlebotomy (PF24RT24), cryoprecipitate-reduced plasma (CRP, cryopoor plasma), thawed plasma liquid plasma (never frozen), solvent detergent (S/D) plasma, and pathogen-inactivated (PI) plasma. FFP, FP24, PF24RT24, and thawed plasma are all considered clinically interchangeable.

Description: Plasma is the acellular, fluid compartment of blood and consists of 90% water, 7% protein and colloids, and 2%–3% nutrients, crystalloids, hormones, and vitamins. The protein fraction contains all soluble clotting factors, including fibrinogen, factor XIII, von Willebrand factor (VWF), and factor VIII primarily bound to its carrier protein VWF, and vitamin K–dependent coagulation factors II, VII, IX, and X. Fibrinolytic proteins are also contained at normal physiologic concentrations. Plasma can be manufactured from whole blood after centrifugation and RBC removal or can be collected by apheresis. Primary indications for plasma transfusion are either general clotting factor replacement or specific factor replacement when purified or recombinant products are not available.

Plasma Products: FFP, FP24, PF24RT24, S/D plasma, PI plasma, and CRP are stored frozen at ≤−18°C for 1 year (except Octaplas [S/D plasma], which has 3 year shelf life). All require thawing, usually in a 37°C water bath for ~20 minutes; however, other thawing devices are FDA-approved. Thawed product expires in 24 hours. FFP, FP24, and PF24RT24 can be stored for 5 days postthaw at 1–6°C and relabeled as thawed plasma, and CRP can be stored for 5 days postthaw at 1–6°C and relabeled as thawed plasma CRP. FFP, FP24, PF24RT24, S/D plasma, and PI plasma have ~1 IU/mL of each clotting factor (S/D and PI plasma have lower amounts of some clotting factors) (Table 34.1).

FFP, FP24, PF24RT24, and Thawed Plasma: FFP is frozen at ≤−18°C within 8 hours of collection (or within 6 hours with the use of some storage bags after apheresis collection), while FP24 and RP24RT24 are frozen within 24 hours. FP24 is considered easier to manufacture than FFP owing to improved logistics.

CRP: CRP is a plasma product consisting of the supernatant expressed during the manufacture of cryoprecipitate from FFP. CRP is therefore deficient in factor VIII, VWF, fibrinogen, cryoglobulin, and fibronectin. CRP is refrozen and stored at ≤−18°C for up to 1 year. CRP is used only for transfusion or plasma exchange (TPE) in patients with thrombotic thrombocytopenic purpura (TTP).

Transfusion Medicine and Hemostasis. https://doi.org/10.1016/B978-0-12-813726-0.00034-9

TABLE 34.1 Comparison of the Different Plasma Products

	Manufacturing Conditions	Storage Conditions	Factor Activities (% Change vs. FFP)	Indications
FFP	Prepared from whole blood or apheresis collection and frozen at ≤−18°C within 8 hours of collection	Kept frozen at ≤−18°C for up to 1 year. May be held for 24 hours at 1–6°C after thawing	Normal factor activities (0.7–1 unit/mL) and fibrinogen (fib) (1–2 mg/mL)	Replacement of deficient or defective plasma proteins (see text)
FP24	Prepared from whole blood or apheresis collection and held at 1–6°C within 8 hours and ≤−18°C within 24 hours of collection	Kept frozen at ≤−18°C for up to 1 year. May be held for 24 hours at 1–6°C after thawing	F2 0%, F5 +1%, F7 −16%, F8 −15%, F9 +6%, F10 0%, VWF Ag +34%, VWF:Risto +22%, fib +29 mg/dL, ATIII 0%, PC −19%, and PS −5%[a]	May be used interchangeably with FFP
PF24RT24	Prepared from apheresis collections and held at room temp (20–24°C) up to 24 hours before freezing at ≤−18°C within 24 hours of collection	Kept frozen at ≤−18°C for up to 1 year. May be held for 24 hours at 1–6°C after thawing	Similar to FFP except F5 −1%, F8 −9%–13%, and PS −11%[a]	May be used interchangeably with FFP
S/D plasma	≤2500 pooled plasma products treated with tri-η-butyl phosphate and triton X-100 to inactivate lipid-enveloped viruses. Available in 200 mL containers	Kept frozen at ≤−18°C for up to 3 years. May be held for 12 hours at 2–4°C or for 3 hours at 20–25°C	fib −9%, F5 −35%, F8 −30%, F11 −14%, ADAMTS13 0%, PS −43%, antiplasmin −80%, and antitrypsin −50%[b]	Generally interchangeable with FFP; contraindicated in severe protein S deficiency
PI plasma	Amotosalen + UVA ([FDA-approved] frozen at ≤−18°C within 24 hours of WB or 8 hours of apheresis collection), riboflavin + UVA, or methylene blue	INTERCEPT kept frozen at ≤−18°C for up to 1 year. May be held for 24 hours at 1–6°C after thawing	INTERCEPT (apheresis-derived): fib −24%, F2 −12%, F5 −4%, F7 −21%, F8 −30%, F9 −21%, F10 −10%, F11 −15%, VWF:risto −1%, ADAMTS13%–8%, ATIII −6%, PC −14%, PS −5%, and antiplasmin −16%[b]	Generally interchangeable with FFP; contraindicated in patients with hypersensitivity to psoralens or in neonates treated with certain phototherapy devices

Continued

TABLE 34.1 Comparison of the Different Plasma Products—cont'd

	Manufacturing Conditions	Storage Conditions	Factor Activities (% Change vs. FFP)	Indications
CRP	The supernatant plasma remaining after cryoprecipitate manufacture refrozen at ≤−18°C within 24 hours	Kept frozen at ≤−18°C for up to 1 year. May be held for 24 hours at 1–6°C after thawing	Significantly reduced fib, F8, F13, VWF, cryoglobulin and fib; retains significant ADAMTS13	Use limited to transfusion or plasma exchange in TTP
LP	Prepared from WB; never frozen	1–6°C for ≤26 days (CPD/CP2D) or ≤35 days (CPDA-1)	At 14 days: fib +1%, F2 = −12%, F5 = −20%, F7 = −37%, F8 = −34%, F11 = −12%, F12 = −2%, PC = −4%, and PS = −45%[c]	Initial treatment of patients undergoing massive transfusion
TP or TPCR	After thaw and 1–6°C storage for 24 hours, FFP, FP24, or PF24RT may be relabeled as TP; Similarly CRP may be relabeled TPCR	1–6°C for up to 4 additional days after initial 24 hours' thaw period has elapsed	In TP made from FFP, F5 −21%, F7 −33%, and F8 −37% at 5 days. In TPCR, F5 −8%, F7 −12%, and F8 0% at 5 days from day 0 baseline[d]	TP is generally interchangeable with FFP; TPCR is contraindicated in patients with isolated factor/protein deficiencies for which TP or TPCR are relatively deficient

ATIII, antithrombin-III; *CPD*, citrate-phosphate-dextrose; *CP2D*, citrate-phosphate-dextrose-dextrose; *CPDA-1*, citrate-phosphate-dextrose-adenine; *CRP*, cryoprecipitate-reduced plasma; *FFP*, fresh frozen plasma; *FP24*, 24 hours after phlebotomy; *LP*, liquid plasma; *PC*, protein C; *PF24RT24*, plasma frozen within 24 hours after phlebotomy held at room temperature up to 24 hours after phlebotomy; *PI*, pathogen-inactivated; *PS*, protein S; *S/D*, solvent detergent; *TP*, thawed plasma; TPCR, thawed plasma cryoprecipitate reduced; *VWF*, von Willebrand factor.

[a]Percent change in mean clotting factor activities and fibrinogen concentration compared with mean of whole blood–derived plasma held at 4°C for 8 hours before frozen storage at −18°C for 1 month.

[b]Percent change in mean clotting factor activities and fibrinogen concentration compared with untreated FFP.

[c]Percent change in mean clotting factor activities and fibrinogen concentration at day 14 compared with day 0.

[d]Percent change in mean clotting factor activities at day 5 of thaw compared with that at day 0.

SD Plasma: SD plasma is FDA-approved (Octaplas, Octapharma USA Inc., Hoboken, NJ, USA; marketed as Octaplas LG in Europe and other markets) and is manufactured from ≤2500 pooled plasma products that have been treated with solvent (tri-η-butyl phosphate)/detergent (triton X-100) to inactivate lipid-enveloped viruses (HIV, hepatitis B, hepatitis C). These products are distributed in 200 mL containers.

PI Plasma: Other PI technologies are available including amotosalen plus UVA light–treated plasma (INTERCEPT plasma, Cerus Corp, Concord, CA), riboflavin plus UVA light–treated plasma (Mirasol Pathogen Reduction Technology, TerumoBCT, Denver, CO), and methylene blue–treated plasma (Therfex System, MacoPharma, Mouvaux, France). Studies in Europe show that these methods are also quite effective at reducing the risk of virus transmission. Only INTERCEPT plasma is FDA-approved (see Chapter 48).

Liquid Plasma: Liquid plasma is defined as plasma that is separated from whole blood at any time during storage at 1–6°C and up to 5 days after the whole blood expiration date. Depending on anticoagulant preservative used to collect the whole blood, expiration date for liquid plasma is 26 or 40 days. Clotting factor activity progressively decreases over time. Activities of some clotting factors and endogenous thrombin potentially fall below 50% after 15 days of refrigerated storage. Liquid plasma also contains viable lymphocytes that could result in transfusion-associated graft-versus-host disease (TA-GVHD); thus, irradiation may be necessary. This product also contains a small amount of RBCs, and sensitization may occur, particularly with RhD incompatibility. Liquid plasma is recommended only for initial trauma resuscitation in massive hemorrhage.

Recovered and Source Plasma (Plasma for Manufacture): Plasma products (liquid plasma and "plasma") that are derived from whole blood and intended for further manufacturing (i.e., fractionation into albumin, intravenous immunoglobulin, and factor concentrates) are sent to a manufacturer from a collection facility through a "short supply agreement." Another product, termed "plasma," is defined as liquid plasma that is frozen at −18°C or a colder temperature with a shelf life of 5 years. Source plasma, an FDA-licensed product, is collected by apheresis and is used only for further manufacturing.

Indications: Plasma transfusions are predominantly indicated for the treatment of bleeding in patients with congenital or acquired coagulation defects (Table 34.2). It is now well established that plasma should not be used as a volume expander or as a source of nutrients. Coagulation defects are routinely determined by measurement of prothrombin time (PT) or international normalized ratio (INR) and partial thromboplastin time (PTT). Plasma is typically indicated when PT and/or PTT are greater than 1.5–1.7 times the normal paired with the presence of bleeding or anticipated risk for significant bleeding. Evidence suggests that plasma transfusions are ineffective in correcting mild to moderate abnormalities. However, marked reductions in substantially elevated coagulation test results can occur with relatively modest transfusion volumes. Variable response to transfusion can be explained due to the nonlinear, exponential relationship between clotting factor activity levels and coagulation test results. There are no evidence-based studies that define a laboratory value that can serve as a transfusion "trigger," although many clinicians will consider transfusion when INR is ≥2.0 depending on presence, or risk, of bleeding.

Liver Disease: Patients with liver failure may develop low levels of the vitamin K-dependent clotting factors (factors II, VII, IX, and X) and prolonged PT/INR, PTT,

TABLE 34.2 Indications for Plasma Transfusion
Indicated
Multiple acquired coagulation factor deficiencies with bleeding or high risk for bleeding
Replacement of an inherited single plasma factor deficiency for which no coagulation factor concentrate exists or is not readily available (e.g., C1 esterase inhibitor)
Liver failure (restricted transfusion strategy)
Massive transfusion
Disseminated intravascular coagulation and bleeding or high risk for bleeding
Rapid reversal of warfarin (if PCC is not readily available)
Plasma infusion or exchange for thrombotic thrombocytic purpura, other thrombotic microangiopathies, diffuse alveolar hemorrhage, catastrophic antiphospholipid syndrome, and for managing clotting factor depletion

and thrombin time (TT). Fibrin split products may also be elevated, and in later stages, fibrinogen level may decrease. Patients undergoing orthotopic liver transplantation for severe liver disease and liver disease with disseminated intravascular coagulation (DIC) may require large volumes of plasma for bleeding control.

A growing body of evidence suggests that use of plasma does not significantly improve outcomes in the context of severe liver disease and perioperatively in liver transplant. Some authorities suggest restrictive plasma use in these settings, and prophylactic plasma product use is currently not recommended before a surgical challenge or liver biopsy in these patients. In the setting of cirrhosis, plasma transfusion may increase portal hypertension and thus risk of bleeding from varices. Consequently, plasma transfusion in patients with severe liver injury should be guided by a combination of clinical assessment and the degree of observed bleeding. (see Chapter 119).

Massive Transfusion: Massive transfusion is generally defined as receiving 10 or more units of RBCs within 24 hours (or one blood volume), and these patients may present or become coagulopathic for a number of reasons. Trauma patients may arrive at the hospital with a prolonged PT (termed acute coagulopathy of trauma [ACT]). ACT is associated with increased mortality and blood product use. Trauma patients can also develop a secondary coagulopathy, termed "lethal triad"—dilutional coagulopathy, acidosis, and hypothermia. Dilutional coagulopathy can develop secondary to administration of crystalloids and RBCs without coagulation factor support.

Early use of plasma and cryoprecipitate in trauma patients undergoing massive transfusion improves survival. Optimal ratio of plasma:RBC in trauma patients undergoing massive transfusion is likely between 1:1 and 1:2. The recently published PROPPR trial did not show a significant 30 day survival advantage for civilian trauma patients who received 1:1 versus 1:2. The need for plasma transfusion may also be determined based on laboratory triggers, such as PT > 1.5 times normal, ROTEM, thromboelastography (TEG), or other point-of-care test results (see Chapter 58).

Rapid Warfarin Reversal: Warfarin inhibits hepatic synthesis of vitamin K–dependent clotting factors by blocking recovery of the form of vitamin K that is active in the carboxylation of these proteins. Factor deficiencies reverse 48 hours after warfarin discontinuation if diet and vitamin K absorption are normal.

In the context of a bleeding patient on warfarin, use of four-factor prothrombin complex concentrate (PCC, Kcentra, CSL Behring) is now preferred to plasma transfusion, if available. In patients on warfarin who have active bleeding, require emergency surgery, or have serious trauma, the deficient clotting factors can be immediately provided by PCC or plasma transfusions. PCC reverses warfarin-induced coagulopathy faster and more effectively than plasma or vitamin K, and with lower volumes than plasma. Neither PCC nor plasma is indicated for nonurgent warfarin reversal, when the patient is not bleeding and has an INR < 9, as vitamin K administration effectively corrects coagulopathy in 12–18 hours (IV vitamin K decreases INR in <4 hours). INR levels need to be closely monitored and vitamin K can be administered should sustained warfarin reversal be desired (see Chapter 54).

DIC: DIC may be secondary to sepsis, liver disease, hypotension, surgery-associated hypoperfusion, trauma, obstetric complications, leukemia (usually promyelocytic), or underlying malignancy. Successful treatment of the underlying cause is paramount. Recent guidelines suggest that plasma should not be initiated based on abnormal laboratory results alone. Rather, patients with DIC and bleeding, patients requiring an invasive procedure where bleeding is expected, or patients at risk for significant bleeding complications should be transfused plasma in amounts sufficient to correct or ameliorate the coagulopathy (up to 30 mL/kg). However, in patients with severe liver disease, bleeding, and DIC, plasma infusions often fail to normalize the PT and PTT (see Chapter 124).

Plasma Exchange Replacement Fluid: Plasma can be used as replacement fluid for patients undergoing TPE either alone or in combination with saline and/or albumin. Patients with TTP require plasma to be used as the sole replacement fluid, as it provides the deficient enzyme, ADAMTS13 (see Chapter 107). For patients with other diagnoses, plasma may be added as a replacement fluid when the patient is actively bleeding secondary to an underlying medical condition, bleeding secondary to the dilutional effects of ongoing TPE, potential bleeding due to a procedure or surgery, or a concomitant coagulopathy exists. Examples include patients with diffuse alveolar hemorrhage or liver failure (see Chapter 75).

Congenital Coagulation Factor Deficiencies: There are rare factor deficiencies for which purified or recombinant plasma concentrates are not available in the United States, such as factors V or XI (available in Europe). Plasma may also be useful for patients deficient in other plasma-containing proteins, such as complement or C1 esterase inhibitor. In these deficiencies, the use of plasma may be indicated in situations of bleeding or for flares of hereditary angioedema when human concentrated C1 esterase inhibitor is not available.

Other Multiple Coagulation Defects: There are other circumstances that result in acquired coagulation defects, such as cardiac surgery or extracorporeal membrane oxygenation where plasma administration may be indicated.

Prophylactic Use: Randomized control trials and meta-analyses have evaluated the efficacy of prophylactic use of plasma products to reduce the risk of bleeding in patients undergoing central venous catheter insertion, lumbar puncture, angiography, liver biopsy, thoracocentesis, abdominal paracentesis, tracheostomy, dental surgery, and vascular surgery. Overall, these studies suggest that bleeding risk is unaltered with prophylactic correction of an abnormal PT or PTT.

Dosage: Plasma derived from whole blood is provided in doses of 200–280 mL (avg. 250 mL). Plasma apheresis collections can be up to 800 mL (termed "jumbo" plasma units) but is usually divided into 200 mL units. On average, there are 0.7–1 unit/mL of activity of each coagulation factor per mL of plasma and 1–2 mg/mL of fibrinogen. Appropriate dose may be estimated from plasma volume, desired increment of factor activity, and expected half-life of the factor being replaced (e.g., factor V has a half-life of 12–36 hours, and thus plasma doses should be repeated 1–2 times daily). Alternatively, plasma dosage may be estimated as 10–15 mL/kg, and with that dose all clotting factor activities will increase by about 30% in absence of rapid and ongoing consumption. Ideally, plasma should be ordered as number of mLs to be infused, but common practice in adults is to transfuse two to four plasma units. Administration frequency depends on clinical response (cessation or reduction of bleeding) or correction of laboratory parameters. Moreover, plasma infusion should be given near the time of a procedure, as this maximizes the hemostatic effect of the product. Lastly, it is important to note that the complete normalization of coagulation laboratory values will likely not be achieved with plasma infusion, and minimal elevations of PT/INR (e.g., INR < 1.5) are usually not corrected with plasma administration.

Compatibility: Plasma is screened for unexpected RBC alloantibodies at the time of collection and, with exception of group A plasma use in trauma, should be ABO-type–compatible for transfusion (Table 21.4). Tests for serologic compatibility, major and/or minor crossmatching, are not performed before administration.

Adverse Events: Plasma transfusion has similar risks as other blood products but is more commonly associated with allergic reactions, transfusion related acute lung injury (TRALI), and transfusion-associated circulatory overload (TACO) (see Chapters 62, 65, and 66).

Alternatives to Plasma Products: Recombinant factors VIII and IX are available to treat patients with hemophilia A and hemophilia B. Recombinant factor VIIa, FEIBA, or emicizumab-kxwh is available for the treatment of patients with factor VIII inhibitors. Plasma-derived factor concentrates that are available in the United States include fibrinogen, factors VII, VIII, IX, X, and XIII, antithrombin, C1 esterase inhibitor, protein C, and VWF, which mitigate need for plasma transfusion for replacement of these specific factors. Thus, in these circumstances, plasma infusion is contraindicated. Outside of the United States, factor XI concentrates are available and should be used in place of plasma product infusion (see Chapters 110–116).

Further Reading

Backholer, L., Green, L., Huish, S., et al. (2017). A paired comparison of thawed and liquid plasma. *Transfusion*, *57*, 881–889.

Chai-Adisaksopha, C., Hillis, C., Siegal, D. M., et al. (2016). Prothrombin complex concentrates versus fresh frozen plasma for warfarin reversal. A systematic review and meta-analysis. *Thromb Haemost*, *116*, 879–890.

Desborough, M., & Stanworth, S. (2012). Plasma transfusion for bedside, radiologically guided, and operating room invasive procedures. *Transfusion, 52*, 20S–29S.

Holbrook, A., Schulman, S., Witt, D. M., et al. (2012). Evidence-based management of anticoagulant therapy: Antithrombotic therapy and prevention of thrombosis, 9th ed: American college of chest physicians evidence-based clinical practice guidelines. *Chest, 141*, e152S–e184S.

McQuilten, Z. K., Crighton, G., Brunskill, S., et al. (2018). Optimal dose, timing and ratio of blood products in massive transfusion: results from a systematic review. *Transfus Med Rev, 32*(1), 6–15.

Neisser-Svae, A., Trawnicek, L., Heger, A., et al. (2016). Five-day stability of thawed plasma: solvent/detergent treated plasma comparable with fresh-frozen plasma and plasma frozen within 24 hours. *Transfusion, 56*, 404–409.

Roback, J. D., Caldwell, S., Carson, J., et al. (2010). Evidence-based practice guidelines for plasma transfusion. *Transfusion, 50*, 1227–1239.

Scott, E., Puca, K., Heraly, J., et al. (2009). Evaluation and comparison of coagulation factor activity in fresh-frozen plasma and 24-hour plasma at thaw and after 120 hours of 1–6°C storage. *Transfusion, 49*, 1584–1591.

Triulzi, D., Gottschall, J., Murphy, E., et al. (2015). A multicenter study of plasma use in the United States. *Transfusion, 55*, 1313–1319.

Yang, L., Stanworth, S., Hopewell, S., Doree, C., & Murphy, M. (2012). Is fresh frozen plasma clinically effective? An update of a systematic review of randomized control trials. *Transfusion, 52*, 1673–1686.

CHAPTER 35

Platelet Products

Melissa M. Cushing, MD and Robert A. DeSimone, MD

Product Names: Platelet products include those derived from whole blood and those collected by apheresis. While the FDA has a nomenclature specific to each method of collection, many terms are in common use. This creates confusion in published papers and with ordering physicians. The FDA calls platelets derived from whole blood "platelets," and these are sometimes also referred to as *whole blood–derived platelets, random donor platelets*, and *platelet concentrates*. Platelets collected by apheresis are called "platelets, pheresis" by the FDA, which are sometimes referred to as single donor platelets, apheresis platelets, and plateletpheresis. The method of collection does not reflect or define a platelet dose. Dose is patient-dependent and can vary given the clinical circumstance but usually approximates 3–4 × 10^{11} platelets for an adult and 10 mL/kg in pediatrics. As whole blood–derived platelet units typically contain 5.5 × 10^{10} platelets, 4–6 units must be "pooled" to make a dose. Many apheresis-derived platelet collections contain 2–3 times the required minimum of 3 × 10^{11} platelets and thus are "split" to make multiple platelet doses from a single collection.

Description: Platelets are an essential component of hemostasis, and deficiencies in platelet number or function can result in bleeding. Thrombocytopenia and/or platelet dysfunction may result from congenital diseases, medications, liver or kidney diseases, sepsis, disseminated intravascular coagulopathy (DIC), hematologic diseases, massive transfusion, and cardiac bypass or extracorporeal membrane oxygenation. Clinical signs of thrombocytopenia or platelet dysfunction include petechiae, easy bruising, or mucosal bleeding. The average in vivo life span of a platelet is ~10 days, but that of a transfused platelet is ~4–5 days. A platelet's life span is shortened by bleeding, DIC, splenomegaly, platelet antibodies, medications, sepsis, endothelial cell or platelet activation, and thrombocytopenia.

Indications: Platelet transfusions are used for prophylaxis to prevent bleeding or for treatment of bleeding in patients who have thrombocytopenia, qualitative defects in their platelet function (inherited or acquired secondary to disease or antiplatelet medications), or in the setting of massive transfusion. AABB guidelines recommend a prophylactic platelet transfusion threshold of 10,000/μL; a higher threshold may be considered in patients with fever, bleeding, or sepsis. A threshold of 10,000/μL was recently validated in a large retrospective analysis of thrombocytopenic hematology/oncology patients undergoing stem cell transplant or chemotherapy, in which a platelet count of <5000/μL was associated with increased bleeding. However, in the setting of autologous hematopoietic cell transplantation for adult patients, recent ASCO guidelines recommend platelet transfusions at the first sign of bleeding rather than prophylactic transfusions. The threshold is increased to >20,000/μL for central venous catheter

Transfusion Medicine and Hemostasis. https://doi.org/10.1016/B978-0-12-813726-0.00035-0

placement and >50,000/μL before lumbar puncture, biopsy, or nonneuraxial surgeries. For procedures involving neuraxial locations, such as the eye or brain, and for major surgery, a threshold of >100,000/μL is suggested. A threshold of >50,000/μL should be considered in actively bleeding patients. Thresholds for neonates are not clearly defined, and practices vary widely; for invasive procedures or bleeding, platelet counts are kept >50,000/μL and >100,000/μL in extremely ill, premature infants. Prophylactic platelet transfusions are generally administered for platelet counts <20,000/μL in neonates and <50,000/μL in extremely ill premature or critically ill neonates.

Relative Contraindications: Platelet transfusions are generally contraindicated in patients with thrombotic thrombocytopenic purpura (TTP) or heparin-induced thrombocytopenia (HIT) unless there is severe or life-threatening hemorrhage because they may increase the risk of thrombosis. In addition, a recent multicenter, randomized controlled trial associated platelet transfusions with greater mortality in patients with acute, spontaneous primary intracerebral hemorrhage on antiplatelet therapies. Thus, platelet transfusions may be contraindicated in this setting.

Whole Blood–Derived Platelets: In the United States using the platelet-rich plasma method (Chapter 9), per CFR and AABB Standards, whole blood–derived platelet units must contain $\geq 5.5 \times 10^{10}$ platelets in 90% of the units tested. Before issuing whole blood–derived platelets, the platelet products must be pooled to make a sufficient adult dose. Transfusion services may pool platelet concentrates, which require bacterial screening and expire in 4 hours postpooling. The FDA has approved a system for prepooling, leukoreducing, and bacterial testing of platelet concentrates. These platelets are referred to as "prestorage pooled platelets." Four to six units of platelets may be pooled using this system to achieve an FDA-approved dose of $2.2–5.8 \times 10^{11}$.

Apheresis: Apheresis platelets are collected into an ACD-A (citric acid, sodium citrate, dextrose) solution as either platelet-enriched plasma or as a platelet pellet that requires resuspension in concurrently collected plasma. Apheresis-derived platelets are collected as leukoreduced.

Buffy Coat–Prepared Platelets: Some countries outside of the United States use buffy coat–prepared platelets derived from whole blood. These buffy coat platelet concentrates can be pooled (prestorage) and stored in a single donor's plasma or platelet additive solution (PAS) (Chapter 9).

Storage: Platelets are stored at 20–24°C with continuous gentle agitation for up to 5 days or at 1–6°C without agitation for up to 3 days for transfusions in patients with active bleeding. Platelets may be stored in plasma or PAS. PAS contains only 35% of the plasma of a standard platelet unit. Platelets must be stored in oxygen-permeable containers because in anoxic conditions platelet metabolism shifts to the anaerobic glycolytic pathway leading to lactic acid production, acidosis, and platelet death. Adequate oxygenation allows aerobic mitochondrial oxidative phosphorylation and the maintenance of pH, as carbon dioxide diffuses as well as oxygen. With the introduction of rapid bacterial testing within 24 hours before issue, the dating of apheresis platelets may be extended up to 7 days (see Chapter 19).

Bacterial Testing: Room temperature storage and the plasma-rich, oxygenated environment of banked platelets may rarely lead to sufficient bacterial levels to cause fever, sepsis, shock, and death in recipients. Most cases have not been life-threatening. As many as 1 in 3000 plateletpheresis collections have evidence of bacterial contamination. Furthermore, late in the 5-day storage period, slow-growing bacteria such as gram-positive cocci may enter an exponential growth phase. To mitigate the risk associated with bacterial contamination, AABB created Standard 5.1.5.2 that requires methods to detect bacteria or use pathogen inactivation technology in all platelet products. Currently there are several methods to meet this Standard (see Chapters 19 and 48).

Pathogen-Reduced Versus Standard Platelets: A recent Cochrane review compared pathogen-reduced (both Mirasol and Intercept) platelets with standard platelets across 15 randomized controlled trials. There was no evidence of a difference in the incidence of clinically significant bleeding complications or serious adverse events. However, participants who received pathogen-reduced platelet transfusions had an increased risk of developing platelet refractoriness with lower 24-hour corrected count increments and required more platelet transfusions, with a mean difference of 1.2 transfusions.

Leukoreduction: Prestorage leukoreduction decreases febrile transfusion reactions by minimizing the level of cytokines released from white blood cells during storage and also reduces the risk of cytomegalovirus (CMV) transmission and human leukocyte antigen (HLA) alloimmunization. Most institutions use only prestorage leukoreduced platelet products (see Chapter 43).

Irradiation: Irradiation of platelet products prevents transfusion-associated graft-versus-host disease (TA-GVHD) (see Chapter 42). Expiration time of platelets does not need to be altered after irradiation.

Washing or Volume Reduction: Volume reduction or washing will remove antibodies contained within the plasma for which the recipient carries the corresponding antigen. Examples include maternal platelets transfused to a neonate with neonatal alloimmune thrombocytopenia (NAIT) or ABO-incompatible platelet products. Patients with recurrent severe allergic reactions may benefit from removal of plasma proteins via washing or volume reduction. Washing or volume reduction results in loss of 5%–30% of the platelets and may also compromise platelet function (see Chapters 46 and 47).

Aliquots: Platelet products are often dispensed in small aliquots for neonatal transfusion. If dispensed in a syringe, these products are acceptable for approximately 4 hours. Syringes do not allow for gas exchange and are usually aliquoted in an open, nonsterile environment.

Quality Control: Per AABB Standards, at least 90% of whole blood–derived platelets must contain $\geq 5.5 \times 10^{10}$ platelets and have a pH of ≥ 6.2 at the end of storage. Ninety-five percent of leukoreduced whole blood–derived platelet units must have $<8.3 \times 10^5$ residual leukocytes. Pooled platelet products must have $<5.0 \times 10^6$ white blood cells. At least 90% of apheresis-derived platelets must contain $\geq 3.0 \times 10^{11}$ platelets and have a pH of ≥ 6.2, and to be considered leukoreduced, 90% of the products tested must have $<5.0 \times 10^6$

white blood cells. Per the CFR, apheresis collections can be split into up to three products: 95% of the products must have $\geq 3.0 \times 10^{11}$ platelets, and 95% must have pH ≥ 6.2. For leukoreduction, 95% of single collections must have $<5.0 \times 10^6$ white blood cells, 95% of double collections must have $<8.0 \times 10^6$ white blood cells (and 95% of product must have $<5.0 \times 10^6$ white blood cells), and 95% of triple collections must have $<12 \times 10^6$ white blood cells (and 95% of products must have $<5.0 \times 10^6$ white blood cells).

Dose: In the setting of hypoproliferative thrombocytopenia, AABB guidelines recommend transfusing 3.0×10^{11} platelets. The PLADO trial randomized patients with hypoproliferative thrombocytopenia to low-dose, medium-dose, or high-dose platelet transfusions (1.1×10^{11}, 2.2×10^{11}, or 4.4×10^{11} platelets per square meter of body surface area, respectively) and found no difference in significant bleeding between the groups, with significantly fewer median platelets transfused in the low-dose group (9.25×10^{11}) relative to the medium-dose (11.25×10^{11}) and high-dose (19.63×10^{11}) groups. However, the low-dose group did require more median transfusions (5) relative to the medium and high-dose groups (3 each).

Product Selection

ABO Compatibility: In general, ABO group–specific platelet transfusions should be administered. As platelet supply is often limited owing to the short shelf life, it may be necessary to select out of group platelets for transfusion. As ABO antigens are present on the surface of platelets, lower recovery of ABO-incompatible platelets (major incompatibility) will be observed compared with compatible platelets (i.e., group A product transfused to a group O recipient versus a group O product to a group O recipient). This difference is not typically of clinical significance.

For minor incompatibility, ABO-incompatible plasma is present in the platelet product. As some group O platelet products have high-titer anti-A or anti-B, often of both IgG and IgM classes, a positive direct antiglobulin test and occasionally immediate hemolysis or rarely death can occur. It is thus recommended that the anti-A and anti-B titer of group O platelets be determined and only those with low titers be administered to group A or B patients.

Studies demonstrate conflicting results about the potential impact of ABO major incompatibility on hematopoietic stem cell transplant (HSCT) outcome. For neonates and infants, ABO-incompatible products should be avoided if possible. If not available, institutions may use PAS platelets or volume-reduced platelets or limit the amount of incompatible plasma per day.

D Compatibility: While not present on platelets, the D antigen is present on residual RBCs within the product. When present in sufficient dose, the D-positive RBCs can result in alloimmunization (anti-D formation) and future risk of hemolytic disease of the fetus and newborn (HDFN). However, this rarely occurs with the minimal volume of RBCs present in today's apheresis platelet products (approximately 0.00043 mL). In a recent study, only 7/485 (1.44%; 95% CI 0.58%–2.97%) recipients had a primary anti-D response after a median serological follow-up of 77 days (range: 28–2111 days). Given the low risk of anti-D formation, D-incompatible transfusions may be given. Hospital policies may include administering RhIg post D-positive platelet transfusion in some at risk patients.

Adverse Events: Platelet products, like other blood components, may result in adverse events that include infectious disease transmission and noninfectious hazards. Bacterial contamination is the most common infectious risk. The noninfectious hazards of transfusion include hemolytic transfusion reactions (usually from ABO-incompatible plasma within the product), allergic reactions, febrile nonhemolytic transfusion reactions (FNHTRs), transfusion-associated circulatory overload (TACO), and transfusion-related acute lung injury (TRALI). Hemolytic transfusion reactions can be mitigated by use of PAS units or units with low-titer anti-A or anti-B. Allergic reactions may be mitigated by the use of PAS units. FNHTRs are mitigated by prestorage leukoreduction but are still common. TACO is mitigated by slowing the rate of transfusion or by splitting units in at risk patients. Lastly, TRALI is mitigated by the use of male-only donors or by excluding donors with HLA or human neutrophil antigens (HNA) antibodies.

Further Reading

Baharoqlu, M. I., Cordonnier, C., Al-Shahi Salman, R., et al. (2016). Platelet transfusion versus standard care after acute stroke due to spontaneous cerebral haemorrhage associated with antiplatelet therapy (PATCH): A randomized, open-label, phase 3 trial. *Lancet, 387*, 2605–2613.

Cid, J., Lozano, M., Ziman, A., et al. (2015). Low frequency of anti-D alloimmunization following D+ platelet transfusion: The anti-D alloimmunization after D-incompatible platelet transfusions (ADAPT) study. *Br J Haematol, 168*, 598–603.

Dunbar, N. M., Katus, M. C., Freeman, C. M., & Szczepiorkowski, Z. M. (2015). Easier said than done: ABO compatibility and D matching in apheresis platelet transfusions. *Transfusion, 55*, 1882–1888.

Estcourt, L. J., Malouf, R., Hopewell, S., et al. (2017). Pathogen-reduced platelets for the prevention of bleeding. *Cochrane Database Syst Rev*. https://doi.org/10.1002/14651858.CD009072.pub3.

Goel, R., Ness, P. M., Takemoto, C. M., et al. (2015). Platelet transfusions in platelet consumptive disorders are associated with arterial thrombosis and in-hospital mortality. *Blood, 125*, 1470–1476.

Kaufman, R. M., Djulbegovic, B., Gernsheimer, T., et al. (2015). Platelet transfusion: A clinical practice guideline from the AABB. *Ann Intern Med, 162*, 205–213.

McQuilten, Z. K., Crighton, G., Brunskill, S., et al. (2018). Optimal dose, timing, and ratio of blood products in massive transfusion: Results from a systematic review. *Transfus Med Rev, 32*, 6–15.

Schiffer, C. A., Bohlke, K., Delaney, M., et al. (2018). Platelet transfusion for patients with cancer: American Society of Clinical Oncology clinical practice guideline update. *J Clin Oncol, 36*, 283–299.

Slichter, S. J., Kaufman, R. M., Assmann, S. F., et al. (2010). Dose of prophylactic platelet transfusions and prevention of hemorrhage. *N Engl J Med, 362*, 600–613.

Solves, P., Carpio, N., Balaguer, A., et al. (2015). Transfusion of ABO non-identical platelets does not influence the clinical outcome of patients undergoing autologous haematopoietic stem cell transplantation. *Blood Transfus, 13*, 411–416.

Uhl, L., Assmann, S. F., Haazma, T. H., et al. (2017). Laboratory predictors of bleeding and the effect of platelet and RBC transfusions on bleeding outcomes in the PLADO trial. *Blood, 130*, 1247–1258.

CHAPTER 36

Cryoprecipitate and Fibrinogen Concentrates

Bruce S. Sachais, MD, PhD and Eric Senaldi, MD

Product: Cryoprecipitated antihemophilic factor (thawed cryoprecipitated AHF; cryoprecipitated AHF, pooled; also called cryoprecipitate and cryo).

Description: Cryoprecipitate is made from human plasma. When fresh frozen plasma (FFP) is thawed in the cold (1–6°C), a precipitate forms (the cryoprecipitate), after which the supernatant (cryosupernatant, cryoprecipitate-poor or cryoprecipitate-reduced plasma) is removed and the plasma is refrozen. Its main constituents are fibrinogen, fibronectin, factor VIII, von Willebrand factor (VWF), and factor XIII. Human fibrinogen concentrates are derived from pooled plasma, purified fibrinogen products labeled with fibrinogen content, and virally inactivated.

Indications: Cryoprecipitate was historically used as a factor VIII replacement for hemophilia A patients, but now they are more purified and available as virally inactivated or recombinant products, making it essentially contraindicated to use this product for this reason in the developed world. Similarly, this product should not be used to treat von Willebrand disease or factor XIII deficiency, as more purified and virally inactivated products containing VWF and factor XIII, respectively, are available. Cryoprecipitate is now used primarily for fibrinogen replacement and in the manufacturing of fibrin sealants and glue (Table 36.1). Outside the United States, fibrinogen concentrates are used instead of cryoprecipitate for all fibrinogen replacement indications. Systematic meta-analysis has not shown any differences in mortality, fibrinogen level increase, bleeding, RBC transfusion, or thromboembolic complications to recommend one product over the other for fibrinogen replacement in acquired hypofibrinogenemia in the bleeding patient.

Fibrinogen Replacement: Hypofibrinogenemia occurs in patients with disseminated intravascular coagulopathy (DIC), with liver failure, after cardiac surgery, during the anhepatic phase of liver transplantation surgery, and during massive transfusion. In an actively bleeding patient or before surgery, fibrinogen product should be given when fibrinogen levels fall below 100 mg/dL, although more evidence and guidelines recommend a trigger of 150–200 mg/dL, as clot strength increases linearly with a minimum of 200 mg/dL required in vitro for optimal rate of clot formation. Early intervention can reduce the amount of red blood cells (RBCs) transfused. When the fibrinogen does fall below 100 mg/dL, there is a prolongation of the prothrombin test (PT) and activated partial thromboplastin time (aPTT) that cannot be corrected by the infusion of plasma products. Once the PT and aPTT are critically abnormal and with significant bleeding in the patient, it can be more important to intervene with the transfusion

Transfusion Medicine and Hemostasis. https://doi.org/10.1016/B978-0-12-813726-0.00036-2

TABLE 36.1 Primary and Secondary Indications, Common Misuses, and Underutilization of Cryoprecipitate

Primary Indications
Acquired/congenital hypofibrinogenemia and dysfibrinogenemia
Massive transfusion with bleeding
Postpartum hemorrhage
As a component of fibrin glue/sealants
Reversal of thrombolytic therapy with bleeding
Uremic coagulopathy
Secondary Indications – For underdeveloped world where recombinant or purified and virally inactivated products may not be available
Hemophilia A
von Willebrand disease
Factor XIII deficiency
Common Misuses
Replacement therapy in patients with normal fibrinogen measurements
Reversal of warfarin therapy
Treatment of bleeding without evidence of hypofibrinogenemia
Treatment of hepatic coagulopathy
Common Underutilization
Massive transfusion with dilutional coagulopathy and bleeding—trauma, cardiac surgery, postpartum hemorrhage

of a fibrinogen product than to await fibrinogen levels. In cardiac surgery, transplant surgery, and trauma, rather than wait for lab testing, the viscoelasticity of the clot is assessed periodically during the procedure to determine the appropriate needs for transfusion. Postoperative fibrinogen levels in cardiac surgery may be an independent predictor of severe bleeding postoperatively. Mathematical modeling would suggest a prophylactic trigger of 115 mg/dL with a target of 280 mg/dL. For those with severe bleeding already, a trigger of 215 mg/dL should be used with a target of 375 mg/dL. Fibrinogen products are also used for fibrinogen replacement in patients with congenital or acquired abnormalities in fibrinogen, such as afibrinogenemia, hypofibrinogenemia, or dysfibrinogenemia. It may also be used preprocedure in plasma exchange to keep the intraprocedure fibrinogen level >50 mg/dL and replenish fibrinogen lost in daily procedures with albumin replacement fluid.

Massive Transfusion: Massive transfusion, defined as the replacement of one blood volume with RBC units (i.e., 10 units in an adult), is often complicated by both a primary and secondary coagulopathy, resulting in thrombocytopenia, hypofibrinogenemia, and low coagulation factor levels. Fibrinogen is the earliest of the factors to be exhausted in the trauma coagulopathy. Low levels of fibrinogen are early predictors of mortality in trauma patients. Fibrinogen levels do not normalize during damage control resuscitation despite high ratios of plasma:RBCs. Therefore, fibrinogen product use should be incorporated into the treatment of massively transfused patients, either as part of a massive transfusion protocol or as replacement once fibrinogen levels reach a predefined threshold through measuring fibrinogen level or by thromboelastography (see Chapter 58).

Women with postpartum hemorrhage (PPH) and fibrinogen levels <200 mg/dL go on to develop severe hemorrhage. Women with levels ≥400 mg/dL rarely develop severe PPH. Given normally elevated fibrinogen levels in pregnancy and the progression to PPH, fibrinogen levels should be kept at or above 200 mg/dL.

Fibrin Glue/Sealant: Fibrin glue/sealant results from the mixture of a fibrinogen source (from plasma or heterologous/autologous cryoprecipitate) with a thrombin source (bovine, human, or recombinant). Fibrin glue is a non–FDA-approved thrombin/ preparation, and it has been widely used in Europe for many years. Fibrin sealants are FDA-approved alternatives to fibrin glue and have some advantages, such as standardization of production, over locally made fibrin glues, but are more expensive. Fibrin glues/ sealants can be used for multiple surgical purposes, including as topical hemostat, sealant, or adhesive. They are not a substitute to stop active arterial bleeding when a suture is required. Multiple fibrin or thrombin-containing products are FDA-approved for use.

The safety profile of each product differs depending on the product components and source. Bovine thrombin has been reported to cause anaphylaxis (due to bovine allergies), coagulopathy through formation of antibodies to factor V or II, and, rarely, death due to severe systemic hypotensive reactions. Consequently, bovine products have an FDA-mandated black box warning on their package inserts. Pooled human plasma sources have the potential risk of viral or prion disease transmission. It is recommended that patients be counseled about this risk although it is greatly reduced via screening donors, testing for viruses, and inactivating or removing viruses. Some human plasma products also contain synthetic aprotinin, which is a potential source of allergic reactions. Recombinant products, although eliminating the risk of infectious transmission or antibody formation, may cause allergic reactions owing to the hamster or snake proteins used to manufacture the product. Lastly, autologous fibrin glue preparations have been used; the infectious risks associated with the use of heterologous fibrin glue are eliminated by replacement with the autologous source.

Factor XIII Deficiency: Factor XIII deficiency is a rare autosomal recessive congenital deficiency. Factor XIII plays an important role in the cross-linking of polymerized fibrin. Patients present with bleeding and delayed wound healing usually first noted at the umbilical stump or after circumcision. They have normal PT and aPTT tests but increased clot solubility. The half-life of factor XIII is 9–15 days. Factor XIII concentrate is the preferred therapy, and if unavailable, cryoprecipitate can be used. Prophylactic replacement is standard of care with factor activity <1%. It may be considered with factor activity <4%–5% in severe bleeding phenotypes.

Bleeding Complications After Thrombolytic Therapy: Approximately 1% and 6% of patients, who receive thrombolytic therapy for either myocardial infarction or stroke, respectively, have intracranial hemorrhage. Cryoprecipitate, as well as antifibrinolytics and plasma, have been used in the algorithm to treat these life-threatening bleeds; cryoprecipitate is especially indicated if the fibrinogen is <100 mg/dL.

Uremic Bleeding: Cryoprecipitate has been reported to shorten the bleeding time in some uremic patients, and it has a variable hemostatic effect. Therefore, in addition to DDAVP (1-deamino-8-D-arginine-vasopressin), platelet transfusion, and dialysis, it may have an adjunctive role.

Processing and Storage: Cryoprecipitate is stored at ≤−18°C with volume of a single unit of cryoprecipitate being 10–15 mL (details of manufacturing are in Chapter 9). For dosing purposes, it is usually pooled postthaw, into a single bag of 10 pooled units for transfusion. Some blood centers supply cryoprecipitate in prepooled units from five donors. Cryoprecipitate must be used within 12 months of collection.

Fibrinogen and factor XIII concentrates are manufactured from pooled human plasma, virally inactivated and lyophilized, and stored at room temperature.

Preparation and Administration

Cryoprecipitate

ABO/D Compatibility: Because cryoprecipitate contains negligible amounts of RBCs and minimal isohemagglutinins, anti-A and/or anti-B, choosing units with ABO or D compatibility is not necessary for most adult and pediatric patients; neonatal transfusion guidelines often recommend the use of ABO-compatible cryoprecipitate.

Thawing and Pooling: Cryoprecipitate takes 10–15 minutes to thaw in a 30–37°C water bath and is usually pooled for easier administration, which takes an additional 10–15 minutes. Cryoprecipitate can be prepooled by the manufacturing blood center, which eases the process of thawing and administration, especially in times of emergency, such as trauma and liver transplantation. After thawing, the cryoprecipitate is maintained between 20 and 24°C.

Expiration: Once thawed, a single unit or a prepooled unit in a closed system of cryoprecipitate expires in 6 hours; pooled units in open system expire in 4 hours.

Fibrinogen and Factor XIII Concentrate: Lyophilized products are mixed with sterile liquid (water). Product is dissolved then aseptically intravenously administered.

Cryoprecipitate Quality Assurance: Per AABB *Standards*, each unit of cryoprecipitate must contain a minimum of 150 mg of fibrinogen, but usually it contains 250–350 mg. Per AABB and FDA standards, it must contain a minimum of 80 IU of factor VIII, but usually it contains 80–120 IU. In addition, cryoprecipitate contains 30–60 mg of fibronectin, 40–60 IU of factor XIII, and ~80 IU of VWF.

Dose

Fibrinogen: In adults, one unit of cryoprecipitate per 10 kg of body weight will increase the fibrinogen concentration by approximately 50 mg/dL. In children, one unit of cryoprecipitate per 5–10 kg of body weight will raise fibrinogen levels by approximately 60–100 mg/dL.

Most clinicians will order a single dose of a 10-unit pool of cryoprecipitate (an equivalent of one blood volume's worth) or 70 mg/kg × body weight (kg) of fibrinogen concentrate and remeasure the fibrinogen level.

Factor XIII: One unit of cryoprecipitate for every 10–20 kg given every 3–4 weeks prophylactically and higher doses more frequently in the setting of active bleeding is a common strategy, with laboratory measurements obtained as indicated. Postoperatively, factor levels should be kept above 5% until wound healing is complete. For factor XIII concentrate, see Chapter 41 for dosing.

Adverse Events: Adverse reactions include fever, chills, and allergic reactions. Large volumes of ABO-incompatible cryoprecipitate may cause a positive direct antiglobulin test and have rarely been reported to cause mild hemolytic transfusion reactions. Severe reactions for fibrinogen concentrate include thrombotic and anaphylactic events. There has been no evidence of increased thromboembolic events with excessive rise in fibrinogen levels. For recombinant Factor XIII, 10% of patients develop nonneutralizing antibodies, which did not lead to bleeding and eventually were undetectable.

International Standards: In the United Kingdom, cryoprecipitate is manufactured from a single unit of FFP by rapid freezing to <−30°C, then thawing slowly to 4°C. Current guidelines require that 75% of the units of cryoprecipitate must include at least 140 mg of fibrinogen and 70 IU/mL of factor VIII with a 24 month storage period.

Further Reading

Bolliger, D., Gorlinger, K., & Tanaka, K. (2010). Pathophysiology and treatment of coagulopathy in massive hemorrhage and hemodilution. *Anesthesiology*, *113*, 1205–1219.

Hedges, S. J., Dehoney, S. B., Hooper, J. S., Amanzadeh, J., & Busti, A. J. (2007). Evidence-based treatment recommendations for uremic bleeding. *Nat Clin Pract Nephrol*, *3*, 138–153.

Levy, J. H., Sziam, F., Tanaka, K. A., & Sniecienski, R. M. (2012). Fibrinogen and hemostasis: A primary hemostatic target for the management of acquired bleeding. *Anesth Analg*, *11*, 261–274.

CHAPTER 37
Granulocyte Products

Suzanne A. Arinsburg, DO

Granulocytes are the immune system's main cellular defense against bacterial and fungal infections. Granulocyte transfusions are used in patients with prolonged neutropenia or functional neutrophil disorders and severe bacterial and fungal infections. Granulocyte transfusion clinical efficacy is not established. However, granulocyte transfusions are still used in clinical management of neutropenic patients with bacterial or fungal infections that are refractory to standard antimicrobial therapy in some institutions.

Granulocyte products

- contain $>1 \times 10^{10}$ granulocytes, 10–30 mL of RBCs, and $1–6 \times 10^{10}$ platelets,
- must undergo crossmatch and emergency release procedure, as results of infectious disease testing may be unknown at time of administration,
- should be irradiated and cytomegalovirus (CMV)–matched to recipient, and
- should be transfused within 24 hours of collection.

Indications: Owing to lack of high-quality data, indications primarily come from case series.

Neonatal Sepsis: Neonates are at risk for bacterial and fungal infection because of diminished storage pool of neutrophils, which can be rapidly depleted; neonates also have a qualitative defect of neutrophil function. In 1992, a clinical trial involving 35 neonates with neutropenia and sepsis randomized to receive either intravenous immunoglobulin (IVIG) (n = 14) or granulocyte transfusion (n = 21) found significantly higher survival in granulocyte transfusion group (100% survival) versus IVIG group (64% survival). Subsequently, more recent studies with newer antimicrobials have not demonstrated improvement in mortality; however, granulocyte dose transfused is the most significant factor predicting favorable outcome. In 2011, the Cochrane Collaboration concluded based on four randomized controlled trials published that there is inconclusive evidence to support granulocyte use in neonatal sepsis.

Neutrophil Function Defects: Patients whose neutrophils have defects in adhesion and motility may develop cutaneous abscesses with common pathogens such as *Staphylococcus aureus,* mucous membrane lesions caused by *Candida albicans*, or sepsis. Disorders of phagocytic microbicidal activity such as chronic granulomatous disease (CGD) are also associated with cutaneous and deep-seated abscesses, lymphadenitis, pulmonary infections, and sepsis. Granulocyte transfusions in CGD are supported by observation that a small number of normal neutrophils may be able to compensate for the metabolic defect in CGD neutrophils. Numerous case reports document the potential benefit of granulocyte transfusions in this patient population.

Transfusion Medicine and Hemostasis. https://doi.org/10.1016/B978-0-12-813726-0.00037-4

Neutropenia and Sepsis: Two randomized multicenter trials failed to demonstrate difference between standard antimicrobial therapy alone versus with granulocyte transfusion; however, both studies had poor accrual. In the European trial, 74 patients were randomized to receive standard antimicrobial therapy either with or without granulocyte transfusions (G-CSF–stimulated donors); no 28-day survival benefit was demonstrated. In the US Resolving Infection on Neutropenia with Granulocytes (RING) trial, 114 patients were randomized to receive standard antimicrobial therapy alone or with granulocytes (G-CSF and dexamethasone–stimulated donors) and demonstrated no survival difference. Although, post hoc analysis comparing subjects receiving high-dose versus low-dose granulocytes did achieve statistical significance. Cochrane review concluded there is low-grade evidence that therapeutic granulocyte transfusions may not increase clinical resolution of an infection.

Processing and Storage

Collection: Granulocytes are collected using continuous flow centrifugation separation apheresis. Final granulocyte yield is dependent on donor white blood cell count and total blood volume processed.

Collection efficiency may be enhanced by the addition of pentastarch, hetastarch, and/or hydroxyethyl starch (HES), all of which increase density gradients in apheresis device, thus improving collection by facilitating separation of leukocytes from RBCs. Donor side effects, such as allergic reactions, edema, and weight gain, may occur with these agents.

Donor stimulation with G-CSF and/or corticosteroid administered 8–16 hours before product collection increases granulocyte product yield. Corticosteroids are not recommended for donors with a history of uncontrolled hypertension, diabetes mellitus, or peptic ulcer disease. Side effects to G-CSF include bone pain, headache, and myalgias, and side effects to corticosteroid stimulation include insomnia, fluid accumulation, and weight gain. A link between development of postcapsular cataracts and repeated use of corticosteroids in granulocyte donors was found in two small studies but not confirmed by two larger studies.

Because granulocytes are usually released for transfusion before receipt of results of donor infectious disease testing, prior testing results are used to decrease transfusion-transmitted infection risk, usually obtained from a previous donation (typically of platelets) within the past 30 days.

Storage: Granulocyte storage should be at 20–24°C without agitation. Granulocyte products have been stored in this manner for up to 8 hours without any reduction of their chemotactic or adhesion properties. Between 8 and 24 hours of storage, ability of transfused granulocytes to migrate to areas of inflammation may be decreased. *In vitro* studies have shown that neutrophils from G-CSF/dexamethasone-stimulated donors have prolonged life span, prolonged function, and changes in expression of genes encoding proteins integral to inflammatory and immune responses. Resultant functional changes, including increase in IL-8 release to Toll-like receptors, are thought to enhance neutrophil function and may benefit patients.

Patient Selection, Dose Preparation, Administration, and Toxicities

Patient Selection: Granulocyte transfusion recipients have neutropenia with absolute neutrophil count (ANC) < 500/μL or proven neutrophil dysfunction and fungemia, bacteremia, and/or proven or probable invasive tissue fungal or bacterial infection.

Dose: Adult dose of granulocytes should be $\geq 4 \times 10^{10}$ (approximately 5.7×10^{8} granulocytes/kg body weight), once daily, either until significant clinical improvement is noted or endogenous ANC is >500/μL, or 42 days. Neonatal granulocyte transfusion dose of 10–15 mL/kg body weight or $1–2 \times 10^{9}$ granulocytes/kg body weight. Typically, 5 days of products are ordered then clinical picture is reassessed, and additional granulocytes are ordered as needed.

Preparation: Granulocyte products are irradiated to prevent transfusion-associated graft-versus-host disease (TA-GVHD). CMV-seronegative granulocyte donors are required for CMV-seronegative recipients, as granulocyte products cannot be leukoreduced. Similarly, they may result in human leukocyte antigen (HLA) alloimmunization. HLA matching is difficult but may be considered in HLA alloimmunized patients. Because granulocytes contain RBCs, granulocytes should be ABO- and RhD-compatible with recipient's plasma and crossmatch should be performed if >2 mL of RBCs are present.

Administration: Granulocytes should be transfused through standard filters (not leukoreduction filters).

Toxicities: Fever and chills are common symptoms during granulocyte transfusions. Pulmonary complications, including decreased oxygen saturation and development or worsening of pulmonary infiltrates, may occur. More severe pulmonary side effects such as respiratory distress with severe hypoxia and hypotension are rare events.

Quality Assurance: Granulocyte products are not licensed by the FDA. AABB *Standards* require that at least 75% of products contain at least 1×10^{10} granulocytes.

International Issues: The UK National Blood Service guidelines for granulocyte transfusion are the same as described earlier but include a method for preparing granulocytes by pooling buffy coats for adults or children: Pool of 10 is usually used for adults; children <50 kg are dosed at 10 mL/kg. Both granulocytes collected by apheresis and pool of 10 buffy coats must contain greater than 5×10^{9} granulocytes.

Further Reading

Estcourt, L. J., Stanworth, S. J., Hopewell, S., et al. (2016). Granulocyte transfusions for treating infections in people with neutropenia or neutrophil dysfunction. *Cochrane Database Syst Rev*, *4*, CD005339.

Gea-Banacloche, J. (2017). Granulocyte transfusions: A concise review for practitioners. *Cytotherapy*, *19*(11), 1256–1269.

Pammi, M., & Brocklehurst, P. (2011). Granulocyte transfusions for neonates with confirmed or suspected sepsis and neutropenia. *Cochrane Database Syst Rev*, *10*, CD003956.

Price, T. H., Boeckh, M., Harrison, R. W., et al. (2015). Efficacy of transfusion with granulocytes from G-CSF/dexamethasone-treated donors in neutropenic patients with infection. *Blood*, *126*(18), 2153–2161.

CHAPTER 38
Albumin and Related Products

Anne M. Winkler, MD

Albumin as compared with nonprotein colloid or crystalloid solutions has not been well addressed in appropriately designed studies. Thus, albumin administration is based on an individual patient's clinical status. Clinical situations where albumin is commonly administered include replacement fluid for therapeutic plasma exchange (TPE), ovarian hyperstimulation syndrome (OHSS), cirrhosis with spontaneous bacterial peritonitis (SBP), large-volume paracentesis (LVP), nephrotic syndrome, and fluid resuscitation in critically ill patients (e.g., sepsis, acute burns). Alternatives to albumin for plasma expansion include crystalloids (e.g., 0.9% sodium chloride [normal saline], Ringer's lactate), alternate protein colloids (e.g., plasma protein fraction [PPF]), and nonprotein colloids (e.g., dextran, gelatin, and starches). Crystalloids and nonprotein colloids have not demonstrated a benefit over albumin but are less expensive; however, nonprotein colloids have been associated with adverse effects.

Albumin: Albumin, the most abundant protein (50%–60%) in human plasma, accounts for 80%–85% of the plasma osmotic pressure and therefore maintains and regulates plasma volume. As a result, albumin infusions draw fluid from the extravascular to intravascular space. Albumin also acts as carrier for other physiologic molecules and administered drugs, provides antioxidant properties, and serves as a buffering molecule with an acidifying effect.

Indications: Albumin was initially used as a treatment for shock during World War II, and its use has expanded; however, advantages and disadvantages of albumin have not been well addressed in appropriately designed studies. The Cochrane Collaboration has published systematic reviews of, and recommendations for, albumin use in specific clinical circumstances. Clinical situations where albumin is commonly administered are described below.

Therapeutic Plasma Exchange: Albumin is the primary replacement fluid used during TPE, except in clinical disorders that require specific factors, such as thrombotic thrombocytopenic purpura (Chapter 73).

Ovarian Hyperstimulation Syndrome: OHSS usually results from iatrogenic administration of human chorionic gonadotrophin (hCG) to induce ovulation. OHSS is typified by enlarged ovaries, which release vascular endothelial growth factor that can increase capillary permeability, leading to a fluid shift from the intravascular compartment to abdominal/pleural spaces. In the most severe form, OHSS is associated with tense ascites, oliguria, dyspnea, hemodynamic instability, and thromboembolism. Treatment includes fluid restriction, analgesics, and monitoring.

Transfusion Medicine and Hemostasis. https://doi.org/10.1016/B978-0-12-813726-0.00038-6

Mild OHSS occurs in ~33% and moderate to severe OHSS in ~5% of women receiving hCG. Increased risks include young age, low body weight, polycystic ovarian syndrome, high-dose hCG, high or rapid rise in estradiol level, and previous history of OHSS. In addition, the risk is proportional to the number of developing follicles and oocytes retrieved. Plasma expanders (albumin, hydroxyethyl starch [HES], and mannitol) reduce rates of moderate and severe OHSS in high-risk women; however, albumin may reduce pregnancy rates.

Cirrhosis With Spontaneous Bacterial Peritonitis: SBP is complicated by type 1 hepatorenal syndrome in 30% of patients. The pathophysiology may be secondary to increased intraperitoneal nitric oxide leading to systemic vasodilation and hypovolemia, resulting in renal vasoconstriction and renal failure. In a randomized trial, albumin infusion with antibiotics versus antibiotics alone decreased renal failure and improved mortality in cirrhotic patients with SBP. Current guidelines recommend albumin administration for all SBP patients at a dose of 1.5 g/kg on day 1 followed by 1 g/kg on day 3 (regimen used in trial).

Large-Volume Paracentesis: Albumin is often used in cirrhotic patients undergoing LVP (removal of >5 L) for diuretic refractory ascites to prevent paracentesis-induced circulatory dysfunction. A recent meta-analysis found insufficient evidence that albumin infusion after LVP lowered mortality in patients without hepatocellular carcinoma but with advanced liver disease. Nevertheless, guidelines currently recommend albumin administration.

Nephrotic Syndrome: Nephrotic syndrome occurs secondary to increased permeability of glomerular capillary basement membranes, leading to increased urinary protein excretion and resultant hypoalbuminemia, edema, renal failure, and hyperlipidemia. Primary treatment is mitigation of the underlying cause, diuretic therapy, and a sodium-restricted diet. Albumin combined with diuretics has been used to increase glomerular vascular pressure to increase diuresis; however, recent studies have shown no benefit or worse outcomes including hypertension, respiratory distress, and electrolyte abnormalities. Thus, albumin replacement in nephrotic patients is currently considered second-line therapy.

Sepsis: Currently, guidelines for fluid resuscitation in patients with sepsis, severe sepsis, and septic shock recommend albumin infusion when patients require substantial amounts of crystalloid; however, these recommendations are based on limited low-quality evidence. Randomized trials and meta-analyses have demonstrated no significant difference in mortality from albumin compared to other fluids.

Burns: Albumin use in resuscitation of acute burn patients remains controversial. A recent meta-analysis failed to demonstrate improved mortality compared with nonalbumin solutions.

Contraindications: Albumin may be contraindicated in any disease state that would be exacerbated by volume expansion, including severe anemia, congestive heart failure, and pulmonary edema. Albumin is also contraindicated in patients with previous anaphylactic reactions to albumin.

Adverse Effects: Adverse effects are rare but include vital sign changes (heart rate, blood pressure, and respiration), nausea, fever/chills, and allergic reactions. Furthermore, because negatively charged albumin binds calcium, administration can lead to hypocalcemia. Albumin solutions may also contain trace amounts of aluminum, which can cause toxicity in infants and chronic renal failure patients. Because albumin acts to increase osmotic pressure, rapid infusion can lead to significant shifts in intravascular volume, and resultant circulatory overload is possible. These shifts can also cause dilutional anemia and electrolyte imbalances.

Cost and Usage: Replacement fluid choice has considerable cost implications. Albumin is substantially more expensive than crystalloid and is often used in situations where randomized trials have shown no improvement in mortality. Albumin is inappropriately used in 50%–75% of cases. Most common inappropriate uses include intradialytic blood pressure support and hypoalbuminemia. Thus, ensuring appropriate use results in decreased cost.

Manufacturing: Albumin products are purified from human sources and must have 96% of protein composition consisting of albumin. The product also contains nonalbumin proteins (<4%), endotoxins, trace metals (e.g., aluminum), prekallikrein activator, bradykinin, sodium, potassium, and stabilizers (e.g., sodium caprylate and/or sodium acetyltryptophanate).

Pathogen inactivation includes heat treatment (e.g., 60°C for 10 hours) and cold ethanol fractionation. To date, no transmission of HIV, HCV, or HBV has been reported due to albumin (single bacterial contamination lot in the 1970s).

Storage: Albumin is stored at room temperature in either glass or specialized plastic containers for up to 2 years. Albumin solutions are inspected for turbidity by the manufacturer and before use to detect potential bacterial contamination.

Preparation and Administration: Albumin is usually provided in 5% and 25% concentrations; the former is slightly hyperoncotic to plasma and therefore may result in dilutional anemia, whereas the latter is significantly hyperoncotic and may result in dilutional anemia, pulmonary edema, and circulatory overload. The concentration used is determined based on the patient's clinical volume status. Typically, 5% expands volume equal to the volume of albumin infused, whereas 25% will expand volume 3.5 times. Thus, if 25% is used in a dehydrated patient, additional fluids are necessary to avoid exacerbating dehydration.

It is possible to dilute 25% albumin, but it must be diluted with normal saline. Dilution with sterile water can lead to hemolysis due to hypotonicity and has resulted in death. Smaller doses may be diluted with 5% dextrose in water (D5W), but large volumes of D5W-diluted albumin may lead to hyponatremia with resultant sequelae, including cerebral edema.

Dosing

Adult: The typical initial adult dose is 25 g, which may be repeated after 15–30 minutes. In a 48-hour period, a maximum of 250 g can be infused; however, no standard dose is applicable to all clinical situations, and clinical parameters must be used to determine appropriateness. Infusion rates vary depending on albumin concentration to prevent complications of rapid volume expansion; as a guideline, 5% solutions are started at 1–2 mL/min and increased to 4 mL/min, while 25% solutions are not infused at rates >1 mL/min. However, infusion rates may be increased in patients with hypoproteinemia.

Pediatric: Dosing depends on the clinical indication, but a typical 25% albumin dose for hypoproteinemia is 0.5–1.0 g/kg, which can be repeated every 1–2 days. For infants and children with hypovolemia, 0.5–1.0 g/kg per dose can be administered up to 6 g/kg per day. For neonates with hypovolemia, a typical dose is 0.25–0.5 g/kg for 5% albumin, and 25% should be avoided. Lastly, dosing for infants and children with nephrotic syndrome varies from 0.25-1.0 g/kg per dose of 25% albumin.

Other Colloid Solutions

Plasma Protein Fraction: PPF is seldom used because of an association with more frequent hypotensive and allergic reactions than albumin. PPF is derived from human plasma, comes as 5% solution, and contains ≥83% albumin and <1% gamma globulin. PPF was associated with a single outbreak of HBV secondary to failure of heating during manufacturing in 1973.

Dextrans: Dextran is a synthetic colloid consisting of mixture of glucose polymers derived from the action of *Leuconostoc mesenteroides* on sucrose and is currently available in 10% dextran 40 (40 kDa) and 6% dextran 70 (70 kDa). Dextrans have high water-binding capacity; for example, 1 g of dextran 40 retains 30 mL of water, while 1 g dextran 70 retains 20–25 mL of water. Dextran is mostly eliminated by the kidneys and should be avoided in patients with impaired renal function. Moreover, renal dysfunction has been reported after dextran infusion. Other side effects of dextrans include allergic reactions and coagulopathy. Dextrans can induce acquired von Willebrand syndrome by decreasing activity of von Willebrand factor and factor VIII and enhancing fibrinolysis, resulting in potential bleeding sequelae. Dextrans were historically used to maintain circulation in shock and reperfusion injury owing to their ability to reduce endothelial cell damage from activated leukocytes; however, dextran use is declining owing to its adverse effects.

Gelatins: Gelatins are synthetic colloids composed of polypeptides produced from degradation of bovine collagen. Gelatins are sterile, pyrogen-free, do not contain preservatives, and have a 3-year expiration date at room temperature. Molecular size varies, and high-molecular-weight products have greater oncotic effect and increased viscosity. However, blood volume increase is generally less than amount administered owing to passage of gelatins into interstitial space and rapid renal clearance. Thus, repeated infusions are necessary to maintain intravascular volume. Similar to dextrans, gelatins also affect hemostasis by interfering with platelet function. In addition, a recent systematic review and meta-analysis demonstrated increased risk of anaphylaxis, mortality, renal failure, and bleeding with gelatin administration. Thus, gelatins should be used with caution.

Hydroxyethyl Starch: HES is a class of synthetic colloids derived from amylopectin, a starch obtained from maize or potatoes, which is similar to glycogen. HES is available in multiple concentrations (3%, 6% and 10%) and molecular weights; it is important to recognize this because of effects on intravascular volume expansion and dose-dependent adverse effects on hemostasis and renal function. In general, HES solutions can retain 20–30 mL of water per g; however, 50% of HES is degraded and renally excreted within 24 hours. HES is used for volume expansion in the critically ill and for fluid resuscitation in trauma; however, HES appears to increase mortality and kidney injury in the critically ill and is no longer recommended by the United States Food and Drug Administration and European Medicines Agency. HES is used for granulocyte collections and in stem cell processing (Chapters 37 and 84).

Further Reading

Bernardi, M., Caraceni, P., & Navickis, R. J. (2017). Does the evidence support a survival benefit of albumin infusion in patients with cirrhosis undergoing large-volume paracentesis? *Expert Rev Gastroenterol Hepatol*, *11*, 191–192.

Chang, R., & Holcomb, J. B. (2016). Choice of fluid therapy in the initial management of sepsis, severe sepsis, and septic shock. *Shock*, *46*, 17–26.

Eljaiek, R., Heylbroeck, C., & Dubois, M. J. (2017). Albumin administration for fluid resuscitation in burn patients: A systemic review and meta-analysis. *Burns*, *43*, 17–24.

Guo, J. L., Zhang, D. D., Zhao, Y., et al. (2016). Pharmacologic interventions in preventing ovarian hyperstimulation syndrome: A systematic review and network meta-analysis. *Sci Rep*, *6*, 19093.

Kütting, F., Schubert, J., Franklin, J., et al. (2017). Insufficient evidence of benefit regarding mortality due to albumin substitution in HCC-free cirrhotic patients undergoing large volume paracentesis. *J Gastroenterol Hepatol*, *32*, 327–338.

Moeller, C., Fleischmann, C., & Thomas-Rueddel, D. (2016). How safe is gelatin? A systematic review and meta-analysis of gelatin-containing plasma expanders vs crystalloids and albumin. *J Crit Care*, *35*, 75–83.

Patel, A., Laffan, M. A., Waheed, U., & Brett, S. J. (2014). Randomized trials of human albumin for adults with sepsis: Systemic review and met-analysis with trial sequential analysis of all-cause mortality. *BMJ*, *349*, g4561.

Pericleous, M., Sarnowski, A., & Moore, A. (2016). The clinical management of abdominal ascites, spontaneous bacterial peritonitis and hepatorenal syndrome: A review of current guidelines and recommendations. *Eur J Gastroenterol Hepatol*, *28*, e10–e18.

Youssef, M. A., & Mourad, S. (2016). Volume expanders for the prevention of ovarian hyperstimulation syndrome. *Cochrane Database Syst Rev* (8), CD001302.

CHAPTER 39

Human Immunoglobulin Preparations

Joseph S.A. Restivo, DO and Matthew S. Karafin, MD, MS

Human Immunoglobulin Preparations

Description: Human immunoglobulins (Ig) are prepared from large pools of whole blood or apheresis-derived plasma. Ig preparations are concentrated (through Cohn fractionation), purified, filtered, and sterilized, making the risk of infectious disease transmission virtually zero (Figure 39.1). Human Ig preparations consist mostly of IgG, with half-life of 21–28 days. Ig preparations can be made for intramuscular (IM), subcutaneous (SC), or intravenous (IV) administration and clinically are used to replenish IgG in patients with hypogammaglobulinemia or for their immunomodulatory properties. Many preparations are available in the United States and throughout the world.

Mechanism of Action: For patients with acquired or congenital Ig deficiency (hypo- or agammaglobulinemia), these products supplement or replace the missing humoral components of their immune system. Diversity of antibody specificities contained in Ig preparations protects patients from increased susceptibility to infection by classical elimination of opsonized infectious organisms via antibody-dependent cell-mediated cytotoxicity or by complement activation. These processes are followed by lysis and/or neutralization of soluble infectious proteins by immunocomplex formation and elimination through the reticuloendothelial system.

In treatment of autoimmune disorders or other diseases associated with antibodies, Ig preparations result in immunomodulation and may alleviate the symptoms of the disease. Mechanisms of Ig-induced immunomodulation are incompletely understood but may include the following:

- Macrophage Fc receptor blockage by immune complexes formed between Ig and native antibodies
- Modulation of complement
- Inhibition of cell-mediated cytotoxicity
- Reduction of apoptosis
- Inhibition of leukocyte adhesion
- Regulation and inhibition of B cells and T cells
- Downregulation of dendritic cells and metalloproteinases
- Suppression of antibody production
- Suppression of inflammatory cytokines and chemokines
- Antiidiotypic regulation of autoreactive B lymphocytes or antibodies

Indications and Dose: There are seven FDA-approved indications for human Ig preparations and an expanding list of off-label uses, many of which are considered first-line therapy. Some of the FDA-approved indications and non–FDA-approved

Transfusion Medicine and Hemostasis. https://doi.org/10.1016/B978-0-12-813726-0.00039-8

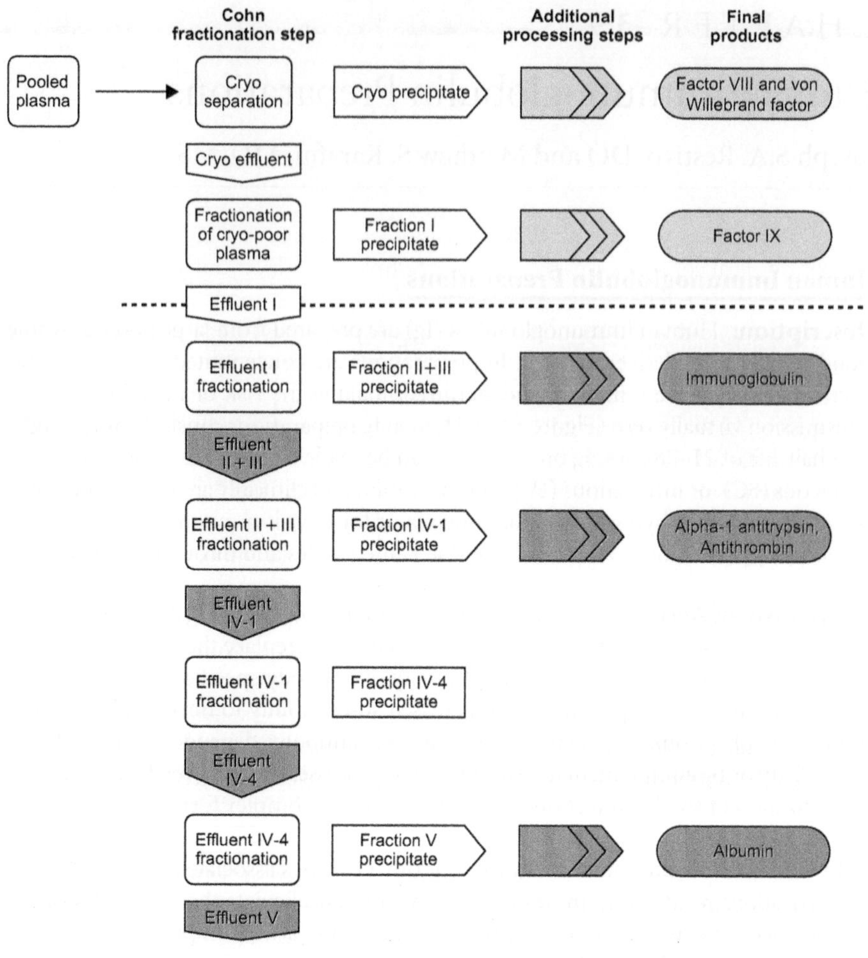

FIGURE 39.1 **Cohn plasma fractionation process.** Modified from Burdick, M. D., Pifat, D. Y., Petteway, S. R., & Cai K. (2006). Clearance of prions during plasma protein manufacture. *Transfus Med Rev, 20*, 57–62.

indications are described below where data support clinical benefit (Table 39.1). There are multiple different products, and not all products have been FDA-approved for each of the listed indications. Moreover, each disease indication has its own distinct recommended therapeutic dosages. Many clinicians use different products interchangeably, and doses are usually spread out over a period of days owing to slow infusion rates and to minimize adverse effects.

FDA-Approved Indications

B-Cell Chronic Leukemia/Lymphoma With Infection: The most common cause of morbidity and mortality in individuals with chronic leukemia/lymphoma (CLL) is infection. As up to 96% of patients with CLL have hypogammaglobulinemia in at least one Ig isotype, administration of immunoglobulin preparations may replace the missing immunoglobulins needed for humoral immunity. A multicenter, double-blind

TABLE 39.1 Some Clinical Indications for Ig Administration

FDA-Approved Indications	Non–FDA-Approved Indications of Proven or Probable Benefit
B-cell chronic leukemia/lymphoma with infection	Antibody-mediated kidney transplant rejection
Chronic inflammatory demyelinating polyradiculoneuropathy	Aplastic anemia secondary to parvovirus
Immune thrombocytopenia	Autoimmune hemolytic anemia
Kawasaki disease	Autoimmune uveitis
Multifocal motor neuropathy	Bullous pemphigoid
Primary/secondary immune deficiency	Birdshot retinochoroidopathy
Passive immunity for certain viruses (e.g., Hep A)	CMV pneumonitis in solid organ transplants
	Dermatomyositis or polymyositis
	Graft-versus-host disease
	Graves ophthalmopathy
	Guillain–Barre syndrome
	Henoch–Shonlein purpura
	Hematopoietic stem cell transplantation
	Hemolytic disease of the fetus/newborn
	Hypogammaglobulinemia associated with multiple myeloma
	IgM antimyelin-associated glycoprotein
	Immune neutropenia
	Lambert–Eaton myasthenic syndrome
	Myasthenia gravis
	Necrotizing fasciitis
	Neonatal alloimmune thrombocytopenia
	Neonatal sepsis
	Paraprotein-associated peripheral neuropathy
	Pemphigus vulgaris and pemphigus foliaceus
	Postinfectious thrombocytopenic purpura
	Relapsing–remitting multiple sclerosis
	Stiff person syndrome
	Toxic epidermal necrolysis and Stevens–Johnson syndrome
	Viral enterocolitis/meningoencephalitis

clinical trial found that those who received intravenous immunoglobulin (IVIG) 400 mg/kg every 3 weeks for 1 year had significantly fewer bacterial infections and experienced longer time to first major infection than those who did not receive this therapy. However, given more recent advances in CLL treatment and management, IVIG use should be restricted to those with recurrent serious bacterial infections and serum IgG levels <500 despite immunization to diphtheria, tetanus, or pneumococcal infection (typical dose 300–500 mg/kg monthly to maintain nadir IgG levels of 500 mg/dL).

Chronic Inflammatory Demyelinating Polyradiculoneuropathy: Chronic inflammatory demyelinating polyradiculoneuropathy (CIDP) is a chronic autoimmune disorder resulting in antibody-mediated demyelination of peripheral nerves that result in weakness and sensory changes. Equivalent outcomes have been observed in the treatment of CIDP with IVIG (typical loading dose 2000 mg/kg, then 1000–2000 mg every three to 4 weeks, with dose divided over 2–5 infusion days) (SC administration under investigation), therapeutic plasma exchange (TPE), or steroids. Decision as to which treatment to use is made on an individual basis, as up to one-third of patients do not respond to IVIG. Balancing risks and benefits of each treatment modality is an important consideration for this disease.

Immune Thrombocytopenia: Immune thrombocytopenia (ITP) is a bleeding disorder characterized by immune-mediated platelet destruction and resultant thrombocytopenia (see Chapters 101 and 102). The most effective pharmacologic therapies for acute ITP include corticosteroids, IVIG, and RhIg. RhIg and IVIG have rapid but temporary response. IVIG (typical dose 2000 mg/kg) is therefore indicated in acute bleeding episodes or before surgery, including splenectomy; in patients at high risk of intracranial hemorrhage; and in those in whose corticosteroids are contraindicated or ineffective. IVIG has also been used to treat ITP during pregnancy, postinfectious thrombocytopenia, ITP associated with HIV infection, and neonatal thrombocytopenia.

Kawasaki Disease: Kawasaki disease is an acute, self-limited childhood disorder manifested by fever, bilateral conjunctivitis, rash, and cervical lymphadenopathy. It is associated with systemic vasculitis, which results in coronary artery aneurysms in 15%–25% of untreated children. IVIG (typical dose 2000 mg/kg) given with aspirin and prednisolone (2 mg/kg) within 7 days of illness presentation reduces this risk from 23% to 4%.

Multifocal Motor Neuropathy: Multifocal motor neuropathy is a chronic progressive disorder resulting in primarily hand weakness. IVIG (SC administration under investigation), at a dose of 2000 mg/kg, is now considered a first-line treatment for this condition.

Primary or Secondary Immune Deficiencies: Patients with primary immunodeficiency syndromes have decreased levels of IgG and increased susceptibility to infections. Prophylactic administration of IVIG reduces number and duration of infections (typical maintenance dose for adults, 400–600 mg/kg every 3–4 weeks to maintain a trough IgG level of at least 500 mg/dL).

Patients with secondary immunodeficiency syndromes have acquired disorders of the immune system, which may be caused by multiple etiologies including hematologic malignancy (e.g., CLL, multiple myeloma, protein-losing enteropathy, nephrotic syndrome, or other severe illnesses). Criteria in support of IVIG use in these conditions include hypogammaglobulinemia (IgG < 200 or total Ig < 400 mg/dL), absent or low natural antibodies, absent or low response to antigenic challenge (i.e., vaccines), and lack of antibody response to the infecting organism.

Passive Immunity for Certain Viruses: Administration of specific formulations of Ig is approved to provide passive immunity for hepatitis A and measles (Rubeola) and can modify or prevent disease manifestations. Prophylactic value is greatest (80%–90% effective for hepatitis A) when given prophylactically before or soon after exposure (within 2 weeks for hepatitis A and within 6 days in unvaccinated person for measles). For hepatitis A, a dose of 0.02 mL/kg is recommended for recently exposed patients and those who will be in a hepatitis A endemic area for <3 months. For measles, a dose of 0.25 mL/kg is generally suggested.

Selected Off-Label Indications

Aplastic Anemia Secondary to Parvovirus: Parvovirus B19 infection can result in severe anemia and reticulocytopenia, especially in immunocompromised individuals or in individuals with sickle cell disease or thalassemia, and IVIG use is considered first-line therapy (typical dose 500 mg/kg weekly for 4 weeks).

Dermatomyositis and Polymyositis: Dermatomyositis is a chronic inflammatory disorder resulting in progressive weakness and rash. IVIG (typical dose 2000 mg/kg monthly) results in improved muscle strength and reduced neuromuscular symptoms.

Guillain–Barre Syndrome (Acute Inflammatory Demyelinating Polyneuropathy): Guillain–Barre syndrome is an acute demyelinating peripheral neuropathy affecting motor and sensory nerves. IVIG (typical dose 2000 mg/kg) is considered equivalent to TPE as a first-line therapy in improving disability and shortening the time to clinical improvement, especially when given within 2 weeks of symptom onset.

Hematopoietic Stem Cell Transplantation: Prophylactic use IVIG or cytomegalovirus (CMV) IVIG in CMV-negative hematopoietic stem cell transplantation (HSCT) recipients during first 100 days posttransplant has been demonstrated to reduce incidence of symptomatic CMV-associated disease, including CMV interstitial pneumonia, in some trials (typical dose 500 mg/kg monthly to maintain normal antibody levels). Because of the high cost of this treatment, and increasing and successful use of prophylactic ganciclovir and other therapies, IVIG is currently not indicated for the routine prophylaxis of HSCT recipients and is limited to allogeneic HCT recipients with IgG<400 mg/dL and with recurrent infections. Its role in preventing severe graft-versus-host disease (GVHD), however, remains less clear, as prolonged IVIG therapy during GVHD prevention may suppress humoral immune recovery. It might be useful in treating selected patients with chronic GVHD, recurrent serious bacterial infections and demonstrable defect in antibody production.

Hemolytic Disease of the Fetus and Newborn: Hemolytic disease of the fetus and newborn (HDFN) results from maternal RBC alloantibodies binding to fetal/neonatal RBCs, which may result in hemolysis, leading to anemia, hydrops fetalis, and possibly death (see Chapter 50). IVIG, usually in conjunction with TPE, has been used to prevent fetal loss before the availability of intrauterine transfusion. After delivery, IVIG has been suggested, but with conflicting results in randomized controlled trials, to significantly reduce the need for exchange transfusions in neonates with HDFN (IVIG treatment of HDFN has been associated with increased risk of necrotizing enterocolitis in the neonate). IVIG is currently recommended at a dose of 500–1000 mg/kg if there is established jaundice and rising total serum bilirubin despite phototherapy.

Incompatible Solid Organ Transplantations: Presence of high-titer reactive antibodies against incompatible graft HLA (human leukocyte antigen) and ABO antigens increases the risk of early humoral graft rejection and mortality, especially in kidney and cardiac transplants. For some patients who have these antibodies, to undergo transplantation, these antibodies must be removed or decreased. IVIG with (100 mg/kg after each TPE) or without TPE (2000 mg/kg each month for 4 months) and with or without rituximab has been shown to decrease sensitization of incompatible antigens in patients awaiting renal and cardiac transplantations. IVIG combined with rituximab and with or without TPE is also used in treatment of biopsy-proven antibody-mediated rejection (typical dose 100–400 mg/kg post TPE).

Lambert–Eaton Myasthenic Syndrome: Lambert–Eaton myasthenic syndrome results from antibodies to the neuromuscular junction, leading to autonomic dysfunction. One randomized control trial demonstrated that IVIG (typical starting dose of 2000 mg/kg) significantly improves generalized central and peripheral muscle strength and decreases serum calcium channel antibody titers for up to 8 weeks.

Multiple Sclerosis: Multiple sclerosis (MS) is a chronic progressive or relapsing and remitting disorder characterized by brain white matter demyelination. IVIG has been demonstrated to reduce the number of exacerbations in patients with relapsing–remitting MS. However, no studies to date have compared IVIG with standard therapies, and one clinical trial (PRIVIG trial) has raised doubt that IVIG is effective as a routine treatment. Consequently, IVIG is considered a second-line option for those patients who fail, decline, or are unable to tolerate standard immunomodulatory therapies such as β-interferon and glatiramer acetate (1000 mg/kg monthly for starting dose for relapsing–remitting MS). Additionally, IVIG may have a role in postpartum relapses, as immunomodulatory drugs are contraindicated in pregnancy and breastfeeding.

Myasthenia Gravis: Myasthenia gravis (MG) is a chronic neurologic autoimmune disorder characterized by weakness and fatigue on repetitive muscle use. IVIG (dose 2000 mg/kg) has been used successfully as a short-term measure for acute severe exacerbations with outcomes similar to TPE.

Neonatal Alloimmune Thrombocytopenia: Neonatal alloimmune thrombocytopenia (NAIT) results from maternal platelet alloantibodies against fetal/neonatal platelets, resulting in neonatal/fetal thrombocytopenia (see Chapter 94). Treatment of NAIT during pregnancy is maternal administration of 1000 mg/kg IVIG weekly beginning at 20–24 weeks of gestational age with or without the use of steroids. Moreover, the neonate may need to receive IVIG (dose 1000 mg/kg) and platelet transfusions after delivery to increase fetal platelet counts and prevent intracerebral hemorrhage.

Posttransfusion Purpura: Posttransfusion purpura (PTP) is a rare complication of transfusion resulting in acute, profound thrombocytopenia, secondary to platelet antibodies that destroy both transfused and autologous platelets (see Chapter 69). Available evidence suggests that IVIG should be a first-line therapy. PTP treatment with IVIG (dose 1000 mg/kg) can result in a rapid increase in platelet count.

Stiff Person Syndrome: Stiff person syndrome is a neurologic disorder associated with truncal and limb rigidity and heightened sensitivity. IVIG (starting dose 2000 mg/kg) is considered a second-line therapy for those who fail or cannot tolerate GABA-ergic medications.

Preparation and Administration

Production: Ig production is strictly regulated by IUIS/WHO (International Union of Immunological Societies/World Health Organization), and multiple requirements are necessary for appropriate IVIG production:

- Source material must be plasma obtained from a minimum pool of 10,000 donors.
- Product must be free of prekallikrein activator, kinins, plasmin, preservatives, or other potentially harmful contaminants.
- IgA content and IgG aggregate levels need to be as low as possible.
- Product must contain at least 90% intact IgG.
- IgG should maintain opsonin activity, complement binding, and other biological activities.
- IgG subclasses should be present in similar proportions to those in normal pooled plasma.
- Antibody levels against at least two species of bacteria (or toxins) and two viruses should be determined.
- Product must demonstrate at least 0.1 IU of hepatitis B antibody per mL and a hepatitis A radioimmunoassay titer of at least 1:1000.
- Manufacturer should specify the contents of the final product, including the diluent and other additives, and any chemical modification of the IgG.

Plasma Collection: IVIG is derived from pools of plasma collected either by whole blood donations as recovered plasma (20%) or by apheresis as source plasma (80%).

Processing: Manufacturers differ in the steps used to fractionate, purify, and stabilize Ig; methods used to inactivate and/or remove viruses; and formulation of the final product. Cold ethanol is commonly used for fractionation, and then the product is purified by filtration, chromatography, and/or precipitation. Viral inactivation is achieved by heat and chemical/enzymatic methods. To limit IgG aggregates, ion exchange chromatography, treatment with pepsin at a pH of 4, polyethylene glycol, and/or stabilizers such as sucrose, glucose, glycine, maltose, sorbitol, and/or albumin are used.

Product Selection: Products are available in liquid or lyophilized forms. Lyophilized forms can be reconstituted to a variety of different concentrations and osmolarities depending on amount, and choice, of liquid used (sterile water, 5% dextrose, or 0.9% saline) and depending on manufacturers' instructions. No other medications or fluids should be mixed with IVIG. Different Ig products have differences in concentration of additives, IgA content, osmolarity, osmolality, and pH. These factors should be considered based on the patient's clinical history (Table 39.2). For instance, IVIG with glucose should be used with care in diabetics. Additionally, antibody titers and biologic function of different products are not typically tested, even though there may be potentially clinically significant differences between products. Most hospital pharmacies stock a limited number of Ig products.

Administration: The FDA has approved IM, IV, and more recently SC Ig products. The IVIG infusion line used should be separate from other infusion lines. Infusion rate

TABLE 39.2 Variables to Consider When Choosing an Ig Product

Variable	Clinical Significance
Sucrose	The FDA issued a warning letter stating that the administration of sucrose-containing products may increase the risk of development of acute renal failure. Patients at increased risk include those with any degree of preexisting renal insufficiency, diabetes mellitus, age >65 years, volume deletion, sepsis, paraproteinemia, and concomitant nephrotoxic drugs.
Sorbitol	Patients with hereditary fructose intolerance who receive sorbitol- or fructose-containing products may develop irreversible multiorgan failure.
Glucose	Glucose-containing products should be used with caution in patients with diabetes or renal dysfunction and the elderly.
Glycine	Glycine-containing products are associated with increased frequency of vasomotor events.
Maltose	Some glucose monitors may interpret maltose as glucose and give falsely elevated results, which may result in iatrogenic insulin overdose.
Sodium	High-sodium products should be cautiously given to patients with heart failure or renal dysfunction, neonates, young children, the elderly, and those at risk for thromboembolism.
pH	Low-pH products should be administered cautiously to those with compromised acid–base compensatory mechanisms, such as neonates or those with renal dysfunction.
Osmolality and osmolarity	The osmolality and osmolarity should be considered in patients with heart disease or renal dysfunction, young children, the elderly, and those at risk for thromboembolism.
Volume	The volume should be considered when administering IVIG to volume-sensitive patients, including patients with renal dysfunction, heart disease, the elderly, neonates, and small children.

for those not previously exposed to IVIG should be low and can be increased gradually as defined by the specific IVIG product used (see manufacturers' instructions). Infusion rates in the elderly, patients at risk for renal dysfunction, or patients at risk for thrombosis should also be low. Vital signs should be monitored every 15 minutes for the first hour and then every 30–60 minutes. SC administration is now FDA-approved for use in primary immunodeficiency treatment. SC administration may be helpful when venous access is difficult or for home/self-administration. Also there is a lower incidence of adverse drug reactions when high doses are given SC.

Adverse Events: Adverse effects occur in 5%–15% of infusions and are more likely to occur in those receiving medication for first time, when increasing the dose, or when switching products. Most of the common adverse reactions, such as headache, nausea, vomiting, chills, fever, and malaise, appear related to the rate and/or dose of infusion (Table 39.3). Other adverse reactions include erythema, phlebitis, eczema, myalgias, flushing, rash, diaphoresis, pruritus, bronchospasm, chest pain, back pain, dizziness, and blood pressure changes.

Adverse reactions that are dose-related may be ameliorated by decreasing infusion rate or administering the total required dose over 2–5 days. Premedicating with IV

TABLE 39.3 Ig Administration

Common	Uncommon	Rare
Anxiety	Anaphylactic/anaphylactoid reactions	Anaphylaxis from anti-IgA
Arthralgia	Aseptic meningitis	Arthritis
Back pain	Chest pain/tightness	Cardiac rhythm abnormalities
Chills	Dyspnea	Hepatitis (noninfectious)
Change in blood pressure	Hemolysis	Hypothermia
Erythema	Pruritus/urticaria	Lymphocytic pleural effusion
Fever	Severe headaches	Neutropenia
Fluid overload	Thromboembolic events	Progressive neurodegeneration
Headache		Pseudohyponatremia
Infusion site pain/swelling		Skin findings (i.e., vasculitis, erythema multiforme)
Malaise/fatigue		Transfusion-related acute lung injury
Myalgia		Uveitis
Nausea/emesis		
Rash/flushing		
Tachycardia		
Tingling		

fluids, antipyretics, antihistamines, antiemetics, or steroids in some may help lessen the severity of adverse events. Additionally, adverse reactions differ among different preparations, such that patients may tolerate one product better than another product. Lastly, improved manufacturing processes render infectious transmission risk negligible. While rare, severe complications can occur. Key complications are reviewed below.

Anaphylactic Reactions: Individuals who are IgA-deficient and have anti-IgA may have anaphylactic reactions. These patients develop severe symptoms including hypotension, wheezing, and shortness of breath. These reactions require halting the infusion and providing epinephrine, antihistamines, steroids, fluids, and oxygen as the clinical situation requires. There are products available with low IgA levels (≤2.2 μg/mL) for use in IgA-deficient individuals.

Aseptic Meningitis: This event is characterized by severe headache, nuchal rigidity, drowsiness, fever, photophobia, painful eye movements, nausea, and vomiting beginning 6–48 hours after infusion. The etiology of this complication is unknown but may be due in part to osmotic changes within the brain due to the IVIG infusion. Patients with a history of migraine and who have received high-dose Ig treatment appear more susceptible. Certain IVIG brands are more prone to this complication. Cerebrospinal fluid demonstrates pleocytosis and elevated protein levels. Symptoms resolve in hours to days and may be prevented prophylactically with premedication of steroids and antimigraine medications, slowing the infusion rate, and/or dividing a dose over more days.

Hemolytic Transfusion Reactions: One study determined a 1.6% incidence of decreased hemoglobin after IVIG administration, especially in non–group O women in an inflammatory state receiving large IVIG doses. Proposed mechanism is from passively transfused anti-A and/or anti-B antibodies within the product.

TABLE 39.4 Hyper-Ig Preparations, Indications, and Dose

Hyper-Ig	Indication	Dose
Botulism Ig	Botulism in infants only	1 mL/kg (50 mg/kg) IV
Cytomegalovirus (CMV) Ig	In CMV-seronegative recipients of a solid organ transplant from a seropositive donor	150 mg/kg IV within 72 hours of transplant. Subsequent single IV doses of 100 mg/kg (renal) or 150 mg/kg (other solid organ) at weeks 2, 4, 6, and 8 posttransplant. Dose is decreased to 50 mg/kg IV (renal) and 100 mg/kg IV (other solid organ) for week 12 and 16 posttransplant
Hepatitis B Ig	Unvaccinated individuals at high risk of infection	0.06 mL/kg IM within 24 hours
	Infants born to hepatitis B–positive mothers	0.5 mL IM within 12 hours of birth
	Prevention of recurrent hepatitis B in liver transplantation recipients	20,000 IU IV concurrent with grafting of the transplanted liver, then 20,000 IU/day IV on days 1–7 postoperatively, then 20,000 IU IV every 2 weeks starting on day 14 postoperatively, then 20,000 IU IV every month starting on month 4 postoperatively; target serum anti-HBs concentration >500 IU/L; if the serum anti-HBs concentration is <500 IU/L within the first week of transplantation, increase the dose to 10,000 IU IV every 6 hours until the target is reached
Rabies Ig	Rabies exposure	20 IU/kg IM up to 7 days after the first vaccine dose
Tetanus Ig	Prophylaxis of tetanus infection in patients with traumatic injuries, other than a minor, clean wound, who have not been immunized within the last 5 years	250 U IM, or for children under 7 years of age either 4 U/kg or 250 U IM; with concomitant vaccination
	Tetanus infection	3000–6000 U IM within 24 hours of infection
Vaccinia Ig	Serious adverse reactions to smallpox vaccine	6000 U/kg IV as soon as symptoms appear. The dose may be repeated depending on response to treatment and severity of symptoms
Varicella-zoster virus Ig	Unvaccinated immunocompromised individuals or neonates who are exposed to varicella	125 U/10 kg IM (minimum dose of 125 U and maximum dose of 625 U within 96 hours of exposure)

Passively Acquired Antibodies: Patients who receive IVIG may passively acquire a variety of antibodies, including antibodies to hepatitis B core antigen (HBc) and anti-CMV, and therefore this may result in false-positive serologic testing. Testing can be repeated at a later time interval, or nonserologic methods can be used to determine presence of the infectious agent. Blood group antibodies may also be passively acquired, particularly anti-A and/or anti-B, resulting in a positive direct or indirect antiglobulin test.

Renal Failure: The FDA issued a warning in 1998 regarding the association of administration of sucrose-containing IVIG products and acute renal failure (Table 39.2).

Thromboembolic Events: IVIG has been associated with deep venous thrombosis, myocardial infarction, cerebrovascular accidents, transverse sinus thrombosis, and pulmonary embolism. These events are possibly related to the increase in blood viscosity after IVIG administration. Patients who received large doses rapidly, and patients with cardiovascular disease as well as elderly, overweight, or immobilized patients are thought to be at highest risk.

Hyperimmune Globulin Products (Hyper-Ig): Hyperimmune globulin (Hyper-Ig) products are manufactured from donors with high titers of the Ig specificity of interest. High titers can be achieved by natural immunity, prophylactic immunizations, or target immunizations. Hyper-Ig products should contain at least fivefold increased titers compared with standard preparations of IVIG according to IUIS/WHO.

Hyper-Ig products transfer a specific passive immunity for a variety of conditions, and are used to prevent the development of specific clinical diseases or alter its symptomatology. For instance, Hepatitis B Ig is used to provide passive immunity to hepatitis B virus associated with needle stick exposure or sexual contact with hepatitis B surface antigen (HBs)-positive individuals, post–liver transplantation for prevention of recurrence, and prevention of hepatitis B vertical transmission. Other proven effective hyperimmunoglobulins include botulism, CMV, rabies, tetanus, vaccinia, variola, and varicella-zoster virus immunoglobulins (Table 39.4).

Adverse Events: Local adverse reactions such as tenderness, pain, soreness, or stiffness of the muscle may occur at an IM injection site. Additionally, adverse events similar to those with IVIG can be associated with hyper-Ig use.

Further Reading

Alejandria, M. M., Lansang, M. A. D., Dans, L. F., & Mantaring, J. B., III. (2013). Intravenous immunoglobulin for treating sepsis, severe sepsis and septic shock. *Cochrane Database Syst Rev*, *16*(9), CD001090.

Eftimov, F., Winer, J. B., Vermeulen, M., et al. (2013). Intravenous immunoglobulin for chronic inflammatory demyelinating polyradiculoneuropathy. *Cochrane Database Syst Rev*, *30*(12), CD001797.

Gajdos, P., Chevret, S., & Toyka, K. (2012). Intravenous immunoglobulin for myasthenia gravis. *Cochrane Database Syst Rev*, *12*, CD002277.

Hughes, R. A., Swan, A. V., & van Doorn, P. A. (2014). Intravenous immunoglobulin for Guillain-Barré syndrome. *Cochrane Database Syst Rev*, *19*(9), CD002063.

Louis, D., More, K., Oberoi, S., et al. (2014). Intravenous immunoglobulin in isoimmune haemolytic disease of newborn: An updated systematic review and meta-analysis. *Arch Dis Child Fetal Neonatal Ed*, *99*, F325–F331.

Perez, E. E., Jordan, S., Orange, J. S., et al. (2017). Update on the use of immunoglobulin in human disease: A review of evidence. *J Allergy Clin Immunol*, *139*(3), S1–S46.

Shehata, N., Palda, V., Bowen, T., et al. (2010). The use of immunoglobulin therapy for patients with primary immune deficiency: An evidence based practice guideline. *Trans Med Rev*, *24*, S28–S50.

Shehata, N., Palda, V., Meyer, R., et al. (2010). The use of immunoglobulin therapy for patients undergoing solid organ transplantation: An evidence based practice guideline. *Trans Med Rev*, *24*, S7–S26.

Van Schaik, I. N., van den Berg, L. H., de Haan, R., et al. (2005). Intravenous immunoglobulin for multifocal motor neuropathy. *Cochrane Database Syst Rev*, *18*, CD004429.

CHAPTER 40

Rh Immune Globulin

Joseph S.A. Restivo, DO and Matthew S. Karafin, MD, MS

Rh immune globulin (RhIg) is a human plasma–derived product consisting of IgG antibodies to the D antigen. First licensed in 1968, it is used to prevent immunization to the D antigen in D-negative individuals and for the treatment of immune thrombocytopenia (ITP). Prevention of anti-D formation in females of childbearing potential is important because anti-D can cause severe, potentially fatal, hemolytic disease of the fetus and newborn (HDFN). Exposure to as little as 0.03 mL of D-positive erythrocytes can result in sensitization. Perinatal administration of RhIg has decreased the risk of forming anti-D in D-negative women carrying D-positive fetuses from approximately 13% to 0.1% and substantially reduced the risk of HDFN. RhIg can also be effective in treating ITP, a bleeding disorder characterized by immune-mediated platelet destruction, in D-positive patients with intact spleen.

Each dose of RhIg suppresses the immune response for up to a certain amount of D-positive red blood cells (RBCs) (Table 40.1). Measuring the amount of D-positive RBCs within the patient is imprecise, and therefore, the dose of RhIg recommended is greater than the exact dose calculated. Half-life of RhIg is ~24 days. Moreover, dosing can also vary slightly by manufacturer. Consequently, only general guidelines are provided in the section below.

The mechanism of action for preventing D immunization is likely due to the anti-D neutralizing the D antigen on RBCs. RhIg is used most commonly in the antenatal/postnatal prevention of anti-D formation in D-negative women and prevention of anti-D formation in D-negative recipients of D-positive RBC-containing products.

Preparation and Administration: RhIg is typically produced from D antigen–negative male donors who are repeatedly exposed to the D antigen. As described for intravenous immunoglobulin (IVIG), the pooled plasma undergoes cold alcohol fractionation, purification, filtration, and infectious disease reduction technologies. There are several RhIg preparations available. Those licensed only for intramuscular (IM) use contain small amounts of IgG aggregates, IgA, IgM, and other proteins that could potentially cause anaphylaxis if administered intravenously. Those products licensed for intravenous (IV) use are prepared by ion exchange chromatography and are purer than the IM preparations. IV injections are generally more comfortable for patients than IM injections if the patient has venous access. IV formulation should be used in ITP patients or other patients with thrombocytopenia to prevent IM bleeding.

Indications to Prevent Anti-D Formation

Perinatal Administration: If a pregnant woman is D-negative and is not sensitized to the D antigen, then she should receive RhIg perinatally to prevent anti-D formation and risk of HDFN in future pregnancies.

In Europe and the United States, serologic weak D variants (0.2%–1.0% prevalence) result in uncertain serologic testing for RhD. Of these women, 80% are classified as

Transfusion Medicine and Hemostasis. https://doi.org/10.1016/B978-0-12-813726-0.00040-4

TABLE 40.1 RhIg Vial Size and Amount of Whole Blood or RBCs Neutralized

Vial Size (μg)	IU	Whole Blood (mL)	RBCs (mL)
50	250	5	2.5
120	600	12	6
300	1500	30	15
1000	5000	100	50

weak D types 1, 2, or 3, who are not at risk of developing anti-D and do not require RhIg administration. Molecular genetic testing is available to classify the mother's weak or partial D type, to determine if RhIg administration can be avoided. Implementation of molecular testing in these women would save ~24,000 vials of RhIg annually and would likely result in cost savings.

Additionally, up to 40% of RhD-negative women may carry RhD-negative fetuses, in which case RhIg administration is unnecessary. Advances in cell-free fetal DNA testing allow for a noninvasive means to determine the baby's RhD status and permit more judicious RhIg prophylaxis strategies. A recent large Dutch study demonstrated 99% sensitivity and 95% specificity of a cell-free fetal DNA RhD assay performed at 27 weeks and 24% reduction in RhIg administration after incorporating the test into the Netherlands' RhIg prophylaxis strategy.

Antepartum and postpartum dosing is usually with 300 μg of RhIg. While some institutions prefer to give a vial of 50 μg at ≤12 weeks' gestational age, there is a risk of inadvertent misadministration of the lower dose. Outside of the United States, varying doses from 100 to 300 μg of RhIg are administered antepartum and postpartum.

RhIg is administered at the following stages:

- 28 weeks gestational age (dose 300 μg)
- At delivery, if the neonate is D-positive, weak D-positive, or D-untested (minimum dose of 300 μg; further dosing determined by fetomaternal hemorrhage [FMH] testing)
- After perinatal events that are associated with FMH (minimum dose of 300 μg; further dosing determined by FMH testing if >20 weeks' gestational age):
  - Therapeutic termination of pregnancy
  - Spontaneous abortion
  - Threatened abortions associated with bleeding
  - Invasive prenatal diagnostics: chorionic villus sampling, amniocentesis, cordocentesis
  - Invasive intrauterine procedures: Evacuation of the uterus due to molar pregnancy, multifetal reduction, fetal therapy (insertion of shunts)
  - Antepartum hemorrhage when bleeding continues intermittently, after 20 weeks' gestation
  - External version of the fetus
  - Abdominal trauma (any type)
  - Ectopic or molar pregnancy
  - Intrauterine fetal death
  - Stillbirth

TABLE 40.2 Calculation of RhIg Dosing

1. % fetal RBCs in maternal circulation is determined by Kleihauer–Betke test or flow cytometry. Maternal blood volume (mL) = 70 mL/kg × maternal weight (kg) or 5000 mL if maternal weight unknown
2. Fetal bleed (mL) = % fetal RBCs × maternal blood volume (mL)
3. Dose of RhIg = fetal bleed (mL)/30 mL per dose (300 μg vial)
4. If the number to the right of the decimal point is <5, round down and add one dose of RhIg (e.g., 2.3→2 + 1 = 3 vials).
   If the number to the right of the decimal point is ≤5, round up and add one dose of RhIg (e.g., 2.6→3 + 1 = 4 vials).

After a perinatal event or delivery beyond 20 weeks of gestational age, when the fetal blood volume exceeds 30 mL, quantification of FMH is recommended (see Chapter 50). If an FMH is detected, then the dose of RhIg should be calculated (Table 40.2). RhIg dose calculators are available.

RhIg should be administered within 72 hours of the event. If a dose is not administered during that time frame, then it should be administered as soon as possible up to 28 days after the event. Despite adequate doses of RhIg, 1–2 in 1000 women will still become D-immunized, and this failure is likely due to FMH before the first RhIg dose at 28 weeks' gestation.

D-Positive Blood Product Transfusion Into D-Negative Recipient: Individuals who receive D-positive RBC-containing products can be given RhIg to prevent D alloimmunization. In patients who do not receive RhIg, ~20% of D-negative patients exposed to D-positive RBCs owing to emergency transfusions form anti-D. Studies report a sensitization rate of <4% in patients exposed to D-incompatible platelet transfusions.

Although the risk is relatively low, reasons to prevent D sensitization are in females with childbearing potential to prevent future risk of HDFN. If an entire RBC product (250 mL RBCs) is administered, then multiple doses of RhIg can be administered, but there is a potential risk of hemolysis as a result of the RhIg. When multiple D-positive RBC products are transfused to a D-negative female with childbearing potential, a combination of RBC exchange and RhIg could be used to prevent D sensitization. A single-volume RBC exchange removes ~65% of D-positive RBCs.

Whole blood–derived platelets contain approximately 0.5 mL of RBCs, and apheresis platelet products contain <0.1 mL (usually much less). Given the small RBC volume in platelet products, a single 300 μg dose (one vial) of RhIg will prevent immunization for up to 30 whole blood–derived platelet and many apheresis platelet products. Granulocyte transfusions typically contain 10–30 mL of RBCs, and therefore to prevent anti-D formation, a dose of at least 600 μg is recommended (2 vials). Alloimmunization resulting from D-incompatible plasma has been rarely reported; thus, use of RhIg is not indicated.

Indications for Use in Immune Thrombocytopenia: ITP is a bleeding disorder characterized by immune-mediated platelet destruction and resultant thrombocytopenia. The most effective pharmacologic therapies for acute ITP include corticosteroids, IVIG, and RhIg. RhIg can be used to treat ITP in D-positive patients with intact spleens. It is thought to work by binding to D-positive RBCs, thus blocking receptor-mediated phagocytosis of platelets in the spleen; additional mechanisms include modulation of Fcγ receptor expression and immunomodulation.

For ITP, dosing is based on the patient's hemoglobin (Hgb): 50 µg/kg if the Hgb is ≥10 g/dL and 25–40 µg/kg when Hgb is 8–10 g/dL. The administration of RhIg results in a decrease in Hgb and therefore should be used with caution in patients with Hgb <8 g/dL. Platelet counts usually rise within 1–2 days and peak within 7–14 days. A second dose (25–60 µg/kg) may be given if there is no response to the first dose, which should be adjusted to the Hgb level (see Chapters 101 and 102).

Adverse Events: Adverse events to low-dose RhIg are rare. However, fever, chills, pain at the injection site, and rarely hypersensitivity (pruritus/rash) reactions have been reported.

Large IV doses of RhIg for the treatment of ITP result in mild hemolytic reactions (fevers, rigors, and headaches) in ~20% of infusions. More severe intravascular hemolytic reactions occur in ~0.7%, resulting in hemoglobinuria, pallor, hypotension, sinus tachycardia, oliguria, anuria, edema, dyspnea, ecchymosis, bleeding, and rarely death. The FDA has issued a "boxed warning" because of this risk. Transfusion related acute lung injury (TRALI) has also been reported.

The risk of disease transmission is limited by donor testing, fractionation, and pathogen inactivation and removal steps (such as ultrafiltration). Owing to improvements in viral reduction, no reported case of infectious disease transmission has occurred since the mid-1990s when hepatitis C was transmitted.

International Standards: There is no universal policy regarding the dosing of perinatal prophylactic RhIg, and dosing varies worldwide: 100 µg in the UK, 100–120 µg in Canada, and 200–250 µg in many European countries.

Further Reading

ACOG Practice Bulletin. (2017). Prevention of RhD alloimmunization. *Obstet Gynecol, 130*(2), 57–70.

Ayache, S., & Herman, J. H. (2008). Prevention of D sensitization after mismatched transfusions for blood components: toward optimal use of RhIG. *Transfusion, 48*, 1990–1999.

BCSH Guidelines. (2008). *Guideline for the use of prophylactic anti-D immunoglobulin.* London, England: British Committee for Standards in Haematology. http://www.bcshguidelines.org/pdf/anti-D_070606.pdf.

Crow, A. R., & Lazarus, A. H. (2008). The mechanisms of action of intravenous immunoglobulin and polyclonal anti-D immunoglobulin in the amelioration of immune thrombocytopenic purpura: what do we really know? *Transfus Med Rev, 22*, 103–116.

De Mattia, D., Del Vecchio, G. C., Russo, G., et al. (2010). Management of chronic childhood immune thrombocytopenic purpura: AIEOP consensus guidelines. *Acta Haematol, 123*, 96–109.

de Haas, M., Thurik, F. F., van der Ploeg, C. P. B., et al. (2016). Sensitivity of fetal RHD screening for safe guidance of targeted anti-D immunoglobulin prophylaxis: Prospective cohort study of a nationwide programme in the Netherlands. *BMJ, 355*, i5789.

Sandler, S. G., Flegel, W. A., Westhoff, C. M., et al. (2015). It's time to phase in RHD genotyping for patients with a serologic weak D phenotype. *Transfusion, 55*, 680–689.

Shaz, B. H., & Hillyer, C. D. (2011). Residual risk of D alloimmunization: Is it time to feel safe about platelets from D+ donors? *Transfusion, 51*, 1132–1135.

CHAPTER 41

Coagulation Factor Products

Scott T. Avecilla, MD, PhD

Patients with coagulation factor deficiencies were first treated with plasma products, but a wide variety of human-, animal-, and laboratory-derived products now exist for safer, more effective treatment. Each coagulation factor product differs in its composition, major indication, and dosage (Table 41.1A–D).

Historically, the major source of coagulation factors for treatment of congenital deficiencies was donated human plasma or its derivative cryoprecipitated antihemophilic factor (cryoprecipitate). Next, the development of plasma fractionation and factor purification became available. In the 1980s, HIV, hepatitis B, and hepatitis C were discovered and resulted in a large proportion of transfusion infection transmission in patients with congenital coagulation factor deficiencies. This led to improved testing, processing, and the development of recombinant products with no human proteins.

Coagulation Factor Concentrate Production: (1) Plasma is collected either through apheresis as source plasma or whole blood as recovered plasma from donors who have undergone a rigorous donor health questionnaire and infectious disease screening similar to which is performed for allogeneic blood donation. (2) Large numbers of individual donor plasma units are pooled, concentrated, and purified to various levels with several processes, which include protein precipitation, chromatographic separation methods, and antibody-antigen capture columns. (3) The concentrated coagulation factors are treated to substantially reduce pathogen-infectious risk with methods such as heat inactivation, organic solvent treatment, in addition to nanofiltration.

Additionally, the pooled, concentrated, and virally inactivated product is tested with nucleic acid amplification tests to many pathogens including HIV, hepatitis C, hepatitis A, hepatitis B, HTLV-I/II, WNV, and parvovirus. While a substantially safer product than those previously offered has resulted, with no case of transfusion-transmitted disease being reported since the mid-1990s, patients still exhibit a high level of apprehension with receiving human-derived factor concentrates, which has led to the development of recombinant products without human proteins.

With the advent of genetic and cellular engineering technologies, it is possible to isolate the genes of specific coagulation factors, such as factor VIII, and to artificially produce properly functioning protein from cell culture, thereby eliminating the infection transmission risk. Another potential benefit of artificially producing coagulation factors was the freedom from reliance on blood donation, thereby ensuring a more predictable supply. While the safety is perceived to be much higher, the trade-off of laboratory-derived products is a substantially higher price. Enhanced purity has also contributed to an increased ability to characterize the biological effect and make it a more uniform product with less lot-to-lot variability than with human-derived factors.

Transfusion Medicine and Hemostasis. https://doi.org/10.1016/B978-0-12-813726-0.00041-6

TABLE 41.1A Coagulation Factor Product Information

Product	Contains	Major Indications	Dosages
Factor VIII $T_{1/2}$ 10–12 hours	Recombinant human factor VIII Advate ($T_{1/2}$ 12 hours) Helixate FS ($T_{1/2}$ 14 hours) Kogenate FS ($T_{1/2}$ 14 hours) Kovaltry ($T_{1/2}$ 14 hours) Novoeight ($T_{1/2}$ 11–12 hours) Nuwiq ($T_{1/2}$ 17 hours) Recombinate ($T_{1/2}$ 14 hours) Xyntha ($T_{1/2}$ 11–12 hours)	*Hemophilia A* Control and prevention of bleeding episodes Perioperative management	Dose (IU) = body weight (kg) × Desired factor VIII rise (IU/dL or % of normal) × 0.5 (IU/kg per IU/dL) The dosage and duration depend on the severity of factor VIII deficiency, the location and extent of the bleeding, and the patient's clinical condition Careful control of replacement therapy is especially important in cases of major surgery or life-threatening bleeding episodes
	Recombinant human factor VIII, long-acting Adynovate ($T_{1/2}$ 14.7 hours) Afstyla ($T_{1/2}$ 14.3 hours) Eloctate ($T_{1/2}$ 19.7 hours)		Target / Maintenance Minor bleed: 20–40 IU/dL / 10–20 IU/kg Mod. bleed: 30–60 IU/dL / 15–30 IU/kg Major bleed: 60–100 IU/dL / 30–50 IU/kg Minor surgery: 60–100 IU/dL / 30–50 IU/kg Major surgery: 80–120 IU/dL / 40–60 IU/kg
	Plasma-derived human factor VIII Hemofil M ($T_{1/2}$ 14.8 hours) Koate-DVI ($T_{1/2}$ 16.1 hours) Monoclate P ($T_{1/2}$ 17.5 hours)		Repeat infusions every 12–24 hours (8–24 hours for patients under the age of 6) for 3 days or more until the bleeding episode is resolved (as indicated by relief of pain) or healing is achieved
	Recombinant human factor VIII (porcine sequence) Obizur ($T_{1/2}$ 2–17 hours)	*Acquired hemophilia A* Treatment of bleeding episodes in adults	Initial dosing of 200 U/kg followed by additional dosing based on factor recovery levels and clinical response

Factor IX	Recombinant human factor XI	*Hemophilia B*			
$T_{1/2}$ 20–24 hours	BeneFIX ($T_{1/2}$ 18.1 hours)	Control and prevention of bleeding episodes	Dose (IU) = body weight (kg) × desired factor VIII rise (IU/dL or % of normal) × 1.0 (IU/kg per IU/dL)		
	Ixinity ($T_{1/2}$ 24 hours)				
	Rixubis ($T_{1/2}$ 26.7 hours)	Perioperative management			
	Recombinant human factor IX, long-acting			Target	Maintenance
			Minor bleed	20–30 IU/dL	10–15 IU/kg
	Alprolix ($T_{1/2}$ 86.5 hours)		Mod. bleed	25–50 IU/dL	15–30 IU/kg
	Idelvion ($T_{1/2}$ 104 hours)		Major bleed/surgery	50–100 IU/dL	30–50 IU/kg
	Plasma-derived human factor XI		Repeat infusions every 12–24 hours for: 1–2 days (minor); 2–7 days (moderate); 7–10 days (major)		
	AlphaNine SD ($T_{1/2}$ 21 hours)				
	Mononine ($T_{1/2}$ 25.3 hours)				
	Plasma-derived human PCC	*Hemophilia B*			
	Profilnine; activity ratio factor II 150, factor VII 35, factor IX 100, factor X 100), no heparin in product	Prevention and control of bleeding Not indicated for use in the treatment of factor VII deficiency	1.0 IU/kg × body weight (kg) × desired increase factor IX (% of normal); mild to moderate bleed may usually be treated with a dose sufficient to raise the plasma factor IX level to 20%–30%; major hemorrhage, factor IX level should be raised to 30%–50%.		
	Bebulin VH; activity ratio factor II 120, factor VII 13, factor IX 100, factor X 139), small amounts of heparin (≤0.15 IU heparin per IU factor IX)	*Hemophilia B* Prevention and control of bleeding	1.2 IU/kg × body weight (kg) × desired increase factor IX (% of normal)		
		Not indicated for use in the treatment of factor VII deficiency		Target	Initial Dose
			Minor bleed	20 IU/dL	25–35 IU/kg, 1 day
			Mod. bleed	40 IU/dL	40–55 IU/kg, 2 days
			Major bleed	>60 IU/dL	60–70 IU/kg, 2–3 days
			Minor surgery	40–60 IU/dL	50–60 IU/kg
			Major surgery	>60 IU/dL	70–95 IU/kg

IU, international units; information derived from current package inserts of respective products.

TABLE 41.1B Coagulation Factor Product Information

Product	Contains	Major Indications	Dosages
Factor VII $T_{1/2}$ 5 hours	Recombinant human factor VII, activated NovoSeven RT ($T_{1/2}$ 2.3–3.9 hours)	***Hemophilia A or B with inhibitors, acquired hemophilia, congenital factor VII deficiency*** Treatment of bleeding episodes	Hemophilia A or B with inhibitors: 90 mcg/kg given every 2 hours by bolus infusion until hemostasis is achieved, or until the treatment has been judged to be inadequate; for severe bleeds, dosing should continue at 3–6 hour intervals after hemostasis is achieved, to maintain the hemostatic plug Doses between 35 and 120 mcg/kg have been used successfully in clinical trials, and both the dose and administration interval may be adjusted based on the severity of the bleeding and degree of hemostasis achieved Acquired hemophilia: 70–90 mcg/kg repeated every 2–3 hours until hemostasis is achieved
		Prevention of bleeding in surgical interventions or invasive procedures	90 mcg/kg body weight given immediately before surgery and repeated at 2-hour intervals for the duration of the surgery • Minor surgery, postsurgical dosing by bolus injection should occur at 2-hour intervals for the first 48 hours and then at 2- to 6-hour intervals until healing has occurred • Major surgery, postsurgical dosing by bolus injection should occur at 2-hour intervals for 5 days, followed by 4-hour intervals until healing has occurred. Additional bolus doses should be administered if required Factor VII deficiency 15–30 mcg/kg body weight every 4–6 hours until hemostasis is achieved Effective treatment has been achieved with doses as low as 10 mcg/kg; dose and frequency of injections should be adjusted to each individual

Activated prothrombin complex concentrate (aPCC)	Plasma-derived human PCC, activated FEIBA NF; contains human factors II, VIIa, IX, X, and VIII; 1 bypass activity unit (arbitrary) shortens aPTT of high titer to 50% of blank value	***Hemophilia A or B with inhibitors*** Indicated for the control of spontaneous bleeding episodes or to cover surgical interventions (>5 Bethesda units)	50 to 100 units of aPCC, per kg of body weight every 6–12 hours with a maximum daily dose of 200 units/kg
von Willebrand factor (VWF) $T_{1/2}$ 11–12 hours	Plasma-derived human VWF Humate-P; 2.4:1 VWF/FVIII Alphanate; 0.6:1 VWF/FVIII Wilate; 1:1 VWF/FVIII	***von Willebrand disease (vWD)*** Treatment of spontaneous and trauma-induced bleeding episodes; it is not indicated for the prophylaxis of spontaneous bleeding episodes	Treatment of bleeding episodes—administer 40–80 IU VWF:ristocetin cofactor (RCo) per kg body weight every 8–12 hours
	Recombinant human VWF Vonvendi; VWF only, may need to supplement with rhFVIII if FVIII <40%; administer in ratio of 1.3:1 VWF/FVIII	Prevention of excessive bleeding during and after surgery. This applies to patients with severe vWD as well as patients with mild to moderate vWD where the use of desmopressin is known or suspected to be inadequate	Emergency surgery, administer a dose of 50–60 IU VWF:RCo/kg body weight; maintenance doses are half of loading dose every 8–12 hours

IU, international units; information derived from current package inserts of respective products.

TABLE 41.1C Coagulation Factor Product Information

Product	Contains	Major Indications	Dosages			
Prothrombin complex concentrate (PCC)	Plasma-derived human PCC KCentra; dose based on FIX content Activity ratio: factor II 76–160, factor VII 40–100, factor IX 80–124, factor X 100–204, protein C 84–164, protein S 48–136, antithrombin III 0.8–6; all human derived; product contains heparin	*Acquired deficiency of the prothrombin complex coagulation factors (e.g., vitamin K antagonist)* Treatment of bleeding and perioperative prophylaxis of bleeding No adequate study in subjects with congenital deficiency is available PCC can be used for the treatment of bleeding and perioperative prophylaxis of bleeding in congenital deficiency of any of the vitamin K–dependent coagulation factors only if purified specific coagulation factor product is not available	Initial INR	2.0–3.9	4.0–6.0	>6.0
			Dose IU FIX/kg	25	35	50
			Max Dose IU (FIX)	2500	3500	5000
			The correction of the vitamin K antagonist-induced impairment of hemostasis is reached at the latest 30 minutes after the injection and will persist for approximately 6–8 hours. However, the effect of vitamin K, if administered simultaneously, is usually achieved within 4–6 hours. Thus, repeated treatment with human prothrombin complex is not usually required when vitamin K has been administered. For patients weighing more than 100 kg the dose calculations have to be based on 100 kg b.w.			
Factor X $T_{1/2}$ 40–60 hours	Plasma-derived human factor X Coagadex ($T_{1/2}$ 30.0 hours)	*Congenital factor X deficiency* On-demand treatment and control of bleeding episodes Perioperative management of bleeding in patients with mild hereditary factor X deficiency (not studied in moderate/severe deficiencies)	For treatment of bleeding episodes: 25 IU per kg body weight, repeated at intervals of 24 hours until the bleeding stops Presurgery, raise plasma FX levels to 70–90 IU/dL Postsurgery, maintain plasma FX levels at a minimum of 50 IU/dL until the patient is no longer at risk of bleeding due to surgery Required dose (IU) = Body weight (kg) × desired dactor X rise (IU/dL or % of normal) × 0.5			

Plasma (FFP, FP24, thawed plasma)	~1 IU/mL each of human factors I, II, V, VII, VIII, IX, X, XI, XII, XIII, ATIII, PC, PS, VWF, and others	*Coagulation factor deficiencies* Treatment of patients with documented coagulation factor deficiencies (congenital or acquired) and active bleeding, or who are about to undergo an invasive procedure Use when factor concentrate does not exist or it is unavailable (e.g., factor V, factor XI, protein S, plasminogen activator inhibitor-1, α_2-antiplasmin)	10–20 mL/kg loading dose with the $T_{1/2}$ of the factor being replaced guiding the frequency of subsequent doses 30 IU/dL coagulation factor levels are typically sufficient to achieve hemostasis; however, this can be difficult to achieve with plasma alone due to the large volumes required
Cryopreciptated antihemophiliac factor (cryoprecipitate)	Plasma-derived human factors I (≥10 mg/mL), VIII (≥5.3 IU/mL), XIII (2.7–4 IU/mL), VWF (5.3–8 IU/mL), and fibronectin	*Hypofibrinogenemia, factor XIII deficiency* Treatment of patients for bleeding or immediately before an invasive procedure in patients with significant hypofibrinogenemia (<100 mg/dL) Factor XIII deficiency if factor concentrate not available	8–10 bag pool (120–150 mL); postinfusion fibrinogen level should be measured to guide further therapy
Fibrinogen (factor I) $T_{1/2}$ 100–150 hours	Plasma-derived human factor I RiaSTAP ($T_{1/2}$ 78.7 hours)	*Congenital fibrinogen deficiency, including afibrinogenemia and hypofibrinogenemia* Treatment of acute bleeding episodes Not indicated for dysfibrinogenemia	$\frac{[\text{Target level} - \text{measured},\ \text{mg/dL}] \times \text{weight (kg)}}{1.7\ (\text{mg/dL per mg/kg body weight})}$ or if baseline unknown, 70 mg/kg
		In Europe, this product has been used as fibrinogen replacement in postpartum hemorrhage and trauma settings	Target fibrinogen level of 100 mg/dL should be maintained until hemostasis is obtained
Factor XIII $T_{1/2}$ 7–12 days	Plasma-derived human factor XIII Corifact ($T_{1/2}$ 6.6 days) Recombinant human factor XIII Tretten (A-subunit, $T_{1/2}$ 5.1 days) Not for B-subunit deficiency	*Congenital factor XIII deficiency* Prophylactic treatment Perioperative management of surgical bleeding	40 IU per kg body weight, loading dose; dosing should be guided by the most recent trough Factor XIII activity level, with dosing every 28 days (4 weeks) to maintain a trough level of approximately 5%–20%. Recommended dosing adjustments of ±5 units per kg should be based on trough FXIII activity levels of <5% or >20%, and the patient's clinical condition.

IU, international units; information derived from current package inserts of respective products.

TABLE 41.1D Coagulation Factor Product Information

Product	Contains	Major Indications	Dosages
Antithrombin III (ATIII) $T_{1/2}$ 2.8–4.8 days	Plasma-derived human ATIII Thrombate III ($T_{1/2}$ 2.5 days) Recombinant human ATIII ATryn ($T_{1/2}$ 11.6–17.7 hours)	*ATIII deficiency* Treatment of patient with congenital ATIII deficiency in connection with surgical or obstetrical procedures or when they suffer from thromboembolism Can be useful in cases of heparin resistance but no clinical trial data supports this usage	$\frac{[\text{Target level} - \text{measured IU}] \times \text{weight (kg)}}{1.4}$ Loading dose and 60% loading every 24 hours for maintenance; Target levels of 80–120 IU/dL; to be maintained 2–8 days
Protein C $T_{1/2}$ 6–10 hours	Plasma-derived human Protein C Ceprotin ($T_{1/2}$ 9.9 hours)	*Protein C deficiency* Prevention and treatment of venous thrombosis and purpura fulminans Replacement therapy	100–120 IU/kg, followed with subsequent dosing of 60–80 IU/kg every 6 hours and maintenance dosing of 45–60 IU/kg every 6 or 12 hours
Thrombin (topical)	Recombinant human factor II, activated Recothrom Plasma-derived human factor II, activated Evitrhom	To aid hemostasis whenever oozing blood and minor bleeding from capillaries and small venules is accessible and control of bleeding by standard surgical techniques (such as suture, ligature, or cautery) is ineffective or impractical; may be used in conjunction with an absorbable gelatin sponge	For TOPICAL USE only, DO NOT INJECT, DO NOT use for the treatment of massive or brisk arterial bleeding Reconstituted solution can be applied on the surface of bleeding tissue; the amount of topical thrombin required depends on the area of tissue to be treated and method of application

IU, international units; information derived from current package inserts of respective products.

Coagulation Factor Therapy Monitoring: While the category of factor concentrate is dictated by the patient's factor deficiency, the selection of a specific product is often a decision made by availability and clinician/patient preferences. Whenever possible, before prescribing a coagulation factor concentrate, the specific details of its dosing, half-life, and other pharmacological characteristics should be reviewed from the most current version of the drug's product insert. A brief outline of laboratory monitoring and dosing strategies is described in this chapter. Clinical presentation, diagnosis, and management are covered in each factor deficiency's chapter (see Chapters 109–116).

The level of coagulation factors present in pooled normal plasma is defined as 100% activity, which can also be designated as 100 international units (IU)/dL or 1 IU/mL (depending on the manufacturer). There are a variety of assays used to make activity determinations, but they mainly fall into two categories: (1) clotting time–based assays and (2) chromogenic assays. Additionally, the level of certain coagulation factors can be determined antigenically (ELISA or latex immunoturbidimetry assays) or via other functional measurements (e.g., VWF:ristocetin cofactor activity). It is often very helpful to discuss the clinical situation with a coagulation laboratory specialist, if available, to select the optimal assay and testing schedule to manage the factor replacement therapy. While the factor concentrate manufacturers have suggested doses that are often weight based, the in vivo recovery of circulating factor activity is only an estimation based on normal test subjects, not actively bleeding patients with possible comorbidities. Thus, it is recommended to collect pre- and postinfusion samples (which can be run as a batch) for the first few administrations to determine the in vivo circulating factor activity recovery for each specific product in a specific patient for optimal dosing.

Coagulation Factor Concentrate Dosing: As indicated above, a weight-based calculation can be performed to determine an initial loading dose. Thereafter, depending on the biological half-life of the factor, subsequent maintenance doses can be given to keep the factor levels at or above a desired threshold, which is typically determined by the current and projected bleeding risk of the patient. An example for generic calculation is as follows:

Required dose (IU) = [body weight (kg) × desired Coagulation Factor Rise (IU/dL or % of normal)]/amount activity recovered (IU/dL per IU/kg)[1]

1. Note that the amount activity recovered is often assumed to be a 1% or 1 IU/dL per IU/kg administered, however, for FVIII, 1 IU/kg dose results in a 2% or 2 IU/dL per IU/kg activity recovered. The activity recovery can be determined from the pre- and postinfusion factor activity measurements as in the below calculation:

Activity recovery (% or IU/dL per IU/kg) = [(Immediate postdose factor level (IU/dL) − predose factor level (IU/dL)) × body weight (kg)]/coagulation factor dose administered (IU)

Further Reading

Key, N. S., & Negrier, C. (2007). Coagulation factor concentrates: Past, present, and future. *Lancet, 370*(9585), 439–448.

Kulkarni, R., Chitlur, M., & Lusher, J. M. (2011). Treatment of congenital coagulopathies. In P. D. Mintz (Ed.), *Transfusion therapy: Clinical principles and practice* (3 ed.) (pp. 167–208). Bethesda: AABB Press.

Shord, S. S., & Lindley, C. M. (2000). Coagulation products and their uses. *Am J Health Syst Pharm, 57*(15), 1403–1417 Quiz 18–20.

Sorensen, B., Spahn, D. R., Innerhofer, P., Spannagl, M., & Rossaint, R. (2011). Clinical review: Prothrombin complex concentrates–evaluation of safety and thrombogenicity. *Crit Care, 15*(1), 201.

CHAPTER 42

Irradiation of Blood Products

Richard O. Francis, MD, PhD

Irradiation (or pathogen inactivation) of blood products is performed to abrogate the risk of transfusion-associated graft-versus-host disease (TA-GVHD), a rare and almost universally fatal complication of blood transfusion with no successful treatment options. Irradiation results in the generation of electrons that damage lymphocyte DNA, and therefore, renders the lymphocytes unable to proliferate. Pathogen inactivation (PI) also prevents lymphocyte replication (Chapter 48). In this chapter, PI will be included under irradiation. A few institutions and countries practice universal irradiation of cellular blood products, but most choose to irradiate cellular blood products only for those patients at increased risk for development of TA-GVHD (Chapter 70).

Indications: Table 42.1 lists the clear indications, the indications deemed appropriate by most authorities, and the indications considered unwarranted by most authorities for irradiated blood products. Even in centers of excellence, some divergence of opinions about indications for irradiation does occur, so each transfusion service should develop its own criteria and indications in concert with ordering physicians.

Guidelines and Standards for Irradiation and Mitigation of Transfusion-Associated Graft-Versus-Host Disease: AABB *Standard* 5.19.3 states:

> *The BB/TS (blood bank/transfusion service) shall have a policy regarding the transfusion of irradiated components. At a minimum cellular components shall be irradiated when: 1) A patient is identified as being at risk for TA-GVHD; 2) The donor of the component is a blood relative of the recipient; 3) The donor is selected for HLA compatibility, by typing or crossmatching.*

The United States currently does not have established or widely adopted guidelines or indications for irradiation of blood components. Therefore, indications vary from institution to institution, and there is heterogeneity of practice. Results of a survey performed by the College of American Pathologists in 2014 demonstrated that 78.6% of institutions provided irradiated components for patients who were receiving products from blood relatives, 68.9% for patients receiving HLA-matched or partially matched products, 66.3% for neonatal exchange transfusions, 63.3% for intrauterine transfusions (IUTs), 62.7% for allogeneic/autologous hematopoietic progenitor cell transplant, and 61.8% for preterm/low-birth-weight infants. These data demonstrate that even for indications that AABB deems irradiation should be performed (e.g., transfusions that are HLA-matched or from blood relatives), there is variation in practice.

Transfusion Medicine and Hemostasis. https://doi.org/10.1016/B978-0-12-813726-0.00042-8

TABLE 42.1 Indications for Irradiated Cellular Blood Products

Indications for which irradiation is considered to be required
Congenital immunodeficiency syndromes (suspected or known)
Allogeneic and autologous hematopoietic progenitor cell transplantation
Transfusions from blood relatives
HLA-matched or partially HLA-matched products (platelet transfusions)
Granulocyte transfusions
Hodgkin disease
Patients treated with purine analogue drugs (fludarabine, cladribine, and deoxycoformycin)
Patients treated with Campath (anti-CD52) and other drugs/antibodies that affect T-lymphocyte number or function
Intrauterine transfusions
Indications for which irradiation is deemed appropriate by most authorities
Neonatal exchange transfusions
Preterm/low-birth-weight infants
Infant/child with congenital heart disease (secondary to possible DiGeorge syndrome)
Acute leukemia
Non-Hodgkin lymphoma and other hematologic malignancies
Aplastic anemia
Solid tumors receiving intensive chemotherapy and/or radiotherapy
Recipient and donor pair from a genetically homogeneous population
Indications for which irradiation is considered unwarranted by most authorities
Solid organ transplantation
Healthy newborns/term infants
Patients with HIV/AIDS

Adapted from Shaz, B. H., Francis, R. O., & Hillyer, C. D. (2017). Transfusion-associated graft-versus-host disease and microchimerism. In M. F. Murphy, D. J. Roberts, M. H. Yazer (Eds.), *Practical transfusion medicine* (5th ed.). Oxford: Wiley-Blackwell.

Universal Irradiation: Several factors related to TA-GVHD and irradiation lead to the adoption of universal irradiation in some centers and countries:

1. TA-GVHD can occur in seemingly immunocompetent patients if the donor has a homozygous HLA haplotype for which the recipient is heterozygous (random donor and recipient partial HLA matching has been estimated to be possible in as many as 1 in 7174 in the United States vs. 1 in 874 in Japan), if the product is received from a relative, or if the recipient's degree of immunocompromise was not known or properly identified before transfusion;
2. TA-GVHD is almost universally fatal;
3. The adverse effects of irradiation on the blood product and its constituents are minimal; and
4. The cost of irradiating products is modest.

Universal irradiation is practiced in some institutions within and outside the United States (e.g., Japan, where the risk of inadvertent, nonrelative partial matching of HLA haplotypes has been estimated as 1:874).

Blood Products Requiring Irradiation: All cellular blood products, defined as RBCs, platelets, granulocytes, whole blood, and liquid plasma (not previously frozen plasma), contain viable T lymphocytes that are capable of causing TA-GVHD. These products should at a minimum be irradiated for patients at increased risk for TA-GVHD (Table 42.1). Three potentially high-risk blood products are granulocytes, products donated from first-degree relatives, and HLA-matched or crossmatched platelets.

Granulocytes are very high-risk products for TA-GVHD because they are given soon after collection, have a high lymphocyte count, and are administered to neutropenic and immunosuppressed patients. Therefore, it is recommended that all granulocyte transfusions undergo irradiation before transfusion.

All cellular blood products transfused to a relative of the donor, and all HLA-matched products (including both HLA-matched/selected and crossmatched platelet products) should be irradiated because the viable donor lymphocytes within the product may be homozygous for an HLA haplotype for which the recipient is heterozygous. This scenario results in the inability of the recipient to recognize the donor lymphocytes as foreign and therefore the donor lymphocytes can proliferate and cause TA-GVHD.

Processing and Storage

Sources of Irradiation: γ-rays, X-rays, and PI technologies can be used to irradiate blood products and cause adequate T-lymphocyte inactivation at the doses described. Usually, γ-rays originate from cesium 137 or cobalt 60 while X-rays are generated from linear accelerators or stand-alone units.

Since the September 11, 2001 attacks, there has been increased regulation of blood irradiators by the US Nuclear Regulatory Commission (NRC). One initiative through the Energy Policy Act is to find alternative technologies that do not use radionuclides, such as electricity, or use lower-risk sources. Therefore, there has been increasing use of X-ray irradiators.

Dose: The dose of irradiation must be sufficient to inhibit lymphocyte proliferation without significantly damaging RBCs, platelets, and granulocytes. Assays to assess the effect of irradiation on T-lymphocyte proliferation include the mixed lymphocyte culture assay and limiting dilution analysis. The recommended dose varies between 15 and 50 Gy (1500 and 5000 cGy) (Table 42.2); the United States requires a dose of 25 Gy (2500 cGy) at the center of the product, and minimum of 15 Gy (1500 cGy) and maximum of 50 Gy (5000 cGy) at any point in, the product.

Quality Assurance: Quality-related measures for blood product irradiation include those focused on the irradiator itself and on the product. AABB *Standard* 5.7.3.2 states:

> *Irradiated blood and blood components shall be prepared by a method known to ensure that irradiation has occurred. A method shall be used to indicate that irradiation has occurred with each batch. Alternate methods shall be demonstrated to be equivalent.*

Quality assurance measures should be performed on the irradiator, including dose mapping, adjustment of irradiation time to correct for isotopic decay, ongoing detection for radiation leakage, timer accuracy, turntable operation, and preventive maintenance. Each batch of irradiated products should have attached a qualitative radiation

TABLE 42.2 Comparison of Irradiated Blood Product Requirements

	United States[a]	United Kingdom[b]	Japan[c]
Techniques	γ-irradiation or X-rays	γ-irradiation or X-rays	γ-irradiation
Dose	2500 cGy at center of product Minimum 1500 cGy at any point Maximum 5000 cGy	Minimum 2500 cGy No part >5000 cGy	Between 1500 and 5000 cGy
Type of product	All cellular products: Whole blood RBCs Platelets Granulocytes	All cellular products: Whole blood RBCs Platelets Granulocytes	All cellular products: Whole blood RBCs Platelets Granulocytes Fresh plasma
Age of product	RBCs—any time platelets—any time granulocytes—before infusion	RBCs < 14 days after collection For risk of hyperkalemia, e.g., exchange or intrauterine transfusion: < 24 h before transfusion platelets—any time during 5-day storage	RBCs: ≤3 days—regardless of recipient ≤14 days—if clinically indicated At any time—if patient immune-compromised
Expiration	RBCs stored for up to 28 days after irradiation or original outdate, whichever is sooner	RBCs stored 14 days after irradiation	Irradiated RBCs—up to 3 weeks after collection
General	All blood from blood relatives All HLA-matched products	All blood from blood relatives All HLA-matched products All granulocytes	All blood from blood relatives All HLA-matched products All granulocytes

[a]Fung, M. K., Eder, A. F., Spitalnik, S.L., & Westhoff, C. M. (Eds.). (2017). *Technical manual* (pp. 753–755) (19th ed.). Bethesda, MD: AABB.

[b]Treleaven, J., Gennery, A., Marsh, J., et al. (2011). Guidelines on the use of irradiated blood components prepared by the British Committee for Standards in Haematology blood transfusion task force. *Br J Hematol, 152*, 35–51.

[c]Asai, T., Inaba, S., Ohto, H., et al. (2000). Guidelines for irradiation of blood and blood components to prevent post-transfusion graft-vs.-host disease in Japan. *Transfus Med, 10*, 315–320.

dosimeter; usually, a label is placed on individual products, and the label physically changes at the triggering dose of radiation.

Storage and Expiration: In the United States, RBC product outdates are shortened to 28 days after irradiation; RBC product outdate is variably shortened to 14–28 days after irradiation in other countries (Table 42.2). These changes in outdating are due to the effects of radiation on erythrocyte membranes leading to increased potassium leakage and accelerated cell damage over time during the storage period. There is no change in outdate of other blood products.

Adverse Events: At recommended doses, radiation causes a very low level of oxidation and damage to lipid components of membranes, which occurs over time. Products, and the constituent cells within, irradiated immediately before transfusion appear to be unaffected and have virtually normal function. The effects of radiation are most significant on RBC products and include increased extracellular potassium and decreased post-transfusion RBC survival. The in vivo viability of irradiated RBCs evaluated by 24-hour posttransfusion recovery is reduced by 3%–10% compared with nonirradiated RBCs.

The increase in extracellular potassium is typically not considered clinically significant because of posttransfusion dilution. However, there may be certain patients for whom attention should be paid to the increased potassium, such as premature infants, infants receiving large RBC volume transfusion, neonates undergoing exchange transfusions, patients receiving intracardiac transfusions via central line catheters, and fetuses receiving IUTs. The high extracellular potassium can be prevented by either irradiating the RBC product shortly before transfusion (usually within 24 hours) or by washing or volume-reducing the RBC product before transfusion. Other methods for potassium removal include filters that specifically remove this element from blood products. At present, these filters are investigational in the United States.

International Considerations: There are differences in dose of γ-irradiation, product requirements, product expiration, and indications among countries (Table 42.2). The United Kingdom has developed indications for receiving irradiated cellular blood components, and Japan irradiates all cellular blood components because of the similarity of HLA haplotypes within the Japanese population. In addition, some countries use PI technologies.

Further Reading

Asai, T., Inaba, S., Ohto, H., et al. (2000). Guidelines for irradiation of blood and blood components to prevent post-transfusion graft-vs.-host disease in Japan. *Transfus Med, 10*, 315–320.

Davey, R. J., McCoy, N. C., Yu, M., Sullivan, J. A., Spiegel, D. M., & Leitman, S. F. (1992). The effect of prestorage irradiation on post-transfusion red cell survival. *Transfusion, 32*, 525–528.

Moroff, G., & Luban, N. L. (1997). The irradiation of blood and blood components to prevent graft-versus-host disease: technical issues and guidelines. *Transfus Med Rev, 11*, 15–26.

Pritchard, A. E., & Shaz, B. H. (2016). Survey of irradiation practice for the prevention of transfusion-associated graft-versus-host disease. *Arch Pathol Lab Med, 140*, 1092–1097.

Treleaven, J., Gennery, A., Marsh, J., et al. (2011). Guidelines on the use of irradiated blood components prepared by the British Committee for Standards in Hematology blood transfusion task force. *Br J Hematol, 152*, 35–51.

CHAPTER 43

Leukoreduction of Blood Products

Theresa Nester, MD

Leukocytes (white blood cells [WBCs]) remain in red blood cell (RBC) and platelet (PLT) components after simple component preparation. Residual leukocytes have the potential to cause harmful effects in the transfusion recipient including febrile nonhemolytic transfusion reactions (FNHTRs), HLA alloimmunization, and transmission of cytomegalovirus (CMV). Other possible effects include transmission of other leukocyte-associated herpesviruses and transfusion-related immunomodulation (TRIM). The likelihood of these effects depend on the number in each component, storage temperature, and recipient clinical status including previous and latent infection with the viruses above, and their immune status. Thus, leukoreduction significantly decreases their incidence.

Before the 1980s, it was recognized that residual leukocytes could be removed by washing components or by some membrane and fiber "filter" methods. These later methods were further developed into "filter leukoreduction." Filter leukoreduction that is performed within 24 hours of collection is termed prestorage leukoreduction, whereas leukoreduction performed after 24 hours, typically just before the product is issued, is termed poststorage leukoreduction. Currently, there are a variety of leukoreduction filters, which remove 3 logs of WBCs. These filters use a combination of barrier filtration (i.e., pore size) and cell adhesion to decrease leukocyte content. Some filters are PLT sparing, which is used to make leukoreduced whole blood, while RBC filters usually also remove PLTs. The RBC product from non-PLT sparing filters will lack PLTs or PLT-derived cytokines. In addition, leukofiltration of RBCs from hemoglobin AS (sickle trait) patients is often less effective secondary to reduced RBC deformability leading to clogging or decreasing the area of the filter. Apheresis-derived PLTs are leukoreduced by the apheresis devices, known as "process leukoreduction." Because of the clear clinical benefit with minimum cost, most blood products are leukoreduced in developed nations.

Definitive Indications

Decreasing Incidence of Febrile Nonhemolytic Transfusion Reactions: The frequency of FNHTRs is significantly decreased (75%–95%) when prestorage leukoreduced products are transfused. There are two mechanisms by which leukocytes in transfused units contribute to FHNTRs. The first mechanism involves leukocyte-derived cytokines and is most commonly associated with PLT products. Leukocyte-derived cytokines (e.g., interleukins [IL-1, IL-6, IL-8] and tumor necrosis factor α [TNF-α]) accumulate in the supernatant during room temperature storage of PLT products. Unlike prestorage leukoreduction, poststorage leukoreduction will be ineffective in addressing this accumulation of cytokines. The second mechanism involves WBC antibodies and is most commonly associated with RBC products. Anti-HLA and anti-HNA antibodies in the transfusion recipient's plasma directed toward the

Transfusion Medicine and Hemostasis. https://doi.org/10.1016/B978-0-12-813726-0.00043-X

transfused WBCs lead to an antigen–antibody complex, resulting in the release of endotoxin, which results in fever and other symptoms (see Chapter 61).

Decreasing Incidence of HLA Alloimmunization: Leukoreduction of RBCs and PLTs has been shown to decrease the incidence of HLA alloantibody formation in transfusion recipients. This is of primary importance in patients who require ongoing PLT transfusion support, as anti-HLA antibodies can lead to PLT refractoriness by binding to the corresponding antigens (major histocompatibility complex) on transfused PLTs. The TRAP study in 1997 assessed reduction in HLA alloimmunization and PLT refractoriness by randomized controlled trial in patients with acute leukemia; in addition, a number of smaller studies had previously addressed the same issue. In aggregate, metaanalysis of these studies showed a relative risk reduction of HLA immunization of approximately one-third for patients with acute leukemia or other hematologic malignancies not previously sensitized to HLAs. These data are then extrapolated to other patient populations. For example, because HLA immunization can contribute to rejection of an organ graft and thus difficulty finding a compatible organ and/or undergoing transplantation, many transfusion physicians recommend leukoreduced cellular products for potential solid organ transplant recipients.

Decreasing Cytomegalovirus Transmission: Blood donors in an active viremia phase (CMV DNA detectable in plasma) appear to be most likely to transmit CMV to a recipient. However, latent CMV can reactivate on transfusion and infect the naïve recipient. Because current leukoreduction filters achieve a 3 log reduction in WBCs in a cellular blood product, leukoreduction substantially decreases the inoculum of latent CMV genomes a recipient receives. Accordingly, transmission by blood transfusion is substantially decreased by leukoreduction (from as high as 30% in susceptible patients to 0%–2.5%). Data indicate that leukoreduced blood has rates of CMV transmission as low as seronegative products (which still have a risk of CMV transmission). However, in patient populations such as fetuses requiring intrauterine transfusion, where monitoring for CMV viremia is not possible, some suggest that these patients may benefit from CMV seronegative products (see Chapter 44).

Potential Indications

Decreasing Other Human Herpesvirus Transfusion-Transmitted Infections: Like CMV, other human herpesviruses can be associated with leukocytes. Viral loads (i.e., EBV and HHV-8) are reduced by leukoreduction.

Prion Disease: Transfusion transmission of variant Creutzfelt–Jakob disease (vCJD) has been reported. As prions can be associated with leukocytes, leukoreduction technology has been assessed for the ability to prevent transfusion transmission of vCJD. Prions can be decreased by passing blood products through existing filters. An additional filter designed to remove prions by selective adsorption has been developed and approved for use in the European Union. Whether such filters will decrease transmission of disease in humans is undetermined; however, a near 50% decrease in transmission was observed in an animal model.

Reduction in the Risk of Transfusion-Associated Graft Versus Host Disease: Evidence suggests that leukoreduction decreases transfusion-associated graft versus host disease (TA-GVHD) risk. However, because no definitive data exists, irradiation or pathogen inactivation is required to prevent TA-GVHD (see Chapter 42).

Controversial Indication

Transfusion-Related Immunomodulation: TRIM is defined as effects of transfusion on the recipient immune system, including potential downregulation of cellular immunity, induction of humoral immunity, and altered inflammatory responses (see Chapter 71). The TRIM effect may not only be secondary to the transfused WBCs but also be the result of the other blood constituents, such as proteins, lipids, and inflammatory mediators. However, as TRIM can be mediated in large part by transfused leukocytes, leukoreduction is predicted to decrease TRIM; nevertheless, the clinical benefit of leukoreduction for this purpose is unestablished.

Contraindication: Bedside leukoreduction should be avoided in patients on ACE inhibitors, as hypotensive episodes may be induced.

Quality Assurance: In the United States, the requirements for leukoreduction of RBCs and apheresis PLTS are $<5 \times 10^6$ leukocytes per product and for whole blood–derived PLTs $<0.83 \times 10^6$ leukocytes per product. At least 95% of the products sampled must meet this specification.

In-Lab Versus Bedside Leukoreduction: Leukoreduction can be performed at the bedside; however, bedside leukoreduction has several substantial problems. These include severe hypotensive reactions for patients taking ACE inhibitors when a bedside filter is used, and lack of quality control to confirm that the resultant product meets the FDA criteria for leukoreduction. For this reason, bedside filtration is infrequently used in current practice.

Universal Versus Diagnosis-Specific Leukoreduction: Whether it is safe, appropriate, and cost-effective to leukoreduce all blood products before storage or to reserve leukoreduced products for particular patient subsets is currently a matter of debate. Practices differ by blood supplier, transfusion service, and region. In the United States, prestorage leukoreduced products are used in 95% of PLTs and 75% of RBCs.

International Differences: In Europe, the requirements for leukoreduction of RBCs and apheresis-derived PLTs are fewer than 1×10^6 leukocytes per product.

Further Reading

AABB Clinical Transfusion Medicine Committee. (2016). Reducing transfusion-transmitted cytomegalovirus infections. *Transfusion, 56*, 1581–1587.

Delaney, M., Maycock, D., Knezevic, A., et al. (2016). Postnatal CMV infection: a pilot comparative effectiveness study of transfusion safety using leukoreduction only transfusion strategy. *Transfusion, 56*, 1945–1950.

Dollard, S. C., Roback, J. D., Gunthel, C., et al. (2013). Measurement of human herpesvirus 8 viral load in blood before and after leukoreduction filtration. *Transfusion*, *53*, 2164–2167.

Douet, J. Y., Bujdoso, R., & Andreoletti (2015). Leukreduction and blood-borne vCJD transmission risk. *Curr Opin Hematol*, *22*, 36–40.

Glenister, K. M., & Sparrow, R. L. (2010). Level of platelet-derived cytokines in leukoreduced red blood cells is influenced by the processing method and type of leukoreduction filter. *Transfusion*, *50*, 185–189.

Mainou, M., Fares, A., Tobian, A., et al. (2016). Reducing the risk of transfusion-transmitted cytomegalovirus infection: a systematic review and meta-analysis. *Transfusion*, *56*, 1569–1580.

Qu, L., Rowe, D. T., Donnenberg, A. D., et al. (2009). Effects of storage and leuoreduction on lymphocytes and Epstein-Barr virusgenomes in platelet concentrates. *Transfusion*, *49*, 1580–1583.

Simancas-Racines, D., Osorio, D., Martí-Carvajal, A. J., & Arevalo-Rodriguez, I. (2015). Leukoreduction for the prevention of adverse reactions from allogeneic blood transfusion. *Cochrane Database Syst Rev* (12):CD009745. https://doi.org/10.1002/14651858.CD009745.pub2.

The Trial to Reduce Alloimmunization to Platelets Study Group. (1997). Leukocyte reduction and ultraviolet B irradiation of platelets to prevent alloimmunization and refractoriness to platelet transfusions. *NEJM*, *337*, 1861–1869.

CHAPTER 44

CMV-Safe Blood Products

Stuart P. Weisberg, MD, PhD

Description: Although typically of little consequence in immunocompetent individuals, primary cytomegalovirus (CMV) infection is associated with considerable morbidity and mortality in at-risk populations. TT-CMV is a serious complication of cellular blood product transfusion.

Indications: Studies suggest that 13%–37% of immunocompromised patients will contract TT-CMV from transfusion of unfiltered cellular blood components that are not mitigated for CMV. Populations at greatest risk for TT-CMV include fetuses (intrauterine transfusion), premature low-birth-weight infants (<1250–1500 g) born to seronegative mothers, seronegative recipients of seronegative allogeneic or autologous hematopoietic stem cell (HSC) transplants, seronegative recipients of seronegative solid organ transplants, and seronegative patients with HIV infection (Table 44.1). Primary infection in these patients can lead to serious end organ damage or failure, including CMV hepatitis, retinitis, colitis, interstitial pneumonitis, esophagitis, polyradiculopathy, transverse myelitis, and subacute encephalitis.

Additional groups that may benefit from CMV-safe transfusion products include pregnant women who are seronegative, to prevent primary infection to the fetus, and seronegative patients who are candidates for HSC transplantation.

Processing and Storage

CMV-Seronegative Products: The standard approach for identifying donors who are not infected with CMV is the detection of CMV antibodies within the serum using various serologic methods, including solid phase fluorescence immunoassay, enzyme immunoassay, latex particle agglutination, and solid phase red cell adherence.

A recent study quantified CMV viral DNA in blood from newly seroconverted donors with short interdonation intervals before and after primary seroconversion. Using a highly sensitive PCR assay, it found only low to barely detectable CMV DNA in 25% of window period donors. The highest levels of CMV DNA were found at the first seropositive donation. Thus, CMV window period donations are likely very rare with very low levels of systemic viremia. The highest risk of TT-CMV may come from primarily seropositive donors. Some studies have also identified donors without detectable levels of CMV antibodies that have CMV-specific cytotoxic T-lymphocytes suggesting they have had a CMV infection previously.

Although the exclusive use of CMV-seronegative components in high-risk individuals markedly decreases the incidence of TT-CMV, it does not completely eliminate it. Multiple studies have shown infection rates in seronegative infants that have been transfused product from seronegative donors vary between 0% and 4%.

Transfusion Medicine and Hemostasis. https://doi.org/10.1016/B978-0-12-813726-0.00044-1

TABLE 44.1 CMV-Seronegative Populations at High Clinical Risk of Morbidity Due to TT-CMV

Pregnant Women
Fetuses (intrauterine transfusion)
Premature infants weighing less than 1250 g
Severe combined/variant immunodeficiency and some other congenital immunodeficiencies
Autologous or allogeneic hematopoietic stem cell recipients or candidates
Solid organ transplant: cytomegalovirus (CMV)–negative recipients of CMV-negative organ donor

However, transfused blood components are not necessarily the sole source of CMV infections seen in these studies. A recent study in very low-birth-weight infants transfused with blood that was both CMV-seronegative and leukoreduced, showed a postnatal CMV infection rate of 6.9%, with no case linked to transfusion and almost all cases linked with CMV-positive breast milk.

CMV-Safe by Leukoreduction: Because white blood cells (WBCs) are the primary vector for TT-CMV, removal of WBCs from the components also markedly reduces the incidence of TT-CMV. Current filtration and apheresis technologies produce blood components well below the maximal limit of 5×10^6 WBCs mandated for leukoreduced products in the United States. Although CMV infection rates in seronegative, recipients receiving standard blood products have been reported to exceed 25%, this rate decreases to 1%–4% with leukoreduction. A prospective randomized trial published in 1995 showed the probability of seronegative recipients of seronegative HSC transplants developing CMV infection was similar with both the filtered components (2.4%) and the seronegative components (1.3%; $P=1.00$). However, because of a statistically greater progression to CMV disease (including lethal CMV pneumonia) in the group receiving filtered compared with seronegative components (2.4% vs. 0%, respectively; $P=.03$), there is a possibility that seronegative components may be slightly safer. Leukoreduction technology has improved considerably since publication of this study, and these results may not be applicable to the modern era. Indeed, recent retrospective studies have not observed any TT-CMV from leukoreduced blood components in seronegative recipients of seronegative HSC transplants.

There does not appear to be a utility for CMV nucleic acid testing to further reduce the risk of TT-CMV, as based on recent large-scale studies, no additional CMV-positive donors were identified that were not already positive by serology.

CMV-Safe by Pathogen Inactivation: PI technologies for platelet (and RBC) components also effectively remove CMV while preserving platelet function. PI platelets have been transfused in Europe for over a decade. In December 2014, the US FDA approved a photochemical treatment for pathogen inactivation in apheresis platelet components. Studies show a 5- to 6-log reduction in infectivity of human CMV in platelet concentrates using this PI technology and prevention of murine CMV transmission in a mouse model. Little clinical data are available to evaluate whether these systems prevent TT-CMV infection as effectively as leukoreduced and CMV seronegative platelets (notably all PI technologies require leukoreduced cellular components). One observational study in 51 thrombocytopenic patients with hematological disease showed no CMV transmission from transfusion of leukoreduced PI platelets from CMV unselected donors. Currently, PI technologies for RBC components are undergoing phase III randomized controlled clinical trials in the United States and Europe.

Practice Guidelines: In an effort to develop clinical practice guidelines for prevention of TT-CMV, the AABB Clinical Transfusion Medicine Committee commissioned a metaanalysis to address the following questions for patients at risk for CMV disease: (1) Does leukoreduction reduce TT-CMV infection? (2) Does CMV seronegativity add additional benefit to leukoreduction? A systematic review of the literature published in 2016 identified only studies of low and very low quality, and only one of the controlled studies was published within the previous decade. Because of the low quality of the published data, the metaanalysis could not provide evidence favoring any particular CMV mitigation strategy, and the committee did not issue clinical practice guidelines.

For prevention of CMV disease in solid organ and HSC transplant patients, evidence-based guidelines support using either CMV prophylaxis or a preemptive approach using CMV antigenemia or PCR monitoring with immediate treatment for evidence of viral replication. Currently, there are no evidence-based practice guidelines recommending universal CMV prophylaxis or monitoring for pregnant women or low-birth-weight infants. Testing and treatment for CMV in these patients is based on clinical suspicion.

According to AABB *Standards*, each institution should weigh the risks and benefits, and must have a policy regarding the selection and processing of blood components to reduce CMV transmission.

Quality Assurance: CMV donor screening assays detect total antibody (IgG and IgM) and are cleared by the FDA. Donors of viable, leukocyte-rich HCT/Ps must be tested for evidence of CMV infection to adequately and appropriately reduce the risk of transmission, and there must be an established and maintained standard operating procedure governing the release of products from donors reactive for CMV (21CFR 1271.85(b)(2)), according to the FDA. A positive test for CMV does not necessarily make the donor ineligible.

Leukoreduction can be performed on whole blood and blood components either pre- or poststorage using any closed system or functionally closed methods, including filtration through an in-line filter integral to the blood collection or apheresis set, filtration through a filter system attached to a component container, or direct in-process leukocyte reduction for certain apheresis collections. Notably, HCT/Ps and granulocyte products cannot be leukoreduced.

Increased dependence on universal leukoreduction as the primary TT-CMV mitigation strategy can leave institutions vulnerable to leukoreduction filter failure. For example, in June of 2016, the Food and Drug Administration issued a large-scale leukoreduction filter recall during which time the supply of leukoreduced blood, and CMV-seronegative blood was temporarily limited. Thus, institutions should maintain policies delineating which patients should receive CMV-safe blood when universal leukoreduction is not possible.

International Practices: The widespread use and availability of universal leukoreduction and lack of compelling evidence favoring CMV-seronegative over leukoreduced blood components to prevent TT-CMV has increased international consensus to use leukoreduced blood components for most adult patients at risk for CMV infection. In 2017, the Canadian National Advisory Committee recommended that Canadian blood services issue CMV-seronegative blood components for the sole purpose of intrauterine transfusion. In 2012, the UK Advisory committee for the safety of tissues and organs recommended that leukoreduced CMV unselected blood components be provided to CMV-at-risk patients. It recommended CMV-seronegative, leukoreduced blood components for intrauterine transfusion, transfusions in neonates up to 28 days and transfusions during pregnancy (but not childbirth).

Despite these recommendations, standardization of practices is lacking. A College of American Pathologists Transfusion Medicine (Comprehensive) Participant Survey in 2015 showed that universal leukoreduction is the primary strategy for TT-CMV mitigation used by 90% of institutions surveyed; however, there was a lack of consensus on which patients should receive CMV-seronegative blood components. For example, 49% of institutions surveyed had policies to give leukoreduced CMV-seronegative blood components to neonates, with much higher rates reported among centers with neonatal intensive care units. A recent survey of UK adult and pediatric transplant centers showed that 22.7% of the centers continued to give CMV-seronegative blood components to HSC transplant patients.

Better standardization of practices for CMV mitigation will require evidence-based guidelines derived from high-quality clinical studies. For example, a dual prevention strategy of transfusing CMV-seronegative leukoreduced blood components is variably used for nonneonatal immunosuppressed patient groups despite being of questionable clinical benefit. High-quality clinical studies should be performed to determine if CMV seronegativity is required to further reduce CMV transmission from leukoreduced blood products. Evaluation of the TT-CMV risk from PI platelets (and RBCs) is also needed. These studies should be carefully designed to discriminate TT-CMV from community-acquired CMV.

Further Reading

AABB, Clinical Transfusion Medicine Committee, Heddle, N. M., Boeckh, M., Grossman, B., et al. (2016). AABB Committee Report: reducing transfusion-transmitted cytomegalovirus infections. *Transfusion*, *56*, 1581–1587.

Bowden, R. A., Slichter, S. J., Sayers, M., et al. (1995). A comparison of filtered leukocyte-reduced and cytomegalovirus (CMV) seronegative blood products for the prevention of transfusion-associated CMV infection after marrow transplant. *Blood*, *86*, 3598–3603.

Hall, S., Danby, R., Osman, H., et al. (2015). Transfusion in CMV seronegative T-depleted allogeneic stem cell transplant recipients with CMV-unselected blood components results in zero CMV transmissions in the era of universal leukocyte reduction: a U.K. dual centre experience. *Transfus Med*, *25*, 418–423.

Mainou, M., Alahdabm, F., Tobian, A. A., Asi, N., Mohammed, K., Murad, M. H., et al. (2016). Reducing the risk of transfusion-transmitted cytomegalovirus infection: a systematic review and meta-analysis. *Transfusion*, *56*, 1569–1580.

Morton, S., Peniket, A., Malladi, R., & Murphy, M. F. (2017). Provision of cellular blood components to CMV-seronegative patients undergoing allogeneic stem cell transplantation in the UK: survey of UK transplant centres. *Transfus Med*, *27*(6), 444–450.

Roback, J. D., Conlan, M., Drew, W. L., Ljungma, P., Nichols, W. G., & Preiksaitis, J. K. (2006). The role of photochemical treatment with amotosalen and UV-A light in the prevention of transfusion-transmitted cytomegalovirus infections. *Transfus Med Rev*, *20*, 45–56.

Roback, J. D., Drew, L. W., Laycock, M. E., Todd, D., Hillyer, C. D., & Busch, M. P. (2003). CMV DNA is rarely detected in healthy blood donors using validated PCR assays. *Transfusion*, *43*, 314–321.

Thiele, T., Kruger, W., Zimmermann, K., et al. (2011). Transmission of cytomegalovirus (CMV) infection by leukoreduced blood products not tested for CMV antibodies: a single-center prospective study in high risk patients undergoing allogeneic hematopoietic stem cell transplantation. *Transfusion*, *51*, 2620–2626.

Weisberg, S. P., Staley, E. M., Williams, L. A., et al. (2017). Survey on transfusion-transmitted cytomegalovirus and cytomegalovirus disease mitigation. *Arch Pathol Lab Med*, *141*(12), 1705–1711.

Ziemann, M., Heuft, H. G., Frank, K., Kraas, S., Görg, S., & Hennig, H. (2013). Window period donations during primary cytomegalovirus infection and risk of transfusion-transmitted infections. *Transfusion*, *53*, 1088–1094.

CHAPTER 45

Frozen Blood Products

Cheryl A. Goss, MD

Cryopreservation of blood products, which maintains their activity, is an effective way to increase their storage time. Plasma products are routinely stored frozen, but RBCs are sometimes cryopreserved (for cryopreserved hematopoietic progenitor cell products, see Chapters 84 and 85). Rarely, platelet products are cryopreserved, but the recovery is low and the product is still under investigation and not FDA licensed. The primary indication for freezing RBCs is the preservation of rare and/or autologous units. Additional indications include inventory management during emergencies or shortages and military contingency operations. RBC products are usually cryopreserved with glycerol. Before use, the product must be thawed and deglycerolized. Frozen RBCs are not routine transfusion practice because of the increased cost (both labor and equipment) and RBC loss during the process.

Cryopreserved RBC Products: Cryopreserving RBCs is a valuable technology to prolong storage of rare units (e.g., RBCs lacking high-frequency antigens). Freezing RBCs increases their storage time to 10 years (and likely even longer) and also provides a product with replenished 2,3-diphosphoglycerate (2,3-DPG) and adenosine triphosphate (ATP) levels. Frozen and deglycerolized units have a decreased number of leukocytes ($\sim 9 \times 10^6$ white blood cell unit) but too many for labeling as leukoreduced. After thawing and deglycerolization up to 20% of the RBCs can be lost. Because of the removal of the supernatant plasma and storage solution, restored ATP and 2,3-DPG, thawed and deglycerolized RBC products may be preferred in some situations, such as neonatal RBC exchange and intrauterine transfusion.

Cryopreservation With Glycerol: RBCs must be protected during freezing to prevent cellular dehydration and mechanical trauma as a result of intracellular ice formation. Glycerol is a penetrating cryoprotective agent, which crosses the cell membrane into the cytoplasm, providing an osmotic force that prevents water from migrating outward as extracellular ice is formed. Glycerol must be introduced slowly, as rapid introduction can itself result in hypertonic damage and hemolysis. RBC units are frozen using either a high- or low-glycerol method. Low, 15%–20%, concentrations of glycerol require the use of liquid nitrogen to achieve rapid cooling rates (−100°C/minutes) and are limited to polyolefin bags. These products are then stored at ≤−120°C. High, 40%–50% concentrations of glycerol are required for slower cooling rates (−1°C/minutes) achieved with mechanical freezers (≤−80°C). They can be frozen in either polyvinyl chloride (PVC) bags or polyolefin (preferred because they are less brittle) and stored at ≤−65°C. Most blood centers use the high-glycerol method.

RBCs collected and stored in citrate phosphate dextrose (CPD) or citrate phosphate dextrose adenine (CPDA-1) for up to 6 days, or up to 3 days postexpiration if

Transfusion Medicine and Hemostasis. https://doi.org/10.1016/B978-0-12-813726-0.00045-3

rejuvenated, can be frozen. RBC units stored in CPD/additive solution-1 (AS-1) or CP2D/AS-3 can be stored up to 42 days before freezing. RBCs stored in CPD/AS-1 can be rejuvenated then frozen but can only be stored frozen for up to 3 years. Frozen RBCs are FDA approved for storage up to 10 years; however, there are reports of successful use of RBCs frozen up to 37 years. An aliquot of the donor's serum or plasma should be frozen and stored for possible use if new donor screening tests are implemented.

Rejuvenation: FDA-approved rejuvenation solutions contain pyruvate, inosine, phosphate, and adenine. They are not intended for intravenous administration; after incubation with the solution, the RBC units are either frozen or washed and stored at 1–6°C for up to 24 hours. The solution may be added at any time between 3 days after collection and 3 days after expiration of the unit. Use of rejuvenation solution for liquid units less than 14 days of age is not recommended because the cells can develop supraphysiologic levels of 2,3-DPG resulting in decreased oxygen affinity.

Thawing and Deglycerolizing: To thaw a frozen product, the unit must be warmed to 37°C in a waterbath or dry warmer. The cells should be gently agitated during thawing to speed up the process, which takes approximately 10 minutes. Glycerol must be gradually removed to prevent hemolysis and be completely removed before it can be infused. Commercial instruments are available for this process. One method uses 12% NaCl to dilute the unit and then washes with 1.6% NaCl until deglycerolization is complete (usually 1–2 L is required). Then RBCs are suspended in isotonic saline (0.9%) with 0.2% dextrose.

Most institutions use an open system for deglycerolization. In 2006, the FDA approved an automated closed system for deglycerolization. The recovered RBCs can be resuspended and stored in AS-3 for up to 14 days or SAG-M for 7 days (not approved for use in the United States). Studies have demonstrated adequate in vivo and in vitro RBC quality and recovery.

Frozen RBCs from donors with sickle cell trait will form a jellylike mass and hemolyze during the deglycerolization process. Modified wash procedures are available to prevent the hemolysis, but many institutions screen donors for hemoglobin S before freezing because specialized wash procedures are not routinely available.

Refreezing: Refreezing may be indicated for extremely rare RBC products that were thawed unintentionally or unexpectedly not used. Products that are deglycerolized, stored ≤20 hours at 1–6°C then refrozen and rethawed when needed, demonstrate no loss in survival and adequate levels of ATP and 2,3-DPG.

Indications

Rare Products: RBC products with a rare phenotype may be frozen for transfusion to patients with antibodies to high-frequency antigens or with multiple antibodies. AABB and American Red Cross maintain the American Rare Donor Program (ARDP), AABB also has centralized database to locate antigen-negative units, the UnCommon Good Program, and International Society Blood Transfusion (ISBT) maintains the International Blood Group Reference Laboratory (IBGRL), which aid in the finding of rare RBC products. These products, which are frequently frozen, can be transported between blood centers for the treatment of patients throughout the world.

Autologous Units: Patients may choose to donate autologous units before a planned surgical intervention. If many units are required or procedure date is changed, units may be frozen to increase their storage time. Additionally, patients with alloantibodies may store frozen autologous RBCs for future use.

Cryopreserved Platelets Products: Platelets can be frozen and stored at ≤80°C for up to 2 years but is not FDA licensed. Several preservative solutions have been described, but the most widely used is dimethyl sulfoxide (DMSO). There are two cryopreservation protocols used to freeze platelets. The traditional method, involves suspension of the platelets in 6% DMSO solution, freezing and storage at ≤−80°C. Before transfusion, the product must be thawed at 37°C and DMSO removed by washing. Up to 25% of the platelets are lost, but the function is preserved. A newer method allows removal of most of the supernatant DMSO before freezing, which eliminates need for washing after thawing and increases platelet recovery. Animal and human studies have shown improved function of cryopreserved platelets over liquid-stored platelets; however, further studies are required. Cryopreservation is time-consuming, complex, and costly. Therefore, this product is not widely available. However, there has been renewed interest in this product over the last several years because of its successful use in military operations since 2001. Phase I clinical trials comparing cryopreserved with standard platelets are underway in the United States and other countries.

Further Reading

Cid, J., Escolar, G., Galan, A., et al. (2016). In vitro evaluation of the hemostatic effectiveness of cryopreserved platelets. *Transfusion*, *56*, 580–586.

Henkelman, S., Lagerberg, J., Reindert, G., Rakhorst, G., & van Oeveren, W. (2010). The effects of cryopreservation on red blood cell rheologic properties. *Transfusion*, *50*, 2393–2401.

Hess, J. (2004). Red cell freezing and its impact on the supply chain. *Transfus Med*, *14*, 1–8.

Rentas, F. (2011). Cryopreserved red blood cells. *J Trauma*, *70*, S45–S46.

CHAPTER 46

Washed Blood Products

Cheryl A. Goss, MD

Washing refers to a process that removes the noncellular fluid in red blood cell (RBC) and platelet products and replaces it, typically with saline. Usually, the process is performed in an open system where storage time is limited to 4–24 hours depending on the storage temperature. Washing removes >99% of plasma proteins (including antibodies) and original supernatant that may contain unwanted substances (e.g., anticoagulant-preservative solution, cytokines, electrolytes). Washing is indicated in a couple of clinical situations including recurrent severe allergic/anaphylactic reactions, and removal of potassium in large volume transfusion in pediatrics. Washing cellular products can take up to 2 hours and therefore limits its utility in emergent clinical situations.

Washed RBC Products: RBC products can be washed with 1–2 L of normal saline using a manual or automated method. The process can result in the loss of up to 20% of the RBC mass. If washing is performed in an open system, the unit can be stored for 24 hours at 1–6°C. More recently, closed system technology enables automated washing and extended storage of RBCs in additive solution for up to 14 days at 1–6°C. However, this technology has been limited mostly to cryopreserved RBCs after deglycerolization and is not FDA approved.

Washed Platelet Products: Platelets can be washed with normal saline, saline buffered with ACD-A or citrate, or platelet additive solutions (PAS) using a manual or automated method. Washing of platelets leads to platelet activation, loss of discoid shape, and reduced granule content. In addition, there is a significant loss of platelets (up to 30%) and reduced aggregation after washing. Recent studies have shown a difference in the functionality of washed platelets based on their storage time and suggest fresher platelets be selected for washing to improve their functionality, if possible. Washed platelets are stored at room temperature and must be used within 4 hours.

Possible Indications (Table 46.1)

Prevention of Severe Allergic/Anaphylactic Transfusion Reactions: The incidence of allergic transfusion reactions (ATRs), which is the most common complication of transfusion, is reported in 1%–3% of transfusions, but may be even higher in the setting of prospective, active surveillance. Most ATRs are mild and consist of urticaria with or without pruritus. Rarely, patients may experience anaphylactic reactions to blood products characterized by dyspnea, wheezing, hypotension, tachycardia, angioedema, shock and, potentially, death. Severe ATRs may be attributed to patients deficient in a particular plasma protein (e.g., IgA, haptoglobin, C3, and C4), food or medication (e.g., peanuts and aspirin) present in donor plasma. Products deficient in

Transfusion Medicine and Hemostasis. https://doi.org/10.1016/B978-0-12-813726-0.00046-5

TABLE 46.1 Indications and Rationale for Washed Blood Products

Indication	Rationale
Severe allergic or anaphylactic reaction	Removes allergenic plasma proteins
Maternal platelets for neonatal alloimmune thrombocytopenia	Removes maternal antibodies from maternal platelet product
Large-volume or rapid transfusion	Decrease the risk of hyperkalemia and cardiac arrhythmia in neonates and small children
Following irradiation and storage	Decreases potassium
Cardiopulmonary bypass	Recent randomized clinical trial demonstrated no clinical benefit
Paroxysmal nocturnal hemoglobinuria	No longer recommended

these plasma proteins are not usually readily available. Therefore, washed cellular products can be used to prevent severe allergic reactions.

A recent retrospective cohort study found that patients receiving washed apheresis platelets had a 95% reduction in the number of ATRs compared with 73% reduction in patients receiving volume-reduced platelets (see Chapter 62). Washing is superior because of more efficient removal of plasma proteins with washing (96% removal) versus volume reduction (51% removal). When cellular blood products are volume reduced, a small amount of plasma (around 50 mL) is left behind to provide glucose and buffering capacity. However, despite the improved removal of plasma proteins achieved by washing, washing has a more severe effect on the platelets than volume reduction.

Recently, the FDA approved the use of PAS. PAS-stored platelet products have 65% PAS and 35% residual donor plasma (about 70 mL). PAS platelets reduce the incidence of acute adverse reactions compared with plasma platelets, notably a 50% decrease in allergic reactions but would not be suitable for patients with a history of severe allergic/anaphylactic reactions.

ABO-Out-of-Group Transfusion: Because of inventory availability, a patient may be given an ABO-incompatible transfusion (i.e., group O platelet to group A recipient). Hemolytic transfusion reactions may be mitigated by washing, plasma reduction, use of PAS platelets, or use of low anti-A titer products (see Chapter 35).

Large Volume or Rapid Transfusion into Neonates and Small Children: Extracellular fluid in cellular blood products contain anticoagulant (citrate), dextrose, and, through storage, potassium, and lactate. RBCs rely on the membrane sodium–potassium pump to maintain higher intracellular potassium than the extracellular environment, and these pumps are energy adenosine triphosphate (ATP) dependent and temperature sensitive. During refrigeration of the RBC unit, potassium leaks out of the RBC. Irradiation of RBCs is known to cause injury to the RBC membrane, leading to increased permeability to potassium, shortening the storage time to maximum of 28 days. If longer stored (with or without irradiation) products are infused rapidly or in large volume (>25 mL/kg) to a neonate or pediatric patient, there is risk of hyperkalemia-induced cardiac arrhythmia, hyperglycemia, and citrate toxicity (see Chapter 33).

Cases of acute cardiac arrhythmia secondary to hyperkalemia leading to death have been reported in pediatric patients undergoing cardiac surgery, especially during rapid infusion through a central line. Washing or volume reduction has been shown to decrease the risk of hyperkalemia.

Neonatal Alloimmune Thrombocytopenia: Neonatal alloimmune thrombocytopenia (NAIT) is the most common cause of moderate to severe thrombocytopenia in an otherwise well newborn. It results from a maternal platelet alloantibody against the fetal platelet antigen (usually anti-HPA-1a). Infants often present with platelet counts <50,000/μL, which can lead to complications, most severely intracranial hemorrhage. Currently, no consensus guidelines are established that indicate what level of thrombocytopenia in the newborn warrants treatment in the absence of hemorrhage or how to treat (platelet transfusion, intravenous immunoglobulin). If platelet transfusion is indicated, the type of platelets used varies, depending on availability and urgency of use, including antigen negative (if the mother had anti-HPA-1a, product would be HPA-1a negative), random platelets (not platelet antigen tested), or maternal platelets, which can serve as an available source of compatible platelets for transfusion to the fetus/neonate. If used, maternal platelets must be washed to remove the antibody and irradiated to prevent transfusion-associated graft-versus-host disease (see Chapter 94).

Cardiopulmonary Bypass: Patients undergoing cardiac surgery may require the use of a cardiopulmonary bypass machine to replace the work of the heart and lungs during the procedure. The tubing of the bypass machine must be primed with fluid to prevent air from entering the patient. Depending on the size of the patient, the machine may require priming with RBCs to prevent excessive hemodilution. Neonates and small children usually need a blood prime, and once the machine is activated they receive a rapid, large volume infusion of RBCs making them at risk for the complications listed above. Therefore, cardiac surgery centers may wash the RBC units before the blood prime. However, the recently published REDWASH study (randomized trial of red cell washing for the prevention of transfusion-associated organ injury in cardiac surgery) demonstrated no benefit with washing; larger studies may be needed in the future.

Paroxysmal Nocturnal Hemoglobinuria: Paroxysmal nocturnal hemoglobinuria (PNH) is a manifestation of complement-mediated autoimmune hemolysis, thrombophilia, and marrow failure. Washing of RBC products for patients with PNH was advocated for many years, and practice remained unchecked for almost 40 years. It is currently accepted practice to provide group-specific RBCs and ABO-compatible plasma containing products to patients with PNH, and washing of RBC units is no longer indicated.

Further Reading

Tobian, A. A., Savage, W. J., Tisch, D. J., et al. (2011). Prevention of allergic transfusion reactions to platelets and red blood cells through plasma reduction. *Transfusion, 51*, 1676–1683.

Tynngard, N., Trinks, M., & Berlin, G. (2010). Platelet quality after washing: the effect of storage time before washing. *Transfusion, 50*, 2745–2752.

Valeri, C. R., Rango, G., Van Houten, P., et al. (2005). Automation of the glycerolization of red blood cells with the high-separation bowl in the Haemonetics APC 215 instrument. *Transfusion*, *45*, 1621–1627.

Weisbach, V., Riego, W., Strasser, E., et al. (2004). The *in vitro* quality of washed, prestorage leukocyte-depleted red blood cell concentrates. *Vox Sang*, *87*, 19–26.

Winkelhorst, D., Oepkes, D., & Lopriore, E. (2017). Fetal and neonatal alloimmune thrombocytopenia: Evidence based antenatal and postnatal management strategies. *Expert Rev Hem*, *10*, 729–737.

Woźniak, M. J., Sullo, N., Qureshi, S., et al. (2017). Randomized trial of red cell washing for the prevention of transfusion-associated organ injury in cardiac surgery. *Br J Anaesth*, *118*, 689–698.

CHAPTER 47

Volume-Reduced Blood Products

Patricia E. Zerra, MD and Cassandra D. Josephson, MD

Volume reduction (also known as hyperpacking or hyperconcentrating), a process performed after a blood component has been manufactured, is the removal of a portion of supernatant of a cellular blood product, such as red blood cell (RBC) or platelet unit. The supernatant contains residual plasma and anticoagulant-preservative solution, and its removal results in a more concentrated cellular product. This procedure is performed more often in pediatric than adult setting. In general, volume reduction requires a centrifugation step followed by expression of supernatant fluid and is performed on request, immediately before issue from transfusion service to the patient for administration. Most transfusion authorities suggest that this process modification should only be used in explicitly indicated circumstances, including prevention of transfusion of unnecessary fluid to volume-sensitive patient, and for removal of the potassium-containing supernatant to prevent hyperkalemia in an at-risk patient.

RBC Products

Methods: Volume-reduced products are manufactured using a modification of the RBC component preparation method described in Chapter 10. Volume-reduced method was described by Strauss et al. and enables manufacturing of small RBC aliquots, with a hematocrit of >90% using single RBC product until expiration on day 42, for neonates. Whole blood collected in CP2D is centrifuged at 5000 g for 5 minutes, the supernatant platelet-rich plasma is removed, and 100 mL of extended storage media, preferably AS-3 (as it does not contain mannitol), is added to and mixed with RBCs. The product is drained by gravity through a leukoreduction filter into a primary storage bag. Attached to this primary storage bag, by way of sterile connecting device, are a cluster of small volume bags. When an aliquot for transfusion is ordered, storage bag is centrifuged in an inverted position to pack the RBCs to a hematocrit of approximately 90%. RBC volume requested flows out into one of the attached small volume bags, which is subsequently disconnected. The remainder of the product is mixed and returned to storage. In addition, the product is mixed thoroughly each week.

A less complicated method of concentrating RBC products is by "inverted gravity sedimentation," which does not require a refrigerated centrifuge. This method stores RBC products in the refrigerator "upside down," which will concentrate an additive solution product to a hematocrit of around 70%–90% within 72 hours.

Indications: Two primary indications for volume-reduced RBCs are prevention of transfusion of unnecessary fluid to a volume-sensitive patient and removal of potassium-containing supernatant to prevent hyperkalemia in an at-risk patient. Volume reduction of RBC products to prevent passive transfusion of potentially hazardous

Transfusion Medicine and Hemostasis. https://doi.org/10.1016/B978-0-12-813726-0.00047-7

concentrations of potassium applies only to large volume transfusions (≥20 mL/kg). Transfusion safety without volume reduction has been established by Strauss. Concentration of extracellular potassium in the supernatant depends on the age of the RBC product and when/if the product was irradiated. Extracellular potassium increases from ~4 mmol/L at day 0 to ~60 mmol/L at day 42 in a nonirradiated additive solution RBC product, or to 80 mmol/L at day 35 in a CPDA-1 RBC product (Table 33.1). Therefore, a 1-kg infant receiving a 15 mL transfusion would only receive 0.3–0.4 mEq of potassium, which is a small amount compared with daily requirement of potassium of 2–3 mEq/kg. There have been several studies confirming lack of hyperkalemia or other untoward events for small volume transfusions of 42-day storage AS RBC products into neonates. There are individual patients, such as those with or at risk of hyperkalemia (i.e., small infants with only one vascular access point and tip of catheter near right atrium) and patients receiving large volume transfusion, who could be at an increased risk of transfusion-related hyperkalemia and may benefit from volume-reduced RBC products.

Adverse Effects: Volume reduction of RBC products does not appear to harm the RBCs. As the removal of supernatant can be performed in a closed system, expiration date of the product does not change.

Platelet Products

Methods: Pooled, individual platelet concentrates, and apheresis platelets may be volume reduced, if necessary. Stored platelets may be centrifuged at 20–24°C either at 580 g for 20 minutes, 2000 g for 10 minutes, or 5000 g for 6 minutes; optimal centrifugation rates and times have not been clearly determined. After centrifugation, platelets must rest without agitation for 20–60 minutes before transfusion. Because volume reduction is performed in an open system, expiration of platelet products is 4 hours, starting at time product was entered for processing.

Indications

ABO Out-of-Group Platelet Transfusion: When ABO-type-specific platelets are unavailable, volume reduction to remove anti-A and/or anti-B in platelet product plasma may be indicated. ABO-incompatible plasma most commonly causes weakly positive direct antiglobulin test but may result in severe acute intravascular hemolysis, which in rare circumstances is fatal. This possibility is especially likely in small children who have a higher "transfusion volume to total blood volume" ratio. The situation with highest risk of hemolysis is group O apheresis product containing high-titer anti-A (usually considered >1:64 IgM or >1:256 IgG) that is transfused to group A or AB recipients. Alternative strategies to mitigate hemolytic transfusion reaction risk include the following: (1) washing product or (2) titering group O platelet products for anti-A and using only those with non–high-titer anti-A for transfusion to group A or group AB individuals.

Prevention of Transfusion-Associated Cardiovascular Overload: Transfusion-associated cardiovascular overload (TACO) may be minimized by using volume reduction to platelet products. This is especially helpful in individuals who are sensitive to volume, such as low-birth-weight newborns and pediatric hematopoietic stem cell transplant patients.

Decrease in Febrile Nonhemolytic Transfusion Reactions: In adults, plasma-reduced platelet components decrease the incidence of febrile nonhemolytic transfusion reactions (FNHTRs). This effect is secondary to a decrease in leukocyte-derived cytokines, which accumulate in the plasma during storage. Decrease in FNHTRs is currently more often attained by prestorage leukoreduction.

Adverse Effects: The centrifugation process has untoward effects on platelets themselves, which may result in platelet loss, clumping, and dysfunction. Thus, AABB Committee on Pediatric Hemotherapy has recommended that "volume reduction of platelets should be reserved for special infants for whom marked reduction of all intravenous fluids is truly needed."

Hematopoietic Progenitor Cell Products: The indication for volume reduction of hematopoietic progenitor cell (HPC) products, particularly bone marrow and peripheral blood HPC-derived products, is ABO incompatibility between the donor and recipient, where donor has antibodies against recipient RBCs. HPC product can be centrifuged and plasma decanted before transfusion.

Further Reading

Josephson, C. D., Castillejo, M. I., Grima, K., & Hillyer, C. D. (2010). ABO-mismatched platelet transfusions: strategies to mitigate patient exposure to naturally occurring hemolytic antibodies. *Transfus Apher Sci*, *42*, 83–88.

Larrsson, L. G., Welsh, V. J., & Ladd, D. J. (2000). Acute intravascular hemolysis secondary to out-of-group platelet transfusion. *Transfusion*, *40*, 902–906.

Sherwood, W. C., Donato, T., Clapper, C., & Wilson, S. (2000). The concentration of AS-1 RBCs after inverted gravity sedimentation for neonatal transfusions. *Transfusion*, *40*, 618–619.

Strauss, R. G. (2000). Data-driven blood banking practices for neonatal RBC transfusions. *Transfusion*, *40*, 1528–1540.

Strauss, R. G. (2008). How I transfuse red blood cells and platelet to infants with anemia and thrombocytopenia of prematurity. *Transfusion*, *48*, 209–217.

Strauss, R. G., Villhauser, P. J., & Cordle, D. G. (1995). A method to collect, store and issue multiple aliquots of packed red blood cells for neonatal transfusions. *Vox Sang*, *68*, 77–81.

CHAPTER 48

Pathogen Reduction Technologies

Susanne Marschner, PhD and Lina Y. Dimberg, PhD

Blood safety is of major importance in transfusion medicine. To decrease safety risks, a combination of donor education, screening, and testing for selected agents has been implemented. Pathogen reduction (PR) technology (PRT, also known as pathogen inactivation) is a proactive approach to reduce contaminating pathogens.

HIV in the blood supply resulted in many infected patients in the 1980's. This spurred development of methods to better safeguard the blood supply, including implementation of enhanced donor screening criteria and sensitive tests based on nucleic acid detection or immunologic methods. These safety measures have drastically reduced risks of transfusion-transmitted infections. However, pathogen screening does not eliminate the risk completely, due to the window period where tests are negative but the unit may be infectious. Furthermore, emerging infectious agents such as chikungunya, West Nile, dengue, Ebola, and Zika viruses have increased through international travel, shipment of goods, and changing climatic conditions. Development of tests and deferral policies may take time. Thus, PR represents a potential proactive rather than reactive approach. Additionally, PR inactivates donor white blood cells (WBCs), thereby eliminating the need for irradiation to prevent TA-GVHD. Several PRT systems are in clinical practice (Table 48.1).

Pathogen Reduction Technologies

Solvent/Detergent Treatment: S/D treatment, which is limited to plasma, includes filtration, pathogen inactivation via membrane disruption with detergents and solvents, followed by extraction of detergents and solvents, and finally additional sterile filtration. Octaplas (Octapharma, Lachen, Switzerland) S/D treatment is performed in pools of ~2500 donors and inactivates lipid-enveloped viruses, while parasites and bacteria are eliminated through the filtration process. A new product, Octaplas LG, has an additional step that eliminates prions. Numerous clinical studies over three decades have demonstrated the safety and efficacy of S/D-treated plasma products. Pooling and extra screening additionally decreases the risk of allergic reactions and TRALI.

THERAFLEX Methylene Blue: THERAFLEX Methylene Blue (MB-Plasma) (Macopharma, Lille, France) treats single plasma units with a photoactive phenothiazine dye, methylene blue, which associates with nucleic acids and, when exposed to visible light, generates a photodynamic reaction that modifies nucleic acids and prevents replication. This technique can inactivate enveloped viruses and has been used in Europe for 20 years. Methylene blue is not effective against intracellular viruses. Methylene blue plasma product has generally been regarded as safe.

Transfusion Medicine and Hemostasis. https://doi.org/10.1016/B978-0-12-813726-0.00048-9

TABLE 48.1 Pathogen Reduction Technologies

Product (Manufacturer)	Mechanism of Action	Pathogen Reduction	Transfusion Component Licensure			
			Plasma	Platelets	Whole Blood	Red Blood Cells
INTERCEPT (Cerus)	Amotosalen + UVA light	• Enveloped viruses • Nonenveloped viruses • Bacteria • Parasites	• CE marked • FDA approved	• CE marked • FDA approved		
INTERCEPT (Cerus)	S-303	• Enveloped viruses • Nonenveloped viruses • Bacteria • Parasites				Phase 3 clinical trials (US, EU)
Mirasol (Terumo BCT)	Riboflavin + UVB light	• Enveloped viruses • Nonenveloped viruses • Bacteria • Parasites	• CE marked	• CE marked • Pivotal clinical trials (US)	• CE marked	Pivotal clinical trials (US)
THERAFLEX (Macopharma)	UV-C light + agitation	• Enveloped viruses • Nonenveloped viruses • Bacteria • Parasites		• CE marked		
THERAFLEX (Macopharma)	Methylene blue + visible light exposure + filtration	• Enveloped viruses • Nonenveloped viruses • Bacteria • Parasites	• CE marked			
Octaplas (Octapharma)	Solvent/detergent	• Enveloped viruses • Bacteria • Parasites • Prions (OctaplasLG)	• CE marked • FDA approved			

THERAFLEX UV-C: The THERAFLEX UV-Platelets (Macopharma, Langen, Germany) uses UV-C light to inactivate bacteria, viruses, and protozoa without a photosensitizer during platelet agitation. However, HIV is resistant to UV-C light. In a dose-escalation trial, repeated transfusion of autologous UV-C-treated platelets was well tolerated. A randomized, double-blind Phase III trial, CAPTURE, comparing clinical efficacy and safety of UV-C-treated platelets and control platelets in thrombocytopenic patients is underway.

Mirasol (Riboflavin/UV Light): Mirasol PRT System (TerumoBCT, Lakewood, CO) uses the natural photosensitizer riboflavin (vitamin B2) plus UV light (280–400 nm) to mediate selective, irreversible damage to nucleic acids, thus impairing replication and repair processes of enveloped and nonenveloped viruses, bacteria, protozoa, and WBCs. More than 700,000 Mirasol plasma and platelet units have been transfused. Mirasol has been evaluated in the clinical setting and demonstrated efficacy to provide adequate support in thrombocytopenic patients requiring platelet transfusion (MIRACLE study) and in patients requiring plasma transfusion (acquired and chronic coagulopathies). Two large noninferiority clinical trials with platelets stored in platelet additive solution (IPTAS) and in plasma (PREPAReS) have recently been completed. The IPTAS study includes separate assessments of both Mirasol-treated and INTERCEPT-treated platelets (discussed below) compared with control platelets. The study closed before completion, thus lacked statistical power; however, incidence of grade 2 ≤ bleeding was similar between study arms. In the PREPAReS trial, Mirasol-treated platelets were compared with control platelets. The primary endpoint of noninferiority in terms of grade 2, 3, and 4 bleeding was met in the intent to treat (ITT) population.

Mirasol is currently under evaluation for the treatment of whole blood. The AIMS study demonstrated a reduction of transfusion-transmitted malaria in patients receiving Mirasol-treated whole blood versus control whole blood. This study demonstrated for the first time that PRT can reduce pathogen transfusion transmission. RBCs derived from Mirasol-treated whole blood have been evaluated in two clinical trials in healthy volunteers, IMPROVE and IMPROVE II. FDA criteria of in vivo recovery were exceeded with no increase of adverse events. A pivotal FDA licensure trial in transfusion-dependent thalassemic patients, PRAISE, is underway.

INTERCEPT (Amotosalen/UVA): INTERCEPT Blood System (Cerus Corp., Concord, CA) uses amotosalen, which binds to nucleic acids and lipids, and intercalates into nucleic acids. Illumination by UV light (320–400 nm) leads to the formation of covalent bonds between pyrimidine bases, preventing replication and transcription. An adsorption step is required to remove amotosalen and its photoproducts. The treatment has been shown to inactivate enveloped and nonenveloped viruses, bacteria, protozoa, and WBCs. Over 2,000,000 products have been transfused. INTERCEPT-treated products have been evaluated in several clinical studies for platelets (SPRITE, SPRINT, TESSI, HOVON-82, IPTAS) in thrombocytopenic patients and plasma in various clinical settings (acquired and chronic coagulopathies) and has been demonstrated to provide adequate transfusion support.

INTERCEPT (Amustaline/Glutathione): INTERCEPT Blood System for RBCs uses amustaline (S-303; Cerus Corporation, Concord, CA) and glutathione, a quenching agent, in a process involving a nucleic-acid-targeted alkylation and no UV light. Treatment with the first-generation S-303 PRT resulted in the formation of antibodies against neo-epitopes on RBCs. To mitigate the immune reactivity and improve RBC storage, the process was modified to include a higher concentration of glutathione with a neutral pH, and the RBCs were resuspended in a standard RBC solution posttreatment. RBCs treated with the modified technology have been evaluated in healthy subjects, demonstrating sufficient in vivo posttransfusion recovery and quality. In a randomized controlled double-blind study in cardiac surgery patients, the frequency of adverse events was similar in patients transfused with S-303-treated RBCs versus control RBCs. Two trials are in progress: one in thalassemia patients in Europe (SPARC) and one in ZIKV endemic regions (RedeS).

PR Blood Components

Plasma: THERAFLEX Methylene Blue, Mirasol, Octaplas, and INTERCEPT are currently used clinically for PR plasma. The two latter are FDA approved. The main drawback to PR plasma is that coagulation factors are reduced (typically up to 30% depending on factor and method). Hemostatic function is maintained, and studies in liver transplant patients demonstrated that blood loss and changes in coagulation factors were not significantly different. Small trials in patients with thrombotic thrombocytopenic purpura confirmed that treated plasma was safe and effective. Some studies indicate a need to transfuse PR plasma more frequently or in higher volumes. Notably, Octaplas has substantially lower risk of allergic reactions and no reported cases of TRALI.

Platelets: THERAFLEX UV-C, INTERCEPT, and Mirasol are available PR platelets. INTERCEPT is FDA approved. A Cochrane metaanalysis included 15 trials and 2075 participants and concluded that in thrombocytopenic hematology/oncology patients, there is no evidence of increased clinically significant bleeding complications with the use of PR platelets, and probably no difference in the incidence of serious adverse events. There is evidence that PR platelet transfusions increase risk of platelet refractoriness, increase platelet transfusion requirement, and decrease corrected count increment. No differences in adverse events or mortalities were reported.

Red Blood Cells: Two PR RBCs are in clinical development: Mirasol (via whole blood treatment) and INTERCEPT amustaline/glutathione. PR may affect the storage lesions, resulting in reduced shelf life (21 days for Mirasol-treated RBCs and 35 days for INTERCEPT-treated RBCs), but FDA criterion for hemolysis and in vivo recovery are met.

Whole Blood: PR whole blood is available with Mirasol. As with componentized blood products, PR has some impact on plasma coagulation factors, platelet efficacy and hemolysis, limiting the shelf life to 14 or 21 days depending on storage condition. Mirasol-treated whole blood is CE marked for transfusion.

Summary: Benefits of using PR products are mitigation of transfusion-transmitted infections, including emergent viruses and residual risks. Additionally, nucleic-acid-targeted PRTs eliminate the need for γ-irradiation of cellular blood components.

However, PR blood components are more expensive than conventional ones; treatment may impact the quality and shelf life; and PRTs are limited in terms of applications and efficiency. Therefore, careful cost-effectiveness analysis and risk analysis (such as risk-based decision making) should precede PRT implementation.

Further Reading

Allain, J. P., Owusu-Ofori, A. K., Assennato, S. M., et al. (2016). Effect of plasmodium inactivation in whole blood on the incidence of blood transfusion-transmitted malaria in endemic regions: The African Investigation of the Mirasol System (AIMS) randomised controlled trial. *Lancet, 387*(10029), 1753–1761.

Drew, V. J., Barro, L., Seghatchian, J., & Burnouf, T. (2017). Towards pathogen inactivation of red blood cells and whole blood targeting viral DNA/RNA: Design, technologies, and future prospects for developing countries. *Blood Transfus*, *15*(6), 512–521.

Estcourt, L. J., Malouf, R., Hopewell, S., et al. (2017). Pathogen-reduced platelets for the prevention of bleeding. *Cochrane Database Syst Rev*, *7*, CD009072.

Irsch, J., & Seghatchian, J. (2015). Update on pathogen inactivation treatment of plasma, with the INTERCEPT Blood System: Current position on methodological, clinical and regulatory aspects. *Transfus Apher Sci*, *52*(2), 240–244.

Liumbruno, G. M., Marano, G., Grazzini, G., Capuzzo, E., & Franchini, M. (2015). Solvent/detergent-treated plasma: A tale of 30 years of experience. *Expert Rev Hematol*, *8*(3), 367–374.

Lozano, M., Cid, J., & Muller, T. H. (2013). Plasma treated with methylene blue and light: Clinical efficacy and safety profile. *Transfus Med Rev*, *27*, 235–240.

Rebulla, P., Vaglio, S., Beccaria, F., et al. (2017). Clinical effectiveness of platelets in additive solution treated with two commercial pathogen-reduction technologies. *Transfusion*, *57*(5), 1171–1183.

CHAPTER 49

Neonatal and Pediatric Transfusion Medicine

Patricia E. Zerra, MD, Jeanne E. Hendrickson, MD and
Cassandra D. Josephson, MD

Red Blood Cell Transfusions

RBC Transfusion Considerations in Neonates: Neonatal RBC transfusion thresholds are not clearly defined, with gestational age, postnatal age, and clinical condition being important considerations. Two randomized control trials (RCTs) (Bell et al. and Kirpalani et al.) comparing liberal to restricted hemoglobin (hb) transfusion thresholds in preterm neonates had conflicting results regarding neurocognitive outcomes, making general recommendations difficult. Two RCTs (one in the United States and one in Europe) are on going to reexamine this question. An additional consideration is that severe anemia (≤8 gm/dL), but not RBC transfusion, was recently shown to be associated with an increased risk of necrotizing enterocolitis.

RBC Product Selection for Neonates: The type of anticoagulant-preservative solution (CPDA-1 with hematocrit ~70% vs. AS with hematocrit ~55%–60%) does not pose significant risk to premature infants and neonates when *small*-volume transfusion (i.e., 10–15 mL/kg of RBCs) is given. Regarding the relative length of storage of products transfused to neonates, a recent RCT demonstrated no difference in outcomes with the use of fresh (≤7 days) versus standard-issue RBCs in premature low-birth-weight infants (age of red blood cells in premature infants trial).

Although the rise in plasma potassium after a *small*-volume transfusion is minimal, a risk of hyperkalemia in neonates receiving *large*-volume transfusions from stored units exists. Fresher, washed, or volume-reduced RBCs have lower potassium loads. Large-volume transfusions also increase the risk of citrate toxicity resulting in hypocalcemia; thus, the transfusion recipient's electrolytes should be carefully monitored.

Leukoreduction: Leukoreduced RBC products are typically provided for neonates in the United States. Leukoreduction decreases the risk of febrile transfusion reactions, human leukocyte antigen (HLA) alloimmunization, and cytomegalovirus (CMV) transmission (leukoreduced products are CMV risk-reduced).

Cytomegalovirus-Seronegative: Some institutions provide blood from CMV-seronegative donors (CMV-negative) for very-low-birth-weight infants (<1500 g) who are at highest risk of contracting CMV. However, given the low risk of transfusion-transmitted cytomegalovirus (TT-CMV) with modern leukoreduction techniques, others have recommended the use of leukoreduced blood from CMV untested donors as an acceptably safe and low-risk alternative. A pilot study determined that this strategy

Transfusion Medicine and Hemostasis. https://doi.org/10.1016/B978-0-12-813726-0.00049-0

was effective, with zero cases of TT-CMV detected in infants (n = 20) transfused with leukoreduced only blood. Notably, most neonatal CMV cases result from transmission from the mother.

Irradiation: Given the potential for an undiagnosed cellular immune deficiency, which would increase the neonate's risk of transfusion-associated graft-versus-host disease (TA-GVHD), cellular blood products given to infants <4–6 months of age should be irradiated. Irradiated RBCs should be provided to neonates and children with congenital immunodeficiencies, hematologic malignancies, undergoing hematopoietic progenitor cell transplantation, requiring treatment with fludarabine, or other purine analogues, having significant immune suppression for other reasons, and receiving products from relatives.

RBC products should be irradiated close temporally to a large-volume transfusion, especially in patients more sensitive to increases in potassium. Alternatively, the potassium level can be decreased through washing or volume reduction.

Neonatal RBC Compatibility Testing: Given that alloimmunization to transfused RBCs in infants under 4 months of age is extremely unusual, AABB *Standards* require limited pretransfusion testing for neonates. ABO and RhD forward typing must be done once, and neonatal plasma/serum must initially be tested for passively acquired maternal anti-A or anti-B using the antiglobulin phase if non-group-O RBCs are to be transfused. One antibody screen must be performed on a sample from either the neonate or mother to ensure maternal RBC alloantibodies are not present. During a single hospitalization, repeat antibody screens are not necessary (until 4 months of age) if the initial screen is negative. Compatibility testing and repeat ABO/D typing are necessary only in certain circumstances, such as an initial positive antibody screen or passively acquired maternal anti-A or anti-B, respectively.

Neonatal RBC Exchange Transfusion Considerations: Exchange transfusion (ET) is used primarily in neonates with hemolytic disease of the newborn, to decrease bilirubin levels and to remove offending antigen-positive RBCs. Cord bilirubin levels >5 mg/dL, bilirubin levels that rise >1 mg/dL per hour, or indirect bilirubin levels >20 mg/dL are all indications for ET. RBC products used for ET should be irradiated, hemoglobin S-negative, volume reduced, and reconstituted with plasma to a hematocrit of about 50%. Postprocedure labs must be checked to ensure hemoglobin, platelet count, prothrombin time, partial thromboplastin time, and fibrinogen levels are appropriate (see Chapter 50).

RBC Transfusions in Children and Adolescents: An RBC transfusion "dose" for neonates and children is 10–15 mL/kg of RBCs, which can increase the hemoglobin approximately 2–3 g/dL. A study of RBC transfusion thresholds (Lacroix et al.) in a pediatric intensive care unit setting found a hemoglobin threshold of 7 g/dL for RBC transfusion can decrease transfusion requirements without increasing adverse outcomes in stable, critically ill children. However, in children with chronic anemias such as thalassemia major, higher hemoglobin levels may be necessary for adequate growth/development and for suppression of erythropoiesis.

RBC Product Selection in Children and Adolescents: Indications for leukoreduction, CMV-seronegative products, and irradiation are similar to those for adults and are discussed in Chapters 42–44.

Platelet Transfusions

Platelet Transfusions in Neonates: Thrombocytopenia in neonates is due to decreased production or accelerated destruction of platelets. Most cases are multifactorial, with prematurity and infection playing a role. The diagnosis of neonatal alloimmune thrombocytopenia (NAIT) or maternal immune thrombocytopenia (ITP) should be considered in otherwise healthy infants with thrombocytopenia. Both NAIT and ITP occur due to passively transferred maternal antibodies against fetal/neonatal platelet antigens.

Platelet transfusion triggers for neonates are not clearly defined and depend on the clinical situation and gestational age. Generally, platelet counts are kept above 50,000/μL in bleeding neonates or during invasive procedures and above 100,000/μL in extremely ill, premature infants. Prophylactic transfusions are generally given for platelet counts below 20,000/μL, with higher thresholds in NAIT. Transfusion of 10 mL/kg of platelets generally leads to an expected increase in platelet count of 50–100,000 μL^{-1}.

Platelet Transfusions in Children: Indications for platelet transfusions in children are similar to those in neonates. Platelets are dosed by volume in children (~10 mL/kg/dose, or approximately 1 unit of whole blood–derived platelets per 10 kg).

Platelet Product Selection

ABO/RhD Compatibility: Platelet components should be ABO-compatible, if possible. The transfusion of group O platelets with high-titer isohemagglutinins to group A or B recipients should be avoided if possible, due to the passive transfer of anti-A or anti-B, which can cause hemolysis. Platelet-additive solution decreases the isohemagglutinin burden, by replacing much of the plasma.

Leukoreduction: Leukoreduced platelet products are typically provided for infants and children in the United States to prevent CMV transmission, HLA alloimmunization, and febrile reactions.

Cytomegalovirus-Seronegative: Leukoreduced, CMV risk-reduced (see RBC recommendations) platelets are generally provided for infants and children.

Irradiation: Irradiation of platelets is indicated for situations similar to those as outlined in the RBC section above. In addition, platelets specifically selected for compatibility (e.g., crossmatched or HLA matched) should be irradiated to prevent TA-GVHD.

Washing or Volume Reducing: Maternal platelets transfused in instances of NAIT must be washed before transfusion, to remove the offending antibody-containing plasma. Rarely, washed products are needed for patients with recurrent allergic reactions. It should be noted, that washing (or volume reducing) results in platelet loss.

Pathogen Inactivation: Scant data exist in the United States on the transfusion of pathogen-inactivated platelets to neonates or children, given the recent FDA approval of this product.

Plasma Transfusions: Reference values of many coagulation factors for children under 6 months of age are different than those for children and adults. Plasma may be transfused for the prevention or treatment of bleeding in patients with acquired or congenital defects (e.g., severe liver disease or massive transfusion). In neonates and children, 10–15 mL/kg of plasma will raise coagulation factor levels by approximately 20%.

Plasma Product Selection: FP24 (plasma frozen within 24 hours of collection) from whole blood and plasma frozen within 24 hours after phlebotomy held at room temperature up to 24 hours after phlebotomy (PF24RT24) prepared from apheresis collections have lower levels of factor VIII than FFP; because plasma is not used to treat factor VIII deficiency, FFP, FP24, and PF24RT24 are typically used interchangeably. Thawed plasma, stored at 1–6°C for up to 5 days, has degradation of factors over time (especially factor VIII); few pediatric studies have evaluated the use of thawed plasma outside of trauma scenarios.

Cryoprecipitate Transfusions: Cryoprecipitate is primarily transfused to increase fibrinogen and is dosed in neonates and children at 1–2 units/10 kg; this equates to approximately 2–5 mL of cryoprecipitate/kg.

Cryoprecipitate Product Selection: It is generally recommended that neonates receive ABO-compatible cryoprecipitate.

Granulocyte Transfusions: Granulocytes may be more efficacious in children than in adults, due to the size of the recipient allowing for larger doses per kg. Indications include bacterial sepsis unresponsive to antibiotics in patients with severe neutropenia, severe infection unresponsive to antibiotics in patients with qualitative neutrophil defects (e.g., chronic granulomatous disease), and refractory fungal sepsis in neutropenic oncology patients.

Granulocyte Product Selection: Granulocytes must be ABO and crossmatch compatible with the recipient and must also be irradiated. CMV-seronegative recipients should receive granulocytes from CMV-seronegative donors if possible.

Further Reading

Bell, E. F., Strauss, R. G., Widness, J. A., et al. (2005). Randomized trial of liberal versus restrictive guidelines for red blood cell transfusion in preterm infants. *Pediatrics, 115*, 1685–1691.

Estcourt, L. J., Stanworth, S., Doree, C., et al. (2015). Granulocyte transfusions for preventing infections in patients with neutropenia or neutrophil dysfunction. *Cochrane Database Syst Rev*, CD005341.

Fergusson, D. A., Hebert, P., Hogan, D. L., et al. (2012). Effect of fresh red blood cell transfusions on clinical outcomes in premature, very low-birth-weight infants. *JAMA, 308*, 1443–1451.

Josephson, C. D., Caliendo, A. M., Easley, K. A., et al. (2014). Blood transfusion and breast milk transmission of cytomegalovirus in very-low-birth-weight infants: A prospective cohort study. *JAMA Pediatr*, *168*, 1054–1062.

Josephson, C. D., Granger, S., Assmann, S. F., et al. (2012). Bleeding risks are higher in children versus adults given prophylactic platelet transfusions for treatment-induced hypoproliferative thrombocytopenia. *Blood*, 748–760.

Kirpalani, H., Whyte, R. K., Andersen, C., et al. (2006). The premature infants in need of transfusion (PINT) study: A randomized, controlled trial of a restrictive (low) versus liberal (high) transfusion threshold for extremely low birth weight infants. *J Pediatr*, *149*, 301–307.

Lacroix, J., Hebert, P. C., Hutchison, J. S., et al. (2007). Transfusion strategies for patients in pediatric intensive care units. *N Engl J Med*, *356*, 1609–1619.

Patel, R. M., Knezevic, A., Shenvi, N., et al. (2016). Association of red blood cell transfusion, anemia, and necrotizing entercolitis in very low-birth-weight infants. *JAMA*, *315*, 889–897.

Sparger, K. A., Assmann, S. F., Granger, S., et al. (2016). Platelet transfusion practices among very-low-birth-weight infants. *JAMA Pediatr*, *170*, 687–694.

Strauss, R. G. (2010). RBC storage and avoiding hyperkalemia from transfusions to neonates and infants. *Transfusion*, *50*, 1862–1865.

Venkatesh, V., Khan, R., Curley, A., et al. (2012). The safety and efficacy of red cell transfusions in neonates: A systematic review of randomized controlled trials. *Br J Haematol*, *158*, 370–385.

CHAPTER 50

Perinatal Transfusion Medicine

Nancy L. Van Buren, MD

Transfusion management of the pregnant woman and fetus requires special consideration. This chapter will address the following related issues: (1) routine prenatal and neonatal transfusion testing in relationship to maternal red blood cell (RBC) alloimmunization, (2) hemolytic disease of the fetus and newborn (HDFN), and (3) transfusion management of HDFN, including maternal, fetal, and neonatal testing and treatment. HDFN occurs when maternal plasma contains an RBC alloantibody against an antigen carried on the fetal RBCs, resulting in hemolytic anemia.

The diagnosis and management of HDFN includes maternal, fetal, and neonatal testing and treatment. HDFN occurs when maternal plasma contains an RBC alloantibody against an antigen carried on the fetal RBCs, resulting in hemolytic anemia. The administration of Rh immune globulin (RhIg) perinatally has dramatically decreased the incidence of HDFN due to anti-D. The primary goals of prenatal testing are to determine which women would benefit from RhIg prophylaxis and which women/fetuses require further monitoring/treatment for HDFN.

Other obstetrical issues associated with transfusion medicine that may occur are discussed in different chapters of this book. These include obstetrical complications resulting in massive transfusion, such as placenta previa, uterine atony or rupture, and disseminated intravascular coagulopathy (see Chapter 117). These obstetrical complications can lead to hysterectomy and loss of future reproductive capacity, and/or loss of the mother, child, or both. In addition, thrombocytopenia can occur in pregnancy and may be secondary to immune thrombocytopenia (ITP); thrombotic thrombocytopenic purpura ; hemolysis, elevated liver enzymes, and low platelets (HELLP syndrome); and acute fatty liver of pregnancy. Transfusion management during pregnancy in patients with hemoglobinopathies, such as transfusion management of sickle cell disease and thalassemia (see Chapter 52). Lastly, neonatal alloimmune thrombocytopenia occurs as a result of maternal platelet alloantibodies directed against an antigen on fetal platelets resulting in thrombocytopenia, which can result in intracranial hemorrhage (see Chapter 94).

Hemolytic Disease of the Fetus and Newborn: HDFN occurs when maternal plasma contains an alloantibody against an antigen carried on the fetal RBCs. The maternal IgG crosses the placenta and coats the fetal RBCs. The sensitized RBCs are removed from circulation by splenic macrophages, which leads to fetal anemia. In an effort to compensate for the RBC loss, bone marrow erythropoiesis is stimulated, and release of immature RBCs results in erythroblastosis fetalis. When the bone marrow fails to produce enough RBCs, then extramedullary erythropoiesis occurs in the spleen and liver. The enlarged spleen and liver (hepatosplenomegaly) results in hepatocellular damage associated with insufficient production of plasma proteins, leading

Transfusion Medicine and Hemostasis. https://doi.org/10.1016/B978-0-12-813726-0.00050-7

to high-output cardiac failure with generalized edema, effusions, and ascites (hydrops fetalis). Hydrops fetalis may develop as early as 17 weeks of gestational age and was previously uniformly fatal. With current management strategies, including intrauterine transfusions (IUTs) and other therapeutic modalities, there is a ~75% survival rate. Severe nonhydropic HDFN, requiring IUT, has a ~90% survival rate. If severe anemia and/or hydrops fetalis develop before the ability to perform IUT (before 18 weeks of gestational age), treatment may include a combination of maternal plasma exchange and intravenous immunoglobulin (IVIG) until IUT is possible.

In utero, the bilirubin released from hemolyzed RBCs is cleared by the placenta. After birth, the neonatal liver has limited capacity to conjugate the bilirubin. When increased levels of unconjugated bilirubin exceed the albumin-binding capacity, the unbound, unconjugated bilirubin crosses the blood–brain barrier and results in neuronal cell death in the basal ganglia and brain stem (known as kernicterus). Treating the neonatal hyperbilirubinemia by phototherapy and RBC exchange, if needed, prevents kernicterus. Some recommend the use of IVIG if the bilirubin level is not sufficiently lowered by phototherapy in an attempt to avoid exchange transfusion. Guidelines for detection, management, and use of phototherapy, and RBC exchange transfusion for hyperbilirubinemia are published by the American Academy of Pediatrics Subcommittee on Hyperbilirubinemia.

Antibody titer and specificity (anti-D, anti-c, and anti-K have the highest likelihood of severe HDFN), immunoglobulin class, and number of antigenic sites on the RBC influence disease severity. In general, the severity of HDFN increases with subsequent pregnancies. Anti-A and/or anti-B are the most common antibodies associated with HDFN, but the disease is usually mild.

Immune sensitization to RBC antigens occurs after fetomaternal hemorrhage (FMH) during pregnancy or delivery, or through previous RBC transfusion. As little as 0.1 mL of D-positive RBCs may result in D sensitization. The incidence of maternal RhD antigen sensitization decreases with ABO-incompatibility between the fetus and mother. Because of the use of RhIg prophylaxis, the incidence of anti-D formation has decreased from 14% to 0.1% of RhD-negative mothers.

Prenatal and Neonatal Transfusion Testing

Testing for RhD Type: At the first prenatal visit, usually at 12 weeks of gestational age, the maternal ABO/RhD type and antibody screen are performed. A challenging area in the laboratory has been properly identifying patients who have a variant D phenotype. These individuals have altered or weakened expression of their D antigen, which often requires a sensitive antiglobulin test or other method to detect, referred to as a serologic weak D phenotype. Depending on the methodology and/or reagents, these patients may test as D-negative, weak D, or D-positive. This leads to confusion about whether these patients should be considered as candidates for RhIg prophylaxis and transfused D-negative blood products.

The three broad categories of D variants include weak D, partial D, and the rare DEL phenotypes, which are most common among those of Asian ancestry and are not detected by conventional serologic typing methods. In particular, individuals with weak D and partial D are often determined to have a serologic weak D phenotype when tested in the transfusion service. Weak D is a quantitative polymorphism resulting in reduced expression of D antigen; whereas, people with partial D can make alloanti-D because of a qualitative polymorphism resulting in an altered D antigen epitope. The majority of individuals with serologic weak D phenotype of European ancestry are weak D types 1, 2, or 3, and will not form anti-D when exposed to D-positive cells, but the percentage varies depending on ethnic background. The remainder, including an estimated 10% of individuals serotyped as weak D, who are partial D variants, can potentially form anti-D and should be managed as if they were D-negative. In 2015, recommendation for *RHD* genotyping for pregnant women and females of childbearing potential with a serologic weak D phenotype, was endorsed by AABB, America's Blood Centers, the American Red Cross, the Armed Forces Blood Program, the College of American Pathologists, and the College of Obstetricians and Gynecologists. Specifically, an algorithm was published for resolving serologic weak D phenotype test results by genotyping to determine candidacy for RhIg and RhD type for transfusions (see Fig. 50.1).

RhIg Prophylaxis: If a pregnant woman is D-negative and is not sensitized to the D antigen, or has a serologic weak D phenotype that should be managed as D-negative, then she should receive RhIg perinatally.

Antepartum and postpartum dosing is usually with 300 μg of RhIg (which neutralizes 30 mL of whole blood or 15 mL of D-positive RBCs), although some institutions prefer to give 50 μg (which neutralizes 5 mL of whole blood or 2.5 mL of D-positive RBCs) at ≤12 weeks of gestational age, particularly with abdominal trauma or threatened

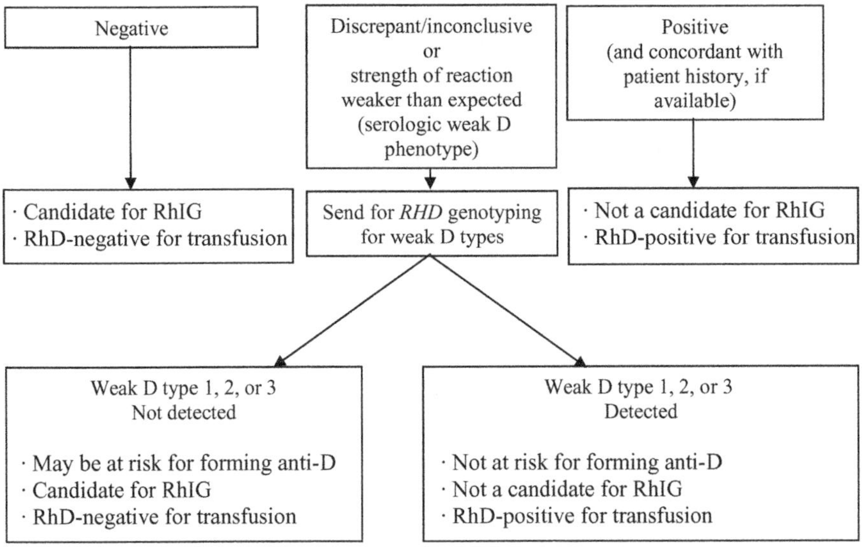

FIGURE 50.1 **Algorithm for Managing a Serologic Weak D Phenotype. The algorithm illustrates how RHD genotyping can determine which individuals with a serologic weak D phenotype can be managed safely as Rh-positive.** (Reproduced from the journal Sandler, S. G., Flegel, W. A., Westhoff, C. M., et al. (2015). It's time to phase in RHD genotyping for patients with a serologic weak D phenotype. *Transfusion, 55*, 680–689 with permission of Wiley Periodicals, Inc.)

miscarriage, but there is a risk of inadvertent misadministration of the lower dose (see Chapter 40). The half-life of passive anti-D is 3–4 weeks and is detectable in over half of women who deliver less than 76 days after administration of RhIg.

RhIg is administered at:

- 28 weeks of gestational age (dose 300 µg);
- at delivery, if the neonate is D-positive, weak D-positive, or D-untested (minimum dose of 300 µg, further dosing determined by FMH testing); and
- after perinatal events associated with FMH, such as abortion, ectopic pregnancy, amniocentesis, chorionic villus sampling, external cephalic version, abdominal trauma, and antepartum hemorrhage (minimum dose of 300 µg, further dosing determined by FMH testing, if >20 weeks of gestational age).

After a perinatal event or delivery beyond 20 weeks of gestational age, when the fetal blood volume exceeds 30 mL, quantification of FMH is recommended. In all situations, RhIg should be administered within 72 hours of the event. If a dose is not administered during that time frame, then it should be administered as soon as possible, even up to 28 days after the event. While passive anti-D may be detectable up to 6 months after administration, very few women (less than 3%) will have demonstrable anti-D after 76–95 days of RhIg administration. Therefore, any woman with detectable anti-D beyond this timeframe may not have had adequate antepartum protection and should be considered suspicious for the development of immune anti-D.

Fetomaternal Hemorrhage Testing: The goal of FMH testing is to determine an adequate dose of RhIg to neutralize fetal D-positive cells in the maternal circulation; thus, preventing maternal alloimmunization to RhD antigen. A sample for FMH testing should preferably be obtained approximately 1 hour after delivery on all D-negative women who deliver a D-positive infant (or 1 hour after an event as described above). FMH testing can be performed by a screening test, which is typically the rosette test. If positive, quantification of FMH is usually performed by an acid elution test (Kleihauer–Betke test) or by flow cytometry. Other FMH detection methodologies are in use, such as gel agglutination and enzyme-linked antiglobulin test.

Rosette Test (Fetal RBC Screen [Qualitative]): The rosette test demonstrates the number of D-positive cells in a D-negative suspension using an anti-D reagent. The anti-D binds to D-positive fetal RBCs, and when indicator D-positive RBCs are added rosettes are formed. This method has an FMH detection limit of about 10 mL. For the test to be valid, fetal cells must be D-positive (not weak D or D unknown—result is false negative) and the maternal cells must be D-negative—result is false positive; in those situations, a test that detects fetal hemoglobin should be used, such as the Kleihauer–Betke or flow cytometry test.

Kleihauer–Betke Test (Fetal RBC Count [Quantitative]): The Kleihauer–Betke test has several limitations, including low sensitivity, poor reproducibility, and tendency to overestimate the FMH volume, yet it is the most commonly used test to quantify FMH in the United States. The Kleihauer–Betke test is performed on a maternal blood smear treated with acid and then stained, so the fetal RBCs remain red, and the maternal RBCs appear as ghosts. 2000 cells are counted, and the percentage of fetal RBCs is determined.

Flow Cytometry (Fetal RBC Count [Quantitative]): Flow cytometry techniques quantitate the amount of hemoglobin F or D-positive RBCs and are simpler, more precise, reliable, and thus may result in lower RhIg administration than the Kleihauer–Betke test. Flow cytometry techniques are not routinely available throughout the United States.

RhIg Dosing in the Presence of Fetomaternal Hemorrhage: The calculation of RhIg dosing used by AABB method (other methods exist) is presented in Table 50.1 (and through an online College of American Pathologist calculator [www.cap.org]).

Other Strategies for the Prevention of Hemolytic Disease of the Fetus and Newborn: Currently, the use of prophylactic RhIg is the only therapy available with proven success for the prevention of alloimmunization against the D antigen. If the mechanism of action of RhIg was better understood, it would assist with the development of other alternative strategies to prevent pregnancy-associated alloimmunization. This could include some novel immunoprophylaxis therapies, as well as recombinant therapies focused on inducing antigen-specific tolerance to other paternally derived RBC antigens.

Other proposed strategies for the prevention of maternal alloimmunization include the provision of antigen-matched RBC products to females of childbearing potential, who require blood transfusion. In fact, several countries in the Western World, including Germany, the Netherlands, Finland, Switzerland, the United Kingdom, and Australia, routinely provide antigen-matched blood to women of childbearing age to prevent alloimmunization. Typically, this includes K, as well as extended Rh antigen matching for C, c, and E. While this is not practical in an emergency situation, it is a worthwhile consideration for those with anticipated transfusion needs as a measure of serologic prevention for future pregnancies. The impact of this strategy may be low because most antigen sensitization in women of childbearing age is pregnancy-related. Nonetheless, prevention of alloimmunization is a worthwhile goal because the development of maternal alloimmunization may result in fetal morbidity or mortality. Antigen matching of RBC products could also be considered for pregnant women requiring IUTs to prevent the development of alloantibodies broadening of maternal antibody specificities.

TABLE 50.1 Calculation of RhIg Dosing

1. % Fetal RBCs in maternal circulation is determined by Kleihauer–Betke test or flow cytometry.
   Maternal blood volume (mL) = 70 mL/kg × maternal weight (kg) or 5000 mL if maternal weight unknown
2. Fetal bleed (mL) = % Fetal RBCs × Maternal blood volume (ml)
3. Dose of RhIg = Fetal bleed (mL)/30 mL per dose (300 μg vial)
4. If the number to the right of the decimal point is <5, round down and add one dose of RhIg.
   (e.g. 2.3→2 + 1 = 3 vials)
   If the number to the right of the decimal point is ≤5, round up and add one dose of RhIg.
   (e.g. 2.6→3 + 1 = 4 vials)

Testing for the Detection of Hemolytic Disease of the Fetus and Newborn

Alloantibodies Currently or Previously Detected: In one series, 1% of pregnancies were complicated by alloimmunization to RBC antigens that were associated with HDFN (40% anti-D, 30% antibodies to other Rh antigens, and 30% were to non-Rh antigens). More than 50 RBC alloantibodies have been implicated in HDFN, with the majority occurring in the Rh, Kell, Duffy, Kidd, and MNS blood group systems. ABO is also a common cause of HDFN but rarely results in severe anemia. The goal of transfusion service prenatal testing is to identify the alloantibody (see Chapter 22) to determine the clinical significance of the alloantibody (Table 50.2) and to perform antibody titrations to assist the clinicians as to when fetal monitoring is required. It should be noted that the outcome for antigen-positive fetuses may vary depending on the characteristics of the maternal alloantibody and the RBC antigen. For example, anti-D may result in severe HDFN with kernicterus; whereas, anti-K is more likely to result in severe fetal anemia with reticulocytopenia due to bone marrow suppression of erythropoiesis. Hence, early detection of maternal alloantibodies is critical for the management of at-risk pregnancies.

Phenotype the Father: If paternity is assured, the father may be phenotyped to assess for risk of HDFN. If the father's RBCs do not carry the antigen, then no further workup needs to be performed. If the father is homozygous, he carries a double dose of the antigen (e.g., a mother has anti-Jk^a and the father types as Jk(a+b−)), then the fetus will express the RBC antigen and is at-risk for HDFN. If the father is heterozygous, he carries a single dose of the antigen (e.g. Jk(a+b+)), then the fetus has a 50% chance of being at-risk. Fetal genotyping or phenotyping may be performed to determine the antigen status of the fetus. Because serologic testing cannot readily predict paternal zygosity for RhD, DNA testing may be useful to predict paternal genotype, if there is maternal anti-D. Fetal DNA can either be obtained through amniocentesis or extracted noninvasively from the maternal plasma, known as cell-free fetal DNA (cffDNA), as early as 12 weeks of gestational age for fetal risk stratification.

Antibody Titration: Antibody titers should be performed using a standard procedure. The Biomedical Excellence for Safer Transfusion Collaborative and the Transfusion Medicine Resource Committee of the College of American Pathologists determined the Uniform procedure for tube and gel card techniques to minimize intralaboratory variation, which differs from the method in the AABB Technical Manual. The current specimen is tested in the antiglobulin phase with a previous sample (frozen at −20°C or colder), if available, and run in parallel to compare the change in titration. Some institutions use RBCs with the strongest dose of the corresponding antigen (R_2R_2 cells in the case of anti-D), and some institutions use RBCs with a single dose of antigen (recommended in the Uniform method); the laboratory should consistently use RBCs of the same phenotype. Antibody titers are performed by serially diluting the patient's plasma and testing with the appropriate RBCs to determine the highest titer where reaction occurs (some laboratories use weak+ and others 1+). The titration end point is reported as a reciprocal of the titer (e.g., 1:16 is reported as 16).

TABLE 50.2 Probability of Causing Severe Hemolytic Disease of the Fetus and Newborn (HDFN) Associated With RBC Antibodies

Blood Group	Highest Likelihood of Severe HDFN	Rare Cases of Severe HDFN	Usually Associated With Mild Disease	Not a Cause of HDFN
MNS		M, S, s, U, Mi^a, Vw, Mur, Mt^a, Hut, Hil, M^v, Far, s^D, En^a, MUT	M, S, s, U, Mt^a, Mit	N
Rh	D, c	C, E, f, Ce, C^w, C^I, E^w, G, Hr_0, Hr, Rh29, Go^a, Rh32, Be^a, Evans, Tar, Rh42, Sec, JAL, STEM	E, e, f, C^I, D^w, Rh29, Riv, LOCR	
Lutheran			Lu^a (rare), Lu^b	
Kell	K	k, Kp^a, Kp^b, Ku, Js^a, Ul^a, K11, K22	Ku, Js^a, K11	K23, K24
Lewis				Le^a, Le^b
Duffy		Fy^a	Fy^b (rare), Fy3 (rare)	
Kidd		Jk^a	Jk^b (rare), Jk3	
Other		Di^a, Wr^a, Rd, Co^a, Co3, PP_1P^kVel, MAMBi, Kg, JONES, HJK, REIT	Di^b, Sc3, Co^b, Ge2 (rare), Ge3, Ls^aLan, At^a, Jr^aJFV, HOFM	P_1, Wr^b, Yt^a, Yt^b, Sc1, Sc2, CH/RG, CROM, KN, JMH, IJr^aHLA, Bg^a, Bg^b, Bg^c

Eder, A. F. (2006). Update on HDFN: New information on long-standing controversies. *Immunhematology, 22*, 188–195.

The critical titer varies between laboratories, but most institutions use 8 to 32 for anti-D. Critical titers for other antibody specificities remain unclear. This is especially true for anti-K, where titers of 4 have been associated with HDFN. Regardless of the maternal alloantibody, some fetuses will have severe anemia with low antibody titers, while others will have minimal evidence of disease with high titers. With the advent of noninvasive measurements of fetal anemia through Doppler assessment of the middle cerebral artery peak systolic velocity, the role of antibody titration, as well as more invasive techniques, such as amniocentesis and cordocentesis, is limited. If monitoring is not performed by middle cerebral artery Doppler, then titrations should be performed every 2–4 weeks after 18 weeks of gestational age. Once the critical titer is reached, no further titration studies are necessary, and the pregnancy should be monitored by middle cerebral artery Doppler. In addition, if the mother had a previously affected fetus or infant, maternal antibody titers are not helpful and the fetus should be monitored for anemia by middle cerebral artery Doppler.

Middle Cerebral Artery Peak Systolic Velocity: Doppler measurements of the peak velocity of systolic blood flow in the middle cerebral artery are equivalent to serial ΔOD_{450} measurements using amniocentesis. Moderate to severe fetal anemia requiring transfusion was predicted by ≥1.5 multiples of the median. Middle cerebral artery blood flow has 88% sensitivity and 82% specificity in predicting fetal anemia.

Amniocentesis: Before the use of middle cerebral artery peak systolic velocity, serial amniocentesis was used to determine the severity of hemolysis. For the majority of antibodies, the most notable exception is anti-K (because anti-K affects early RBC precursors, which results in both hemolytic anemia and suppression of erythropoiesis); amniotic fluid bilirubin levels correlate with the severity of hemolysis. The bilirubin level is quantified by spectrophotometry and measured as the change in the optical density at the wavelength 450 nm (ΔOD_{450}). The ΔOD_{450} is then plotted on the Liley chart, which is most accurately used after 27 weeks of gestational age, to determine the severity of hemolysis (zone I predicts mild or no disease, zone II predicts moderate disease, and zone III predicts severe disease). Serial measurements are used to follow the severity of disease; when values are unchanging or increasing then the disease is worsening. Currently, amniocentesis is performed to determine the phenotype/genotype of the fetus and risk for HDFN and to determine fetal lung maturity. Although rare, amniocentesis can result in fetal injury or death.

Periumbilical Cord Blood Sampling: Fetal blood can be sampled through percutaneous umbilical blood samplings (also known as cordocentesis), which has a 1%–2% risk of fetal loss. This sample can be tested for hemoglobin level, hematocrit, blood type, and direct antiglobulin test (DAT).

Perinatal Transfusion Management of Hemolytic Disease of the Fetus and Newborn

Intrauterine Transfusions: IUT is performed after 18–20 weeks of gestational age. With advances of ultrasonography, allogeneic blood is directly infused into the umbilical cord vein. If the vein is not accessible, an alternative method is injection of RBCs into the fetal peritoneal cavity, after which the RBCs are absorbed into circulation through the lymphatic system. A combination of intravascular and intraperitoneal transfusions can also be used to prolong the interval between transfusions. There is a 1%–2% risk of fetal mortality with this procedure. Neonates who have received IUT can generally avoid exchange transfusion after birth but may require additional small volume transfusions because of suppression of erythropoiesis and the presence of residual maternal alloantibodies.

RBC Transfusions: The indications for intrauterine RBC transfusion are fetal anemia secondary to parvovirus infection, HDFN, twin-to-twin transfusion, large-volume fetal hemorrhage, and homozygous β-thalassemia. Fetal anemia can be detected by amniotic fluid ΔOD_{450} or Doppler middle cerebral artery measurements, cordocentesis blood sample (hemoglobin <10 g/dL or hemoglobin 4–6 standard deviations below the mean for gestational age), or fetal hydrops noted on ultrasound examination. For HDFN, the most common antibodies requiring the use of IUT are anti-D (44%), anti-D with other antibodies (42%), anti-c (6%), anti-K (5%), and anti-E (2%), as reported at one institution. Fetal transfusion will help correct the anemia and also suppress fetal erythropoiesis, consequently diminishing the production of RBCs with antigen corresponding to the offending maternal alloantibody. The suggested transfusion volume is between 20% and 50% of the fetoplacental blood volume. The desired fetal hematocrit after the transfusion is usually 40%–45%. IUT volume can be calculated by a variety of methods, Table 50.3 provides one of the calculation methods used. Alternatively, the transfused volume can be estimated by multiplying the estimated fetal weight by 0.02 for each 10% increase in hematocrit desired. The normal fetus can tolerate rapid infusion of large blood volumes (5–7 mL/minute), but this is not the case with a hydropic fetus where the transfusion volume should be limited and the posttransfusion hematocrit should be lower for the first procedure. A hydropic fetus requires more frequent transfusions of smaller volumes. Typically, IUT is repeated every 2–4 weeks to maintain a fetal hematocrit at 27%–30% (fetal hematocrit decreases 1–2% per day), until 35 weeks of gestational age with delivery at 37–38 weeks after confirmation of fetal lung maturity by amniocentesis. Intravascular exchange transfusions can be performed instead of simple intravascular transfusions, but the simple transfusion is preferred because of the shorter procedure time and being technically easier.

Neonates who have received IUT may have suppressed erythropoiesis, which can result in hypoproliferative anemia (low levels of erythropoietin and low reticulocyte counts). This may require RBC transfusions several weeks after birth and/or the use of exogenous erythropoietin. Additionally, IUT for anti-D HDFN has been associated with increased risk of cholestatic liver disease in neonates.

TABLE 50.3 Volume of Blood for Intrauterine Transfusion

1. Determine fetal and placental total blood volume
   - Multiply the ultrasound-estimated fetal weight (gm) × 0.14 mL/gm
2. Multiply the result by the difference in posttransfusion (desired) and pretransfusion (initial) Hct
3. Divide the result by the Hct of the RBC product

RBC Product Selection: The RBCs should be group O, D-negative, CMV-safe (some centers may decide to provide CMV-seronegative, especially if the mother is CMV-seronegative), leukoreduced, irradiated, and hemoglobin S-negative. RBC products should be antigen-negative for the corresponding antibody(ies) and crossmatch compatible with the maternal plasma. Some recommend products that are less than 7 days old because they have improved levels of 2,3-DPG and lower levels of potassium. An alternative is to use frozen, deglycerolized RBCs, which also have near normal levels of 2,3-DPG. If an RBC unit stored for >7 days needs to be used, then it should be washed to remove potassium. In rare circumstances, especially when antigen-compatible allogeneic blood is unavailable, maternal RBCs may be used, even if the blood is ABO-mismatched for the fetus. The maternal RBC product should be washed to remove the antibody-containing plasma. The RBC product should have a high hematocrit, 75%–85%, to minimize volume overload.

Postnatal Transfusion Management of Hemolytic Disease of the Fetus and Newborn

Neonatal Testing: The cord blood sample should be tested for ABO/D, including weak D, and the corresponding antigen if the mother has an RBC antibody. A false-negative D type can occur if the fetal RBCs are coated with IgG anti-D, known as "blocked D." In this situation, an eluate will demonstrate anti-D. A DAT should be performed using anti-IgG on the cord blood if there is a maternal RBC antibody or suspected HDFN due to ABO incompatibility between the mother and infant. The DAT contributes to the diagnosis of HDFN, but the strength does not correlate with the severity of disease.

Treatment of the Neonate With Hemolytic Disease of the Fetus and Newborn: Phototherapy with ultraviolet light can be used in infants with mild to moderate HDFN. In severe cases, RBC exchange transfusion may be necessary to prevent kernicterus. Indications for exchange include cord hemoglobin ≤10 g/dL, cord bilirubin ≥5.5 mg/dL, rising bilirubin despite phototherapy ≥0.5 mg/dL per hour, bilirubin >20 mg/dL in a term infant (lower bilirubin levels in premature infants or infants with sepsis, hypoxemia, acidosis, and hypothermia). The benefits of RBC exchange transfusion include removal of the bilirubin, removal of sensitized RBCs, removal of maternal antibody, provision of antigen-negative RBCs, and suppression of erythropoiesis. IVIG (dose 0.5 g/kg) may decrease the need for RBC exchange in neonates with HDFN. While these therapies have shown to be effective, their use is not without risk and side effects. Additionally, discontinuation of maternal breast milk feedings should be considered for infants with HDFN because persistent HDFN due to passive transfer of anti-D in breast milk has been reported.

RBC Exchange Transfusion: Exchange transfusion is needed for the prevention of kernicterus, which may develop with bilirubin levels as low as 8–12 mg/dL in very-low-birth-weight neonates but rarely develops at levels <25 mg/dL in full-term infants. RBC units selected for exchange transfusion should be group O, antigen-negative for the corresponding maternal antibody (i.e., RhD-negative, if anti-D is present), leukoreduced (generally considered to be adequate to prevent transfusion-transmitted CMV;

TABLE 50.4 Volume and Preparation of Blood for Exchange Transfusion

1. Exchange volume (mL) = 2 × TBV
2. TBV = 100 mL/kg premature infant = 85 mL/kg term infant
3. Volume of RBC product mL = $\frac{\text{Exchange volume} \times \text{reconstituted Hct}}{\text{Hct of RBC product}}$

   Volume of FFP (mL) = Exchange volume − volume of RBC product

however, some centers also choose to provide CMV-seronegative, if mother is CMV-negative), irradiated, and hemoglobin S-negative, which are less than 5–7 days old (alternately, frozen and deglycerolized RBC units, especially if rare RBC units are necessary). The RBC units should be crossmatch-compatible with the maternal plasma, if available for testing. If a maternal sample is not available, then the infant's plasma and/or an eluate from the infant's RBCs can be used. The additive solution should be removed and replaced with group AB frozen plasma to dilute the RBCs to an appropriate hematocrit, 40%–50%. An additional 75–100 mL may be added to the volume requested for exchange transfusion to account for the volume lost in tubing and blood warmer. In rare circumstances, particularly when antigen-compatible allogeneic blood is unavailable, maternal RBCs can be used (with the plasma removed), even if they are ABO-mismatched. Typically the volume exchanged is twice the blood volume of the infant (Table 50.4). Of note, the infant should be closely monitored postnatally for hemolysis or anemia because clinically significant maternal antibody may persist in the infant's circulation after exchange transfusion.

Further Reading

American Academy of Pediatrics (AAP). (2004). Management of hyperbilirubinemia in the newborn infant 35 or more weeks gestation. *Pediatrics*, *114*, 297–316.

College of American Pathologists online RhIG dose calculator. http://www.cap.org/web/home/involved/council-committees/transfusion-medicine-topic-center.

Cure, P., Bembea, M., Chou, S., et al. (2017). 2016. Proceedings of the National Heart, Lung, and Blood Institute's scientific priorities in pediatric transfusion medicine. *Transfusion*, *57*, 1568–1583.

Delaney, M., Wikman, A., van de Watering, L., & for the BEST Collaborative, et al. (2017). Blood group antigen matching influence on gestational outcomes (AMIGO) study. *Transfusion*, *57*, 525–532.

Eder, A. F. (2006). Update on HDFN: New information on long-standing controversies. *Immunohematology*, *22*, 188–195.

Engelfriet, C. P., Reesink, H. W., Judd, W. J., et al. (2003). Current status of immunoprophylaxis with anti-D immunogloblulin. *Vox Sang*, *85*, 328–337.

Fung, M. K., Eder, A. F., Spitalnik, S. L., et al. (Eds.). (2017). *Technical manual* (19th ed.) Bethesda: American Association of Blood Banks.

Giannina, G., Moise, K. J., & Dorman, K. (1998). A simple method to estimate the volume for fetal intravascular transfusion. *Fetal Diagn Ther*, *13*, 94.

Hendrickson, J. E., & Delaney, M. (2016). Hemolytic disease of the fetus and newborn: Modern practice and future investigations. *Transfus Med Rev*, *30*(4), 159–164.

Judd, W. J., & for the Scientific Section Coordinating Committee of the AABB (2001). Practice guidelines for prenatal and perinatal immunohematology, revisited. *Transfusion, 41*, 1445–1452.

Kennedy, M. S., McNanie, J., & Waheed, A. (1998). Detection of anti-D following antepartum injections of Rh immune globulin. *Immunohematology, 14*(4), 138–140.

Li, M., & Blaustein, J. C. (2017). Persistent hemolytic disease of the fetus and newborn (HDFN) associated with passive acquisition of maternal anti-D in breast milk. *Transfusion, 57*(9), 2121–2124.

Oepkes, D., Seaward, G., Vandenbussche, F. P., et al. (2006). Doppler ultrasonography versus amniocentesis to predict fetal anemia. *N Engl J Med, 355*, 156–164.

Patel, R. M., Meyer, E. K., & Widness, J. A. (2016). Research opportunities to improve neonatal red blood cell transfusion. *Transfus Med Rev, 30*(4), 165–173.

Rath, M. E., Smits-Wintjens, V. E., Walther, F. J., & Loprior, E. (2011). Hematological morbidity and management in neonates with hemolytic disease due to red cell alloimmunization. *Early Hum Dev, 87*, 583–588.

Sandler, S. G. (2010). Effectiveness of the RhIG dose calculator. *Arch Pathol Lab Med, 134*, 967–968.

Sandler, S. G., Flegel, W. A., Westhoff, C. M., et al. (2015). It's time to phase in *RHD* genotyping for patients with a serologic weak D phenotype. *Transfusion, 55*, 680–689.

Sandler, S. G., & Gottschall, J. L. (2012). Postpartum Rh immunoprophylaxis. *Obstet Gynecol, 120*, 1428–1438.

Sandler, S. G., Roseff, S. D., Domen, R. E., et al. (2014). Policies and procedures related to testing for weak D phenotypes and administration of Rh immune globulin. *Arch Pathol Lab Med, 138*, 620–625.

Yu, A., Morris, E., Adams, R., & Fung, M. K. (2017). Obstetrics and gynecology physician knowledge of Rh immune globulin prophylaxis. *Transfusion, 57*, 1385–1390.

CHAPTER 51

Autoimmune Hemolytic Anemias

Nancy L. Van Buren, MD

Autoimmune hemolytic anemia (AIHA) refers to a group of disorders where autoantibodies are directed against red blood cell (RBC) membrane antigens resulting in shortened RBC survival (normally 100–120 days) through activation of the complement system and/or RBC removal within the reticuloendothelial system (RES). Classification of AIHA includes warm autoimmune hemolytic anemia (WAIHA), cold agglutinin disease (CAD), mixed-type AIHA due to wide thermal amplitude of the autoantibodies, drug-induced hemolytic anemia, and paroxysmal cold hemoglobinuria (PCH). The causes may include medications, autoimmune disease, lymphoma, or hematopoietic malignancy associated with monoclonal gammopathy, or recent viral infection. AIHA is classified as idiopathic if no underlying cause is identified.

In the AIHAs, the degree of hemolysis depends on antibody class and characteristics, such as concentration, "antigen" affinity, ability to fix complement, and thermal amplitude. These antibodies typically react as "panagglutinins," meaning they show in vitro reactivity with all tested RBCs. These antibodies bind to a ubiquitous antigen or have nonspecific binding. In WAIHA, IgG autoantibody most commonly reacts with Rh group proteins or glycophorins on the RBC surface as panagglutinin antibodies. These IgG autoantibodies "coat" RBCs, and these coated RBCs are removed by phagocytosis by Fcγ receptors primarily within the spleen, thus achieving extravascular hemolysis. Drugs most commonly implicated in AIHA include antibiotics, such as penicillin, ceftriaxone, cefotetan, and piperacillin, as well as certain nonsteroidal antiinflammatory drugs, quinine, purine nucleoside analogues (e.g., fludarabine), and platinums (e.g., cisplatinin).

In contrast to IgG alloantibodies, IgM molecules efficiently activate the complement cascade (starting with C1q) and can result in intravascular hemolysis. The degree of intravascular hemolysis is affected by antibody concentration, thermal amplitude, and amount of inactivation by complement regulatory proteins, such as decay-accelerating factor (DAF or CD55) and membrane inhibitor of reactive lysis (MIRL, HRF, or CD59). Cold agglutinins (CAs) are autoantibodies present in most individuals, but typically have no clinical significance, unless they are sufficiently reactive at peripheral body temperatures. Both benign and pathologic conditions are caused by IgM autoantibodies, usually of I or i antigen specificity. CAD is caused by autoantibodies that are of higher titer (>512 at 4°C and >128 at 22°C) and are associated with agglutination or complement fixation at 30–37°C. Patients undergoing cardiac surgery with CA, in the absence of CAD, can safely be managed by undergoing normothermic cardiopulmonary bypass (avoiding cold cardioplegia). In contrast, hematology should be consulted for patients with CAD before cardiac surgery, and additional laboratory testing should be performed, including CA titers. The underlying causes of CAD include *Mycoplasma pneumoniae*, Waldenstrom macroglobulinemia, and IgM monoclonal gammopathy. Severe cases of CAD often result in intravascular hemolysis due to complement activation, but

Transfusion Medicine and Hemostasis. https://doi.org/10.1016/B978-0-12-813726-0.00051-9

can result in extravascular hemolysis because RBCs are in the cool peripheral circulation for a short time period, and when the RBCs are warmed in the central circulation, the IgM dissociates. The short time period may only be sufficient to activate the complement cascade to the C3b stage and not to the membrane attack complex stage. The C3b-coated RBCs are cleared by hepatic macrophages with receptors specific for C3b, if there is a sufficient quantity of C3b molecules on the RBCs (>500–800/RBC). C3b-coated RBCs, which are not removed from the hepatic circulation, are unharmed, and the C3b is degraded to C3dg; C3dg-coated RBCs have near-normal survival.

Clinical Presentation: The presence of erythrocyte autoantibodies does not always cause hemolysis and thus anemia. In fact, most patients with low concentrations of IgG autoantibodies will have laboratory findings such as a positive DAT but will not have anemia or related symptoms. Clinically significant hemolysis results in anemia, jaundice, and splenomegaly. Laboratory testing should include a complete blood count, reticulocyte count, serum bilirubin level, serum lactic dehydrogenase (LDH) level, serum haptoglobin value, and peripheral blood smear. Hemolytic anemia results in anemia, elevated reticulocyte count, increased bilirubin levels (especially the indirect fraction), elevated LDH level, and decreased haptoglobin level. The peripheral smear may demonstrate spherocytes (WAIHA) or RBC agglutination (CAD), which may aid in the diagnosis of hemolytic anemia. Urinalysis will show hemoglobinuria when intravascular hemolysis is present.

Warm Autoimmune Hemolytic Anemia: WAIHA represents 60%–80% of AIHA cases, including those that are drug-induced. WAIHA is secondary to IgG (rarely IgA or IgM) warm autoantibodies often with Rh specificity, which typically react at 37°C and results in primarily extravascular hemolysis. Extravascular hemolysis is when the hemolysis occurs in the RES, resulting in increased serum bilirubin, but not necessarily hemoglobinemia or hemoglobinuria. IgG-sensitized RBCs are destroyed or damaged within the RES, predominantly by splenic macrophages.

Pathogenesis: While many cases of WAIHA are primary (idiopathic), approximately 50% of WAIHAs are secondary to underlying disease states, including lymphoma and various lymphoproliferative disorders, autoimmune diseases, and drugs. For this reason, it is important to assess patients diagnosed with WAIHA for underlying disease, along with medication history. Lymphoma and other lymphoproliferative disorders are the most commonly associated condition, representing nearly half, while autoimmune diseases account for 27% in one study. The underlying clinical condition is frequently present preceding the diagnosis of WAIHA. Several studies have documented that the diagnosis of lymphoma is made subsequent to the clinical presentation of AIHA in a significant number of cases, which highlights the importance of evaluating patients for lymphoid malignancies at the time of presentation to avoid treatment delays and improve response. A detailed description of the pathogenesis and mechanisms of drug-induced hemolytic anemia follows at the end of this chapter. IgG-coated RBCs are detected serologically by a positive DAT with IgG; in addition, the RBCs may sometimes be coated with complement, and thus the DAT may also be positive for complement (see Chapter 21). IgG, when eluted from RBCs, typically reacts as a panagglutinin and does not demonstrate specificity. 80% of individuals with WAIHA have autoantibody present in the plasma as well as on the RBC membrane when sufficient antibody titers are present (see Chapter 20). The plasma-phase autoantibody usually reacts as a panagglutinin in the indirect antiglobulin test (i.e., antibody screen/antibody identification panel). Occasionally, antibody identification demonstrates specificity to antigens in the Rh system (usually c or e) or other RBC antigens and is said to have "apparent specificity."

Treatment: Treatment for WAIHA depends on the clinical severity of the disease. First-line treatment is usually corticosteroids using high-dose prednisone, with or without rituximab, which may offer a higher rate of complete remission and relapse-free survival. Splenectomy is considered to be an effective second-line therapy. Other treatments may include IVIG and alternate immunosuppressive medications. A daily maintenance dose of prednisone is sometimes required to control hemolysis. Supplemental folic acid is also customary because chronic hemolysis may lead to folate deficiency. Rituximab has been found to be an effective treatment for both primary and secondary WAIHA, with favorable results in both adult and pediatric patients. It may also be useful in the treatment of cases of relapsed WAIHA. In secondary WAIHA, treatment should also target the underlying illness. For patients with chronic lymphocytic leukemia (CLL), AIHA often diminishes in parallel with response to therapy: treatment decisions are based on the International Workshop on CLL guidelines for active disease.

Severe anemia with cardiac or cerebral dysfunction requires urgent management, which may include RBC transfusion. Patients with severe disease, who are not responsive to RBC transfusion and other immunomodulatory treatments secondary to rapid RBC destruction, may occasionally benefit from plasma exchange for removal of the pathogenic antibodies (see below). Preliminary data suggest that hematopoietic stem cell transplantation may have a role in treatment of WAIHA; however, more data are required to assess its efficacy. The diagnosis of primary WAIHA may also precede the development of non-Hodgkin lymphoma, sometimes by many years, so it is important to follow these patients on a continuing basis after treatment of the AIHA.

Transfusion Management: During an acute presentation in a patient newly diagnosed with AIHA, finding the appropriate RBC product for transfusion can be a challenge and close communication between the transfusion service and the treating physician is necessary. Because of the presence of a strong autoantibody, both the direct and indirect antiglobulin tests (and thus the antibody screening tests and panels) will be positive as the autoantibody, reacting as a panagglutinin, will react with all tested cells. The presence of positive reactions (i.e., in vitro agglutinin) in all tested cells makes the serologic identification of coexisting *allo*antibodies difficult, as their identification typically requires patterns of differential reactivity. Absorption techniques using either donor or patient RBCs are available at some hospital laboratories, but usually reference laboratories perform these specialized time-consuming tests. As 12%–40% of patients with warm autoantibodies also have clinically significant alloantibodies, it is imperative that appropriate methods for determining the presence of coexisting alloantibodies be used. It is also helpful to recognize that the presence of a panagglutinin will create positive crossmatch results; thus, the crossmatch will not be helpful in determining compatibility with underlying alloantibodies.

Performing RBC phenotyping/genotyping in patients with autoantibodies can be helpful because it focuses the antibody workup on the possible alloantibodies the patient is capable of forming. In addition, if a complete phenotype can be determined, then the transfusion service can provide phenotype-matched RBCs, which may prevent future alloimmunization and delayed hemolytic transfusion reactions as well as circumvent the need for exhaustive absorption studies (see Chapter 20). This is a worthwhile consideration to enhance transfusion safety, especially because individuals with WAIHA have higher rates of alloimmunization (12%–40%).

The Johns Hopkins Hospital published their approach to patients with warm autoantibodies (Shirey et al.). If possible, this included a phenotype for C, E, c, e, K, Jk^a, Jk^b, Fy^a, Fy^b, S, and s, allowing them to provide phenotype-matched, as well as antigen-negative units for any identified alloantibodies. During analysis of subsequent samples, if the serologic results were consistent with previous findings, phenotype-matched products were provided. Twelve of the 20 patients studied could be fully phenotyped and 8 patients could be partially phenotyped or phenotyping was indeterminate. The patients received between 2 and 39 products, and none developed new alloantibodies during the study period of 13 months.

Because of the presence of the warm, IgG autoantibody in these patients, it is typically not possible to identify crossmatch-compatible units. Restrictive transfusion strategies for hemodynamically stable patients should be followed per AABB RBC transfusion guidelines. Blood should not be withheld from patients if they are experiencing cardiopulmonary symptoms due to anemia. Because the autoantibodies target Rh or Rh-related antigens or glycophorin, which are nearly universally present on RBCs, administration of "incompatible" blood is often necessary. Most blood banks and clinicians will select so-called "least incompatible" blood, meaning that they will choose those units for which the crossmatch demonstrates the least reactivity after using special compatibility test procedures to rule out underlying alloantibodies. The predicative value of this approach for in vivo hemolysis is not known. When the autoantibody does demonstrate apparent specificity (e.g., anti-e), selection of antigen-negative units (e.g., e−) seems like an appropriate decision, although these antibodies will still react with the patient's RBCs. Arguments against this approach include the following: there are scant data to support this approach, and this practice increases the likelihood that an alloantibody will be induced to the antithetical antigen (e.g., E), if it is not carried on the patient's RBCs. In addition, the apparent antigen specificity is usually to a high-frequency antigen making it difficult to obtain antigen-negative RBCs (e.g., 98% of the population is e positive), which may delay transfusion while acquiring antigen-negative RBCs.

Risks of transfusion in patients with WAIHA are increased hemolysis, resulting from the increase in RBC mass or inability to recognize and respect underlying alloantibodies, and congestive heart failure secondary to circulatory overload. There are also risks to withholding transfusion when a patient has a justifiable need, even when the compatibility test is strongly positive, particularly in patients with cardiopulmonary symptoms or rapidly developing anemia with reticulocytopenia. It is recommended in severe cases that transfusion commence using small volumes of RBCs with close clinical observation.

Therapeutic plasma exchange (TPE) has been used in the management of WAIHA in severe cases unresponsive to RBC transfusion and other immunomodulatory therapies.

WAIHA is an ASFA category III indication for TPE secondary to the rare and conflicting reports of its successfulness (see Chapter 75). Theoretically, TPE would be less effective in WAIHA because the IgG intravascular distribution is 45% (IgM is 80%). The pathogenic IgG antibody coating the RBCs is less available in the plasma resulting in inefficient antibody removal by TPE. However, in patients with severe hemolysis, TPE may be used as a temporizing measure until immunosuppressive therapy can take effect.

Autoimmune Hemolytic Anemia Associated With a Negative Direct Antiglobulin Test: A patient may have AIHA with a negative DAT secondary to either RBC-bound IgG being below the threshold of detection (standard DAT can detect >300–500 bound IgG molecules/RBC), RBC-bound IgA and IgM being responsible (these are not detected by most routine reagents), or low-affinity IgG, which is washed off the RBCs. This accounts for approximately 3%–11% of cases, and the typical clinical presentation is consistent with WAIHA. Because physicians may reject the diagnosis based on the negative DAT, it is important to recognize that this entity exists and perform additional diagnostic testing to minimize delays in treatment.

Alternative Testing Techniques: Reference laboratories may have alternative techniques for evaluating patients with WAIHA associated with a negative DAT. Flow cytometry can be used to detect small amounts of RBC-bound IgG, IgA, and IgM. In addition, nonlicensed anti-IgA and anti-IgM reagents used to detect RBC-bound IgA or IgM, respectively, are available. Cold low ionic strength saline (LISS) wash DAT may be helpful in detecting low-affinity autoantibodies, as well as 4°C LISS wash IAT, PEG IAT, and Donath–Landsteiner (DL) antibody testing, if PCH is suspected.

Cold Agglutinin Disease: CAD accounts for 16%–32% of cases of AIHA. It may be primary or secondary to an underlying condition. Of the infectious causes, *M. pneumoniae* or mononucleosis (Epstein–Barr virus) infection are the most common precipitating factors. In adults, CAD is a well-recognized entity that is often caused by a bone marrow clonal B-cell lymphoproliferative disorder, such as lymphoma, CLL, IgM monoclonal gammopathy, or Waldenstrom macroglobulinemia. Regardless of the cause, CAD results in AIHA and cold-induced circulatory symptoms. When associated with an infection, CAD typically presents with acute onset and is usually transient. When associated with lymphoproliferative disorders, it is usually chronic. Secondary CAD is often seen in children or young adults, while primary CAD is seen most often in elderly patients. The clinical severity depends on the antibody thermal amplitude, ability to fix complement, and titer. The typical clinical manifestations are moderate chronic hemolytic anemia in a middle-aged or elderly person, which is exacerbated by the cold and is associated with a good prognosis. In severe cases, the IgM autoantibody can result in life-threatening intravascular hemolysis.

CAD is characterized by a high-titer (>10,000 at 4°C) IgM autoantibody that is reactive at >30°C (high thermal amplitude) with I or i specificity (rarely P^r) (Table 51.1). The DAT is positive for C3 only (see Chapter 23). CAD secondary to *M. pneumoniae* is associated with an IgM autoantibody with anti-I specificity. The specificity of the antibody associated with infectious mononucleosis is anti-i. The pathogenesis of the

TABLE 51.1 Typical Serologic Findings in AIHA

	WAIHA	CAD	Mixed-Type AIHA	PCH	Drug-Induced
DAT (routine)	IgG IgG + C3 C3	C3 only	IgG + C3 C3	C3 only	IgG IgG + C3
Ig type	IgG	IgM	IgG, IgM	IgG	IgG, IgM
Eluate	IgG antibody	Nonreactive	IgG antibody	Nonreactive	Reactive with drug-treated RBCs
Serum	By IAT, 35% agglutinate untreated red cells at 20°C	IgM agglutinating antibody, titer ≥1000 (60%) at 4°C, reactive at 30°C	IgM IAT-reactive antibody plus IgM agglutinating antibody reactive at 30°C	Negative routine IAT result, IgG biphasic hemolysin in Donath–Landsteiner test	Antibody reactivity with drug-treated RBCs
Specificity	Broadly reactive, multiple specificities reported	Usually anti-I	Usually unclear	Anti-P	See Figure 51.1

AIHA, autoimmune hemolytic anemia; *CAD*, cold agglutinin disease; *DAT*, direct antiglobulin test; *IAT*, indirect antiglobulin test; *IgG*, immunoglobulin G; *IgM*, immunoglobulin M; *PCH*, paroxysmal cold hemoglobinuria; *WAIHA*, warm AIHA.

Modified from Fung, M. K., Eder, A. F., Spitalnik, S. L., Westhoff, C. M. (Eds.) (2017). *Technical manual* (19th ed.). Bethesda, MD: AABB Press.

autoantibody formation is unknown; theories include (1) immune dysfunction, (2) similar antigens shared by the infectious agent and RBC (known as antigen mimicry), and (3) infection-induced antigenic changes resulting in increased antigenicity.

The CAs involved in primary CAD have been found to be monoclonal (rather than polyclonal) IgM antibodies with κ light chain restriction in the majority of cases. Flow cytometry studies also demonstrate a κ-clonal B-lymphocyte population in bone marrow aspirates of patients with primary CAD. In addition, a clonal lymphoproliferative disorder, usually characterized as lymphoplasmacytic lymphoma, has been identified through immunohistochemical staining of the bone marrow of these patients. These studies suggest that there may be an overlap between Waldenstrom macroglobulinemia and primary CAD.

Pathogenesis: CAD results from IgM cold autoantibodies (also known as CAs), which lead to RBC agglutination at cold temperatures (4–18°C) in vitro and hemolysis in vivo when the antibody is present in high titer (>1:10,000) and reactive at warm temperatures (~37°C). The IgM autoantibodies, when present in sufficient quantity and able to react at near 37°C, activate the classic complement pathway on the RBC membrane beginning with calcium-dependent C1q binding, under appropriate conditions, and ending with full assembly of the membrane (or terminal) attack complex.

The membrane attack complex breaches the RBC membrane and leads to intravascular hemolysis resulting in hemoglobinemia, hemoglobinuria, and hemosiderinuria. Plasma-free hemoglobin binds haptoglobin and thus free haptoglobin measurements are low. In some clinical situations, complement activation is not sufficient to activate full assembly of the membrane attack complex, due to antibody concentration and reactivity at required temperatures and complement regulatory proteins, and the complement activation is only taken through the C3 stage, resulting in C3b-coated RBCs. C3b-positive RBCs can be phagocytized by macrophages with C3b receptors in the liver Kupffer cells resulting in extravascular hemolysis. Because the RBCs must be coated with at least 500–800 C3b molecules for RBC clearance, many C3b-coated RBCs survive normally. In addition, C3b is degraded to iC3b or Cdg by factor I; the former is formed if factor H, membrane cofactor protein, or complement receptor CR1 is present.

Treatment: Treatment of CAD depends on the severity and rapidity of onset of the symptoms, and the underlying etiology for the formation of the CA. The typical treatments for WAIHA are less effective in CAD (i.e., corticosteroids and splenectomy). In the majority of cases the anemia is mild; thus cold avoidance is the sole treatment used to prevent exacerbations. Cold avoidance includes wearing weather-appropriate clothing, avoiding cold drinks, keeping the thermostat turned up, as well as the use of prewarmed IV fluids and warming blankets to minimize exacerbation of hemolysis during hospitalization. Hemolysis will increase during acute infection, due to increased production of C3 and C4 and subsequent complement-mediated destruction of RBCs. Therefore, it is recommended that patients with chronic CAD receive vaccination against influenza or pneumococcal infections, as well as rapid treatment of other febrile illnesses. Additional treatment options for chronic CAD may include rituximab, prednisone, fludarabine, chlorambucil, and cyclophosphamide. Rituximab monotherapy or in combination with fludarabine may be of particular use in treating primary CAD because it targets the clonal B-cell population. Recently, rituximab in combination with bendamustine has also shown promise as a highly effective therapy for patients with CAD. Treatment of secondary CAD should target underlying disease. Patients with transient CAD secondary to infection (e.g., mycoplasma pneumonia) usually require supportive measures only (transfusion and cold avoidance).

Transfusion Management: Prewarming techniques may be used to mitigate the autoantibody reactivity and thus can aid in the identification of underlying alloantibodies in patients with CAD (see Chapter 22). In less than 10% of patients with CAD, an alternate technique, such as cold adsorption, will be required.

Severe anemia can result from CAD and cardiovascular compromise and death can follow; such cases may require RBC transfusions. These patients can usually be transfused with RBCs crossmatch compatible at 37°C. Some suggest using a blood warmer, especially in severe cases, in addition to keeping the patient warm and transfusing the product slowly. Life-threatening CAD is an ASFA category II indication for TPE (see Chapter 75). TPE may be beneficial in severe cases because TPE efficiently removes the primarily intravascular IgM pathogenic antibody. However, the effect of TPE is usually temporary and should therefore be combined with immunosuppressive therapy.

In cases of CAD where agglutination occurs at room temperature, it may be necessary to increase the temperature of both the room and the TPE equipment to 37°C.

Combined Cold and Warm Autoimmune Hemolytic Anemia (Mixed AIHA): Combined cold and warm AIHA occurs when a patient has serologic findings characteristic of WAIHA and has a CA of high titer and thermal amplitude and therefore has both WAIHA and CAD (Table 51.1). The IgG warm antibody is usually the more pathogenic antibody; therefore, treatments are similar to those for WAIHA. Depending on the titer and thermal amplitude of the cold autoantibody, a combination of serologic techniques used for patients with WAIHA and CAD can be used. Similar to WAIHA, it may be idiopathic or due to autoimmune disease, drug, or lymphoproliferative disease (e.g., lymphoma).

Paroxysmal Cold Hemoglobinuria: PCH is rare in adults, but accounts for 32% of cases of immune hemolytic anemia in children. It is a disorder that occurs due to the formation and activity of the DL antibody, which is a weak-affinity "biphasic" IgG cold-reactive, complement-binding autoantibody most frequently shown to have P specificity. This biphasic antibody binds to autologous P-positive RBCs at low temperature and initiates complement activation, but intravascular hemolysis does not occur until the RBCs with the antigen–antibody complex warm to 37°C. Patients with DL antibodies generally have a negative antibody screen because the antibodies do not react with RBCs under usual laboratory testing conditions. Because the DL antibody does not cause RBC agglutination above 20°C, it does not interfere with pretransfusion testing. PCH may be idiopathic or secondary to viral infections or syphilis. The clinical manifestation of PCH is an acute intravascular hemolytic anemia. In severe cases, shaking chills, abdominal pain, and high fever may be observed. It is predominantly a pediatric disease, typically occurring after a viral illness, such as an upper respiratory or gastrointestinal infection; however, rare causes may include vaccination, autoimmune disorders, and hematopoietic malignancies. Usually, it is self-limited, resolving completely within several weeks. There are only rare reported cases of chronic PCH, which were more common historically with advanced syphilis. The DAT is usually positive with complement only (Table 51.1). Along with the clinical presentation, the diagnosis is supported by a positive DL test and the absence of warm autoantibodies or CAs. Spherocytosis or rouleaux may be seen in up to 50% of patients on the peripheral blood smear. Although erythrophagocytosis by neutrophils is a pathognomonic finding, this is observed in less than 10% of cases. Treatment for PCH is primarily supportive, including cold avoidance. Corticosteroids have not been shown to improve outcome. Rare cases with severe or recurrent hemolysis have been treated with IVIG, azathioprine, or rituximab. Syphilis treatment usually eliminates syphilis-associated PCH.

There is some suggestion that transfused p or P^k RBCs will have better in vivo survival, but the patient is likely to require transfusion before these RBC products becoming available. Therefore, the use of crossmatch-compatible RBCs of common P types should be provided, which usually provides adequate RBC support. In addition, RBCs should be transfused through a blood warmer, and the patient should be maintained at a warm temperature.

Drug-Induced Hemolytic Anemia: Administration of a number of commonly used medications can lead to hemolysis typically occurring secondary to the mechanisms described below. Drug-induced hemolytic anemia may occur acutely as intravascular hemolysis soon after the patient receives the drug, or it may have a milder presentation as extravascular hemolysis, sometimes months after drug administration. Drugs may induce the formation of antibodies, either against the drug itself or against RBC antigens that may result in a positive DAT with or without hemolysis. Some of the antibodies produced require the presence of the drug (drug-dependent) for their detection and/or destruction, while others do not (drug-independent). If a DAT is positive and an eluate from the patient's RBCs demonstrates drug-independent antibodies, then primary WAIHA should also be considered as this is more common than drug-induced hemolytic anemia. It is impossible to distinguish drug-induced hemolytic anemia due to drug-independent antibodies from primary WAIHA using serological tests; therefore, the only way to confirm the diagnosis is by hematological improvement after discontinuation of the drug. If an eluate does not react during an antibody screen and the patient has recently received a drug, then it is important to test for drug-dependent antibodies. The clinical manifestations of drug-induced hemolytic anemia are variable. Patients who have an acute presentation of hemolysis after receiving a drug often have a history of prior administration of the same drug and therefore have previously formed antibodies. The disease has an excellent prognosis, and treatment is usually to discontinue the medication. Drug-induced hemolytic anemia is most often associated with an IgG antibody with Rh specificity. The drug is adsorbed onto the RBC membrane, and antibodies are made to the drug, membrane components, or part-drug, part-membrane (Fig. 51.1).

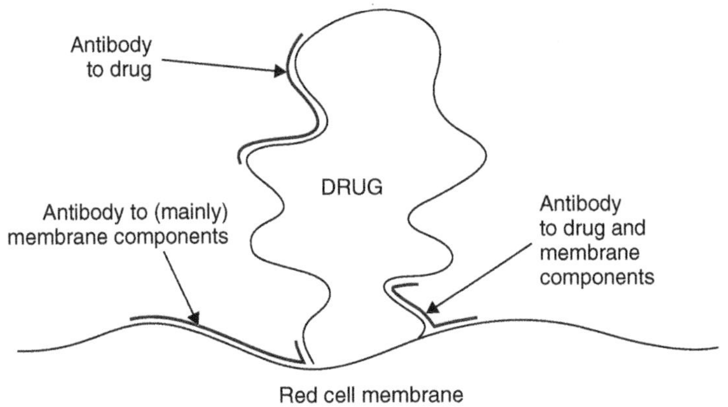

FIGURE 51.1 **Proposed unifying theory of drug-induced antibody reactions.** The *blue lines* represent antigen-binding sites on the F(ab) region of the drug-induced antibody. Drugs (haptens) bind loosely, or firmly, to cell membranes, and antibodies may be made to (1) the drug (producing in vitro reactions typical of a drug adsorption [penicillin-type] reaction); (2) membrane components or mainly membrane components (producing in vitro reactions typical of autoantibody); or (3) part-drug, part-membrane components (producing an in vitro reaction typical of the so-called immune complex mechanism). (Based on a cartoon by Habibi as cited by Garratty. From Fung, M. K., Eder, A. F., Spitalnik, S. L., Westhoff, C. M. (Eds.) (2017). *Technical manual* (19th ed.). Bethesda, MD: AABB Press.)

From a blood bank and transfusion service perspective, it is imperative to consider drug-induced hemolytic anemias under the appropriate clinical circumstances. Serologically, drug-induced hemolytic anemia may be difficult to demonstrate until the requisite medication is introduced into the test system. Specialized referral laboratories are able to perform these assays.

Drug Adsorption (Hapten) Hypothesis: In the drug adsorption hypothesis, the drug binds to the RBC membrane, and the antibody is largely directed against the drug itself or against a metabolite of the drug. The DAT is positive with IgG, and possibly complement, and occasionally results in extravascular hemolysis. The antibody in serum/plasma and elute only react with drug-treated RBCs (not untreated RBCs). Approximately, 3% of patients receiving large doses of penicillin, and 4% of patients receiving cephalosporins develop a positive DAT. The dose of the drug required to cause hemolysis varies between drugs: millions of units daily for weeks in the case of penicillin. Treatment of the hemolytic anemia is discontinuation of the medication.

Immune Complex Hypothesis: The hemolysis occurs as a result of the drug/anti-drug immune complex binding to the RBC. Cephalosporins may react by this mechanism. Usually, the DAT is positive for complement only, but IgG can be present. The drug must be present to demonstrate the antibody in the patient's plasma. Only a small amount of drug is required to result in hemolysis. This mechanism results in intravascular hemolysis with hemoglobinemia, hemoglobinuria, renal failure, and occasionally in death. Treatment of the hemolytic anemia is discontinuation of the medication and transfusion support.

Autoantibody Induction by Drugs: Drugs may also cause a positive DAT by inducing a drug-independent immune response. Alternatively, a drug-initiated immune response may occur. The proposed mechanism is through drug effect on immunoregulatory T cells. Serologically, this autoantibody is indistinguishable from a non-drug-induced warm autoantibody. Methyldopa is the classic example, but fludarabine, cephalosporins, procainamide, and nonsteroidal antiinflammatory drugs have also been associated with this mechanism. Treatment of the hemolytic anemia is discontinuation of the drug. The DAT may remain positive for months, but clinical improvement should be seen within 2 weeks. If the hemolytic anemia is severe, it may be treated like a WAIHA.

Nonimmunologic Protein Adsorption (Membrane Modification): Some drugs result in nonimmunologic adsorption of proteins and result in a positive DAT without hemolysis. Cephalosporins, diglycoaldehyde, suramin, cisplatin, clavulanate, sulbactam, and tazobactam are associated with this mechanism.

Further Reading

Barbara, D. W., Mauermann, W. J., Neal, J. R., et al. (2013). Cold agglutinins in patients undergoing cardiac surgery requiring cardiopulmonary bypass. *J Thorac Cardiovasc Surg*, *146*, 668–680.

Berentsen, S., Randen, U., Oksman, M., et al. (2017). Bendamustine plus rituximab for chronic cold agglutinin disease: Results of a Nordic prospective multicenter trial. *Blood*, *130*(4), 537–541.

Berentsen, S., & Tjonnfjord, G. (2012). Diagnosis and treatment of cold agglutinin mediated autoimmune hemolytic anemia. *Blood Rev, 26*, 107–115.

Birgens, H., Frederiksen, H., Hasselbalch, H. C., et al. (2013). A phase III randomized trial comparing glucocorticoid monotherapy versus glucocorticoid and rituximab in patients with autoimmune haemolytic anaemia. *Br J Haematol, 163*(3), 393–399.

Denomme, G. A. (2013). Prospects for the provision of genotyped blood for transfusion. *Br J Haematol, 163*, 3–9.

Eder, A. F. (2005). Review: Acute Donath-Landsteiner hemolytic anemia. *Immunohematology, 21*(2), 56–62.

Garratty, G. (2010). Immune hemolytic anemia associated with drug therapy. *Blood Rev, 24*, 143–150.

Garratty, G., & Arndt, P. A. (2014). Drugs that have been shown to cause drug-induced immune hemolytic anemia or positive direct antiglobulin tests: Some interesting findings since 2007. *Immunohematology, 30*, 66–79.

Garratty, G., & Petz, L. D. (2002). Approaches to selecting blood for transfusion to patients with autoimmune hemolytic anemia. *Transfusion, 42*, 1390–1392.

Go, R. S., Winters, J. L., & Kay, N. E. (2017). How I treat autoimmune hemolytic anemia. *Blood, 129*, 2971–2979.

Johnson, S. T., Fueger, J. T., & Gottschall, J. L. (2007). One center's experience: The serology and drugs associated with drug-induced immune hemolytic anemia – a new paradigm. *Transfusion, 47*, 697–702.

Michel, M. (2011). Classification and therapeutic approaches in autoimmune hemolytic anemia: An update. *Expert Rev Hematol, 4*, 607–618.

Prince, S. D., Winestone, L. E., Nanc, S. J., et al. (2017). Recurrent Donath-Landsteiner hemolytic anemia: A pediatric case report. *Transfusion, 57*, 1401–1406.

Schwartz, J., Padmanabhan, A., Aqui, N., et al. (2016). Guidelines on the use of therapeutic apheresis in clinical practice–evidence-based approach from the Apheresis Applications Committee of the American Society for Apheresis: The sixth special issue. *J Clin Apher, 31*, 149–162.

Shanbhag, S., & Spivak, J. (2015). Paroxysmal cold hemoglobinuria. *Hematol Oncol Clin N Am, 29*, 473–478.

Shirey, R. S., Boyd, J. S., Parwani, A., Tanz, W. S., Ness, P. M., & King, K. E. (2002). Prophylactic antigen-matched donor blood for patients with warm autoantibodies: An algorithm for transfusion management. *Transfusion, 42*, 1435–1441.

Zanella, A., & Barcellini, W. (2014). Treatment of autoimmune hemolytic anemias. *Haematologica, 99*(10), 1547–1554.

CHAPTER 52

Transfusion Management of Patients With Sickle Cell Disease and Thalassemia

Yvette C. Tanhehco, PhD, MD, MS and Patricia A. Shi, MD

Introduction: Patients with hereditary hemoglobinopathies such as sickle cell disease (SCD) and thalassemia may require lifelong red blood cell (RBC) transfusion support, warranting special consideration by blood centers and transfusion services.

Sickle Cell Disease: SCD is usually caused not only by homozygosity for the hemoglobin (Hb) S mutation (sickle cell anemia) but is also caused by heterozygosity of HbS with a β-thalassemia or HbC. It affects not only African Americans (AA) but also people of Hispanic, Mediterranean, Middle Eastern, South Asian, and Caribbean descent. Approximately 100,000 Americans are affected. Despite improved childhood survival because of early detection and interventions, the median overall survival is ~43 years.

Pathophysiology: Deoxygenated sickle Hb forms polymers within the erythrocyte that distort its shape and decrease its deformability, leading to increased blood viscosity, vasoocclusion, intravascular hemolysis, and subsequent anemia. Patients may have recurrent episodes of severe pain and other complications, such as acute chest syndrome and stroke. Organs with slow flow through sinusoids, such as spleen, liver, bone marrow, and penis, are especially vulnerable to occlusion. Repeated vasoocclusion/hemolysis over time causes widespread end-organ damage, including the brain, heart, lungs, and kidneys. Patients are also at increased risk of thrombotic events.

Red Blood Cell Transfusion: RBC transfusion improves oxygen delivery, suppresses endogenous erythropoiesis, and also reduces vasoocclusion, hemolysis, and blood viscosity by decreasing the fraction of HbS-containing RBCs. RBC transfusion may be indicated either acutely or chronically, with simple (allogeneic RBCs transfused without autologous RBC removal) or exchange (autologous RBCs removed and replaced with allogeneic RBCs) transfusion performed depending on clinical presentation and feasibility. The end hematocrit (Hct) goal of transfusion should generally be ≤30% (≤36% appropriate in chronic stroke prophylaxis and possibly other chronic transfusion indications) to minimize increased blood viscosity.

With simple transfusion, achieving a therapeutic HbS reduction to ≤30%–50% may not be possible without increasing the patient Hct to a point that risks circulatory overload or significantly increases blood viscosity. In contrast, red cell exchange (RCE) allows the achievement of a low HbS level without increasing total blood volume or Hct, if desired. The risk of transfusional iron overload is also less with RCE compared with simple transfusion. RCE is usually performed using an automated apheresis instrument but can be performed manually when apheresis is unavailable or in infants

Transfusion Medicine and Hemostasis. https://doi.org/10.1016/B978-0-12-813726-0.00052-0

with small total blood volumes. Drawbacks of RCE compared to simple transfusion include an increased number of RBC units transfused and thus possibly increased risk of alloimmunization, venous access requirements (two large-bore peripheral needles or a dialysis-type central venous catheter), and increased cost. A modified automated RCE, where RBC depletion by isovolemic hemodilution is performed before RCE, can be performed to decrease the number of RBC units needed or alternatively increase the time interval between procedures because of a greater reduction in HbS. Table 52.1 describes recommended transfusion management for various complications that arise in SCD patients.

For automated RCE procedures, the apheresis instrument will calculate the required volume of RBC replacement based on patient and target parameters. Use of the RBCX Calculation Tool app developed by TerumoBCT is recommended (see Chapter 76). If unavailable, a simple way to estimate the number of units to order, which assumes that the patient has 100% HbS-containing RBCs, is provided in Table 52.2. Performance of manual RCE is outlined in Table 52.3. Chapter 49 includes the dosing calculation for simple RBC transfusion in pediatric patients.

Red Blood Cell Product Selection: RBC products are typically matched for DCE and K RBC antigens, leukocyte-reduced, and HbS-negative to decrease the risk of RBC and HLA alloimmunization and accurately assess HbS percentage.

Antigen Matched: All SCD patients should have an extended antigen genotype or phenotype (ABO, Rh, Kell, Kidd, Duffy, Lewis, and MNS system antigens) on record to better manage alloimmunization risk. Molecular genotyping is preferable to serologic phenotyping due to likely history of recent transfusion, its ability to detect Rh variants (D, CEe), and type for antigens for which there are no serologic reagents.

Prophylactic RBC matching for C, E, K antigens decreases the risk of RBC alloimmunization, and more extensive prophylaxis that includes Duffy, Kidd, and S antigens can be initiated, especially if alloantibodies to Kidd, Duffy, S, or Dombrock antigens have formed. Limitations to prophylactic matching include increased cost, inventory management, and procurement delays, so urgently needed RBC transfusion should not be delayed for the purpose of prophylactic matching (in contrast to matching for already formed RBC alloantibodies).

Risks of Transfusion

Alloantibody Formation: SCD patients are prone to forming RBC alloantibodies because of differences in RBC antigen allele frequency between patients of predominantly AA ethnicity and Caucasian donors, high rates of RhCE variant (i.e., partial C and partial e) expression among AA, and postulated altered immune regulation. The RBC alloimmunization rate is reported to decrease from as high as ~50% without prophylactic matching to ~10% with limited matching to the most commonly formed alloantibodies (C, E, K) and to ~5% with extended matching (also Duffy, Kidd, S).

Autoantibody Formation: Autoantibodies are more frequently detected in SCD patients who are RBC alloimmunized (~50%–70%) compared with nonalloimmunized (5%–10%). Some autoantibodies may actually be alloantibodies due to RhCE variants. SCD patients may be predisposed to autoantibody formation due to

TABLE 52.1 Transfusion Management for Sickle Cell Disease (SCD) Complications

Indication	Description	Usual Method of Transfusion
Acute		
Acute symptomatic anemia	• Worsening of chronic anemia due to blood loss, increased hemolysis, viral suppression of erythropoiesis, or sequestration • Treat if symptomatic	Simple
Aplastic crisis	• Decreased Hb with reticulocytopenia • Usually caused by parvovirus B19 infection • Erythropoietic arrest for >5 days may lead to marked anemia with compensatory expanded plasma volume and secondary heart failure	Simple
Acute sequestration	• Splenic or hepatic sequestration occurs due to RBC trapping within the sinusoids • Signs and symptoms include pain, symptomatic anemia, hypotension, organ enlargement • Laboratory findings for hepatic sequestration include mild liver enzyme elevation and direct bilirubinemia. • Splenic sequestration has a recurrence rate of ~50%	• Simple in most cases, but exchange transfusion may be indicated if hemodynamically stable and hyperviscosity is a concern • With simple transfusion, hyperviscosity from the release of sequestered autologous RBCs may be avoided by transfusing RBCs in small aliquots (5 mL/kg) over 4 hours
Preoperative (moderate- to high-risk surgery using general anesthesia)	• Pre-op transfusion to a Hb goal of 10 g/dL reduces the risk of perioperative SCD-related complications • Perioperative treatment should also include hydration, body temperature maintenance, adequate oxygenation, and incentive spirometry.	• Simple sufficient in many cases, but exchange transfusion is recommended with high-risk surgery or with baseline Hb >9 g/dL[a,b]
Acute multiorgan failure	• Life-threatening complication resulting from widespread tissue infarction secondary to vasoocclusion • Signs and symptoms include rapid drop in platelet counts, highly elevated, LDH, fever, and acute dysfunction in at least two organs, most commonly the lungs, liver, and kidney	Exchange preferable to simple, especially in a rapidly decompensating patient

Continued

TABLE 52.1 Transfusion Management for Sickle Cell Disease (SCD) Complications—cont'd

Indication	Description	Usual Method of Transfusion
Acute stroke	• The incidence rate of ischemic stroke is greatest in children and older patients, whereas the incidence rate of hemorrhagic stroke, which has a high case-fatality rate, is greatest in patients in their twenties • Stroke treatment should also follow standard recommendations for non-SCD stroke treatment	• Exchange to a target HbS level of ≤30% is recommended for both ischemic and hemorrhagic stroke • Simple transfusion to raise the Hb to 10g/dL before RCE may be warranted if RCE cannot occur soon
Fat embolism syndrome	• Caused by extensive bone marrow infarction and necrosis due to vasoocclusion in bone marrow sinusoids • Signs and symptoms include hypoxemia, nonfocal encephalopathy, thrombocytopenia, and skin and mucosal petechiae	Exchange
Severe hepatic crisis/intrahepatic cholestasis	• Massive vasoocclusion in the sinusoids leads to hepatic ischemia and dysfunction • Signs and symptoms include tender hepatomegaly, liver enzyme elevation, coagulopathy, and markedly elevated bilirubinemia (often >50mg/dL) • May be recurrent and progress to chronic liver disease	Exchange
Acute chest syndrome	• Develops in >30% of patients with SCD in their lifetime • Leading cause of death • Signs and symptoms include pulmonary infiltrate on chest X-ray, shortness of breath, chest pain, and/or fever • May not be the presenting diagnosis; often develops during admission for pain crisis or following anesthesia • Precipitants include fat embolism from bone marrow infarction, pulmonary infarction, pneumonia, asthma	Simple or exchange[c] • Simple transfusion may be effective if instituted promptly • RCE is recommended with rapid clinical deterioration or progressive hypoxemia/respiratory distress despite simple transfusion
Preoperative high-risk surgery	• Transfusions to a Hb goal of 10g/dL before surgery reduces the risk of perioperative sickle cell–related complications in surgeries requiring general anesthesia • Perioperative treatment should also include hydration, body temperature maintenance, adequate oxygenation, and incentive spirometry	Simple or exchange[d] • RCE may be warranted in high-risk surgeries such as major cardiothoracic, vascular, or neurosurgical procedures and in patients with HbSC to prevent hyperviscosity

Chronic prophylactic		
Pregnancy	• With fetal or maternal complications: indicated in patients with previous fetal/neonatal mortality or morbidity, current risk factors (e.g., multiple gestations), obstetric complications (e.g., preeclampsia), or SCD-related complications (pain crisis or acute chest syndrome; more common in the third trimester) • Universal 3rd trimester prophylaxis is controversial	Simple or exchange • Decision of which trimester to initiate chronic transfusion should be individualized per history and risk factors
Stroke prevention (overt and silent cerebral infarct)	• Primary stroke prophylaxis: Prophylactic chronic transfusion in children with high stroke risk by transcranial Doppler (TCD) ≥200 cm/s reduces a first stroke risk of 10% of patients per year by 92% (STOP trial)[e]. Hydroxyurea is noninferior to transfusions for stroke prevention in high-risk (by TCD) children who have received at least 1 year of transfusions and have no MRA-defined severe vasculopathy (TWITCH trial)[f]. Transfusions should be continued indefinitely, however, in the absence of hydroxyurea due to an otherwise high risk of TCD reversion, stroke, and silent infarcts (STOP II trial)[g] • Secondary stroke prevention: chronic transfusion is effective for prevention of recurrent stroke, decreasing overt stroke recurrence from ~70% in the absence of chronic transfusion therapy to 20% for new overt stroke and 27% for silent cerebral infarcts[h]. Hydroxyurea is not as effective as transfusion for prevention of recurrent stroke (SWITCH trial)[i] • Silent cerebral infarcts: in patients with silent cerebral infarct alone (normal TCD and no history of overt stroke), chronic transfusion therapy reduces the incidence of overt stroke and new/enlarging silent cerebral infarcts, which are associated with neurocognitive disability and poor academic achievement (SIT trial)[j]	Simple or exchange • The goal of chronic transfusion therapy for both primary and secondary stroke prophylaxis is to maintain HbS ≤30% and Hb at 10 g/dL • Some reports suggest that the HbS target can be raised to <50% after 3 years of stability, but strong evidence is lacking
Controversial indications		
Recurrent pain episodes	• STOP trial showed a significant decrease in frequency of severe pain episodes with chronic transfusion	Simple or exchange
Recurrent acute chest syndrome	• Multiple episodes of ACS may contribute to chronic lung disease such as pulmonary fibrosis and pulmonary hypertension • STOP trial showed a decrease in frequency of ACS with chronic transfusion	Simple or exchange
Recurrent splenic sequestration	• Chronic transfusion to prevent recurrence or delay splenectomy	Simple

Continued

TABLE 52.1 Transfusion Management for Sickle Cell Disease (SCD) Complications—cont'd

Indication	Description	Usual Method of Transfusion
Priapism	• Acute transfusion reasonable if symptoms persist despite initial treatment with hydration, analgesia, and urologic intervention • Reports of ASPEN syndrome (Association of sickle cell disease, Priapism, exchange transfusion, and neurologic events) are related to acute increases in blood viscosity avoidable by targeting an end Hct close to the patient's baseline.	Simple or exchange avoiding an end Hct > 30%
Leg ulcers	• Chronic transfusion therapy reasonable in conjunction with local measures (surgical debridement or grafts) until ulcer healed	Simple or exchange
Possible future indications		
Pulmonary hypertension[k]	• Prevalent in 6%–10% of adult patients when confirmed by right heart catheterization • Associated with increased mortality • Nitric oxide depletion by hemolysis may be causal factor • Agents approved for pulmonary hypertension in other disease populations are not yet proven beneficial in SCD patients	
Chronic kidney disease[l]	• Microalbuminuria is a risk factor for chronic kidney disease and is associated with low hemoglobin.	
Congestive heart failure/ diastolic dysfunction	• Appears to be related to anemia for which transfusion may therefore assuage[m,n]	
Nonindications		
Acute pain crisis	Main treatment is hydration, analgesia, and incentive spirometry	
Avascular necrosis	Causative factors of AVN are still unclear; treatment is focused on local interventions such as core decompression.	

[a]Howard, J., Malfroy, M., Llewelyn, C., Choo, L., Hodge, R., Johnson, T., et al. (2013). The transfusion alternatives preoperatively in sickle cell disease (TAPS) study: A randomized, controlled, multicentre clinical trial. *Lancet (Lond Engl), 381*(9870), 930–938.

[b]Vichinsky, E. P., Haberkern, C. M., Neumayr, L., Earles, A. N., Black, D., Koshy, M., et al. (1995). A comparison of conservative and aggressive transfusion regimens in the perioperative management of sickle cell disease. The Preoperative Transfusion in Sickle Cell Disease Study Group. *N Engl J Med, 333*(4), 206–213.

[c]Adams, R. J., McKie, V. C., Hsu, L., Files, B., Vichinsky, E., Pegelow, C., et al. (1998). Prevention of a first stroke by transfusions in children with sickle cell anemia and abnormal results on transcranial Doppler ultrasonography. *N Engl J Med, 339*(1), 5–11.

[d]Ware, R. E., Davis, B. R., Schultz, W. H., Brown, R. C., Aygun, B., Sarnaik, S., et al. (2016). Hydroxycarbamide versus chronic transfusion for maintenance of transcranial doppler flow velocities in children with sickle cell anemia-TCD with transfusions changing to hydroxyurea (TWiTCH): A multicentre, open-label, phase 3, noninferiority trial. *Lancet (Lond Engl), 387*(10019), 661–670.

[e]Abboud, M. R., Yim, E., Musallam, K. M., & Adams, R. J. (2011). Discontinuing prophylactic transfusions increases the risk of silent brain infarction in children with sickle cell disease: data from STOP II. *Blood, 118*(4), 894–898.

[f]Hulbert, M. L., McKinstry, R. C., Lacey, J. L., Moran, C. J., Panepinto, J. A., Thompson, A. A, et al. (2011). Silent cerebral infarcts occur despite regular blood transfusion therapy after first strokes in children with sickle cell disease. *Blood, 117*(3), 772–779.

[g]Estcourt, L. J., Fortin, P. M., Hopewell, S., Trivella, M., Hambleton, I. R., & Cho, G. (2016). Regular long-term red blood cell transfusions for managing chronic chest complications in sickle cell disease. *Cochrane Database Syst Rev*, (5), Cd008360.

[h]Hirst, C., & Williamson, L. (2012). Preoperative blood transfusions for sickle cell disease. *Cochrane Database Syst Rev, 1*, Cd003149.

[i]Ware, R. E., Schultz, W. H., Yovetich, N., Mortier, N. A., Alvarez, O., Hilliard, L., et al. (2011). Stroke with transfusions changing to hydroxyurea (SWiTCH): A phase III randomized clinical trial for treatment of children with sickle cell anemia, stroke, and iron overload. *Pediatr Blood Cancer, 57*(6), 1011–1017.

[j]DeBaun, M. R., Gordon, M., McKinstry, R.C., Noetzel, M. J., White, D.A., Sarnaik, S. A., et al. (2014). Controlled trial of transfusions for silent cerebral infarcts in sickle cell anemia. *N Engl J Med, 371*(8), 699–710.

[k]Detterich, J. A., Kato, R. M., Rabai, M., Meiselman, H. J., Coates, T. D., & Wood, J. C. (2015). Chronic transfusion therapy improves but does not normalize systemic and pulmonary vasculopathy in sickle cell disease. *Blood, 126*(6), 703–710.

[l]Alvarez, O., Montane, B., Lopez, G., Wilkinson, J., & Miller, T. (2006). Early blood transfusions protect against microalbuminuria in children with sickle cell disease. *Pediatr Blood Cancer, 47*(1), 71–76.

[m]Westwood, M. A., Shah, F., Anderson, L. J., Strange, J. W., Tanner, M. A., Maceira, A. M., et al. (2007). Myocardial tissue characterization and the role of chronic anemia in sickle cell cardiomyopathy. *J Magn Reson Imaging, 26*(3), 564–568.

[n]Niss, O., Fleck, R., Makue, F., Alsaied, T., Desai, P., Towbin, J. A., et al. (2017). Association between diffuse myocardial fibrosis and diastolic dysfunction in sickle cell anemia. *Blood, 130*(2), 205–213.

TABLE 52.2 Estimation for Number of Red Blood Cell (RBC) Units to Order for Automated Red Cell Exchange (RCE)[a]

1. RBC volume (mL) = current Hct × total blood volume
2. Number of RBC units to maintain current Hct = RBC volume/180 mL
3. Number of RBC units to raise current Hct to desired Hct = (desired Hct − current Hct)/3
4. Total number of RBC units to order = Equation 3 + Equation 4 (roundup)

[a]Assumes the patient's RBC Volume is 100% HbS-containing RBCs.

TABLE 52.3 Manual RCE Procedure[a]

Adults

1. Phlebotomize 500 mL whole blood, infuse 500 mL normal saline
2. Phlebotomize 500 mL whole blood, infuse 2 RBC units
3. Repeat Steps 1 and 2 until the total blood volume exchanged equals 1.5 times the patient's RBC volume

Infants

1. Calculate RBC product volume to prepare for transfusion based on formula in Table 50.4
2. Phlebotomize 5–10 mL/kg, infuse an equivalent volume of RBC product
3. Repeat Step 2 until 1–2 times the patient's total blood volume has been exchanged

[a]Utilizing stopcocks for whole blood removal and infusion of RBC product is helpful.

disease-related inflammation and RBC membrane changes, leading to neoantigen exposure. Autoantibodies in SCD are not associated with development of rheumatologic disease but may cause clinically significant hemolysis and complicate underlying alloantibody identification and crossmatching (see Chapter 51).

Infection: SCD patients are susceptible to sepsis from bacterially contaminated products, particularly with encapsulated organisms and babesia, due to splenic autoinfarction and dysfunction. Additionally, SCD patients on iron chelation or with iron overload are susceptible to serious infection with *Yersinia enterocolitica*, which, being cold-tolerant and heme-avid, contaminates ~1 in 500,000 units RBC units.

Iron Overload: Iron overload can occur with repeated simple transfusion and also with repeated RCE when the target end Hct is above the starting Hct because the body has no mechanism for excreting excess iron. Serum ferritin >1000 μg/L along with a history of at least 10 transfusions may be useful as a screening test to determine which patients should undergo magnetic resonance imaging (MRI) of the liver ± heart to determine the need for iron chelation therapy. SCD patients tend to be spared cardiac iron overload compared with thalassemia patients, with about 2.5% of SCD patients affected.

Hyperhemolytic Transfusion Reaction: Hyperhemolytic transfusion reactions (HHTRs) are reported in 1%–4% of transfused SCD patients and can be potentially life-threatening. The patient presents with anemia below pretransfusion levels, hemolysis, and often reticulocytopenia approximately 7–10 days after a RBC transfusion. Some patients may have newly detected allo- or autoantibodies, but others have no detectable antibodies (i.e., negative antibody screen and direct antiglobulin test). Possible mechanisms for HHTRs include bystander hemolysis, erythropoiesis suppression, contact hemolysis via activated macrophages, and complement activation. Further RBC transfusions during the event, even if extensively antigen matched, may exacerbate hemolysis to life-threatening severity. Acute management may include erythropoietin, steroids, IVIG, limiting activity to minimize oxygen consumption, and possibly eculizumab and/or rituximab. Patients with a history of HHTR are at risk of future HHTR, so transfusions should be avoided if possible, but if absolutely necessary, rituximab prophylaxis may be helpful before future transfusion.

Thalassemia: Thalassemia mutations are the most common inherited gene mutations in the world. An estimated 1000 patients with β-thalassemia major live in the United States, and the prevalence of thalassemia may be increasing because of immigration. Thalassemia primarily affects people of Mediterranean, Middle Eastern, and Asian descent, with Asians and Caucasians making up the majority in the United States. Chronic lifelong transfusion is needed in those severely affected to prevent life-threatening organ damage. The only curative treatment is hematopoietic stem cell transplantation or more recently gene therapy.

Pathophysiology: Thalassemia results from mutations in the β-globin gene (β-thalassemia) or α-globin gene (α-thalassemia) that reduce the synthesis of that globin, leading to ineffective erythropoiesis, iron hyperabsorption, RBC membrane damage, and extravascular hemolysis. The resulting anemia, extramedullary hematopoiesis, hypermetabolic state, and iron overload lead to life-threatening organ damage. Patients are also at increased risk of thrombotic events.

β-Thalassemia: Thalassemias are phenotypically classified by their clinical severity. Thalassemia major (Cooley's anemia) patients have anemia and ineffective erythropoiesis severe enough to be chronically transfusion-dependent, usually within the first year of life. Thalassemia intermedia patients are also anemic but do not require chronic transfusions to prevent organ damage. Some patients initially classified as thalassemia intermedia may eventually need chronic lifelong transfusion. Both phenotypes result from homozygosity or compound heterozygosity for mutations (including HbE) that abolish or severely reduce β-globin expression. Thalassemia minor patients inherit only a single β-globin mutation and have a mild anemia that requires no specific treatment.

α-Thalassemia: Three or more of the four α-globin genes (HbH disease [3/4 genes] or Hb Bart's [4/4 genes]) need to be mutated to result in thalassemia severe enough to require chronic transfusions. With HbH disease, transfusion dependence is variable; however, the HbH-Constant Spring genotype is often transfusion-dependent.

Red Blood Cell Transfusion: Indications for transfusion include growth retardation, failure to thrive, symptomatic anemia, Hb < 7 g/dL, progressive hypersplenism, and facial bone or other skeletal changes. Transfusions increase oxygen delivery and suppress endogenous erythropoiesis, thereby decreasing extramedullary hematopoiesis, hypersplenism, erythroid bone marrow expansion, bone fractures/osteopenia, and gastrointestinal absorption of iron. Patients with thalassemia major are managed with chronic simple transfusions to maintain their pretransfusion Hb at 9–10 g/dL, requiring transfusion intervals of 2–5 weeks. A higher pretransfusion Hb target of 11–12 g/dL may be appropriate for patients with heart disease or continued inadequate suppression of erythropoiesis. The patient's transfusion volume should be monitored annually, and if >200 mL/kg of RBC per year is required to maintain Hb > 10 g/dL, splenectomy should be considered to decrease transfusion requirements and iron overload. The extended RBC antigen phenotype or genotype (ABO, Rh, Kell, Kidd, Duffy, Lewis, and MNS systems) should be determined before the initiation of transfusion to better manage alloimmunization risk.

Leukoreduced, prophylactically C, E, K antigen matched RBC products are recommended. Notably, routine washing and irradiation of RBC units is unnecessary in the absence of severe allergic reactions and immune deficiency, respectively.

Risks of Transfusion

Alloantibody and Autoantibody Formation: The prevalence of RBC alloimmunization in the US thalassemia population is 15%–20%, with the most common antibodies being to C, E, K, and Kidd antigens; anti-Mi^{a+} is frequent in Chinese patients. Alloimmunization risk increases with years of transfusion, although transfusion before 1 year of age may induce some immune tolerance. RBC autoantibodies, for which RBC alloimmunization is the main risk factor, are present in ~5%, of patients. Although these do not typically cause hemolysis, they complicate underlying alloantibody identification and crossmatching. Transfused patients are also at risk of HLA alloimmunization, with HLA antibodies being present in about 50% of patients.

Transfusion Reactions: About 50% of patients experience a transfusion reaction, with allergic and febrile nonhemolytic transfusion reactions most commonly; allergic reactions may be more common than in the general population. Transfusion-transmitted infections are a possibility. Although less frequent that in SCD, thalassemia patients are also at risk of a hyperhemolytic transfusion reaction.

Iron Overload: Chronic transfusion in the absence of iron chelation results in iron overload, with cardiomyopathy, liver fibrosis, and endocrine dysfunction. Efficacy of iron chelation should be followed by serial ferritin at least every 3 months and cardiac/liver T2 MRI at least annually. Patients with thalassemia intermedia, even in the absence of transfusions, also need to be monitored for iron overload due to disease-induced gastrointestinal hyperabsorption (see Chapter 75).

Further Reading

Buchanan, G. R., & Yawn, B. P. (2014). *Evidence-based management of sickle cell disease*. U.S. Department of Health and Human Services National Institutes of Health, National Heart, Lung, and Blood Institute.

Chou, S. T., & Fasano, R. M. (2016). Management of patients with sickle cell disease using transfusion therapy: Guidelines and complications. *Hematol Oncol Clin N Am.*, *30*(3), 591–608.

Coates, T. D., & Wood, J. C. (2017). How we manage iron overload in sickle cell patients. *Br J Haematol, 177*(5), 703–716.

Davis, B. A., Allard, S., Qureshi, A., et al. (2017). Guidelines on red cell transfusion in sickle cell disease. Part I: Principles and laboratory aspects. British Committee for Standards in Haematology. *Br J Haematol, 176*(2), 179–191.

Davis, B. A., Allard, S., Qureshi, A., et al. (2017). Guidelines on red cell transfusion in sickle cell disease Part II: Indications for transfusion. *Br J Haematol, 176*(2), 192–209.

Estcourt, L. J., Fortin, P. M., Hopewell, S., et al. (2017). Blood transfusion for preventing primary and secondary stroke in people with sickle cell disease. *Cochrane Database Syst Rev, 1*, CD003146.

Howard, J. (2016). Sickle cell disease: When and how to transfuse. *Hematol Am Soc Hematol Educ Program, 2016*(1), 625–631.

Nsimba, B., Habibi, A., Pirenne, F., et al. (2017). Delayed hemolytic transfusion reaction and hyperhemolysis syndrome without detectable alloantibodies or autoantibodies in a patient with sickle cell disease: A fatal case report and literature review. *Transfus Clin Biol* S1246-7820(17)30537-2.

Okusanya, B. O., & Oladapo, O. T. (2016). Prophylactic versus selective blood transfusion for sickle cell disease in pregnancy. *Cochrane Database Syst Rev, 12*, CD010378.

CHAPTER 53

Transfusion Management of Patients Undergoing Hematopoietic Stem Cell and Solid Organ Transplantation

Sara Rutter, MD and Jeanne E. Hendrickson, MD

Transfusion in Patients Undergoing Hematopoietic Stem Cell Transplantation: Human leukocyte antigen (HLA) compatibility is of greater importance than ABO compatibility in donor selection for HSC transplantation (HSCT). As such, ABO incompatibility may be present in up to 50% of donor/recipient HSCT pairs. Major ABO incompatibility occurs when the recipient expresses an antibody to A or B antigen present on the donor cells. Minor ABO incompatibility occurs when the donor has an antibody to A or B antigen present on recipient cells. The type of ABO incompatibility must be considered for transfusion product selection in the pretransplant, peritransplant, and posttransplant periods. Other considerations for predicting the duration of needed transfusion support include the source of donor HSCs and the type of recipient preparative conditioning.

Product Selection

Leukoreduction: Leukocyte reduction of RBCs and platelets decreases febrile nonhemolytic transfusion reactions, transmission of viruses such as cytomegalovirus (CMV), and HLA alloimmunization. Therefore, it is recommended that HSCT recipients receive leukoreduced products.

Cytomegalovirus Safe Products: HSCT recipients are susceptible to multiple viruses, and thus strategies to minimize transfusion transmitted viral transmission are recommended. Leukoreduced products are considered "CMV safe" for transplant recipients, and the approach of selecting both leukoreduced and CMV seronegative products is not necessary.

Irradiation: HSCT recipients are considered at risk for transfusion-associated graft versus host disease (TA-GVHD) and should receive irradiated cellular blood products (i.e., red blood cell, platelet, and granulocyte transfusions). Unlike GVHD associated with transplantation, TA-GVHD is nearly uniformly fatal due to bone marrow involvement producing pancytopenia, as well as skin, liver, and intestinal involvement.

ABO/RhD Compatibility: Careful selection of compatible blood components is required in the pre-, peri-, and posttransplant period (Table 53.1). For RhD mismatched HSCTs, some centers provide RhD negative RBCs regardless of the direction of the mismatch to prevent anti-D alloimmunization.

Transfusion Medicine and Hemostasis. https://doi.org/10.1016/B978-0-12-813726-0.00053-2

TABLE 53.1 Transfusion Support for Patients Undergoing Hematopoietic Stem Cell Transplantation According to the ABO Incompatibility and the Stage of the Transplant

Type of Incompatibility	Transplant Stage	ABO Blood Group Selection	
		RBCs	Platelets[a]/Plasma
Major incompatibility	Preparative regimen	Recipient	Donor
	Transplantation	Recipient	Donor
	Recipient antibodies detected	Recipient	Donor
	Recipient antibodies no longer detected	Donor	Donor
Minor incompatibility	Preparative regimen	Donor	Recipient
	Transplantation	Donor	Recipient
	Recipient cells circulating	Donor	Recipient
	Recipient cells no longer circulating	Donor	Donor
Bidirectional incompatibility	Preparative regimen	Group O	Group AB
	Transplantation	Group O	Group AB
	Recipient antibodies detected/recipient cells circulating	Group O	Group AB
	Recipient antibodies no longer detected/ recipient cells no longer circulating	Donor	Donor

[a]Because of a short shelf life, and the generally limited availability of group AB platelets, it may not always be possible to provide identically matched ABO platelet products, as outlined in this table. In some instances, blood banks and transfusion services may consider providing either a limited number of units of ABO mismatched platelets per day or ABO mismatched units that have been volume reduced to diminish plasma content.

Modified from Fung, M. K., Eder, A. F., Spitalnik, S. L., & Westhoff, C. M. (Eds.). (2017). *Technical manual* (19th ed.). Bethesda, MD: AABB Press, with permission.

Red Blood Cell Transfusion: Many HSCT recipients are transfused at threshold hematocrits of 21%–25%, with few studies to date investigating the optimal transfusion threshold. A recent subanalysis of oncology patients in the platelet dosing trial showed that a hematocrit <25% was associated with an increased risk of bleeding (OR 1.29). The end goals of transfusion must be taken into consideration while awaiting future trials to refine best practices for RBC transfusion thresholds in HSCT recipients.

Platelet Transfusion: Platelet transfusion triggers of 10,000/μL are typically used for HSCT recipients with uncomplicated thrombocytopenia to prevent bleeding, and this threshold is recommended by the AABB's Clinical Practice Guidelines. A subset of HSCT recipients, including those with a higher bleeding risk (based on past personal history or based on risk factors for CNS bleeding, for example), may be transfused at higher platelet thresholds. Platelet refractoriness, which may be immune or nonimmune mediated, can result in a lower corrected count increment. Immune-mediated platelet refractoriness may require the use of crossmatched or HLA-matched platelets.

Limited platelet product availability may make location of the "first-choice" product difficult, and a limited number of ABO mismatched products may be substituted in cases of necessity. Platelets stored in platelet additive solution decrease the likelihood of allergic reactions, and pathogen reduced (PR) platelets decrease the likelihood of bacterial contamination but also decrease posttransfusion platelet recovery.

Plasma Transfusion: Plasma may rarely be needed for HSCT recipients who are coagulopathic to treat or prevent bleeding. Ideally, plasma products should also be compatible with both donor and recipient ABO types (Table 53.1).

Granulocyte Transfusion: Granulocytes are selectively transfused to some HSCT recipients with neutropenia (absolute neutrophil count <500/μL) and severe bacterial or fungal infections that are nonresponsive to antibiotic therapy. The efficacy of granulocyte therapy continues to be debated. Special considerations for granulocytes include the need for them to be ABO/RhD compatible with the recipient due to RBC contamination, irradiated, and ideally collected from CMV seronegative donors, if applicable, due to inability for leukoreduction.

Special Considerations Based on Donor/Recipient ABO Mismatch: There is a risk in *major* mismatched HSCTs of delayed erythropoiesis due to persistent recipient production of anti-A or anti-B against the antigens of engrafting donor RBCs. In some instances, RBC engraftment is delayed 40 or more days after transplantation, with RBC transfusion dependency typically persisting until engraftment occurs. There is a risk in *minor* mismatched HSCTs for the production of anti-A or anti-B by donor lymphocytes, with these antibodies targeting residual recipient RBCs. This "passenger lymphocyte" syndrome typically occurs 1–2 weeks after HSCT and subsides once recipient RBCs cease to exist. Of note, incompatibility may also be seen against RBC antigens of non-ABO blood group systems, with delayed RBC engraftment or hemolysis being possible.

Transfusion in Patients Undergoing Solid Organ Transplantation

Description: Transfusion requirements for solid organ transplantation vary largely by the patient's underlying disease and the organ being transplanted. Whereas patients undergoing liver or intestinal transplantation may require transfusion of large volumes of RBCs, plasma, and other blood products because of underlying coagulopathy, transfusions during renal transplantation are rare. Transfusion requirements during cardiac transplantation are variable, with at least 10% of patients having massive bleeding.

ABO compatibility of donor and recipient blood types is a priority for selection of solid organs for transplantation, unlike for HSCT. A number of strategies have been utilized in recent years, however, to improve the survival of ABO-incompatible solid organ transplants. These strategies include the selection of recipients with low titers of relevant isohemagglutinins (anti-A and/or anti-B), the utilization of A_2 organ donors with lower levels of antigen expression than A_1 donors, recipient plasma exchange, IVIG, rituximab, and other immunosuppression to achieve lower titers of isohemagglutinins in the peritransplant period.

Product Selection

Leukoreduction: Leukocyte reduction of RBCs and platelets decreases febrile nonhemolytic transfusion reactions, transmission of viruses such as CMV, and HLA alloimmunization. Therefore, solid organ recipients should receive leukoreduced products.

Cytomegalovirus Safe Products: Solid organ recipients are susceptible to multiple viruses, and thus strategies to minimize transfusion transmitted viral transmission are recommended. Leukoreduced products are considered "CMV safe" for transplant recipients, and the approach of selecting both leukoreduced and CMV seronegative products is not necessary.

Irradiation: Irradiation of cellular blood products is not recommended for solid organ transplant recipients.

ABO/RhD Compatibility: Blood product selection for ABO identical solid organ transplants is intuitive. For ABO minor mismatched solid organ transplants, some centers transfuse RBCs of the organ donor ABO type to prevent hemolysis from passenger lymphocyte–derived antibodies; see "special considerations" section. Massive transfusion scenarios in patients with the need for a relatively rare type of blood (e.g., RhD negative recipients) may require the provision of RhD positive RBCs in the middle of a case when the bleeding is most brisk, with a switch back to RhD negative RBCs toward the end of the case. When possible, platelets and plasma that are ABO identical or compatible with the recipient should be provided for solid organ transplant recipients.

Red Blood Cell Transfusion: RBC transfusion thresholds for solid organ transplant recipients have not been studied extensively. Preoperative strategies to increase hemoglobin values, in combination with initiatives for intraoperative blood conservation, improved surgical techniques, thromboelastography, and medications have decreased RBC transfusions in some transplant settings.

Platelet Transfusion: Solid organ transplant recipients may require platelet transfusions in the intra- or perioperative period, especially in the setting of massive transfusion and dilutional coagulopathy. The need for platelets depends on bleeding and on laboratory evidence of thrombocytopenia or platelet dysfunction; point-of-care viscoelastic testing such as rotational thromboelastrography or thromboelastrography can help to guide product selection. Platelet transfusion support for patients undergoing liver transplantation is complicated by the tendency of these patients to bleed while simultaneously being at an increased risk of thrombosis.

Platelets stored in platelet additive solution decrease the likelihood of allergic reactions but have decreased plasma, which may be disadvantageous in the massive transfusion setting. PR platelets decrease the likelihood of bacterial contamination, but also decrease posttransfusion platelet recovery. Given the low risk of septic transfusion reactions in this patient population and the high risk of bleeding, PR platelets may not be beneficial in this setting.

Plasma Transfusion: Plasma may be needed intraoperatively for solid organ transplant recipients who are coagulopathic due to their underlying disease or who develop dilutional coagulopathy from massive transfusion. Patients undergoing liver transplantation or small bowel transplantation are particularly likely to require plasma transfusion, with many centers preparing massive transfusion boxes containing a near 1:1 ratio of RBCs:plasma to have in the operating room for such transplants; these products are then supplemented as needed with platelets and cryoprecipitate.

Special Considerations Based on Donor/Recipient ABO Mismatch: A special consideration for solid organ transplantation includes the possibility of delayed RBC hemolysis 1–2 weeks posttransplant due to the production of antibodies from passenger lymphocytes in the transplanted organ. Organs from group O donors transplanted into ABO mismatched recipients (especially group A recipients) have the highest risk of passenger lymphocyte syndrome. As described previously, some centers preferentially transfuse donor-type (in this case, group O) RBC products early in the transplant period to minimize this risk, and others rely on immunosuppressive medications for risk mitigation. Donor-derived antibodies against minor blood group antigens in recipient can also mediate passenger lymphocyte syndrome in solid organ transplant recipients. Plasma exchange and/or RBC exchange with RBCs lacking the corresponding antigen(s) have been used in patients with severe hemolysis.

Further Reading

Carson, J. L., Grossman, B. J., Kleinman, S., et al. (2012). Red blood cell transfusion: A clinical practice guideline from the AABB. *Ann Intern Med, 157*(1), 49–58.

Cid, J., Lozano, M., Klein, H. G., & Flegel, W. A. (2014). Matching for the D antigen in haematopoietic progenitor cell transplantation: Definition and clinical outcomes. *Blood Transfus, 12*(3), 301–306.

Dahl, D., Hahn, A., Koenecke, C., et al. (2010). Prolonged isolated red blood cell transfusion requirement after allogeneic blood stem cell transplantation: Identification of patients at risk. *Transfusion, 50*(3), 649–655.

Estcourt, L. J., Stanworth, S. J., Doree, C., Hopewell, S., Trivella, M., & Murphy, M. F. (2015). Comparison of different platelet count thresholds to guide administration of prophylactic platelet transfusion for preventing bleeding in people with haematological disorders after myelosuppressive chemotherapy or stem cell transplantation. *Cochrane Database Syst Rev, 11*, Cd010983.

Hall, S., Danby, R., Osman, H., et al. (2015). Transfusion in CMV seronegative T-depleted allogeneic stem cell transplant recipients with CMV-unselected blood components results in zero CMV transmissions in the era of universal leukocyte reduction: A UK dual centre experience. *Transfus Med, 25*(6), 418–423.

Madisetty, J., & Wang, C. (2017). Transfusion medicine and coagulation management in organ transplantation. *Anesthesiol Clin, 35*(3), 407–420.

Price, T. H., Boeckh, M., Harrison, R. W., et al. (2015). Efficacy of transfusion with granulocytes from G-CSF/dexamethasone-treated donors in neutropenic patients with infection. *Blood, 126*(18), 2153–2161.

Roback, J. D., Caldwell, S., Carson, J., et al. (2010). Evidence-based practice guidelines for plasma transfusion. *Transfusion, 50*(6), 1227–1239.

Staley, E. M., Schwartz, J., & Pham, H. P. (2016). An update on ABO incompatible hematopoietic progenitor cell transplantation. *Transfus Apher Sci, 54*(3), 337–344.

Stanworth, S. J., Hudson, C. L., Estcourt, L. J., Johnson, R. J., & Wood, E. M. (2015). Risk of bleeding and use of platelet transfusions in patients with hematologic malignancies: Recurrent event analysis. *Haematologica, 100*(6), 740–747.

Treleaven, J., Gennery, A., Marsh, J., et al. (2011). Guidelines on the use of irradiated blood components prepared by the British Committee for Standards in Haematology blood transfusion task force. *Br J Haematol, 152*(1), 35–51.

CHAPTER 54

Transfusion Management of Patients Receiving Antithrombotic Therapy

Scott T. Avecilla, MD, PhD

Hemostasis is the process by which blood components work in concert to halt bleeding as a consequence of injury. A blood clot forms in two general phases called primary and secondary hemostasis. Primary hemostasis involves the exposure of collagen in the subendothelium after a vascular injury. The collagen and blood flow shear forces stimulate von Willebrand factor to unfurl over the exposed collagen and provide a surface on which platelets can preferentially adhere. Adherent platelets undergo further activation and aggregate in response to physiologic agonists, which include collagen, adenosine diphosphate (ADP), arachidonic acid, thrombin, and epinephrine. Once an initial platelet plug is formed, it requires further stabilization in the form of a fibrin cap to continue to halt bleeding. It is on the surface of the aggregated platelets where the coagulation cascade occurs and results in the generation of cross-linked fibrin strands. While the prothrombotic processes are engaged to halt bleeding from the site of injury, simultaneously anticoagulant/fibrinolytic processes are also activated to keep the resultant thrombus from occluding the vessel. A brief overview of transfusion management of patients who are undergoing antithrombotic therapies will be discussed here.

Venous Thrombosis Prophylaxis: Patients with a prior history of spontaneous or provoked (prolonged periods of immobility, smoking, oral contraceptive use) venous thrombosis often undergo extensive laboratory workups to determine if a specific etiology can be found. These etiologies include activated protein C resistance (via the factor V Leiden mutation or others), prothrombin gene mutation, antiphospholipid syndrome, persistently elevated factor VIII, and homocysteinemia. Deficiencies in protein C, protein S, or antithrombin have also been implicated in pathological prothrombotic states. Prevention of further thromboses can be achieved with drug therapy, which includes warfarin, heparin (unfractionated or low molecular weight), fondaparinux (pentasaccharide), direct anti-Xa inhibitors, and direct thrombin inhibitors (DTIs) (Table 54.1).

Warfarin: Warfarin (via PO) is a small molecule inhibitor of vitamin K epoxide reductase (VKOR), which prevents the in vivo regeneration of vitamin K and results in a vitamin K deficiency, which in turn reduces the amount of the properly carboxylated vitamin K–dependent coagulation factors (factors II, VII, IX, X and proteins C, S, and Z). The drug has a long half-life of 40 hours and variable activity in individuals secondary to genetic polymorphisms in VKOR or metabolic enzymes such as CYP2C9, all of which contribute to difficulty in proper dosing for individual patients. The therapeutic efficacy of the drug is monitored via a normalized prothrombin time (PT)–based ratio or international normalized ratio (INR) with a typical target range of 2–3.

Transfusion Medicine and Hemostasis. https://doi.org/10.1016/B978-0-12-813726-0.00054-4

TABLE 54.1 Antithrombotic Therapies

Drug Class	Agent	Mechanism of Action	Route of Admin	$T_{1/2}$	Reversal
Vitamin K antagonists	Warfarin	Small molecule inhibitor of vitamin K epoxide reductase (VKOR), which prevents the in vivo regeneration of vitamin K	PO	40 hours	4-factor PCC
Heparins	Unfractionated heparin	Increases antithrombin activity to primarily deactivate factors IIa and Xa	IV, SC	1.5 hours	1 mg protamine sulfate (IV) per 100 U UFH administered over the last 4 hours.
	Low-molecular-weight heparin	Increases antithrombin activity to primarily deactivate factors Xa	SC	Variable, 4.5 hours for enoxaparin	1 mg protamine sulfate (IV) to 1 mg LMWH administered over the previous 4 hours (maximum dose 100 mg protamine sulfate over 2 hours).
Pentasaccharide	Fondaparinux	Increases antithrombin activity to primarily deactivate factors Xa	SC	17–21 hours	No antidote[a]
Direct Xa inhibitors	Rivaroxaban	Small molecule inhibitor of factor Xa	PO	5–9 hours	No antidote[a]
	Apixaban		PO	12 hours	Low-dose aPCC (20–40 IU/kg) may have some efficacy in reversal but awaiting confirmation with clinical trials.
	Edoxaban		PO	10–14 hours	
Direct thrombin inhibitors (DTI)	Dabigatran	Small molecule inhibitor of factor IIa	PO	12–17 hours	Idarucizumab
	Argatroban		IV	50 minutes	No antidote[a]
	Bilvalrudin		IV	25 minutes	No antidote[a]
	Lepirudin		IV	1.3 hours	No antidote[a]

Continued

TABLE 54.1 Antithrombotic Therapies—cont'd

Drug Class	Agent	Mechanism of Action	Route of Admin	$T_{1/2}$	Reversal
Cyclooxygenase inhibitors	Aspirin	Irreversible inhibitor of cyclo-oxygenase-1 (COX-1)	PO	2–5 hours	No antidote[a] Possible benefit from platelet administration 8–20 hours after last dose of aspirin.
ADP receptor inhibitors	Clopidogrel Prasugrel Ticlodipine	Irreversible $P2Y_{12}$ inhibitors	PO	7–8 hours 7 [2–15] hours 12 hours	No antidote[a] Possible benefit from platelet administration 4 half-lives time after last dose of agent.
	Ticagrelor	Reversible $P2Y_{12}$ inhibitors		7–8.5 hours	No antidote[a]
PAR-1 inhibitors	Vorapaxar	Reversible PAR-1 inhibitor (effectively irreversible)	PO	8 days	No antidote[a] Neither platelets nor dialysis are expected to be beneficial for bleeding episodes.
GPIIb/IIIa Inhibitors	Abciximab Eptifibatide Tirofiban	Reversible inhibitor of GPIIb/IIIa	IV IV IV	10–30 minutes 2.5 hours 2 hours	No antidote[a] Recovery of platelet function expected 4 half-lives time after last dose of agent.

[a]The patient can be supported with appropriate component therapy to replete hemorrhaged blood until the antithrombotic effect of the drug is no longer present.

However, reversal of warfarin therapy is needed to recover clotting ability in patients who have any of the following: (1) overdosed (e.g., supratherapeutic INR), (2) active bleeding, or (3) a hemostatic challenge, such as surgery or other procedure. Depending on the rapidity and magnitude of correction of the warfarin effect needed, differing therapeutic options exist. First and foremost, vitamin K repletion via oral (PO), subcutaneous (SC), or intravenous (IV) routes is available. Alternatively, prothrombin complex concentrates (PCC: factors II, VII, IX, X and proteins C and S) has been recommended for rapid reversal in cases of severe hemorrhage or intracranial bleeding.

Of important note, as indicated in Table 41.1, products in the PCC category vary markedly in composition such that they are likely not equally suitable for warfarin reversal (Kcentra [4 factor PCC] is FDA approved for warfarin reversal). When PCCs are not available, plasma can be used, often requiring higher transfusion volumes (10–20 mL/kg) to achieve similar corrections in clotting ability, typically with a goal of INR < 1.5 (see Chapters 34 and 41).

Heparins: Unfractionated heparin (UFH) (via IV, SC) is a group of highly sulfated glycosaminoglycans, which bind to antithrombin and cause a 1000-fold kinetic increase in antithrombin's activity, primarily deactivating factors IIa and Xa. UFH levels are usually monitored by either the activated partial thromboplastin time (PTT) or by an anti-Xa assay (see Chapter 156). UFH has a typical biological half-life of 1.5 hours, and it can be neutralized with protamine sulfate (via IV). Outside of the controlled, high-dose UFH therapy given during cardiopulmonary bypass, UFH therapeutic effect is infrequently actively reversed. Because UFH has a relatively short half-life, it is typically sufficient to stop therapy and monitoring. If deemed clinically necessary, protamine sulfate can be used to rapidly neutralize UFH; however, because protamine sulfate is an irreversible anticoagulant if given in excess, dosing should be carefully calculated. Reversal can be achieved with a 1-mg protamine sulfate per 100 U UFH administered over the last 4 hours. A general rule of thumb is to neutralize 80% of the estimated UFH dose (i.e., estimating residual dose since last administration taking into account a $T_{1/2}$ 1.5 hours) as to ensure no overdose.

Low-molecular-weight heparin (LMWH) (via SC) and fondaparinux (via SC) are FDA approved without the need for routine monitoring, unlike UFH. Because of their smaller molecular size, when bound to antithrombin, the complex has preferential targeting to factor Xa and therefore the PTT is not a suitable test to monitor levels. LMWH ($T_{1/2}$ variable, 4.5 hours for enoxaparin) can be neutralized with protamine sulfate with 1-mg protamine sulfate to 1-mg LMWH administered over the previous 4 hours (maximum dose 100 mg protamine sulfate over 2 hours). Fondaparinux ($T_{1/2}$ 17–21 hours), however, does not have an antidote and overdoses should be managed with clinical supportive measures, i.e., red cell transfusion if resultant hemorrhage is excessive. Administration of plasma products cannot reverse the effects of UFH/LMWH/fondaparinux. Similarly, because the targets of antithrombin are factors Xa and IIa, recombinant factor VIIa is not expected to reverse anticoagulation.

Direct Xa Inhibitors: Rivaroxaban (via PO), apixaban (via PO), and edoxaban (via PO) are direct factor Xa inhibitors (DXIs), which have been FDA approved for use without routine laboratory monitoring. Should monitoring been needed, an anti-Xa assay with an appropriate dose curve can be used (see Chapter 156). Alternative methods, including a modified PT test, are also being explored for monitoring direct factor Xa inhibitors. Rivaroxaban has a half-life of 5–9 hours. Apixaban's apparent $T_{1/2}$ is 12 hours due to a prolonged absorption time. Edoxaban also has a longer $T_{1/2}$ of 10–14 hours. Hemorrhage while anticoagulated with direct factor Xa inhibitors should be managed with supportive measures. Because of the highly protein-bound nature of the drugs, dialysis is not expected to be useful in removing either drug. Recent data in healthy human subjects treated with rivaroxaban indicate that treatment with a 4 factor PCC (4-PCC; factors II, VII, IX, X, protein C/S) can correct rivaroxaban-induced PT prolongation and more importantly, it can normalize endogenous thrombin potential (ETP) values. More data have emerged, which show that activated PCC (FEIBA, aPCC) at a low dose (20–40 IU/kg) ex vivo can restore ETP and thrombin generation lag times in subjects treated with rivaroxaban to close to normal levels. While these data are limited by the fact that no bleeding episodes occurred in the study, that the subjects were healthy individuals, and that correction of PT and ETP values are not validated markers for bleeding risk while taking rivaroxaban, there at least is some rationale to attempt 4-PCC or aPCC use as a final effort after supportive blood component therapy has failed to stabilize the bleeding patient. As indicated earlier (Warfarin section), PCCs vary in composition, and they will likely differ in efficacy as a reversal agent for direct Xa inhibitor therapy. No antidote is currently available for DXIs, although there are several under clinical study. Of note, a recombinant catalytically inactive human factor Xa is being investigated as a potential antidote for DXIs as well as for LMWH and fondaparinux.

Direct Thrombin Inhibitors: DTIs are small molecule inhibitors of factor IIa (thrombin), which include the following: dabigatran (via PO, $T_{1/2}$ 12–17 hours), argatroban (via IV, $T_{1/2}$ 50 minutes), bivalrudin (via IV, $T_{1/2}$ 25 minutes), and lepirudin (via IV, $T_{1/2}$ 1.3 hours). With the recent FDA approval of dabigatran, for use to prevent thrombosis in patients with nonvalvular atrial fibrillation, it is expected that a much larger patient population will be treated with this DTI as a replacement for warfarin. Dabigatran has been approved for use without routine monitoring, and no FDA-approved laboratory monitoring test yet exists (see Chapter 156). Several institutions have adopted a plasma-diluted thrombin time as a reliable method to measure the pharmacological levels of DTIs, but until there is widespread adoption of a standard measurement method, data to correlate drug levels with risk of bleeding do not exist. Because DTIs (other than dabigatran) also do not have an antidote currently, cessation of the drug and supportive clinical measures are all that are available to manage a bleeding patient while under DTI therapy. The dabigatran product insert indicates that it can be removed by hemodialysis with removal levels of up to 60% reported over 2–3 hours; however, no clinical data exist to support this intervention. Practically speaking, a DTI overdose is unlikely to be treated by hemodialysis due to the reluctance of practitioners to place an appropriate IV access in a bleeding patient with highly abnormal coagulation test results (PT, aPTT, TT). Recombinant factor VIIa, PCCs, or plasma are not expected to reverse DTI anticoagulation. Data from the same study, which investigated the use of 4-PCCs for reversal of rivaroxaban (Direct Xa Inhibitors section), showed that 4-PCCs were ineffective in correcting the dabigatran-induced PT and ETP value abnormalities. Low-dose aPCC (20–40 IU/kg) ex vivo was found in a new study to correct the ETP and thrombin generation lag time. After conventional measures to support the bleeding patient have failed, consideration of nonspecific bleeding reduction strategies may be warranted. Antifibrinolytic agents such as ε-amino caproic acid or tranexamic acid, while not specific antidotes or procoagulant agents per se, can stabilize residual clot formation that may occur in the presence of antithrombotic therapy. No established clinical data support their use at this time. A neutralizing monoclonal antibody against dabigatran (idarucizumab) has been recently FDA approved as a reversal agent for dabigatran anticoagulation for emergency surgery/urgent procedures for life-threatening or uncontrolled bleeding. The availability of idarucizumab may be limited at the time an emergent reversal is required, and the above alternative strategies can be considered.

Arterial Thrombosis Prophylaxis: Patients with a prior history or risk of arterial thrombosis, the major example being coronary artery disease and myocardial infarction, are often treated with drugs that target platelet receptors and enzymes. Antiplatelet drugs currently include inhibitors of cyclooxygenase 1 (COX-1), ADP receptor, protease-activated receptor 1 (PAR-1), and GPIIb/IIIa (Table 54.1).

Aspirin: Aspirin (via PO, $T_{1/2}$ 2–5 hours; acetylsalicylic acid, [ASA]) is an irreversible inhibitor of COX-1 and a mainstay of coronary artery disease prophylaxis and therapy. COX-1 is necessary in the platelet activation pathway to generate thromboxane-A2 (TxA2), a bioactive signaling lipid from arachidonic acid. TxA2 binds to its cognate receptor and stimulates Ca^{2+} influx to result in platelet aggregation. Before a hemostatic challenge such as surgery, the patient should stop taking ASA 7–10 days prior, taking into account the platelet life span of 7–10 days and the patient's absolute platelet counts. One nonrandomized observational study indicated that ASA-treated patients bled less if allowed at least 3 days to recover since the last ASA dose. However, in cases of hemorrhage that require immediate treatment, a platelet transfusion may help by introducing functional platelets into circulation. No guidelines exist for this practice however.

Adenosine Diphosphate Receptor Inhibitors: ADP receptor inhibitors are small molecule inhibitors of the $P2Y_{12}$ ADP receptor, which include the following: clopidogrel (via PO, $T_{1/2}$ 7–8 hours), prasugrel (via PO, $T_{1/2}$ 7 [2–15] hours), ticlodipine (via PO, $T_{1/2}$ 12 hours), and ticagrelor (via PO, $T_{1/2}$ 7–8.5 hours). Clopidogrel, prasugrel, and ticlopidine are all irreversible $P2Y_{12}$ (ADP receptor) inhibitors, which inhibit platelet function for the lifetime of the platelet (7–10 days). Ticagrelor, which is an allosteric reversible inhibitor of the ADP receptor, can be expected to have recovery of platelet function 28–34 hours after cessation of therapy. There is great variability in patient response (i.e., resistance) to clopidogrel due to a two-step metabolism in which there is polymorphic variability in the required activation step, which makes estimation of platelet inhibition by pharmacological parameters alone unreliable. Prasugrel, however, does not exhibit such variable response because it bypasses the activation step, which makes it a more uniformly effective platelet function inhibitor. Both clopidogrel and prasugrel are expected to be mostly cleared after 24–28 hours since last dose and it would be expected that platelet transfusions could theoretically be helpful after that washout period since before then, the transfused platelets would likely be deactivated immediately in circulation. No currently accepted, evidence-based guidelines exist to help direct support in bleeding patients on ADP receptor inhibition therapy. There are recent data, which indicate that platelet transfusion therapy given to patients on antiplatelet therapy who suffer from intracranial hemorrhage have an increased risk of death or other severe adverse event and as such platelet transfusion may not be beneficial.

Protease-Activated Receptor 1 Inhibitors: PAR-1 is the cognate receptor for thrombin, the most potent activator of platelet aggregation. vorapaxar (via PO, $T_{1/2}$ 8 days) is a reversible antagonist of PAR-1; however, due to its markedly long half-life, it is effectively an irreversible platelet function inhibitor. While it is quite effective (>80%

aggregation inhibition) after a week of therapy, there is no practical reversal strategy available to regain platelet activity in a patient who is bleeding. Even after 4 weeks of discontinuation of therapy, platelet inhibition can be detected and neither dialysis nor platelet transfusion are expected to be beneficial to stop bleeding episodes.

GPIIb/IIIa Inhibitors: GPIIb/IIIa (CD41) inhibitors are molecules that block the platelet fibrinogen receptor, which include the following: abciximab (via IV, $T_{1/2}$ 10–30 minutes), eptifibatide (via IV, $T_{1/2}$ 2.5 hours), and tirofiban (via IV, $T_{1/2}$ 2 hours). Because the GPIIb/IIIa blockers do not bind irreversibly to their targets, normal platelet function is expected to return within 24–48 hours after last dose. Without platelet transfusion, 50% functional return can be observed from 4 to 12 hours from last dose, depending on the specific agent. Platelet transfusion can be used in emergency cases of hemorrhage with expected variable efficacy.

Lab-Guided Platelet Transfusion: With the availability of rapid platelet functional assays, an assessment of platelet inhibition level coupled with a platelet count could be performed to calculate the estimated number of functional circulating platelets. The patient can thus be transfused in a lab-guided fashion to achieve recommended functional platelet counts the procedure needs. No rigorous clinical data exist for the transfusion of platelets to patients treated with antiplatelet therapy, so the above is only a proposal that has not been validated but it offers a rational, possibly reproducible method to treat. Of note, while the above tests have been used to predict antiplatelet therapy resistance, they cannot at this time predict bleeding tendency.

Further Reading

Avecilla, S. T., Ferrell, C., Chandler, W. L., & Reyes, M. (2012). Plasma-diluted thrombin time to measure dabigatran concentrations during dabigatran etexilate therapy. *Am J Clin Pathol*, *137*, 572–574.

Baharoglu, M. I., Cordonnier, C., Al-Shahi Salman, R., de Gans, K., Koopman, M. M., Brand, A., et al. (2016). Platelet transfusion versus standard care after acute stroke due to spontaneous cerebral haemorrhage associated with antiplatelet therapy (PATCH): A randomised, open-label, phase 3 trial. *Lancet*, *387*(10038), 2605–2613.

Chai-Adisaksopha, C., Hillis, C., Siegal, D. M., Movilla, R., Heddle, N., Iorio, A., et al. (2016). Prothrombin complex concentrates versus fresh frozen plasma for warfarin reversal. A systematic review and meta-analysis. *Thromb Haemost*, *116*(5), 879–890.

Eerenberg, E. S., Kamphuisen, P. W., Sijpkens, M. K., Meijers, J. C., Buller, H. R., & Levi, M. (2011). Reversal of rivaroxaban and dabigatran by prothrombin complex concentrate: A randomized, placebo-controlled, crossover study in healthy subjects. *Circulation*, *124*(14), 1573–1579.

Marlu, R., Hodaj, E., Paris, A., Albaladejo, P., Crackowski, J. L., & Pernod, G. (2012). Effect of non-specific reversal agents on anticoagulant activity of dabigatran and rivaroxaban: A randomised crossover ex vivo study in healthy volunteers. *Thromb Haemost*, *108*, 217–224.

CHAPTER 55

Blood Transfusion in Economically Restricted and Developing Countries

Heather A. Hume, MD, FRCPC

Over the past 10–20 years there have been major improvements in blood safety in low and lower middle-income countries (LIC, LMIC), largely driven by the human immunodeficiency virus (HIV) epidemic. Unfortunately, improvements in blood supply adequacy have not been as marked, and many challenges to the provision of an adequate supply of safe blood in these countries remain, particularly in countries that have relied or still rely heavily on external funding. Indeed, these challenges could increase if external funding decreases without concomitant increase in country-based funding. This chapter reviews some of these challenges with particular reference to sub-Saharan Africa (SSA) outside of southern Africa, where most countries are LIC or LMIC.

Income groups referred to are those defined by the World Bank (LIC, LMIC, and high-income country [HIC] being countries with gross national income per capita in 2017–18 of ≤US$1005, US$1006–3995 and ≥US$12,236, respectively) and, unless stated otherwise, data concerning transfusion activities are taken from the World Health Organization (WHO) 2016 Global Status on Blood Safety and Availability (GDBS), which is based on data reported from 180 countries, primarily data collected in 2013.

Blood Suppliers

Blood Systems: WHO recommends that all activities related to blood collection, testing, processing, storage, and distribution be coordinated at the national level and governed by a national blood policy and legislative framework. In 2013, 79% of HIC and 41% of LIC reported having specific legislation covering safety and quality of blood transfusion. However, only 15 (33%) countries in Africa reported having a system of regular inspection of blood transfusion services by a national regulatory authority.

Supply: Whole blood (WB) donations/1000 population is an indicator of general blood availability. WHO has estimated that at least 10 donations/1000 population are required to adequately fulfill any country's transfusion requirements. Median WB donation rate in HIC in 2013 was 32.1 donations/1000 population, whereas in LMIC it was 7.8 and LIC 4.6. Within WHO's Africa region, 38 countries reported collecting <10 donations/1000 population—almost unchanged from the 2011 report when it was 39. A review published in 2008 concluded that 26% of in-hospital maternal deaths due to hemorrhage in SSA were directly due to lack of blood, and more recently a study of severely anemic children (Hb < 5 g/dL) found that, at 8 hours after admission, 52% (54/103) of children who had not received a transfusion versus 4% (39/899) who had received a transfusion died. Inadequate supply of blood remains one of the most pressing problems in LMIC and LIC.

Transfusion Medicine and Hemostasis. https://doi.org/10.1016/B978-0-12-813726-0.00055-6

Donors: In contrast to HIC where most donors are voluntary, nonremunerated donors (VNRD), in many LMIC and LIC a significant proportion of donations are collected from family/replacement donors (FRDs), i.e., donors who are recruited by a patient and/or his/her family to donate blood to replace blood that he/she may require or have required (without those donations being specifically reserved for that particular patient).

To promote the safest donors possible, WHO has set a target for all member states to collect 100% of their blood from VNRD by 2020. However, the 2016 GDBS report shows that 71 countries, 22 of them in Africa, still collect <50% of their blood from VNRD and depend on FRD (and in a few cases some paid donors) for the majority of their blood supply. Reasons for the large number of FRD are cultural, practical, and financial. Prior to the creation of centralized transfusion services, most blood was collected locally in hospitals and donors were FRD; therefore, this system is well known to communities and culturally corresponds to the strong family/community bonds present in many LIC. In addition, donations obtained from VNRD in a centralized system are two- to fivefold more expensive than those obtained from FRD in a localized center. Thus, there is debate in the medical literature about eliminating FRD in countries where this practice remains prevalent. In some locations, studies have shown that positive marker rate for HIV, hepatitis B virus (HBV), and hepatitis C virus (HCV) are not statistically different in FRD versus first-time VNRD. A major consideration in determining the optimal donor recruitment strategy is ability to convert first-time donors, whether VNRD or FRD, into repeat donors. In countries with a high prevalence of HIV, HBV, and HCV in the general population repeat donors are, by a significant margin, the safest donors.

Testing for Transfusion-Transmissible Infections: Great improvements have been made over the past decades in testing for HIV, HBV, and HCV. In the 2016 GDBS report, 176/180 responding countries reported having a policy of screening all donations for HIV (the majority testing for antibodies only or together with antigen testing) and HBsAg, and 174/180 for HCV antibodies. However, in spite of these policies, 13 countries, including 6 in Africa, reported not being able to test 100% of collections, possibly because of an irregular supply of test kits, and only 66% of LIC reported that testing was performed in a quality-assured manner (i.e., with standard operating procedures and participation in external quality control program). In several African countries, particularly in settings without centralized testing laboratories and/or limited resources, rapid diagnostic tests are used for HIV, HBV, and HCV detection. These tests have variable and often low sensitivity for HBV and HCV, especially when used in the absence of rigorous quality assurance programs. The reported median proportion of collections with positive or reactive results on screening tests in HIC versus LIC, respectively, are as follows: HIV 0.003 versus 1.08, HBV 0.03 versus 3.70, HCV 0.02 versus 1.03 (although in LIC, unlike HIC, these results are often reactive but not confirmed positive results). One study, using mathematical modeling, estimated the residual risk for transmission of HIV, HBV, and HCV in SSA to be 1, 4.3, and 2.5/1000 units, respectively. Another study, using data from repeat donors in five SSA countries, estimated residual risk of HIV transmission as 0.03/1000 units.

Malaria is endemic in many LIC/LMIC countries, but donations are not tested for malaria and so potentially carry malaria transmission risk. In a review of 17 published studies from SSA, median prevalence of malaria among donors was 10%. However, little is currently known about rate and severity of transfusion-transmitted malaria. LIC and LMIC are also areas where emerging transfusion-transmissible diseases, such as Zika virus, may appear and add to transfusion-transmissible infection (TTI) risk. Finally, bacterial testing is not performed in the majority of these countries.

Processing: The 2016 GDBS survey indicates that 97% of WB donations collected in HIC versus 50% (31% in the 2008 survey) in LIC is separated into components. Although WHO encourages component preparation, it can be argued that complete conversion of a LIC's blood supply from WB to components is not always the best option. It is more expensive to prepare components and much of the blood is used to treat acute hemorrhage where WB is often an appropriate therapeutic option. As well given the current residual risks of TTI, it is unlikely that plasma not used for transfusion would be acceptable to plasma fractionators - thus much of the plasma would either be outdated or used inappropriately. However, some access to components is important for treatment of clotting factor deficiencies and for support of more sophisticated medical and surgical treatments that are gradually being introduced.

Large numbers of transfusions in SSA are for treatment of severely anemic young children. Accordingly, one important processing consideration is the availability of partial WB or RBC units. This is a major reason for RBC unit preparation in some African blood centers, where WB is collected into a multiple bag system, plasma is removed, and two or three pediatric RBC doses are prepared.

Processing measures that are routinely used in HIC to reduce transfusion complications are neither available nor affordable in most LIC and LMIC. These include prestorage leukoreduction, irradiation, and transfusion-related acute lung injury mitigation measures.

Transfusion Services: Somewhat analogous to the situation in HIC in the 1980s–90s, most of the attention and funding to improve transfusion safety in LIC has been directed to blood suppliers, with little evaluation of, or funding for improving transfusion activities within transfusing facilities. However, it is likely (although not documented) that this relatively neglected component of the transfusion process represents an important weak link in the provision of safe transfusions in LIC and LMIC.

Clinical Use: According to the 2016 GDBS survey, in HIC 79% of transfusion recipients are >60 years while in LIC 67% of transfusion recipients are <5 years, followed by females aged 15–45 years. Transfusion rates of WB and blood components by World Bank income groups are shown in Table 55.1.

Although WHO and many publications state that blood is often prescribed inappropriately in SSA, there are little evidence-based data examining to what extent this is true. WHO has published guidelines addressing appropriate transfusion indications; in the 2016 GDBS, 126 countries, including 34 in Africa, reported the existence of national transfusion guidelines. However, little information is available on the extent of guideline implementation. Systematic implementation of alternate therapies to transfusion, such as treatment of iron-deficiency anemia, use of tranexamic acid early in traumatic or postpartum hemorrhage, and delayed cord clamping in newborns, is also lacking.

TABLE 55.1 Blood Transfusion Rates per 1000 Population Reported in the WHO 2016s

World Bank Income Group	Whole Blood/RBCs	Plasma	Platelets (Adult Dose)
HIC	32.0 (7–49)	7.6 (2.2–18.9)	3.5 (0.8–11.8)
UMIC	12.5 (6.1–37.4)	3.3 (0.1–22)	0.8 (0.01–4.4)
LMIC	5.4 (0.4–17.4)	0.8 (0.1–14.8)	0.02 (<0.01–1.9)
LIC	3.4 (0.3–10.0)	0.06 (0.01–1.0)	–

HIC, high-income country; *LIC*, low-income country; *LMIC*, lower middle-income country; *UMIC*, upper middle-income country.

Pretransfusion Testing: In many SSA hospital blood banks pretransfusion testing consists only of recipient ABO/D phenotyping plus room-temperature saline cross-matching. Antibody screening is not performed, compatibility testing using indirect antiglobulin technique or equivalent is rarely available, and most transfusion services do not have units phenotyped beyond ABO/D. However, 6% of transfused patients in SSA do reportedly develop clinically significant RBC alloantibodies. Therefore, patients are at risk for hemolytic transfusion reactions (HTRs).

Requesting, Issuing, and Administering Blood: Specific forms for requesting and issuing blood are not consistently available or used throughout SSA, nor are written procedures for identifying patients, administering transfusions or monitoring patients during transfusion. Challenges include frequent inconsistencies in spelling of patients' names, unknown birth dates, lack of patient identification bands, and "stock out" of supplies. Owing to the lack of monitoring and reporting, many acute transfusion reactions, including HTRs due to ABO incompatibility, are unrecognized and/or unreported.

Transfusion Medicine Research: Because evidence is lacking, several groups have developed recommendations concerning research priorities for SSA and/or LMIC/LIC to improve transfusion safety. Priorities include the development of strategies that are culturally, geographically and economically appropriate to safely increase blood supply, decrease TTI risk, ensure the equitable distribution of blood, increase transfusion medicine education, and ensure sustainable funding.

Further Reading

Ala, F., Allain, J. P., Bates, I., Boukef, K., Boulton, F., Brandful, J., et al. (2012). External financial aid to blood transfusion services in sub-Saharan Africa: A need for reflection. *PLoS Med*, *9*(9), e1001309.

Allain, J. P., & Smit Singa, C. T. (2016). Family donors are critical and legitimate in developing countries. *Asian J Trans Sci*, *10*, 5–11.

Bates, I., Hassall, O., & Mapako, T. (2017). Transfusion research priorities for blood services in sub-Saharan Africa. *Br J Haematol*, *177*, 855–863.

Bloch, E. M., Vermeulen, M., & Murphy, E. (2012). Blood transfusion safety in Africa: A literature review of infectious disease and organizational challenges. *Transfus Med Rev*, *26*, 164–180.

Custer, B., Zou, S., Glynn, S. A., et al. (2018). Addressing gaps in international blood availability and transfusion safety in low- and low-middle income countries: an NHLBI workshop. *Transfusion*, *58*, 1307–1310.

Jayaraman, S., Chalabi, Z., Perel, P., Guerriero, C., & Roberts, I. (2010). The risk of transfusion-transmitted infections in sub-Saharan Africa. *Transfusion*, *50*, 33–42.

Lefrère, J. J., Dahourou, H., Dokekias, A. E., et al. (2011). Estimate of the residual risk of transfusion-transmitted human immunodeficiency virus infection in sub-Saharan Africa: A multinational collaborative study. *Transfusion*, *51*, 486–492.

Owusu-Ofori, A. K., Parry, C., & Bates, I. (2010). Transfusion-transmitted malaria in countries where malaria is endemic: A review of the literature from sub-Saharan Africa. *Clin Infect Dis*, *51*, 1192–1198.

Tagny, C. T., Murphy, E. L., & Lefrère, J. J. (2014). Le groupe de recherches transfusionnelles d'Afrique francophone: bilan des cinq premières années [The francophone Africa blood transfusion research network: A five-year report]. *Transfus Clin Biol*, *21*, 37–42.

World Health Organization. http://www.who.int/bloodsafety/global_database/en/.

CHAPTER 56

Management of Patients Who Refuse Blood Transfusion

Burak Bahar, MD and Jeanne E. Hendrickson, MD

A doctrine introduced in 1945 by Jehovah's Witnesses teaches that the Bible prohibits the consumption, storage, and transfusion of human blood (Genesis 9:3, 4 and Acts 15:19, 20). The Watch Tower Bible and Tract Society of Pennsylvania have issued many doctrines since that time, citing that "blood is sacred to God," and "even in the case of an emergency, it is not permissible to sustain life with transfused blood." These beliefs stem from the interpretation of Biblical scriptures. Many Jehovah's Witnesses carry medical directive "No Blood" cards, stating that blood transfusions are unacceptable. The use of blood derivatives, however, is not specifically prohibited, and the Watch Tower encourages members to personally decide whether accepting these component fractions violates the doctrine(s). Examples of potentially accepted blood product derivatives include cryoprecipitate, albumin, immunoglobulin therapy, human- derived clotting factor concentrates, and interleukins. Recombinant proteins (e.g., recombinant factor VIIa) are generally accepted by Jehovah's Witnesses, as are blood substitutes (hemoglobin-based oxygen carriers) (Fig. 56.1).

While standard transfusions are unacceptable, there are some related procedures that are not specifically prohibited. These include plasma exchange, dialysis, intraoperative blood salvage, hemodilution, blood donation strictly for the purpose of further fractionation of components, and transfusion of autologous blood as long as a continuous circuit with the patient remains. Transfusion of preoperatively donated autologous blood is, however, typically prohibited, due to the belief that blood should not be taken out of the body and stored (Table 56.1).

The right of a competent adult to refuse consent for medical treatment is accepted, and documentation of refusal for transfusion should be placed in the medical record. Electronic medical record and laboratory information systems may be useful in preventing blood from being inadvertently ordered or transfused to a non-consenting patient. Worst-case scenario discussions should be held with patients who refuse blood products, and documentation to this effect should be included in the medical record; some clinicians opt to have patients sign the notes stating that these discussions were held. Forcing a non-consenting patient to receive a transfusion unwillingly can be viewed as battery, and a Jehovah's Witness who accepts a transfusion can be spiritually cut off from a community of family and friends.

Trauma: Situations of trauma are difficult, in that medical directive cards may not be immediately available. If there is any doubt in a clinician's mind as to the wishes of the patient or as to what is legally appropriate, it is recommended that the clinician treat per the accepted standards of care until legal documentation is available.

Transfusion Medicine and Hemostasis. https://doi.org/10.1016/B978-0-12-813726-0.00056-8

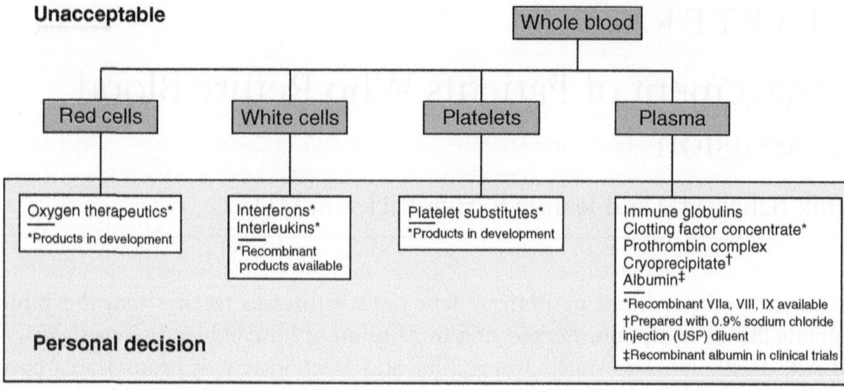

FIGURE 56.1 Blood components and fractions. (Reproduced from Bodnaruk, Z. M., Wong, C. J., & Thomas, M. J. (2004). Meeting the clinical challenge of care for Jehovah's Witnesses. *Transfus Med Rev, 18*(2), 105–116, with permission from Elsevier.)

Pregnant Women and Children: The treatment of pregnant women and children deserves special attention. Given that minor children are not considered capable of informed consent, it is recommended that the clinician seek legal intervention in cases where the fetus or child is placed at risk by parental refusal for transfusion. Laws governing the locality in which the patient is treated determine whether a treating physician can emergently transfuse a non-consenting pregnant woman or a child, with legal experts often determining that a life-saving blood transfusion is in the best interest of the child.

Blood Management: If a patient does not allow transfusion, the treating physician should attempt to find an alternative therapy within the boundaries of the patient's religious beliefs. Such options include blood conservation, optimizing coagulation, and use of non-blood adjunctive therapies. If time allows, adequate preoperative planning and incorporation of strategies to maximize erythropoiesis may decrease the likelihood of severe anemia intraoperatively or postoperatively. Blood conservation can also be achieved by reducing blood loss through decreased phlebotomy and meticulous surgical care, intraoperative blood salvage, and/or acute normovolemic hemodilution.

Depending on what products are deemed to be acceptable by the individual patient and what the clinical situation is, coagulation may be optimized with vitamin K, cryoprecipitate, 1-deamino-8-D-arginine vasopressin (DDAVP), fibrinogen concentrates, antifibrinolytic drugs, prothrombin complex concentrates, or recombinant factor VIIa. Potential adjunctive therapies to increase RBC mass and improve oxygenation include iron, folate, vitamin B12, erythropoietin, alternative oxygen carriers, or hyperbaric oxygen. The potential complications of each of these treatments must be weighed against their potential beneficial effects.

The Hospital Information Services department of the Watch Tower Society has established more than 1700 Hospital Liaison Committees worldwide to support health-care providers in their care of Jehovah's Witnesses. These committees can be reached at 718-560-4300 and https://www.jw.org/en/medical-library/hospital-liaison-committee-hlc-contacts/united-states/.

TABLE 56.1 Jehovah's Witness Religious Position on Medical Therapy

Acceptable Treatment

- Most surgical and anesthesiological blood conservation measures (e.g., hemostatic surgical instruments, controlled hypotension regional anesthesia, minimally invasive surgery, meticulous surgical hemostasis)
- Most diagnostic and therapeutic procedures (e.g., phlebotomy for laboratory testing, angiographic embolization)
- Pharmacologic agents that do not contain blood components or fractions such as the following:
  - Drugs to enhance hemostasis (e.g., tranexamic acid, epsilon-aminocaproic acid, aprotinin, desmopressin, recombinant factor VIIa, conjugated estrogens)
  - Hematopoietic growth factors and hematinics (e.g., albumin-free erythropoietin, iron)
  - Recombinant products (e.g., albumin-free coagulation factors)
  - Synthetic oxygen therapeutics (e.g., perfluorochemicals)
  - Non–blood volume expanders (e.g., saline, lactated Ringer's, hydroxyethyl starches)

Personal Decision (Acceptable to Some, Declined by Others)

- Blood cell salvage[a] (intraoperative or postoperative autotransfusion)
- Acute normovolemic hemodilution[a]
- Intraoperative autologous blood component sequestration[a] (including intraoperative plateletpheresis, preparation of fibrin gel, platelet gel, platelet-rich plasma)
- Cardiopulmonary bypass[b]
- Apheresis[b]
- Hemodialysis[b]
- Plasma-derived fractions (e.g., immune globulins, vaccines, albumin, cryoprecipitate[c])
- Hemostatic products containing blood fractions (e.g., coagulation factor concentrates, prothrombin complex concentrate, fibrin glue/sealant, hemostatic bandages containing plasma fractions, thrombin sealants)
- Products containing plasma-derived blood fractions such as human serum albumin (e.g., some formulations of erythropoietin, streptokinase, G-CSF, vaccines, recombinant clotting factors, nuclear imaging products)
- Products containing a blood cell–derived fraction
- Epidural blood patch
- Blood cell scintigraphy (e.g., radionuclide tagging for localization of bleeding)
- Peripheral blood stem cell transplantation (autologous or allogeneic)
- Other transplants (organ, HSC [hematopoietic stem cell], bone)

Unacceptable Treatment

- Transfusion of allogeneic whole blood, red blood cells, white cells, platelets, or plasma
- Preoperative autologous blood donation

[a]Patients might request that continuity is maintained with their vascular system.
[b]Circuits not primed with allogeneic blood.
[c]Cryoprecipitate suspended in 0.9% sodium chloride injection (USP) diluent.
Bodnaruk, Z. M., Wong, C. J., & Thomas, M. J. (2004). Meeting the clinical challenge of care for Jehovah's Witnesses. *Transfus Med Rev, 18*(2), 105–116, with permission from Elsevier.

Summary: An adult patient with decision-making capacity may defer some or all treatment modalities for themselves, which may include refusing blood products. Optimal management of patients who refuse blood transfusion requires close communication with the patient and their support network, early preoperative planning, and a thorough assessment of possible alternatives to transfusions.

Further Reading

Bodnaruk, Z. M., Wong, C. J., & Thomas, M. J. (2004). Meeting the clinical challenge of care for Jehovah's Witnesses. *Transfus Med Rev*, *18*(2), 105–116.

Booth, G. S., & Pilla, M. A. (2016). Preventing the preventable: How the blood bank laboratory information system fails to protect patients that refuse blood. *Transfus Apher Sci*, *55*(2), 245.

Jassar, A. S., Ford, P. A., Haber, H. L., Isidro, A., Swain, J. D., Bavaria, J. E., et al. (2012). Cardiac surgery in Jehovah's Witness patients: Ten-year experience. *Ann Thorac Surg*, *93*(1), 19–25.

Lawson, T., & Ralph, C. (2015). Perioperative Jehovah's Witnesses: A review. *Br J Anaesth*, *115*(5), 676–687.

Rogers, D. M., & Crookston, K. P. (2006). The approach to the patient who refuses blood transfusion. *Transfusion*, *46*(9), 1471–1477.

Scharman, C. D., Shatzel, J. J., Kim, E., & DeLoughery, T. G. (2017). Treatment of individuals who cannot receive blood products for religious or other reasons. *Am J Hematol*. https://doi.org/10.1002/ajh.24889. Epub 8/18/17.

Spahn, D. R., & Goodnough, L. T. (2013). Alternatives to blood transfusion. *Lancet*, *381*(9880), 1855–1865.

Zeybek, B., Childress, A. M., Kilic, G. S., Phelps, J. Y., Pacheco, L. D., Carter, M. A., et al. (2016). Management of the Jehovah's witness in obstetrics and gynecology: A comprehensive medical, ethical, and legal approach. *Obstet Gynecol Surv*, *71*(8), 488–500.

CHAPTER 57

Platelet Transfusion Refractory Patients

Christopher A. Tormey, MD and Jeanne E. Hendrickson, MD

Platelet refractoriness is defined as an inappropriately low platelet count increment after platelet transfusion. There are nonimmune and immune causes for platelet refractoriness (Table 57.1), with nonimmune causes being responsible for the majority of cases. Evaluation of the potentially platelet refractory patient includes determining the etiology of the underlying thrombocytopenia, reviewing the patient's underlying illness and current medications, determining sites and severity of active bleeding, and ascertaining future platelet threshold goals.

Calculation of Refractoriness: Determination of platelet refractoriness is based on evaluating the increase in platelet count after transfusion. Ideally, such an evaluation will examine platelet increases within 1 hour after transfusion of ABO identical or compatible platelets stored for less than 48 hours, on two different occasions. The most commonly used calculation in the United States is the corrected count increment (CCI), although other equations may also be used. The CCI is calculated as follows:

$$\frac{\text{Posttransfusion platelet count} - \text{pretransfusion platelet count}\ (/L) \times \text{BSA}\ \left(\text{m}^2\right)}{\text{Number of platelets transfused}\ \left(10^{11}\right)}$$

A typical body surface area (BSA) for adult is 2 m^2, and it can be assumed that 3–4 × 10^{11} platelets are in an apheresis unit or that 0.5–0.7 × 10^{11} platelets are in a whole blood–derived random donor platelet concentrate. Generally, CCIs >7500 are considered satisfactory, while a CCI less than 5000 on two occasions is typically considered to be consistent with platelet refractoriness. It is important to note that while a poor CCI indicates that a patient may be platelet refractory, the value itself provides no insight as to the underlying cause of refractoriness; low CCI values prompt further clinical and laboratory investigation.

Nonimmune Refractoriness: Nonimmune causes of platelet refractoriness are listed in Table 57.1, with two-thirds of platelet refractoriness thought to be due to nonimmune causes.

Immune Refractoriness: The most common immune-mediated cause of platelet refractoriness is antibodies directed against HLA class I antigens (HLA-A or HLA-B), occurring in 30%–40% of cases. These alloantibodies typically form after exposure to the corresponding HLA class I antigens on either platelets or white blood cells; such exposure may occur through transfusion or pregnancy. However, the presence of HLA class I antibodies alone does not mean that these antibodies are responsible for platelet refractoriness. In the Trial to Prevent Alloimmunization to Platelets (TRAP), for example, the incidence of HLA alloantibodies was twofold higher than the incidence of platelet refractoriness.

Transfusion Medicine and Hemostasis. https://doi.org/10.1016/B978-0-12-813726-0.00057-X

TABLE 57.1 Etiologies of Platelet Refractoriness

Nonimmune Causes: 60%–70% of Causes	Immune Causes: 30%–40% of Causes
Fever	Human leukocyte antigen antibodies
Infection	Human glycoprotein platelet antigen antibodies
Disseminated intravascular coagulopathy	Anti-A or anti-B antibodies
Medications	Drug-induced antibodies
Bleeding	Platelet autoantibodies
Hematopoietic stem cell transplant	Possible: cellular-mediated immunity
Graft-versus-host disease	
Veno-occlusive disease	
Splenomegaly	

Antibodies against human glycoprotein platelet antigens (HPAs) have been reported to cause platelet refractoriness on occasion. Patients with Bernard–Soulier syndrome and Glanzmann Thrombasthenia may become broadly immunized to the platelet glycoproteins GPIb/IX/V and GPIIb/IIIa. Patients with *auto*antibodies against these and similar platelet antigens (e.g., in the setting of immune thrombocytopenia or post-transfusion purpura) will also demonstrate transfusion refractoriness.

Drug-induced antibodies and major mismatched ABO-incompatible transfusions can also lead to antibody-mediated platelet refractoriness. In general, a 20% increase in platelet recovery may be seen after transfusion with ABO identical or compatible units, given the presence of A and B antigens on platelets (e.g., a platelet unit from a group O donor may survive 20% better in a group O recipient than a platelet unit from a group A donor will, given the recipient's naturally occurring anti-A antibodies).

Recent animal studies suggest that cellular-mediated immune etiologies of platelet refractoriness may exist, although they would not be identified by antibody-based clinical assays.

Laboratory Testing for Immune-Mediated Platelet Transfusion Refractoriness

Human Leukocyte Antigen Antibody Detection: Testing for HLA antibodies can be performed by lymphocytotoxicity testing (known as panel-reactive antibody [PRA]) in which serum is reacted with a panel of HLA-typed lymphocytes; a PRA >20% suggests probable HLA alloimmunization. HLA antibodies can also be detected by enzyme-linked immunosorbent assay (ELISA) or flow cytometry.

Human Glycoprotein Platelet Antigen Antibody Detection: Although HPA antibodies are a relatively uncommon cause of platelet refractoriness, testing for these antibodies may be considered in platelet refractory patients. Multiple platforms for testing are available and include ELISA and indirect platelet immunofluorescence testing.

Platelet Transfusion Support for the Human Leukocyte Antigen Alloimmunized Patient: There are three main methods for selecting matched platelets (Table 57.2): (1) HLA antigen matched, (2) crossmatched, and (3) HLA antibody-specific prediction.

TABLE 57.2 Methods for Managing Immune-Mediated Platelet Refractoriness

	Human Leukocyte Antigen (HLA) Matched	Crossmatched	Antibody Specificity Prediction
Method	HLA type the patient and provide platelets collected from an HLA-matched donor	Combine donor platelets with patient's serum to determine crossmatch compatibility	Identify HLA antibodies in patient and then provide platelets without those specific HLA antigens
Pros	Prevents future alloimmunization if high grade match	Useful for anti–human glycoprotein platelet antigen (anti-HPA) and anti-HLA Rapid availability HLA typing not required	Larger donor pool Patient HLA typing not required
Cons	Not useful for anti-HPA Patient and donor HLA typing required Must recruit HLA-matched donors Limited donor pool for rare HLA types	Difficult to find suitable crossmatch in highly sensitized patients Risk of alloimmunization against mismatched donor HLA antigens Frequent crossmatching necessary	Not useful for anti-HPA Potential risk of alloimmunization against mismatched donor HLA antigens Must type donor HLA

From Forest and Hod, Management of the Platelet Refractory Patient, Hematology Oncology Clinics of North America 2016: 665–677, with permission from Elsevier.

Human Leukocyte Antigen Antigen-Matched Platelets: With this strategy, products are selected based on compatibility between donor and recipient HLA-A and HLA-B antigens. Matches are graded from "A to D," based on the degree of matching. "A" matches are optimal but difficult to locate, with all four donor/recipient HLA-A and HLA-B antigens being identical. "B" matches are easier to locate, with the donor either being homozygous at HLA-A and HLA-B antigens and having no antigens not present in the recipient (BU) or with the donor having one or more mismatches to the recipient but all within a cross-reactive group (CREG) (BX). "C" and "D" matches are less desirable and may not yield platelet increments any different than randomly selected units. One recent study showed that although "A" and "B" matches are more likely to result in a CCI >5000 than random platelets, 70% of such "A" and "B" matched platelets still failed to result in an adequate CCI. These data highlight the fact that nonimmune causes are likely present in alloimmunized patients and raise the question of whether recipient cellular-mediated immunity may also be involved in platelet refractoriness.

Crossmatched Platelets: Crossmatching, using the solid phase red cell adherence assay, flow cytometry, or other techniques, can detect compatible platelets without requiring knowledge of HLA or HPA type or antibody identification. In general, crossmatch-compatible platelets are thought to be as good a product as HLA-matched platelets in terms of posttransfusion platelet increase.

Human Leukocyte Antigen Antibody-Specific Prediction: This method, which is rarely used alone or used for long term, simply avoids HLA-A or HLA-B donor antigens against which the recipient is alloimmunized against.

Bleeding, Platelet Refractory Patients: Transfusion of large numbers of platelets is often attempted in thrombocytopenic or bleeding, platelet refractory patients, but may not result in a detectable increase in platelet count. A hematocrit below 25% was recently shown to be a potential risk factor for bleeding in thrombocytopenic patients. A continuous slow infusion of platelets may be used in some instances, although the literature to support such a practice is scarce. Alternative potential therapies include IVIG, plasma exchange, rituximab, splenectomy, corticosteroids, other immunosuppressive medications, antifibrinolytic agents, thrombopoietin mimetics, and potentially recombinant factor VIIa. Except for antifibrinolytic agents, these therapies have generally not proven very beneficial.

Summary: Platelet refractoriness is a complex problem encountered by many patients with hypoproliferative thrombocytopenia. Optimal care for platelet refractory patients requires a team-based approach, both in terms of the workup, the products selected for transfusion, and the goals of transfusion therapy.

Further Reading

Leukocyte reduction and ultraviolet B irradiation of platelets to prevent alloimmunization and refractoriness to platelet transfusions. *N Engl J Med*, *337*(26), (1997), 1861–1869.

Forest, S. K., & Hod, E. A. (2016). Management of the platelet refractory patient. *Hematol Oncol Clin N Am*, *30*(3), 665–677.

Nahirniak, S., Slichter, S. J., Tanael, S., Rebulla, P., Pavenski, K., Vassallo, R., et al. (2015). Guidance on platelet transfusion for patients with hypoproliferative thrombocytopenia. *Transfus Med Rev*, *29*(1), 3–13.

Narvios, A., Reddy, V., Martinez, F., & Lichtiger, B. (2005). Slow infusion of platelets: A possible alternative in the management of refractory thrombocytopenic patients. *Am J Hematol*, *79*(1), 80.

Rioux-Masse, B., Cohn, C., Lindgren, B., Pulkrabek, S., & McCullough, J. (2014). Utilization of cross-matched or HLA-matched platelets for patients refractory to platelet transfusion. *Transfusion*, *54*(12), 3080–3087.

Shehata, N., Tinmouth, A., Naglie, G., Freedman, J., & Wilson, K. (2009). ABO-identical versus nonidentical platelet transfusion: A systematic review. *Transfusion*, *49*(11), 2442–2453.

Slichter, S. J., Davis, K., Enright, H., Braine, H., Gernsheimer, T., Kao, K. J., et al. (2005). Factors affecting posttransfusion platelet increments, platelet refractoriness, and platelet transfusion intervals in thrombocytopenic patients. *Blood*, *105*(10), 4106–4114.

Stanworth, S. J., Navarrete, C., Estcourt, L., & Marsh, J. (2015). Platelet refractoriness–practical approaches and ongoing dilemmas in patient management. *Br J Haematol*, *171*(3), 297–305.

Triulzi, D. J., Assmann, S. F., Strauss, R. G., Ness, P. M., Hess, J. R., Kaufman, R. M., et al. (2012). The impact of platelet transfusion characteristics on posttransfusion platelet increments and clinical bleeding in patients with hypoproliferative thrombocytopenia. *Blood*, *119*(23), 5553–5562.

CHAPTER 58

Massive Transfusion

Nancy M. Dunbar, MD, Jansen N. Seheult, MB, BCh, BAO and
Mark H. Yazer, MD

Introduction: Massive transfusion for an adult recipient is traditionally defined as the transfusion of ≥10 red blood cell (RBC) units < 24 hours (approximately, 1 total blood volume [TBV]). Some definitions that are more applicable in real time include the transfusion of ≥4 RBC units in 1 hour with anticipation of a continued need for blood product support, and replacement of 50% of the TBV within 4 hours. In pediatric recipients, massive transfusion is usually based on the percentage of the TBV transfused compared with the patient's age-based estimated blood volume.

Much of the existing evidence informing the current approach to massive transfusion is based on traumatically injured patients. However, massive transfusion may also be required in medical and surgical settings and in cases of obstetric hemorrhage, the most common cause of maternal mortality worldwide.

Massive Transfusion in Trauma

Coagulopathy of Trauma: Up to 5% of civilian trauma patients require massive transfusion and as much as 40% of trauma-related mortality in patients admitted to trauma centers is due to uncontrolled or refractory hemorrhage. These patients can present with a coagulopathy known as early trauma-induced coagulopathy or acute traumatic coagulopathy, which is characterized by dysregulation of procoagulant, anticoagulant, and fibrinolytic pathways and platelet dysfunction and is associated with increased mortality (Fig. 58.1).

Hemostatic Resuscitation in Trauma: Current recommendations for the initial management of massively bleeding trauma patients include damage control resuscitation, a multimodal resuscitation paradigm that includes damage control surgery, permissive hypotension, rapid rewarming, limited crystalloid administration, early blood component resuscitation, and correction of hyperfibrinolysis. Blood components are transfused using fixed ratio component therapy (FRCT) until laboratory test results, including thromboelastography (TEG) or rotational thromboelastometry (ROTEM), are available to guide resuscitation.

The first study that suggested a survival benefit from FRCT in trauma was in military combat patients. Additional retrospective studies observed similar findings in civilian trauma. The Prospective, Observational, Multicenter, Major Trauma Transfusion (PROMMTT) study showed that a plasma:RBC transfusion ratio less than 1:2 was associated with increased mortality in the first 6 hours of resuscitation. Although these studies have shifted the practice of trauma resuscitation toward FRCT, some authors

Transfusion Medicine and Hemostasis. https://doi.org/10.1016/B978-0-12-813726-0.00058-1

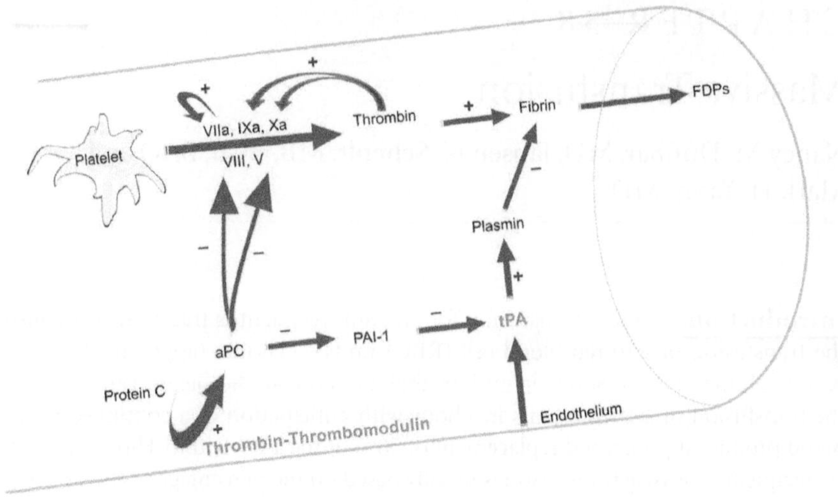

FIGURE 58.1 Mechanisms of trauma-induced coagulopathy. Thrombin–thrombomodulin complexes stimulate activation of protein C, which inactivates factors V, VIII, and plasminogen activator inhibitor-1 (PAI-1). Damage to the endothelium also stimulates release of tissue-type plasminogen activator (tPA), which is responsible for hyperfibrinolysis. *aPC*, activated protein C; *FDPs*, fibrin degradation products.

have pointed out limitations and confounding factors that limit their generalizability and the strength of their conclusions.

The Pragmatic Randomized Optimal Platelet and Plasma Ratios (PROPPR) trial, the largest randomized trial to date on FRCT, showed no differences in the primary outcomes of 24-hour and 30-day all-cause mortality between patients who received 1:1:1 versus 1:1:2 FRCT, although there were some differences in secondary outcomes favoring the patients who received the 1:1:1 strategy. This study was also limited by the absence of a control arm where the patients were resuscitated using a strategy other than FRCT, for example, using goal-directed component therapy based on laboratory results. Thus, the optimal method of blood component provision for trauma patients remains unknown.

Massive Transfusion in Nontrauma-Related Hemorrhage: The need for massive transfusion also occurs nontrauma settings including surgery (cardiac surgery, liver transplantation, ruptured abdominal aortic aneurysm repair), medicine/critical care (gastrointestinal bleeding), and obstetrics (postpartum hemorrhage [PPH]). Bleeding in nontrauma situations may have differences in underlying pathophysiology. For example, cardiac surgery with cardiopulmonary bypass causes severe derangements in the hemostatic system due to dilution, hypothermia, acidosis, consumption due to coagulation factor activation on foreign surfaces, acquired von Willebrand factor deficiency, hyperfibrinolysis, anticoagulation with unfractionated heparin, and platelet inhibition. Patients undergoing liver transplantation may have underlying hypo- or dysfibrinogenemia requiring cryoprecipitate transfusion; coagulopathy may be worsened by dilution and consumption of coagulation factors caused by blood loss in the hepatectomy and anhepatic stages, followed by primary hyperfibrinolysis and a global reduction in coagulation factors at the time of graft reperfusion. Patients with obstetric hemorrhage may need more aggressive cryoprecipitate support based on studies suggesting a link between PPH and fibrinogen <200 mg/dL. Adjunct therapies, including uterotonics and antifibrinolytics, may also be helpful in PPH.

Component Therapy During Massive Transfusion

Red Blood Cells: Patients presenting with life-threatening hemorrhage should receive uncrossmatched group O RBCs until their blood group is determined. Rh(D)-negative group O uncrossmatched RBCs are generally used for the initial resuscitation of females of childbearing age due to the risk of Rh(D)-alloimmunization, which could result in hemolytic disease of the fetus and newborn during a future pregnancy. The risk of anti-D formation after Rh(D)-positive RBC transfusion to an Rh(D)-negative recipient is approximately 20% and after platelet transfusion the risk is <2%.

Samples for type and screen testing should be sent to the blood bank as soon as possible to permit the transition to ABO-group-specific RBCs. Crossmatched RBCs can usually be issued within an hour of the transfusion service receiving the recipient's samples if an unexpected antibody is not detected on the antibody detection test. The evaluation of an unexpected antibody may take hours to days to complete. Even in patients with unexpected antibodies there is a low risk of hemolysis after receipt of uncrossmatched RBCs.

Plasma, Platelets, and Cryoprecipitate: Plasma generally refers to a plasma product that is frozen with 8 hours of collection and transfused within 24 hours of thawing (i.e., thawed fresh frozen plasma [FFP]). Thawed plasma (which can be stored for up to 5 days after thawing) and liquid plasma (which is stored in the liquid state at 1–6°C for up to 26 days) are alternative options as thawing frozen plasma for immediate use can delay availability by up to 30 minutes. Both thawed and liquid plasma have decreased coagulation factor activity compared with thawed FFP, as levels decline during storage. The clinical efficacy of these products compared with thawed FFP has not been well studied.

While group AB plasma is generally used until blood group is determined, the use of group A plasma has also been used to supplement the often-limited group AB plasma inventory.

Platelets and cryoprecipitate are generally transfused without regard recipient ABO group during massive transfusion in adults.

In the military combat setting, warm fresh whole blood from walking donors has been used for trauma resuscitation, as it provides RBCs, functional platelets, and plasma in one component. Based on this experience, there has been recent interest in the use of low-titer group O uncrossmatched cold-stored whole blood in civilian trauma patients.

Massive Transfusion Protocols: A massive transfusion protocol (MTP) is a formal system for rapidly providing large quantities of blood products to patients requiring massive transfusion. MTPs are typically developed and implemented by a multidisciplinary team that includes representation from the hospital transfusion service, emergency department, anesthesia, surgery, critical care, and trauma services.

As outlined in the American College of Surgeons' Trauma Quality Improvement Program, the protocol should specifically address the trigger(s) for MTP activation, blood component provision including delivery method(s), frequency and type of laboratory monitoring, resuscitation goals, process for MTP termination, and ongoing monitoring for performance improvement.

Several scoring systems have been developed to predict which patients are likely to require massive transfusion in both the trauma and nontrauma setting. These include the Trauma Associated Severe Hemorrhage (TASH) score, the Assessment of Blood Consumption (ABC) score, scoring systems developed for liver transplant recipients and cardiac surgery patients, and the obstetric shock index. While these scoring systems may be helpful, the judgment of the clinical team remains the major trigger for MTP activation.

Although the contents of MTP package rounds vary from institution to institution, they generally provide RBCs, plasma, and platelets in a 1:1:1 to 1:1:2 ratio, for example, six units of RBCs, six units of group AB or A plasma and one dose of apheresis or whole blood platelets. The role of the MTP is to provide a large quantity of blood products to the clinical team at the beginning of the resuscitation that can be used while awaiting the results of laboratory and/or point of care testing, such as TEG or ROTEM.

While much of the research into MTPs has been conducted in civilian trauma patients, in general, there are no unique disadvantages of implementing MTPs in the nontrauma setting from the resource allocation perspective, with the potential advantage that blood components are issued more quickly and with less variability. Based on the limited data available in these patient populations, however, there appears to be no significant difference in survival between patients receiving higher ratios of plasma:RBCs or platelets:RBCs compared with those receiving lower ratios.

Pharmacologic Agents: Tranexamic acid (TXA) inhibits fibrinolysis, a complication seen in about one-third of traumatically injured patients requiring massive transfusion, by reversibly antagonizing the lysine-binding site on plasminogen and plasmin. The Clinical Randomisation of an Antifibrinolytic in Significant Hemorrhage 2 (CRASH-2) trial demonstrated the efficacy of early administration of TXA (within 3 hours) in reducing mortality in the setting of traumatic injury. An international, randomized, double-blind, placebo-controlled trial has also shown that TXA reduced death due to bleeding in women with PPH, especially if given early, with no significant increase in adverse effects.

Activated recombinant factor VIIa (rfVIIa) has been widely used off-label to treat bleeding. There is no current evidence that the use of this product confers any survival benefit in the treatment of trauma-related hemorrhage. Furthermore, the high cost of the product and increased risk of arterial thromboembolic events mean that rfVIIa is no longer recommended as an adjunctive therapy in MTPs.

Outside of the United States, fibrinogen concentrate and prothrombin complex concentrate products, which consist of various combinations and concentrations of factors II, VII, IX, and X, and proteins C and S, are being used in massive transfusions to replace cryoprecipitate and plasma, but there is currently limited evidence to recommend their routine use in MTPs.

Further Reading

Camazine, M. N., Hemmila, M. R., Leonard, J. C., et al. (2015). Massive transfusion policies at trauma centers participating in the American College of Surgeons Trauma Quality Improvement Program. *J Trauma Acute Care Surg*, *78*, S48–S53.

Diab, Y. A., Wong, E. C., & Luban, N. L. (2013). Massive transfusion in children and neonates. *Br J Haematol*, *161*, 15–26.

Dobson, G. P., Letson, H. L., Sharma, R., et al. (2015). Mechanisms of early trauma-induced coagulopathy: The clot thickens or not? *J Trauma Acute Care Surg*, *79*, 301–309.

Dudaryk, R., Hess, A. S., Varon, A. J., et al. (2015). What is new in the blood bank for trauma resuscitation. *Curr Opin Anaesthesiol*, *28*, 206–209.

Dunbar, N. M., Yazer, M. H., & on behalf of the BEST collaborative (2017). Safety of the use of group A plasma in trauma - the STAT Study. *Transfusion*, *57*, 1879–1884.

Etchill, E. W., Myers, S. P., McDaniel, L. M., et al. (2017). Should all massively transfused patients be treated equally? An analysis of massive transfusion ratios in the nontrauma setting. *Crit Care Med*, *45*, 1311–1316.

Holcomb, J. B., del Junco, D. J., Fox, E. E., et al. (2013). The prospective, observational, multicenter, major trauma transfusion (PROMMTT) study: Comparative effectiveness of a time-varying treatment with competing risks. *JAMA Surg*, *148*, 127–136.

Holcomb, J. B., Tilley, B. C., Baraniuk, S., et al. (2015). Transfusion of plasma, platelets, and red blood cells in a 1:1:1 vs a 1:1:2 ratio and mortality in patients with severe trauma: The PROPPR randomized clinical trial. *JAMA*, *313*, 471–482.

Rossaint, R., Bouillon, B., Cerny, V., et al. (2016). The European guideline on management of major bleeding and coagulopathy following trauma: Fourth edition. *Crit Care*, *20*, 100.

Simmons, J. W., & Powell, M. F. (2016). Acute traumatic coagulopathy: Pathophysiology and resuscitation. *Br J Anaesth*, *117*, iii31–iii43.

Spinella, P. C., Frazier, E., Pidcoke, H. F., et al. (2015). All plasma products are not created equal: Characterizing differences between plasma products. *J Trauma Acute Care Surg*, *78*, S18–S25.

Treml, A. B., Gorlin, J. B., Dutton, R. P., et al. (2017). Massive transfusion protocols: A Survey of Academic Medical Centers in the United States. *Anesth Analg*, *124*, 277–281.

CHAPTER 59

Patient Blood Management

Ruchika Goel, MD, MPH and Patricia A. Shi, MD

Patient blood management (PBM) is the term given to the appropriate use of blood transfusion to optimize patient care. Patient-centric approach is its core principle (Fig. 59.1). PBM is "transfusing the right product in the right dose to the right patient at the right time for the right reason." PBM has five main tenets (Fig. 59.2). In addition to improving patient safety, PBM programs decrease hospital costs and conserve clinical resources. Cost can range from $522–1183/RBC unit when accounting for storage, testing, and labor.

Implementing Blood Management Program: Hospital-wide, comprehensive, and multidisciplinary PBM program can optimize patient care, avoid unnecessary transfusions of blood products, and limit adverse effects. Successful PBM programs are rooted in evidence-based medicine and enhance patient safety and outcomes through

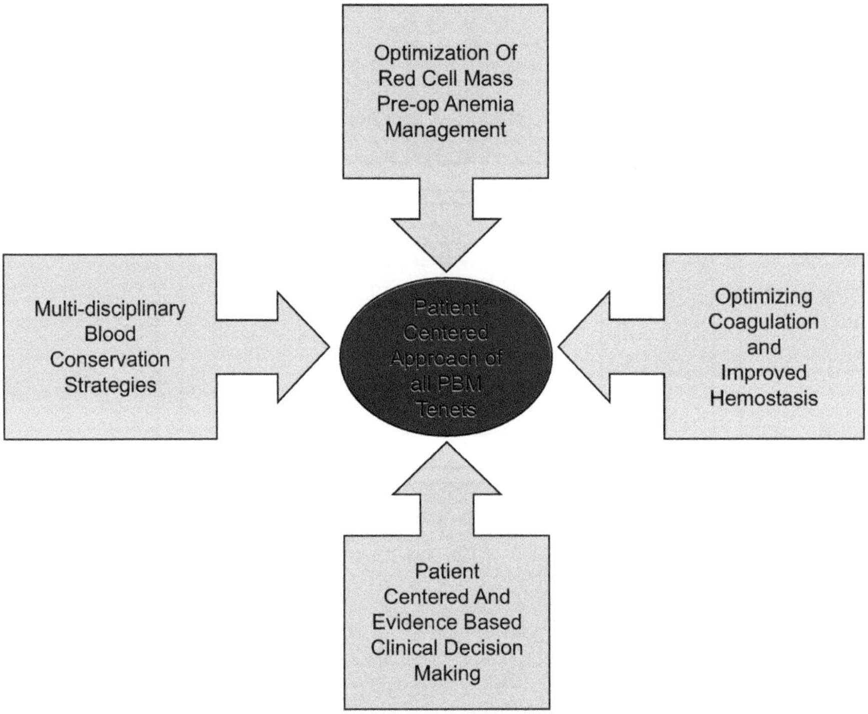

FIGURE 59.1 Key principles of patient blood management. (Adapted from Society of Blood Management 2012 learning resources.)

Transfusion Medicine and Hemostasis. https://doi.org/10.1016/B978-0-12-813726-0.00059-3

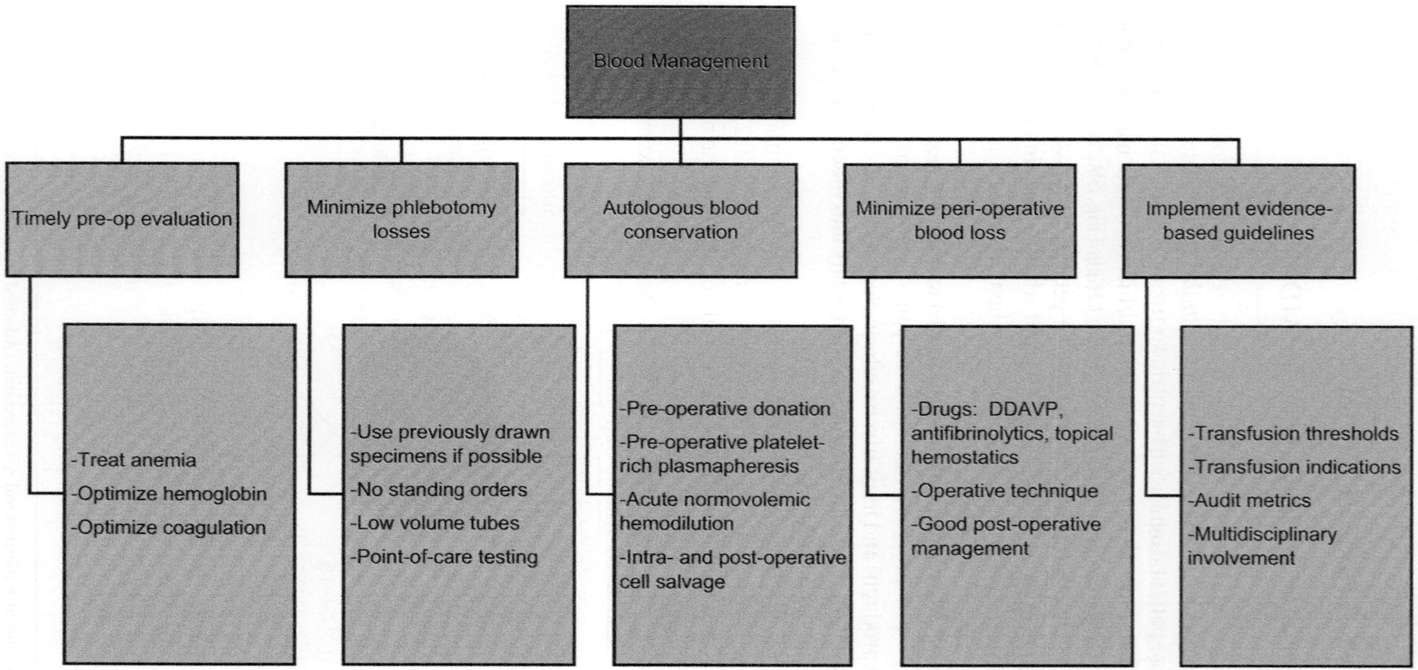

FIGURE 59.2 Patient blood management (PBM) matrix: key tenets of PBM.

measurable improvements. An integral component is engagement and education of ordering providers. Thus, successful PBM require active input, commitment, and leadership from medical, surgical, anesthesia, intensive care, and hospitalist services, with operational support from administration, finance, and information technology.

Paramount to the success of the program is an initial audit of current transfusion practice, ideally organized by medical specialty and even by individual physicians within that specialty. Appropriate use of products should be audited, including modified (irradiated or washed) and autologous products. Evidence-based transfusion guidelines must then be developed by a multidisciplinary team, usually through transfusion committee. Creation of specific guidelines for certain populations, scenarios, or diseases, such as pediatric, massive transfusion, or sickle cell, is also typically developed.

Transfusion guidelines are ideally incorporated into the ordering process, and physicians and other hospital staff must be educated to them to ensure understanding and compliance. Appropriate audit metrics to determine whether guidelines are being followed must be developed, and data collected, analyzed, and acted on. Auditing can be performed prospectively, concurrently, or retrospectively, and institutional information systems should be harnessed for automated data collection and analytic ease. Possible audit metrics include blood use per diagnosis-related group, pretransfusion hemoglobin and INR levels, blood product use and blood recovery use for specific surgical procedures, and clinically relevant outcomes. Feedback through repeat audits that normalize for severity of illness, compare physicians with their peers, include patient outcomes, and use of transfusion alternatives should be presented at all levels of the hospital hierarchy to stimulate staff awareness and motivation for improvement.

Minimizing Inappropriate Blood Use Through Evidence-Based Guidelines: Ideally, physiologic indicators of tissue oxygen delivery and ischemia should guide RBC transfusion therapy. Randomized controlled studies in critical care populations, both adult and pediatric, show no difference in functional outcome, morbidity, or survival between restrictive (transfusing for hemoglobin <7 g/dL) and liberal (transfusing for hemoglobin <10 g/dL) transfusion strategies. Creating restrictive transfusion criteria reduces the likelihood of transfusion with an average savings of a single RBC unit per transfused patient, without harming the patient.

In other hospitalized patients, the RBC transfusion threshold should be individually assessed for each patient depending on their comorbidities and anemia tolerance. A recent trial in hip fracture repair patients >50 years old and with cardiovascular disease showed no difference in survival or functional outcome between liberal and restrictive (hemoglobin <8 g/dL) transfusion strategy.

Prophylactic platelet transfusion in the absence of bleeding, fever, or sepsis, or an interventional procedure is not indicated with platelet count >10,000/μL. Although not based on randomized controlled trials, many centers transfuse to platelet count of 50,000/μL for interventional procedures, but recent data suggest that, at least for relatively minor-risk surgeries such as central venous catheter placement, platelet count of >20,000/μL is sufficient. Institutions typically use platelet threshold of 20,000/μL in the presence of fever, sepsis, or other bleeding risks and threshold of 50,000/μL in the presence of active bleeding.

In regard to plasma transfusion, a recent metaanalysis of randomized controlled trials showed no definite benefit to plasma transfusion in reducing blood loss in either prophylactic or therapeutic setting. Typically, corrections in international normalized ratio (INR) are small unless pretransfusion INR is >2.5 in the setting of bleeding.

Minimizing Phlebotomy-Related Blood Loss: Phlebotomy-related losses can significantly contribute to anemia and transfusion requirements in hospitalized patients, especially in the intensive care unit setting. Both frequency and volume of blood draws should be limited. Strategies to limit blood draw frequency include the following: limiting orders to only medically necessary tests, embedding disease-specific clinical guidelines into order forms or computer screens, and eliminating "standing" orders. Strategies to limit volume collected include the following: use of low-volume tubes, reducing discard volume from indwelling lines, and point-of-care testing. Point-of-care testing requires smaller blood volumes and has quicker turnaround time than standard laboratory testing. Point-of-care instrumentation is available for blood counts, standard coagulation parameters, and thromboelastography/rotational thromboelastometry measurements.

Treating Preoperative Anemia and Coagulopathy: Preoperative anemia is the strongest predictor of perioperative transfusion and increases risk of perioperative complications. Often presurgical laboratory testing is performed or reviewed too late to treat anemia (iron, B12, folate, or erythropoietin). Erythropoietin can also be used preoperatively to facilitate preoperative autologous donation (PAD) or acute normovolemic hemodilution (ANH). Concomitant intravenous iron supplementation can be used as feasible. Ideally, there is a multidisciplinary preoperative anemia clinic to prepare these patients.

Evaluation includes medical history, including personal and family history of bleeding and use of anticoagulant or antiplatelet medications. Patients should be provided a schedule for switching long-acting anticoagulant or antiplatelet drugs to shorter acting agents, to be stopped immediately before surgery. Herbal or vitamin supplements may also affect bleeding risk and should be discontinued 1–2 weeks preoperatively.

Autologous Blood Conservation: Autologous blood transfusion minimizes, including PAD, ANH, and perioperative blood recovery, allogeneic blood risks such as transfusion-transmitted disease.

Preoperative Autologous Donation: PAD was used frequently in elective procedures with high likelihood of significant blood loss, but its role has been substantially decreased due to improved surgical techniques and its costs and risks. These costs and risks include wastage of ~50% of autologous blood collected (unused); administrative costs from extra handling of such units; risk of inducing preoperative anemia, which increases the likelihood of transfusion of both autologous and allogeneic RBCs; and risk of clerical errors with transfusion of incorrect product. Additionally, the number of PAD units that can be collected and transfused is limited by 42-day RBC shelf life. RBC units can be frozen for up to 10 years to allow more collections: thawing and deglycerolizing takes a few hours, and thawed products only have a 24-hour shelf life. Furthermore, delays in sending the blood expeditiously to the requesting hospital and then thawing and deglycerolizing are often encountered.

PAD is most useful in individuals with anticipated elective surgery who have high-frequency or multiple RBC alloantibodies, or IgA deficiency (particularly with plasma or platelet products). Long-term frozen storage without anticipated surgery, however, is of dubious value; significant autologous blood would have to be collected to meet emergency needs. With the advent of molecular testing, blood centers have increased ability to identify rare units, thus decreasing the use and need for autologous frozen blood.

PAD results in decreased hemoglobin, which takes time to recover. Adequate iron stores must be ensured through oral or intravenous iron replacement therapy as needed. Intravenous compared with oral iron acts faster and more effectively, and newer formulations have little risk of anaphylaxis. Erythropoietin can also be used to increase number/frequency of PAD products collected, up to one RBC unit per week of treatment. PAD can be well tolerated by high-risk donors, such as the elderly, children, pregnant women, and patients with atherosclerotic coronary artery disease.

Acute Normovolemic Hemodilution: ANH involves removal of one or more whole blood units into standard collection bags containing citrate anticoagulation and replacement of lost volume with crystalloids or colloids. ANH reduces RBC loss because blood shed subsequently has lower hematocrit: for example, 2-L blood loss at hematocrit of 20% results in 400 mL rather than 900 mL RBC loss at hematocrit of 45%.

Cardiovascular status should be closely monitored during hemodilution procedure. The degree of anemia can affect oxygen transport, although concomitant drop in blood viscosity and compensatory increased cardiac output tend to offset this risk. Collected blood is stored and reinfused in the operating room during surgery, in reverse order of collection to transfuse bags with highest hematocrit toward end of surgery, after blood loss is controlled. When stored at room temperature (up to 8 hours), it will contain functional platelets and clotting factors. Recent metaanalyses suggest that ANH reduces allogeneic blood transfusion, particularly in cardiac surgery, but studies are complicated by significant heterogeneity and publication bias.

Intraoperative Blood Recovery: This is when blood shed into operative field is recovered, mixed with anticoagulant, and stored in a sterile reservoir until enough blood is collected for reinfusion. Approximately, 50% of blood lost during surgical procedure can be recovered. Intraoperative blood recovery should be considered for surgeries where mean transfusion rate is at least 1 unit. Devices are available that collect blood for reinfusion, either with or without washing before transfusion. Devices that wash blood (by centrifugation with normal saline) are preferable, to remove free hemoglobin, activated clotting and complement factors, inflammatory cytokines, fat particles, and contaminants. Large volumes can be processed quickly in the event of rapid blood loss. The processed product contains RBCs diluted in saline to a >45% hematocrit and can be stored at room temperature up to 8 hours from completion of processing. Unwashed products usually have hematocrit of 20%–30% (due to dilution with irrigation fluids), must be filtered to remove clots and tissue debris, and can be stored at room temperature up to 8 hours from start of collection, due to risk from other contaminants.

Postoperative Blood Recovery: This is when blood shed from surgical drains and/or wounds is recovered and reinfused. This is predominantly used in cardiac and orthopedic surgery, where reduction in allogeneic blood use has been demonstrated; drainage sites include mediastinum or knee/hip sites. Blood salvaged from a serosal cavity has little residual fibrinogen or platelets to allow clotting; therefore, anticoagulant addition is usually unnecessary. Product can be washed or not. Unwashed blood may contain undesirable constituents as previously described but is usually well tolerated.

Adverse Events: Blood recovery from a contaminated site, such as with spilled intestinal contents, is relatively contraindicated, but bacterial contamination may still occur with environmental or skin organisms such as coagulase-positive and coagulase-negative *Staphylococcus*, *Propionibacterium*, and *Corynebacterium* species. Tumor cell contamination of recovered blood is a theoretical concern with malignancies, but increased risk of recurrence or metastasis with reinfusion is unproven and use of leukoreduction filters removes most tumor cells. Aspiration of pharmacologic or hemostatic agents, fat, heavily bacterially contaminated fluids, amniotic fluid, gastric fluid, or bone fragments should be avoided through use of a double suction setup. Care must be taken to vent air from the final product bag for infusion, to prevent potentially fatal air embolism. Adverse effects including respiratory distress, hypotension with anaphylaxis, and fever are more likely when product is collected over long-time interval (maximum of 6 hours from start of postoperative collection allowed).

Minimizing Perioperative Blood Loss: Perioperative blood loss can be minimized by the following: (1) drugs to reduce surgical bleeding, (2) operative techniques to reduce surgical bleeding, and (3) adequate postoperative management (Table 59.1).

Drugs

Antifibrinolytics: Tranexamic acid (TXA) and α-aminocaproic acid, lysine analog inhibitors of plasminogen activation and plasmin activity, are the primary antifibrinolytics used in the United States, with TXA more commonly studied. TXA has been shown to reduce death from bleeding in trauma and postpartum hemorrhage. TXA also reduces blood loss and allogeneic RBC transfusion in orthopedic and on-pump cardiac surgery in adults, with no evidence of increased thromboembolic or mortality risk. TXA in off-pump cardiac surgery, however, has been associated with a possible risk of seizures. TXA is also used topically in orthopedic surgery.

Topical Hemostatic Agents: Topical hemostatic agents can be divided into three basic categories: thrombin-based fibrin sealants; autologous platelet gel; and commercial derivatives of collagen, gelatin, and cellulose that provide a matrix for endogenous coagulation. Thrombin-based agents and autologous platelet gel contain active clotting factors, while collagen, gelatin, and cellulose are nonactive agents forming physical matrix over bleeding site. These mechanical agents may be sufficient for bleeding patients with intact coagulation, whereas active hemostatic agents are appropriate for coagulopathic patients with bleeding. Fibrin sealant, combination of a solution of human thrombin and calcium with a solution of human fibrinogen and factor XII to form fibrin, appear to be safe and may reduce allogeneic transfusion and perioperative blood loss in various surgical procedures.

TABLE 59.1 Potential Perioperative Patient Blood Management Strategies

Preoperative Strategies to Augment Hemoglobin Levels	Intraoperative Blood Management Strategies	Postoperative Blood Management Strategies
Restrictive transfusion threshold strategies guided by appropriate evidence base	Restrictive transfusion threshold strategies guided by appropriate evidence base	Restrictive transfusion threshold guided by appropriate evidence base
Minimize phlebotomy-related blood loss preoperatively	Minimize phlebotomy-related blood loss intraoperatively	Minimize phlebotomy-related blood loss postoperatively
Treating preoperative anemia: • Oral and Intravenous iron replacement (with or without concomitant folate)	Intraoperative blood recovery and cell salvage	Postoperative blood recovery
Treating preoperative anemia: • Erythropoiesis-stimulating agents e.g., erythropoietin (with or without concomitant iron and folate)	Improved surgical techniques, e.g., robotic surgeries	Repleting vitamin K postoperatively
Autologous blood conservation (decreasing evidence base)	Acute Normovolemic hemodilution	Use of antifibrinolytic agents: TXA, EACA as appropriate, e.g., postpartum hemorrhage
	Prophylactic use of fibrinogen concentrates	
	Use of antifibrinolytic agents: tranexamic acid (TXA), epsilon aminocaproic acid (EACA)	
	Use of point-of-care testing (e.g., Thromboelastography or Rotational Thromboelastometry to guide transfusion decisions	
	Fresh whole blood or reconstituted whole blood	

Operative Techniques: Certain operative principles apply to all surgeries. Maintenance of normal body temperature and pH is critical for normal platelet and coagulation function. Body temperature can be maintained through blood/intravenous fluid warmers, air-warming devices, and physical coverage. Acidosis is avoided by preventing hypovolemia, hypotension, and excessive normal saline administration.

Careful planning can improve pace and efficiency of surgery and thus minimize blood loss. Surgical techniques that reduce bleeding include laparoscopic, robotic, or endovascular approaches; or preoperative embolization. Surgical instruments such as ultrasonic scalpels, bipolar vessel sealers, and argon beam coagulators can improve hemostasis at incision sites. Other techniques include patient positioning to elevate blood loss site; decreasing pressure on abdominal contents (which can obstruct inferior vena cava flow); tourniquet use; reduction of central venous pressure; infusion of local vasoconstrictive agents; and controlled hypotension.

Adequate Postoperative Management: Postoperative blood salvage should be considered. Body temperature, cardiac output, and ventilation/oxygenation should continue to be monitored. Rapid diagnosis and control of surgical hemorrhage, with reexploration as needed, is crucial. Postoperative anemia should be evaluated. Patients may have absolute or functional iron deficiency, for which intravenous iron and/or erythropoietin treatment may be appropriate. Drug side effects or lack of stress gastritis prophylaxis can also contribute to anemia.

Further Reading

AABB. (2015). *AABB White Papers: Building a better patient blood management program identifying tools, solving problems and promoting patient safety*.

Carless, P. A., Henry, D. A., Moxey, A. J., O'Connell, D., Brown, T., & Fergusson, D. A. (2010). Cell salvage for minimizing perioperative allogeneic blood transfusion. *Cochrane Database Syst Rev, 4*, CD001888.

Carson, J. L., Guyatt, G. H., Heddle, N. M., et al. (2016). Clinical practice guidelines from the AABB: Red blood cell transfusion thresholds and storage. *JAMA, 316*(19), 2025–2035.

Ellingson, K. D., Sapiano, M. R. P., Haass, K. A., et al. (2017). Continued decline in blood collection and transfusion in the United States-2015. *Transfusion, 57*(Suppl. 2), 1588–1598.

Henry, D. A., Carless, P. A., Moxey, A. J., et al. (2011). Anti-fibrinolytic use for minimizing perioperative allogeneic blood transfusion. *Cochrane Database Syst Rev, 3*, CD001886.

Kaufman, R. M., Djulbegovic, B., Gernsheimer, T., et al. (2015). Platelet transfusion: A clinical practice guideline from the AABB. *Ann Intern Med, 162*(3), 205–213.

Levy, J. H., & Sniecinski, R. M. (2012). Prohemostatic treatment in cardiac surgery. *Semin Thromb Hemost, 38*, 237–243.

Liumbruno, G. M., Bennardello, F., Lattanzio, A., Piccoli, P., & Rossetti, G. (2011). Recommendations for the transfusion management of patients in the peri-operative period. I. The pre-operative period. *Blood Transfus, 9*, 19–40.

Markowitz, M. A., Waters, J. H., & Ness, P. M. (2014). Patient blood management: A primary theme in transfusion medicine. *Transfusion, 54*(10 Pt 2), 2587.

Yang, L., Stanworth, S., Hopewell, S., Doree, C., & Murphy, M. (2012). Is fresh-frozen plasma clinically effective? An update of a systematic review of randomized controlled trials. *Transfusion, 52*, 1673–1686.

CHAPTER 60

Overview of Adverse Events and Outcomes After Transfusion

Ruchika Goel, MD, MPH and Aaron A.R. Tobian, MD, PhD

Transfusion of blood products can lead to number of adverse events and outcomes in recipients, ranging from subclinical infection with a virus that can remain undiagnosed for decades, to acute immune hemolysis or acute septic reaction resulting in rapid onset of hypotension, shock, and ultimately death. A number of classification schemes exist to categorize adverse events and outcomes, including groupings by pathogenesis (immune vs. nonimmune; infectious [termed transfusion-transmitted infections, TTI] vs. noninfectious [termed noninfectious serious hazards of transfusion, NiSHOTs]), reaction type (febrile or allergic), and time to development. Acute transfusion reactions occur within 24 hours of transfusion, and delayed reactions occur more than 24 hours after transfusion, and both can range from mild to life-threatening (Table 60.1).

Serious Hazards of Transfusion: As serious infectious hazards of transfusion including human immunodeficiency virus (HIV), hepatitis C virus (HCV), and hepatitis B virus (HBV) are increasingly rare, attention has turned more to NiSHOTs.

NiSHOTs include transfusion-related acute lung injury (TRALI), transfusion-associated circulatory overload (TACO), and hemolytic transfusion reactions (HTRs). Occurrence of such reactions is not uncommon, as evidenced by FDA data ranking these three entities as the leading causes of transfusion-associated death in the United States in 2015. The latest information (2016) from the UK hemovigilance scheme, serious hazards of transfusion (SHOT) with 20 years of follow-up regarding adverse events and reactions showed that acute transfusion reactions together with failure to administer anti-D globulin and handling and storage errors (HSEs) ranked highest followed by incorrect blood component transfused (IBCT; equals "mistransfusion"). Importantly, longitudinal analysis of hemovigilance data from the United Kingdom also suggests that as focused steps are taken to mitigate SHOTs, such as TRALI and bacterial contamination of blood components, somewhat less-well-defined entities such as HSE are now taking lead in transfusion-associated morbidity.

Leading Causes of Transfusion-Associated Death: As stated above, the three leading causes of transfusion-associated death in the United States are TRALI, TACO, and HTR (Chapters 63, 65, and 66). Significant strides have been made to develop prevention strategies for these NiSHOTs in the past few years.

Transfusion-Related Acute Lung Injury: TRALI is defined as "new episode of ALI that occurs during or within 6 hours of completed transfusion which is not temporally related to any competing etiology for ALI." The diagnosis is considered a clinical

Transfusion Medicine and Hemostasis. https://doi.org/10.1016/B978-0-12-813726-0.00060-X

TABLE 60.1 Transfusion Reactions

Name	Temporal Relationship	Severity
Acute Transfusion Reactions		
Acute hemolytic	0–24 hours	Mild–severe
Anaphylactic	0–1 hours	Severe
Febrile	0–4 hours	Mild
Hypotensive	0–1 hours	Mild–moderate
Metabolic complications	0–4 hours	Mild–moderate
Septic	0–6 hours	Mild–severe
TACO	0–6 hours	Mild–severe
TRALI	0–6 hours	Mild–severe
TAD	0–24 hours	Mild
Urticarial/allergic	0–4 hours	Mild–moderate
Delayed Transfusion Reactions		
Alloimmunization	Days–months	None–severe
Delayed hemolytic	Days	Mild–severe
Iron overload	Years	Mild–severe
PTP	Week–weeks	Moderate–severe
TA-GVHD	Week–weeks	Severe
TA-MC	Months–years	Unknown
TRIM	Week–weeks	Mild–moderate
TTD	Days–years	None–severe

and radiographic diagnosis, not a laboratory-defined one. The pathogenesis is not fully understood but likely requires several "hits"; including donor HLA antibodies.

In the 2015 fatalities reported to FDA, TRALI continued to be responsible for highest number of reported fatalities (38%). In contrast, a review of cumulative (2010–16) UK SHOT data for deaths shows that while pulmonary complications are the leading cause of death (61/115 cases, 53.0%) (Fig. 60.1), TACO was the leading reason in ~87%, and TRALI was only responsible for 8% of deaths. Indeed, TRALI mitigation schemes, such as male-only plasma, in the United Kingdom, United States, and elsewhere have substantially decreased its risk. Close evaluation of TRALI prevention strategies and better understanding of recipients at risk and TRALI pathophysiology are important to continue reducing its incidence.

Transfusion-Associated Circulatory Overload: Transfusion-associated volume/circulatory overload is a result of infusion of blood product volume that exceeds recipient's circulatory capacity, either due to high volume or excessive infusion rate. In 2015, the FDA reported TACO as the second most common cause of transfusion-related fatality being responsible for 24% of reported deaths. The UK SHOT report 2016 reported pulmonary complications being the leading cause of death (53.0% cases) of which TACO was the leading reason ~87% of deaths.

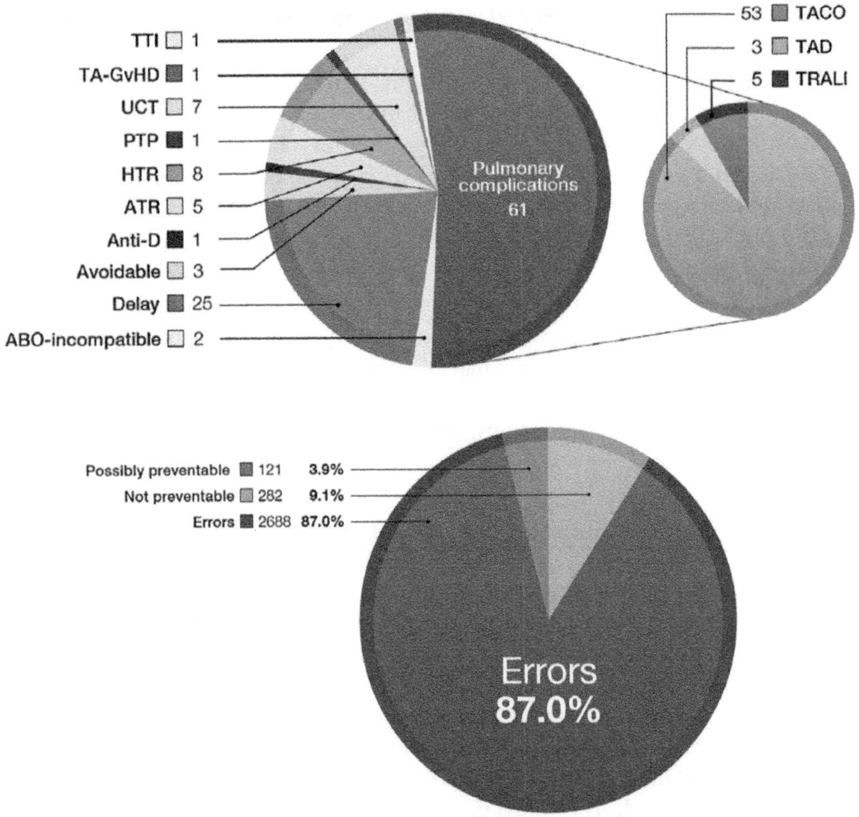

FIGURE 60.1 Data from 2016 UK SHOT Annual Report. Transfusion-related deaths, cumulative data 2010–16 (n = 115) showing that pulmonary complications are the leading cause of death in 61/115 (53.0%) cases and that errors accounted for majority of reports. *Anti-D*, failure to give anti-D globulin; *ATR*, acute transfusion reaction; *HTR*, hemolytic transfusion reaction; *PTP*, posttransfusion purpura; *TACO*, transfusion-associated circulatory overload; *TAD*, transfusion-associated dyspnea; *TA-GvHD*, Transfusion Associated Graft versus Host Disease; *TRALI*, transfusion-related acute lung injury; *TTI*, transfusion-transmitted infection; *UCT*, unclassifiable complication of transfusion.

TACO more frequently occurs in elderly, children, and patients with impaired cardiac function. Symptoms include dyspnea, orthopnea, tachycardia, and widening of pulse pressure often associated with increase in blood pressure and decreased oxygen saturation within 6 hours from initiation of transfusion. Several strategies are available to decrease TACO risk, including appropriate medical assessment of patients receiving transfusion followed by restriction of transfusion rate in at-risk patients to 2.0–2.5 mL/kg per hour. Additionally, avoidance of overtransfusion by proper assessment and optimization of patient's transfusion needs by posttransfusion evaluation will likely help in decreasing risk.

Hemolytic Transfusion Reaction and Mistransfusion: Both ABO and non-ABO antibody-related HTR are significant cause of transfusion-related mortality and morbidity. In 2015 FDA report, mistransfusion was responsible for ~20%

of transfusion-related mortalities: HTRs due to non-ABO (14%) and ABO (7.5%) incompatibilities.

Mistransfusion, when the patient receives the wrong blood, occurs for a variety of reasons, including improper identification of intended recipient during initial sample collection for blood typing (so-called "wrong blood in tube" [WBIT]), improper typing or pretransfusion testing of blood component or recipient, or misidentification of recipient and/or blood product at time of initiation of transfusion. Mistransfusion has an estimated incidence of 1:14,000 units with fatality rates approaching 1:500,000 units transfused. A number of technologies exist to reduce mistransfusion risk, including barcoding of patient and blood unit identifiers, radio frequency identification, combination locked pouches that require special bedside code, and bedside ABO tests that would allow an operator to "retype" patient and "match" patient blood type with blood unit label. Most of these methods have not had widespread implementation to date, and each of these methods involves human actions, interpretation, or decision, thus introducing a scope of human error.

Developing Noninfectious Serious Hazards of Transfusion: Significant concerns remain on the horizon and may become future NiSHOTs, including transfusion-associated morbidity and mortality.

Transfusion-Associated Morbidity and Mortality: A significant body of data associates blood transfusion itself with increased morbidity and mortality in various clinical settings. However, increasing the number of randomized controlled trials to date demonstrates an equivalent outcome in patients transfused at higher and lower hemoglobin thresholds thus showing safety of restrictive transfusion thresholds. Additionally, earlier studies proposed the association between age of blood and clinical outcomes (worse outcome with older age blood products). However, all recent randomized controlled trials have refuted this association. While RBC storage lesion per se is a well-accepted phenomenon, there has been no conclusive RCT evidence supporting any significant clinical effects of transfusing younger versus older RBCs in adults.

Hemovigilance Systems: Hemovigilance systems are reporting systems that collect data on adverse events and errors of transfusion; these data are analyzed to determine methods to improve transfusion standards, assist in guideline development and ultimately increase safety and quality of transfusion. This effort to minimize errors and adverse outcomes of transfusion has proved successful. The UK SHOT, established in 1996, is the best example; its development resulted in detection of correlation between TRALI incidence and transfusion of plasma from donors with history of multiple pregnancies and subsequent decrease in TRALI after implementation of male-only plasma.

The National Heart Lung and Blood Institute (NHLBI) Recipient Epidemiology and Donor Evaluation Study-III (REDS-III) is focused on improving transfusion safety. Its Phase 2 study titled *Severe Transfusion Reactions Including Pulmonary Edema (STRIPE)* focuses on strategies that will prevent or reduce complications related to TACO and other transfusion complications.

In addition to tracking known adverse outcomes of transfusion and resulting in development of mitigation strategies, an important additional benefit of fully implemented national and international hemovigilance networks is early detection of novel or developing infectious and noninfectious hazards of transfusion.

Further Reading

Centers for Disease Control and Prevention. (National Safety Network (NHSN): Biovigilance Component. Available from: https://www.cdc.gov/nhsn/pdfs/biovigilance/bv-hv-protocol-current.pdf; https://www.fda.gov/downloads/BiologicsBloodVaccines/SafetyAvailability/ReportaProblem/TransfusionDonationFatalities/UCM598243.pdf; https://www.shotuk.org/wp-content/uploads/SHOT-Report-2016-Summary.pdf.

CHAPTER 61

Febrile Nonhemolytic Transfusion Reactions

Irina Maramica, MD, PhD, MBA

Febrile nonhemolytic transfusion reactions (FNHTRs) are common, occurring with 1–3% of transfusions. FNHTR manifests as fever and/or chills without hemolysis occurring in the patient during or within 4 hours of transfusion cessation. Diagnosis is made by excluding other causes of fever. Most common causes are passively transfused proinflammatory cytokines or recipient antibodies reacting with donor leukocytes. Most reactions are easily managed. FNHTRs are mitigated through prestorage leukoreduction.

Clinical Presentation: FNHTR is defined as occurring during or within 4 hours of cessation of transfusion, with fever (≥38°C and change of ≥1°C) from pretransfusion or chills/rigors. Fever may be absent due to antipyretic premedication. One-third of reactions are severe, resulting in rigors, temperature elevation of >2°C, dyspnea, headache, nausea, and vomiting.

Diagnosis: FNHTR is diagnosis of exclusion, as no specific tests are available. Differential diagnosis includes hemolysis, sepsis and TRALI, or fever due to medications or medical conditions. Hemolytic transfusion reaction is excluded thorough transfusion reaction workup, including clerical check, ABO confirmation of product and recipient, and serological workup. Transfusion-associated sepsis usually presents with high increase in fever and hypotension. Gram stain and culture of patient and unit support excluding septic transfusion reactions. TRALI typically presents with fever, dyspnea, and hypotension. Review of patient's record can rule out fevers due to underlying disease or medications, such as beta-lactam antibiotics, procainamide, isoniazid, barbiturates, quinidine, and diphenylhydantoin.

Incidence: FNHTRs are the most common acute transfusion reactions. Prestorage leukocyte reduction has dramatically decreased its incidence to <0.2% for both RBCs and platelets, with higher incidence using active versus passive reporting. Certain patient populations are considered to be at a higher risk, including patients with hematologic diseases, who maybe HLA alloimmunized from frequent transfusions, or more sensitive to infused cytokines due to baseline inflammatory state.

Platelets: FNHTRs' incidence varies with platelet product type; higher (0.4%–2.2% of transfusions) with nonleukoreduced or bedside leukocyte-reduced products versus 0.1%–0.15% with prestorage leukoreduced platelet products. More reactions are observed when nonprestorage leukoreduced 4–5 days old platelet concentrates are transfused, due to increase in cytokine levels: 1.1% with ≤3 day versus 4.6% > 3 day platelets. Use of platelet additive solution decreased FNHTR rate with apheresis platelets from 0.50% to 0.17%.

Transfusion Medicine and Hemostasis. https://doi.org/10.1016/B978-0-12-813726-0.00061-1

Red Blood Cells: FNHTR incidences also vary with leukoreduction: nonleukoreduced 0.33%–0.34% to leukoreduced 0.18%–0.19%.

Plasma: FNHTRs rarely occur with plasma products: incidence of 0.02% of transfusions. Recent systematic review showed that methylene blue–treated Fresh Frozen Plasma (FFP) led to fewer FNHTRs than FFP.

Pathophysiology: Fever in FNHTRs is triggered by several different mechanisms, resulting in release of cytokines, including IL-1β, IL-6, and TNF- α from activated monocytes and macrophages. These cytokines induce production of prostaglandin E2, which acts on the hypothalamus to increase body temperature. Immune and nonimmune mechanisms have been implicated its pathophysiology:

1. One immune-mediated mechanism is a result of recipient white blood cell alloantibodies reacting with donor white blood cells to release cytokines. Both granulocyte and HLA antibodies can stimulate donor's white blood cells. Immune-mediated FNHTRs are most commonly seen in previously transfused or pregnant patients who are receiving nonleukocyte-reduced products, thus are prevented with leukoreduction.
2. Second immune mechanism is when cytokines are produced and released by recipient's macrophages, stimulated by immune complexes formed between recipient's antibody and donor cell (leukocyte or platelet) antigens. Leukoreduction is effective only in preventing reactions mediated by antibody interaction with donor leukocytes but not platelets. Antibody interaction with donor platelets provides possible explanation for febrile reactions in alloimmunized recipients of prestorage leukocyte-reduced platelets. Repeat FNHTRs may indicate development of refractoriness to platelet transfusions, as these same antibody interactions with transfused platelets can lead to their clearance.
3. Nonimmune mechanism is due to passive transfer of proinflammatory cytokines accumulated in plasma portion of product during storage at room temperature. This mechanism is not dependent on leukocyte antibody presence. Leukocyte-derived cytokines have been demonstrated in supernatant plasma of stored platelet products, and their levels correlate with storage duration and prestorage leukocyte number. Cytokines do not accumulate in sufficient quantities during refrigerated RBC storage.

Bedside leukocyte reduction does not prevent these reactions. Prestorage leukocyte reduction is effective, although it does not eliminate these reactions completely, possibly due to platelet-derived mediators that are not removed by leukoreduction filters. Proinflammatory cytokines released by storage-induced platelet microparticles have been demonstrated in supernatants of leukoreduced platelet concentrates and apheresis platelets, including platelet factor 4 (PF-4), transforming growth factor β1 (TGF-β1), β-thromboglobulin (β-TG) and soluble CD40 ligand (sCD40L). Notably, these are removed by nonplatelet-sparing RBC leukoreduction filters.

Treatment: Transfusion should be discontinued immediately and antipyretics administered. Acetaminophen (adult dose: 325–650 mg orally, pediatric dose: 10–15 mg/kg orally or rectally) is preferred because aspirin and nonsteroidal antiinflammatory agents are contraindicated in patients receiving platelet transfusions. Patients who develop rigors are commonly given meperidine (25–50 mg intravenously) to control severe shaking that can significantly increase oxygen demand. This drug should be avoided in patients with renal failure or on MAO inhibitor therapy. These are self-limited reactions without long-term consequences.

Recent retrospective analysis of FNHTR management and outcomes demonstrated postreaction clinical activity represents substantial burden on hospital and patient care: >40% of implicated products were incompletely transfused, one-fourth of patients underwent chest imaging and majority had microbial cultures, patients had exposure to unplanned medications, and 15% had disposition escalation. Based on these considerations, authors provided estimates of FNHTR management of $160 per patient.

Prevention: Both pre- and poststorage leukoreduction, are effective in preventing antibody-mediated reactions to RBCs and platelets. Poststorage leukoreduction is not effective in preventing reactions to platelets that are mediated by accumulation of cytokines released from leukocytes during storage. These reactions can be prevented by prestorage filtration, which is not effective in removing platelet-derived mediators, and some patients experience repeat reactions to prestorage leukoreduced blood products. Decreasing cytokine levels in platelet products by removal of stored supernatant plasma effectively lowers FNHTR rate to 0.6% compared with poststorage (20%) leukoreduction. However, this method is not superior to prestorage leukoreduction and required additional centrifugation step compromises platelet function. Using leukoreduced apheresis, platelets stored in PAS decreased FNHTRs with relative risk of 0.336 in relation to platelets stored in 100% plasma. Prestorage leukoreduction remains most practical and efficient way of preventing FNHTRs.

Routine premedication with antipyretics is commonly used to prevent FNHTRs in high-risk patients and those with recurrent reactions. Two prospective, randomized, controlled trials examined effectiveness of acetaminophen and diphenhydramine in preventing reactions. 51 patients with hematological malignancies experienced 12 febrile and 3 allergic reactions when transfused with 98 prestorage leukoreduced apheresis platelets. Overall rate of reactions was similar in the treatment arm (15.4%) and the placebo arm (15.2%), but FNHTRs were more common in the treatment arm (15.4%) compared with placebo arm (8.7%). The second study of 315 patients receiving poststorage leukoreduced products also failed to demonstrate effectiveness of premedication on preventing NHTR, with the overall rate of reactions of 1.44% of transfusions in treatment group and 1.51% in placebo group (FNHTR rate: 0.35% vs. 0.64%, $P = .8$). Use of premedication to prevent FNHTRs should be evaluated with consideration of its costs and potential toxicities, potential risk that antipyretics could mask fever due to more serious reasons, and significantly reduced FNHTR rates associated with leukoreduced components.

Further Reading

Cohen, R., Escorcia, A., Tasmin, F., et al. (2017). Feeling the burn: The significant burden of febrile nonhemolytic transfusion reactions. *Transfusion*, *57*, 1674–1683.

Cohn, C. S., Stubbs, J., Schwartz, J., et al. (2014). A comparison of adverse reaction rates for PAS C versus plasma platelet units. *Transfusion*, *54*, 1927–1934.

Dzik, W. H., Anderson, J. K., O'Neill, E. M., et al. (2002). A prospective, randomized clinical trial of universal WBC reduction. *Transfusion*, *42*, 1114–1122.

Heddle, N. M., Blajchman, M. A., Meyer, R. M., et al. (2002). A randomized controlled trial comparing the frequency of acute reactions to plasma-removed platelets and prestorage WBC-reduced platelets. *Transfusion*, *42*, 556–566.

Kennedy, L. D., Case, L. D., Hurd, D. D., Cruz, J. M., & Pomper, G. J. (2008). A prospective, randomized, double-blind controlled trial of acetaminophen and diphenhydramine pretransfusion medication versus placebo for the prevention of transfusion reactions. *Transfusion*, *48*(11), 2285–2291.

Pagliano, J. C., Pomper, G. J., Fisch, G. S., et al. (2004). Reduction of febrile but not allergic reactions to RBCs and platelets after conversion to universal leukoreduction. *Transfusion*, *44*, 16–24.

Rogers, M. A. M., Rohde, J. M., & Blumberg, N. (2016). Haemovigilance of reactions associated with red blood cell transfusion: Comparison across 17 countries. *Vox Sang*, *110*, 266–277.

Sadaah, N., van Hout, F. M. A., Schipperius, M. R., et al. (2017). Comparing transfusion reaction rates for various plasma types: A systemic review and meta-analysis/regression. *Transfusion*, *57*, 2104–2114.

CHAPTER 62

Allergic Transfusion Reactions

William J. Savage, MD, PhD

Allergic reactions are common reactions to blood transfusions. On one end of the spectrum, typical mild allergic transfusion reactions (ATRs) consist of isolated, pruritic/urticarial lesions and occur during or within 2 hours of transfusion. On the other end of the spectrum, anaphylaxis is an acute, systemic allergic reaction that is characterized most significantly by hypotension and/or respiratory compromise. Anaphylaxis typically occurs early after transfusion has started. The term "anaphylactoid" is no longer favored but has been used historically to describe either systemic hypersensitivity reactions not mediated by IgE or moderate severity ATRs that involve allergic signs or symptoms other than cutaneous manifestations.

Clinical Features: ATRs occur in 0.03%–0.61% of red blood cell (RBC) transfusions, 0.3%–6% of platelet transfusions, and 1%–3% of plasma transfusions. About 90% of these reactions are mild. Anaphylactic reactions occur in approximately 1 in 20,000–47,000 components transfused. The incidence of any allergic reaction to factor VIII, factor IX, and von Willebrand factor concentrates is approximately 1 in 5000 doses, with a roughly similar incidence for concentrates and recombinant factors. Anaphylaxis accounts for 5% of transfusion-related mortality reported to the FDA (eight cases reported from 2011 to 2015). Leukoreduction does not reduce the incidence of ATRs. Atopic disease in blood donors does not contribute to ATRs in most cases; the rare exception is passively transfused IgE that has specificity to an allergen exposure in the transfusion recipient, e.g., peanuts. Atopic disease in the recipient, particularly hay fever, is associated with a risk of ATRs.

Symptom onset of anaphylactic reactions is between seconds and 45 minutes after the start of transfusion, while symptom onset for mild ATRs occurs up to 2–3 hours after completing transfusion. Generalized pruritus may precede urticarial eruptions. There may be generalized erythema (flushing) of the skin or angioedema. Symptoms of severe reactions evolve quickly. Upper and/or lower airway obstruction is caused by angioedema and results in hoarseness, stridor, and/or the complaint of a "lump" in the throat. Lower airway obstruction results in audible wheezing, feeling of chest tightness, substernal pain, dyspnea, or cyanosis. There may be profound hypotension, possibly leading to loss of consciousness, tachycardia, cardiac arrhythmias, or cardiac arrest. Severe gastrointestinal symptoms (abdominal cramps/pain, nausea, vomiting, diarrhea) may also be present.

Diagnosis: Some symptoms that are consistent with ATRs overlap with other types of reactions, particularly dyspnea and shock. Immediate hemolytic transfusion reactions can be differentiated by the presence of fever and absence of cutaneous manifestations. Hypotensive reactions have marked drops in blood pressure without cutaneous

Transfusion Medicine and Hemostasis. https://doi.org/10.1016/B978-0-12-813726-0.00062-3

symptoms or other signs of anaphylaxis. Transfusion-related acute lung injury can be differentiated from ATRs by the presence of fever, chest X-ray findings of pulmonary edema, and absence of cutaneous symptoms. Bacterial contamination is associated with hypotension and shock, but the presence of rigors and fever and the absence of urticaria and angioedema differentiate it from ATRs. Flushing and hypotension may be seen in patients taking angiotensin-converting enzyme inhibitors but can be differentiated from ATRs by the absence of pruritus or pulmonary findings. Other causes of the patient's allergic reaction should be ruled out, such as coincidental administration of a drug or food.

Patients with a severe ATR may be tested for IgA deficiency and the presence of anti-IgA, although the prevalence of IgA deficiency among patients with severe reactions is low. Screening for IgA deficiency can be performed with routine IgA testing. Reference laboratories can perform IgA tests that are more sensitive and can distinguish severe IgA deficiency from IgA concentrations that are present, but below the limit of detection of routine assays. Anti-IgA testing is currently only performed in reference laboratories, and a common diagnostic approach is to screen for anti-IgA antibodies only after severe IgA deficiency is diagnosed.

Pathophysiology: ATRs manifest clinically as type I hypersensitivity responses, and biomarkers of immediate hypersensitivity reactions, e.g., tryptase, are elevated after severe ATRs. In IgE-mediated hypersensitivity responses, IgE binds to mast cells and basophils by means of IgE Fc receptors, causing activation and release of mediators (such as histamine, heparin, leukotrienes, platelet-activating factor, cytokines, and chemokines). These mediators instigate changes in smooth muscle tone and extravasation of cells and fluid into the tissues, resulting in signs and symptoms of the skin, respiratory tract, and cardiovascular and gastrointestinal systems. In addition, IgG may also mediate ATRs through complement fixation, one of the results of which is the formation of C3a and C5a anaphylatoxins.

IgA Deficiency and Anti-IgA: IgA deficiency is the most extensively studied predisposition to ATRs, although IgA deficiency underlies the pathophysiology of only a minority of cases. In the setting of transfusion, IgA deficiency typically refers to a severe deficiency (<0.05 mg/dL). Severe IgA deficiency occurs in approximately 1 in 900 blood donors in the United States and United Kingdom, 1 in 500 in Finland, and 1 in 93,000 in Japan. Of IgA-deficient individuals who are otherwise healthy, 20%–30% form anti-IgA, while approximately 80% of IgA-deficient individuals with autoimmune diseases (such as rheumatoid arthritis or systemic lupus erythematosus) form anti-IgA. Anti-IgA is usually IgG but may be IgM or IgE, and it can be of broad or limited (anti-IgA subclass or allotype) specificity.

In a study of 32,376 blood donors, the frequency of severe IgA deficiency (<0.05 mg/dL) and class-specific anti-IgA was 1 in 1,200, which is greater than the observed frequency of anaphylactic reactions of 1 in 20,000–47,000 transfusions. Therefore, it is likely that most individuals with IgA deficiency and anti-IgA do not have anaphylactic reactions. Thus, IgA-deficient individuals without a history of anaphylaxis must be monitored during transfusion, as any transfusion recipient should be. It may not be necessary to restrict IgA-deficient patients to washed or IgA-deficient products without a trial of unmodified blood products.

Patients with IgA deficiency, anti-IgA antibodies, and a history of ATRs need to receive blood components and immunoglobulin products that lack IgA. Plasma components must be obtained from IgA-deficient donors. Cellular components must be washed to remove as much plasma as is feasible.

Anaphylactic reactions have been reported in individuals who are deficient or lack allotypic forms of other normal serum proteins, such as haptoglobin, C3 and C4, and who form IgG or IgE antibodies against these proteins.

Treatment

Mild Allergic Transfusion Reactions: Treatment of mild ATRs consists of temporarily stopping the transfusion and administration of antihistamines. If the symptoms resolve promptly, then the transfusion can be continued. Mild ATRs are the only transfusion reaction for which it is routinely permissible to restart a transfusion after a reaction manifests. If the symptoms recur or persist, then the transfusion should be discontinued.

Severe Allergic Transfusion Reactions: Epinephrine is the most effective initial medication for treating anaphylaxis, and the initial dose(s) of epinephrine (0.01 mg/kg; maximum dose 0.5 mg) may be given intramuscularly in the lateral thigh as frequently as every 5 minutes. In addition, antihistamines, H2-receptor antagonists, and intravenous glucocorticoids can be used. Glucocorticoids are most effective at treating late phase allergic responses, which can manifest up to 24 hours later. Observation of the patient for up to 24 hours may be required for the resolution of symptoms.

Prevention: The majority of clinical trials demonstrate that the routine use of premedication does not decrease the incidence of ATRs. Several studies indicate that plasma reduction, via washing or platelet additive solution, reduces the incidence of ATRs. Evidence does not support a relationship between infusion rate and incidence or severity of ATRs.

Prophylactic Premedication in Patients Without Prior History: Although premedication with diphenhydramine to prevent ATRs is a common practice, reported in up to 80% of transfusions, there are no data to demonstrate that routine premedication before transfusions decreases the incidence of ATRs. Premedication without a reasonable expectation of benefit simply exposes a patient to undue risk.

For diphenhydramine, which has anticholinergic effects and crosses the blood–brain barrier, the side effects include dry mouth, urinary retention, drowsiness, decreased alertness (impairing driving abilities), delirium (especially in elderly hospitalized patients), and impaired cognitive performance, memory, and attention. It can alternatively cause restlessness and nervousness. Nonsedating antihistamines, e.g., cetirizine or loratadine, are usually preferable to diphenhydramine.

Prophylactic Premedication in Patients With a Prior History of Allergic Transfusion Reactions: In patients who have had previous ATRs, it may be beneficial to premedicate with antihistamines, H2-receptor antagonist, or intravenous steroids, depending on the severity of the previous reaction. These medications most likely do not prevent reactions; rather, they mitigate symptoms in patients with a higher likelihood to experience reactions. Thus, in patients with moderate ATRs, premedication may be more warranted as a method to decrease the risks of transfusion. In patients with moderate to severe reactions, washing or plasma-reducing RBC or platelet products may be warranted to. For patients with recurrent ATRs to platelets, concentrating platelets reduces the incidence by approximately 75%, and washing reduces the incidence by approximately 95%.

Further Reading

Hirayama, F., Yasui, K., Matsuyama, N., & Okamura-Shiki, I. (2018). Possible utility of the basophil activation test for the analysis of mechanisms involved in allergic transfusion reactions. *Trans Med Rev, 32*, 43–51.

Marti-Carvajal, A. J., Sola, I., & Gonzalez, L. E. (2010). Pharmacological interventions for the prevention of allergic and febrile non-haemolytic transfusion reactions. *Cochrane Database Sys Rev, 16*, CD007539.

Savage, W. J., Hamilton, R. G., Tobian, A. A. R., et al. (2014). Defining risk factors and presentations of allergic reactions to platelet transfusion. *J Allergy Clin Immunol, 133*, 1772–1775.

Savage, W. J., Tobian, A. A. R., Savage, J. H., et al. (2015). Transfusion and component characteristics are not associated with allergic transfusion reactions to apheresis platelets. *Transfusion, 55*, 296–300.

Tobian, A. A., Savage, W. J., Tisch, D. J., et al. (2011). Prevention of allergic transfusion reactions to platelets and red blood cells through plasma reduction. *Transfusion, 51*, 1676–1683.

CHAPTER 63

Acute Hemolytic Transfusion Reactions

Ruchika Goel, MD, MPH and Aaron A.R. Tobian, MD, PhD

Acute hemolytic transfusion reactions (AHTRs) may occur when either incompatible RBCs or large amounts of incompatible plasma are transfused, which can lead to antibody–antigen binding in the recipient. AHTRs can lead to minimal hemolysis with no clinical sequelae, or can result in brisk hemolysis, induction of disseminated intravascular coagulation (DIC), hypotension, and shock, followed by renal failure and/or death.

Incidence: Studies from the REDS III database showed that in 2013–14, AHTRs comprised 0.41% of all reported adverse effects to transfusion. Per the National Blood Collection and Utilization Surveys, 2013 and 2015, AHTRs were estimated to have occurred less than 100 times annually at a national level. Between fiscal year (FY) 2002 and 2015, the number of AHTR resulting in fatalities has decreased. In FY2015, there were two reported ABO hemolytic transfusion fatalities (5% of confirmed transfusion-associated fatalities), and four non-ABO hemolytic transfusion fatalities (11% of confirmed transfusion-associated fatalities).

Mistransfusion may range from being asymptomatic in approximately 50% of recipients to death in 2-7% of ABO-incompatible recipients. Volume of incompatible blood transfused and antibody titer likely play a role in recipient signs/symptoms. While, transfusion of as little as 30 mL of incompatible RBCs has been reported to result in fatality, in other cases, large volumes have been transfused with no or minor symptoms.

Transfusion of ABO-Incompatible RBCs: Transfusion of ABO-incompatible RBCs is most associated with mistransfusion event, which is the "transfusion of a unit of blood that has been 'incorrectly typed, labeled crossmatched, or issued,' or the transfusion of 'correctly typed, labeled, crossmatched, and issued' unit of blood that is administered to the incorrect patient." Thus, transfusion service errors (typing, labeling, crossmatching, and issuing), and patient identification errors (patient sampling for blood typing, or patient and unit identification at the time of transfusion [at the bedside]), can lead to mistransfusion.

Transfusion errors are a sentinel event (an unexpected occurrence resulting in death or serious adverse event or risk thereof) according to the Joint Commission that must be reported. A sentinel event requires root cause analysis to identify what happened, why it happened, and what factors (human, equipment, and/or environmental) contributed to its occurrence. In addition, an action plan must be developed to reduce risk of event reoccurring.

Transfusion of Non-ABO-Incompatible RBCs to Alloimmunized Recipients: Recipients may develop non-ABO antibodies when exposed to RBC antigens (also known as alloimmunization). These antibodies may make future RBC transfusions incompatible. In extremely rare circumstances, non-ABO-incompatible blood

Transfusion Medicine and Hemostasis. https://doi.org/10.1016/B978-0-12-813726-0.00063-5

is administered knowingly. These include emergent clinical situations where the best available RBC product is incompatible with recipient's plasma, and the product is either not available or there is not enough time to identify antibody specificity and/or obtain antigen negative product. In other circumstances, AHTR due to non-ABO-incompatible blood occurs because the patient's antibody was not identified before transfusion. This may be due to false-negative antibody identification test (indirect antiglobulin test [IAT]), lack of antibody presence detection by IAT, or not enough time to perform IAT.

Transfusion of Significant Amounts of Incompatible Plasma: Transfusion of incompatible plasma most often occurs with administration of out-of-group platelets, most commonly group O platelet to group A recipient, which may result in AHTR. Passive ABO antibodies can also occur in administration of non-ABO identical whole blood or plasma product, which often result in a positive DAT and rarely in a severe AHTR. Non–group O platelet units are rarely associated with AHTR. Increased use of single-donor apheresis platelets in the United States, coupled with low availability of type-specific apheresis platelets, has likely led many adult hospital transfusion services to issue ABO-mismatched platelets that contain large plasma volumes and potentially high titers of isohemagglutinins. In the United Kingdom, all blood donations undergo anti-A IgM testing and those with titers >100 are labeled accordingly, and hospital transfusion services are aware to transfuse to only group O recipients. Most US hospital transfusion services do not routinely measure isohemagglutinin titers (IgM or IgG) before transfusing ABO-mismatched platelets or plasma, although this practice is becoming more standard and blood centers are performing titers and labeling products as high titer accordingly. The majority use "critical titer" of 100 for anti-A/B at room temperature. Notably, direct correlation between hemoglobin loss and rising anti-A/B titers was demonstrated: larger decrease in hemoglobin value was observed as titers increased from undetectable to low (32 or less) to high (512 or more). Thus, high-titer products should be limited to ABO-compatible recipients.

Clinical Manifestations, Diagnosis, and Evaluation: AHTR occurs within 24 hours of transfusion with any of the signs/symptoms and supporting laboratory evidence as listed in Table 63.1. The clinical and laboratory picture with ABO or other RBC antigen compatibility or evidence of mistransfusion confirms the diagnosis. Notably, fever and chills/rigors are usually an early manifestation of AHTR and thus it is important that transfusion recipients are monitored during initial minutes of transfusion, transfusion is discontinued immediately if symptoms arise, and transfusion reaction evaluation should be initiated.

Transfusion Service Evaluation: Transfusion service should be immediately informed when AHTR occurs or is suspected. The blood product with all attached tubing should be returned to transfusion service. First, clerical check of patient and product is performed as to determine if incorrect blood type, product type, or patient was transfused. Blood samples are drawn from recipient to visually inspect specimen for hemolysis, perform DAT, and confirm ABO typing. In addition, sample from the product is also retyped. Patient's post-transfusion DAT may be positive for IgG and C3 (when IgM binds to RBCs it fixes complement, which remains on RBC and is detected in DAT). If DAT is positive for IgG, eluate may reveal IgG anti-A or anti-B. Repeat ABO testing on pretransfusion specimen should also be performed. Clinical laboratory evaluation for hemolysis and DIC is listed in Table 63.1.

TABLE 63.1 Signs/Symptoms and Laboratory Manifestations of Acute Hemolytic Transfusion Reactions (AHTRs)

Signs/Symptoms of AHTR	Laboratory Manifestations of AHTR
Chills/rigors	Positive direct agglutinin test and eluate
Back/flank pain	Decreased fibrinogen
Hematuria	Decreased haptoglobin
Oliguria/anuria	Elevated bilirubin
Disseminated intravascular coagulopathy	Elevated LDH
Fever	Hemoglobinemia
Hypotension	Hemoglobinuria
Renal failure	Plasma discoloration consistent with hemolysis
Pain and/or oozing at IV site	Spherocytes on blood film

Other Causes of Acute Hemolysis: Nonimmune hemolysis can occur from transfusion of improperly stored RBC product resulting in thermal injury (exposed directly to ice or placed on heater), mechanical injury (constricted access lines or pressurized infusions), osmolar injury (hypotonic solutions or medications infused with product), improper deglycerolization, and bacterial contamination. Patients with underlying hemolytic anemia, such as those with sickle cell disease, thrombotic thrombocytopenic purpura, or implanted circulatory devices, may have acute hemolysis, which is unrelated to transfusion.

Pathophysiology

Intravascular Hemolysis: Hemolysis likelihood depends on class and subclass of the antibody, thermal range over which this antibody can fix complement, number of RBCs transfused (or volume of incompatible plasma transfused), density of corresponding antigen on RBCs, efficiency of complement system, and complement control proteins in circulation.

Most AHTRs are due to transfusion of ABO-incompatible RBCs into patients with preexisting, naturally occurring anti-A and/or anti-B antibodies. These antibodies are typically mixture of IgM and IgG. IgM can efficiently bind C1q and initiate the complement cascade, leading to the formation of the membrane attack complex and erythrocyte lysis. Some IgG antibodies can fix complement efficiently as well and are usually of IgG1 and IgG3 subclasses, and these can cause mild to fatal AHTRs. Thus, antibodies to Kell, Rh, Duffy, and Kidd systems can cause AHTRs, and these can be fatal as well.

Clinical Sequelae: Intravascular hemolysis results in release of cell free hemoglobin, RBC stroma, and nonstroma proteins. Free hemoglobin binds nitric oxide (NO) at rate 1000 times that of RBC. Hemoglobin scavenging leads to decreased bioavailability of NO and thus vasoconstriction and alterations in capillary response to hypoxia. RBC stroma, which is the cytoskeletal framework supporting hemoglobin, can also contribute to DIC pathogenesis via activation of platelets and coagulation cascade. RBC stroma has also been shown to increase blood pressure and is toxic to the glomerulus and renal tubule and thus can cause acute renal failure. Ultimately, increased cytokines and hypotension stimulate a compensatory sympathetic nervous system response contributing to renal, splanchnic, and cutaneous vasoconstriction that, in combination with pathophysiology described above, leads to shock and circulatory collapse.

Management: Once the possibility of AHTR is recognized and transfusion is discontinued, supportive care is immediately instituted to manage hypotension and maintain adequate renal perfusion. These measures include intravenous crystalloid (10–20 mL/kg of normal saline), diuretic to maintain urine output between 30 and 100 mL/hour or greater (>1 mL/kg/hour), administration of low-dose dopamine (1–5 μg/kg/minute) for hypotension as needed, and for active bleeding and DIC, transfusion of plasma, platelets, and/or cryoprecipitate as needed.

Prevention: Strict adherence to pretransfusion bedside patient identification procedures to assure proper specimen collection will aid in preventing clerical errors and misidentification. AABB *Standards* require that blood container label and records shall be examined to detect errors in identifying patient, blood, or blood component. In addition, labeling process should include second check to ensure accuracy of affixed labels, including correct donation identification number, ABO/Rh, expiration date (as appropriate), and product name and code. Two determinations of recipient's ABO group are made—one on a current sample and second by one of following methods: (1) testing a second current sample, (2) comparison with previous records, or (3) retesting same sample if patient identification was verified using an electronic identification system or another process validated to reduce misidentification risk.

Other prevention strategies include technological innovations aimed at reducing human error, including barcoding blood components, patient identification systems, and barrier systems. In addition, hemovigilance and/or quality assurance programs to monitor transfusion practices, investigate and analyze root cause of adverse events, and reinforce physician and nurse transfusion practices improve transfusion safety. Prevention is through administering ABO-compatible products, and if incompatible products are given to administer low-titer products and potentially those with less plasma (e.g., platelets stored in platelet-additive solutions). While there is no universal policy for preventing hemolytic transfusion reactions, AABB *Standards* mandate that each transfusion service establish "a policy concerning transfusion of components containing significant amounts of incompatible ABO antibodies or unexpected red cell antibodies."

Further Reading

Fatalities Reported to FDA Following Blood Collection and Transfusion Annual Summary for Fiscal Year 2015. https://www.fda.gov/BiologicsBloodVaccines/SafetyAvailability/ReportaProblem/TransfusionDonationFatalities/default.htm (accessed January 2018).

Karafin, M., et al. (2017). Demographic and epidemiologic characterization of transfusion recipients from four US regions: Evidence from the REDS-III recipient database. *Transfusion*, *57*, 2903–2913.

Sapiano, M., et al. (2017). Supplemental findings from the national blood collection and utilization Surveys, 2013 and 2015. *Transfusion*, *57*, 1599–1624.

CHAPTER 64

Delayed Hemolytic Transfusion Reactions

Patricia E. Zerra, MD and Cassandra D. Josephson, MD

Delayed hemolytic transfusion reactions (DHTRs) typically occur 3–10 days after red blood cell (RBC) transfusion that appear to be serologically compatible. These reactions occur in patients who have been alloimmunized to minor RBC antigens during previous transfusions and/or pregnancies; pretransfusion testing fails to detect these alloantibodies due to their low titer. After reexposure to antigen-positive RBCs, an anamnestic response occurs, with rapid rise in antibody titer. Decreased survival of transfused RBCs may result, primarily due to extravascular hemolysis. In the majority of cases, however, anamnestic antibody production does not cause detectable hemolysis. The term delayed *serologic* transfusion reaction (DSTR) defines reactions in which anamnestic antibody is identified serologically, with absence of clinical evidence of accelerated RBC destruction. Antigens implicated most often in DHTRs and DSTRs are in the Kidd, Duffy, Kell, and MNS systems, in order of decreasing frequency.

Incidence: In combination, DHTRs and DSTRs occur in approximately 1 in 1500 transfusions, with DSTRs occurring up to four times more often than DHTRs. Data from Mayo Clinic calculated an incidence rate of 1:6944 patients transfused for DHTR and 1:3146 patients transfused for DSTR using gel technique and 1:1200 patients transfused for DHTR and 1:611 patients transfused for DSTR using PEG technique. Antibodies to E, Fy(a), and Jk(a) were most frequently identified.

Clinical Manifestations: DHTRs are characterized clinically by an unexpected drop in hemoglobin or less than expected posttransfusion increment in hemoglobin (Table 64.1). This diagnosis should be considered days to weeks after transfusion, although hemolysis may be more prolonged. Symptoms of extravascular hemolysis include fever, chills, jaundice, malaise, back pain and, uncommonly, renal failure. Symptoms may mimic vasoocclusive pain crisis in patients with sickle cell disease (SCD), and thus the underlying diagnosis of DHTR is often missed.

Pathophysiology: A minority of patients exposed to foreign RBCs, during either previous transfusions or pregnancies, may become sensitized to minor RBC antigens. The primary immune response, occurring over weeks to months, is typically not clinically significant. Alloantibody titers to these RBC antigens may decrease over time in the patient's plasma without continued exposure to inciting antigen(s), resulting in negative pretransfusion antibody screening. However, transfusion of offending antigen-positive RBCs may lead to an anamnestic response, resulting in the rapid production of IgG antibody and primarily extravascular hemolysis. Thermal range, antibody specificity, and IgG subclass are the three factors that influence DHTR severity.

Transfusion Medicine and Hemostasis. https://doi.org/10.1016/B978-0-12-813726-0.00064-7

TABLE 64.1 Timeline of Delayed Hemolytic Transfusion Reaction

Time (Days)	Event	Explanation
0	Pretransfusion antibody screening negative	Antibody titer below detectable levels
1	RBC transfusion	
3–10	Clinical signs of hemolysis may appear	Accelerated destruction of transfused donor RBCs
10–21	Posttransfusion sample: positive direct antiglobulin (DAT) and positive antibody screen due to newly detected antibody	Antibody titer increases
>21	DAT may become negative	Antibody-sensitized donor RBCs removed from circulation
21–300	DAT may persist as positive; eluates may reveal alloantibody specificity or panagglutination	Alloantibody binding nonspecifically to autologous RBC, or development of a warm autoantibody

Modified from Hillyer, C. D., Silberstein, L. E., Ness, P. M., Anderson, K. C., Roback, J. D. (Eds.), (2007). *Blood banking and transfusion medicine: Basic principles & practice,* (2nd ed.). Philadelphia: Elsevier.

Diagnosis: To investigate suspected DHTR, posttransfusion specimens should be evaluated for antibody identification and direct antiglobulin (DAT) studies. Positive antibody screen with newly identified alloantibody and/or positive DAT will help confirm the reaction; DAT may show mixed field reaction, with transfused cells but not autologous cells being coated with antibody. If DAT is positive for IgG, then eluate testing should be pursued to identify RBC-coating antibody specificity. Supplemental serologic testing may be necessary to confirm diagnosis including repeat antibody screen and DAT on pretransfusion specimen (retained in transfusion service), to ensure previous results were not erroneous. Establishing patient's phenotype on pretransfusion specimen may also be helpful. Additionally, RBCs from transfused product (using stored "segment") should be phenotyped for antigen corresponding to newly identified antibody. Failure to identify new antibody or negative DAT does not rule out DHTR, however, as hemolysis can occur in the absence of detectable antibody in an antibody-negative DHTR.

Other laboratory findings suggestive of DHTR include reticulocytosis, unconjugated hyperbilirubinemia, and urine urobilinogen. In patients with SCD, unexpected decrease in the relative proportion of transfused (Hgb A) RBCs provides additional support of potential DHTR. Although DHTRs primarily involve extravascular hemolysis, some antibodies may fix complement and cause intravascular hemolysis. In such instances, hemoglobinuria may also be present, along with elevated serum lactate dehydrogenase and decreased haptoglobin.

DSTRs are often first identified in the blood bank, when posttransfusion antibody screen is positive. Close communication with clinical team is indicated to confirm reaction is solely serological and not hemolytic.

Differential Diagnosis: Alloantibody formation may lead to development of RBC autoantibodies, which are often transient, but may result in hemolytic anemia.

DAT positivity beyond 3–4 months after transfusion (and thus beyond the lifespan of the transfused RBCs) supports the presence of autoantibodies.

In patients with SCD, other causes of increased RBC destruction must also be included in the differential diagnosis. These causes include fever, underlying disease, or hypersplenism. The serologic evaluation, including antibody screen and DAT, would be expected to be negative.

Management: Specific treatment is usually not necessary, and RBC transfusion should be avoided except in cases of severe symptomatic anemia. In those cases, additional RBC transfusions with RBC products that lack antigen corresponding to newly developed alloantibody may be necessary. However, if transfusion is ordered before specific identification of newly formed antibody, risk of hemolysis must be weighed against transfusion benefits. If decision is made to transfuse, sufficient hydration and close monitoring should be performed in light of risk of ongoing hemolysis.

Additional treatments remain controversial due to lack of randomized controlled studies and incomplete understanding of exact mechanisms underlying DHTRs. IVIG, recombinant erythropoietin, and corticosteroids have been used off-label as successful treatment strategies. As complement activation may play a role in DHTRs, eculizumab, an anti-C5 monocolonal antibody, has been used successfully as management for life-threatening episodes of DHTRs in patients with SCD.

Prevention: To prevent DHTRs and DSTRs, AABB *Standards* mandates permanent preservation of all records of potentially clinically significant antibodies and review of previous records before RBCs being issued for transfusion. Once clinically significant antibody has been identified, the patient should receive offending antigen-negative units for all future RBC transfusions. The 3-day interval requirement for RBC type and screening of recently transfused or pregnant patients is based on the finding that anamnestic antibody responses may occur within 3 days of transfusion.

As SCD-related complications such as acute chest syndrome and vasoocclusive crisis at time of transfusion have been associated with alloimmunization, judicious use of RBC transfusions during these proinflammatory states is important for prevention of alloimmunization and risk for subsequent DHTRs.

Some patients may require additional antigen matching to prevent further alloantibody formation once RBC alloimmunization has occurred. By prospectively avoiding incompatibility to C, E, and K antigens, alloimmunization rate among chronically transfused SCD patients was reduced from 3% to 0.5% per transfused RBC unit in the Stroke Prevention Trial in Sickle Cell Anemia, and hemolytic transfusion reactions were reduced by 90%.

Administration of rituximab before subsequent RBC transfusions can be considered in patients with history of alloimmunization and severe DHTR for prevention of further antibody development and to minimize the risk of DHTR recurrence.

Hyperhemolytic Transfusion Reaction: Hyperhemolytic transfusion reactions (HHTRs) have been reported in patients with SCD, thalassemia, and other diseases. In these serious reactions, both donor and recipient RBCs are destroyed, leading to more severe anemia than was present before transfusion. In addition, such patients may have reticulocytopenia. Some patients may have alloantibodies or autoantibodies, but others have no detectable antibodies to RBC antigens. Similar reactions may occur after subsequent transfusion even if the RBCs are extensively phenotypically matched. Possible mechanisms include bystander hemolysis, erythropoiesis suppression, and RBC destruction secondary to contact lysis via activated macrophages. Bystander hemolysis is an immune-mediated hemolysis, where the RBC does not carry the antigen for which the antibody is directed. Treatments for patients with HHTRs have included erythropoietin, IVIG, steroids, and plasma exchange.

Further Reading

Dumas, G., Habibi, A., Onimus, T., Merle, J., Razazi, K., Dessap, A. M., et al. (2016). Eculizumab salvage therapy for delayed hemolysis transfusion reaction in sickle cell disease patients. *Blood, 127*, 1062–1064.

Fasano, R. M., Booth, G. S., Miles, M., Du, L., Koyama, T., Meier, E. R., et al. (2014). Red blood cell alloimmunization is influenced by recipient inflammatory state at time of transfusion in patients with sickle cell disease. *Br J Haematol, 168*, 291–300.

Garratty, G. (2004). Autoantibodies induced by blood transfusion. *Transfusion, 44*, 5–9.

Habibi, A., Mekontso-Dessap, A., Guillaud, C., Michel, M., Razazi, K., Khellaf, M., et al. (2016). Delayed hemolytic transfusion reaction in adult sickle-cell disease: Presentations, outcomes, and treatments of 99 referral center episodes. *Am J Hematol, 91*, 989–994.

Vamvakas, E. C., Pineda, A. A., Reisner, R., Santrach, P. J., & Moore, S. B. (1995). The differentiation of delayed hemolytic and delayed serologic transfusion reactions: Incidence and predictors of hemolysis. *Transfusion, 35*, 16–32.

Vichinsky, E. P., Luban, N. L., Wright, E., & Stroke Prevention Trial in Sickle Cell Anemia (2001). Prospective RBC phenotype matching in a stroke-prevention trial in sickle cell anemia: A multicenter transfusion trial. *Transfusion, 41*, 1086–1092.

Win, N., Doughty, H., Telfer, P., Wild, B. J., & Pearson, T. C. (2001). Hyperhemolytic transfusion reaction in sickle cell disease. *Transfusion, 41*, 323–328.

Winters, J. L., Richa, E. M., & Bryant, S. C. (2010). Polyethylene glycol antiglobulin tube versus gel microcolumn: Influence on the incidence of delayed hemolytic transfusion reactions and delayed serologic transfusion reactions. *Transfusion, 50*, 1444–1452.

CHAPTER 65

Transfusion-Associated Circulatory Overload

Amit Gokhale, MD and Jeanne E. Hendrickson, MD

Incidence: TACO is estimated to occur in up to 1% of transfusions, with higher rates reported in studies using active surveillance methodologies. For several reasons, TACO is likely one of the most underreported transfusion complications to hospital transfusion. A retrospective Medicare database review of over 2 million transfusions showed that patient characteristics such as age (greater than 85 years), history of heart failure, female sex, white race, and a history of chronic pulmonary disease inferred a greater risk of developing TACO. Epidemiologic data suggest that additional risk factors for TACO include being at the extremes of age, having a positive fluid balance, undergoing orthopedic surgery, having an acute myocardial infarction, and having renal failure. Higher rates of infusion, larger transfusion volumes, and transfusion of plasma are additional risk factors for the development of TACO.

Clinical Manifestations: Symptoms of TACO include dyspnea, orthopnea, cough, chest tightness, cyanosis, hypertension, and congestive heart failure. Tachycardia and a widened pulse pressure may also occur. TACO generally occurs toward the end of a transfusion but may occur up to 6 hours afterward.

Pathophysiology: TACO is caused by the inability of the cardiopulmonary system in the recipient to tolerate the volume or rate of transfusion. Thus, TACO represents cardiogenic pulmonary edema.

Diagnosis: TACO is diagnosed using a combination of clinical, radiographic, electrocardiographic, laboratory, hemodynamic, and echocardiographic findings. Physical examination of patients with TACO may reveal lung crackles and rales, elevated jugular venous pressure, and possibly an S3 gallop from volume overload. Chest X-ray may show alveolar and interstitial edema, Kerley B-lines, pleural effusions, or cardiomegaly. There are currently no confirmatory laboratory tests available, although brain natriuretic peptide (BNP) and N-terminal pro-BNP (NT-proBNP) may be elevated in TACO. Although nonspecific, these laboratory results may be helpful when combined with the patient's overall clinical symptoms.

TACO is defined by the National Healthcare Safety Network Manual (NHSN) Biovigilance Component as the new onset or exacerbation of three or more of the following, within 6 hours of the end of transfusion:

- Acute respiratory distress dyspnea, orthopnea, cough
- Evidence of positive fluid balance
- Elevated BNP

Transfusion Medicine and Hemostasis. https://doi.org/10.1016/B978-0-12-813726-0.00065-9

- Radiographic evidence of pulmonary edema
- Evidence of left heart failure
- Elevated central venous pressure

The NHSN also categorizes adverse events, including TACO, by severity (nonsevere, severe, life-threatening, or deadly) and by imputability (definite, probable, or possible). TACO cases with definite imputability are described as those in which no other explanations for circulatory overload are possible. Probable cases are described as those in which the transfusion was likely a contributing factor to the circulatory overload and *either* the patient received other fluids *or* has a history of cardiac insufficiency that could explain the circulatory overload. Possible cases are described as those in which the patient's history of preexisting cardiac insufficiency likely explains the circulatory overload.

Differential Diagnosis: The diagnoses of TACO and TRALI should be considered in any patient who develops respiratory difficulty during or soon after a transfusion (see Chapter 66 for further details on TRALI; Table 65.1 compares TACO and TRALI). Patients with TRALI typically do not have evidence of volume overload, nor do they respond to diuretics. Bacterial contamination, anaphylaxis, acute intravascular hemolysis, and transfusion-associated dyspnea (TAD) should also be in the differential diagnosis of respiratory distress that occurs after transfusion. TAD is characterized by acute respiratory distress occurring within 24 hours of transfusion, not meeting the criteria for TACO, TRALI, or allergic reactions.

Management: It is critical that the transfusion be stopped if the patient develops respiratory distress. A transfusion reaction should be reported to the transfusion service, where an investigation will be undertaken. Diuretic therapy, including intravenous furosemide, is the main symptomatic treatment. Other supportive care measures include supplemental oxygen, as necessary, and sitting the patient upright. Pharmacologic therapies to minimize left ventricular afterload may also be beneficial.

Prevention: Avoiding unnecessary transfusions is the single most prudent strategy for prevention. If transfusion is necessary, selection of the smallest number of products needed to obtain the transfusion goal is recommended. For example, the provider could order 1 unit instead of 2 units, and then reassess the patient; this example of patient blood management is part of AABB's Choosing Wisely campaign. Attention should also be paid to the infusion rate. Optimal infusion rates for blood products have not been studied extensively, but it is likely that slower infusion rates (e.g., 1 unit infused over 3–4 hours) are safer than faster infusion rates in patients at risk of circulatory overload. Patients at very high risk of TACO may be transfused with a split unit, with each aliquot infused over 3–4 hours. Preemptive diuretic therapy may also help to prevent TACO in at risk patients.

Summary: TACO, one of the most common non-infectious serious hazards of transfusion, causes significant morbidity and mortality. An increased awareness of patient and transfusion-specific risk factors, in combination with judicious use of transfused blood products, will improve transfusion safety.

TABLE 65.1 Comparison Between TACO and TRALI

	TACO	TRALI
Clinical manifestations	Dyspnea, respiratory distress, hypoxia, orthopnea, hypertension, jugular venous distention, congestive heart failure	Dyspnea, respiratory distress, hypoxia, pulmonary edema, fever, tachycardia, hypotension
Pathophysiology	Volume overload	Primarily antibody mediated (anti-human leukocyte antigen (HLA) or anti-human neutrophil antigen (HNA))
Chest X-ray and laboratory findings	Chest X-ray shows alveolar and interstitial edema, Kerley B-lines, pleural effusions, or cardiomegaly Elevated brain natriuretic peptide may be present.	Chest X-ray shows bilateral infiltrates in interstitial and alveolar spaces, lack of cardiomegaly. Anti-HLA or HNA antibodies in the blood donor (and cognate antigen(s) in the transfusion recipient)
Treatment	Stop transfusion. Supportive care, including diuretics. Sit patient upright.	Stop transfusion. Supportive care, including oxygen and possibly ventilatory support. Diuretics typically are not effective.
Reporting	Report to transfusion service for transfusion reaction evaluation.	Report to transfusion service for transfusion reaction evaluation. Transfusion service will report reaction to blood center for further evaluation (including donor HLA/HNA antibody screening and recipient HLA typing)
Future transfusion considerations	Judicious use of blood products. Transfuse future blood products more slowly, and in high-risk cases consider splitting units with each split to be transfused over 3–4 hours. Consider preemptive diuretic therapy.	Judicious use of blood products. Avoid further transfusions from the implicated donor.

Further Reading

Alam, A., Lin, Y., Lima, A., Hansen, M., & Callum, J. L. (2013). The prevention of transfusion-associated circulatory overload. *Transfus Med Rev*, *27*(2), 105–112.

Andrzejewski, C., Jr., Casey, M. A., & Popovsky, M. A. (2013). How we view and approach transfusion-associated circulatory overload: Pathogenesis, diagnosis, management, mitigation, and prevention. *Transfusion*, *53*(12), 3037–3047.

Clifford, L., Jia, Q., Subramanian, A., Yadav, H., Schroeder, D. R., & Kor, D. J. (2017). Risk factors and clinical outcomes associated with perioperative transfusion-associated circulatory overload. *Anesthesiology*, *126*(3), 409–418.

Centers for Disease Control and Prevention. *National hemovigilance safety network biovigilance component, hemovigilance module surveillance protocol* [9/1/17]. Available from: www.cdc.gov/nhsn.

Raval, J. S., Mazepa, M. A., Russell, S. L., Immel, C. C., Whinna, H. C., & Park, Y. A. (2015). Passive reporting greatly underestimates the rate of transfusion-associated circulatory overload after platelet transfusion. *Vox Sanguinis*, *108*(4), 387–392.

Roubinian, N. H., Looney, M. R., Keating, S., Kor, D. J., Lowell, C. A., Gajic, O., et al. (2017). Differentiating pulmonary transfusion reactions using recipient and transfusion factors. *Transfusion*, *57*(7), 1684–1690.

CHAPTER 66

Transfusion-Related Acute Lung Injury

Amit Gokhale, MD and Jeanne E. Hendrickson, MD

Incidence: The incidence of transfusion-related acute lung injury (TRALI) and deaths due to TRALI has decreased substantially over the past decade, as TRALI mitigation strategies have been implemented. In 2006, 35 TRALI-related fatalities were reported to the FDA. These accounted for more than 50% of all transfusion-related fatalities, with 60% of the cases being attributed to plasma products. In contrast, only 12 TRALI fatalities that met definite or probable criteria were reported to the FDA in 2015, comprising 29% of all transfusion-related fatalities. The current incidence of nonfatal TRALI cases is estimated by National Blood Collection and Utilization Surveys to be 1 per 60,000 blood components transfused.

Clinical Manifestations: Typically occurring during or soon after an allogeneic transfusion, signs and symptoms of TRALI include the acute onset of respiratory distress with hypoxemia, dyspnea, and tachypnea; these symptoms may be accompanied by fever, tachycardia, and/or hypotension. Physical examination may reveal rales and diminished breath sounds, without other evidence of fluid overload.

Pathophysiology: The end result of TRALI is an increased permeability of the pulmonary microcirculation, such that high protein fluid enters the interstitium and alveolar air spaces. Both immunologic and non-immunologic mechanisms likely play a role, with two "hits" being necessary to induce TRALI. The first "hit" is related to the patient's underlying illness status, and the second "hit" involves neutrophil activation induced by transfusion. The majority of cases are immunologic. Some recipient risk factors include recent surgery, active infection, chronic alcohol abuse, shock, higher peak airway pressure while being ventilated, current smoking, and a positive fluid balance.

The primary proposed *immunologic* mechanism is that human leukocyte antigen (HLA) or human neutrophil antigen (HNA) antibodies in donor plasma react with antigens on recipient leukocytes, which may need to be primed by a coexisting process. This ultimately leads to lung injury, with fluid accumulation in the alveoli. It is also possible that immune complexes of antibody and soluble HNA or HLA may be recognized by neutrophil Fc receptors, that antibodies may induce monocytes or platelets to secrete inflammatory mediators, which then activate neutrophils, or that antibodies may be directed against HLAs on recipient pulmonary interstitial cells.

In some TRALI cases, donor HLA or HNA alloantibodies are not detected; this may be secondary to assay limitations or to a *non-immunologic* mechanism of TRALI. Non-immunologic mechanisms may include co-transfusion of elements that accumulate during blood storage, including biologically active lipids and/or cytokines.

Transfusion Medicine and Hemostasis. https://doi.org/10.1016/B978-0-12-813726-0.00066-0

Diagnosis: A high index of suspicion is necessary when considering TRALI, with the diagnosis being based on clinical and radiographic findings in a patient with no prior lung injury. The acute lung injury must occur within 6 hours of transfusion. Chest X-ray may show bilateral infiltrates involving alveolar and interstitial spaces, without cardiomegaly. If intubation is required, frothy pink secretions may be seen. There are no classic laboratory findings associated with TRALI, but transient leukopenia has been reported. At least one study suggests that a significantly elevated brain natriuretic peptide may help to rule out TRALI as a cause of respiratory distress after transfusion.

The National Healthcare Safety Network Hemovigilance Surveillance Protocol defines TRALI as

- no evidence of acute lung injury before transfusion,
- acute lung injury onset during or within 6 hours of cessation of transfusion,
- hypoxemia defined as either PaO_2/FiO_2 less than or equal to 300 mm Hg or oxygen, saturation less than 90% on room air, and
- radiographic evidence of bilateral infiltrates, and no evidence of left atrial hypertension (i.e., circulatory overload).

Imputability criteria for TRALI range from definite (with no alternative risk factors for acute lung injury present) to possible (with evidence for other causes of acute lung injury, either direct or indirect, being present). Possible TRALI tends to be less likely due to an *immunologic* mechanism and more likely due to a recipient-specific etiology.

Differential Diagnosis: Respiratory compromise may occur around the time of transfusion in patients with transfusion-associated circulatory overload (TACO), transfusion-associated dyspnea (TAD), or acute respiratory distress syndrome (ARDS). Patient history and fluid overload status, among other things, can be used to help differentiate TRALI from TACO, TAD, or ARDS. TACO is more likely to be associated with hypertension and a response to diuretics, unlike TRALI (see Chapter 65).

Management: Transfusion should be stopped if TRALI is suspected, and a transfusion reaction reported to the transfusion service. The management of TRALI is primarily supportive, with severe cases requiring mechanical ventilation. Although TRALI may be fatal, most patients improve clinically within 48–96 hours.

Donor/Recipient Investigation: After a potential case of TRALI, the transfusion service should notify the blood collection facility. This facility then investigates the pregnancy and/or transfusion history of the implicated donor(s) and determines if the donor has antibodies to HLAs or HNAs. In cases where donor antibodies are discovered, testing of the patient is typically performed to see whether an antibody/antigen interaction could possibly have occurred. AABB *Standards for Blood Banks and Transfusion Services* requires that donors implicated in TRALI are evaluated regarding their continued eligibility to donate.

Prevention: From a recipient perspective, judicious transfusion of blood products is recommended as a general strategy to prevent transfusion-associated adverse events;

particular attention should be paid to patients with risk factors for TRALI. From a donor perspective, much progress has been made in the past 10–15 years in terms of identifying and implementing strategies to prevent plasma-containing products at highest risk of causing TRALI in recipients from being transfused. Pregnancy history is one risk factor for donor HLA alloimmunization, with the Leukocyte Antibody Prevalence Study reporting pregnancy/HLA alloimmunization rates as: 1 pregnancy, 11.2%, 2, 22.3%, 3, 27.5%, and 4 or more, 32.2%. The 30th edition of the AABB *Standards for Blood Banks and Transfusion Services* states that "Plasma, Apheresis Platelets, and Whole Blood for allogeneic transfusion shall be from males, females who have not been pregnant, or females who have been tested since their most recent pregnancy and results interpreted as negative for HLA antibodies." With the implementation of strategies such as these, the risk of TRALI has decreased significantly in the United States, Canada, and other countries.

The use of pooled solvent detergent plasma (S/D plasma) instead of other sources of frozen plasma is another potential TRALI-mitigating strategy. To date, no cases of TRALI have been reported after S/D plasma transfusion. This has been attributed to the fact that the pooling of many units of plasma results either in dilution of anti-HLA antibodies or in binding of these antibodies to soluble HLAs. An alternative TRALI mitigation strategy for platelets may be to use platelet additive solution, which reduces the plasma volume by two-thirds.

The relative percentage of TRALI cases due to RBCs has increased in the past decade, due in part to the success of TRALI mitigation strategies targeted at plasma products. The utility of other RBC modifications to mitigate TRALI risk remains under investigation.

Summary: TRALI incidence and mortality has decreased substantially over the past decade, due to the implementation of donor center–based mitigation strategies. Despite the fact that TRALI mortality is now relatively uncommon, continued awareness of this potentially fatal transfusion complication is needed.

Further Reading

AABB. *Association bulletin #14-02-TRALI risk mitigation update* [9/1/17]. Available from: www.aabb.org/programs/publications/bulletins/Documents/ab14-02.pdf.

Centers for Disease Control and Prevention. *National hemovigilance safety network biovigilance component, hemovigilance module surveillance protocol* [9/1/17]. Available from: www.cdc.gov/nhsn.

Muller, M. C., van Stein, D., Binnekade, J. M., van Rhenen, D. J., & Vlaar, A. P. (2015). Low-risk transfusion-related acute lung injury donor strategies and the impact on the onset of transfusion-related acute lung injury: A meta-analysis. *Transfusion*, *55*(1), 164–175.

Roubinian, N. H., Looney, M. R., Keating, S., Kor, D. J., Lowell, C. A., Gajic, O., et al. (2017). Differentiating pulmonary transfusion reactions using recipient and transfusion factors. *Transfusion*, *57*(7), 1684–1690.

Schmickl, C. N., Mastrobuoni, S., Filippidis, F. T., Shah, S., Radic, J., Murad, M. H., et al. (2015). Male-predominant plasma transfusion strategy for preventing transfusion-related acute lung injury: A systematic review. *Crit Care Med*, *43*(1), 205–225.

Toy, P., Bacchetti, P., Grimes, B., Gajic, O., Murphy, E. L., Winters, J. L., et al. (2015). Recipient clinical risk factors predominate in possible transfusion-related acute lung injury. *Transfusion*, *55*(5), 947–952.

Toy, P., Gajic, O., Bacchetti, P., Looney, M. R., Gropper, M. A., Hubmayr, R., et al. (2012). Transfusion-related acute lung injury: Incidence and risk factors. *Blood*, *119*(7), 1757–1767.

Vlaar, A. P., & Juffermans, N. P. (2013). Transfusion-related acute lung injury: A clinical review. *Lancet*, *382*(9896), 984–994.

CHAPTER 67

Septic Transfusion Reactions

Irina Maramica, MD, PhD, MBA

Transfusion of bacterially contaminated blood products may result in no symptoms, bacterial infection, sepsis, or death in the recipient. Bacterial pathogens have emerged as most common cause of transfusion-transmitted infections. Incidence of septic transfusion reactions (STRs) varies with type of platelet product. STR is under recognized and underreported, and high level of suspicion should be maintained in certain patient populations. Significant improvement in bacterial safety of platelets has been achieved since 2004 through implementation of multiple technologies.

Clinical Presentation: Clinical consequences of transfusing bacterially contaminated blood products are influenced by virulence, concentration (STR with ≥5 CFU/mL) and growth rate of bacteria, and immune status of recipient (most fatalities occur in neutropenic patients). Most organisms isolated from contaminated platelet and RBC units tend to be skin-associated contaminants, usually not associated with STR. Historically, gram-negative organisms and high levels of endotoxin, most often achieved with storage times longer than 21 days for RBCs or 3 days for platelets have been associated with more severe reactions and shorter delay between symptom onset and transfusion. In recipient, endotoxin stimulates macrophages and endothelial cells, leading to massive release of proinflammatory cytokines (interleukin [IL]-1β, IL-6, and IL-8), TNFα, and nitric oxide, and also activates complement and coagulation cascades, thereby causing inflammation, intravascular coagulation, hemorrhage, and death. Gram-positive bacteria such as *Bacillus cereus*, *Staphylococcus aureus*, and coagulase-negative staphylococci have also been implicated in severe or fatal reactions with *S. aureus* causing one-third of fatalities over the last 5 years.

Most patients develop fever of ≥39°C (102.2 F) or an increase of ≥2°C (3.9 F) from pretransfusion value, during or within 4 hours of transfusion. Fever might be accompanied by hypotension, rigors, tachycardia, dyspnea, nausea/vomiting, and in rare cases septic shock. Differential diagnosis includes noninfectious transfusion reactions such as febrile nonhemolytic, hemolytic, and transfusion-related acute lung injury.

In 2014, AABB published clinical criteria for investigation of suspected bacterial contamination. When used as part of active surveillance, these criteria demonstrated high diagnostic sensitivity. Active surveillance includes culturing platelets at time of issue and evaluating patients as soon as positive culture results are obtained and with any change in clinical condition during or within 24 hours of transfusion. High level of suspicion should be maintained in patients who are immunosuppressed, neutropenic, transfused during surgery, febrile, or receiving antipyretics.

Management: If sepsis is suspected, transfusion should be stopped immediately and open port capped or tubing clamped. Two sets of blood cultures should be drawn

Transfusion Medicine and Hemostasis. https://doi.org/10.1016/B978-0-12-813726-0.00067-2

from patient's arm opposite from transfusion site even if the patient is on antibiotics. Remaining product bag should be returned to transfusion service after placing in plastic-sealed bag, to minimize leakage and contamination risk. Gram stain and optional FDA-approved rapid assay should be performed on remaining product immediately to help guide initial antibiotic therapy. Cultures on remaining product should be done both aerobically and anaerobically. If there is no blood remaining in bag, 10–20 mL of sterile broth should be aseptically injected into bag and mixed thoroughly before sampling. Alternatively, culture can be performed on residual sample from index component prepared at time of issue. Culturing segments yields high rate of false-negative results and should be avoided. Blood supplier should be immediately notified and cocomponents from same donation quarantined or recalled if contamination is likely.

Incidence and Implicated Bacteria

RBC Products: The estimated incidence of RBC contamination is ~1:30,000 units transfused, with approximate STR rate of 1:250,000 units. Low prevalence of contamination is due to poor viability of most bacteria in blood stored at 1–6°C. Exception is gram-negative, psychrophilic bacteria that grow at cold temperatures and cause infections with high mortality rate. Bacteria most commonly implicated are *Yersinia enterocolitica* that grows well in iron-rich environment of stored RBCs, *Pseudomonas* species, and *Serratia* species. Spiking experiments have demonstrated that following a 2 weeks of lag-phase, concentration of this bacterium reaches 10^9/mL, with parallel rise in endotoxin level, after 4 weeks of storage.

Platelets (Apheresis and Whole Blood–Derived): Because platelets are stored at room temperature, they provide favorable environment for bacterial growth. Bacterial contamination of platelets is leading infectious risk of transfusion therapy, with bacterial contamination being detected by current screening methods in ~1:6000 apheresis platelet components, depending on culture methods and apheresis technology. Residual risk of STR with apheresis platelets similarly vary between 1:100,000–0:1,000,000 distributed platelet products with 0–4 FDA-reported fatalities per year. In ~70% of cases, platelet contamination occurs with gram-positive bacteria found on the donor skin. Most frequently isolated organisms include *S. aureus*, coagulase-negative staphylococci, streptococci, aerobic and anaerobic diphtheroid bacilli, and gram-positive bacilli. Most commonly implicated bacteria in fatalities include *S. aureus*, *Staphylococcus epidermidis*, *Klebsiella pneumoniae*, *Escherichia coli*, and *Serratia marcescens*.

Plasma and Cryoprecipitate: Rare reports of patients developing endocarditis or mediastinal wound infection after transfusion of cryoprecipitate and plasma contaminated by environmental bacteria, *Burkholderia cepacia* and *Pseudomonas aeruginosa*. These bacteria grow optimally at 37°C, and contamination of product occurs through contaminated bath water use.

Mitigation Strategies

RBCs: Culturing RBCs after collection is not practical because cold storage temperatures prolong the initial lag phase to ~10 days for most bacteria, keeping their numbers below detection level. Currently, RBCs are screened for bacterial contamination by visual inspection for hemolysis before transfusion.

Platelets: Initial inoculum size is <10 CFU/mL and STRs occur with bacterial concentrations of $\geq 10^5$ CFU/mL that are usually achieved on storage days 4 and 5. Methods for mitigation bacterial contamination are discussed in Chapter 19.

Summary: Bacterial contamination detection, development, and implementation of pathogen reduction methodology and optimization of storage and handling of blood products have substantially reduced STR risk due to platelet transfusion. New products such as cold storage, lyophilized, and cryopreserved platelets will also have mitigated risk.

Further Reading

AABB. *Clinical recognition and investigation of suspected bacterial contamination of platelets.* Association Bulletin #14–04. Available from: http://www.aabb.org/programs/publications/bulletins/Documents/ab14-04.pdf.

Hong, H., Xiao, W., Lazarus, H. M., Good, C. E., Maitta, R. W., & Jacobs, M. R. (2016). Detection of septic transfusion reactions to platelet transfusions by active and passive surveillance. *Blood*, *127*, 496.

Tomasulo, P. A. (2017). Reducing the risk of septic transfusion reactions from platelets. *Transfusion*, *57*(5), 1099–1103.

CHAPTER 68

Metabolic, Hypotensive, and Other Acute Reactions and Complications

Nancy L. Van Buren, MD

Metabolic complications may occur when large volumes of blood products are transfused. The so-called lethal triad of massive transfusion, including acidosis, hypothermia, and coagulopathy will be discussed in this chapter. Other complications of massive transfusion will be reviewed, including hyperkalemia, hypoglycemia, hypocalcemia from citrate in transfused products, and other metabolic abnormalities associated with hypothermia. Hypotensive reactions and transfusion-associated dyspnea will also be discussed.

Metabolic Complications of Transfusion: Metabolic complications of blood transfusion are most often seen in neonates or in circumstances in which large volumes of blood products are transfused, such as massive transfusion (see Chapter 58). The category of "metabolic complications" typically includes acidosis, citrate toxicity (hypocalcemia), hyperkalemia, hypokalemia, and hypothermia.

Pathophysiology: Metabolic complications are secondary to blood product storage in the cold (1–6°C), citrate within the anticoagulant/preservative solution, and the RBC storage lesion; together these three elements may result in hypothermia, hypocalcemia, acidosis, and hyperkalemia. Changes in RBCs during storage include a decrease in ATP and 2,3-DPG, and an increase in hemolysis. In the supernatant fluid, increases in potassium and decreases in pH, sodium, and glucose are observed. In addition, irradiation of RBC products increases the amount of potassium in the supernatant over time.

Hyperkalemia and Hypokalemia: Hyperkalemia can result in cardiac arrhythmias and potentially death. Potassium levels increase during storage (27 mmol/L at day 0–78.5 mmol/L at day 35 in CPDA-1 and 45–50 mmol/L at day 42 in AS; Table 31.1); irradiation increases supernatant potassium levels (see Chapter 42).

Patients at particular risk for complications from hyperkalemia include infants and neonates who receive large volumes of irradiated and/or older units and those with higher pretransfusion potassium levels. Use of central lines increases the risk of cardiac arrhythmias. The Society for Pediatric Anesthesia released recommendations to reduce the incidence of transfusion-associated hyperkalemia and associated cardiovascular risks. These recommendations include use of fresh RBC products in cases where massive transfusion is anticipated and use as soon as possible after irradiation; and if RBCs with high potassium levels are the only readily available option, the unit can be washed. Other recommendations include use of AS units, supernatant removal, monitoring of potassium levels, transfusion through peripheral infusion line, and management of pretransfusion potassium levels. Potassium filters are in development and use of perioperative salvage technology to wash allogeneic RBCs has been reported.

Transfusion Medicine and Hemostasis. https://doi.org/10.1016/B978-0-12-813726-0.00068-4

Other patients at risk for clinical problems due to hyperkalemia include those with severe tissue injury or underlying renal insufficiency or failure. Rapid rates of blood transfusion (100–150 mL/minute or greater) commonly develop transient hyperkalemia, and rapid transfusion through a central venous catheter has been associated with hyperkalemic cardiac arrest, particularly in vulnerable populations of patients.

While hyperkalemia from blood transfusion is a well-recognized, hypokalemia may also occur in association with transfusion. This can occur after blood transfusion, because donor RBCs are often potassium-depleted. Citrate metabolism causes further movement of potassium into cells, particularly in those receiving large amounts of plasma. Patients at risk include pediatric liver transplant patients and massively transfused trauma patients receiving other potassium-poor solutions, such as crystalloid, platelets, and plasma. Metabolic alkalosis may occur when renal function is impaired due to the release of bicarbonate when citrate is metabolized, which results in a lower serum potassium level.

Hypothermia: Hypothermia has a range of effects, including the following:

1. Decreases in tissue oxygenation due to increases in hemoglobin's affinity for oxygen, which can result in a metabolic acidosis.
2. Increases in the metabolic rate, resulting in increased oxygen consumption.
3. Impairment of the metabolism of citrate and some medications.
4. Inhibition of coagulation factor enzymatic reactions and disruption of platelet function leading to a bleeding diathesis.
5. Induction of ventricular arrhythmias when large volumes of cold blood are infused through a central catheter in close proximity to the cardiac conducting system. This risk is exacerbated by coexisting hypocalcemia and hyperkalemia.

Hypothermia (cold toxicity) results in increased blood loss and transfusion requirements. Normothermia is defined as a core body temperature of 36–38°C. Severe hypothermia with temperatures <32°C is associated with significantly increased mortality, which has prompted the development of guidelines for maintaining normothermia during surgery. Management and prevention of hypothermia in the operative setting may include use of active warming (e.g., forced air warming both preoperatively and intraoperatively), minimizing heat loss through warming of IV fluids, irrigation fluids, and blood warmers for transfusion.

Hypocalcemia: Hypocalcemia is a result of citrate toxicity (i.e., excess citric acid) in the anticoagulant solution, as citrate chelates calcium to prevent blood from clotting. Citrate (an acid) is metabolized primarily in the liver through the Krebs cycle, which results in the release of bicarbonate (a base), which is subsequently excreted by the kidneys. Recipients are at increased risk of hypocalcemia if the amount of citrate is large, such as in massive transfusion or apheresis, or if the recipient is unable to metabolize the citrate adequately secondary to hepatic impairment, which may be from hypothermia, or liver failure. In addition, metabolic alkalosis may result if the bicarbonate cannot be excreted, such as in renal failure.

Clinical Features: Hypothermia causes increased oxygen requirements, impaired metabolism of citrate and lactate, release of potassium, increased affinity of hemoglobin for oxygen, and ventricular arrhythmias. Hypothermia commonly occurs in patients who require massive transfusion from hemorrhagic shock. Infusion of unwarmed blood products is a well-known contributing factor. The consequences of hypothermia include metabolic acidosis, coagulopathy, peripheral vasoconstriction, cardiac arrhythmias, and other associated morbidities.

Citrate toxicity results in hypocalcemia, which causes paresthesia, nausea, hyperventilation, and depressed cardiac function. Hyperkalemia may result in cardiac arrest.

Contributing to the hemostatic abnormalities, which often develop during massive transfusion, including dilutional and consumptive coagulopathy, hypothermia impacts platelet function and impairs the coagulation cascade with resultant decreased ability to form stable clots. This is due to reduction in coagulation factor activity for each 1°C drop in temperature, which underscores the importance of avoiding hypothermia in patients requiring massive transfusion.

Diagnosis: Laboratory testing of ionized calcium, potassium, and pH can be used for diagnosis of hypocalcemia, hyperkalemia, and acidosis.

Treatment: Hypothermia can be treated with a blood warmer and other methods to keep the recipient warm. Ionized calcium concentrations should be monitored and calcium replacement given as needed, usually calcium gluconate or calcium citrate. Treatment of hyperkalemia might include insulin, calcium gluconate, and furosemide or possibly Kayexalate (sodium polystyrene).

Prevention: Use of blood warmers and other patient-warming devices can minimize hypothermia. Neonates and other at-risk patient populations for hyperkalemia, especially patients receiving large RBC volumes, can receive RBC products, which are <7 days-10 days old or RBC units with less supernatant (washed or plasma reduced), and avoidance of RBC units that have been irradiated more than 12 hours before transfusion. Slower infusion rates will decrease the risk of citrate toxicity and hyperkalemia. Use of a peripheral line will decrease risk of cardiac arrhythmia.

Hypotensive Reactions: Acute hypotensive transfusion reactions present as a sudden drop in blood pressure occurring during or within 1 hour of cessation of transfusion. Typically, it is the sole manifestation of the reaction; however, other symptoms may also be present, including facial flushing, dyspnea, and abdominal pain. It is a rare reaction with a reported incidence of 1.3 per 10,000 transfusions and has been implicated with red blood cells, platelets, and plasma. For adults, this is defined as a decrease in systolic blood pressure of ≥30 mmHg and systolic blood pressure of ≤80 mmHg. For infants, children, and adolescents (1 year to <18 years) this is defined as a blood pressure drop of ≥25% systolic blood pressure, and for neonates (<1 year old OR <12 kg body weight) there must be >25% drop in baseline blood pressure value based on whichever method of measurement is being recorded. These reactions respond rapidly to transfusion cessation and supportive care. The FDA issued a letter in May 1999 to alert the public about hypotensive reactions and bedside leukocyte reduction filters after receiving more than 80 reports of these events. Hypotensive reactions are most commonly seen with the use of negatively charged bedside leukoreduction filters for platelet transfusions in patients receiving angiotensin-converting enzyme (ACE) inhibitors, but not exclusively so. In addition, hypotensive reactions are seen in patients undergoing extracorporeal blood processing procedures such as hemodialysis and therapeutic apheresis. Notably, most transfusion recipients taking ACE inhibitors do not have hypotensive reactions.

Pathophysiology: Hypotensive reactions are thought to be due to bradykinin and des-Arg-bradykinin, which are two vasoactive kinins that are generated by activation of the contact system. Activated factor XII (Hageman factor) converts prekallikrein to kallikrein, which cleaves kininogen and releases bradykinin, which causes hypotension and edema by activating B2-kinin receptors on the vascular endothelium through the release of endothelium-derived nitric oxide, prostacyclin, and endothelium-derived hyperpolarizing factor. Bradykinin is rapidly inactivated by ACE; however, ACE inhibitors hinder this process. For this reason, hypotensive reactions are more likely to occur in patients with hypertension who are on ACE inhibitors. Originally, hypotensive reactions were reported with the use of negatively charged bedside leukoreduction filters, which causes kinin activation. Subsequently, hypotensive reactions have been reported with positively charged bedside leukoreduction filters. In the absence of bedside leukoreduction filters, hypotensive transfusion reactions have also been reported during cardiopulmonary bypass and radical prostatectomy, the pathophysiology of which is yet to be defined. Other causes postulated are blood warmers and other patient-specific factors. Lastly, cases have been reported without ACE inhibitor therapy.

Clinical Features: Abrupt hypotension as defined above is usually the sole manifestation of this transfusion reaction. In addition to a sudden decrease in blood pressure, patients may have facial flushing, dyspnea, and abdominal cramps or pain with or without nausea, loss of consciousness, and symptoms of shock. It is noteworthy that most cases of acute hypotensive transfusion reactions spontaneously resolve when the transfusion is discontinued.

Diagnosis: The diagnosis is based on the clinical symptoms. Hypotensive reactions should be differentiated from other transfusion reactions where hypotension is present, such as transfusion-related acute lung injury (TRALI), which would have respiratory findings consistent with pulmonary edema and fever; anaphylactic reaction, which would have respiratory findings, cutaneous findings, and possibly gastrointestinal symptoms; acute hemolytic transfusion reaction, septic hypotensive transfusion reactions, and severe hypotensive reaction triggered by some other cause, such as an anaphylactic drug reaction, hemorrhage, pulmonary embolism, or acute cardiac event.

Treatment: Hypotensive reactions require immediate intervention, including stopping the infusion and providing supportive care. Administration of emergency medications such as vasopressors, antihistamines, fluids, and steroids may be ordered by the physician, along with resuscitation and treatment of respiratory distress and/or cardiovascular shock, if symptoms do not resolve immediately after cessation of transfusion. If the patient is on ACE inhibitor, stopping the drug should be considered. Once the transfusion is discontinued, the symptoms typically promptly disappear. Supportive care may be necessary until the patient recovers fully.

Prevention: Transfusions for patients on ACE inhibitors that cannot be discontinued should be administered at slow rates of infusion and carefully monitored. In the past, the majority of hypotensive reactions were associated with the use of bedside leukoreduction filters. Because most reactions are idiosyncratic to a specific unit, patients without risk factors are unlikely to have subsequent hypotensive reactions to transfusion. Consideration of washing cellular components to reduce accumulated bradykinin is seldom needed but may be given to patients who experience recurrent reactions. No other routine preventative measures have been identified. As the majority of blood components within the United States are prestorage leukoreduced, this complication is currently seen infrequently, although the importance of recognizing and reporting this reaction to the transfusion service should be emphasized.

Transfusion-Associated Dyspnea: Transfusion-associated dyspnea (TAD) is defined as respiratory distress within 24 hours of transfusion that does not meet criteria for TRALI, transfusion-associated circulatory overload (TACO) or allergic reaction. The respiratory distress is not explained by a patient's underlying or preexisting medical condition. TAD is used primarily in the hemovigilance system.

Clinical Features: Patient develops acute respiratory distress that occurs within 24 hours of transfusion, and other transfusion reactions are ruled out.

Diagnosis: The diagnosis is based on the clinical symptoms after allergic reaction, TACO, and TRALI are excluded. Air embolism should also be considered in patients presenting with acute respiratory distress in the appropriate setting, as it has rarely been reported with intraoperative and perioperative blood recovery systems when air is allowed into the blood infusion bag. Of note, fatal outcome rarely result from <100 mL air embolization.

Treatment: Discontinue transfusion.

Further Reading

Association for Surgical Technologists. (2015). *AST standards of practice for maintenance of normothermia in the perioperative patient*. Available at: http://www.ast.org/uploadedFiles/Main_Site/Content/About_Us/SOP_For_Normothermia.pdf.

Centers for Disease Control and Prevention. (2016). *National healthcare safety network (NHSN): Blood safety surveillance* Atlanta, GA. Available at: http://www.cdc.gov/nhsn/acute-care-hospital/bio-hemo/index.html.

Cyr, M., Eastlund, T., Blais, C., et al. (2001). Bradykinin metabolism and hypotensive transfusion reactions. *Transfusion*, *41*, 136–150.

Delaney, M., Wendel, S., Bercovitz, R. S., et al. (2016). Transfusion reactions: Prevention, diagnosis, and treatment. *Lancet*, *388*, 2825–2835.

Kaira, A., Palaniswamy, C., Patel, R., et al. (March 2012). Acute hypotensive transfusion reaction with concomitant use of angiotensin-converting enzyme inhibitors: A case report and review of the literature. *Am J Ther*, *19*(2), e90–e94.

Pagano, M. B., Ness, P. M., & Chajewski, O. S. (2015). Hypotensive transfusion reactions in the era of prestorage leukoreduction. *Transfusion*, *55*(7), 1668–1674.

Sapiano, M. R. P., Savinkina, A. A., Ellingson, K. D., et al. (2017). Supplemental findings from the national blood collection and utilization surveys, 2013 and 2015. *Transfusion*, *57*, 1599–1624.

Sihler, K. C., & Napolitano, L. M. (2010). Complications of massive transfusion. *Chest*, *137*(1), 209–220.

Society for Pediatric Anesthesia. (January 27, 2015). *Wake up safe. The pediatric anesthesia quality improvement initiative*. Available at: http://www.wakeupsafe.org/Hyperkalemia_statement.pdf?201501300915.

Vossoughi, S., Perez, G., Whitaker, B. I., et al. (2017). Analysis of pediatric adverse reactions to transfusions. *Transfusion*, *57*(2), 489–490.

CHAPTER 69

Posttransfusion Purpura

Emily K. Storch, MD[a] and Melissa M. Cushing, MD

Posttransfusion purpura (PTP) is a rare complication of transfusion that most commonly occurs in previously pregnant women. PTP typically occurs 2–14 days after a transfusion of a blood product (most commonly a red blood cell [RBC] product), resulting in acute, profound thrombocytopenia (platelet count <10,000/μL). The incidence is thought to be 1 in 50,000–100,000 components transfused. The thrombocytopenia is secondary to high-titer platelet alloantibodies.

Pathophysiology: PTP is an immune thrombocytopenia resulting from platelet alloantibodies, most often anti-HPA-1a alone or in combination with antibodies to other platelet antigens almost invariably on the GPIIb/IIIa receptor complex. The reaction is usually due to transfusion of RBCs, but other products (e.g., platelets and plasma) have been implicated. The transfused product contains the immunogenic platelet glycoprotein, which induces an anamnestic response. PTP occurs most frequently in women who were previously sensitized during pregnancy, yet sensitization from transfusion does occur. About 2% of the general population is homozygous HPA-1b; however, the incidence of PTP is lower than that would be expected given this frequency. In 2015, 305 cases were reported in the United States to the National Blood Collection and Utilization Survey (NBCUS), at a rate of 1:57,823 components transfused. This represents an increase from 2013 (259 cases or 1:78,014 components transfused) and a further increase from 2011 (209 cases or 1:100,158 components transfused). Patients who develop anti-HPA-1a often share certain human leukocyte antigen (HLA) genotypes, such as HLA-B8 or HLA-DRB3*0101, similar to that seen in neonatal alloimmune thrombocytopenia (NAIT). Interestingly, the increased risk of PTP in women who have previously had children with NAIT has not been reported. Women with a history of PTP may be at increased risk for having a pregnancy affected by NAIT. The antibody involved in PTP destroys both transfused and autologous platelets. The mechanism of destruction of autologous platelets is unknown; theories include the following:

- The antibody produced cross-reacts with autologous platelets.
- Donor-derived, soluble platelet glycoprotein is adsorbed on to the autologous platelets.
- The immune response includes an autoimmune component.

Clinical Manifestations: The patient with PTP presents with unexplained purpuric rash, bruising, or mucosal bleeding 2–14 days after transfusion. The disease can be

a. This chapter reflects the views of the author and should not be construed to represent FDA's views or policies.

Transfusion Medicine and Hemostasis. https://doi.org/10.1016/B978-0-12-813726-0.00069-6

self-limited and the platelet count usually recovers within 21 days in the absence of treatment. However, approximately 30% of patients have major hemorrhage and the mortality rate approaches 10%. The diagnosis may be delayed secondary to the interval between transfusion and disease onset. A febrile nonhemolytic transfusion reaction often occurs with the implicated transfusion.

Diagnosis: The differential diagnosis includes other diseases with rapid onset of severe thrombocytopenia such as autoimmune thrombocytopenia, drug-induced thrombocytopenia, disseminated intravascular coagulation, heparin-induced thrombocytopenia, and thrombotic thrombocytopenic purpura. The diagnosis is confirmed by the clinical presentation and the detection of platelet-specific alloantibodies. Large platelets are seen on blood smear, and the bone marrow shows increased megakaryocytes. The majority of cases have HPA-1a antibodies, but antibodies against many other HPA antigens have been reported, and occasionally multiple antibodies are present. Enzyme-linked immunosorbent assays (ELISA), monoclonal antibody immobilization of platelet antigens assay, and other assays are used to diagnose platelet antibodies. Documenting the patient's platelet phenotype can be difficult, due to the adsorption of antibody onto endogenous platelets and the severe thrombocytopenia. Molecular HPA genotyping is very helpful in confirming the diagnosis, especially in cases with no detectable antibodies (see Chapter 31).

Treatment: The primary treatment is high-dose intravenous immunoglobulin (IVIG) (400–500 mg/kg per day for 5 days or 1 g/kg for 2 days) alone or in combination with corticosteroids resulting in a platelet count >100,000/μL in 3–5 days with an 85% response rate. Corticosteroids alone appear to be ineffective. The mechanism of action of IVIG is thought to be due to antiidiotypic antibodies, Fc receptor blockade, nonspecific binding of Ig to the platelet surface, and/or acceleration of IgG catabolism. Plasma exchange was previously used before discovering the efficacy of IVIG and may still be helpful in patients who do not respond to IVIG. For those who do not respond, splenectomy has been performed. If the antibody specificity has been determined, platelets lacking the offending antigen (usually HPA-1a negative) have been used in bleeding patients with modest clinical improvement although transfused antigen-negative platelets also have decreased survival. However, serologically compatible platelets are recommended in the acute setting to avoid reexposure to the implicated antibody. Coadministration of IVIG may improve the survival of the transfused platelets.

Prevention: Recurrence of PTP after a subsequent RBC transfusion is uncommon but has been reported. Some authorities have advocated the use of washed or leukoreduced RBC or RBC and platelet products from corresponding antigen-negative donors for subsequent transfusions, but the value of this practice is unclear. A report from the United Kingdom described a decrease in cases of PTP after implementation of universal leukoreduction, likely secondary to decrease in platelets and platelet microvesicles in the leukoreduced RBC product.

Further Reading

Curtis, B. R. (2008). Genotyping for human platelet alloantigen polymorphisms: Applications in the diagnosis of alloimmune platelet disorders. *Semin Thromb Hemost*, *34*(6), 539–548.

Curtis, B. R., & McFarland, J. G. (2014). Human platelet antigens - 2013. *Vox Sang*, *106*(2), 93–102.

Delaney, M., Wendel, S., Bercovitz, R. S., Cid, J., Cohn, C., Dunbar, N. M., et al. (2016). Transfusion reactions: Prevention, diagnosis, and treatment. *Lancet*, *388*(10061), 2825–2836.

Gonzalez, C. E., & Pengetze, Y. M. (2005). Post-transfusion purpura. *Curr Hematol Rep*, *4*(2), 154–159.

Landau, M., & Rosenberg, N. (2011). Molecular insight into human platelet antigens: Structural and evolutionary conservation analyses offer new perspective to immunogenic disorders. *Transfusion*, *51*(3), 558–569.

Sapiano, M. R. P., Savinkina, A. A., Ellingson, K. D., Haass, K. A., Baker, M. L., Henry, R. A., et al. (2017). Supplemental findings from the national blood collection and utilization surveys, 2013 and 2015. *Transfusion*, *57*(Suppl. 2), 1599–1624.

Williamson, L. M., Stainsby, D., Jones, H., Love, E., Chapman, C. E., Navarrete, C., et al. (2007). The impact of universal leukodepletion of the blood supply on hemovigilance reports of posttransfusion purpura and transfusion-associated graft-versus-host disease. *Transfusion*, *47*(8), 1455–1467.

CHAPTER 70

Transfusion-Associated Graft-Versus-Host Disease

Richard O. Francis, MD, PhD

Transfusion-associated graft-versus-host disease (TA-GVHD), a rare (1 fatality reported to FDA 2011–15) and almost universally fatal complication of blood product transfusion, is due to the cotransfusion of viable lymphocytes in cellular blood products, such as whole blood, red blood cells (RBCs), platelets, granulocytes, and liquid (not previously frozen) plasma. If the immune system of the recipient does not recognize and produce an immune response against the cotransfused lymphocytes, they can engraft and mount an immune response against the host; thus the term graft-versus-host disease. The recipient's immune system can be limited in responding to the cotransfused lymphocytes because of the following: (1) significant immunosuppression and (2) planned or inadvertent human leukocyte antigen (HLA) matching between donor and recipient. Irradiation or pathogen inactivation (PI) of the product can mitigate TA-GVHD. Both of these methods cause DNA damage in the cotransfused lymphocytes and render them incapable of cell division and engraftment. It is critical to identify recipients who are at risk of TA-GVHD and ensure that blood products are treated to prevent this lethal transfusion complication.

Normal Clearance of Transfused Lymphocytes: In a study investigating the clearance of cotransfused lymphocytes in immunocompetent recipients, three phases were found: (1) 99.9% of lymphocytes were cleared over the first 2 days, (2) there was a 1-log *increase* in the number of circulating donor lymphocytes on days 3–5, and (3) there was a second clearance event leading to the small number and percent of chimeric cells. It was postulated that the transient increase in donor lymphocytes represents one arm of an in vivo mixed lymphocyte reaction with activated donor T-lymphocytes proliferating in response to exposure to HLA-incompatible recipient cells. The second clearance step results from the recipient's immune system mounting an augmented response against the donor cells. Irradiation of blood components eliminates the expansion of donor cells and thus abrogates transfusion-associated microchimerism (TA-MC) and TA-GVHD.

Pathophysiology: The mechanism of TA-GVHD is similar to that of acute GVHD after hematopoietic progenitor cell (HPC) transplantation, where donor lymphocytes attack host tissues. There are three separate phases of development: phase 1 is conditioning, phase 2 is the afferent phase, and phase 3 is the efferent phase. The conditioning regimen results in host tissue damage and activation, which leads to the production of inflammatory cytokines. The afferent phase results in donor T-cell activation through antigen presentation, followed by proliferation, and then differentiation of activated

Transfusion Medicine and Hemostasis. https://doi.org/10.1016/B978-0-12-813726-0.00070-2

T-cells. Inflammatory cytokines are released during the efferent phase, leading to the damage of host tissues, and finally cell death and host tissue destruction.

Clinical Manifestations: Once coadministered viable lymphocytes have been infused with the blood products, they can engraft and produce an immune response against HLA-rich tissues and organs in the recipient. Symptoms include erythema, liver dysfunction, GI symptoms, and profound pancytopenia. Fever accompanies this process due to release of inflammatory cytokines. Pancytopenia of TA-GVHD differentiates it from HPC-associated GVHD and gives it a near 100% mortality.

TA-GVHD in adults results in death within 3 weeks from symptom onset in >95% of cases. In neonates, the clinical manifestations are similar, but the time between transfusion and symptom onset is longer than for adults: fever occurs around 28 days, rash around 30 days, and death around 51 days. In both groups, fever is usually the presenting symptom, followed by an erythematous maculopapular rash, which typically begins on the face and trunk and spreads to the extremities. Liver dysfunction usually manifests as obstructive jaundice or acute hepatitis. GI complications include nausea, anorexia, or diarrhea. Leukopenia and pancytopenia develop later and progressively become more severe, often times leading to sepsis and candidiasis, multiorgan failure, and death.

Diagnosis: Diagnosis is based on clinical findings in conjunction with laboratory and biopsy results. However, TA-GVHD is often not suspected, the symptoms, findings, and laboratory tests can easily be interpreted as being due to severe viral infection or adverse reaction to administered medication. Thus TA-GVHD is potentially underdiagnosed. Discovery of donor lymphocytes or DNA in the patient's peripheral blood or tissue biopsy with the appropriate clinical picture confirms the diagnosis. Donor-derived DNA is usually detected by polymerase chain reaction (PCR)-based HLA typing, but other methods include use of restriction fragment length polymorphism analysis, variable number tandem repeat analysis, microsatellite marker analysis, fluorescence in situ hybridization, and cytogenetics.

The National Healthcare Safety Network Manual, Biovigilance Component, case definition criteria includes

- A clinical syndrome occurring from 2 days to 6 weeks after transfusion characterized by the following:
  - Rash (erythematous, maculopapular eruption centrally that spreads to extremities and may progress to generalized erythroderma and hemorrhagic bullous formation)
  - Diarrhea
  - Fever
  - Hepatomegaly
  - Liver dysfunction
  - Marrow aplasia
  - Pancytopenia
  - Characteristic histological appearance of skin or liver biopsy
  - WBC chimerism

Treatment: Treatment is largely palliative and aimed at attempting to improve the function of, or render the recovery of, the recipient's immune system and bone marrow. This is largely unsuccessful. Approaches include corticosteroids, antithymocyte globulin, and cyclosporin used with hematopoietic growth factors. There are a few reports of spontaneous resolution, successful treatment with a combination of cyclosporin, steroids and OKT3 (anti-CD3 monoclonal antibody) or antithymocyte globulin, and treatment with autologous or allogeneic hematopoietic stem cell transplantation. Transient improvement has been seen with nafamostat mesilate, a serine protease inhibitor that inhibits cytotoxic T-lymphocytes.

Prevention: Patients at increased risk must be identified and transfused with lymphocyte-inactivated products, usually through irradiation or PI methods (see Chapter 42).

Risk Factors (Table 42.1)

Blood Product Factors

Age of Blood: Fresh blood increases the risk of TA-GVHD because over time lymphocytes undergo apoptosis and fail to stimulate a mixed lymphocyte culture response during the storage period. A series of cases of TA-GVHD in Japan in immunocompetent patients found that 62% of patients had received blood less than 72 hours old and a US series found about 90% of cases received blood less than 4 days old. In addition, a case of fatal TA-GVHD was recently reported, which involved a previously healthy US Army soldier who was transfused with fresh whole blood that was less than 24 hours old.

Leukocyte Dose: Leukocyte reduction of blood products may decrease the risk of TA-GVHD but does not eliminate it. The serious hazards of transfusion data reported a decrease in the number of TA-GVHD cases after universal leukocyte reduction of blood components in the United Kingdom in 1999.

Blood Products: All cellular blood products, including RBCs, platelets, granulocytes, whole blood, and fresh (not frozen) plasma, contain viable T-lymphocytes that are capable of causing TA-GVHD. Granulocyte transfusions are the highest risk products because they are given fresh, have a high lymphocyte count and are administered to neutropenic and immunosuppressed patients.

Patients at Increased Risk

Congenital Immunodeficiency Patients: The first reported cases of TA-GVHD occurred in the 1960s, in children with T-lymphocyte congenital immunodeficiency syndromes. TA-GVHD has occurred in children with severe combined immunodeficiency syndromes and with variable immunodeficiency syndromes, such as Wiskott–Aldrich and DiGeorge syndromes. Because of the possibility of pediatric patients having an undiagnosed immunodeficiency, some institutions irradiate all blood components for children under a certain age, and some pediatric hospitals irradiate all cellular blood components. This is particularly necessary for infants undergoing cardiac surgery who may have unrecognized DiGeorge syndrome. It is recommended that all patients with suspected or confirmed congenital immunodeficiency receive irradiated products.

Allogeneic and Autologous Hematopoietic Progenitor Cell Recipients: Both allogeneic and autologous HPC transplant recipients are at increased risk of TA-GVHD. Patients who undergo allogeneic HPC transplantation have received irradiated blood products routinely for more than 40 years. Multiple organizations, including The European School of Haematology (ESH), European Group for Blood and Marrow Transplantation (EBMT), and Foundation for the Accreditation of Cellular Therapy (FACT), recommend irradiated blood products for allogeneic and autologous HPC recipients, but it is unclear for how long before and after transplantation these patients require irradiated products.

Hematologic Malignancies: Patients with hematologic malignancies, especially Hodgkin disease, are at increased risk. It is recommended that these patients receive irradiated products; however, it is less clear if this requirement should be only during active treatment and how long it should be continued.

Recipients of Fludarabine, Other Purine Analogues and Drugs/Antibodies That Affect T-Lymphocyte Number or Function: TA-GVHD was initially reported in patients with chronic lymphocytic leukemia (CLL) receiving fludarabine, a purine analogue that results in profound lymphopenia. There are nine cases of TA-GVHD in CLL, acute myeloid leukemia, and patients with non-Hodgkin lymphoma who received fludarabine up to 11 months before transfusion. Additionally, TA-GVHD occurred in a patient who received fludarabine for treatment of autoimmune disease. Other purine analogues, including deoxycoformycin (pentostatin) and chlorodeoxyadenosine (cladribine), have been associated with TA-GVHD. It is recommended that all patients who have received fludarabine or other purine analogues, and alemtuzumab (anti-CD52) or other drugs/antibodies that affect T-lymphocyte function or number be transfused with irradiated products; however, it is unclear for how long these patients should receive such products. The current recommendation is for at least a year and until recovery from the resulting lymphopenia.

Fetuses and Neonates: Fetuses and neonates have immature immune systems and may be at increased risk. In neonates, most cases of TA-GVHD reported are in those with congenital immunodeficiency or those who received products from related donors. At least 10 cases were reported after neonatal exchange transfusions, of which 4 occurred in infants who had previously received intrauterine transfusion (IUT). Seven cases were in preterm infants (excluding those who received a product from a relative). A single case report involved a full-term infant receiving extracorporeal membrane oxygenation (ECMO). The use of irradiated products for fetal and neonatal transfusions is therefore recommended for exchange transfusions and IUT, preterm infants, infants with congenital immunodeficiency, and those receiving products from relatives; its need is unclear for other neonatal transfusions.

Aplastic Anemia: Because patients with aplastic anemia are usually treated with intensive chemotherapy regimens and possible HPC transplantation, some authorities recommend that they receive irradiated products, especially during myelosuppressive therapy.

Patients Receiving Chemotherapy and Immunotherapy: TA-GVHD has occurred in patients with solid tumors, including neuroblastoma, rhabdomyosarcoma, urothelial carcinoma, and small cell lung cancer during intensive myeloablative therapy. Therefore, it is recommended that patients with solid tumors receive irradiated products, especially during myelosuppressive therapy.

Solid Organ Transplantation Recipients: GVHD is a rare complication of solid organ transplantation, which usually results from the passenger lymphocytes contained within the solid organ and not from transfusion, even though these individuals are highly immunosuppressed and transfused. There have been four cases of TA-GVHD in solid organ transplant recipients including a liver transplant recipient with preexisting pancytopenia, a heart transplant recipient, and two inconclusive cases in kidney recipients. TA-GVHD risk in solid organ transplant recipients appears to be low and the use of irradiated products is generally considered to be unwarranted.

Human Immunodeficiency Virus and Acquired Immune Deficiency Syndrome patients: HIV/AIDS is not considered a risk factor for TA-GVHD, as there is only a single case report of a child with AIDS developing transient TA-GVHD. It is postulated that HIV infects the transfused T-lymphocytes preventing the development of TA-GVHD. The use of irradiated blood products in HIV/AIDS patients is not currently indicated. Some institutions do provide irradiated blood products for HIV/AIDS patients likely because of the immunosuppressive nature of HIV/AIDS, the high degree of fatality of TA-GVHD, and the ability of HAART to decrease the burden of circulating HIV particles thus theoretically preventing the virus from infecting the cotransfused lymphocytes.

Cardiovascular Surgery: Before the adoption of universal irradiation, the reported incidence of TA-GVHD after cardiovascular surgery was 0.15%–0.47% in Japan. Fifty-six of the 122 cases of TA-GVHD reported in Japan from 1985 to 1993 were in patients after cardiovascular surgery; 28% used blood from a relative and 72% used blood less than 72 hours old. A lower risk was reported for women than men, possibly secondary to women having previous exposure to leukocytes during pregnancy and childbirth. There are six cases of TA-GVHD after cardiovascular surgery reported in the United States and United Kingdom. Possible reasons for this increased risk are that the RBC products transfused are usually less than 72 hours old, and cardiac surgery may result in reduced cell-mediated immunity. The recommendation for irradiation of blood products for cardiac surgery patients is warranted in Japan but not in the United States at this time.

Immunocompetent Patients: TA-GVHD is reported in immunocompetent patients; the majority of cases occurred with the use of fresh whole blood from a close relative. In a review of 122 cases of TA-GVHD in immunocompetent patients in Japan, 67% had not received products from a related donor, and of the 66 noncardiovascular surgery patients, 39 had solid tumors and 27 had other conditions. The risk of receiving a blood product from a homozygous donor is greatest in populations with limited HLA haplotype polymorphisms, such as Japan. The frequency of reported cases is substantially lower than these estimates, which may be a result of unrecognized and/or unreported cases, lymphocytes in blood products that are either nonviable or insufficient to cause disease, and/or recipients being able to destroy the donor lymphocytes because of minor HLA differences between donor and recipient. In addition, a systematic review examined 348 cases of TA-GVHD and found that the majority of cases were transfusions of immunocompetent patients, with cellular blood components that were stored for ≤10 days, and a donor HLA profile lacking identifiably foreign antigen types compared with the recipient. Thus, these data demonstrate the importance of viable lymphocytes that are able to evade immune detection in causing TA-GVHD. Irradiation of products from close relatives and HLA-matched products is recommended for immunocompetent patients. The risk is otherwise minimal for other immunocompetent patients, and irradiation is not warranted.

Transfusion-Associated Microchimerism: TA-MC is defined as the presence of transfused donor leukocytes comprising up to 5% of the recipient's peripheral blood leukocytes, which can remain for long periods of time (maximum length unknown). A proposed hypothesis for this phenomenon is that there are genetic polymorphisms between the donor and recipient linked to cytokine production, where some of the polymorphisms lead to immune suppression in the recipient that allows for tolerance to the transfused allogeneic cells and the development of TA-MC.

Irradiation of blood products prevents TA-MC, while leukocyte reduction of blood products does not appear to affect its incidence. This suggests that even a low number of residual lymphocytes can cause TA-MC, but those residual lymphocytes are not capable of engraftment (which would cause TA-GVHD). TA-MC has been reported most extensively in trauma patients, occurring in up to 25%–50% of transfused trauma patients. It has also been reported in sickle cell disease and thalassemia patients but has not been shown as a sustained event in HIV-positive individuals. In trauma patients, age, gender, injury severity score, splenectomy, and number of products transfused do not appear to correlate with TA-MC establishment. As TA-MC has no known clinical correlation or adverse effects, there are no recommendations to modify transfusion guidelines or products. TA-MC can be prevented by irradiation; however, this is not an indication for irradiation.

Further Reading

Cid, J. (2017). Prevention of transfusion-associated graft-versus-host disease with pathogen-reduced platelets with amotosalen and ultraviolet A light: A review. *Vox Sang*, *112*, 607–613.

Kopolovic, I., Ostro, J., Tsubota, H., et al. (2015). A systematic review of transfusion-associated graft-versus-host disease. *Blood*, *126*, 406–414.

Utter, G. H., Reed, W. F., Lee, T. H., & Busch, M. P. (2007). Transfusion-associated microchimerism. *Vox Sang*, *93*, 188–195.

Williamson, L. M., Stainsby, D., Jones, H., et al. (2007). The impact of universal leukodepletion of the blood supply on hemovigilance reports of posttransfusion purpura and transfusion-associated graft versus host disease. *Transfusion*, *47*, 1455–1467.

CHAPTER 71

Transfusion-Related Immunomodulation

Theresa Nester, MD

Transfused blood products may have effects on recipient immunity. Broadly, these effects have been called transfusion-related immunomodulation (TRIM). Some TRIM effects are generally accepted, while others are a matter of debate. The TRIM effects can be categorized as beneficial or deleterious (Table 71.1). The following effects are more evident with blood that has not been leukoreduced.

Accepted Transfusion-Related Immunomodulation Effects

Solid Organ Transplantation: Since the 1960s, it has been appreciated that transfusion of whole blood before organ transplantation results in significantly improved allograft survival in both humans and experimental animals. This effect was most notable for transplanted cadaveric kidneys. This phenomenon, or "transfusion effect," occurs to the greatest extent with the transfusion of whole blood, and to a lesser extent with RBC products. Leukocytes appear to play a central role in this process, as there is little effect with washed RBC products (~80%–90% leukocyte reduced) and essentially no transfusion effect when giving leukoreduced products (99.9% leukoreduced). The transfusion effect appears to be antigen specific, leading to tolerance to HLA antigens. There is a strong correlation between improved allograft survival and coincidence of HLA antigens on the donor organ and on the transfused white blood cells. This is especially true for the HLA-DR locus; a mismatch results in increased sensitization, and a partial mismatch results in the transfusion effect. However, transfusion also carries the risk of sensitization to HLA antigens. The formation of HLA antibodies can cause acute humoral rejection of the allograft and therefore the majority of transplants do not use crossmatch-incompatible allografts (i.e., HLAs present on the allograft for which the recipient has HLA antibodies). Current immunosuppressive medications have greatly reduced the benefits of the transfusion effect. Thus, current practice is to provide potential transplant recipients (particularly kidney and heart transplantation) with leukoreduced products to prevent HLA alloimmunization.

Decreased Likelihood of Recurrent Spontaneous Abortions: It has been observed that transfusion of allogeneic blood, from paternal or other sources, has a beneficial effect in preventing recurrent spontaneous abortion in women possibly by decreasing the T-cell response and generating suppressor T cells. Although the effect is mild, it is reproducible.

Debated Transfusion-Related Immunomodulation Effects

Cancer, Infection, and Autoimmunity: Subtle TRIM effects have been reported in situations in which mild impairment of cellular immunity may be predicted to result in

Transfusion Medicine and Hemostasis. https://doi.org/10.1016/B978-0-12-813726-0.00071-4

TABLE 71.1 Established and Proposed Transfusion-Related Immunomodulation Effects
Decreased renal allograft rejection
Improvement in autoimmune diseases (e.g., Crohn's disease)
Decreased repetitive spontaneous abortions
Increased tumor recurrence
Increased postoperative infection

alterations in pathophysiology. For example, published data suggest that transfusion may increase cancer recurrence and/or metastasis, presumably due to decreased antitumor immunity. Likewise, published data demonstrate that increased postoperative infections correlate with transfusion, suggesting TRIM inhibition of antimicrobial immunity. In addition, studies suggest that transfusion may have a therapeutic benefit in certain autoimmune states (such as Crohn's disease). In each of these situations, an equally large number of reports show no statistically significant effects. Thus, at this point, it is reasonable to conclude clinically significant TRIM effects in cancer, postoperative wound infections and autoimmune disorders *may* or *may not* occur and thus remain disputed.

Proposed Mechanisms of Transfusion-Related Immunomodulation

Traditional TRIM: At first glance, the combination of alloimmunization (which is a positive immune response) and other TRIM effects (which are generally immunosuppressive) may seem contradictory. However, humoral and cellular immunity represent distinct response pathways, which can be mutually antagonistic (i.e., Th1- vs. Th2-type responses). Thus, leukocytes in transfused RBCs may simultaneously enhance humoral immunity while suppressing cellular immunity. In this way, enhanced antibody responses and cellular immunosuppression may simultaneously occur. In the current era, where many blood inventories are completely leukoreduced using prestorage leukoreduction techniques, the TRIM effect from white cells is predicted to be significantly smaller.

Recent TRIM: If a TRIM effect still exists after leukoreduction, it is even more difficult to elucidate. It may depend on specific methods of storage and processing of units, as well as donor characteristics and patient immune state. Research has focused on extracellular vesicles left in a unit after processing; these and residual platelets may release inflammatory cytokines. Additionally, extracellular vesicle-bound hemoglobin from stored red cells may enhance inflammation. If an immunosuppressive effect still exists, it may be an outcome of the macrophage activities of erythrophagocytosis (which can suppress the oxidative burst) and iron recycling. Any immunomodulatory effect from the transfusion may depend on the current immune state of the patient, as well as the quantity of blood transfused. Thus, much more research is needed to clarify the effect for a given patient population.

Potential Problems With Transfusion-Related Immunomodulation Science: The clinical studies referred to above seldom actually measured immunity per se, but rather rely on secondary outcomes assumed to be affected by immunologic alterations. Much of the animal data suffers the same problem; in particular, that the effects being observed may or may not be due to alterations in immunity (which is not itself tested). An example of this has recently been reported by Hod et al. Some pathogenic bacteria use iron as a growth factor, and transfusion of stored RBCs induces profound expansion of the bacteria. Traditional TRIM research would attribute increased bacterial infection to suppressed immunity; however, in this case, it appears that while the effect of transfusion on bacteria is quite real, its mechanism is independent of the immune system. Thus, while secondary outcomes of transfusion may be real and reproducible, considerable scrutiny must be applied to determine whether the mechanism of such outcomes is due to effects on the immune system. Nevertheless, the observed TRIM correlations are consistent with a hypothesis of immunomodulation being causative.

Further Reading

Goubran, H., Sheridan, D., Radosevic, J., et al. (2017). Transfusion-related immunomodulation and cancer. *Transfus Apher Sci*, *56*, 336–340.

Hod, E. A., Zhang, N., Sokol, S. A., et al. (2010). Transfusion of red blood cells after prolonged storage produces harmful effects that are mediated by iron and inflammation. *Blood*, *115*, 4284–4292.

Muszynski, J. A., Spinella, P. C., Cholette, J. M., et al. (2017). Transfusion–related immunomodulation: Review of the literature and implications for pediatric critical illness. *Transfusion*, *57*, 195–206.

Vamvakas, E. C., & Blajchman, M. A. (2007). Transfusion-related immunomodulation (TRIM): An update. *Blood Rev*, *21*, 327–348.

CHAPTER 72

Iron Overload

Yelena Z. Ginzburg, MD and Francesca Vinchi, PhD

Iron overload is an excess of systemic iron, leading to its progressive accumulation in vital organs (e.g., liver, heart, pancreas, and endocrine organs). When untreated, iron overload increases the risk of liver cirrhosis, heart failure, diabetes mellitus, osteoporosis, hypogonadism, and neurodegenerative symptoms. These potentially fatal complications are preventable by iron depletion therapies.

Iron overload results from pathologically increased intestinal iron absorption or as a side effect of clinical interventions (i.e., recurrent red blood cell [RBC] transfusions and parenteral iron administration to treat anemia).

200 billion RBCs are produced every day requiring ample iron to maintain adequate erythropoiesis. Iron transported in the circulation is mostly used for hemoglobin (Hb) synthesis in RBCs. Iron supply to the bone marrow is largely maintained by iron recycling in reticuloendothelial macrophages, which engulf senescent RBCs, releasing iron back into circulation. Small amounts of iron are absorbed by duodenal enterocytes to replace ordinary iron losses. When iron absorption is uncontrolled or parenteral iron is supplied, a significant iron accumulation may occur, leading to parenchymal iron deposition and overload.

Pathophysiology: Iron overload can be primary, when genetically inherited as disorders of enhanced iron absorption (e.g., hereditary hemochromatosis [HH]), or secondary, when acquired through the administration of recurrent RBC transfusions (e.g., β-thalassemia major). Thus, patients with HH continually absorb iron, despite excess iron stores. Transfusion-requiring β-thalassemia major patients often receive up to 4 RBC units per month and may accumulate 10 g of iron per year, threefold more than total normal body content in a healthy 70-kg male. These two conditions both lead to pathological tissue iron overload.

Because no physiological mechanism exists for active iron excretion in humans, iron absorption and recycling are tightly regulated to maintain iron homeostasis (Fig. 72.1). The peptide hormone hepcidin orchestrates systemic iron flux by binding to the iron exporter ferroportin on the surface of iron-releasing cells (macrophages and enterocytes), blocking cellular iron release into the circulation. Therefore, inherited and acquired disorders that prevent normal hepcidin production result in iron overload.

Hepcidin production is regulated by multiple and opposing signals. (1) Elevated iron stores and inflammation (IL-1, IL-6, activin-B) induce hepcidin to prevent iron overload and deprive invading microorganisms of growth-essential iron, respectively. (2) Conversely, erythropoietic demand and hypoxia suppress hepcidin production, enabling further iron supply for Hb synthesis. Recently, erythroferrone has been identified as a key erythropoietin-responsive erythroid regulator of hepcidin. Additional factors, including GDF15, may also be important.

Transfusion Medicine and Hemostasis. https://doi.org/10.1016/B978-0-12-813726-0.00072-6

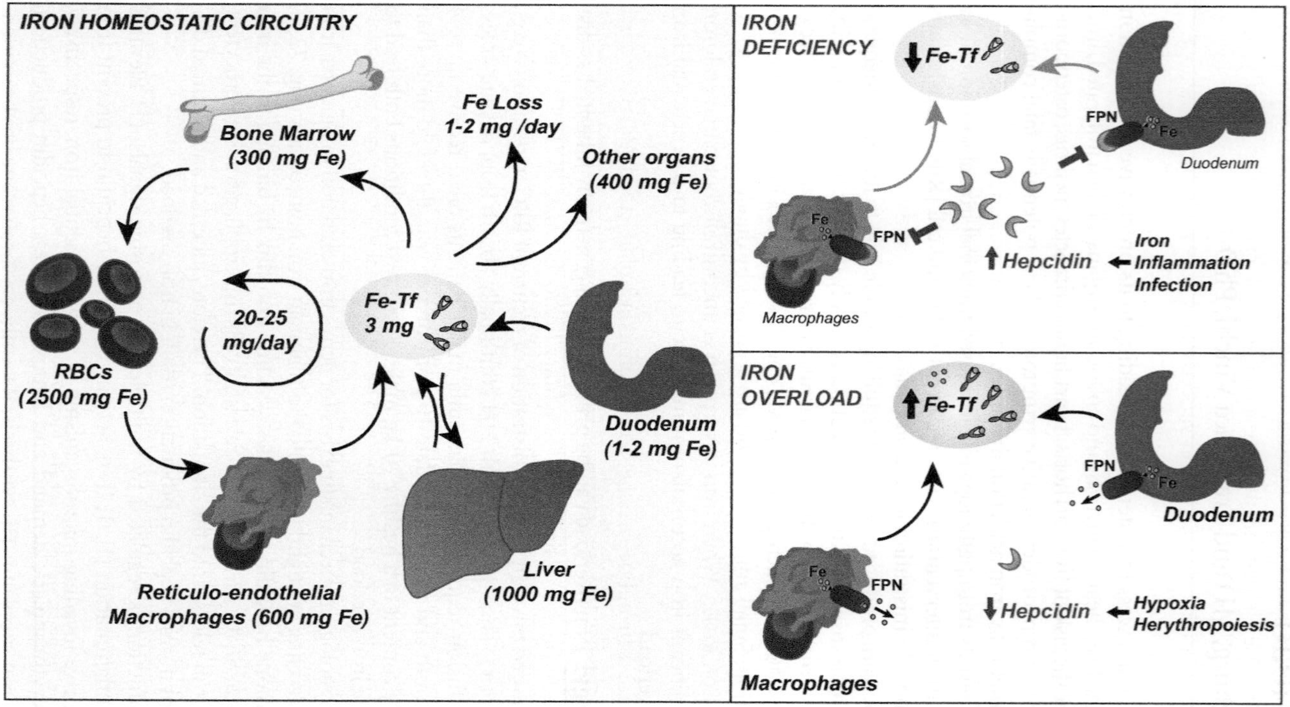

FIGURE 72.1 **Schema of mechanisms regulating iron hemeostasis.** *Iron cycling* involved transferrin-bound iron in circulation and within hemoglobin inside red blood cells. Iron recycling from senescent red blood cells in macrophages provides the largest proportion of iron available for erythropoiesis. Iron is also used for multiple proteins in other tissues. Small amounts of iron are absorbed daily to offset losses, and the liver is the site of iron storage. *Iron deficiency* results from increased hepcidin (induced by circulating iron, inflammation, or infection), which prevents iron absorption and recycling by blocking iron egress from cells through ferroportin. *Iron overload* results from decreased hepcidin (induced by hypoxia and expanded erythropoiesis), which enables increased iron absorption and recycling, eventually overwhelming transferrin iron binding capacity, leading to nontransferrin-bound iron and dysfunctional parenchymal iron deposition. *Fe*, iron; *Fe-Tf*, transferrin bound iron (holo-transferrin); *FPN*, ferroportin; *RBCs*, red blood cells.

Insufficiently elevated hepcidin is implicated as a cause of primary and secondary iron overload. Primary iron overload diseases (i.e., HH) result from mutations in hepcidin or its regulators (HFE, HJV, TfR2), leading to insufficient hepcidin production, inappropriately high iron absorption, and iron overload. Secondary iron overload diseases (e.g., transfusion-dependent β-thalassemia major and myelodysplastic syndrome) result in iron overload predominantly from excess iron acquisition through repeated RBC transfusions.

Clinical Manifestations: Symptoms of iron overload are often limited, until organ damage occurs. Clinical features include hepatic dysfunction, cardiomyopathy, arrhythmias, diabetes mellitus, impotence, arthropathy, fatigue, and predisposition to infections. Increasing evidence implicates nontransferrin-bound iron (NTBI) as the cause of organ iron deposition. Normally, iron circulates bound to transferrin, which is saturated in iron-overloaded patients. When circulating iron exceeds transferrin's iron-binding capacity (55%–60%), NTBI is generated and translocated across cell membranes, leading to iron accumulation in parenchymal cells. Recently, the ion channel SLC39A14 (Solute Carrier Family 39 Member 14; ZIP14) was identified as the specific importer of NTBI in hepatocytes and pancreatic cells. Labile plasma iron (LPI), the pathological NTBI fraction that permeates cells, is redox active and potentially chelatable. Thus, through the generation of reactive oxygen species, which oxidize lipids, proteins, and nucleic acid, LPI accumulation, may result in organ damage. Liver and heart demonstrate the highest propensity to accumulate iron.

Diagnosis: The diagnosis of iron overload requires clinical correlation of the sign and symptoms with an estimation of total body iron accumulation/stores. Estimating total body iron stores is imprecise and can be accomplished via direct and indirect methods. Indirect methods include serum ferritin concentration, transferrin saturation, heart and liver MRI (R2 Ferriscan or T2* method), and superconducting quantum interface device (SQUID). Serum ferritin is widely used as marker of body iron stores despite its acute phase reactant properties. SQUID provides the greatest accuracy at high cost and is not commonly available. Liver MRI estimation of iron accumulation is not predictive of cardiac iron overload, and cardiac function requires monitored by echocardiography and MRI. Direct iron measurements can be accomplished by liver and/or heart biopsy with increased accuracy albeit as a consequence of an invasive procedure not without risk.

Although subclinical iron overload on MRI or liver biopsy may be evident after 20 RBC units transfused (4–5 g of iron), collagen formation and portal fibrosis in the liver and symptomatic heart disease can be observed in 2 and 10 years after initiating chronic RBC transfusions, respectively.

In primary iron overload diseases, testing for known gene mutations causing HH enable diagnosis and some degree of predictability of disease severity (depending on the specific mutation). Other diseases may also be associated with iron overload, including liver diseases, porphyrias, and aceruloplasminemia (rare).

Management: Patients with primary iron overload often develop complications in their 40s (depending on specific mutation), and early implementation of therapeutic phlebotomy (see Chapter 82) confers normal life expectancy. Secondary iron overload patients often develop complications of iron overload in their 20s if untreated. With iron chelation therapy, patients have a prolonged survival and a delayed onset of cardiac iron overload. Iron chelation should be applied once serum ferritin exceeds 1000 μg/L, typically once 120 mL of RBCs/kg of body weight have been transfused. Significant progress has been made in validating the utility of MRI screening to identify pre-clinical liver and heart iron overload. In fact, Ferriscan has been FDA approved for clinical diagnosis and has enabled more homogenous decision-making regarding when to initiate iron chelation therapy.

Iron-chelating agents directly bind iron and excrete the complex in the urine or feces. Although iron chelation and excretion is slow and inefficient, the effectiveness of chronic use has been demonstrated in β-thalassemia major patients. In iron-chelated patients, therapy significantly reduces the complications of iron overload and leads to an improved quality of life. The therapeutic goal is to maintain liver iron concentration at below 5 mg/g liver dry weight and serum ferritin below 1000 μg/L.

Both parenteral (**deferoxamine**) and oral (**deferiprone** or **deferasirox**) iron-chelating agents are available for use in the United States. Deferoxamine is typically administered via daily overnight subcutaneous infusion, deferiprone orally three times a day, and deferasirox, due to its longer half-life, as a once-daily oral monotherapy. The novel deferasirox formulation, Jadenu, has further reduced gastrointestinal side effects. These chelating agents exhibit differences in organ-specific efficacy (e.g., deferiprone more effective in removing cardiac iron) and their varying side-effect profiles require different patient follow-up recommendations. Combination therapy or switching between chelators may be required to improve response to individual chelators.

Chronic RBC transfusion therapy is also used to treat complications of sickle cell disease (SCD). Exchange transfusion may be used instead of simple transfusion to reduce the overall transfused iron burden (see Chapter 76). Chelation therapy might be needed for SCD patients.

Currently, research studies are ongoing to provide novel therapeutic options for iron overload. For example, hepcidin-like molecules have been shown to ameliorate iron overload in preclinical models of HH and β-thalassemia (alone and in combination with chelators) and are ongoing clinical evaluation.

Further Reading

Bassett, M. L., Hickman, P. E., & Dahlstrom, J. E. (2011). The changing role of liver biopsy in diagnosis and management of haemochromatosis. *Pathology, 43*, 433–439.

Cappellini, M. D., Cohen, A., Porter, J., Taher, A., & Viprakasit, V. (2014). *Guidelines for the management of transfusion dependent thalassaemia (TDT)* (3rd ed.). Nicosia, CY: Thalassaemia International Federation.

Coffey, R., & Ganz, T. (2017). Iron homeostasis: An anthropocentric perspective. *J Biol Chem, 292*, 12727–12734.

Ganz, T. (2013). Systemic iron homeostasis. *Physiol Rev, 93*, 1721–1741.

Muckenthaler, M. U., Rivella, S., Hentze, M. W., & Galy, B. (2017). A red carpet for iron metabolism. *Cell, 168*, 344–361.

Shenoy, N., Vallumsetla, N., Rachmilewitz, E., Verma, A., & Ginzburg, Y. (2014). Impact of iron overload and potential benefit from iron chelation in low-risk myelodysplastic syndrome. *Blood, 124*, 873–881.

CHAPTER 73

Transfusion-Transmitted Diseases

Louis M. Katz, MD and Roger Y. Dodd, PhD

Introduction: Transfusion-transmitted diseases (TTDs) are caused by viruses, bacteria, protozoa, and prions. Examples of broad spectrum of infections of contemporary interest to transfusion medicine community include *Babesia*, *Plasmodia*, dengue, and Zika viruses in addition to historically important transfusion-transmitted agents—human immunodeficiency virus (HIV), hepatitis C virus (HCV), and hepatitis B virus (HBV). Bacterial sepsis from contaminated platelets is considered elsewhere.

The minimal prerequisites for a pathogen to require consideration as a transfusion-transmitted infection include the following:

- The occurrence of *asymptomatic* chronic or acute blood-borne phase, during which individual is clinically well enough to qualify as a donor.
- Survival of the agent in contemporary components.
- Infectiousness by parenteral route.
- Reservoir of susceptible recipients.
- Occurrence of clinical morbidity in recipients.

Level of concern afforded to agent should be predicated on that agent's

- Clinical severity after transfusion transmission.
- Incidence and prevalence.
- Rate of emergence in a potential donor population.

TTD mitigation is based on careful donor education and recruitment, donor interview and physical examination, donor testing, and application of good manufacturing practices. These combined strategies result in much lower donor prevalence and incidence rates of relevant pathogens than in general population (Table 73.1) and have lowered residual risks of transfusion-transmitted HBV, HCV, and HIV, compared with the general population, by 3–5 orders of magnitude.

Hepatitis Viruses

Hepatitis B: HBV is a DNA virus from the family *Hepadnaviridae.* It is transmitted parenterally, sexually, and perinatally. In the United States and other low-prevalence countries, most infections are horizontal from adult to adult, while in countries with high prevalence, both horizontal and vertical (i.e., perinatal) infections are common. There are an estimated 800,000–1.4 million chronic infections in the United States, but in this era of universal prenatal screening and universal immunization, fewer than 20,000 acute infections are estimated to occur annually.

Transfusion Medicine and Hemostasis. https://doi.org/10.1016/B978-0-12-813726-0.00073-8

TABLE 73.1 Prevalence and Incidence Rates/100,000 for HBV, HCV, and HIV—United States

Agent	Donor Prevalence	Population Prevalence	Donor Incidence	Population Incidence
HBV	1760	4900	3.0	111
HCV	299	1800	1.9	13
HIV	20	136	1.6	15

Acute Infection: Incubation period is 60–150 days. Incidence of symptomatic hepatitis varies by age. Most children <5 years are asymptomatic. Of the infected persons, 30%–50% of ≥5 years of age will have signs and symptoms clinically suggestive of hepatitis, and 0.5%–1% of infected individuals, usually ≥60 years of age, will have fulminant acute infection resulting in death from liver failure. Testing in acutely infected adults demonstrates the presence of viral DNA, surface antigen (HBsAg), and antibody to the core antigen (anti-HBc IgM). Development of antibody to HBsAg (anti-HBs) and disappearance of HBsAg indicates development of immunity generally without further risk of clinical disease. Asymptomatic infection is common and reactivation can occur with immunosuppression and during treatment of hepatitis C with direct-acting antivirals.

Chronic Infection: 90% of infected infants, but only ~6% of persons infected after age 5, develop chronic HBV infection. Approximately 25% of those chronically infected during childhood and 15% of those who become chronically infected later die prematurely from cirrhosis or liver cancer if untreated, but most remain asymptomatic unless affected by onset of end-stage liver disease. Patients with chronic infection are DNA, HBsAg, IgG anti-HBc positive, and generally IgM anti-HBc and anti-HBs negative. Occult HBV infection, characterized by the presence of anti-HBc and HBV DNA in the absence of HBsAg, appears to result in transfusion transmission only rarely where anti-HBc donor screening is not used.

Risk of Transfusion Transmission: In the United States, with testing for HBsAg, anti-HBc, and minipool HBV DNA, current residual risk estimates are ≤1:1,000,000 and expected to fall further due to universal immunization.

Hepatitis C: HCV is a parenterally transmitted RNA *Hepacivirus* in flavivirus family, distinguished by low rate of symptomatic acute infection but high proportion of chronic infections that result in substantial morbidity and mortality from chronic liver injury over long periods of observation. In the United States, there are estimated 3.2 million chronic infections. Injecting drug use remains predominant mode of transmission in the United States, but nonsterile medical injections have been important in the developing world. New infections in the United States declined from 240,000 annually in the 1980s to 19,000 in 2006. Seroprevalence in the general US population is 1.8% compared with as high as 79% in intravenous drug users.

Acute infection with HCV is asymptomatic in 80% of patients. Approximately 20% will have clinical hepatitis after incubation period of 7–8 weeks. Chronicity develops in 75%–85% of infections, with cirrhosis developing in 20%–30% after an average of ≥20 years. Hepatocellular carcinoma develops in cirrhotic patients an average of 30 years after infection. End-stage liver disease from HCV infection is the most common indication for liver transplantation in this country, and mortality attributed to HCV is now higher than from HIV. Very high effectiveness (sustained viral response rates of >95%) of direct acting antiviral agents for HCV has been a remarkable recent development.

Risk of Transfusion Transmission: During the 1970s, 10% or more of US transfusion recipients developed evidence of hepatitis, now known to have been largely HCV. With the use of serology since 1990 and nucleic acid testing (NAT) since 1999, residual risk has fallen to 1:1,148,000 products.

Hepatitis A: Hepatitis A virus (HAV), an RNA virus in the family *Picornaviridae*, usually causes asymptomatic infection or self-limited acute disease with a mortality rate of <0.2%. It does not result in chronic infection. HAV is transmitted during close person-to-person contact with infected individual or after ingestion of contaminated food or drink. Seropositive rate in the United States is ~33%. Incidence in the general population is <7/100,000. Viremia occurs 7–14 days before clinical onset, and infectious virus can survive in banked blood for its storage duration. HAV transmission by blood is rare, but can be amplified in neonatal intensive care units, where multiple infants may be infected by receipt of aliquots of blood components from an individual infected donor. Its rarity is attributed to short infectious viremia, low incidence of acute infection in US donors, absence of chronic infection, preexisting immunity in recipients, and neutralization of virus from a concurrent blood product containing specific antibody. Solvent–detergent process used to make pathogen-reduced plasma fails to inactivate this nonenveloped virus, and transmission has occurred. Subsequently, donor NAT for source plasma was implemented, and it is recommended that recipients of plasma derivatives be vaccinated. Universal vaccination in childhood is now recommended by the Centers for Disease Control and Prevention, and continuing decline in new infection are expected.

Hepatitis E: Hepatitis E virus (HEV) is an RNA virus endemic to large areas of the developing world, where specific genotypes (1 and 2) cause epidemics associated with fecal contamination of potable water. Sporadic infections in developed countries are generally caused by zoonotic genotypes (3 and 4). While usually self-limited, HEV can be lethal in pregnant women, their fetuses and patients with chronic liver disease. Chronic infections with genotypes 3 and 4 are recognized in the developed world in immunocompromised, e.g., transplant recipients. Transfusion transmission remains rare, but has been reported from endemic areas, and more rarely in nonendemic regions, resulting from asymptomatic donor viremia. Donor studies in the United States and Canada demonstrate low prevalence of RNA positivity. Rates in the EU are considerably higher, generally attributed to dietary exposure to poorly cooked pork products. Because of the latter, donor NAT screening has been implemented in the United Kingdom and the

Netherlands. Transfusion-transmitted HEV has not been reported in the United States and need, or not, for mitigation strategies is subject of chronic controversy.

A variety of "non–A–E" hepatitis agents have been associated with residual reports of posttransfusion hepatitis but on further study have not proven to be causal. Examples include SEN-V, TTV, and human *Pegivirus* or HPgV (formerly hepatitis G or GBV-C). While infections with these agents do not qualify as TTDs, identification of putative pathogens will continue in the era of sensitive molecular "pathogen" discovery, and the role of such agents in transfusion adverse outcomes will require careful investigation. Metagenomic next-generation sequencing and other studies do not support the existence of additional hepatitis viruses.

Retroviruses: These RNA viruses are characterized by the enzyme reverse transcriptase and a unique replication cycle. Viral RNA is reverse transcribed to DNA and integrated in host genome where it persists as a provirus. Subsequent transcription and translation of viral genes and processing of proteins results in progeny viruses that bud from plasma membrane and can infect other cells. They can also spread by fusion of infected and uninfected cells and replication of integrated proviral DNA during mitosis or meiosis.

Human Immunodeficiency Virus(es): HIV-1 and 2 are lentiviruses that cause acquired immunodeficiency syndrome (AIDS) via depletion of CD4-positive T-lymphocytes, producing profound cellular immunodeficiency. Transmission is most commonly sexual, vertical, or by breastfeeding, but HIV transmission by injection drug use and transfusion of infected blood products is highly efficient. HIV-1 circulates as both cell-associated and cell-free virions at levels of $\geq 10^5$/mL with high replication rate.

Risk of Transfusion Transmission: Since 1999 with sequential application of donor education and deferral strategies, serological donor screening, and NAT, residual risk for HIV is estimated at 1:1,470,000 transfusions. Window period with current tests is ≤9 days from infection. Only five reported transmissions by blood in the United States since 1999, all of which appear to have been from very early donor infections in window period.

Human T-Cell Lymphotropic Retrovirus: Deltaretroviruses HTLV-1 and -2 have 60%–70% RNA sequence homology and shared tropism for T-lymphocytes: HTLV-1 preferentially for CD4+ lymphocytes, and HTLV-2 for CD8+ lymphocytes and less so for CD4+ lymphocytes, B lymphocytes, and macrophages. Unlike HIV, HTLV is very highly cell-associated, rarely detected in cell-free plasma and replicates slowly in infected humans. HTLV-1 is distributed worldwide, with endemic foci in southern Japan, the Caribbean, and certain parts of South America, Africa, the Middle East, and Melanesia. HTLV-2 is endemic among Amerindians in both North and South America and among African Pygmies. An HTLV-2 epidemic has occurred over past 40–50 years among injection drug users in the United States, Brazil, and Europe. Transmission of both HTLV-1 and -2 occurs by parenteral exposure, sexual contact, and vertical transmission from mother to child during pregnancy and breastfeeding. In the United States, risk factors for HTLV-1 include a link to an endemic area and multiple sex partners. Injection drug use or sex with a drug user is significantly associated with HTLV-2 infection.

Most HTLV-1 and -2 infections are asymptomatic, but there is a 1%–5% lifetime risk that disease will develop as long as decades after infection. Diseases associated with HTLV-1 infection include adult T-cell leukemia/lymphoma, HTLV-associated myelopathy/tropical spastic paraparesis (HAM/TSP), lymphocytic pneumonitis, uveitis, polymyositis, and arthritis. HTLV-2 does not appear to cause hematologic malignancy but has been associated rarely with HAM/TSP. HTLV-2 infection in blood donors is associated with higher rates of common infections (acute bronchitis, pneumonia, and urinary infection) than in general population, suggesting subtle immunomodulatory effect.

Risk of Transfusion Transmission: In the United States, seroconversion rates after receipt of seropositive cellular components have ranged from 14% to 30%. RBCs, platelets, and whole blood, but not frozen plasma, have resulted in seroconversion of transfusion recipients. Products stored for >7 days or leukoreduced are less likely to transmit the viruses.

Seroprevalence in blood donors in the United States is approximately 10–20/100,000 donors. Incident infections are rare in repeat donors, an observation that has led to one-time testing of donors in some European countries. With serological testing, residual risk of transfusion transmission is estimated at 1:2,700,000 products transfused.

Herpesviruses: Human herpesviruses (HHVs) are double-stranded DNA viruses characterized by establishment of lifetime latency in permissive cells after primary infection. Cytomegalovirus (CMV) is the herpesvirus most relevant to transfusion medicine. Other leukotropic herpesviruses (Epstein–Barr virus [EBV] and human herpesvirus 8 [HHV-8]) may also be present in donors and blood products, but specific interventions have not generally been proposed or required.

Cytomegalovirus: Community CMV transmission is usually through close contact to a person shedding infectious virus. Transplacental and perinatal infections also occur. Recent studies show that infant infections, some of which were previously attributed to transfusion transmission, are more probably result of maternal infection and via breast milk. CMV is transmitted by transfusion, hematopoietic stem cell (HSC), and solid organ transplantation. In the United States, in the population-based NHANES III study, 59% of individuals >5 years old were seropositive indicating persistent CMV infection. This is age dependent, rising to more than 90% in octogenarians. Age of acquisition is associated with race/ethnicity and other socioeconomic and demographic correlates. More than 50% of US blood donors are CMV seropositive.

Primary infection in immunocompetent individuals is usually unrecognized but may involve generally mild "mononucleosis-like" syndrome. Latency is established, and persists for life in cells of the monocyte/macrophage lineage, from which reactivation can occur. Transplacental infection after primary infection during pregnancy can result in fetal wastage, intrauterine growth retardation, deafness, mental retardation, blindness, and thrombocytopenic bleeding. Primary infection in premature infants and recipients of solid organ or HSC transplantation can result in pneumonitis, hepatitis, retinitis, marrow suppression, and multisystem organ failure that can be lethal. Reactivation of latent infection during late-stage immunodeficiency in AIDS is associated with blinding retinitis and other serious morbidity and mortality from infection in several organ systems. CMV infection can be detected with anti-CMV serology (of limited diagnostic use due to high seroprevalences), CMV antigen testing, and amplification of CMV DNA.

Risk of Transfusion Transmission: Those at greatest risk for transfusion transmission of CMV are seronegative recipients with cellular immune defects (including premature infants) who may develop primary infection after receipt of cellular components from seropositive donors. Latently infected leukocytes are primary vector of transfusion transmission and leukoreduction greatly decreases transmission. Expansion of routine leukoreduction has been associated with decreased transfusion transmission and is an argument for universal leukoreduction given imprecision of identifying at risk recipients. Furthermore, use of active CMV surveillance with antigen and NAT in high-risk patient populations and availability of effective antiviral therapy has attenuated clinical impact of CMV infection. Requirement for CMV seronegative units has become increasingly controversial in the face of these observations and, for example, in Canada use of serological testing is being restricted to components intended for intrauterine transfusion.

Epstein–Barr Virus: EBV is associated with a variety of diseases including infectious mononucleosis, Burkitt lymphoma, nasopharyngeal carcinoma, and posttransplantation lymphoproliferative disease (PTLD). Natural EBV transmission is via infected saliva. Acute infection in children is generally not recognized. In adolescents and adults, it results in infectious mononucleosis. EBV is latent in B-lymphocytes.

Occasional cases of posttransfusion EBV infection, including PTLD, have been reported. Serological screening of blood donors is impractical due to the high population prevalence (~90%). Donation age individuals with active infection (infectious mononucleosis) are usually symptomatic and would not qualify to donate. Contemporary leukoreduction processes reduce EBV genetic material to undetectable levels and may effectively reduce transmission risk.

Human Herpesvirus 8: HHV-8 is associated with Kaposi sarcoma, primary effusion lymphoma, and multicentric Castleman disease. It is transmitted primarily by direct person-to-person contact in the developing world, and by sexual contact among gay men. Seroprevalence in US blood donors is 3%–4%. Among gay men without HIV, this is as high as 25%. The latter, and concern about transfusion transmission, has been among reasons cited for maintenance of stringent deferrals of men having sex with other men. In Uganda, where the HHV-8 seroprevalence is 40%, seroconversion after transfusion of seropositive product was seen in 2.3% of seronegative recipients. One study in the United States demonstrated a 0.082% risk of seroconversion per transfused component. There is no documentation of transfusion transmission of HHV-8 associated with clinical illness. An association with increased posttransfusion mortality was suggested in an Ugandan observational study where fresh, nonleukoreduced products are commonly used.

Other Viruses

Parvovirus B19: This is a nonenveloped, single-stranded DNA virus from the family *Parvoviridae*. B19 has at least three genotypes: genotype 1 infections predominate in the United States and Europe; genotype 2 infections are confined to those born before 1973; and genotype 3a and 3b infections occur in West Africa. Infection with B19 is very common, with 50% of US high school students being seropositive, increasing to 90% in older adults.

Natural transmission is respiratory and transplacental to the fetus from up to 30% of women experiencing primary infection during pregnancy. Transfusion of blood and blood components, plasma derivatives, and organ transplantation are infrequent routes.

B19 causes erythema infectiosum or "fifth disease" (generally benign febrile rash), oligoarthritis, and neurologic and myocardial infections. Acute infection features viremia up to 10^{12} genome equivalents/mL until seroconversion occurs at ~9 days. B19 may persist in the bone marrow, liver, tonsils, and skin of healthy persons with no recognized clinical significance.

B19 is highly tropic for erythroid progenitor cells where P blood group antigen (globoside) is its receptor. Acute infection impairs erythropoiesis for 7–10 days, with complete cessation for 3–7 days, causing aplastic crises in patients with sickle cell disease, other hemolytic diseases, and conditions associated with decreased RBC survival. In women infected during weeks 9–20 of pregnancy, fetal death occurs in 10%–15% from hydrops fetalis. Persistent high-titer viremia and sustained RBC aplasia occur infrequently, primarily in those who fail to develop neutralizing antibodies, including those receiving intense chemotherapy and immunosuppression, organ transplant recipients, and those with HIV.

Rare transmissions by labile components and plasma derivatives are recognized. They result from donations during the 1–2 week high-titer viremic period. Among volunteer whole blood donors, 1/6000–1/16,000 have titers $\geq 10^6$ IU/mL. In a study involving more than 12,100 B19 DNA tested blood donations given to surgical patients (who had 78% pretransfusion B19 IgG seroprevalence), no B19 transmissions occurred from units containing $<10^6$ IU/mL. Hence, most transfusion recipients are at low risk of an infectious exposure. Plasma-derivative manufacturers use "in process" NAT to identify and remove infected source (generally paid donors) and recovered (untransfused plasma from volunteer whole blood donations) plasma units to produce pools with $\leq 10^4$ IU/mL. Results are not generally available until after transfusion or expiration of recovered plasma and donors are not notified.

Arboviruses

West Nile Virus: West Nile virus (WNV) is a flavivirus that was endemic to Africa, Middle East, and Eastern Europe. After introduction to the United States in New York City in 1999, it spread coast-to-coast in only 3 years. It is transmitted by mosquitoes to birds and mammals. Incubation period in humans is 3–14 days. Infection is subclinical in 80% of cases. Symptoms range from fever and headache (West Nile fever) to meningoencephalitis (~1/150 infections). Acute flaccid paralysis is well recognized and source of chronic disability. There is no human vaccine, and prevention is by avoiding mosquito bites.

Risk of Transfusion Transmission: Since NAT and refined strategies were implemented to switch from pool to individual NAT after reaching predetermined triggers of WNV activity in blood regions, 14 transmissions have been reported from subsequent 14 transmission seasons. These were associated with the use of insensitive triggers, suboptimal communication about donor infections in overlapping blood regions and, in a single case, an untested, urgently released granulocyte product.

Dengue Virus: Mosquito-borne infection is of current interest as an emerging pathogen with potential for transfusion transmission. Dengue is caused by four related flaviviruses spread person-to-person by *Aedes aegypti* and *Aedes albopictus*, vectors that are potentially present in 16 and 35 states in the United States, respectively. Forty percent of the world's population lives in areas with dengue risk, including many areas frequently visited by US travelers. It has spread rapidly in Latin America and the Caribbean since the 1980s. Dengue is endemic in Puerto Rico, US Virgin Islands, and American Samoa, and there have been outbreaks in Hawaii, Texas, and Florida during the last 10 years.

Most infections are asymptomatic, but illness ranges from an undifferentiated fever to classic "break bone fever" and severe dengue (dengue hemorrhagic fever and dengue shock syndrome). A ~7-day viremia is a feature of both asymptomatic and symptomatic infection. Asymptomatic blood donors from Hong Kong, Singapore, and Puerto Rico have transmitted dengue to recipients. These reports are limited, compared with high rates of vector-borne infection, but there is no systematic surveillance for transfusion-transmitted dengue and its recognition during intense arthropod-borne epidemics is problematic. Viremic, asymptomatic donors have been identified in Brazil, Central America, and Puerto Rico using nucleic acid and antigen detection tests. Rates of donor dengue viremia in Puerto Rico have been comparable with those of WNV viremia found in US donors during the most active WNV seasons. Retrospective NAT in a Brazilian study of 39,134 donor samples identified 16 transfusions of RNA-positive blood, and with five probable and one possible transmissions. Chart review found no evidence of morbidity attributable to these infections.

Current risk of dengue from transfusion in the United States relates mainly to donations from asymptomatic or presymptomatic travelers returning to the United States from endemic areas. Deferral for travel to malarious areas offers some protection, but proportion of dengue affected areas visited are malaria-free, and travelers to those areas could conceivably introduce dengue into the blood supply. Transfusion-transmitted dengue was identified by the AABB's Transfusion Transmitted Diseases (TTD) Committee, as one of three highest priority emerging infections posing a potential threat to transfusion recipients in the United States and Canada and has been the subject of discussions at the FDA's Blood Products Advisory Committee.

Chikungunya Virus: This alphavirus, that caused geographically limited, cyclic outbreaks mainly in sub-Saharan Africa after its recognition in the 1950s, spread explosively into the Indian Ocean and beyond beginning in 2004. This was, in part, attributable to single-nucleotide polymorphism associated with adaption to its mosquito vector(s). While the large bulk of infections have occurred in the developing world, small clusters of autochthonous infections have occurred in Europe and elsewhere. Chikungunya causes nonspecific acute febrile illness that is generally short-lived but can be associated with prolonged arthritis and disability. Mortality and major morbidity are unusual. Transfusion transmission has not been observed despite millions of infections during this pandemic, while modeling exercises using conservative inputs suggest that asymptomatic, viremic donors should be common during substantial outbreaks. During extensive epidemic on in La Reunion, measures to protect blood supply from its theoretical risk included suspension of RBC collections, their import from the mainland, and urgent implementation of pathogen reduction for platelets. The virus reached the Western Hemisphere in the Caribbean in 2013 with subsequent wide distribution, and more than 2,400,000 cases suspected or confirmed through March 2017. Local epidemics persist, with substantial activity in the past year in Brazil and the Southern Cone. In the United States, limited local transmission occurred in Florida, but imported cases continue to be recognized (175 reported in 2016). No action has been taken by the blood community or the FDA. Travel surveys by the AABB's TTD committee, conceived to address potential risks from dengue, chikungunya, and other acute "tropical" infections, estimate that a 28-day deferral for travel outside the United States and Canada would result in deferral of 2.64% and 4.02% of current donors in summer and winter, respectively.

Zika Virus: Zika, a flavivirus identified in the late 1940s in nonhuman primates, emerged from Africa, spread to South Asia and the Pacific during the 2000s and to the Americas, with an explosive epidemic starting in 2013–14. A signal development was its association with severe neurodevelopmental congenital Zika syndrome and Guillain–Barre syndrome in adults. Zika virus transmitted by transfusion has been described in Brazil in four patients, but recipients did not demonstrate any relevant symptoms. It is very likely that Zika can also be transmitted by transplanted organs. Sensitivity to transfusion risk was magnified by occurrence of severe neurologic damage (especially microcephaly) in the fetuses of mothers infected during pregnancy. Suggested strategies for dealing with transfusion risk were varied. In nonendemic areas, risks may be reduced by excluding donors who have exposure through travel or sexual contact with someone at risk. In both endemic and nonendemic areas, risk can be further reduced by NAT of donors (serology being of essentially no value because antibody becomes detectable after most of the viremic phase and further, is not specific for Zika) or by pathogen reduction of platelet and plasma products. Initial FDA guidance followed onset of explosive epidemic in the Caribbean. It required suspension of collections in areas with local transmission unless pathogen reduction and/or testing were implemented. Donor travel and sexual contact deferrals were included. The FDA then mandated the emergency implementation of universal individual NAT under IND when autochthonous transmission appeared in Florida, beginning with states most at risk of local transmission. Fifty-one confirmed positive donors from 11.6 million screened were identified through September 2017. Infection was predominantly acquired during travel to areas with local transmission outside the continental United States. CDC has estimated cost of this approach at $137 million per year, with clinical burden addressed being speculative. While incidence of new vector-borne infections fell precipitously between 2016 and 2017, the situation locally and globally is clearly fluid. As the epidemic continues to evolve, current approach is being revisited, with active discussion of a switch to minipool screening and ID-NAT conversion using an approach similar to WNV.

Anaplasma phagocytophilum: This is the agent of human granulocytic anaplasmosis (HGA), a gram-negative obligate intracellular bacterial pathogen, transmitted primarily by *Ixodes* ticks. Clusters, called morulae, of the organism can be seen in neutrophils on examination of peripheral blood smears. Incubation after tick exposure is 5 days to 3 weeks. Recognition and incidence of HGA appear to be increasing in the United States, especially in Northeast and North Central states, and in the EU.

Most vector-borne infections are not recognized clinically, but symptoms include malaise, fever, myoarthralgia, and headache. In the elderly and immunocompromised, it can cause acute kidney injury, sepsis syndrome, disseminated intravascular coagulation, and respiratory distress. Laboratory abnormalities are nonspecific but include cytopenias and elevated transaminase levels. Therapy with tetracyclines is effective. In addition to direct visualization of *Anaplasma phagocytophilum* on blood smears, serology and PCR are available for diagnosis.

Transfusion transmission is well documented with ~10 cases published, with single recipient fatality. Transmission has been reported from RBCs and platelets, and the organism survives at least 3 weeks at refrigerator temperatures. Leukoreduction reduces titers by >2 orders of magnitude but transmission from both leukoreduced RBCs and platelets, suggests that the infectious dose is quite small.

No donor screening assay are FDA-approved, leukoreduction is insufficient to prevent transmission, and, while HGA is effectively inactivated by pathogen reduction processes in development, these are not (yet) available for all blood components. Whether current risk of HGA from blood justifies specific interventions is controversial, and FDA-sponsored workshop in 2017 recommended continued surveillance for the time being.

Protozoa

***Plasmodium* Spp. (Malaria):** Five *Plasmodium* species, *Plasmodium falciparum*, *Plasmodium malariae*, *Plasmodium vivax*, *Plasmodium ovale*, and *Plasmodium knowlesi*, caused more than 214,000,000 malaria cases and almost 438,000 deaths in humans in 2015. *P. falciparum* results in most serious disease and large fraction of malaria mortality that concentrates in children ≤5 years old in sub-Saharan Africa. Transmission to humans is by female anopheline mosquitoes. Infection usually persists for 1–2 years, but can last decades, especially in the case of *P. malariae*.

Infection: Malarial infection results in fever, chills, headache, hemolytic anemia, and splenomegaly. More severe manifestations include cerebral malaria, acute kidney injury, and cardiac involvement. When individuals have been repeatedly infected they develop partial immunity and can be asymptomatic but parasitemic, resulting in transmission risk from donors to transfusion recipients. Diagnosis is classically by microscopic examination of blood smears, which is technically demanding and insensitive in donors of greatest interest. Rapid diagnostic tests that detect parasite proteins are being used increasingly in endemic areas where high-quality microscopy may be operationally difficult. Antibody assays have been developed, but in endemic areas are nonspecific for transmissible parasitemia and do not detect all five species with optimal sensitivity. NAT has generally been considered research tool rather than routine diagnostic, and may not, in fact, be sensitive enough for parasitemia levels associated with infectivity. A randomized study of whole blood pathogen reduction in Ghana has demonstrated decreased risk of transmission from 22% to 4% among recipients of untreated versus treated units.

Risk of Transfusion Transmission: Approximately 100 cases of transfusion-induced malaria have been recognized and reported in the United States since 1963. Since 2007, there have been a total of 9 cases reported (0-3 per year). Generally, these cases result from infected donors who have lived in endemic regions for prolonged intervals and are asymptomatic due to partial immunity; casual travelers are rarely, if ever, the source. Prevention of malaria transfusion transmission uses donor criteria based on natural histories of these infections. They require travelers from nonendemic countries to malaria endemic areas, who have no immunity to attenuate symptoms, to be deferred for 1 year. Those who have lived in and/or emigrated from such areas, and may be semiimmune, are deferred for 3 years, as are those who have had, and recovered from, malaria. Many US cases would have been prevented had extant donor deferral criteria been successfully applied. Even with these criteria, risk persists because transfusion transmissions of *P. falciparum*, *P. vivax*, and *P. ovale* have been reported 13, 27, and 7 years, respectively, after departure from malarious areas. Because *P. malariae* infection can persist for >70 years without symptoms, elimination of transfusion-induced malaria using donor history alone is practically impossible. Licensed in vitro screening for at-risk donors using antibody tests, antigen detection or NAT is not available in the United States. Donor deferral guidelines have been revised by FDA to reduce annual loss of more than 60,000 potential donors with travel to areas of Mexico where malaria risk is minimal.

Trypanosoma cruzi: The zoonotic agent of Chagas disease is transmitted mainly in South and Central America and Mexico, primarily by bite of infected triatomine (reduviid or "kissing") bugs. In Latin America, vertical transmission from infected mothers, oral transmission from foodstuffs contaminated with infected arthropod feces and transfusion (for which donor serological screening is long-established) are alternative routes. Parasitemic reservoir hosts (wide variety of mammals) and infected triatomine vectors are present in southern half of the US, but autochthonous, vector-borne infection is rare. In the US, most cases are associated with prior extended residence in Latin America, and less so to apparent vertical transmission in immigrant families. Acute vector-borne infections are generally asymptomatic or demonstrate transient inflammatory signs at bug bite site, although younger children may develop acute myocarditis or meningoencephalitis. Signs of acute infection resolve without treatment, but lifelong low level parasitemia persists, giving rise to transfusion transmission risk. 20%–40% of chronically infected individuals develop cardiac or gastrointestinal symptoms years to decades later.

Seven reported cases of transfusion transmission from the US and Canada and prevalence of estimated 100,000 infected immigrant prospective donors resulted in enthusiasm for donor screening, and the FDA licensed donor test for antibodies to

Trypanosoma cruzi in donors in 2006. Voluntary testing from 2007 onward demonstrated infection rate of ~1/30,000 donors, concentrated, as expected, in areas with large Latin American immigrant populations. Lookback studies of >250 prior recipients from infected donors identified by testing have found only two infections (0.79%), both from a single platelet donor. This contrasts with much higher historical rates from Latin America. Explanation is likely complex but includes historic transfusion of fresh whole blood in Latin America, inactivation of the parasite under current component processing, and storage conditions in the United States and possibly parasite strain differences in the United States compared with Latin America where transfusion transmission was prevalent. Follow-up of test-negative donors in the United States did not identify a single incident infection during more than 6 million person-years of follow-up. In light of these data on transmission and lack of incident infections, the FDA permits a unique selective donor screening algorithm, whereby a donor is indefinitely qualified using a single negative test. This results in testing <20% of current donations.

***Babesia* spp.:** Babesiosis is caused by intraerythrocytic protozoans *Babesia microti*, *Babesia duncani*, *Babesia divergens* (limited primarily to Europe, where transfusion transmissions are not recognized), and *B. divergens*–like (MO-1 and EU-1) strains. *B. microti* causes the large majority of human infections. In the United States, white-footed mouse is the reservoir and deer tick, *Ixodes scapularis* (also vector of Lyme, borreliosis, and HGA), is the vector. Transmission follows bites from infected ticks, primarily nymphs, in spring and summer. Although reported from >20 states, the large majority of infections are acquired in seven states (MA, RI, CN, NY, NJ, WI, and MN). Geographic range is expanding, attributed to expansion of deer populations, increased incursion of humans into endemic habitat and improving awareness.

Incubation period varies from 1 to 9 weeks. Babesiosis severity relates primarily to patient's clinical status. Very young, elderly, splenectomized and those with hereditary hemolytic anemias have historically been considered at greatest risk for morbidity and mortality. Analysis of reported transfusion-transmitted infections concluded that risk is more generalized among transfusion recipients and argued those at highest risk for morbidity receive "*Babesia*-safe" components. Approximately one-third of vector-transmitted infections remain asymptomatic, parasitemia may persist for >2 years, and transfusion cases are identified year-round, so spring-summer tick-borne season has little relevance for transfusion transmission mitigation.

Typical babesiosis symptoms resemble malaria (fever, fatigue, malaise, myalgia, anorexia, nausea, vomiting, and diarrhea) with laboratory findings consistent with hemolysis (anemia, reticulocytosis, and hemoglobinuria). Case fatality rates ~5% after vector transmission.

More than 200 cases of transfusion-transmitted *Babesia* infections have been reported with >2/3 cases from past 15 years. All-cause (i.e., not attributable) mortality rate approaches 19% after transfusion transmission. Case reports of transfusion-associated *Babesia* infections implicate RBC products, cryopreserved RBC products, and several cases involve whole blood–derived platelet units, secondary to RBC contamination. Cases have occurred primarily in endemic states, and those in nonendemic states largely reflect interstate and interregional movement of both blood and donors. Risk is likely underestimated due to failure to recognize infection and/or its relationship to recent transfusion. Babesiosis is now nationally notifiable and included in biovigilance network of the National Healthcare Safety Network.

Diagnostic laboratory evaluation includes blood smear examination, which requires differentiation of *Babesia* sp. from *Plasmodium* sp. infections. Molecular and serologic tests are available. Indirect immunofluorescence antibody testing and immunoassays are available from commercial and public health laboratories, and both serological and NAT are under IND for donor screening. Diagnostic and donor screening antibody tests may remain positive after resolution or cure of babesiosis and are thus somewhat nonspecific with regard to transfusion risk. PCR testing is more sensitive for detecting transmissible infection before seroconversion but may miss donors with very low parasitemias. Approval of donor screening tests, followed by FDA guidance for their use, is expected. Absent investigational donor screening or an FDA-licensed donor test, mitigation of transfusion transmission involves deferral of potential donors with history of *Babesia* infection. The growing number of transfusion transmissions demonstrates this is not optimal. Asking about tick exposures has no predictive value for serostatus when used as donor screening question.

Nucleic acid and serological testing used under investigational exemptions in highly endemic blood regions have been shown to essentially eliminate transfusion transmission in context of passive surveillance. Whether PCR alone or PCR with serology is an optimal donor screening strategy, and its regional extent is a subject of current controversy.

Transmissible Spongiform Encephalopathies: Transmissible spongiform encephalopathies (TSEs) (prion diseases) are rare, lethal neurodegenerative diseases caused by prion proteins called PrP^{TSE}, as distinguished from normal cellular protein, PrP^{C}. Consensus supports etiologic role of infectious proteins, devoid of nucleic acids. PrP^{TSE} is an abnormal conformation of PrP^{C} that recruits transformation of additional PrP^{C} to PrP^{TSE}, resulting in deposits of aggregates in central nervous system (CNS) tissue followed by progressive dementia and other characteristic neurological findings.

Transfusion transmission of sporadic, familial, or iatrogenic Creutzfeldt–Jakob disease (CJD) has not been observed and cohort studies of intensively transfused groups have failed to establish any epidemiologic association. In an ongoing lookback effort, no cases of CJD have been observed among 826 recipients (of whom 154 were still alive as of December 31, 2014) of blood from 65 donors subsequently diagnosed with CJD after 3934 person-years of follow-up. As a result of iatrogenic transmissions with long incubation periods (e.g., from human pituitary-derived growth hormone), concern arose in the mid-1990s that CJD transmission might occur from asymptomatic donors to transfusion or plasma-derivative recipients. This resulted in lifetime donor deferral for CJD risk (cadaveric human pituitary growth hormone receipt and dura mater grafts or family history of CJD).

Human variant CJD (vCJD) was recognized in 1996 in the United Kingdom. It occurs in younger patients than CJD and has distinctive clinical, radiographic, histopathologic, and biochemical features compared with sporadic CJD. Etiologic agent is the prion that causes bovine spongiform encephalopathy (BSE or "mad cow disease"), transmitted to by consumption of bovine products contaminated with infectious tissue after massive UK BSE epidemic of the 1980s and 1990s. Both the UK BSE and vCJD outbreaks have subsided, attributed to control of the food chain, after peak of 28 vCJD cases in 2000. As of October 2017, total of 178 definite or probable cases and deaths from vCJD have been reported in the United Kingdom. Fifty-three cases have been diagnosed elsewhere through January 2017, mostly in France and several other European countries. Four cases have been seen in the United States (associated with exposure in endemic countries), and two, in Canada. Most patients with extra-European cases had >6 months exposure in the United Kingdom during BSE epidemic.

While vCJD outbreak is waning (one case has been reported since 2013 in the United Kingdom), there is concern about further waves due to delayed onset of the disease in infected individuals. Abnormal prion protein accumulation has been documented in surgical tissue samples in the United Kingdom. Two large studies of non-CNS tissues archived after routine surgery in the United Kingdom have provided estimates of a 1/2000–4000 prevalence of subclinical vCJD. Negative controls were not included, and specificity of these results has been questioned. A more recent survey of appendices archived outside the recognized vCJD outbreak, found that cases with immunohistochemical evidence of infection were clustered immediately before and after the epidemic lending some credibility to existence of subclinical infection at a similar (statistically) rate of 1/4200 and mandating continued careful surveillance "second wave."

Polymorphism at codon 129 of the PrP^C gene leads to variation in susceptibility to and incubation period of human TSEs. All vCJD cases except the last UK case reported to date are methionine (MM) homozygotes at codon 129, compared with only 37% of general British population, and maximally susceptible to vCJD. A larger population (52%) of MM–valine heterozygotes and (11%) valine homozygotes may be at risk for infection with a longer incubation.

Four transfusion-transmitted infections with vCJD prion have been reported in the UK. Three of four were diagnosed with clinical vCJD 6.5, 7.8, and 9 years after transfusion of blood from two different donors who developed clinical vCJD 40 and 21 months after donating. All three were homozygous for MM at codon 129. The fourth case was heterozygous (MM/valine), had no clinical signs or symptoms of vCJD at death, but had abnormal prion protein aggregates in lymphoid tissues at autopsy, and was considered to have preclinical infection 5 years after transfusion from donor who developed clinical vCJD 18 months after donating. These four cases represent 6% of 66 UK recipients who received blood from 18 donors subsequently diagnosed with vCJD (and 23% of exposed MM homozygotes). Single case report epidemiologically links transfusion of plasma derivatives to potential transmission.

On the basis of vCJD prion distribution in lymphoid tissue, early observation that nearly all cases of vCJD were associated with potential exposure in the United Kingdom or to UK bovine products, and with subsequent reports of animal and human infection by transfusion, the FDA recommended vCJD donor deferrals in 1999. The first called for indefinite deferral of donors who had spent more than 6 months (subsequently reduced to 3 months) in the United Kingdom from 1980 to 1996, after control of food chain for BSE risk, and of recipients of bovine insulin from the United

Kingdom. Models predicted that this would remove 90% of the risk at a "cost" of deferring <5% of otherwise eligible donors. Deferrals were expanded to include US military personnel and their dependents who spent certain durations on military bases in the EU where UK beef was imported during the BSE epidemic, and recipients of transfusions in the United Kingdom and France. Experience at US blood centers has largely confirmed donor loss predictions of FDA models. With apparent waning of vCJD epidemic, current US donor deferrals have been reviewed by FDA, resulting in addition of transfusion in Ireland from 1980 to the present causing deferral, and with limitation of deferrals for potential exposure in France to that lasting 5 years or more from 1980 to 2001.

Prion removal strategies (e.g., affinity filters) remain under evaluation, especially in the United Kingdom, but performance characteristics are not suitable for consideration for use in the immediate future, especially as epidemic wanes.

There are no donor screening tests for vCJD; their use would be fraught with operational and ethical concerns, particularly among low-risk donors, where most positive results will be false-positive results.

Chronic wasting disease (CWD) of deer and elk is prevalent at rates as high as 15% in cervid populations in multiple areas of the United States and Canada and appears to be spreading. Concern has been expressed that, given hunting popularity, there is a risk of exposure to and infection with the CWD prion during handling or consumption of infected animals. Apparent clusters of classic CJD in hunters have been alleged, but after evaluation are not related to the CWD agent. There is currently no plan to intervene for this theoretical risk in hunters who donate blood beyond hygienic measures when handling cervids or their tissues.

Future Emerging Infections: This chapter highlights only selected, well-characterized TTDs. The transfusion medicine community's experiences with HIV, WNV, and most recently Zika virus provides sufficient evidence that continuous vigilance for new transfusion-transmitted infections is needed. That new agents will emerge is a given, and our community must make every effort to anticipate which of many pathogens may represent a threat. AABB's TTD Committee has published, and periodically updated, a list of more than 70 agents of interest with the potential to emerge in North America, and established on ongoing review aimed at early recognition, evaluation, and understanding of agents with potential to impact transfusion safety (Figure 73.1).

The most critical activity is "horizon scanning" that *daily* surveils a spectrum of online and print resources to identify new and emerging human pathogens. These include media reports, professional meetings, organizational and peer-reviewed publications, open source and subscription websites (e.g., ProMED, CDC.gov, WHO.int, PAHO.org), and personal networks. The most difficult element is effective hemovigilance requiring non–transfusion medicine clinicians to routinely elicit transfusion history when constructing differential diagnosis in patients with apparent infections, reviewing the prerequisites proposed in the introduction to this chapter, imputing blood as a "vector" and bring cases to blood bankers' attention.

In the future, pathogen reduction of all blood components may further mitigate the risk as well as advances in molecular methodologies for blood screening.

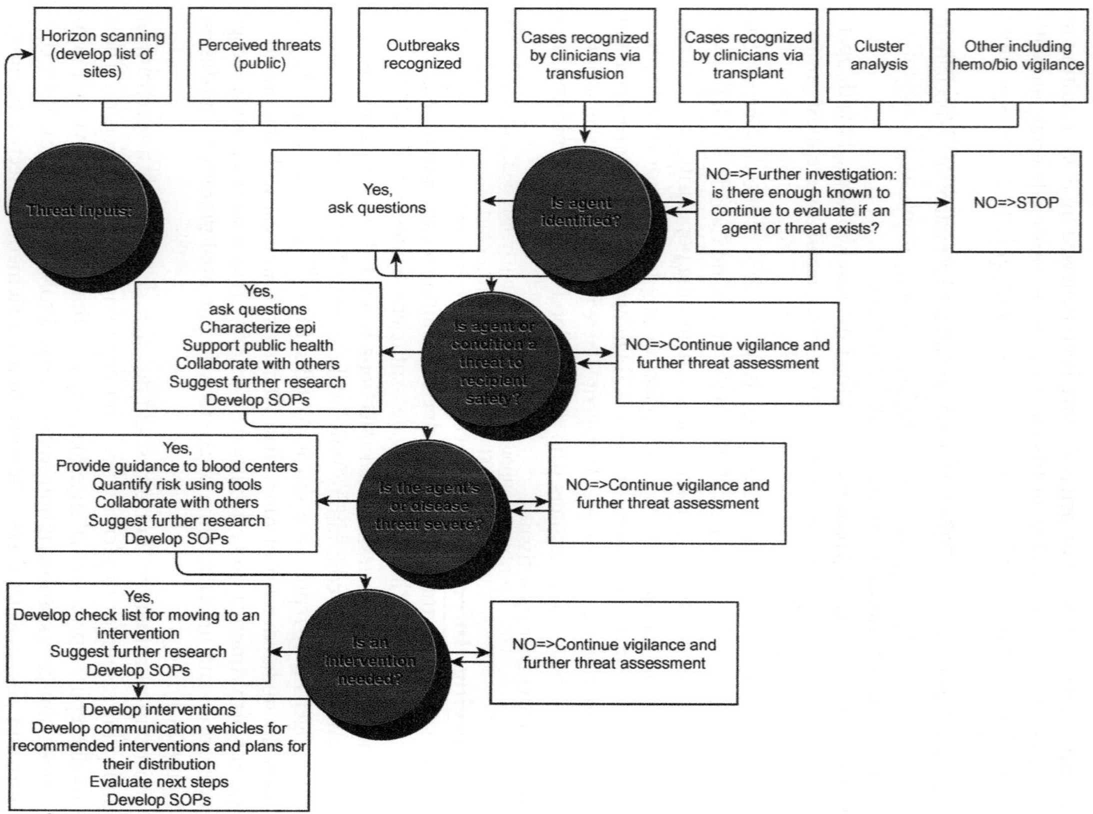

FIGURE 73.1 A framework for the recognition, assessment, and management of emerging infectious agents posing risks of transfusion transmission and disease. *SOP*, standard operating procedure. (From Stramer, S. L., Dodd, R. Y., & AABB Transfusion-Transmitted Diseases Subgroup (2013). Transfusion-transmitted emerging infectious diseases: 30 years of challenges and progress. *Transfusion*, 53, 2375–2383.)

Further Reading

Bern, C., Kjos, S., Yabsley, M. J., & Montgomery, S. P. (2011). *Trypanosoma cruzi* and Chagas' disease in the United States. *Clin Microbiol Rev, 24*, 655–681.

Herwaldt, B. L., Linden, J. V., Bosserman, E., et al. (2011). Transfusion-associated babesiosis in the United States: a description of cases. *Ann Intern Med, 155*, 509–519.

Katz, L. M., & Rossmann, S. N. (2017). Zika and the blood supply: a work in progress. *Arch Path Lab Med, 141*, 85–92.

Oei, W., Janssen, M. P., van der Poel, C. L., et al. (2013). Modeling the transmission risk of emerging infectious diseases through blood transfusion. *Transfusion, 53*, 1421–1428.

Perkins, H. A., & Busch, M. P. (2010). Transfusion-associated infections: 50 years of relentless challenges and remarkable progress. *Transfusion, 50*, 2080–2099.

Stramer, S. L., Dodd, R. Y., & AABB Transfusion-Transmitted Diseases Subgroup (2013). Transfusion-transmitted emerging infectious diseases: 30 years of challenges and progress. *Transfusion, 53*, 2375–2383.

Stramer, S. L., Hollinger, F. B., Katz, L. M., et al. (2018). *Emerging infectious disease agents supplement fact sheets*. Available at: http://www.aabb.org/tm/eid/Pages/appendix2.aspx.

Zou, S., Stramer, S. L., & Dodd, R. Y. (2012). Donor testing and risk: current prevalence, incidence, and residual risk of transfusion-transmissible agents in US allogeneic donations. *Trans Med Rev, 26*, 119–128.

CHAPTER 74

Overview of Therapeutic Apheresis

Elizabeth A. Godbey, MD Joseph Schwartz, MD, MPH and Huy P. Pham, MD, MPH

Apheresis is derived from a Greek word "aphairesis," which means "to remove forcibly." Whole blood is removed from a subject, separated into components extracorporeally (RBCs, white blood cells, platelets, and plasma), desired blood component is removed, and remaining components are returned with or without replacement fluid(s). Therapeutic apheresis (TA) is used to remove pathogenic substances in the plasma (termed therapeutic plasma exchange [TPE]) or pathogenic cells (termed cytapheresis).

Principle: Successful use of TA as a treatment modality requires that a disease results from a substance found in plasma (e.g., antibody to acetylcholine receptor in myasthenia gravis) or by a blood component (e.g., hemoglobin S RBCs in sickle cell disease), and that the pathogenic substance or component can be removed efficiently enough to permit resolution of the disease.

McLeod's criteria to determine TA success in disease treatment consist of three components: (1) disease pathogenesis suggests clear rationale for TA, (2) abnormality meaningfully corrected by TA, and (3) strong evidence TA confers meaningful clinical benefit. American Society for Apheresis (ASFA) publishes an evidence-based guideline for TA use, which is updated every 3 years. It provides an ASFA category (Table 74.1) and grade of recommendation (Table 74.2) for each disease entity. The newest edition (7th edition, 2016) describes 87 diseases and 179 clinical indications (Table 74.3).

Methods: Apheresis devices separate whole blood into component fractions, allow removal of the desired fraction, and return the remaining components. This separation can be achieved either by centrifugation or membrane filtration. Centrifugation separates based on differential density; filtration devices separate based on size. In the United States, most TA procedures are performed using centrifugation.

Vascular Access: Apheresis procedures require high blood flow rates (up to 150 mL/minutes). Peripheral venous access or double-lumen dialysis/apheresis catheter for adults are typically utilized. For pediatric patients, access size depends on the weight of the patient. Central and femoral catheters each pose their own risks and benefits. Patients who require prolonged TA may need a tunneled catheter or other long-term access device (e.g., arteriovenous fistula, graft, or port).

Volume Exchanged and Frequency: Determination of exchanged volume is based on a model of an isolated one-compartment intravascular space (Fig. 74.1). This model works best for components located predominantly in the intravascular

Transfusion Medicine and Hemostasis. https://doi.org/10.1016/B978-0-12-813726-0.00074-X

TABLE 74.1 American Society for Apheresis Category Indications

Category	Description
I	Disorders for which apheresis is accepted as first-line therapy, either as a primary standalone treatment or in conjunction with other modes of treatment.
II	Disorders for which apheresis is accepted as second-line therapy, either as a standalone treatment or in conjunction with other modes of treatment.
III	Optimum role of apheresis therapy is not established. Decision-making should be individualized.
IV	Disorders in which published evidence demonstrates or suggests apheresis to be ineffective or harmful. Institutional review board approval is desirable if apheresis treatment is undertaken in these circumstances.

From Schwartz, J., et al. (2016). Guidelines on the use of therapeutic apheresis in clinical practice – evidence-based approach from the apheresis applications committee of the American Society of Apheresis. *J Clin Apher, 31*, 149–338.

compartment (e.g., IgM, RBCs), and less well for IgG (which is 45% extravascular). Frequency is determined by time for reequilibration into intravascular space and the need to minimize the risk of bleeding as a result of depletion of coagulation factors, especially fibrinogen. Reequilibration of the intravascular IgG with extravascular IgG typically occurs within 2 days. Five to six one-PV exchanges over 14 days when combined with immunosuppressive medications achieve a 70%–85% reduction in IgG. Typically, plasma exchange processes 1–1.5 total blood volume (TBV) every other day.

Calculations: Calculations commonly used in TA include calculation of TBV, plasma volume, and extracorporeal volume (ECV). All approaches to calculating TBV overestimate in obese patients and underestimate in muscular patients. Nonetheless, they provide reasonable approximation. TBV is ~70 mL/kg. Plasma volume = TBV × (1 − Hct). Total RBC volume = TBV × Hct.

ECV is the amount of blood outside the patient filling the apheresis set and tubing. ECV varies by system, type of procedure, and ancillary equipment. ECV should not exceed 15% of TBV to avoid intraprocedural volume depletion and/or anemia. Typically, ECV is not a problem in adults. If ECV exceeds 10%–15% of TBV, then blood priming with RBCs or 5% albumin is performed.

Replacement Solutions: Replacement fluids may be used to replace the volume removed during TA (primarily to maintain intravascular volume). Additional functions include replacement of plasma proteins (e.g., immunoglobulins, coagulation factors) or blood components (e.g., RBCs). Using plasma or albumin/saline is based on disease, patient's risk of bleeding, and/or comorbidities. Plasma products are used only in the treatment of thrombotic microangiopathies and in patients with underlying coagulopathy (such as liver disease), actively bleeding, and/or immediately pre- or postinvasive procedures. Plasma can be given at the end of the procedure to prevent coagulopathy or treat procedure-related coagulopathy.

Anticoagulation: Anticoagulant solution is necessary to maintain flow through plastic tubing and devices. Heparin and/or citrate may be used. Most devices use citrate anticoagulant. If a device requires heparin, but patient has a history of heparin-induced

TABLE 74.2 Grading Recommendations

Recommendation	Description	Methodological Quality of Supporting Evidence	Implications
Grade 1A	Strong recommendation, high-quality evidence	Randomized controlled trials (RCTs) without important limitations or overwhelming evidence from observational studies	Strong recommendation, can apply to most patients in most circumstances without reservation
Grade 1B	Strong recommendation, moderate-quality evidence	RCTs with important limitations (inconsistent results, methodological flaws, indirect, or imprecise) or exceptionally strong evidence from observational studies	Strong recommendation, can apply to most patients in most circumstances without reservation
Grade 1C	Strong recommendation, low-quality or very low-quality evidence	Observational studies or case series	Strong recommendation but may change when higher quality evidence becomes available
Grade 2A	Weak recommendation, high-quality evidence	RCTs without important limitations or overwhelming evidence from observational studies	Weak recommendation, best action may differ depending on circumstances or patients' or societal values
Grade 2B	Weak recommendation, moderate-quality evidence	RCTs with important limitations (inconsistent results, methodological flaws, indirect, or imprecise) or exceptionally strong evidence from observational studies	Weak recommendation, best action may differ depending on circumstances or patients' or societal values
Grade 2C	Weak recommendation, low-quality or very low-quality evidence	Observational studies or case series	Very weak recommendations; other alternatives may be equally reasonable

From Schwartz, J., et al. (2016). Guidelines on the use of therapeutic apheresis in clinical practice – evidence-based approach from the apheresis applications committee of the American Society of Apheresis. *J Clin Apher, 31*, 149–338.

thrombocytopenia, then citrate should be considered. Citrate is rapidly metabolized by the liver. Therefore, citrate has lower risk of systemic bleeding.

Citrate: Citrate prevents coagulation by binding ionized calcium, which is required in clot formation. Citrate may result in hypocalcemia, which is usually mild, but can be more severe depending on infusion rate and patient's hepatic and renal function. Calcium can be infused during procedure to mitigate adverse effects.

Heparin: Heparin acts by potentiating activity of antithrombin. It has a half-life of ~90 minutes and results in systemic anticoagulation. Heparin is not typically used for TA because of its systemic anticoagulation and side effects.

TABLE 74.3 Clinical Indications for Therapeutic Apheresis

Disease Group/Name/Condition	Therapeutic Apheresis Modality	Category	Recommendation Grade
Acute Disseminated Encephalomyelitis, Refractory to Steroids			
	Therapeutic plasma exchange (TPE)	II	2C
Acute Inflammatory Demyelinating Polyradiculoneuropathy (Guillain–Barre Syndrome)			
Primary Treatment	TPE	I	1A
After IVIG	TPE	III	2C
Acute Liver Failure			
	TPE	III	2B
	Plasma exchange, high volume (not in the United States)	I	1A
Age-related macular degeneration, dry	Rheopheresis	I	1B
Amyloidosis, systemic			
	B_2 microglobulin column	II	2B
	TPE	IV	2C
Antineutrophil Cytoplasmic Antibodies (ANCA)-Associated Rapidly Progressive Glomerulonephritis (Granulomatosis With Polyangiitis and Microscopic Polyangiitis)			
Dialysis dependence	TPE	I	1A
Diffuse alveolar hemorrhage (DAH)	TPE	I	1C
Dialysis independence	TPE	III	2C
Antiglomerular Basement Membrane Disease (Goodpasture's Syndrome)			
Dialysis dependence	TPE	III	2B
DAH	TPE	I	1C
Dialysis independence	TPE	I	1B
Aplastic Anemia; Pure Red Cell Aplasia			
Aplastic anemia	TPE	III	2C
Pure red cell aplasia	TPE	III	2C
Atopic (Neuro-) Dermatitis (Atopic Eczema), Recalcitrant			
	ECP	III	2C
	IA	III	2C
	TPE	III	2C
Autoimmune Hemolytic Anemia: Warm Autoimmune Hemolytic Anemia; Cold Agglutinin Disease			
Warm autoimmune hemolytic anemia (severe)	TPE	III	2C
Cold agglutinin disease (severe)	TPE	II	2C
Babesiosis, Severe			
	RBC exchange (REX)	II	2C

TABLE 74.3 Clinical Indications for Therapeutic Apheresis—cont'd

Disease Group/Name/Condition	Therapeutic Apheresis Modality	Category	Recommendation Grade
Burn Shock Resuscitation			
	TPE	III	2B
Cardiac Neonatal Lupus			
	TPE	III	2C
Cardiac Transplantation			
Cellular/recurrent rejection	ECP	II	1B
Rejection Prophylaxis	ECP	II	2A
Desensitization	TPE	II	1C
Treatment of antibody-mediated rejection	TPE	III	2C
Catastrophic Antiphospholipid Syndrome			
	TPE	II	2C
Chronic Focal Encephalitis (Rasmussen Encephalitis)			
	TPE	III	2C
Chronic Inflammatory Demyelinating Polyradiculoneuropathy			
	TPE	I	1B
Coagulation Factor Inhibitors			
Alloantibody	TPE	IV	2C
Autoantibody	TPE	III	2C
Alloantibody	Immunoadsorption (IA)	III	2B
Autoantibody	IA	III	1C
Complex Regional Pain Syndrome			
Chronic	TPE	III	2C
Cryoglobulinemia			
Symptomatic/severe	TPE	II	2A
Symptomatic/severe	IA	II	2B
Cutaneous T-Cell Lymphoma; Mycosis Fungoides; Sezary Syndrome			
Erythrodermic	ECP	I	1B
Nonerythrodermic	ECP	III	2C
Dermatomyositis or Polymyositis			
	TPE	IV	2B
	ECP	IV	2C
Dilated Cardiomyopathy, Idiopathic			
NYHA II-IV	IA	II	1B
NYHA II-IV	TPE	III	2C
Erythropoietic Porphyria, Liver Disease			
	TPE	III	2C

Continued

TABLE 74.3 Clinical Indications for Therapeutic Apheresis—cont'd

Disease Group/Name/Condition	Therapeutic Apheresis Modality	Category	Recommendation Grade
	REX	III	2C
Familial Hypercholesterolemia			
Homozygotes	Low-density lipoprotein (LDL) apheresis	I	1A
Heterozygotes	LDL apheresis	II	1A
Homozygotes with small blood volume	TPE	II	1C
Focal Segmental Glomerulosclerosis			
Recurrent in transplanted kidney	TPE	I	1B
Steroid resistant in native kidney	LDL apheresis	III	2C
Graft-Versus-Host Disease			
Skin (chronic)	ECP	II	1B
Nonskin (chronic)	ECP	II	1B
Skin (acute)	ECP	II	1C
Nonskin (acute)	ECP	II	1C
Hashimoto's Encephalopathy: Steroid-Responsive Encephalopathy Associated With Autoimmune Thyroiditis			
	TPE	II	2C
HELLP Syndrome			
Postpartum	TPE	III	2C
Antepartum	TPE	IV	2C
Hematopoietic Progenitor Cell (HPC) Transplantation, ABO Incompatible			
Major HPC incompatibility, Marrow	TPE	II	1B
Major HPC incompatibility, Apheresis	TPE	II	2B
Minor HPC incompatibility, Apheresis	REX	III	2C
HPC Transplantation, HLA Desensitization			
	TPE	III	2C
Hemophagocytic Lymphohistiocytosis; Hemophagocytic Syndrome; Macrophage Activating Syndrome			
	TPE	III	2C
Henoch–Schönlein Purpura			
Crescentic	TPE	III	2C
Severe extrarenal disease	TPE	III	2C
Heparin-Induced Thrombocytopenia and Thrombosis			
Precardiopulmonary bypass	TPE	III	2C
Thrombosis	TPE	III	2C
Hereditary Hemochromatosis			
	Erythrocytopheresis	I	1B

TABLE 74.3 Clinical Indications for Therapeutic Apheresis—cont'd

Disease Group/Name/Condition	Therapeutic Apheresis Modality	Category	Recommendation Grade
Hyperleukocytosis			
Symptomatic	Leukocytapheresis	II	1B
Prophylactic or secondary	Leukocytapheresis	III	2C
Hypertriglyceridemic Pancreatitis			
	TPE	III	2C
Hyperviscosity in Monoclonal Gammopathies			
Symptomatic	TPE	I	1B
Prophylactic or secondary	TPE	I	1C
Immune Thrombocytopenia			
Refractory	TPE	III	2C
Refractory	IA	III	2C
Immunoglobulin A Nephropathy			
Crescentic	TPE	III	2B
Chronic progressive	TPE	III	2C
Inflammatory Bowel Disease			
Ulcerative colitis	Adsorptive cytapheresis	III/II	1B/2B
Crohn's Disease	Adsorptive cytapheresis	III	1B
Crohn's Disease	ECP	III	2C
Lambert–Eaton Myasthenic Syndrome			
	TPE	II	2C
Lipoprotein (a) Hyperlipoproteinemia			
	LDL apheresis	II	1B
Liver Transplantation			
Desensitization, ABOi LD	TPE	I	1C
Desensitization, ABOi DD	TPE	III	2C
Antibody-mediated rejection (ABOi and HLA)	TPE	III	2C
Lung Transplantation			
Bronchiolitis obliterans syndrome	ECP	II	1C
Antibody-mediated rejection	TPE	III	2C
Desensitization	TPE	III	2C
Malaria, Severe			
	REX	III	2B
Multiple Sclerosis			
Acute CNS inflammatory demyelinating disease	TPE	II	1B

Continued

TABLE 74.3 Clinical Indications for Therapeutic Apheresis—cont'd

Disease Group/Name/Condition	Therapeutic Apheresis Modality	Category	Recommendation Grade
Acute CNS inflammatory demyelinating disease	IA	III	2C
Chronic progressive	TPE	III	2B
Myasthenia Gravis			
Moderate–severe	TPE	I	1B
Prethymectomy	TPE	I	1C
Myeloma Cast Nephropathy			
	TPE	II	2B
Nephrogenic Systemic Fibrosis			
	ECP	III	2C
	TPE	III	2C
Neuromyelitis Optica Spectrum Disorders			
Acute	TPE	II	1B
Maintenance	TPE	III	2C
***N*-Methyl D-Aspartate Receptor Antibody Encephalitis**			
	TPE	I	1C
Overdose, Envenomation, and Poisoning			
Mushroom poisoning	TPE	II	2C
Envenomation	TPE	III	2C
Drug overdose/poisoning	TPE	III	2C
Paraneoplastic Neurologic Syndromes			
	TPE	III	2C
	IA	III	2C
Paraproteinemic Demyelinating Neuropathies/Chronic Acquired Demyelinating Polyneuropathies			
Anti-MAG neuropathy	TPE	III	1C
Multifocal motor neuropathy	TPE	IV	1C
IgG/IgA	TPE	I	1B
IgM	TPE	I	1C
Multiple myeloma	TPE	III	2C
IgG/IgA/IgM	IA	III	2C
Pediatric Autoimmune Neuropsychiatric Disorders Associated With Streptococcal Infections and Sydenham's Chorea			
PANDAS (exacerbation)	TPE	II	1B
Sydenham's chorea, severe	TPE	III	2B
Pemphigus Vulgaris			
Severe	TPE	III	2B
Severe	ECP	III	2C

TABLE 74.3 Clinical Indications for Therapeutic Apheresis—cont'd

Disease Group/Name/Condition	Therapeutic Apheresis Modality	Category	Recommendation Grade
Severe	IA	III	2C
Peripheral Vascular Diseases			
	LDL apheresis	II	1B
Phytanic Acid Storage Disease (Refsum's Disease)			
	TPE	II	2C
	LDL apheresis	II	2C
Polycythemia Vera; Erythrocytosis			
Polycythemia vera	Erythrocytapheresis	I	1B
Secondary erythrocytosis	Erythrocytapheresis	III	1C
Posttransfusion Purpura			
	TPE	III	2C
Prevention of RhD Alloimmunization After RBC Exposure			
Exposure to RhD(+) RBCs	REX	III	2C
Progressive Multifocal Leukoencephalopathy Associated With Natalizumab			
	TPE	I	1C
Pruritis Due to Hepatobiliary Diseases			
Treatment resistant	TPE	III	1C
Psoriasis			
	ECP	III	2B
Disseminated pustular	Adsorptive cytoapheresis	III	2C
	Lymphocytapheresis	III	2C
	TPE	IV	2C
Red Cell Alloimmunization in Pregnancy			
Before intrauterine transfusion availability	TPE	III	2C
Renal Transplantation, ABO Compatible			
Antibody-mediated rejection	TPE/IA	I	1B
Desensitization, LD	TPE/IA	I	1B
Desensitization, DD	TPE/IA	III	2C
Renal Transplantation, ABO Incompatible			
Desensitization, LD	TPE/IA	I	1B
Antibody-mediated rejection	TPE/IA	II	1B
A_2/A_2B into B, DD	TPE/IA	IV	1B
Scleroderma (Systemic Sclerosis)			
	TPE	III	2C

Continued

TABLE 74.3 Clinical Indications for Therapeutic Apheresis—cont'd

Disease Group/Name/Condition	Therapeutic Apheresis Modality	Category	Recommendation Grade
	ECP	III	2A
Sepsis With Multiorgan Failure			
	TPE	III	2B
Sickle Cell Disease, Acute			
Acute stroke	REX	I	1C
Acute chest syndrome, severe	REX	II	1C
Priapism	REX	III	2C
Multiorgan failure	REX	III	2C
Splenic/hepatic sequestration; intrahepatic cholestasis	REX	III	2C
Sickle Cell Disease, Nonacute			
Stroke prophylaxis/iron overload prevention	REX	I	1A
Recurrent vasoocclusive pain crisis	REX	III	2C
Preoperative management	REX	III	2A
Pregnancy	REX	III	2C
Stiff Person Syndrome			
	TPE	III	2C
Sudden Sensor.ineural Hearing Loss			
	LDL apheresis	III	2A
	Rheopheresis	III	2A
	TPE	III	2C
Systemic Lupus Erythematosus			
Severe	TPE	II	2C
Nephritis	TPE	IV	1B
Thrombocytosis			
Symptomatic	Thrombocytapheresis	II	2C
Prophylactic or secondary	Thrombocytapheresis	III	2C
Thrombotic Microangiopathy, Coagulation Mediated			
THBD mutation	TPE	III	2C
Thrombotic Microangiopathy, Complement Mediated			
Complement factor gene mutations	TPE	III	2C
Factor H autoantibodies	TPE	I	2C
MCP mutations	TPE	III	1C
Thrombotic Microangiopathy, Drug-Associated			
Ticlopidine	TPE	I	2B
Clopidogrel	TPE	III	2B
Calcineurin inhibitors	TPE	III	2C
Gemcitabine	TPE	IV	2C

TABLE 74.3 Clinical Indications for Therapeutic Apheresis—cont'd

Disease Group/Name/Condition	Therapeutic Apheresis Modality	Category	Recommendation Grade
Quinine	TPE	IV	2C
Thrombotic Microangiopathy; Hematopoietic Stem Cell Transplant–Associated			
	TPE	III	2C
Thrombotic Microangiopathy, Shiga Toxin Mediated			
Severe neurologic symptoms	TPE/IA	III	2C
Streptococcus pneumonia	TPE	III	2C
Absence of severe neurological symptoms	TPE	IV	1C
Thrombotic Thrombocytopenic Purpura			
	TPE	I	1A
Thyroid Storm			
	TPE	III	2C
Toxic Epidermal Necrolysis			
Refractory	TPE	III	2B
Vasculitis			
HBV–PAN	TPE	II	2C
Idiopathic PAN	TPE	IV	1B
EGPA	TPE	III	1B
Behcet's disease	Adsorption granulocytapheresis	II	1C
Behcet's disease	Plasma exchange	III	2C
Voltage-Gated Potassium Channel Antibodies			
	TPE	II	2C
Wilson's Disease, Fulminant			
Fulminant	TPE	I	1C

From Schwartz, J., et al. (2016). Guidelines on the use of therapeutic apheresis in clinical practice – evidence-based approach from the apheresis applications committee of the American Society of Apheresis. *J Clin Apher, 31*, 149–338.

Adverse Events

Allergic Reactions: Allergic reactions (mild urticaria to anaphylaxis) are mostly associated with use of plasma as replacement fluid but can occur rarely with albumin. Future procedures may require premedication with antihistamines and/or corticosteroids. Atypical allergic reactions are associated with ethylene oxide gas sterilization of tubing sets, which are characterized by periorbital edema, chemosis, and tearing. Recurrence can be minimized by double priming of the tubing.

Citrate Toxicity: Mild hypocalcemia (tingling, oral paresthesia, nausea, vomiting, abdominal pain [most common symptom in pediatric patients], or chest discomfort) due to citrate is the most common adverse reaction. Severe hypocalcemia can cause

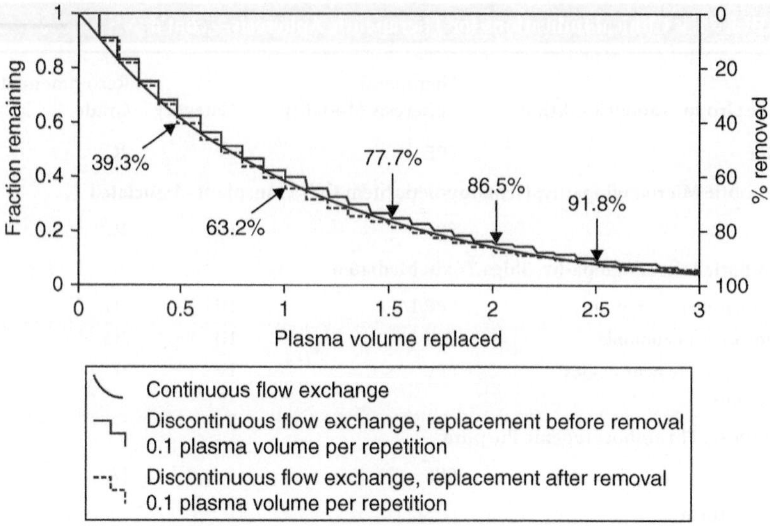

FIGURE 74.1 Relationship of removal of plasma constituents to plasma volume processed in therapeutic plasma exchange, assume the constituent is an ideal solute. (From Hillyer, C. D., Silberstein, L. D., Ness, P. M., et al. (Eds.). (2007). *Blood Banking and Transfusion Medicine: Basic Principles & Practice* (2nd ed.). Philadelphia, San Diego: Churchill Livingstone Elsevier.)

tetany, gastrointestinal symptoms, hypotension, cardiac dysrhythmias, or seizures. The majority of citrate reactions can be managed by slowing citrate infusion rate, adjusting whole blood to anticoagulant ratio, or administering calcium (either intravenously or orally). For severe toxicity, stopping the procedure and giving intravenous calcium are recommended. Citrate toxicity risk increases in patients with small size, hepatic or renal dysfunction, or with TPE using plasma as replacement fluid.

Hypotension: Hypotension reaction may result from hypovolemia, vasovagal reaction, and angiotensin-converting enzyme (ACE) inhibitors. Symptoms include lightheadedness, dizziness, sweating, and nausea. Treatment includes Trendelenburg's position and administration of fluid boluses.

ACE Inhibitor Reactions: ACE inhibitors may cause facial flushing or hypotension during TA, which is due to generation and accumulation of excess bradykinin (a potent vasodilator). Some experts prefer that ACE inhibitor therapy be discontinued 24–48 hours before starting apheresis procedures, if possible. If procedure must be done, decision of how to proceed is based on emergent nature of procedure and risks/benefits for patient. If albumin replacement is used, slow infusion rates are recommended.

Evaluation of a New Patient for Therapeutic Apheresis Initiation:

- Is there a clear rationale for TA use, will this correct the abnormality, and is there clinical evidence that TA benefits the patient (McLeod's criteria)?—Will TA affect the patient's comorbidities and medications?
- What anticoagulation, replacement fluid, vascular access, and volume of blood processed should be used?

- How many procedures total and at what frequency should be performed?
- What endpoints will be used to determine TA efficacy? What are the criteria for discontinuation?
- When and where should TA be initiated?

Further Reading

Mokrzycki, M. H., & Balogun, R. A. (2011). Therapeutic apheresis: A review of complications and recommendations for prevention and management. *J Clin Apher, 26*, 243–248.

Pham, H. P., & Schwartz, J. (2015). How to approach an apheresis consultation using the American Society for Apheresis guidelines for therapeutic apheresis procedures. *ISBT Sci Ser, 10S*, 79–88.

Schwartz, J., Padmanabhan, A., Aqui, N., Balogun, R. A., Connelly-Smith, L., Delaney, M., et al. (2016). Guidelines on the use of therapeutic apheresis in clinical practice – evidence based approach from the Writing Committee of the American Society for Apheresis: The Seventh Special Issue. *J Clin Apher, 31*, 149–338.

CHAPTER 75

Therapeutic Plasma Exchange

Elizabeth A. Godbey, MD, Joseph Schwartz, MD, MPH and
Huy P. Pham, MD, MPH

Therapeutic plasma exchange (TPE) is a procedure in which whole blood of the patient is passed through an apheresis device, which separates and removes plasma. Other components are returned to the patient together with replacement fluid.

Pathophysiology: TPE is used to treat diseases that are thought to be caused by a substance in plasma whose removal can help with disease resolution. TPE is mostly used for antibody (and rarely immune complex) removal from circulation. It can be used to remove other molecules, such as drugs, paraproteins, and low-density lipoproteins (LDLs), but other apheresis devices are more efficient at LDL removal (Chapter 80).

Volume Exchanged: Typically, unless otherwise indicated, 1 to 1.5 plasma-volumes are exchanged approximately every other day based on the apheresis kinetics (Figure 75.1). In addition to removing pathogenic substances, TPE also removes normal plasma constituents, such as coagulation factors, immunoglobulins, and platelets (Table 75.1). In patients with normal bone marrow and liver function, endogenous synthesis replete most coagulation factors and platelets within 2–4 days. It is not advisable to check routine labs, such as coagulation (PT/aPTT and fibrinogen) and chemistry panels immediately post-TPE; it takes ~24 hours for equilibrium to be established between intra- and extravascular spaces. Furthermore, different replacement fluids have different advantages and disadvantages (Table 75.2). Regarding the type of replacement fluid used in TPE procedures, albumin and/or normal saline are usually given (the ratio of albumin to saline varies among institutions, but albumin is the predominant fluid given to prevent hypovolemic reactions) if plasma is not indicated. Plasma products are typically used only in the treatment of thrombotic microangiopathies (TMAs) and in patients with underlying coagulopathy, such as liver disease, and/or who are actively bleeding. In addition, plasma can be given at the end of the procedure to prevent coagulopathy in patients who need an invasive procedure immediately after TPE or to treat procedure-related coagulopathy.

Drug Removal During Apheresis: TPE has been used to treat acute drug toxicity when other modalities such as gastric lavage, dialysis, hemoperfusion, and forced diuresis are ineffective. Drugs that are highly protein-bound and have small volumes of distribution are most effectively removed by TPE. More commonly, therapeutic drug clearance by TPE is a concern. A drug is most likely to be removed during distribution phase, and therefore, it is prudent to dose drugs after TPE and not immediately before. Data suggest that prednisone, digoxin, cyclosporine, ceftriaxone, valproic acid, and phenobarbital might not be removed by TPE. Salicylates and tobramycin should be supplemented after TPE, and phenytoin should be monitored.

Transfusion Medicine and Hemostasis. https://doi.org/10.1016/B978-0-12-813726-0.00075-1

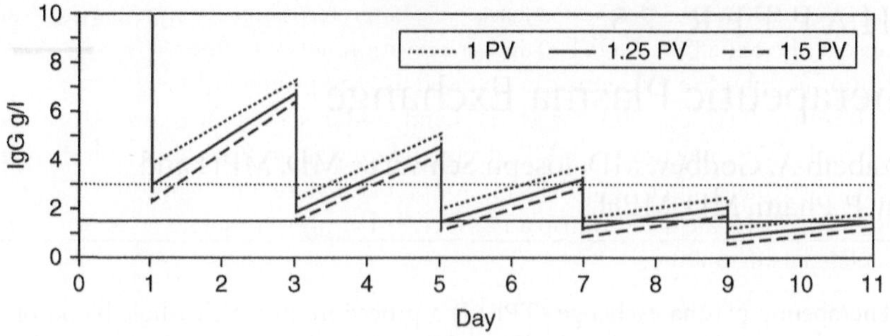

FIGURE 75.1 Apheresis kinetics for IgG. (Modified from Hillyer, C. D., Silberstein, L. E., Ness, P. M., Anderson, K. C., & Roback, J. D. (Eds.). (2007). *Blood Banking and Transfusion Medicine: Basic Principles & Practice* (2nd ed.). Philadelphia, San Diego: Churchill Livingstone Elsevier.)

TABLE 75.1 Plasma Exchange Removal Following 1 Plasma Volume Therapeutic Plasma Exchange

	% Removal	% Recovery at 48 Hours
Coagulation factors	25–50	80–100
Fibrinogen	63	65
Immunoglobulins	63	45
Paraproteins	20–30	Variable
Liver enzymes	55–60	100
Bilirubin	45	100
C3	63	60–100
Platelets	25–30	75–100

From Weinstein, R., in McLeod, B. C., Weinstein, R., Winters, J. L., et al. (Eds.). (2010). *Apheresis Principles and Practice* (3rd ed.). Baltimore: AABB Press.

TABLE 75.2 Comparison of Replacement Fluid

Replacement Fluid	Advantage	Disadvantage
Crystalloid	Low cost Hypoallergenic No infectious risk	Hypooncotic No immunoglobulins No coagulation factors
Albumin	Iso-oncotic Minimal infectious risk	Higher cost No immunoglobulins No coagulation factors
Plasma	Immunoglobulins Coagulation factors Iso-oncotic	Infection risk Citrate toxicity Allergic reactions ABO compatibility

From Chhibber, V. and King, K. E., in McLeod, B. C., Weinstein, R., Winters, J. L., et al. (Eds.). (2010). *Apheresis Principles and Practice* (3rd ed.). Baltimore: AABB Press.

Category I Indications: American Society for Apheresis recently published guidelines on TPE indications (Table 74.3).

Acute Inflammatory Demyelinating Polyradiculoneuropathy (Guillain–Barre Syndrome): TPE and intravenous immunoglobulin (IVIG) have demonstrated equal clinical efficacy in the treatment of this disease. Volume of exchange in TPE = (200–250 mL of plasma) × (body weight measured in kilograms), spread over 10–14 days.

ANCA-Associated Rapidly Progressive Glomerulonephritis (Granulomatosis With Polyangiitis; and Microscopic Polyangiitis) (Category I for Dialysis Dependent and Diffuse Alveolar Hemorrhage): ANCA (antineutrophil cytoplasmic antibody) small vessel vasculitis is treated with corticosteroids and immunosuppressive drugs. TPE should be added in cases of severe renal disease or pulmonary hemorrhage. Daily TPE with plasma replacement fluid should be performed for diffuse alveolar hemorrhage (DAH); otherwise, every other day with albumin is recommended.

Antiglomerular Basement Membrane Disease (Goodpasture Syndrome): This syndrome stems from antiglomerular basement membrane antibodies, which result in damage to alveolar and renal basement membranes. Patients are treated with cyclophosphamide, steroids, and TPE. TPE is performed daily or every other day with albumin as replacement fluid (unless DAH is present, in which case plasma is recommended).

Chronic Inflammatory Demyelinating Polyradiculoneuropathy: Chronic inflammatory demyelinating polyradiculoneuropathy exhibits proximal and distal symmetric muscle weakness with or without numbness that becomes more severe for ≥2 months. Corticosteroids, TPE, and IVIG have similar efficacy. TPE is performed 2–3 times per week until the patient improves, and then frequency is tapered.

Focal Segmental Glomerulosclerosis: TPE is used in the management of patients with recurrent focal segmental glomerulosclerosis in a renal allograft. TPE is performed daily or every other day, and multiple regimens have been reported. Tapering the frequency of TPE is individually done based on patient's proteinuria severity. Some patients require long-term exchanges as maintenance therapy.

Hyperviscosity in Monoclonal Gammopathies: Hyperviscosity, which may occur in disorders such as Waldenström macroglobulinemia and multiple myeloma, may result in a multitude of symptoms. Each patient has an individual viscosity for which he/she becomes symptomatic, most commonly around 6–7 cP. TPE removes paraproteins, thereby decreasing viscosity. TPE is initiated as soon as possible after diagnosis, followed by disease-specific treatments to prevent future accumulation of paraproteins. Daily 1–1.5 TPV with albumin as replacement fluid, typically 1–3 total. Patient status, serum viscosity, and paraprotein may all be considered when determining duration of TPE treatment.

Liver Transplantation: TPE may be used to reduce isoagglutinin titers in ABO-incompatible liver transplants. The replacement fluid for TPE is plasma (compatible with both donor and recipient), or a mixture of plasma and albumin. For desensitization, TPE will continue until a goal titer is achieved, but for rejection, the treatment is based on liver function. Individual institutions will determine their titer goals. Typically, 1–1.5 TPV is processed daily or every other day.

Myasthenia Gravis: Cholinesterase inhibitors, thymectomy, immunosuppression, and TPE or IVIG are the mainstays of myasthenia gravis therapy. Patients typically improve quickly with TPE but may need maintenance therapy. Albumin is used as replacement fluid, and the procedure is performed daily or every other day.

N-Methyl D-Aspartate Receptor Antibody Encephalitis: N-methyl D-aspartate receptor (NMDAR) encephalitis results from antibodies directed against the NMDAR GluN1 subunit. It results in a severe neurologic disease with many pronounced symptoms. Most patients are female, about half of whom have a neoplasm (this is most commonly an ovarian teratoma). When NMDAR encephalitis is diagnosed, a tumor should be ruled out and removed if discovered. Immunosuppression medications, IVIG, and TPE are primary treatment options. TPE should be performed with albumin every other day for 5–6 treatments.

Paraproteinemic Demyelinating Neuropathies/Chronic Acquired Demyelinating Polyneuropathies (Category I for IgG/IgA and IgM): These are chronic progressive illnesses, which present with a variety of neurologic symptoms and are caused by several different proteins. TPE can remove the disease-causing antibodies. When TPE is used, 5–6 (replacement fluid albumin) are completed over about 2 weeks.

Progressive Multifocal Leukoencephalopathy Associated With Natalizumab: TPE and discontinuation of natalizumab are first-line treatments for progressive multifocal leukoencephalopathy associated with natalizumab. TPE may be performed every other day using albumin. Immune reconstitution inflammatory syndrome (IRIS) may occur several weeks after TPE and may be fatal. IRIS should be treated with corticosteroids (not with TPE).

Renal Transplantation, ABO Compatible: TPE may be used to remove HLA antibodies before transplant during desensitization or after transplant during antibody mediated rejection (AMR) episodes. TPE is used in conjunction with immunosuppressive medications. TPE is performed five times, daily or every other day, using albumin or plasma as replacement fluid. Antibody titers and/or renal function can also be used to guide therapy decisions.

Renal Transplantation, ABO Incompatible (Category I for Desensitization, Category II for AMR): TPE may be used in desensitization to prepare for ABO-incompatible renal transplantation or in treatment of episodes of AMR. Treatments are continued until antibody titers fall below a specific threshold (goal titers are defined at each institution). TPE may be completed daily or every other day, using albumin or plasma. If plasma is used, it must be compatible with donor and recipient.

Thrombotic Microangiopathy, Complement Mediated (Only Factor H Autoantibody Is Category I): Atypical hemolytic syndrome (aHUS), or complement-mediated thrombotic microangiopathy (TMA), is thought to be due to the alternative complement pathway. Primary therapy is with TPE, until Thrombotic thrombocytopenic purpura (TTP), HUS, and TMA secondary to drugs or hematopoietic stem cell transplant are eliminated from the differential. Eculizumab is the treatment of choice for complement-mediated TMA. TPE could be continued until eculizumab can be obtained. Daily TPE is performed with plasma.

Thrombotic Microangiopathy, Drug Associated: TMA associated with ticlopidine usually has very low ADAMTS13 levels and inhibitors, resembling idiopathic TTP. For this reason, daily TPE with plasma as replacement fluid is effective in ticlopidine-associated TMA. TMA associated with other medications does not resemble idiopathic TTP—typically, they lack ADAMTS13 inhibitors and do not have severe ADAMTS13 deficiency. TPE is not as effective in nonticlopidine drug-associated TMAs and hence, TPE is a category I treatment for ticlopidine-associated TMA but is category III or IV treatment for others. The primary treatment in the setting of other drugs is to stop administering drug, provide supportive therapy, and provide defined interventions according to each drug in question.

Thrombotic thrombocytopenic purpura (TTP): TTP is classically defined by thrombocytopenia, microangiopathic hemolytic anemia (MAHA), neurologic changes, renal failure, and fever (Chapter 107). Thrombocytopenia and MAHA without known cause are sufficient to initiate emergent treatment. Remember to rule out other causes of TMA (e.g., disseminated intravascular coagulopathy [DIC], malignant hypertension [atypical], HUS, pernicious anemia). Acquired TTP is associated with autoantibody to ADAMTS13 resulting in severe ADAMTS13 deficiency. TPE is a life-saving therapy and has decreased the overall mortality from >90% to <20%. TPE should be initiated as soon as possible. If TPE cannot be initiated right away, plasma infusions should be started at approximately 30–40 mL/kg per day, avoiding volume overload, until TPE can be initiated. Additionally, corticosteroids may be given. Medications that are sometimes added to therapy include rituximab, cyclosporine, azathioprine, vincristine, and other immunosuppression drugs. Congenital TTP has decreased ADAMTS13 activity, but does not have inhibitors; therefore, infusions of plasma (10–15 mL/kg) or cryoprecipitate are used. TPE is typically completed daily, using plasma, until platelet count is $>150{,}000 \times 10^9/L$ with normalized LDH for several days.

Wilson Disease, Fulminant: Wilson disease is an autosomal recessive inherited disorder that impairs biliary copper processing. Consequently, copper builds up in the liver, brain, cornea, and kidneys. Diets low in copper, zinc acetate, chelation therapy, and liver transplant are the primary modes of treatment. TPE may be used to reduce very high copper levels and toxins while an acutely ill patient is waiting for a liver transplant. The use of plasma as replacement fluid can also treat coagulopathy resulting from liver failure.

Category II Indications

Acute Disseminated Encephalomyelitis: Acute disseminated encephalomyelitis is an acute demyelinating disease. The disease is thought to result from an autoimmune attack of antigens of the CNS. Steroid therapy is the first-line treatment, and TPE is pursued in patients who do not respond to steroid therapy or for whom steroids are contraindicated. Most commonly, five every other day procedures with albumin as replacement fluids are used, but 3–6 every other day procedures have been reported.

Autoimmune Hemolytic Anemia (Only Severe Cold Agglutinin Disease Is Category II): Avoidance of cold is the treatment for cold agglutinin disease, but in severe cases, rituximab is used as well. TPE can be completed daily or every other day with albumin.

Cardiac Transplantation (Only Desensitization Is Category II): TPE may be performed daily or every other day using albumin or plasma as replacement fluid. A specific recommendation for evaluating efficacy of treatment has not been endorsed, but groups have used cardiac function, cardiac biopsy, and donor specific antibody (DSA) levels to monitor TPE efficacy. If TPE is performed immediately before or after the transplantation, then plasma may be used as part of the replacement fluid to reduce the risk of perioperative bleeding.

Catastrophic Antiphospholipid Syndrome: Catastrophic antiphospholipid syndrome is characterized by thrombosis in ≥3 organ systems in days or weeks, along with serologic evidence of antiphospholipid antibodies. Use of anticoagulation, glucocorticosteroids, and TPE/IVIG all together is recommended. TPE is performed daily or every other day with plasma replacement fluid (discontinuation based on patient status).

Cryoglobulinemia: Cryoglobulins are comprised of immunoglobulins that reversibly precipitate at cool temperatures and can deposit in small vessels, resulting in vascular damage. Cryoglobinemia is associated with many conditions. TPE removes cryoglobulins; it may be used alone or with immunosuppressive agents and results in improvement of up to 80% of patients. Cryoglobulin may precipitate in the extracorporeal circuit. To prevent this, one may consider warming the room, lines, and replacement fluid. Typically, 3–8 procedures are performed every 1–3 days, and the patient is then reevaluated. Maintenance treatments are sometimes needed.

Familial Hypercholesterolemia (Only Homozygous With Small Blood Volume Are TPE Category II): TPE lowers serum cholesterol in patients with familial hypercholesterolemia (FH) who are unresponsive or intolerant of medical management, but LDL apheresis is more effective. TPE may be used in small children if they are too small for LDL apheresis devices. TPE is performed once every 1–2 weeks and is continued until a better treatment modality can be initiated.

Hashimoto's Encephalopathy: Steroids are the primary treatment, but other medications can be added should steroid treatment fail. Clinical improvement has been described with TPE in patients who did not respond to steroid therapy (case reports).

Hematopoietic Stem Cell Transplantation, ABO Incompatible (Major Hematopoietic Progenitor Cell Marrow or Apheresis): TPE can decrease isoagglutinins. In major ABO incompatibility (the recipient has isoagglutinins against the donor A/B RBC antigens), TPE should be performed daily before infusion of hematopoietic progenitor cells to reduce the IgM/IgG titers to <16 before transplant. Replacement fluid is albumin or albumin and plasma (plasma must be compatible with both donor and recipient).

Lambert–Eaton Myasthenic Syndrome: Lambert–Eaton myasthenic syndrome (LEMS) is a disruption of neuromuscular transmission that is caused by an autoimmune attack on the presynaptic neuromuscular junction. In the setting of LEMS, any malignancy should be identified and treated, immunosuppression should be initiated to decrease creation of autoantibodies, and medical support of acetylcholine neurotransmission should be initiated. 3,4-Diaminopyridine is the primary treatment for symptoms of LEMS. TPE can be added for patients with severe or quickly worsening symptoms, or for patients for whom IVIG is not an option. A 1–1.5 TPV TPE using albumin, as replacement fluid is recommended daily or every other day until clinical response, EMG improvement, or 2–3 weeks of TPE treatment.

Multiple Sclerosis (Acute, Category II; Chronic Progressive, Category III): Primary treatment for multiple sclerosis (MS) includes a variety of immunomodulating medications. If TPE is pursued, 1–1.5 TPV are exchanged with albumin. For acute patients, 5–7 TPE across 2 weeks.

Myeloma Cast Nephropathy: Diuresis, chemotherapy, immune modulation, and supportive care are the mainstays of treatment. TPE is a second-line treatment and works by decreasing circulating light chains. If TPE is pursued, it is completed daily or every other day using albumin.

Neuromyelitis Optica Spectrum Disorders (Acute, Category II; Disease Maintenance, Category III): Neuromyelitis optica spectrum disorders are demyelinating diseases, which cause pathology of the spinal cord and optic nerve. It is thought to result from an autoantibody to aquaporin-4 (found on astrocyte foot processes). Immunosuppressant medications are used in acute settings, and TPE may be added as a second-line therapy.

Amanita Mushroom Poisoning: TPE may be an added as a second-line treatment. If used, daily TPE with albumin or plasma should be continued until the toxin's symptoms have resolved.

Pediatric Autoimmune Neuropsychiatric Disorders Associated With Streptococcal Infections: Cognitive behavioral therapy, psychiatric medications, antibiotics, or corticosteroids may be used are first-line treatments. In very severe cases, IVIG or TPE may be used. 3–6 TPE with albumin, spread over 7–14 days, is recommended.

Phytanic Acid Storage Disease (Refsum Disease): Phytanic acid (PA) storage disease (Refsum disease) is an autosomal recessive disease caused by a deficiency of phytanoyl-CoA hydrolase, causing buildup of PA, and resulting neurologic symptoms. The primary method of avoiding symptoms is to avoid PA in the diet. TPE may reduce PA if needed. TPE is performed using albumin; the schedule depends on PA measurements and symptoms. LDL apheresis may also be used (category II).

Systemic Lupus Erythematosus, Severe: Systemic lupus erythematosus (SLE) is a chronic disease, which is treated with immunosuppressing medications. TPE is not typically used for SLE patients (category II for severe SLE, category IV for nephritis). In the setting of lupus cerebritis or DAH, TPE may be completed daily or every other day. Plasma is used as replacement fluid if DAH is present.

Vasculitis: HBV-Associated Polyarteritis Nodosa: HBV-PAN (polyarteritis nodosa) is treated with glucocorticoids, hepatitis B antivirals, and TPE. Because TPE works well to remove immune complexes, it is a category II. In contrast, idiopathic PAN is category IV.

Further Reading

Schwartz, J., Padmanabhan, A., Aqui, N., Balogun, R. A., Connelly-Smith, L., Delaney, M., et al. (2016). Guidelines on the use of therapeutic apheresis in clinical practice – evidence based approach from the Writing Committee of the American Society for Apheresis: The seventh special issue. *J Clin Apher*, 3, 149–341.

CHAPTER 76

Therapeutic Erythrocytapheresis and Red Cell Exchange

Elizabeth M. Staley, MD, PhD, Joseph Schwartz, MD, MPH and
Huy P. Pham, MD, MPH

Therapeutic erythrocytapheresis (ET) is a procedure in which patients' RBCs are selectively removed to reduce excessive RBC mass. It has been used for treatment of polycythemia vera, reactive erythrocytosis, and hereditary hemochromocytosis.

RBC exchange (RBCx) is a procedure in which the patient's RBCs are replaced with allogeneic RBCs. RBCx is mostly used to treat patients with sickle cell disease (SCD). RBCx is safe; however, use of allogenic RBCs as replacement fluid places patient at risk for transfusion-associated adverse events. Recently, a modified RBCx procedure, termed isovolemic hemodilution (IHD) or depletion RBCx, has been introduced for patients with SCD to reduce number of RBC units used during procedure and/or lengthen time between procedures.

ET and RBCx may be performed manually or with automated system (this chapter addresses automated procedures).

Exchange Volume and Replacement Fluids: Fluid exchange volume is automatically calculated by the device as function of clinical parameters (e.g., patient's gender, height, weight, and Hct), type of replacement fluid used (including replacement fluid Hct), desired postprocedure Hct, and desired fraction of cells remaining (FCR, which is defined as ratio of target patient's RBCs remaining over initial RBCs). Typically, the goal in patients with SCD is final hemoglobin S (HbS) ≤30%; therefore, if the patient's initial HbS is 100%, FCR would be set at 30%. FCR is dictated by preexchange HbS, and desired final HbS percentage; for example, if preexchange HbS is 60% with HbS goal of 30%, then FCR would be 50%. Most apheresis devices additionally require entry of goal end Hct. For patients with SCD, this value should be ≤30% to prevent hyperviscosity.

Because IHD involves RBC depletion followed by exchange, IHD RBCx should only be performed in stable patients because hypotension may occur during the depletion phase. During depletion phase, the patient's Hct is lowered to safe target Hct (typically the higher of 8% less than initial Hct or 22%) using either 5% albumin or 0.9% normal saline to maintain blood pressure. Once target Hct is reached, device begins exchange procedure using RBC units as replacement fluid.

For all RBCx procedures, leukoreduced RBC products are recommended as replacement fluid, to mitigate potential alloimmunization and febrile reactions. RBC unit Hct varies with anticoagulant-preservative solution. In children, Hct of each unit is typically determined, and subsequently averaged to obtain replacement fluid Hct, enabling accurate prediction of postprocedure Hct.

Transfusion Medicine and Hemostasis. https://doi.org/10.1016/B978-0-12-813726-0.00076-3

In clinically unstable patients, priming exchange the set with 5% albumin before starting the procedure is advisable. Using 5% albumin conserves RBCs, simplifies procedure, and returns the patient's blood mixed with albumin as opposed to saline. However, in severely anemic patients, priming with RBCs is recommended. For ET, 5% albumin and/or 0.9% saline may be used to maintain intravascular volume and hemodynamic parameters.

Indications

ABO Incompatible (Minor Incompatibility): HPC ABO minor incompatibility occurs when the donor's plasma contains antibodies against the recipient's RBCs. To prevent or treat passenger lymphocyte syndrome, where donor lymphocytes secrete antibodies resulting in hemolysis, RBCx can be performed replacing recipient RBCs with group O RBCs. Goal is recipient RBCs <35%, 1 RBC volume exchange.

Babesiosis: Babesiosis is caused by RBC intracellular parasite that can result in hemolysis. Most cases are subclinical. RBCx is indicated in patients with high parasite burden (parasitemia >10%), or in the presence of severe hemolysis, disseminated intravascular coagulation, or pulmonary, renal, or hepatic failure. RBCx's goal is residual parasitemia <5% and/or symptom amelioration.

Erythropoietic Porphyria: Erythropoietic protoporphyria (EPP) is a hereditary disorder caused by genetic mutations occurring in one of two mitochondrial enzymes (ferrochelatase or 5-aminolevulinate synthase) that are critical to heme biosynthesis. Enzyme functional alterations ultimately result in protoporphyrin accumulation in RBCs, plasma, skin, hepatocytes, and stool. RBCx has been used in patients with EPP presenting in acute liver failure, to rapidly decreasing RBC and plasma protophyrin levels, and preventing further liver deposition/accumulation. RBCx target is FCR of 20%–30% and final Hct of 35%.

Hereditary Hemochromatosis: Hereditary hemochromatosis is an inherited disorder resulting in iron accumulation/deposition in the liver, heart, pancreas, and other organs. Primary therapy is iron removal via phlebotomy or erythrocytapheresis, which is performed biweekly as tolerated until ferritin level remains <50 ng/mL. Target final Hct for each procedure is >30%, and threshold to perform ET is Hct >34%.

Malaria: Malaria is a vector-borne protozoal disease that is transmitted to humans via mosquito bites or blood transfusion. Intraerythrocytic phase of protozoal life cycle is responsible for malarial symptoms. In severely ill patients (parasitemia >10%), RBCx functions to improve blood rheological properties and reduce pathogenic mediators, such as parasite-derived toxins, hemolytic metabolites, and cytokines. However, RBCx benefit remains uncertain. RBCx goal is residual parasitemia of <5%.

Erythrocytosis: Erythrocytosis (polycythemia) is increase in RBC mass, which is either primary, resulting from myeloproliferative disorder, or secondary, resulting from a hemoglobinopathy, chronic hypoxia, malignancy, or dysregulated erythropoietin production. Blood viscosity increases significantly at Hct >50%, resulting in headache, dizziness, slow mentation, confusion, fatigue, myalgia, angina, dyspnea, and thrombosis.

Treatment of secondary erythrocytosis should target treatment of underlying disorder. Additionally, patients with symptomatic hyperviscosity may be treated with therapeutic phlebotomy. Therapeutic end point for phlebotomy varies in accordance with underlying etiology; target Hct is <45% in patients with MPD, 50%–52% in patients with pulmonary hypoxia or high oxygen affinity hemoglobins, and 55%–60% in patients with cyanotic congenital heart disease. Albumin and/or saline may be used during procedure as needed to maintain hemodynamic parameters. For chronic ET or diseases of increased RBC mass or iron, frequency of procedures is determined by Hct or iron levels, respectively.

Rh Alloimmunization: D alloimmunization is a potential transfusion complication after transfusion of D-positive RBCs into D-negative recipient, placing recipient at risk for anti-D formation, and subsequently hemolytic disease of the fetus and newborn. RBCx has been used after large volume D-positive RBC transfusion (>20% total RBC volume), to reduce number of D-positive RBCs. Postprocedure residual D-positive RBC volume should be covered with RhIg administration. One red cell volume (RCV) is typically performed.

Sickle Cell Disease: RBCx has been preformed for SCD with acute life- or organ-threatening complications such as stroke, multiorgan failure, severe acute chest, priapism or splenic sequestration (Chapter 52). RBCx is preferable to simple transfusion because HbS concentration is reduced rapidly without increasing blood viscosity or volume overload. One procedure is sufficient to treat acute complications of SCD resulting in HbS ≤30%.

For patients receiving chronic transfusion therapy (e.g., for stroke prevention), long-term RBCx is advantageous, as it reduces transfusion-associated iron accumulation. However, chronic RBCx has other risks including establishment of venous access and increased RBC product exposure. No evidence supports one method of chronic transfusion over other. When being used as chronic transfusion therapy, RBCx is performed at intervals to maintain target HbS level.

Further Reading

Calvo-Cano, A., Gómez-Junyent, J., & Lozano, M. (2016). The role of red blood cell exchange for severe imported malaria in the artesunate era: A retrospective cohort study in a referral centre. *Malar J*, *15*, 215–220.

Poullin, P., Sanderson, F., Bernit, E., et al. (2016). Comparative evaluation of the depletion-red cell exchange program with the Spectra Optia and the isovolemic hemodilution-red cell exchange method with the COBE Spectra in sickle cell disease patients. *J Clin Apher*, *31*, 429–433.

Schwartz, J., Padmanabhan, A., Aqui, N., et al. (2016). Guidelines on the use of therapeutic apheresis in clinical practice-evidence-based approach from the Writing Committee of the American Society for Apheresis: The seventh special issue. *J Clin Apher*, *31*, 149–338.

CHAPTER 77

Therapeutic Thrombocytapheresis

Sierra C. Simmons, MD, MPH, Joseph Schwartz, MD, MPH and
Huy P. Pham, MD, MPH

Therapeutic thrombocytapheresis (or commonly referred to as plateletpheresis or platelet depletion) is used in primary and sometimes secondary thrombocytosis to rapidly remove platelets for prevention or treatment of hemorrhage and/or thrombosis. Reduction of platelet count achieved by this procedure is short-lived; thus, other definitive and longer-term therapies are also needed.

Volume Exchanged: 1.5–2 total blood volumes (TBVs) are typically processed, resulting in approximately 30%–60% reductions in platelet count.

Replacement Fluid: Replacement fluid is not routinely used because in many cases, <15% of the patient's TBV is removed. However, if >15% of the TBV is expected to be removed, then normal saline and/or 5% albumin can be used for fluid replacement. Normal saline is used to maintain the patient's blood pressure and volume as necessary throughout the procedure.

Anticoagulant: Citrate should be used as the anticoagulant, and the anticoagulant-to-whole-blood ratio should range between 1:6 and 1:12. Heparin is not recommended as an anticoagulant because it might cause platelet clumping in the circuit.

Indications

Thrombocytosis: Thrombocytosis, generally defined as a platelet count >450,000/μL, can occur from a reactive process (secondary thrombocytosis, accounting for ~80%–90% of the cases) or underlying hematologic malignancy (primary thrombocytosis), including polycythemia vera, essential thrombocythemia, chronic myelogenous leukemia, primary myelofibrosis, and myelodysplastic syndromes/myeloproliferative neoplasms with ring sideroblasts and thrombocytosis. Risks factors for development of thrombotic or hemorrhagic complications in primary thrombocytosis include the following: >60 years of age, pregnancy, erythrocytosis, leukocytosis, a JAK2 V617F mutation, and a prior history of thrombosis. The differential diagnosis for secondary thrombocytosis includes acute bleeding, hemolysis, infection, inflammation, iron deficiency, and postoperative stressors including postsplenectomy. Secondary thrombocytosis has a significantly lower risk of life-threatening thrombosis or hemorrhage in comparison with primary thrombocytosis. The rationale for the thrombotic/hemorrhagic risk for patients with primary thrombocytosis includes shorter platelet survival and increased circulating markers of platelet activation, such as platelet factor 4, thrombomodulin, and β-thromboglobulin, as well as endothelial dysfunction and platelet–leukocyte aggregation.

Transfusion Medicine and Hemostasis. https://doi.org/10.1016/B978-0-12-813726-0.00077-5

There is an increased risk of thromboembolic events with high platelet counts, although the risk is not determined exclusively by the circulating number. Hemorrhagic risk is greatest with platelet counts >1,000,000/μL, due to the development of acquired von Willebrand syndrome associated with a decrease in the functional amount of von Willebrand factor.μ

In patients with acute thrombotic or hemorrhagic complications, the goal is to decrease the platelet count to <350,000–400,000/μL until the effect of a cytoreductive agent (e.g., hydroxyurea), given concurrently, takes place, thereby preventing the rapid reaccumulation of platelets. Thrombocytapheresis may be appropriate for selected high-risk patients whose platelet-lowering agents are contraindicated or intolerable (e.g., in pregnant women to prevent recurrent fetal loss) or when the onset of pharmacologic therapy would be too slow (e.g., before urgent surgery). Platelet depletion used to treat rebound thrombocytosis postsplenectomy has been described in case reports; however, the rationale is undefined, and the efficacy is unclear.

Therapeutic trials of platelet depletion could be performed in patients with complications from secondary thrombocytosis; however, it is likely that there are other factors contributing to the complications, such as preexisting atherosclerosis, hypercoagulable states, or malignancy, and the underlying cause should be addressed.

Frequency of thrombocytapheresis in general should be performed daily or as needed.

Further Reading

Piccin, A., Steurer, M., Mitterer, M., Blöchl, E. M., Marcheselli, L., Pusceddu, I., et al. (2015). Role of blood cells dynamism on hemostatic complications in low-risk patients with essential thrombocythemia. *Intern Emerg Med*, *10*(4), 451–460.

Schwartz, J., Padmanabhan, A., Aqui, N., Balogun, R. A., Connelly-Smith, L., Delaney, M., et al. (2016). Guidelines on the use of therapeutic apheresis in clinical practice—evidence-based approach from the Writing Committee of the American Society for Apheresis: The seventh special issue. *J Clin Apher*, *31*(3), 149–338.

CHAPTER 78

Therapeutic Leukocytapheresis and Adsorptive Cytapheresis

Elizabeth M. Staley, MD, PhD, Joseph Schwartz, MD, MPH and Huy P. Pham, MD, MPH

Therapeutic leukocytapheresis (or leukapheresis) is a procedure in which white blood cells (WBCs) are selectively removed from patient's circulation, generally with the aim of treating hyperleukocytosis and/or hyperviscosity. Additionally, leukocytapheresis has been performed prophylactically, such as to prevent tumor lysis syndrome before initiation of chemotherapy. Unfortunately, neither therapeutic nor prophylactic leukocytapheresis appear to effect long-term survival in leukemic patients.

Selective leukocyte apheresis incorporates adsorptive columns into extracorporeal stage of apheresis procedure with the goal of removing circulating leukocytes and immune system modulation. This procedure has been used for the treatment of inflammatory bowel disease (IBD), systemic lupus erythematosus, psoriasis, Behçet's disease, rheumatoid arthritis, and exacerbations of idiopathic interstitial pneumonias.

Processing Volume: Therapeutic leukocytapheresis processes between 1.5 and 2 total blood volumes (TBVs) resulting in 30%–60% reduction in WBC count. Postprocedure WBC count is often difficult to predict, as leukocytes are mobilized from extramedullary sites into intravascular space during procedure. Procedural efficacy may be improved through use of erythrocyte-sedimenting agent, such as 6% hydroxyethyl starch (HES). HES enhances separation of WBCs from other blood components, thereby, improving leukocyte removal efficiency. Addition of HES is recommended when treatment is intended for removal of mature myeloid cells. However, each patient's renal and cardiovascular status must be assessed before use of HES due its renal excretion and function as volume expander.

Replacement Fluid: If leukocytapheresis procedures result in <15% of TBV removal, then it is sufficient to use normal saline as needed to maintain patient's blood pressure. When volume removal >15% of TBV, then replacement with colloid solution, such as 5% albumin, is recommended. Furthermore, apheresis circuit may be primed with irradiated RBCs for selected anemic patients. This is often preferable to RBC transfusion, which can result in increase in viscosity.

Leukocyte Adsorption Devices: There are currently two methods of selective apheresis available in Europe and Japan, leukocytapheresis (LCAP) using the Cellsorba (Asahi Medical, Japan) or granulocyte/monocyte apheresis (GMA) using the Adacolumn (JIMRO, Japan). The Adacolumn consists of a column containing cellulose acetate beads in isotonic saline. It selectively retains monocytes and granulocytes through binding of FCγR. Meanwhile, the Cellsorba column consists of cylindrical

Transfusion Medicine and Hemostasis. https://doi.org/10.1016/B978-0-12-813726-0.00078-7

nonwoven polyester fibers, which removes leukocytes through filtration and adhesion. In addition to effectively removing 90%–100% of monocytes and granulocytes, it eliminates 30%–60% of circulating lymphocytes.

Indications

Hyperleukocytosis

Symptomatic Leukostasis: Correlation between WBC count and symptomatic hyperviscosity is poor. Typically, symptomatic leukostasis is observed in acute myeloid leukemia (AML) when WBC count >100,000/μL and acute lymphoblastic leukemia (ALL) when the WBC count >400,000/μL. However, in AML with monocytic or monoblastic subtype, symptoms have been described at WBC counts as low as 50,000/μL. It has been postulated that symptomatic leukostasis occurs at lower counts in AML because of large, inflexible nature of myeloid blasts, in addition to their ability to elaborate cytokines known to increase expression of cell adhesion molecules and induce inflammation. Symptoms of hyperleukocytosis have also been described in patients with chronic myelogenous leukemia in association with an acute increase in circulating immature myeloid cells (WBC counts >100,000–200,000/μL) and in patients with chronic lymphocytic leukemia with WBC counts > 400,000/μL.

Cytoreduction with leukocytapheresis rapidly reverses hyperviscosity manifestations of leukostasis. However, due to variation and poor correlation between WBC count and clinical symptoms, defining WBC or blast count goal for postprocedure is impractical. In general, leukocytapheresis can be repeated in persistently symptomatic patients until clinical symptoms are resolved. Fig. 78.1 contains a flowchart assisting with the clinical evaluation and technical aspect for leukopheresis. Concurrent chemotherapy is essential to prevent rapid reaccumulation of circulating blasts and treat patients' underlying disease.

Prophylactic Leukocytapheresis: Leukocytapheresis has been performed prophylactically in the setting of elevated WBC counts before leukostatic symptom onset. Prophylactic leukocytapheresis was historically believed to reduce early mortality; however, it was not shown to improve long-term survival in patients with AML and blast counts >100,000/μL. A recent study examining outcomes in AML patients presenting with hyperleukocytosis, reported that leukocytapheresis did not affect early mortality. Similarly, a recent retrospective analysis of propensity score matched leukemic patients presenting with hyperleukocytosis found leukocytapheresis did not affect rates of early disease complications (disseminated intravascular coagulation or tumor lysis syndrome) or patient outcome. Additionally, a retrospective analysis of pediatric AML patients suggested that leukocytapheresis confers no survival benefit to patients whose WBC count <200,000/μL. Lastly, prophylactic leukocytapheresis offers no advantage compared with aggressive induction chemotherapy and supportive care in adults with ALL and WBC counts >400,000/μL. However, as pulmonary and CNS complications occur in 50% of children with ALL and WBC counts >400,000/μL, prophylactic leukocytapheresis may be considered in this group of patients. Additionally, there are reports of prophylactic leukocytapheresis being used to maintain blast counts below 100–150,000/μL in pregnant patients with CML. This is believed to be effective at

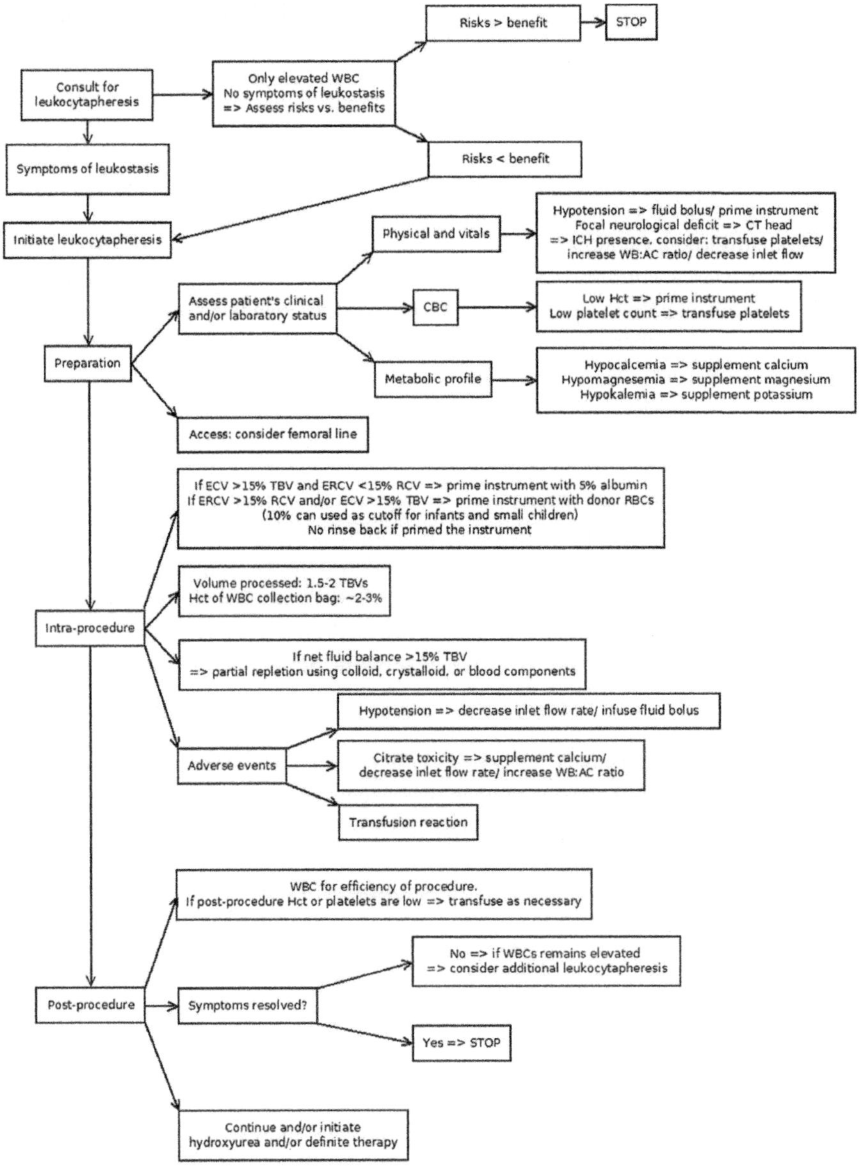

FIGURE 78.1 Leukocytapheresis flowchart. *CBC*, complete blood count; *ECV*, extracorporeal volume; *ERCV*, extracorporeal red cell volume; *Hct*, hematocrit; *ICH*, intracranial hemorrhage; *RBC*, red blood cell; *RCV*, red cell volume; *TBV*, total blood volume; *WB:AC ratio*, whole blood:anticoagulant ratio; *WBC*, white blood count. (Adapted from Pham, H. P., & Schwartz, J. (2015). How we approach a patient with symptoms of leukostasis requiring emergent leukocytapheresis. *Transfusion, 55*(10), 2306–2311.)

preventing placental insufficiency; however, it should ideally be performed in conjunction with treatment of the patients underlying disease.

Inflammatory Bowel Disease: IBD refers to ulcerative colitis (UC) and Crohn disease, chronic inflammatory disorders affecting gastrointestinal tract mediated by immune dysregulation/dysfunction and predicated by genetic, environmental, and physiological factors. Primary treatment is immunosuppression, usually corticosteroids, and aminosalicylates, although azathioprine, 6-mercaptopurine, and infliximab. Results from a large, randomized, sham-controlled, clinical trial suggest that GMA may only be effective in patients with severe UC. Other studies showed potential benefit of selective leukocyte apheresis in IBD, including more rapid remission of disease and fewer adverse reactions as well as endoscopic evidence of healing and diminished leukocyte infiltration in the bowel mucosa. Treatment regimens consist of weekly treatments for 5 weeks (LCAP) and 5–10 weeks (GMA).

Disseminated Pustular Psoriasis: Psoriasis is a chronic immune-mediated disorder characterized by spontaneous skin lesions. Disease is mediated by dysregulation of immune response resulting from genetic predisposition in association with environmental and physiological stimuli. Several small studies report symptomatic/clinical improvement when GMA was performed in conjunction with standard therapy. Treatment regimen consists of five sessions performed weekly for five consecutive weeks.

Vasculitis-Behcet Disease: Behçet's disease (BD) is an immune-mediated vasculitis affecting all vessels in arterial and venous systems. It manifests as recurrent ocular, mucocutaneous, gastrointestinal, articular, and CNS symptoms. Treatment is immunosuppression. A small study demonstrated improvement in 64% of patients with refractory ocular BD who underwent GMA. Treatment consists of five sessions performed weekly for five consecutive weeks.

Further Reading

Choi, M. H., Choe, Y. H., Park, Y., et al. (2018). The effect of therapeutic leukapheresis on early complications and outcomes in patients with acute leukemia and hyperleukocytosis: A propensity score-matched study. *Transfusion*, *1*, 208–216.

Pham, H. P., & Schwartz, J. (2015). How we approach a patient with symptoms of leukostasis requiring emergent leukocytapheresis. *Transfusion*, *10*, 2306–2311.

Schwartz, J., Padmanabhan, A., Aqui, N., et al. (2016). Guidelines on the use of therapeutic apheresis in clinical practice-evidence-based approach from the Writing Committee of the American Society for Apheresis: The seventh special issue. *J Clin Apher*, *31*, 149–338.

Staley, E. M., Simmons, S. C., Feldman, A. Z., et al. (2018). Management of chronic myeloid leukemia in the setting of pregnancy: When is leukocytapheresis appropriate? A case report and review of the literature. *Transfusion*, *2*, 456–460.

Zeller, B., Glosli, H., Forestier, E., et al. (2017). Hyperleucocytosis in paediatric acute myeloid leukaemia - the challenge of white blood cell counts above 200 × 10(9)/l. The NOPHO experience 1984-2014. *Br J Haematol*, *3*, 448–456.

CHAPTER 79

Extracorporeal Photopheresis

Sierra C. Simmons, MD, MPH, Joseph Schwartz, MD, MPH and Huy P. Pham, MD, MPH

Extracorporeal photopheresis (ECP) involves the ex vivo exposure of peripheral blood mononuclear cells (MNCs), including pathogenic or autoreactive T-lymphocytes, to photoreactive 8-methoxypsoralen (8-MOP) and ultraviolet A (UVA) light, followed by reinfusion of these MNCs. ECP was first successfully used for the treatment of cutaneous T-cell lymphoma (CTCL), its only FDA-approved indication. It is also used in the treatment of cell-mediated immunity disorders and autoimmune diseases, such as graft-versus-host disease (GVHD) and solid organ graft rejection.

Extracorporeal Photopheresis Principles: ECP is an immunomodulation procedure that results in an antigen-specific immune response directed to autoreactive or pathogenic T cells without causing generalized immunosuppression. The precise mechanisms of ECP are unknown, but the desired therapeutic effects are believed to result from multiple synergistic actions (Fig. 79.1).

Methods: In the United States, two ECP systems are FDA-approved, UVAR XTS (Mallinckrodt Pharmaceuticals, Bedminster, NJ) and CellEx (Mallinckrodt Pharmaceuticals, Bedminster, NJ). The UVAR XTS system, which is soon phasing out, uses single-needle venous access, operating in a discontinuous flow. The newer device, the CellEx system, allows for single- or double-venous access. The extracorporeal volume (ECV) in the CellEx is ~255–865 mL, depending on the patient's hematocrit, single- or double-access, and the return bag threshold volume. The manufacturer recommends the procedure be performed only in patients with a hematocrit >27%; however, many institutions allow procedures to be done in patients with a hematocrit >25%, as long as the ECV is tolerated. If the device's ECV exceeds 10%–15% of the patient's total blood volume, then procedure modification may be required, which includes priming the apheresis system with RBCs or giving a fluid (normal saline or 5% albumin). 8-MOP is added after MNC collection, then the MNC-psoralen product is exposed to UVA light to activate psoralen, intercalate into DNA, and trigger photooxidation and cell death. This mixture is then reinfused into the patient. The CellEx takes about ~1.5 hours to complete. ECP is typically performed on 2 consecutive days, every 2–4 weeks, depending on the indication and clinical status of the patient.

Anticoagulant: ECP typically requires heparin; citrate may be used if heparin is contraindicated.

Adverse Effects: ECP is a safe and well-tolerated procedure with limited side effects. A low-grade fever may occur within 2–12 hours after MNC reinfusion, most likely due to cytokine release. Psoralen compounds are contraindicated in patients with aphakia

Transfusion Medicine and Hemostasis. https://doi.org/10.1016/B978-0-12-813726-0.00079-9

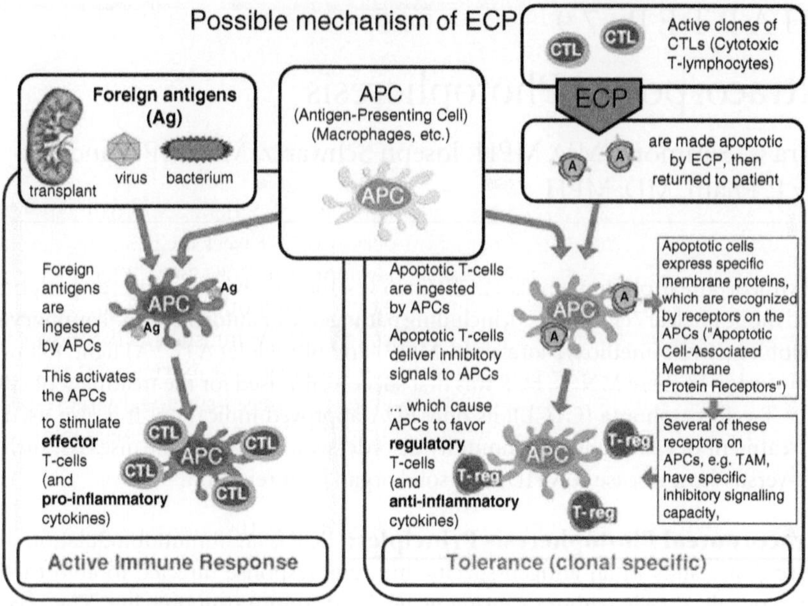

FIGURE 79.1 **Possible mechanisms by which extracorporeal photopheresis (ECP) downregulates T-cell activity.** On the left side above are depicted the events that lead to an active cellular immune response. Uptake of a foreign antigen by antigen-presenting cells (APCs) is followed by antigen presentation to T-cell clones that have antigen-binding sites that are specific for that antigen. Proinflammatory cytokines are secreted, which stimulate clonal expansion of specific cytotoxic T-lymphocytes (CTLs). These then go on to mount a cellular immune attack on the foreign antigen. In the right panel are the effects of ECP. Active cytotoxic T cells are rendered apoptotic by ECP, which when reinfused, are taken up by APCs. The APCs have receptors that recognize apoptosis-specific proteins on the T cell. This signal causes the APCs to produce antiinflammatory cytokines and to promote the development of T regulatory cells, which suppress the active cellular immune response. The effect is clonally specific. (From Ward, D. M. (2011). Extracorporeal photopheresis: how, when, and why. *J Clin Apher, 26*(5), 276–285.)

and in those who have exhibited reactions to psoralen compounds or who have a history of photosensitive disease (e.g., porphyria cutanea tarda). Patients should avoid sun exposure for 24+ hours postprocedure and a high fat diet preprocedure that renders opacity to the plasma, which can potentially interfere with establishing the MNC interface during the separation process and during penetration with the UVA in the photoactivation process. Of note, based on a recent FDA warning, ECP may increase the risk of venous thromboembolism, including pulmonary embolism.

Indications

Cardiac Allograft Rejection (for Prophylaxis or Treatment of Rejection): ECP has been used to treat and prevent cellular rejection and vasculopathy in cardiac allografts. Patients treated prophylactically after cardiac transplantation may exhibit fewer rejection and infection episodes, and they have better survival. Additionally, a decrease in coronary artery intimal thickening and panel reactive antibody levels has been demonstrated versus those treated with immunosuppressants alone. ECP is usually performed on 2 consecutive days weekly or every 2–8 weeks for several months, but this can vary. There is no definitive data for ECP duration or discontinuation point as of yet.

Atopic Recalcitrant Dermatitis: Atopic dermatitis is a common, chronic skin disease among infants and children often with allergic rhinitis and/or asthma due to an exaggerated T-cell response to allergens and other irritants. Initial treatment consists of antiinflammatory therapy. One series every 2 weeks for 12 weeks with tapering is recommended.

Cutaneous T-Cell Lymphoma; Mycosis Fungoides; Sézary Syndrome (for Erythrodermic or Nonerythrodermic Involvement): CTCL refers to a group of lymphoproliferative disorders caused by clonal, malignant CD3+/CD4+ T-lymphocytes involving the skin, including mycosis fungoides and Sézary syndrome (SS). Response rates from ECP range from 36% to 73% with a complete response in 14%–26% of patients. Some patients require long-term ECP for disease control, while others may tolerate fewer procedures. Treatment every 2–4 weeks for 6 or more months should be planned. When a maximal response is achieved, the frequency may be reduced to 1 series every 6–12 weeks with discontinuation if no relapses occur. If CTCL recurs in >25% of the skin, two daily procedures once or twice monthly should be reinstituted. If there is evidence of disease progression after 6 months with ECP alone, then combination therapy should be attempted. If there is minimal or no clinical response after 3 months with combination therapy, ECP should be discontinued. For SS, 2 monthly series are recommended.

Graft-Versus-Host Disease (for Acute or Chronic Skin GVHD or for Acute or Chronic Nonskin GVHD): GVHD results from activation of donor T cells after allogeneic hematopoietic progenitor cell transplantation and may be acute (aGVHD) or chronic (cGVHD) or may present as an overlap syndrome. The typical regimen for aGVHD is 1 series (2–3 treatments) weekly until disease response (usually within 4 weeks), tapered to biweekly, then to twice every 2 weeks before discontinuation. The typical regimen for cGVHD is 1 series (2 consecutive days) every 1–2 weeks until disease response or for 8–12 weeks, followed by a taper to every 2–4 weeks until a maximal response is achieved.

Inflammatory Bowel Disease: Inflammatory bowel disease, including ulcerative colitis and Crohn's disease, leads to chronic inflammation (leukocyte migration) within the GI tract. Effective lines of therapy include antiinflammatory agents, steroids, and other immunosuppressants, followed by thiopurines and other options. ECP has been evaluated in the setting of patients with Crohn's disease in two case series and was found to allow the discontinuation or reduction of steroid dose in a steroid-dependent population.

Lung Allograft Rejection: The initial treatment of bronchiolitis obliterans syndrome (BOS) from acute lung allograft rejection is usually high-dose corticosteroids; if the patient does not respond, then alternative immunosuppressive therapies have been used. ECP was initially used in the context of refractory BOS and demonstrated a beneficial effect. Additionally, ECP may be effective in patients with persistent acute rejection and early BOS, thus preventing further loss of pulmonary function. A common regimen includes 1 series (2 consecutive days) 5 times during the first month, followed by 1 series biweekly for the next 2 months, then 1 series monthly for 3 months for a total of 12 series over 6 months. The ideal duration is unknown, and long-term therapy may be beneficial.

Nephrogenic Systemic Fibrosis: Nephrogenic systemic fibrosis is a systemic disorder seen in patients with acute or typically chronic kidney disease after an average of 2 days after administration of gadolinium. Because of the lack of an effective therapy, ECP was attempted with the schedule ranging from a minimum of 1 series (2 consecutive days) every 2–4 weeks to rounds of 5 procedures performed every other day with an increasing number of weeks (1–4) between rounds for a total of 4 rounds. The ECP time to response ranged from 4 to 16 months.

Pemphigus Vulgaris: Pemphigus vulgaris is a potentially fatal but rare autoimmune mucocutaneous blistering disease with an associated antibody to desmoglein 1 and 3. CD4+ T cells reactive to desmoglein 1 and 3 have been encountered. If ECP is incorporated into the treatment regimen, it should be continued daily or every other day until a clinical response is observed.

Psoriasis: Psoriasis, a chronic cutaneous disorder, causes thickened plaques and papules from dysregulation and hyperproliferation of the epidermis. There are data to support the use of ECP from one study. One series (2 consecutive days) per week for 4 months with tapering was used.

Scleroderma (Systemic Sclerosis): Accumulation of collagen and extracellular matrix proteins in the skin and other organs in systemic sclerosis (SSc) is thought to be the result of activated T cells, macrophages, and cytokines. ECP may be attempted with treatment occurring on 2 consecutive days every 4–6 weeks for 6–12 months.

Further Reading

Baskaran, G., Tiriveedhi, V., Ramachandran, S., Aloush, A., Grossman, B., Hachem, R., et al. (2014). Efficacy of extracorporeal photopheresis in clearance of antibodies to donor-specific and lung-specific antigens in lung transplant recipients. *J Heart Lung Transplant*, *33*, 950–956.

Boshell, M. N., Peavey, D. B., Simmons, S. C., Williams, L. A., III, & Pham, H. P. (2018). Blood prime for patients with single-needle access requiring extracorporeal photopheresis: How to do and why it may be useful? *J Clin Apher*, *33*(1), 121–123.

Colvin, M. M., Cook, J. L., Chang, P., Francis, G., Hsu, D. T., Kiernan, M. S., et al. (2015). Antibody-mediated rejection in cardiac transplantation: Emerging knowledge in diagnosis and management: A scientific statement from the American Heart Association. *Circulation*, *131*, 1608–1639.

Knobler, R., Berlin, G., Calzavara-Pinton, P., Greinix, H., Jaksch, P., Laroche, L., et al. (2014). Guidelines on the use of extracorporeal photopheresis. *J Eur Acad Dermatol Venereol*, *28*(Suppl. 1), 1–37.

Schwartz, J., Padmanabhan, A., Aqui, N., Balogun, R. A., Connelly-Smith, L., Delaney, M., et al. (2016). Guidelines on the use of therapeutic apheresis in clinical practice-evidence-based approach from the Writing Committee of the American Society for Apheresis: The seventh special issue. *J Clin Apher*, *31*, 149–338.
Spratt, E. A. G., Gorcey, L. V., Soter, N. A., & Brauer, J. A. (2015). Phototherapy, photodynamic therapy and photopheresis in the treatment of connective tissue diseases: A review. *Br J Dermatol*, *173*, 19–30.
Ward, D. M. (2011). Extracorporeal photopheresis: How, when, and why. *J Clin Apher*, *26*(5), 276–285.

CHAPTER 80

LDL Apheresis

Monika Paroder-Belenitsky, MD, PhD and Huy P. Pham, MD, MPH

Low-density lipoprotein (LDL) apheresis (LA) removes apo-B containing lipoprotein (LDL and lipoprotein(a) [Lp(a)]). The primary indication for LA is familial hypercholesterolemia (FH), an autosomal dominant disorder of cholesterol metabolism, resulting in elevated plasma LDL-C levels.

Numerous LDL-C removal methodologies are used by various systems available worldwide. Available technologies include heparin-induced extracorporeal LDL-C precipitation, dextran sulfate cellulose adsorption, double filtration plasmapheresis, polyacrylate full blood adsorption, and immunoadsorption—all have similar efficacy (Table 80.1). Only Liposorber LA-15 (Kaneka Pharma America LLC, New York, NY) and Plasmat Futura (previously Plasmat Secura) (B. Braun Medical, Bethlehem, PA) systems are FDA-approved.

Anticoagulant: LA instruments use heparin, either alone or in combination with citrate, as the anticoagulant. Citrate cannot be used in the Liposorber LA-15 system because this system removes cholesterol selectively based on the electrostatic interaction between the column and apoB100 lipoproteins and citrate can interrupt this interaction. In a patient with heparin-induced thrombocytopenia, lepirudin was reportedly used as the anticoagulant for the direct adsorption of lipoprotein (DALI) system. A small case series describing the use of citrate, as the only coagulant for this DALI system was also published.

Adverse Events and Contraindications: LA has an overall adverse event rate of 11%, and without significant difference between different LA systems. Most of these reactions are mild. The most common side effects are postprocedure bleeding (3.5%), vomiting (2.5%), hypoglycemia (2.4%), and hypotension (2.2%). The use of angiotensin-converting enzyme inhibitors (ACEIs) is contraindicated in adsorption-based LA, and discontinuation is recommended 24 hours before procedure. Angiotensin receptor blockers can be used as alternatives. Recently, the use of turmeric herbal therapy has been associated with symptoms similar to those observed in patients taking ACEI.

Indications

Familial Hypercholesterolemia: FH is an autosomal dominant disorder of lipoprotein metabolism mostly caused by a mutation in the LDL receptor (LDL-R) that results in elevated levels of LDL cholesterol (LDL-C), thereby increasing the risk of premature cardiovascular consequences. FH is not an uncommon inherited disorder—approximately 1 in 500 Americans are heterozygotes, and 1 in 1,000,000 homozygotes. Typically, FH homozygotes have cholesterol in the range of 700–1200 mg/dL, and heterozygotes have cholesterol in the range of 350–500 mg/dL.

Transfusion Medicine and Hemostasis. https://doi.org/10.1016/B978-0-12-813726-0.00080-5

TABLE 80.1 Comparison of Currently Available Low-Density Lipoprotein (LDL) Apheresis Systems

System/Commercial Instrument	Mechanism of LDL Removal	Substances Removed[a]	Advantages	Disadvantages
Dextran Sulfate Liposorber LA-15[b]	Binding to dextran sulfate on the basis of electrical charge	LDL: 56%–65% HDL: 9%–30% TG: 34%–40% Lp(a): 52%–61%	Column can be regenerated	System requires plasma separation, which may be interfered by a high hematocrit
Heparin-induced extracorporeal LDL precipitation. Plasmat Secura and Plasmat Futura[b]	Precipitation of LDL by heparin at an acidic pH	LDL: 67% HDL: 15% TG: 41% Lp(a): 62%	System removes fibrinogen	System is complicated. High hematocrit may interfere with plasma separation
Double filtration plasmapheresis	Separation based on size by filtering plasma with a second filter	LDL: 56% HDL: 25% TG: 49% Lp(a) 53%	System removes fibrinogen	High hematocrit may interfere with plasma separation. System causes loss of some albumin, HDL, and IgG
Plasmaselect	Immobilized sheep apolipoprotein B-100 antibodies	LDL: 64% HDL: 14% TG: 42% Lp(a): 64%	Column can be regenerated	System causes exposure to animal proteins
Lipoprotein hemoperfusion DALI	Binding to polyacrylate-coated polyacrylamide beads on the basis of electrical charge	LDL: 61% HDL: 30% TG: 42% Lp(a): 64%	Plasma separation is not necessary	Column cannot be regenerated or reused
Liposorber D	Binding to dextran sulfate covalently bonded to cellulose, on the basis of electrical charge	LDL: 62% HDL: 2.5% TG: 38%–68% Lp(a): 56%–72%	Plasma separation is not necessary. Procedure time is shorter than Liposorber LA-15	Column cannot be regenerated

[a]Percentage removed in a typical treatment.
[b]FDA-cleared system in the United States.
Modified from Winters, J. L., in McLeod, B. C., Weinstein, R., Winters, J. L., et al. (Eds.). (2010). *Apheresis Principles and Practice* (3rd ed.). Baltimore: AABB Press.

If untreated, persistently elevated LDL-C typically leads to tendinous xanthomatosis and premature atherosclerotic coronary artery disease. While statins remain the mainstay of therapy for hypercholesterolemia in general, they have a limited effect due to the need for production of a functional LDL-R for their effect. Given the well-established role of LDL-C in cardiovascular disease, all therapies are aimed at reducing intravascular LDL-C. Targeted lipid goals are often not possible to achieve despite maximal medical therapy and lifestyle adjustments, thereby necessitating LA to extracorporeally remove LDL-C. LA is an FDA-approved indication for the following high-risk patient populations for whom diet has been ineffective and maximum drug therapy has either been ineffective or not tolerated:

- Functional homozygotes with LDL-C >500 mg/dL
- Functional heterozygotes with LDL-C ≥300 mg/dL
- Functional heterozygotes with LDL ≥200 mg/dL and documented cardiovascular disease

In addition, although it is not an FDA-approved indication, patients without FH but with elevated LDL-C or Lp(a) levels who cannot tolerate or are unresponsive to conventional therapy (drug and diet) can also potentially be treated by LA.

In children with homozygous FH, it is recommended that LA should begin by 7 years of age to prevent the development of aortic stenosis. Long-term outcomes studies have demonstrated significant reduction in the number of cardiovascular events as well as stabilization or regression of coronary stenosis.

The goal in the treatment of hypercholesterolemia by LA is to reduce the LDL-C by >60% from baseline as well as to reduce the time-average cholesterol. Procedures are typically performed every 1–2 weeks, but the frequency may be adjusted to meet the LDL-C goal. Regular and lifelong apheresis therapy is required to achieve and maintain LDL-C levels at target ranges.

Focal Segmental Glomerulosclerosis Refractory to Standard Therapies: LA using the Liposorber LA-15 system is FDA-approved for use in patients with nephritic syndrome associated with focal segmental glomerulosclerosis (FSGS) refractory to standard therapies (i.e., steroids, calcineurin inhibitors) or when standard therapy is not well tolerated and in pediatric patients after renal transplant with recurring primary FSGS. The exact mechanism of how LA works to improve nephritic syndrome is largely unknown.

Peripheral Vascular Disease: Peripheral vascular disease is characterized by narrowing of blood vessel lumens secondary to atherosclerosis, thus, leading to insufficient blood flow. LA-induced reductions in LDL-C, CRP, and fibrinogen are thought to enhance microcirculation and reduce blood viscosity and adhesion molecules.

Refsum Disease: Phytanic acid storage disease (also known as Refsum disease) is an autosomal recessive disease resulting from the inability to metabolize phytanic acid due to a defect in the phytanoyl-CoA hydrolase. The accumulation of this branched chain fatty acid may manifest clinically as retinitis pigmentosa, peripheral neuropathy, cerebellar ataxia, sensorineural deafness, and anosmia. Elevated blood levels of phytanic acid can be rapidly reduced by TPE. Because phytanic acid is also bound to lipoproteins and triglycerides, LA can be considered.

Sudden Sensorineural Hearing Loss: Sudden sensorineural hearing loss is defined as the loss of at least 30 dB in three sequential frequencies. Elevated fibrinogen and LDL-C are considered risk factors; thus, their reduction with LA may be beneficial.

Further Reading

Grützmacher, P., Öhm, B., Szymczak, S., Dorbath, C., Brzoska, M., & Kleinert, C. (March 2017). Primary and secondary prevention of cardiovascular disease in patients with hyperlipoproteinemia (a). *Clin Res Cardiol Suppl*, *12*(Suppl. 1), 22–26.

Schwartz, J., Padmanabhan, A., Aqui, N., Balogun, R. A., Connelly-Smith, L., Delaney, M., et al. (June 2016). Guidelines on the use of therapeutic apheresis in clinical practice-evidence-based approach from the Writing Committee of the American Society for Apheresis: The seventh special issue. *J Clin Apher*, *31*(3), 149–162.

Winters, J. L. (2011). Lipid apheresis, indications, and principles. *J Clin Apher*, *26*, 269–275.

CHAPTER 81

Immunoadsorption

Monika Paroder-Belenitsky, MD, PhD and Huy P. Pham, MD, MPH

Immunoadsorption (IA) selectively removes certain plasma proteins. IA can be specific (only removing antibody specific for single antigen) or nonspecific (removing all antibodies). Although several IA systems are available worldwide, none are currently used in the United States.

This chapter discusses only IA using staphylococcal protein A (SPA) columns, which remove immunoglobulin, as an example of a selective system. SPA is a cell wall component of certain strains of *Staphylococcal aureus* and has high affinity for Fc portion of IgG and for aggregated IgG and IgG-containing immune complexes. SPA binds strongly to IgG1, IgG2, and IgG4 and variably with IgG3, IgM, and IgA. Additionally, IA has immunomodulatory affects, including activating complement by the alternate pathway and reducing circulating immune complexes by altering antigen–antibody ratios.

Protein A Immunoadsorption: Two extracorporeal SPA devices, Prosorba and Immunosorba, use highly purified SPA linked to a solid matrix to selectively remove immunoglobulins from plasma. The Prosorba column (Fresenius Kabi, Redmond, WA) contained SPA immobilized on a silica matrix, which removes free immunoglobulins and immune complexes. As of 2006, Prosorba column is no longer commercially available. Immunosorba column (Fresenius Kabi AG, Bad Homburg, Germany), consists of SPA linked to a sepharose matrix. The process involves separation of plasma in a cell separator, passing 7–9 L of plasma alternatively through one of two parallel columns, and returning processed plasma to the patient. Treatment of 2.5 plasma volumes reduces plasma immunoglobulin concentrations: IgG1 (97%), IgG2 (98%), IgG3 (40%), IgG4 (77%), IgM (56%), and IgA (55%).

Adverse Effects: Most procedures result in minor, but not major, complications. Common adverse effects include chills, low-grade fever, musculoskeletal pain, hypotension, nausea, and vomiting. Short-term flare in joint pain and swelling may appear within 1 hour of the procedure and last up to 2 hours. Severe respiratory and cardiovascular toxicities can occur, rarely resulting in fatality. Additionally, the use of IA is contraindicated in patients taking angiotensin converting enzyme inhibitor (ACEI).

Indications: IA, or therapeutic plasma exchange (TPE), can be used to treat various immune-mediated disorders. However, unlike IA, TPE removes pathogenic and nonpathogenic (i.e., coagulation factors) substances from plasma nonselectively.

Atopic Dermatitis, Recalcitrant: Atopic dermatitis (AD) is the most common chronic relapsing skin disease, frequently affecting children. AD-affected individuals have elevated levels of IgE and are predisposed to allergic rhinitis and asthma later in life.

Transfusion Medicine and Hemostasis. https://doi.org/10.1016/B978-0-12-813726-0.00081-7

IA has been used to reduce serum IgE; however, short-term decreases are followed by rebound elevation within weeks after discontinuation of IA.

Cryoglobulinemia: Cryoglobulins are immunoglobulins that reversibly precipitate and deposit in small vessels at temperature below 37°C, activate complement, and recruit leukocytes, thereby causing vascular damage. Cryoglobulinemia is associated with viral infections (especially hepatitis C), autoimmune diseases, and lymphoproliferative disorders and can cause vasculitis, neuropathy, and glomerulonephritis. TPE or IA can be used as adjunct therapy for severe symptomatic cryoglobulinemia in conjunction with the treatment of the underlying disease(s). It is critical to warm apheresis system and room to avoid precipitation of cryoglobulins. A single randomized controlled trial in patients with hepatitis C–associated cryoglobulinemia who had not responded to previous conventional medications, reported greater clinical improvement (in severity of organ involvement) in patients who underwent IA (three times a week [45 mL/kg processed for 12 weeks]) in combination with medical therapy (80%) versus those who did not (33%).

Dilated Cardiomyopathy With NYHA II–IV: Dilated cardiomyopathy (DCM) is characterized by left ventricular enlargement with impaired systolic function and can ultimately result in cardiac failure. It accounts for ~50% of heart transplantations in the United States. DCM can be idiopathic, rarely secondary to mutations in cytoskeletal proteins, or more commonly sequela of viral myocarditis. Approximately 80% of patients with DCM have autoantibodies to myocardial antigens, and IA has demonstrated short- and long-term patient improvement. No outcome differences between treatment schedules have been noted. Repeated IA procedures can be helpful in patients with worsening cardiac function. Generally, patients who undergo IA were less likely to require transplant or left ventricular assist device (LVAD) than those who did not.

Coagulation Factor Inhibitors: Patients with congenital factor deficiencies (i.e., hemophilia A or B) can develop alloantibodies against the deficient factor, known as inhibitors. Some patients develop autoantibodies to coagulation factors, resulting in acquired deficiency. In patients with factor inhibitors, therapy should be individualized, depending on clinical status and inhibitor titer. Perioperative treatment options for bleeding are to replace factor, bypass factor (e.g., using recombinant factor VIIa in patients with strong inhibitors to factor VIII or IX), or remove and/or suppress inhibitor. Inhibitor suppression therapy is usually achieved through immunosuppressants and inhibitor removal through either IA or TPE; IA is more effective than TPE. IA is performed daily (three plasma volumes processed) until antibody titer decreases and bleeding is easily controlled with other therapeutic modalities. Anticoagulant use in apheresis system should be minimized.

Refractory Immune Thrombocytopenia: Immune thrombocytopenia (ITP) is caused by autoantibodies to platelet surface antigens, which results in accelerated platelet destruction. Risk of fatal bleeding from thrombocytopenia increases with age. Therapeutic modalities used include corticosteroids, IVIG, RhIg, splenectomy, and thrombopoietin receptor agonists. IA can be considered in patients with refractory ITP and life-threatening bleedings or in patients in whom splenectomy is contraindicated. Studies of IA have yielded highly variable outcomes, from no improvement to complete remission.

Multiple Sclerosis: Multiple sclerosis (MS) is a demyelinating disorder of the central nervous system, often progressive and after relapsing course. While MS pathophysiology remains not entirely understood, it involves humoral and cellular arms of the immune system. Immune-modifying pharmacologic agents are mainstay of therapy. Limited studies suggest that TPE or IA may benefit patients with acute attacks who fail initial treatment with high-dose steroids. In patients with steroid-refractory MS relapses, four studies reported significant clinical improvement in 73%–85% of patients. However, in MS patients with nonactive relapsing–remitting or secondary progressive course, no clinical improvement was observed. Most studies of IA in MS have used single-use tryptophan adsorbers.

Paraneoplastic Neurologic Syndromes: Paraneoplastic neurologic syndromes affect ~1% of cancer patients and may precede cancer diagnosis in 50% of the cases. It associates with specific CSF and serum antibodies. A series of 13 patients with paraneoplastic opsoclonus/myoclonus or paraneoplastic cerebellar degeneration were treated with IA (six total treatments performed twice weekly for 3 weeks); three achieved complete remission and three achieved partial neurological remissions, but all patients subsequently relapsed.

Paraproteinemic Polyneuropathies Caused by IgG, IgA, or IgM: Paraproteinemic polyneuropathies (PPs) are chronic illness resembling chronic inflammatory demyelinating polyradiculoneuropathy. PP is most commonly seen with monoclonal gammopathy of undetermined significance. TPE and IA can be used, although the clinical experience with IA is extremely limited.

Pemphigus Vulgaris: Pemphigus vulgaris (PV) is a rare mucocutaneous blistering disease, caused by autoantibodies to desmosomal proteins desmoglein 1 and 3. Treatment options include corticosteroids and other immune modulatory agents. IA has primarily been used in treating patients with PV in Europe. Decrease in circulating levels of antibodies correlate with skim lesion improvement.

Renal Transplant: TPE or IA can be used in conjunction with immunosuppressive agents before transplantation in patients with antibodies to donor human leukocyte antigens. These procedures can be continued postoperatively if antibody-mediated rejection occurs to remove donor-specific antibodies, as this can result allograft loss.

Thrombotic Microangiopathy, Shiga Toxin Mediated: Hemolytic uremic syndrome (HUS) is characterized by Coombs-negative hemolytic anemia, thrombocytopenia, and acute kidney injury. Most commonly, HUS is caused by *Escherichia coli* O157:H7 infection and Shiga-like toxin-induced damage to renovascular endothelium. HUS treatment is supportive care. While TPE was performed in 251 patients with HUS in the 2011 German outbreak with no clinical benefit, improvement in neurological deficits was reported with IA in 12 patients with *E. coli* O104:H4-associated HUS unresponsive to TPE or eculizumab.

Further Reading

Huestis, D. W., & Morrison, F. (1996). Adverse effects of immune adsorption with staphylococcal protein A columns. *Transfus Med Rev*, *10*, 62–70.

Meyersburg, D., Schmidt, E., Kasperkiewicz, M., & Zillikens, D. (2012). Immunoadsorption in dermatology. *Ther Apher Dial*, *16*(4), 311–320.

Oji, S., & Nomura, K. (October 2017). Immunoadsorption in neurological disorders. *Transfus Apher Sci*, *56*(5), 671–676.

Sanchez, A. P., Cunard, R., & Ward, D. M. (2013). The selective therapeutic apheresis procedures. *J Clin Apher*, *28*(1), 20–29.

Schwartz, J., Padmanabhan, A., Aqui, N., Balogun, R. A., Connelly-Smith, L., Delaney, M., et al. (2016). Guidelines on the use of therapeutic apheresis in clinical practice-evidence-based approach from the Writing Committee of the American Society for Apheresis: The seventh special issue. *J Clin Apher*, *31*(3), 149–162.

CHAPTER 82

Therapeutic Phlebotomy

Yelena Z. Ginzburg, MD and Francesca Vinchi, PhD

Therapeutic phlebotomy entails the removal of blood to treat diseases in which decreasing red blood cell (RBC) mass, hematocrit, and blood viscosity, or inducing iron restriction, enabling management of disease-associated symptoms and complications. Standardized indications for therapeutic phlebotomy for each disease are currently unavailable. Therefore, specific regimens are individually tailored to patient's needs.

Therapeutic phlebotomy is a clinical procedure commonly used to treat diseases associated either with elevated iron stores (i.e., hereditary hemochromatosis [HH]) or RBC mass (i.e., polycythemia vera [PV]). Because RBCs contain high amounts of the iron-containing hemoglobin, the removal of circulating RBCs decreases total body iron, preventing or reversing the adverse effects of excess iron, including liver, cardiac, and endocrine dysfunction (see Chapter 72). In patients with PV, therapeutic phlebotomy results in reduced RBC mass by iron-restricting erythropoiesis and is associated with decreased risk of thrombosis.

Indications

Hereditary Hemochromatosis: HH is a common inherited disorder characterized by progressive accumulation of dietary iron. In the United States, the most common hemochromatosis genotype is the homozygous mutation in the *HFE* gene (C282Y or H63D; type 1 HH), involved in the regulation of hepcidin, the peptide hormone central to the regulation of iron metabolism. Autosomal recessive mutations in hepcidin itself and its other regulators (HJV, TfR2) as well as autosomal dominant mutations in its receptor (ferroportin) result in juvenile, type III, and type IV HH, respectively.

Serum transferrin saturation and ferritin levels are typically used for screening. Genetic testing for the most common HH-associated mutations is routinely available. Liver biopsy and noninvasive quantitative techniques (i.e., MRI) enable diagnosis of parenchymal iron overload. Life-threatening complications of iron overload include cirrhosis, hepatocellular carcinoma, diabetes, and heart disease. The severity of liver disease closely reflects the magnitude of hepatic iron deposition and is often reversible with therapy.

Therapeutic phlebotomy is the treatment of choice in patients with HH. However, the benefit of therapeutic phlebotomy has not been established for HH individuals without elevated serum ferritin or evidence of tissue iron load. In cases with evidence of frank or impending iron overload, complications and symptoms can be prevented or reversed by reducing iron overload for which therapeutic phlebotomy is an established treatment. Response rate to therapeutic phlebotomy is dependent on the individual severity of iron overload and no evidence-based guidelines are available to direct the end point of therapeutic phlebotomy. The recommendations that exist are based on (1) a theoretical argument that achieving iron deficiency is necessary to normalize tissue iron and (2) that a stated target prevents variable interpretation and practice.

Transfusion Medicine and Hemostasis. https://doi.org/10.1016/B978-0-12-813726-0.00082-9

Thus, practice guidelines propose goals of therapy as reducing and maintaining low-normal iron stores (ferritin 50–100 ng/mL). The typical regimen is weekly phlebotomy until ferritin approximates 50 ng/mL followed by lifelong maintenance phlebotomy. Although therapeutic phlebotomy is not a cure for HH, effectively phlebotomized patients exhibit normalized tissue iron, decreased risk of liver cirrhosis, and a significantly longer median survival.

Secondary Iron Overload: Iron overload can result from chronic RBC transfusion in patients with acquired and inherited anemias. These patients typically do not tolerate therapeutic phlebotomy. Iron chelation therapy is a standard first-line treatment to prevent and treat iron overload in these patients (see Chapter 72).

Other diseases associated with acquired iron overload include chronic hepatitis C, fatty liver disease, porphyria cutanea tarda, insulin resistance-associated hepatic iron overload, African dietary iron overload (Bantu siderosis), parenteral iron overload in renal failure patients on hemodialysis, and oral iron ingestion. Therapeutic phlebotomy is indicated in some individuals with these conditions and decisions about treatment are made on an individual basis.

Polycythemia Vera: PV is a clonal progressive myeloproliferative disorder with significant erythrocytosis. Diagnosis is based on WHO criteria with elevated hemoglobin (>18.5 g/dL in men or >16.5 g/dL in women) or RBC mass (>25% predicted) with splenomegaly, bone marrow hypercellularity, low serum erythropoietin levels, normal arterial oxygen saturation, and mutated JAK2 (V617F in approximately 95% of cases, or exon 12 mutations in approximately 4% of cases). Patients with PV have increased risk of arterial or venous thromboses. Depending on the presenting hematocrit, alternate day to weekly therapeutic phlebotomy effectively lowers hematocrit to <45%. Recommendations for phlebotomy are based on the observation of reduced cardiovascular morbidity and decreased incidence of thrombotic events in patients with PV with hematocrit <45%. Despite this, hematocrit response is not associated with improved survival or reduction in thrombotic complications; in fact, leukocytosis and thrombocytosis are associated with increased risk of death and thrombosis, respectively. Additional work is needed to determine specific benefit of therapeutic phlebotomy in patients with PV.

Secondary Polycythemia: Polycythemia can be secondary to increased erythropoietin production either directly due to pathologically increased serum erythropoietin (e.g., impaired renal perfusion from renal artery stenosis), or erythropoietin-producing tumors, or indirectly via response to chronic hypoxia, as in lung or cardiac disease, in smokers, at high altitudes, and in carbon monoxide poisoning. Therapeutic phlebotomy is recommended for patients with hypoxemic lung disease or cyanotic congenital heart disease who exhibit blood hyperviscosity and hematocrit >55%. Polycythemia can also occur after kidney transplantation and in response to testosterone replacement therapy, especially in patients using intramuscular testosterone. In these cases, decisions regarding need for therapeutic phlebotomy should be made on an individual basis.

Recommended Approach: A unit (450–500 mL) of whole blood containing 200–250 mg of iron is removed during each phlebotomy. A half unit is removed for patients with lower body mass, anemia, or cardio/pulmonary disorders. Patients with normal body mass may tolerate alternate day therapeutic phlebotomy. Suggested criteria for initiating therapeutic phlebotomy include ferritin concentration: (1) >200 ng/mL for patients of age <18 years (regardless of gender); (2) >500 ng/mL for women of child-bearing age; (3) >300 ng/mL for men >18 years. Patients undergoing frequent therapeutic phlebotomy (e.g., alternate day to weekly) require prephlebotomy hemoglobin evaluation and serum ferritin measurements every 2–3 months. In patients with polycythemia, phlebotomy goal is hematocrit <45% with an expected decrease of approximately 3% with each 450–500 mL phlebotomy.

Units drawn via phlebotomy can be used for allogeneic transfusion if they meet the following criteria: (1) they are collected from HH individuals only, (2) they are collected at no expense to the individual, (3) the individual meets FDA allogeneic blood donor eligibility criteria, and (4) the center has an FDA variance for use of these units.

Erythrocytapheresis represents an alternative approach enabling the removal of more RBCs and therefore more iron per single treatment (~400–500 mg) in qualified patients (see Chapter 76). The application of this intervention has anyway several limitations, including high costs, long treatment time, and treatment qualification restrictions relative to standard therapeutic phlebotomy. Oral or parenteral iron chelator therapy is a treatment option for patients who cannot tolerate phlebotomy and iron-overloaded patients with severe anemia precluding therapeutic phlebotomy.

Adverse Events: Adverse events of therapeutic phlebotomy are similar to those of whole blood donation (see Chapter 8) and include vasovagal reactions and bruising or hematoma at the site of venipuncture. Symptoms of hypovolemia can occur in a minority of patients (more typically if prephlebotomy hemoglobin <11 g/dL) and can usually be prevented with adequate hydration and avoidance of strenuous exercise within 24 hours following the procedure. In addition, patients undergoing therapeutic phlebotomy for erythrocytosis are at risk for developing iron deficiency and therefore accompanying symptoms of fatigue.

Further Reading

Adams, P. C., & Barton, J. C. (2010). How I treat hemachromatosis. *Blood, 116*, 317–325.

Arber, D. A., et al. (2016). The 2016 revision to the World Health Organization classification of myeloid neoplasms and acute leukemia. *Blood, 127*, 2391–2405.

Assi, T. T., & Baz, E. (2014). Current applications of therapeutic phlebotomy. *Blood Transfus, 12*(Suppl. 1), s75–s83.

Bacon, B. R., et al. (2011). Diagnosis and Management of Hemochromatosis: 2011 practice guideline by the American Association for the Study of Liver Diseases. *Hepatology, 54*, 328–343.

Buzzetti, E., et al. (2017). Interventions for hereditary haemochromatosis: An attempted network meta-analysis. *Cochrane Database Syst Rev, 3*, CD011647.

European Association For The Study Of The Liver. (2010). EASL clinical practice guidelines for HFE hemochromatosis. *J Hepatol, 53*, 3–22.

Kim, K. H., & Oh, K. Y. (2016). Clinical application of therapeutic phlebotomy. *J Blood Med, 7*, 139–144.

Marchioli, R., et al. (2013). Cardiovascular events and intensity of treatment in polycythemia vera. *N Engl J Med, 368*, 22–33.

CHAPTER 83

Overview of Cellular Therapy

Rona S. Weinberg, PhD

Cellular therapy is the use of viable cells and tissues for the treatment of disease. Cell and tissue donors can be autologous, syngeneic, or allogeneic. Current good tissue and manufacturing practices must be followed to prevent the introduction, transmission and spread of communicable diseases. Both require donor qualification, controlled environments, use of sterile supplies and reagents, and assays to ensure the purity and potency of products. When selecting a cell type for the development of a cellular therapy, the goals and purpose of the therapy must be considered. The starting cells should be readily available in sufficient quantities to provide the desired effects. Either stem cells or adult tissue-specific differentiated cells can be used. Stem cells are undifferentiated cells capable of unlimited self-renewal; under appropriate conditions they differentiate into specific cell type(s). Stem cells are most useful for repair or repopulation of damaged tissues and can be derived from embryonic or adult tissues. Adult tissue-specific differentiated cells may be used when the function of a specific differentiated cell type is desired.

The concept that all cellular therapy product development and manufacture must address safety, quality, identity, potency, and purity (SQUIPP) has its roots in the development of blood transfusions as a cellular therapy.

Donors: Autologous donors provide their own cells. Autologous cells are readily available, do not cause immunologic complications such as graft-versus-host disease (GVHD) or graft rejection and transmission of infectious agents is not relevant. However, autologous cells may contain tumor or other abnormal cells and do not provide tumor-suppressing immunologic effects (graft-versus-tumor effects, GVT).

Syngeneic donor cells are genetically identical to those of the recipient, i.e., from an identical twin, and therefore immunologic complications such as rejection, hemolysis, or GVHD should not occur. However, immunologic effects that help to eradicate malignancy and prevent tumor relapse (GVT) are also absent. Transmission of infectious agents from donor to recipient can occur.

Allogeneic donors, either related or unrelated to the recipient, are genetically distinct from the recipient. Depending on cell type and planned use of cells, it may be necessary for the donor and the recipient to be human leukocyte antigen (HLA) matched to prevent immunologic complications such as GVHD and/or graft rejection. HLA-matched donor availability may be limited. Allogeneic cells may provide immunologic effects such as GVT. Transmission of infectious agents from donor to recipient can occur.

Xenogeneic cells or tissues are obtained from nonhuman animals and are immunologically distinct from human cells; therefore, graft rejection is a major concern. Tissue culture media used to produce cellular therapy products often contain xenogeneic material such as fetal bovine serum. Animal-derived cells and reagents may be

Transfusion Medicine and Hemostasis. https://doi.org/10.1016/B978-0-12-813726-0.00083-0

contaminated with potentially infectious agents that cannot be detected by currently available assays, for example, prions that cause bovine spongiform encephalopathy.

Donor eligibility must be determined to prevent the spread of communicable diseases. Donors are tested for relevant infectious diseases for which licensed tests are available and outlined in the 21 CFR 1271, including HIV types 0, 1, and 2, hepatitis B and C viruses, human T-cell lymphoproliferative viruses I and II, syphilis, and cytomegalovirus (CMV). Testing for other relevant diseases, which are also licensed, include West Nile virus and *Trypanosoma cruzi*. Physical examination and review of health history and medical records evaluate risk factors and symptoms of communicable diseases, including those for which no licensed tests are available, such as Zika virus (one platform FDA approved), *Babesia microti*, and human transmissible spongiform encephalopathy.

Sources of Cells: When selecting a cell type for the development of a cellular therapy, the goals and purpose of the therapy must be considered. The starting cells should be readily available in sufficient quantities to provide the desired effects. Cellular therapies have been developed using either stem cells or adult tissue-specific differentiated cells (Table 83.1). Stem cells are undifferentiated cells that are capable of unlimited self-renewal and under the appropriate conditions can be induced to differentiate into specific cell type(s). Stem cells are most useful for repair or repopulation of damaged tissues and can be derived from embryonic or adult tissues. The dogma surrounding the use of adult stem cells is that adult stem cells have limited differentiation capacity, usually restricted to the tissue from which they are derived. However, this has been blurred by reports of bone marrow (BM) stem cells being used to repair other tissues, such as in cardiac repair. Adult tissue-specific differentiated cells may be used when the function of a specific differentiated cell type is desired. Homologous cells perform the same basic function in the recipient as in the donor (e.g., hematopoietic progenitor cell [HPC] product for BM reconstitution) while a nonhomologous product performs a different function in the recipient than the donor (e.g., HPC product used to enhance cardiac muscle repair). To prevent immune reactions in the recipient and/or graft rejection, the immunogenicity of allogeneic cells must be considered. Large-scale production of "off the shelf" cellular therapies that are available for distribution to patients may require the development of banks of cells with multiple HLA types, or the production of cells that are universally accepted by recipients with varying HLA types.

Stem Cells: Stem cell therapies that are currently considered to be standard of care are HPCs for hematopoietic stem cell transplant (HSCT), epithelial stem cell treatments for burns and corneal tissue for corneal replacement.

Hematopoietic Progenitor Cell Products: HPC products contain multipotential stem cells that can differentiate into red blood cells, white blood cells, and megakaryocytes to make platelets and are used to reconstitute a recipient's BM. HPC products can be harvested from BM, peripheral blood (PB) by apheresis, and umbilical cord blood (CB). HSCT is a life-saving treatment for hematologic malignancies, BM failure syndromes, radiation sickness, genetic disorders such as hemoglobinopathies, and in conjunction with high-dose chemotherapy for a variety of malignancies. HPC products

TABLE 83.1 Sources of Cells

Cell Type	Characteristics	Uses	Pros and Cons
Stem cells	Undifferentiated cells. Capable of unlimited self-renewal. Can be induced to differentiate into specific cell type(s).	Replace or repair damaged cells or tissue.	Pro: Can differentiate into multiple cell types. Con: Rare cells difficult to obtain from solid tissues; expansion may lead to metastatic transformation.
Hematopoietic stem cells	Derived from the bone marrow. "Adult" stem cells. Multipotent: differentiate into red blood cells, white blood cells, and megakaryocytes. Differentiation limited to blood cells.	Hematopoietic stem cell transplant (HSCT) to reconstitute bone marrow.	Pro: Can cure hematologic malignancies and genetic disorders. Con: Appropriate allogeneic donor may not be available; autologous cells may contain malignancy.
Embryonic stem cells	Derived from the inner cell mass of an embryonic blastocyst 5 days after fertilization. Pluripotent: capacity to differentiate into any cell type.	Potential to repair or replace any damaged cells or tissue.	Pro: Potential to differentiate into all cell types. Con: Requires destruction of an embryo; may form teratomas when transplanted.
Induced pluripotent stem cells	Derived from adult somatic cells. Human embryonic stem cell–like cells. Pluripotent: capacity to differentiate into any cell type.	Potential to repair or replace any damaged cells or tissue.	Pro: Autologous cells are readily available; can differentiate into all cell types; diseases caused by a single gene mutation can be studied in cell lines derived from patients. Con: Risk of mutation; may form teratomas if transplanted.
Mesenchymal stem cells	Derived from bone marrow, cord blood, placenta, and other tissues. Differentiate into bone, cartilage, and fat (mesodermal origin tissues) and also neural, hepatic, and renal tissue.	Potential to repair or replace bone, cartilage, fat, and neural, hepatic, and renal tissue. Immunomodulatory: prevent graft-versus-host disease, enhance immune reconstitution after HSCT. Secrete cytokines that enhance tissue repair.	Pro: Readily available in numerous tissues; differentiation capacity; paracrine function for tissue repair. Immune tolerance of transplanted cells. Con: Poorly characterized; tissue source may affect functional characteristics; mutation risk; malignant transformation.

Continued

TABLE 83.1 Sources of Cells—cont'd

Cell Type	Characteristics	Uses	Pros and Cons
Differentiated cells	No self-renewal	Provide specific cellular function.	Pro: Functional characteristics. Con: No engraftment, short term effects.
Immunomodulatory cells: T-, natural killer, and dendritic cells	Derived from blood. Ex vivo incubation with viruses or tumor cells enhances function.	Tumor vaccines. Ameliorate infectious complications of HSCT.	Pro: Functional characteristics. Con: No engraftment, short-term effects.
Adoptive cell transfer: genetically modified T cells	Derived from blood mononuclear cells. Ex vivo gene transduction with T-cell receptor or chimeric antigen receptor genes.	Immunotherapy for cancer.	Pro: Functional characteristics. Con: No engraftment, short-term effects

replace abnormal and/or damaged BM stem cells and when used to treat malignancies and may also provide immunologic effects that help to eradicate the tumor cells and prevent relapse (see Chapters 84 and 85).

Human Embryonic Stem Cells: Because creation of human embryonic stem cell (hESC) lines requires destruction of human embryos, hESC research has been hindered by moral and political debates. hESCs are derived from the inner cell mass of an embryonic blastocyst 5 days after fertilization. hESCs are undifferentiated pluripotent stem cells that have the capacity to either self-renew indefinitely or to differentiate into any cell in the body when exposed to the correct combination of signaling and growth factors. hESC lines are clonal populations of undifferentiated hESC maintained indefinitely in vitro. Knowledge of the regulatory mechanisms that induce lineage commitment in embryos has been used to systematically obtain highly enriched populations of differentiated cells from hESC lines. Differentiation of these cells is a multistep process. Initially, regulatory signaling pathways are manipulated to induce the formation of mesoderm, endoderm, or ectoderm germ layers in vitro. Germ layer–specific tissues can be produced from each. Hematopoietic, vascular, and cardiac cells can be obtained from mesoderm; hepatocytes and pancreatic cells from endoderm; and neural cells from ectoderm. When injected into immune-deficient mice, hESCs form teratomas-containing cells derived from all germ layers. This has raised safety concerns regarding the tumorigenicity of hESC-derived therapies. To improve safety, development of therapies from hESC has focused primarily on their development into specific differentiated cell populations. Purity and potency assays for these products must include safeguards to ensure that the final cellular therapy product does not contain undifferentiated tumorigenic stem cells.

Induced Pluripotent Stem Cells: Induced pluripotent stem cells (iPSCs) are hESC-like cells derived from adult somatic cells. Adult somatic cells are treated with transcription factors (Oct4, Sox2, c-Myc, and KLF4) to create embryonic stem cell–like "stemness" and pluripotency. In 2006, Takahashi et al. created iPSCs by introducing these transcription factors into mouse fibroblasts using retroviral transduction. The following year, they introduced the same transcription factors into human fibroblasts and succeeded in creating human iPSCs. These iPSCs resemble hESCs with regard to their cell surface markers, telomerase activity, ability to remain undifferentiated in culture indefinitely, capacity to form teratomas containing cell lineages derived from all three embryonic germ layers in immunodeficient mice, and to differentiate into cell lineages derived from all three embryonic germ layers in vitro. Because multiple copies of the transgenes appear to be inserted randomly in donor cell DNA, and c-Myc is a known oncogene, numerous questions remain regarding the tumorigenicity of iPSCs and the cells and tissues derived from them. Subsequently, to alleviate safety concerns, several other groups have used varying combinations of these and other transcription factors, vectors that are not incorporated into the host cell genome, and small molecules to successfully develop iPSC lines from a variety of somatic cell types.

iPSCs from patients with diseases caused by a single gene mutation can be differentiated into cell lineages that express the disorder and have proven to be outstanding "disease in a dish" models. Proof of this principle was provided by use of fibroblast-derived iPSCs from a patient with type 1 spinal muscular atrophy (SMA). SMA is caused by a single mutation in the *survival motor neuron 1 (SMN1)* gene resulting in decreased SMN1 protein and reduced motor neuron survival. Treatment of motor neurons differentiated from SMA iPSCs with valproic acid, and tobramycin induced expression of SMN1 protein and in vitro motor neuron survival. Similarly, iPSC-derived neurons have been used to model Alzheimer and Parkinson diseases. In addition, CRISPR-CAS9 gene editing technology has been used to correct monogenic disease mutations in iPSC lines in vitro, thus increasing the utility of iPSC disease models by providing isogeneic study controls. iPSCs are being used for large-scale drug screening for neurologic diseases such as familial dysautonomia, Alzheimer disease, motor neuron disease, amyotrophic lateral sclerosis (ALS), and Fragile X syndrome, and liver and cardiac drug toxicities. Of note, large-scale screening of approved drugs using iPSC disease models has identified statins as a potential therapy for achondroplasia with mutations in fibroblast growth factor receptor 3. The long-term goal is also to provide donor-specific cells able to replace defective cells and tissues.

The first such clinical trial involved transplantation of sheets of autologous iPSC-derived retinal pigment epithelial cells for macular degeneration. The therapy prevented further macular deterioration and modest vision improvement. However, the study was halted when mutations were observed in another patient's iPSCs (see Chapter 86).

Mesenchymal Stem Cells: Mesenchymal stem cells (MSCs) are nonhematopoietic multipotent stem cells that were first identified in BM stroma and their pivotal role in maintenance of the hematopoietic microenvironment and homing as well as bone formation and remodeling are well documented. MSCs have also been identified in and recovered from a variety of tissues, including umbilical CB, placenta, adipose tissue, endometrium, dental tissues, amniotic fluid, and Wharton's jelly. In vitro cultures are used to enrich and recover MSC. Because there are no quantitative assays for MSC, the International Society for Cellular Therapy proposed minimal criteria for identifying MSC. MSC should adhere to plastic culture dishes, express CD73, CD90, and CD105 cell surface markers, lack CD14, CD34, CD45 cell surface markers, lack HLA-DR, and have the ability to differentiate into adipocytes, chondrocytes, and osteoblasts. Although derived from mesoderm, when exposed to the appropriate factors in vitro, MSC can differentiate into cell lineages derived not only from mesoderm (bone, cartilage, fat) but also from ectoderm (neural) and endoderm (hepatocyte). MSCs are also inflammation and immuno-modulatory, presumably through the secretion of cytokines that modulate dendritic and NK, T- and B-cell functions. Preclinical animal studies and human clinical trials are ongoing to investigate the capacity of MSC to populate and regenerate tissue and to modulate immunity and inflammation. Obtaining large numbers of cells needed for transplantable products has been a challenge because long-term culture leads to telomere shortening and senescence, diminished differentiation capacity, and malignant transformation of MSC lines. Improved neurologic function and survival after MSC transplantation in mouse models of neurodegenerative disorders has led to ongoing clinical trials for ALS, Parkinson disease, and Alzheimer disease. MSCs are also being tested for treatment of myocardial infarction with limited success in clinical trials. The mechanism by which MSCs participate in cardiac repair are complex because even when cardiac tissue regeneration occurred in animal models, direct tissue repopulation by MSC and long-term persistence of injected MSC could not be confirmed. Suggested mechanisms of action include differentiation into and/or fusion with cardiocytes and endothelial cells; paracrine effects of MSC-secreted vascular endothelial growth factor, IL-6, platelet-derived growth factor, and fibroblast growth factor resulting in mobilization of autologous cardiomyocytes and angiogenesis; and inhibition of inflammatory responses to protect the myocardium. Similarly, the mechanisms by which MSCs inhibit GVHD after HSCT are multifaceted and most likely include inhibition of T-, NK, B-, and antigen-presenting cells. These immunologic mechanisms are also being investigated for the treatment of autoimmune disorders such as type 1 diabetes and rheumatoid arthritis.

Differentiated Cells

Immunomodulatory Cells (See Chapter 87): T-, natural killer, and dendritic cells are essential to the functioning of immune system. These cells can be incubated with tumor cells or antigens to make vaccines comprised of activated immune cells that recognize and kill tumor cells. The FDA licensed a cellular therapy product derived from adult mononuclear cells for the immunomodulatory treatment of prostate cancer. Similarly, viruses such as CMV and Epstein–Barr virus (EBV) that cause serious complications in immune compromised patients are incubated in vitro with T cells, which can be infused in patients to ameliorate infectious complications of treatments such as HSCT.

Adoptive Cell Transfer: Adoptive cell transfer (ACT) of genetically modified T cells is an innovative and effective immunotherapy for the treatment of cancer. T-cells are genetically engineered to express either a T-cell receptor (TCR) or chimeric antigen receptor (CAR-T cell) against the patient's tumor. Although the structure and mechanisms of action of these receptors differ, both are designed to interact with antigens on the tumor cells resulting in targeted tumor cell death. Manufacture of genetically modified T cells is a complex procedure. Autologous PB mononuclear cells are collected by apheresis. Mononuclear cells or enriched T cells are incubated with viral vectors containing engineered receptor genes, and transduced T cells are expanded. The procedure involves "open" manufacturing steps and requires 2 to 3 weeks in a clean room laboratory. As a result, there are significant constraints on the number of patients who can be treated. Quality control assays must be performed to ensure purity, potency, and absence of gene-modified viral vectors. Two pharmaceutical companies have recently received FDA approval for the treatment of acute lymphoblastic leukemia in patients up to 25 years old and refractory aggressive non-Hodgkin lymphoma.

Release Testing: To ensure safety and efficacy cellular therapy products must be tested for purity and potency. Assays must be developed to determine that the HCT/P provides the expected effect. Assays can be appropriate laboratory tests or controlled clinical data obtained by administering the product. Assays should identify the cell types and viability of the cells and should also rule out contamination with foreign material and or microbial agents. Products that depend on function of living cells cannot be sterilized. Use of aseptic technique is essential. The final product should be assayed for bacterial and fungal contamination, and for more than minimally manipulated products endotoxin and mycoplasma.

Development of New Cellular Therapies: Stem cell and other cellular therapies hold promise of cures for diseases. The only stem cell therapies that are currently considered to be standard of care are HPCs for HSCT, epithelial stem cell treatments for burns and corneal tissue for corneal replacement. The FDA-licensed HPC products derived from CB for use in allogeneic HSCT. Seven other cellular therapies have been licensed by the FDA: two autologous cultured chondrocytes for cartilage repair, autologous fibroblasts for improvement of appearance of moderate to severe nasolabial folds, allogeneic-cultured keratinocytes and fibroblasts scaffolding for gingival wounds, autologous cellular immunotherapy indicated for the treatment of asymptomatic or minimally symptomatic metastatic castrate resistant (hormone refractory) prostate cancer, autologous CAR-T cells for treatment of acute lymphoblastic leukemia in patients up to 25 years old, and autologous CAR-T cells for refractory aggressive non-Hodgkin lymphoma. Unfortunately, some clinics offer cellular therapies with unsubstantiated claims, which are expensive and sometimes harmful treatments to desperate patients.

To be safe and effective, development of stem cell therapies must include carefully controlled in vitro studies followed by preclinical studies in animals to evaluate the in vivo behavior and safety of cells that cannot be predicted by in vitro studies, and, ultimately, carefully controlled clinical trials that are vigilantly monitored for safety and efficacy. The International Society for Stem Cell Research recognized the uniqueness of stem cell therapies and published strict guidelines for the development of stem cell products. Unique qualities of stem cells addressed in these guidelines include the following: difficulties controlling stem cell self-renewal, differentiation, and tumorigenicity; limitations of animal models; recognition that effects may be unpredictable, irreversible and persistent for many years requiring long-term safety and efficacy monitoring.

Included are recommendations for strict oversight of product development, extensive preclinical animal testing models that provide proof of principle and safety data, clinical trials, and safety monitoring by individuals with stem cell–specific scientific and ethical expertise; informed consent for allogeneic donors; donor-infectious disease testing; establishment of surrogate markers of cellular identity and potency of the products; elimination of xenogeneic components of culture media; and harmonization of international standards to improve accessibility to cellular therapies.

The National Institutes of Health (NIH) website (http://www.clinicaltrials.gov) lists more than 6000 cellular therapy clinical trials. Of these, more than 3000 are for HPC use for diseases as diverse as hematologic malignancies to cardiac repair, autoimmune diseases, and neurologic disorders. More than 700 trials are for MSC therapy for diseases including spinal cord injury, ischemic stroke, limb ischemia, liver failure, and GVHD; and ACT such as CAR T and TCR trials account for more than 700 trials. Approximately 60 hESC trials for macular dystrophy, macular degeneration, and other retinal diseases are in progress. Sixty trials are recruiting the donation of disease-specific somatic cells for the development of iPSC to be used as experimental models of genetic and other disorders are also listed.

Further Reading

Kaiser, A. D., Assenmacher, M., Schröder, B., Meyer, M., Orentas, R., Bethke, U., et al. (2015). Towards a commercial process for the manufacture of genetically modified T cells for therapy. *Cancer Gene Ther*, *22*, 72–78.

Miao, C., Lei, M., Hu, W., Han, S., & Wang, Q. (2017). A brief review: The therapeutic potential of bone marrow mesenchymal stem cells in myocardial infarction. *Stem Cell Res Ther*, *8*, 242–247.

Murry, C. E., & Keller, G. (2008). Differentiation of embryonic stem cells to clinically relevant populations: Lessons from embryonic development. *Cell*, *132*, 661–680.

Pontikoglou, C., Deschasseaux, F., Sensebe, L., & Papdaki, H. A. (2011). Bone marrow mesenchymal stem cells: Biological properties and their role in hematopoiesis and hematopoietic stem cell transplantation. *Stem Cell Rev*, *7*, 569–589.

Shi, Y., Inoue, H., Wu, J. C., & Yamanaka, S. (2017). Induced pluripotent stem cell technology: A decade of progress. *Nat Rev Drug Discov*, *16*, 115–130.

Sipp, D., Caulfield, T., Kaye, J., Barfoot, J., Blackburn, C., Chan, S., et al. (2017). Marketing of unproven stem cell–based interventions: A call to action. *Sci Transl Med*, *9*, 1–5.

Takahashi, K., Tanabe, K., & Ohnuki, M. (2007). Induction of pluripotent stem cells from adult human fibroblasts by defined factors. *Cell*, *131*, 861–872.

Ullah, I., Subbarao, R. B., & Rho, G. J. (2015). Human mesenchymal stem cells—current trends and future prospective. *Biosci Rep*, *35*, 1–18.

Volkman, R., & Offen, D. (2017). Concise review: Mesenchymal stem cells in neurodegenerative diseases. *Stem Cells*, *35*, 1867–1880.

CHAPTER 84

Hematopoietic Progenitor Cell Products Derived From Bone Marrow and Peripheral Blood

Rona S. Weinberg, PhD

Hematopoietic stem cells (HSCs) are blood-forming multipotent stem cells that produce all blood cell lineages and maintain hematopoiesis. Hematopoiesis occurs in bone marrow (BM), but a small number of HSCs circulate in peripheral blood. Like other stem cells, HSCs have the capacity of self-renewal by means of asynchronous cell division, which produces an identical daughter stem cell and another cell, the progeny of which become terminally differentiated blood cells. Blood differs from other tissues in that it contains a multiplicity of cells that due to relatively short life spans (red blood cells [RBCs], 120 days; platelets, 1 week; granulocytes, less than 1 day) must be replaced daily. It is estimated that 10×10^{12} cells are lost and replaced daily. Most HSCs remain quiescent in the BM. Their enormous proliferative capacity is accomplished by progeny destined to terminal differentiation. Progeny cells go through a series of divisions and maturation steps, each of which results in an amplification of the number of proliferating cells. HSCs are not morphologically identifiable but are identified by the glycoprotein cell surface antigen CD34 and their ability to "rescue" lethally irradiated recipients by engrafting and reconstituting the BM. Progeny of HSCs differentiate into Hematopoietic progenitor cells (HPCs). HPCs cannot self-renew, can differentiate and become restricted to a single lineage, are not morphologically identifiable, and are identified by the glycoprotein cell surface antigen CD34 and in vitro colony formation (colony-forming units [CFUs]). HPCs differentiate into morphologically identifiable precursors, which ultimately become mature blood cells that are released into the circulation. In general, adult stem cells, such as HSCs, compared with embryonic stem cells, have a limited capacity to differentiate; meaning they only differentiate into the tissues in which they reside and cannot dedifferentiate and become committed to other cell lineages. However, this distinction is becoming blurred by the use of HSCs to repair cardiac and neural tissue and the development of induced pluripotent stem cells (see Chapter 83).

Hematopoietic Stem Cell Transplantation: Hematopoietic stem cell transplantation (HSCT) is the replacement of a recipient's BM with donor HSCs. Historically, HSCT was referred to as bone marrow transplantation (BMT) because BM was the source of transplanted HSCs. HSCT is more accurate because HSCs can also be harvested from peripheral blood by apheresis and umbilical cord blood. Experimental HSCT began in the 1960s to treat BM failure related to lethal radiation exposure. Indications for HSCT include hematologic malignancies, BM failure syndromes, genetic disorders, such as hemoglobinopathies, and in conjunction with high-dose

Transfusion Medicine and Hemostasis. https://doi.org/10.1016/B978-0-12-813726-0.00084-2

chemotherapy for a variety of malignancies. HSCT replaces abnormal and/or damaged HSCs to restore hematopoiesis and immunologic function and may also provide immunologic effects that help to eradicate malignancies and prevent relapse.

Hematopoietic Progenitor Cell Products: HPC products contain both HSC and HPC. Although HPCs may play a role in early engraftment, HSCs are required to sustain long-term engraftment. HPC products can be collected from BM (HPC, Marrow), peripheral blood by apheresis (HPC, Apheresis), and umbilical cord blood (HPC, Cord Blood; see Chapter 85). The HPC product's total nucleated cell count (TNC) is used to determine the adequacy of the collection. Similarly, because HSCs and HPCs are both identified by the CD34 cell surface antigen, enumeration of CD34+ cells by flow cytometry, according the ISHAGE protocol, is used as a quality indicator for HPC products.

HPC, Marrow: BM is the physiologic site of hematopoiesis, and high concentrations of HSCs/HPCs can be collected from it. BM harvest is an invasive procedure involving aspiration from multiple sites on the iliac crest, performed in an operating room, requiring regional or general anesthesia. HPC, Marrow contains bone spicules, fat, and clots that should be filtered out during collection and/or processing. The weight of the recipient determines the number of cells required and volume collected. Generally, 10–15 mL of BM/kg recipient weight is collected, resulting in volumes of 1000–1500 mL for an average adult. Lymphocytes and granulocytes are the predominant cells with ~0.5%–2% being CD34+ cells. HPC, Marrow hematocrits range from 20% to 30%, resulting in volumes of RBCs as great as 450 mL. When the donor and recipient are ABO/Rh incompatible it may be necessary to remove plasma and RBCs to prevent hemolytic reactions on infusion. Adverse events associated with BM harvesting are those associated with anesthesia, pain, bruising, and rarely infection at collection sites.

HPC, Apheresis: HSCs are mobilized (released) from BM into peripheral blood by treatment with hematopoietic growth factors with or without prior chemotherapy, enabling collection by apheresis. The number of CD34+ cells in HPC, Apheresis provides an indication of the collected HSC/HPCs number. Granulocyte colony–stimulating factor (G-CSF) is the primary hematopoietic growth factor used for HSC mobilization. G-CSF stimulates stromal cells to release proteolytic enzymes, particularly metalloproteinase-9, which cleave adhesion of HSCs to BM stroma. In patients treated with myeloablative chemotherapy its effects are synergistic. In healthy donors, peak mobilization is predictable and occurs approximately 5 days after initiation of G-CSF. In patients who have received high doses of chemotherapy, peak mobilization is less predictable. Monitoring increases in white blood cells (WBCs), platelets, and circulating CD34+ cells are predictive of peak mobilization. Circulating CD34+ cell counts of ≥10/μL correspond with optimal apheresis collections of CD34+ cells. Some patients who have received high doses of chemotherapy have poor mobilization with G-CSF, resulting in poor recovery of CD34+ cells by apheresis. The interaction of the CXCR4 ligand on HSCs and CXCR12 also known as stromal-derived factor-1 (SDF-1) on stromal cells modulates the retention, migration, and homing of HSCs. Plerixafor is an antagonist of CXCR4 and acts by disrupting the interaction of CXCR4 and CXCR12, thereby releasing HPCs into the circulation within several hours. Use of plerixafor has been limited by its cost. However, transplant centers are finding it to be cost-effective when used in conjunction with G-CSF in poor mobilizers because it reduces the total number of apheresis collections required to reach target CD34+ cell dosages. Adverse events related to mobilization include bone pain, fever, headache, nausea, and flu-like symptoms. Health and safety concerns for donors treated with G-CSF were allayed by a prospective NMDP study, reported in 2009, in which no cases of leukemia were observed in 2408 G-CSF mobilized normal allogeneic donors with a median follow-up of 49 months.

HPC, Apheresis volumes range from 300 to 600 mL. TNCs of HPC, Apheresis are 10 times that of HPC, Marrow and products are comprised of 40%–50% lymphocytes and 30%–50% granulocytes. Mobilized HPC, Apheresis yields 10- to 20-fold and 40- to 50-fold more CD34+ cells than HPC, Marrow and non-mobilized HPC, Apheresis, respectively. HPC, Apheresis hematocrits are approximately 5% and therefore generally contain <30 mL of RBCs.

HPC, Marrow versus HPC, Apheresis (Table 84.1): National Marrow Donor Program/Be the Match (NMDP) reported Center for International Blood and Marrow Transplant Research (CIBMTR) 2016 data indicating that BM remains the primary mode of collection for allogeneic pediatric transplants, while apheresis collections predominate for adult recipients. Each type of collection has advantages and disadvantages, for donors and patients, to be considered when selecting the appropriate collection method. BM harvest is an invasive procedure requiring regional or general anesthesia. HPC, Apheresis is easier for the donor because it does not require anesthesia

TABLE 84.1 Comparison of HPC, Marrow and HPC, Apheresis

Characteristic	HPC, Marrow	HPC, Apheresis
Collection	Requires general anesthesia; painful recovery	Donor mobilization using G-CSF ± Plerixafor may result in bone pain, flu-like symptoms; Ease of collection; Multiple collections to achieve CD 34 + cell target dosage is possible
Volume	Large volume, >1000 mL	Smaller volume, 300–600 mL; Multiple collections
Hematocrit	20%–30%, requires RBC depletion for ABO/Rh incompatibility and in some instances cryopreservation	5%–10%. May not require RBC depletion for ABO/Rh incompatibility
CD34+ cell count	Lower than HPC, Apheresis	Higher than HPC, Marrow; Multiple collections possible to achieve target dosage
CD3+ cells	Fewer than in HPC, Apheresis	Greater than in HPC, Marrow
GVHD	Lower risk than HPC, Apheresis	Greater risk than HPC, Marrow
GVT	Less than HPC, Apheresis	Greater than HPC, Marrow
Recipients	Preferred for pediatric patients, hemoglobinopathies	Preferred for adult patients, malignancies

and recovery is quicker. HPC, Apheresis products contain more CD34+ cells, multiple collections can be performed to meet target cell dosages, and engraft more quickly reducing peritransplant morbidity and mortality. HPC, Apheresis contains more lymphocytes and carries not only a greater risk of chronic graft-versus-host disease (GVHD) but also a greater probability of advantageous immunomodulatory graft-versus-tumor (GVT) effects. In HPC, Marrow transplants comparatively delayed engraftment results in greater peritransplant morbidity and mortality, and GVT is less effective resulting in a greater risk of relapse. Therefore, HPC, Marrow may be better suited to reduce the risk of GVHD in non-malignant disorders (e.g., hemoglobinopathies) or when there are donor safety concerns about administration of G-CSF (e.g., sickle cell trait donors). In contrast, GVT effects of HPC, Apheresis are advantageous for malignancies, particularly when non-myeloablative preparative regimens are used. Long-term survival is similar regardless of the mode of collection, probably because each has advantages and disadvantages that affect survival. Compared with HPC, Marrow, HPC, Apheresis contains fewer RBCs reducing the risk of hemolytic reactions when donor and recipient are ABO incompatible.

Processing HPC Products: Processing laboratories perform procedures to customize and ensure the safety of HPC products. HPC, Marrow and Apheresis each have unique qualities to be considered when determining appropriate processing procedures. Cryopreservation methods enable advance collection and storage of HPC products.

Plasma and RBC Reduction: Plasma is easily removed by centrifugation, which separates cellular and plasma components of the product, followed by removal of the plasma by aspiration. Loss of cells is minimal. Even when ABO incompatibilities are not a consideration, plasma reduction may be desirable to reduce infusion volumes and, if cryopreservation is planned, to reduce volume of cryoprotectant.

RBCs can be removed by sedimentation through hydroxyethyl starch (HES) with or without centrifugation. HES causes RBCs to clump, increasing their density and sedimentation rate. WBCs including CD34+ cells remain in the plasma and can be recovered by centrifugation. Alternatively, density gradient centrifugation on ficoll-hypaque can be used to obtain a buffy coat containing WBCs including CD34+ cells. For large volume products such as HPC, Marrow, cell washing devices such as the COBE 2991 may be used. RBC reduction can result in significant losses (up to 40%) of WBCs. Methods should be validated, and acceptable cell loss should be determined before implementing these procedures.

Positive and Negative Cell Selection: HSCs and HPCs express CD34, while lymphocytes such as those that cause GVHD and many tumor cells do not. CD34+ cell enrichment (positive selection) can therefore be used to remove lymphocytes and tumor cells (negative selection) from HPC products, thus reducing the risk of GVHD and relapse, respectively. When added to HPC products, anti-CD34 monoclonal antibodies (mABs) bind only to CD34+ cells. Magnetic particles attached to the anti-CD34 mABs allow CD34-positive and negative cells to be separated by passage over a magnetic column. Alternatively, mABs can be labeled with fluorescent dyes such that fluorescent activated cell sorting separates positive and negative cells.

T-lymphocytes are immunomodulatory. In instances such as HLA haploidentical donor and recipient HSCTs, T cells are removed to prevent GVHD. Yet, these cells are necessary to modulate GVT effects, prevent infectious complications, and prevent graft rejection. Risks and benefits must be weighed. CD34+ cell enrichment results in depletion of T cells. T-cell depletion can also be accomplished by positive selection and removal of T cells using anti-T-cell mABs as described above. T-cell depletion with mABs against specific T-cell subsets may be preferred to CD34+ cell enrichment because of the reduction of GVHD and preservation of GVT effects.

Donor lymphocyte infusions can be saved as a by-product of CD34+ cell enrichment or other T-cell depletion procedures and cryopreserved for future use. Infusion of T-lymphocytes may provide GVT effects, and prevent graft rejection, infection, and relapse, particularly after nonmyeloablative preparative regimens and/or CD34+ cell enriched or T cell–depleted transplants. Specific lymphocyte subsets can be enriched or depleted using mABs.

Purity and Potency: Quality factors that can affect the purity, potency, and engraftment include the following: donor health, mobilization protocol, collection procedure, processing including use of aseptic technique, clinical grade reagents, sterile materials and skill, cryoprotectant and freezing rate, storage temperature, and thawing rate and temperature. TNC/kg recipient body weight, CD34+ cells/kg recipient body weight and viability are potency parameters commonly used to assess each HPC product. CD34+ cell enumeration and viability are assessed by flow cytometry using the standardized ISHAGE protocol. In vitro colony forming units (CFU) may also be used as a surrogate assay to assess potency and viability. However, CFU assays are difficult to standardize, and data are not available for up to 2 weeks. Transplant centers establish acceptable targets for each parameter. Generally, TNC dosage $\geq 3\times10^8$/kg recipient weight, CD34+ cell dosage $\geq 2\times10^6$/kg recipient weight, and viability ≥90% are targeted to ensure neutrophil engraftment (absolute neutrophil count ≥500) within 2–3 weeks depending on the type of donor (autologous, syngeneic, or allogeneic). Time to engraftment is also affected by the recipient's disease and prior exposure to chemotherapy and radiation. To ensure safety, HPC products should be free of microbial contamination. Use of aseptic technique is essential during collection and processing procedures to prevent microbial contamination. The final HPC product must be assayed for bacterial and fungal contamination. The ultimate gold standard is engraftment.

Cryopreservation and Storage: Allogeneic and syngeneic products may be collected just before infusion and stored short term at either room temperature (20–24°C) or refrigerated (2–8°C). Cell viability decreases, and risk of microbial contamination increases during room temperature or refrigerated storage and therefore storage at these temperatures should not exceed 72 hours. All autologous and some allogeneic and syngeneic HPCs are collected far in advance of infusion and must be cryopreserved to maintain viability. Dimethyl sulfoxide (DMSO) is added to HPCs at concentrations between 5% and 10% to prevent cellular dehydration and formation of ice crystals in the cells. Immediately after the addition of DMSO, HPCs are cooled at a rate of −1 to −2°C per minute. Once frozen, HPCs can be stored in mechanical freezers (≤−80°C), vapor phase liquid nitrogen (≤−150°C), or liquid phase liquid nitrogen (−196°C).

No temperature fluctuations occur in liquid phase liquid nitrogen; however, infectious substances have been transmitted between products. Vapor phase liquid nitrogen storage is most often selected, because the storage temperature facilitates indefinite viability, infectious substances cannot be transmitted from one product to another, and temperature is maintained for several days during power failures.

Thawing, Infusion, and Adverse Events: HPCs are typically thawed at 37°C at the patient's bedside and infused immediately at a rate of 5–20 mL/minute. HPCs can be filtered to remove cellular debris; however, leukoreduction filters must never be used. Most adverse events that occur during infusion can be attributed to DMSO, lysed granulocytes, or ABO incompatibility. Less than 1 g DMSO/kg recipient weight should be infused per day. Although removal of DMSO could reduce complications, this procedure is usually avoided because it can significantly compromise cell number and viability and ultimately engraftment. Infusion adverse events are usually mild and include nausea, vomiting, abdominal cramps, cough, hypertension, hypotension, and occasionally cardiac symptoms (see Chapter 88).

Further Reading

Bashey, A., Zhang, M.-J., McCurdy, S. R., St Martin, A., Argall, T., Anasetti, C., et al. (2017). Mobilized peripheral blood stem cells versus unstimulated bone marrow as a graft source for T-cell–replete haploidentical donor transplantation using post-transplant cyclophosphamide. *J Clin Oncol*, *35*, 3002–3012.

Doulatov, S., Notta, F., Laurenti, E., & Dick, J. E. (2012). Hematopoiesis: A human perspective. *Cell Stem Cell*, *10*, 120–136.

Mielcarek, M., Storer, B., & Martin, P. J. (2012). Long-term outcomes after transplantation of HLA-identical related G-CSF-mobilized peripheral blood mononuclear cells versus bone marrow. *Blood*, *119*, 2657–2678.

National Marrow Donor Program/Be the Match, Current uses and outcomes of hematopoietic cell transplantation (HCT). (2016). Summary Slides. Available at: https://bethematchclinical.org/resources-and-education/hct-presentation-slides/#901/.

Pulsipher, M. A., Chitphakdithai, P., & Miller, J. P. (2009). Adverse events among 2,408 unrelated donors of peripheral blood stem cells: Results of a prospective trial from the National Marrow Donor Program. *Blood*, *113*, 3604–3611.

CHAPTER 85

Cord Blood Banking

Andromachi Scaradavou, MD

Cord blood (CB), the blood remaining in the placenta and umbilical cord after birth is rich in hematopoietic stem cells (HSCs) and can be used for bone marrow reconstitution of patients undergoing myeloablative or non-myeloablative therapy that do not have matched related donors. CB is also being evaluated for tissue repair and regeneration. CB banking refers to the systematic procurement, processing, testing, and storage of CB collections (units; CBU) and organization of all relevant data CB banks have large repository of CBU, fully tested and stored that can be provided for transplantation when needed.

Cord Blood as a Hematopoietic Stem Cell Source for Unrelated Transplantation:

Advantages: CB is a rich source of HSC and can be collected easily, without risks to the mother or the newborn baby.

- There are large numbers of donors from all ethnic backgrounds increasing the chances of minority patients to find a suitable donor.
- There are advanced techniques for processing and storing the CBU in large Inventories, with currently no established "expiration date". Since all testing has been completed, cryopreserved CB is an "off-the-shelf" product and can be sent to the transplant centers upon demand, so that patients can be treated without any delays.
- There is a lower risk of viral infections of the newborn babies, particularly those relevant to transplantation (such as cytomegalovirus – CMV), in comparison to adult volunteer donors.
- CB contains immunologically "naïve" or "different" T cells; as a result, transplants can be performed without "perfect" HLA matching between donor and recipient with acceptable incidence of graft-versus-host disease. The ability to use partially matched CB grafts increases significantly the probability of finding suitable donors for patients.

Considerations: The volume and number of total nucleated cells (TNC) contained in a CBU is defined at the time of collection. TNC dose (TNC/kg body weight of recipient) may not be adequate for larger patients. The donor cannot be approached for a second donation.

- There is a potential risk for transmission of (unknown) genetic diseases.

Types of Cord Blood Banking: *Private/family banks* are for-profit companies that facilitate storage of CBU for personal or family use. They advertise widely to the public and to physicians and have established a presence in many obstetrical waiting rooms

Transfusion Medicine and Hemostasis. https://doi.org/10.1016/B978-0-12-813726-0.00085-4

and hospital delivery areas. Many family banks are also storing cord tissue. *Directed donation* refers to banking of a CBU for a particular person in the family that already has a medical condition (malignant or genetic) that could potentially benefit from CB transplantation. *Public banks* store CBU that can be used for any suitable patient, unrelated to the baby donor. Several public CB banks have also implemented programs for family donation (*hybrid* banks).

Recommendations: The American Academy of Pediatrics (AAP) policy in 2017 states that "*public CB banking is the preferred method for collecting, processing and using CB cells for transplantation.*" Additionally, "*quality-assessment reviews by several national and international accrediting bodies show private CB banks to be underused for treatment, less regulated for quality control, and more expensive for the family than public CB banks.*" The American Society of Blood and Marrow Transplantation (ASBMT) presented similar recommendations in 2008, encouraging directed donations in certain clinical situations and public cord blood donations, where possible.

Public Cord Blood Banking

Collection: The collection process involves harvesting of CB, obtaining maternal informed consent for donation and maternal blood sample for testing, review of maternal medical records, and completion of the maternal questionnaire to evaluate donor risk factors.

CB can either be collected in utero, during the third stage of labor, after the delivery of the baby, a method performed usually by the obstetrician or obstetrical staff, or ex utero, after the third stage of labor and delivery of the placenta, with the placenta suspended from a stand, usually performed by trained collection personnel. With either method, careful cleaning of the venipuncture site on the umbilical cord is important to avoid bacterial or maternal blood contamination. CB flows by gravity into a blood collection bag containing anticoagulant (CPDA). To maximize collection volume and avoid clotting, timing of the CB collection in relation to the delivery of the placenta is critical. Further, the volume of CB that can be collected for banking is influenced by several variables, such as the infant's birth weight and gestational age, as well as the amount of placental transfusion, i.e. how much blood is allowed to go to the newborn before clamping the umbilical cord. The recent practice of "delayed" cord clamping after birth has imposed challenges in the collection of sufficient CB volumes for banking and clinical transplantation.

Processing: The aim of processing (minimal manipulation) is to reduce the cell volume to be frozen, by removing the bulk of RBCs and plasma. Today automated processing systems are used, which require less time and processing occurs in a closed system reducing the potential for product contamination. The two systems most commonly used by CB banks are AXP (AutoExpress, Thermogenesis, CA, USA) and Sepax (Biosafe, Switzerland). Automated processing results in products with lower hematocrit and predefined volume. The final mononuclear cell product is in the cryopreservation (freezing) bag. The bags can have one or two compartments and "segments" i.e., internally attached tubing, that contains small amounts of product. Cells from the segment can be used for testing prior to the release of the CBU for transplantation.

Cryopreservation: Cryopreservation is performed by addition of the cryoprotectant dimethyl sulfoxide (DMSO) to a final concentration of 10% with continuous mixing while maintaining the product temperature between 0 and 4°C. According to NETCORD-FACT standards the time interval from the CBU collection to processing must not exceed 48 hours.

Storage: Cryopreserved CBU are stored in liquid nitrogen freezers (liquid phase). The most commonly used freezing device is the BioArchive System (Thermogenesis, CA), which combines controlled rate freezing of the product and long-term storage, thereby avoiding the transfer of the frozen CBU to another device and exposure to higher temperature ("transient warming effect"). This computerized system allows for automated freezing and individually controlled storage of each CBU: the system positions each CBU in its slot, records the position and also retrieves it without affecting other products. One BioArchive freezer can store 3600 CBU.

Testing: Most public CB banks use a cutoff collection volume or cell count to select CBU for processing and cryopreservation storage. Additional manufacturing or testing criteria are also used to identify CBU, which will be banked.

Potency Evaluation:

- TNC: Enumeration of nucleated cells (includes white blood cells and nucleated RBCs), as well as complete blood count with differential, is performed before and after processing.
- CD45+/CD34+ cell counts and viability: Analysis is performed by flow cytometry to count the cells and assess their viability before cryopreservation. Significant variations exist between laboratories regarding this assay.
- CFU (colony-forming units): Although not mandated by the FDA, several banks perform CFU to assess hematopoietic progenitor cell (HPC) function in the CBU. The traditional assay is not yet been standardized.
- In addition, other assays, such as expression of aldehyde dehydrogenase are used by some banks to assess CBU potency.

Bacteriology: Bacterial and fungal cultures of the CBU are performed to exclude microbial contamination. Most banks are using bacteriology screening assays. CBU with positive results are not allowed for transplantation.

Infectious Disease Markers: Infectious disease markers (IDMs) testing of the CBU is performed according to the FDA regulations. The maternal blood sample is tested for all relevant communicable diseases with FDA-approved screening assays, in laboratories that have been certified by CLIA, as required for all stem cell donors. Testing includes serologic evaluation for HIV, HCV, HBV, HTLV, syphilis, Chagas, and NAT for HIV, HCV, HBV and West Nile Virus. Donor eligibility is assigned according to the FDA regulations after review of the risk factors captured in the maternal questionnaire and maternal medical records, and the results of IDM tests.

Identity Testing: Identification of blood group (ABO) and Rh antigens is performed. HLA typing is performed, using DNA-based methods, for HLA-A, -B, and -DRB1 at minimum; typing for HLA-C and -DQB1 is recommended. Initially, HLA typing is performed on the preprocessing sample of the CBU. Prior to release of a CBU for transplantation, confirmatory HLA typing must be performed from cells extracted from the attached segment of the cryopreservation bag to confirm the CBU identity.

Hemoglobinopathy Screening: Screening is performed to identify sickle hemoglobin (HbS). Additionally, molecular testing for hemoglobinopathy can be performed to provide precise diagnosis, e.g., thalassemia.

Search Inventory: CBU with complete evaluation that meet FDA regulations, FACT/AABB criteria as well as additional bank-specific manufacturing criteria can be "released" to search inventory; therefore, they become available for search for transplant centers and transplant registries. Because all testing has been completed, cryopreserved CBU is "off-the-shelf" product.

Cord Blood Unit Release for Clinical Transplantation: Before release for transplantation, the CBU identity is verified by performing HLA typing from the bag's attached segment. Banks also perform quality control studies (TNC, CD45+/CD34+ cell count and viability, CFU assays) on cells obtained from the segment to assess the cryopreservation effect on the final product.

Transportation to the Transplant Centers: Products are shipped frozen in special portable liquid nitrogen shipping containers designed and validated to maintain a controlled environment and very low temperature (T = −196°C) for the entire transit time (max: 5–7 days). Temperature is continuously monitored during transportation and must remain ≤−150°C (NETCORD-FACT Standards). On receipt at the transplant center, the shipping container is inspected for integrity, and opened, to verify product integrity and product-specific information. The CBU in its metal canister is removed from the container and placed in a liquid nitrogen or mechanical freezer (storage temperature <−150°C) until the day of transplant.

Thawing and Preparation for Infusion: There are two methods of preparing CBU for infusion: (1) thawing of frozen CBU at 37°C, reconstitution (diluting the thawed product with albumin/dextran solution) and washing (centrifugation and removal of supernatant plasma and DMSO), or (2) thawing and reconstitution without washing and removal of DMSO, a procedure used to minimize cell losses. Importantly, dilution has to be 8–10-fold. Most banks provide specific instructions for thawing and preparation of the CBU. Bed-side thaw procedures are not recommended because they have been associated with serious infusion reactions; preparation of the products for infusion in stem cell laboratories under controlled conditions is the preferred approach.

The CBU is infused through a central venous catheter; it should not be administered through a leukocyte reduction filter and should not be irradiated. CBU that have not undergone RBC depletion during processing (i.e., RBC-replete CBU or whole blood frozen CBU) should undergo washing to remove the RBC fragments and free hemoglobin, and DMSO, before administration, to decrease the potential severe infusion reactions.

National Cord Blood Inventory: The Stem Cell Therapeutic and Research Act of 2005 established the C.W. Bill Young Cell Transplantation Program. Under this program, a network of US CB banks provides high-quality CBU (***NCBI units***) to facilitate the use of CB for transplantation purposes. The Act also specified a Single Point of Access (SPA), i.e., a single electronic system that is used to search the registry for unrelated marrow donors and donated CBU, as well as the Stem Cell Therapeutic Outcomes Database (SCTOD), administered by the Center for International Blood and Marrow Transplant Research (CIBMTR). All transplant centers are required to report posttransplant outcome information to CIBMTR, after patients grant informed consent. The centrally collected information is distributed to CB banks by monthly reports.

Oversight and Regulatory Aspects of Public Cord Blood Banking

Food and Drug Administration: The FDA considers CB to be a biological drug. The FDA Guidance of October 2009 (updated in 2011) outlined requirements so that CB banks apply for biological licensure of unrelated, allogeneic, minimally manipulated CB units. CBU meeting the FDA requirements are considered *licensed (hematopoietic progenitor cells—HPCs, CB) for use in unrelated donor hematopoietic progenitor cell transplantation procedures in conjunction with an appropriate preparative regimen for hematopoietic and immunologic reconstitution in patients with disorders affecting the hematopoietic system that are inherited, acquired, or result from myeloablative treatment.* As of February 2018, there are seven licensed US CB banks.

CBU that were collected before implementation of license requirements, or do not meet all FDA requirements, are considered "investigational" products and are distributed under an IND (investigational new drug).

Current Good Manufacturing Practices: The FDA regulations are contained in the CFR. With respect to CB banking, the CFR focuses on current good manufacturing practices (cGMPs) that ensure safety, quality, identity, potency, and purity of the product.

NETCORD-FACT and AABB: The Foundation for the Accreditation of Cellular Therapy and the American Association of Blood Banks have issued standards for CB banking and release of CB products for transplantation. The agencies inspect public CB banks to evaluate compliance with the Standards and ensure their optimal function.

Unrelated Cord Blood Grafts in Clinical Transplantation: *CBU selection* is based primarily on cell dose (TNC/kg and CD34/kg) and HLA match with the patient. "Traditional" HLA match involves matching at 4-6/6 HLA-A, -B antigens and -DRB1 alleles; most centers now consider HLA matching at allele level for four loci (HLA-A, -B, -C, -DRB1). Other CBU characteristics that impact quality and safety, such as CB bank of origin, availability of attached segment for identity testing, as well as infectious disease testing and hemoglobinopathy results must also be considered. *Diseases that can be treated with CB transplantation*: Practically, every disease that can be treated with allogeneic HSC transplantation, can be treated (and has been treated) using CB as the source of stem cells. *Optimal clinical practices*: Targeted care strategies for CB transplant recipients were recently summarized in a publication based on the practices of six experienced US transplant centers.

Ex vivo Expansion of Cord Blood Hematopoietic Progenitor Cells to Improve Engraftment: With the increasing use of CB as alternative donor source, efforts are made to address the CBU cell dose limitation. Low cell dose for large/adult patients leads to prolonged post-transplant pancytopenia, with median time to an Absolute Neutrophil Count of 500/μl (ANC>500) in the peripheral blood of 21–25 days, increases the risk of infections and bleeding, and the overall Transplant-Related Mortality. Consequently, only large (or very large) CBU, with median TNC $>150 \times 10^7$ are being utilized for transplantation, and these represent only a small fraction of the stored CB inventories. To shorten the post-transplant cytopenia, and achieve engraftment times comparable to other stem cell sources, *ex vivo* expansion of CB-derived hematopoietic progenitor cells is being actively studied. Using a combination of cytokines, additional molecules, culture media and various culture times, many groups have succeeded in multi-fold expansion of the CB-derived stem cells in vitro. Some of these strategies have been implemented in clinical trials (Table 85.1). In addition, studies have addressed improving the homing of CB-derived cells by co-culture with mesenchymal cells or by a short incubation with agents prior to infusion.

The promising results of the ex vivo expansion studies imply that cell dose may not be a limitation in the CBU graft selection in the near future, and this way, better HLA-matched CBU can be prioritized without concern for delayed recovery. Such an approach would improve the clinical outcomes of CB transplantation, as well as the utilization of current CB inventories that contain thousands of "small" or "medium" CBU, currently not being used for clinical transplantation. *Engineering CB derived T cells*: Given the risk of post-transplant viral infections in immunocompromised recipients, adoptive T cell therapy using donor-derived, *ex vivo* expanded, pre-existing Virus-Specific T cells (VST) has emerged as an effective way to treat viral infections. To transfer this approach to CB requires generating virus-protective T cells from naïve CB lymphocytes, but it has been successful under certain conditions. Despite the lengthy manufacturing periods, several studies have shown clinical efficacy for the VST. Additionally, recent work has involved generation of tumor-directed T cells and Chimeric Antigen Receptor (CAR) T cells originating from CB that could be used to treat disease relapse.

Cord Blood in Regenerative Medicine: CB contains, in addition to HSCs, many other cell types with antiinflammatory and repair-inducing properties and are being actively investigated in various diseases, such as traumatic brain injury, cerebral palsy, spinal cord injury, autism, hypoxic ischemic encephalopathy, stroke, hearing loss, diabetes, and tissue regeneration/repair. In most cases, the initial studies involve only autologous CBU for safety. For diseases where safety and some efficacy can be seen, related CBU (mostly from sibling donors) are used. However, since these diseases affect large numbers of patients that do not have autologous or sibling CBU, unrelated,

TABLE 85.1 Cord Blood Manipulation Strategies

Compound	Time in Culture (Days)	Phase I/II Completed (# pts)	Infusion of T-Cell Fraction	Persistence of Manipulated Cells	Lead Institution/ Corporate Partner	Median Days to ANC >500
Ex vivo Expansion Strategies to Increase the Total Cell Dose of the Cord Blood Graft						
Notch Ligand (NLA101)	16	Yes (15)	No	No	FHCRC/NOHLA	16
Copper Chelation (StemEx)	21	Yes (101)	No	No	Gamida-Cell/Teva	21
MSC Co-culture	14	Yes (32)	No	No	MDACC/Mesoblast	15
StemRegenin 1 (MGTA-456)	15	Yes (17)	Yes	Yes	UMinn/Magenta	15
NiCord	21	Yes (11)	Yes	Yes	Duke/Gamida-Cell	11
UM171	7	No	Yes	Yes	Univ. of Montreal/ Excellthera	19
Enhancement of the Bone Marrow Homing Ability of the Cord Blood Cells						
PGE-2 analogue	Incubation 2 hours	Yes (12)	CBU	Yes	DFCI/Fate Therapeutics	17.5
Fucosylation	Incubation 30 minutes	Yes (22)	CBU	Yes	MD Anderson Cancer Center	17
DPPIV inhibitor	Oral administration	Yes (24)	CBU	N/A	Univ. Indiana	21

already stored CBU will probably be utilized also. These studies pose different challenges in quantifying improvements and defining endpoints. Further, the regulatory framework needs to be further developed. Nonetheless, they create very exciting opportunities in cell therapy and regenerative medicine.

Further Reading

American Academy of Pediatrics. (2017). Cord blood banking for potential future transplantation (policy statement). *Pediatrics, 140*(5), e20172695.

Ballen, K. (2017). Umbilical cord blood transplantation: Challenges and future directions. *Stem Cell Transl Med*, 6, 1312–1315.

Barker, N., Byam, C., & Scaradavou, A. (2011). How I treat: The selection and acquisition of unrelated cord blood grafts. *Blood, 117*, 2332–2339.

Barker, J. N., Kurtzberg, J., Ballen, K., Boo, M., Brunstein, C., Cutler, C., et al. (2017). Optimal practices in unrelated donor cord blood transplantation for hematologic malignancies. *Biol Blood Marrow Transplant, 23*, 882–896.

Mehta, R. S., Dave, H., Bollard, C. M., & Shpall, E. J. (2017). Engineering cord blood to improve engraftment after cord blood transplant. *Stem Cell Investig*, 4, 41.

CHAPTER 86

Regenerative Medicine

Marie Csete, MD, PhD

Regenerative medicine is a multidisciplinary field focused on repair, replacement and/or regeneration of diseased cells or organs, or in many cases on protecting vulnerable diseased cells from death. Stem cell (SC) therapies are at the core of many regenerative therapies, although most SC therapies are still at the stage of clinical trial testing. Growth of the field is highlighted in the clinicaltrials.gov registry, which contains >6000 SC listings, and the Alliance for Regenerative Medicine estimates that $5.2 billion dollars was raised for this industry in 2016. Regulation of SC therapies continues to be challenging. Although FDA has yet to approve a mesenchymal stem cell (MSC) therapy, clinics that administer these cells (usually autologous from lipoaspirates) are numerous, prompting California to pass legislation designed to inform and protect patients versus Texas passing a Right to Try Act to ease access to unapproved stem cell therapies.

Definitions

Stem Cells: Regardless of donor source tissue or donor age, SCs have the potential to self-renew and generate more than one differentiated daughter cell type. Adult SCs are generally limited in differentiation potential compared with embryonic SCs (ESCs).

Embryonic Stem Cells: These pluripotent (meaning able to differentiate into many cell types) SCs were first isolated from blastocyst stage embryos but can now be isolated without destruction of the embryo at the 8-cell stage. Under ideal conditions, undifferentiated ESCs can proliferate indefinitely, whereas adult SCs undergo proliferative senescence. Expansion of ESCs without generating mutations remains a challenge.

Induced Pluripotent Stem Cells: Induced pluripotent stem cells (iPSCs) are created from more differentiated cells by transient overexpression of genes normally expressed only in pluripotent SCs. The result of this transient overexpression is demethylation of the genome, an epigenetic reprogramming that opens up the iPSC to development programs. The first iPSC-based clinical trial in Japan, using autologous iPSC-derived retinal pigment epithelial cells targeting age-related macular degeneration was abruptly stopped after a single patient in 2014, when the second patient's reprogrammed cells were found to contain mutations. Because most people with macular degeneration are older, their cells may be more mutation prone under the stress of cellular reprogramming. Thus, the strategy for cell sourcing changed to creating allogeneic SC generated iPSC bank covering the immunologic heterogeneity of the Japanese population. The allogeneic iPSC-based trial for macular degeneration resumed in 2017.

iPSC technology continued to evolve, which now enables generation of a therapeutic cell product from another differentiated cell type by transient overexpression of key

Transfusion Medicine and Hemostasis. https://doi.org/10.1016/B978-0-12-813726-0.00086-6

developmental genes in the lineage of the desired cell type. To date, cells created from direct somatic reprogramming are under development but not in clinical trials.

Parthenogenic Pluripotent Stem Cells: Parthenogenic pluripotent stem cells (PPSCs) are derived from stimulated unfertilized eggs, and so, contain only maternal genome. The homozygosity of these cells is a potential advantage for immunologic matching. The first clinical trial taking place in Australia uses neural progenitors derived from PPSC injected into the striatum and substantia nigra of patients with Parkinson disease.

Trends in Stem and Progenitor Cell–Based Therapies

Mesenchymal Stem Cell Therapies: MSCs are found in many organs including dental pulp but clinically are derived most often from bone marrow or adipose tissue. Although the identity of undifferentiated MSC is defined fairly well by expression of surface markers, the populations used in clinical studies are heterogeneous and also differ considerably from person to person. MSCs are also defined by their ability to differentiate down adipogenic, chondrogenic, and osteogenic lineages. Autologous MSCs or stromal vascular fraction cells (and related multipotent adult progenitor cells) continue to be widely studied in clinical trials, and MSCs are administered in hundreds of clinics and offices without regulatory approval. FDA recently issue guidances suggesting that they will impose stronger oversight of these office-based SC treatments, but the mechanics of regulating this widespread practice is daunting. Autologous MSCs have been administered to thousands of patients worldwide and are generally well tolerated and do not elicit a systemic immune response. Nonetheless, a recent report of three patients administered autologous fat-derived SC bilaterally into vitreous points to the danger of their unregulated use: these patients developed serious eye pathology and vision loss. The only allogeneic MSC therapy with regulatory approval for marketing in North America is for treatment of medication-resistant graft-versus-host disease in pediatric patients.

MSC mechanisms of action include secretion of trophic (survival) factors and angiogenic factors, factors that control vascular tone to mediate vasodilatation, and complex immunomodulatory properties. MSCs can suppress T cell, NK cell, B cell, and dendritic cell proliferation and activation and promote phenotype switching of macrophages from pro- to anti-inflammatory type. MSCs can also increase production of T regulatory cells, which mediate immune tolerance. In addition to immunomodulatory properties, MSCs secrete exosomes containing active enzymes and other proteins, lipids, and miRNA, which mediate similar functions as parent MSCs. Thus, exosomes are now being investigated independently for therapeutic potential. Exosome internalization by host cells may in part explain the relatively prolonged therapeutic effect of MSCs beyond grafted MSC residence time in host tissue. MSCs also have capacity to transfer mitochondria to host cells, which may improve survival of vulnerable diseased or ischemic host cells (i.e., mediate antiapoptotic effects).

Lessons learned from clinical studies and knowledge of MSC functions. First, it is difficult without significant adjunctive engineering approaches to use MSC to (re)generate structure. Second, deeper phenotyping of patients in MSC clinical trials is needed to be able to better predict which patients will and will not respond to a particular therapy. Nonetheless, their immunomodulatory properties are being evaluated in efficacy phase clinical trials for multiple diseases, including systemic lupus erythematosus, multiple sclerosis, decompensated liver disease, Crohn's disease, and cardiomyopathies, as well as in solid organ transplantation to lower doses of pharmacologic immunosuppressive drugs. MSCs are also being studied in *acute* diseases including adult respiratory distress syndrome, sepsis, and burn wound healing. Because a single injection of MSC is generally safe, MSC therapies easily pass phase I of clinical trial testing. But potency of these therapies across disease types is still largely unknown and awaits the results of many ongoing phase II trials.

Nonmesenchymal Stem Cell Adult Stem Cell Therapies: A few important examples within this large field are presented here. Limbal SC therapy for replacement of superficial cornea continues to be a success story worldwide. Autologous limbal SC can be used in the case of unilateral corneal injury; allogeneic limbal SCs are used in the case of bilateral injury.

A phase I study of adult neural SC used in the treatment of chronic stroke was reported in 2016; although safety was the major end-point, the therapy also showed efficacy in a patient population traditionally thought to have fixed lesions. Of note, the cells used in this trial were engineered to express an oncogene under a controllable promoter, allowing their in vivo expansion to numbers needed for adequate dosing. Another phase I study in chronic stroke reported positive results in 2016, using marrow-derived MSCs treated with a plasmid carrying another gene that promoted proliferation for manufacturing a cell bank, and the gene was lost after significant expansion.

Results of neural progenitor therapy for treatment of amyotrophic lateral sclerosis are difficult to evaluate because the slope of anticipated clinical decline is a difficult end-point; although some patients appeared to benefit in early clinical trials, initial results suggest inadequate potency. Retinal progenitor cells are being studied by two different groups for treatment of retinitis pigmentosa. These cells have the unique capacity to differentiate into photoreceptors, which may make them an optimal cell type for diseases characterized by photoreceptor degeneration.

Although many centers continue to evaluate a variety of SC therapies for treatment of acute myocardial infarction, generally clinical trials in this area have yielded disappointing results, with potency limited. A delivery problem exacerbated by the beating heart is that most cells in suspension delivered into the parenchyma are rapidly washed away; in all organs grafted/injected cells that do not anchor quickly enough will undergo anoikis. Integration of the transplanted cells to achieve coordinated contraction is another hurdle unique in heart applications. Consequently, use of SC therapies alone for myocardial infarction is declining, and moving toward therapies that capture or enhance endogenous regenerative capacity, continuing with optimization of tissue-engineered patch grafts, instructive matrices, exosomes, and autologous mitochondria grafts.

Embryonic Stem Cell–Derived Cell Therapies: The first ES-derived clinical trial in 2010 conducted using preoligodendrocytes generated from an ESC master cell bank, and applied to acute thoracic spinal cord injury subjects, was halted prematurely although no safety issues were encountered. These cells are currently being studied again for cervical spinal cord injury using higher doses than in the original trial. ES-derived pancreatic endoderm cells (that develop into more mature beta cells in vivo) are also being tested for treatment of type 1 diabetes. Initiation of this therapy required years of effort devoted to finding an encapsulation device that limited access of immune cells to the graft, prevented fibrin deposition, but allowed biochemical signals to cross between host and graft. The capsules containing the ES-derived cells for this trial are implanted under the dermis. ES-derived retinal pigment epithelial cells are being studied for treating macular degeneration, delivered both as cell suspensions and on engineered grafts. ES-derived cardiovascular progenitor cells are being evaluated for treating chronic cardiomyopathy.

The lessons learned from these earliest trials are invaluable: In general, although there is a risk of pluripotent cells present in the first master cell banks, no teratomas have resulted from injection of the ES-derived progenitors. The importance of finding the optimal dose (cell number) is underscored by the first hESC-preoligodendrocyte spinal cord injury trial, in which subjects were likely given too few cells. The importance of very long-term monitoring is also raised by these early trials, not only for safety monitoring, but because therapeutic effects may not manifest until years after the cells are delivered. The relative immune privilege of the central nervous system and the eye is supported by the first ES-derived (allogeneic) cell therapy trials.

Combination Stem Cell and Gene Therapy: Given recent advances in gene editing and funding, we can anticipate combination SC and gene therapy product development. The FDA has already approved combination cell and gene therapy (cells are manipulated to overexpress a growth factor) for ALS. A dramatic success of gene editing to correct an inherited mutation, followed by expansion of corrected keratinocyte SC to generate "normal" skin grafts was reported to treat severe epidermolysis bullosa.

CrispR was used to inactivate endogenous retroviruses in pigs, removing one of the major roadblocks to use of pig organs for human transplantation. Given the continued shortage of optimal human organs for transplantation, epidemic of nonalcoholic steatohepatitis, expense and difficulty of generating large organs from SC, these advances combined with progress in editing the pig genome to limit immunogenicity are important advances for future regenerative therapies.

Tissue Engineering Approaches to Regenerative Therapy: Several groups are making advances in tissue-engineered trachea and larynx, as well as esophagus. Advances in tissue-engineered cartilage and bone are also being seen in the clinic, as well as engineered blood vessels including a hemodialysis prosthesis. For more complex organs, decellularization–recellularization methods are still a focus of intensive study; a clinical trial of decellularized heart valves was recently registered. Similarly, 3D bioprinted organs are not yet in clinical trial although mini organs (rather than monolayer SC cultures) are increasingly used to understand development (and regeneration) and as disease models for drug screening. Nanoparticles used to deliver small RNA species with SC also represent a future promising combined therapeutic approach.

Conclusion: Progress in understanding the mechanisms of action of SC in mediating a therapeutic response may lead to noncellular therapies that recapitulate these responses. Furthermore, because insufficient potency is a common point of failure of SC therapies in development, these noncellular therapies arising from SC research can be used to augment cell-mediated regenerative effects in combined therapies. Similarly, tissue-engineered matrices, scaffolds, and other instructive factors should and will be combined with SC to increase disease-modifying effects. Finally, recent progress in gene therapy and editing, cellular reprogramming, and cell manufacturing methods means regenerative medicine is poised to adapt long-awaited combination gene and cell therapies.

Further Reading

Cahill, T. J., Choudhury, R. P., & Riley, P. R. (2017). Heart regeneration and repair after myocardial infarction: Translational opportunities for novel therapeutics. *Nat Rev Drug Discov, 16*, 699–717.

Ellis, C., Ramzy, A., & Kieffer, T. J. (2017). Regenerative medicine and cell-based approaches to restore pancreatic function. *Nat Rev Gastroenterol Hepatol, 14*, 612–628.

Higuchi, A., Kumar, S. S., Benelli, G., et al. (2017). Stem cell therapies for reversing vision loss. *Trends Biotechnol, 35*, 1102–1117.

Hirsch, T., Torhoeft, T., Teig, N., et al. (2017). Regeneration of the entire human epidermis using transgenic stem cells. *Nature, 551*, 327–332.

Islam, M. N., Das, S. R., Emin, M. T., et al. (2012). Mitochondrial transfer from bone-marrow-derived stromal cells to pulmonary alveoli protects against acute lung injury. *Nat Med, 18*, 759–765.

Kalladka, D., Sinden, J., Pollock, K., et al. (2016). Human neural stem cells in patients with chronic ischaemic stroke (PISCES): A phase 1, first-in-man study. *Lancet, 388*, 787–796.

Kuriyan, A. E., Albini, T. A., Townsend, J. H., et al. (2017). Vision loss after intravitreal injection of autologous "stem cells" for AMD. *N Engl J Med, 376*, 1047–1053.

Marquardt, L. M., & Heilshorn, S. C. (2016). Design of injectable materials to improve stem cell transplantation. *Curr Stem Cell Rep, 2*, 207–220.

Niu, D., Wei, H. J., Lin, L., et al. (2017). Inactivation of porcine endogenous retrovirus in pigs using CRISPR-Cas9. *Science, 357*, 1303–1307.

Phinney, D. G., & Pittenger, M. F. (2017). Concise review: MSC-derived exosomes for cell-free therapy. *Stem Cells, 35*, 851–858.

CHAPTER 87

T-Cell Immune Therapies

Philip Norris, MD and Shibani Pati, MD, PhD

Cellular therapies designed to modulate or enhance immune responses are gaining acceptance as viable treatments for a variety of diseases, most notably in the oncology field. Some therapies such as granulocyte infusion have been a tool in transfusion medicine for more than 40 years, while novel therapies such as chimeric antigen receptor (CAR) T cells have only very recently been licensed for clinical use. This chapter will focus on T cell–based immunotherapy. Mechanistically, T-cell immune therapies range in function from replacing missing immune cells after chemotherapy (antiviral T-cell therapy), to augmenting or redirecting T-cell responses against tumors (tumor-infiltrating lymphocytes, or TILS, and CAR T cells), to modulating the systemic immune response in the case of regulatory T cells (Tregs). While the promise of T cell–based immune therapies is clear, the final determination of their benefit and the role transfusion medicine will play in providing these novel therapies is not defined and will depend on institutional resources and expertise.

T Cells as Immune Therapeutics: T cells are key members of the adaptive immune response, with effector functions that include cytokine secretion, promotion of antibody responses, and killing virus-infected or tumor cells. In addition, some T cells play a role in damping down immune responses, such as Tregs that can inhibit both T-cell and B-cell function. As adoptive immune therapy, T cells have been used to treat a variety of cancers, as well as to treat or prevent viral infections (Table 87.1). As opposed to transfer of effector/killer T cells, Treg cells can counteract a variety of inflammatory conditions. The best studied of these is as prevention or treatment of graft-versus-host disease (GVHD), and trials have been performed for the treatment of autoimmune disease or to build tolerance in organ transplantation.

Antiviral Cellular Therapy and Prophylaxis: Viral infections normally controlled by the immune system are frequently lethal in settings of profound T-cell depletion, such as after allogeneic hematopoietic stem cell transplantation (HSCT). Examples of problematic viruses include infection with the chronic viruses, cytomegalovirus and Epstein–Barr virus and the acute virus adenovirus. One solution to preventing or treating viral infection after HSCT is to infuse donor lymphocytes. Efficacy of infusion of unselected donor lymphocytes is limited by the typical low frequency of virus-specific T cells in the population and the side effect of increased alloreactive T cells and frequency of GVHD. The therapeutic index of donor lymphocytes can be improved through depletion of alloreactive T cells and enrichment for virus-specific T cells. Techniques used to enrich for these cells include selecting for cells of known specificity using multimers of peptide-MHC, stimulating cells with viral peptide and capturing those that secrete interferon-γ, or expanding virus-specific cells in vitro.

Transfusion Medicine and Hemostasis. https://doi.org/10.1016/B978-0-12-813726-0.00087-8

TABLE 87.1 T-Cell Immunotherapies

Disease	Therapy	Progress
Tumor eradication	CAR T cells	FDA approved
Tumor eradication	Tumor-infiltrating lymphocytes	Phase 1 through 3 trials
Tumor eradication	Enhanced affinity T-cell receptor	Phase 1 and 2 trials
Viral infections posthematopoietic stem cell transplantation	Effector T-cell transfer	Phase 1 through 3 trials
Graft-versus-host disease	Treg transfer	Phase 1 and 2 trials
Autoimmunity	Treg transfer	Phase 1 and 2 trials
Organ transplantation	Treg transfer	Phase 1 and 2 trials

Chimeric Antigen Receptor T Cells: CAR T cells are patient-derived T cells that are removed and genetically engineered to recognize specific targets found on cancer cells. The engineered T-cell receptor typically has an antibody on the extracellular surface coupled to an activation domain intracellularly. When the antibody recognizes a target, such as CD19 found on B-cell acute lymphocytic leukemia (ALL), the T cell becomes activated and kills the target cell bearing the antigen. CAR T-cell therapies have proven lifesaving in cases of refractory to conventional therapyand based on this efficacy two firms to date have received FDA licensure for these novel drugs. There are several limitations to first-generation clinical CAR T-cell therapies. T cells that recognize a given target kill all the cells bearing that target, and the range of cancers expressing unique antigens is relatively small, precluding general use of the therapy. Furthermore, patients treated for B-cell ALL become deficient in B cells, requiring replacement immunoglobulin therapy to treat the acquired immune deficiency. Finally, the robust, unregulated antitumor activity of the CAR T cells results in rapid tumor lysis and cytokine release, which can be treated with anticytokine therapy, but in severe cases can be fatal. Overall, CAR T-cell therapy represents a breakthrough in cellular immunotherapy, and further refinement and modification of their generation and use will expand the scope of their use.

Tumor-Infiltrating Lymphocytes: To date, CAR T cells have not had the same success in treating solid tumors, as they have in treating hematological malignancies. It has long been recognized that solid tumors contain T cells that recognize cancer cells and have infiltrated the tumor. However, the tumor microenvironment is immune suppressive and prevents tumor clearance by these TILs. In fact, these cells form the basis of many successful immune therapies that have been developed, including the immune checkpoint inhibitors such as anti-CLTA4 and anti-PD1 antibodies, which unleash the antitumor activity of these cells. Exogenous expansion and reinfusion of TILs may be even more effective than immune checkpoint inhibitors, and this is an active area of research. TIL therapy requires large numbers of cells to be effective, which requires a 5–6 week expansion period and represents a bottleneck in the production of these cells. Current strategies to optimize TIL therapy are focused on improving large-scale expansion of the cells, selecting TILs with a functional phenotype to augment in vivo persistence and function, and optimizing infusion of TILs, such as use of a lympho-depleting conditioning regimen before TIL infusion. While not currently practical for use beyond a few specialized clinical centers, TILs hold the potential to develop into an important personalized cellular immune therapy.

Regulatory T Cells: The concept of Tregs remained elusive and ill-defined until the identification of the forkhead box P3 (FOXP3) gene as a master transcriptional regulator in the development of T cells with a regulatory phenotype. Treg cells express high levels of the α-chain of the interleukin (IL)-2 receptor, and expansion of Tregs is highly IL-2 dependent. Tregs have also been found to be more plastic in nature than previously appreciated and can revert to proinflammatory, IL-17 producing cells after adoptive transfer. Finally, clinical trials have been conducted with polyclonal Treg cells, and research is underway to improve antigen specificity of Tregs, such as through development of CAR Tregs.

In spite of the fact that Treg biology is still being elucidated, these cells have been shown to be beneficial in a number of clinical conditions in small trials. Treg infusion after HSCT has been shown to decrease the incidence of GVHD compared with historical controls, and importantly, does not appear to impair the graft-versus-tumor response. In the field of organ transplantation, Treg-based therapies hold the promise of long-term tolerance of the transplanted organ without exogenous immune suppression, and a recently published trial in liver transplant patients showed an increased rate of organ tolerance in Treg-treated subjects compared with historical controls. Of the autoimmune diseases, type 1 diabetes mellitus has received the most attention as a target of Treg cell therapy, but multiple other diseases such as lupus, rheumatoid arthritis, and multiple sclerosis make attractive targets for Treg therapy.

Transfusion Medicine Interface With T-Cell Therapies: Transfusion medicine professionals will potentially interact with T-cell immune therapies at multiple points from their production to use in clinical settings. Transfusion services and blood centers possess expertise in many of the areas that are key to production and distribution of cellular therapies, including medical, quality, and regulatory programs. Given widely distributed donor collection centers, blood banks can play a role in the collection of mononuclear cells from donors to be used in production of T-cell therapies. More sophisticated blood centers with good manufacturing practice clean room facilities may be able to perform manipulations of the cells, such as T-cell isolation or more extensive manipulation using standardized operating procedures. Finally, given their proximity to hospitals, storage and distribution of T-cell therapies are another possibility as T cells move beyond proof of concept clinical trials to mainline medical therapies.

Further Reading

Brunstein, C. G., Miller, J. S., McKenna, D. H., Hippen, K. L., DeFor, T. E., Sumstad, D., et al. (2016). Umbilical cord blood-derived T regulatory cells to prevent GVHD: Kinetics, toxicity, profile, and clinical effect. *Blood*, *127*, 1044–1051.

Jain, M. D., & Davila, M. L. (2017). Emerging principles from the clinical application of chimeric antigen receptor (CAR) T cell therapies for B cell malignancies. *Stem Cells*. https://doi.org/10.1002/stem.2715.

Lee, S., & Margolin, K. (2012). Tumor-infiltrating lymphocytes in melanoma. *Curr Oncol Rep*, *14*, 468–474.

Leen, A. M., Heslop, H. E., & Brenner, M. K. (2014). Antiviral T-cell therapy. *Immunol Rev*, *258*, 12–29.

Suhoski Davis, M. M., McKenna, D. H., & Norris, P. J. (2017). How do I participate in T-cell immunotherapy? *Transfusion*, *57*, 1115–1121.

Tang, Q., & Vincenti, F. (2017). Transplant trials with Tregs: Perils and promises. *J Clin Investig*, *127*, 2505–2512.

CHAPTER 88

Adverse Events Associated With Hematopoietic Progenitor Cell Product Infusion

Yen-Michael S. Hsu, MD, PhD and Melissa M. Cushing, MD

Adverse events associated with hematopoietic progenitor cell (HPC) product infusion can vary depending on the type of product infused (Table 88.1). The Center for Disease Control National Health Safety Network provides a Biovigilance Component Hemovigilance Module Surveillance Protocol that provides definitions for blood transfusion reactions. Similarly, the *Circular of Information for the Use of Cellular Therapy Products* also provides descriptions of the known side effects and hazards associated with HPC infusions. This chapter focuses on the common types of clinically significant reactions that can occur during or within hours after HPC infusion, and the strategies to mitigate and treat adverse reactions.

Incidence: The incidence of adverse reaction to HPC infusion varies depending on the sources of graft obtained (e.g., bone marrow, peripheral blood, or umbilical cord), HPC product condition (e.g., fresh/thawed/washed), and the infusion volume. Complex host factors (e.g., patient demographics, primary diagnoses, infection status, or responses to conditional regiments) can also modulate the recipients' immune responses that lead to abnormal clinical signs and symptoms during the HPC infusion. Recently, Mulay et al. reported that roughly 50% of HPC infusions are associated with infusion-related adverse reactions. In the pediatric cohort, another study with over 200 patients described an HPC infusion adverse event incidence of 55%–62% within 24 hours of infusion.

Cryopreserved products have different risks than fresh products. Although the HPCs survive the freeze–thaw process and the effect of DMSO, granulocytes and RBCs lyse because of their particular susceptibility to osmotic stress. Postthaw washing eliminates the cell debris and DMSO. With postthaw nonwashed HPC product infusions (i.e., bedside thaws), the frequency of adverse events is associated with the number of granulocytes in the product at the time of graft harvest. Host factors, such as recipient age and sex, have been associated with infusion-related toxicity, with more side effects occurring in female patients. It was also observed that pediatric patients with lower body weight (<10 kg) might have a slightly higher incidence of adverse events. The exact etiology of these findings has not yet been elucidated. It is hypothesized that recipients with smaller body surface area and lower intravascular volume per body weight may receive relatively higher DMSO and/or cellular debris concentrations during HPC infusion.

With the exception of umbilical cord blood HPC transplants, allogeneic transplants are generally infused to the recipient shortly (within 48 hours) after collection from the donors. In the interim, the allogeneic grafts are maintained either at room temperature (bone marrow HPC product) or refrigerated temperature (peripheral blood HPC product). The rate of adverse events during the infusion of noncryopreserved products

Transfusion Medicine and Hemostasis. https://doi.org/10.1016/B978-0-12-813726-0.00088-X

TABLE 88.1 Adverse Reactions to Stem Cell Infusions

Nonimmunologic Complications	Immunologic Complications
Dimethyl sulfoxide toxicity	Acute/delayed hemolytic reaction
Septic transfusion reaction	Febrile, nonhemolytic reaction
Fat emboli	Allergic reaction
Bleeding from excessive anticoagulant	Transfusion-related acute lung injury
Circulatory/volume overload	Graft-versus-host disease
Hypothermia	
Nonimmunologic hemolysis	
Transmission of infectious disease or disease agent	

is significantly lower than that of cryopreserved products, with reported rates ranging from 0% to 26.8%.

Nonimmunologic Complications

Circulatory Overload: Circulatory or volume overload occurs when the intravenous infusion of an HPC product acutely raises the central venous hydrostatic pressure, resulting in fluid extravasation and subsequent pulmonary edema. Female, pediatric/elderly, and renal insufficiency patients are reported to be more susceptible. For those recipients with increased susceptibility, slowing the infusion rate, splitting the total HPC infusion volume into multiple infusions, or using diuretics before and in between infusions should be considered.

Septic Transfusion Reaction: Bacterial contamination can occur during any step in the process, including HPC collection, processing, and infusion. The reported HPC product contamination rate varies among institutions and by product type, ranging from 1% to 6%. Skin flora and environmental organisms are the predominant bacteria identified in contaminated HPCs, with coagulase negative *Staphylococcus* species accounting for the majority of positive cultures. Rarely, fungal contamination is detected. Fortunately, in vitro microbial growth in the HPC product may not correlate with a clinical septic reaction. The risk can be extrapolated from studying human transfusable platelet products where at least 100,000 CFU/mL microbial activity in a product is associated with clinical sepsis. Fortunately, bacterial contamination can be significantly reduced by the cryopreservation and thawing/washing processes.

Unlike the more common blood components and products, microbial-contaminated HPCs are not readily replaceable. When an HPC product is determined to be culture-positive, the clinical transplant team must be immediately notified. The possibility of replacing or discarding the contaminated HPC product should be considered, especially when the recipient is receiving a fully myeloablative conditioning regimen or when additional noncontaminated aliquots of HPC are available. Alternatively, the transplant team may decide to infuse the product with prophylactic antimicrobial therapy targeting the bacterial species identified. Before HPC infusion, many of the immune-compromised HPC recipients would already be receiving a broad-spectrum prophylactic antimicrobial regimen. Overall, patients who receive culture-positive products do not have increased mortality compared with patients receiving culture-negative products. When a septic reaction is suspected, it is important to evaluate the infectious etiologies both for the recipient (e.g., underlying infection or infected vascular access) and HPC product (e.g., donor or processing contamination) level during the treatment and investigation.

Dimethyl Sulfoxide Toxicity: HPCs require the addition of a cell membrane–permeable cryoprotectant, most commonly DMSO for cryopreservation. Most autologous and cord blood products are cryopreserved. DMSO toxicity is the most common reaction associated with the thawed nonwashed HPC product infusions and commonly manifests in abdominal cramping, nausea/vomiting, hypotension, and bradycardia (more rarely hypertension and tachycardia). Interestingly, in the pediatric cohort, the DMSO content does not appear to have significant association with HPC-associated adverse events.

DMSO can cause an increased histamine release and mast cell degranulation, resulting in an urticarial rash, pruritus, facial flushing, or periorbital swelling. DMSO has been implicated as a potential neurotoxin, but total DMSO dosage of less than 1 g/kg recipient body weight has been correlated with reduced neurological toxicity. Severe DMSO toxicity resulting in cardiac arrhythmias can be fatal. The DMSO toxicity can be mitigated by albumin- or dextran-based dilution of the HPC product, with or without washing. Although washing the product can effectively remove DMSO without significantly impacting the CD34+ cellular content, one study reported delayed platelet engraftment with washed compared with nonwashed products.

Fat Emboli and Nonimmune Hemolysis: It is not uncommon to detect cellular aggregates or debris in thawed HPC products. The loss of anticoagulant from the harvested graft is sometimes associated with significant fibrin formation that leads to cell aggregation. The common HPC cryopreservation strategy is not optimized to prevent destruction of red blood cells and granulocytes during either freezing or thawing phases. Therefore, washing the thawed HPC product and infusing it through a 170 μm blood filter can mitigate such adverse events. Thawed HPC cellular debris and hemolytic by-products can also be reduced with RBC removal before product cryopreservation.

Immunologic Complications

Immune Hemolysis: Thirty to fifty percent of allogeneic transplants are ABO incompatible. Major ABO incompatibility (e.g., Group O recipient receiving Group A HPC graft) is associated with immediate destruction of donor RBCs. However, the extent of clinical hemolysis depends on a combination of factors, including recipient's isohemagglutinin titer and the infused donor RBC volume. Only up to 1% of HPC recipients experience immediate clinical hemolysis related to red cell antibodies. Several studies have shown that 10%–15% of HPC transplants with minor ABO incompatibility are associated with delayed immune hemolysis. HPC manufacturing facilities are required to monitor and limit the amount of major incompatible donor RBC content in the HPC products as per Foundation for Accreditation of Cellular Therapy (FACT) or AABB (previously known as American Association of Blood Banks) standards. According to common practice, the red cell volume is typically reduced to less than 30 mL or 0.4 mL/kg. For minor ABO incompatibility, plasma reduction of the HPC product to remove hemagglutinins can be considered. Treatment for hemolysis requires active hydration. Resuming the infusion of remaining product(s) after the resolution of the acute adverse event may be necessary. As per the American Society for Apheresis (ASFA) therapeutic apheresis guidelines published in 2016, therapeutic plasma exchange could be considered in patients with severe hemolysis due to major ABO incompatibility (i.e., category II indication).

Febrile Nonhemolytic Reactions: Similar to blood products, allogeneic HPC products can cause febrile nonhemolytic transfusion reactions with a shared pathogenesis and identical clinical features. The clinical manifestations include fever, chills, or rigors. Recently, Chen et al. described a new type of febrile nonhemolytic reaction and coined the term "infusion-related febrile reactions" (IRFR) in the haploidentical peripheral blood stem cell product recipients. Defined by a temperature of 38°C observed within 24 hours of infusion and elevated C3 complement and C-reactive protein (CRP) levels, IRFR is associated with proinflammatory cytokine release and the infusion of large doses of mononuclear, CD34+, and CD3+ cells. If a significant febrile reaction is detected, the infusion must be stopped to rule out other more severe etiologies, such as hemolysis or septic reactions.

Allergic Reactions: Allogeneic HPC products contain similar cellular and plasma constituents to allogeneic blood components. Thus, allergic reactions also manifest with common signs and symptoms, including pruritus, urticaria, or orbital swelling. Upper and lower airway obstruction and frank anaphylaxis may occur during more severe and even life-threatening reactions. The allergens may be proteins in an allogeneic donor's plasma or chemical agents added during processing, such as dextran or DNAse. Recipients are typically administered antihistamines and steroids to prevent or treat allergic reactions; however, the efficacy of the prophylactic regimen has not been proven.

Transfusion-Related Acute Lung Injury: Transfusion-related acute lung injury (TRALI) is characterized by noncardiogenic pulmonary edema and associated with recipient neutrophils activated by donor-derived antibodies targeting human leukocyte antigens (HLA) or human neutrophil antigens (HNA) in many cases. Non–antibody-mediated TRALI occurs and may be mediated by biologic response modifiers, such as bioactive lipids, present in the transfused blood products. Recipient-derived HLA or HNA antibodies directly reacting with donor WBC are uncommonly described as a cause of TRALI.

HPC-associated TRALI is rarely observed. If TRALI is suspected during infusion, the infusion should be immediately stopped and followed with supportive care. In response, the laboratory may plasma-reduce or wash the remaining HPC products before reinitiating the infusion. Preexisting or concurrent underlying pulmonary conditions, such as respiratory infections, aspiration, or volume overload, should be investigated and ruled out before confirming a diagnosis of TRALI.

Documentation and Reporting: Adverse reactions must be documented in the patient's medical record and reported in accordance with a facility's policies and the applicable laws and regulations. Each event must be reported to the patient's physician and the medical director of the processing facility. In addition, reactions may need to be reported to the appropriate hospital quality and safety committees, as well as outside institutions (such as NMDP, the FDA, and local government regulating bodies). If the patient participates in a clinical trial, the Institutional Review Board and clinical trial sponsor should also be notified. Transplant programs and cell processing laboratories accredited by FACT or AABB are also required to establish a quality assurance program that routinely monitors these adverse events for practice improvement. Recently, the field of cellular therapy, involving genetically modified adoptive immune therapy (i.e., chimeric antigen receptor T cells, CAR T cells), has rapidly advanced. The severe adverse events associated with these products, such cytokine release syndrome, neurologic toxicity, on-target off-tumor effect, anaphylaxis, and graft-versus-host disease, should also be documented and reported to further ensure the safe use of these novel cellular therapy products in the general population.

Conclusion: The great majority of cellular therapy infusion-related adverse events are mild to moderate in severity and self-limited. Various mitigation strategies have been proposed. First, for cryopreserved HPC products, reducing DMSO before product freezing or postthaw may reduce the incidence of DMSO toxicity. Second, infusion premedication may be helpful in some patient populations, but efficacy remains to be proven. Third, for apheresis HPC products, careful monitoring of the collection interface during leukapheresis can further reduce the introduction of donor granulocytes and red blood cells to mitigate infusion reactions. Fourth, adverse reactions must be carefully documented and reported to provide adequate monitoring of new infusion practices or the use novel cell therapy products.

Recommended Reading

Castillo, N., Garcia-Cadenas, I., Garcia, O., Barba, P., Diaz-Heredia, C., Martino, R., et al. (2015). Few and nonsevere adverse infusion events using an automated method for diluting and washing before unrelated single cord blood transplantation. *Biol Blood Marrow Transplant, 21*, 682–687.

Chen, Y., Huang, X. J., Wang, Y., Liu, K. Y., Chen, H., Chen, Y. H., et al. (2015). Febrile reaction associated with the infusion of haploidentical peripheral blood stem cell: Incidence, clinical features, and risk factors. *Transfusion, 55*, 2023–2031.

Hong, H., Xiao, W., Lazarus, H. M., Good, C. E., Maitta, R. W., & Jacobs, M. R. (2016). Detection of septic transfusion reaction to platelet transfusion by active and passive surveillance. *Blood, 127*, 496–502.

Jacobs, M. R., Good, C. E., Fox, R. M., Roman, K. P., & Lazarus, H. M. (2013). Microbial contamination of hematopoietic progenitor and other regenerative cells used in transplantation and regenerative medicine. *Transfusion, 53*, 2690–2696.

Mulay, S. B., Greiner, C. W., Mohr, A., Bryant, S. C., Lingineni, R. K., Padely, D., et al. (2014). Infusion technique of hematopoietic progenitor cells and related adverse events (CME). *Transfusion, 54*, 1997–2003.

Otrock, Z. K., Sempek, D. S., Carey, S., & Grossman, B. J. (2017). Adverse events of cryopreserved hematopoietic stem cell infusions in adults: A single-center observational study. *Transfusion*, *57*, 1522–1526.

Schwartz, J., Padmanabhan, A., Aqui, N., Balogun, R. A., Connelly-Smith, L., Delaney, M., et al. (2016). Guidelines on the use of therapeutic apheresis in clinical practice-evidence-based approach from the writing committee of the American Society for apheresis: The seventh special issue. *J Clin Apher*, *31*, 149–162.

Shu, Z., Heimfeld, S., & Gao, D. (2014). Hematopoietic SCT with cryopreserved grafts: Adverse reactions after transplantation and cryoprotectant removal before infusion. *Bone Marrow Transplant*, *49*, 469–476.

Staley, E. M., Schwartz, J., & Pham, H. P. (2016). An update on ABO incompatible hematopoietic progenitor cell transplantation. *Transfus Apher Sci*, *54*, 337–344.

Truong, T. H., Moorjani, R., Dewey, D., Guilcher, G. M. T., Prokopishyn, N. L., & Lewis, V. A. (2016). Adverse reactions during stem cell infusion in children treated with autologous and allogeneic stem cell transplantation. *Bone Marrow Transplant*, *51*, 680–686.

CHAPTER 89

Quality and Regulatory Issues in Cellular Therapy

Eva D. Quinley, MS, MT(ASCP)SBB, CQA(ASQ) and
Joseph Schwartz, MD, MPH

Cellular therapy is the use of viable cells and tissues for the treatment of disease. Modern cell-based therapies include regenerative medicine, which is the process of replacing or regenerating human cells, tissues, or organs to restore or establish normal function. With these advances, a need existed to incorporate quality systems and regulatory oversight. While the basics of quality and quality systems are the same for blood banking/transfusion services and cellular therapies, there are some distinct differences. This is seen both in standards created for cellular therapy and in the applicable US Food and Drug Administration (FDA) regulations. Accreditation of cellular therapy laboratories is through the Foundation for the Accreditation of Cellular Therapies (FACT), AABB, the College of American Pathologists (CAP) and other nongovernmental agencies, which are primarily in the United States but also accredit internationally. Since 2005, the FDA has regulated cellular therapies based on 21 CFR 1271, which contains the current good tissue practice (cGTP) regulations and the regulations pertaining to use of an investigational new drug (IND). Certain laboratory tests performed on cells or tissue intended for transplantation are regulated under the Clinical Laboratory Improvement Amendments (CLIA) under the auspices of Centers for Medicare and Medicaid Services (CMS). State health departments may also have local regulations relevant for cellular therapy. As additional uses for cell therapies are found, it is expected that there will be further expansion of regulatory oversight (Table 89.1). Individuals and organizations or institutions involved in cellular therapy processing must be familiar with the requirements of these agencies.

There are many potential forms of cell therapy. Human cells, tissues, or cellular or tissue-based products (HCT/Ps) are defined as articles containing or consisting of human cells or tissues that are intended for implantation, transplantation, infusion, or transfer into a human recipient. Examples of HCT/Ps include, but are not limited to, bone, ligament, skin, dura mater, heart valve, cornea, hematopoietic stem/progenitor cells derived from peripheral and cord blood and bone marrow, manipulated autologous chondrocytes, epithelial cells on a synthetic matrix, and semen or other reproductive tissue.

Accreditation Standards: FACT, AABB, and CAP have created sets of standards for quality systems applicable to organizations that manufacture HCT/Ps and accredit organizations that meet these standards.

Foundation for the Accreditation of Cellular Therapy: FACT Standards promote improvement and progress in cellular therapy. Because cellular therapy products often are exchanged internationally, the requirements help to ensure consistency in

Transfusion Medicine and Hemostasis. https://doi.org/10.1016/B978-0-12-813726-0.00089-1

TABLE 89.1 Overview of Key Events in the Regulations of Cellular Therapies

1912	Public Health Service Act—blood defined as a biologic
1938	Federal Food, Drug, and Cosmetic Act—products must be safe
1972	FDA granted jurisdiction over human biological products, therapeutics, and blood banking diagnostics
1986	First clinical trials with nonhematopoietic cells
1997	First FDA license for autologous chondrocytes
1998	Regulation of unrelated allogeneic peripheral and umbilical cord blood hematopoietic progenitor cells
2001	Establishment registration and product listing finalized
2004	Donor eligibility requirements finalized
2005	21 CFR 1271 in effect
2010	First licensed cellular immunotherapy product
2011	First FDA license for cord blood granted
2017	First CAR T-cell product licensed to treat children and young adults with B-cell acute lymphoblastic leukemia—first licensed immunotherapy
2017	First CAR T-cell product a licensed to treat non-Hodgkin lymphoma

product quality worldwide. The Standards are the cornerstone of the FACT accreditation program. The Standards address both clinical and laboratory practices and are applicable to HCT/Ps and therapeutic cells (TCs), which are obtained from the bone marrow, peripheral blood, or umbilical cord blood. FACT accreditation is awarded to organizations following successful documentation of compliance with the Standards. On-site inspections are conducted by inspectors who are qualified by training and experience and who have working knowledge of the industry. The inspection is quality-oriented and in line with the FDA current good tissue practice (cGTP) regulations. FACT inspections focus on all areas of manufacturing, including donor eligibility, collection, processing, storage, distribution, administration, clinical outcomes, and other clinical parameters. With the expansion of cellular therapies, FACT has developed different sets of standards to address different needs of cellular therapy facilities to include the following:

1. *FACT-JACIE International Standards for Hematopoietic Cellular Therapy Product Collection, Processing, and Administration.* For these standards, FACT joined with the Joint Accreditation Committee (JACIE) of the International Society for Cellular Therapy (ISCT), and the European Group for Blood and Bone Marrow Transplantation (EBMT) to publish standards for hematopoietic progenitor–derived cellular therapy. These Standards cover all aspects of collection, processing, storage, and administration of the bone marrow and cells derived from peripheral blood. For umbilical cord blood products, these Standards only apply to the administration of the cellular product.
2. *Common Standards for Cellular Therapies* that represent the basic fundamentals of cellular therapy that can be applied to any cell source or therapeutic application and are intended to be used throughout product development and clinical trials. These Standards apply to cells collected from nonhematopoietic sources (e.g., adipose tissue) or cells collected from hematopoietic sources for nonhomologous use (e.g., mesenchymal stromal cells for cardiac repair).

TABLE 89.2 AABB Quality System Essentials for Cellular Therapy
1. Organization
2. Resources
3. Equipment
4. Agreements
5. Process controls
6. Documents and records
7. Deviations and nonconforming products or services
8. Internal and external assessments
9. Process improvements
10. Safety and facilities

3. *Immune Effectors Standards*, which apply to immune effector cells, are used to modulate an immune response for therapeutic intent, such as dendritic cells, natural killer cells, T cells, and B cells. This includes, but is not limited to, genetically engineered chimeric antigen receptor T cells (CAR-T cells) and therapeutic vaccines.
4. *NetCord-FACT International Cord Blood Standards* that cover all phases of cord blood collection, processing, testing, selection and release, and distribution. These NetCord-FACT Standards are organized into sections, which cover the following:

- Operations, including quality management; Donor management and collection; Processing; Listing, search, selection, reservation, release, and distribution.

American Association of Blood Banks: The first edition of AABB *Standards for Cellular Therapy Product Services* was published in 2004. The Standards for cellular therapy products are similar to those for blood banks and transfusion services with an emphasis on a quality system composed of 10 elements (Table 89.2). The accreditation process is also similar in that accreditation is voluntary and involves assessments by trained AABB assessors. Accreditation is based on compliance with the Standards and is focused on donor eligibility, collection, processing, storage, distribution, and certain clinical components.

College of American Pathologists: CAP accredits laboratories and has deemed authority by CMS. It also offers an accreditation service to help laboratories earn accreditation under ISO 15189 Medical Laboratories.

Regulation: The number of uses of HCT/Ps has increased rapidly, and as the field has expanded, regulations have been promulgated, which govern the procurement, manufacture, and transplant of HCT/Ps. Regulation is focused on safety and efficacy: safety issues include sterility, purity, identity, segregation, and tracking, whereas efficacy includes potency and stability.

Unmanipulated bone marrow cell transplants are regulated by the Health Research and Services Administration (HRSA). This organization also oversees the marrow and cord blood (HPC, Bone Marrow and HPC, Cord Blood, respectively) donations under the National Marrow Donor Program (NMDP). Other HCT/Ps, such as HPC, Apheresis, were regulated under 21 CFR 1270 until May 2005 when 21 CFR 1271 came into effect and superseded 1270. The purpose of these regulations is to create a unified registration and listing system for establishments that manufacture HCT/Ps and to establish donor eligibility, cGTP, and other procedures to prevent the introduction, transmission, and spread of communicable diseases by HCT/Ps. The Office of Cellular, Tissue and Gene Therapies (OCTCT) within the Center for Biologics Evaluation and Research (CBER) regulates HCT/Ps under section 361 of the Public Health Service Act (PHS Act/42 U.S.C. 262). Regulation solely under section 361 (designation as a 361 HCT/P) requires establishments to adhere to regulations designed to prevent the transmission of communicable disease but do not require premarket review or notification for such products. The criteria for regulation solely under section 361 are as follows:

1. HCT/P is minimally manipulated;
2. HCT/P is intended for homologous use only;
3. Manufacture of the HCT/P does not involve the combination of the cell or tissue component with a drug or device, except for a sterilizing, preserving, or storage agent, if the addition of the agent does not raise new clinical safety concerns with respect to the HCT/Ps; and either
4. HCT/P does not have a systemic effect and is not dependent on the metabolic activity of living cells for its primary function; or
5. HCT/P has a systemic effect or is dependent on the metabolic activity of living cells for its primary function, and
   a. is for autologous use,
   b. is for allogeneic use in a first or second degree relative, or
   c. is for reproductive use.

Any cell-based product that contains cells or tissues that "are highly processed, are used for other than their normal function, are combined with nontissue components, or are used for metabolic purposes" would also be subject to the Public Health Safety Act, Section 351 (designation as a 351 HCT/P), which regulates the licensing of biologic products and requires the submission of an IND application to the FDA before studies involving humans are initiated. 21 CFR Parts 210 and 211 apply to the manufacture of these cellular products as well as the applicable parts of 21 CFR Part 600. The FDA is allowed to inspect facilities, announced or unannounced, where HCT/Ps are manufactured at reasonable times and in a reasonable manner.

The Clinical Laboratory Improvement Amendments: Laboratories that perform testing on cells and/or tissue intended for implantation, transplantation, or infusion are subject to CLIA regulations under CMS. The tests regulated are FDA-required tests for communicable diseases, FDA-required testing for emerging infectious diseases, and sterility testing in cases where donor notification is necessary (e.g., stem cells).

Investigational New Drug: With certain exceptions, clinical investigations in which a drug is administered to human subjects must be conducted under an IND as required in 21 CFR Part 312. The definition of the term *drug* in section 201(g)(1) of the Food, Drug, and Cosmetic Act (FD&C Act) includes, among other things, "articles intended for use in the diagnosis, cure, mitigation, treatment, or prevention of disease…" and "articles (other than food) intended to affect the structure or any function of the body of man or other animals." Biological products subject to licensure under

TABLE 89.3 Regulations Applicable to Investigational New Drugs (INDs)

Regulation	Content
21 CFR 312.23	IND content and format
21 CFR 312.42	Clinical holds
21 CFR 312.50–21 CFR 312.69	Responsibilities of sponsors/investigators
21 CFR 50 and 56	IRB ad consents
21 CFR 58	Good laboratory practice
21 CFR 1271	Human cells, tissues, and cellular and tissue-based products (includes current good tissue practice)

section 351 of the PHS Act may also be considered drugs within the meaning of the FD&C Act (Table 89.3).

Each IND has a sponsor, which is the individual or organization that takes responsibility for and initiates a clinical investigation, and one or more investigators, who conduct the clinical investigation. The sponsor and investigator may be the same person. An IND submission contains a specified set of documents and information (Table 89.3). IND submissions are reviewed, approved or disapproved, and monitored by the FDA. There are four types of IND studies: phase I evaluates safety of the drug; phase II studies efficacy and dose; phase III confirms efficacy and provides statistical evidence of effectiveness; and phase IV includes postmarketing studies designed (1) to compare a drug with other drugs already in the market; (2) to monitor a drug's long-term effectiveness and impact on a patient's quality of life; and (3) to determine the cost-effectiveness of a drug therapy relative to other traditional and new therapies.

Current Good Tissue Practice: The core requirements of the cGTPs are most directly related to preventing the introduction or transmission or spread of communicable disease (Table 89.4).

Donor Eligibility: Donor eligibility determination, based on donor screening and testing for relevant communicable disease agents and diseases, is required for all donors of cells used in HCT/Ps, unless they are autologous. Donor eligibility determination should be performed according to written procedures and is required before transplant. The review must be conducted by a qualified and responsible individual. Only in cases of documented urgent medical need can an HCT/Ps be transplanted without completion of donor eligibility. Documentation of donor eligibility must accompany the product at all times.

Quality Program: Any organization that performs steps in the manufacture of HCT/Ps must have a quality program that addresses the cGTPs and ensures that procedures exist for all steps in manufacturing as well as those required for the quality system, including management review, periodic internal audits, and a mechanism to evaluate and document information pertaining to deviations and complaints.

TABLE 89.4 Current Good Manufacturing Practices (cGMPs) Versus Core Current Good Tissue Practices (cGTPs)

cGMPs	cGTPs
• Organization and personnel	• Facilities
• Buildings and facilities	• Environmental control
• Equipment	• Equipment
• Control of components, containers, and closures	• Supplies/reagents
• Production and process controls	• Processing/process controls
• Holding and distribution	• Labeling controls
• Laboratory controls	• Storage
• Records and reports	• Receipt, predistribution, and distribution
	• Donor eligibility

Personnel: The number of personnel must be adequate to perform the work. Personnel should have the education, experience, and training to perform their assigned tasks and should only be assigned tasks for which they are trained and deemed competent.

Procedures: Well-written procedures that are compliant with regulations are required for all steps in the manufacturing process. These procedures must be reviewed and approved by a responsible person and must be available and easily accessible to those who are doing the work. Procedures from other organizations may be used if they meet the cGTP requirements and are appropriate for the task.

Facilities: Facilities must be of suitable size and orientation to prevent contamination or mix-ups; facilities must be maintained in a clean and sanitary condition. The manufacturing operations should be performed in clearly defined areas with adequate lighting, ventilation, plumbing, drainage, and access to sinks and toilets. Sewage, trash, or refuse should be removed in a timely manner. Written procedures governing maintenance and cleaning should be in place and followed, and cleaning records should be retained for a defined period of time no less than 3 years.

The cGTPs require strict environmental monitoring. Controls may include temperature, humidity, ventilation and air filtration and flow, cleaning and disinfection of rooms, and maintenance of equipment used to control conditions. The environment should be monitored regularly using validated methods.

Equipment: Equipment used in the manufacture of HCT/Ps must be of appropriate design to facilitate cleaning and operations. Equipment should be qualified for its intended use, including calibration as appropriate. The equipment must be well maintained, and there should be documentation of maintenance, cleaning, and calibration activities readily available. Traceability of equipment used in the manufacture of each HCT/P is also a requirement.

Supplies and Reagents: The supplies and reagents used in manufacturing of HCT/Ps must be qualified and sterile, if necessary. Supplies and reagents must be stored according to manufacturer's instruction. Documented verification of each supply or reagent is required, i.e., test results or certificates of analysis. Records must be maintained to allow traceability of supply or reagent use, including lot numbers used in the manufacturing of each HCT/P.

Recovery: HCT/Ps must be recovered in a way that minimizes the risk of contamination. Written procedures should describe the manner in which the HCT/Ps are obtained from the donor.

Processing and Process Controls: Procedures should prevent contamination or cross-contamination, and pooling of donors is not allowed in the manufacturing process. There should be inspection and/or testing during the manufacturing process and sampling for product testing must be representative of the state of the product.

Process Changes and Validation: Changes to processes must be documented and validated before implementation. Any changes made must be well communicated. Validation is required for those changes to processes which cannot be completely verified.

Labeling: Labeling must be performed under strict controls. Labels must be verified for accuracy, legibility, integrity, and insurance that they meet all requirements. Procedures should exist to instruct how products are to be labeled, and individuals who apply labels should be trained accordingly. Donor eligibility determination documentation must accompany each HCT/P. For licensed products, a circular of information is also necessary as part of labeling.

Storage: Storage areas must be designed to prevent mix-ups and contamination or cross-contamination. Products should be stored to prevent improper release of HCT/Ps. When there are temperature requirements for storage, temperatures must be monitored to assure they remain in acceptable range.

Receipt and Distribution: Each incoming HCT/P must be inspected and accepted, rejected, or quarantined. Before making an HCT/P available for distribution all manufacturing records must be reviewed with verification that all release criteria have been met, including eligibility requirements. Shipping containers must be validated for their intended use and designed to protect the product from contamination. Records must be maintained to allow complete traceability and trackability of the product.

Records: Records must be maintained concurrently with the work being performed. Records must be in line with good documentation practices (legible, indelible, accurate, and complete), and they must be maintained in a well-established records management system according to a documented record retention schedule. Records should allow review of the product history before distribution and afterward for follow-up purposes.

Complaint File: Manufacturers must maintain procedures for the handling of complaints, including their review and evaluation. It must be determined if the complaint represents a deviation or an adverse reaction. Certain deviations and adverse reactions must be reported to the FDA (Table 89.5).

TABLE 89.5 Reportability of Adverse Reactions and Events to the FDA

Adverse reactions	• Fatal • Life-threatening • Results in permanent impairment of a body function or permanent damage to a body structure • Necessitates medical or surgical intervention, including hospitalization
361 HCT/Ps	• Deviation forms current good tissue practices • Unexpected or unforeseeable event that may lead to the transmission or potential transmission of a communicable disease or HCT/P contamination if it meets the criteria of an adverse reaction as above
351 IND HCT/Ps	• No specific investigational new drug (IND) reporting requirements • Report deviations in the manufacture of IND in annual and safety reports
351 Licensed HCT/Ps	• Report any event in which the safety, purity, or potency of a distributed product may be affected

Licensure: In 2010, the FDA issued a license for a cellular product, representing the first product in a new therapeutic class known as active cellular immunotherapies. The product (Provenge, Dendreon Corporation) is an autologous cellular immunotherapy, designed to stimulate a patient's own immune system to respond against cancer and is used in treating advanced prostate cancer. The product is made using autologous immune cells collected by apheresis. In August 2017, the FDA approved a CAR T-cell therapy, Kymriah (CTL019), to treat children and young adults with B-cell lymphoma that was refractory to treatment or had relapsed two times. Two months later, the FDA cleared another CAR T-cell therapy, Yescarta, the first licensed product to treat non-Hodgkin's lymphoma.

HPC, Cord Blood: Effective October 20, 2011, HPC, Cord Blood products were subject to FDA licensure requirements as a biologic drug under 21 CFR Part 600, 601, and 610. The first license for an HPC, Cord Blood product was granted in November 2011. As with all drug products, licensure of HPC, Cord Blood requires strict adherence to all applicable regulations. Of note, is the requirement for a complete batch record for each HPC, Cord Blood unit. This batch record must be reviewed and approved by quality as part of the release of an HPC, Cord Blood for licensure. HPC, Cord Blood products that do not meet licensure criteria may still be used under IND.

Future: In the future, it is almost certain that more HCT/Ps will become licensed products. Research is already in progress to improve and enhance the efficacy of these products in the health-care setting. It is also certain that there will be additional novel uses for HCT/Ps. As new applications for cellular therapies continue to be developed and introduced for clinical use, there will be changes to regulations, including addition of regulations, to accommodate these new products. The new regulations and standards will need to ensure the continued safety, purity, and potency of cellular therapies for patients.

Further Readings

AABB Standards for Cellular Therapy (8th ed.) (2017). AABB, Bethesda, MD.

FACT-JACIE international standards for hematopoietic cellular therapy product collection, processing, and administration (6th ed.) (2015). Foundation for the Accreditation of Cellular Therapy (FACT).

Halme, D., & Kessler, D. (2006). FDA regulation of stem-cell-based therapies. *N Engl J Med, 355*(16), 1730–1735.

NetCord-FACT international standards for cord blood collection, banking, and release for administration (6th ed.) (2016). International NetCord Foundation and Foundation for Accreditation of Cellular Therapy.

Philippidis, A. (May 2, 2011). A clearer regulatory path for cell therapy may help improve investment. *Genet Eng Biotechnol News* Insight and Intelligence Report.

CHAPTER 90

Tissue Banking in the Hospital Setting

Robert A. DeSimone, MD and Cassandra D. Josephson, MD

In many hospitals, management of human tissue, specifically tissue allografts, has been centralized. Centralized tissue services coordinate the supply, dispensing, and accounting of products among numerous clinical services and have an infrastructure to efficiently receive, process, store, issue, and manage products.

Tissue Suppliers: Tissue suppliers aseptically recover tissue for transplantation from deceased human donors. Suppliers then process and store the recovered tissue. Some of the suppliers use patented processing methods to further refine the tissue through lyophilization and washing/sterilization methods. In compliance with FDA regulations and American Association of Tissue Banks (AATB) Standards, allograft donors should be screened through a review of their medical and social history to identify medical conditions and risk factors associated with infectious diseases, such as Zika virus, that make them ineligible for donation. Allograft donors are thus screened similarly to blood donors. Additionally, microbiological cultures are performed. Currently, disease transmission from transplanted tissue is nearly nonexistent because of advances in donor screening, donor testing, and tissue culture and treatment methods.

Tissue Types: Tissues can be allogeneic or autologous. Most transplanted tissue is allogeneic and does not require HLA or ABO compatibility, because of a lack of abundant residual cellular material in processed grafts. Tissues include bones, cartilage, tendons, meniscus, skin, fascia lata, corneas, heart valves, pericardium, dura mater, and blood vessels. Allogeneic bone and corneas are the most frequently transplanted tissues in the United States. Bone allografts are most commonly used for orthopedic surgeries to decrease infection risk, decrease surgical time, provide structural support, and promote healing. Corneal allografts are used to correct opacities. Tendons, cartilage, and ligaments are used in orthopedic and reconstructive surgery. Skin allografts are used to treat deep burn wounds. Amniotic/chorionic membranes produce an acellular matrix that serves as a scaffold for cellular incorporation and revascularization in soft tissue reconstructive surgery and wound care. The pericardium and dura mater are used for cerebrospinal leaks and dural defects. The pericardium and fascia lata are used in soft tissue reconstruction. Veins and arteries are used in coronary artery bypass grafting, tissue revascularization, dialysis shunt, and aneurysm repair.

Autologous tissues include endocrine (e.g., parathyroid gland, pancreatic tissue) and connective tissue, skin, and bone (e.g., skull flaps and iliac crests) that are reimplanted. Storage conditions are dictated by the tissue type. Skull flaps recovered from craniotomies in the setting of traumatic brain injury are the most common autologous tissue stored in hospital tissue services. Autologous tissues should not be collected from patients with systemic infections, or if the tissue is in close proximity to an infected site.

Transfusion Medicine and Hemostasis. https://doi.org/10.1016/B978-0-12-813726-0.00090-8

Regulatory Oversight: Transmission of HIV, hepatitis C virus, *Clostridium* spp., group A streptococcus, *Candida albicans,* and prions have been reported after tissue transplantation. These infections provided an impetus for regulatory agencies to provide oversight and accreditation through regulations and guidelines. Depending on the extent of processing and its intended use and effects, human tissue allografts can be regulated as human tissue, biological products, drugs, or medical devices. Specific agencies responsible for administration and oversight of tissue banks and hospital tissue services include AATB, the Joint Commission (TJC), AABB, College of American Pathologists (CAP), Association of periOperative Registered Nurses (AORN), Eye Bank Association of America (EBAA) and the FDA, namely the Code of Federal Regulations (21 CFR 1270 and 1271). Three FDA rules related to operation and handling of tissues includes registration of tissue establishments, donor eligibility, and compliance with current good tissue practices (cGTP) requirements. In 2005, TJC released tissue storage and issuance standards, specifying requirement of written procedures, traceability methods, and investigative plans that examine adverse events. The FDA also generated written regulations to guide tissue suppliers and tissue processors. In 2006, the AABB issued "Guidelines for Managing Tissue Allografts in Hospitals."

Decentralized Tissue Services: Most US hospitals use a decentralized model, where the transplanting physician and/or service is ultimately responsible for the proper handling of tissues from receipt to implantation. Over time, this system has been efficient for surgeons because it eliminates administrative bureaucracy and allows for tailored ordering of tissues. Most often in this system, the supplier delivers the tissue and takes responsibility for its proper handling and storage, including transport to the operating room. Regulatory agencies such as TJC have recognized the nonstandard practices of handling tissues in the decentralized model. TJC does not prohibit decentralization, but requires that a single person keep track of each individual service, i.e., ophthalmologic corneas, orthopedic tendons, cartilage, and ligaments.

Centralized Tissue Services: The TJC's standards released in 2005 emphasized centralization of tissue management in transfusion services.

Centralized tissue services can provide the following:

- Guidance about necessary equipment
- Recommendations for and supervision of staffing and personnel resources
- Direction about physical space and location of tissue services vis-à-vis surgical suites
- Evaluation of the specific tissue needs of physicians in varying specialties
- Leadership in specialized training of medical technologists
- Supportive assistance to laboratory infrastructure and balance with transfusion service

Further Reading

Basha, M. A. (2015). Guidline implementation: Autologus tissue management. *AORN J*, *102*, 271–280.

Eastlund, D. T., Eisenbrey, A. B., & for the Tissue Committee (2006). *Guidlines for managing tissue allografts in hospitals*. Bethesda, MD: AABB.

Hillyer, C. D., & Josephson, C. D. (2007). Tissue oversight in hospitals: The role of the transfusion services. *Transfusion*, *47*, 185–187.

Hinsenkamp, M., Muylle, L., Eastlund, T., Fehily, G., Noël, L., & Strong, D. M. (2012). Adverse reactions and events related to musculoskeletal allografts: Reviewed by the World Health Organisation Project NOTIFY International Orthopaedics (SICOT), *36*, 633–641.

Schlueter, A. J., Josephson, C. D., & Brubaker, S. A. (2017). Human tissue allografts and the hospital transfusion service. In M. K. Fung, A. F. Eder, S. L. Spitalnik, & C. M. Westhoff (Eds.), *Technical manual* (19th ed.) (pp. 695–708). Bethesda, MD: AABB Publisher.

Strong, D. M., & Shinozaki, N. (2010). Coding and traceability for cells, tissues and organs for transplantation. *Cell Tissue Bank*, *11*, 305–323.

WHO. Aide Memoire on key safety requirements for essential minimally processed human cells and tissues for transplantation. Available from: http://www.who.int/transplantation/AM-SafetyEssential%20HCTT.pdf.

WHO guiding principles on human cell, tissue, organ transplantation. Available from: http://www.who.int/transplantation/Guiding_PrinciplesTransplantation_WHA63.22en.pdf.

PART II

Hemostasis

CHAPTER 91

Overview of the Coagulation System

Morayma Reyes Gil, MD, PhD

Hemostasis means stopping blood loss or bleeding. Hemostasis is an orchestrated, balanced, and tightly regulated process. Hemostasis can be subdivided into three sequential processes: primary hemostasis, secondary hemostasis and tertiary hemostasis. In primary hemostasis the interaction of the injured endothelium with von Willebrand factor (VWF) and platelets is crucial for the formation of a platelet plug at the injury site. In secondary hemostasis, the coagulation factors are activated on the surface of injured endothelium and activated platelets which ultimately form a fibrin mesh that stabilizes the platelet plug to allow wound healing. And last, in tertiary hemostasis, fibrinolysis is activated to dissolve the platelet plug and return the normal architecture of the endothelium, smooth endothelial lining and normal lumen size.

Primary Hemostasis: During injury or damage to a blood vessel, the first response to prevent blood loss is vasoconstriction, leading to activation of endothelial cells (ECs). Activated EC secrete VWF, which major role is to recruit platelets to the wounded endothelium. VWF first binds to exposed collagen in the injured endothelium and recruits circulating platelets by recognizing glycoprotein Ib-V-IX (GPIb) on platelets. The binding of VWF to platelets is the first step in the platelet adhesion process. The role of VWF in hemostasis is discussed in details in Chapter 109.

Platelets are essential for normal hemostasis, and their functions can be summarized with the triple A mnemonic: adhesion, activation, and aggregation (Fig. 91.1).

Platelet Adhesion: VWF plays a major role in platelet adhesion by recruiting platelets to the injured endothelium and subendothelial matrix, which is abundant in collagens and other extracellular matrix proteins. Under high shear conditions, the A3 domain of VWF first bind to subendothelial collagen. This process induces uncoiling of VWF and exposes the A1 domain of VWF, which in turns bind to platelet-GPIb-V-IX receptor (GP1b). Binding of the platelet glycoprotein GPIb to collagen-bound VWF anchors the platelets at the site of injury. In turn, platelets have other glycoproteins that directly bind collagens, including glycoprotein Ia/IIa and glycoprotein VI. Signaling via these glycoproteins, adhesion proteins and soluble ligands induce activation from the intracellular compartment to the extracellular compartment, a process called "inside-out" activation. A very important protein that undergoes "inside-out" activation is glycoprotein IIb/IIIa, wherein phosphorylation of its cytoplasmic domain induces changes in the extracellular domain resulting in increased affinity for VWF and fibrinogen. These changes induced by adhesion and signaling initiate the activation of platelets.

Transfusion Medicine and Hemostasis. https://doi.org/10.1016/B978-0-12-813726-0.00091-X

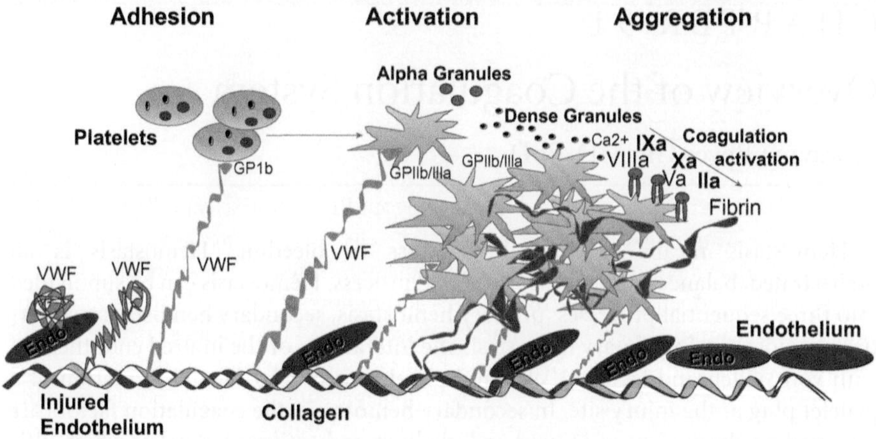

FIGURE 91.1 Overview of platelet functions.

Platelet Activation: Platelet activation results from a series of events that are orchestrated to happen in sequence and sometimes in synchrony leading to synergism. One of the first steps during activation is the shape change. Platelets are discoid in shape when resting, but during activation their cytoplasm expands and flattens and the cytoskeleton rearranges to develop fingerlike extensions called lamellipodia and filopodia. The adhesion receptors are redistributed in the lipid rafts and concentrate to the filopodia ends (sticky ends). These shape change is essential to make platelets more adherent to the injured endothelium and to other activated platelets thus increasing the surface area of platelets protecting the injured endothelium and physically preventing further blood loss.

Another important step during platelet activation is the redistribution of phospholipids to the outer membrane. Platelets use flippases, floppases, and scramblases to move phospholipids to the outer membrane. This process is very important for the recruitment and activation of coagulation factors on the surface of activated platelets, which is crucial for the initiation of secondary hemostasis.

Platelet Secretion: Platelets contain three types of granules, alpha granules and dense granules, and lysosomes (Fig. 91.2). The largest and most abundant type of granule is the alpha granule, which contains large proteins important for adhesion, angiogenesis, and wound healing. For example, VWF, fibrinogen, and vitronectin are matrix proteins contained within the alpha granule, which contribute to platelet adhesion, clot formation, expansion, and stabilization. Platelet alpha granules also contain growth factors that induce and stabilize new vessels. Alpha granules contain not only factors that promote new blood vessel growth (angiogenic factors) but also anti-angiogenic factors that induce vessel stability and maturation. Among the most important angiogenic factors in alpha granules are vascular endothelial growth factor, epidermal growth factor, and platelet-derived growth factor, whereas angiogenesis inhibitors include angiostatin, thrombospondin, and endostatin. In addition, platelet alpha granules contain immune modulators, such as platelet factor 4 (PF4), CCL5 (RANTES), and interleukin-8 (IL-8).

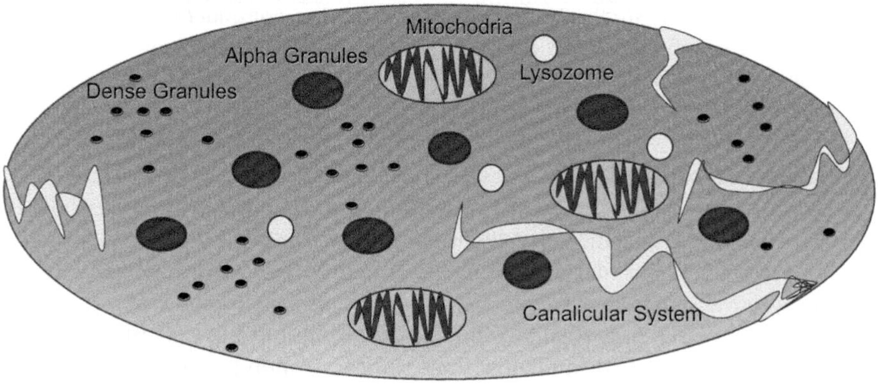

FIGURE 91.2 Overview of platelet structure.

Dense granules are smaller granules that contain ATP, ADP, serotonin, histamine, calcium, inorganic diphosphate, inorganic phosphate, and calcium. Mitochondria are another important organelle for platelets, as they produce ATP and may also participate in the regulation of the platelet activation response.

Platelets contain a network of interconnected channels, the open canalicular system, that extends from the inside of the platelet to the outside environment and may facilitate the rapid release of the constituents of platelet granules and also the uptake of circulating proteins that are stored in the alpha granules. For instance, coagulation factor V and fibrinogen are not made by platelets but are absorbed from the blood and stored in the alpha granules, whereas, VWF, is made de novo by platelets. The secretion of alpha and dense granules is not necessarily simultaneous and could be selective.

Microparticle Secretion: During activation, platelets also actively form and secrete microparticles, which can be differentiated from other intracellular granules and vesicles, as platelet derived microparticles contain a membrane rich in phospholipid, platelet surface glycoproteins and may play a role during platelet aggregation and thrombus formation (Chapter 162).

Platelet Aggregation: Glycoprotein IIb/IIIa (GPIIb/IIIa, also known as integrin $\alpha_{IIb}\beta_3$) is the major receptor on the platelet surface, with an average of 80,000 GPIIb/IIIa receptors per platelet. In resting platelets, GPIIb/IIIa exists in an inactive conformation, unable to bind to its primary ligands, fibrinogen and VWF. During activation, platelets undergo a process called "inside-out" activation that leads to a change in the conformation of the extracellular domain of GPIIb/IIIa to an active form with much higher affinity for its ligands. In addition, the surface density of active GPIIb/IIIa increases during platelet activation. Functional GPIIb/IIIa is essential for platelet aggregation, and it is considered the final step in platelet activation leading to aggregation, as binding of platelets to fibrinogen and/or VWF leads to platelet aggregation irrespectively of the activator or agonist. Not surprisingly, mutations that impair GPIIb/IIIa function manifest as a severe platelet disorder named Glanzmann thrombocytopenia (Chapter 97). Once a platelet clot has been formed, activation of the coagulation cascade on the surface of platelets is necessary to form a fibrin mesh to stabilize the clot under high shear stress. Extracellular calcium secreted by platelets and exposure of phospholipids (phosphatidylserine) on the surface of platelets, and the endothelial wall is essential for the activation and recruitment of coagulation factors.

Secondary Hemostasis: The classic coagulation cascade first proposed by Macfarlane more than 50 years ago was a breakthrough, as it led to the generation of in vitro models of coagulation that today we know as the routine coagulation tests, prothrombin time (PT) and partial thromboplastin time (PTT). It described a cascade of serine proteases and other factors interacting in a stepwise fashion defined as the extrinsic pathway (TF-factor VII [FVII]) and the intrinsic pathway (surface-contact factors). These in vitro tests imply that these two coagulation pathways are independent, hierarchical, and somewhat redundant, as they both merge into the common pathway. However, there are limitations and discrepancies of these coagulation tests that could not explain certain clinical presentations, such as why hemophilia patients bleed, when a normal PT implies hemophiliacs have the necessary amount of extrinsic factors and common pathways factors to form a normal clot and prevent bleeding. On the other hand, patients with contact factor deficiencies have very prolonged PTT but do not bleed. In addition, patients with vascular disorders bleed, but they exhibit normal coagulation tests (PT and PTT) and normal platelet count and function. These discrepancies highlight a major flaw of these routine coagulation tests: they lack the cellular contribution/regulation provided by the vessel wall and endothelial cells.

In an effort to better represent in vivo coagulation the cellular coagulation model has been proposed. In this model, there is only one in vivo coagulation cascade, and it is initiated and regulated by vascular cells. In vivo coagulation is best described by "stages" of plasma protein interaction with vascular cells rather than "pathways" (Fig. 91.3).

Tissue factor (TF) is a transmembrane protein found on the surface of many extravascular cells, including vascular smooth muscle cells and adventitial cells, which form

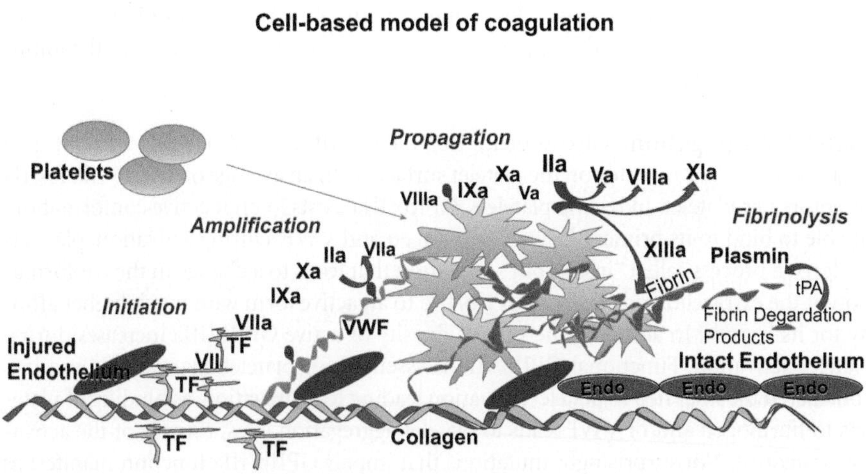

FIGURE 91.3 Cell-based model of coagulation.

integral part of the blood vessel wall. When the blood vessel is damaged, platelets and coagulation factors leak into the subendothelial compartment of the vessel wall where membrane-bound TF can directly activate FVII and initiate the coagulation cascade on the surface of activated platelets and endothelial cells with exposed phospholipids, especially phosphatidylserine, where the coagulation factors bind and undergo activation.

In the initiation stage, TF on the surface of extravascular cells binds activated FVII (FVIIa), which circulates in small amounts (1%–2% of total FVII) in the plasma. The enzymatic activity of the TF/FVIIa complex cleaves small amounts of both FIX and FX, which then activates the common pathway. aFIX can activate FV, which binds aFX to form the prothrombin complex. The initiation stage generates small amounts of thrombin, which is not sufficient to sustain a large clot but necessary to induce amplification of the clot.

The presence of VWF bound to subendothelial collagen and activated platelets is important for the amplification stage. The small amount of thrombin generated during the initiation stage can dissociate FVIII from VWF, thus activating FVIII. Activated FIX derived from the initiation stage can now bind aFVIII on the surface of activated platelets.

In the propagation stage, a large amount of thrombin is formed through the actions of the "tenase" (FIXa and FVIIIa) and "prothrombinase" (FXa and FVa) complexes. The tenase complex converts FX to FXa, and the prothrombinase complex converts prothrombin to thrombin. Thrombin is a master positive regulator of the coagulation cascade. Thrombin continues activating FV and FVIII and also activates FXI. aFXI then maintains the activation of FIX fueling the tenase complex. FXIa is important for maintenance of thrombin generation during the propagation phase. The balance and availability of the other factors necessary to fuel the tenase complex (aFVII and/or aFVIII bound to aFIX) may explain the variable bleeding phenotype in patients with FXI deficiency. Once large amount of thrombin is generated then thrombin activates FXIII, which is necessary for cross-linking of fibrin and stabilization of the platelet clot. Small amounts of FXIII are sufficient to stabilize the clot as only severe cases of FXIII deficiency (FXIII $< 1\%$) present with delayed bleeding due to lack of cross-linking of the fibrin clot. Thrombin directly contributes to enlargement of the platelet plug by directly activating platelets, via binding to PAR receptors on platelets. In addition, thrombin inhibits fibrinolysis by activating thrombin-activatable fibrinolytic inhibitor (TAFI).

Role of Endothelial Cells in Regulating Coagulation: Endothelial cells are key regulators of coagulation and counterbalance thrombin by several mechanisms. Heparin and dermatan sulfates secreted by the injured endothelium accelerate the action of the anticoagulants antithrombin and heparin cofactor II. Endothelial cells sense how much thrombin has been formed locally by the thrombomodulin receptor, which binds thrombin and converts protein C to activated protein C facilitated by the endothelial protein C receptor (EPCR). Endothelial cells also play a major role during the activation of the fibrinolytic system by releasing tissue plasminogen activator (tPA). Endothelial cells express both isoforms of tissue factor pathway inhibitor (TFPI), which downregulates thrombin formation by directly inhibiting TF/FVIIa and aFX. Endothelial cells also activate the contact factor pathway and modulate inflammation during clot remodeling and wound healing.

Tertiary Hemostasis: Fibrinolysis: Once clot formation is completed the process of clot dissolution (fibrinolysis), and wound remodeling is initiated. Matrix metalloproteinases (MMPs) are activated largely by the contact system of the coagulation cascade (kallikrein), as well as by the fibrinolytic system (plasmin), and specific enzymes (trypsin). MMPs, such as collagenase, degrade collagen, and other matrix proteins. Endothelial cells express the receptor for prekallikrein, which activate FXII facilitated by the cofactor high-molecular-weight kininogen (HMWK). aFXII and prekallikrein can cleave plasminogen into plasmin. As mentioned above, tPA is the major activator of plasmin. Plasmin cleaves the fibrin polymers and generates fibrin degradation products, which interfere with platelet aggregation and fibrin polymerization, destabilizing the platelet clot, which dissolves under high shear stress.

The cell-based model of coagulation provides a better explanation for the various pattern of bleeding observed in patients with bleeding disorders: patients with platelet disorders and von Willebrand disease present with mucocutaneous bleeding, petechial, and bruising; patients with coagulation factor deficiencies such as hemophiliacs or factor VII deficiency present with joint bleeds, muscle bleeds, and if severe intracerebral hemorrhage or intracranial hemorrhage in neonates; and patients with increased fibrinolysis (e.g., FXIII or PAI-1 deficiency) present with delayed bleeding after trauma or surgery. In addition, the cell-based model of coagulation explains why patients with contact factor deficiencies despite a prolonged PTT do not bleed but may exhibit impaired fibrinolysis. Finally, the cell-based model of coagulation recognizes the major role of endothelial cells in regulating hemostasis and explains how inflammatory, infectious, or prothrombotic disorders can alter the hemostasis balance and lead to excessive thrombosis and/or bleeding (DIC).

Further Reading

Brass, L. F., Wannemacher, K. M., Ma, P., et al. (2011). Regulating thrombus growth and stability to achieve an optimal response to injury. *J Thromb Haemost*, *9*(Suppl. 1), 66–75.

Curnow, J., Pasalic, L., & Favaloro, E. J. (2016). Why do patients bleed? *Surg J (NY)*, *2*(1), e29–e43.

Furie, B., & Furie, B. C. (2007). In vivo thrombus formation. *J Thromb Haemost*, *5*(Suppl. 1), 12–17.

Heemskerk, J. W., Bevers, E. M., & Lindhout, T. (2002). Platelet activation and blood coagulation. *Thromb Haemost*, *88*, 186–193.

Hoffman, M., & Monroe, D. M. (2001). A cell-based model of hemostasis. *J Thromb Haemost*, *85*, 958–965.

Jackson, S. P. (2007). The growing complexity of platelet aggregation. *Blood*, *109*, 5087–5095.

Whiteheart, S. W. (2011). Platelet granules: Surprise packages. *Blood*, *118*, 1190–1191.

CHAPTER 92

Approach to the Pediatric Patient With a Bleeding Disorder

Ahmad Al-Huniti, MD, Anjali Sharathkumar, MD
and Deepa Manwani, MD

Basic Considerations for Evaluation of a Pediatric Patient With a Bleeding Disorder: Children with bleeding disorders generally present with mild bleeding symptoms including easy bruising, epistaxis, gum bleeding, abnormal clotting tests from presurgical evaluation, or known family history of bleeding disorder. Rarely, they can present with life-threatening bleeding in vital organs or serious postsurgical bleeding. Clinical history is the key to further characterize the bleeding phenotype. The history should focus on answering following three questions before performing extensive laboratory workup:

Question #1: Am I dealing with a child with bleeding disorder?
Question #2: Is it a congenital or acquired bleeding disorder?
Question #3: Is it a disorder of primary hemostasis or secondary hemostasis?

The subsequent sections will highlight how to address these questions.

Am I Dealing With Bleeding Disorder?: This question helps characterize clinically significant bleeding. History should include the most common site, type and duration of bleeding, medication and herbal history, and bleeding on exposure to hemostatic challenge (procedures such as tonsillectomy, wisdom teeth extraction, circumcision, menstrual history, postpartum bleeding if appropriate). Family history of bleeding should be obtained along the same lines. Objective characterization of the bleeding phenotype can be done using standardized bleeding questionnaires to quantitate bleeding symptoms. Pediatric Bleeding Questionnaire (PBQ) provides a comprehensive assessment of bleeding symptoms. PBQ has high negative predictive value (99%) and low positive predictive value (14%) for von Willebrand disease (VWD). Thus, PBQ is a valuable tool to rule out VWD. Physical exam should include the pattern of bruising and assessment for joint hypermobility. The Beighton scoring system is easy method to assess joint hypermobility. It is worth noting that bleeding disorders can be silent in children who have not yet faced a hemostatic challenge and family history, specifically pattern of inheritance (X-linked recessive vs. autosomal dominant disorders), plays a key role in these cases. The common bleeding symptoms are elaborated below:

- **Bruising:** The first step in evaluation of bruises in children is to rule out nonaccidental trauma. Inflicted trauma is most likely to manifest over the head, chest, back, and the long bones and may retain the outlines of the instrument. Bruises associated with primary defects of hemostasis are usually located over areas of typical childhood trauma, such as the bony protuberances of extremities

Transfusion Medicine and Hemostasis. https://doi.org/10.1016/B978-0-12-813726-0.00092-1

or spinous processes. Bruises that are not limited to distal extremities, are larger than a quarter coin, and are associated with hematomas and bruising out of proportion to the mechanism of injury are likely suggestive of underlying bleeding disorder. Intramuscular hematomas usually present with large bruises along with swelling and pain with use of the affected muscle.

- **Epistaxis:** Epistaxis is usually due to local factors such as trauma, allergic rhinitis, or dry nasal mucosa but is also a common presenting symptom of an underlying bleeding disorder. Bleeding disorders are diagnosed in 25%–33% of patients referred to pediatric hematology for recurrent epistaxis. The clinical significance of epistaxis is enhanced by requiring emergency department visit, occurring in both nostrils, and occurring in association with other bleeding signs and a family history of similar bleeding. Hereditary hemorrhagic telangiectasia may manifest as mucosal bleeding especially epistaxis.
- **Hemarthrosis:** Joint effusion, warmth, and pain are indicative of acute joint bleed while limited range of motion, synovial thickening, and chronic pain are indicative of chronic bleeds. Younger children can present with only refusal to walk or use of affected joint. There is growing awareness about limitation of range of motion of joints among carriers of hemophilia and other bleeding disorders secondary to recurrent microbleeds.
- **Menorrhagia:** Menorrhagia is defined by either of menses lasting more than 7 days, frequent pad changes (<2 hours), or more than one period per month. Menorrhagia can result in anemia and decrease in quality of life. A pictorial blood flow assessment chart can be used to provide semiobjective assessment of menstrual blood loss. Menorrhagia in setting of underlying bleeding disorder often occurs with first cycle at menarche. The American College of Obstetrics and Gynecology recommends evaluation of VWD in women presenting with menorrhagia. Women with platelet dysfunction disorders and other coagulation disorders can also present with menorrhagia.
- **Surgical bleeding:** Circumcision, tonsillectomy, or wisdom teeth extraction are common surgical procedures performed in pediatric population. Surgical bleeding in bleeding disorder patients is characterized by uncontrolled bleeding during or after the procedure, bleeding that extends beyond the surgical site, unexpected need for blood transfusion, or delayed bleeding after procedure. Posttonsillectomy bleeding is usually delayed until 7–10 days after procedures in patients with underlying bleeding disorder.
- **Bleeding in infancy:** History of bleeding during newborn period should include cephalhematoma, umbilical cord bleeding, bleeding after heel stick, and postcircumcision bleeding.

Is Bleeding Congenital or Acquired?: In general, early presentation during newborn period or infancy points toward more severe bleeding disorder and congenital cause. In this clinical setting, the diagnosis of bleeding disorder is urgent to guide appropriate clinical management.

Family history is critical for differentiating congenital bleeding disorders from acquired disorders. X-linked inheritance pattern suggests diagnosis of hemophilia either factor VIII or factor IX deficiency. It is essential to keep in mind that approximately one-third of patients with hemophilia have a negative family history. Hereditary hemorrhagic telangiectasia and common bleeding disorders such as VWD and mild platelet function abnormalities have autosomal dominant inheritance. Common bleeding disorders can have variable clinical phenotype and can be more symptomatic in women because of the additional hemostatic challenges of menstruation and child birth.

Acquired bleeding disorder are generally due to underlying systemic illness, medication, secondary structural or functional VW deficiency, or coagulation factor inhibitor. As coagulation factors are mainly synthesized in the liver, liver dysfunction is a common cause of acquired bleeding disorder. Generally, it presents with multiple clotting factor deficiencies. Accumulation of urea and acidosis in renal failure can interfere with platelet function. Vitamin K in essential for the synthesis of multiple coagulation factors and deficiency secondary to malabsorption syndromes or antibiotics can lead to bleeding. Mechanical heart valve can lead to cleavage of VW factor by mechanical shear forces and results in acquired VWD. Adsorption of VWF on malignant cells of Wilms tumor causes acquired VWD. Medication such as aspirin, nonsteroidal antiinflammatory drugs, and others can have significant effect on platelet function. Disseminated intravascular coagulation is caused by widespread activation of coagulation cascade, which results in thrombocytopenia and consumptive coagulopathy.

Am I Dealing With Defect in Primary Hemostasis or Secondary Hemostasis?: Primary hemostasis defects are either due to thrombocytopenia, platelet dysfunction disorders or vessel wall defects. Secondary hemostasis defects are caused by inherited or acquired deficiency of coagulation proteins or rarely due to presence of inhibitor of coagulation protein (Refer to Chapter 91). Table 92.1 provides the clinical difference between primary and secondary hemostasis defects. VWD is the most common disorders of primary hemostasis affecting up to 1 in 1000 population. Hemophilia (deficiency of factor VIII or IX) is the most common coagulation disorder affecting up to 1 in 5000 boys. Deficiencies of other coagulation proteins, i.e., fibrinogen, factor II, V, X, VII, XI, XIII, etc. are autosomal recessive, rare and present with variable clinical symptoms.

Laboratory Evaluation: Laboratory evaluation should be guided by clinical suspicion for a primary or secondary hemostasis defect. Results of initial screening tests may direct more specific investigations. Fig. 92.1 summarizes an algorithm for the evaluation of a patient with suspected bleeding disorder.

Clinical Assays for Evaluating Primary Hemostasis: Complete blood count and peripheral blood smear are the first step of evaluation. It helps rule out conditions such as hematological malignancy, bone marrow failure syndrome, MYH9 disorders, and systemic disorders such as liver pathology.

Platelet function assay (PFA-100): Platelet function assay is an equivalent of in vitro bleeding time and is now widely available. It attempts to simulate in vivo platelet adhesion and aggregation. The use of PFA as a screening tool remains controversial and is generally not recommended anymore. Normal PFA does not rule out mild–moderate

TABLE 92.1 Is It a Disorder of Primary Hemostasis Versus Secondary Hemostasis

Clinical Presentation	Disorders of Primary Hemostasis (Vessel Wall, Platelets, VWD)	Disorders of Secondary Hemostasis (Clotting Protein Deficiency)
Common site of bleeding	Mucocutaneous bleeding	Deep tissue bleeds: muscles, soft tissue, joints
Bruising	Small (<dime size), superficial	Large, deep, palpable
Epistaxis	Yes	Rare
Gum bleeding	Yes	No
Subcutaneous hematoma	Rare	Common
Bleeding following minor cuts	Yes	No
Postoperative/posttraumatic bleeding	Immediate	Generally delayed
Menorrhagia	Yes	Yes
Abnormal blood tests	Low hemoglobin, low platelets, prolongation of bleeding time/PFA-100; abnormal results of VWD panel	Prolongation of PT, PTT, thrombin time, low factor levels

platelet dysfunction. Thus, in cases of clinical suspicion for primary hemostasis defects more specific tests should be obtained such as platelet aggregation studies and VWD evaluation. VWD evaluation includes VWF: antigen, VWF: ristocetin cofactor assay, factor VIII activity, and VWF multimer analyses.

Clinical Assays for Evaluating Secondary Hemostasis: Commonly used screening assays for disorders of secondary hemostasis include the following: prothrombin time (PT), partial thromboplastin (PTT) time, and thrombin time (TT).

PT evaluates "extrinsic pathway" (TF, FVII) and common pathway (FX, FV, FII, fibrinogen). PTT evaluates contact system (prekallikrein, FXII), intrinsic pathway (FXI, FIX, FVIII), and common pathway. PT and PTT results should be compared with age-specific laboratory reference interval. PT and PTT are prolonged in newborns indicating immature vitamin K pathways. In cases of mild coagulation factor, deficiency PT and PTT may be normal. Therefore, specific factor assays should be obtained based on clinical suspicion. As the PT and PTT terminal event is conversion of fibrinogen into fibrin, these tests do not evaluate the function of FXIII. When initial laboratory evaluation is negative, FXIII assay should be obtained to detect FXIII deficiency. PTT could be prolonged in the absence of bleeding symptoms with deficiencies of contact system.

Elevated PT and PTT could be due to coagulation factor deficiency or presence of coagulation factor inhibitor. Lupus anticoagulant (LA) antibodies arise transiently in setting of infection and are responsible for PTT elevation in the absence of bleeding symptoms. LA has been associated with thrombotic events in childhood especially if persistent. Rarely, LA antibodies can be directed against prothrombin and results in acute bleeding and elevation of PT and PTT. In most cases, LA has benign course and is usually transient.

Mixing study, where patient plasma is mixed 1:1 with normal pooled plasma and PTT or PT is measured, is used to differentiate between coagulation factor deficiency versus

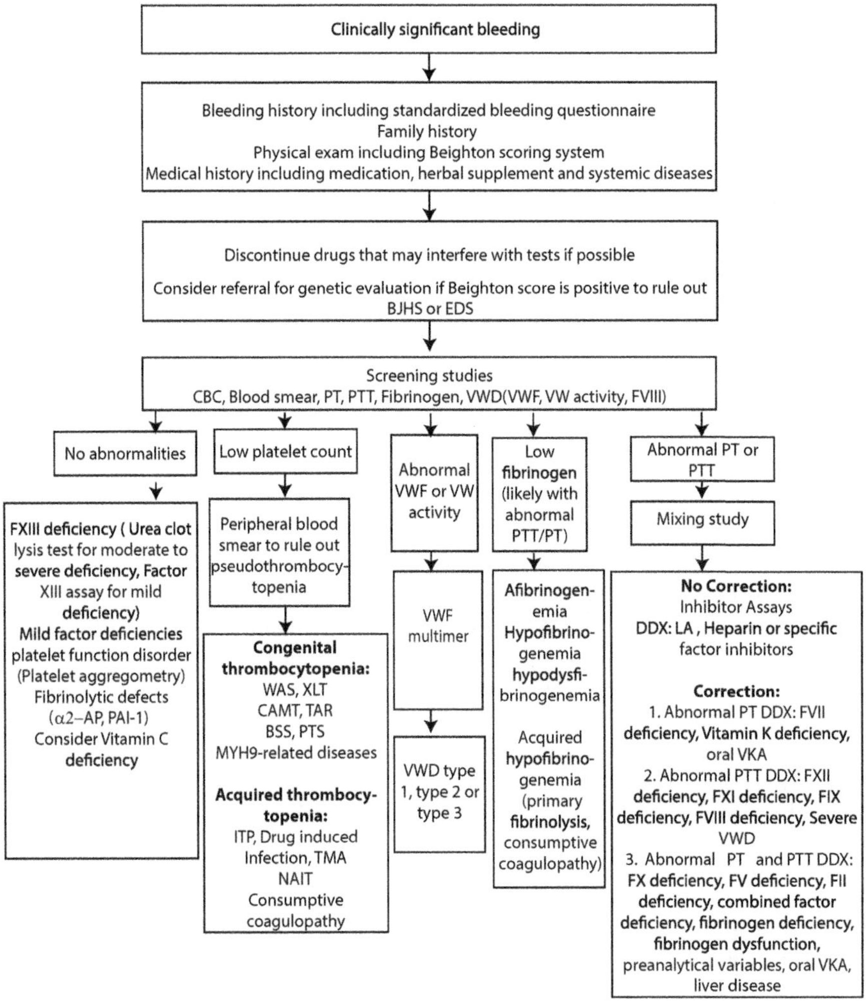

FIGURE 92.1 Laboratory evaluation of bleeding disorder. *α2-AP*, antiplasmin; *BJHS*, benign joint hypermobility syndrome; *BSS*, Bernard-Soulier syndrome; *CAMT*, congenital amegakaryocytic thrombocytopenia; *CBC*, complete blood count; *DDX*, differential diagnosis; *EDS*, Ehlers-Danlos syndrome; *ITP*, idiopathic thrombocytopenic purpura; *LA*, lupus anticoagulants; *NAIT*, neonatal alloimmune thrombocytopenia; *PAI*, plasminogen activator inhibitor; *PT*, prothrombin time; *PTS*, Paris-Trousseau syndrome; *PTT*, partial thromboplastin time; *TAR*, thrombocytopenia with absent radius; *TMA*, thrombotic microangiopathy; *TT*, thrombin time; *VKA*, vitamin K antagonist; *VWD*, von Willebrand disease; *VWF*, von Willebrand factor; *WAS*, Wiskott Aldrich syndrome; *XLT*, X linked thrombocytopenia.

coagulation factor inhibitor or LA. PTT or PT completely corrects in cases of factor deficiency. PTT or PT remains prolonged in cases of LA or presence of factor inhibitor.

The TT (or thrombin clotting time) is not necessarily a part of the basic screening profile. It tests conversion of fibrinogen to fibrin. Prolongation of the TT is consistent with the presence of heparinlike anticoagulants, hypofibrinogenemia, afibrinogenemia, dysfibrinogenemia, fibrin degradation products, and inhibitors of thrombin.

Currently available screening tests do not diagnose deficiency of fibrinolysis inhibitors such as α2-antiplasmin or plasminogen activator inhibitor (PAI) 1 and levels should be determined if suspected. These deficiencies can result in bleeding due to hyperactive fibrinolysis and require global coagulation assays or a specialized laboratory to measure the levels of these proteins.

Global coagulation assays: These tests include both the cellular and enzymatic components of coagulation. Currently, thromboelastography (TEG) and thrombin generation assay (TGA) are being explored for clinical practice. TEG can serve as point of care test, which makes it appealing to be used in acute setting such as surgery, trauma, and intensive care unit. TEG use in cardiac surgery has led to decrease in transfusion requirement, but it has several pitfalls that limit its use for evaluation of bleeding disorder. TGA evaluates thrombin generation and decay and is able to investigate bleeding diathesis and thrombophilia. The time required to run the test and lack of standardized protocol are major drawbacks for TGA, and it is currently a research tool.

Finally, all coagulation assays are susceptible to preanalytical variables. Heparin contamination and improper collection of blood sample are the common causes of abnormal coagulation assays. Sample should be collected properly, filled to at least 90% of expected fill volume, and tested within 2 hours of collection to prevent unreliable results due to activated specimen or disproportionately high citrate level. Patients with polycythemia can have falsely abnormal coagulation results due to decrease plasma volume which, in turn, results in disproportionately high citrate level.

In summary, approach to diagnosis of bleeding disorder in a child relies heavily on a detailed clinical history and thorough clinical examination. Laboratory testing should be used to confirm clinical suspicion and guide the management. Fig. 92.1 provides a comprehensive algorithm about evaluation of a child with suspected bleeding disorder.

Further Reading

Anderst, J. D., Carpenter, S. L., Abshire, T. C., & Section on Hematology/Oncology; Committee on Child Abuse and Neglect of the American Academy of Pediatrics (2013). Evaluation for bleeding disorders in suspected child abuse. *Pediatrics*, *131*(4), e1314–1322.

Branchford, B., & Di Paola, J. (2015). Approach to the child with a suspected bleeding disorder. In D. G. Nathan, F. A. Oski, & S. H. Orkin (Eds.), *Nathan and Oskis hematology and oncology of infancy and childhood* (pp. 999–1009).

Lancé, M. D. (2015). A general review of major global coagulation assays: Thrombelastography, thrombin generation test and clot waveform analysis. *Thromb J*, *13*(1), 1.

Nowak-Göttl, U., Limperger, V., Bauer, A., Kowalski, D., & Kenet, G. (2015). Bleeding issues in neonates and infants – update 2015. *Thromb Res*, *135*, S41–S43.

O'Brien, S. H. (2012). An update on pediatric bleeding disorders: Bleeding scores, benign joint hypermobility, and platelet function testing in the evaluation of the child with bleeding symptoms. *Am J Hematol*, *87*(Suppl. 1), S40–S44.

Revel-Vilk, S. (2011). Clinical and laboratory assessment of the bleeding pediatric patient. *Semin Thromb Hemost*, *37*(7), 756–762.

Sharathkumar, A. A., & Pipe, S. W. (April 2008). *Pediatr Rev*, *29*(4), 121–130.

CHAPTER 93
Congenital Thrombocytopenia

Michele P. Lambert, MD, MSTR

Introduction: Inherited thrombocytopenia was once considered a rare disease, but more recent studies using next-generation sequencing (NGS) and genome-wide association studies suggest that the prevalence of inherited or familial thrombocytopenia may be much higher than originally appreciated and that some individuals who were previously thought to have immune thrombocytopenia (ITP) may, in fact, have a defect in platelet or megakaryocyte development. Further characterization and investigation of individuals and families will most certainly lead to expansion of the number of disorders and modify phenotypes. This will allow us to further improve management and avoid incorrect diagnosis and treatment of individuals who present for the first time to a clinician with a low platelet count.

Diagnosis: Patients with congenital thrombocytopenia may present with mild-to-moderate mucocutaneous bleeding (epistaxis, petechiae and bruising, gastrointestinal [GI] or genitourinary tract bleeding) (Table 93.1). Severe bleeding is rarer but is possible depending on the degree of thrombocytopenia and whether or not there is associated platelet dysfunction. Some disorders may have associated physical or laboratory characteristics that suggest the diagnosis and so a thorough history, physical examination, and evaluation of the peripheral smear may guide the remainder of the diagnostic evaluation and limit unnecessary testing. A few well-characterized inherited thrombocytopenias have clinically available genetic testing that allows for complete evaluation of the entire family of an affected individual even in the absence of thrombocytopenia in other family members.

Differential Diagnosis: Initial evaluation must include a careful personal and family history with particular attention to ethnicity and consanguinity, bleeding history, including onset of bleeding and prior hemostatic challenges (surgery or trauma), and associated medical complications seen in the inherited disorders, including hearing loss, renal disease, cataracts, pigment changes, eczema, skeletal abnormalities, cardiac anomalies, frequent infections, or hematologic malignancy. As the most likely causes of thrombocytopenia are acquired (ITP or drug-induced thrombocytopenia), the onset of bleeding is as important as a careful evaluation of any current or recent medication exposures. Careful evaluation of the peripheral smear and physical examination to evaluate for hepato/splenomegaly or lymphadenopathy may suggest myeloproliferative, myelophthisic, or aplastic bone marrow disease that could account for the low platelet count. Inherited disorders should be suspected in any patient with longstanding, refractory thrombocytopenia; particularly if associated with any of the medial complications discussed above. In patients with mild thrombocytopenia, consideration

Transfusion Medicine and Hemostasis. https://doi.org/10.1016/B978-0-12-813726-0.00093-3

TABLE 93.1 Inherited Thrombocytopenias Classified by Platelet Size

Inherited Condition	Gene (Location)	Inheritance	Key Features
Microthrombocytic			
Wiskott–Aldrich syndrome/ X-linked thrombocytopenia	*WAS* (Xp11)	X-linked	Thrombocytopenia, eczema, severe immunodeficiency, small platelets
Congenital autosomal recessive small-platelet thrombocytopenia	*FYB* (5p13.1)	AR	Thrombocytopenia with small platelets
Normothrombocytic			
Congenital amegakaryocytic thrombocytopenia	*MPL* (1p34)	AR	Hypomegakaryocytic thrombocytopenia with eventual development of bone marrow failure
Thrombocytopenia with absent radii	*RBM8A* (1q21.1)	AR	Thrombocytopenia that improves with age, limb anomalies (but relatively normal thumbs)
Amegakaryocytic thrombocytopenia with radio/ulnar synostosis	*HOXA11* (7p15) *MECOM* (3q26.2)	AD	Severe thrombocytopenia that improves with age, skeletal abnormalities (radio/ulnar synostosis, clinodactyly, syndactyly, hip dysplasia), hearing loss
Familial platelet disorder with predisposition to acute myeloid leukemia (FPD/ AML)	*RUNX1* (21q22)	AD	Thrombocytopenia, myelodysplasia or even AML, platelet dysfunction
Paris-Trousseau/Jacobsen syndrome	*FLI1* (11q23.3–24.2)	AR	Thrombocytopenia with large granules; may have associated cardiac anomalies, mental retardation, abnormal facies
Familial thrombocytopenia 2	*ANKRD26* (10p12.1)	AD	Mild-to-moderate thrombocytopenia with mild bleeding symptoms
Macrothrombocytopenic			
Bernard–Soulier syndrome (BSS)	GPIbα (17), GPIbβ (22), GPIX (3)	AR	Platelet dysfunction with large platelets
Velocardiofacial syndrome	22q11	AD	Cardiac anomalies, cleft palate, hypocalcemia, thymic aplasia, and typical facies. BSS-like thrombocytopenia ±autoimmune
Benign Mediterranean macrothrombocytopenia	GPIb/IX/V; ITGB3	AD	May represent as heterozygous BSS mutations
Platelet type von Willebrand disease	GPIbα (17)	AD	Decreased high-molecular-weight VWF multimers with thrombocytopenia because of increased platelet affinity for VWF

TABLE 93.1 Inherited Thrombocytopenias Classified by Platelet Size—cont'd

Inherited Condition	Gene (Location)	Inheritance	Key Features
MYH9-related disease	*MHY9* (22q11.2)	AD	Large platelets, leukocyte inclusions may have sensorineural hearing loss, cataracts, glomerulonephritis, or renal failure
Gray platelet syndrome	*NBEAL2* (3p21.31), *GFI1B* (9q34.13)	AD, AR	Large, pale platelets with absence of α granules
GATA-1 mutation of X-linked thrombocytopenia with thalassemia (GATA-1)	*GATA1* (Xp11.23)	X-linked	Thrombocytopenia with variable anemia
ACTN1-related thrombocytopenia	*ACTN1* (14q24.1)	AD	Macrothrombocytopenia
TUBB1-related thrombocytopenia	*TUBB1* (20q13.1)	AD	Macrothrombocytopenia
FLNA-related thrombocytopenia	*FLNA* (Xq28)	X-linked	
TRPM7	*TRPM7* (15q21.2)	AD	Atrial fibrillation and macrothrombocytopenia in one family

should also be given to the diagnosis of congenital thrombocytopenia, particularly if there are no prior blood counts with normal platelet counts. The advent of NGS has led the identification of several forms of familial thrombocytopenia with predisposition to malignancy and therefore establishing a molecular diagnosis, whenever possible, may have important implications for other family members and may help inform risk prediction. Targeted therapies for some of the inherited thrombocytopenias, particularly those with associated other medical complications, are being developed, as are gene therapies. Some of the more severe disorders, such as Wiskott–Aldrich syndrome (WAS) or Bernard–Soulier Syndrome, may benefit from bone marrow transplant in the appropriate setting.

Management: Many patients with congenital thrombocytopenia have mild-to-moderately low platelet counts and therefore have no or moderate bleeding. Treatment may be necessary in the setting of trauma or surgical intervention (including dental procedures) and consists of supportive therapy: physical tamponade, local/topical glue or fibrin products, DDAVP, or antifibrinolytics. In disorders with more severe bleeding phenotype or in settings of significant bleeding, platelet transfusion may be necessary to control bleeding. Management of specific disorders is detailed below. For a few disorders with severe thrombocytopenia such as MYH9-related disease (MYH9-RD) and WAS, treatment with the novel thrombopoietin (TPO) receptor agonists has been tried with varying success.

Congenital Thrombocytopenias: There are multiple ways to organize the inherited thrombocytopenias, including by manner of inheritance, by size of platelets, or by pathologic defect. Clinically, size of platelets is a useful differentiation schema as it allows for the easy subcategorization of patients into useful groups that can then be systematically evaluated.

Disorders With Small Platelets (Microthrombocytopenia)

Wiskott–Aldrich Syndrome and X-Linked Thrombocytopenia: WAS and X-linked thrombocytopenia (XLT) are rare congenital thrombocytopenias caused by mutations in the *WAS* gene that encodes WASp, which plays an important role in signal transduction from cell surface receptors to the actin cytoskeleton. These patients usually have significantly small platelets (mean platelet volume [MPV] 3.5–5 fL) and may have associated immune dysfunction, resulting in eczema, frequent infections, or autoimmune disease. Recent reports have suggested that platelet size may be more variable than originally thought, and some patients with documented *WAS* mutations may have somewhat higher MPV. There is an increased risk of lymphoma and solid tumors in adolescents and young adults (particularly those with significant immunodeficiency). Mutations with essentially no protein expression result in the more severe WAS: immunodeficiency, autoimmunity, microthrombocytopenia, and eczema. Mutations with single amino acid substitutions and more normal levels of protein expression tend to result in isolated thrombocytopenia (XLT). Thorough evaluation of patient with small platelets includes sequencing of the *WAS* gene and evaluation of WASp levels and interrogation of the immune system. Splenectomy has been used in the past to improve platelet counts; however, asplenia in the setting of underlying immunodeficiency is a concern and thus should not be done without careful consideration. Management focuses on control of bleeding in the acute setting and monthly IVIG and vigilance regarding infections. Platelet transfusion can be used to control more significant hemorrhage. Thrombopoietin receptor agonists have been used to improve platelet counts in patients with XLT and ITP associated with WAS with varying success. Patients with more severe disease are more often offered with hematopoietic stem cell (HSC) transplant, and outcomes have been generally good.

Congenital Autosomal Recessive Microthrombocytopenia Associated With FYB Mutations: Recently, two families with congenital microthrombocytopenia and significant bleeding symptoms without variants in *WAS* were found to harbor mutations in *FYB*, which encodes adhesion- and degranulation-promoting adaptor protein, a protein involved in platelet activation. The affected individuals have platelet counts between 5–69 × 10^9/L with a low MPV (5.8–7.2 fL) and normal white blood cells with associated mild iron deficiency anemia that responded to iron supplementation. The patients had mucosal bleeding and menorrhagia, petechiae, and intraabdominal bleeding due to ruptured corpus luteum without any associated syndromic features and without eczema or predisposition to infection. Knockout mice demonstrated similar platelet defects to those shown in this family except for the degree of thrombocytopenia, which was more pronounced in humans than in murine models.

Disorders With Normal-Sized Platelets (Normocytic Thrombocytopenia): Most of these disorders are caused by defects in transcription factors important in megakaryocyte differentiation or defects in TPO signaling. Platelets are normal in size but may have decreased function because of abnormal production of granule contents because of transcription factor dysregulation. These disorders are generally associated with predisposition to bone marrow failure or leukemic transformation.

Congenital Amegakaryocytic Thrombocytopenia: These very rare patients often present with severe thrombocytopenia at birth, which is recognized early because of significant petechiae, bruising, and mucosal bleeding. Occasionally, the thrombocytopenia is attributed to maternal ITP or neonatal alloimmune thrombocytopenia, but persistent thrombocytopenia beyond the second month of life, negative workup for these entities and lack of response of platelet counts to appropriate therapies lead to further evaluation. Patients have markedly elevated serum TPO levels (in the thousands of pg/mL range) compared with the modestly elevated TPO levels seen in most other thrombocytopenias. A relatively large percentage of patients may present with intracranial hemorrhage. CAMT results (most often) from mutations in the TPO receptor (c-MPL), and many patients are compound heterozygotes in this autosomal recessive disorder. Because TPO is critical for megakaryopoiesis, patients have profound thrombocytopenia and consequent significant bleeding symptoms. However, TPO is not only required for megakaryopoiesis but also for stem cell renewal and results in a gradual decline in HSCs causing eventual bone marrow failure. Treatment is supportive in the first few months. Bone marrow examination will reveal absent or markedly decreased megakaryocytes and may show a hypocellular marrow. Once the diagnosis is established, the only curative therapy is HSC transplant.

Amegakaryocytic Thrombocytopenia With Radio/Ulnar Synostosis: Amegakaryocytic thrombocytopenia with radio/ulnar synostosis (ATRUS) is an autosomal dominant congenital thrombocytopenia characterized by skeletal abnormalities (radio/ulnar synostosis, hip dysplasia, clinodactyly/syndactyly) and sensorineural hearing loss. Patients present in the neonatal period with severe thrombocytopenia may improve with time, but some patients may develop progressive cytopenias and eventual bone marrow failure. Mutation analysis of two affected kindreds demonstrated a mutation in the *HOXA11* gene. Recent analysis of additional kindreds demonstrated that variants in EVI1 encoded by the MDS1 and EVI1 complex locus *(MECOM)* located on chromosome 3q26.2 may be responsible for this disorder in some families. EVI1 is a zinc-finger transcription factor, interacts with other transcriptions factors such as RUNX1 and GATA1, and has been shown to be important in HSC renewal. EVI1 downregulates *MPL*. The risk of malignancy with *MECOM* mutations is not known, but patients with severe thrombocytopenia or bone marrow (BM) failure often undergo HSCT at a young age. Interestingly, in sporadic acute myeloid leukemia (AML), high expression of *MECOM* occurs in approximately 5%–10% of patients and is associated with poor prognosis.

Thrombocytopenia Absent Radii: Thrombocytopenia absent radii (TAR) syndrome is usually considered an autosomal recessive disorder with incomplete penetrance. Patients present with neonatal thrombocytopenia, which can be severe and result in significant bleeding. This thrombocytopenia is in association with classic upper extremity abnormalities, which almost invariably affect the radius but may range from more mild with isolated radial abnormalities to severe with phocomelia. Most patients have intact hands, and thumbs are always present but may be hypoplastic. There is also a high incidence of cardiac and renal abnormalities, as well as milk protein allergy. The milk protein allergy may result in life-threatening GI hemorrhage requiring platelet transfusion. Abnormalities of the central nervous system, including dysgenesis of the corpus callosum, have also been reported. The thrombocytopenia is generally severe initially, with platelet counts <30,000/μL but gradually improves so that many patients will have low normal platelet counts by about 1 year of age. NGS techniques identified the molecular cause of to be complex. Most patients have a mutation in the *RBM8A* gene, which encodes a protein Y14, the function of which is not clearly understood. The second mutation, in many individuals with TAR syndromes, occurs in an EVI1 regulatory region of the *RBM8A* gene on the other allele. Treatment is supportive as the thrombocytopenia often improves. Early diagnosis of milk protein allergy (if present) to prevent GI bleeding is important. Patients may present with pulmonary hemorrhage associated with undiagnosed milk protein allergy, and recurrent respiratory symptoms with iron deficiency should be evaluated promptly with consideration for pulmonary hemosiderosis. Evaluation for other associated abnormalities and appropriate orthopedic, cardiac, renal, and neurology followup is critical for outcomes.

Familial Platelet Disorder With Predisposition to Acute Myelogenous Leukemia: Familial platelet disorder with predisposition to acute myelogenous leukemia (FPD/AML) is an autosomal dominant disorder characterized by thrombocytopenia and a strong predisposition (20%–60% of patients with high variability within families) to the development of leukemia, most often AML of the M0 subtype, although lymphosarcoma, myelodysplasia, and other tumors have also been reported. Patients generally present early in life with thrombocytopenia and a mild bleeding diathesis, but platelet count may be normal making complete blood count an inadequate screening test for affected family members. The bleeding may be out of proportion to the degree of thrombocytopenia, and some patients have variable aggregation defects to a number of agonists. FPD/AML is caused by mutations in the *RUNX1* gene (*AML1, CBFA2*), which is a transcription factor critical for normal thrombopoiesis. Haploinsufficiency of *RUNX1* results in abnormal expression of downstream platelet targets, which may contribute to the platelet dysfunction and thrombocytopenia. *RUNX1* is also involved in sporadic AML/myelodysplastic syndrome (MDS) with translocations 8;21 and 12;21 such that its role in cancer predisposition is clear if poorly understood. Treatment is supportive with counseling and screening of family members about the risk of malignancy. A few patients with germline deletions of *RUNX1* have also been described, and these patients have congenital thrombocytopenia with or without other features (cardiac abnormalities, mental retardation, seizures) and early presentation with MDS/AML.

Familial Thrombocytopenia 2 or ANKRD26-Related Thrombocytopenia: A recent addition to the list of autosomal dominant thrombocytopenias is thrombocytopenia 2 (THC2). Clinically, these patients present with mild-to-moderate thrombocytopenia and normal-sized platelets associated with mild bleeding symptoms. Bone marrow examinations reveal small, hypolobated megakaryocytes. Initially, this disorder was shown by linkage analysis to be located on chromosome 10p11–12, and then mutations in *MASTL, ACBD4*, and *ANKRD26* were found in pedigrees and, in the case of *ANKRD26*, a large series of 78 patients from 12 families. Subsequent studies support *ANKRD26* as the most likely causative genetic variant responsible for THC2.

ANKRD26 encodes for an inner cell membrane protein that interacts with signaling proteins. The variants in *ANKRD26* occur in the 5′untranslated region and abrogate RUNX1 and FLI1 binding, resulting in loss of ANKRD26 silencing and hyperactivation of mitogen-activated protein kinase pathway, and impaired proplatelet formation. There may be a mild decrease in alpha-granule content and associated mild defects in platelet function. Several large cohort studies have demonstrated a predisposition to leukemia (about 23-fold increased risk), MDS (about 12-fold risk), and chronic myelogenous leukemia (CML; 21-fold risk) in these families.

ETV6-Associated Thrombocytopenia With Predisposition to Acute Lymphoblastic Leukemia: Heterozygous *ETV6* germline mutations have been identified in several families with inherited thrombocytopenia, variable red cell macrocytosis, and multiple family members with hematologic malignancies, primarily B-cell acute lymphoblastic leukemia (ALL). Germline *ETV6* mutations have been implicated in ~1% of childhood ALL patients, and therefore this form of thrombocytopenia may be underrecognized. Patients with heterozygous germline *ETV6* mutations have variable thrombocytopenia with platelet counts from 8 to 132×10^9/L and BM that may show hypolobated MKs and mild dyserythropoiesis or may be normal. Platelet ultrastructure and MPV are normal with few elongated alpha-granules reported. It is not known if platelet function is impaired.

Paris-Trousseau/Jacobsen Syndrome: Paris-Trousseau/Jacobsen syndrome (PT/JS) is due to a terminal deletion of 11q and patients have multiple associated other anomalies, including cardiac defects (which are often the major problems at birth), dysmorphic facies, mental retardation, and congenital thrombocytopenia with abnormal platelet granules. The thrombocytopenia is due to deletion of FLI1, which encodes Fli-1, a transcription factor important in megakaryopoiesis. Bone marrow examination shows micromegakaryocytes, and electron microscopy of the platelets shows abnormally large platelet granules. There is often a platelet function defect with the thrombocytopenia, which may persist even as the platelet count improves, beginning in infancy and extending into adolescence. In PT, the platelet defect predominates. In JS, there is a larger deletion and associated cardiac anomalies, facial dysmorphism, mental retardation, and trigonocephaly predominate. NGS has uncovered novel *FLI1* variants in individuals with platelet dysfunction (with mild thrombocytopenia), and homozygous *FLI1* missense mutations have been reported in two patients with a PTS phenotype. Treatment is supportive.

Disorders With Large Platelets (Macrothrombocytopenia): These disorders are generally associated with variants affecting the cytoskeleton or cytoskeletal interactions resulting in changes in platelet size and shape. Some of these disorders have very mild platelet dysfunction, whereas others have a more severe phenotype.

Bernard–Soulier Syndrome, 22q11.2 Deletion Syndrome, Mediterranean Macrothrombocytopenia, and Platelet-Type von Willebrand Disease: Several disorders with varying degrees of bleeding result from mutations in genes encoding for the von Willebrand factor (VWF) receptor on the surface of platelets, the GPIb/IX/V complex. Patients with homozygous mutations or compound heterozygous deletions/mutations in *GPIBA* or *GPIBB* or *GPIX* resulting in absent expression of the receptor complex have a severe bleeding disorder, Bernard–Soulier syndrome (BSS). Heterozygous deletions or missense mutations result in somewhat decreased expression of the complex on platelet surfaces, fairly normal platelet function, and mild thrombocytopenia without significant bleeding previously called Mediterranean thrombocytopenia. In patients with the classic deletion of 22qDS, the *GPIBB* gene is also deleted, resulting in essentially Mediterranean thrombocytopenia in some patients. A few patients have been reported with 22qDS and BSS resulting from deletion of *GPIBB* on one chromosome and mutations on the other chromosome. Patients with platelet-type von Willebrand disease (a subset of whom were called Montreal platelet disorder) have specific mutations in *GPIBA*, resulting in an increased affinity of the receptor for VWF. As a result, high-molecular-weight multimers of VWF spontaneously bind to the circulating platelets causing intermittent thrombocytopenia, which may be a macrothrombocytopenia, associated with a decrease in high-molecular-weight VWF multimer as in type 2B von Willebrand disease. These topics are covered in more detail in Chapters 96 and 109.

MYH9-Related Diseases: The MYH9-RD (formerly May–Hegglin anomaly, Sebastian syndrome, Fechtner syndrome, and Epstein syndrome) are a group of autosomal dominant disorders with associated macrothrombocytopenia resulting from mutations in *MYH9*, which codes for nonmuscle myosin heavy chain IIA. Clinically, patients present with varying degrees of macrothrombocytopenia (some may have severe thrombocytopenia), leukocyte inclusions, sensorineural hearing loss, cataracts, and glomerulonephritis. The thrombocytopenia may result in a bleeding diathesis, which is generally mild, although a few patients have severe thrombocytopenia and significant bleeding. There may be a correlation between genotype and phenotype, in that patients with mutations in the ATPase region within the head domain have more systemic disease with associated hearing loss and/or nephropathy, whereas patients with C-terminal region mutations have predominantly the hematologic manifestations. Because hearing loss is limited initially to high frequencies and renal disease may be relatively silent initially (with asymptomatic proteinuria), patients with identified MYH9 mutations should be thoroughly evaluated for other manifestations of disease. Treatment with the TPO receptor agonists has been tried in patients with MYH9-related diseases and severe thrombocytopenia with some success.

Autosomal Dominant Macrothrombocytopenia Associated With Hearing Loss: A second gene associated with hearing loss and macrothrombocytopenia was recently identified again using NGS and variants in several independent families were described. These individuals have variants in *DAIPH1* and present with congenital hearing loss (in contrast to the MYH9-related disorders with later onset hearing loss) and macrothrombocytopenia. Platelet counts were 63–147 × 10^9/L, and MPV was 11.2–14.1 fL and there does not appear to be associated renal or ocular findings as have been reported in the MYH9-related disorders. The macrothrombocytopenia may be progressive as was recently described in a Japanese cohort with DIAPH1 variants.

Gray Platelet Syndrome: The Gray platelet syndrome (GPS) is an autosomal dominant or autosomal recessive disorder causing mild-to-moderate thrombocytopenia with platelet dysfunction because of the absence of platelet alpha granules, giving the characteristic gray appearance on May–Grünwald–Giemsa staining. Because of the abnormal packaging of alpha granule contents, proteins are instead redirected to the demarcation membrane lumen and are secreted into the extracellular marrow space, resulting in bone marrow reticular fibrosis. The lack of platelet alpha granule contents results in mild, variable platelet dysfunction and bleeding manifestations. Mutations in *NBEAL2* encoding for a BEACH domain protein related to LYST (mutated in Chediak–Higashi syndrome) have recently been identified as the cause of an autosomal recessive form of GPS in some families.

In some families, pathologic variants in *GFI1B*, which encodes GFI1B, a zinc-finger transcription factor located on chromosome 9q34.13, appear to have a gray platelet–like syndrome with mild-to-moderate macrothrombocytopenia, red cell anisopoikilocytosis and alpha-granule deficiency with platelet dysfunction. This is inherited in an autosomal dominant fashion and appears to be due to a dominant negative effect exerted by mutant GFI1B proteins.

GATA-1-Related Thrombocytopenia: Another form of XLT, but with macrothrombocytes, results from mutations in *GATA1*, a gene that encodes for a 2 zinc-finger protein essential for both erythroid- and megakaryocyte-specific gene transcription. Mutations in *GATA1* can cause macrothrombocytopenia with or without an associated microcytic anemia resembling thalassemia. Because GATA-1 is an important transcription factor in many platelet-associated genes, there may be platelet dysfunction. Patients present in early childhood with bleeding diathesis, which results from the thrombocytopenia and concomitant platelet dysfunction. Because of the variable thrombocytopenia and anemia, clinical diagnosis may be difficult and patients may be misdiagnosed with WAS/XLT (if attention is not paid to the platelet size and genetic testing is not performed). Treatment is supportive although patients with severe thrombocytopenia or bleeding diathesis may benefit with HSC transplantation.

Other Macrothrombocytopenias: Several other macrothrombocytopenias have been described, including macrothrombocytopenia with mutations in platelet β1 tubulin (*TUBB1*) or actinin 1 (*ACTN1*) and macrothrombocytopenia associated with Filamin A mutations (*FLNA*). A recent report described a family with a variant in a magnesium channel (TRPM7) that results in macrothrombocytopenia and atrial fibrillation. However, many patients who present with macrothrombocytopenia still have no identifiable explanation for the disorder. New genome-wide assessment techniques may help to define these disorders and increase our knowledge about which polymorphisms affect platelet count and size. Physicians should consider referring patients with these disorders to a center specializing in platelet disorders that has the ability to do genome-wide studies so that we may learn more about these diseases. Treatment of the undefined thrombocytopenia that is thought to be congenital is supportive. Trials with the novel TPO receptor agonists are underway in some settings.

Further Reading

Babushok, D. V., Bessler, M., & Olson, T. S. (2016). Genetic predisposition to myelodysplastic syndrome and acute myeloid leukemia in children and young adults. *Leuk Lymphoma*, *57*, 520–536.

Balduini, C. L., Pecci, A., & Savoia, A. (2011). Recent advances in the understanding and management of MYH9-related inherited thrombocytopenias. *Br J Haematol, 154*, 161–174.

Bariana, T. K., Ouwehand, W. H., Guerrero, J. A., Gomez, K., & on behalf of BRIDGE (2016). Dawning of the age of genomics for platelet granule disorders: improving insight, diagnosis and management. *Br J Haematol, 176*, 705–720.

Geddis, A. E. (2009). Congenital amegakaryocytic thrombocytopenia and thrombocytopenia with absent radii. *Hematol Oncol Clin N Am, 23*, 321–331.

Grossfeld, P. D., Mattina, T., Lai, Z., et al. (2004). The 11q terminal deletion disorder: A prospective study of 110 cases. *Am J Med Genet A, 129A*, 51–61.

Gunay-Aygun, M., Zivony-Elboum, Y., Gumruk, F., et al. (2010). Gray platelet syndrome: Natural history of a large patient cohort and locus assignment to chromosome 3p. *Blood, 116*, 4990–5001.

Horvat-Switzer, R. D., & Thompson, A. A. (2006). HOXA11 mutation in amegakaryocytic thrombocytopenia with radio-ulnar synostosis syndrome inhibits megakaryocytic differentiation in vitro. *Blood Cells Mol Dis, 37*, 55–63.

Levin, C., Koren, A., Pretorius, E., Rosenburg, N., Shenkman, B., Hauschner, H., et al. (2015). Deleterious mutation in the *FYB* gene is associated with congenital autosomal recessive small-platelet thrombocytopenia. *J Thromb Haemostasis, 12*, 1285–2192.

Millikan, P. D., Balamohan, S. M., Raskind, W. H., & Kacena, M. A. (2011). Inherited thrombocytopenia due to GATA-1 mutations. *Semin Thromb Hemost, 37*, 682–689.

Owen, C. (2010). Insights into familial platelet disorder with propensity to myeloid malignancy (FPD/AML). *Leuk Res, 34*, 141–142.

Pippucci, T., Savoia, A., Perrotta, S., et al. (2011). Mutations in the 5′ UTR of ANKRD26, the ankirin repeat domain 26 gene, cause an autosomal-dominant form of inherited thrombocytopenia, THC2. *Am J Hum Genet, 88*, 115–120.

Songdej, N., & Rao, A. K. (2017). Hematopoietic transcription factor mutations: Important players in inherited platelet defects. *Blood, 129*, 2873–2881.

Thrasher, A. J. (2009). New insights into the biology of Wiskott-Aldrich syndrome (WAS). *Hematol Am Soc Hematol Educ Program*, 132–138.

Toriello, H. V. (2012). Thrombocytopenia-absent radius syndrome. *Semin Thromb Hemost, 37*, 707–712.

CHAPTER 94

Fetal and Neonatal Alloimmune Thrombocytopenia

Catherine E. McGuinn, MD, William B. Mitchell, MD and James B. Bussel, MD

Fetal and neonatal alloimmune thrombocytopenia (FNAIT) is characterized by the presence of transient isolated neonatal thrombocytopenia secondary to maternal antibodies against paternally inherited antigens expressed on the fetal platelets. FNAIT is the most common cause of severe thrombocytopenia (<50,000/μL) in term neonates as well as the most common cause of intracranial hemorrhage (ICH). While the incidence varies depending on the definition of thrombocytopenia and in particular how it is identified (clinically vs. by screening), it is estimated that approximately 1 in 1000 to 1 in 5000 live births are affected.

Estimates of the incidence, severity, and other features of cases of FNAIT are highly dependent on how cases are accrued (Table 94.1). One way is the summation of clinically detected neonates, with cases identified because the neonate is sufficiently affected to merit a blood count secondary to bleeding and bruising prompting diagnosis. Additional cases can be identified by a platelet count done for other reasons (i.e., sepsis investigation). Cases can also be identified by screening of either all mothers (for antigen status, e.g., human platelet antigen 1 [HPA-1]) or of all neonates (by platelet count). Cases accrued by maternal platelet antigen screening are substantially more frequent but clinically milder than those accrued by clinical detection of the neonate (Table 94.1).

TABLE 94.1 FNAIT Identified by Screening Versus Clinical Diagnosis

	Screening	Clinical Diagnosis
Incidence	1:500–1:1000	1:3000–1:10,000
Thrombocytopenia (platelet count/μL)	50–100,000	90% < 50,000; 50% < 20,000
ICH	2:100,000	10%–20% of cases (10% of 1:5000 = 2:100,000)
Development	Not studied	Mildly abnormal even without ICH
Findings associated with FNAIT	Not studied	1/3 but mainly minor
Antenatal treatment of next pregnancy	C/S at 37 weeks	IVIG ± steroids
Maternal disease associated	Not studied	High incidence of autoimmunity (RA and DM)
Prediction of severity	Antibody titers	Previous fetus had an ICH
Worsening of next pregnancy	Sometimes	99 + % at least as severe
Timing of sensitization	75% at birth	60% prior to birth of 1st neonate

Transfusion Medicine and Hemostasis. https://doi.org/10.1016/B978-0-12-813726-0.00094-5

FNAIT with severe thrombocytopenia has a substantial risk for in utero and perinatal ICH with resulting long-term neurological disability or death. Prompt investigation and management of neonatal (or rarely fetal) thrombocytopenia is critical for the neonate, future pregnancies, and possibly pregnancies in female siblings of the mother. Usually, results of parental platelet antigen and antibody testing are not immediately available. Therefore, treatment needs to be instituted empirically pending these results, while recognizing that diagnostic testing to explore the possibility of FNAIT is important. Significant improvement in subsequent fetal/neonatal outcomes have occurred with maternal treatment with intravenous immune globulin and/or steroids in the next pregnancy.

FNAIT is usually characterized by the presence of transient isolated thrombocytopenia secondary to the production and transplacental passage of maternal antibodies directed against paternally inherited antigens expressed on the fetal/neonatal platelet. FNAIT is the platelet equivalent of hemolytic disease of the fetus and newborn (HDFN) in which the same type of incompatibility occurs, only in the case of FNAIT involving platelet instead of red cell antigens (Table 94.2).

Pathophysiology: HPAs are platelet-specific antigens that reside on platelet surface membrane glycoproteins (see Chapter 31). HPAs are expressed on fetal platelets from the 16th week of gestation and are the antigenic source for alloimmunization in NAIT.

The most frequently involved alloantigen in NAIT is HPA-1a in the Caucasian population, where incompatibility at the HPA-1 polymorphism accounts for ~70%–80% of severe NAIT. Ethnic variation in allele distribution is important to consider, in particular that HPA-4 incompatibility accounts for 80% of FNAIT in the Asian population in whom the HPA-1a/1b polymorphism is extremely rare.

Additionally, incompatibilities at HPA-5b, HPA-3a and 3b, and HPA-9b have been identified. HPA-5b incompatibility is the second most common cause of FNAIT but is usually associated with a less severe phenotype. One possible explanation is the high density of GPIIIa molecules on the platelet surface (~26,000 potential targets for anti-HPA-1a). In contrast, HPA-5 is expressed on the platelet surface at much lower density (~1000 potential targets of anti-HPA5b). HPA-3 incompatibility represents only 1%

TABLE 94.2 Differences and Similarities Between NAIT and Hemolytic Disease of the Fetus and Newborn (HDFN)

Differences with HDFN	Similarities With HDFN
Primiparous pregnancy can be affected in NAIT	Pathophysiology is similar: antigen incompatibility of the parents and maternal antibody production of which the IgG crosses the placenta and affects the fetus by destroying fetal cells
No routine screening exists for NAIT	EPO and TPO receptors are in the fetal brain
No preventative treatment is available for NAIT	Neonatal cytopenias are transient until the maternal antibodies disappear from neonatal circulation
IVIG is relatively effective for in utero disease in NAIT	No evidence for or against a role for breast feeding in perpetuating neonatal disease exists
	Severe cytopenia is associated with significant clinical consequences

of cases although 10% of pregnancies have an incompatibility of HPA-3a or 3b, and most cases are severe, similar to HPA-1a incompatibility. In addition, anti-HPA3a or 3b may be hard to detect. HPA-9b is more recently described and is now reported to be the second most common cause of severe cases of FNAIT in Caucasians. HPA-9b is found only in 1:500–1:1000 people, and antibodies against it are extremely difficult to measure. Therefore, if the father has HPA-9b in a thrombocytopenic fetus who clinically appears to have FNAIT and in whom the rest of the workup does not reveal an alternative etiology, the diagnosis of FNAIT secondary to HPA-9b incompatibility is made even in the absence of detectable anti-HPA-9b antibody.

In the Caucasian population, 97%–98% of individuals express HPA-1a on their platelets (75% HPA-1a/1a homozygous or 25% 1a/1b heterozygous); the remaining 2%–2.5% are HPA-1a negative. It is these HPA-1b/1b homozygous mothers who are at risk of the development of NAIT in each pregnancy, including the first, if their partner is HPA-1a heterozygous or homozygous.

In addition to HPA alloantigen incompatibilities and the presumptive role of antigen exposure to the mother, in the development of NAIT, additional genetic modifiers are involved in the development of alloantibodies. Only approximately 12% of mothers who are HPA-1a-negative produce antibodies; the additional presence of HLA DRB3*0101 in the mother confers a 10–20-fold increased relative risk of antibody production.

It is believed that the exposure of incompatible fetal platelets to the maternal circulation results in the production of maternal IgG anti-HPA antibodies. However, it has been proposed that these antibodies could arise from expression of GPIIIa and thus HPA-1a on the surface of the syncytiotrophoblast from the maternal side of the placenta. The standard explanation of pathophysiology is that these IgG antibodies are transported across the placenta and bind to fetal platelets. These antibody-coated platelets are removed from fetal circulation by the reticuloendothelial system with resultant fetal thrombocytopenia. Recent work has suggested that these alloantibodies also damage fetal megakaryocytes and thus contribute to fetal thrombocytopenia. Furthermore, because GPIIIa is expressed on vascular endothelial cells, it may also contribute to fetal ICH via an alternative mechanism in addition to the creation of severe fetal thrombocytopenia.

Clinical Presentation: FNAIT can occur in any pregnancy with most cases evident in full-term neonates. NAIT should be suspected when a neonate presents with a platelet count <50,000/μL, with or without bleeding symptoms in the perinatal period. Bleeding symptoms in the neonate with NAIT can include petechiae, hematoma, melena, hematuria, hematemesis, hemoptysis or, most significantly, ICH. Additional cases of NAIT can be identified in neonates without bleeding symptoms who have thrombocytopenia detected incidentally on blood counts obtained for sepsis or other nonhemorrhagic evaluations.

As we incorporate additional information from screening programs for FNAIT, it has been identified that 75% of sensitization occurs after the birth of the first child, similar to Rh disease, attributed to delivery-related transplacental hemorrhage. However, in more than 60% of clinically severe cases the first affected pregnancy is caused by prior sensitization. In those cases, sensitization must occur antenatally (Table 94.1).

Thrombocytopenia with FNAIT can occur as early as the 13th week of gestation, with up to 75% of clinical ICHs occurring prenatally. In untreated cases, ICH occurs in 10%–20% of affected fetuses/neonates with a 15% fatality rate, and may present with fetal distress, encephalomalacia, intracranial cysts, and focal neurological exam including a full fontanelle, anemia, or poor feeding. The presentation can include fetal anemia or porencephaly; the most common time of hemorrhage seems to be approximately 32 weeks of gestation.

Diagnosis: Diagnosis of NAIT should be explored when a neonate presents with severe thrombocytopenia (<50,000/μL) especially in the first 24 hours of life. First, pregnancies can be affected, and in most cases there is no family history. However, the identification of NAIT has important clinical implications for the mother's future pregnancies.

Consideration should be given to alternative diagnoses through review of maternal history with specific attention to maternal infections, hypertension, platelet count, and autoimmune disorders. In addition, neonatal clinical course and associated abnormalities need to be carefully explored. Initial evaluation should include confirmation of thrombocytopenia and review of peripheral smear. Especially in the neonate, platelet count may be inaccurate due to difficult sample collections. Even if alternative etiologies of thrombocytopenia are identified, testing for FNAIT should be performed because a third of proven cases of FNAIT have comorbid conditions that contribute to thrombocytopenia in addition to FNAIT.

Diagnosis is confirmed by demonstrating a platelet antigen incompatibility *and* the presence of maternal antibody directed against that specific antigen (vs. nonspecific platelet alloantibodies). Testing should only be performed by an experienced reference laboratory, and diagnosis can usually be made with maternal and paternal samples (as a surrogate for the neonatal/fetal sample). Maternal and paternal samples can be genotyped via polymerase chain reaction or high-throughput methods to identify incompatibilities at HPA (see Chapter 31). Testing should include HPA-1, 3, 5, 9b, and 15 with inclusion of HPA-4 if the family is of Asian descent. The demonstration of a maternal alloantibody that is specific to the paternal platelet is essential to confirm the diagnosis (except in the setting of HPA-9b). This serological testing is performed by enzyme immunoassay with monoclonal-antibody-based identification of platelet glycoprotein assay (MACE or MAIPA) of flow cytometry, which allows for identification and quantification of glycoprotein-specific antibodies (see Chapter 31).

Because specialized testing will not typically produce results immediately, management needs to be based on clinical suspicion.

Differential Diagnosis: Thrombocytopenia is nonspecific and can be a manifestation of many disease processes within the neonatal period. Hence the differential is very broad and should include infection, maternal hypertension, chronic hypoxia, necrotizing enterocolitis, thrombosis, maternal autoimmune thrombocytopenia, and inherited thrombocytopenia including bone marrow failure syndromes, such as the rare congenital amegakaryocytic thrombocytopenia (CAMT) (see Chapter 90). Full maternal and neonatal history and physical exam should be included in the assessment of thrombocytopenia, including review of timing of the presentation of thrombocytopenia, and also careful review of the peripheral smear. Early onset (<72 hours) severe thrombocytopenia in an otherwise full term neonate should prompt a diagnostic work for and clinical treatment of FNAIT. CAMT is a mimicker of FNAIT and should be suspected if there is response to platelet transfusions but not intravenous immunoglobulin (IVIG) and if the thrombocytopenia does not improve within 1–2 weeks.

Management: Therapy should be initiated in the newborn with severe thrombocytopenia before confirmation of the diagnosis. Initial evaluation should include a physical exam and head ultrasound. Head ultrasound should be obtained as close to birth as feasible to facilitate management and document the timing of ICH if present, i.e., antenatal versus peripartum or postnatal. Transfusion parameters differ widely according to institutional protocol, but a general guideline might be to recommend intervention at the platelet count of ≤30,000/μL in a well neonate, 50,000/uL in a sick premature infant, and ≤100,000/μL in a neonate with documented ICH.

In a recent change, the cornerstone of initial therapy is infusion of random donor platelets (e.g., platelets that are not antigen-negative for the corresponding platelet antibody). Random donor platelets are effective at raising the platelet count in >70% of cases, despite immunological incompatibility, and are readily available. Certain centers may have matched platelets on hand that are HPA-1b1b and HPA-5a5a. If available, antigen-negative platelets from a donor without antibodies produce greater posttransfusion platelet count increases; however, this is uncommon in the United States giving the time required for donor activation, screening, collection, and processing. In some European countries, the United Kingdom, France, and Netherlands, who have less geographic constraints, arrangement are in place to have HPA-matched platelets available with time-sensitive transportation arrangements. Currently in the United States, random donor platelets are the frontline therapy because testing, processing, and the mother's ability to donate may lead to unacceptable delays in therapy. Antigen-matched or maternal platelets are reserved for patients who do not respond or may require additional transfusions. Processing of maternal platelets by definition will match the fetal type. However, the mother will also have antibody to her neonate's platelet antigen in the plasma. Previously, maternal platelets had to be washed, which often lead to damage of the platelets, but current technology allows the donor to provide more concentrated platelets and thus less plasma and less antibody. However, there are IgG antibodies in platelet alpha granules, so even prefect elimination of plasma would not fully eliminate maternal anti-HPA antibody transmission, albeit it would be at a very low titer. Response to platelet therapy should be monitored within 1 hour of platelet administration.

Frontline complementary options for neonatal management of NAIT include the addition of IVIG 1 gm/kg/day for 1–3 days with the concept of inhibiting the peripheral immune-mediated platelet destruction and protecting the transfused incompatible platelets; this is our practice. Corticosteroids can also be considered in conjunction with IVIG administration, for example, 1 mg IV q8 hours for 1–3 days with the IVIG but are not recommended at higher doses and/or more prolonged usage because this increases the risk of fungal infection in the neonate.

Platelets produced by neonates with FNAIT may exhibit accelerated clearance until the disappearance of platelet antibody, and additional transfusions of IVIG and/or platelets may be required in the first 2–4 weeks of life. Platelet counts should be monitored until a normal count is observed off therapy to exclude the possibility of an inherited thrombocytopenia.

In neonates without ICH, complications of FNAIT are rare, but patients should be followed up after 4–8 weeks to ensure normal platelet count and development.

Antenatal: The goal of management in FNAIT is to prevent ICH, the most devastating complication of severe thrombocytopenia. Because the majority of ICH occurs before the onset of labor, antenatal treatment must be considered in the management of subsequent pregnancies at risk for NAIT. Because screening is not available at this time, there is no plan in place if a case is identified before delivery of the first affected neonate other than possibly elective delivery at 37 weeks.

The first issue with antenatal management is noninvasive fetal diagnosis, which is based on paternal zygosity. Three quarters of HPA-1a-positive fathers are homozygous HPA-1a. If paternity is assured and the father is a homozygote for the corresponding platelet antigen, then the fetus can be assumed to be at least as severely affected as the previous fetus. If the father is unknown, unavailable, or a heterozygote, the usual procedure in the past would be to do an amniocentesis for DNA-based fetal HPA typing (i.e., HPA-1a). Currently, the goal would be to use free fetal DNA for diagnosis, which is possible with HPA-1a but not other platelet antigens at this time.

However, if it is known that the fetus is antigen-positive, then the primary question is if and when an ICH occurred in a previous sibling. Recently, to better understand FNAIT and to monitor therapy, serial fetal blood sampling (FBS) was used. However, our understanding of FNAIT has increased, and it is also clear that FBS, especially in a thrombocytopenic fetus, has well-described risks. Therefore, alternative noninvasive treatment strategies have been advocated for the empiric antenatal management of at-risk pregnancies. In affected pregnancies with a previous sibling with ICH, maternal IVIG 2 g/kg/week, in two infusions beginning at 12 weeks of gestation, with augmentation of therapy by adding prednisone 0.5 mg/kg and then increasing to 1 mg/kg at 20–28 weeks has been recommended. In less severely affected families starting with IVIG 1 g/kg/week at ~20 weeks combined with prednisone 0.5 mg/kg/day and adding a second dose of 1 g/kg/week IVIG at 30–32 weeks is effective. The goal of antenatal management is to minimize treatment side effects and invasive procedures but maximize effective platelet count in the fetal and perinatal period. Elective cesarean section is usually the standard recommendation at 37–38 weeks of gestation after documentation of fetal lung maturity. Vaginal delivery has been explored after using FBS to document a response to antenatal therapy, although this practice is empiric rather than based on study results relating delivery outcome to the fetal platelet count.

There is considerable current debate about how much IVIG and/or steroids are required. This is partly dependent on the desire to have a protocol of treatment that is as close to 100% effective as possible. Using 1 g/kg/week may not substantially increase the platelet count in all patients but nonetheless may help to avoid ICH until delivery. The issue is that ICH is a rare outcome, and the balance of preventing ICH while minimizing both cost and maternal side effects is difficult. This debate highlights the need for a noninvasive marker of severity similar to the MCA Doppler used in FNHDN so that treatment can proceed from empirical to rational.

Screening: Screening programs for maternal platelet antibodies and/or HPA typing have been preliminarily explored to identify at-risk pregnancies and thereby minimize the risk of fetal/neonatal ICH in the first affected pregnancy. The challenges to the development of a successful prenatal screening program include lack of biological markers for severity of disease, invasive risk of confirmatory testing with FBS, limited consensus on antenatal treatment in cases in which there is no previous affected fetus/neonate to guide management, lack of prophylactic medication, and little clarity regarding cost-effectiveness.

Overall goals for the future for FNAIT thus would be the following:

(a) Develop a noninvasive marker of disease severity
(b) Expand the use of free fetal DNA to all relevant human platelet antigens
(c) Institute routine screening of mothers
(d) Develop prophylaxis similar to anti-D for HDFN
(e) Develop (if safe and effective) FcRn inhibition to prevent transfer of maternal IgG antibody to the fetus
(f) Determine risks for ICH to guide antenatal treatment of a thrombocytopenic fetus

The cost effective implementation of an effective screening program incorporating new laboratory methods and noninvasive treatment algorithms has been considered, and their role remains to be clarified in prospective studies.

Further Reading

Arinsburg, S. A., Shaz, B. H., Westhoff, C., & Cushing, M. M. (2012). Determination of human platelet antigen typing by molecular methods: Importance in diagnosis and early treatment of neonatal alloimmune thrombocytopenia. *Am J Hematol*, *87*, 525–528.

Birchall, J. E., Murphy, M. F., Kaplan, C., Kroll, H., & European Fetomaternal Alloimmune Thrombocytopenia Study Group (2003). European collaborative study of the antenatal management of feto-maternal alloimmune thrombocytopenia. *Br J Haematol*, *122*, 275–288.

Bussel, J. (2009). Diagnosis and management of the fetus and neonate with alloimmune thrombocytopenia. *J Thromb Haemost*, *7*(Suppl. 1), 253–257.

Bussel, J. B., Berkowitz, R. L., & Hung, C. (2010). Intracranial hemorrhage in alloimmune thrombocytopenia: Stratified management to prevent recurrence in the subsequent affected fetus. *Am J Obstet Gynecol*, *203*(135), e1–14.

Bussel, J. B., & Primiani, A. (2008). Fetal and neonatal alloimmune thrombocytopenia: Progress and ongoing debates. *Blood Rev*, *22*, 33–52.

Bussel, J. B., Zacharoulis, S., & Kramer, K. (2005). Clinical and diagnostic comparison of neonatal alloimmune thrombocytopenia to non-immune cases of thrombocytopenia. *Pediatr Blood Cancer*, *45*, 176–183.

Chakravorty, S., & Roberts, I. (2012). How I manage neonatal thrombocytopenia. *Br J Haematol*, *156*, 155–162.

Davoren, A., Curtis, B. R., Aster, R. H., & McFarland, J. G. (2004). Human platelet antigen-specific alloantibodies implicated in 1162 cases of neonatal alloimmune thrombocytopenia. *Transfusion*, *44*, 1220–1225.

Kamphuis, M. M., Paridaans, N., & Porcelijn, L. (2010). Screening in pregnancy for fetal or neonatal alloimmune thrombocytopenia: Systematic review. *BJOG, 117*, 1335–1343.

Kiefel, V., Bassler, D., & Kroll, H. (2006). Antigen-positive platelet transfusion in neonatal alloimmune thrombocytopenia (NAIT). *Blood, 107*, 3761–3763.

Peterson, J. A., Gitter, M., Bougie, D. W., Pechauer, S., Hopp, K. A., Pietz, B., Szabo, A., Curtis, B., McFarland, J., & Aster, R. H. (2014). Low frequency human platelet antigens as triggers for neonatal alloimmune thrombocytopenia. *Transfusion, 54*(5), 1286–1293.

Pacheco, L. D., Berkowitz, R. L., & Moise, K. J. (2011). Fetal and neonatal alloimmune thrombocytopenia: A management algorithm based on risk stratification. *Obstet Gynecol, 118*, 1157–1163.

Schoot, C. E., Thurik, F. F., Veldhuisen, B., & Haas, M. (2013). Noninvasive prenatal blood group genotyping. *Transfusion, 53*, 2834–2836.

CHAPTER 95

Acquired Neonatal Thrombocytopenia

Catherine E. McGuinn, MD, William B. Mitchell, MD and
James B. Bussel, MD

This chapter will focus on the acquired etiologies of neonatal thrombocytopenia as congenital and alloimmune thrombocytopenias are discussed in Chapters 93 and 94. Neonatal thrombocytopenia occurs in 1%–2% of healthy term neonates but is common in the neonatal intensive care unit where thrombocytopenia occurs in up to one-third of all admissions. Fetal platelet counts increase during gestation, reaching adult values by the 22nd week. At term, more than 98% of healthy neonates have platelet counts >150,000/μL. Thrombocytopenia in the fetus and neonate is defined as a platelet count <150,000/μL while acknowledging this reference range is not well-standardized for preterm infants and that mild thrombocytopenia may overlap with normal values. Thrombocytopenic neonates may present symptomatically with petechiae, hematomas, and gastrointestinal, umbilical, or intracranial bleeding (ICH). The increased risk of life-threatening bleeding that accompanies severe thrombocytopenia and prematurity makes the accurate diagnosis and intervention of acquired thrombocytopenia critical.

Pathophysiology: The mechanisms responsible for neonatal thrombocytopenia can be broadly categorized as follows:

1. increased platelet consumption,
2. decreased platelet production, and
3. hypersplenism.

In practice, determining the precise mechanisms leading to thrombocytopenia in a critically ill neonate may be difficult due to overlapping clinical presentations and especially the limited ability to perform any testing including bone marrow aspiration and biopsy, which remain the diagnostic gold standard.

Increased platelet consumption can be further categorized into immune- and non-immune-mediated etiologies. Immune thrombocytopenia reflects an increased rate of immunoglobulin or complement-mediated platelet clearance, which may be allo-immune or autoimmune. Alloimmune neonatal thrombocytopenia, as discussed in Chapter 94, is secondary to transplacental transfer of maternal antibody to a nonshared platelet antigen. Alternatively, maternal immune thrombocytopenia (ITP) can result in neonatal thrombocytopenia from transplacental passage of a maternal auto-anti-platelet antibody reactive with all platelets including fetal and neonatal ones. Rarely, neonates can develop an autoimmune-mediated ITP from systemic lupus erythematosus (SLE) (see Chapters 100–103). Nonimmune-mediated pathology that contributes to platelet consumption includes viral, bacterial, or fungal sepsis, necrotizing enterocolitis (NEC), disseminated intravascular coagulation (DIC), vascular malformations with Kasabach–Merritt syndrome (KMS), congenital thrombotic thrombocytopenia purpura (TTP), type 2B von Willebrand disease, and thrombosis.

Transfusion Medicine and Hemostasis. https://doi.org/10.1016/B978-0-12-813726-0.00095-7

Decrease in platelet production can be seen in the perinatal period with chronic fetal hypoxia, congenital infections, and bone marrow infiltration. Chronic fetal hypoxia is secondary to placental insufficiency from pregnancy-induced hypertension, preeclampsia, HELLP syndrome (hemolytic anemia, elevated liver enzymes, low platelet count) or intrauterine growth restriction (IUGR) may have a depressive effect on megakaryocytic platelet production. Congenital infections, which can decrease bone marrow platelet production, include typical TORCH infections which include Toxoplasmosis, Other (syphilis, varicella-zoster, parvovirus), Rubella, Cytomegalovirus (CMV) and Herpes Infections (Herpes Simplex Virus (HSV) and Human Herpesvirus 6 (HHV-6)), and HIV. Bone marrow infiltration may result from congenital leukemia, neuroblastoma, histiocytosis, transient myeloproliferative disease in Down syndrome, or osteopetrosis. Especially, the latter may have accompanying hepatosplenomegaly. Fetal and neonatal megakaryocytes are smaller with lower ploidy producing less platelets per megakaryocyte. This means a newborn may have limited response to increased demand for platelets. This inability to respond to thrombocytopenia may be more pronounced in the preterm neonate, where the relatively low level of thrombopoietin also limits platelet production.

Lastly, the sequestration with increased entrapment of platelets in the spleen may occur with hemolytic anemia, congenital hepatitis, congenital viral infections, portal vein thrombosis, or virtually any cause of substantial splenomegaly.

Clinical Presentation: The clinical presentation of thrombocytopenia ranges from asymptomatic (detected on a routine laboratory test) to life-threatening bleeding. Symptomatic presentations may include petechiae, ecchymoses, melena, cephalohematoma, hematuria, endotracheal bleeding, umbilical stump bleeding and oozing from venipuncture/heelstick sites, or most seriously focal neurological symptoms.

Clinical history should include the age of neonate, gestational age at birth, birth weight, Apgar scores, timing of onset of thrombocytopenia, and details of prenatal care/delivery including fetal monitoring records. It is important to incorporate the maternal history in clinical decision-making, specifically any history of hypertension, HELLP, preeclampsia, ITP, and maternal infectious serologies and cultures (TORCH infections, and Group B streptococcus status). Additionally, it is important to note any familial history of previous thrombocytopenia, especially neonatal, with attention to severity, duration, and clinical manifestations.

Physical examination should include signs of bleeding, presence of hepatosplenomegaly, survey for dysmorphic features, e.g., abnormalities of thumbs, radius, inability to rotate the forearm and the presence of any features suggestive of trisomies 21, 18, 13, Turner syndrome, or Jacobsen syndrome. Laboratory evaluation should include complete blood count with review of the peripheral smear with special attention to platelet agglutination and confirmation of maternal platelet count. Additional testing and monitoring for signs of bleeding (head sonogram and also urine, stool, and gastric aspirate) will be dictated by the clinical situation.

Differential Diagnosis: The approach to thrombocytopenia in this neonatal period can be categorized based on the timing of presentation into early (<72 hours) and late onset (>72 hours), with additional focus on the severity of presentation and the prenatal/postnatal age of the infant.

Early-onset thrombocytopenia is most often associated with pregnancy/placental complications, infection, sepsis, DIC (see Chapter 120), or Fetal and Neonatal Alloimmune Thrombocytopenia (FNAIT) (see Chapter 91). With a marked, <50K platelet count on day 1 of life, early-onset thrombocytopenia, an immediate diagnostic evaluation is indicated and intervention with consideration of infection and FNAIT for appropriate therapeutic management. In less severe presentations of early-onset thrombocytopenia, maternal placental insufficiency, IUGR, preeclampsia, HELLP, maternal diabetes, and fetal hypoxia are the relevant etiologies for consideration. Overlapping an early and late timeframe of identification the TORCH infections, thrombosis, vascular consumption from KMS or autoimmune thrombocytopenia (ITP/SLE) should be included in the differential diagnosis.

In late-onset thrombocytopenia, the vast majority are secondary to postnatal bacterial or fungal sepsis, liver disease, and/or NEC. In a premature infant, subsequent hypogammaglobulinemia may contribute to infections. If one of the foregoing is not definite, differential diagnosis for late-onset thrombocytopenia should also include inborn errors of metabolism and drug-induced thrombocytopenia. Finally, it is important to consider genetic thrombocytopenia in the etiology of neonatal presentation, including thrombocytopenia-absent radii syndrome, congenital amegakaryocytic thrombocytopenia, Wiskott–Aldrich syndrome, Fanconi anemia, MYH-9-related thrombocytopenia, Chediak–Higashi syndrome, Bernard–Soulier syndrome, Jacobsen syndrome, trisomies 21, 18, and 13, congenital TTP, and von Willebrand type 2B (see Chapter 93).

Diagnosis of Acquired Thrombocytopenia

Chronic Fetal Hypoxia: In premature infants, thrombocytopenia is often secondary to placental insufficiency leading to chronic hypoxia, which is associated with pregnancy-induced hypertension, preeclampsia, HELLP syndrome, maternal diabetes, and IUGR. This form of thrombocytopenia often occurs in clinically well preterm infants, is present at birth or <72 hours from delivery, and in most cases is benign and self-limited. Platelet counts <50,000 are infrequent. Resolution is often seen within 10 days without specific intervention. These neonates can also have additional hematologic abnormalities including transient neutropenia, increased nucleated red blood cells, increased erythropoietin levels, and hyposplenism, which often resolve spontaneously in 2–3 days.

Infections: Prenatal viral and perinatal bacterial infections in the term or preterm infant can present with early or late onset of mild–severe thrombocytopenia. Viral etiologies that have been implicated include toxoplasmosis, rubella, CMV, HSV, HHV-6, enterovirus, parvovirus B19, and varicella. Usually, these infants present with moderate thrombocytopenia and additional findings, which may include low birth weight, microcephaly, hepatosplenomegaly, chorioretinitis, and hearing impairments. Up to 60% of neonates with serious bacterial infection have platelet counts <100,000/μL. Common organisms causing sepsis and DIC in ill-appearing neonates include Group B *Streptococcus*, *Escherichia coli*, and fungus.

Necrotizing Enterocolitis: NEC is a gastrointestinal disorder presenting with feeding intolerance, abdominal distension, bilious emesis, bloody stools, and pneumatosis intestinalis. Diagnosis can be confirmed by the preceding clinical and radiographic criteria. Etiology is multifactorial with contributions in preterm infants from ischemia, infection, enteral feeding, and alterations in the immunological barrier. NEC can result in significant morbidity and mortality, and severe thrombocytopenia has been associated with worse clinical outcomes in these infants.

Thrombocytopenia Secondary to Maternal Immune Thrombocytopenia: Thrombocytopenia can rarely occur (3% of neonatal thrombocytopenia cases) when there is a maternal history of ITP or SLE during pregnancy. Clinical presentation is usually relatively mild with a <1% rate of intracranial hemorrhage. The etiology of thrombocytopenia is the transplacental passage of maternal IgG platelet antibodies and their binding to neonatal platelets with resultant thrombocytopenia. Neonates are only severely affected in ~9%–15% of at-risk pregnancies, but fetal/neonatal platelet count cannot be reliably predicted by maternal platelet count, level of platelet antibody or maternal history of treatment, and prenatal sampling is not recommended. A family history of an affected or unaffected sibling presentation, including severity and timing of thrombocytopenia, can be very helpful in predicting outcomes in the current pregnancy provided that the mother's ITP has not changed, e.g., increased treatment before or during the second pregnancy.

Vascular Consumption: Thrombocytopenia can be secondary to consumption of platelets as sign of arterial/venous thrombosis, KMS, von Willebrand type 2B, or a thrombotic microangiopathy. Thrombosis has been described in the neonate most commonly at the sites of portal vein, renal vein, and inferior vena cava and could be especially suspected if hematuria is present.

KMS is associated with the presentation of thrombocytopenia purpura with a rapidly enlarging Kaposiform hemangioendothelioma or Tufted angioma in the newborn infant. Thrombocytopenia in KMS is secondary to platelet trapping and activation by the abnormally proliferating endothelium and consumptive coagulopathy and thus often presents at several months of age. The clinical presentation can be diverse with multiple lesions, visceral involvement, DIC, and/or severe thrombocytopenia.

Rare genetic etiologies of platelet consumption can include von Willebrand type 2B, which results in platelet consumption from an increased affinity interaction at the von Willebrand factor (VWF) and clearance of platelet and highest-molecular-weight VWF multimers. Upshaw Schulman syndrome, congenital TTP from a genetic mutation in ADAMTS-13 can also present with thrombocytopenia from a consumptive microangipathy process often accompanied by hemolytic anemia and hyperbilirubemia.

Management: Platelet transfusions are the only specific therapy for the management of neonatal thrombocytopenia, and currently there is limited consensus on transfusion thresholds in this age group. Given the lack of evidence-based guidelines and the challenges and inaccuracies of extrapolating from adult literature secondary to all of the increased risk of ICH in the neonate and major differences in platelet function and coagulation factor levels, there are no clear recommendations. Therefore, the clinical status of the neonate, risk of bleeding/intervention, and platelet count must be considered in making recommendations for platelet transfusions. Additionally, some centers have incorporated the platelet mass (mean platelet volume × platelet count) into transfusion parameters to account for the size and number of platelets available for functional hemostasis and, possibly more in the future, platelet reticulocyte counts.

In practice, it is commonly accepted to maintain the platelet count >50,000/μL in a neonate with bleeding symptoms. A recent survey showed a significant portion of neonatologists in the United States would maintain the platelet count >100,000/μL in a bleeding neonate, while the same survey undertaken in Europe had a more restrictive pattern to platelet transfusion. The same survey showed that in a clinically stable preterm neonate the trigger for platelet transfusion ranged from 20 to 50,000/μL, and was 30,000/μL in a term neonate. In a clinically ill or unstable neonate, the majority of practitioners aimed to maintain a higher platelet count in the range of 50–100,000/μL. PlaNeT-2, a ongoing multicenter randomized clinical trial compares a liberal (>50K) versus a restrictive (>25K) prophylactic platelet transfusion strategy in preterm infants may provide evidence to guide practice in the future [http://www.planet-2.com Website accessed 12/27/17] (see Chapter 49).

Etiology-Specific Treatment

Thrombocytopenia Due to Maternal Immune Thrombocytopenia: A neonate born to a mother with ITP should have a cord blood platelet count at delivery and avoidance of intramuscular injections until the result is available. An infant with any degree of thrombocytopenia should be followed clinically and with serial platelet measurements because the nadir usually occurs 2–5 days postnatally. If the platelet count is <50,000/μL, head ultrasound should be obtained to assess for ICH. If treatment of the neonate is indicated in the setting of clinical bleeding or platelet count <20–30,000/μL, IVIG 1 gm/kg can be administered. If there is identification of life-threatening bleeding, IVIG can be combined with a platelet transfusion.

Vascular Consumption: Management of KMS syndrome can include supportive therapy with antifibrinolytics, and plasma (e.g., fresh frozen plasma [FFP]), cryoprecipitate, and platelet transfusions. Additional specific therapy depends on clinical presentation including surgical excision, vascular embolization, compression therapy, immunosuppressive therapy, chemotherapy, and anticoagulation. Interferon therapy should be avoided.

In the setting of thrombosis, intervention should focus on clot resolution. Treatments may include removal of central catheters, thrombectomy, thrombolysis, and anticoagulation. Platelet transfusions may be given as needed to maintain hemostatic parameters. The platelet count may also increase in conjunction with successful anticoagulation.

For von Willebrand type 2B therapy may include von Willebrand concentrate and combination with platelet transfusions to address hemostasis. While for congenital TTP replacement of ADAMTS-13 can be accomplished through FFP or potentially recombinant ADAMTS-13, currently in clinical trials.

Further Reading

Christensen, R. D. (2011). Platelet transfusion in the neonatal intensive care unit: Benefits, risks, alternatives. *Neonatology*, *100*, 311–318.

Cremer, M., Sola-Visner, M., & Roll, S. (2011). Platelet transfusions in neonates: Practices in the United States vary significantly from those in Austria, Germany, and Switzerland. *Transfusion, 51*, 2634–2641.

Curley, A., Venkatesh, V., Stanworth, S., Clarke, P., Watts, T., New, H., Willoughby, K., Khan, R., Muthukumar, P., & Deary, A. (2014). Platelets for Neonatal Transfusion - Study 2: A Randomised Controlled Trial to Compare Two Different Platelet Count Thresholds for Prophylactic Platelet Transfusion to Preterm Neonates. *Neonatology, 106*, 102–106.

Del Vecchio, A., & Motta, M. (2011). Evidence-based platelet transfusion recommendations in neonates. *J Matern Fetal Neonatal Med, 24*(Suppl. 1), 38–40.

Hall, G. W. (2001). Kasabach-merritt syndrome: Pathogenesis and management. *Br J Haematol, 112*, 851–862.

Holzhauer, S., & Zieger, B. (2011). Diagnosis and management of neonatal thrombocytopenia. *Semin Fetal Neonatal Med, 16*, 305–310.

Josephson, C. D., Su, L. L., & Christensen, R. D. (2009). Platelet transfusion practices among neonatologists in the United States and Canada: Results of a survey. *Pediatrics, 123*, 278–285.

Kenton, A. B., Hegemier, S., & Smith, E. O. (2005). Platelet transfusions in infants with necrotizing enterocolitis do not lower mortality but may increase morbidity. *J Perinatol, 25*, 173–177.

Liu, Z. J., & Sola-Visner, M. (2011). Neonatal and adult megakaryopoiesis. *Curr Opin Hematol, 18*, 330–337.

Provan, D., Stasi, R., & Newland, A. C. (2010). International consensus report on the investigation and management of primary immune thrombocytopenia. *Blood, 115*, 168–186.

Roberts, I., Stanworth, S., & Murray, N. A. (2008). Thrombocytopenia in the neonate. *Blood Rev, 22*, 173–186.

Scully, M., Knöbl, P., Kentouche, K., et al. (2017). Recombinant ADAMTS-13: first-in-human pharmacokinetics and safety in congenital thrombotic thrombocytopenic purpura. *Blood, 130*(19), 2055–2063.

Sola-Visner, M., Saxonhouse, M. A., & Brown, R. E. (2008). Neonatal thrombocytopenia: What we do and don't know. *Early Hum Dev, 84*, 499–506.

CHAPTER 96

Bernard–Soulier Syndrome and Other GPIb-IX-Related Receptor Defects

Shawn M. Jobe, MD, PhD

GPIb-IX is the major platelet receptor for von Willebrand factor (VWF). Normal GPIb-IX function is essential both for normal megakaryocyte/platelet maturation and for normal platelet function, most prominently adhesion. Mutations in GPIb-IX result in a number of different syndromes. Clinical manifestations of these syndromes range from a severe platelet function defect in patients with a homozygous deficiency of one of the molecular components of GPIb-IX, to mild thrombocytopenia in patients with heterozygous deficiency. Platelet-type VWD, a syndrome similar to type 2B VWD, also is caused by mutations in the GPIb-IX receptor.

Bernard–Soulier Syndrome: Bernard–Soulier syndrome (BSS) is an autosomal recessive platelet disorder characterized by thrombocytopenia, the presence of giant platelets, and defective ristocetin-induced platelet agglutination.

BSS occurs when a mutation is present in both alleles of one of the components of the platelet receptor GPIb-IX. GPIb-IX is the receptor that is predominantly responsible for the VWF-dependent adhesion of platelets in damaged blood vessels. The absence of GPIb-IX results in greatly diminished platelet accumulation at sites of vessel injury and bleeding. GPIb-IX is a multimeric receptor complex composed of the products of three separate glycoprotein (GP) subunits (Ibα, Ibβ, IX), which are encoded by three different genes. Homozygous or compound heterozygous mutations responsible for BSS have been identified within each of these GPIbα, GPIbβ, and GPIX components. The absence of any one of component results in markedly decreased cell surface expression of the entire GPIb-IX complex because intracellular association of the GPIbα, GPIbβ, and GPIX subunits is necessary for the appropriate trafficking of GPIb-IX. Although classically associated with GPIb-IX, BSS caused by mutations in the gene-encoding Glycoprotein V has not been identified.

Giant platelets are a characteristic feature of BSS, and it is not uncommon to see platelets as large as 20 μm in diameter (RBCs are 8 μm). Platelet counts in individuals with BSS range from 20,000/μL to near normal. A normal interaction between GPIbα and the cytoskeletal protein filamin is an important determinant of platelet size, and its disruption leads to the production of abnormally large platelets. Platelet aggregation studies in patients with BSS demonstrate a poor agglutination response to ristocetin and botrocetin. Both ristocetin and botrocetin bind and activate the GPIb-IX receptor. In vivo the GPIb-IX receptor only binds VWF once the A1 domain of VWF has been functionally presented due to tensile forces imposed by shear within the vasculature. In vitro, the normally shear-induced conformational change in VWF that results in functional presentation of the A1 domain can be mimicked by allowing VWF to bind to ristocetin. In platelet-rich plasma absence of ristocetin-induced agglutination could

Transfusion Medicine and Hemostasis. https://doi.org/10.1016/B978-0-12-813726-0.00096-9

either be due to a defect in GPIb-IX or decreased or dysfunctional VWF. To discern between these two possibilities the patient's platelets are mixed with normal platelet-poor plasma and ristocetin. If the defect in ristocetin-induced agglutination persists, this would indicate a defect in the patient's GPIb-IX receptor, and not absent or dysfunctional VWF. GPIb-IX levels determined by flow cytometry typically demonstrate the absence of the GPIb-IX receptor complex from the platelet surface. However, variant forms of BSS have been identified that have expression of a dysfunctional GPIb-IX complex on the platelet surface. Sequencing of the genes encoding for the molecular components of GPIb-IX may also assist in the diagnosis.

BSS patients most often have a clinically significant bleeding disorder. Mild to moderate bleeds might be adequately treated using desmopressin (DDAVP) and antifibrinolytic agents, and platelet transfusions are recommended for more severe bleeding manifestations. BSS patients with an absence of GPIb-IX can become refractory to platelet transfusion due to the development of GPIb-IX antibodies.

Benign Mediterranean Macrothrombocytopenia: Mediterranean macrothrombocytopenia is an autosomal-dominant disorder characterized by moderate thrombocytopenia, large platelets, and a mild bleeding diathesis. The bleeding symptoms can usually be adequately treated using DDAVP and antifibrinolytic agents. A causative mutation (Ala156Val) in the gene encoding GPIbα (*GPIBA*) is present in approximately 50% of affected individuals.

Mediterranean macrothrombocytopenia was first described in 1975 in a study that compared 145 asymptomatic Mediterranean and 200 Northern European subjects. In this study, an increased incidence of individuals with thrombocytopenia and large platelets was noted in the Mediterranean population. Recently, a heterozygous missense mutation (Ala156Val) of GPIbα was identified as a potentially important cause of this disorder. Sequencing of consecutive patients who presented to an Italian clinic with symptoms of Mediterranean macrothrombocytopenia identified the Ala156Val mutation in 6 of the 12 patients. This relatively high incidence suggests that Ala156Val mutations in *GPIBA* may account for a large proportion of cases of macrothrombocytopenia in Mediterranean populations. The remaining gene mutations responsible for Mediterranean macrothrombocytopenia are unknown. However, the recent identification of a heterozygous mutation in GPIBA (pVal31Leu) in a similar Indian pedigree points toward additional causative mutations in *GPIBA*.

Thrombocytopenia and Velocardiofacial (DiGeorge) Syndrome: Macrothrombocytopenia often occurs in patients with hemizygous deletion of the chromosomal region 22q11. This deletion results in a constellation of syndromes called the velocardiofacial syndrome (VCF). Characteristics of patients with VCF include a typical facies, hypocalcemia, thymic aplasia, cardiac anomalies, and learning disabilities.

More than half of the 22q11 deletions in patients with VCF include the region encoding GPIbβ. Heterozygous absence of the GPIbβ component of the GPIb-IX complex accounts for the mild thrombocytopenia observed in approximately 40% of patients with VCF. Clinically significant bleeding is rarely reported. However, deficiencies of other genes within the deleted region may result in immune dysregulation, and patients with VCF have an increased tendency for autoimmune thrombocytopenia.

Platelet-Type Von Willebrand Disease (Gain of Function Mutation of GPIb-IX): Patients with platelet-type von Willebrand disease (PT-VWD) present clinically with thrombocytopenia and low VWF levels. Significant mucocutaneous bleeding can occur in patients with PT-VWD. Platelet transfusion is typically used as therapy when other supportive measures have been attempted. DDAVP administration or transfusion with plasma-derived VWF concentrates may provide some benefit but should be used with caution, as they may aggravate the thrombocytopenia.

PT-VWD is caused by a unique set of mutations in the GPIbα subunit of the GPIb-IX complex. Causative mutations for platelet-type VWD have been identified in the VWF-binding domain of GPIbα and a region of GPIbα called the macroglycopeptide region. These mutations cause GPIbα to bind with VWF with increased affinity. Spontaneous binding of circulating VWF to the platelet GPIb-IX receptor results in a low-plasma VWF level. Thrombocytopenia might result either from clearance of VWF-bound platelets, or from impaired thrombopoiesis due to increased megakaryocyte GPIb-IX receptor activation.

Laboratory studies in PT-VWD and type 2B VWD are similar, and distinguishing these two clinical entities can be difficult. Both PT-VWD and type 2B VWD result in mildly decreased levels of plasma VWF and a disproportionate decrease in high-molecular-weight VWF multimers. Both result in thrombocytopenia, and both show an enhanced platelet agglutination response to ristocetin that is particularly observable at low ristocetin concentrations. In PT-VWD, the enhanced ristocetin agglutination response is due to GPIbα mutations that increase the binding affinity of the platelet GPIb-IX receptor to VWF. In type 2B VWD, the enhanced agglutination response is due to specific mutations in VWF in its GPIbα-binding domain (A1 domain) that increase its binding affinity to the platelet GPIb-IX receptor. The two can be distinguished by mixing studies characterizing the binding of normal VWF to the patient's platelets, and the binding of the patient's VWF to normal platelets. Alternatively, DNA sequencing of the gene encoding the GPIb-α subunit and exon 28 of VWF can also be done to distinguish PT-VWD and type 2B VWD, respectively (see Chapter 134).

Further Reading

Kanaji, T., Ware, J., Okamura, T., & Newman, P. J. (2012). GPIbα regulates platelet size by controlling the subcellular localization of filamin. *Blood, 119*, 2906–2913.

Lambert, M. P., Arulselvan, A., Schott, A., Markham, S. J., Crowley, T. B., Zackai, E. H., et al. (September 22, 2017). The 22q11.2 deletion syndrome: Cancer predisposition, platelet abnormalities and cytopenias. *J Med Genet A*.

Lopez, J. A., Andrews, R. K., Afshar-Kharghan, V., & Berndt, M. C. (1998). Bernard-Soulier syndrome. *Blood, 91*, 4397–4418.

Savoia, A., Kunishima, S., De Rocco, D., Zieger, B., Rand, M. L., Pujol-Moix, N., et al. (2014). Spectrum of the mutations in Bernard-Soulier syndrome. *Hum Mutat, 35*, 1033–1045.

CHAPTER 97

Glanzmann Thrombasthenia

Shawn M. Jobe, MD, PhD

Glanzmann thrombasthenia (GT) is an autosomal-recessive platelet function disorder in which platelet appearance and platelet number are unaffected. Patients with GT present with platelet-type bleeding, which may be severe, such as purpura, epistaxis, oral mucosal bleeding, menorrhagia, or gastrointestinal bleeding.

Pathophysiology: The most common genetic cause of GT is mutations in both alleles of one of the genes that encode for the α_{IIb} or β_3 polypeptide of the platelet integrin receptor $\alpha_{IIb}\beta_3$. $\alpha_{IIb}\beta_3$ is the integrin receptor that mediates platelet adhesion to fibrinogen and von Willebrand factor and is normally expressed at high levels on the platelet surface. In the absence of functional $\alpha_{IIb}\beta_3$, both platelet adhesion and platelet recruitment, or aggregation, are markedly impaired.

The majority of causative mutations in patients with GT result in severely decreased platelet surface expression of $\alpha_{IIb}\beta_3$. Mutations in the α_{IIb} or β_3 component result in a similar phenotype. Reported mutations include point mutations, splice defects, and small deletions and gene inversions.

Variant Glanzmann Thrombasthenia: Variant forms of GT have been identified in which normal or modestly decreased levels of $\alpha_{IIb}\beta_3$ are expressed on the platelet surface. In most cases, the phenotype in patients with variant GT is due to the expression of a dysfunctional $\alpha_{IIb}\beta_3$ protein on the platelet surface. Mutations in either the extracellular or intracellular domain of $\alpha_{IIb}\beta_3$ can result in variant GT. Intermolecular events that occur at the $\alpha_{IIb}\beta_3$ intracellular cytoplasmic tail on platelet activation are required for normal platelet function. These platelet activation events result in a change of $\alpha_{IIb}\beta_3$ structure from an inactivated form to an activated ligand binding conformation, a process known as inside-out signaling. Mutations in $\alpha_{IIb}\beta_3$ that interfere with this inside-out signaling process have been identified. Another group of patients with variant GT have mutations in the MIDAS (metal-ion-dependent adhesion site) extracellular domain of β_3. MIDAS domain mutations result in defective receptor activation, impaired ligand binding, or instability of the receptor complex on the platelet surface. Constitutive activation of $\alpha_{IIb}\beta_3$ can also result in variant GT. Platelets from a patient with GT with an extracellular mutation (Cys560Arg) in β_3 spontaneously bound fibrinogen, perhaps preventing appropriate engagement of the platelet receptor at sites of injury. Another variant form of GT, due to a mutation in kindlin-3, occurs together with a leukocyte adhesion defect, type III. Members of the kindlin family of proteins are essential for normal integrin activation. Because kindlin-3 is the primary kindlin expressed in the hematopoietic lineage, all integrins present on leukocytes and platelets, including β_1, β_2, and β_3, are dysfunctional.

Transfusion Medicine and Hemostasis. https://doi.org/10.1016/B978-0-12-813726-0.00097-0

Diagnosis: Platelet aggregation studies in patients with GT demonstrate absent or greatly diminished responses to all agonists. Response to the agglutinating agent, ristocetin, is unaffected as platelet agglutination that occurs in response to ristocetin is mediated by the GP Ib-IX receptor complex and is not $\alpha_{IIb}\beta_3$ dependent. Flow cytometry studies for CD41 (α_{IIb}) or CD61 (β_3) can be used to distinguish patients with absent $\alpha_{IIb}\beta_3$ expression from patients with variant GT. $\alpha_{IIb}\beta_3$ can be identified on the surface of platelets of patients with variant GT. Most patients with GT will have greatly decreased or absent surface levels of $\alpha_{IIb}\beta_3$.

Management: Patients with GT frequently require platelet transfusions for bleeding episodes often starting from infancy. In patients with absent $\alpha_{IIb}\beta_3$, $\alpha_{IIb}\beta_3$ on the surface of the transfused platelet may be recognized as a foreign protein. Consequently, alloimmunization to $\alpha_{IIb}\beta_3$ or HLA antigens may occur and result in the development of a platelet refractory state. Recombinant factor VIIa has been used with some effectiveness in patients with GT who are unresponsive to platelet transfusions. Hematopoietic stem cell transplantation may also be an effective therapeutic option in severe cases.

Further Reading

Buitrago, L., Rendon, A., Liang, Y., Simeoni, I., Negri, A., Thrombo Genomics, C., et al. (2015). alphaIIbbeta3 variants defined by next-generation sequencing: predicting variants likely to cause Glanzmann thrombasthenia. *Proc Natl Acad Sci USA*, *112*, E1898–E1907.

Nurden, A. T., Pillois, X., & Wilcox, D. A. (2013). Glanzmann thrombasthenia: State of the art and future directions. *Semin Thromb Hemost*, *39*, 642–655.

Poon, M. C., Di Minno, G., d'Oiron, R., et al. (2016). New insights into the treatment of Glanzmann Thrombasthenia. *Transfus Med Rev*, *30*, 92–99.

Svensson, L., Howarth, K., McDowall, A., et al. (2009). Leukocyte adhesion deficiency-III is caused by mutations in KINDLIN3 affecting integrin activation. *Nat Med*, *15*, 306–312.

CHAPTER 98

Platelet Storage Granule Defects

Shawn M. Jobe, MD, PhD

Platelets contain two major types of granules, dense and α-granules. Granule contents include numerous factors essential to the normal formation and stabilization of a hemostatic plug. Either a decrease in the number of platelet granules or a defect in their release can result in a bleeding diathesis. Several clinical syndromes are associated with storage granule defects. Bleeding manifestations in patients with storage granule defects are typically mild.

Characteristics of the δ-granules and α-granules are as follows:

Platelet-dense granules: contain calcium, serotonin, adenosine diphosphate (ADP) and adenosine triphosphate (ATP), and have a dark, electron-dense appearance in electron microscopy. Dense granule contents contribute to the activation of neighboring platelets.

Platelet α-granules: contain proteins with a number of different functions. Generally, these contents promote the adhesion of platelets and the stimulation of inflammatory and vascular cells. Adhesion proteins, procoagulant molecules, pro- and anti-angiogenic proteins, inflammatory cytokines, platelet receptors, and bactericidal proteins have been identified within the α-granule. Release of different α-granule components may be differentially regulated depending on the agonist used.

Either a decrease in number of platelet granules or a defect in their release can result in a bleeding diathesis. Several clinical syndromes, such as Hermansky–Pudlak, Chediak–Higashi, gray platelet, arthrogryposis–renal dysfunction–cholestasis (ARC) and others, can be associated with platelet granule deficiency in addition to defects in other cells and organs. Bleeding manifestations in patients with storage granule defects are typically mild.

Diagnosis: Dense granule content and release can be evaluated using luminometry. Luminometry is an assay that is often performed in concert with aggregometry, which measures the release of platelet-dense granule ATP into the plasma. To distinguish defects in δ-granule release and formation, electron microscopy can be used. Because individual platelets have only 3–5 δ-granules, assessment of multiple platelets should be performed by an experienced laboratory. Platelets from patients with no α-granules appear gray when evaluated by light microscopy, due to the absence of the basophilic α-granule. Expression of the P-selectin (CD62p) adhesive protein on the platelet surface is frequently used as a marker of α-granule membrane fusion. Fusion of the α-granule membrane with the external platelet membrane occurs on platelet activation. As a result of this membrane fusion, P-selectin is expressed on the platelet surface. The presence of additional systemic manifestations or hematologic findings may also aid in diagnosis.

Transfusion Medicine and Hemostasis. https://doi.org/10.1016/B978-0-12-813726-0.00098-2

Management: Bleeding manifestations of platelet storage defects are typically mild and can be treated with supportive agents, such as desmopressin (DDAVP) and antifibrinolytic agents. Platelet transfusion may be necessary in cases of severe bleeding. Care of these patients is often more focused on the accompanying systemic manifestations.

Hermansky–Pudlak Syndrome: Hermansky-Pudlak syndrome (HPS) is an autosomal recessive disease characterized by the absence of platelet-dense granules and oculocutaneous albinism. Ophthalmologic complications include nystagmus and cataracts. In addition, pulmonary fibrosis, typically noted clinically in the fourth decade of life, occurs in 60% of patients, and granulomatous colitis affects approximately 15% with *HPS1*.

Mutations in 10 different genes have been reported to result in HPS. All mutations result in defective lysosome packaging and formation. *HPS1* is the most common of the nine HPS subtypes (*HPS1*, *AP3B1*, *HPS3*, *HPS4*, *HPS5*, *HPS6*, *BLOC1S3*, *Palin* and *DTNBP1*, *AP3D1*) and is responsible for approximately 90% of the cases. Most cases of *HPS1* occur in members of a pedigree residing on the Caribbean island of Puerto Rico, secondary to a founder effect. Characterization of the causative molecular defect is recommended in patients with suspected HPS, as it may contribute to the management of associated syndromic characteristics and familial counseling.

Chediak–Higashi Syndrome: Chediak–Higashi syndrome (CHS), like HPS, is characterized by a paucity of dense granules and variable degrees of oculocutaneous albinism. The gene defect responsible for CHS occurs in the *LYST* gene, which is presumed to play a role in vesicular trafficking. The blood smear of patients with CHS is remarkable for the presence of granulocytes with huge cytoplasmic granules. Neutropenia and pronounced immunodeficiency occur in individuals with CHS. Patients with CHS often develop a potentially life-threatening complication in the first decades of life; a nonmalignant lymphohistiocytic infiltration referred to as the accelerated phase of CHS. Hematopoietic stem cell transplantation cures the hematologic manifestations of the illness, but neurologic manifestations of the illness, such as ataxia and decreased cognitive abilities, will continue to progress and may present later in life.

Gray Platelet Syndrome(s): Patients with gray platelet syndrome (GPS) have a gray appearance when evaluated on a blood smear, due to a paucity of basophilic α-granules. In some pedigrees, GPS is associated with a mild to severe myelofibrosis and extramedullary hematopoiesis. In another pedigree, the granule deficiency affects other hematopoietic cells, and gray-appearing neutrophils with absent granules are observed. Macrothrombocytopenia has been reported in some pedigrees.

The gene defect responsible for the classical GPS has recently been identified. Mutations in the *NBEAL2* gene were identified in multiple pedigrees with GPS. Interestingly, *NBEAL2*, like *LYST*, encodes for a protein domain known as a BEACH domain implicating an important potential role of this protein structure in granule biogenesis. Myelofibrosis in patients with *NBEAL2* deficiency may be related to premature ectopic secretion of α-granules into the bone marrow by developing megakaryocytes.

Other syndromes may also result in a paucity of α-granules and a gray platelet phenotype. An X-linked form of GPS occurs in individuals with a mutation in the

DNA-binding region of the megakaryocyte and erythroid transcription factor GATA-1. Because GATA-1 is also essential in erythropoiesis, these patients often will also have an associated thalassemia. Gray platelets are also one of the defining characteristics of ARC syndrome. Patients with ARC syndrome have a molecular defect in either the *VPS33B* gene or *VP16B* gene, which encodes a closely associated protein. Both *VPS33B* and *VP16B* encode proteins essential for normal vacuolar fusion. Recently, an autosomal-dominant GPS was identified in association with mutations in *GFI1B*.

White Platelet Syndrome: White platelet syndrome is an autosomal-dominant disorder characterized by macrothrombocytopenia and decreased α-granule content. This syndrome was described in a large multigenerational pedigree from Minnesota. The α-granule content of the platelets is low, giving some of the platelets a gray appearance. The unique characteristic of platelets in patients with white platelet syndrome is the presence of large, fully developed Golgi complexes. Typically, Golgi bodies are only present during thrombopoiesis. The gene defect(s) responsible is unknown.

Other Granule Defects: Quebec platelet syndrome is an autosomal dominant bleeding disorder. α-granules are present, but α-granule protein levels are decreased. Degradation of α-granule proteins is due to overexpression of urokinase plasminogen activator in the platelet α-granule. The causative genetic defect is a tandem duplication of the *PLAU* gene, which encodes for urokinase plasminogen activator. Release of the ectopically expressed urokinase in the formed clot may also result in accelerated clot lysis and delayed bleeding manifestations. Antifibrinolytic agents may substantially benefit these patients.

Patients with the absence of both α- and δ-granules have also been reported in the literature. These patients typically had a mild to moderate bleeding diathesis.

Further Reading

Blavignac, J., Bunimov, N., Rivard, G. E., & Hayward, C. P. (2011). Quebec platelet disorder: Update on pathogenesis, diagnosis, and treatment. *Semin Thromb Hemost*, 6, 713–720.

Chen, C. H., Lo, R. W., Urban, D., Pluthero, F. G., & Kahr, W. H. (2017). alpha-granule biogenesis: From disease to discovery. *Platelets*, *28*, 147–154.

Monteferrario, D., Bolar, N. A., Marneth, A. E., et al. (2014). A dominant-negative GFI1B mutation in the gray platelet syndrome. *N Engl J Med*, *370*, 245–253.

Paterson, A. D., Rommens, J. M., Bharaj, B., et al. (2010). Persons with Quebec platelet disorder have a tandem duplication of PLAU, the urokinase plasminogen activator gene. *Blood*, *115*, 1264–1266.

Seward, S. L., Jr., & Gahl, W. A. (2013). Hermansky-pudlak syndrome: Health care throughout life. *Pediatrics*, *132*, 153–160.

White, J. G., Key, N. S., King, R. A., & Vercellotti, G. M. (2004). The white platelet syndrome: A new autosomal dominant platelet disorder. *Platelets*, *15*, 173–184.

CHAPTER 99

Failure to Release and Aspirin-Like Defects

Shawn M. Jobe, MD, PhD

Multiple platelet signaling and synthetic pathways are necessary for the optimal propagation and stabilization of the forming thrombus. Mutations in platelet receptors, signaling proteins, and synthetic pathways have been identified in patients with granule release and aspirin-like defects. Patients with platelet granule release and aspirin-like defects typically have mild mucocutaneous bleeding. Rare patients have been identified with an isolated defect in platelet procoagulant activity.

Defects in pathways mediating thromboxane A2 synthesis, ADP release and responsiveness, and platelet procoagulant activity have been identified in patients with congenital disorders of hemostasis (Fig. 99.1). Representative disorders are described below.

Examples of Failure of Granule Release and Aspirin-Like Defects

Thromboxane Pathway Defects: Thromboxane (TxA2) is an agonist released by activated platelets, which supports both autocrine (self) and paracrine (adjacent) platelet activation. The importance of this pathway in thrombus formation is illustrated by the clinical effectiveness of aspirin. Aspirin's antithrombotic effects are mediated through irreversible inhibition of cyclooxygenase, a key enzyme in the TxA2 synthetic pathway. Aspirin-like platelet defects are observed in patients with cyclooxygenase deficiency, and a similar phenotype is observed in patients with a deficiency of another enzyme necessary for TxA2 synthesis, thromboxane synthase. TxA2's effectiveness depends on its ability to bind and activate the G-protein-coupled TxA2 receptor, and mutations in the TxA2 receptor have been found in kindred with a mild-platelet function disorder.

ADP/ATP Receptor Defects: ADP and ATP are soluble platelet agonists released from the dense granules of activated platelets. ADP binding to the receptor P2Y12 on adjacent platelets stabilizes the thrombus. Two clinically important antithrombotics, clopidogrel and ticlopidine, function by inactivating P2Y12. Patients with dysfunctional or absent P2Y12 receptor present clinically with a mild bleeding diathesis and have decreased aggregation in response to ADP. Dominant negative mutation of another ADP/ATP receptor P2X1 results in a similar clinical phenotype.

Defects in Platelet Intracellular Signaling Pathways: The soluble platelet agonists ADP, TxA2, and thrombin, bind G-protein-coupled receptors. These receptors depend on the actions of several G-proteins for activation of downstream signaling pathways. A mutation in the G-protein, Gαq, has been found in a patient with a mild bleeding diathesis and decreased aggregation to multiple agonists. Alterations in other intracellular signaling pathways, including calcium mobilization, tyrosine kinase activity, and phospholipase activity, have also been identified in patients with bleeding

Transfusion Medicine and Hemostasis. https://doi.org/10.1016/B978-0-12-813726-0.00099-4

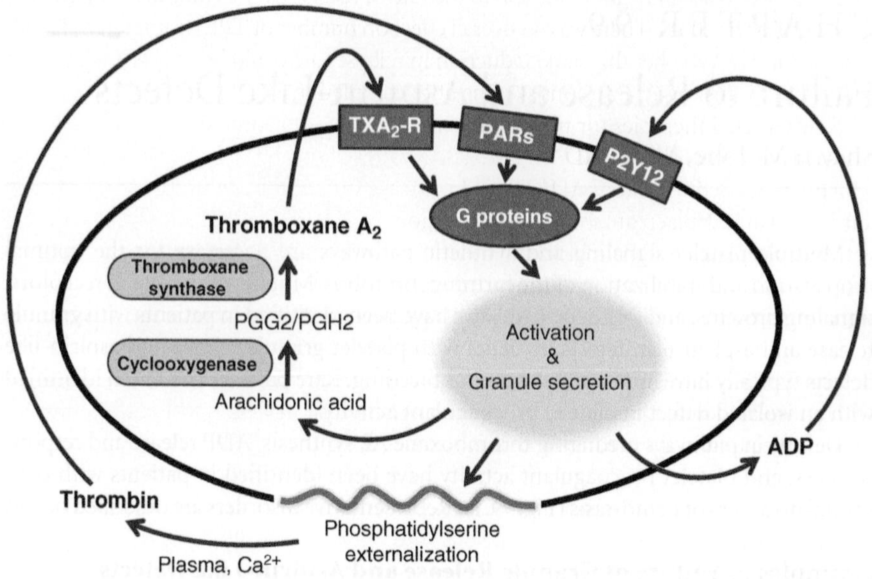

FIGURE 99.1 Paracrine and autocrine pathways support platelet activation. A simplified schematic of key pathways supporting platelet function is presented. Illustrated pathways have either been identified in individuals with abnormal platelet function or have been pharmacologically targeted in the clinic (P2Y12, clopidogrel; cyclooxygenase, non-steroidal anti-inflammatory drugs (NSAIDs), and aspirin). Not shown here for the purpose of simplification are numerous additional intracellular signaling pathways that contribute to activation and granule secretion.

disorders and abnormal platelet aggregation responses. The genes responsible for many of these alterations remain unidentified.

Defects in Platelet Procoagulant Activity: Disorders of platelet procoagulant activity have been demonstrated in several bleeding disorder patients with normal platelet aggregation activity and normal platelet count. One such disorder, Scott syndrome, is characterized by normal platelet aggregation and platelet counts, and decreased platelet procoagulant activity. Decreased procoagulant activity occurs due to impaired phosphatidylserine externalization, a critical factor accelerating the assembly and activity of the procoagulant tenase and prothrombinase enzyme complexes on the platelet surface. Causative mutations in the gene *TMEM16F*, which encodes for a membrane protein, have been identified in patients with Scott syndrome. Interestingly, another bleeding disorder, Stormorken syndrome, is characterized by increased platelet phosphatidylserine externalization and platelet procoagulant activity on circulating platelets. Defects in platelet adhesiveness have been suggested to account for some of the bleeding manifestations in these patients.

Diagnosis: Typically, aggregation studies indicate an impaired second wave of aggregation and decreased ATP-mediated luminescence when luminometry is performed. Aggregation studies will not identify a patient with a procoagulant defect, and specialized testing is required to examine for a defect in platelet phosphatidylserine externalization (see Chapters 141 and 142). Alternatively, targeted sequencing of genes can assist in the diagnosis and subsequent management of symptomatic patients with suspect platelet function defect due to a congenital mutation affecting one of these pathways.

Management: Adequate hemostasis can usually be achieved using DDAVP or an antifibrinolytic agent. More severe bleeding may require platelet transfusion.

Further Reading

Castoldi, E., Collins, P. W., Williamson, P. L., & Bevers, E. M. (2011). Compound heterozygosity for 2 novel TMEM16F mutations in a patient with Scott syndrome. *Blood, 117*, 4399–4400.

Cattaneo, M. (2011). The platelet $P2Y_{12}$ receptor for adenosine diphosphate: Congenital and drug-induced defects. *Blood, 117*, 2102–2112.

Rao, A. K. (2003). Inherited defects in platelet signaling mechanisms. *J Thromb Haemost, 1*, 671–681.

Stormorken, H., Holmsen, H., Sund, R., et al. (1995). Studies on the haemostatic defect in a complicated syndrome. An inverse Scott syndrome platelet membrane abnormality? *Thromb Haemost, 74*, 1244–1251.

Suzuki, J., Umeda, M., Sims, P. J., & Nagata, S. (2010). Calcium-dependent phospholipid scrambling by TMEM16F. *Nature, 468*, 834–838.

CHAPTER 100

Other Platelet Abnormalities

Grace F. Monis, MD, PhD

There are countless mechanisms for causing platelet abnormalities, many of which are likely yet to be discovered. This chapter is meant to touch on additional platelet abnormalities not completely discussed in other chapters of this book—but is not by any means all inclusive. It is important to note that some platelet abnormalities may incidentally be detected in laboratory assays without an associated manifestation of clinical signs or symptoms. Such results can lead to an inappropriate (and harmful) action when interpreted without clinical context. With this in mind, consider these categories in the differential diagnosis of a platelet-like disorder workup to create a more comprehensive evaluation and continue to look further.

Platelet Abnormality as a Side Effect of Medications and Supplements: The effects of antiplatelet medications, aspirin, and other nonsteroidal antiinflammatory medications (NSAIDs) have already been described in other chapters. In addition to these, many drugs and herbal supplements have a platelet disorder listed in their side effect profile. Such drugs can provoke a thrombocytopathy and/or thrombocytopenia. Examples of these drugs include antiepileptics, antipsychotics, endocrine replacements, over-the-counter supplements and food ingredients (Table 100.1). It is recommended to review all medications and supplements for side effects, and if clinically safe, recommend that the patient refrain from taking these medications in a reasonable time before platelet testing and/or surgical procedures to clear them from the patient's system.

Platelet Abnormalities Related to Shear Stress and Artificial Surfaces: The interaction of blood with nonbiological surfaces and/or conditions that increase turbulent flow is associated with platelet abnormalities. A resulting acquired von Willebrand disease and activation of coagulation pathways are also contributing factors. Examples of devices with artificial surfaces are often used during cardiopulmonary bypass and extracorporeal membrane oxygenation. Increased shear stress and turbulent flow is seen in heart valve disease (e.g., aortic stenosis), cavernous hemangiomata (Kasabach–Merritt syndrome), and disseminated intravascular coagulation. It is well known that thrombocytopenia ensues due to consumption, but platelets also may show signs of dysfunction. This phenomenon has been described as an "exhausted platelet syndrome" where the platelets are initially hyperactivated by injured endothelium releasing ADP, then subsequently, become unresponsive. Other studies describe mechanisms of thrombocytopathy involving alteration of platelet membrane lipid organization, loss of platelet surface integrins and receptors (integrin a2bB3, GPIbα, and GPVI shedding), and oxidative stress.

Transfusion Medicine and Hemostasis. https://doi.org/10.1016/B978-0-12-813726-0.00100-8

TABLE 100.1 Examples of "Other" Drugs and Supplements That Can Induce Platelet Abnormalities

Drug Category	Common Examples
Neurological/antipsychotics	Selective serotonin reuptake inhibitors, valproic acid, levetiracetam (Keppra), levodopa
Endocrine	Levothyroxine
Antibiotics	Beta-lactam antibiotic
Methylxanthines	Caffeine, theobromine, theophylline
Supplements, herbs, and foods	Fish oil, garlic, ginseng, *Ginkgo biloba*, ginger, bilberry, dong quai, feverfew, turmeric, meadowsweet, willow, black tree fungus, alcohol

Thrombocytopathy Associated With Systemic Derangements

Uremia: Patients with uremia during renal failure have an increased risk of bleeding and thrombosis. The platelet dysfunction is likely multifactorial involving the effects of several accumulated toxins that result in both prothrombotic and platelet inhibitory mechanisms that contribute to hemostatic imbalance. Renal disease often has an associated anemia, which adds to ineffective platelet function and platelet contact along endothelial surfaces (observed in vitro). Alterations to platelet membranes and signaling may occur as well. Some studies point to reduced integrin response to stimulation, reduced thromboxane A2, decreased dense granules, elevated cyclic AMP and cyclic GMP (molecules that inhibit platelet signaling). Treatment with dialysis and erythropoietin can help mitigate the bleeding and thrombosis of renal failure.

Trauma and Resuscitation: Platelet dysfunction during trauma and resuscitation is due to many reasons. Hemodilution causes thrombocytopenia, anemia, and decreased coagulation factors. In vitro studies found that a 40% hemodilution alone caused a 40%–50% decrease of both platelet adhesion and aggregation. In the trauma setting, this is combined with acidosis, endothelial disruption, dysfibrinogenemia, and hypothermia to alter normal platelet functioning. Obstetrical emergencies can also present a similar profile of hemostatic imbalance to trauma patients with hemodilution of pregnancy.

Acquired Inhibitors of Platelet Function: Heparin-induced thrombocytopenia and immune thrombocytopenia are discussed in other chapters. It is important to consider that other antibodies or paraproteins formed toward specific platelet receptors (such as VWF receptor GP1b and collagen receptor GPVI) or nonspecific platelet coating may also cause acquired thrombocytopathies. This can be seen in oncology patients with monoclonal gammopathies or Waldenstrom macroglobulinemia. A case has also been reported implicating the formation of an atypical IgM platelet cold agglutinin.

Further Reading

Bedi, H. S., Tewarson, V., & Negi, K. (2016). Bleeding risk of dietary supplements: A hidden nightmare for cardiac surgeons. *Indian Heart J*, *68*(Suppl. 2), S249–S250.

Casari, C., & Bergmeier, W. (2016). Acquired platelet disorders. *Thromb Res, 141*(Suppl.2), S73–S75.

Cowman, J., Müllers, S., Kenny, D., et al. (2017). Platelet behaviour on von Willebrand factor changes in pregnancy: Consequences of haemodilution and intrinsic changes in platelet function. , *7*(1).

Djunic, I., Elezovic, I., Ilic, V., Milosevic-Jovcic, N., Bila, J., Suvajdzic-Vukovic, N., et al. (2014). The effect of paraprotein on platelet aggregation. *J Clin Lab Anal, 28*(2), 141–146.

Hacquard, M., Richard, S., Lacour, J., Lecompte, T., & Vespignani, H. (2009). Levetiracetam-induced platelet dysfunction. *Epilepsy Res, 86*(1), 94–96.

Horacek, J., Maly, I., Svilias, L., Smolej, P., Cepkova, Vizda, Sadilek, Fatorova, & Zak (2015). Prothrombotic changes due to an increase in thyroid hormone levels. *Eur J Endocrinol, 172*(5), 537–542.

Jalaer, I., Tsakiris, D., Solecka-Witulska, B., & Kannicht, C. (2017). The role of von Willebrand factor in primary haemostasis under conditions of haemodilution. *Thromb Res, 157*, 142–146.

Konkle, B. (2011). Acquired disorders of platelet function. *Hematol Am Soc Hematol Educ Program, 2011*, 391–396.

Mondal, N., Chen, Z., Trivedi, J., Sorensen, E., Pham, S., Slaughter, M., et al. (2017). Oxidative stress induced modulation of platelet integrin $\alpha2b\beta3$ expression and shedding may predict the risk of major bleeding in heart failure patients supported by continuous flow left ventricular assist devices. *Thromb Res, 158*, 140–148.

Munday, A. D., Mersha, M. S., Bolgiano, D., Konkle, B. A., & Lopez, J. A. (2011). Intake of Omega-3 fatty acids suppresses shear-induced platelet aggregation and reduces lipid raft localization of the platelet glycoprotein Ib-IX-V complex. *Blood, 118*(21), 1128.

Ranucci, M., Baryshnikova, E., Ciotti, S., & Silvetti (2017). Hemodilution on cardiopulmonary bypass: Thromboelastography patterns and coagulation-related outcomes. *J Cardiothorac Vasc Anesth, 31*(5), 1588–1594.

Shenkman, B., Budnik, I., Einav, Y., Hauschner, H., Andrejchin, M., & Martinowitz, U. (2016). Model of trauma-induced coagulopathy including hemodilution, fibrinolysis, acidosis, and hypothermia: Impact on blood coagulation and platelet function. *J Trauma Acute Care Surg, 82*(2), 287–292.

CHAPTER 101

Acute (Childhood) Immune Thrombocytopenia

Deepa Manwani, MD and William B. Mitchell, MD

Immune thrombocytopenia (ITP) is a bleeding disorder characterized by immune-mediated platelet destruction and resultant thrombocytopenia. Two forms of ITP had been described previously: acute ITP that resolves within 6 months and chronic ITP that persists beyond 6 months. Recently, an international working group on ITP has modified these designations to reflect the relatively common occurrence of spontaneous remission in children between 6 months and 1 year of diagnosis. Newly diagnosed ITP now denotes those within the first 3 months of diagnosis, persistent ITP refers to those between 3 months and 1 year of diagnosis, and chronic ITP now represents more than 1 year of disease persistence. ITP is one of the most common acquired bleeding disorders in the pediatric population, with an estimated prevalence of 4–8 per 100,000 children per year. Children typically present with severe thrombocytopenia but rarely have serious bleeding. In the majority of children, the disease resolves spontaneously over a period of weeks, irrespective of treatment. Only 10%–20% of children with ITP go on to have chronic disease. In contrast, most adults who are present with ITP develop chronic disease with thrombocytopenia persisting beyond 1 year, and spontaneous recovery occurs in less than 10% of such cases. Acute and chronic ITP are approached differently and therefore are discussed in two separate chapters; acute childhood ITP will be reviewed in this chapter, whereas chronic ITP in adults will be presented in Chapter 102.

Pathophysiology: Although ITP is an autoimmune disease, the precise pathophysiologic mechanisms responsible for the disorder are poorly understood. A major feature of ITP is Fc receptor–mediated clearance of IgG autoantibody–coated platelets by macrophages in the reticuloendothelial system. Most affected children produce both IgG and IgM autoantibodies reactive to multiple platelet antigens, usually platelet surface glycoproteins such as GPIIb/IIIa and GPIb/IX. However, not all patients have measurable antibodies, and the mechanisms that trigger autoantibody production are not clear. Children with acute ITP show increased expression of gamma interferon–dependent genes in the early stages of the disease, indicating the presence of a global proinflammatory state. T-cell abnormalities are also involved in the pathophysiology of acute ITP. Imbalances in the mutually inhibitory Th1 and Th2 patterns of cytokine response in T-helper lymphocytes have been measured in patients with ITP. Platelet autoreactive T-cell clones have been identified in the peripheral blood of children with ITP, and cytotoxic T cells from ITP patients can destroy platelets with a high effector cell to target ratio. In addition, abnormalities in the numbers and function of regulatory T cells ($CD4^+$, $CD25^+$ T cells) have been shown to be important for the development of many autoimmune diseases, including ITP. Ultimately, the normal mechanisms of

Transfusion Medicine and Hemostasis. https://doi.org/10.1016/B978-0-12-813726-0.00101-X

self-tolerance and autoantibody suppression are reestablished in the majority of children with ITP, and the thrombocytopenia resolves.

Clinical Manifestations: The physical findings of ITP are usually limited to those associated with thrombocytopenia and include petechiae, bruising, and mucocutaneous bleeding. Patients with acute ITP usually have a rapid onset of bleeding symptoms, which appear over a few hours or days. Typically, children with ITP have surprisingly few other complaints. There may be a history of a recent upper respiratory infection, febrile illness, or immunization, but the history is otherwise unremarkable. In the majority of children with ITP, there is no history of persistent fevers, recurrent infection, weight loss, bone pain, fatigue, or other constitutional symptoms. Despite severe thrombocytopenia in the majority of cases, there is a surprisingly low incidence of overt or severe bleeding. Occasionally children with ITP develop nasal or oral bleeding, but more serious or life-threatening bleeding is unusual. The most serious complication of acute ITP is intracranial hemorrhage (ICH). Although ICH (and other serious bleeding) does occur in childhood ITP, the risk is low (approximately 1:600). Some children with ITP will have other findings, such as fever, lymphadenopathy, or splenomegaly, but these findings are atypical and deserve further scrutiny for a more serious condition such as leukemia, myelodysplasia, rheumatologic disorder, or infection.

Diagnosis: ITP is a clinical diagnosis that is characterized by the acute onset of thrombocytopenia and associated symptoms in an otherwise well child, the absence of other physical or laboratory findings, and the exclusion of other causes of thrombocytopenia. Careful clinical and laboratory assessments are necessary to exclude other causes of thrombocytopenia, particularly in patients with atypical features. The most compelling diagnostic test is the presence of a robust platelet response to standard therapy.

Clinical Evaluation: The clinical evaluation of a patient presenting with possible ITP must include a detailed medical history (including past medical history, family history, medication history, review of systems, growth and development, etc.) and physical examination. A history of arthritis, diabetes, inflammatory bowel disease, or other autoimmune disease should be investigated further. A history of poor growth, frequent infections, family history of blood disease, or other unusual findings is also not consistent with ITP, and further investigation may be warranted. Full physical examination is extremely important for diagnosis. Physical findings such as jaundice, lymphadenopathy, and hepatosplenomegaly are atypical of patients with acute ITP and should be investigated further to rule out a more serious malignant, genetic, or autoimmune disorder.

Laboratory Evaluation: The most important laboratory testing for the diagnosis of ITP is the complete blood count and the examination of peripheral blood smear. Usually, the platelet count is severely low in children with ITP, often under 20,000/μL, but the white blood count and differential, hemoglobin, and red blood cell indices should be normal. The presence of a few large platelets is the characteristic of ITP, but numerous giant platelets are more typical of Bernard–Soulier syndrome or MYH-9 mutation–related disorders. Irregularities in the red blood cell or white cell number or morphology are not the characteristic of ITP and should trigger further investigation. The presence of white blood cell blasts is indicative of leukemia, and bone marrow evaluation is necessary.

Other laboratory testing may be performed as indicated, but no additional tests are part of the routine evaluation for the diagnosis of ITP. Bone marrow evaluation is usually unnecessary in the majority of typical cases and should be reserved for children presenting with atypical features. Rheumatologic studies (antinuclear antibody, lupus anticoagulant, complement, antiphospholipid assays, lupus anticoagulant, etc.) and blood chemistries are not essential for diagnosis unless specifically indicated as in a patient with an older age of presentation or atypical clinical findings. Serum immunoglobulins and rheumatologic studies should be considered before administering therapeutic immune globulin. Blood type and a Coombs test are helpful in treatment planning. If there is concern for an immune disorder, lymphocyte subsets should also be considered before the administration of steroids. Assays for platelet antibodies have poor specificity and sensitivity for ITP and are therefore not recommended for routine evaluation. Platelet survival and/or function studies likewise are also not recommended.

Differential Diagnosis: The differential diagnosis in children presenting with thrombocytopenia is broad, and patients presenting with thrombocytopenia and atypical features require more intensive evaluation. In infants presenting with thrombocytopenia, congenital thrombocytopenic disorders such as amegakaryocytic thrombocytopenia and thrombocytopenia-absent radius should be considered. Wiskott–Aldrich syndrome should be ruled out in infant boys presenting with low platelets in the setting of eczema, failure to thrive, chronic infections, and small platelets on peripheral smear. Patients with skeletal anomalies should be evaluated for Fanconi anemia or other bone marrow failure disorders. It is important to distinguish between the primary acute presentation and ITP secondary to other illnesses such as HIV, hepatitis C, systemic lupus erythematosus, autoimmune lymphoproliferative syndrome, or other autoimmune disease, as treatment in these cases is different. Leukemia, myelodysplasia, and aplastic anemia may present with thrombocytopenia but rarely as the only laboratory finding. These children usually have other presenting signs or symptoms, such as hepatosplenomegaly, limb pain, lymphadenopathy, weight loss, persistent fever, anemia, leukocytosis, or neutropenia. Bone marrow evaluation should be performed in all patients with atypical features.

Management: There is an ongoing debate among hematologists regarding treatment of children with uncomplicated acute ITP. Recent evidence suggests that serious, life-threatening bleeding is exceedingly rare in children with acute ITP, even when the platelet count is under 20,000/μL, and may not be prevented by therapy directed at raising the platelet count. The most serious complication of acute ITP is ICH. Prevention of ICH is the major motivation for treating children with acute ITP, presenting with a severely low platelet count (under 20,000/μL). A recent prospective international study of children with acute ITP showed the incidence of ICH in children with ITP to be very low (about 0.17%). Additionally, all of the first-line treatments for ITP have potentially serious side effects. Therefore, some children with typical acute ITP with no or mild bleeding symptoms may be managed with close observation alone without specific pharmacologic therapy, providing that close follow-up and frequent reevaluations can be provided. Patients must be followed closely for bleeding symptoms or for the development of a more serious condition such as aplastic anemia. For children who are present with bleeding, or who are unable to comply with close follow-up for geographic or other reasons, or in whom the risk of ICH is increased, treatment to raise the platelet count to over 20,000/μL may be indicated.

First-line pharmacologic therapies for acute ITP include intravenous immunoglobulin (IVIG), Rh immune globulin (RhIg), and corticosteroids (Table 101.1). There is no evidence that any one of these treatments is more effective than the other, and all have side effects, therefore treatment choice is somewhat arbitrary. Most patients (70%) will respond to any of the therapies by 72 hours. IVIG and RhIg are costly and must be given via intravenous infusion but are associated with a more rapid platelet response (platelets over 20,000/μL often within 24 hours). RhIg is less expensive than IVIG but is only effective in patients who are D positive with an intact spleen. RhIg also causes hemolysis and should not be given to patients with anemia or a positive Coombs' test, as the hemoglobin decreases 0.5–2 g/dL in most patients. Corticosteroids may be given orally and are relatively inexpensive but may take 48–72 hours to have a therapeutic effect.

In the majority of children, ITP resolves spontaneously in weeks to months, with or without treatment. Occasionally children will have recurrent episodes of thrombocytopenia in the months following presentation, particularly in the setting of infection. Although the risk of late relapse is unknown, it is thought to be quite low, and in general, children with ITP can expect complete resolution of symptoms without evidence of an underlying autoimmune or other hematologic or malignant process.

Some special treatment considerations include the patient who presents with acute major bleeding and acute refractory ITP. In patients with acute bleeding that is significant, combination treatments such as high-dose intravenous methylprednisolone (30 mg/kg) with IVIG are indicated. Adjunct therapies include antifibrinolytic agents for mucosal bleeding and hormonal therapy for uterovaginal bleeding. Platelet transfusions are important for life-threatening bleeding and may need to be given continuously.

TABLE 101.1 Treatment Options for Acute (Childhood) Immune Thrombocytopenia

Treatment	Dose	Course	Side Effects	Response (Platelets > 20,000/μL)
Corticosteroid: prednisone, prednisolone, solumedrol	1–4 mg/kg/day	1–14 days	Hyperglycemia, irritability, weight gain, hypertension, anxiety, dysphoria,	60% by 72 hours
IVIG	1 g/kg/day	1–2 days	Headache, chills, fever, rarely aseptic meningitis, anaphylaxis	24–48 hours
RhIg	75 μg/kg	1 dose	Headache, chills, fever, fatigue, decrease in hemoglobin, vomiting, hemolysis, anemia	24–48 hours

IVIG, intravenous immunoglobulin; *RhIg*, Rh immune globulin.

In refractory patients, subsequent approaches include combination therapy, rituximab, thrombopoietin-receptor agonists, or splenectomy. Alhough there are numerous second- and third-tier agents (including immunosuppressive and chemotherapeutic agents) because of paucity of data, there are no clear guidelines on the sequence in which these should be used.

Further Reading

Blanchette, V., & Bolton-Maggs, P. (2008). Childhood immune thrombocytopenic purpura: Diagnosis and management. *Pediatr Clin North Am*, *55*, 393–420.

Bolton-Maggs, P., Tarantino, M. D., Buchanan, G. R., et al. (2004). The child with immune thrombocytopenic purpura: Is pharmacotherapy or watchful waiting the best initial management? A panel discussion from the 2002 meeting of the American Society of Pediatric Hematology/Oncology. *J Pediatr Hematol Oncol*, *26*, 146–151.

Cooper, N. (April 2017). State of the art - how I manage immune thrombocytopenia. *Br J Haematol*, *177*(1), 39–54.

Cooper, N., & Bussel, J. (2006). The pathogenesis of immune thrombocytopaenic purpura. *Br J Haematol*, *133*, 364–374.

Cuker, A., Cines, D. B., & Neunert, C. E. (September 2016). Controversies in the treatment of immune thrombocytopenia. *Curr Opin Hematol*, *23*(5), 479–485.

Johnsen, J. (2012). Pathogenesis in immune thrombocytopenia: New insights. *Hematology Am Soc Hematol Educ Program*, 306–312.

Kuhne, T., Buchanan, G. R., Zimmerman, S., et al. (2003). Intercontinental childhood ITPSG. A prospective comparative study of 2540 infants and children with newly diagnosed idiopathic thrombocytopenic purpura (ITP) from the Intercontinental Childhood ITP Study Group. *J Pediatr*, *143*, 605–608.

Neunert, C., Lim, W., Crowther, M., et al. (2011). The American Society of Hematology 2011 evidence-based practice guideline for immune thrombocytopenia. *Blood*, *117*, 4190–4207.

Nugent, D. J. (2006). Immune thrombocytopenic purpura of childhood. *Hematol Am Soc Hematol Educ Program*, 97–103.

Rodeghiero, F., Stasi, R., Gernsheimer, T., et al. (2009). Standardization of terminology, definitions and outcome criteria in immune thrombocytopenic purpura of adults and children: Report from an international working group. *Blood*, *113*, 2386–2393.

Tarantino, M. D., Young, G., Bertolone, S. J., et al. (2006). Single dose of anti-D immune globulin at 75 microg/kg is as effective as intravenous immune globulin at rapidly raising the platelet count in newly diagnosed immune thrombocytopenic purpura in children. *J Pediatr*, *148*, 489–494.

CHAPTER 102
Chronic Immune Thrombocytopenia

William B. Mitchell, MD

Introduction: Immune thrombocytopenia (ITP) is a bleeding disorder characterized by immune-mediated platelet destruction with resultant thrombocytopenia and mucocutaneous bleeding. Chronic ITP is defined by ITP persistence beyond 12 months, with spontaneous recovery occurring in less than 10% of adults. The estimated incidence of ITP is ~100 cases per 1 million persons per year, with about half in adults. Approximately twice as many women are affected as men. Patients with chronic ITP often have problematic bleeding that requires ongoing therapy, and hemorrhagic deaths are not uncommon. Chronic ITP in adults is presented in this chapter. Newly diagnosed/childhood ITP is reviewed in Chapter 101.

Pathophysiology: Chronic ITP is a heterogeneous disease that may be the common end-stage phenotype of numerous immunologic insults. The initiating event is a complex immunologic process, involving platelets, megakaryocytes, B cells, T cells, and other components of the immune system. A major feature of ITP is the Fc receptor–mediated clearance of IgG autoantibody–coated platelets by macrophages in the reticuloendothelial system. Platelet-associated IgG autoantibodies can be measured in about 50%–60% of patients with ITP and are usually reactive to platelet surface glycoproteins such as GPIIb/IIIa, GPIa/IIa, and GPIb/IX. However, approximately 40%–50% of patients do not have measurable antibodies, and therapies aimed at reducing autoantibody production and autoantibody-mediated platelet destruction are effective in only 50%–70% of patients.

T cells and their secretory factors are important for the stimulation of antibody-producing B-cell clones and may be important in the development of antiplatelet antibody production. Cytotoxic T cells from ITP patients can also destroy platelets with a high effector cell to target ratio. Platelet autoreactive T-cell clones have been identified in patients with chronic ITP, and genes involved in T cell–mediated cytotoxicity are upregulated in many ITP patients. Regulatory T cells that normally suppress autoreactive B-cell clones are defective in some patients with chronic ITP.

The pathophysiology of ITP also involves decreased or inadequate production of platelets. Thrombopoietin levels can be inappropriately low in patients with ITP. Additionally, antiplatelet antibodies and possibly antiplatelet T cells suppress production and maturation of bone marrow megakaryocytes.

Clinical Manifestations: The physical findings associated with chronic ITP are usually limited to those associated with thrombocytopenia and include petechiae, ecchymoses, and purpura. Mucocutaneous bleeding can be widespread and may include epistaxis, gingival bleeding, hematuria, and menorrhagia. Rarely, patients manifest severe bleeding, such as intracranial hemorrhage. The clinical evaluation of a

Transfusion Medicine and Hemostasis. https://doi.org/10.1016/B978-0-12-813726-0.00102-1

patient with chronic ITP must also include a detailed past medical and family history, medication history, and review of systems.

Laboratory Evaluation: The most important laboratory tests for the management of chronic ITP are the complete blood count and the examination of peripheral blood smear. The white blood cell count and differential, hemoglobin, and red blood cell indices are typically normal. The presence of large platelets is a distinguishing feature of the peripheral smear of patients with ITP, but irregularities in the red blood cell or white blood cell morphology are atypical and should trigger further investigation. Other laboratory testing may be performed as indicated, but no additional tests are necessary in the routine evaluation of chronic ITP. Bone marrow evaluation is not needed in the majority of cases but should be reserved for patients presenting with atypical features and for patients over 60 years of age. Assays for antigen-specific platelet autoantibodies have poor specificity and sensitivity for ITP and are therefore not recommended as routine testing. Additional laboratory studies, such as coagulation tests, platelet survival or function studies, rheumatologic studies, blood chemistries, and serum immunoglobulins, may be helpful in eliminating alternative diagnoses but are usually unnecessary in typical cases. Myelodysplasia and other bone marrow infiltrative or failure disorders rarely present without other cellular or physical findings; therefore, careful attention to the history, physical exam, and laboratory findings is important so that subtle findings are not overlooked.

Differential Diagnosis: ITP is the most common cause of acquired severe thrombocytopenia in an otherwise healthy individual, but the diagnosis is one of the exclusion, and other causes of thrombocytopenia must be excluded. Bone marrow failure or infiltrative disorders such as aplastic anemia, leukemia, and myelodysplasia should be considered in patients with atypical physical or laboratory findings. Patients presenting with splenomegaly should be evaluated for diseases of abnormal platelet distribution, such as congestive splenomegaly from portal hypertension. Many drugs can be responsible for thrombocytopenia, and careful drug history is important. Drug-induced thrombocytopenia is reviewed in Chapter 103. Other diseases that result in thrombocytopenia, such as sepsis, thrombotic thrombocytopenic purpura, and disseminated intravascular coagulopathy, rarely present without associated presenting features.

Management

First-Line Therapies: Treatment of chronic ITP must be tailored to the individual's platelet count and bleeding symptoms. Patients presenting with platelets >50,000/μL often have no or few symptoms and frequently do not require therapy. Patients with moderate thrombocytopenia (platelets 20,000–50,000/μL) can usually be followed closely on an outpatient basis and treated only to relieve symptomatic bleeding. Patients with severe thrombocytopenia (platelets <20,000/μL) may have overt bleeding and require therapy. Management decisions should not only be based exclusively on the platelet count but also must depend on age, lifestyle, and other medical conditions because these factors contribute to the overall risk of serious bleeding.

The most common first-line therapy for adults with chronic ITP is corticosteroid, usually prednisone (1–2 mg/kg per day for 2–4 weeks) with a taper if there is a platelet response. Pulsed high-dose dexamethasone at 40 mg per day for 4 days without taper is an effective alternative oral regimen and may be repeated at 2–4 week intervals. Approximately 50%–75% of patients will respond to corticosteroids, irrespective of dosage. Intravenous immunoglobulin (IVIG) (1 g/kg/dose for 1–2 doses) may be given alone or added to steroid if a more rapid response is required. Rh immune globulin (RhIG) (50–75 μg/kg) is also effective for adults with ITP. However, both IVIG and RhIG require IV administration and are considerably more expensive than steroids. RhIG is effective only in patients who are D positive with an intact spleen. RhIG also causes hemolysis and should not be given to patients with anemia, as the hemoglobin decreases 0.5–2 g/dL in most patients.

Splenectomy: Despite first-line measures, many patients will continue to have moderate to severe thrombocytopenia with bleeding symptoms and require alternative therapies. Splenectomy is associated with significant morbidity and an increased risk of *Streptococcus pneumonia* sepsis but is an effective and durable treatment for adults with chronic ITP who have not had a sustained response to initial therapy. Approximately 70% of patients will achieve sustained remission after splenectomy. Laparoscopic splenectomy appears to be preferable to open splenectomy, as the recovery time is shorter and the outcome is similar.

Refractory Immune Thrombocytopenia: Refractory ITP is defined as the persistence of thrombocytopenia after splenectomy. Approximately 10% of patients will be refractory to all standard therapy, including splenectomy. Although the overall prognosis of ITP is good, the mortality in this refractory and unresponsive group of patients is high—between 6% and 15%. As the likelihood of durable and complete remission in these patients is much lower, the goal is to balance the bleeding risk with lifestyle issues and toxicity of therapy. The American Society of Hematology (ASH) guidelines suggest that maintaining a platelet count of 30,000–50,000/μL in chronic ITP patients without other risk factors and without overt bleeding is a generally acceptable level for reducing the risk of spontaneous hemorrhage. An individualized approach is important for disease management, as no single protocol is appropriate for all patients. No randomized clinical trials have been performed comparing the numerous therapies for chronic ITP, and no trials have assessed the outcomes of bleeding and death. Therefore, no evidence-based algorithm can be proposed as the standard of care for patients with chronic refractory ITP. However, several agents have been shown to be useful for the treatment of chronic ITP and, while none offers a cure, some result in durable remission and are useful in alleviating bleeding symptoms.

Thrombopoietin-receptor agonists (TPO-RA) have emerged as effective agents in refractory/unresponsive ITP. Romiplostim, a thrombopoietin-mimetic peptibody, and eltrombopag, a small molecule, have been FDA approved for the treatment of chronic ITP in patients who have failed first-line therapy. The ASH 2011 *Clinical Practice Guideline on ITP* recommends these agents as an alternative to splenectomy for second-line therapy. Both agents have a 70%–80% response rate in adult patients with chronic, severe, refractory/unresponsive ITP. Patients who do not respond to one TPO-RA may respond to the other.

Rituximab, an anti-CD20 monoclonal antibody, is an effective drug for the treatment of severe, refractory ITP. The most common dosage regimen is 375 mg/m^2/week for four consecutive weeks, but other dosage regimens have been utilized. Up to 50% of patients achieve sustained platelet response with rituximab. Dexamethasone pulses added to IVIG may increase the response rate.

Many other agents have been used to treat chronic refractory ITP, but there are few randomized clinical trials to support their use. Azathioprine, cyclosporine, and mycophenolate mofetil are immunosuppressive agents commonly used to treat chronic refractory ITP and are generally well tolerated. Azathioprine, a purine analogue, may take up to 3–6 months to be effective, but the long-term efficacy is up to 45%. Cyclosporine, a calcineurin inhibitor, also has a response rate over 40%. Mycophenolate mofetil, a purine nucleotide synthesis inhibitor, has shown a favorable response in up to 75% of patients. Sirolimus is an mTOR inhibitor that has been used successfully to treat both chronic refractory ITP and multiple cytopenias (Evans syndrome). Other treatments that have been used with minimal toxicity and varying efficacy include danazol, dapsone, and 6-mercaptopurine. Chemotherapy agents such as cyclophosphamide, vinca alkaloids, combination therapies, and hematopoietic stem cell transplantation have more significant toxicities and are usually reserved for highly refractory patients with severe thrombocytopenia and bleeding. Newer investigational agents, such as monoclonal immunosuppressive agents and a Syk inhibitor, are currently under investigation.

Further Reading

Bussel, J. B., Cheng, G., Saleh, M. N., et al. (2007). Eltrombopag for the treatment of chronic idiopathic thrombocytopenic purpura. *N Engl J Med, 357*, 2237–2247.

Cines, D. B., & McMillan, R. (2007). Pathogenesis of chronic immune thrombocytopenic purpura. *Curr Opin Hematol, 14*, 511–514.

Cooper, N., & Bussel, J. (2006). The pathogenesis of immune thrombocytopaenic purpura. *Br J Haematol, 133*, 364–374.

Ghanima, W., Godeau, B., Cines, D. B., et al. (2012). How I treat immune thrombocytopenia: The choice between splenectomy or a medical therapy as a second-line treatment. *Blood, 120*, 960–969.

Godeau, B., Provan, D., & Bussel, J. (2007). Immune thrombocytopenic purpura in adults. *Curr Opin Hematol, 14*, 535–556.

Neunert, C., Lim, W., Crowther, M., et al. (2011). The American Society of Hematology 2011 evidence-based practice guideline for immune thrombocytopenia. *Blood, 117*, 4190–4207.

Rodeghiero, F., Stasi, R., Gernsheimer, T., et al. (2009). Standardization of terminology, definitions and outcome criteria in immune thrombocytopenic purpura of adults and children: Report from an international working group. *Blood, 113*, 2386–2393.

CHAPTER 103

Drug-Induced Thrombocytopenia

William B. Mitchell, MD

Introduction: Drug-induced thrombocytopenia (DIT) is a common clinical problem, and numerous drugs have been implicated in the development of thrombocytopenia. The risk of thrombocytopenia after any drug is low, and only a small number of patients taking a suspected medication will develop the problem. However, many patients who develop DIT are taking multiple medications and are critically ill, making the diagnosis difficult. Rapid recognition of thrombocytopenia in affected patients and identification and removal of the offending agent before clinically significant bleeding occurs is imperative. Available laboratory tests have limited specificity and sensitivity and lack a sufficiently rapid turn around time to make a diagnosis in the acute setting. Thus, DIT is usually a clinical diagnosis that can be supported only by resolution of thrombocytopenia after cessation of drug. Certain drugs have a well-documented association with thrombocytopenia (Table 103.1). Heparin-induced thrombocytopenia is discussed in Chapter 104.

Pathology: More than a 100 drugs have been implicated in DIT. Drugs cause thrombocytopenia by two basic mechanisms: decreased platelet production or increased platelet destruction. Decreased production is usually due to marrow suppression. It is an expected complication of chemotherapy agents but can be seen with other agents. More commonly, DIT is due to targeted immune destruction of platelets brought on by the drug.

Decreased Production: DIT resulting from decreased production is usually the result of generalized dose-dependent myelosuppression. Many chemotherapeutic agents suppress the bone marrow and cause leukopenia, anemia, and often severe thrombocytopenia. The cytopenias resolve with cessation of chemotherapy. Other nonchemotherapeutic agents, such as the antiepileptic drug valproate, can also cause myelosuppression. The dose-dependent association between valproate and neutropenia and thrombocytopenia is well documented. Selective depression of megakaryocyte production has been associated with drugs, such as thiazide diuretics, ethanol, and tolbutamide, and can lead to isolated thrombocytopenia.

Increased Destruction: Platelet destruction in DIT is usually immune-mediated, and the mechanism involves the drug-induced production of antiplatelet antibodies that bind to platelets and trigger their consumption by macrophages in the reticuloendothelial system.

Hapten-Induced Antibody: Certain small-molecule drugs may act as haptens and can become covalently linked to platelet surface antigens to form hapten–protein

Transfusion Medicine and Hemostasis. https://doi.org/10.1016/B978-0-12-813726-0.00103-3

TABLE 103.1 Drugs Commonly Implicated as Triggers of Drug-Induced Thrombocytopenia[a]

Drug Category	Drugs Implicated in Five or More Reports	Other Drugs
Heparins	Unfractionated heparin, low-molecular weight heparin	
Cinchona alkaloids	Quinine, quinidine	
Platelet inhibitors	Abciximab, eptifibatide, tirofiban	
Antirheumatic agents	Gold salts	D-penicillamine
Antimicrobial agents	β-lactam antibiotics, linezolid, rifampin, sulfonamides, vancomycin	
Sedatives and anticonvulsant agents	Carbamazepine, phenytoin, valproic acid	Diazepam
Histamine receptor antagonists	Cimetidine	Ranitidine
Analgesic agents	Acetaminophen, diclofenac, naproxen	Ibuprofen
Diuretic agents	Chlorothiazide	Hydrochlorothiazide
Chemotherapeutic and immunosuppressant agents	Fludarabine, oxaliplatin	Cyclosporine, rituximab, ABT-263

[a]For a more extensive list, see the University of Oklahoma website: (http://moon.ouhsc.edu/platelets/ditp.html).
Modified from Aster, R. H., & Bougie, D. W. (2007). Drug-induced immune thrombocytopenia. *N Engl J Med, 357*, 580–587.

complexes. Antibodies directed at these cell surface protein complexes are produced and result in platelet destruction and thrombocytopenia. This phenomenon is responsible for the thrombocytopenia observed with penicillin and penicillin derivatives. Hapten-induced immune hemolytic anemia occurs much more commonly with penicillins, whereas DIT by this mechanism is less common.

Drug-Dependent (Quinine Type) Antibody Formation: This form of DIT is characterized by the formation of an antibody that binds to platelets only in the presence of the drug in the soluble state. These antibodies are usually directed at glycoprotein GPIIb/IIIa or GPIb/V/IX complexes. The drug binds noncovalently to the glycoprotein to produce a "compound" epitope or to induce a conformational change that is recognized as a novel epitope. Quinine, sulfonamide antibiotics, and nonsteroidal antiinflammatory drugs trigger thrombocytopenia by this mechanism.

GPIIb/IIIa Inhibitors: Specific antiplatelet medications, the GPIIb/IIIa inhibitors such as tirofiban, eptifibatide, and abciximab, cause DIT by antibody-dependent mechanisms. Acute thrombocytopenia often occurs within hours of drug exposure and may suggest a nonimmune mechanism. However, in the presence of tirofiban and eptifibatide, conformational changes in the GPIIb/IIIa molecule occur, exposing neoepitopes termed ligand-induced binding sites (LIBS). Antibodies are produced that are specific to these LIBS, resulting in antibody-mediated platelet destruction that can lead to severe thrombocytopenia and bleeding. It is thought that anti-LIBS autoantibodies may be normally present in the circulation, accounting for the acute onset of thrombocytopenia in previously untreated patients. In the case of abciximab, antibody formation may also be against the murine component of the chimeric Fab fragment specific for GPIIIa.

Drug-Induced Antibody: In rare instances, drugs induce the production of true antiplatelet autoantibodies that cause platelet destruction even in the absence of the sensitizing agents. In these cases, the autoantibodies may persist indefinitely, leading to chronic DIT that is indistinguishable from immune thrombocytopenia or ITP. Gold salts and procainamide can induce this form of DIT.

Immune Complex: In this form of DIT, antibodies causing thrombocytopenia are produced as a result of drug–antibody immune complex formation, leading to platelet destruction and consumption. Heparin-induced thrombocytopenia results from antibody-bound complexes of platelet factor 4 (secreted from platelets) and heparin binding to the Fc gamma RIIA receptors on platelets. This leads to both platelet activation and clearance. This is discussed in detail in Chapter 100.

Apoptosis: The BH3 mimetic antineoplastic agents, such as ABT-263, cause a novel form of thrombocytopenia through apoptosis. Platelet life span is determined by the amount of the antiapoptotic protein BCL-X_L within each platelet. BH3 mimetic agents pharmacologically inactivate BCL-X_L, resulting in acute and dose-dependent thrombocytopenia.

Clinical Manifestations: Patients with DIT usually present with moderate to severe thrombocytopenia (platelet count $\leq$ 50,000/μL) and evidence of bleeding. Usually the thrombocytopenia becomes clinically apparent 1–2 weeks after starting the drug, but rarely it can occur in patients who have been taking a drug for months or years. Thrombocytopenia may occur more quickly if the drug has been administered previously or with drugs such as quinine or abciximab. Bleeding symptoms are variable and range from mild cutaneous findings (such as petechiae and ecchymoses) to more significant overt hemorrhage. When the thrombocytopenia is severe, patients can present with severe bleeding from the nose, gums, gastrointestinal tract, or uterus. Life-threatening spontaneous intracranial hemorrhage, while rare, can occur in patients with severe thrombocytopenia.

Diagnosis: The diagnosis of DIT must often be made on clinical findings alone, as the laboratory assays do not have rapid turn around time and may not have sufficient sensitivity to detect drug-induced antibodies.

Clinical Evaluation: DIT should be considered in any patient taking medication who presents with sudden, unexplained thrombocytopenia. It is important to obtain a complete and accurate medical history and list of medications, including all prescription and nonprescription medications, nonconventional therapies, and herbal remedies. Specific attention should be given to higher-risk medications such as β-lactam antibiotics, quinines, and sulfonamides. Often, the diagnosis is confirmed only after platelet recovery with discontinuation of a suspected, sensitizing drug.

Laboratory Evaluation: Various in vitro methods exist to detect drug-induced binding of IgG to platelets. Radiolabeled or fluorescein-labeled anti-IgG can be used to detect platelet-bound immunoglobulin, but these tests may not help distinguish between primary immune thrombocytopenia and DIT. Enzyme-linked immunosorbent assay, flow cytometry, and immunoprecipitation Western blotting have more specificity for detecting platelet-reactive antibodies induced by drugs. Unfortunately, these tests are not widely available, may take several days to complete, and have low sensitivity; therefore, they may not be supportive of the acute care of patients with suspected DIT.

Differential Diagnosis: Other causes of thrombocytopenia should be considered, particularly in hospitalized patients taking multiple medications. Thrombocytopenia from sepsis, disseminated intravascular coagulopathy, or thrombotic thrombocytopenic purpura may be difficult to distinguish from DIT, particularly in critically ill patients. Heparin-induced thrombocytopenia should be considered in patients on heparin. Patients with immune thrombocytopenia (nondrug-induced), either primary or secondary, can present with acute thrombocytopenia. Patients with HIV infection can present with thrombocytopenia that may be immune-mediated or infection-induced. Patients who present with other physical or laboratory findings should be evaluated for bone marrow failure or infiltrative disorder. Patients with thrombocytopenia and hepatosplenomegaly should be evaluated for congestive disorders, such as portal hypertension, malignancy, or hepatitis.

Management: Discontinuation of the sensitizing drug is the necessary treatment. As many drugs as possible should be discontinued in patients with suspected DIT. If a particular therapy is medically necessary, a nonimmunologic, non–cross-reacting alternative should be considered. Platelet transfusions should be used for life-threatening bleeding or patients at particularly high risk for hemorrhage but may not be effective because of the presence of antibody. In patients with severe DIT who are acutely ill, intravenous immunoglobulin (IVIG), plasma exchange, and corticosteroids have been used, but the clinical benefit is variable. Usually, the thrombocytopenia improves within the first few days after discontinuing the drug. However, sensitivity to the drug persists indefinitely, and further use of the suspected causative agent should be avoided.

Further Reading

Arnold, D. M., Kukaswadia, S., Nazi, I., et al. (2013). A systematic evaluation of laboratory testing for drug-induced immune thrombocytopenia. *J Thromb Haemostat, 11*, 169–176.

Aster, R. H., & Bougie, D. W. (2007). Drug-induced immune thrombocytopenia. *N Engl J Med, 357*, 580–587.

Li, X., Swisher, K. K., Vesely, S. K., & George, J. N. (2007). Drug-induced thrombocytopenia: An updated systematic review. *Drug Saf, 30*, 185–186.

Mason, K. D., Carpinelli, M. R., Fletcher, J. I., et al. (2007). Programmed a nuclear cell death delimits platelet life span. *Cell, 128*, 1173–1186.

Visentin, G. P., & Liu, C. Y. (2007). Drug-induced thrombocytopenia. *Hematol Oncol Clin North Am, 21*, 685–696 (vi).

CHAPTER 104

Heparin-Induced Thrombocytopenia

Allyson Pishko, MD and Adam Cuker, MD, MS

Epidemiology: The reported incidence of heparin-induced thrombocytopenia (HIT) after heparin exposure is 0.2%–7%, though the accuracy of such estimates is limited by the challenges of disease recognition and diagnosis. Risk factors may be host- or heparin-related.

Heparin-Related Risk Factors: The most important heparin-related risk factors are type of heparin formulation and length of exposure. Among heparin formulations, the incidence is lower with smaller, less charged species. Large metaanalyses suggest that low-molecular weight heparins (LMWHs) carry a 5- to 10-fold lower risk of HIT than unfractionated heparin (UFH). The risk of HIT is nearly negligible with the pentasaccharide fondaparinux. A study implementing an "Avoid Heparin" campaign has shown that replacing UFH with LMWH for prophylaxis and treatment can reduce the incidence of HIT within an institution.

Duration of heparin exposure also contributes to risk. A metaanalysis reported UFH thromboprophylaxis exposure for ≥6 days was associated with an overall incidence of HIT of 2.6%. In contrast, briefer courses were associated with an incidence of 0.2% in a hospital database review.

The dose and route of administration of heparin may contribute to the risk of HIT, but this is less clear. Several studies demonstrate a greater incidence of HIT in patients treated with intravenous therapeutic-intensity UFH than among those receiving subcutaneous thromboprophylaxis. A substantial limitation of these studies is confounding in treatment indication between the two groups, which makes the true risk differences difficult to interpret.

Host-Related Risk Factors: Host-related risk factors for HIT include patient population and demographic variables. There is a reported threefold greater risk among surgical than medical patients. Type of surgery also appears to be important. Cardiac surgery patients have a reported incidence of HIT of 1%–2% versus ~5% of patients who receive UFH thromboprophylaxis after major orthopedic surgery. The mechanistic basis of these risk differences is unknown but may in part relate to differences in platelet activation and PF4 release associated with the respective surgeries. Similarly, in trauma patients, the severity of trauma affects HIT risk with a greater incidence of HIT after major trauma compared with minor trauma. The risk of HIT is <1% in critically ill patients and those undergoing chronic hemodialysis and <0.1% in obstetrical patients and children.

Demographic factors such as age and gender also contribute to risk. HIT occurs almost exclusively in middle-aged and older individuals. As with some other immune-mediated disorders, women may be at increased risk for HIT.

Transfusion Medicine and Hemostasis. https://doi.org/10.1016/B978-0-12-813726-0.00104-5

Prevalent infections such as periodontitis predispose to anti-PF4/heparin antibody formation and may increase disease risk. Infections with gram-negative organisms were shown in a case–control study to be associated with a higher incidence of antiheparin/PF4 antibody seroconversion.

Small genetic studies suggest association of Fc receptor and platelet glycoprotein polymorphisms with the risk of developing HIT and HIT-associated thrombosis but require confirmation. A retrospective, genome-wide association study suggested TDAG8 and HLA-DRA single-nucleotide polymorphisms (SNPs) were potentially associated with development of HIT.

Pathophysiology: Unlike most drug-induced thrombocytopenias, which predispose to hemorrhage, platelet activation and predisposition to thrombosis are central to HIT pathogenesis (Fig. 104.1). On exposure to heparin, susceptible hosts mount an antibody response to complexes of PF4, a chemokine stored in the α-granules of platelets, and heparin. The targets of this response are epitopes on PF4, which become exposed as a result of a conformational change in the molecule when it complexes with heparin. Although PF4/heparin complexes range in size, large multimolecular complexes, which form optimally over a narrow molar ratio of PF4 to heparin, show the greatest capacity to promote platelet activation. The antibody response to these complexes is polytypic and may consist of IgM, IgA, and/or IgG. However,

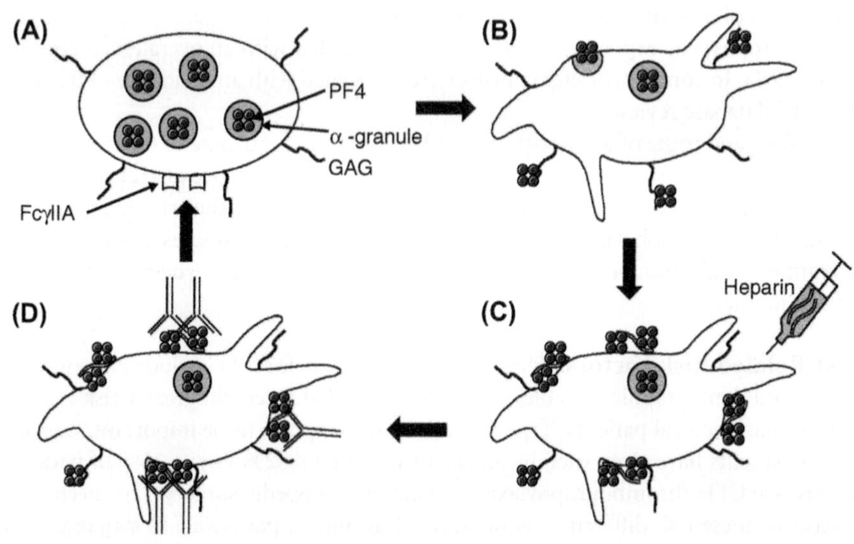

FIGURE 104.1 A simplified model of heparin-induced thrombocytopenia pathophysiology. A resting platelet with platelet factor 4 (PF4)–containing α-granules is shown (A). On platelet activation, positively charged PF4 tetramers are released from the platelet and bind to negatively charged glycosaminoglycans (GAGs) on the cell surface (B). When negatively charged heparin molecules are administered, they displace the GAGs, forming large multimolecular complexes with PF4 tetramers on the platelet surface (C). PF4 undergoes a conformational change on complexing with heparin, exposing a cryptic epitope. The Fab region of HIT IgG binds to this epitope, forming HIT immune complexes. The Fc portion engages the activating FcγIIA receptors on the surface of the same or adjacent platelets, thereby establishing a positive feedback loop of platelet activation, PF4 release, and immune complex formation (D).

anti-PF4/heparin molecules of the IgG class carry the primary potential for pathogenicity. When these molecules bind PF4/heparin complexes on the platelet surface, their Fc portion engages activating FcγIIA receptors on the same or adjacent platelets, leading to additional platelet activation and release of PF4. Activation of coagulation on the surface of the activated platelets results in thrombin generation. HIT antibodies also activate monocytes to promote thrombosis through release of tissue factor and other procoagulant molecules and may further promote coagulation by inhibiting activation of the anticoagulant, protein C, through disruption of its interaction with thrombomodulin. In addition to their effects on platelets, PF4/heparin immune complexes activate endothelial cells, resulting in enhanced elaboration of proadhesive molecules such as E-selectin and von Willebrand factor.

Diagnosis: HIT is a clinicopathologic disorder, the diagnosis of which rests on a compatible clinical picture and laboratory demonstration of anti-PF4/heparin antibodies that induce heparin-dependent platelet activation.

Clinical Diagnosis: The cardinal clinical manifestation of HIT is a fall in platelet count in the setting of a proximate heparin exposure. However, this scenario is common among hospitalized patients, occurring in 36% of hospitalized patients who received heparin for ≥4 days in a multicenter study, and has poor specificity for HIT. Therefore, attention to other characteristic clinical features, including the timing and degree of thrombocytopenia, the presence of thrombosis or hemorrhage, and the plausibility of alternative explanations for thrombocytopenia, is mandatory for assessing the clinical probability of HIT.

Timing of Platelet Count Fall: A platelet count fall beginning 5–10 days after initial heparin exposure is typical for HIT. Patients with previous exposure to heparin may develop *rapid-onset HIT* within a few hours of exposure to heparin due to preexisting anti-PF4/heparin antibodies. Owing to the transient nature of these antibodies, such patients invariably have a history of recent heparin exposure, usually within the preceding month. Rarely, manifestations of HIT may develop after heparin is discontinued. This phenomenon, referred to as *delayed-onset HIT*, occurs a median of 10–14 days after heparin withdrawal and is often associated with disseminated intravascular coagulation (DIC).

Degree of Platelet Count Fall: Degree of platelet count fall, measured from the peak platelet count following heparin initiation to the nadir platelet count, is ≥50% in most patients. The nadir platelet count need not meet the traditional laboratory definition of thrombocytopenia ($<150 \times 10^9$/L). This is particularly relevant in surgical patients who experience an initial postoperative thrombocytosis followed by a significant decline in platelet count that does not fall below this threshold. The median nadir platelet count in HIT is ~60×10^9/L and seldom falls below 20×10^9/L unless concomitant DIC is present.

Thrombosis and Hemorrhage: In contrast to most other thrombocytopenic disorders, significant bleeding events are rare in HIT regardless of the degree of thrombocytopenia. The most frequent complication is thrombosis, which may be limb- or life-threatening. New thromboembolism is the presenting feature in approximately half of individuals. Patients without thrombosis at presentation have up to a 40% risk of developing thrombosis in the 10 days following cessation of heparin if an alternative anticoagulant is not initiated promptly. Lower extremity deep vein thrombosis (DVT), pulmonary embolism, and central venous catheter-related DVT are the most common thrombotic manifestations of HIT, outnumbering arterial events by ~2:1. Arterial thromboembolism most frequently involves the peripheral arteries, but stroke and myocardial infarctions may also occur in association with HIT. Thrombosis of the cerebral sinuses, visceral vessels, and the bilateral adrenal veins with resultant hemorrhagic necrosis is well documented.

Unusual Clinical Sequelae: Rare sequelae of HIT, which may curiously occur in the absence of thrombocytopenia, include anaphylactoid reactions following intravenous heparin bolus, transient global amnesia, and skin necrosis at subcutaneous injection sites.

Alternative Causes of Thrombocytopenia: Just as important as remaining vigilant for characteristic clinical features of HIT is a deliberate search for alternative etiologies of thrombocytopenia. HIT must be distinguished from a nonimmune form of thrombocytopenia (formerly called HIT type I) characterized by a lesser fall in the platelet count that typically occurs within 48 hours of heparin exposure and resolves with continued administration. This condition, due to a direct platelet-activating effect of heparin, is of no known clinical consequence. Other thrombotic thrombocytopenias such as antiphospholipid syndrome and malignancy-associated microangiopathy may masquerade as HIT. Other causes of hospital-acquired thrombocytopenia include infection, medications other than heparin, DIC, hemodilution, posttransfusion purpura, and intravascular devices such as balloon pumps, ventricular assist devices, and cardiopulmonary bypass. After CPB, the platelet count falls by a mean of 42% and typically begins to recover on postoperative day 2 or 3. Recovery in platelet count after CPB followed by a second fall between days 5 and 10 is highly suggestive of HIT. A platelet count fall immediately post-CPB that persists beyond 4 days without recovery should trigger the clinician to explore alternate etiologies of thrombocytopenia such as infection.

Clinical Scoring Systems: Several scoring systems have been developed to assist clinicians in synthesizing the complex data necessary to estimate the clinical probability of HIT. The most extensively studied of these, the 4Ts, incorporates four typical features of HIT (Table 104.1). The model yields an integer score between 0 and 8 with scores of 0–3, 4–5, and 6–8 classified as low, intermediate, and high pretest probability, respectively. In a metaanalysis, a low probability score was associated with a negative predictive value of 99%. Intermediate and high probability scores were associated with positive predictive values of 14% and 64%, respectively. The 4Ts has neither been directly compared to a gestalt approach to diagnosis nor has it been subjected to a formal impact analysis on clinician behavior and outcomes. Another scoring system, the HIT Expert Probability Score, exhibited improved operating characteristics in a single institution series. This scoring system is currently undergoing prospective evaluation.

Laboratory Diagnosis: Given the challenges of judging the probability of HIT on clinical grounds, physicians rely heavily on laboratory testing to assist in diagnosis. This topic is reviewed in detail in Chapter 156.

TABLE 104.1 The 4Ts Scoring System

Category	2 Points	1 Point	0 Points
Thrombocytopenia	Platelet count fall >50% and platelet nadir ≥20 × 10^9/L	Platelet count 30%–50% or platelet nadir 10–19 × 10^9/L	Platelet count fall <30% or platelet nadir <10 × 10^9/L
Timing of platelet count fall	Clear onset between days 5–10 and platelet fall ≤1 day (prior heparin exposure within 30 days)	Consistent with days 5–10 fall, but not clear (e.g., missing platelet counts); onset after day 10; or fall ≤1 day (prior heparin exposure 30–100 days ago)	Platelet count fall ≤4 days without recent exposure
Thrombosis or other sequelae	New thrombosis (confirmed); skin necrosis; acute systemic reaction postintravenous unfractionated heparin bolus	Progressive or recurrent thrombosis; nonnecrotizing (erythematous) skin lesions; suspected thrombosis (not proven)	None
Other causes of thrombocytopenia	None apparent	Possible	Definite

The 4Ts score is the sum of the values in each of the four categories. Scores of 0–3, 4–5, and 6–8 are considered to correspond to a low, intermediate, and high probability of heparin-induced thrombocytopenia, respectively.

Management: Treatment of HIT requires immediate withdrawal of all heparin, including heparin flushes and heparin-bonded catheters. Heparin cessation alone, however, is often insufficient to prevent thrombosis. Historical series of untreated patients document a 5%–10% daily risk of thrombosis in the first several days after discontinuation of heparin. Prompt initiation of an alternative anticoagulant reduces the risk of thrombosis and constitutes standard of care for treatment of HIT with or without thrombosis. Several such agents are available (Table 104.2).

Argatroban: Argatroban is a hepatically cleared direct thrombin inhibitor (DTI) approved in the United States and elsewhere for management of HIT. The pivotal studies leading to its approval (ARG-911 and ARG-915) reported a hazard ratio for thrombosis of 0.33 for argatroban-treated subjects compared with untreated historical controls. A major limitation of these studies is that they did not require serologic confirmation of HIT and 36% of enrollees tested negative for anti-PF4/heparin antibodies on post hoc analysis. Argatroban is monitored with the activated partial thromboplastin time (APTT). Reduction in initial dose is recommended in patients with liver or multiorgan dysfunction, heart failure, anasarca, or recent cardiac surgery (Table 104.2).

Bivalirudin: Unlike argatroban, bivalirudin undergoes partial enzymatic inactivation and has a briefer half-life (~25 minutes), rendering it potentially attractive in settings such as coronary angiography, cardiac surgery, and critical illness with multiorgan failure in which rapid dose titration is desired. Bivalirudin has been studied extensively in coronary angiography where it is licensed for patients with and without HIT. Validated protocols are also available for its use in cardiac surgery with and without CPB.

TABLE 104.2 Anticoagulants Available for the Treatment of Heparin-Induced Thrombocytopenia (HIT)

Agent	Initial Dosing	Monitoring of Anticoagulant Effect	Clearance (Half-Life)
Argatroban[a]	**Bolus** None **Continuous infusion** Normal organ function → 2 mcg/kg/minute Liver dysfunction (bilirubin > 1.5 mg/dL), heart failure, multiple organ failure, postcardiac surgery, anasarca → 0.5–1.2 mcg/kg/minute	Adjust to APTT of 1.5–3.0 times patient baseline	Hepatobiliary (40–50 minutes)
Bivalirudin[b]	**Bolus** None **Continuous infusion** Normal organ function → 0.15–0.2 mg/kg/hour Renal or hepatic dysfunction → dose reduction may be appropriate	Adjust to APTT of 1.5–2.5 times patient baseline	Enzymatic and renal (25 minutes)
Danaparoid	**Bolus** <60 kg → 1500 U 60–75 kg → 2250 U 75–90 kg → 3000 U >90 kg → 3750 U **Accelerated initial infusion** 400 U/hour × 4 hours, then 300 U/hours × 4 hours **Maintenance infusion** Normal renal function → 200 U/hour Renal insufficiency → 150 U/hour	Adjust to anti-FXa of 0.5–0.8 U/mL	Renal (24 hours)
Fondaparinux[b]	<50 kg → 5 mg SC daily 50–100 kg → 7.5 mg SC daily >100 kg → 10 mg SC daily Cl_{cr} 30–50 mL/minutes → use caution Cl_{cr} < 30 mL/minutes → contraindicated	May not be necessary; some authorities recommend monitoring by anti-FXa	Renal (17–20 hours)
DOACs	Optimal dosing for HIT has not been established. Refer to package inserts for standard VTE dosing.	Routine monitoring not indicated	
Apixaban[b]			Biliary/direct intestinal excretion and renal (12 hours)
Rivaroxaban[b]			Hepatobiliary and renal (7–13 hours)
Dabigatran[b]			Renal (12–17 hours)
Edoxaban[b]			Biliary/intestinal excretion and renal (9–11 hours)

APTT, activated partial thromboplastin time; Cl_{cr}, creatinine clearance; *DOAC*, direct oral anticoagulants; *VTE*, venous thromboembolism.

[a]These dosing recommendations are lower than the dosing guidelines on the US FDA label due to concerns about bleeding.

[b]Not approved for treatment of HIT.

Published evaluation of bivalirudin in critical illness and multiorgan failure is limited to case series. A retrospective analysis proposes dose reduction for renal dysfunction or hemodialysis. In the operating room and catheterization laboratory, dosing is monitored with the activated clotting time. Elsewhere, the APTT is used (Table 104.2).

Danaparoid: Danaparoid is an antithrombin-dependent inhibitor of FXa. It is labeled for treatment of HIT in multiple jurisdictions, although it is no longer marketed in the United States, and drug shortages have limited its availability elsewhere. Efficacy was demonstrated in an open-label randomized trial of intravenous danaparoid versus dextran-70 in 42 patients with HIT complicated by thrombosis. Significantly more danaparoid-treated subjects were judged to have clinical recovery from thrombosis at discharge (56% vs. 14%). Danaparoid should be given intravenously by bolus followed by accelerated initial infusion and then maintenance infusion (Table 104.2).

Fondaparinux: Fondaparinux is an indirect FXa inhibitor that is administered subcutaneously once daily. In the United States, it is licensed for the prevention and treatment of venous thromboembolism. A retrospective propensity-matched study of 239 patients treated for HIT with fondaparinux showed similar rates of thrombosis, thrombosis-related death, and major hemorrhage as argatroban and danaparoid. Although the incidence of anti-PF/heparin seroconversion after fondaparinux exposure is similar to LMWH, the incidence of fondaparinux-induced HIT appears to be negligible. Several cases of HIT induced or exacerbated by fondaparinux have been described, though the attribution of fondaparinux in these cases remains uncertain and spontaneous HIT may be an alternative explanation. Overall, a favorable risk/benefit profile, ease of once daily subcutaneous administration, and lack of need for routine anticoagulant monitoring have contributed to the growing use of this agent in stable patients. In some countries, fondaparinux is now used more frequently than intravenous agents. Fondaparinux has been shown to be more cost-effective than conventional agents because of lower drug costs, decreased hospital length of stay, and less intense monitoring compared with parenteral alternative anticoagulants.

Direct Oral Anticoagulants: Apixaban, rivaroxaban, edoxaban, and betrixaban are orally administered direct factor Xa-inhibitors, and dabigatran is an orally administered DTI. These agents, collectively referred to as direct oral anticoagulants (DOACs), have several theoretical advantages in the treatment of HIT. They are given orally in fixed doses, do not require routine laboratory monitoring of anticoagulant effect, and do not cross-react with HIT antibodies in vitro. Emerging data suggest that they may be safe and effective therapies in patients with acute HIT. A case series and literature review identified 69 patients with acute HIT treated with rivaroxaban (n = 46), apixaban (n = 12), or dabigatran (n = 11). Only two patients experienced thrombosis (one each in the rivaroxaban and dabigatran groups) and there were no major bleeds. More data are needed to confirm these promising results, which are subject to selection bias and reporting bias.

Limitations of Current Therapies: Despite improvements in the clinical detection and treatment of HIT, approved therapies for HIT (i.e., argatroban and danaparoid) are associated with important limitations, including incomplete antithrombotic efficacy

TABLE 104.3 Measures for Preventing Venous Limb Gangrene in Patients With Acute Heparin-Induced Thrombocytopenia (HIT)

1. If a patient is receiving a vitamin K antagonist (VKA) at the time HIT is diagnosed, the VKA should be stopped and reversed with vitamin K.
2. A VKA should not be initiated until the platelet count has recovered to a stable plateau.
3. When a VKA is initiated, large loading doses (e.g., warfarin >5 mg) should be avoided.
4. The VKA should be overlapped with a parenteral anticoagulant for at least 5 days and until the INR has reached its intended target.

TABLE 104.4 Algorithm for Transitioning Patients From Argatroban to a Vitamin K Antagonist

Argatroban Dose	Management
≤2 mcg/kg/minutes	1. Stop argatroban when INR combined argatroban and warfarin is >4 2. Repeat INR in 4–6 hours 3. If INR is <2, restart argatroban at 10% increased dose 4. Repeat procedure until INR ≥2 off argatroban is achieved
>2 mcg/kg/minutes	1. Reduce argatroban dose to 2 mcg/kg/minutes 2. Repeat INR in 4–6 hours 3. Stop argatroban when INR con combined argatroban and warfarin is >4 4. Repeat INR in 4–6 hours 5. If INR is <2, restart argatroban at 10% increased dose 6. Repeat procedure until INR ≥2 off argatroban is achieved

and narrow therapeutic indices. They decrease thrombosis by approximately one-half to two-thirds but do not reduce amputation or mortality and are associated with a ~1% rate of major bleeding per treatment day. Additionally, administration of these agents is complex and prone to error. They require continuous intravenous infusion, serial laboratory monitoring, and frequent dose adjustments. Monitoring of argatroban by APTT may also be confounded by HIT-associated consumptive coagulopathy, in which protocol-driven reduction in the argatroban dose due to a "supratherapeutic" APTT may result in subtherapeutic dosing and thrombus progression. Fondaparinux and the DOACs offer hope of reducing therapeutic complexity.

Transitioning to Oral Therapy: Patients with acute HIT who are receiving a vitamin K antagonist (VKA) are at risk of venous limb gangrene, a thrombotic complication of the microvasculature due to depletion of protein C. Appropriate precautions should be taken to guard against this limb-threatening complication (Table 104.3). Another potential challenge in transitioning patients from a parenteral anticoagulant, particularly argatroban, to a VKA is posed by the International Normalized Ratio (INR)-raising effect of argatroban. If this effect is ignored, on discontinuation of argatroban, patients may be inadequately anticoagulated during a period of great thrombotic risk. Algorithms have been established to guide the transition from argatroban to warfarin (Table 104.4).

Patients with HIT are being increasingly transitioned to DOACs rather than warfarin after platelet count recovery. Because of their rapid onset of action, bridging with a parenteral agent is not required in patients transitioning to a DOAC. A retrospective study of 22 patients demonstrated transition from a short course of argatroban to DOAC to be a safe approach with no bleeding or recurrent thrombosis. In a single center report of 6 patients with confirmed or probable HIT who transitioned to a DOAC after recovery of platelet count, none had recurrent thrombosis.

Duration of Anticoagulation: As with other provoked thromboses ascribed to a transient risk factor, patients with HIT-associated thrombosis generally require anticoagulation for 3–6 months. Bilateral lower extremity compression ultrasonography should be considered in all patients with HIT, even in the absence of signs and symptoms because silent DVT is common and its presence may influence the recommended duration of anticoagulation. The optimal length of anticoagulation in individuals presenting with HIT without thrombosis is unknown. In a historical series of 62 untreated patients with isolated HIT, the cumulative incidence of thromboembolism at 30 days was 53%. Most events occurred within 10 days of heparin discontinuation, a period corresponding to platelet recovery. Therefore, anticoagulation should be continued at least until the platelet count has recovered to a stable plateau with consideration for extending therapy to a month in patients with additional thrombotic risk factors.

Intravenous Immunoglobulin: Case reports have described patients with severe and refractory HIT who responded to intravenous immunoglobulin (IVIG) transfusion. We consider adding IVIG in patients with acute HIT and severe, persistent thrombocytopenia despite treatment with an alternative anticoagulant.

Platelet Transfusion: There is a long-held concern that platelet transfusion may precipitate thrombosis. Published evidence is conflicting. A case series reported no thrombosis in 41 platelet-transfused patients with suspected HIT. In contrast, platelet transfusion in patients with HIT was associated with higher odds of arterial thrombosis and mortality in an administrative database study. In general, routine prophylactic platelet transfusion in patients with confirmed or strongly suspected HIT should be avoided, particularly because HIT is a prothrombotic rather than a prohemorrhagic diathesis. Platelet transfusion may be considered in situations of diagnostic uncertainty, high bleeding risk, or clinically significant bleeding.

Heparin Reexposure in Patients With a History of Heparin-Induced Thrombocytopenia: The HIT immune response wanes over time in a predictable manner after an episode of acute HIT. Functional assays become negative at a median of 50 days after heparin cessation, whereas anti-PF4/heparin antibody titers begin to decline after heparin is discontinued and are no longer detectable in most patients by day 100. Identification of a patient's position along this sequence by HIT laboratory testing may be used to determine the safety of heparin reexposure for cardiovascular surgery or coronary angiography (Table 104.5). The successful use of preoperative plasmapheresis to remove circulating anti-PF4/heparin antibodies and enable heparin use during cardiac surgery in patients with HIT has been reported in a small number of cases.

TABLE 104.5 Recommendations for Intraoperative Anticoagulation in Patients With a History of Heparin-Induced Thrombocytopenia (HIT)

	Laboratory Profile		Intraoperative anticoagulation[a]	
Clinical Picture	Immunologic Assay	Functional Assay	Cardiovascular Surgery	Coronary Angiography
Remote HIT	Negative	Negative	Use unfractionated heparin (UFH).	Use an alternative anticoagulant such as bivalirudin. If an alternative anticoagulant is not available, use UFH.
Subacute HIT	Positive	Negative	Delay surgery, if possible, until immunologic assay becomes negative. If surgery cannot be delayed, use UFH or an alternative anticoagulant such as bivalirudin.	Use an alternative anticoagulant such as bivalirudin.
Acute HIT	Positive	Positive	Delay surgery, if possible, until functional and immunologic assays become negative. If surgery cannot be delayed, use an alternative anticoagulant such as bivalirudin.	

[a]If heparin is used, it should be limited to the intraoperative setting and strictly avoided before and after the surgery.

Further Reading

Agnelli, G., Buller, H. R., Cohen, A., Curto, M., Gallus, A. S., Johnson, M., et al. (2013). Oral apixaban for the treatment of acute venous thromboembolism. *N Engl J Med*, *369*(9), 799–808.

Aljabri, A., Huckleberry, Y., Karnes, J. H., Gharaibeh, M., Kutbi, H. I., Raz, Y., et al. (2016). Cost-effectiveness of anticoagulants for suspected heparin-induced thrombocytopenia in the United States. *Blood*, *128*(26), 3043–3051.

Amiral, J., Bridey, F., Dreyfus, M., Vissoc, A. M., Fressinaud, E., Wolf, M., et al. (1992). Platelet factor 4 complexed to heparin is the target for antibodies generated in heparin-induced thrombocytopenia. *Thromb Haemost*, *68*(1), 95–96.

Cai, Z., Yarovoi, S. V., Zhu, Z., Rauova, L., Hayes, V., Lebedeva, T., et al. (2015). Atomic description of the immune complex involved in heparin-induced thrombocytopenia. *Nat Commun*, *6*, 8277.

Cuker, A. (2016). Management of the multiple phases of heparin-induced thrombocytopenia. *Thromb Haemost*, *116*(5), 835–842.

Cuker, A., Arepally, G., Crowther, M. A., Rice, L., Datko, F., Hook, K., et al. (2010). The HIT expert probability (HEP) score: A novel pre-test probability model for heparin-induced thrombocytopenia based on broad expert opinion. *J Thromb Haemost*, *8*(12), 2642–2650.

Cuker, A., & Cines, D. B. (2012). How I treat heparin-induced thrombocytopenia. *Blood*, *119*(10), 2209–2218.

Cuker, A., Gimotty, P. A., Crowther, M. A., & Warkentin, T. E. (2012). Predictive value of the 4Ts scoring system for heparin-induced thrombocytopenia: A systematic review and meta-analysis. *Blood, 120*(20), 4160–4167.

Davidson, S. J., Wadham, P., Rogers, L., & Burman, J. F. (2007). Endothelial cell damage in heparin-induced thrombocytopenia. *Blood Coagul Fibrinolysis, 18*(4), 317–320.

Dudek, A. Z., Pennell, C. A., Decker, T. D., Young, T. A., Key, N. S., & Slungaard, A. (1997). Platelet factor 4 binds to glycanated forms of thrombomodulin and to protein C. A potential mechanism for enhancing generation of activated protein C. *J Biol Chem, 272*(50), 31785–31792.

Dyke, C. M., Smedira, N. G., Koster, A., Aronson, S., McCarthy, H. L., 2nd, Kirshner, R., et al. (2006). A comparison of bivalirudin to heparin with protamine reversal in patients undergoing cardiac surgery with cardiopulmonary bypass: The EVOLUTION-on study. *J Thorac Cardiovasc Surg, 131*(3), 533–539.

Flaker, G., Lopes, R. D., Al-Khatib, S. M., Hermosillo, A. G., Hohnloser, S. H., Tinga, B., et al. (2014). Efficacy and safety of apixaban in patients after cardioversion for atrial fibrillation: Insights from the ARISTOTLE trial (apixaban for reduction in stroke and other thromboembolic events in atrial fibrillation). *J Am Coll Cardiol, 63*(11), 1082–1087.

Girolami, B., Prandoni, P., Stefani, P. M., Tanduo, C., Sabbion, P., Eichler, P., et al. (2003). The incidence of heparin-induced thrombocytopenia in hospitalized medical patients treated with subcutaneous unfractionated heparin: A prospective cohort study. *Blood, 101*(8), 2955–2959.

Goel, R., Ness, P. M., Takemoto, C. M., Krishnamurti, L., King, K. E., & Tobian, A. A. (2015). Platelet transfusions in platelet consumptive disorders are associated with arterial thrombosis and in-hospital mortality. *Blood, 125*(9), 1470–1476.

Greer, I. A., & Nelson-Piercy, C. (2005). Low-molecular-weight heparins for thromboprophylaxis and treatment of venous thromboembolism in pregnancy: A systematic review of safety and efficacy. *Blood, 106*(2), 401–407.

Greinacher, A., Eichler, P., Lubenow, N., Kwasny, H., & Luz, M. (2000). Heparin-induced thrombocytopenia with thromboembolic complications: Meta-analysis of 2 prospective trials to assess the value of parenteral treatment with lepirudin and its therapeutic aPTT range. *Blood, 96*(3), 846–851.

Greinacher, A., Holtfreter, B., Krauel, K., Gatke, D., Weber, C., Ittermann, T., et al. (2011). Association of natural anti-platelet factor 4/heparin antibodies with periodontal disease. *Blood, 118*(5), 1395–1401.

Jay, R. M., & Warkentin, T. E. (2008). Fatal heparin-induced thrombocytopenia (HIT) during warfarin thromboprophylaxis following orthopedic surgery: Another example of 'spontaneous' HIT? *J Thromb Haemost, 6*(9), 1598–1600.

Junqueira, D. R., Zorzela, L. M., & Perini, E. (2017). Unfractionated heparin versus low molecular weight heparins for avoiding heparin-induced thrombocytopenia in postoperative patients. *Cochrane Database of Syst Rev, 4*, Cd007557.

Kang, M., Alahmadi, M., Sawh, S., Kovacs, M. J., & Lazo-Langner, A. (2015). Fondaparinux for the treatment of suspected heparin-induced thrombocytopenia: A propensity score-matched study. *Blood, 125*(6), 924–929.

Karnes, J. H., Cronin, R. M., Rollin, J., Teumer, A., Pouplard, C., Shaffer, C. M., et al. (2015). A genome-wide association study of heparin-induced thrombocytopenia using an electronic medical record. *Thromb Haemost, 113*(4), 772–781.

Kowalska, M. A., Krishnaswamy, S., Rauova, L., Zhai, L., Hayes, V., Amirikian, K., et al. (2011). Antibodies associated with heparin-induced thrombocytopenia (HIT) inhibit activated protein C generation: New insights into the prothrombotic nature of HIT. *Blood, 118*(10), 2882–2888.

Krauel, K., Hackbarth, C., Furll, B., & Greinacher, A. (2012). Heparin-induced thrombocytopenia: In vitro studies on the interaction of dabigatran, rivaroxaban, and low-sulfated heparin, with platelet factor 4 and anti-PF4/heparin antibodies. *Blood, 119*(5), 1248–1255.

Lewis, B. E., Wallis, D. E., Leya, F., Hursting, M. J., & Kelton, J. G. (2003). Argatroban anticoagulation in patients with heparin-induced thrombocytopenia. *Arch Intern Med, 163*(15), 1849–1856.

Linkins, L. A., Dans, A. L., Moores, L. K., Bona, R., Davidson, B. L., Schulman, S., et al. (2012). Treatment and prevention of heparin-induced thrombocytopenia: Antithrombotic therapy and prevention of thrombosis, 9th ed: American College of Chest Physicians evidence-based clinical practice guidelines. *Chest, 141*(2 Suppl.), e495S–e530S.

Lubenow, N., Eichler, P., Lietz, T., & Greinacher, A. (2005). Lepirudin in patients with heparin-induced thrombocytopenia - results of the third prospective study (HAT-3) and a combined analysis of HAT-1, HAT-2, and HAT-3. *J Thromb Haemost, 3*(11), 2428–2436.

Lubenow, N., Hinz, P., Thomaschewski, S., Lietz, T., Vogler, M., Ladwig, A., et al. (2010). The severity of trauma determines the immune response to PF4/heparin and the frequency of heparin-induced thrombocytopenia. *Blood, 115*(9), 1797–1803.

Martel, N., Lee, J., & Wells, P. S. (2005). Risk for heparin-induced thrombocytopenia with unfractionated and low-molecular-weight heparin thromboprophylaxis: A meta-analysis. *Blood, 106*(8), 2710–2715.

McGowan, K. E., Makari, J., Diamantouros, A., Bucci, C., Rempel, P., Selby, R., et al. (2016). Reducing the hospital burden of heparin-induced thrombocytopenia: Impact of an avoid-heparin program. *Blood, 127*(16), 1954–1959.

Newman, P. M., Swanson, R. L., & Chong, B. H. (1998). Heparin-induced thrombocytopenia: IgG binding to PF4-heparin complexes in the fluid phase and cross-reactivity with low molecular weight heparin and heparinoid. *Thromb Haemost, 80*(2), 292–297.

Noel, E., Abbas, N., Skaradinskiy, Y., & Schreiber, Z. (2015). Heparin-induced thrombocytopenia in a patient with essential thrombocythemia: A case based update. *Case Rep Hematol, 2015*, 985253.

Oliveira, G. B., Crespo, E. M., Becker, R. C., Honeycutt, E. F., Abrams, C. S., Anstrom, K. J., et al. (2008). Incidence and prognostic significance of thrombocytopenia in patients treated with prolonged heparin therapy. *Arch Intern Med, 168*(1), 94–102.

Padmanabhan, A., Jones, C. G., Pechauer, S. M., Curtis, B. R., Bougie, D. W., Irani, M. S., et al. (2017). IVIg for treatment of severe refractory heparin-induced thrombocytopenia. *Chest, 152*(3), 478–485.

Pongas, G., Dasgupta, S. K., & Thiagarajan, P. (2013). Antiplatelet factor 4/heparin antibodies in patients with gram negative bacteremia. *Thromb Res, 132*(2), 217–220.

Rauova, L., Poncz, M., McKenzie, S. E., Reilly, M. P., Arepally, G., Weisel, J. W., et al. (2005). Ultralarge complexes of PF4 and heparin are central to the pathogenesis of heparin-induced thrombocytopenia. *Blood, 105*(1), 131–138.

Refaai, M. A., Chuang, C., Menegus, M., Blumberg, N., & Francis, C. W. (2010). Outcomes after platelet transfusion in patients with heparin-induced thrombocytopenia. *J Thromb Haemost*, *8*(6), 1419–1421.

Rosenberger, L. H., Smith, P. W., Sawyer, R. G., Hanks, J. B., Adams, R. B., & Hedrick, T. L. (2011). Bilateral adrenal hemorrhage: The unrecognized cause of hemodynamic collapse associated with heparin-induced thrombocytopenia. *Crit Care Med*, *39*(4), 833–838.

Schmitt, B. P., & Adelman, B. (1993). Heparin-associated thrombocytopenia: A critical review and pooled analysis. *Am J Med Sci*, *305*(4), 208–215.

Schulman, S., Kakkar, A. K., Goldhaber, S. Z., Schellong, S., Eriksson, H., Mismetti, P., et al. (2014). Treatment of acute venous thromboembolism with dabigatran or warfarin and pooled analysis. *Circulation*, *129*(7), 764–772.

Selleng, S., Malowsky, B., Strobel, U., Wessel, A., Ittermann, T., Wollert, H. G., et al. (2010). Early-onset and persisting thrombocytopenia in post-cardiac surgery patients is rarely due to heparin-induced thrombocytopenia, even when antibody tests are positive. *J Thromb Haemost*, *8*(1), 30–36.

Sharifi, M., Bay, C., Vajo, Z., Freeman, W., Sharifi, M., & Schwartz, F. (2015). New oral anticoagulants in the treatment of heparin-induced thrombocytopenia. *Thromb Res*, *135*(4), 607–609.

Smythe, M. A., Koerber, J. M., & Mattson, J. C. (2007). The incidence of recognized heparin-induced thrombocytopenia in a large, tertiary care teaching hospital. *Chest*, *131*(6), 1644–1649.

Stein, P. D., Hull, R. D., Matta, F., Yaekoub, A. Y., & Liang, J. (2009). Incidence of thrombocytopenia in hospitalized patients with venous thromboembolism. *Am J Med*, *122*(10), 919–930.

Stone, G. W., White, H. D., Ohman, E. M., Bertrand, M. E., Lincoff, A. M., McLaurin, B. T., et al. (2007). Bivalirudin in patients with acute coronary syndromes undergoing percutaneous coronary intervention: A subgroup analysis from the acute catheterization and urgent intervention triage strategy (ACUITY) trial. *Lancet (Lond Engl)*, *369*(9565), 907–919.

Tsu, L. V., & Dager, W. E. (2011). Bivalirudin dosing adjustments for reduced renal function with or without hemodialysis in the management of heparin-induced thrombocytopenia. *Ann Pharmacother*, *45*(10), 1185–1192.

Tutwiler, V., Madeeva, D., Ahn, H. S., Andrianova, I., Hayes, V., Zheng, X. L., et al. (2016). Platelet transactivation by monocytes promotes thrombosis in heparin-induced thrombocytopenia. *Blood*, *127*(4), 464–472.

Walenga, J. M., Prechel, M., Jeske, W. P., Hoppensteadt, D., Maddineni, J., Iqbal, O., et al. (2008). Rivaroxaban–an oral, direct Factor Xa inhibitor–has potential for the management of patients with heparin-induced thrombocytopenia. *Br J Haematol*, *143*(1), 92–99.

Wallis, D. E., Workman, D. L., Lewis, B. E., Steen, L., Pifarre, R., & Moran, J. F. (1999). Failure of early heparin cessation as treatment for heparin-induced thrombocytopenia. *Am J Med*, *106*(6), 629–635.

Warkentin, T. E. (1998). Clinical presentation of heparin-induced thrombocytopenia. *Semin Hematol*, *35*(4 Suppl. 5), 9–16 discussion 35–6.

Warkentin, T. E., Elavathil, L. J., Hayward, C. P., Johnston, M. A., Russett, J. I., & Kelton, J. G. (1997). The pathogenesis of venous limb gangrene associated with heparin-induced thrombocytopenia. *Ann Intern Med*, *127*(9), 804–812.

Warkentin, T. E., Greinacher, A., Koster, A., & Lincoff, A. M. (2008). Treatment and prevention of heparin-induced thrombocytopenia: American College of chest physicians evidence-based clinical practice guidelines (8th edition). *Chest, 133*(6 Suppl.), 340s–380s.

Warkentin, T. E., & Kelton, J. G. (2001a). Delayed-onset heparin-induced thrombocytopenia and thrombosis. *Ann Intern Med, 135*(7), 502–506.

Warkentin, T. E., & Kelton, J. G. (2001b). Temporal aspects of heparin-induced thrombocytopenia. *N Engl J Med, 344*(17), 1286–1292.

Warkentin, T. E., Levine, M. N., Hirsh, J., Horsewood, P., Roberts, R. S., Gent, M., et al. (1995). Heparin-induced thrombocytopenia in patients treated with low-molecular-weight heparin or unfractionated heparin. *N Engl J Med, 332*(20), 1330–1335.

Warkentin, T. E., Pai, M., & Linkins, L. A. (2017). Direct oral anticoagulants for treatment of HIT: Update of Hamilton experience and literature review. *Blood, 130*(9), 1104–1113.

Warkentin, T. E., Roberts, R. S., Hirsh, J., & Kelton, J. G. (2005). Heparin-induced skin lesions and other unusual sequelae of the heparin-induced thrombocytopenia syndrome: A nested cohort study. *Chest, 127*(5), 1857–1861.

Warkentin, T. E., Sheppard, J. A., Chu, F. V., Kapoor, A., Crowther, M. A., & Gangji, A. (2015). Plasma exchange to remove HIT antibodies: Dissociation between enzyme-immunoassay and platelet activation test reactivities. *Blood, 125*(1), 195–198.

Warkentin, T. E., Sheppard, J. A., Horsewood, P., Simpson, P. J., Moore, J. C., & Kelton, J. G. (2000). Impact of the patient population on the risk for heparin-induced thrombocytopenia. *Blood, 96*(5), 1703–1708.

Warkentin, T. E., Sheppard, J. A., Sigouin, C. S., Kohlmann, T., Eichler, P., & Greinacher, A. (2006). Gender imbalance and risk factor interactions in heparin-induced thrombocytopenia. *Blood, 108*(9), 2937–2941.

Welsby, I. J., Um, J., Milano, C. A., Ortel, T. L., & Arepally, G. (2010). Plasmapheresis and heparin reexposure as a management strategy for cardiac surgical patients with heparin-induced thrombocytopenia. *Anesth Analg, 110*(1), 30–35.

CHAPTER 105

Autoimmune Lymphoproliferative Syndrome

Marcus A. Carden, MD and Michael A. Briones, DO

Introduction: Autoimmune lymphoproliferative syndrome (ALPS) was first described in 1967 when Canale and Smith reported on five children with massive lymphadenopathy and splenomegaly mimicking lymphoma. In 1992 Sneller et al. similarly described two patients with a progressive lymphoproliferative disorder associated with autoimmunity and whose blood and lymph nodes demonstrated an increase in an atypical population of CD41992$^-$ and CD8$^-$ T cells (double-negative T cells, DNTs). We now know an elevation of DNTs in the peripheral blood results from a primary defect in Fas-mediated T-cell apoptosis and is a requirement to diagnose ALPS. Although the role of DNTs in ALPS is not completely understood, recent evidence suggests a hyperactive mammalian target of rapamycin (mTOR) pathway within T-cell subsets may be responsible, and sirolimus, an mTOR inhibitor, shows significant promise in the treatment of symptomatic patients with ALPS. Because of significant morbidity associated with ALPS and its association with lymphoma, clinicians should have a heightened suspicion of this disease in appropriate clinical circumstances. With modern treatments, patients with ALPS can expect to have a near-normal life span.

Pathophysiology: ALPS is a primary immunodeficiency defined by a defect in the Fas-mediated external apoptosis pathway, which plays a pivotal role in lymphocyte homeostasis. During the process of downregulating the body's normal immune response, activated B and T lymphocytes increase the expression of the cell-surface protein Fas (also termed CD95/APO1). During this process, activated T lymphocytes increase expression of Fas ligand. Fas and Fas-ligand interact through the Fas-activating death domain, triggering the recruitment and intracellular cascade of procaspases 8 and 10, whose downstream effects initiate DNA degradation, proteolysis, and cellular apoptosis. As such, Fas maintains lymphocyte homeostasis and peripheral immune tolerance, and therefore, defects in this external pathway of apoptosis can result in aberrant lymphoproliferation, autoimmunity, and cancer.

Genetics: ALPS is a genetic disorder of apoptosis, and its research has informed us about the critical homeostatic mechanisms that regulate the development and death of lymphocytes. Most patients with ALPS have an identified genetic defect in the Fas-mediated extrinsic apoptosis pathway. The majority (60%–70%) harbor a germline mutation in the FAS gene (classified as ALPS-FAS), which is typically inherited in an autosomal dominant manner and codes for a cell-surface protein responsible for the initial receptor–ligand interactions in this pathway. Approximately 10% of patients have somatic mutations in this gene (ALPS-sFAS). Less than 1% possess Fas ligand mutations (ALPS-FASLG), whereas 2%–3% of patients contain CASP10 gene mutations

Transfusion Medicine and Hemostasis. https://doi.org/10.1016/B978-0-12-813726-0.00105-7

(ALPS-CASP10), which codes for an intracellular enzyme that is part of the downstream pathway in Fas-mediated lymphocyte apoptosis. Approximately 20%–30% of patients have no identifiable mutation (ALPS-U). Most ALPS mutations arise from autosomal dominant germline inheritance, whereas a minority arise from somatic mutations. Variable penetrance and expressivity suggest other modifiers can impact phenotypic expression.

Clinical Manifestations: The median age at presentation is 24 months. Pathologic lymphoproliferation most commonly manifests as lymphadenopathy and/or splenomegaly. More pronounced in infancy, the lymphadenopathy often regresses during adolescence and may resolve spontaneously. Splenomegaly is present in 95% of patients. Because of associated autoimmune cytopenias, ALPS patients may present with fatigue, pallor, jaundice/icterus, spontaneous bruising/bleeding, or recurrent infections. Extranodal lymphoproliferation may also result in uveitis, hepatitis, glomerulonephritis, pulmonary lesions, encephalitis, and myelitis. Dermatologic diatheses are common and may manifest as urticarial or polymorphic maculopapular skin rashes. Autoimmune-related alopecia, osteopenia, and vasculitis have been reported.

The risk of autoimmunity is lifelong and becomes greater with age. Compared to the general population, there is a significant risk of developing non-Hodgkin and Hodgkin lymphoma in patients with ALPS. The relative risks are estimated to be as high as 61 and 149, respectively, in some reports.

Diagnosis: The diagnostic criteria for ALPS are based on international expert consensus (see Table 105.1). Increased circulating vitamin B12 levels are commonly found and result from increased haptocorrin synthesized and released from ALPS-related lymphocytes. The classification of ALPS is based on which genetic mutation is present (see Genetics above).

Differential Diagnosis: Many hematologic disorders, involving antibody-mediated cytopenias and/or lymphoproliferation, should be considered in the workup for ALPS. Systemic lupus erythematosus or Evans Syndrome may be the result of unrecognized ALPS, and anticardiolipin and rheumatoid factor antibodies may be found circulating in patients with ALPS. Nonmalignant, immune disorders that can mimic ALPS include Rosai–Dorfman disease, Castleman's disease, and Kikuchi–Fujimoto disease. ALPS-related disorders (and their associated genetic mutations) include caspase-8 deficiency state, RAS-associated autoimmune lymphoproliferative disease (NRAS), Dianzani autoimmune lymphoproliferative disease (unknown), X-linked lymphoproliferative syndrome (SH2D1A), STAT3 gain-of-function mutations causing autoimmune disease, CTLA-4 haploinsufficiency with autoimmune infiltration disease, and LRBA deficiency disease.

Management: As symptoms of autoimmune cytopenias and lymphadenopathy may fluctuate over time in ALPS, many patients require no systemic therapy and can be serially monitored with routine clinical and radiologic (when necessary) surveillance. Immunosuppression may not be required until symptomatic cytopenias are present.

TABLE 105.1 : Diagnostic Criteria for Autoimmune Lymphoproliferative Syndrome (ALPS)

Required

1. Chronic nonmalignant, noninfectious lymphadenopathy, and/or splenomegaly (>6 months)
2. Elevated peripheral blood $CD3^+$ DNTs

Primary

1. Defective in vitro lymphocyte apoptosis (demonstrated in two separate assays)
2. Somatic or germline mutation in ALPS causative gene (FAS, FASL, CASP10)

Secondary

1. Elevated biomarkers (any of following)
   a. Plasma sFASL > 200 pg/mL
   b. Plasma IL-10 > 20 pg/mL
   c. Plasma or serum vitamin B12 > 1500 ng/L
   d. Plasma IL-18 > 500 pg/mL
2. Immunohistochemical findings consistent with ALPS (as determined by an experienced hematopathologist)
3. Autoimmune cytopenias (anemia, thrombocytopenia, neutropenia) and elevated IgG levels (polyclonal hypergammaglobulinemia)
4. Family history of ALPS or nonmalignant lymphoproliferation with or without autoimmunity

Definitive diagnosis: required plus one primary accessory criteria
Probable diagnosis: required plus one secondary accessory criteria
Probable and definitive ALPS should be treated the same clinically

Based on consensus from the First International ALPS Workshop 2009, see Oliveira, J. B., Bleesing, J. J., Dianzani, U., et al. (2010) Revised diagnostic criteria and classification for the autoimmune lymphoproliferative syndrome (ALPS): Report from the 2009 NIH international workshop. *Blood, 116*,e35–e40.

Similarly, patients who develop organ compromise from lymphoproliferation may require medical intervention. All patients with familial ALPS should undergo genetic counseling.

The successful use of sirolimus to treat patients with symptomatic ALPS has changed the landscape of treatment. Rapid improvement in autoimmune cytopenias, lymphadenopathy, and splenomegaly has been observed within 1–3 months of starting sirolimus in patients refractory to other treatments, although it is now often used as up-front therapy. After loading with 3 mg/m, patients are started at 2–2.5 mg/m^2 daily dosing with a goal of trough levels of 5–15 ng/mL. Routine monitoring of blood counts, liver enzymes, and triglycerides is recommended.

Other potential immunomodulatory agents used for symptomatic patients include corticosteroids, intravenous immunoglobulin, rituximab, and mycophenolate mofetil. Splenectomy should be reserved for patients with life-threatening or symptomatic splenomegaly or cytopenias refractory to all other treatments, as the risk of postsplenectomy thrombosis and overwhelming sepsis in these cases is well documented. Similarly, the use of hematopoietic stem cell transplantation is reserved for severe, refractory cases of ALPS. All patients with ALPS should be appropriately immunized.

Practitioners should be vigilant about screening for lymphoma transformation in patients with ALPS. Computed tomography or positron emission tomography scans should be considered in the appropriate clinical setting, as routine screening remains controversial. All suspicious nodes should be biopsied for definitive diagnosis.

Further Reading

Bowen, R. A. R., Dowdell, K. C., Dale, J. K., et al. (2012). Elevated vitamin B12 levels in autoimmune proliferative syndrome attributable to elevated haptocorrin in lymphocytes. *Clin Biochem*, *45*, 490–492.

Bride, K. L., Vincent, T., Smith-Whitley, K., et al. (2016). Sirolimus is effective in relapsed/refractory autoimmune cytopenias: Results of a prospective multi-institutional trial. *Blood*, *127*, 17–28.

Canale, V. C., & Smith, C. H. (1967). Chronic lymphadenopathy simulating malignant lymphoma. *J Pediatr*, *70*, 891–899.

Magerus-Chatinet, A., Stolzenberg, M. C., Loffredo, M. S., et al. (2009). FAS-L, IL-10, and double-negative CD4-CD8-TCR alpha/beta T cells are reliable markers of autoimmune lymphoproliferative syndrome (ALPS) associated with FAS loss of function. *Blood*, *113*, 3027–3030.

Neven, B., Magerus-Chatinet, A., Florkin, B., et al. (2011). A survey of 90 patients with autoimmune lymphoproliferative syndrome related to TNFRSF6 mutation. *Blood*, *118*, 4798–4807.

Oliveira, J. B., Bleesing, J. J., Dianzani, U., et al. (2010). Revised diagnostic criteria and classification for the autoimmune lymphoproliferative syndrome (ALPS): Report from the 2009 NIH international workshop. *Blood*, *116*, e35–e40.

Price, S., Shaw, P. A., Seitz, A., et al. (2014). Natural history of autoimmune lymphoproliferative syndrome associated with FAS gene mutations. *Blood*, *123*, 1989–1999.

Rao, V. K., Dugan, F., Dale, J. K., et al. (2005). Use of mycophenolate mofetil for chronic, refractory immune cytopenias in children with autoimmune lymphoproliferative syndrome. *Br J Haematol*, *129*, 534–538.

Rao, V. K., & Oliveira, J. B. (2011). How I treat autoimmune lymphoproliferative syndrome. *Blood*, *118*, 5741–5751.

Seidel, M. G. (2014). Autoimmune and other cytopenias in primary immunodeficiencies: Pathomechanisms, novel differential diagnosis, and treatment. *Blood*, *124*, 2337–2344.

Sneller, M. C., Straus, S. E., Jaffe, E. S., et al. (1992). A novel lymphoproliferative/autoimmune syndrome resembling lpr/gld disease. *J Clin Investig*, *90*, 334–341.

Teachey, D. T. (2012). New advances in the management and treatment of autoimmune lymphoproliferative syndrome. *Curr Opin Pediatr*, *24*, 1–8.

Teachey, D. T., Greiner, R., Seif, A., et al. (2009). Treatment with sirolimus results in complete responses in patients with autoimmune lymphoproliferative syndrome. *Br J Haematol*, *145*, 101–106.

Volkl, S., Rensing-Ehl, A., Allgauer, A., et al. (2016). Hyperactive mTOR pathway promotes lymphoproliferation and abnormal differentiation in autoimmune lymphoproliferative syndrome. *Blood*, *128*, 227–238.

CHAPTER 106

Hemolytic Uremic Syndrome

Michael White, MD and Michael A. Briones, DO

Hemolytic uremic syndrome (HUS) is a thrombotic microangiopathy (TMA) characterized by the triad of microangiopathic hemolytic anemia, thrombocytopenia, and acute kidney injury (AKI). This disease can be divided into three major subtypes: typical HUS, atypical HUS (aHUS), and DEAP-HUS (deficiency of CFHR plasma proteins and autoantibody positive HUS). Typical HUS makes up the majority of cases in children (90%) and is associated with bloody diarrhea caused by infection with Shiga-like toxin-producing *Escherichia coli* (typical HUS; STEC-HUS). aHUS (non-STEC HUS) accounts for 5%–10% of cases and is a heterogeneous disorder that often occurs in the absence of diarrhea and is associated with a wide variety of conditions, namely genetically defined complement dysregulation. A third type is DEAP-HUS, which accounts for about 1% of cases, and is characterized by the presence of autoantibodies directed to complement factor H (CFH).

Pathophysiology: TMA in HUS results in endothelial cell damage, accumulation of platelet thrombi, and vessel obstruction. Red blood cells (RBC) are then destroyed in these vessels, causing schistocyte formation (Fig. 106.1). These vascular abnormalities are most evident in the renal glomeruli.

Hemolytic Uremic Syndrome: Approximately 15% of children with *E. coli* O157:H7 infections develop HUS within 1 week of onset. STEC infections occur after ingestion of contaminated fruits/vegetables, undercooked meat, or nonpasteurized milk. Bacteria invade colonic mucosal epithelial cells, and then the toxin migrates into the bloodstream and binds to platelets, monocytes, and neutrophils. These aggregates circulate to target organs, such as the kidneys, and induce cytotoxic effects on glomerular cells.

Renal glomerular endothelial cells express glycolipid Gb3, the predominant membrane receptor for the Shiga toxin, making them primary targets. The injured cells expose collagen and ultralarge VWF, causing platelets aggregation. Additionally, exposure of tissue factor activates the coagulation cascade, leading to vessel microthrombosis.

Atypical Hemolytic Uremic Syndrome: aHUS is not associated with STEC infection and is caused by many etiologies that lead to TMA. The pathophysiology of complement-mediated HUS is that a trigger causes uncontrolled continuous activation of the alternative complement pathway in a susceptible individual with either gene mutations or antibodies to complement proteins, which results in renal endothelial damage and leads to TMA. The genetic causes for complement alternative pathway dysregulation are identified in about 60% of cases. The most common mutations affect CFH, membrane cofactor protein (MCP, CD46), factor I, factor B, C3 convertase, thrombomodulin, and other CFH-related proteins.

Transfusion Medicine and Hemostasis. https://doi.org/10.1016/B978-0-12-813726-0.00106-9

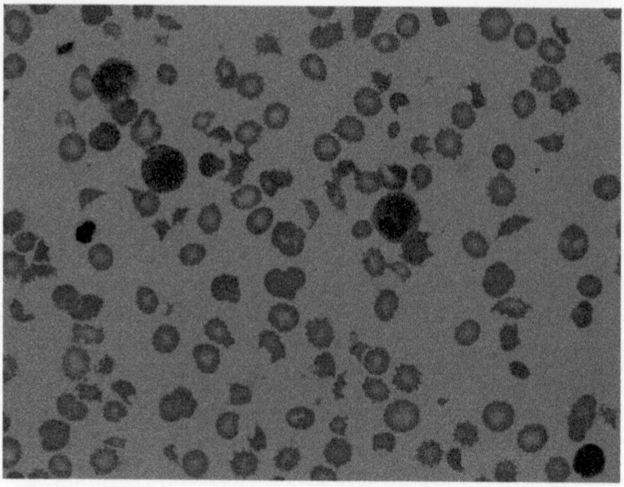

FIGURE 106.1 Note the absence of platelets and the presence of a nucleated erythrocyte and schistocytes, consistent with a microangiopathic process. (From Hoffman, R., Benz, E. J., Shattil, S. J., et al. (Eds.). (2005). *Hematology: Basic principles and practice* (4th ed.). Philadelphia, PA: Elsevier.)

aHUS also occurs secondary to infections with *Streptococcus pneumoniae* or HIV. Children with *Streptococcus*-associated aHUS usually present with pneumonia and/or bacteremia. Neuraminidase is proposed to expose the Thomsen–Friedenreich antigen and lead to complement fixation.

Drug-induced TMA has been reported with exposure to numerous medications, including antineoplastic (cisplatin, bleomycin, gemcitabine), immunosuppressive (cyclosporine, tacrolimus), antiplatelet (ticlopidine), and other agents (quinidine, interferon).

DEAP-HUS: DEAP-HUS is a rare and newly identified form of HUS characterized by deletion of genes coding for CFH-related proteins and the presence of anti-CFH antibodies. These autoantibodies result in a dysregulation of the alternative complement pathway and cell damage. These patients have poorer outcomes compared with patients with STEC-HUS, being more likely to progress to end-stage renal disease (ESRD) and death.

Clinical Manifestation: HUS is characterized by hemolytic anemia, thrombocytopenia, and AKI. The anemia can be profound due to hematochezia and RBC destruction from hemolysis and may require transfusion. Despite thrombocytopenia, serious bleeding complications are rare. Oliguria or anuria is somewhat common due to AKI. Stool cultures often identify *E. coli* in the case of bloody diarrhea. In aHUS, the presentation may be varied because of different underlying causes.

Laboratory Features and Diagnosis

Complete Blood Count and Coagulation Markers: The hallmark findings are hemolytic anemia and thrombocytopenia. Schistocytes are seen on peripheral blood smear, reflecting fragmentation of RBCs as they traverse vessels partially occluded by microthrombi. Reticulocytosis, elevated lactate dehydrogenase and indirect bilirubin, and low haptoglobin levels all reflect intravascular hemolysis. The direct antiglobulin test is typically negative. Thrombocytopenia is mild to moderately severe (40–140,000/mm^3). The anemia and thrombocytopenia do not reflect the severity of the renal dysfunction. However, hemolysis and thrombocytopenia extending past 10 days is associated with long-term renal sequelae. Leukocytosis is often present. The prothrombin time and partial thromboplastin time are normal, differentiating HUS from DIC, but the fibrinogen and D-dimer may be elevated.

Renal Function: Blood urea nitrogen and creatinine are typically markedly elevated. Urinalysis shows varying degrees of nephritis with protein, RBCs, white blood cells, and casts. Persistent proteinuria may be associated with an increased risk of progressive renal dysfunction.

Complement Levels: Complement levels are not needed in patients with typical HUS. However, all patients with aHUS should be screened for genetic forms because this may impact management decisions. This includes C3, C4, CFH, CFI, and CFB levels, leukocyte CD46 expression, and presence of anti-CFH antibodies should be evaluated. Confirmatory gene testing can be performed in specific cases. Because DEAP-HUS predominates in children from 7 years through adolescence, screening should be performed in this group.

Differential Diagnosis: The differential diagnosis of HUS includes other TMAs: thrombocytopenic purpura (TTP), DIC, glomerulonephritis, malignant hypertension, posttransplantation, Kasabach–Merritt syndrome, hemolysis with prosthetic cardiac devices, vasculitis (particularly lupus), antiphospholipid antibody syndrome, and HELLP syndrome. No laboratory test clearly distinguishes between these disorders, making the history and medical workup paramount in determining the etiology and appropriate therapy (plasma therapy vs. other).

Management/Prognosis

Hemolytic Uremic Syndrome: The mainstay of therapy for diarrhea-associated HUS is supportive care and early management of intravascular volume loss, anemia, and renal failure. This may prevent historic complications such as hyponatremia, pulmonary edema, myocardial insufficiency, gangrenous colitis, and central nervous system catastrophe.

Antibiotics to treat bacterial colitis should be avoided unless the patient has signs of sepsis. Platelet transfusion is not recommended unless active bleeding accompanies thrombocytopenia. RBC transfusion is appropriate in symptomatic patients. All dialysis modalities appear to be effective in supporting patients with acute renal failure, and choice depends on the expertise of the treating team. Plasma exchange has not been proven to positively affect outcomes. However, when there is question whether HUS is due to STEC infection or to nonspecific diarrhea in a patient with possible aHUS, plasma exchange should be implemented.

Intravenous immunoglobulin, fibrinolytic and antiplatelet agents, and corticosteroids are generally ineffective during the acute phase of disease. Renal injury progressing to ESRD requiring kidney transplantation is rare, occurring in less than 5% of patients. Neurological complications (seizures, altered mental status, and stroke) may occur in up to 25% of children with HUS. Death rarely occurs (2%–4%).

aHUS and DEAP-HUS: In aHUS, 50% of patients develop ESRD and 25% die in the acute phase of the disease. Plasma exchange or plasma infusion therapy remains the first-line treatment for aHUS. Recommendations to start plasma therapy within 24 hours of presentation are justified by the often rapid deterioration of renal function in patients with CFH, combined CFI, C3, and CFB mutations. The goal is to exchange 1–1.5 times the plasma volume (60–75 mL/kg) per session, with plasma replacement. When plasma exchange cannot be performed within 24 hours of presentation, plasma infusion (10–20 mL/kg) can be given. Differentiating between aHUS and TTP can be difficult, but the initial treatment for both is plasma therapy. Renal transplantation is a limited option because of the high recurrence rate (50%).

Complement pathway blockade with eculizumab, a humanized anti-C5 monoclonal antibody, has shown efficacy in several trials. This targeted medication inhibits terminal complement activation, which prevents terminal complex formation and cell lysis. Eculizumab may eventually replace plasma therapy in plasma-sensitive patients and control the disease in plasma-resistant patients. In patients with complement autoantibodies, the addition of rituximab therapy to the immunosuppressive regimen has been shown to be effective in patients with CFH antibodies and deletion of CFHR1 and CFHR3 genes. Patients with mutations in genes for CFH, CFI, or C3 who are unresponsive to plasma therapy or eculizumab and/or have relapsing disease are likely to progress to ESRD.

Further Reading

Ariceta, G., Besbas, N., Johnson, S., et al. (2009). European Paediatric Study Group for HUS. Guideline for the investigation and initial therapy of diarrhea-negative hemolytic uremic syndrome. *Pediatr Nephrol, 24*, 687–696.

Caprioli, J., Noris, M., Brioschi, S., et al. (2006). International registry of recurrent and familial *HUS/TTP*. Genetics of HUS: The impact of MCP, CFH, and IF mutations on clinical presentation, response to treatment, and outcome. *Blood, 108*, 1267–1279.

Kose, O., Zimmerhackl, L. B., Jungraithmayr, T., Mache, C., & Nurnberger, J. (2010). New treatment options for atypical hemolytic uremic syndrome with the complement inhibitor eculizumab. *Semin Thromb Hemost, 36*, 669–672.

Loirat, C., & Fremeaux-Bacchi, V. (2011). Atypical hemolytic uremic syndrome. *Orphanet J Rare Dis, 6*, 60.

Loirat, C., Garnier, A., Sellier-Leclerc, A., & Kwon, A. T. (2010). Plasmatherapy in atypical hemolytic uremic syndrome. *Semin Thromb Hemost, 36*, 673–681.

Taylor, C. M., Machin, S., Wigmore, S. J., & Goodship, T. H. (2010). Clinical practice guidelines for the management of atypical haemolytic uraemic syndrome in the United Kingdom. *Br J Haematol, 148*, 37–47.

Zheng, X. L., & Sadler, J. E. (2008). Pathogenesis of thrombotic microangiopathies. *Ann Rev Pathol, 3*, 177–249.

CHAPTER 107

Thrombotic Thrombocytopenic Purpura

Christine L. Kempton, MD, MSc and Ana G. Antun, MD, MSc

Thrombotic thrombocytopenic purpura (TTP) is a syndrome consisting of microangiopathic hemolytic anemia (MAHA), thrombocytopenia, and end-organ damage secondary to microvascular thrombi. Anemia, thrombocytopenia, fever, neurological signs, and renal dysfunction makeup the classic pentad; however, TTP can present without the full pentad; up to 35% of patients do not have neurological signs at presentation, and renal abnormalities and fever are not prominent features.

TTP should be suspected and treatment initiated in the setting of unexplained MAHA and thrombocytopenia. Treatment of TTP with therapeutic plasma exchange (TPE) has drastically decreased the mortality rate from greater than 90% to less than 20%.

Pathophysiology: The first description of TTP was by Moschowitz in 1924. These descriptions were of a disease presenting as a pentad of signs and symptoms (thrombocytopenia, fever, anemia, hemiparesis and hematuria). Postmortem examination revealed widespread thrombi in the terminal circulation of organs. In 1982, the presence of ultralarge von Willebrand factor (VWF) multimers was described by Moake et al. in a patient with relapsing TTP. These ultralarge VWF multimers are thought to adhere to platelets and contribute to the thrombotic occlusion of small vessels. The presence of ultralarge VWF provided significant insight into the pathophysiology of small vessel thrombi, but the reason for their accumulation was unclear until the late 1990s, when a deficiency of VWF cleaving protease was identified in a patient with chronic relapsing TTP. The cleaving protease was defined as a disintegrin and metalloprotease with thrombospondin type 1 motif, 13 (ADAMTS13). Currently, the predominant mechanism believed to be responsible for TTP is ADAMTS13 deficiency, resulting in the accumulation of platelet-hyperadhesive ultralarge VWF multimers, which bind platelets, and leads to both thrombi in the microvasculature and thrombocytopenia. The mechanism for severe ADMATS13 deficiency in acquired TTP is via autoantibodies directed against ADAMTS13, as demonstrated by anti-ADAMTS13 IgG (rare IgA and IgM) in ~75% of patients with TTP during the acute phase. Anti-ADAMTS13 antibodies can alter ADAMTS13 activity by several mechanisms: (1) inhibit the VWF proteolytic activity of ADAMTS13, (2) form ADAMTS13-specific immune complex, and (3) increase clearance of ADAMTS13. The mechanism by which ADAMTS13 becomes deficient or dysfunctional in those patients without detectable anti-ADAMTS13 IgG (20%–25% of acute patients TTP) remains unclear, though several potential mechanisms have been hypothesized and include (1) lack of sensitivity of anti-ADAMTS13 IgG assays in detecting IgG trapped within immune complexes, (2) involvement of other Ig isotypes, or (3) other nonimmune mechanisms. Congenital TTP (Upshaw-Schulman syndrome) is caused be recessive inherited biallelic mutations

Transfusion Medicine and Hemostasis. https://doi.org/10.1016/B978-0-12-813726-0.00107-0

of the *ADMATS13* gene (homozygous or compound heterozygous), causing congenital TTP. Worldwide, there are 150 distinct mutations of ADAMTS13 gene located in the *N*-terminal region consisting in ~70% of missense mutations and ~30% of truncating mutations with variable clinical penetrance.

Secondary forms of TTP can also be associated with drugs (clopidogrel, mitomycin C, cyclosporine, quinine, ticlopidine), bacterial infection and HIV, pregnancy and postpartum, cancer and organ transplantation, and autoimmune disease (connective tissue disease). The pathophysiology of these secondary types of TTP is varied and depends on the underlying cause; some disorders are associated with ADAMTS13 autoantibody formation, whereas other disorders are driven by endothelial cell activation and damage. Despite the link between ADAMTS13 and TTP, ADAMTS13 deficiency is not pathognomonic for TTP. Low ADAMTS13 activity can be seen in healthy individuals and in individuals with disseminated intravascular coagulopathy, immune thrombocytopenia, sepsis, hepatic dysfunction, malignancy, and pregnancy. Furthermore, patients with TTP have been reported to have normal or minimally reduced ADAMTS13 activity. Therefore, the diagnosis of TTP remains a clinical one.

TTP can at times be difficult to distinguish from hemolytic uremic syndrome (HUS). Typical HUS is secondary to bacterial enterocolitis, most commonly *Escherichia coli* 0157:H7. Atypical HUS (aHUS), also called diarrhea-negative HUS, is the result of uncontrolled complement activity (see Chapter 106).

Epidemiology: The annual prevalence is ~10 cases/million people and an annual incidence of 1 new case/million. Approximately 90% of all TTP cases occur during adulthood and 10% occur during childhood and adolescents. There is a predilection for women of African ancestry to be affected. Specifically, the incidence of acquired TTP is 2.5 times greater in women than in men and 4.9 times greater in African Americans compared with other races. TTP can affect all ages, though the median age at presentation of acquired TTP is 49 years (interquartile range [IQR] 35–63 years) and in those with ADAMTS13 activity of <10%, the median age is 40 years (IQR 33–50 years). TTP remains life-threatening disease with a mortality rate of 10%–20% in spite of appropriate therapeutic management.

Clinical Manifestations: Although part of the early descriptions of TTP, the classic pentad of anemia, thrombocytopenia, fever, neurological signs, and renal failure is present in only 10% of patients. Severe thrombocytopenia ($<30 \times 10^9$/L) and microangiopathic hemolytic anemia characterized by schistocytes on the blood smear are the constant signs of TTP associated with skin and mucosal hemorrhage, weakness, and dyspnea. The most common organ ischemia/infarction manifestations are in the brain, heart, and mesenteric ischemia. Approximately 60% have neurological symptoms at presentation: headache, confusion, stroke, seizures, and coma. Twenty five percent have heart ischemia manifest as isolated EKG abnormalities to myocardial infarction, and 35% present with abdominal pain and diarrhea due to mesenteric ischemia. Overall, fever is present in 24% of patients and renal abnormalities in 59% and may manifest as proteinuria or increasing serum creatinine level. Importantly, the absence of fever, renal dysfunction, neurologic findings or other end-organ damage does not exclude the diagnosis of TTP. Patients with congenital TTP related to a mutation of the *ADAMTS13* gene, and resulting ADAMTS13 deficiency, may present any time from the neonatal period through young adulthood. In two-thirds of patients, the interval between relapses is every 2–3 weeks, and in one-third of patients the interval between relapses may be as long as years.

Diagnosis: TTP should be considered in the differential diagnosis of all patients presenting with MAHA and thrombocytopenia in the absence of underlying disease. MAHA is defined as schistocytes on peripheral smear and hemolytic anemia. Hemolytic anemia is diagnosed by an elevated lactate dehydrogenase (LDH), indirect bilirubin, and reticulocyte count. A microangiopathic cause for the hemolytic anemia is supported by the presence of schistocytes on the peripheral smear (>2 per field at 100× magnification). Given that MAHA and thrombocytopenia are not specific to TTP, other causes should also be evaluated (Table 107.1). Prothrombin time (PT), activated partial thromboplastin time (aPTT), direct antiglobulin test (DAT), renal and liver function tests may help in diagnosis. PT and aPTT typically are within normal limits in TTP, and DAT is typically negative. These standard investigations are not specific for TTP and should be complemented by analysis of ADAMTS13 activity levels and the presence of inhibitor antibodies of ADAMTS13. Although low activity levels of ADAMTS13 and the presence of inhibitory antibodies to ADAMTS13 may clarify the prognosis, the diagnostic value of these tests remains unclear (see Chapter 157).

Differential Diagnosis: The patient's history and physical examination are most useful in determining the cause of MAHA and thrombocytopenia (Table 107.1). The distinction between aHUS and TTP may be difficult; patients with aHUS typically have significantly worse renal function than those with TTP and lack neurologic symptoms. In the adult patient, aHUS is not readily distinguishable from TTP and may be initially managed similarly. Children with diarrhea-associated HUS are not routinely treated with TPE (see Chapter 106).

Management: Left untreated, the mortality associated with TTP is greater than 90%. TPE is the primary therapy for TTP. TPE is hypothesized to work by removing ultralarge VWF multimers and anti-ADAMTS13 autoantibodies, and concurrently restoring

TABLE 107.1 Differential Diagnosis of Microangiopathic Hemolytic Anemia
Autoimmune diseases (antiphospholipid antibody syndrome, systemic lupus erythematosus)
Disseminated intravascular coagulopathy
Hematopoietic stem cell transplantation
Kasabach–Merritt syndrome
Malignancy
Malignant hypertension
Medications:
Calcineurin inhibitors: cyclosporine and tacrolimus
Cancer chemotherapy: mitomycin C, cisplatin, and gemcitabine
Pregnancy (hemolysis-elevated liver enzymes and low platelets [HELLP] syndrome, eclampsia)
Prosthetic heart valves
Radiation nephritis
Scleroderma renal crises

ADAMTS13 protease activity. However, the benefit of TPE even in the absence of severe deficiency calls this hypothesis into question. TPE should be instituted as soon as possible after TTP is suspected, based on the findings of an unexplained MAHA and thrombocytopenia. If TPE is not available, plasma infusion (15–30 mL/kg) should be initiated while arrangements for TPE are made (see Chapter 75). Although some patients may have had signs or symptoms of TTP for weeks, instituting TPE should not be delayed on the basis of apparent clinical stability. There are no clinical prognostic factors that can accurately predict the onset of potentially catastrophic end-organ damage. There are clinical situations of secondary TTP, such as pregnancy and systemic lupus erythematosus, where TPE should be instituted as in acquired TTP because these disorders are associated with ADAMTS13 autoantibody formation (i.e., they have pathophysiology identical to acquired TTP). In contrast, TPE use in thrombotic microangiopathy secondary to malignant hypertension, prosthetic intravascular devices, or hematopoietic stem cell transplantation is typically not indicated. For the remainder of cases of nonacquired TTP, the utility of TPE is uncertain; typically, a trial of TPE is initiated.

TPE can be initiated using plasma products (fresh-frozen plasma [FFP], plasma frozen within 24 hours, or thawed plasma) or cryoprecipitate reduced plasma (CRP). Although underpowered to be definitive, a small randomized study comparing CRP and plasma found no difference in patient outcome, and nonrandomized study suggested that CRP is at least as effective as FFP as replacement for plasma exchange in TTP. The typical TPE regimen is 1.0 total plasma volume exchanged daily until the platelet count is above 150,000/μL. Most authorities will continue TPE for 2 days after the platelet count reaches >150,000/μL. An alternative approach is after 2 days of a platelet count of >150,000/μL to taper TPE by performing five TPE in over a 2-week period. Following cessation of TPE, the patient should be closely followed for evidence of recurrence over the next few weeks. Durable treatment response is a complete response at least for 30 days after discontinuation of TPE. Exacerbation is defined as a recurrent disease within 30 days after achieving complete response. Exacerbations of TTP typically become evident within 1–2 weeks of stopping TPE. Ideally the central catheter used for TPE can remain in place for 1–2 weeks while the durability of remission is being confirmed.

During treatment and in the immediate period after discontinuing TPE, daily laboratory monitoring should include hematocrit/hemoglobin, platelet count, reticulocyte count, and LDH. The platelet count is the most reliable measure of response to TPE. LDH will decrease with TPE but will less predictably normalize and define a complete response. If the platelet count decreases or the LDH increases after discontinuing TPE, the patient should be reevaluated for reinitiating TPE. Infection is associated with worsening or relapses of TTP. Accordingly, the central venous catheter should always be assessed for signs of infection in patients with clinical deterioration despite ongoing TPE or early relapse after cessation of TPE.

All patients should receive folic acid to facilitate RBC production secondary to the increased RBC destruction. Given the potential for an underlying autoimmune disorder leading to ADAMTS13 inhibitory antibodies, corticosteroids are typically used in the treatment of acquired TTP. Regimens include 1–2 mg/kg of prednisone daily until remission is achieved then slowly tapered, or 10 mg/kg day of methylprednisolone for 3 days and then 2.5 mg/kg day.

Rituximab is a chimeric monoclonal antibody against the B-cell marker CD20, an effective adjunct to TPE therapy in both refractory and relapsing acquired TTP. In the front-line therapy, a phase 2 trial of rituximab initiated within 3 days onset of diagnosis of TTP, patients received 375 mg/m^2 weekly for 4 weeks. Rituximab was infused at least 4 hours before TPE. Compared with historical controls that received a similar protocol of TPE, rituximab led to a significant reduction in the rate of relapse over 49 months of follow-up, 57% to 10% (P=.001). There was no overall effect on number of TPE treatments or length of hospital stay. Whether this same reduction in relapse rate could be achieved with lower doses, fewer infusions or infusions starting after completion of TPE are unknown.

New targeted therapies for treatment acute episodes of acquired TTP are emerging. These include recombinant ADAMTS13 (BAX930) (ClinicalTrials.gov: NCT02216084), which is intended to restore ADAMTS13 protease function by saturating ADAMTS13 antibodies and caplacizumab (ClinicalTrial.gov: NCT02553317), a nanobody that targets the A1 domain of VWF blocking the interaction of VWF and platelets.

Historically, it has been recommended that platelet transfusions should be avoided during an acute episode of TTP for fear of "fueling the fire." However, in a systematic review of the Oklahoma registry, platelet transfusions were not associated with adverse outcomes. Therefore, although platelets should not be transfused for management of severe thrombocytopenia alone, they should be part of strategies to treat bleeding or prevent bleeding in the setting of invasive procedures.

Refractory Disease: Refractory disease has been defined in various ways; definitions include a lack of improvement in platelet count after seven daily TPE procedures, lack of normalization of platelet count after 3 weeks of treatment, and worsening of disease despite treatment. In a small study, those patients that had ADAMTS13 deficiency and an inhibitor to ADAMTS13 had a more prolonged clinical course as compared to those with ADAMTS13 deficiency without an inhibitor (see Chapter 157). Those with an inhibitor had a longer time to platelet recovery (23 days vs. 7 days). A variety of treatment approaches have been used for refractory TTP but none in a randomized clinical trial. One approach is to perform two exchanges per day with 1.0 plasma volume per each exchange and add steroids, rituximab, or other immunosuppressive agents if not already utilized. Bortezomib is a proteasome inhibitor that induces apoptotic cell death mainly in immature dendritic cells and depletion of residual autoreactive B cells and plasma cells. Based on seven publications, 12 patients with relapsed/refractory acquired TTP were treated with Bortezomib in combination with other therapeutic modalities, 11 patients survived the acute episode and maintained remission.

Exacerbation: Recurrence of active TTP before 30 days after the last TPE procedure should be considered an exacerbation of the initial episode rather than a relapse (see below) and treated with reinitiation of TPE. Overall, 20%–45% of patients will experience an exacerbation. Rituximab can be considered in this setting if not used as part of the initial treatment.

Relapse: Relapse is typically defined as disease recurrence (i.e., thrombocytopenia and elevated LDH) after 30 days from the last TPE procedure from the previous episode. The cumulative estimated risk of recurrent TTP in patients with severe ADAMTS13 deficiency (<10%) is 41% at 7.5 years, and nearly half of those occur within the first year. This is compared with a 4% cumulative risk of recurrence at 7.5 years in patients with ADAMTS13 activity >10%. Ten-year survival did not differ between those with <10% or >10% ADAMTS13 activity but was lower in those with an anti-ADAMTS13 inhibitor titer >2 BU/mL (see Chapter 157). In the event of a relapse, TPE should be reinitiated. Multiple other therapies have been employed to prevent additional relapses in a patient that has demonstrated a predisposition to recurrent TTP. These approaches include rituximab, bortezomib, splenectomy, and other immunosuppressive medications (e.g., cyclosporine), although mixed reports exist with respect to the effectiveness of these approaches.

Congenital Thrombotic Thrombocytopenic Purpura: For patients with congenital TTP secondary to a congenital deficiency of ADAMTS13, the treatment of choice is replacement of the missing cleaving protease by prophylactic plasma infusions. This can be achieved by infusions of plasma (10 mL/kg) approximately every 2–3 weeks. The interval between infusions should be adjusted according to clinical signs and symptoms. Prophylactic plasma infusion is also required in pregnant women to prevent relapses. Recombinant ADAMTS13 (rADAMTS13) is currently under investigation (ClinicalTrial.gov: NCT02216084) and if effective may replace plasma infusion as standard of care.

Further Reading

Berangere, S. J., Coppo, P., & Veyradier, A. (2017). Thrombotic thrombocytopenic purpura. *Blood, 129*, 2837–2845.

George, J. (2010). How I treat patients with thrombotic thrombocytopenic purpura. *Blood, 116*, 4060–4469.

Moake, J. L., Rudy, C. K., Troll, J. H., et al. (1982). Unusually large plasma factor VIII:von Willebrand factor multimers in chronic relapsing thrombotic thrombocytopenic purpura. *N Engl J Med, 307*, 1432–1435.

Rock, G. A., Shumak, K. H., Buskard, N. A., et al. (1991). Comparison of plasma exchange with plasma infusion in the treatment of thrombotic thrombocytopenic purpura. Canadian Apheresis Study Group. *N Engl J Med, 325*, 393–397.

Rock, G., Shumak, K. H., Sutton, D. M., et al. (1996). Cryosupernatant as replacement fluid for plasma exchange in thrombotic thrombocytopenic purpura. *Br J Haematol, 94*, 383–386.

Sadler, J. E. (2017). Pathophysiology of thrombotic thrombocytopenic purpura. *Blood, 130*, 1181–1188.

Scully, M., McDonald, V., Cavenagh, J., et al. (2011). A phase 2 study of the safety and efficacy of rituximab with plasma exchange in acute acquired thrombotic thrombocytopenia purpura. *Blood, 118*, 1746–1753.

Terrell, D. R., Williams, L. A., Vesely, S. K., et al. (2005). The incidence of thrombotic thrombocytopenic purpura-hemolytic uremic syndrome: All patients, idiopathic patients, and patients with severe ADAMTS-13 deficiency. *J Thromb Haemost, 3*, 1432–1436.

Zeigler, Z. R., Shadduck, R. K., Gryn, J. F., et al. (2001). Cryoprecipitate poor plasma does not improve early response in primary adult thrombotic thrombocytopenic purpura (TTP). *J Clin Apheresis, 16*, 19–22.

CHAPTER 108

Antiphospholipid Syndrome

Lucia R. Wolgast, MD

Introduction: The antiphospholipid (aPL) syndrome (APS) is an autoimmune thrombophilic condition that is defined by a combination of clinical and laboratory criteria. In general terms, APS patients have developed circulating antibodies against plasma proteins that bind to phospholipids (i.e., aPL antibodies) with subsequent clinical morbidity of thrombosis and/or pregnancy complications. The investigational criteria for APS (often referred to as the Sydney Investigational Criteria) are detailed in Table 108.1 and require that patients have documented evidence for vascular thrombosis and/or obstetric complications, such as unexplained recurrent miscarriages, intrauterine growth restriction, intrauterine fetal demise, preeclampsia/toxemia, placental abruption, and preterm labor. The laboratory criteria require identifying the persistent aPL antibodies (i.e., at least two abnormal measurements at least 12 weeks apart), including elevated medium to high tiers of anticardiolipin (aCL) IgG or IgM antibodies, anti-β_2-glycoprotein I (anti-β2GPI) IgG or IgM antibodies, and/or positive lupus anticoagulant (LA). The laboratory diagnosis of APS is discussed in Chapter 158.

It is important for the reader to understand that these criteria were not designed to be requirements for the clinical diagnosis of APS. Rather, they were intended to provide

TABLE 108.1 Sydney Investigational Criteria for the Diagnosis of the Antiphospholipid Syndrome[a]

Clinical
• **Vascular thrombosis (one or more episodes of arterial, venous, or small vessel thrombosis). For histopathologic diagnosis, there should be no evidence of inflammation in the vessel wall.**
• **Pregnancy morbidities attributable to placental insufficiency, including (a) three or more otherwise unexplained recurrent spontaneous miscarriages, before 10 weeks of gestation, (b) one or more fetal losses after the 10th week of gestation, (c) stillbirth, and (d) episode of preeclampsia, preterm labor, placental abruption, intrauterine growth restriction, or oligohydramnios that are otherwise unexplained.**
Laboratory
• **Medium- or high-titer aCL or anti-β2GPI IgG and/or IgM antibody present on two or more occasions, at least 12 weeks apart, measured by standard ELISAs.**
• **Lupus anticoagulant in plasma, on two or more occasions, at least 12 weeks apart, detected according to the guidelines of the ISTH SSC Subcommittee on Lupus Anticoagulants and Phospholipid-Dependent Antibodies.**

β2GPI, β2-glycoprotein I; *aCL*, anticardiolipin; *aPL*, antiphospholipid; *ELISA*, enzyme-linked immunosorbent assay; *Ig*, immunoglobulin.

[a]"Definite APS" is considered to be present if at least one of the clinical criteria and one of the laboratory criteria are met.

Modified from Miyakis, S., Lockshin, M. D., Atsumi, T., et al. (2006). International consensus statement on an update of the classification criteria for definite antiphospholipid syndrome (APS). *J Thromb Haemost*, *4*, 295–306.

Transfusion Medicine and Hemostasis. https://doi.org/10.1016/B978-0-12-813726-0.00108-2

TABLE 108.2 Proposed Criteria for the Classification of Catastrophic Antiphospholipid Syndrome (APS)

1. Evidence of involvement of three or more organs, systems, and/or tissues[a]
2. Development of manifestations simultaneously or in less than a week
3. Confirmation by histopathology of small vessel occlusion in at least one organ or tissue[b]
4. Laboratory confirmation of the presence of antiphospholipid antibodies (lupus anticoagulant and/or anticardiolipin antibodies)[c]

Definite Catastrophic APS

- All four criteria

Probable Catastrophic APS

- All four criteria, except for only two organs, systems and/or tissues involvement
- All four criteria, except for the absence of laboratory confirmation at least 6 weeks apart due to the early death of a patient never previously tested for aPL before the catastrophic APS event
- Criteria 1, 2, and 4
- Criteria 1, 3, and 4 and the development of a third event in more than a week but less than a month, despite anticoagulation

[a]Usually, clinical evidence of vessel occlusions, confirmed by imaging techniques when appropriate. Renal involvement is defined by a 50% rise in serum creatinine, severe systemic hypertension (N180/100 mm Hg), and/or proteinuria (N500 mg/24 hours).
[b]For histopathological confirmation, significant evidence of thrombosis must be present, although, in contrast to Sydney criteria, vasculitis may coexist occasionally.
[c]If the patient had not been previously diagnosed as having an APS, the laboratory confirmation requires that the presence of antiphospholipid antibodies must be detected on two or more occasions at least 6 weeks apart (not necessarily at the time of the event), according to the proposed preliminary criteria for the classification of definite APS.
Modified from Asherson, R. A., Cevera, R., de Groot, P. G., et al. (2003). Catastrophic antiphospholipid syndrome: international consensus statement on classification criteria and treatment guidelines. *Lupus, 12*, 530–534.

a uniformly rigorous definition of APS for the purpose of standardizing research on the disorder. In "real-world" clinical practice, some patients may be appropriately diagnosed for presumptive APS without meeting the strict investigational criteria. Some APS patients may be positive for other "noncriteria" clinical laboratory tests—see Chapter 158—that have not been included by consensus panels as diagnostic criteria for the disorder, but these positive noncriteria laboratory tests may indicate a clinical risk and warrant treatment. Furthermore, APS patients may have positive APS criteria assays but not the typical criteria manifestations. These "noncriteria manifestations" include thrombocytopenia, livedo reticularis, skin ulcers, nephropathy, migraine, cognitive defects, diffuse alveolar hemorrhage, and valvular heart disease (Libman–Sacks endocarditis), and these clinical manifestations may need to be managed differently. Occasional patients may even test entirely negative for the APS criteria assays but have typical clinical manifestations of the disorder—a situation referred to as seronegative APS (termed SNAPS).

At present, APS may be divided into the following subcategories: (1) **Primary APS** is the "stand alone" disorder, in the absence of systemic lupus erythematosus (SLE), (2) **secondary APS** occurs in the presence of APS, (3) **catastrophic APS (CAPS)** manifests as disseminated thrombosis in large and small vessels with resulting multiorgan failure (Table 108.2), and (4) **SNAPS** includes patients whose diagnostic tests are entirely negative but who, on clinical grounds, are still suspected to have the disorder.

Antigenic Specificities and Pathophysiology of Antiphospholipid Antibodies

β2-glycoprotein I: β2-glycoprotein I (β2GPI), a 50-kd glycoprotein member of the complement control protein superfamily 5, is a major antigenic target for aPL antibodies. It is a scavenger protein that binds and clears apoptotic cells, microparticles, and other subcellular elements in circulating blood, including von Willebrand factor (VWF). Thromboprotective roles include binding and clearance of low-density lipoprotein (oxLDL), binding to VWF to reduce platelet adhesion to collagen, acting as a cofactor for tissue-type plasminogen activator (t-PA), and annexin A2 allowing increased fibrinolytic activity.

β2GPI contains five homologous complement control domains, with a fifth domain that contains a phospholipid binding site (Fig. 108.1). This binding site has cationic residues with an affinity for anionic phospholipids and a hydrophobic loop, which can insert into the lipid bilayer. All five domains of β2GPI have been reported to be recognized by anti-β2GPI antibodies derived from patients. However, the most significant correlation is IgG antibodies against an epitope on domain I (Gly40-Arg43), which has a strong association with thrombosis. The antibodies against the domain I recognize an epitope, which is cryptic in the soluble, unbound protein, but becomes exposed after β2GPI binds to phospholipid bilayers. Transmission electron microscopy and X-ray scattering studies of stained β2GPI molecules indicate that the free unbound protein has a circular confirmation, whereas β2GPI bound to phospholipid has either a "J" or "S" shape conformation, exposing hidden domains to the immune system.

Direct injury to endothelial surfaces causes β2GPI to bind to exposed phospholipids. Complexes of β2GPI bound to exposed aPLs, such as cardiolipin, create antigenic targets for aPL antibodies. These aPL antibodies then bind to the β2GPI/phospholipid complexes and interfere with anticoagulant and fibrinolytic pathways. Through cellular signaling pathways, the bound aPL antibodies create a prothrombotic state—upregulating tissue factor expression and adhesion molecules. aPL antibodies can also interfere with β2GPI cofactor functions for t-PA and annexin A2 inhibiting fibrinolytic activity. aPL antibodies reduce the effectiveness of β2GPI in clearing atherogenic proteins such as oxLDL and procoagulant microparticles. aPL antibodies can reduce the effectiveness of β2GPI in regulating VWF-mediated platelet adhesion. In the placenta, aPL antibodies cause placental insufficiency by activating trophoblasts and endometrial cells, resulting in abnormal trophoblast proliferation, increased trophoblast apoptosis, disruption in differentiation of decidual endometrial cells, and interference in maternal spiral artery maturation. The presence of aPL antibodies can also disrupt protective anticoagulant mechanisms to endothelial and trophoblastic injury, such as the binding of the anticoagulant annexin A5 to exposed phospholipids.

aPL antibodies have been shown to activate platelets and monocytes and interfere with β2GPI's role as a regulator of the complement cascade. β2GPI complexed with HLA-DR7 on cell surfaces can trigger complement-mediated cytotoxicity, which has been shown to play a significant role in obstetric APS.

Other Antigenic Targets of Antibodies Associated With Antiphospholipid Syndrome: Targets other than β2GPI that have been identified for aPL antibodies

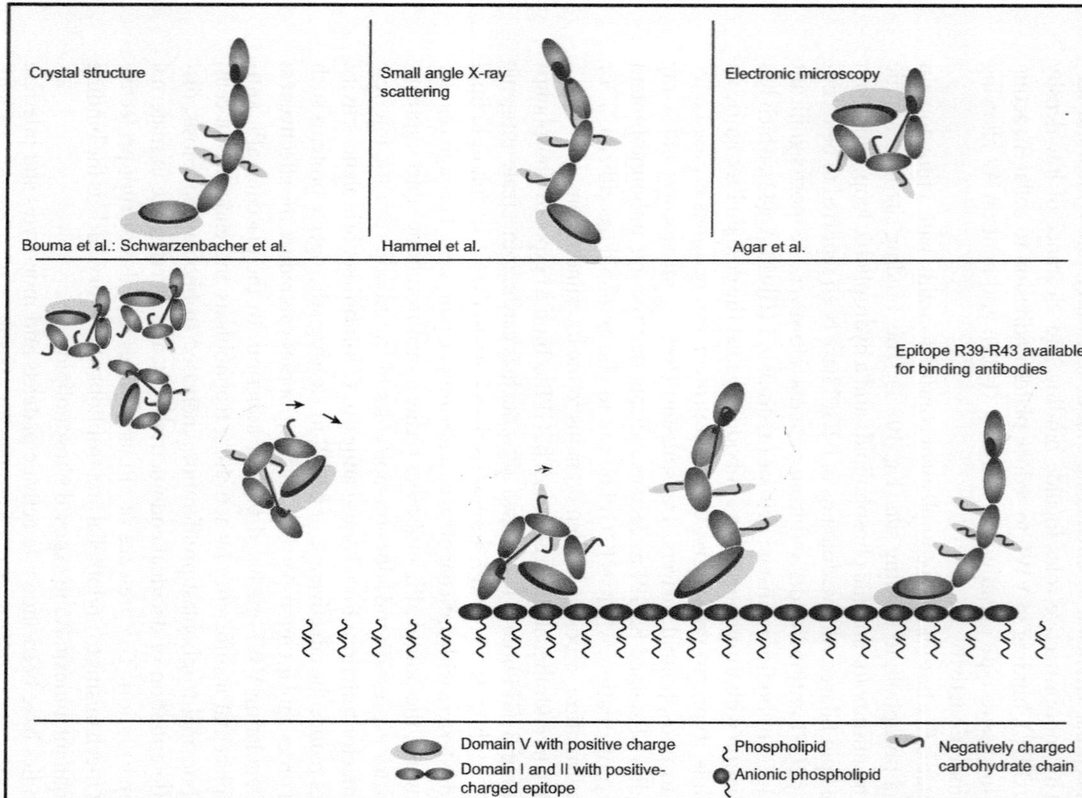

FIGURE 108.1 Models for proposed β2GPI structures. By X-ray crystallography, the structure of β2GPI showed a "J" shape. However, small angle X-ray scattering (SAXS) of β2GPI in solution revealed an S-shape conformation, with a carbohydrate chain on top of the interface between domains I and II. In contrast to the SAXS structure, transmission electron microscopy (TEM) of negatively stained unbound β2GPI mounted on grids showed a circular conformation, but a "J" shape for β2GPI bound to phospholipid. The arginine 39-arginine 43 epitope on domain I that is recognized by thrombogenic aPL antibodies is cryptic in the free-β2GPI proposed conformations. This epitope then becomes exposed and available for antibody recognition after the protein is bound to phospholipid. (Reprinted From de Laat, B., & de Groot, P. G. (February 2011). Autoantibodies directed against domain I of β2-glycoprotein I *Curr Rheumatol Rep 13*(1), 70–76.)

include other anionic phospholipids, including phosphatidylserine, phosphatidylethanolamine, and phosphatidylinositol, which have all been associated with increased thrombosis and pregnancy morbidity. Other antibodies associated with APS clinical manifestations include prothrombin (coagulation factor II), coagulation factor V, protein C, protein S, annexin A2, annexin A5, high- and low-molecular-weight kininogens, heparin, factor VII/VIIa, plasmin, vimentin, VWF, thrombospondin, and P-selectin.

Clinical Manifestations of the Antiphospholipid Syndrome: The clinical symptoms of APS can be categorized into "criteria manifestations" and "noncriteria manifestations." The "criteria manifestations" include vascular thrombotic events found in thrombotic APS, pregnancy complications found in obstetric APS, and multiorgan damage found in CAPS. The "noncriteria manifestations" include other pathologic conditions that have been found in APS patients but have not been included in the definition of "definite APS" by the expert consensus groups. These noncriteria manifestation in conjunction with positive laboratory criteria findings may warrant patient management and treatment.

Criteria Manifestations of Thrombotic and Obstetric Antiphospholipid Syndrome

Venous and Arterial Thromboembolism: Criteria manifestations of thrombotic APS include both venous and arterial thrombosis with 60% of APS patient presenting with venous thrombosis, 30% presenting with arterial thrombosis, and 10% presenting with both venous and arterial thrombosis. The most common manifestations are deep vein thrombosis of the lower extremity, pulmonary embolism, and stroke/transient ischemic attacks. Other venous manifestations include thrombosis of thoracic, abdominal, pelvic, renal and/or retinal veins. Other arterial manifestations include myocardial infarction, nonatherosclerotic coronary artery occlusion, pulmonary hypertension attributed to thrombotic occlusion, and occlusions of abdominal, renal, and retinal arteries. The diagnosis of APS should be considered in patients spontaneously presenting with thrombosis particularly in the absence of other risk factors. The diagnosis of thrombotic APS is especially suspected when transient ischemic attacks or strokes occur in younger patients, particularly those without the usual risk factors for cerebrovascular disease—e.g., cigarette smoking, hypertension, and diabetes mellitus. In addition, thrombotic APS may manifest in the presence of a predisposing provocative factors such as estrogen hormone replacement therapy, oral contraceptives, pregnancy, the postpartum state, vascular stasis, surgery, or trauma. Some patients with venous thrombosis may also have a concurrent genetic thrombophilic conditions such as factor V-Leiden polymorphism.

Pregnancy Complications: Criteria manifestations of obstetric APS include recurrent fetal loss, preeclampsia, placental abruption, prematurity, intrauterine fetal demise, intrauterine growth restriction, and oligohydramnios. Women with obstetric APS generally present with a history of recurrent (i.e., three or more) miscarriages. In approximately half of patients, the pregnancy losses occur in the first trimester; other patients present with later losses, most in the second trimester, but some even later, including stillbirth. The specific pregnancy complications that define obstetric APS include the absence of another explanation for the complications, three or more recurrent spontaneous first trimester miscarriages, or one mid-trimester loss, stillbirth, episode of preeclampsia, preterm labor, placental abruption, intrauterine growth restriction, or oligohydramnios. Pregnant patients with APS are also more prone to develop deep vein thrombosis during pregnancy or the puerperium. Interestingly, early reproductive failure (i.e., infertility) is not associated with APS.

Noncriteria Clinical Manifestations Associated With Antiphospholipid Syndrome: The most common noncriteria clinical manifestations include cardiac, neurological, dermatological, and hematological symptoms, including coronary artery disease in the absence of thrombotic occlusion, peripheral vascular disease, valvular abnormalities, migraines, seizures, livedo reticularis (Fig. 108.2), skin ulcerations/necrosis, thrombocytopenia, and immune thrombocytopenic purpura. Many of the noncriteria manifestations occur in patients with SLE or other autoimmune disorders and may warrant additional management to prevent APS-associated thrombosis and/or pregnancy complications. One or two early spontaneous miscarriages (less than 10 weeks gestation) with a low positive aCL or anti-β2GPI are also considered a noncriteria manifestation that may warrant additional management. Other rarer noncriteria manifestations include glomerulonephritis, APS nephropathy, primary biliary cirrhosis, pancreatitis, acute sensorineural hearing loss, osteonecrosis, and bleeding due to acquired factor deficiencies or von Willebrand disease.

Catastrophic Antiphospholipid Syndrome: Rare patients present with a *catastrophic* form of APS, which is characterized by severe widespread vascular occlusions, and a high mortality. The formal diagnostic criteria include evidence of involvement of at least three organs, systems and/or tissues, development of manifestations simultaneously or within 1 week, histopathologic confirmation of small vessel occlusion, and laboratory confirmation of the presence of aPL antibodies (Table 108.2).

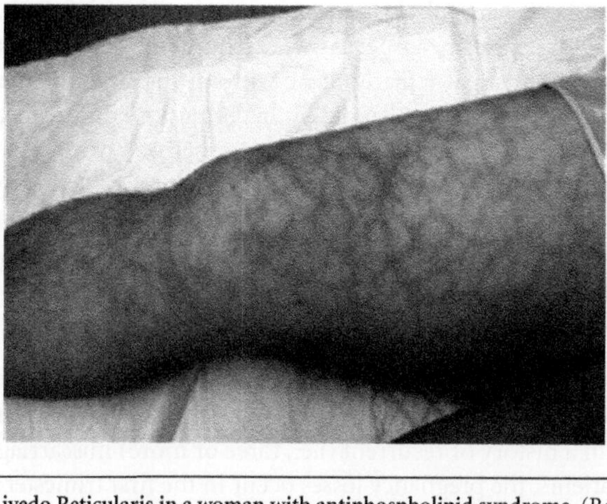

FIGURE 108.2 **Livedo Reticularis in a woman with antiphospholipid syndrome.** (Reprinted with permission from Elsevier, Ruiz-Irastorza, G., Crowther, M., Branch, W., & Khamashta, M. A. (October 30, 2010). Antiphospholipid syndrome. *Lancet 376*(9751),1498–1509.)

In contrast to the diagnostic criteria for APS, which exclude the diagnosis when histopathology shows evidence for vasculitis, the diagnostic criteria for CAPS do permit the presence of histological evidence for vasculitis together with thrombosis. These patients present with evidence for severe multiorgan ischemia/infarction, usually with concurrent microvascular thrombosis. Patients with CAPS can present with massive venous thromboembolism, along with respiratory failure, stroke, abnormal liver enzymes, renal impairment, adrenal insufficiency, and areas of cutaneous infarction. The respiratory failure is usually due to acute respiratory distress syndrome and diffuse alveolar hemorrhage. Laboratory evidence for disseminated intravascular coagulation is frequently present. According to the CAPS Registry, a web-based database of 433 patients with CAPS (https://ontocrf.costaisa.com/en/web/caps/), the majority of CAPS patients are female (69%), in their late thirties (mean age of 38.5 years), but patients can be present at any age (range 0–85 years). In half of the CAPS cases, the patients' catastrophic event was their first APS manifestation. Precipitating factors of CAPS include infections, drugs (sulfur-containing diuretics, captopril, and oral contraceptives), surgical procedures, and cessation of prior anticoagulant therapy. In 26.9% of cases, the patients also had SLE. The most frequently affected organ was the kidney (73% of episodes), followed by lungs (58.9%), the brain (55.9%), the heart (49.7%), and the skin (45.4%). Other organs were also affected, including the peripheral vessels, intestines, spleen, adrenal glands, pancreas, retina, and bone marrow. Improved aggressive treatment has reduced mortality from ~50% to ~20% and relapse is rare in survivors.

Pediatric Antiphospholipid Syndrome: APS has become increasingly recognized to be a significant cause of thrombosis in the pediatric population. A European registry report of an initial 121 cases described thrombotic manifestations similar to those seen with adults with APS. However, there was a significant and interesting difference between children with primary APS and the secondary APS, who had other autoimmune disease. The patients with primary APS tended to be younger and had a higher frequency of arterial thrombotic events, whereas the secondary APS patients were older and had a higher frequency of venous thrombotic events associated with hematologic and skin manifestations. The catastrophic form of the syndrome has been reported in children but occurs more rarely than in adults. Thrombosis is rare in newborns delivered from mothers with APS, and only a few cases are reported, mostly associated with other prothrombotic factors. aPL antibodies have been found in up to 30% of offspring of mothers with APS. Other complications in neonates born to APS mothers include prematurity, neurodevelopment abnormalities, and learning disabilities.

Treatment of Patients With Antiphospholipid Syndrome: There is general agreement that patients with spontaneous and/or recurrent thrombosis require long-term anticoagulant therapy and that pregnant patients with a history of recurrent spontaneous pregnancy losses require antithrombotic therapy for most of the gestational period and for the puerperium. There have been uncertain approaches to treatment of patients with single thrombotic events, patients with provoked thrombotic events (i.e., trauma, surgery, stasis, airplane travel, pregnancy, or estrogens), patients with 1–2 early miscarriages, patients with noncriteria clinical manifestations, and patients with low-titer laboratory results. In addition, asymptomatic patients with multipositivity for criteria laboratory tests may be at increased risk for a clinical event and may warrant preventative treatment. Table 108.3 provides a summary of current recommended guidelines and treatment considerations.

Thrombotic Antiphospholipid Syndrome Treatment: In general, patients with thrombotic APS based on the diagnostic criteria should be treated with warfarin for the long term, at a therapeutic INR of 2.0–3.0. This is regardless of single or multiple positivity for aPL antibodies. In addition, patients with an arterial thrombotic event (i.e., stroke and/or myocardial infarction) should either be placed on a higher warfarin dose with INR 3.0–4.0 or be placed on long-term warfarin at a therapeutic INR of 2.0–3.0 and 100 mg daily dose of aspirin. APS patients with myocardial infarction and a stent placement should also be treated with clopidogrel (75 mg per day).

Direct-acting oral anticoagulants (DOACs), either direct factor-Xa or thrombin inhibitors, have been proven effective for treatment of venous thromboembolism. However, their use specifically in APS patients has not been thoroughly evaluated. A recent randomized, open-label trial with 54 patients receiving rivaroxaban and 56 patients on warfarin showed the same safety profiles with no recurrent thrombosis or bleeding. Clinical trials are currently underway to confirm the safety and efficacy of DOACs in APS patients. DOACs can be considered in patients with poor anticoagulant compliance with warfarin or vitamin K antagonist allergy. DOACs are not recommended for APS patients with arterial thrombosis or with recurrent thrombosis while on therapeutic warfarin anticoagulation.

Obstetric Antiphospholipid Syndrome Treatment: The current approach to treating pregnant women with obstetric APS includes daily low-dose aspirin (75–81 mg per day) and either unfractionated heparin (UFH) or low-molecular-weight heparin (LMWH). Although clinical studies have shown efficacy with UFH, most clinicians treat with LMWH because it has a better pharmacokinetic profile and lower risk of heparin-induced thrombocytopenia and osteopenia. Heparin is then withheld when labor begins or 24 hours before a cesarean section. Anticoagulation is then resumed 6 weeks postpartum because of the increased risk of venous thromboembolism (VTE) in this period. With this management, the likelihood of a good pregnancy outcome in women with APS has been estimated to be about 75%–80%

Management of Noncriteria Clinical Manifestations: Treatment for noncriteria clinical manifestations of APS includes rituximab, glucocorticoids, vitamin D supplementation, and hydroxychloroquine. Rituximab may be effective in controlling noncriteria manifestations of APS such as persistent thrombocytopenia, autoimmune hemolytic anemia, cardiac valve disease, chronic skin ulcers, aPL-nephropathy, and/or cognitive dysfunction. Because low vitamin D levels correlated with noncriteria APS manifestations, it is recommended that vitamin D deficiency (<10–20 ng/mL) and insufficiency (<30 ng/mL) be corrected as a preventative strategy.

Catastrophic Antiphospholipid Syndrome Treatment: Patients with CAPS require aggressive treatment because of the high mortality. Treatment for CAPS is directed toward the thrombotic events and suppression of the cytokine cascade.

TABLE 108.3 Antiphospholipid Syndrome (APS) Treatment Strategy

Clinical Manifestation	Laboratory Result[a]	Treatment
Thrombotic APS		
Venous thromboembolism	Positive aCL and/or anti-β2GPI (IgG and/or IgM) immunoassays with medium or high titers[b] or Positive LA (confirm positive using two different assays)	Long-term vitamin K antagonist, INR 2.0–3.0
	Low-titer positivity for aCL (IgG or IgM < 40 PL) or anti-β2GPI (IgG or IgM < 99th percentile)	Anticoagulation treatment and duration determined by additional risk factors at time of thrombosis.
Arterial thromboembolism (including stroke and myocardial infarction)	Positive aCL and/or anti-β2GPI (IgG and/or IgM) immunoassays with medium or high titers[b] or Positive LA (confirm positive using two different assays)	Long-term vitamin K antagonist, INR 2.0–3.0 plus aspirin 100 mg per day or Long-term vitamin K antagonist, INR 3.0–4.0 Myocardial Infarction with percutaneous coronary interventions and stent placement: Long-term vitamin K antagonist, INR 2.0–3.0, aspirin 100 mg per day, and clopidogrel 75 mg per day.
	Low-titer positivity for aCL (IgG or IgM < 40 GPL) or anti-β2GPI (IgG or IgM < 99th percentile)	Anticoagulation treatment and duration determined by additional risk factors at time of thrombosis.
Noncriteria manifestations	Positive aCL and/or anti-β2GPI (IgG and/or IgM) immunoassays with medium or high titers[b] or Positive LA (confirm positive using two different assays)	Consider rituximab and/or vitamin D supplementation
Noncriteria manifestations	Low-titer positivity for aCL (IgG or IgM < 40 GPL) or anti-β2GPI (IgG or IgM < 99th percentile)	Consider rituximab and/or vitamin D supplementation
Refractory thrombotic APS (recurrent thrombosis despite treatment)		Higher intensity vitamin K antagonist, INR 3.0–4.0. First line add-on treatments: Antiplatelet agents or low-molecular-weight (LMW) heparin Second line add-on treatments: Glucocorticoids, hydroxychloroquine, statins, and/or rituximab
Obstetric APS		
Pregnant women with history of criteria pregnancy complications[c] or thrombotic events	Positive aCL and/or anti-β2GPI (IgG and/or IgM) immunoassays with medium or high titers[b] or Positive LA (confirm positive using two different assays)	Unfractionated or LMW heparin plus low-dose aspirin

Continued

TABLE 108.3 Antiphospholipid Syndrome (APS) Treatment Strategy—cont'd

Clinical Manifestation	Laboratory Result[a]	Treatment
Pregnant women with history of criteria pregnancy complications[c] or thrombotic events	Low-titer positivity for aCL (IgG or IgM < 40 GPL) or anti-β2GPI (IgG or IgM < 99th percentile)	Unfractionated or LMW heparin plus low-dose aspirin[d]
Noncriteria manifestation—one or two early miscarriage (<10 weeks of gestation)	Positive aCL and/or anti-β2GPI (IgG and/or IgM) immunoassays with medium or high titers[b] or Positive LA (confirm positive using two different assays)	Unfractionated or LMW heparin plus low-dose aspirin[d]
Noncriteria manifestation—one or two early miscarriage (<10 weeks gestation)	Low-titer positivity for aCL (IgG or IgM < 40 GPL) or anti-β2GPI (IgG or IgM < 99th percentile)	No treatment recommendation, however, assess pregnant patients for treatment due to additional risk factors.
Refractory obstetric APS		Consider add-on treatment with glucocorticoids, intravenous IgG (IVIG), and/or hydroxychloroquine
Asymptomatic Antiphospholipid Positive Carriers		
None	Triple positive for LA, aCL, and anti-β2GPI with medium or high titers[b]	Long-term prophylaxis with low-dose aspirin. Consider anticoagulant prophylaxis for high-risk situations—e.g., immobilization, surgery, air travel, etc. Consider unfractionated or LMW heparin for pregnant patients
	Double positive for LA, aCL, or anti-β2GPI with medium or high titers or Persistently positive for isolated aCL IgG or IgM with medium or high titers or Single positive for LA	Long-term prophylaxis with low-dose aspirin. Consider anticoagulant prophylaxis for high-risk situations—e.g., immobilization, surgery, air travel, etc. Consider unfractionated or LMW heparin for pregnant patients
	Low-titer positivity for aCL (IgG or IgM < 40 GPL) or anti-β2GPI (IgG or IgM < 99th percentile) and/or transiently positive.	No treatment, however, considers prophylaxis treatment with LMW heparin in situations with increased risk—surgery, prolonged immobilization, pre-/peripartum, ovarian stimulation, thalidomide therapy, and the patient is currently positive[d]
Patients with SLE or autoimmune disease	Single positive LA or persistently positive aCL with medium or high titers	Low-dose aspirin and hydroxychloroquine

[a]Laboratory tests should be deferred until at least 12 weeks after the clinical event to avoid interferences of the acute phase of the disease. Earlier testing may yield false positive results. LA, aCL, and anti-β2GPI tests must be positive on two or more occasions at least 12 weeks apart to be considered "Positive."

[b]Medium or high titers include aCL IgG, or IgM > 40 GPL and/or anti-β2GPI IgG or IgM > 99th percentile.

[c]Sydney Investigational Criteria pregnancy complications include (a) >3 unexplained recurrent spontaneous miscarriages at <10 weeks of gestation, (b) one or more fetal losses after the 10th week of gestation, (c) stillbirth, and (d) episode of preeclampsia, preterm labor, placental abruption, intrauterine growth restriction, or oligohydramnios.

[d]From recent recommendations by international consensus in *Autoimmunity Reviews* 16 (2017) 911–924 and 1103–1108.

A triple-therapy strategy of anticoagulation with heparin, high-dose glucocorticoids, and either intravenous immunoglobulin (IVIG) or plasma exchange or both has improved outcomes. This triple-therapy strategy has resulted in the significant reduction in mortality rate to 30% in CAPS. Cyclophosphamide is recommended for patients with CAPS and inflammatory features of SLE or high-titer aPL antibodies. Rituximab may be useful in refractory or relapsing cases of CAPS.

Refractory Antiphospholipid Syndrome Treatment: For thrombotic APS patients who are refractory to standard warfarin anticoagulant therapy, an increase in warfarin intensity is recommended with an INR 3.0–4.0. If APS patients have a recurrence on high-intensity warfarin, then add-on treatment with an antiplatelet agent (i.e., clopidogrel) or LMWH may be considered, as well as a DOAC for recurrent venous thromboembolism. If anticoagulation fails, combinations of anticoagulation with glucocorticoids, hydroxychloroquine, statins, and/or rituximab may be considered. In patients with refractory obstetric APS, glucocorticoids, IVIG, and hydroxychloroquine have been shown to be effective. In addition, glucocorticoids should be considered in obstetric APS patients who have a severe immune thrombocytopenia or a significant contraindication to heparin therapy.

Management of Patients With Clinical Manifestations and Low-Titer Antiphospholipid Antibody Positivity: Currently there are no treatment recommendations for patients with low-titer positivity for aCL (IgG or IgM < 40 GPL) or anti-β2GPI (IgG or IgM < 99th percentile) or transiently positive aPL antibodies. For patients with venous or arterial thromboembolism, anticoagulation treatment and duration should be determined by additional risk factors at time of thrombosis. In pregnant patients with history of greater than three early spontaneous miscarriages, one or more late miscarriages, stillbirth, an episode of preeclampsia, preterm labor, placental abruption, intrauterine growth restriction or oligohydramnios, and unfractionated or LMW heparin plus low-dose aspirin is recommended. No treatment is recommended in patients with low-titer antibodies and one or two miscarriages; however, they should be assessed for other risk factors.

Management of Asymptomatic Antiphospholipid Antibody-Positive Patients: Asymptomatic patients with no previous thrombosis but a high-risk aPL antibody profile (i.e., triple positive for APS criteria assays, persistently positive medium to high titers of aCL antibodies, or persistent LA positive) be given long-term prophylaxis with low-dose aspirin, especially if there are other thrombotic risk factors. Furthermore, aPL antibody carriers should receive thromboprophylaxis with usual doses of low-molecular-weight heparin in high-risk situations, such as surgery, prolonged immobilization, and the puerperium. In addition to low-dose aspirin, aPL-antibody carriers with SLE should receive hydroxychloroquine as primary prophylaxis. In patients with low-titer positivity for (IgG or IgM < 40 GPL) or anti-β2GPI (IgG or IgM < 99th percentile) or transiently positive aPL antibodies, no treatment is recommended. However, if a patient is found to be currently positive during situations with increased risk—i.e., surgery, prolonged immobilization, pre-/peripartum, ovarian stimulation, thalidomide therapy—consider prophylaxis treatment with LMW heparin. No prophylaxis treatment is recommended in pregnant women with elevated low titer or transient aPL antibodies and without any history of spontaneous pregnancy losses, other attributable pregnancy complications, thrombosis, or embolism.

Antiphospholipid Syndrome Treatment Duration and Monitoring: Current treatment for APS involves long-term, indefinite management with anticoagulation. It is important to note that when treating APS patients, anticoagulation may interfere with LA testing and cause false positive results. Therefore diagnosing APS and monitoring for persistently positive criteria assays can be a challenge. In addition, it is uncertain how to manage APS patients whose aPL antibody tests become persistently negative. At this point, it is unclear if the prothrombotic state induced by aPL antibodies fully disappears when patients are negative on aPL antibody tests. Because of inadequate data, there are no current recommendations on ceasing anticoagulant therapy in APS patients who become seronegative.

Future Management of Antiphospholipid Syndrome: APS is a heterogenous disorder requiring future efforts in improving the diagnosis through additional novel assays and evolving guidelines to identify patients with a propensity for thrombosis and/or pregnancy complications. Future targeted therapies for APS are still experimental and include tissue factor inhibitors, complement inhibitors, intracellular signaling inhibitors, and antagonists to platelet GP receptors and β2GPI receptors such as annexin A2, TLR4, and ApoE2. These potential treatments will require evaluation in well-designed prospective clinical trials.

Further Reading

Arnaud, L., Conti, F., Massaro, L., et al. (2017). Primary thromboprophylaxis with low-dose aspirin an antiphospholipid antibodies: Pros and cons. *Autoimmun Rev, 16*, 1103–1108.

Asherson, R. A., Cevera, R., de Groot, P. G., et al. (2003). Catastrophic antiphospholipid syndrome: International consensus statement on classification criteria and treatment guidelines. *Lupus, 12*, 530–534.

Breen, K. A., Sanchez, K., Kirkman, N., et al. (2015). Endothelial and platelet microparticles in patients with antiphospholipid antibodies. *Thromb Res, 135*, 368–374.

Cervera, R. (2017). Antiphospholipid syndrome. *Thromb Res, 151*, S43–S47.

Cervera, R., Rodriguez-Pinto, I., Colafrancesco, S., et al. (2014). 14th international congress on antiphospholipid antibodies task force report on catastrophic antiphospholipid syndrome. *Autoimmun Rev, 13*, 699–707.

de Laat, B., & de Groot, P. G. (2011). Autoantibodies directed against domain I of β2-glycoprotein I. *Curr Rheumatol Rep, 13*, 70–76.

Erkan, D., Aguiar, C. L., Andrade, D., et al. (2014). 14th international congress on antiphospholipid antibodies: Task force report on antiphospholipid syndrome treatment trends. *Autoimmun Rev, 13*, 685–696.

Miyakis, S., Lockshin, M. D., Atsumi, T., et al. (2006). International consensus statement on an update of the classification criteria for definite antiphospholipid syndrome (APS). *J Thromb Haemost, 4*, 295–306.

Ruiz-Iraastorza, G., Cuadrado, M. J., Ruiz-Arruza, I., et al. (2011). Evidence-based recommendations for the prevention and long-term management of thrombosis in antiphospholipid antibody positive patients: Report of a task force at the 13th international congress on antiphospholipid syndrome. *Lupus*, *20*, 206–218.

Sciascia, S., Coloma-Bazan, E., Radin, M., et al. (2017). Can we withdraw anticoagulation in patients with antiphospholipid syndrome. *Autoimmun Rev*, *16*, 1109–1114.

CHAPTER 109

von Willebrand Disease

Morayma Reyes Gil, MD, PhD

von Willebrand disease (VWD) is the most common hereditary bleeding disorder affecting as much as 1% of the general population. The VWF (von Willebrand factor) protein has three essential hemostatic functions: (1) binding to FVIII thereby prolonging its half-life, (2) binding to collagen in the underlying subendothelial matrix, and (3) binding to platelets; thus, VWF recruits platelets to the injury site by functioning as a bridge between platelets and the subendothelial matrix. Therefore, deficiency in VWF highlights the indispensable role of VWF in primary and secondary hemostasis and explains the unique features of bleeding manifestations of VWD. Patients who are symptomatic for this disease have mucosal bleeding symptoms and bleeding immediately after invasive procedures or surgery. Severe VWD may also present as gastrointestinal bleed and bleeding in muscles and joints. The most common presentation in males is epistaxis, whereas in females is menorrhagia. The bleeding history is the most important criteria to seek and establish a diagnosis of VWD. PTT is often normal and thus specific testing for VWD antigen and function is required. VWD workups should only be performed for patients with bleeding history. A family history of bleeding is useful to establish the autosomal dominant inheritance. Minor mucosal bleeding does not require treatment, whereas minor invasive procedures (e.g., dental procedures) can be treated with antifibrinolytics or desmopressin (DDAVP) (if proven beneficial). For major surgeries or trauma, a VWF-containing product (Humate-P or Wilate) is recommended.

In 1926 Erik Adolf von Willebrand characterized a bleeding disorder in a family in the Aland archipelago off the coast of Finland. 23 of 66 members of the family, mainly females, had bleeding problems. He first called the disease "hereditary pseudohemophilia" but renamed it "constitutional thrombopathy" to emphasize involvement of platelet. VWD is caused by a qualitative or quantitative deficiency of the VWF protein and is the most common inherited bleeding disorder. Common bleeding manifestations include mucosal bleeding, menorrhagia, and bleeding immediately after invasive procedures or surgery.

The Protein: VWF is a huge glycoprotein composed of many individual monomers that are linked together to form multimers. Each monomer contains the sites of VWF hemostatic functions: binding to collagen, binding to platelets, and binding to coagulation FVIII (Fig. 109.1). However, the larger the multimers the more efficient they are at carrying their hemostatic functions.

VWF is synthesized as a large pro-VWF and undergoes several posttranslational modifications necessary for proper function and secretion. Monomers are dimerized in the endoplasmic reticulum and then glycosylated in the Golgi apparatus. Following glycosylation, the C terminal dimers are then N-terminal multimerized up to 20 million

Transfusion Medicine and Hemostasis. https://doi.org/10.1016/B978-0-12-813726-0.00109-4

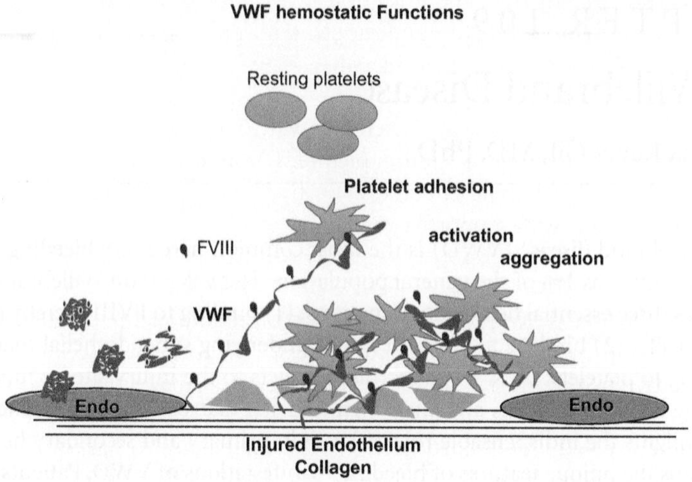

FIGURE 109.1 Depiction of von Willebrand factor (VWF) hemostatic functions: adhesion to collagen, recruitment of platelets, and coagulation FVIII carrier. VWF is secreted in a globular form. VWF uncoils under high shear stress after it binds collagen in the subendothelium, and the large uncoiled multimers bind to GP1b on platelets and recruit them to the wounded endothelium. VWF carries coagulation FVIII and delivers it to the site of activated platelets where coagulation factors are recruited and activated.

daltons in size. The VWF propeptide is cleaved and packed together with mature VWF in specialized compartments called Weibel–Palade Bodies (WPB) in endothelial cells. The VWF propeptide (VWFAgII) is important for proper VWF multimerization and storage. WPB store other proteins that modulate coagulation, angiogenesis, and inflammation, such as coagulation FVIII, tissue plasminogen activator, P-selectin, P-selectin cofactor (CD63), osteoprotegerin, angiopoetin-2, endothelin-1, and interleukin 8. Interestingly, the formation of WPBs is defective in VWD. On activation, endothelial cells secrete VWF along with the other content of WPB. Secreted VWF circulates in a globular form until it gets exposed to subendothelial collagen, where it binds and, under high shear stress, uncoils into a linear chain. The largest multimeric forms of the protein play an important role in recruiting platelets to the injury sites. The third hemostatic function of VWF is to carry coagulation FVIII. It protects FVIII from degradation and brings it in close proximity to phospholipids on the surface of activated platelets and injured endothelium, where coagulation is initiated.

Platelet-Derived von Willebrand Factor: VWF is also produced in megakaryocytes and stored as large VWF multimers in alpha granules within the platelets. Platelet-derived VWF is important for platelet adhesion, aggregation, and thrombus formation. Although, platelet-derived VWF is encoded by the same gene in chromosome 12, its glycosylation, which differs from endothelial-derived VWF, renders it resistance to ADAMTS13 cleavage. It is estimated that ~20% of circulating VWF is derived from platelets and they represent the largest multimers in the multimer assay. Recent studies using transgenic mouse models and transplantations demonstrate

	Normal	Type I	Type 3	Type 2A	Type 2B	Type 2N	Type 2M	PT-VWD
VWF:Ag	NL	↓	AB	↓	↓	NL or ↓	NL or ↓	↓
VWF:RCo	NL	↓	AB	↓↓↓	↓↓	NL or ↓	↓ or ↓↓	↓↓
VWF:CBA	NL	↓	AB	↓↓↓	↓↓	NL or ↓	↓↓ or ↓	↓
FVIII	NL	NL or ↓	<5%	NL or ↓	NL or ↓	↓↓	NL	NL or ↓
RIPA	NL	often NL	AB	↓	NL or ↑	NL	NL or ↓	often NL
LD-RIPA	AB	AB	AB	AB	↑↑↑	AB	AB	↑↑↑
PFA	NL	NL/Abn	Abn	Abn	NL/Abn	NL	NL/Abn	NL/Abn
Platelet count	NL	NL	NL	NL	↓ or -NL	NL	NL	↓ or -NL
Usual Tx		DDAVP	VWFc	VWFc	VWFc	VWFc	VWFc	platelets
Response to DDAVP		Good	Poor	Poor	↓ platelets	Poor	Poor	↓ platelets
Response to VWFc		Good	Good	Good	Good	Good	Good	↓ platelets
Frequency in population		1-2%	very rare	rare	rare	rare	rare	rare
VWF multimers			AB					

FIGURE 109.2 Summary of von Willebrand disease (VWD) types (types 1, 3, 2A, 2B, 2M, 2N) and absence of or defective platelet GPIb receptor for VWF, platelet-type VWD (PT-VWF). The lower portion illustrates the VWF multimers that are identified by SDS-agarose gel electrophoresis of VWF for each of these variants. *AB*, absent; *Abn*, abnormal; *CBA*, collagen binding assay; *LD-RIPA*, ristocetin-induced platelet aggregation to low dose ristocetin; *NL*, normal; *PFA*, platelet function analyzer; *PT-VWD*, platelet-type VWD; *Tx*, treatment; *VWF:Ag*, von Willebrand factor antigen; *VWF:RCo*, VWF activity by ristocetin cofactor assay; *VWFc*, VWF concentrate. Modified from RR Montgomery.

that endothelial-derived VWF is necessary for hemostasis (prevention of blood loss), whereas platelet-derived VWF is more important for clot formation and stability. Thus, platelet-derived VWF may represent a new target for stroke and arterial thrombosis.

Pathophysiology

Overview of Molecular Variants and Prevalence: See Fig. 134.1 for protein domains and encoding exons; see Fig. 109.2 for VWD subtype characteristics.

Type 1 von Willebrand Disease: Type 1 VWD is the most common type of VWD (75%–80% of all VWD), with a prevalence of 1% of the population, but symptomatic disease (bleeding) affects only a fraction of this population. Type 1 VWD is a quantitative defect, causes mainly by missense mutations that lead to decreased amount of VWF protein and thus decreased function. The ratio of VWF antigen to activity approximates 1. Western blot analysis of VWF in plasma of patients with type I VWD shows normal pattern of all multimers but in decreased amount. Only in severe cases of type 1 VWD, factor VIII levels are proportionally low, as each VWF monomer can bind a molecule of FVIII, and it is estimated that only ~20% of VWF binding sites are saturated. Thus, the ratio of FVIII/VWF antigen tends to be greater than 1.5 in most cases of type I VWD. A defect in the *VWF* gene can be found in about 63%–70% of patients, particularly if the VWF level is <35%. Thus, mutational analysis is not recommended for diagnosis of type I VWD as it will miss nearly 40% of the cases. The majority of type

1 VWD mutations result in decreased synthesis and/or secretion, with the exception of mutations that result in increased clearance, classified as type 1C. VWF Vicenza is an example of type IC VWD. Type 1C can be differentiated from other type I VWDs by increased levels of VWF propeptide in comparison to VWF antigen. In addition, DDAVP challenge may be useful for diagnosis as it shows increased peak level of VWF (5–10 × baseline levels) 15–30 minutes after DDAVP challenge that rapidly decrease (>50% decrease after 4 hours).

Type 2 von Willebrand Disease: Type 2 VWD (15%–20% of VWD) by definition is a qualitative defect in which one of its functions is affected, whereas the antigen is borderline to normal. There are four subclassifications of type 2 VWD based on the defect and the multimer pattern.

Type 2A von Willebrand Disease: This is the most common subtype among the type 2 VWD (10%–12% of VWD). It can be further subclassified into two types based on the type of mutation. Type 2A VWD type I results from mutations in the terminal portions (domain D2, D3, and C terminal) thus impairing multimerization and resulting in absence of high- and intermediate-molecular-weight multimers and sometimes decreased low-molecular-weight multimers. Type 2A VWD type II results from mutations in the A2 domain that increase susceptibility to proteolysis by ADAMTS13 metalloprotease. Although large multimers are released from the endothelium, the rapid proteolysis of ADAMTS13 results in a multimer pattern of absent high- and intermediate-molecular weight multimers and increased low-molecular-weight multimers sometimes with abnormal fragments. Absence of high- and intermediate-molecular-weight multimers results in decreased VWF activity as measured by VWF ristocetin cofactor or VWF collagen binding assay compared with VWF antigen.

Type 2B von Willebrand Disease: This is the rarest type among type 2 VWDs (3%–5%) that results from a gain of function increasing the binding affinity to GP1b on platelets. About 30% of patients present with mild thrombocytopenia, and a lack of the large-molecular-weight multimers as the largest multimers spontaneously bind platelets inducing their activation and clearance. Because of the lack of large-molecular-weight multimers, type 2B presents as decreased VWF activity as measured by VWF:RCo, which seems paradoxical. A more physiologically relevant laboratory assay to diagnose type 2B is ristocetin-induced platelet agglutination. Ristocetin is an antibiotic that induces platelet agglutination at high dose, but patients with type 2B exhibit platelet agglutination at low concentration of ristocetin. DNA sequencing of exon 28, which encodes for the A1 domain where type 2B mutations are found, can be used for confirmation. Further details regarding testing will be discussed in the VWD laboratory chapter (see Chapter 134).

Type 2M von Willebrand Disease: Type 2M VWD is characterized by either decreased binding to GP1b on platelets (mutations in the A1 domain), the molecular opposite of type 2B VWD, or decreased binding to collagen (mutations in the A3 domain). Thus, the two activity assays are necessary and indispensable. Mutations that result in decreased binding to GP1b exhibit decreased VWF ristocetin cofactor activity

(VWF:RCo), whereas mutations that lead to decreased binding to collagen exhibit relatively normal VWF:RCo but decreased collagen binding activity. Functional activity of the VWF molecule is approximately one half of the antigenic activity. Type 2M VWD affects 1%–2% of the VWD population. Multimer patterns are typically normal in type 2M VWD, although cases of either decreased or increased high-molecular-weight multimers and/or decreased resolution of the triplets have been described.

Type 2N von Willebrand Disease: This rare type of VWD accounts for 1%–2% of the VWD population and is also known as the Normandy variant, where it was first described. Despite normal VWF:RCo and collagen binding activity, this type is also classified as qualitative (type 2) because the binding of VWF to FVIII is severely decreased, leading to rapid clearance of unbound FVIII. Mutations in the FVIII binding domain of VWF in the D′ or D3 domain prevent the normal stabilization of FVIII by VWF. Type 2N may resemble mild hemophilia A as FVIII levels are mildly low (>5%), but a differentiation can be made based on a more predominant mucosal bleeding presentation and a consistent ratio of FVIII/VWF antigen (~0.5%) even in acute phase reaction, whereas failure of FVIII increase along VWF during acute phase reaction is more characteristic of hemophilia A. Type 2N VWD may be inherited in a recessive or compound heterozygous fashion (i.e., one allele is Type 2N and other allele is another type of VWD). Confirmation of Type 2N VWD diagnosis is made by a VWF-FVIII binding assay.

Type 3 von Willebrand Disease: This is rarest (~1 in 1×10^6 individuals) and most severe type of VWD with complete absence of VWF protein, undetectable activity, and antigen levels. It is the autosomal recessive form of VWD, most commonly inherited as compound heterozygosity (two parents who are heterozygous for type 1 or 2 VWD). Clinical bleeding manifestations in type 3 VWD resemble more a severe/moderate hemophilia presentation as FVIII levels tend to be in the moderate hemophilia range (3%–5% factor VIII activity), but the absence of VWF accentuates bleeding.

Platelet-Type or Pseudo-von Willebrand Disease: The name pseudo-VWD illustrates the fact that the gain of function mutations causing this disease is found on the platelet GPIb receptor and not in the VWF protein. From a laboratory perspective, pseudo-VWD resembles type 2B VWD, featuring loss of high-molecular-weight multimers, mild thrombocytopenia, and increased aggregation to low-dose ristocetin. Pseudo-VWD may be distinguished from type 2B VWD by ristocetin aggregation with washed platelets from affected patients, flow cytometry, or GPIb sequencing (see more in Chapter 134).

Diagnosis: The clinical history is the single most important criteria in establishing a diagnosis of VWD. VWD workups should not be pursued in patients without a history of bleeding or history of family members diagnosed with VWD. Details of the bleeding history are found in Chapter 92. Bleeding score may be helpful, although somewhat unreliable in children, as children may be difficult historians or may not have been hemostatically challenged. Mucocutaneous bleeding (dental and nosebleeds) and purpura are the most prevalent bleeding symptoms among children, and later in life heavy menses is a common presentation in adolescent women. Many undiagnosed mild VWD patients present as adults with unexplained or exaggerated bleeding after a surgery or trauma.

Routine coagulation testing (prothrombin time [PT], activated partial thromboplastin time [PTT], thrombin time, fibrinogen, complete blood count, and platelet count) is often normal in a patient with VWD, therefore specific testing for VWD is required. PTT is only prolonged if the FVIII activity is less than 40%. The PFA-100 became popular as a replacement for the bleeding time, but it is only sensitive to severe VWD type 1 (VWF levels are less than 30%) and some severe VWD type 2. The specific laboratory tests utilized in making a diagnosis and in differentiating the types of VWD are summarized in Fig. 109.2 (see more details in Chapter 134).

von Willebrand Disease Screening Tests: A panel, including VWF antigen, VWF:RCo, VWF collagen binding activity, and VWF multimers, is recommended to diagnose the specific VWD type. Most hematologists will usually perform the VWF antigen, VWF:RCo, and the FVIII activity on initial screening and will perform the multimer testing, collagen binding activity, and blood type for further evaluation and VWD classification. Diagnostic levels for VWF antigen and/or VWF activity should be less than 35% (0.35 IU/dL). Importantly, VWF:RCo is often lower than VWF antigen in African Americans due to a common polymorphism (~40%) that affects VWF binding to ristocetin but does not affect VWF hemostatic functions (*VWF* Exon 28 1472). The VWF propeptide and rapid clearance following DDVAP challenge may help in the diagnosis of VWD type 1C. The VWF collagen binding assay can be considered when the other VWF testings (VWF:RCo, VWF antigen, and multimers) are relatively normal, and bleeding symptoms are present.

There are important physiological factors that regulate VWF levels and need to be considered during VWD workups. It is well known that VWF clearance is in part modulated by blood type, and blood type O is associated with lower VWF levels because of accelerated clearance. As blood type O is the most common blood type, it is extremely important to only test individuals with high clinical suspicion of a hereditary bleeding disorder to establish true disease. Other physiologic factors that transiently increase VWF level are pregnancy, the menstrual cycle, and exercise. Because VWF levels in some women can fluctuate with their menstrual cycle, it is recommended to test in the last day of their period, when VWF levels are the lowest. VWF levels also increase with age. Most importantly, VWF is an acute phase reactant that substantially increases in response to physiological stress. Thus, it is often recommended to retest in patients with high clinical suspicion and borderline normal results. VWF levels less than 35% in patients with strong clinical history are diagnostic of VWD. Patients with borderline VWF levels (35%–50%) require careful consideration of bleeding history and blood type. Patients with definite bleeding symptoms but VWF levels above 50% may need to be retested if any of the physiological factors described above applies. Alternatively, other causes of bleeding should be considered in patients with high clinical suspicion and consistently normal VWD testing (repeated over a period of several months). Other bleeding disorders that can mimic VWD include platelet disorders, plasminogen activator deficiency 1, and vascular disorder such as Ehlers–Danlos syndrome.

Acquired von Willebrand Syndrome: Acquired von Willebrand syndrome (AvWS) manifests as new onset of bleeding in patients with an underlying disorder that affects VWF levels or its functions such as autoimmune disorders, monoclonal gammopathies/paraproteinemias, certain malignancies (lymphomas/lymphoproliferative diseases, Wilms tumor, hemangiomas), and cardiovascular conditions with aberrant turbulence (e.g., ventricular septal defect, aortic stenosis, pulmonary stenosis, mitral valve prolapse, hypothyroidism). In addition, prosthetic heart valves and left ventricular assist devices can result in AvWS. The mechanisms involved in AvWS include increased clearance, absorption by tumor cells or abnormal tissue (e.g., amyloidosis), antibodies that decrease or inhibit VWF activity (paraproteins), decreased production (hypothyroidism), or destruction (cardiovascular conditions). Most AvWS manifest a lower VWF activity compared with the VWF antigen (see Chapter 135). The most common multimer pattern in acquired VWD is selective loss of high-molecular-weight multimers, but other patterns can be seen. AvWS can be cured by eliminating the underlying disorder. For antibody-mediated AvWS, intravenous immunoglobulin plasma exchange or immunosuppressive agents such as corticosteroids or cyclophosphamide are utilized.

Management: Most minor bleeding in VWD does not require treatment. However, persistent nosebleeds or mucosal bleeds can be treated by antifibrinolytic therapy and/or DDAVP in patients with good response to DDVAP (see below). Menorrhagia is commonly treated with hormonal therapy. Invasive surgeries may require VWF concentrates or platelets (for PT-VWD). AvWS tends to respond better to VWF concentrates.

Antifibrinolytics: Aminocaproic acid (50–100 mg/kg per dose, max 6 g per dose) can be given every 6–12 hours for 3–7 days. Tranexamic acid (25 mg/kg per dose, max of 1.5 g per dose) can be given every 8 hours for 3–7 days. Frequently a single dose of an antifibrinolytic is sufficient to stop epistaxis. The topical nasal saline gel can also be helpful in the prevention of epistaxis. Other topical agents such as gel foam, collagen, and topical thrombin and fibrin glue can be used to stop nose and dental bleeds. Severe nosebleeds may require topical vasoconstricting agents and cauterization.

Desmopressin: Intravenous and/or nasal DDAVP is an inexpensive and common treatment for most VWD types 1 and 2. DDAVP induces massive release of large multimeric forms of VWF from endothelial cells but is not effective in cases of abnormal production of VWD (severe type 1, type 2A VWD type I, and type 3) and is contraindicated for type 2B, as it may exacerbate thrombocytopenia and potentially promote thrombosis. A therapeutic trial with DDAVP should be carried out, and VWF antigen and activity should be monitored after 1 and 4 hours to establish DDAVP responsiveness and VWF clearance. Intravenous DDAVP is given as 0.3 mcg/kg in 30–50cc of saline, whereas intranasal DDAVP is administered as one spray (150 mcg for patients less than 50 kg) or two sprays (300 mcg for patients over 50 kg). Both the intranasal and the intravenous form of DDAVP may be given once or twice a day for up to three doses with careful monitoring of fluid intake. DDAVP should not be given more than three doses within a few days to avoid tachyphylaxis.

For invasive and long surgeries, a VWF-containing product is recommended (see Chapter 41). VWF levels should be monitored during surgery and 1–2 days postoperatively.

Further Reading

Favaloro, E. J. (2017). Diagnosis or exclusion of von Willebrand disease using laboratory testing. *Methods Mol Biol, 1646*, 391–402.

Federici, A. B., & Mannucci, P. M. (2007). Management of inherited von Willebrand disease in 2007. *Ann of Med, 39*, 346–358.

Federici, A. B., Mannucci, P. M., Castaman, G., et al. (2009). Clinical and molecular predictors of thrombocytopenia and risk of bleeding in patients with von Willebrand disease type 2B: A cohort study of 67 patients. *Blood, 113*, 526–534.

Flood, V. H., Gill, J. C., Morateck, P. A., et al. (2010). Common VWF exon 28 polymorphisms in African Americans affecting the VWF activity assay by ristocetin cofactor. *Blood, 116*, 280–286.

Gill, J. C., Endres-Brooks, J., Bauer, P. J., Marks, W. J., Jr., & Montgomery, R. R. (1987). The effect of ABO blood group on the diagnosis of von Willebrand disease. *Blood, 696*, 1691–1695.

Haberichter, S. L., Balistreri, M., Christopherson, P., et al. (2006). Assay of the von Willebrand factor (VWF) propeptide to identify patients with type 1 von Willebrand disease with decreased VWF survival. *Blood, 108*, 3344–3351.

James, P. D., Notley, C., Hegadorn, C., et al. (2007). The mutational spectrum of type 1 von Willebrand disease: Results from a Canadian cohort study. *Blood, 109*, 145–154.

Mannucci, P. M., & Ghirardini, A. (1997). Desmopressin: Twenty years after. *Thromb Haemost, 78*, 958.

Peake, I., & Goodeve, A. (2007). Type 1 von Willebrand disease. *J Thromb Haemost, 5*, 7–11.

Rodeghiero, F., Tosetto, A., Abshire, T., et al. (2010). ISTH/SSC bleeding assessment tool: A standardized questionnaire and a proposal for a new bleeding score for inherited bleeding disorders. *J Thromb Haemost, 8*, 2063–2065.

Sadler, J. E., Budde, U., Eikenboom, J. C., et al. (2006). Update on the pathophysiology and classification of von Willebrand disease: A report of the subcommittee on von Willebrand factor. *J Thromb Haemost, 4*, 2103–2114.

Tosetto, A., Rodeghiero, F., Castaman, G., et al. (2006). A quantitative analysis of bleeding symptoms in type 1 von Willebrand disease: Results from a multicenter European study (MCMDM-1 VWD). *J Thromb Haemost, 4*, 766–773.

CHAPTER 110

Hemophilia A

Surbhi Saini, MBBS and Amy L. Dunn, MD

Hemophilia A (also known as classical hemophilia) results from congenital deficiency of factor VIII (FVIII). It is an X-linked recessive disorder that results in decreased or absent circulating FVIII activity, leading to lifelong bleeding tendency. Hemophilia A has an incidence of approximately 1:5000 male births and accounts for approximately 85% of cases of hemophilia. It affects all racial and ethnic groups equally.

Pathophysiology: FVIII is a plasma glycoprotein consisting of six domains, A1–A2–B–A3–C1–C2 (Fig. 110.1). The encoding gene is found on the long arm of the X chromosome (Xq.28). The mature protein is a heterodimer with a light chain consisting of domains A3–C1–C2 and a heavy chain with the domains A1–A2–B. The majority of FVIII is thought to be synthesized in hepatic endothelial cells, but it may also be produced in endothelial cells in general (e.g., elevated FVIII levels during liver failure). On release into the circulation, it is noncovalently linked to von Willebrand factor (VWF), which prevents enzymatic degradation of nascent FVIII. During coagulation, the tissue factor (TF)–FVIIa complex activates FX and FIX, leading to conversion of prothrombin to thrombin. The initial thrombin cleavage of the FVIII light chain causes FVIII to be released into the circulation and which is then activated to FVIIIa by further thrombin-mediated proteolysis. FVIIIa, along with FIXa, in the presence of calcium, then act as cofactors on a phospholipid surface during activation of factors X, V and, ultimately, thrombin. This is also known as the tenase complex. Patients with hemophilia A are unable to generate adequate thrombin due to lack of FVIII and become dependent on the TF pathway. Circulating tissue factor pathway inhibitor (TFPI) efficiently downregulates the TF–FVIIa pathway as well as FXa, leading to decreased thrombin and bleeding. Thrombin-activated fibrinolysis inhibitor (TAFI) production is also decreased in hemophilia, leading to more rapid dissolution of the fibrin clot. A large number of molecular defects have been described in hemophilia A, including large gene deletions, inversions, single gene rearrangements, deletions, and insertions. A list of mutations leading to hemophilia A can be found at http://hadb.org.uk/ as well as https://www.cdc.gov/ncbddd/hemophilia/champs.html.

Clinical Manifestations: The hallmark of hemophilia-related bleeding is delayed bleeding along with joint and muscle bleeding. As hemophilia A is X-linked, the vast majority of affected patients are male. Females, however, can be affected by extreme X-chromosome lyonization, or with gene abnormalities such as Turner syndrome.

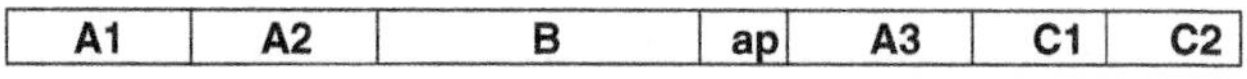

FIGURE 110.1 Domain structure of factor VIII.

Transfusion Medicine and Hemostasis. https://doi.org/10.1016/B978-0-12-813726-0.00110-0

Heterozygous female carriers of hemophilia A may exhibit a bleeding tendency, most commonly heavy menstrual bleeding and postsurgical bleeding. There is a high rate of spontaneous mutation within the *F8* gene, and approximately 30% of newly diagnosed patients will have no family history of hemophilia.

In general, the severity of bleeding depends on the percentage of residual clotting factor activity. Patients with levels of >5%–40% are classified as having mild hemophilia, patients with levels of 1%–≤5% as moderate and those with less than 1% activity as having severe disease. Approximately 60% of persons with hemophilia have severe disease. They suffer from spontaneous bleeding while those with mild to moderate disease typically bleed only when challenged with trauma or surgery. In the newborn period, the most common findings are bleeding and bruising after venipuncture, heel sticks, immunizations, and circumcision. Intracranial hemorrhage remains the most dreaded complication of hemophilia in the first 2 years of life and occurs in about 7%–10% of infants. Infants born of known carrier mothers should not undergo instrumented birth and should not be circumcised until testing for FVIII rules out hemophilia. Older children and adults may experience excessive bruising, epistaxis, soft tissue hematomas, intracranial bleeding, and hemarthrosis. The single largest preventable cause of morbidity is degenerative joint disease due to recurrent hemarthrosis. Females who carry hemophilia may also have bleeding symptoms such as menorrhagia, oral bleeding, bleeding with childbirth, surgical, and trauma-related bleeding.

Diagnosis: An X-linked inheritance pattern, elevated partial thromboplastin time (PTT), and decreased plasma FVIII levels confirm the diagnosis. The most commonly used PTT reagents may not detect mild deficiency of FVIII. Prenatal diagnosis can be performed in the case of a known family history. Cord blood testing of FVIII levels can also be performed at the time of delivery.

Differential Diagnosis: Hemophilia A and B are clinically indistinguishable, and individual factor levels must be used to clarify the diagnosis. Patients with mildly low FVIII levels and an autosomal inheritance pattern may have type 3 von Willebrand disease (vWD). Patients with type 3 vWD will have moderately low FVIIII and absent VWF multimers, along with essentially absent VWF antigen and ristocetin cofactor activity. In this case, the low FVIII levels are a result of increased proteolysis, not decreased production. vWD type 2N should also be considered in the setting of mildly low FVIII levels, autosomal inheritance, and poor response to recombinant FVIII therapy. In vWD type 2N, the pathophysiology involves decreased FVIII binding to VWF, leading to rapid proteolysis of FVIII. This type of vWD can be evaluated via a VWF to FVIII binding assay. Acquired low FVIII levels can also result from autoantibody formation.

Management

Comprehensive Care: A series of federally funded comprehensive hemophilia treatment centers (HTCs) exist to care for persons with hemophilia. They are typically staffed with hematologists, orthopedists, physical therapists, nurses, genetic counselors, psychologists, and social workers who specialize in the care of patients with bleeding disorders. It has been shown that patients who receive their care in an HTC setting have a longer life expectancy.

Factor Concentrates: The mainstay of hemophilia care is FVIII replacement with intravenously delivered FVIII concentrates. Concentrates are either plasma derived, containing varying amounts of VWF, or recombinant products, and both undergo multiple viral and pathogen attenuation steps. No infectious complications have been reported because these steps were incorporated into the manufacturing process; however, the possibility of contamination with new infectious agents such as prions cannot be excluded with plasma-derived products.

Infusions can be delivered in response to bleeding episodes ("demand therapy") or to prevent bleeding ("prophylaxis"). To prevent or minimize long-term sequelae, demand therapy should be given as soon as possible after a bleeding episode is recognized. Because the need for urgent treatment is so important, many patients affected by hemophilia are proficient in self-infusion techniques. During a bleeding episode, factor replacement therapy should never be delayed to perform imaging or laboratory studies.

The dose and frequency of factor delivery is calculated based on the half-life of the product, the intravascular volume of distribution (1 unit of FVIII per kilogram raises the plasma concentration by about 2%) and the desired clotting factor activity. For example, to raise the factor level of a 20 kg child with severe hemophilia to 100%, the dose should be 20 kg × 50 IU = 1000 IU. Correction to FVIII levels of 40% activity is considered hemostatic in most cases; however, in the setting of surgery or life-/limb-threatening hemorrhage, higher FVIII levels (80%–100%) are recommended. Postsurgical hemostasis should maintain FVIII levels above 50%–70% for the first week and above 30% for the second week. Ancillary measures such as compressive dressings, cauterization, packing, and splinting should also be implemented when appropriate. Additionally, antiplatelet agents should be avoided. Table 110.1 illustrates a suggested approach to factor replacement therapy for commonly encountered bleeding events. Standard plasma and recombinant products have a half-life of 8–12 h. Recently, novel techniques such as conjugation of the recombinant FVIII molecule to neonatal Fc receptors (rFVIIIFc) or polyethylene glycol (PEGylated FVIII), as well as development of single-chain forms with increased VWF binding, has made it possible to extend the product half-life to 18–24 h, primarily by inhibition of degradation of the infused product in the plasma. In some situations, this allows for decreased frequency of infusions for patients on prophylaxis, and/or achievement of higher trough levels improving their overall quality of life.

Prophylaxis: In developed countries, prophylactic therapy delivered one to four times per week is considered the standard of care and is the only therapy proven to prevent the long-term complication of degenerative joint disease. It is common practice to begin prophylaxis before the onset of recurrent joint bleeding, typically before the age of 3 years. Dosage can be individualized based on IV access, bleeding phenotype, and pharmacokinetics.

Desmopressin: Desmopressin, or DDAVP (1-deamino-8-D-arginine vasopressin), is a synthetic form of the hormone vasopressin. The product may be given intravenously, subcutaneously, or via nasal delivery. DDAVP causes release of FVIII and VWF from their endothelial storage sites. Patients with mild hemophilia A and some with

TABLE 110.1 Suggested Approaches to Treatment of Bleeding Episodes

Bleed Site	Desired Activity	Length of Therapy	Ancillary Measures
Central nervous system	100%	7–14 days, then strongly consider prophylactic therapy for a minimum of 6 months	Continuous infusion FVIII; antiepileptic prophylaxis; surgical intervention
Oral cavity or mucosal bleeding	30%–60%	3–7 days	Antifibrinolytic therapy; custom mouthpiece; topical thrombin powder
Retropharynx	80%–100%	7–14 days	Continuous infusion FVIII; antifibrinolytic therapy
Nose	30%–60%	1–3 days	Packing, cautery; saline nose spray/gel; nasal vasoconstrictor spray; antifibrinolytic therapy
Gastrointestinal tract	40%–80%	3–7 days	Antifibrinolytic therapy; endoscopy with cautery
Genitourinary tract	40%–60%	1–3 days	Vigorous hydration; evaluation for stones/urinary tract infection; avoid antifibrinolytic therapy; glucocorticoids
Muscle	40%–80%	Every other day until pain-free movement	Rest, ice, compression, elevation; physical therapy
Iliopsoas muscle	80%–100%	Until radiographic evidence of resolution	Continuous infusion FVIII; bedrest; physical therapy
Joint	40%–80%	1–2 days	Rest, ice, compression, elevation; physical therapy
Target joint	80% day 1, 40% days 2 and 4	3–4 days	Rest, ice, compression, elevation; physical therapy

moderate disease can be tested with DDAVP, and if they show a response by manifesting hemostatic levels or at least a threefold increase in FVIII, then DDAVP is often sufficient to treat mild bleeding symptoms such as nose, mouth, and soft tissue bleeding. FVIII storage pools become depleted after multiple doses, so this treatment is not adequate for lengthy therapy, and fluid intake must be monitored closely as hyponatremia may result, particularly in children less than 2 years of age and in the elderly. In most cases, life- or limb-threatening bleeding episodes require FVIII replacement.

Antifibrinolytic Therapies: Antifibrinolytic medications such as aminocaproic acid or tranexamic acid are used to prevent excessive fibrinolysis and are particularly useful in diminishing bleeding symptoms in locations with prominent fibrinolytic activity, such as the mouth, gastrointestinal tract, and uterus.

Liver Transplant: Liver transplantation has been performed in several patients with hemophilia as a result of severe liver disease. The transplant effectively cures the hemophilia.

Complications

Infectious Complications: In the late 1970s and early 1980s, before incorporation of viral attenuation steps (mid-1980s for HIV and late 1980s for HBV and HCV) in factor manufacturing, many patients became infected with HCV and/or HIV from contaminated concentrates. Although many patients succumbed to these infections, there is a large cohort of long-term survivors as a result of highly active antiretroviral therapies for HIV and combination therapy for HCV.

Inhibitors: A serious complication of congenital hemophilia A is the development of inhibitory alloantibodies to FVIII. These inhibitory antibodies occur in approximately 20%–30% of patients with severe hemophilia, and to a lesser extent in those with mild and moderate disease (2%–3%). They are more likely to occur in the setting of a family history of inhibitors, in patients having large gene disruptions and in non-white patients. Nongenetic risk factors such as factor replacement during inflammatory states (known as the "danger theory") and delivery via continuous infusion may play a role. To date, it is still debated whether antibody formation is influenced by the type of product used; however, some data suggest that plasma-derived products that contain VWF may be less immunogenic.

Alloantibodies to FVIII in congenital hemophilia are most commonly directed against the A2 and C2 domains and often develop within the first 10–20 exposures to exogenous FVIII but can develop at any age. They typically neutralize both endogenous and exogenous FVIII, an important aspect in patients with mild–moderate disease who develop inhibitors because it changes their phenotype to severe disease. The antibody titer is measured using a Bethesda assay and is expressed in units (BU). One BU is the amount of antibody, which lowers the plasma factor level by 50%. Low-titer inhibitors (<5 BU) are often transient but may be persistent and carry clinical significance, while high-titer inhibitors (>5 BU) significantly impact patient care and quality of life. Autoantibodies can develop in patients without hemophilia, leading to a condition known as acquired hemophilia. These autoantibodies occur most commonly in the setting of pregnancy, malignancy, and autoimmune conditions; however, 50% of cases are idiopathic.

Immune tolerance therapy (ITT) with repeated exposure to FVIII concentrate over a period of months to years may eradicate the antibody. ITT has been accomplished using both high-dose (100–200 IU/kg per day) and low-dose standard half-life concentrates (50 IU/kg thrice weekly) of FVIII. More recently, there is an increasing interest in the concept of decreased immunogenicity with the use of extended half-life products, particularly rFVIIIFc. While randomized controlled trials in previously untreated patients are awaited, these products could be considered in patients with refractory inhibitors. Patients treated with more frequent dosing have not only demonstrated less frequent breakthrough bleeding during ITT but also more frequently require central venous access device placement to ensure IV access. Both recombinant and vWF-containing plasma-derived products have been used successfully. Additionally, immunosuppressive agents such as cyclosporine and rituximab have been used, with some success. Bleeding in the setting of a high-titer inhibitor often requires bypassing therapy with either high dosage of recombinant FVIIa or an activated prothrombin complex concentrate (see Chapter 41).

Joint Disease: Degenerative joint disease due to recurrent hemarthrosis is the single largest preventable cause of morbidity for patients with hemophilia A. The pathogenesis of this arthropathy is multifactorial, with iron and free radical formation being the most likely culprit triggering the degenerative changes. Recurrent hemarthrosis causes an inflammatory reaction resulting in a thickened, hyperemic synovium. This inflamed synovium participates in a vicious cycle of further bleeding, and ultimately the destruction of bone and articular cartilage. Unfortunately, the physical findings are often subtle in the early stages of joint disease, and investigation with magnetic resonance imaging may be required to demonstrate hemosiderin deposition and or hypertrophic synovium. More recently, point-of-care ultrasound protocols have proved time-efficient and user-friendly in detecting early joint bleeds and/or abnormalities and guiding treatment. Synovectomy, arthroscopic, or radionuclide has been used to address recurrent bleeding in these target joints, but degenerative it remains to be seen whether it can halt articular changes are common even in the setting of bleed reduction.

Recent Advances in Hemophilia Care

Gene Therapy: Gene therapy has the potential to change the landscape in hemophilia care. For hemophilia A, progress in gene therapy has been slow, largely due to the large size of the *F8* genome that requires transfer to a viral cassette. Currently, a codon-optimized *F8* gene transfer strategy using an adeno-associated virus has shown promise in early clinical trials. Preclinical animal models suggest that liver-directed *F8* gene transfer may prevent, and even eradicate preexisting inhibitors in patients with hemophilia.

Nonfactor Replacement Products: Development of a recombinant, monoclonal bispecific antibody that mimics the actions of the intrinsic tenase complex (Hemlibra ®, Emicizumab-kxwh), and hence obviates the need for FVIII replacement, has been a significant milestone in the treatment of patients with hemophilia A and inhibitors. A recently concluded phase III trial of this molecule has demonstrated safety and efficacy with once-weekly subcutaneous dosing.

The concept of "rebalancing hemostasis" and thus increasing thrombin generation has been applied to the development of another category of products to treat hemophilia. Fitusiran (ALN-AT3SC), a small interfering RNA that decreases natural antithrombin levels, is currently undergoing an open label extension trial. Concizumab (monoclonal anti-TFPI antibody), inhibits TFPI leading to increase in thrombin generation via the TF-FVIIa pathway. Preliminary clinical trials are being conducted.

Further Reading

Abshire, T., & Kenet, G. (2004). Recombinant factor VIIa: Review of efficacy, dosing regimens and safety in patients with congenital and acquired factor VIII or IX inhibitors. *J Thromb Haemost*, *2*(6), 899–909.

Dunn, A. L., & Abshire, T. C. (2006). Current issues in prophylactic therapy for persons with hemophilia. *Acta Haematol*, *115*, 162–171.

Franchini, M. (2007). The use of desmopressin as a hemostatic agent: A concise review. *Am J Hematol*, *82*, 731–735.

Gouw, S. C., van der Bom, J. G., & van den Berg, M. (2007). Treatment-related risk factors of inhibitor development in previously untreated patients with hemophilia A: The CANAL cohort study. *Blood*, *109*, 4648–4654.

Hay, C. R. M., & DiMichele, D. M. (2012). The principal results of the International Immune Tolerance Study: A randomized dose comparison. *Blood*, *119*, 1335–1344.

Hooiveld, M., Roosendaal, G., Vianen, M., van den Berg, M., Bijlsma, J., & Lafeber, F. (2003). Blood-induced joint damage: Longterm effects *in vitro* and *in vivo*. *J Rheumatol*, *30*, 339–344.

Manco-Johnson, M. J., Abshire, T. C., Shapiro, A. D., Riske, B., Hacker, M. R., Kilcoyne, R., et al. (2007). Prophylaxis versus episodic treatment to prevent joint disease in boys with severe hemophilia. *N Engl J Med*, *357*, 535–544.

Soucie, J. M., Nuss, R., Evatt, B., Abdelhak, A., Cowan, L., Hill, H., et al. (2000). Mortality among males with hemophilia: Relations with source of medical care. The Hemophilia Surveillance System Project Investigators. *Blood*, *96*, 437–442.

Peyvandi, F., Mannucci, P. M., Garagiola, I., El-Beshlawy, A., Elalfy, M., Ramanan, V., et al. (2016). A randomized trial of factor VIII and neutralizing antibodies in hemophilia A. *N Engl J Med*, *374*, 2054–2064.

Pipe, S. W. (2016). New therapies for hemophilia. *Hematol Am Soc Hematol Educ Program*, *2016*, 650–656.

CHAPTER 111

Hemophilia B

Surbhi Saini, MBBS and Amy L. Dunn, MD

Hemophilia B, also known as Christmas disease, results from a congenital deficiency or absence of coagulation factor IX (FIX). It is an X-linked recessive disorder with an incidence of approximately 1:25,000 live male births and accounts for 15%–20% of hemophilia cases. It is thought to affect around 3300 patients in the United States. The term "Christmas disease" comes from Stephen Christmas, who was the first patient in whom the distinction between FVIII and FIX deficiency was made. While less common than hemophilia A, hemophilia B has been made famous as the royal disease. Through Queen Victoria and her progeny, hemophilia B spread to the Spanish, Russian, and German royal families of the 19th and 20th century.

Pathophysiology: FIX is a single-chain vitamin K–dependent glycoprotein and has a molecular weight of 57,000 Da. Its gene is 33 kb long and is positioned on the long arm (q) of the X chromosome (Xq27.1). FIX is a serine protease believed to be synthesized in the liver and released into the circulation in its inactive form. During coagulation, FIX is activated by both the intrinsic and extrinsic pathways. In the extrinsic pathway, tissue factor (TF), along with FVIIa, activates FIX. In the first step of the intrinsic or contact pathway, FIX is directly activated by FXIa. This step is nonphospholipid dependent. FIXa in turn activates FVIII. FVIIIa along with FIXa and calcium then act as cofactors on a phospholipid surface (typically a platelet) during activation of factors X and V and, ultimately, thrombin. Patients with hemophilia B are unable to generate adequate thrombin and become dependent on the less robust TF pathway. Tissue factor pathway inhibitor (TFPI) quickly downregulates the TF pathway and, as a result, a bleeding diathesis ensues.

A number of mutations have been described in the *F9* gene, including small and large deletions and additions, rearrangements, and missense mutations. A list of mutations can be found at http://www.factorix.org and https://www.cdc.gov/ncbddd/hemophilia/champs.html. A unique variant, hemophilia B Leyden, occurs when point mutations in the promoter region disrupt transcription factor binding sites. During puberty, androgen effects on this promoter region lead to rising FIX levels.

Clinical Manifestations: The hallmark of hemophilia-related bleeding is delayed bleeding along with joint and muscle bleeding. In comparison, patients with von Willebrand disease more commonly manifested immediate and mucocutaneous bleeding. As hemophilia B is X-linked, the vast majority of affected patients are male. Females can be affected in cases of extreme X-chromosome lyonization, or with gene abnormalities such as Turner syndrome.

In general, the severity of bleeding depends on the residual circulating FIX activity. Patients with levels of ≥5%–40% are historically classified as having mild hemophilia, patients with levels of 1%–<5% as moderate and those with less than 1% activity as

Transfusion Medicine and Hemostasis. https://doi.org/10.1016/B978-0-12-813726-0.00111-2

having severe disease. Approximately 60% of patients have severe disease. Commonly, patients with severe disease will suffer from spontaneous bleeding while those with mild to moderate disease typically bleed with trauma or surgery. In the newborn period, the most common findings are bleeding from blood draws, heel sticks, immunizations, and circumcision. Neonates may also have birth-related intracranial hemorrhage. Infants born of known carrier mothers should not undergo instrumented birth and should not be circumcised until testing for FIX rules out hemophilia. Older children and adults may experience excessive bruising, hematomas, intracranial bleeding, joint bleeding, muscle bleeding, and mouth bleeding. The single largest preventable cause of morbidity is degenerative joint disease due to recurrent hemarthrosis. Carrier females may experience bleeding symptoms such as menorrhagia, oral bleeding, surgical, and trauma-related bleeding.

Diagnosis: Hemophilia B presents with an X-linked inheritance pattern and an elevated partial thromboplastin time (PTT), normal prothrombin time (PT) and decreased plasma FIX activity. Commonly used PTT reagents may not be sensitive enough to diagnose mild FIX deficiencies, in which case residual plasma FIX activity would be diagnostic. Prenatal diagnosis can be performed in the case of a known family history. Cord blood testing of FIX levels should be performed at the time of delivery in cases of known carrier mothers. It should be noted, however, that FIX levels in term neonates are reduced to around 50% of normal adult values and normalize at around 6 months of age. Neonates with moderate and severe disease will still have distinctly low FIX levels, but diagnosis of mild hemophilia B in a neonate is complicated by the low normal levels at birth.

Differential Diagnosis: Hemophilia A and B are clinically indistinguishable, and individual factor levels must be used to clarify the diagnosis. Low FIX levels can be seen in advanced vitamin K deficiency; however, the other vitamin K–dependent proteins will be affected as well and the PT will be elevated. Acquired low FIX levels are rarely a result of autoantibody formation.

Management

Comprehensive Care: Comprehensive hemophilia treatment centers (HTCs) are federally funded centers that exist to care for patients with hemophilia. Typically, staffing includes hematologists, orthopedists, infectious disease specialists, physical therapists, nurses, psychologists, and social workers who specialize in the care of patients with bleeding disorders. Patients who receive their care in an HTC setting have a longer life expectancy than those cared for outside of this setting.

Factor Concentrates: The cornerstone of hemophilia care is FIX replacement with intravenously delivered FIX concentrates. Concentrates are either plasma-derived or recombinant products, and both currently undergo multiple viral and pathogen attenuation steps. No infectious complications have been reported since these steps were incorporated into the manufacturing process (mid-1980s for human immunodeficiency virus [HIV] and late 1980s for hepatitis B virus [HBV] and hepatitis C virus [HCV]); however, the possibility of contamination with new infectious agents such as prions cannot be excluded with plasma-derived products. Additionally, cryoprecipitate contains sufficient levels of FIX to treat bleeding episodes but due to concerns of viral safety should not be used when FIX concentrates are readily available.

Factor infusions can either be delivered in response to bleeding episodes ("on-demand therapy") or to prevent bleeding ("prophylaxis"). To prevent or minimize long-term sequela, on-demand therapy should be given as soon as possible after a bleeding episode is recognized. Because the need for urgent treatment is so important, many patients affected by hemophilia are educated in home infusion techniques. In the emergency department or office setting, factor replacement therapy should never be delayed to perform imaging or laboratory studies. On-demand therapy can be used to effectively treat bleeding episodes, but it is not effective in preventing the most common complication of hemophilia, which is blood-induced joint disease.

In developed countries, prophylactic therapy delivered 1–3 times per week dosed to keep trough levels above 1% is considered standard of care. If prophylactic infusions are started at a young age before the development of arthropathy, it is termed primary prophylaxis and is the only therapy proven to prevent the long-term complication of degenerative joint disease. While there is no universally accepted starting age, dose or frequency of infusions for primary prophylaxis, it is common practice to begin prophylaxis before the onset of recurrent joint bleeding. This is because the risk of irreversible joint damage increases if the onset of prophylaxis is delayed. Secondary prophylaxis refers to long-term factor replacement after a hemophilia complication has been experienced. This practice is commonly used after intracranial hemorrhage or once joint disease has become established. While secondary prophylaxis effectively reduces the frequency of bleeding episodes, it has not been shown to halt the progression of joint disease.

The dose and frequency of factor delivery is calculated based on the half-life of the product (typically 20–24 hours for FIX), the intravascular volume of distribution (one international unit [IU] of FIX per kilogram raises the plasma concentration by about 1%) and the desired clotting factor activity. For example, to raise the factor level of a 30 kg child with severe hemophilia B to 100% the dose should be 30 kg × 100 IU = 3000 IU. Forty percent activity is considered hemostatic in most cases; however, in the setting of surgery or life/limb threatening hemorrhage higher levels are necessary. Ancillary measures such as compressive dressings, cautery, packing, and splinting should also be implemented when appropriate. Agents that affect platelet function should be avoided. Table 111.1 illustrates an approach to factor replacement therapy of commonly encountered bleeding events.

More recently, development of strategies to extend the half-life of recombinant clotting factors has led to the approval of four extended half-life FIX products. Bioengineered recombinant FIX coupled to neonatal Fc receptors (rFIXFc), polyethylene glycol (PEGylated FIX), or albumin (rFIX-FP) allows for three- to fivefold increase in half-life, significantly decreasing the dosing frequency. For patients on prophylaxis, this could increase compliance and decrease the number of bleeding episodes, with the ultimate goal of improvement in long-term outcomes, most specifically, degenerative joint disease.

Antifibrinolytic Therapies: Antifibrinolytic medications such as aminocaproic acid or tranexamic acid are used to prevent excessive fibrinolysis and are particularly useful in diminishing bleeding symptoms in locations with prominent fibrinolytic activity, such as the mouth, gastrointestinal tract, and uterus.

TABLE 111.1 Suggested Approaches to Treatment of Bleeding Episodes

Bleed Site	Desired Activity	Length of Therapy	Ancillary Measures to Consider
Central nervous system	100%	7–14 days then strongly consider prophylactic therapy for a minimum of 6 months	Continuous infusion FIX; antiepileptic prophylaxis; surgical intervention
Persistent oral/ mucosal	30%–60%	3–7 days	Antifibrinolytic therapy; custom mouthpiece; topical thrombin powder
Retropharyngeal	80%–100%	7–14 days	Continuous infusion FIX; antifibrinolytic therapy
Nose	30%–60%	1–3 days	Packing, cautery; saline spray/gel; vasoconstrictor spray; antifibrinolytic therapy
Gastrointestinal	40%–80%	3–7 days	Antifibrinolytic therapy; endoscopy with cautery
Persistent gross urinary	40%–60%	1–3 days	Vigorous hydration; evaluation for stones/urinary tract infection; avoid antifibrinolytic therapy; glucocorticoids
Muscle	40%–80%	Every third day until pain-free movement	Rest, ice, compression, elevation; physical therapy
Iliopsoas	80%–100%	Until radiographic evidence of resolution	Continuous infusion FIX; bedrest; physical therapy
Joint	40%–80%	1–2 days	Rest, ice, compression, elevation; physical therapy
Target joint	80% day 1, 40% day 3	3–4 days	Rest, ice, compression, elevation; physical therapy

Gene Therapy: Gene therapy using an adeno-associated virus (AAV) vector to deliver human *F9* cDNA to the liver has shown promising results as a potential cure of hemophilia B. Three-year follow-up of 10 patients enrolled in the Phase 1 trial showed long-term therapeutic FIX expression without late toxic side effects. Currently, various trials using wild-type *F9* gene, codon-optimized F9 gene, and gain-of-function Padua variant F9 gene transduced via AAV or lentiviral vectors are ongoing.

Complications

Infectious Complications: In the late 1970s and early 1980s, before incorporation of viral attenuation steps in factor manufacturing, many patients became infected with HCV and/or HIV from contaminated factor concentrates. Thanks to highly active antiretroviral therapies for HIV and the combination therapy for HCV there is a large cohort of long-term survivors. There have been no cases of viral transmission related to clotting factor concentrates in nearly two decades.

Inhibitors: Occurrence of inhibitors in patients with severe hemophilia B is much rarer than hemophilia A (1%–3% vs. 20%–30%). Anaphylactic reactions to exogenous FIX, which can be life-threatening, occur in roughly 50% of patients with hemophilia B inhibitors and have been reported before recognition of a positive antibody titer. Most patients with an anaphylactic reaction will subsequently develop an inhibitory antibody, most of which are high titer in nature. Because of this potential rare complication, the first several (10–20) exposures to exogenous FIX concentrate should occur in the medical setting. Successful desensitization to FIX has been achieved in some patients and immune tolerance induction (ITI) with repeated exposure to FIX concentrate may eradicate the antibody. However, it is quite difficult to achieve immune tolerization in hemophilia B, with a reported success rate of only 15%–30%. Bleeding in the setting of a high-titer inhibitor often requires bypassing therapy with recombinant FVIIa or FEIBA. In patients who have had a severe allergic reaction to an FIX product, FEIBA is not recommended.

Nephrotic Syndrome: Nephrotic syndrome is a reported complication of ITI with FIX products and typically occurs 8–9 months into therapy. The precise mechanism of this complication is yet unknown, although the deposition of FIX-antibody immune complexes within the renal basement membrane has been postulated. Immunosuppressive agents such as cyclosporine, mycophenolate mofetil, and rituximab have been used with some success. It is commonly steroid-unresponsive and requires cessation of exposure to FIX.

Joint Disease: Degenerative joint disease due to recurrent hemarthrosis is the single largest preventable cause of morbidity for patients with hemophilia B. The most commonly affected joints are the elbows, knees, and ankles. The pathogenesis of this arthropathy is multifactorial, with iron and free radical formation being the most likely trigger of the degenerative changes. Recurrent hemarthrosis causes joint capsule stretching as well as an inflammatory reaction within the synovium. The synovium becomes thickened and highly vascularized, vulnerable to further recurrent bleeding. Osteopenia and proteoglycan loss causes osseous and cartilage destruction, leading to end stage joint disease. The physical findings in the early stages of joint disease are subtle, and investigation with MRI or power Doppler ultrasound may be required to demonstrate hemosiderin deposition and/or hypertrophic synovium. Initial treatment is to reduce or prevent further bleeding with clotting factor. Arthroscopic and radionuclide synovectomy can address recurrent bleeding into target joints, but it remains to be seen whether they can halt articular destruction.

Further Reading

Christophe, O. D., Lenting, P. J., Cherel, G., Boon-Spijker, M., Lavergne, J. M., Boertjes, R., et al. (2001). Functional mapping of anti-factor IX inhibitors developed in patients with severe hemophilia B. *Blood*, *98*, 1416–1423.

Collins, P. W., Chalmers, E., Hart, D. P., Liesner, R., Rangarajan, S., Talks, K., et al. (2013). Diagnosis and treatment of factor VIII and IX inhibitors in congenital haemophilia: (4th edition). UK Haemophilia Centre Doctors Organization. *Br J Haematol*, *160*, 153–170.

DiMichele, D. M. (2007). Inhibitor development in haemophilia B: An orphan disease in need of attention. *Br J Haematol, 138*, 305–315.

Dunn, A. L., & Abshire, T. C. (2006). Current issues in prophylactic therapy for persons with hemophilia. *Acta Haematol, 115*, 162–171.

Jansen, N. W., Roosendaal, G., & Lafeber, F. P. (2008). Understanding haemophilic arthropathy: An exploration of current open issues. *Br J Haematol, 143*, 632–640.

Manco-Johnson, M. J., Abshire, T. C., Shapiro, A. D., Riske, B., Hacker, M. R., Kilcoyne, R., et al. (2007). Prophylaxis versus episodic treatment to prevent joint disease in boys with severe hemophilia. *N Engl J Med, 357*, 535–544.

Mannucci, P. M., & Franchini, M. (2014). Emerging drugs for hemophilia B. *Expert Opin Emerg Drugs, 19*, 407–414.

Nathwani, A. C., Reiss, U. M., Tuddenham, E. G., Rosales, C., Chowdary, P., McIntosh, J., et al. (2014). Long-term safety and efficacy of factor IX gene therapy in hemophilia B. *N Engl J Med, 371*, 1994–2004.

Warrier, I. (1998). Management of haemophilia B patients with inhibitors and anaphylaxis. *Haemophilia, 4*, 574–576.

CHAPTER 112

Factor XI Deficiency

Jennifer Davila, MD

Factor XI (FXI) deficiency, was first described in 1953. It is characterized by a highly variable bleeding phenotype. Some patients have no apparent excessive bleeding while others have more substantial bleeding. In contrast to other coagulation factor deficiencies, excessive bleeding in FXI deficiency is also highly variable in an individual patient, who might hemorrhage after one hemostatic challenge but not after another. Homozygotes or compound heterozygotes with <15% FXI levels usually have more severe bleeding tendencies than heterozygotes who have 25%–70% FXI levels (and who have little or no excessive bleeding). Although FXI deficiency is classified among rare inherited coagulation disorders (estimated prevalence of severe deficiency in most populations is 1:1,000,000), its frequency is remarkably higher among Ashkenazi Jews (1:450 individuals).

Biochemistry: FXI is synthesized in both the liver and perhaps to a minor extent in megakaryocytes. FXI is a 160-kDa glycoprotein, which separates into two 80-kDa subunits linked by disulfide bonds. It comprises heavy chains with four repeats that have binding sites for high-molecular-weight kininogen (HK), thrombin, platelets, FIX, and FXII. It also comprises a light chain where the serine protease is located. It is converted to the active serine protease form, FXIa. This conversion is accelerated by calcium, platelets (a source of phospholipids), and thrombin. Originally, FXI was thought to predominantly be activated by FXII or HK, due to contact activation and the so-called intrinsic pathway. Contact activation is not thought to be as important in the physiologic activation of FXI, and instead the activation of FXI is predominately mediated by thrombin. Thrombin's role in the activation of FXI is the result of a "feed-forward loop" to promote stable clot formation and protection against fibrinolysis by thrombin activatable fibrinolytic inhibitor (TAFI). This latter relationship between FXI and fibrinolysis might explain why, in contrast to other coagulation factor deficiencies, FXI deficiency leads to excessive mucosal bleeding.

Genetics: The *F11* gene is located on chromosome 4q34–35, near the *KLKB1* gene, which encodes for prekallikrein (PK). Most mutations produce an inability to make adequate protein. Mutations that cause qualitative defects associated with a dysfunctional protein are unusual. Accordingly, in most cases, the amount of functional protein correlates with the protein antigen level. There are three main mutations:

- Type I occurs in the last intron of the gene, which results in a disruption of the mRNA splicing or a premature translation termination. The type I mutation is rarely found in the Jewish patient population (accounts for only 1%).
- Type II in exon 5 resulting in a premature stop codon.
- Type III in exon 9 resulting in a nucleotide substitution (missense mutation).

Transfusion Medicine and Hemostasis. https://doi.org/10.1016/B978-0-12-813726-0.00112-4

The type II mutation is the earliest FXI founder mutation, which can be traced to early patriarchal times. Accordingly, it exists not only in patients of Jewish descent but also among subjects who can trace their lineage to other Middle Eastern populations. Ashkenazi Jews with FXI deficiency typically have either type II or type III mutations, whereas other Middle Eastern populations only harbor the type II mutation. Homozygotes for the type II mutation have the lowest FXI levels (approximately 1%). Type III FXI deficiency displays the mildest deficiency of FXI levels, often approximately 10% of normal. Compound heterozygotes for the type II/III mutations usually have moderate deficiencies of FXI with levels of 2%–5% of normal. Persons of Arab and Iraqi Jewish descent who are FXI deficient usually have the type II mutation. FXI deficiency is also seen in other ethnic groups, and in these patients the defect can be attributable to mutations within other exons. Given the existence of founder effects, information on the ethnic background of the patient can guide mutation screening.

Clinical Manifestations: FXI deficiency is a rare bleeding disorder where bleeding usually occurs after surgery or trauma. The bleeding can occur immediately or sometimes only after scar tissues have detached. Severe FXI deficiency is defined as a plasma activity level of <1%-15U/dL. The bleeding phenotype is not correlated with the genotype but frequently with site of injury. When a site of injury with high fibrinolysis is involved, the risk of bleeding is increased (49%–67%) in comparison with sites without fibrinolysis (1.5%–40%). Areas of high fibrinolytic activity are found in the nose, mouth, and the genitourinary tract. The bleeding in these sites is thought to be related to impaired activation of TAFI with the concomitant increase in fibrinolysis. Bleeding from orthopedic and abdominal surgery is less common, and spontaneous bleeding is rare altogether. Excessive bleeding has been documented in heterozygotes with borderline normal FXI level between 50 and 70 IU/dL. Therefore, other determinants such as bleeding history or bleeding score should be used to predict and assess bleeding potential in these patients. FXI deficiency is a common cause of menorrhagia. Consequently, postpartum hemorrhage can occur in approximately 20% of women with FXI deficiency.

Laboratory Diagnosis: There is variability in the sensitivity of the partial thromboplastin time (PTT) to detect heterozygous FXI deficiency, and consequently it should not be considered a reliable screening tool. In a patient with normal PTT and history of bleeding, FXI plasma activity levels should be measured to assess for deficiency, but it bears repeating that FXI levels do not always predict who will bleed excessively. The use of global coagulation assays to determine bleeding risk in patients with FXI deficiency has not been proven. Development of inhibitory antibodies to FXI is encountered in approximately 5% of patients who are homozygous for the type II mutation. The incidence increases in patients who are homozygous for the type II mutation and have received plasma.

Management: Given the variability in bleeding manifestations, bleeding assessment and management must be individualized. The lack of obvious genotype–phenotype correlation often leads to over or undertreatment of FXI deficiency. Minor and major surgery should be treated with prophylactic factor replacement with FFP (15–20 mL/kg). Although there is no consensus, some clinicians recommend that patients undergoing major or high-risk surgery have their nadir FXI level targeted closer to >20% or as close to 50% of normal for 7 days or until healing is complete. Afterward, they can be kept at a trough above 30% for an additional 3–7 days. Less invasive surgery can be managed by maintaining lower FXI levels above 20%–30% for a shorter period of time (3–5 days). The half-life of FXI is approximately 40–70 hours. Therefore, plasma products can usually be given every 1–2 days to keep hemostatic trough levels of 15%–20%. Nose- and mouth-related bleeding as well as menorrhagia can often be managed with antifibrinolytic agents, such as aminocaproic acid or tranexamic acid.

Purified FXI concentrates are not currently available in the United States but are used in Europe. In the United States, patients usually receive plasma products, as recombinant FXI is not available in the United States. Fibrin glue can be used to help with hemostasis. DDAVP has also been used for mucosal bleeding or minor surgery. The efficacy of recombinant FVIIa (rFVIIa) to treat patients with FXI has been shown in several case reports. The optimal dosing regimen is unknown and requires clinical data. Doses ranging between 12 and 90 mcg/kg, some with concurrent use of tranexamic acid, have been used to manage various bleeding challenges, including from postpartum bleeding, in patients with and without inhibitors. Concern for catastrophic thrombotic events, however, suggests the use of lower doses to manage bleeding.

Further Reading

Davies, J., & Kadir, R. (2016). The management of factor XI deficiency in pregnancy. *Semin Thromb Hemost*, *42*(7), 732–740.

Duga, S., & Salomon, O. (2013). Congenital factor XI deficiency: An update. *Semin Thromb Hemost*, *39*(6), 621–631.

Livnat, T., Tamarin, I., Mor, Y., Winckler, H., Horowitz, Z., Korianski, Y., et al. (2009). Recombinant activated factor VII and tranexamic acid are haemostatically effective during major surgery in factor XI-deficient patients with inhibitor antibodies. *Thromb Haemost*, *102*(3), 487–492.

O'Connell, N. M. (2003). Factor XI deficiency–from molecular genetics to clinical management. *Blood Coagul Fibrinolysis*, *14*, S59–S64.

Salomon, O., Zivelin, A., Livnat, T., et al. (2003). Prevalence, causes, and characterization of factor XI inhibitors in patients with inherited factor XI deficiency. *Blood*, *101*, 4783–4788.

CHAPTER 113

Factor VII Deficiency

Glaivy Batsuli, MD and Shannon L. Meeks, MD

Factor VII (FVII) deficiency is the most common autosomal recessive rare bleeding disorder. It was first described in 1951 and has an estimated prevalence of 1 in 500,000. The manifestation of bleeding symptoms in FVII deficiency is clinically heterogeneous. Some individuals with FVII deficiency may have mild symptoms, whereas others may have severe and potentially life-threatening bleeding symptoms. Moreover, a proportion of patients remain asymptomatic. Even in one individual with FVII deficiency, there can be a discrepancy between the plasma FVII level and bleeding risk, which can make the prediction of bleeding risk and management challenging.

Pathophysiology: FVII is a 50-kDa, single-chain, vitamin K–dependent serine protease that is synthesized in the liver. FVII is a unique factor in which a small portion of FVII circulates in the plasma as activated FVII (FVIIa) even in the absence of activation during coagulation. On vessel injury, exposed tissue factor (TF) on the vascular lumen interacts with both FVII and FVIIa. This complex of TF and FVII/FVIIa is able to activate factor IX to factor IXa and factor X to factor Xa, which ultimately results in the formation of a fibrin clot. Additionally, FVII can be proteolytically cleaved to FVIIa by several enzymes, including thrombin, factor IXa, factor Xa, and the TF/FVIIa complex. The lipoprotein tissue factor pathway inhibitor is the primary inhibitor of FVIIa.

Genetics: FVII deficiency is inherited in an autosomal recessive pattern. The *F7* gene is encoded on chromosome 13 (13q34) and consists of 9 exons encoding 406 amino acids. The majority of *F7* gene mutations causing FVII deficiency are missense mutations; splice site mutations, nonsense mutations, and deletions are less common. More than 200 mutations have been identified and can be found in the International Society on Thrombosis and Haemostasis (ISTH) international database for mutations causing rare bleeding disorders (http://www.isth.org/?MutationsRareBleedin).

Clinical Manifestations: The most common bleeding symptoms are epistaxis, easy bruising, gum bleeding, menorrhagia in females, and bleeding after trauma or surgery. Individuals with FVII deficiency, particularly those with severe phenotypes, may develop hemarthrosis, hematomas, or GI bleeding similar to individuals with classical hemophilia. Newborns with severe FVII deficiency may present with intracranial hemorrhage (ICH). Approximately one-third of individuals will remain asymptomatic throughout their lifetime.

Homozygous patients with severe FVII deficiency typically have plasma FVII activity levels <2%. These patients tend to present with severe and sometimes life-threatening bleeding. Complete absence of FVII is considered incompatible with life. Patients with heterozygous FVII deficiency tend to have levels approaching 50%. Based on prior analyses, FVII activity levels of 15%–20% are considered adequate for hemostasis; thus

Transfusion Medicine and Hemostasis. https://doi.org/10.1016/B978-0-12-813726-0.00113-6

patients with FVII levels >20% tend to be asymptomatic. However, there can be a discrepancy between the plasma FVII levels and bleeding risk.

Diagnosis: An isolated prolonged prothrombin time (PT) that corrects with a mixing study is found on initial evaluation of FVII deficiency. In heterozygotes the PT is 1–3 seconds prolonged, while the PT may be prolonged more than 20 seconds in patients with severe disease. The activated partial thromboplastin time (aPTT) and thrombin time will be normal. Subsequent testing of plasma FVII procoagulant activity levels often confirms the diagnosis.

Measurement of plasma FVII activity can vary depending on the thromboplastin reagent used. In general, recombinant thromboplastin derived from the human tissue factor gene is considered the most reliable in predicting bleeding risk. Samples should not be stored on ice, as this may induce cold activation of coagulation factors and underestimation of FVII in the plasma sample. Similarly, the presence of recombinant FVIIa in the plasma leads to an overestimation of the FVII level. FVII antigen levels can be used to identify patients with dysfunctional proteins. Whole gene sequencing of the *F7* gene is recommended for a molecular diagnosis but is not mandatory.

Differential Diagnosis: An inherited FVII deficiency must be differentiated from the more common acquired deficiencies of FVII. Acquired FVII deficiency may be seen in vitamin K deficiency, liver disease, and disseminated intravascular coagulation. Inherited FVII deficiency can be differentiated from acquired deficiencies by normal levels of other vitamin K–dependent proteins (i.e., factors II, IX, and X, proteins C and S). FVII has the shortest half-life (~3–6 hours) of the coagulation factors; thus deficiency of FVII occurs earlier during the course of an acquired disease process than the other vitamin K–dependent factors.

Management: A recent report from the European Network of Rare Bleeding Disorders (EN-RBD) recommends an FVII trough of >20% to prevent bleeding. Several different products can be used for FVII replacement depending on availability, including fresh frozen plasma (FFP), prothrombin complex concentrates (PCCs), plasma-derived FVII concentrates, and recombinant FVIIa. Recombinant FVIIa is the most frequently used agent for management of bleeds and is considered first-line therapy. Recombinant FVIIa is recommended at a dose of 15–20 mcg/kg for FVII deficiency. This dose is much lower than the 90–120 mcg/kg dose used for treatment in patients with hemophilia and inhibitors or Glanzmann Thrombasthenia. In mild/moderate bleeds or after a minor surgical procedure, a single dose of recombinant FVIIa is often adequate for treatment. For severe bleeding or for major surgery, dosing can be repeated every 4–6 hours until hemostasis is achieved; the dose and frequency should be individualized to the patient. Long-term prophylaxis with FVII replacement using recombinant FVIIa or PCCs is recommended for individuals with a severe phenotype, particularly after ICH and for children with recurrent episodes of hemarthrosis.

The FVII concentration in FFP and PCCs are low and may be inadequate for treatment depending on the clinical scenario. In countries where it is available, plasma-derived FVII concentrates is used for treatment. Plasma-derived FVII concentrates often require high infusion volumes and have a short half-life of 4–6 hours. However, 1 IU of plasma-derived

FVII concentrate reportedly increases the plasma FVII level by 1.9%. Dosing of 15–20 units/kg for minor bleeds and 30–40 units/kg for severe bleeds are recommended.

Patients with FVII deficiency have an intact intrinsic coagulation pathway and are able to amplify the thrombin burst to generate significantly more thrombin and a more stable fibrin clot. Deep vein thrombosis has been reported in 3%–4% of individuals with FVII deficiency in association with surgery and FVII replacement therapy. Spontaneous thromboses may also occur. Inhibitors to FVII, though rare, have been reported in a few cases after exposure to recombinant FVIIa.

The most limiting aspect of FVII replacement remains its short half-life. There has been development of newer bioengineered recombinant FVII products that have shown promise in improving this barrier. Specifically, a recombinant FVIIa and albumin fusion protein showed a 3–4 fold increase in half-life (6–10 hours) in a phase I randomized, placebo-controlled clinical trial. Unfortunately, lack of a dose response and development of cross-reactive anti-FVII antibodies in patients with hemophilia and inhibitors have halted further testing of PEGylated recombinant FVIIa agents in prior clinical trials.

Antifibrinolytic agents, such as tranexamic acid and aminocaproic acid, can be helpful for mucosal bleeding. Antifibrinolytics can be used alone or in combination with FVII replacement for bleeding events depending of severity. Hormonal agents such as combined oral contraceptives and intrauterine devices can be used to augment treatment of menorrhagia.

Further Reading

Golor, G., Bensen-Kennedy, D., Haffner, S., Easton, R., Jung, K., Moises, T., et al. (2013). Safety and pharmacokinetics of a recombinant fusion protein linking coagulation factor VIIa with albumin in healthy volunteers. *J Thromb Haemost*, *11*(11), 1977–1985.

Koh, P. L., Ng, H. J., Lissitchkov, T., Hardtke, M., & Schroeder, J. (2013). The TRUST trial: Anti-drug antibody formation in a patient with hemophilia with inhibitors after receiving the activated factor VII product Bay 86-6150. *Blood*, *122*(21), 573.

Ljung, R., Karim, F. A., Saxena, K., Suzuki, T., Arkhammar, P., Rosholm, A., et al. (2013). 40K glycoPEGylated, recombinant FVIIa: 3-month, double-blind, randomized trial of safety, pharmacokinetics and preliminary efficacy in hemophilia patients with inhibitors. *J Thromb Haemost*, *11*(7), 1260–1268.

Mariani, G., Herrmann, F. H., Dolce, A., Batorova, A., Etro, D., Peyvandi, F., et al. (2005). Clinical phenotypes and factor VII genotype in congenital factor VII deficiency. *Thromb Haemost*, *93*(3), 481–487.

Napolitano, M., Siragusa, S., & Mariani, G. (2017). Factor VII deficiency: Clinical phenotype, genotype and therapy. *J Clin Med*, *6*(4).

Palla, R., Peyvandi, F., & Shapiro, A. D. (2015). Rare bleeding disorders: Diagnosis and treatment. *Blood*, *125*(13), 2052–2061.

Peyvandi, F., Di Michele, D., Bolton-Maggs, P. H., Lee, C. A., Tripodi, A., & Srivastava, A. (2012). Classification of rare bleeding disorders (RBDs) based on the association between coagulant factor activity and clinical bleeding severity. *J Thromb Haemost*, *10*(9), 1938–1943.

Peyvandi, F., Palla, R., Menegatti, M., Siboni, S. M., Halimeh, S., Faeser, B., et al. (2012). Coagulation factor activity and clinical bleeding severity in rare bleeding disorders: Results from the European Network of Rare Bleeding Disorders. *J Thromb Haemost*, *10*(4), 615–621.

CHAPTER 114

Factor II, Factor V, and Factor X Deficiencies

Karen L. Zimowski, MD and Shannon L. Meeks, MD

Introduction: Inherited deficiencies of factors II, V, and X (FII, FV, and FX) are rare (estimated frequencies are FII, 1 in 2,000,000; FV, 1 in 1,000,000; and FX, 1 in 1,000,000). Patients who are homozygous or compound heterozygous for defects in the *F2, F5*, or *F10* genes can have moderate to severe bleeding symptoms, with patients who have FX deficiency being more likely to manifest severe symptoms. The European Network of Rare Bleeding Disorders analyzed the correlation of factor activity and clinical bleeding in patients with rare bleeding disorders, observing a strong association for FX and combined FV and FVIII deficiencies and a weak association for FV and FII deficiencies. In general, however, the lower the factor levels, the more severe the bleeding. Overall, a spectrum of bleeding symptoms have been reported, including easy bruising, mucosal bleeding, menorrhagia, and surgical/trauma-induced bleeding, while hemarthroses and intracranial hemorrhage are less common. Heterozygotes are typically asymptomatic; however, easy bruising, mucosal bleeding, and bleeding with surgery/trauma have been reported.

Pathophysiology: FII (also known as prothrombin) is a vitamin K–dependent glycoprotein with a plasma concentration of approximately 100 µg/mL and a half-life of 3 days. It is synthesized in the liver and is the inactive zymogen of thrombin. FV is a nonenzymatic cofactor that is synthesized in both hepatocytes and megakaryocytes, with 80% being secreted and 20% being stored in platelet α-granules. The half-life of FV is 36 hours. FX is a vitamin K–dependent glycoprotein that is synthesized in the liver as an inactive zymogen and circulates at a concentration of 8–10 µg/mL. The half-life of FX is 40 hours.

Role in Coagulation: In the coagulation cascade, FX is activated by FVIIa from the tissue factor (extrinsic) pathway or FIXa from the contact factor (intrinsic) pathway. FXa associates with FVa and phospholipid to form the prothrombinase complex, which, in the presence of calcium, cleaves prothrombin to its active form, thrombin. Thrombin in turn cleaves fibrinogen generating fibrin, which polymerizes into a clot. Thrombin also promotes platelet activation, activates factor XIII, enhances clot stability by activating thrombin-activatable fibrinolysis inhibitor, and upregulates its own production by activating FXI, FVIII, and FV. Thrombin's procoagulant activity is downregulated by antithrombin, heparin cofactor II, and protease nexin I. Thrombin also binds to thrombomodulin and activates protein C, which in turn inactivates FVIIIa and FVa.

Genetics: Deficiencies of FII, FV, and FX are all inherited in an autosomal recessive fashion, and as such consanguinity increases the incidence of these deficiencies.

Transfusion Medicine and Hemostasis. https://doi.org/10.1016/B978-0-12-813726-0.00114-8

The prothrombin gene (*F2*) is found on chromosome 11p11.2. The hemostatic level of prothrombin is between 20% and 40%. More than 40 mutations leading to prothrombin deficiency have been described. Combined deficiencies of the vitamin K–dependent proteins (FII, FVII, FIX, and FX) occur when there is an abnormality in the γ-glutamyl carboxylase gene or the vitamin K epoxide reductase complex. The *F5* gene is located on chromosome 1q24.2. The hemostatic level for FV is between 10% and 30%. More than 100 mutations in *F5* have been identified to date. A combined deficiency of FV and FVIII occurs due to a mutation in the *LMAN1* (lectin mannose-binding protein 1) gene found on chromosome 18. *LMAN1* encodes a transmembrane protein that acts as a chaperone in the intracellular transport of both FV and FVIII. Mutations in the *MCFD2* (multiple coagulation factor deficiency 2) gene, which encodes a protein that acts as a cofactor for LMAN1, have also been described. The *F10* gene is found on chromosome 13q34. The hemostatic level for FX is 10%–20%. More than 100 mutations have been found in the *F10* gene. FX deficiency, like prothrombin deficiency, can be associated with a combined deficiency of the vitamin K–dependent proteins.

Diagnosis: FII, FV, and FX are essential parts of the common pathway of coagulation. The finding of a prolonged prothrombin time (PT) and activated partial thromboplastin time (PTT), which correct when mixed with normal plasma, suggest a factor deficiency, and individual factor levels should be checked. However, PT and PTT prolongation may be minimal; thus, if the clinical index of suspicion for a factor deficiency is high, factor levels should be measured. Most clinical labs use functional assays to test factor levels. Antigen testing via immunoassay can be useful in separating a type I and type II deficiency. Type I or hypoproteinemia occurs when there is a decreased level of normally functioning protein, characterized by a proportional decrease in antigen and activity. Type II or dysproteinemia occurs when there is a normal antigen level but a decreased level of activity.

Differential Diagnosis: Both FII and FX are vitamin K–dependent proteins, and therefore their levels decrease in vitamin K deficiency and with warfarin therapy. FII, FV, and/or FX deficiencies may be seen in liver disease. Acquired prothrombin deficiency may occur in the presence of inhibitors similar to those seen with the lupus anticoagulant. Amyloidosis may be associated with low FX levels due to adsorption of FX to amyloid.

Management: Acute bleeding episodes for these factor deficiencies can be treated with fresh frozen plasma (FFP) at 15–20 mL/kg, which should achieve hemostatic factor levels in most cases. For surgical procedures or more severe bleeding episodes, a loading dose of 15–20 mL/kg of FFP followed by 3–6 mL/kg every 12–24 hours is usually adequate. For FII or FX deficiency another option is nonactivated prothrombin complex concentrates (PCCs) that are typically given at a dose of 20–30 IU/kg based on FIX units and contain FII, FIX, FX, and small amounts of FVII. They vary in the amounts of each factor based on the specific product and lot—for example, Bebulin VH (Baxter, Deerfield, IL) is a PCC with more FX and Profilnine SD (Grifols, Los Angeles, CA) has more FII. However, these products are labeled based on FIX units, and the exact amount of FII or FX in the product is unknown. Four-factor PCCs (4F-PCCs) contain physiologic amounts of FVII and protein C and S in addition to FII, FIX, and FX. Kcentra (CSL-Behring, Germany), currently the only Federal drug administration (FDA)-approved 4F-PCC, has been used as an antidote for bleeding associated with vitamin K antagonism and could be considered for use in factor II and X deficiency although it has not been studied in this patient population.

Since approximately 20% of FV is contained within platelet α-granules, platelets can also be given for FV deficiency. In theory, the transfusion of platelets may aid in hemostasis by delivering FV to the site of bleeding. In rare cases, alloantibodies to FV develop in patients with congenital FV deficiency, rendering FFP ineffective; bleeding in these cases typically responds to platelet transfusions. No pure FII or FV concentrates are available, although efforts are underway at developing a plasma-derived FV concentrate. A high-purity plasma-derived FX concentrate (pdFX, Coagadex) was FDA-approved in 2015 for on-demand treatment of acute bleeding episodes and perioperative bleeding in patients ≥12 years old with hereditary FX deficiency at a dose of 25 IU/kg. Ongoing studies are assessing the use of this product in younger patients and as prophylaxis to prevent bleeding episodes. In general, bleeding symptoms and factor levels should be followed with an aim of maintaining trough levels of FX >20%, FII >30%, and FV >20%. Antifibrinolytic therapies such as ε-aminocaproic acid or tranexamic acid may be administered for mucosal bleeding and hormonal therapy with estrogens and/or progesterones may help reduce menstrual bleeding. Prophylaxis with once-weekly PCCs for FII or FX deficiency or FFP for FII, FV, or FX deficiency has been reported for patients with severe disease.

Further Reading

Kinard, T. N., & Sarode, R. (2014). Four factor prothrombin complex concentrate (human): Review of the pharmacology and clinical application for vitamin K antagonist reversal. *Expert Rev Cardiovasc Ther*, *12*(4), 417–427.

Lancellotti, S., Basso, M., & Cristofaro, R. (2013). Congenital prothrombin deficiency: An update. *Semin Thromb Hemost*, *39*(06), 596–606.

Meeks, S. L., & Abshire, T. C. (2008). Abnormalities of prothrombin: A review of the pathophysiology, diagnosis, and treatment. *Haemophilia*, *14*(6), 1159–1163.

Menegatti, M., & Peyvandi, F. (2009). Factor X deficiency. *Semin Thromb Hemost*, *35*(04), 407–415.

Mumford, A. D., Ackroyd, S., Alikhan, R., Bowles, L., Chowdary, P., Grainger, J., et al. (2014). Guideline for the diagnosis and management of the rare coagulation disorders: A United Kingdom Haemophilia Centre Doctors' Organization guideline on behalf of the British Committee for Standards in Haematology. *Br J Haematol*, *167*(3), 304–326.

Palla, R., Peyvandi, F., & Shapiro, A. D. (2015). Rare bleeding disorders: Diagnosis and treatment. *Blood*, *125*(13), 2052–2061.

Peyvandi, F., & Menegatti, M. (2016). Treatment of rare factor deficiencies in 2016. *Hematol Am Soc Hematol Educ Program*, *2016*(1), 663–669.

Peyvandi, F., Palla, R., Menegatti, M., Siboni, S. M., Halimeh, S., Faeser, B., et al. (2012). Coagulation factor activity and clinical bleeding severity in rare bleeding disorders: Results from the European Network of Rare Bleeding Disorders. *J Thromb Haemost*, *10*(4), 615–621.

Shapiro, A. (2016). Plasma-derived human factor X concentrate for on-demand and perioperative treatment in factor X-deficient patients: Pharmacology, pharmacokinetics, efficacy, and safety. *Expert Opin Drug Metab Toxicol*, *13*(1), 97–104.

Thalji, N., & Camire, R. (2013). Parahemophilia: New insights into factor V deficiency. *Semin Thromb Hemost*, *39*(06), 607–612.

Zheng, C., & Zhang, B. (2013). Combined deficiency of coagulation factors V and VIII: An update. *Semin Thromb Hemost*, *39*(06), 613–620.

CHAPTER 115

Congenital Disorders of Fibrinogen

Glaivy Batsuli, MD and Shannon L. Meeks, MD

Congenital fibrinogen disorders include a spectrum of defects that fall into two categories: quantitative (type I) and qualitative (type II) fibrinogen disorders. Quantitative (type I) disorders include the absence of fibrinogen (afibrinogenemia) or low fibrinogen activity and antigen levels typically <150 mg/dL (hypofibrinogenemia). Qualitative (type II) disorders entail low fibrinogen activity but discordant normal antigen (dysfibrinogenemia) or reduced antigen (hypodysfibrinogenemia). The incidence of afibrinogenemia is estimated at 1 in 1 million; however, it is higher in areas where consanguinity is common. Hypofibrinogenemia, dysfibrinogenemia, and hypodysfibrinogenemia are thought to occur more frequently; however, estimating the incidence of these subtypes is challenging, as most patients are asymptomatic. Patients with these disorders can present with a spectrum of clinical manifestations including no symptoms, bleeding, thrombosis, or both hemorrhagic and thrombotic complications.

Pathophysiology: Fibrinogen is a 340-kDa hexamer synthesized by hepatocytes that consists of two sets of three homologous polypeptide chains: Aα, Bβ, and γ (Fig. 115.1). These chains are joined at their respective amino terminus by five disulfide bridges forming a central E region (E domain). The E domain also contains fibrinopeptide A (FPA) and fibrinopeptide B (FPB) on the α and β chains, respectively, and is flanked by two D regions (D domains) on either side. Fibrinogen circulates in the plasma at 150–350 mg/dL and has a half-life of 3–4 days. Platelet alpha granules also contain an internalized pool of fibrinogen.

Role in Coagulation: Fibrinogen plays a critical role in both primary and secondary hemostasis. Fibrinogen released from platelet alpha granules serves as a bridge for the platelet glycoprotein IIb/IIIa receptor on the platelet surface during platelet aggregation. After the formation of a platelet plug, the coagulation cascade is activated and thrombin is generated. Thrombin subsequently cleaves fibrinogen to fibrin inducing release of FPA and FPB from the central E domain to form a fibrin clot. The fibrin units are assembled into an organized polymeric structure and cross-linked by factor XIIIa to further stabilize the fibrin clot. Antithrombin is an important regulator of coagulation; it inhibits thrombin formation and sequesters circulating thrombin within fibrin. Fibrinogen also interacts with plasminogen, tissue plasminogen activator, and α2-antiplasmin to help regulate fibrin deposition and fibrinolysis.

Genetics: Fibrinogen is encoded by three genes: Aα (*FGA*), Bβ (*FGB*), and γ (*FGG*) on chromosome 4. Each gene is transcribed and translated separately to produce proteins containing 644 (Aα), 491 (Bβ), and 437 (γ) amino acids. Fibrinogen mutations that result in congenital fibrinogen disorders have been described in all three encoding

Transfusion Medicine and Hemostasis. https://doi.org/10.1016/B978-0-12-813726-0.00115-X

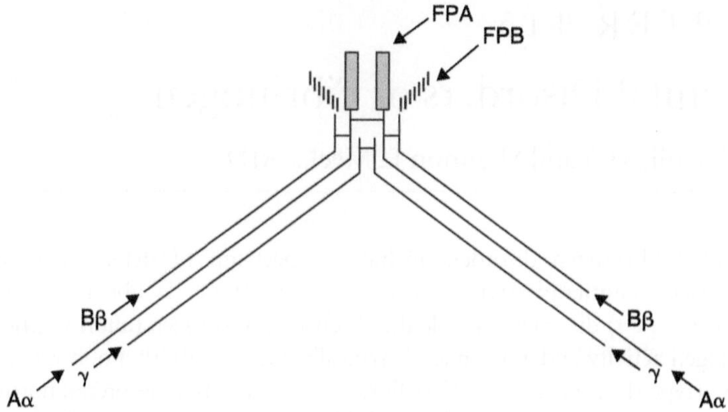

FIGURE 115.1 Fibrinogen is composed of three chains, the Aα, Bβ, and γ chains, arranged as a heterodimer, Aα2Bβ2γ2. The conversion of fibrinogen to fibrin, α2β2γ2, requires the cleavage of peptide bonds to release FPA and FPB. (From Hoffman, R., Benz, E. J., Shattil, S. J., et al. (Eds.). (2005). *Hematology: Basic Principles and Practice* (4th ed.). Philadelphia, PA: Elsevier.)

genes of fibrinogen; however, the majority of mutations are found in the *FGA* gene. Individuals with afibrinogenemia tend to have severe null mutations that result in the absence of fibrinogen. These mutations are inherited in an autosomal recessive fashion as homozygous or compound heterozygous mutations. Heterozygotes with one normal fibrinogen allele and one mutated allele tend to have decreased amounts of fibrinogen as seen in individuals with hypofibrinogenemia. Dysfibrinogenemia and hypodysfibrinogenemia have an autosomal dominant inheritance, primarily of a single missense mutation, that results in the production of a dysfunctional protein.

Clinical Manifestations: Individuals with afibrinogenemia may present with mucosal bleeding, subcutaneous bleeding including muscle hematomas, and in rare cases life-threatening intracranial hemorrhage. The majority of these patients (85–90%) typically present in the neonatal period with umbilical cord bleeding. Menometrorrhagia may occur but is less common. Additionally, these patients may have poor wound healing, painful bone cysts, first trimester miscarriages, and an unexplained increased risk of spontaneous splenic rupture. Individuals with hypofibrinogenemia are often asymptomatic or have a less severe bleeding phenotype. Major bleeding after trauma or surgery has been reported in patients with significantly decreased fibrinogen activity (<50 mg/dL). The fibrinogen activity level tends to correlate well with bleeding severity. Patients with quantitative defects of fibrinogen are still at risk for venous and arterial thrombosis with or without fibrinogen replacement. This has been attributed to the formation of unstable and loosely packed thrombi that are prone to embolize.

Patients with qualitative fibrinogen disorders can present with a heterogeneous range of clinical symptoms. Patients may have bleeding after surgery, thrombosis, or both. However, the majority of patients are asymptomatic. Bleeding symptoms are generally mild but may occur frequently. First trimester miscarriage is less common in dysfibrinogenemia compared with afibrinogenemia. Chronic thromboembolic pulmonary hypertension and amyloidosis have been associated with dysfibrinogenemia.

Diagnosis: The initial evaluation for a fibrinogen disorder will reveal a prolonged prothrombin time (PT), prolonged activated partial thromboplastin time (aPTT), and a prolonged thrombin time (TT), primarily when the fibrinogen activity is <100 mg/dL. If there is concern for heparin contamination a reptilase time (RT), an assay that uses snake venom as opposed to thrombin used in the thrombin time as the reagent to measure the conversion of fibrinogen to fibrin, may be used. The RT will remain prolonged in fibrinogen disorders but normalize in heparin contamination. The fibrinogen activity and antigen level (measured via immunoassays using antifibrinogen antibodies) can help differentiate between quantitative and qualitative disorders. In afibrinogenemia and hypofibrinogenemia, the fibrinogen activity and antigen level will be similarly low. A low fibrinogen activity with a normal antigen suggests dysfibrinogenemia. Fibrinogen genotype analysis is important for disease management, as some mutations are predictive of clinical phenotype, particularly in cases of dysfibrinogenemia.

Differential Diagnosis: It is important to rule out more common acquired deficiencies of fibrinogen such as disseminated intravascular coagulation, liver disease, and administration of L-asparaginase therapy when considering the diagnosis of congenital fibrinogen deficiency. The presence of heparin or fibrin/fibrin degradation products may alter initial screening labs (i.e., PT, aPTT, and TT). It is important to note that fibrinogen is an acute phase reactant that is elevated during periods of stress or infection. Fibrinogen is also affected by age, gender, race, obesity, and pregnancy.

Management: Fibrinogen replacement with cryoprecipitate or plasma-derived fibrinogen concentrates is readily used in afibrinogenemia and hypofibrinogenemia. The first-line treatment is plasma-derived fibrinogen concentrate that has a mean half-life of ~3 days. There are four plasma-derived fibrinogen products commercially available that varies by country. In the United States, plasma-derived fibrinogen concentrate Riastap (CSL Behring) is available for treatment at a recommended dose of 70 mg/kg. Plasma-derived fibrinogen concentrates can be used for long-term prophylaxis in patients with afibrinogenemia or severe hypofibrinogenemia at a frequency of every 7–14 days to maintain a goal fibrinogen activity of >100 mg/dL. Cryoprecipitate, which contains 250 mg of fibrinogen per cryoprecipitate unit, can be used as an alternative for treatment and prophylaxis. Infusion of 1 unit of cryoprecipitate per 5 kg of patient's weight followed by 1 unit/15 kg to maintain hemostasis is recommended. Risk of thrombosis, allergic reactions, anaphylaxis, and inhibitor development are rare but have been reported with fibrinogen replacement. As a result, replacement is often performed and monitored in a medical facility. Antifibrinolytic agents such as tranexamic acid or aminocaproic acid are useful for mucosal bleeding, minor procedures, or menorrhagia.

Management of patients with dysfibrinogenemia is more challenging due to the lack of strong evidence-based recommendations and the clinical heterogeneity among patients. Treatment should be tailored to the individual taking into consideration the patient's personal history, family history of bleeding and thrombosis, and the molecular defect.

Further Reading

Bornikova, L., Peyvandi, F., Allen, G., Bernstein, J., & Manco-Johnson, M. J. (2011). Fibrinogen replacement therapy for congenital fibrinogen deficiency. *J Thromb Haemost*, *9*(9), 1687–1704.

de Moerloose, P., Casini, A., & Neerman-Arbez, M. (2013). Congenital fibrinogen disorders: An update. *Semin Thromb Hemost*, *39*(6), 585–595.

de Moerloose, P., Schved, J. F., & Nugent, D. (2016). Rare coagulation disorders: Fibrinogen, factor VII and factor XIII. *Haemophilia*, *22*(Suppl. 5), 61–65.

Mosesson, M. W. (2005). Fibrinogen and fibrin structure and functions. *J Thromb Haemost*, *3*(8), 1894–1904.

Palla, R., Peyvandi, F., & Shapiro, A. D. (2015). Rare bleeding disorders: Diagnosis and treatment. *Blood*, *125*(13), 2052–2061.

CHAPTER 116

Factor XIII, α_2-Antiplasmin, and Plasminogen Activator Inhibitor-1 Deficiencies

Glaivy Batsuli, MD and Shannon L. Meeks, MD

Factor XIII (FXIII), α_2-antiplasmin (α_2-AP), and plasminogen activator inhibitor type 1 (PAI-1) deficiencies are all very rare bleeding disorders. These proteins play a critical role in stabilizing a fibrin clot (FXIII) and regulating fibrinolysis through the inhibition of plasmin (α_2-AP) or the inhibition of plasminogen conversion to plasmin (PAI-1). FXIII affects 1 in 2 to 3 million individuals worldwide; the true incidence of α_2-AP and PAI-1 deficiencies is unknown.

Pathophysiology: FXIII circulates in plasma as a transglutaminase heterotetramer that consists of two catalytic A subunits and two carrier B subunits. The A subunit homodimers are present in platelets and monocytes; the B subunits are synthesized in the liver and circulate in excess of the A subunit. FXIII has a half-life of 9–10 days.

α_2-AP is a serine protease inhibitor that is synthesized in the liver. It circulates in the plasma in free form or bound to plasminogen. α_2-AP has a plasma concentration of 0.7 mg/mL and a half-life of 2–6 days.

PAI-1 is a serine protease inhibitor synthesized and released from hepatocytes, endothelial cells, adipocytes, and megakaryocytes. Although the majority of PAI-1 exists in an active form, there is a portion of PAI-1 that exists in an inactive (or latent) form that can be activated with denaturants or phospholipids in vitro. PAI-1 has an exceptionally short half-life of 10 minutes.

Role in Coagulation: Formation of a fibrin clot followed by fibrinolysis for repair of damaged endothelium after vessel injury is tightly regulated pathways that must be balanced to avoid pathologic thrombosis from excess clot formation and bleeding from excess clot degradation. Thrombin generated during the activation of the coagulation cascade catalyzes the conversion of fibrinogen to fibrin and the activation of FXIII to FXIIIa that exposes the A subunits and releases the B subunits. In turn, FXIIIa cross-links fibrin to α_2-AP and enables incorporation of antifibrinolytic proteins within fibrin to form a stable clot that is resistant to early degradation by fibrinolytic proteins. FXIII/FXIIIa also cross-links fibrinogen, fibronectin, and collagen to form a clot resistant to fibrinolysis. Additionally, FXIII/FXIIIa plays an important role in wound healing, tissue repair, and embryonic implantation.

After the formation of the fibrin clot, the zymogen plasminogen is able to bind to the fibrin surface. Plasminogen is proteolytically cleaved to form plasmin, which is the primary protease responsible for the initiation of fibrinolysis. Plasmin formation is catalyzed by tissue-plasminogen activator (t-PA). α_2-AP is a competitive inhibitor of plasminogen and prevents its absorption within fibrin. When cross-linked to fibrin, α_2-AP

Transfusion Medicine and Hemostasis. https://doi.org/10.1016/B978-0-12-813726-0.00116-1

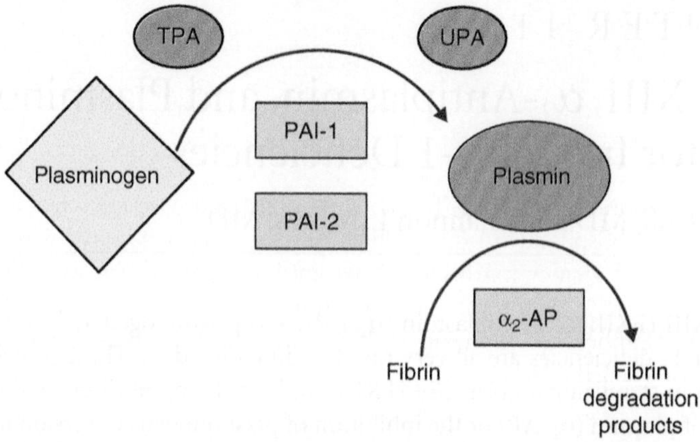

FIGURE 116.1 Schematic representation of the fibrinolytic system. *PAI-1 and PAI-2*, plasminogen activator inhibitor-1 and -2; *t-PA*, tissue-type plasminogen activator; *UPA*, urokinase-type plasminogen activator; *α_2-AP*, α_2-antiplasmin, ▭, inhibitors; ⬬, activators. (From Hoffman, R., Benz, E. J., Shattil, S. J., et al. (Eds.). (2005). *Hematology: Basic principles and practice* (4th ed.). Philadelphia, PA: Elsevier.)

can inactivate free plasmin and induce increased fibrin resistance to local plasmin via FXIIIa. The plasminogen activator t-PA is also inhibited by α_2-AP. PAI-1 is the main inhibitor of t-PA (Fig. 116.1).

Genetics: FXIII deficiency is an autosomal recessive disorder. It is caused by defects in either the *F13A* gene that encodes the A subunit on chromosome 6 or the *F13B* gene that encodes the B subunit on chromosome 1. The majority of disease-causing mutations are missense mutations in the A subunit. Generally, patients with severely low FXIII levels have homozygous or compound heterozygous mutations, whereas patients with levels approaching 30 IU/dL tend to have heterogyous mutations.

The α_2-AP and PAI-1 genes are located on chromosomes 17 and 7, respectively. Both deficiencies are inherited in an autosomal recessive manner. Homozygous and compound heterozygous mutations in the PAI-1 gene associated with bleeding have been described. There is a slightly increased incidence of PAI-1 deficiency reported in Amish populations of Eastern and Southern Indiana.

Clinical Manifestations

FXIII Deficiency: Individuals with FXIII deficiency may present with easy bruising, mucocutaneous bleeding, impaired wound healing, umbilical cord bleeding, postsurgical or trauma-induced bleeding, menorrhagia, recurrent spontaneous miscarriages, and spontaneous intracranial hemorrhage. The majority of neonates with FXIII deficiency will have a history of umbilical cord bleeding. There is a strong correlation between bleeding symptoms and FXIII activity levels. Patients with FXIII activity levels <1 IU/dL tend to have a severe phenotype and severe bleeding episodes, whereas those with FXIII levels >30 IU/dL are usually asymptomatic.

α_2-AP Deficiency: Heterozygous individuals with α_2-AP deficiency may be asymptomatic or have mild to moderate bleeding symptoms that can manifest as delayed bleeding after trauma, dental procedures, or invasive surgery. However, individuals with homozygous α_2-AP deficiencies may present with severe bleeding symptoms similar to classical hemophilia, intramedullary hematomas, or umbilical cord bleeding.

PAI-1 Deficiency: Although the association between elevated PAI-1 levels and increased risk of arterial thrombosis has been well established, reports of PAI-1 deficiency and a bleeding diathesis remain a very rare phenomenon. Bleeding symptoms in PAI-1 deficiency are similar to those with α_2-AP deficiency; however, spontaneous bleeding is not common. Complete absence of PAI-1 has been associated with cardiac fibrosis.

Diagnosis: Routine hemostasis screening laboratories in these disorders are usually normal, including prothrombin time, activated partial thromboplastin time, and thrombin time. A high index of suspicion is necessary to obtain additional studies that would lead to a diagnosis. Studies may show decreased plasminogen activity or a shortened euglobulin lysis time. Rapid clot dissolution in 5M urea or 1% monochloroacetic acid solution in the clot solubility test can be used as a qualitative screening test in patients with severe FXIII deficiency. Characterizing the FXIII molecular defect is important for determining whether an A or B subunit defect is present due to the potential treatment implications.

α_2-AP deficiency can be detected via specific activity and antigen assays that can help differentiate a type 1 quantitative defect versus a type 2 qualitative defect. Diagnosis of PAI-1 deficiencies can be challenging because of limitations in the PAI-1 activity assays. The current assays are primarily designed to detect elevated PAI-1 activity, and the normal reference ranges often include levels down to zero. Immunoassays detecting PAI-1 antigen levels are commercially available. The diagnosis should be considered in a patient with a shortened euglobulin lysis time, PAI-1 activity level <1 IU/dL, and a clinical history of bleeding.

Differential Diagnosis: Acquired FXIII deficiency has been described in inflammatory bowel disease, Henoch–Schönlein Purpura, and patients who have undergone cardiac surgery. Acquired α_2-AP deficiency can be seen in end-stage liver disease, renal disease, and disseminated intravascular coagulation. Acquired deficiency of PAI-1 is rare.

Management

Factor XIII Deficiency: There are two FXIII products currently available for patients with FXIII deficiency. A plasma-derived product, Fibrogammin/Corifact (Europe and the United States, respectively, CSL Behring), can be used in patients with A or B subunit defects. Every 28-day dosing at 40 units/kg is recommended for long-term prophylaxis in patients with severe deficiency or a bleeding history. The recombinant FXIII-A subunit product, NovoThirteen/Tretten (Europe and the United States, respectively,

Novo Nordisk), only contains the A subunit and is contraindicated in patients with FXIII-B subunit deficiency. For long-term prophylaxis, infusion of 35 units/kg every 28 days is recommended. A target FXIII activity trough >15 IU/dL is suggested for prevention of severe bleeding. Fresh-frozen plasma (FFP) and cryoprecipitate are alternative sources of FXIII, if FXIII concentrates are unavailable. FFP contains larger amounts of FXIII per volume compared with cryoprecipitate, but its use may be limited in cases where fluid overload is a concern. Antifibrinolytic agents are useful in cases of minor or mucosal bleeding.

α_2-AP and PAI-1 Deficiencies: The primary treatment modality for bleeding control and prevention in α_2-AP and PAI-1 deficiencies is the use of antifibrinolytic agents such as tranexamic acid and ε-aminocaproic acid given either orally or intravenously. These agents compete for binding to lysine residues on plasminogen and prevent binding of plasminogen to fibrin thus inhibiting fibrin degradation. Antifibrinolytics and hormonal agents are useful for menorrhagia. In rare cases, FFP has been used in the preoperative management of these deficiencies.

Further Reading

Carpenter, S. L., & Mathew, P. (2008). Alpha2-antiplasmin and its deficiency: Fibrinolysis out of balance. *Haemophilia, 14*(6), 1250–1254.

Caudill, J. S., Nichols, W. L., Plumhoff, E. A., Schulte, S. L., Winters, J. L., Gastineau, D. A., et al. (2009). Comparison of coagulation factor XIII content and concentration in cryoprecipitate and fresh-frozen plasma. *Transfusion (Paris), 49*(4), 765–770.

Heiman, M., Gupta, S., Khan, S. S., Vaughan, D. E., & Shapiro, A. D. (1993). Complete plasminogen activator inhibitor 1 deficiency. In M. P. Adam, H. H. Ardinger, R. A. Pagon, S. E. Wallace, L. J. H. Bean, H. C. Mefford, et al. (Eds.), *GeneReviews*. Seattle, (WA): University of Washington, Seattle. University of Washington, Seattle. GeneReviews is a registered trademark of the University of Washington, Seattle. All rights reserved.

Iwaki, T., Tanaka, A., Miyawaki, Y., Suzuki, A., Kobayashi, T., Takamatsu, J., et al. (2011). Life-threatening hemorrhage and prolonged wound healing are remarkable phenotypes manifested by complete plasminogen activator inhibitor-1 deficiency in humans. *J Thromb Haemost, 9*(6), 1200–1206.

Mehta, R., & Shapiro, A. D. (2008). Plasminogen activator inhibitor type 1 deficiency. *Haemophilia, 14*(6), 1255–1260.

Menegatti, M., Palla, R., Boscarino, M., Bucciarelli, P., Muszbek, L., Katona, E., et al. (2017). Minimal factor XIII activity level to prevent major spontaneous bleeds. *J Thromb Haemost, 15*(9), 1728–1736.

de Moerloose, P., Schved, J. F., & Nugent, D. (2016). Rare coagulation disorders: Fibrinogen, factor VII and factor XIII. *Haemophilia, 22*(Suppl. 5), 61–65.

Mumford, A. D., Ackroyd, S., Alikhan, R., Bowles, L., Chowdary, P., Grainger, J., et al. (2014). Guideline for the diagnosis and management of the rare coagulation disorders: A United Kingdom Haemophilia Centre Doctors' Organization guideline on behalf of the British Committee for Standards in Haematology. *Br J Haematol, 167*(3), 304–326.

CHAPTER 117

Bleeding Disorders in Pregnancy

Christine L. Kempton, MD, MSc and Kalinda Woods, MD

Postpartum hemorrhage (PPH) is a leading cause of maternal death worldwide. Although the majority of PPH occurs as a result of uterine atony or other anatomical defect, the risk is increased when a congenital or acquired hemostatic defect is present. PPH can be categorized as primary, if there is abnormal bleeding during the first 24 hours (>500 or >1000 mL following vaginal or Caesarean delivery, respectively, or results in hemodynamic instability), and secondary, if abnormal bleeding from 24 hours until 6 weeks postpartum. This chapter will review the more common congenital and acquired hemostatic disorders affecting pregnancy.

Congenital Disorders

Preconception and Prenatal Counseling: Management of pregnancy in women with congenital bleeding disorders is best started before conception. Preconception counseling allows for discussion with a hematologist, obstetrician, and genetic counselor regarding risks to the mother and fetus. Strategies to reduce the risks, including prenatal testing, can be reviewed. Ultimately a treatment plan for delivery should be developed by a hematologist and shared with the patient, obstetrician, and anesthesiologist involved in the patient's care. The location of delivery will depend in part on the severity of the women's bleeding disorder and the severity of the known or potential bleeding disorder in the fetus.

von Willebrand Disease: During pregnancy, von Willebrand factor (VWF) and factor VIII (FVIII) levels begin to increase in the 6th week, peak between the 29th and 35th weeks of gestation, and return to near baseline within 7–10 days following delivery. Because of this rise, most patients with von Willebrand disease (VWD) will not have significant bleeding during the antepartum period; however, women with more severe disease, who experience little or no increase in their VWF and FVIII levels, may have bleeding complications that require active management (see Chapter 109). It is unknown whether women with VWD are more likely to miscarry compared with women without VWD, however, there is observational evidence to suspect this may be the case. In the event of miscarriage, bleeding can be severe and treatment to raise VWF levels is appropriate.

Management of Labor and Delivery: Because FVIII and VWF increase during pregnancy in patients with type 1 and some type 2 VWD, FVIII, ristocetin cofactor activity (VWF:RCo) and VWF antigen (VWF:Ag), levels should be checked at approximately 34–36 weeks of gestation to plan appropriately for delivery. Both the FVIII and VWF:RCo should be ≥50% at the time of delivery and for neuraxial anesthesia. Most patients with mild type 1 disease will have a rise of FVIII and VWF:RCo above this

Transfusion Medicine and Hemostasis. https://doi.org/10.1016/B978-0-12-813726-0.00117-3

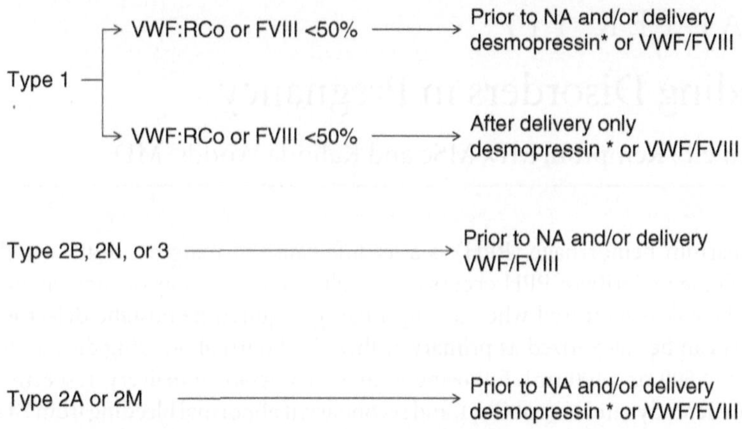

FIGURE 117.1 Treatment of von Willebrand disease during labor and delivery. *For use in patients known to respond to DDAVP. *FVIII*, factor VIII; *NA*, neuraxial anesthesia; *VWF:RCo*, von Willebrand factor ristocetin cofactor activity.

threshold naturally during pregnancy, and treatment can be deferred until after delivery, when levels will begin to return to baseline (Fig. 117.1). In patients with third trimester FVIII and VWF:RCo levels that are <50%, treatment should be initiated before neuraxial anesthesia and delivery. If with treatment, levels ≥50% are achieved, then neuraxial anesthesia can be used safely. Neuraxial anesthesia without normalization of FVIII and VWF:RCo puts the patient at risk of epidural hematoma and irreversible paralysis. In many patients with type 2 and all patients with type 3 VWD, VWF and FVIII levels are not expected to appreciably change during pregnancy; therefore, most pregnant women with types 2 or 3 VWD will need treatment before neuraxial anesthesia and delivery. The best VWF and FVIII levels to target are a matter of debate. Some argue that treatment should mimic the typical supraphysiological levels of pregnancy; however, the benefit of this approach over a VWF/FVIII target of 100% has not been studied in a prospective fashion.

In all patients, treatment should maintain FVIII levels >50% for 3–4 days following vaginal delivery and 4–5 days following Caesarean section, or until abnormal bleeding has ceased. Some centers may be able to monitor their patient's FVIII and VWF:RCo activity following delivery, allowing therapy to be withheld in patients with FVIII and VWF:RCo activity ≥50% until levels reach <50%; however, many hospitals are not able to perform this type of testing in real time and therefore empiric treatment should be employed. Although peripartum treatment reduces the risk of early PPH, it may occur up to 2–3 weeks after delivery. Accordingly, postpartum treatment plans should include strategies to promote identification and management of delayed bleeding.

Options for treatment to raise VWF levels include desmopressin (1-deamino-8-arginine-vasopressin, DDAVP) or VWF concentrates (see Chapter 41). Desmopressin can be given intranasally (1.5 mg/mL, total dose equals 300 μg) or intravenously (0.3 μg/kg). Desmopressin increases FVIII and VWF in most patients with type 1 VWD. Levels typically increase by about threefold, but not all patients respond adequately. Therefore, it is preferable to have confirmed a response to desmopressin at both 1 and 4 hours after administration before its use at the time of a hemostatic challenge. Tachyphylaxis to this drug occurs, so test administration should be done at least several weeks before its intended use. Because hyponatremia is a potential adverse event with desmopressin, fluid intake should be reduced to 75% of normal and drinks with higher NaCl content should replace free water for 24 hours after its use. Desmopressin does not cross the placenta in significant amounts; therefore, direct effects on the fetus are not of concern. There is concern regarding neonatal hyponatremia if the mother develops hyponatremia before delivery. For patients who do not respond or are intolerant to desmopressin, or have disease types where desmopressin is not effective, i.e., type 3 and 2B, VWF concentrates can be used to target trough VWF:RCo and FVIII levels of 40%–50%.

Management of Bleeding and Breastfeeding: Desmopressin is minimally excreted into breast milk and poorly absorbed by infants; therefore, it is thought to be generally safe for use while breastfeeding. Concentrations of tranexamic acid found in the breast milk are low (1% of maternal serum concentrations) and are thought to be a safe alternative for use while breastfeeding. There are no data on excretion of aminocaproic acid in breast milk. VWF concentrates are safe for use while breastfeeding.

Carriers of Hemophilia A or B: In general, no therapy is needed for women with baseline FVIII or factor IX (FIX) levels ≥50%. For those with levels ≤50% at the time of delivery, treatment is necessary before neuraxial anesthesia and delivery. Women with baseline FVIII levels <50% that rise to >50% at the end of pregnancy do not need treatment before neuraxial anesthesia and delivery but will need postpartum treatment as their FVIII levels will decline after deliver and return to baseline ~1 week postpartum. Treatment should maintain FVIII and FIX levels ≥50% for 3–4 days postpartum after vaginal delivery and 4–5 days postpartum after Caesarean section. Treatment options to raise FVIII levels in women with FVIII deficiency include desmopressin and FVIII products. Treatment of FIX carriers is with FIX products only (see Chapter 41). Tranexamic acid can be used as an adjunct to desmopressin and factor replacement therapy.

Mode of Delivery in Carriers of Severe Hemophilia: For women who are pregnant with a child who is known or could possibly have severe hemophilia A, the risk of intracranial hemorrhage (ICH) is approximately 1%–2% when born via vaginal delivery or unscheduled Caesarean section. This increases to up to 8% in women who are not known to be carriers, likely due to use of assistive devices during delivery. The risk of ICH is reduced, though never zero, when an elective Caesarean section is used for delivery. It is important for the patient and obstetrician to discuss the risks and benefits of elective Caesarean section.

Factor XI Deficiency: Only 30% of a cohort of pregnant women with severe factor XI (FXI) deficiency experienced PPH. PPH was more likely to occur if there was a past history of surgical bleeding. Accordingly, the decision to utilize prophylaxis during labor and delivery should be based on the baseline FXI level, history of bleeding, and planned mode of delivery. In women with FXI deficiency and a history of bleeding, or with FXI levels ≤15% and undergoing Caesarean section, the goal of therapy, as with major surgery, is to maintain an FXI level of ≥30%–50% using plasma product transfusion, ~15 mL/kg, or FXI concentrates, if available. As the half-life of FXI is

≥50 hours, additional plasma products can be transfused every 48 hours to maintain FXI levels above 40% for 3–4 days following a vaginal delivery or 4–5 days following a Caesarean section. For women without a history of bleeding despite severely reduced FXI levels who are undergoing vaginal delivery, hemostatic treatment can be withheld until there is evidence of clinical bleeding. However, before neuraxial anesthesia, prophylactic plasma product transfusion is required in women with severe FXI deficiency (<15%) even in the absence of a history of bleeding. Recombinant factor VIIa (rFVIIa), as reported in small case series, has been effectively used for management of women with FXI deficiency undergoing elective Caesarean section and neuraxial anesthesia. Tranexamic acid has also been used for up to 2 weeks postpartum to reduce PPH.

Platelet Function Defects: Management of platelet function defects at the time of labor and delivery should be based on the history of bleeding, patient preferences regarding neuraxial anesthesia, and the response to desmopressin during other hemostatic challenges. Decisions regarding management in pregnancy are similar to those during major surgery (see Chapters 96–99). In general, options for treatment include desmopressin, platelet transfusions, and tranexamic acid. In many cases where the platelet defect is mild, desmopressin alone will be sufficient. In more severe cases, such as Glanzmann's thrombasthenia or Bernard–Soulier patients who have received prior platelet transfusions, the risk of alloimmunization is increased. The pregnant mother may become sensitized to fetal platelet antigens inherited from the father. If this occurs, the mother may become refractory to platelet transfusions and the antibodies may cross the placenta causing fetal/neonatal alloimmune thrombocytopenia.

Other Inherited Defects: Deficiencies of fibrinogen and factor XIII (FXIII) may result in recurrent pregnancy loss and peripartum hemorrhage (see Chapters 115 and 116). Early and regular replacement of fibrinogen or FXIII is necessary for a successful pregnancy. Plasma-derived fibrinogen concentrates are available (see Chapter 41); however, they have not been formally tested in pregnant women. Nonetheless, fibrinogen concentrates are preferred over cryoprecipitate as they are heat-treated and therefore considered safer for repeated infusions. During pregnancy, it has been suggested to maintain the fibrinogen level at >60 mg/dL and increase to >150 mg/dL at the time of labor and delivery. FXIII concentrate is preferred over plasma products and cryoprecipitate for pregnant women with FXIII deficiency. Regular infusions are best started early in pregnancy and continued to target a trough FXIII level >10 IU/dL during pregnancy and increased to >20 IU/dL during labor and delivery.

Factor VII (FVII) levels rise during normal pregnancy and levels of 15%–25% are considered adequate for hemostasis (see Chapter 113). rFVIIa, 15–30 μg/kg, should be used for those women with FVII deficiencies who have persistently low FVII levels after reaching the third trimester (see Chapter 41). A first dose can be given before neuraxial anesthesia and repeated every 4–6 hours, depending on the patient's bleeding history and severity of PPH. For many patients, a single dose will suffice. Tranexamic acid can be used for management of heavy PPH.

Delivery of a Fetus With a Potential Bleeding Disorder: At the time of delivery, if prenatal testing has not excluded a bleeding disorder in the fetus, then the fetus should be handled as though one is present. Specifically, assisted deliveries using forceps and vacuum extraction should be avoided. The use of forceps and vacuum extraction increases the risk of ICH and cephalohematoma. Additionally, scalp-monitoring electrodes should not be used. Medications and immunizations should be delivered subcutaneously rather than intramuscularly, and circumcision should be avoided or delayed until adequate workup can be performed.

To detect an ICH at the earliest state after vaginal or attempted vaginal delivery, an ultrasound of the head should be performed within 24–48 hours of birth of an infant that might have a severe bleeding disorder (i.e., severe hemophilia, type 3 VWD, FXIII deficiency, etc.). If a hemorrhage is detected, this is a potential medical emergency, and a pediatric hematologist should be consulted immediately. Factor replacement should be given at the direction of a pediatric hematologist.

Acquired Disorders

Acquired Factor VIII Inhibitors: FVIII inhibitors occur rarely in pregnancy. They occur most commonly within the first 3 months postpartum, and only ~5% antepartum. Bleeding is variable, ranging from asymptomatic to fatal (mortality 5.6%). Although pregnancy-related FVIII inhibitors often disappear spontaneously, resolution may take a year or longer. Corticosteroid therapy should be considered in all patients to facilitate inhibitor eradication and reduce the risk of bleeding complications. Cyclophosphamide and other therapies can be reserved for those who fail to respond adequately to corticosteroids (see Chapter 126).

Thrombocytopenia: When evaluating gravid women with thrombocytopenia, the same general approach is utilized as in nongravid patients. However, several additional disorders and treatment options should be considered.

Gestational Thrombocytopenia: Gestational thrombocytopenia is the most common cause of thrombocytopenia in pregnancy, occurring in 5%–8% of all pregnancies. It is typically mild (platelet count ≥70,000/μL), occurs after the mid-second trimester, resolves following delivery, does not affect the fetus, and occurs in women without a history of thrombocytopenia (excluding a history of gestational thrombocytopenia). No treatment is needed.

Immune Thrombocytopenia: Immune thrombocytopenia (ITP) can lead to more profound thrombocytopenia and occur earlier in pregnancy than is seen with gestational thrombocytopenia. The platelet count may progressively decline during pregnancy, reaching a nadir in the third trimester. A platelet count ≥50,000/μL is generally considered safe for vaginal and Caesarean delivery, but some physicians prefer a platelet count ≥80,000/μL for neuraxial anesthesia. These thresholds are based on limited observational data and the most recent version of the American Society of Hematology guidelines does not endorse a specific platelet count for labor and delivery.

Treatment of Immune Thrombocytopenia in Pregnancy: If treatment is required, options include intravenous immunoglobulin (IVIG) (0.4 g/kg per day for 5 days or 1 g/kg per day for 2 days) or corticosteroids; however, IVIG is generally preferred over corticosteroids, particularly when longer durations of therapy are required.

With longer corticosteroid exposure, there are increased risks of hypertension, hyperglycemia, osteoporosis, and premature rupture of the membranes. Prednisone, prednisolone, or methylprednisolone are favored over dexamethasone because the former are well metabolized by the placenta, exposing the fetus to only 10% of the maternal dosage. Regardless of the specific corticosteroid used, one should target the lowest possible dosage for the shortest duration. Splenectomy has been successfully performed during pregnancy and can be considered in severe and refractory ITP. The second trimester is the optimal time to perform this intervention because it can frequently lead to miscarriage or premature delivery in the other trimesters. Immunosuppressive agents such as cyclophosphamide, vincristine, and danazol are contraindicated in pregnancy. Azathioprine has been used in women following organ transplantation; however, it remains category D because of the potential for fetal harm. A number of case reports have detailed the safe use of rituximab for treatment of lymphoma during pregnancy. The infants in these cases have had abnormal B-cell development that normalized after 6 months without reported infectious complications. Rituximab for treatment of ITP during pregnancy is limited even further to a few case reports. Given the limited experience and likely bias to report positive outcomes, rituximab is generally not recommended for treatment of ITP during pregnancy. Thrombopoietic agents are increasingly used for management of ITP in nonpregnant patients; however, it is considered category C. Despite several published case reports of successful use, more clinical studies are needed before their routine use in pregnancy.

Management of the Fetus or Neonate: Maternal antibodies can affect the fetal platelet count, leading to thrombocytopenia in the newborn. The maternal platelet count does not predict the infant's platelet count, but splenectomy and a history of neonatal thrombocytopenia at delivery in a prior sibling are likely risk factors for neonatal thrombocytopenia. Because of the lack of reliability of platelet counts obtained through fetal scalp vein sampling and the risk associated with percutaneous umbilical blood sampling, fetal platelet count monitoring is not recommended. Despite the potential for thrombocytopenia in up to 20% of infants, platelet counts <20,000/μL occur in 4% and ICH occurs in ≤1.5% of infants born to mothers with ITP. The likelihood of ICH or other fetal complications is not influenced by the method of delivery; therefore, obstetric indications alone should determine the method of delivery.

The neonate's platelet count should be measured from a cord blood sample at delivery and serially in those that are thrombocytopenic, as the nadir may not occur until 2–5 days following delivery. It is generally recommended that infants with a platelet count ≤20,000/μL or those with hemorrhage receive treatment (IVIG or corticosteroids). Some experts recommend imaging the brain of infants with platelet counts ≤50,000/μL to exclude occult ICH.

Disseminated Intravascular Coagulation: Disseminated intravascular coagulation (DIC) is associated with major obstetrical complications such as placental abruption, amniotic fluid emboli, retained fetus syndrome, and eclampsia. The diagnosis and treatment of DIC in pregnancy is similar to that in the nonpregnant patient (see Chapter 124).

Hemolysis, Elevated Liver Enzymes, and Low Platelets: This is a syndrome that is part of a spectrum associated with preeclampsia and eclampsia. Mild thrombocytopenia may precede other signs and symptoms, such as right upper quadrant pain and elevated liver transaminases. There are no standard laboratory values to use as diagnostic criteria. Prompt delivery is the treatment of choice. Clinical improvement is anticipated soon after delivery and if is not realized 72 hours after delivery, alternative diagnosis such as thrombotic thrombocytopenia purpura (TTP) and atypical hemolytic uremic syndrome should be considered.

Acute Fatty Liver of Pregnancy: Acute fatty liver of pregnancy is a rare disorder, with liver failure dominating the clinical picture. Thrombocytopenia is typically mild, and hemolysis is usually not present unless DIC has developed. Delivery is the mainstay of management along with supportive care.

Thrombotic Thrombocytopenia Purpura: TTP in pregnancy has been reported to occur at any period during gestation but most commonly later in pregnancy or after delivery. Signs and symptoms are similar to those of TTP not associated with pregnancy (see Chapter 107). In contrast to the other pregnancy-related disorders listed above, the clinical course of TTP is not altered by delivery. The presence of preeclampsia, right upper quadrant pain, and elevated liver enzymes may point toward other pregnancy-related microangiopathies, although there may be times that this distinction is difficult and the response to therapy, delivery, and/or plasma exchange may be the only means to distinguish between these disorders.

Subsequent pregnancies carry a risk of recurrent TTP. In the Oklahoma TTP/HUS registry, 19 women with a history of TTP became pregnant and ~25% of these women developed recurrent TTP. Of those with a history of pregnancy/postpartum-related TTP, 18% had recurrent TTP during a subsequent pregnancy. If the cause of TTP was idiopathic, it was more likely for these women to develop TTP during a subsequent pregnancy (43%). There were no maternal deaths associated with pregnancy. However, the pregnancies were often complicated and fetal death occurred in 40%, although not all fetal deaths appeared related to maternal TTP.

Further Reading

Asahina, T., Kobayashi, T., Takeuchi, K., & Kanayama, N. (2007). Congenital blood coagulation factor XIII deficiency and successful deliveries: A review of the literature. *Obstet Gynecol Surv*, *62*, 255–260.

Cines, D. B., & Levine, L. D. (2017). Thrombocytopenia in pregnancy. *Blood*. https://doi.org/10.1182/blood-2017-05-781971.

Demers, C., Derzko, C., David, M., & Douglas, J. (2006). Gynaecological and obstetric management of women with inherited bleeding disorders. *Int J Gynaecol Obstet*, *95*, 75–87.

Hauser, I., Schneider, B., & Lechner, K. (1995). Post-partum factor VIII inhibitors. A review of the literature with special reference to the value of steroid and immunosuppressive treatment. *Thromb Haemost*, *73*, 1–5.

Kujovich, J. L. (2005). Von Willebrand disease and pregnancy. *J Thromb Haemost*, *3*, 246–253.

Kulkarni, A. A., Lee, C. A., & Kadir, R. A. (2006). Pregnancy in women with congenital Factor VII deficiency. *Haemophilia, 12*, 413–416.

Mannucci, P. M. (2005). Use of desmopressin (DDAVP) during early pregnancy in factor VIII deficient women. *Blood, 105*, 3382.

Neunert, C., Lim, W., Crowther, M., et al. (2011). The American Society of Hematology 2011 evidence-based practice guideline for immune thrombocytopenia. *Blood, 117*, 4190–4207.

Salomon, O., Steinberg, D. M., Tamarin, I., et al. (2005). Plasma replacement therapy during labor is not mandatory for women with severe factor XI deficiency. *Blood Coagul Fibrinolysis, 16*, 37–41.

Stavrou, E., & McCraie, K. R. (2009). Immune thrombocytopenia in pregnancy. *Hematol Oncol Clin N Am, 23*, 1299–1316.

Vesely, S. K., Li, X., McMinn, J. R., et al. (2004). Pregnancy outcomes after recovery from thrombotic thrombocytopenic purpura-hemolytic uremic syndrome. *Transfusion, 44*, 1149–1158.

CHAPTER 118

Vascular Bleeding Disorders

Patricia E. Zerra, MD and Michael A. Briones, DO

Hereditary Vascular Malformations

Hereditary Hemorrhagic Telangiectasia (Osler–Weber–Rendu Disease): Hereditary hemorrhagic telangiectasia (HHT) is a systemic autosomal dominant (AD) vascular disease that is characterized by mucocutaneous telangiectasias and multiple arteriovenous malformations (AVMs). It is the most common inherited vascular bleeding disorder with an estimated incidence of 1 in 5000–10,000 people and affects all ages, races, and sexes equally.

There are three identified pathologic gene variants that cause HHT, which are thought to play a role in regulating angiogenesis. The most common mutation (HHT type 1), is of the *ENG* gene on chromosome 9q for the glycoprotein endoglin. HHT type 2 is due to variants in *ACVRL1* on chromosome 12 encoding activin receptorlike kinase 1. Roughly, 1% of cases are due to mutation in the *MADH4* gene (*SMAD4*), resulting in a combined syndrome of HHT and juvenile polyposis.

HHT has a wide variety of clinical manifestations. Telangiectasias (dilated blood vessels with thin fragile walls and defective endothelial cell junctions, leading to rupture and bleeding) are present at birth, but enlarge and become visible in childhood. Bleeding occurs spontaneously or after minor trauma and may be prolonged due to defective vessel contraction. Slow progressive bleeding from epistaxis (the most common manifestation) or lesions in the gastrointestinal (GI) tract can lead to iron deficiency anemia. Up to half of patients have multiple AVMs in the lungs, causing decreased oxygenation and/or hemorrhage. Telangiectasis in the liver and central nervous system can increase mortality. In addition to hemorrhagic complications, some patients may suffer from pulmonary hypertension or thromboembolic events.

Diagnosis is primarily clinical; however, molecular genetic testing is available. Platelet counts and coagulation studies are normal. Management is primarily symptomatic, although recurrent GI or pulmonary bleeding may require surgical intervention and/or interventional radiology interventions. In some instances, antifibrinolytic therapy may be beneficial.

Hereditary Connective Tissue Disorders

Ehlers–Danlos Syndromes: Ehlers–Danlos syndrome (EDS) is a clinically heterogeneous group of connective tissue disorders with abnormal and decreased synthesis of subendothelial collagen. The synthesis of collagen is very complex, and an abnormality in any one of 20 genes may result in decreased synthesis of collagen or production of collagen with a variant amino acid sequence, resulting in the EDS phenotype. At least 11 syndromes are included in EDS based on clinical characteristics and inheritance

Transfusion Medicine and Hemostasis. https://doi.org/10.1016/B978-0-12-813726-0.00118-5

pattern. Most EDS subtypes are caused by mutations in genes encoding the fibrillar collagens type I, III, and V, or in genes coding for enzymes involved in the posttranslational modification of these collagens.

The principal clinical features in EDS are skin hyperextensibility, delayed wound healing with atrophic scarring, joint hypermobility, easy bruising, and generalized connective tissue fragility. Cutaneous findings include thin skin and a tendency to develop nonpalpable purpuric lesions as a result of dermal blood vessel fragility.

There are six recognized subtypes of EDS based on clinical presentation, inheritance pattern, and molecular defects; the classic, hypermobility, and vascular subtypes are most common. In the classic subtype (EDS type I/II), inherited in an AD pattern, bruising is accompanied by soft, fragile, hyperextensible skin that splits easily after minor trauma. Wound healing is delayed, and scars have a papyraceous or "cigarette-paper-like" aspect. In hypermobility EDS (type III), also AD, patients have joint and spine hypermobility with frequent joint dislocations, smooth skin, and autonomic dysfunction. Vascular-type EDS (type IV or arterial-ecchymotic type) is an AD disorder caused by structural defects type III collagen, encoded by the *COL3A1* gene, causing vascular fragility of both large and small arteries, veins, and capillaries. Skin is not hyperelastic, but rather thin and translucent, showing a visible venous pattern over the chest, abdomen, and extremities. Excessive bruising is common along with gum bleeding and profuse bleeding after tooth extraction. Bleeding can present as acute abdominal pain, cerebral stroke, hemoptysis, hematemesis, renal colic and hematuria, retroperitoneal bleeding, muscular swelling, shock, or sudden death. This subtype has the worst prognosis due to potentially fatal vascular and GI complications.

Diagnosis of EDS is clinical although molecular genetic testing for some EDS subtypes is available.

Pseudoxanthoma Elasticum: Pseudoxanthoma elasticum, an autosomal recessive disorder, is characterized by mineralization and fragmentation of elastin in the skin, retina, and blood vessels. It is caused by mutations in the *ABCC6* gene on chromosome 16.

Mineralization can affect elastic fibers in the skin, eyes, blood vessels, and, less frequently, the digestive tract. Characteristic skin findings may be subtle in children; however, 80% of patients develop progressive skin lesions by age 20. Affected individuals may develop yellow to orange papules with a pebbly appearance called pseudoxanthomas on their necks, underarms, and other flexor areas. They may also have eye abnormalities, such as changes in the pigmented cells of the retina known as peau d'orange or angioid streaks, caused by tiny breaks Bruch's membrane underlying the retina. Bleeding and scarring of the retina may also occur, causing vision loss. Mineralization of the arteries may cause accelerated arteriosclerosis. Rarely, bleeding from blood vessels in the digestive tract may occur.

Acquired Vascular Disorders

Purpura Simplex (Idiopathic Purpura): Purpura simplex is spontaneous easy bruising on the trunk and the lower extremities that typically occurs following minor trauma. The disorder, usually affecting women, is likely secondary to increased fragility of skin vessels, and is one of the most frequent reasons for increased bruising.

If investigation does not reveal additional bleeding history, or a family history of bleeding, and screening coagulation laboratory tests are normal, the patient should be reassured and advised to avoid aspirin or other medications causing platelet dysfunction. Serious bleeding does not occur.

Senile Purpura: Senile purpura typically affects elderly patients as dermal tissues atrophy with progressive loss of collagen in skin and blood vessels, causing them to become more fragile. Usually occurring in sun-exposed areas on the dorsal side of the hands and wrists or on the forearms, these lesions may be large and well demarcated. Venipuncture may result in rapidly spreading purpura, and thus prolonged pressure should be applied after venipuncture in elderly patients. No treatment hastens lesion resolution and although cosmetically displeasing, the disorder has no health consequences.

Purpura Due to Infections: Infections may cause purpura and mucous membrane bleeding through direct vessel damage by microorganisms, toxins, vasculitis, thrombocytopenia, or disseminated intravascular coagulopathy (DIC). DIC from infections, especially meningococcal disease, may be associated with renal failure, rapidly spreading purpura with skin necrosis, and circulatory failure. Treatment of the underlying infection is critical, with supportive care for the coagulopathy. Heparin treatment to counteract excessive thrombin generation may be helpful in rare instances if there is associated thrombosis, although the risk of life-threatening hemorrhage is extremely high with anticoagulant administration.

Scurvy: Scurvy is a result of insufficient dietary intake of vitamin C, which is required for collagen synthesis. Clinical manifestations include mucous membrane bleeding, orbital hemorrhage, and subperiosteal hemorrhage (especially in infants). Rarely, hematuria, hematochezia, and melena are noted; costochondral beading (scorbutic rosary) may also occur. Low-grade fever, anemia, coiled fragmented hair, and poor wound healing are typical signs of scurvy. Vitamin C administration is effective in curing infantile scurvy. Symptoms of irritability, fever, tenderness to palpation and hemorrhage generally resolve within 7 days of initiating treatment.

Henoch–Schönlein Purpura: Henoch–Schönlein purpura (HSP), also known as anaphylactoid purpura, purpura rheumatic, or immunoglobulin A (IgA) vasculitis, is a leukocytoclastic vasculitis predominantly affecting small blood vessels. It is the most common systemic vasculitis in children with an annual incidence of approximately 10–20 per 100,000 and a peak age of onset between 4 and 6 years. Caucasians have the highest incidence and unlike most vasculitides, males are affected more commonly than females (2:1). HSP is most prevalent during the winter/spring, and this seasonal distribution supports the hypothesis that an infectious agent triggers this condition. Group A β-hemolytic streptococcus, *Staphylococcus aureus*, influenza, parainfluenza, Epstein–Barr virus, adenovirus, parvovirus, and mycoplasma have all been reported as triggers for HSP. Other potential triggers include medications (nonsteroidal antiinflammatory drugs, angiotensin-converting enzyme inhibitors, and antibiotics), food allergies, and insect bites. Pathogenesis has been linked to deposition of IgA1 in the glomerulus, skin, and blood vessels of the GI tract.

HSP classically presents with lower-extremity palpable purpura without thrombocytopenia or coagulopathy, arthritis, abdominal pain, and renal disease. The purpuric rash, appearing as bullae, necrotic lesions, or deep bruising, is usually on dependent areas but may be seen on the arms, face, and ears. Purpura may be preceded by a maculopapular or urticarial rash that usually disappears within 24 hours. GI manifestations, affecting 50%–75% of children, include bleeding, intussusception, and abdominal pain and may precede purpura by up to 2 weeks. Renal disease with microscopic hematuria with or without proteinuria, affects 20%–60% of children. The risk of chronic renal impairment and end-stage renal disease is 2%–15% and less than 1%, respectively. HSP is most often self-limited with symptom resolution within 4–6 weeks of onset, and treatment is supportive. Steroids may be used for patients with renal involvement, persistent purpura, or severe abdominal pain, however, have not been shown to alter long-term clinical outcomes and are not used routinely.

Further Reading

Bristow, J., Carey, W., Egging, D., et al. (2005). Tenascin-X, collagen, elastin, and the Ehlers-Danlos syndrome. *Am J Med Genet C Semin Med Genet*, *139*, 24–30.

De Paepe, A., & Malfait, F. (2004). Bleeding and bruising in patients with Ehlers-Danlos syndrome and other collagen vascular disorders. *Br J Haematol*, *127*, 491–500.

Fuchizaki, U., Miyamori, H., Kitagawa, S., et al. (2003). Hereditary haemorrhagic telangiectasia (Rendu-Osler-Weber disease). *Lancet*, *362*, 1490–1494.

Geisthoff, U. W., Nguyen, H. L., Röth, A., & Seyfert, U. (2015). How to manage patients with hereditary haemorrhagic telangiectasia. *Br J Haematol*, *171*, 443–452.

Malfait, F., & De Paepe, A. (2014). The Ehlers-Danlos syndrome. *Adv Exp Med Biol*, *802*, 129–143.

Mao, J. R., & Bristow, J. (2001). The Ehlers-Danlos syndrome: On beyond collagens. *J Clin Investig*, *107*, 1063–1069.

Parambil, J. G. (2016). Hereditary hemorrhagic telangiectasia. *Clin Chest Med*, *37*, 513–521.

Sobey, G. (2015). Ehers-Danlos syndrome: How to diagnose and when to perform genetic tests. *Arch Dis Child*, *100*, 57–61.

Trapani, S., Micheli, A., Grisolia, F., et al. (2005). Henoch Schönlein purpura in childhood: Epidemiological and clinical analysis of 150 cases over a 5-year period and review of literature. *Semin Arthritis Rheum*, *35*, 143–153.

Weiss, P. F. (2012). Pediatric vasculitis. *Pediatr Clin North Am*, *59*, 407–423.

CHAPTER 119

Hemostasis in Liver Disease

Margarita Kushnir, MD and Henny H. Billett, MD

Liver disease is associated with multiple hemostatic defects leading to both hemorrhagic and thrombotic manifestations. Its effect on coagulation factors, endogenous anticoagulants, and platelets results in rebalanced hemostasis and creates challenging clinical scenarios requiring careful management.

Decreased Synthesis of Pro- and Anticoagulant Factors: All coagulation factors (except FVIII and von Willebrand factor [VWF]) are synthesized in hepatocytes and can be deficient in chronic liver disease. FVIII and VWF are often elevated due to their increased production in endothelium. In the setting of acute liver failure, the rate of development of a coagulation factor deficiency is directly proportional to its half-life in plasma. Because FVII has the shortest half-life (approximately 4–5 hours) of all coagulation factors, patients with acute liver failure develop FVII deficiency before levels of other coagulation factors fall. Components of the common pathway of coagulation also have relatively short half-lives and become deficient next. Because of more pronounced deficiency of FVII and the factors involved in the common coagulation pathway, liver failure tends to have a greater effect on the prothrombin time (PT) compared with the partial thromboplastin time (PTT). FVIII, VWF, and fibrinogen are all acute phase reactants that are usually elevated early in liver failure, but as it progresses, fibrinogen falls due to decreased hepatic production while FVIII and VWF levels remain normal or high.

The liver also synthesizes proteins that inhibit coagulation, such as protein C, protein S, and antithrombin. Deficiency of these proteins likely counteracts shortage of procoagulant factors. This helps maintain hemostasis but can also lead to inappropriate thrombosis.

Vitamin K Deficiency: In patients with cirrhosis and obstructive jaundice, decreased bile salt production and/or delivery can cause decreased absorption of fat-soluble vitamins in the small intestine, including vitamin K, an essential cofactor for the production of biologically active factors II, VII, IX, and X, as well as protein C and S. Prolonged antibiotic use, which inhibits the gut flora that produces Vitamin K, and severe dietary deficiency, as may occur in cases of starvation (usually medically induced) or alcoholic cirrhosis with continued alcohol abuse, may also contribute to a deficiency of vitamin K.

Thrombocytopenia: Thrombocytopenia in liver disease is also often multifactorial. Splenomegaly, caused by portal hypertension, results in sequestration of platelets and leads to their reduction in the circulation. Thrombocytopenia is further exacerbated by deficiency of thrombopoietin, a growth factor that stimulates the production and differentiation of megakaryocytes in the bone marrow, and is predominantly

Transfusion Medicine and Hemostasis. https://doi.org/10.1016/B978-0-12-813726-0.00119-7

produced in the liver. Furthermore, underlying disease may shorten platelet survival due to hepatitis B and C viremia or antiplatelet antibodies, as can be seen in autoimmune hepatitis. Alcohol causes direct megakaryocyte toxicity and can also contribute to thrombocytopenia.

Fibrinogen: Fibrinogen levels tend to remain normal to only slightly depressed (usually >100 mg/dL) in chronic liver disease. However, the fibrinogen that is produced has increased amounts of sialic acid, which renders it less functional and prolongs the thrombin time. Despite this alteration, in most cases, the dysfibrinogen of liver disease does not contribute to a clinically meaningful coagulopathy.

Clinical Manifestations: Although bleeding has, for a long time, been considered a common manifestation of liver disease because of abnormal coagulations tests, it is increasingly recognized that cirrhosis is, to a greater extent, a prothrombotic condition. Many hemorrhagic complications of cirrhosis are caused by hemodynamic derangements of portal hypertension, such as gastropathy and esophageal varices, rather than an underlying hypocoagulable state. Because PT and PTT do not reflect levels of endogenous anticoagulant proteins (protein C, S, ATIII), standard coagulation tests are poorly predictive of bleeding in patients with liver disease. In fact, no association has been found between PT and bleeding in past studies.

To account for the effect of reduced anticoagulant levels, assessments of global coagulation function using thrombin generation assays have been performed in several studies. There was no significant difference in endogenous thrombin potential when plasmas of patients with chronic cirrhosis and controls were compared. Despite significantly prolonged PT and PTT, patients with acute liver failure have also been found to have similar endogenous thrombin generation, in the presence of thrombomodulin, compared with healthy controls. Preserved thrombin generation is thought to be due to rebalanced coagulation attributed to reduction of protein C production and increase in factor VIII in liver disease.

Patients with cirrhosis usually have mild–moderate thrombocytopenia until the later stages of the disease when it can worsen. Bleeding rarely occurs with platelet counts >50,000/μL. It is thought that an increase in VWF released from endothelial cells compensates for the thrombocytopenia to maintain normal platelet function until the platelet count is very severe. A 2006 study showed that thrombin generation in patients with cirrhosis correlates with platelet numbers and is diminished with severe thrombocytopenia. It was found that platelet count of 56,000 × 10/μL is required to generate thrombin amounts equivalent to lower limit of the normal range.

Monitoring Laboratory Parameters in Liver Disease: In patients with chronic liver disease, it is common to see prolonged PT and, to a lesser extent, PTT. Worsening in these parameters can indicate progressive impairment of synthetic liver function and is often accompanied by other lab abnormalities, such as hypoalbuminemia and hyperbilirubinemia.

In an acutely ill patient with cirrhosis or new onset liver failure, it can be difficult to distinguish coagulopathy of advanced liver disease from disseminated intravascular coagulation (DIC) as both can cause hypofibrinogenemia, prolonged PT/PTT, and thrombocytopenia. D-dimer levels, like in DIC, are often elevated in liver disease, although elevations are generally more pronounced in DIC. Because FVIII, along with all other factors, is consumed in DIC, obtaining FVIII level may be helpful in differentiating between DIC and coagulopathy of liver disease, where it is normal or elevated. A helpful aide to distinguishing these is that in DIC, as the coagulopathy gets worse, the fibrinogen and VIII first increase and then decrease together whereas in liver disease, as the disease progresses and the fibrinogen decreases, the factor VIII will increase—an inverse relationship. However, coagulopathy and DIC may also coexist and clinical judgment is important in distinguishing these two entities. DIC is usually (but not always) acute and generally has an obvious underlying cause such as sepsis.

Management: As coagulation parameters do not reflect patient's true hemostasis, it is not necessary to correct them in asymptomatic patients with cirrhosis. Before surgery, patients are often given products to lower their INR to less than 1.5 and increase fibrinogen to >100 mg/dL. Clinical outcome data do not support this approach as prophylactic transfusion of blood products has not been shown to reduce procedure-related bleeding in this population. There is also concern, particularly with administration of multiple units of fresh frozen plasma (FFP), for expansion of intravascular volume, which may actually increase the bleeding risk. A recent study used thromboelastography (TEG) to guide blood product transfusions before invasive procedures in patients with severe coagulopathy due to cirrhosis and showed that TEG significantly decreased the amount of blood products used compared with standard protocols, without an increase in bleeding complications.

Patients are generally transfused with platelets to achieve platelet counts >50,000/μL before surgery (75,000/μL in ocular or neurosurgery). Response to platelet transfusions may be blunted due to splenomegaly, and this goal may be difficult to attain. Eltrombopag, a thrombopoietin-receptor agonist, has been used in patients with chronic liver disease before procedures; a significant decrease in platelet transfusion requirements among patients on eltrombopag was observed, but there was a higher rate of portal vein thrombosis.

In acute bleeding, patients are usually treated with cryoprecipitate (for low fibrinogen levels) and platelets to keep platelet count above 50,000/μL. FFP is sometimes given, but volume is a concern, as volume overload can worsen portal hypertension and exacerbate bleeding. As there can be an element of vitamin K deficiency in any form of liver disease, vitamin K should be given. It is most reliably administered intravenously, given the edematous state and poor oral absorption in decompensated cirrhosis. In patients with concomitant renal failure, it is important to address uremia. Depending on the site of bleeding, local control is often the most important aspect of management as is the case with variceal bleeds, which can be treated endoscopically.

If bleeding is difficult to control with these standard measures, other treatment modalities may be considered. Desmopressin (DDAVP) may transiently increase VWF to help maximize platelet efficiency, although high-quality data for this maneuver are lacking. The use of antifibrinolytics, such as aminocaproic acid or tranexamic acid, can also be tried as fibrinolysis is often abnormal in liver disease. Finally, prothrombin complex concentrates can be considered in life-threatening refractory bleeding, especially if FVII or other vitamin K–dependent factors are particularly low (e.g., FII and FX), and there is a dire concern about volume overload. However, the benefit of these products must be weighed against their relative contraindication in patients with liver failure because of their inherent thrombotic risk.

Anticoagulation in Liver Disease: Treatment of venous thromboembolism in patients with cirrhosis is complicated by concern for bleeding and baseline PT derangements. Anticoagulation may be precluded by large esophageal varices, which carry considerable bleeding risk. Extent of thrombosis is often an important determinant in assessing the risk: benefit ratio. Warfarin may be used in patients with normal PT. In patients with prolonged PT, low-molecular-weight heparin, which does not require INR monitoring, may be a better option. Direct oral anticoagulants have not been well studied in patients with chronic liver disease and are not commonly used. In the absence of active bleeding and high-risk varices, thromboprophylaxis is generally indicated in hospitalized patients with cirrhosis.

Further Reading

Afdhal, N. H., Giannini, E. G., Tayyab, G., et al. (2012). Eltrombopag before procedures in patients with cirrhosis and thrombocytopenia. *N Engl J Med*, *367*(8), 716–724.

Blonski, W., Siropaides, T., & Reddy, K. (2007). Coagulopathy in liver disease. *Curr Treat Options Gastroenterol*, *10*(6), 464–473.

Caldwell, S. H., Hoffman, M., Lisman, T., et al. (2006). Coagulation disorders and hemostasis in liver disease: Pathophysiology and critical assessment of current management. *Hepatology*, *44*(4), 1039–1046.

De Pietri, L., Bianchini, M., Montalti, R., et al. (2016). Thrombelastography-guided blood product use before invasive procedures in cirrhosis with severe coagulopathy: A randomized, controlled trial. *Hepatology*, *63*(2), 566–573.

Gatt, A., Riddel, A., Calvaruso, V., et al. (2010). Enhanced thrombin generation in patients with cirrhosis-induced coagulopathy. *J Thromb Haemost*, *8*(9), 1994–2000.

Habib, M., Roberts, L. N., Patel, R. K., et al. (2014). Evidence of rebalanced coagulation in acute liver injury and acute liver failure as measured by thrombin generation. *Liver Int*, *34*(5), 672–678.

Kor, D. J., Stubbs, J. R., & Gajic, O. (2010). Perioperative coagulation management–fresh frozen plasma. *Best Pract Res Clin Anaesthesiol*, *24*(1), 51–64.

Kujovich, J. L. (2005). Hemostatic defects in end stage liver disease. *Crit Care Clin*, *21*(3), 563–587.

Northup, P. G., & Caldwell, S. H. (2013). Coagulation in liver disease: A guide for the clinician. *Clin Gastroenterol Hepatol*, *11*(9), 1064–1074.

Rai, R., Nagral, S., & Nagral, A. (2012). Surgery in a patient with liver disease. *J Clin Exp Hepatol*, *2*(3), 238–246.

Shah, N. L., Intagliata, N. M., & Northup, P. G. (2014). Procoagulant therapeutics in liver disease: A critique and clinical rationale. *Nat Rev Gastroenterol Hepatol*, *11*(11), 675–682.

Stellingwerff, M., Brandsma, A., Lisman, T., et al. (2012). Prohemostatic interventions in liver surgery. *Semin Thromb Hemost*, *38*(3), 244–249.

Tripodi, A., Chantarangkul, V., Primignani, M., et al. (2012). Thrombin generation in plasma from patients with cirrhosis supplemented with normal plasma: Considerations on the efficacy of treatment with fresh-frozen plasma. *Intern Emerg Med*, *7*(2), 139–144.

Tripodi, A., Primignani, M., Chantarangkul, V., et al. (2006). Thrombin generation in patients with cirrhosis: The role of platelets. *Hepatology*, *44*(2), 440–445.
Tripodi, A., & Manucci, P. (July 14, 2011). The coagulopathy of chronic liver disease. *N Engl J Med*, *365*(2), 147–156.
Trotter, J. (2006). Coagulation abnormalities in patients who have liver disease. *Clin Liver Dis*, *10*(3), 665–678.

CHAPTER 120

Bleeding Risks With Vitamin K Deficiency

Shilpa Jain, MD, MPH and Suchitra S. Acharya, MD

Introduction: In 1894, the condition known as hemorrhagic disease of the newborn was first discovered to be related to a deficiency of vitamin K (VK), which is an essential cofactor for the synthesis of clotting factors, FII, FVII, FIX, FX, as well as the endogenous inhibitors of coagulation, proteins C and S in the liver. Deficiency of VK can lead to a quick decline in the levels of VK-dependent clotting factors because of their short half-life and can present with bleeding manifestations known as vitamin K deficiency bleeding (VKDB). VK deficiency occurs most commonly during the neonatal period and early infancy but can also be present in individuals with malabsorption syndromes, decreased production of bile salts, warfarin therapy, insufficient intake, or decreased intestinal flora related to antibiotic use.

Pathophysiology: VK is a lipid-soluble vitamin that is essential for the posttranslational modification of glutamic acid residues in the VK-dependent clotting factors, FII, FVII, FIX, FX, to γ-carboxyglutamic acid residues. This modification leads to the activation of these clotting proteins involved in the formation of the coagulation complexes (tenase and prothrombinase) to generate thrombin. Carboxylation of the factors is catalyzed by the enzyme γ-glutamyl carboxylase (GGCX) using the reduced form of VK (VK hydroquinone, $KH_{2)}$ as a cofactor, which in the process gets oxidized to VK epoxide (KO). This is then converted back to KH_2 by the enzyme KO reductase complex subunit 1 (VKORC1), which completes the VK cycle as illustrated in Fig. 120.1.

The major source of dietary VK is in the form of K_1 (phylloquinone), which is primarily derived from yellow or green plants and oils and the daily requirement is 1 μg/kg. Optimal absorption of VK from food requires an intact intestine and is highly dependent on bile salts, which can be impaired with hepatic or intestinal disease. Another form of VK is K_2 (menaquinones-n), which is endogenously synthesized from microbial gut flora and is found in animal and soy protein, which is less readily absorbed from the intestine. Decreased VK synthesis in the gut can occur either from the lack of adequate gut flora such as in immature gastrointestinal (GI) tract of newborns or with altered microbia from prolonged antibiotic use. Newborns are especially susceptible to VKDB, as they are born with limited VK stores because placental transfer of VK is low resulting in an almost 50% reduction in the VK-dependent clotting factors levels compared with adults. This gets further compounded in exclusively breastfed infants, as breast milk has very low levels of VK (5–15 μg/L) compared with infant formulas that are fortified (50–60 μg/L).

Etiology: One must have a high degree of suspicion to entertain a diagnosis of VKDB at all ages. During early infancy, it is generally classified into three distinct groups based

Transfusion Medicine and Hemostasis. https://doi.org/10.1016/B978-0-12-813726-0.00120-3

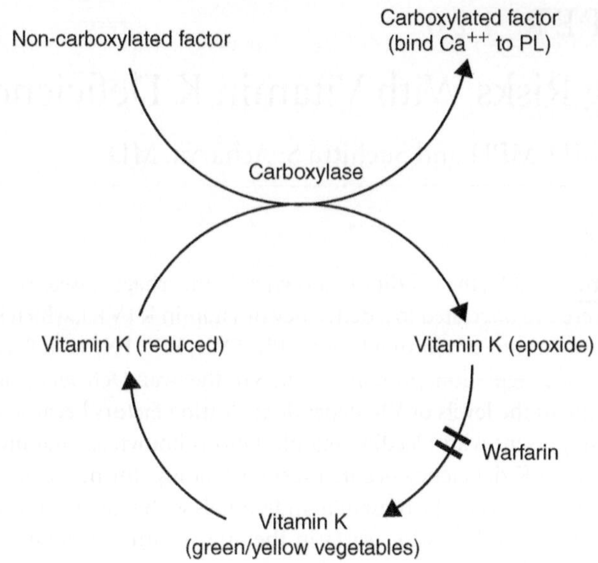

FIGURE 120.1 The vitamin K cycle and its role in the production of carboxylated clotting factors.

on the timing of clinical presentation. (1) **Early VKDB** typically occurs within 24 hours of birth and is related to maternal use of warfarin, antiseizure, or antituberculosis drugs, which affect the storage and function of VK. (2) **Classical VKDB** presents in the first week of life. Common risk factors include maternal use of medications, exclusive breastfeeding (low VK), and lack of microbial flora in the newborn gut. (3) **Late onset VKDB** presents between a week and 6 months of age and is seen primarily in exclusively breastfed infants with a history of diarrhea. With the increasing desire for a natural birthing process and parental refusal of VK prophylaxis at birth, there has been a surge in VKDB in infancy.

Other disorders interfering with VK intestinal absorption and synthesis such as alpha-1 antitrypsin deficiency, biliary atresia, celiac disease, chronic diarrhea, cystic fibrosis, and cholestatic liver disease can present with VK deficiency during infancy or later. Deficiency of VK can also occur secondary to drugs that decrease cholesterol absorption, such as cholestyramine and metabolic disorders that involve dysregulated lipid metabolism (i.e., Gaucher disease). Table 120.1 reviews the age of presentation, bleeding site, cause, and treatment for different clinical manifestations of VK deficiency in children. A rare cause for severe clinical presentation in the perinatal period is the presence of mutations in the enzymes involved in the VK cycle, which leads to an autosomal recessive bleeding disorder, referred to as VK-dependent coagulation factor deficiencies (VKCFDs). Two subtypes of VKCFD have been identified: VKCFD1 for mutations in GGCX and VKCFD2 for mutations in VKORC1.

In the adolescent and adult population, VK deficiency is usually associated with prolonged fasting, excessive vomiting, and medications. It is unclear whether a nutritionally poor diet alone can induce significant VK deficiency as gut flora produced VK_2 compensates for the lack of VK_1 in the diet. Consequently, a clinically meaningful VK

TABLE 120.1 Vitamin K Deficiency in Infants

Syndrome	Age at Presentation	Bleeding Site	Cause	Treatment
Early hemorrhagic disease of the newborn (HDN) (5%, at-risk mothers)	1 day	Intracranial hemorrhage (ICH), GI, umbilical, bony abnormalities	Anticonvulsants; anti-tuberculosis medicines	10 mg vitamin K daily × 2 weeks
Classic HDN (0.01%–1%)	1–7 days	ICH, GI, umbilical, ENT	Idiopathic; maternal drugs	0.5–1 mg vitamin K at birth
Late HDN (0.02%)	2–24 weeks	ICH, GI, skin, ENT	Idiopathic; malabsorption; liver disease; breastfeeding	Vitamin K at birth; repeat as needed orally/IM

deficiency that is induced by poor diet is usually only seen in patients who also have inadequate gut flora because of recent antibiotic use. Warfarin use is the most common cause of VK deficiency in adults, as it inhibits VKORC1 and interrupts the cycling of KH_2 from KO. One must also consider accidental or intentional ingestion of warfarin when evaluating patients who present with mucocutaneous bleeding observed to be selectively deficient in the VK-dependent clotting factors. Long-acting rat poison (a warfarin-like compound) ingestion should also be considered when reversal of presumed warfarin overdose has been attempted, only to have its bleeding manifestations recur within 24 hours due to its accumulation in the subcutaneous fat and slow release.

Clinical Features: In particular, one should consider the diagnosis of VKDB in the context of unexplained intracranial or GI hemorrhage especially seen in the first year of life. Other sites of bleeding during neonatal/infant period can be umbilical, skin, nasal, or after circumcision. Clinical features specific to the etiology of VK deficiency may also be present.

Diagnosis: Diagnosis of VKDB is suspected if both screening tests for coagulation (prothrombin time [PT] and activated partial prothrombin time [aPTT]) are prolonged and associated with concomitantly low (below 30%–35%) clotting factor activity levels (FII, FVII, FIX, FX). Because the PT is more sensitive than the aPTT to abnormalities in the coagulation cascade related to the short half-life of FVII (~4 hours), VK deficiency causes a disproportionately elevated PT compared with the aPTT. Accordingly, one of the diagnostic criteria to confirm a case of VK deficiency bleeding (VKDB) is a PT ≥ 4s the control value. While this approach may miss subclinical deficiency, it still assesses one of the main pathological outcomes of decreased VK. Another test to support the diagnosis is response of the PT to empirical VK treatment but may not be possible if other blood products have been concomitantly administered.

Accurate measurements of VK in the form of serum levels of VK1 before treatment can be difficult. However, if obtained accurately, adult plasma levels less than 100 pg/mL or cord blood levels less than 50 pg/mL are suggestive of VK deficiency. A useful marker of early or subclinical VK deficiency in infants is the presence of increased levels of under- or partially carboxylated VK-coagulation factors, the so-called "*p*roteins *i*nduced by *v*itamin *K a*bsence" (PIVKA). The most sensitive assay done in laboratories involves measuring PIVKA-II (noncarboxylated prothrombin) using enzyme immunoassays even on a posttreatment clotted sample. Moreover, the long half-life of PIVKA-II also enables a retrospective diagnosis of VKDB long after the original bleeding event.

Differential Diagnosis: When excessive mucosal bleeding is encountered (nose, mouth, GI/genitourinary tract, and bruising), screening laboratory testing such as complete blood count, PT/a PTT, fibrinogen, thrombin time and platelet function analyzer (PFA) will help direct the focus toward a correct diagnosis. It is important to note that apart from a prolonged PT/aPTT, the rest of these hemostatic labs are normal in VKDB. Both disseminated intravascular coagulation and liver disease will display abnormalities of all the other laboratory screening tests. Heparin contamination will disproportionately elevate the aPTT compared with the PT in contrast to VK deficiency. In addition, heparin contamination and dysfibrinogenemia can be ruled out by a normal reptilase time. Fibrinogen deficiency is unlikely given a normal fibrinogen activity and reptilase time. Acquired deficiencies of VK-dependent clotting factors (autoimmune disorders such as lupus) can occur, and acute onset of bleeding could be attributed to antibodies to these factors causing a reduction in one or more of the VK-dependent factors most commonly FII deficiency.

Management: After a diagnosis of VK deficiency is established, it is imperative to rapidly correct the coagulopathy to minimize ongoing hemorrhage. Management varies by the clinical manifestations and the underlying cause. Acute bleeding, which is not severe, can be treated with parenteral administration of VK, which starts to normalize the PT and aPTT within 4–6 hours. If a history of maternal medication use is present, risk of early hemorrhagic disease of the newborn can be reduced by the administration of 5 mg of oral VK daily during the third trimester. Classic hemorrhagic disease of the newborn has been effectively eliminated by the prophylactic administration of 0.5–1 mg dose of VK intramuscularly at birth. Parents who refuse VK at birth may be placing their infant at risk of VKDB, particularly if they are exclusively breastfed and need to be educated against it. Parenteral administration of high-dose VK is also effective in complete or partial correction of the coagulation factor defects associated with inherited VKCFD.

In older children, intramuscular VK is given at a dose of 5–10 mg, whereas the preferred administrative route is 5–10 mg orally in adults, if bleeding manifestations are not severe. For the patient who cannot swallow oral medications, VK can be administered subcutaneously; however, this route is associated with inconsistent drug absorption. Finally, it can also be administered intravenously, which has occasionally induced anaphylaxis. If there is central nervous system, GI or extensive mouth and nose bleeding, transfusion with fresh frozen plasma (FFP) at 15–20 mL/kg should be considered, which should raise most factor levels 20%–25% above their baseline. If there is a concern about volume overload with FFP, prothrombin complex concentrates, which include most of the VK-dependent factors may be useful (see Chapter 41).

Further Reading

Clarke, P., & Shearer, M. J. (2007). Vitamin K deficiency bleeding: The readiness is all. *Arch Dis Child*, *92*, 741–743.

Cranenburg, E. C., Schurgers, L. J., & Vermeer, C. (2007). Vitamin K: The coagulation vitamin that became omnipotent. *Thromb Haemost*, *98*, 120–125.

Czogalla, K. J., Watzka, M., & Oldenberg, J. (2016). VKCFD2- form clinical phenotype to molecular mechanism. *Hamostaseologie*, *36*(Suppl. 2), S13–S20.

Doneray, H., Tan, H., Buyukavci, M., & Karakelleoglu, C. (2007). Late vitamin K deficiency bleeding: 16 cases reviewed. *Blood Coagul Fibrinolysis*, *18*, 529–530.

Flood, V. H., Galderisi, F. C., Lowas, S. R., Kendrick, A., & Boshkov, L. K. (2008). Hemorrhagic disease of the newborn despite vitamin K prophylaxis at birth. *Pediatr Blood Cancer*, *50*, 1075–1077.

Laposata, M., Van Cott, E. M., & Lev, M. H. (2007). Case records of the Massachusetts General Hospital. Case 1–2007. A 40-year-old woman with epistaxis, hematemesis, and altered mental status. *N Engl J Med*, *11*, 174–182.

Marcewicz, L. H., Clayton, J., Maenner, M., Odom, E., et al. (2017). Parental refusal of vitamin K and neonatal preventive services: A need for surveillance. *Matern Child health J*, *21*(5), 1079–1084.

Pavlu, J., Harrington, D. J., Voong, K., et al. (2005). Superwarfarin poisoning. *Lancet*, *628*, 12–18.

Spahr, J. E., Maul, J. S., & Rodgers, G. M. (2007). Superwarfarin poisoning: A report of two cases and review of the literature. *Am J Hematol*, *82*, 656–660.

van Hasselt, P. M., de Koning, T. J., Kvist, N., et al. (2008). Prevention of vitamin K deficiency bleeding in breastfed infants: Lessons from the Dutch and Danish biliary atresia registries. *Pediatrics*, *121*, e857–e863.

CHAPTER 121

Bleeding Risks With Cardiac Disease

Margarita Kushnir, MD and Henny H. Billett, MD

Most of the hematologic complications that occur in cardiac disease are linked to the disturbances in the dynamics of flow. Although there are fewer hemorrhagic complications due to intrinsic cardiac disease than occur with cardiac surgeries and implantation devices, those that do occur are linked to the same pathophysiologies that are seen in the operating rooms and the cath labs.

Cardiac Disease: Cardiac disease can result in congestive heart failure and congestive hepatopathy (cardiac cirrhosis). This is associated with a decrease in hepatic synthetic function, thrombocytopenia, and splenomegaly (see Chapter 119 on Liver Disease). Specific myocardial diseases are rare but can occur. Amyloid infiltration can cause myocardial dysfunction and may be associated with a decrease in factor X and dysfibrinogenemia. Although these may cause a coagulopathy and abnormal coagulation testing, amyloidosis can also cause impaired left ventricular function, leading to aneurysm formation and mural thrombi; these may require anticoagulation.

Cyanotic heart diseases may lead to polycythemia. Increased hematocrit can cause rheologic sludging issues while also lowering production of clotting factors. More commonly, the increased hematocrit may cause a factitious elevation of the prothrombin time (PT) and partial thromboplastin time (PTT) as a result of excessive citrate:plasma ratio in the tube used for testing. This laboratory artifact can be overcome by drawing the blood into a syringe containing less citrate to compensate for the decreased plasma.

Certain cardiac conditions such as a ventricular septal defect (VSD), severe aortic or pulmonic stenosis, and some artificial valves may cause vascular turbulence and contribute to an acquired von Willebrand syndrome (AVWS, see Chapter 109). In this condition, there is loss of the high-molecular-weight multimers secondary to a shearing effect on the large von Willebrand protein. This loss causes an inappropriately low VWF activity compared with the VWF antigen level with a laboratory picture similar to congenital type 2A VWD. The severity of the AVWS correlates well with the severity of the stenosis. Paravalvular leaks have also been reported to induce the same hemostatic phenomenon. Correction of the VSD or abnormal heart valve will correct the AVWS.

The high shear rates can also lead to coagulopathies by increasing hemolysis. The increased hemolysis may lead to severe anemia and an increased bleeding diathesis while the free hemoglobin may cause an increase in thrombotic potential. If the shear stress is sufficient, thrombocytopenia and thrombocytopathies can complicate an already complicated situation.

Cardiac Surgery: The most common disturbances in hemostasis occur during cardiac surgery or with cardiac device implantation. Cardiopulmonary bypass (CPB), extracorporeal membrane oxygenation (ECMO), left ventricular access devices (LVADs) all severely stress the hemostatic system. Bleeding diathesis related to CPB surgery or

Transfusion Medicine and Hemostasis. https://doi.org/10.1016/B978-0-12-813726-0.00121-5

LVADs is usually evident from persistent bleeding from nasogastric, chest, and endotracheal tubes, urinary catheters, IV and phlebotomy sites, and surgical wounds. This bleeding from multiple sites helps to differentiate the systemic bleeding sometimes associated with CPB from localized surgical bleeding. Coping with hematologic complications of cardiac surgeries requires an intense interaction between the cardiologist, the cardiothoracic surgeon, the transfusion medicine team, and the hematologist.

With the increasing frequency of coronary artery repair, left ventricular assist devices, heart transplants and valve repairs, the use of CPB and ECMO are becoming commonplace. The complexity of these systems and the ability to manipulate the blood supply are remarkable and so it is understandable that there might be a significant toll taken on hemostatic mechanisms, both from the bleeding and the thrombosis standpoint. The problem often starts with the patient who is typically quite ill when the procedures starts and may have been hypotensive, in shock, with features of congestive hepatosplenomegaly. In the operating room, numerous vascular access devices are placed and the patient is connected to CPB, usually after priming the pump. Priming generally occurs with crystalloid but colloids, albumin, and rarely plasma can be used. These will cause the dilution of the red cells, the coagulation factors, and platelets. Once the circuit is established, the prime is used to lower the patient's body temperature to protect vital organs. At the typical temperatures of 28–32°C, the chemical interactions needed for the normal coagulation pathway are severely hindered. Heparin is infused to anticoagulate the patient, but heparin binding can be erratic as it nonselectively binds to many other plasma proteins leading to wide variations in bioavailability. Failure to adequately anticoagulate the patient can lead to a clot in the system and sudden death so anti-Xa levels, TEG or activated clotting times are measured to ensure optimal heparin effect. Still under- and overanticoagulation occur. In some procedures, and particularly in patients who experience heparin-induced thrombocytopenia, bivalirudin is used. Bivalirudin has fewer difficulties in that it has much better bioavailability and does not require antithrombin binding for full effect. However, unlike heparin that is easily reversed with protamine, there is currently no reversal agent for bivalirudin.

The pump itself can cause hemolysis and thrombocytopathy. Thrombocytopathy occurs because of the depletion of platelet alpha (and some dense) granules and some glycoproteins (GP1bα and GPVI), which are particularly vulnerable to the turbulent pump action (although these are very quickly restored once the patient is off-pump). The dysfunctional platelets are gradually removed from the circulation, so the nadir of the platelet count often is not reached until 1–3 days after the operation. Hemolysis may have the opposite effect. Free hemoglobin has been implicated in thrombosis and in nitric oxide depletion, which is necessary for good blood flow and vasodilation. AVWS can also occur during turbulent flow and/or increased shear stress and may increase the bleeding diathesis for these patients.

Fibrinolysis is increased in CPB as a result of increased tissue plasminogen activator (tPA), decreased plasminogen activator inhibitor-1 (PAI-1) and decreased α_2 antiplasmin (α_2AP). However, antifibrinolytics are used more often than in the past so these problems are minimized. Intraoperative red cell salvage procedures allow for autotransfusion of patient red cells and may decrease transfusion need but the process washes out platelets and plasma, causing further hemodilution.

Left Ventricular Assist Devices: LVADs provide a life-saving bridge to heart transplantation for patients with refractory heart failure and are being used with increased frequency in patients with advanced heart failure as a sole (destination) therapy. Newer LVADs have a lower risk of thrombosis than previous models, but bleeding continues to be a prevalent complication. The cited hemorrhagic complications are as high as 17%–63% per year. Bleeding appears to be secondary to AVWS that develops in these patients as well as to the anticoagulation required for LVADs.

Screening Laboratory in Cardiopulmonary Bypass Bleeding Assessment: A CBC, PT, PTT, fibrinogen and, if necessary, mixing studies are initial screening tests. The CBC will easily allow for the detection of a thrombocytopenia, which can then be corrected with platelet transfusions as needed. DDAVP releases VWF from its endothelial and platelet storage sites and may be used in conjunction with platelet transfusions to correct the associated thrombocytopathies and increase VWF. The PT, PTT, and fibrinogen levels will be useful screening laboratory tests to determine a specific diagnosis and to guide further management. If these tests are abnormal and there is no inhibitor, dilutional coagulopathy may be present. Dilution is best treated with plasma products (e.g., fresh frozen plasma), and cryoprecipitate will remediate hypofibrinogenemia. If there is evidence of an inhibitor, there may still be excess heparin in the system, which can be confirmed by heparin and/or reptilase assays. Rarely, patients will present with a significant bleeding diathesis, a prolonged PT and PTT that is not normalized with mixing studies and is not due to heparin. This may point to an inhibitor.

Increasingly, fibrin glue with human thrombin is used as a local tissue sealant, but topical bovine thrombin may also still be used. Because bovine thrombin preparations are often contaminated with bovine FV, antibovine FV antibodies can be formed, which sometimes cross-react with human FV. Antibodies may also be formed against the vitamin K-dependent clotting factors (FII, VII, IX, and X) but cross-reactivity to FV is more common. Late bleeding, several weeks after the surgery, may occur. These antibodies are best treated with appropriate blood product support and with immunosuppressive agents, corticosteroids, intravenous immunoglobulin, or plasma exchange. If patients are bleeding heavily, factor concentrates may be used: prothrombin complex concentrates, fibrinogen concentrates, VIIa and XIII concentrates have all been used although the evidence of better outcomes for these is unclear.

In summary, recognition of the varied hematologic sequela of heart disease, interventions, and devices is important for optimal management of cardiac patients, who are often already in intricate clinical situations that put them at risk for thrombosis and hemorrhage.

Further Reading

Besser, M. W., Ortmann, E., & Klein, A. A. (2015). Haemostatic management of cardiac surgical haemorrhage. *Anaesthesia*, *70*(Suppl. 1), 87–95.

Davidson, S. (2014). State of the art - how I manage coagulopathy in cardiac surgery patients. *Br J Haematol*, *164*(6), 779–789.

Grottke, O., Fries, D., & Nascimento, B. (2015). Perioperatively acquired disorders of coagulation. *Curr Opin Anaesthesiol, 28*(2), 113–122.
Hillegass, W. B., & Limdi, N. A. (2016). Valvular heart disease and acquired type 2A von Willebrand syndrome: The "hemostatic" waring blender syndrome. *JAMA Cardiol, 1*(2), 205–206.
Jakobsen, C. J. (2014). Transfusion strategy: Impact of haemodynamics and the challenge of haemodilution. *J Blood Transfus, 2014*, 627141.
Nascimbene, A., Neelamegham, S., Frazier, O. H., Moake, J. L., & Dong, J. F. (2016). Acquired von Willebrand syndrome associated with left ventricular assist device. *Blood, 127*(25), 3133–3141.
Paparella, D., & Whitlock, R. (2016). Safety of salvaged blood and risk of coagulopathy in cardiac surgery. *Semin Thromb Hemost, 42*(2), 166–171.
Proudfoot, A. G., Davidson, S. J., & Strueber, M. (2017). von Willebrand factor disruption and continuous-flow circulatory devices. *J Heart Lung Transplant, 36*, 1155–1163.
Ranucci, M. (2015). Hemostatic and thrombotic issues in cardiac surgery. *Semin Thromb Hemost, 41*(1), 84–90.
Thiele, R. H., & Raphael, J. (2014). A 2014 update on coagulation management for cardiopulmonary bypass. *J Semin Cardiothorac Vasc Anesth, 18*(2), 177–189.

CHAPTER 122

Bleeding Risks With Renal Disease

Susmita N. Sarangi, MBBS, MD and Suchitra S. Acharya, MD

Introduction: Renal disease has long been associated with a bleeding diathesis; Morgagni described the association of uremia and bleeding as early as 1764. Although dialysis and erythropoietin-stimulating agents (ESAs) have greatly decreased this bleeding tendency, bleeding is still reported in 24%–50% of patients with chronic renal failure with a fivefold increased risk of intracranial hemorrhage. The etiology of bleeding is multifactorial, as the complex interplay of components of the coagulation system get interrupted by uremia, dialysis, medications, and even the underlying anemia can tip the balance to the prothrombotic state in underlying renal disease. This chapter will focus on the hemostatic disorders associated with chronic renal disease, the laboratory findings, and the management of bleeding in these patients.

Pathophysiology: Uremic patients demonstrate disordered platelet/vessel interaction and disturbances in various stages of platelet function, which unfortunately is not reflected in the commonly used hemostatic parameters such as routine PT/aPTT testing.

Vessel Wall–Related Factors: There are abnormal platelet–vessel wall interactions due to decreased functional GpIIb/IIIa complex and decreased binding of fibrinogen and von Willebrand factor (VWF) to GpIIb/IIIa receptors, thereby impairing platelet adhesion. This could be related to uremic toxins, as it usually improves with dialysis.

Biochemical Changes Due to Uremia: Guanidinosuccinic acid (GSA) (a derivative of L-arginine) and L arginine, are both precursors of nitric oxide (NO) and tend to accumulate in uremic plasma. NO production is thereby increased in uremic patients, which leads to increased cyclic guanosine monophosphate (GMP), producing vascular relaxation and decreased platelet aggregation. GSA also impairs the secondary wave of ADP-induced platelet aggregation. Uremic toxins are also thought to impair the release and synthesis of thromboxane A2 due to dysregulated prostaglandin metabolism in platelets in advanced renal disease, which improves with dialysis.

Role of Red Cell Mass: Chronic renal disease results in decreased erythropoietin production leading to decreased red blood cell (RBC) mass. The use of erythropoietin and RBC transfusions historically has shown improvement in the bleeding time and resultant bleeding manifestations, which may be related to increased availability of ADP from the RBC's and increased NO scavenging by available hemoglobin. There are no clinical outcome studies correlating with more sensitive assays such as the platelet functional analyzer (PFA) and bleeding risk. The improved rheology of flowing blood

Transfusion Medicine and Hemostasis. https://doi.org/10.1016/B978-0-12-813726-0.00122-7

with the increased RBC mass allows platelets and clotting factors to effectively migrate toward the subendothelium and thus more efficiently form plugs over breaches within the vascular wall.

Role of the Fibrinolytic Pathway: Chronic renal disease can also result in decreased fibrinolysis secondary to low tissue plasminogen activator (TPA) and increased plasminogen activator inhibitor-1 (PAI-1). Occasionally, this may contribute to clotting within the shunt placed for dialysis and more systemic clotting problems.

Role of Platelets (Platelet Dysfunction): Although platelet numbers are usually normal, intrinsic platelet dysfunction in the form of altered secretion of ADP and serotonin from platelet granules (acquired storage pool defect), impaired platelet contractility and defective assembly of cytoskeletal proteins contribute to defective platelet activation.

External Factors: There is also continuous platelet activation at the dialyzer membrane and risk of heparin-induced thrombocytopenia with the use of heparin in the circuit. Recently, the use of heparin-free dialysis such as citrate, saline flushes, and direct antithrombin inhibitors has ameliorated this risk. Additionally, medications, including various antibiotics and antiinflammatory drugs may affect platelet function and also anticoagulants (those requiring renal excretion) can accumulate in the body and should be dosed according to renal function.

Nephrotic Syndrome: Nephrotic syndrome is described separately as it is a well-known hypercoagulable state, with an annual incidence of venous thromboembolic events (VTEs) reported at 1.02% per year. This rate is higher at 10% in the first 6 months of diagnosis. Commonly reported events are renal vein thrombosis, deep vein thrombosis, and pulmonary embolism, which are multifactorial in etiology. There is ongoing urinary loss of small anticoagulant proteins, such as antithrombin and plasminogen. In addition, elevated fibrinogen levels are seen, likely due to a secondary compensatory mechanism leading to increased protein loss, which then provides more substrate for thrombin and increased platelet aggregation. Dyslipidemia can contribute to accelerated atherosclerosis in these patients. Hypoalbuminemia is an independent risk factor with albumin levels <2.8 mg/dL associated with increased risk of thrombotic events, with every 1 mg/dL drop in albumin increasing VTE risk by 2.13-fold. Prophylactic anticoagulation should therefore be judiciously considered in these patients during times of increased albumin loss, especially with added VTE risk factors (prolonged immobilization, obesity, vascular access devices) while minimizing bleeding risk.

Clinical Manifestations: Bleeding as a result of platelet dysfunction can manifest as mucocutaneous bleeding, such as epistaxis, oral bleeding, bruising, genitourinary and gastrointestinal hemorrhage and increased bleeding from venipuncture sites or after invasive procedures. The incidence of ischemic stroke and subdural hematoma are higher in patients receiving hemodialysis hypothesized to be due to a sudden drop in intravascular volume and inadequate compensatory vascular responses. Bleeding from the gastrointestinal tract is also frequently seen in dialysis patients due to gastritis secondary to hypergastrinemia and intestinal telangiectasia. Spontaneous retroperitoneal hemorrhage, hepatic subcapsular hematomas, and hemorrhagic pericarditis leading to cardiac tamponade have also been rarely reported.

Diagnosis: Standard coagulation tests such as the prothrombin time (PT) and activated partial thromboplastin time (aPTT) are usually normal, although platelet counts can sometimes be lower than normal but are usually >80,000/μL. Dysfibrinogenemia is characterized by a low fibrinogen activity, low normal to normal fibrinogen antigen, and a prolonged thrombin time and reptilase time. Classically, bleeding time was used to evaluate the bleeding risk in uremic patients; however, it is a cumbersome test that cannot predict true bleeding risk. The PFA can be used to detect disturbances in VWF and platelet adhesion but is less sensitive to pick up some platelet secretion or activation defects. It is also known to give false-positive results in anemic states. Peripheral smear review can reveal acanthocytes and sometimes fragmented schistocytes. Herein lies the importance to not rely solely on tests to predict bleeding risk in surgery or during invasive procedures because there is a poor correlation between the laboratory coagulation abnormalities and clinically significant bleeding. In nephrotic states, appropriate testing for ATIII levels should be performed as well.

Management: Typically, the coagulopathy of renal failure is corrected by dialysis, with peritoneal dialysis being more effective than hemodialysis. A target serum creatinine level below 6 mg/dL has been shown to alleviate uremic platelet dysfunction. Patients who have not begun dialysis, but still have significant renal dysfunction (such as encountered in the patient who requires a kidney biopsy to establish a diagnosis), will benefit in maintaining an RBC mass in a more normal range, aiming for a hematocrit of approximately 30%. This level can be achieved more rapidly by RBC transfusions, or given time with the use of ESAs. The effects of ESAs can take weeks, although there is an increase in the number of reticulated more metabolically active platelets within a week. The use of ESAs carries a black box warning, due to increased mortality seen from myocardial infarction, stroke, heart failure, and thrombosis when hemoglobin levels rose >11 g/dL and as such, all correctable causes of anemia should be treated before considering their use. When more acute control of bleeding is desired, one may use intravenous desmopressin (DDAVP) (0.3–0.4 mcg/kg—maximum dose 20 mcg), which acts within 30 minutes, but effects are short lived. DDAVP asserts its action by inducing the release of VWF from endothelial cells, leading to more efficient platelet adhesion; however, tachyphylaxis can develop shortly when the VWF stores have all been used up. In addition, the need for fluid restriction to avoid hyponatremic seizures can pose a barrier to its use. In more urgent scenarios, cryoprecipitate (rich in factor VIII, VWF, and fibrinogen) has been used to correct the platelet dysfunction, with beneficial effects seen within 4–12 hours. However, the clinical response can be variable, and the risk of transfusion-related adverse events makes this a less attractive treatment modality. Conjugated estrogens are a good alternative in patients where a prolonged hemostatic effect is desired. Intravenous doses of 0.6 mg/kg have been used both in males and females for five consecutive days, with maximum effects seen in the first 5–7 days, although effects can last for 14–21 days. Estrogen exerts its action by increasing the synthesis of VWF and possibly by blocking the effect of NO in its multiple influences, both on the blood vessel wall and on platelet activation and aggregation. Platelet transfusions have also been used in the face of acute bleeding but are only transiently effective, as the transfused platelets will become ineffective in the uremic environment.

Further Reading

Jalal, D. I., Chonchol, M., & Targher, G. (2010). Disorders of hemostasis associated with chronic kidney disease. *Semin Thromb Hemost*, *36*(1), 34–40.

Lionaki, S., Derebail, V. K., Hogan, S. L., Barbour, S., Lee, T., Hladunewich, M., et al. (2012). Venous thromboembolism in patients with membranous nephropathy. *Clin J Am Soc Nephrol*, *7*(1), 43–51.

Lutz, J., Menke, J., Sollinger, D., Schinzel, H., & Thurmel, K. (2014). Haemostasis in chronic kidney disease. *Nephrol Dial Transplant*, *29*(1), 29–40.

Ozkan, G., & Ulusoy, G. (2013). Bleeding diathesis in hemodialysis patients. In P. H. S (Ed.), *Hemodialysis*. InTech.

Pavord, S., & Myers, B. (2011). Bleeding and thrombotic complications of kidney disease. *Blood Rev*, *25*(6), 271–278.

CHAPTER 123

Bleeding Risks With Cancer

Michael A. Briones, DO

Etiology and Pathogenesis

Vascular Etiology: The tumor microenvironment of many malignancies is rich with angiogenesis-stimulating growth factors; the most well-known is vascular endothelial growth factor (VEGF). The neoangiogenesis are typically unorganized and structurally dissimilar to the nonneoplastic vasculature and are more prone to bleeding. Malignant cells can also directly invade the vasculature, such as with lymphomatous involvement of the gastrointestinal (GI) tract leading to hematochezia. Bleeding may also occur within the body of the tumor, such as in hepatocellular carcinoma. Agents targeting angiogenesis have emerged and are in widespread use in both solid and hematologic malignancies, which can affect vessel and cause bleeding. VEGF inhibitors have been shown to cause predisposition to thrombosis and bleeding and after inhibition of VEGF signaling, have shown multiplicity of actions on vascular walls and interaction on the coagulation system. It is thought that inhibition of VEGF diminishes the regenerative capacity of endothelial cells and cause defects that expose procoagulant phospholipids on the luminal plasma membrane or underlying matrix, leading to thrombosis or hemorrhage. However, endothelial cell defects alone are unlikely to explain life-threatening hemorrhage in patients on anti-VEGF therapy and likely multifactorial with the complex interaction of weakening of the wall of major vessels by tumor erosion, necrosis, cavitation, and other concurrent pathological conditions are likely to play a central role. Other agents targeting the VEGF pathway such as sorafenib and sunitinib have also been linked to an increased incidence in bleeding complications.

Platelet Disorders: Quantitative and qualitative platelet defects may occur as a result of the malignancy and/or its treatment. Underproduction of platelets can be seen with bone marrow infiltration by the malignancy such as leukemias and metastatic solid tumors. Most patients with acute leukemias have significant thrombocytopenia at the time of diagnosis due to leukemic involvement of the marrow. Thrombocytopenia (along with anemia and leukopenia) is a common complication of cytotoxic chemotherapeutic agents. The degree and duration of myelosuppression varies greatly depending on the chemotherapeutic agent(s) used as well as patient factors, such as age and prior treatment(s). Radiation therapy, especially when targeting the pelvis, may also cause myelosuppression and subsequent thrombocytopenia.

Hypersplenism is seen in a variety of hematologic malignancies but can also result from other cancer-related causes, such as portal hypertension due to hepatic metastases. Immune thrombocytopenia can also be seen in lymphoproliferative disorders, chronic leukemia such as chronic lymphocytic leukemia (CLL), Hodgkin's lymphoma,

Transfusion Medicine and Hemostasis. https://doi.org/10.1016/B978-0-12-813726-0.00123-9

and occasionally solid tumors. Thrombocytopenia secondary to sepsis/infections, which are relatively common occurrences in patients with cancer, may also lead to thrombocytopenia or worsen existing thrombocytopenia.

Qualitative platelet defects may occasionally occur in patients with cancer. Defects in the platelet membrane, abnormal platelet granules, acquired storage pool disorders, and decreased prostaglandin synthesis have been described. Patients with myeloproliferative neoplasms (MPNs), such as essential thrombocytosis, commonly have bleeding complications with platelet counts >1 million/μL. The specific abnormality leading to platelet impairment has not been well established but may be related to an acquired von Willebrand syndrome.

In patient with multiple myeloma, there is associated platelet dysfunction. The paraproteins in these disorders are thought to interact with platelet surface glycoproteins and impair platelet adherence, activation, and/or aggregation. Cancer patients with impaired renal function from a number of causes may have impaired platelet function due to uremia. The mechanism is thought to be secondary intrinsic platelet defects, abnormalities in the interaction between the platelet and endothelium, or the buildup of toxins normally cleared by the kidney.

Coagulation and Fibrinolysis: Numerous factors related to malignancy can disrupt the normal coagulation cascade. Vitamin K deficiency, due to malnutrition or malabsorption, is common and results in deficiencies of vitamin-dependent factors. Hepatic dysfunction is also a common cause of quantitative abnormalities in coagulation proteins and can lead to a decrease in coagulation factors synthesized in the liver.

The development of specific coagulation factor inhibitors is a rare but potentially serious event. The most frequent inhibitor is directed against FVIII and has been reported in a number of solid and hematologic malignancies, including CLL, lung cancer, and prostate cancer. The pathogenesis is felt to involve the development of a polyclonal IgG autoantibody targeting a tumor antigen that resembles FVIII. Rarely, autoantibodies may also occur against factor V (FV).

Acquired von Willebrand disease is not only a rare but also potentially serious disorder that may arise in the setting of malignancy and can be seen in monoclonal gammopathy of uncertain significance as well as other plasma cell disorders, MPNs, CLL, non-Hodgkin lymphoma, and solid tumors such as Wilms tumor have been associated with this condition. Mechanisms are thought to be development of specific autoantibody or a nonspecific antibody that bonds to von Willebrand factor (VWF), resulting in destruction and removal of the VWF-FVIII complex from circulation, adsorption of the VWF-FVIII complexes onto malignant cells is seen in plasma cell disorders and in MPNs with significant thrombocytosis, the excess platelets bind the large von Willebrand multimers leading to lose of VWF protein. FX deficiency is seen in approximately 9% of patients with AL amyloidosis, and of these patients, 50% will develop significant bleeding. The mechanism is thought to be due to adsorption of FX to the AL fibrils, resulting in increased clearance from the circulation.

Disseminated intravascular coagulation (DIC) is one of the most frequently encountered conditions in malignancy that leads to bleeding. The pathogenesis of DIC involves excessive thrombin generation from release of tissue factor, cancer procoagulant and promotes annexin II receptor expression resulting in massive activation of the clotting cascade. Widespread fibrin formation and deposition occurs leading to thrombotic microangiopathy leading to consumption of clotting factors and platelets as well as secondary fibrinolysis predispose to bleeding. DIC has been reported in multiple solid tumors as well as hematologic malignancies and is especially prominent in acute promyelocytic leukemia (APL); most patients will manifest with DIC and/or hyperfibrinolysis, typically at diagnosis or once cytotoxic chemotherapy is initiated. Morbidity and mortality can be high despite early recognition and therapy. Hyperfibrinolysis, typically in conjunction with DIC is associated with excessive fibrinolysis and significant bleeding. The pathogenesis is likely due to overproduction of plasminogen activators by the malignant cells or depletion of the fibrinolytic inhibitors, such as α2-antiplasmin.

Thrombotic thrombocytopenic purpura–hemolytic uremic syndrome (TTP-HUS) is another microangiopathic hemolytic anemia that is observed in patients with malignancies. The pathogenesis is thought to be due to tumor microemboli, which in turn lead to fragmentation of erythrocytes and not associated with autoantibodies against ADAMTS13. Medications including gemcitabine, mitomycin C, cisplatin, and cyclosporin have also been implicated with the development of TTP. In these cases, the etiology may involve damage to the endothelium, although the pathogenesis remains poorly understood.

Clinical Manifestations: Local bleeding from the primary tumor itself may cause symptoms such as hemoptysis in lung cancer, hematochezia or melena in colorectal cancer, or hematuria in bladder or prostate cancer. The most common type of bleeding in patients treated with bevacizumab is epistaxis; severe bleeding most often manifests as pulmonary and GI hemorrhage. With quantitative or qualitative platelet defects, bleeding tends to be mucocutaneous bleeding such as epistaxis, gingival bleeding, menorrhagia, petechiae, or ecchymoses (generally small and superficial). Patients with quantitative coagulation factor deficiencies due to vitamin K deficiency or hepatic dysfunction rarely develop deep tissue bleeding and have mucocutaneous bleeding.

Approximately 25% of patients with an FVIII inhibitor can have bleeding in cutaneous, muscular, soft tissue, and/or mucosal surfaces. Unlike patients with congenital hemophilia, these patients rarely experience hemarthroses. However, they may experience retroperitoneal, retropharyngeal, or CNS bleeding, which explains the high morbidity and mortality from this disorder. Patient with AvWDs typically experience mucosal or cutaneous bleeding. More severe bleeding from other sites including the GI tract can be seen, and severe bleeding may occur after invasive procedures. Bleeding manifestations in DIC and hyperfibrinolysis is generalized systemic bleeding with ecchymoses, petechiae, mucosal bleeding, injection sites, line site, and internal bleeding such as CNS bleeding may occur (more common in APL).

The classic pentad of anemia, thrombocytopenia, fever, renal dysfunction, and neurologic changes is generally not observed with malignancy-associated TTP or medication-associated thrombotic microangiopathic anemia. Bleeding manifestations are most commonly seen on cutaneous or mucosal surfaces.

Diagnosis: A thorough history and physical examination with importance to the bleeding history is utmost important with emphasis on the type of cancer, location, chemotherapy received, medications taken. A thorough physical examination may yield information about the origin of bleeding. Laboratory testing should include a complete blood count, peripheral blood smear, prothrombin time (PT), and partial thromboplastin time (PTT); fibrinogen and thrombin time can indicate possible disorders in platelets or the clotting cascade, thus narrowing the differential diagnosis. For anatomic or vascular causes of bleeding are generally self-evident, and the diagnostics used are dependent on the location of the bleeding such as endoscopy is used for bleeding of the GI or genitourinary (GU) tract just as bronchoscopy may be used for patients presenting with hemoptysis. Focused radiographic imaging may also be useful, as in the case of retroperitoneal or CNS bleeding. Diagnosis of thrombocytopenia is readily evident on complete blood count. The cause may be obvious from the clinical scenario, such as in a patient receiving cytotoxic chemotherapy. Evaluation of the peripheral blood smear is also important, and attention should be paid to the platelet and erythrocyte morphology.

When hypersplenism is present, splenomegaly may be apparent on physical examination. Radiographic imaging (e.g., ultrasound) may also show an enlarged spleen, though some patients may not have significant splenic enlargement. Qualitative platelet abnormalities require more specialized testing. When multiple coagulation factors are deficient, the prothrombin time (PT) and/or partial thromboplastin time (PTT) may be prolonged. The diagnosis of hepatic dysfunction is generally straightforward based on evaluation of liver function tests. The diagnosis of FVIII inhibitors, AvWS, DIC/hyperfibrinolysis are discussed in previous chapters.

Management: A wide variety of therapeutic options are available, and the individual treatment depends on a multitude of factors including the etiology, site, and degree of bleeding. Local therapies may be used depending on the degree of bleeding and the patient's clinical status. Options include hemostatic dressings (e.g., fibrin sealants) and agents (e.g., topical aminocaproic acid) that can be directly applied to the bleeding site. Palliative surgery or radiation therapy can be used to control bleeding from a primary or metastatic lesion. Endoscopic interventions may be used for localized GI, GU, or pulmonary lesions. Interventional radiology procedures such as arterial embolization (e.g., in renal cell carcinoma) may be an option in selected cases.

When severe bleeding results from thrombocytopenia, platelet transfusions should be given, with a goal platelet count >50,000/μL. In patients who do not have any apparent bleeding, prophylactic platelet transfusions may be given when the platelet count is ≤10,000/μL. Antifibrinolytic agents, including aminocaproic acid or tranexamic acid, can be used as an adjunct.

The use of exogenous thrombopoietic growth factors to offset thrombocytopenia due to chemotherapy has been evaluated in multiple settings using recombinant human thrombopoietin (TPO) and pegylated recombinant human megakaryocyte growth and development factor (PEG-rHuMGDF), both of which stimulate the TPO receptor. However, the use of these agents has not gained widespread. Over the past few years, eltrombopag and romiplostim, both TPO receptor agonists, have gained approval by the Food and Drug Administration (FDA) for the treatment of chronic immune thrombocytopenia but have not been routinely used, and further trials are ongoing to evaluate the utility of romiplostim and eltrombopag in improving chemotherapy-induced thrombocytopenia in patients with malignancies.

When a quantitative platelet defect is due to hypersplenism, splenectomy or splenic irradiation can be considered depending on the etiology and the degree of thrombocytopenia. Autoimmune thrombocytopenia due to CLL should be treated in a similar fashion to immune thrombocytopenia in patients who do not have CLL. Among the standard treatment options are high doses of corticosteroids, intravenous immunoglobulin (IVIG), and rituximab.

The management of qualitative platelet disorders is highly contingent on the underlying etiology. Thrombocytosis associated with MPNs should be addressed with cytoreduction and plateletpheresis if needed for severe bleeding. Similarly, plasmapheresis and/or cytoreduction may be required in patients with lymphoproliferative disorders and a concomitant qualitative platelet abnormality (or hyperviscosity). In other situations, adjunctive desmopressin or aminocaproic acid may be helpful.

For patients with bleeding associated with vitamin K deficiency, exogenous vitamin K should be given, preferably via the enteral route unless there is concern for malabsorption. Intravenous vitamin K can be used in the setting of significant bleeding. When bleeding is due to liver dysfunction, use of plasma products such as fresh frozen plasma and cryoprecipitate (to raise fibrinogen >100 mg/dL) may be necessary depending on the severity of bleeding, and concomitant thrombocytopenia may require platelet transfusion.

For patients with an FVIII inhibitor, treatment of the underlying disease is essential. Management of the bleeding disorder is contingent on the degree of bleeding and level of inhibitor. For patients with low inhibitor levels (<5 BU/mL) and/or mild bleeding, treatment with desmopressin will increase FVIII levels by increasing VWF levels. Recombinant FVIII concentrates can also be used in these patients. On the other hand, for those with more severe bleeding or higher inhibitor titers (≥5 BU/mL), a bypassing agent is necessary to restore hemostasis, which include recombinant FVIIa or plasma-derived activated prothrombin complex concentrates (aPCCs). In patients who have persistent inhibitors, immunosuppressive agents (e.g., corticosteroids, cyclophosphamide, and rituximab) are the cornerstones for eradicating the inhibitor. IVIG, plasma exchange, and immune tolerance induction are options for patients with persistent or very-high-titer inhibitors.

The goals of treatment in AvWS are to control acute bleeding symptoms and minimize bleeding episodes. In general, treatment of the underlying disorder, where possible, should be undertaken, as this will sometimes lead to long-term resolution of the AvWS. For all patients, acute bleeding should be managed using desmopressin, antifibrinolytic agents (aminocaproic acid and tranexamic acid), and/or VWF-containing concentrates. The management of DIC entails treating the underlying malignancy. This is especially true for APL, where treatment with all-transretinoic acid or arsenic trioxide should be given promptly when the diagnosis is suspected, given the potential for catastrophic outcomes. For all patients, coagulation factors and fibrinogen should be replaced using plasma or cryoprecipitate as needed. Platelets should also be given, especially if bleeding is present. When hyperfibrinolysis accompanies DIC, as in prostate cancer, the optimal management strategy is treatment of the underlying malignancy. Acute bleeding can be treated with platelet, plasma, or cryoprecipitate transfusions along with antifibrinolytic agents, although these have not been well studied and are generally only temporizing measures. While the role of plasma exchange for idiopathic TTP is well established, TTP associated with malignancy or chemotherapeutic agents does not respond to plasma exchange. Unfortunately, treatment is limited to supportive measures, treatment of the underlying disorder and/or cessation of the offending agent(s).

Further Reading

Escobar, M. A. (2012). Bleeding in the patient with a malignancy: Is it an acquired factor VIII inhibitor? *Cancer, 118*, 312–320.

Falanga, A., Russo, L., & Milesi, V. (2014). The coagulopathy of cancer. *Curr Opin Hematol, 21*, 423–429.

George, J. N. (2011). Systemic malignancies as a cause of unexpected microangiopathic hemolytic anemia and thrombocytopenia. *Oncology (Williston Park), 25*, 908–914.

Green, D. (2007). Management of bleeding complications of hematologic malignancies. *Semin Thromb Hemost, 33*, 427–434.

Korte, W. (2011). Thrombosis and bleeding in cancer patients. In I. Oliver (Ed.), *The MASCC textbook of cancer supportive care and survivorship* (pp. 174–178). New York: Springer.

Pereira, J., & Phan, T. (2004). Management of bleeding in patients with advanced cancer. *Oncologist, 9*, 561–570.

Thachil, J., Falanga, A., Levi, M., Liebman, H., & Di Nisio, M. (2015). Management of cancer-associated disseminated intravascular coagulation: Guidance from the SSC of the ISTH. *J Thromb Haemost, 13*, 671–675.

Tiede, A., Rand, J. H., Budde, U., Ganser, A., & Federici, A. B. (2011). How I treat the acquired von Willebrand syndrome. *Blood, 117*, 6777–6785.

Valent, J., & Schiffer, C. A. (2011). Thrombocytopenia and platelet transfusions in patients with cancer. *Cancer Treat Res, 157*, 251–265.

Zangari, M., Elice, F., Tricot, G., & Fink, L. (2009). Bleeding disorders associated with cancer dysproteinemias. *Cancer Treat Res, 148*, 295–304.

CHAPTER 124

Disseminated Intravascular Coagulopathy

Surbhi Saini, MBBS and Amy L. Dunn, MD

Disseminated intravascular coagulopathy (DIC) is defined as:

> *An acquired syndrome characterized by the intravascular activation of coagulation with loss of localization arising from different causes. It can originate from and cause damage to the microvasculature, which if sufficiently severe, can produce organ dysfunction.*
>
> International Society on Thrombosis and Haemostasis (ISTH).

DIC is a clinicopathological syndrome with variable clinical severity that can range from isolated laboratory abnormalities to massive hemorrhage and/or thrombosis. The systemic activation of dysregulated coagulation, the hallmark of DIC, occurs secondary to a multitude of underlying conditions. Treatment is focused on treating the underlying disease with concurrent replacement of coagulation factors by blood product transfusion.

Pathophysiology: To maintain normal hemostasis, three primary components of the coagulation system—vascular integrity, coagulation factors, and platelets—are very well coordinated to result in a delicate balance between bleeding (anticoagulation) and clotting (procoagulation). Any disruption of this balance can cause DIC. The most common conditions underlying DIC are sepsis and cancer; other less common causes include significant trauma and burns, obstetrical complications (e.g., retained products of conception, placental abruption), large vascular malformations, severe allergic reactions, and envenomation. Any of these conditions can cause a generalized, unregulated activation of the coagulation system leading to widespread thrombin generation and deposition of fibrin-rich microthrombi throughout the circulation. Resultant obstruction of microvasculature causes hypoxic end-organ damage, mainly to the vital organs such as liver, kidney, and brain. Additionally, excessive consumption of endogenous anticoagulant proteins (protein C, protein S, and antithrombin [AT]), and disruption of hemostatic anticoagulant pathways, contributes to the coagulation disarray. This is aggravated by the relative insufficiency of the fibrinolytic system to clear the microthrombi from the circulation.

During normal hemostasis, tissue factor (TF) expression is restricted to the surface of cells not exposed to flowing blood (subendothelium, fibroblasts). Microvascular injury exposes TF to circulating FVII, initiating coagulation. In DIC, inflammatory cytokines (TNF-alpha, IL-1, IL-6) induce intravascular monocytes and macrophages to express TF. This intravascular expression of TF initiates exaggerated, pathologic coagulation, ultimately leading to the depletion of coagulation factors, which in turn can predispose to hemorrhage. It is this simultaneous development of pathologic thrombosis and hemorrhage that is the hallmark of DIC. The generation of thrombin also leads

Transfusion Medicine and Hemostasis. https://doi.org/10.1016/B978-0-12-813726-0.00124-0

to the exposure of negatively charged phospholipids on the surface of platelets, which plays a role in sustaining thrombin generation, in addition to the platelet microparticles generated by cell damage.

Clinical Manifestations: Clinical presentation of DIC is heterogeneous and to some extent, reflective of the underlying pathophysiology. Severity ranges from isolated laboratory derangements (often referred to as compensated or nonovert DIC) to multi-organ failure and death, indicative of the dynamic nature of DIC. Microvascular occlusion leads to renal, cardiac, central nervous system, and/or pulmonary failure. Digital gangrene and purpura fulminans can occur as a result of widespread skin necrosis. A subset of patients present with primarily hemorrhagic symptoms, most commonly, generalized mucocutaneous bleeding. Mortality rates are high, ranging from 34% to 86%.

Diagnosis: There is no gold standard single test for the diagnosis of DIC. The presence of a triggering condition, clinical signs/symptoms, and supportive laboratory data aids in the diagnosis, and this can be supported by the use of a five-step scoring algorithm developed by the ISTH (Fig. 124.1). This scoring system was designed to assist in diagnosis and to be utilized as a reference standard. Since its inception, it has proven both sensitive and specific for the diagnosis of overt DIC, and increasing scores strongly correlate with patient mortality.

The complete blood count is used to assess platelet number and the degree of anemia associated with the underlying illness. In more advanced DIC, the peripheral smear will show evidence of thrombotic microangiopathy (schistocytes, thrombocytopenia, anemia). Varying degrees of thrombocytopenia may be present in DIC; however, thrombocytopenia may also be present as a result of many of the underlying causes of DIC such as malignancy or sepsis.

The activated partial thromboplastin time (aPTT) and prothrombin time (PT) are in vitro measurements of the coagulation pathway and can be used to assess factor activation and consumption during DIC. In nearly half of the patients, aPTT and PT would be normal or shortened, reflecting the accelerated thrombin generation phase of DIC. The PT is typically more affected than the PTT in patients with DIC.

Fibrinogen levels below 150 mg/dL are typically the result of consumption (e.g., DIC) or decreased production (e.g., liver failure). Levels less than 100 mg/dL significantly increase the risk of bleeding. Fibrinogen is also an acute-phase reactant, and caution must be used when interpreting fibrinogen levels in early DIC, where plasma levels can remain normal or even be elevated.

The thrombin time measures the time to conversion of fibrinogen to fibrin. Decreased concentrations of fibrinogen, decreased clearance of fibrin degradation products, and the presence of heparin prolong the thrombin time.

Fibrin degradation products (FDPs) are formed when fibrinogen, cross-linked fibrin, and non–cross-linked fibrin are degraded by plasmin. The extent of fibrin formation in DIC can be assessed indirectly by the generation of FDPs. D-dimers are a result of plasmin breakdown of cross-linked fibrin and are proposed by the ISTH as the FDP marker of choice when diagnosing DIC. The D-dimer test has a high negative predictive value but is limited by a low positive predictive value.

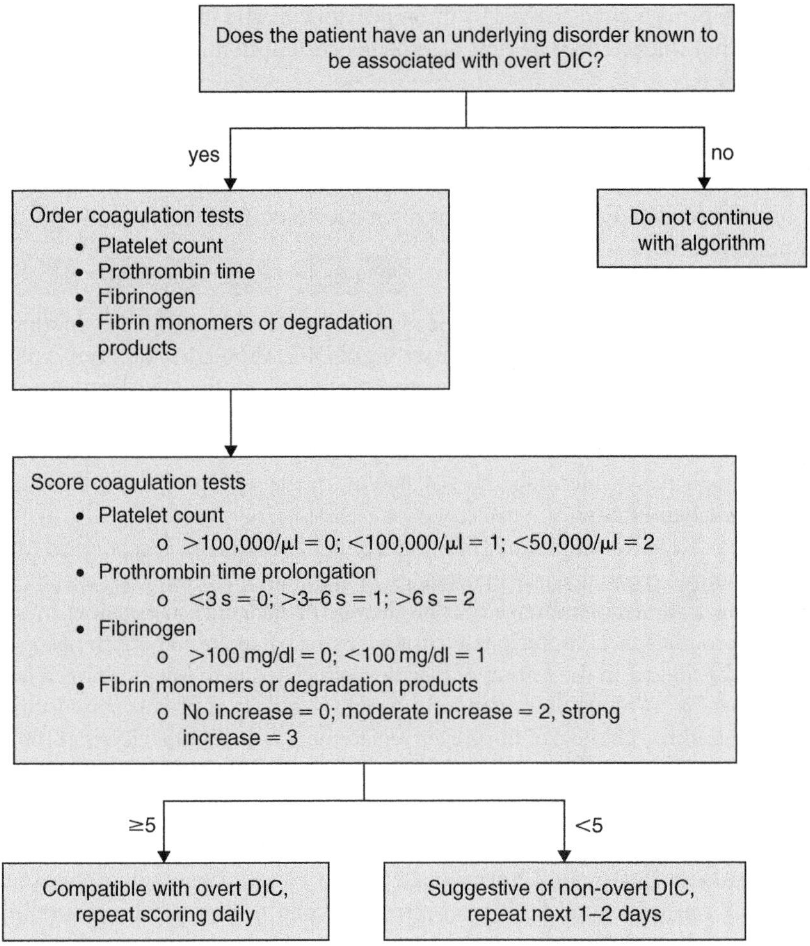

FIGURE 124.1 Diagnostic algorithm for the diagnosis of disseminated intravascular coagulopathy (DIC).

Levels of the endogenous coagulation inhibitor AT are typically decreased in DIC. Levels less than 80% may be associated with thrombosis. Protein C and S may also be reduced in patients with DIC. The current ISTH criteria concluded that routine testing of these parameters showed no additional value to the 4 test scoring system. However, in selected circumstances, testing may prove beneficial.

Differential Diagnosis: Liver failure and severe vitamin K deficiency can mimic DIC. To distinguish between liver disease and vitamin K deficiency, it is helpful to measure factor V and factor VII levels. In liver disease both factors will be decreased, but in vitamin K deficiency only factor VII will be low. Thrombotic thrombocytopenic purpura and hemolytic uremic syndrome are also thrombotic microangiopathies but rarely show evidence of consumption of clotting proteins. Heparin-induced thrombocytopenia is characterized by thrombocytopenia and increased risk of thrombosis but is not characterized by clotting protein consumption.

Management: The cornerstone of DIC management is identification and appropriate treatment of the underlying trigger with ongoing supportive care. Management strategies should be individualized and reassessed frequently (Table 124.2). Replacement of consumed coagulation components is often necessary. Table 120.1 shows potential treatment options.

Blood Product Support: There are no randomized clinical trials (RCT) on which to base guidelines for platelet therapy in the setting of DIC. Although it has been hypothesized, there is currently no evidence to support the hypothesis that platelet transfusions may worsen DIC by providing a phospholipid surface for ongoing thrombin generation. Within our practice, we aim to keep the platelet count greater than 20,000/μL in patients at high risk of bleeding and >50,000/μL in the setting of active bleeding or preprocedural hemostasis.

The role of fresh frozen plasma (FFP) replacement in DIC has been studied in one small randomized controlled trial in neonates. In that study, plasma and platelet therapy did not show a survival advantage over no therapy or exchange transfusion. Infusion of plasma products or cryoprecipitate (in the setting of severe hypofibrinogenemia) should be considered in the setting of bleeding associated with low fibrinogen levels; however, there is currently no evidence for prophylactic use of plasma support in the absence of bleeding. The goal of fibrinogen replacement is to maintain levels at 100 mg/dL to prevent or treat bleeding. The use of recombinant factor VIIa is contraindicated in DIC because of its prothrombotic risk.

Anticoagulant Pathway Therapy: Depletion of anticoagulant proteins such as protein C, protein S, and AT has been postulated to play a role in sustaining DIC, and therefore replacement of these proteins has been explored as a therapeutic option. Over the past decade, evidence has been slowly mounting to support this approach. Subsequently, in a large RCT of recombinant-human activated protein C (rh-APC), adult patients with sepsis treated with rh-APC infusions for 96 hours demonstrated decreased overall mortality at 28 days compared with controls; however, there was a trend toward higher bleeding rates in the treated arm. Subsequent studies have failed to establish a benefit of this therapy in pediatric or adult patients with DIC in less severe sepsis and septic shock, including those with lowered Protein C levels. Smaller, pilot studies have suggested that normalization of serum protein C levels using protein C concentrates (plasma derived, zymogen form) is safe and potentially beneficial.

AT concentrates have also been shown, in animal models of sepsis, to improve coagulation and survival rates. However, a double-blind, placebo-controlled, multicenter phase III clinical trial in patients with severe sepsis failed to demonstrate any survival advantage of high-dose AT and was associated with an increased risk of hemorrhage when administered with heparin. A *post hoc* analysis suggested improved organ failure and mortality rates in patients without heparin and treated early in severe sepsis. Additional trials are needed to clearly define the indications and dosing strategies.

Recombinant human soluble thrombomodulin (rhTM) has been shown to decrease thrombin formation by binding to thrombin and acting in complex with activated protein C. This agent has also shown antiinflammatory properties and has been studied in a randomized, double-blind controlled trial of patients with DIC due to malignancy or infection who were randomized to rhTM or low-dose heparin. 66% of patients in the rhTM arm resolved their DIC compared with 50% of patients in the heparin arm. In a subsequent placebo-controlled randomized controlled trial, rhTM failed to show any benefit, and currently, this drug is not available in the United States.

Anticoagulants: Although there are no data to support the routine use of anticoagulants in DIC, with evidence of thromboembolic disease, anticoagulation with heparin, either unfractionated or low-molecular weight, should be considered. Microvascular thrombosis can generally be treated with low-dose heparin, whereas large vessel disease may require full therapeutic dosage. Due consideration should be given to venous thromboembolism (VTE) prophylaxis in acutely ill patients without bleeding and at low risk for bleeding. To date, there are no trials evaluating the clinical usefulness of direct oral anticoagulants in patients with DIC.

Antifibrinolytic Therapy: Hyperfibrinolysis can occur in patients with DIC, particularly in the setting of acute promyelocytic leukemia. This can be recognized by an abnormally short euglobulin clot lysis time, and antifibrinolytic agents such as aminocaproic acid or tranexamic acid may be of use in such a scenario. These agents should be used with caution in the setting of hematuria because of the risk of renal collecting system thrombosis. Suggested dosage and indications for all of the above agents are included in Table 124.1.

TABLE 124.1 International Society on Thrombosis and Hemostasis Diagnostic Algorithm for the Diagnosis of Overt Disseminated Intravascular Coagulopathy (DIC)

1. **Risk assessment: Does the patient have an underlying disorder known to be associated with overt DIC?**
   If yes: proceed
   If no: do not use this algorithm
2. **Order global coagulation tests** (platelet count, prothrombin time (PT), fibrinogen, D-dimer levels)
3. **Score global coagulation test results**
   (i) Platelet count: >100 = 0; <100 = 1; <50 = 2
   (ii) Elevated fibrin-related marker (D-dimer, FDPs): no increase = 0; moderate increase = 2; strong increase = 3
   (iii) Prolonged prothrombin time: <3 second = 0; 3–6 second = 1; >6 second = 2
   (iv) Fibrinogen level: >1.0 g/L = 0; <1.0 g/L = 1
4. **Calculate score**
5. If >5: compatible with overt DIC; repeat scoring daily
   If <5: suggestive (not affirmative) for nonovert DIC; repeat next 1–2 days

TABLE 124.2 Management Guidelines for Disseminated Intravascular Coagulopathy (DIC)

Product	Dose	Indication
Aminocaproic acid	100 mg/kg PO every 6 hour 33.3 mg/kg/hour IV	DIC associated with hyper fibrinolytic states
Cryoprecipitate or fibrinogen concentrate	1 unit per 10 kg of body weight	Clinically relevant bleeding; fibrinogen < 100 mg/dL unresponsive to FFP
Fresh frozen plasma	15–30 cc/kg	Clinically relevant bleeding; INR ≥2; 1.5-fold prolongation of aPTT; fibrinogen < 100 mg/dL; peri-op hemostasis for invasive procedures
Heparin	500 IU/hour (no loading dose) 5–10 IU/kg/hour low dose	DIC with evidence of ongoing thrombin generation unresponsive to conventional measures
Platelet product	Per domestic transfusion guidelines	Maintain plts > 20×10^9/L; plts > 50×10^9/L with major bleeding; periop hemostasis for invasive procedures

Further Reading

Abraham, E., Laterre, P. F., Garg, R., Levy, H., Talwar, D., Trzaskoma, B. L., et al. (2005). Drotrecogin alfa (activated) for adults with severe sepsis and a low risk of death. *N Engl J Med*, *353*(13), 1332–1341.

Baratto, F., Michielan, F., Meroni, M., Dal Palu, A., Boscolo, A., & Ori, C. (2008). Protein C concentrate to restore physiological values in adult septic patients. *Intensive Care Med*, *34*(9), 1707–1712.

Bernard, G. R., Vincent, J. L., Laterre, P. F., LaRosa, S. P., Dhainaut, J. F., Lopez-Rodriguez, A., et al. (2001). Efficacy and safety of recombinant human activated protein C for severe sepsis. *N Engl J Med*, *344*(10), 699–709.

Decembrino, L., D'Angelo, A., Manzato, F., Solinas, A., Tumminelli, F., De Silvestri, A., et al. (2010). Protein C concentrate as adjuvant treatment in neonates with sepsis-induced coagulopathy: A pilot study. *Shock*, *34*(4), 341–345.

Eid, A., Wiedermann, C. J., & Kinasewitz, G. T. (2008). Early administration of high-dose antithrombin in severe sepsis: Single center results from the KyberSept-trial. *Anesth Analg*, *107*(5), 1633–1638.

Gando, S., Saitoh, D., Ogura, H., Mayumi, T., Koseki, K., Ikeda, T., et al. (2008). Natural history of disseminated intravascular coagulation diagnosed based on the newly established diagnostic criteria for critically ill patients: Results of a multicenter, prospective survey. *Crit Care Med*, *36*(1), 145–150.

Hoffman, M., & Dargaud, Y. (2012). Mechanisms and monitoring of bypassing agent therapy. *J Thromb Haemost*, *10*(8), 1478–1485.

Levi, M., de Jonge, E., & van der Poll, T. (2001). Rationale for restoration of physiological anticoagulant pathways in patients with sepsis and disseminated intravascular coagulation. *Crit Care Med*, *29*(7 Suppl.), S90–S94.

Levi, M., & Meijers, J. C. (2011). Dic: Which laboratory tests are most useful. *Blood Rev*, *25*(1), 33–37.

Ranieri, V. M., Thompson, B. T., Barie, P. S., Dhainaut, J. F., Douglas, I. S., Finfer, S., et al. (2012). Drotrecogin alfa (activated) in adults with septic shock. *N Engl J Med*, *366*(22), 2055–2064 Epub 2012/05/24.

Singh, B., Hanson, A. C., Alhurani, R., Wang, S., Herasevich, V., Cartin-Ceba, R., et al. (2013). Trends in the incidence and outcomes of disseminated intravascular coagulation in critically ill patients (2004-2010): A population-based study. *Chest*, *143*(5), 1235–1242.

Squizzato, A., Hunt, B. J., Kinasewitz, G. T., Wada, H., Ten Cate, H., Thachil, J., et al. (2016). Supportive management strategies for disseminated intravascular coagulation. An international consensus. *Thromb Haemost*, *115*(5), 896–904.

Taylor, F. B., Jr., Toh, C. H., Hoots, W. K., Wada, H., & Levi, M. (2001). Towards definition, clinical and laboratory criteria, and a scoring system for disseminated intravascular coagulation. *Thromb Haemost*, *86*(5), 1327–1330.

Toh, C. H., & Downey, C. (2005). Back to the future: Testing in disseminated intravascular coagulation. *Blood Coagul Fibrinolysis*, *16*(8), 535–542.

Umemura, Y., Yamakawa, K., Ogura, H., Yuhara, H., & Fujimi, S. (2016). Efficacy and safety of anticoagulant therapy in three specific populations with sepsis: A meta-analysis of randomized controlled trials. *J Thromb Haemost*, *14*(3), 518–530.

Warren, B. L., Eid, A., Singer, P., Pillay, S. S., Carl, P., Novak, I., et al. (2001). Caring for the critically ill patient. High-dose antithrombin III in severe sepsis: A randomized controlled trial. *JAMA*, *286*(15), 1869–1878.

Yamakawa, K., Aihara, M., Ogura, H., Yuhara, H., Hamasaki, T., & Shimazu, T. (2015). Recombinant human soluble thrombomodulin in severe sepsis: A systematic review and meta-analysis. *J Thromb Haemost*, *13*(4), 508–519.

CHAPTER 125

Coagulopathy in Sickle Cell Disease

Deepa Manwani, MD, Jennifer Davila, MD and Caterina P. Minniti, MD

Sickle cell disease (SCD) results from a single base pair change in the β-globin subunit, yet the downstream effects and complex manifestations that result are protean and variable. The abnormal HbS is insoluble when deoxygenated and polymerizes causing deformation of and damage to the red blood cell (RBC) membrane resulting in cells that are less deformable, prone to hemolysis, and more adhesive with increased phosphatidylserine surface exposure. Heterotypic cellular interactions between the sickle RBCs, endothelium, leukocytes, and platelets, predominantly in the postcapillary venules, result in the pathognomonic vasoocclusion and ischemia reperfusion injury. Over-activation of the innate immune system secondary to ischemia-re perfusion and damage-associated molecular pattern molecules (DAMS) such as heme, result in tissue damage and contribute to activation of coagulation cascade and platelets.

The chronic activation of coagulation and platelets results in a clinically significant but underappreciated prothrombotic state with increased incidence and recurrence of venous thromboembolic events (VTEs—deep venous thrombosis and PE's-pulmonary embolism). Up to 25% of individuals with SCD will have developed a VTE by adulthood, with the median age of first VTE occurring at 29.99 years, a significantly younger age than the general population. The age at which VTE occurs is comparable with the age observed in individuals with high-risk thrombophilia. The risk of VTE in SCD is compounded by additional risk factors such as by recurrent hospitalizations, increased periods of immobility, frequent use of central venous catheters, and infection, and in these situations may be classified as a provoked rather than an unprovoked VTE. Increased mortality has been observed in adults with SCD and thrombosis, and the recurrence rate in non–catheter-associated VTE is 25%. There is an increased prevalence of pulmonary embolism found in SCD patients at autopsy, especially those with sudden death, and it has been suggested that pulmonary embolism may underlie development of some cases of pulmonary hypertension. Additionally, both retrospective and prospective analyses of patients with acute chest syndrome report increased pulmonary embolism. Use of administrative discharge databases further corroborates the increased incidence of pulmonary embolism in adults with SCD when compared with age- and race-matched controls. In the pediatric SCD population, central venous catheter placement increases the risk of deep vein thrombosis. Pregnancy-related VTE is also increased in women with SCD with an odds ratio of 6.7 (95% CI, 4.4–10.1) in one series. The Royal College of Obstetricians and Gynecologist guidelines recommend low-molecular-weight heparin be administered to women with SCD while in hospital and 7 days postdischarge following vaginal delivery or for a period of 6 weeks following caesarean section (https://www.rcog.org.uk/en/guidelines-research-services/guidelines/gtg61).

Transfusion Medicine and Hemostasis. https://doi.org/10.1016/B978-0-12-813726-0.00125-2

Alteration of Coagulation: Nearly every component of coagulation, including platelets, is affected by SCD with an increase in procoagulant activity and a reduction in some naturally occurring anticoagulants. Tissue factor (TF) is an essential component of the factor VIIA-TF complex enzyme, the initiator of blood coagulation in vivo. TF is expressed by endothelial cells and monocytes, and increased levels are reported in SCD. The number of circulating TF laden cells and microparticles increases during painful crises, as compared with steady state. In general, increased numbers of TF-expressing endothelial cells, monocytes, red blood cells, and their associated microparticles influence the coagulation cascade. There is an association between increased markers of hemolysis in SCD and whole-blood TF procoagulant activity.

An overall increased state of thrombin generation in SCD is evidenced by chronic elevation of procoagulant proteins such as thrombin–antithrombin complexes, prothrombin fragments (F1.2) and D-dimers, and other markers of thrombin generation. Moderately decreased levels of the anticoagulant proteins C and S are observed in patients with SCD in steady state, and these may be further decreased during acute pain episodes. These decreases may occur secondary to consumption due to chronic activation of the clotting cascade, inflammation, and/or liver disease, and levels are higher than those seen with inherited deficiencies. Factor V Leiden and prothrombin G20210A are rare in African Americans and in patients with SCD. Decreased levels of factor V have also been reported, suggesting chronic consumption of procoagulant factors due to the increase in TF expression and thrombin generation.

von Willebrand factor (VWF) has also been implicated in the hypercoagulable state of SCD. Extracellular hemoglobin binds with high affinity to VWF, thus preventing VWF from being cleaved by ADAMTS-13 leading to accumulation of ultralarge, extremely adhesive VWF multimers. Plasma-free heme also induces exocytosis of VWF from Weibel–Palade bodies, and total activity of VWF has been shown to directly correlate with hemolysis. This pathophysiology is demonstrated clinically by the description of a thrombotic thrombocytopenic purpura-like syndrome in SCD patients. Overall it is safe to conclude that the balance of the coagulation system in SCD is tipped toward thrombosis. This system is a potential target for disease-modifying interventions with anticoagulants or antiplatelets agents.

Targeting Coagulation in Sickle Cell Disease

Heparin: In addition to their anticoagulant effect, the antiadhesive effect of heparins is mediated via blockade of P-selectin. A single-center, randomized, double-blind clinical trial of tinzaparin versus placebo demonstrated reduction in the severity and duration of painful events. A study of the effects of unfractionated heparin in acute chest syndrome in SCD (NCT02098993) is ongoing, with the primary outcome being time to hospital discharge.

Direct Thrombin and Factor X Inhibitors: Current studies of new oral anticoagulants and their potential role in SCD are ongoing, aimed at evaluating the impact of prophylactic dosing of apixaban on daily pain scores (NCT02179177) and of rivaroxaban on soluble vascular adhesion molecule and interleukin-6 (NCT02072668).

TABLE 125.1 Ongoing Clinical Trials of Anticoagulants and Antiplatelet Agents in Sickle Cell Disease

Study Title	Intervention	Clinical Trials/Phase	Status
Apixaban in patients with sickle cell disease	Apixaban	NCT02179177 Phase 3	Ongoing
The effect of rivaroxaban in sickle cell disease	Rivaroxaban	NCT02072668 Phase 2	Ongoing
Feasibility study or unfractionated heparin in acute chest syndrome	Unfractionated heparin	NCT02098993 Phase 2	Ongoing
A pharmacokinetic and pharmacodynamic dose-ranging phase II study of ticagrelor followed by a 4 weeks extension phase in pediatric patients with sickle cell disease	Ticagrelor	NCT02214121 Phase 2	Complete
A study to evaluate the effect of ticagrelor in reducing the number of days with pain in patients with sickle cell disease (Hestia2)	Ticagrelor	NCT02482298 Phase 2	Complete

Alteration of Platelets in Sickle Cell Disease: Platelets have been shown to circulate in SCD patients in an activated state in both "steady state" and during painful crisis. This is evidenced by elevated platelet expression of CD62, CD63, PAC-1, P-selectin, activated glycoprotein IIb/IIIa, plasma soluble factor-4, and β-thromboglobulin. It is also proposed that platelets directly contribute to the inflammatory milieu of SCD via manufacture and release of pro- and antiinflammatory molecules on activation. Platelets are well known to form aggregates in SCD by binding erythrocytes, monocytes, and neutrophils. Overall, evidence suggests that the chronic activation of platelets in SCD contributes to the vasculopathy and thromboinflammatory state described in SCD. Accordingly, these alterations have been targeted by antiplatelet therapies with the goal of ameliorating the SCD phenotype (Table 125.1).

Targeting Platelets in Sickle Cell Disease: Numerous phase 1 and 2 trials with antiplatelet agents, most underpowered, have demonstrated safety and biomarker efficacy without a positive impact on pain events or rates. A phase 3 randomized, double-blind, placebo controlled study of prasugrel for prevention of volatile organic compound also did not demonstrate a significant improvement in the number of painful crises in the treatment arm. A phase 2 study, using ticagrelor (NCT02482298), to determine whether the P2Y12 inhibitor can reduce the number of days of pain, pain intensity, and analgesic use has recently been completed and results are not yet available. It is possible that different aspects of SCD pathophysiology may be positively affected by antiplatelet therapy.

Futher Reading

Kumar, R. S. J., Creary, S. E., & O'Brien, S. (2016). Risk factors for venous thromboembolism in children with sickle cell disease: A multicenter cohort study. *Blood*, *128*(7), 7 Abstract.

Lim, M. Y., Ataga, K. I., & Key, N. S. (2013). Hemostatic abnormalities in sickle cell disease. *Curr Opin Hematol, 20*(5), 472–477.

Naik, R. P., Streiff, M. B., Haywood, C., Jr., Nelson, J. A., & Lanzkron, S. (2013). Venous thromboembolism in adults with sickle cell disease: A serious and under-recognized complication. *Am J Med, 126*(5), 443–449.

Naik, R. P., Streiff, M. B., & Lanzkron, S. (2013). Sickle cell disease and venous thromboembolism: What the anticoagulation expert needs to know. *J Thromb Thrombolysis, 35*(3), 352–358.

Naik, R. P., Streiff, M. B., Haywood, C., Jr., Segal, J. B., & Lanzkron, S. (2014). Venous thromboembolism incidence in the cooperative study of sickle cell disease. *J Thromb Haemost, 12*(12), 2010–2016.

Wun, T., & Brunson, A. (2016). Sickle cell disease: An inherited thrombophilia. *Hematol Am Soc Hematol Educ Program, 2016*(1), 640–647.

CHAPTER 126

Acquired Coagulation Factor Inhibitors

Christine L. Kempton, MD, MSc and Duc Q. Tran, MD, MSc

Acquired coagulation factor inhibitors are autoantibodies directed against native clotting factor in persons without an underlying bleeding disorder. Although rare, these disorders can result in life-threatening hemorrhage, which can be difficult to manage. The following factor inhibitors will be discussed: factor VIII (FVIII), von Willebrand factor (VWF), factor V (FV), prothrombin, and thrombin.

Inhibitors of Factor VIII

Pathophysiology and Epidemiology: Acquired FVIII inhibitors are autoantibodies that bind to native FVIII in a person without congenital hemophilia A. Antibody binding leads to functional FVIII deficiency. The incidence is 1.48 per million person years, although this rate increases with age. In those over the age of 85, the incidence is 14.66 per million person years; in those under the age of 16, it is only 0.045 per million person years. In approximately 40%–50% of cases, the acquisition of the FVIII inhibitor is related to an underlying condition such as autoimmune diseases (e.g., rheumatoid arthritis and systemic lupus erythematosus), malignancy, dermatological disorders, and pregnancy. The association between FVIII inhibitor and underlying disorder decreases with age. In a large observational study, an acquired inhibitor was secondary to an associated condition in all patients less than 40 years but in only 25% of those over 85 years. Pregnancy-associated FVIII inhibitor development is discussed in Chapter 117.

Clinical Manifestations: Approximately 6% of patients may initially present for evaluation of a prolonged activated partial thromboplastin time (aPTT) without personal history of bleeding. Although 94% will have bleeding at presentation, approximately one-third will have only minor bleeding that does not require any treatment. The most common sites of bleeding include the subcutaneous tissue, muscle, gastrointestinal tract, and genitourinary tract. In contrast to congenital hemophilia A, hemarthroses are rare. Most patients will have only one bleeding event, though one-third may have two or more. Approximately one-third of patients with an acquired FVIII inhibitor have spontaneous resolution; however, this may not occur for over 12 months (range 10–21 months) after diagnosis. In one report, fatal bleeding occurred in 9%–22% of patients, occurring a median of 19 days after diagnosis, but sometimes much later. In the European registry (EACH2), fatal bleeding occurred in only 3%. Overall, higher inhibitor titers were associated with more severe bleeding. However, in a retrospective cohort, the inhibitor titer in those with fatal hemorrhage was identical to that in those who did not require any hemostatic therapy. Therefore, at this time there are no parameters that are adequately predictive of inhibitor resolution or fatal bleeding.

Transfusion Medicine and Hemostasis. https://doi.org/10.1016/B978-0-12-813726-0.00126-4

Diagnosis: These patients have a prolonged aPTT, with a normal prothrombin time (PT) and thrombin time. aPTT mixing studies may correct immediately after mixing with normal pooled plasma (NPP) but will again prolong after incubation consistent with the time and temperature dependence of acquired FVIII inhibitors. Factor VIII, IX, and XI activity assays that are performed to investigate the prolonged aPTT typically demonstrate a markedly reduced FVIII activity level, whereas other clotting factor activities will be normal. Most patients will have an FVIII activity <5%. An inhibitor to FVIII, measured using the Bethesda assay (see Chapter 132), will be present.

Management: Two major aspects of management of acquired FVIII inhibitors include treatment of acute bleeding and eradication of the inhibitor.

Hemostatic Treatment: Minor bleeding such as ecchymoses or epistaxis that is self-limited does not require active treatment, although close monitoring and follow-up is necessary. For bleeding requiring treatment, several options are available: desmopressin, human (h) FVIII infusions, bypassing agents (recombinant factor VIIa [rFVIIa] and activated prothrombin complex concentrate [aPCC], such as FEIBA), and recombinant porcine (rp) FVIII (Table 126.1).

Desmopressin: Desmopressin is only effective when FVIII activity is >5%. The effect of desmopressin decreases with subsequent dosing, and thus its use is limited to bleeding that requires a short duration of therapy. Additionally, the half-life of the released FVIII will be shortened secondary to binding of the inhibitory antibody, but the exact duration is variable from patient to patient.

TABLE 126.1 Treatment Options of Acquired FVIII Inhibitors

	Dose	Patient Population	Duration of Response	Monitoring	Potential Side Effect
Desmopressin	0.3 μg/kg	FVIII >5% Inhibitor titer <2 BU/mL	Variable	FVIII activity	Hyponatremia, seizures
Human FVIII concentrates	100 or 20 IU/kg for each Bethesda titer + 40 IU/kg	Inhibitor titer <5 BU/mL	Variable	FVIII activity	
Recombinant factor VIIa	70–90 μg/kg	All	2–3 hours	None	Thrombosis
FEIBA	75–100 U/kg Max: 200 U/kg/day	All	8–12 hours	None	Thrombosis
Recombinant porcine FVIII	200 U/kg (alternative 50–100 U/kg if no porcine FVIII titer is negative)	Porcine inhibitor titer <20 BU/mL	4–12 hours	FVIII activity	Inhibitor to porcine FVIII

Human Factor VIII: Infusion of human factor VIII can be utilized when the inhibitor titer is <5 BU/mL. Regardless of the initial dosage, FVIII activity should be measured 15–30 minutes after the bolus. Because the half-life of the factor will be reduced, levels should be monitored carefully to determine the dosing interval.

Porcine Factor VIII: Recombinant porcine FVIII has been FDA-approved for treatment of bleeding in patients with acquired FVIII inhibitor and can be used when the antiporcine FVIII inhibitor titer is <20 BU/mL. The manufacturer recommends an initial administration dose of 200 U/kg; however, lower doses have been used successfully and can be considered when an antiporcine FVIII inhibitor titer is not detected or the bleeding event for which treatment is being prescribed is not severe. FVIII activity levels should be monitored regularly, and intermittent rpFVIII should be administered to maintain target FVIII activity until the resolution of bleeding.

Bypass Therapy: When the inhibitor titer is ≥5 BU/mL, a bypassing agent should be used; either rFVIIa or aPCC. Both agents are effective or partially effective in approximately 80%–90% of bleeding episodes. Thrombotic events have been reported with both rFVIIa and aPCCs, and are more likely to occur in elderly populations who also have underlying vascular disease. There is no available laboratory testing to monitor efficacy or toxicity. Laboratory parameters of disseminated intravascular coagulation (see Chapter 124) should be monitored if therapy is given for more than 5 days. With both agents, the dose and frequency should be adjusted to the clinical severity and the patient's response. The duration of therapy will be determined by bleeding severity and the risk of toxicity secondary to hemostatic therapy. With severe bleeds, hemostatic therapy should continue for at least several doses beyond the cessation of bleeding.

Eradication of the Inhibitor: Because of the inability to predict who may have a fatal hemorrhage or who will spontaneously remit, immune suppression should be considered in all patients. Primary treatment options include corticosteroids, cyclophosphamide, and rituximab.

Corticosteroids and Cyclophosphamide: Corticosteroids should be included in the initial therapy. Cyclophosphamide (oral or intravenous) can be used in combination with corticosteroids initially or following several weeks of corticosteroid monotherapy if an adequate response is not seen. Currently, data suggest that the combination of corticosteroids and cyclophosphamide increases the likelihood and durability of remission compared with corticosteroids alone. Pulse intravenous cyclophosphamide (500–750 mg/m^2) given every 4 weeks will lead to less bladder exposure to acrolein, and therefore, a lower risk of hemorrhagic cystitis than with oral cyclophosphamide. In clinical practice, pulse intravenous dosing of cyclophosphamide has been safe and effective. If oral cyclophosphamide is utilized, a daily dosage of 1–2 mg/kg is recommended.

Rituximab: The role of rituximab during first-line therapy for FVIII inhibitor eradication is less clear. Although high response rates have been reported, in the European registry (EACH2), only 61% of patients treated with a rituximab-containing regimen, and 42% of patients treated with rituximab alone, achieved a stable remission. In those that did achieve a stable complete response after receiving rituximab, the relapse rate was low (3%, compared with 18% and 12% in those who received corticosteroids alone and corticosteroids plus cyclophosphamide). Thus if rituximab is to be used, it is currently recommended that it is used in combination with corticosteroids and/or cyclophosphamide.

An overall approach to inhibitor eradication in those with an acquired FVIII inhibitor should include corticosteroids as a first-line approach in all patients. It appears that the addition of cyclophosphamide to corticosteroids improves response rates. However, in the absence of prospective randomized clinical trials, data are inconclusive and treatment decisions should be individualized for the patient by the treating clinician. Regardless of the initial approach, if after 3 weeks the inhibitor titer is not declining, treatment should be modified.

Inhibitors of von Willebrand Factor

Pathophysiology and Epidemiology: The exact incidence and prevalence of acquired von Willebrand syndrome (AvWS) is unknown. Kumar et al. estimated, based on a retrospective review, that 0.04% of the population has AvWS. Although AvWS can occur by mechanisms other than an antibody-mediated inhibition of VWF activity (Table 126.2), this section will focus on antibody-mediated mechanisms. Lymphoproliferative disorders and monoclonal gammopathy of undetermined significance (MGUS) are associated with nearly half of all AvWS.

Clinical Manifestations: AvWS typically manifests similarly to congenital von Willebrand disease (VWD) (see Chapter 109) with mucocutaneous or postsurgical bleeding. However, one-fourth of patients may be asymptomatic. In contrast to congenital VWD, patients with AvWS will have previously normal hemostasis without a personal or family history of abnormal bleeding.

Diagnosis: Testing for AvWS is similar to that for congenital VWD and includes measurement of VWF activity, VWF antigen, FVIII assay, and multimer analysis. A

TABLE 126.2 Diseases Associated With Acquired von Willebrand Syndrome

Mechanism	Associated Disease(s)
Antibodies: Inhibition of von Willebrand factor (VWF) function or increased clearance by non-specific binding leading to clearance of VWF	Lymphoproliferative/MGUS/Autoimmune disorders
Adsorption on malignant cells	Lymphoproliferative disorders, Myeloproliferative disorders
Degradation by increased shear stress	Cardiac disease such ventricular septal defect, aortic stenosis,or severe coronary artery disease
Decreased production	Hypothyroidism
Proteolysis	Myeloproliferative disorders, uremia, ciprofloxacin, hyperfibrinolysis, fibrinolytic therapy
Idiopathic	Valproic acid, amyloidosis, mixed cryoglobulinemia

VWF propeptide is also useful to demonstrate normal production of VWF antigen and point to inhibition or clearance as the cause of low VWF activity. A type 2 VWD discordant pattern of VWF activity to antigen is most commonly seen (activity: antigen ratio <0.6). Multimer analysis may demonstrate the loss of large molecular weight multimers in 50%–80% of cases (see Chapter 135). Detection of an inhibitor can be attempted by performing a VWF activity assay after mixing the patient's plasma with equal parts of NPP. Unfortunately, this method will only detect the less commonly encountered antibodies that directly impair the function of VWF. In vitro mixing studies will not detect the more common, nonneutralizing antibodies that accelerate VWF clearance. Because of these limitations, testing for an inhibitor is infrequently positive but is slightly more likely to occur in the setting of lymphoproliferative disorders, neoplasia, and immunological disease (~30% of these cases).

Management: As with other autoimmune coagulation defects, the two major components of treatment include hemostatic therapy for acute bleeds and antibody eradication.

Hemostatic Therapy: Desmopressin and VWF concentrates (see Chapter 41) are the primary treatments. Both desmopressin and VWF-containing products are effective in one-third of patients. However, as the recovery and half-life of individual patient response may be reduced, careful monitoring is required. The choice of agent will depend on baseline VWF activity, anticipated duration of therapy, and ability to tolerate the potential side effects of desmopressin. Desmopressin is preferred over a VWF product in the setting of (1) mildly reduced VWF activity, (2) therapy that is anticipated to be of 3 days duration or less, and (3) the patient not having uncontrolled hypertension or cardiovascular disease. Antifibrinolytic therapies such as aminocaproic acid or tranexamic acid may also be used as adjunctive therapies, particularly in the setting of mucocutaneous bleeding.

Antibody Eradication: In the setting of an associated condition for which therapy exists, treatment of the underlying condition is the most appropriate route for antibody eradication. If adequate treatment for the underlying disease is not available, or a quick response to therapy is not anticipated, intravenous immunoglobulin (IVIG) or plasma exchange may be considered. IVIG has been utilized in patients with AvWS and has been reported to be effective in approximately one-third of patients, although responses were more frequent in the setting of lymphoproliferative disorders, IgG MGUS, neoplasia, and immunological diseases. Plasma exchange has been used infrequently and is usually ineffective in patients with the typical IgG-mediated disorder. It is more likely to be effective in patients with IgM MGUS.

An overall approach to the treatment of AvWS suspected to be secondary to an autoantibody is to treat acute bleeding with desmopressin or a VWF-containing product and monitor response. If the response is inadequate, the addition of IVIG (typically 2 g/kg infused in divided doses daily over 2–5 days) or plasma exchange can be considered. Because both IVIG and plasma exchange may improve the recovery and half-life of the VWF-containing product in addition to potentially improving the patient's native VWF antigen level and activity, these agents could be used concurrently to treat severe bleeding.

Inhibitors of factor V: The development of an anti-FV inhibitor is rare. It may occur in association with the use of topical bovine thrombin, aminoglycosides, β-lactam antibiotics, and malignancy. Patients may develop an immune response to bovine thrombin after use in cardiovascular surgery; these antibodies cross-react to human thrombin and FV. Patients may be asymptomatic or display severe bleeding weeks after the exposure. Anti-FV inhibitory antibodies should be suspected in the setting of abnormal bleeding with a newly prolonged PT (and rarely a prolonged aPTT) but normal thrombin time. The PT will fail to correct on mixing with NPP, and the FV activity will be low. An inhibitor titer can be measured utilizing the Bethesda assay. Plasma and platelet (30% of FV is normally contained within platelet α-granules) products can be used to support hemostasis. When associated with topical bovine thrombin or antibiotics, the inhibitor is typically short-lived and immunosuppressive therapy is not warranted. Conversely, those associated with malignancy or autoimmunity are more persistent and may require immunosuppressive treatment such as corticosteroids.

Inhibitors of Prothrombin and Thrombin: Antibodies that bind thrombin may occur after exposure to bovine thrombin. Not all antibodies will inhibit the function of thrombin. Antibodies that neutralize the function of thrombin will lead to prolongation of the PT, aPTT, and thrombin time. Prolongation of the thrombin time helps to distinguish inhibitors to thrombin from those directed against FV and prothrombin. Antiprothrombin antibodies occur most commonly in association with antiphospholipid antibody syndrome (see Chapter 108). Although antiprothrombin antibodies have been purported to be associated with lupus anticoagulant activity and a prothrombotic state, rarely these antibodies lead to increased clearance of prothrombin resulting in a bleeding diathesis. On laboratory testing, both the PT and aPTT will be prolonged and factor II activity will be reduced, whereas the thrombin time will be unaffected.

Further Reading

Collins, P. W. (2007). Acquired hemophilia A in the United Kingdom: A 2-year national surveillance study by the United Kingdom Haemophilia Center Doctors' Organisation. *Blood, 109*, 1870–1877.

Favaloro, E. J., Posen, J., Ramakrishna, R., et al. (2004). Factor V inhibitors: Rare or not so uncommon? A multi-laboratory investigation. *Blood Coagul Fibrinolysis, 15*, 637–647.

Knoebl, P., Marco, P., Baudo, F., et al. (2012). Demographic and clinical data in acquired hemophilia A: Results from the European Acquired Haemophilia (EACH2) Registry. *J Thromb Haemost, 10*, 622–631.

Kruse-Jarres, R., Kempton, C. L., Baudo, F., et al. (2017). Acquired hemophilia A: Updated review of evidence and treatment guidelines. *Am J Hematol, 92*, 695–705.

Lollar, P. (2005). Pathogenic antibodies to coagulation factors. Part II. Fibrinogen, prothrombin, thrombin, factor V, factor XI, factor XII, factor XIII, the protein C system and von Willebrand factor. *J Thromb Haemost, 3*, 1385–1391.

Tiede, A., Rand, J. H., Budde, U., Ganser, A., & Federici, A. B. (2011). How I treat the acquired von Willebrand syndrome. *Blood, 117*, 6777–6785.

CHAPTER 127

Overview of Purposes of Hemostasis Testing and Common Sources of Error

Mikhail Roshal, MD, PhD and Morayma Reyes Gil, MD, PhD

The human hemostasis system is complex and tightly regulated. The maintenance of adequate blood flow, interrupted only by formation of temporary thrombi localized to places of injury, requires coordination between endothelium, platelets, white blood cells, and numerous soluble plasma elements. Disruptions of the hemostasis system can lead to hemorrhage, pathological thrombosis, or both. The pattern and location of the pathological event frequently provide valuable information about the components of the system, which are most affected and the severity of the lesion that may be present. However, owing to the interdependence of the network, there is a limit to information that can be gained by clinical observation alone. Laboratory studies are frequently needed to provide additional information for clinical decision-making. This chapter provides an introduction to testing according to purpose and lists some common diagnostic pitfalls that may lead to erroneous or clinically unhelpful results. Clinical testing in the coagulation laboratory may be broadly subdivided according to purpose.

Screening to Assess Risk of Hemorrhage in Otherwise Asymptomatic Patients Before an Invasive Procedure: A structured bleeding history is much more powerful than laboratory tests in predicting bleeding risk in patients undergoing invasive procedures. Thus, patients who did not show hemorrhagic tendencies during prior hemostatic stress (e.g., dental procedures, trauma, or childbirth) are unlikely to have a significant congenital bleeding diathesis. The prevalence of such disorders in the general population is low; indiscriminate conventional testing does little to predict hemorrhage because of its relatively low sensitivity and specificity. On the other hand, abnormal tests are frequently predictable based on the clinical history. Nonetheless, prothrombin time (PT), activated partial thromboplastin time (PTT), fibrinogen, and platelet count are often performed. Additional screening tests that assess platelet function, von Willebrand factor (VWF), and (rarely) the fibrinolytic system are occasionally performed. It should be noted that none of the tests have a well-documented value in predicting bleeding in the absence of supportive clinical history. Global hemostasis testing platforms such as thromboelastography/thromboelastometry are occasionally used for presurgical evaluation, but the value of these tests in predicting hemorrhage in this situation is yet to be proven. When abnormalities are found, the screening tests are usually followed up by more specific tests.

Measuring the Effect of Hemostasis System-Modifying Drugs: Physicians have an ever-expanding arsenal of medications that inhibit various components of hemostasis, in an effort to prevent unwanted clot formation or hemorrhage. Some, such as warfarin and unfractionated heparin, have a variable dose response and must

Transfusion Medicine and Hemostasis. https://doi.org/10.1016/B978-0-12-813726-0.00127-6

be closely monitored to prevent under- and overanticoagulation. Others have a more predictable dose response and need to be measured only in special situations, such as unexpected bleeding or thrombosis while on medication. Unlike many therapeutic drug monitoring tests in other areas of medicine, the drug monitoring in hemostasis is frequently performed by measuring the effect rather than the concentration of a medication. This is done for two major reasons. First, owing to complexity of the coagulation system, some drugs may have large patient-to-patient variation in their effect at similar concentrations. For instance, infants have ~50% of the adult concentration of antithrombin, which serves as a target for heparin therapy. Thus, infants and others with low antithrombin levels may show less anticoagulation at the same plasma concentration of heparin. Second, measuring the effect of the drug on the coagulation system is frequently more rapid and more accessible than the concentration-based approaches. The simplest tests often provide enough information to monitor drug effect. When the medications have a predictable effect on the factors in the clotting cascade, PT, PTT, and thrombin time–based tests, they offer convenient, inexpensive, and widely available approaches. The tests may not be entirely specific but frequently provide sufficient information. However, for many (mostly newer) drugs, the response of either PT or PTT is not satisfactory, either due to poor dose response or due to presence of interferences. In those cases, specific tests that directly measure the effects of drugs on their physiologic targets with minimal interference from other factors have been developed.

Numerous platforms are now offered to monitor effects of aspirin and ADP-receptor antagonists on platelets. Unfortunately, the lack of agreement between methods in detecting poor response, and the absence of conclusive evidence that such monitoring improves patient outcomes outside of limited circumstances, makes the indications for testing controversial. The field is rapidly evolving, thus it is possible that clear clinical benefit from the platelet-based therapy response determination may be demonstrated in future studies.

Monitoring Hemostasis System Under Stress and Assessing the Need for Replacement Products: Surgery and severe trauma challenge the hemostasis system in numerous ways. Because of the reductive nature of much of the current coagulation testing, no individual standard test can measure the numerous perturbations associated with these challenges. Numerous tests measuring plasma elements (PT, PTT, and fibrinogen), platelet number and function, and occasionally fibrinolysis may be needed to sufficiently assess the status of the hemostasis system under stress. More integrative tests, such as thrombin generation assays that are capable of capturing thrombin activity past the point of clot formation and thromboelastography/thromboelastometry that measure clot formation and stability in whole blood, offer attractive alternatives to the more specific tests. However, the usage of these tests is still evolving and their exact place in the arsenal of testing is yet to be determined. At present, it is not clear whether the global hemostasis tests improve patient outcomes compared with the standard screening tests.

Outside of trauma and surgery, various disease states can have numerous effects on the hemostasis system either as primary manifestation (e.g., heparin-induced thrombocytopenia) or as secondary manifestation (disseminated intravascular coagulation in sepsis). In such cases testing may be targeted to specific suspected abnormalities.

Diagnostic Workup to Assess Risk of Recurrent Thrombosis or Bleeding in Patients With Unexpected Thromboembolic or Hemorrhagic Event: This category of tests is often the most challenging. Although research studies and astute clinical observations have given us a fairly detailed understanding of most of the involved factors and many of their interactions and perturbations, we cannot yet quickly convert much of the knowledge into directly actionable clinical information in the laboratory. Frequently, coagulation workups in search of a lesion involve multistep processes that start with screening tests and proceed to multiple individual tests to pinpoint a lesion. Often, once a potential lesion is localized, it must be confirmed at a separate time point to assess its persistence and to improve diagnostic accuracy.

Such workups may be expensive and slow and should only be undertaken if there is sufficient clinical suspicion. Owing to relative rarity of any individual defect and multiple tests that are frequently required to find them, the rate of false positivity can become unacceptably high if the pretest probability of finding a lesion is similar to that of the general population. Routine, extensive, screening workups for hemorrhagic states or thrombophilia are therefore rarely, if ever, indicated. Use of reflex testing algorithms and rational testing panels clearly improves the diagnostic yield and cost efficiency of complex coagulation testing.

The foregoing makes it clear that knowing when to order tests is frequently just as important as knowing how the test is performed or how to interpret it. Whenever relevant, such indications and interpretive information are listed in the next chapters.

Sources of Diagnostic Errors in Hemostasis Testing: Modern automated analyzers, improved external quality control programs, and sophisticated coagulation reagents have clearly reduced analytical errors in the hemostasis laboratory. Despite that, interferences due to numerous interactions of hemostasis system components still present a significant challenge in measuring specific analytes appropriately. Whenever possible, the analytical challenges are highlighted in the subsequent chapters. However, as is true for most areas of laboratory medicine, preanalytical and postanalytical problems remain major contributors to the relatively high rate of erroneous or diagnostically unhelpful results. There are numerous points where preanalytical and postanalytical errors can occur.

Wrong Timing of the Test: This source of errors is potentially underappreciated, and, in the authors' experience, accounts for a large portion of diagnostically unhelpful results in specialty hemostasis testing. For instance, thrombophilia testing is frequently ordered too close to acute thrombotic events, while patients are on medications or have physiologic states that strongly influence the analytes of interest. Thus protein C (PC), protein S (PS), and antithrombin (AT) tests in the setting of recent or ongoing thrombosis often lead to low results because of consumptive decrease in these proteins. These values are not indicative of congenital deficiency. Tests for PS or von Willebrand disease may not be diagnostic during pregnancy or during acute phase response. Even subtle issues like circadian rhythms affecting analytes such as plasminogen activator inhibitor-1, and therefore tissue plasminogen activator activity, may also have an effect on the diagnostic yield. Whenever possible, these issues are highlighted in the chapters that cover testing for individual analytes within this section.

Wrong Test: Thanks in part to the ever-increasing menu of coagulation tests, ordering clinicians may not be familiar with appropriate testing algorithms for the diagnosis of hemostasis disorders. For instance, while PC antigen result may appropriately indicate a normal level, the test does not rule out a qualitative PC deficiency, which would be detected by a PC activity test. To avoid potential misdiagnosis, it is imperative for laboratories to provide guidance for appropriate test ordering and interpretation of results.

Wrong Patient: Errors in patient sample identification occasionally occur at both the point of blood draw, usually due to the wrong label attached to the sample or within the laboratory. The error risk in hemostasis testing is increased by the frequent requirement for plasma aliquots for later testing. This requirement creates an additional opportunity for error when numerous secondary tubes are generated. Requirement for two unique matching identifiers, both on the order and on the sample, automated aliquoting systems, and barcode tracking has a strong potential to reduce these types of serious errors.

Wrong Sample Type: This is also a potentially underrecognized error, particularly in the reference laboratory setting where aliquots rather than primary tubes are received. For example, samples drawn with inappropriate or no anticoagulant such as ethylenediaminetetraacetic acid or serum separators are clearly unacceptable for most activity-based coagulation tests; however, this error may not be readily picked up when an aliquot is received for testing. Occasionally, samples with too much citrate can be received. This occurs either because 3.8% rather than the recommended 3.2% citrate tubes are used for coagulation factor testing or because the tubes were underfilled (less than 90% of the fill volume) or had significantly elevated hematocrit (above 55%). These events produce spurious results because of insufficient recalcification during testing. Strict sample requirement guidelines should be observed to minimize such relatively common sources of error.

Sample Collection: The vast majority of preanalytical errors in coagulation arise from inappropriate sample collection. Appropriate selection of needle size is important as blood flowing through small needles may result hemolysis and activation of coagulation. It is indicated to use needle size 21 gauge (green needle) for adults, whereas needle size 23 gauge (blue needle) has been traditionally used for children, although validation studies using this needle size or smaller are lacking. If multiple specimens are collected, the coagulation sample (blue tubes) should be the first or second tube collected following blood cultures. Appropriate mix of blood with anticoagulant is extremely important to prevent clot formation or partial activation of coagulation factors. Thus each blue top tube should be mixed by inverting gently three or four times immediately after venipuncture to ensure proper mixing of blood and anticoagulant. Tourniquet pressure should not be longer than 1 minute as venous stasis increases the concentration of large molecules such as VWF and factor VIII (FVIII). Drawing from a vascular access device (VAD) should always be discouraged but if required, the line should be flushed with 5.0 mL saline and the first 5.0 mL of blood or six dead space volumes of the VAD discarded.

Table 127.1 summarizes the most common errors during sample collection, an explanation of the effects of these errors and how to avoid them.

TABLE 127.1 Common Preanalytical Errors in Coagulation Testing

Preanalytical Error That Should Be Avoided	Effect on Laboratory Results	Explanation	Possible Consequences on Patients
Poor mixing of blue top tubes *Please invert the tube gently three or four times immediately after blood draw*	Prolongation of clotting times (e.g., prolonged PT and PTT)	Blood clots form when the anticoagulant (citrate) and blood do not mix rapidly	Misdiagnosis of factor deficiencies, von Willebrand disease, and false supertherapeutic levels of anticoagulants (e.g., warfarin and heparin)
Slow collection *Please use green needle for adults. Blue needle is only acceptable in children*	Shortening of clotting times (e.g., short PT and PTT)	Hemolysis activates coagulation factors and platelets	Missing diagnosis of factor deficiencies, von Willebrand disease, and false undertherapeutic levels of anticoagulants (e.g., warfarin and heparin)
Refrigerating blood tubes *Keep blood tubes at RT*	Short PT	Cold temperatures activate Factor VII	False undertherapeutic INR for patients on warfarin
Excessive duration of tourniquet application *Apply tourniquet for no more than 1 minute*	Short PTT	Venous stasis increases the concentration of large molecules such as VWF and FVIII	Missing diagnosis of von Willebrand disease and hemophilia. False undertherapeutic heparin level
Unfilled blue top tubes *Fill blood collection tubes to 3 mL*	Prolonged PT/PTT	1:9 ratio of citrate: plasma is needed for appropriate antiocoagulation. Too much citrate prolongs clotting times	Misdiagnosis of factor deficiencies, von Willebrand disease, and false supertherapeutic levels of anticoagulants

Italic, list the correct measures to avoid those errors; **Bold**, common errors.
INR, international normalized ratio; *PT*, prothrombin time; *PTT*, partial thromboplastin time; *RT*, room temperature; *VWF*, von Willebrand factor.

Compromised Sample: Clotted, hemolyzed, or lipemic samples are usually not appropriate for testing. Unfortunately, clots may not always be detected when automated processing is used. Although modern photooptical instruments tend to be less sensitive to hemolysis and lipemia, and electromechanical methods are not affected by optical properties of the plasma, the physiologic effect of these conditions may still cause problematic results. Release of tissue factor through cell lysis may activate coagulation, increasing D-dimer concentration, decreasing fibrinogen, and causing spurious PT and PTT results. High fat content in the plasma may activate factor VII and lower other clotting factor activities. Such samples should be redrawn if possible. If hemolysis is determined to occur in vivo, use of electromechanical detection systems is preferable.

Sample Handling: A lot can go wrong between the sample collection and testing. Samples should be transported at temperatures between 15 and 22°C (see Table 127.1).

This is sometimes difficult to control when samples are received from distant locations. Temperatures that are too low can activate platelets, whereas temperatures that are too hot may cause factor degradation. For plasma-based tests, plasma should be separated from cells as soon as possible and definitely within 4 hours of blood draw. For heparin activity testing, there is a 1-hour requirement for plasma separation.

Separation of plasma from platelet components is critical for most coagulation tests. Platelets can serve as activators of coagulation and source of factors and can change optical properties of the plasma. Centrifugation should be done at 15–22°C at 1500 g. Single centrifugation is acceptable for most testing, whereas double centrifugation is required for lupus anticoagulant (LA) testing. This is done by transferring plasma to a new tube and performing a second spin to remove residual platelets. If the second spin is not done, there is a risk of a false negative LA result, particularly if testing is done on a previously frozen sample.

Once samples are frozen, specific storage and thawing guidelines must be observed. Repeat freeze–thaw events may result in loss of FVIII and factor V activity.

Postanalytic Issues: Once the sample analysis is complete, the opportunities for error and misinterpretation do not necessarily end. Values "flagged" as abnormal may neither necessarily represent clinically actionable information nor do normal values necessarily exclude disease states. For instance, given that an incidence of AT deficiency in the general population is very low, and reference ranges generally exclude 2.5% of values from presumed normal subjects, does a slightly low value represent congenital AT deficiency? Similarly, does a patient who has both low PC and PS test values require additional anticoagulation? Does VWF activity of 55% completely rule out von Willebrand disease? These and many other questions may need to be addressed by providing not just a value and a reference range but an interpretative comment that integrates available laboratory and clinical data whenever possible.

Further Reading

Adcock, D. M., Hoefner, D. M., Kottke-Marchant, K., et al. (2008). *Collection, transport, and processing of blood specimens for testing plasma-based coagulation assays and molecular hemostasis assays: Approved guideline-fifth edition*. CLSI document H21–A5. Wayne, PA: Clinical Laboratory Standards Institute.

Favaloro, E. J., Lippi, G., & Adcock, D. M. (2008). Preanalytical and postanalytical variables: The leading causes of diagnostic error in hemostasis? *Semin Thromb Hemost, 34*, 612–613.

Funk, D. M. (2012). Coagulation assays and anticoagulant monitoring. *Hematol Am Soc Hematol Educ Program, 2012*, 460–465.

Lippi, G., Plebani, M., & Favaloro, E. J. (2012). Interference in coagulation testing: Focus on spurious hemolysis, icterus, and lipemia. *Semin Thromb Hemost, 39*, 258–266.

Magnette, A., Chatelain, M., Chatelain, B., et al. (2016). Pre-analytical issues in the haemostasis laboratory: Guidance for the clinical laboratories. *Thromb J, 14*, 49.

Salvagno, G. L., Lippi, G., Montagnana, M., et al. (2009). Influence of temperature and time before centrifugation of specimens for routine coagulation testing. *Int J Lab Hematol, 31*, 462–467.

Watson, H. G., & Greaves, M. (2008). Can we predict bleeding? *Semin Thromb Hemost, 34*, 97–103.

Woodhams, B., Giradot, O., Blanco, M. J., Colesse, G., & Gourmelin, Y. (2001). Stability of coagulation proteins in frozen plasma. *Blood Coagul Fibrinolysis, 12*, 229–236.

CHAPTER 128

Prothrombin Time

Mikhail Roshal, MD, PhD and Morayma Reyes Gil, MD, PhD

Prothrombin time (PT) was one of the first generated in vitro tests of the hemostasis system. It measures the time to form a fibrin clot in platelet-poor plasma stimulated with high amount of tissue factor (TF) and anionic phospholipid at the optimal concentration of calcium. The clotting is initiated by phospholipid-bound TF binding factor VII or VIIa (activated form of factor VII [FVII]). Under the test conditions, FVIIa can directly activate FX, which in turn initiates conversion of prothrombin to thrombin with slow kinetics. The small amount of thrombin in the initiation phase generates activated FV. FVa greatly accelerates FXa protease activity toward prothrombin, allowing the production of enough thrombin to rapidly convert fibrinogen into fibrin, forming a clot. Clot formation can be detected either by measuring changes in plasma viscosity via an electromechanical device or plasma absorbance via photooptical means. The time to clot is reported in units of seconds.

Since the test's invention over 75 years ago, its significance has much expanded, as it has found new uses in anticoagulant drug monitoring. Today it is likely the most widely ordered laboratory coagulation test, with over 800 million PT tests performed annually in the world. The test is sensitive to multiple plasma coagulation factors and suffers from fewer interferences than activated partial thromboplastin time (PTT). This makes it the test of choice for assessing plasma coagulation factor status in situations where all the factors are affected similarly, including trauma, disseminated intravascular coagulopathy (DIC), and surgery. Additionally, high levels of FVIII that shorten PTT do not affect PT. Because FVIII elevations are often present in liver disease, PT is more sensitive to compromise synthetic function of the liver. PT is still the test of choice to monitor treatment with vitamin K synthesis inhibitors (warfarin). Modified PT-based tests are used to measure concentrations of factors II, V, VII, and X. Some commercial tests use PT-type activation for measuring anticoagulants protein C and S activities, as well as activated protein C resistance.

Quick and Owren Prothrombin Time: Two variants of the PT test are currently in use. In most countries, the earlier version of the test based on the Quick method dominates. In the Quick-type PT test, TF and phospholipid are added to decalcified plasma with a final test plasma dilution of 1:3. This variant of the test assesses concentrations of vitamin K-dependent factors II, VII, and X, and additionally, non–vitamin K-dependent FV and fibrinogen.

The Owren-type PT test is now used predominantly in the Nordic countries, Benelux, and Japan. This test was designed specifically to monitor vitamin K-dependent factors and is insensitive to FV and fibrinogen concentration in the test plasma. In the Owren-type PT method, bovine plasma depleted of vitamin K-dependent factors is added together with calcium, phospholipid, and TF. The bovine plasma provides a source of fibrinogen and FV, making the test sensitive only to concentration of vitamin K-dependent factors II, VII, and X. The test plasma is diluted 1:21.

Transfusion Medicine and Hemostasis. https://doi.org/10.1016/B978-0-12-813726-0.00128-8

International Normalized Ratio: Many different sources of TF and phospholipid used in commercial PT reagent preparations pose a significant challenge to standardizing the monitoring of oral anticoagulants using PT. There is considerable variation in the sensitivity of various PT formulations to the concentration of the vitamin K-dependent factors. Therefore, taking a simple ratio of the test PT to the mean of normal PT would not produce comparable results when different reagents are used. To harmonize the treatment monitoring of oral anticoagulants, each reagent lot is assigned an international sensitivity index (ISI) based on an international standard. The ISI is calculated by comparing PTs obtained from a given reagent and instrument combination to that of a reference reagent obtained by using a manual tilt-tube method. This comparison is done using plasmas deficient in vitamin K-dependent factors and is usually performed by the reagent vendor. It is preferable that the ISI calibration is performed for specific combinations of reagent and instrument. If generic ISI (reagent only) is used, local ISI calibration verification is now mandatory (see Clinical Laboratory Standards Institute [CLSI] document H54-A). The international normalized ratio (INR) can be calculated as follows:

$$\text{INR} = (\text{test PT/geometric mean of normal subjects' PT})^{\text{ISI}}$$

It should be noted that ISI calibration is performed using only vitamin K-deficient plasmas; the sensitivity determination is only relevant to detection of vitamin K deficiency or to monitoring warfarin therapy. Other conditions that prolong PT, such as liver failure or DIC, involve decreases in FV and alterations in quantity and/or quality of fibrinogen. Sensitivity to these changes is not assayed during ISI calibration. It has been shown that the use of INR is the single largest contributor to interinstitutional variation of the model of end-stage liver disease score used for liver transplant priority list. Determining ISI using plasma from liver disease patients has been shown to dramatically improve INR variation in liver failure. Unfortunately, liver disease ISI determination is not routinely performed. Further details of PT use for monitoring oral anticoagulant therapy are provided in Chapter 156.

Test Limitations: Common preanalytical and analytical variables that can limit the usefulness of PT for monitoring coagulation factor levels are listed below.

Temperature and Time of Sample Storage: Samples stored for over 24 hours can show prolonged PT because of factor degradation. There is no consensus on refrigerated versus room temperature plasma storage. Cold activation of FVII can theoretically affect the PT, particularly if bovine thromboplastin is used. This effect is less evident with human thromboplastins and no effect on PT was seen during cold storage for 24 hours in one study. Some PT reagent manufacturers provide more restrictive guidelines for plasma storage. These guidelines should be followed or the individual lab can establish stability and storage conditions using samples from their patient population.

Effect of Collection Tubes, High Hematocrit, Underfilled Tubes, and Anemia: Samples collected in anticoagulants other than sodium citrate or without anticoagulant (serum) are not acceptable. The samples are collected in plastic tubes containing 3.2% sodium citrate. The final blood to citrate ratio is intended to be 9:1.

This ratio is optimal for samples with typical adult hematocrit. In practice, it is the final concentration of citrate in the plasma that plays an important role in PT determination. As citrate distributes to plasma and not cell compartment, patients who have high hematocrits (above 55%) will have significantly greater citrate concentration if standard citrate tubes are used. This can lead to inadequate recalcification of samples during PT testing and falsely prolonged results. Citrate concentration should be adjusted to the expected plasma volume before blood draw. A nomogram has been developed to help with adjusting citrate amount in samples with high hematocrit (see CLSI document H21-A5). Similarly, underfilled tubes can also produce overcitrated samples. There is currently no evidence that patients with anemia (hematocrits below 20%) need adjustment of citrate concentration unless clotting occurs.

Plasma Discoloration: Discoloration of the plasma due to hemolysis, lipemia, or hyperbilirubinemia can cause problems with photooptical detection of the clot. The extent of such interference varies from instrument to instrument. In addition, there is a concern that membrane phospholipids in hemolyzed red blood cells may cause increased or decreased activation of the coagulation system. Empirical evidence for either is lacking. Plasma discoloration does not have an effect on end point determination if electromechanical devices are used.

Effect of Medications: Treatment with recombinant FVIIa can produce very short PTs and mask deficiencies of other factors. Direct thrombin inhibitors variably and mildly prolong PT with greater sensitivity in Quick-type PT than in Owren-type PT. New oral anti-Xa inhibitors prolong PT. Heparin concentrations below 2U generally do not affect the PT.

Lupus Anticoagulants: Lupus anticoagulants (LA) can inhibit clotting reactions in vitro but are usually not associated with increased risk of bleeding. PT is less sensitive to LA than PTT. Nevertheless, some cases of LA can prolong PT. In those patients, INR should not be used for warfarin monitoring and chromogenic Factor X levels are recommended to monitor warfarin.

Factor VII R304Q Variant: Some polymorphic substitutions in FVII show highly variable interaction with TF depending on the source. For instance, mutation leading to R304Q (FVII Padua) almost totally abolishes interaction of FVII with rabbit TF but has about 30% activity when human recombinant TF is used. Significantly prolonged PT due to suspected FVII deficiency with no history of severe bleeding should be evaluated using a human TF source.

Test Interpretation: PT is slightly prolonged in newborns because of lower coagulation factor levels and reaches the adult reference range within the first week of life. Quick-type PT will be prolonged with significant combined or individual factor deficiencies in factors II, VII, X, and V and fibrinogen or inhibitors of these factors. High concentration of fibrin split products can also prolong the PT. Notably, PT demonstrates an exponential response to changes in factor level concentration. Thus, a change in factor level from 5% to 10% produces a much greater decrease in PT than a change

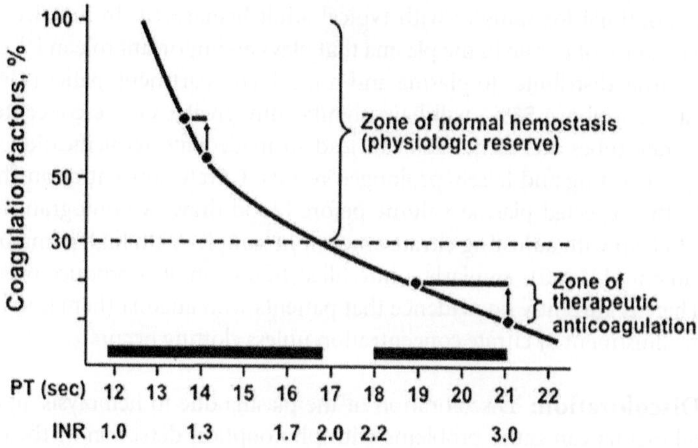

FIGURE 128.1 **The general relationship between the concentration of coagulation factors and the result of prothrombin time (PT) and international normalized ratio (INR) studies.** The normalization of modest elevations in the INR required much larger volumes of plasma than would be expected, and modest doses of plasma can result in marked changes in the INR when markedly elevated. The cause of this phenomenon can be explained by the nonlinear, exponential, relationship between coagulation factor concentration and standard coagulation test results. As shown above, small increases of coagulation factors correlate with marked changes in coagulation studies when coagulation factors are depleted. The opposite is true when the coagulation factors are at higher concentrations. (From Kor, D. J., Stubbs, J. R., & Gajic, O. (2010). Perioperative coagulation management – fresh frozen plasma. *Best Pract Res Clin Anaesth, 24*, 51–64.)

from 50% to 60%. The response, illustrated in Fig. 128.1, should be kept in mind when plasma products are considered for PT correction. Owren PT will only be prolonged if the activities of factors II, X, and/or VII are decreased. Like Quick-type PT, it is sensitive to inhibitors of factors II, VII, X, and V. If the cause of the significant prolongation of the PT is unclear, factor deficiency can be distinguished from inhibitor effect by performing a 1:1 mix with normal plasma. Lack of correction of PT argues for the presence of inhibitor, whereas complete correction is suggestive of factor deficiency as a cause of PT prolongation. It should be noted that severe combined factor deficiencies, such as those in massive transfusion and DIC, may not correct into the normal range as concentrations of multiple factors below 60% may keep the PT above the reference range. To avoid this issue, some laboratorians recommend performing a 4:1 mix of normal and patient plasma to always bring factor levels above 75%. This approach could, in principle, mask weak inhibitors due to greater dilution effect. On the hand, new oral Xa inhibitors that prolong PT will most times correct on a 1:1 mix. Thus, PT mixes should always be interpreted in light of clinical information. PT is usually prolonged in patients on rivaroxaban, whereas patients on apixaban may have a normal PT. The prolongation effects of new oral Xa inhibitors on PT is not linear, thus, it is difficult to estimate the drug level using a routine PT assay. However, a new lab-developed test, named modified PT, utilizes higher amounts of calcium chloride to increase assay dynamic range and can estimate levels of direct Xa inhibitors. However, this is a lab-developed test that is variable depending on the PT reagents used, and thus appropriate reference ranges need to be established.

Further Reading

Adcock, D., Kressin, D., & Marlar, R. A. (1998). The effect of time and temperature variables on routine coagulation tests. *Blood Coagul Fibrinolysis*, *9*, 463–470.

Barrett, Y. C., Wang, Z., & Knabb, R. M. (2013). A novel prothrombin time assay for assessing the anticoagulant activity of oral factor Xa inhibitors. *Clin Appl Thromb Hemost*, *19*, 522–528.

Clinical and Laboratory Standards Institute Document H21-A5. *Collection, transport, and processing of blood specimens for testing plasma-based coagulation assays and molecular hemostasis assays; approved guideline* (5th ed.). Wayne, PA, US.

Clinical and Laboratory Standards Institute Document H47-A2. *One-stage prothrombin time (PT) test and activated partial thromboplastin time (APTT) test; approved guideline* (2nd ed.). Wayne, PA, US.

Clinical and Laboratory Standards Institute Document H54-A. *Procedures for validation of INR and local calibration of PT/INR systems; approved guideline*. Wayne, PA, US.

Girolami, A., Bertozzi, I., de Marinis, G. B., Bonamigo, E., & Fabris, F. (2011). Activated FVII levels in factor VII Padua (Arg304Gln) coagulation disorder and in true factor VII deficiency: A study in homozygotes and heterozygotes. *Hematology*, *16*, 308–312.

Horsti, J. (2002). Has the Quick or the Owren prothrombin time method the advantage in harmonization for the International Normalized Ratio system? *Blood Coagul Fibrinolysis*, *13*, 641–646.

Porte, R. J., Lisman, T., Tripodi, A., Caldwell, S. H., & Trotter, J. F. (2010). The International Normalized Ratio (INR) in the MELD score: Problems and solutions. *Am J Transplant*, *10*, 1349–1353.

CHAPTER 129

Activated Partial Thromboplastin Time

Mikhail Roshal, MD, PhD and Morayma Reyes Gil, MD, PhD

Along with prothrombin time (PT), activated partial thromboplastin time (PTT) serves as a common screening test for coagulation factor deficiencies. The test is sensitive to significantly decreased activities of so-called intrinsic (FXI, FIX, FVIII), common (FX, FV, FII, and fibrinogen), and contact (kallikrein, high-molecular-weight kininogen [HMWK] and FXII) system components. Clotting is initiated with a combination of anionic surface activator, a phospholipid source, and calcium. During PTT testing, platelet-poor plasma is initially incubated with a surface activator (celite, kaolin, micronized silica, ellagic acid, etc.). The timing of the reaction starts on subsequent recalcification of the plasma. The initial incubation leads to induction of the contact system, resulting in activation of FXII. FXIIa in turn activates FXI that converts FIX into FIXa. FIXa slowly cleaves and activates FX. FXa converts a small amount of prothrombin (FII) into thrombin. Thrombin then generates FVIIIa and FVa. These cofactors greatly accelerate the activities of FIXa and FXa, respectively, leading to generation of enough thrombin to convert fibrinogen to fibrin and hence to form a clot. Clot formation can be detected either by measuring changes in plasma viscosity or absorbance. The time to clot is reported in units of seconds, with normal PTTs typically falling in the range of 20–35 seconds.

Test Utility: PT and PTT are complementary tests used in the assessment of factors involved in the clotting cascade in vivo. They are often used together in routine presurgical evaluation. While sensitivities of various PTT formulations to specific factors vary, under most circumstances the PTT will fall outside the normal range if single intrinsic or common factor activities fall below 30%–40% or if milder deficiencies, involving multiple factors, are present. Fibrinogen concentrations significantly below 100 mg/dL also prolong the PTT.

In addition to deficiencies, specific inhibitors of the factors involved in PTT-initiated clotting cascade can also cause an abnormally prolonged test result. Most PTT test formulations show sensitivity to lupus anticoagulants (LA), which allows the use of the test in the initial screening for LA.

Besides screening for factor inhibitors of deficiencies, PTT has found numerous other uses. While not ideal for these purposes, PTT is also currently utilized to monitor treatment with unfractionated heparin and some direct thrombin inhibitors in most laboratories in the United States because of its wide availability and low cost (see Chapter 159). PTT is a suboptimal test for assessing therapy with newer oral direct thrombin inhibitor dabigatran due to a shallow response curve (see Chapter 159). Use of PTT in the detection of LA is covered elsewhere (see Chapter 158).

Test Limitations: Although PTT is sometimes ordered in trauma, disseminated intravascular coagulation, and liver disease, it has several disadvantages for screening for

Transfusion Medicine and Hemostasis. https://doi.org/10.1016/B978-0-12-813726-0.00129-X

multifactor deficiencies compared with PT. First, high FVIII concentration in inflammation and liver disease can mask factor deficiencies by shortening the PTT. Second, PTT measures some factors that are not associated with bleeding risk. Deficiencies in FXII, kallikrein, and HMWK are not associated with increased risk of hemorrhage as they do not seem to play a significant role in clotting cascade activation in vivo. Third, LA and heparin contamination from line draws can lead to falsely high results more frequently than in PT determination. Fourth, in the setting of trauma, the additional incubation required for contact activation makes PTT a slower test to perform than PT. Although PTT will be increased in vitamin K deficiency and warfarin use, it is not used for monitoring therapy.

Interlaboratory Variation: Large variations exist in commercial reagent preparations, instrument detection protocols, and sensitivities to clotting inhibitors. For these reasons, each laboratory establishes its own reference range specific to their reagent/instrument combination and assesses sensitivity to relevant anticoagulants and factor deficiencies. This means that PTT values in seconds are not comparable between different systems. Moreover, PTT reagents vary in their sensitivity to LA. Some reagents are formulated with increased phospholipid content or lupus anticoagulant insensitive activators to neutralize most LA, whereas others are specifically designed to be more sensitive. In case of LA, these modifications frequently result in PTT being normal in one laboratory and substantially prolonged in another.

Preanalytical and Analytical Considerations

Storage Requirements: If heparin effect is to be measured, plasma must be separated from cellular compartment within 1 hour or tested immediately. Release of platelet factor 4 from platelets neutralizes heparin, leading to underestimation of its effect during prolonged storage. For other uses, PTT samples should be tested within 4 hours of collection. Either room temperature or refrigerated storage is acceptable during this time. Platelet-poor plasma can be frozen for longer-term storage without affecting the PTT.

Effect of Collection Tubes, High Hematocrit, Underfilled Tubes, and Anemia: The PTT test has identical requirements to PT (see Chapter 128).

Plasma Discoloration: Plasma discoloration due to lipemia, hyperbilirubinemia, and hemolysis has a similar effect on PTT to that seen with PT (see Chapter 124).

Effect of Medications: Unfractionated heparin has a greater effect on PTT than low-molecular-weight heparin. A single pentasaccharide drug, fondaparinux, has no or very little effect on PTT depending on formulation effect on PTT. Direct thrombin inhibitors and direct anti-FXa agents prolong PTT. Vitamin K-dependent factor deficiency due to warfarin therapy prolongs PTT.

Lupus Anticoagulant: Lupuslike anticoagulants can inhibit clotting reactions in vitro but are usually not associated with increased risk of bleeding. In general, PTT is more sensitive to LA than PT, although a wide range of sensitivities is seen with different reagent formulations. These depend on specific type and amount of phospholipid and the type of surface activator used. Some atypical lupus inhibitors may have little or no effect on PTT but prolong PT. Detailed information on LA detection is provided elsewhere (see Chapter 161).

Effect of High FVIII: FVIII is an acute phase reactant that shows significant amplitude of physiologic variation. In addition, some patients have congenitally elevated FVIII levels (Chapter 158). Significantly elevated FVIII levels can shorten the PTT and mask mild deficiencies in other factors.

Test Interpretation: Prolonged PTT indicates a reduced activity of one or more of FII, FV, FVIII, FIX, FXI, FXII, kallikrein, HMWK, or reduced concentration of fibrinogen or a nonspecific inhibitory effect. Reduced activity can either be due to factor deficiencies or inhibition. If the cause is not clear, a 1:1 mix with normal plasma can be tested. If PTT prolongation is due to factor deficiency, a correction of PTT is expected. It should be noted that severe multiple factor deficiency may not correct into the normal range, whereas single factor deficiencies should. Clinical information about a probable cause of the deficiency should be used to guide the expectation about the extent of the expected correction. Inhibitors will generally prevent PTT correction into the normal range. However, weak or low titer inhibitors that have a slight effect on PTT may be diluted enough to allow correction and cause confusion with factor deficiencies. Slow acting inhibitors, such as those directed against FVIII, can result in partial or near complete correction of 1:1 PTT mix when measured immediately. On 1 or 2 hour incubation, however, a substantial prolongation of the mix's PTT will be seen. When the difference between the immediate and incubated mix reaches 8–12 seconds, a possibility of specific inhibitor of FVIII should be investigated (see Chapter 132).

Mild factor deficiencies can be masked by very high FVIII levels. Short PTTs are generally caused by high concentration of FVIII, fibrinogen, or sample activation before testing.

Not all factor deficiencies cause increased risk of hemorrhage. Contact factor deficiencies are not associated with bleeding risk. A significant elevation of PTT with complete correction on mixing and no history of bleeding are suggestive of FXII, HMWK, or kallikrein deficiencies.

Further Reading

Kershaw, G. (2017). Performance of activated partial thromboplastin time (APTT): Determining reagent sensitivity to factor deficiencies, heparin, and lupus anticoagulants. *Methods Mol Biol*, *1646*, 75–83.

Clinical and Laboratory Standards Institute Document H21-A5. *Collection, transport, and processing of blood specimens for testing plasma-based coagulation assays and molecular hemostasis assays; approved guideline* (5th ed.). Wayne, PA, US.

Fritsma, G. A., Dembitzer, F. R., Randhawa, A., et al. (2012). Recommendations for appropriate activated partial thromboplastin time reagent selection and utilization. *Am J Clin Pathol*, *137*, 904–908.

Turi, D. C., & Peerschke, E. I. (1986). Sensitivity of three activated partial thromboplastin time reagents to coagulation factor deficiencies. *Am J Clin Pathol*, *85*, 43–49.

Uprichard, J., Manning, R. A., & Laffan, M. A. (2010). Monitoring heparin anticoagulation in the acute phase response. *Br J Haematol*, *149*, 613–619.

CHAPTER 130

Mixing Studies

Connie H. Miller, PhD

In mixing studies, patient plasma with a prolonged activated partial thromboplastin time (PTT) or prothrombin time (PT) is mixed with normal pool plasma (NPP). PTT or PT is measured after mixing the two samples. Mixing studies are used to distinguish among potential causes for prolonged screening test results—in particular, to distinguish between a factor deficiency and the presence of an inhibitor.

Methods: The standard procedure for PTT or PT is performed on a 1:1 mixture of patient plasma and NPP, usually immediately and after 1 hour incubation at 37°C.

Interpretation: Mixing study outcomes are summarized in Table 130.1. Complete correction of a 1:1 mix suggests a factor deficiency, either congenital or acquired. Failure to correct completely suggests an inhibitor interfering with one or more coagulation factors or a lupus anticoagulant. If repeating the PTT mix after incubation for 1 hour at 37°C gives a longer time, a FVIII inhibitor, which is time-dependent, may be present. Rarely, time-dependent inhibitors may be directed against FV. A PT mix that corrects most often indicates a multifactor deficiency due to the presence of liver disease, vitamin K deficiency, or warfarin; however, it may represent FVII deficiency alone.

TABLE 130.1 Interpretation of Activated Partial Thromboplastin Time (PTT) and Prothrombin Time (PT) Mixing Studies in Specimens Without Heparin

Observation	Possible Cause	Tests to Perform
PTT 1:1 Mix of Patient and NPP[a]		
Corrected immediately and at 1 hour	Factor deficiency	Factor VIII, IX, XI, XII activities (factors II, V, X if PT also prolonged)
Corrected immediately and prolonged at 1 hour	Factor VIII inhibitor	Factor VIII activity; if low, FVIII inhibitor
	Weak LA[b]	LA tests
Incompletely corrected	LA	LA tests
	Factor inhibitor	Factor VIII, IX, XI, XII activities
PT 1:1 Mix of Patient and NPP		
Corrected immediately	Factor deficiency	Factor II, V, VII, X activities
Incompletely corrected	LA	LA tests
	Factor inhibitor	Factor II, V, VII, X activities

[a]Normal pool plasma.
[b]Lupus anticoagulant.

Transfusion Medicine and Hemostasis. https://doi.org/10.1016/B978-0-12-813726-0.00130-6

The definition of "correction" varies from laboratory to laboratory because of differences in the PTT and PT methods used. Some thromboplastin reagents are more sensitive than others. Attempts to standardize across laboratories have been generally unsuccessful. Consultation with the particular laboratory involved or someone experienced in interpreting its results may be helpful, particularly when partial correction occurs.

Test Performance: In clinical situations where specimen volume is limited, as in pediatrics, mixing studies may be skipped and appropriate diagnostic tests performed instead. With automated analyzers, performing factor assays may be quicker and use less plasma than mixing studies. A lupus anticoagulant or nonspecific inhibitor may be indicated by the reduction of multiple factors.

Sources of Error

Heparin: In a patient who is hospitalized or has a central venous access device, the specimen should be treated first with a heparin-neutralizing agent (such as heparinase) and the PTT or PT repeated. The tests should normalize if slight heparin contamination or a therapeutic level of heparin is present. This step is often omitted in outpatients, particularly in children unlikely to have heparin exposure, to conserve the specimen.

Therapeutic Agents: Treatment with antithrombin or anti-Xa drugs, such as lepirudin, dabigatran, or rivaroxaban can result in noncorrection of PTT or PT mixes.

Weak Inhibitors: Mixing studies may be misleading if the PTT or PT is only slightly prolonged because a weak inhibitor can be diluted sufficiently in a 1:1 mix to disappear.

Quality Assurance: This is the same as for standard PTT and PT. NPP should be checked for the presence of all coagulation factors and the absence of inhibitors.

Further Reading

Forte, K., & Abshire, T. (2000). The use of Hepzyme in removing heparin from blood samples drawn from central venous access devices. *J Pediatr Oncol Nurs, 17*, 179–181.

Kershaw, G., & Orellana, D. (2013). Mixing tests: Diagnostic aides in the investigation of prolonged prothrombin times and activated partial thromboplastin times. *Semin Thromb Hemost, 39*, 283–290.

Monis, G., Ferrell, C., & Reyes, M. (2012). Lupus anticoagulant increases activated partial thromboplastin time (PTT) prolongation in incubated 1:1 mix. *Clin Lab Sci, 25*, 165–169.

CHAPTER 131

Coagulation Factor Testing

Connie H. Miller, PhD

Coagulation factors may be measured by methods assessing both their presence, as antigens, and their ability to function, as activity. Inherited coagulation factor deficiencies may be of two types. Type I defects (quantitative) have decreased absolute amounts of the factor, resulting in both decreased activity and antigen levels. They result from lack of production or increased clearance of the gene product. Type II defects (qualitative) have a defective gene product, resulting in decreased activity but normal or slightly decreased levels of antigen. Type II defects result from a mutation changing the gene product or altered posttranslational modification. Acquired defects may mimic either type by decreased or faulty production, increased clearance or inactivation, or the presence of a circulating inhibitor, usually an antibody, which reacts with the factor to mask its activity or remove it from the circulation. Fig. 131.1 illustrates the relationship among functional and immunologic tests and how they reveal underlying structure and function. This chapter reviews both the coagulation factor activity (functional) and antigen (immunologic) assays for factors II, V, VII, VIII, IX, X, XI, XII, and XIII, which are indicated in the diagnosis of congenital and acquired coagulation factor deficiencies, characterization of coagulation factor defects, and monitoring of coagulation factor therapy.

Methods: Activity assays may be clot-based or chromogenic (amidolytic). Clot-based factor assays are modifications of the screening tests to allow a single factor to be rate-limiting in clot formation. Each compares the relative abilities of a test plasma and a control plasma to correct the defect of plasma deficient in a given factor. Correction may be measured in an activated partial thromboplastin time (PTT)–based or prothrombin time (PT)–based assay. Chromogenic assays also provide all components needed except for the factor to be measured; the endpoint is the cleavage of a synthetic substrate through an enzymatic reaction, detected colorimetrically. Immunologic tests for measuring coagulation factors include enzyme-linked immunosorbent assay, latex immunoassay, and other standard methods. The measurement of most clinical relevance is activity; immunologic methods are used primarily to characterize the type of defect present as quantitative or qualitative.

FVIII Activity: FVIII activity (VIII:C) can be measured by a one-stage assay based on the PTT, a two-stage assay (rarely used now), or a chromogenic method. Results are comparable when testing patient plasma, except for a small number of mild hemophilia A patients who may give higher or lower values with the two-stage and chromogenic assays than with one-stage assays and those treated with certain hemophilia treatment products. Recombinant FVIII lacking the B domain, whether or not it is modified to last longer, may be more accurately measured with a chromogenic assay. For accurate quantitation of low levels of VIII:C, a second "low" curve may be generated using

Transfusion Medicine and Hemostasis. https://doi.org/10.1016/B978-0-12-813726-0.00131-8

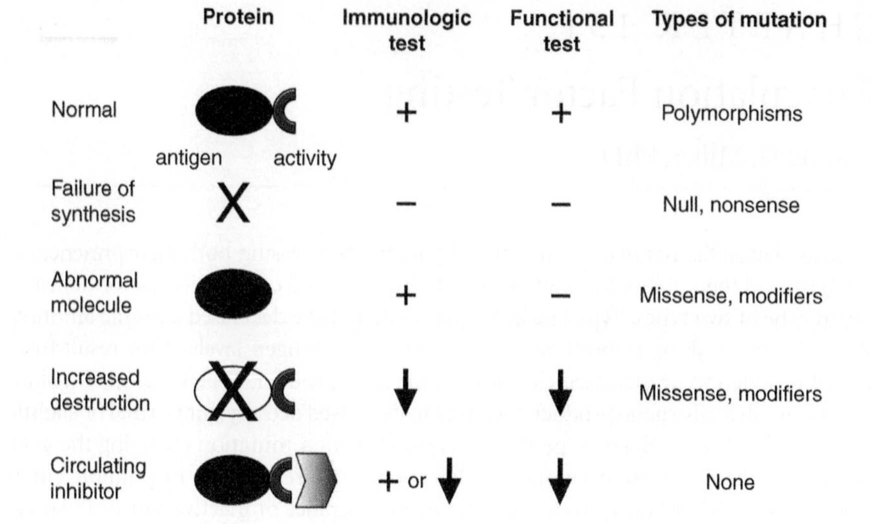

Protein
Immunologic test
Functional test
Types of mutation
Normal
+
+
Polymorphisms
antigen
activity
Failure of synthesis
X
–
–
Null, nonsense
Abnormal molecule
+
–
Missense, modifiers
Increased destruction
Missense, modifiers
Circulating inhibitor
+ or
None

FIGURE 131.1 Functional and immunologic tests of coagulation factors.

VIII:C reference plasma at concentrations of 0.5–20 U/dL. Measurement of VIII:C concentrates requires use of a concentrate standard.

FVIII Antigen: FVIII antigen (VIII:Ag) can be measured using only human or monoclonal antibodies and does not precipitate. It is not routinely measured in clinical practice. The old term "factor VIII–related antigen" or VIII:RAg seen in the literature actually refers to von Willebrand factor antigen (VWF:Ag).

FIX Activity: FIX activity (IX:C) may be measured by one-stage assay based on the PTT or by chromogenic assay. A low curve may also be used to quantitate more accurately the levels of IX:C seen in hemophilia patients.

FIX Antigen: FIX antigen may be detected with heterologous or monoclonal antibodies.

FXI and FXII Activities: FXI and FXII activities are usually measured by one-stage PTT-based assay. FXII deficiency is often the cause of a prolonged PTT in a nonbleeding patient.

FII, FV, FVII, and FX Activities: FII, FV, FVII, and FX activities are usually measured in a one-stage PT-based method, although other methods are available. All can also be detected immunologically.

FXIII Activity: FXIII is not detected by tests with the endpoint of clot formation or thrombin generation. Laboratories have screened for FXIII function by assessing stability of a clot in either 5 M urea or 1% monochloroacetic acid; however, this method detects only severe FXIII deficiency. Actual quantitation of FXIII activity requires measurement of amine incorporation or ammonia release.

FXIII Antigen: Immunologic assays specific for the FXIII-A and FXIII-B subunits are used for diagnosis and classification of the deficiency state.

Test Performance: For quantitation in most functional assays, clotting time or optical density is plotted against concentration using a curve generated from multiple dilutions of a reference plasma of known concentration. Results are expressed as units per milliliter (U/mL) or, more commonly, as units per deciliter (U/dL), which is equivalent to percent of "normal" (%). A result of 100 U/dL or 1.00 U/mL is equivalent to 100%. Results can be expressed in international units if they are derived from a standard, which is defined by the manufacturer or calibrated in-house against an international standard.

Deficient plasmas may be obtained from individuals with a known factor deficiency; however, because of viral risk, artificially depleted plasmas have been developed. They appear to be satisfactory, except in the case of FVIII-deficient plasma depleted by monoclonal antibodies, which lacks von Willebrand factor and performs differently in FVIII inhibitor assays.

Reference ranges for coagulation factors vary with the population to be tested and methods used, and should be derived locally whenever possible. Some factors, such as FVIII, do not follow a normal distribution in the population. Their ranges must be calculated in transformed units or as 5th–95th percentiles using nonparametric methods. It is not appropriate to remove outlying values to produce normality. Some coagulation factors show differences among populations that affect the reference ranges. FVIII activity levels vary significantly by ABO blood group, with group O individuals having lower levels. FVIII is significantly higher and FXII significantly lower in African Americans than in other populations.

Sources of Error: Factor assays performed with clotting techniques are subject to interference from heparin, other anti-Xa drugs, and antithrombin drugs. Heparinase may be used to remove heparin before assay. Lupus anticoagulants will influence clot-based tests for multiple factors. Chromogenic assays are not influenced by heparin, lupus anticoagulants, or fibrin degradation products. Measurement of antigen alone may give false assurance that a factor is present, because it may be nonfunctional.

Measurement of factor activity using only one dilution of patient plasma may mask the presence of an inhibitor. Multiple dilutions may show that the patient's plasma has a dilution curve that does not parallel the normal, usually indicating presence of an inhibitor, often a lupus anticoagulant.

Quality Assurance: Factor assays require a reference plasma, used to produce a curve against which the quantity of a factor can be measured. Commercial reference plasmas should be used as the source of the quantitation to set the calibration for each factor. Positive and negative controls should be run with each assay.

International Standards: Use of a national or international standard provides consistency over time and allows results to be compared across laboratories. Such calibrated standards are preferable to use of pooled plasma from a small number of individuals. International standards are available for FII, FV, FVII, FVIII, FIX, FX, FXI, and FXIII in plasma (National Institute for Biological Standards and Control, Potters Bar).

Further Reading

Kohler, H. P., Ichinose, A., Seitz, R., Ariens, A. S., & Muszbek, L. (2011). Diagnosis and classification of factor XIII deficiencies. *J Thromb Haemost*, *9*, 1404–1406.

Miller, C. H. (2006). Laboratory tests for the diagnosis of thrombotic disorders. *Clin Obstet Gynecol*, *49*, 844–849.

Monagle, P., Barnes, C., Ignjatovic, V., Furmedge, J., Newall, F., Chan, A., et al. (2006). Developmental hemostasis: Impact for clinical hemostasis laboratories. *Thromb Haemost*, *95*, 362–372.

Peyvandi, F., Duga, S., Akhavan, S., & Mannucci, M. (2002). Rare coagulation deficiencies. *Haemophilia*, *8*, 308–321.

Peyvandi, F., Oldenburg, J., & Friedman, K. D. (2016). A critical appraisal of one-stage and chromogenic assays of factor VIII activity. *J Thromb Haemost*, *14*, 248–261.

Pruthi, R. K. (2016). Laboratory monitoring of new hemostatic agents for hemophilia. *Semin Hematol*, *53*, 28–34.

CHAPTER 132

Specific Factor Inhibitor Testing

Connie H. Miller, PhD

Coagulation factor inhibitor assays are used to identify inhibitors occurring in individuals with inherited factor deficiencies (alloantibodies) and those not congenitally deficient (autoantibodies). Alloantibodies are sought when congenitally deficient patients fail to respond to appropriate therapy. Autoantibodies are suspected when low factor levels occur in previously unaffected patients, particularly when multiple dilutions of patient plasma give different factor activity levels, suggesting that the patient's dilution curve does not parallel the normal.

Coagulation factor inhibitors are detected in the laboratory primarily through their ability to neutralize specific coagulation factors and must be distinguished from nonspecific inhibitors, such as the lupus anticoagulant. Recognition of specific inhibitors may be complicated by their presentation as multifactor deficiencies (Table 132.1). At dilutions used in most clinical factor assays, these antibodies may interfere with tests of other factors in their pathway. When patient plasma is tested in higher dilution, accurate measurement of the noninhibited factors is obtained.

Factor VIII (FVIII) inhibitor quantitation was standardized in 1974, when investigators meeting in Bethesda, Maryland, agreed on a standard method and a unit of measure: the Bethesda unit (BU), defined as the amount of inhibitor, which destroys one half of the FVIII activity (VIII:C) in 1 mL of normal plasma within 2 hours when incubated at 37°C. Modifications to increase the sensitivity and specificity of the Bethesda method have been adopted, particularly the Nijmegen-Bethesda (NB) method and use of chromogenic measurement of FVIII. Preanalytical heat inactivation of patient plasma to allow testing with FVIII present has also been introduced. The Bethesda method has been extended to the quantitation of inhibitors to other coagulation factors.

Antibody detection methods, including enzyme-linked immunosorbent assays (ELISA) and fluorescence immunoassays, detect most inhibitors that neutralize factor activity and those that bind to the factor but fail to neutralize its activity in vitro.

TABLE 132.1 Spurious Effects of Factor Inhibitors on Factor Activity Assays

	Dilution	Factor VIII	Factor IX	Factor XI	Factor XII
Factor VIII inhibitor	1:5	<1	23	19	40
	1:20	<1	45	38	55
	1:100	<1	78	65	82
Factor IX inhibitor	1:5	<1	<1	<1	–
	1:20	110	<1	70	–

Transfusion Medicine and Hemostasis. https://doi.org/10.1016/B978-0-12-813726-0.00132-X

Methods and Test Interpretation: A factor inhibitor is detected by performing the relevant coagulation factor assay on a mixture of patient plasma and normal pool plasma (NPP) and comparing the result to a control mixture. Inhibitors to FVIII and some inhibitors to factors V and XI are time-dependent and require incubation. Inhibitors to other factors usually react immediately.

Antibodies with type 1 or "simple" kinetics show high affinity and can be saturated. In antibody excess, all activity is neutralized. Antibodies with type 2 or "complex" kinetics show lower affinity and can dissociate from antigen. Both free antigen and free antibody may be present. Most hemophilic inhibitors have type 1 kinetics; type 2 is more common among autoantibodies.

FVIII Inhibitors: The Bethesda assay is performed by incubating a 1:1 mixture of patient plasma and NPP and a control 1:1 mixture of NPP with imidazole buffer at 37°C for 2 hours and measuring the VIII:C remaining. In the NB method, FVIII-deficient plasma or 4% bovine serum albumin is substituted for buffer in patient dilutions and the control mixture, and NPP is buffered with imidazole to pH 7.4 to maintain protein concentration and pH of the mixtures during incubation. In either method, the activity in the patient mixture is divided by the activity remaining in the control mixture and multiplied by 100 to give the % residual activity (RA). The RA is converted to BU using a graph plotting the logarithm of RA against BU or the equation:

$$BU = (2 - \log RA)\,(0.301)^{-1}$$

If RA = 100%, BU = 0. For RA 25%–100%, BU of the lowest dilution in that range is reported. If no dilution falls above 25% RA, patient plasma is tested serially diluted with imidazole buffer, and the first dilution falling between 25% and 75% RA is multiplied by the dilution factor and reported. Examples of tests on inhibitors with type 1 and type 2 kinetics are shown in Table 132.2. For Type 2 inhibitors, which do not respond

TABLE 132.2 Calculation of Results for Type 1 and Type 2 Inhibitors Measured in Nijmegen-Bethesda Units (NBU)

	Patient Dilution	Patient Mix VIII:C[a]	Control Mix VIII:C[a]	% Residual Activity[b]	Calculated NBU[c]	Total NBU[d]
Type 1 inhibitor	Undiluted	19	45	42.2	1.24	1.2
	1:2	30	45	66.7	0.58	1.2
	1:4	35	45	77.8	0.36	1.4
Type 2 inhibitor	Undiluted	27	45	60.0	0.74	0.7
	1:2	27	45	60.0	0.74	1.5
	1:4	27	45	60.0	0.74	3.0

[a]Units per deciliter of FVIII activity.
[b]Patient mix VIII:C/control mix VIII:C X 100.
[c]NBU read from graph or calculated as $NBU = (2 - \log \%\ \text{residual activity})\,(0.301)^{-1}$.
[d]Calculated NBU X dilution factor.

linearly to dilution, the dilution closest to an RA of 50% is often reported, but quantitation is inexact. Some laboratories report the first dilution falling within the 25%–75% RA range. It is useful to compare the same dilution in subsequent assays to document rise or fall. Results are reported as BU or Nijmegen-Bethesda units. To test patients with measurable VIII:C, such as recently treated hemophilia patients or nonhemophilic patients, plasma may be heated to 56°C for 30 minutes and centrifuged before testing to destroy VIII:C. A chromogenic FVIII assay may also be used in the Bethesda or NB assay.

An ELISA for FVIII inhibitors is commercially available, and other immunologic methods have been described. Because they also detect nonneutralizing antibodies, quantitation using these methods is not directly comparable to that of clot-based tests. They may be used to confirm the presence of specific antibodies.

FIX Inhibitors: FIX inhibitors, which occur infrequently in patients with severe hemophilia B, may be measured by the Bethesda method without incubation, as they are not time-dependent.

FXI Inhibitors: FXI inhibitors occur in up to one-third of congenitally deficient patients homozygous for the type II mutation Glu117Stop and other FXI-deficient patients. They may be time-dependent and are detected by the Bethesda method.

FV, FVII, FX, and FXIII Inhibitors: Inhibitors are seen rarely in congenital deficiencies of FV, FVII, and FX and may be detected by Bethesda-type assays. FXIII inhibitors may be detected by inhibition of specific assays of FXIII activity.

von Willebrand Factor Inhibitors: Inhibitors directed against von Willebrand factor (VWF) in type 3 von Willebrand disease (VWD) may be detected by mixing patient and NPP and performing a VWF activity assay, such as ristocetin cofactor or collagen-binding assay. Many VWF inhibitors, however, are nonneutralizing. VWF inhibitors in non-VWD patients will be detected only rarely in a Bethesda-type assay. VIII:C may also be inhibited in a Bethesda assay, presumably as a secondary phenomenon.

Sources of Error: Lupus anticoagulants, nonspecific inhibitors, and heparin can give false positive results in inhibitor assays. Presence of circulating FVIII may lead to underestimation of the inhibitor titer. Heating plasma to 56°C, to destroy endogenous and infused FVIII without affecting the antibody, may increase the inhibitor titer in some patients. Clot-based inhibitor assays detect only those antibodies, which neutralize factor activity in vitro. In vivo factor recovery studies are required for detection of inhibitors that bind to the factor and are cleared from the circulation.

Further Reading

Franchini, M., Castaman, G., Coppola, A., Santoro, C., Zanon, E., Di Minno, G., et al. (2015). Acquired inhibitors of clotting factors: AICE recommendations for diagnosis and management. *Blood Transfus*, *13*, 498–513.

Kershaw, G., Jayakodi, D., & Dunkley, S. (2009). Laboratory identification of factor inhibitors: The perspective of a large tertiary hemophilia center. *Semin Thromb Haemost*, *35*, 760–768.

Manco-Johnson, M. J., Nuss, R., & Jacobson, L. J. (2000). Heparin neutralization is essential for accurate measurement of Factor VIII activity and inhibitor assays in blood samples drawn from implanted venous access devices. *J Lab Clin Med, 136*, 74–79.

Miller, C. H. (2018). Laboratory testing for factor VIII and IX inhibitors. *Haemophilia, 24*, 186–197.

Miller, C. H., Platt, S. J., Rice, A. S., Soucie, M. J., & The Hemophilia Inhibitor Research Study Investigators (2012). Validation of Nijmegen-Bethesda assay modifications to allow measurement of hemophilic inhibitors during replacement therapy and facilitate inhibitor surveillance. *J Thromb Haemost, 10*, 1055–1061.

Salomon, O., Zivelin, A., Livnat, T., Dardik, R., Loewenthal, R., Avishai, O., et al. (2003). Prevalence, causes, and characterization of factor XI inhibitors in patients with inherited factor XI deficiency. *Blood, 101*, 4783–4788.

CHAPTER 133

Thrombin Time and Fibrinogen Evaluation

Mikhail Roshal, MD, PhD and Morayma Reyes Gil, MD, PhD

Decreased fibrinogen concentration or impaired fibrinogen function can lead to hemorrhage. Thus fibrinogen testing is frequently utilized in the setting of trauma, surgery, disseminated intravascular coagulopathy, and fibrinolytic treatment to determine the need for replacement product. Congenital decreases in fibrinogen such as congenital hypofibrinogenemia and afibrinogenemia are rare. In addition, congenital qualitative deficiencies in fibrinogen activity (dysfibrinogenemia) are rare, whereas acquired dysfribrinogenemia is more common. Clauss fibrinogen and prothrombin time (PT)–based clottable fibrinogen methods allow for rapid quantitative determination of clottable fibrinogen in a clinical laboratory. Additionally, point-of-care whole-blood tests utilizing thromboelastography or rotational thromboelastometry can be utilized to monitor fibrinogen in major surgery and trauma situations. Thrombin time (TT) is methodologically related to Clauss fibrinogen and provides a useful tool for screening for heparin, thrombin inhibitors, and dysfibrinogenemia.

Clottable fibrinogen determination is one of the basic screening tests in a coagulation laboratory. Decreased fibrinogen concentration or impaired function can lead to hemorrhage. Thus fibrinogen testing is frequently utilized in the setting of trauma and surgery to determine the need for replacement product. In addition to trauma and surgery settings, fibrinogen concentration may also be reduced in consumptive coagulopathies, fibrinolytic therapy, and due to compromised fibrinogen synthesis. Inherited disorders of fibrinogen such as congenital hypofibrinogenemia, afibrinogenemia, and dysfibrinogenemia are rare. Clauss fibrinogen and PT-based clottable fibrinogen methods allow for rapid quantitative determination of clottable fibrinogen in a clinical laboratory. In addition, point-of-care whole-blood tests utilizing thromboelastography or rotational thromboelastometry can be utilized to monitor fibrinogen in major surgery and trauma situations. TT determination is methodologically related to Clauss fibrinogen but uses lower thrombin concentration. It provides a useful tool for screening for heparin, thrombin inhibitors, and dysfibrinogenemia.

Physiologic Role of Fibrinogen: Fibrinogen is a plasma glycoprotein with a multitude of activities in the hemostasis system. The protein is a product of three genes *FGA*, *FGB*, and *FGG* and is primarily synthesized by hepatocytes, although extrahepatic fibrinogen synthesis has been observed in lung, kidney, and other tissues. Fully assembled fibrinogen is a hexamer of three dimers Aα, Bβ, and γ chains. The nomenclature refers to the small polypeptides A and B that are cleaved from Aα and Bβ chains by thrombin during conversion of fibrinogen to fibrin (see Chapter 115).

In addition to the plasma fibrinogen pool, the protein is also stored in the platelet alpha granules. The platelet fibrinogen pool provides a localized boost in fibrinogen concentration at the site of platelet activation. Fibrinogen serves as a scaffold for platelet

Transfusion Medicine and Hemostasis. https://doi.org/10.1016/B978-0-12-813726-0.00133-1

aggregation via the activated form of integrin αIIbβ3 (also known as glycoprotein IIb/IIIa). Platelet aggregation via fibrinogen cross-linking provides an initial hemostatic barrier following blood vessel injury as part of the rapid primary hemostatic response. Subsequently, thrombin activation on the platelet surface leads to conversion of fibrinogen to fibrin and the formation of a more durable hemostatic barrier consisting of platelets and fibrin. During the conversion of fibrinogen to fibrin, thrombin cleaves fibrinopeptides A and B from fibrinogen Aα and Bβ chains, respectively, forming so-called fibrin monomers. Fibrin monomers polymerize into a noncovalently linked, staggered, linear network that stabilizes the initial platelet plug. Fibrin fibers are subsequently covalently cross-linked by activated factor XIII, a step that prevents premature dissolution of the fibrin clot. Subsequent degradation of fibrin requires action of plasmin. Paradoxically, fibrinogen also possesses antithrombotic activity, as it sequesters thrombin at nonsubstrate sites. Thus lack of fibrinogen in afibrinogenemia can lead to hemorrhagic and thrombotic complications. Dysfibrinogenemia can also present as either hemorrhagic and/or thrombotic complications, as abnormal fibrin results not only in defective clots but also may be resistance to plasmin cleavage and lysis.

Clinical Significance: Functional fibrinogen concentration is an important physiologic parameter. The normal concentration of clottable fibrinogen is approximately 150–400 mg/dL of plasma, although the normal ranges may vary somewhat from laboratory to laboratory and from method to method. A substantial decrease in clottable fibrinogen below 100 mg/dL can lead to hemorrhage, as fibrinogen becomes a limiting reagent in the formation of a hemostatic plug. Functional fibrinogen concentrations above 150–200 mg/dL may be needed for optimal hemostasis in the setting of trauma resuscitation and major surgery.

Hemostatic function of fibrinogen can be compromised by either quantitative or qualitative changes in the molecule. Substantial quantitative abnormalities of fibrinogen are most often acquired. Dramatic declines in fibrinogen concentrations are frequently seen in conditions of hemostatic stress, such as significant trauma, hemorrhage, and disseminated intravascular coagulation (DIC). Low fibrinogen level in the setting of trauma is an independent predictor of poor survival. Severely compromised liver synthetic function due to synthesis inhibitor antileukemic drug L-asparaginase, and occasionally liver failure can also significantly reduce fibrinogen concentrations. Congenital decrease in fibrinogen concentration due to mutations in fibrinogen genes leading to hypofibrinogenemia or afibrinogenemia is rare.

Increased fibrinogen concentration is most often seen in inflammation. Fibrinogen is a classic acute phase reactant. Hepatic synthesis of fibrinogen can increase up to 20-fold from baseline levels under conditions of severe stress. Genetic and environmental factors can also play a role in a subset of patients with high fibrinogen concentration. Persistently elevated fibrinogen levels have been linked to atherosclerosis risk.

Qualitative fibrinogen defects (dysfibrinogenemia) can be either congenital or acquired. Acquired dysfibrinogenemia is frequently seen in liver disease due to aberrant fibrinogen with increased amount of sialic acid. This modification is similar to fetal fibrinogen and likely has only modest, if any effect on bleeding risk. Dysfibrinogenemia associated with kidney disease such as nephrotic syndrome and renal cell carcinoma has also been reported. Congenital disorders of fibrinogen are discussed in Chapter 115.

Testing Overview

Functional Fibrinogen Determination: The gold standard method for measuring clottable fibrinogen relies on conversion of all fibrinogen into fibrin using thrombin, and then measuring the total protein content of the fibrin clot. Such measurements are cumbersome and impractical in day-to-day operations of a clinical laboratory. Thus, commercially available clinical tests rely on measuring the rate of fibrin polymerization, rather than the total fibrin amount, as an estimate of clottable fibrinogen concentration.

Clauss Fibrinogen Test: The Clauss fibrinogen method allows isolation of fibrinogen measurements from other coagulation factors and phospholipids. This method utilizes a very high (50–100 NIH units) concentration of thrombin in diluted patient plasma. Under these circumstances, fibrinogen becomes a limiting reagent for the rate of clot formation, and the time to clot formation becomes proportional to the clottable fibrinogen concentration. The values are read from a standard curve made with dilutions of a fibrinogen calibrator. Either electromechanical or absorbance-based measurements can be used to detect clot formation. Electromechanical methods generally produce better precision. Nonetheless, both methodologies yield clinically acceptable results under most circumstances.

Prothrombin Time–Based Tests: PT-based methods measure the rate of change in plasma turbidity in diluted plasma stimulated with PT reagent containing tissue factor and phospholipid. The rate of change should be proportional to the rate of fibrin polymerization. This test is inexpensive, as it provides "free" fibrinogen estimation when PT is run. It is, however, subject to a wider variety of interferences, which can produce an inaccurate result under some circumstances.

Thromboelastography/Thromboelastometry: Thromboelastography (TEG 5000, Haemonetics, Braintree, MA) and related rotational thromboelastometry (ROTEM, TEM Innovations, Munich, Germany) are designed as point-of-care tests to assess global hemostasis in whole blood in the setting of trauma or surgery. The tests measure the rotational shear force applied by the clotted whole-blood sample to a pin linked to a detector. This force is minimal in an unclotted sample but increases with the formation of the clot and is proportional to clot strength. Under standard activation conditions, the magnitude of the shear force is correlated with both platelet count and fibrinogen activity. Modifications that block platelet aggregation can mostly isolate fibrin formation from platelet aggregation and thus provide an estimate of clottable fibrinogen concentration. Both TEG 5000 and ROTEM offer specific fibrinogen modules. The usual goal of the study is to assess the need for blood products during procedures. Thus fibrinogen modulus results are treated as semiquantitative or qualitative estimates to direct the use of cryoprecipitate. Overall the results of the clottable fibrinogen modulus correlate well with Clauss fibrinogen measurements and may produce more physiologically relevant results when volume expanders are used during resuscitation. The tests are also useful for detection of fibrinolysis if extended beyond the point of maximum clot formation.

TABLE 133.1 Interpretation of Fibrinogen Activity Results

Common causes of low clottable fibrinogen	**Age-related:** Slightly lower clottable fibrinogen in neonates due to presence of fetal fibrinogen that has slower activation kinetics **Acquired:** DIC, liver failure, hemorrhage, increased fibrinolysis due to treatment with fibrinolytic agents (e.g. tPA, streptokinase, etc.), high doses of direct thrombin inhibitors, high concentration of heparin, acquired dysfibrinogenemia **Congenital (mutations in *FIB* genes):** afibrinogenemia, hypofibrinogenemia, dysfibrinogenemia
Common causes of elevated fibrinogen	**Age-related:** Slightly higher values postmenopause **Acquired:** Inflammation, poststrenuous exercise, smoking, slight seasonal changes **Congenital:** G-455A polymorphism, other polymorphisms, multifactorial

Functional Fibrinogen Test Interpretation: Common physiologically relevant causes of changes in fibrinogen concentration are listed in Table 133.1. In the context of hemostatic stress, fibrinogen values below 100 mg/dL are frequently considered critical.

Test Limitations: The Clauss fibrinogen test has relatively few known interferences due to high thrombin concentrations that are used to initiate fibrin polymerization. However, treatment with volume expanding colloids such as hydroxyethyl starch may lead to falsely elevated fibrinogen values when photooptical instruments are used. The fibrinogen test is insensitive to heparin at therapeutic concentrations, but very high concentrations of heparin can variably falsely decrease fibrinogen values. Similarly, direct thrombin inhibitors may yield a falsely low fibrinogen value at high direct thrombin inhibitor levels. Commercial calibrators for fibrinogen preparations vary and may introduce substantial lab-to-lab variation in fibrinogen values.

PT-based tests frequently give elevated values compared with the Clauss fibrinogen method, particularly in DIC, liver disease, and treatment with oral anticoagulants. In addition, the test is not sensitive to inhibition by fibrin split products that likely has physiologic relevance. The test produces mostly normal values in patients with dysfibrinogenemia. In general, the PT-based test seems to correlate better with antigenic (total) rather than clottable fibrinogen when such values significantly diverge. Differences in turbidity between the calibrator plasma and the test plasma may introduce additional variables. The test is more sensitive to heparin and direct thrombin inhibitors.

Thromboelastography-based fibrinogen estimation is a relatively new technology. Limitations of this type of testing are not yet entirely clear. Because the testing is usually performed outside the laboratory, it raises unique issues with quality control, training, and maintaining proficiency. Outside proficiency testing by the United Kingdom National External Quality Assessment Service (UK NEQAS) has indicated a wide variation of results between centers, to the point that patient management could be significantly altered in some cases.

Total Fibrinogen Determination: Total plasma fibrinogen concentration can be determined by a variety of methodological approaches, including heat and salt precipitation from a known amount of plasma, and immunologic approaches, such as radial immune diffusion and enzyme-linked immunoadsorbent method. These methods do not distinguish between functional and nonfunctional fibrinogen.

Test Interpretation: The total fibrinogen determination is useful to evaluate for the possibility of dysfibrinogenemia. A normal ratio (approximately 0.8–1.7) of clottable to total fibrinogen argues against qualitative fibrinogen defect.

Test Limitations: Immune testing is subject to interferences from heterophile antibodies that cause either increased or decreased values because prozone or high-dose Hook effect.

Thrombin Time: TT determination is closely methodologically related to the Clauss fibrinogen method. TT utilizes undiluted plasma and 1–5 NIH units of thrombin. Under the condition of lower concentration of added thrombin and normal to high fibrinogen concentrations, the rate of clot formation becomes sensitive to therapeutic concentrations of thrombin inhibitors and less sensitive to mild changes in fibrinogen concentration above 100 mg/dL. TT is reported in seconds. The usual upper limit of the normal range is between 11 and 25 seconds depending on thrombin concentration, source of thrombin, and instrumentation used.

Test Interpretation: In a hospital-based clinical laboratory, the presence of pharmacologic thrombin inhibitors, e.g., heparin or direct thrombin inhibitors, in the sample accounts for the vast majority of prolonged TT results. Thus TT is frequently utilized to detect the presence of heparin and direct thrombin inhibitors as a cause of prolonged partial thromboplastin time and/or PT and to exclude interference of these drugs with specialized coagulation testing. Unlike the Clauss fibrinogen test, TT is sensitive to even very low doses of unfractionated heparin. It is exquisitely sensitive to direct thrombin inhibitors. For both classes of drugs, TT often becomes prolonged beyond a useful measurable range at therapeutic doses depending on test conditions. Low-molecular-weight heparin has a modest effect on some TT reagents at therapeutic concentrations. Often, inadequate heparin clearance from venous or arterial puncture lines leads to heparin contamination. If heparin is suspected as a cause of elevated TT, the drug can be removed by either an enzymatic or absorbent treatment. In addition, TT may be prolonged because of low or very high fibrinogen levels or by increased nonpolymerizable fibrin fragments in conditions such as DIC and fibrinolytic therapy and in many forms of dysfibrinogenemia. Elevated TTs have also been reported in children with nephrotic syndrome. The cause of the elevation is likely multifactorial. Antibodies directed at thrombin may also prolong TT. Autoantibodies can arise in autoimmune disorders, due to plasma cell dyscrasia or spontaneously. Occasionally these antibodies can lead to bleeding and even to fatal hemorrhages. This situation is unlike the alloantibody that is usually directed against bovine thrombin in the setting of recent bovine thrombin administration and is commonly benign. Prolonged TT is also seen in a large proportion of amyloidosis patients. The mechanism behind TT prolongation is not fully elucidated and may vary from patient to patient. Prolonged TT in amyloidosis has not been proven to increase the bleeding risk.

Thrombin Time 1:1 Mix: Performing a 1:1 mix study on the TT may be useful to elucidate the cause of prolonged TT. Afribrinogenemia and hypofibrinogenemia (fibrinogen < 100 mg/dL) should show correction on a 1:1 mix. Most weak inhibitors such as fibrin degradation products, low levels of heparin, and mild dysfibrinogenemia may also show correction. However, in some cases of acquired dysfibrinogenemia that exhibits a normal fibrinogen activity to antigen ratio despite a very prolonged TT, the 1:1 mix usually does not correct. However, if the patient plasma is defibrinated by heating at 56°C for 10 minutes, then the 1:1 mix will be correct, as 50% of fibrinogen (~150 mg/dL) provided by the normal pool plasma should be sufficient to obtain a normal TT. The control for this assay should be a 1:1 mix of the normal pool plasma with either 1 part buffer or 1 part defibrinated normal pool plasma.

Test Limitations: Test systems that utilize bovine rather than human thrombin can become elevated because of antibovine thrombin antibodies. These antibodies are known to arise in the setting of recent topical "thrombin glue" administration. Usually such antibodies do not cross-react with human thrombin and are physiologically unimportant. If a human thrombin–based testing system is available, it will usually show normal results in such cases. Recent administration of radiocontrast can inhibit fibrinopeptide A release in vitro, prolonging TT. The physiologic relevance of this phenomenon is unclear. TT should be repeated after the substance has cleared. Common interferences, such as hemolysis or other discoloration of the sample, can impact photooptical clot-detection systems. Additionally, underfilled tubes (increased citrate to plasma ratio) or clotted samples should not be used for testing.

Further Reading

Cunningham, M. T., Brandt, J. T., Laposata, M., et al. (2002). Laboratory diagnosis of dysfibrinogenemia. *Arch Pathol Lab Med*, *126*, 499–505.

Cunningham, M. T., Olson, J. D., Chandler, W. L., et al. (2012). External quality assurance of fibrinogen assays using normal plasma: Results of the 2008 College of American Pathologists proficiency testing program in coagulation. *Arch Pathol Lab Med*, *136*, 789–795.

Fenger-Eriksen, C., Moore, G. W., Rangarajan, S., Ingerslev, J., & Sørensen, B. (2010). Fibrinogen estimates are influenced by methods of measurement and hemodilution with colloid plasma expanders. *Transfusion*, *50*, 2571–2576.

Kitchen, D. P., Kitchen, S., Jennings, I., Woods, T., & Walker, I. (2010). Quality assurance and quality control of thrombelastography and rotational thromboelastometry: The UK NEQAS for blood coagulation experience. *Semin Thromb Hemost*, *36*, 757–763.

Miesbach, W., Schenk, J., Alesci, S., & Lindhoff-Last, E. (2010). Comparison of the fibrinogen Clauss assay and the fibrinogen PT derived method in patients with dysfibrinogenemia. *Thromb Res*, *126*, e428–e433.

Rourke, C., Curry, N., Khan, S., et al. (2012). Fibrinogen levels during trauma hemorrhage, response to replacement therapy, and association with patient outcomes. *J Thromb Haemost*, *10*, 1342–1351.

CHAPTER 134

Laboratory Diagnosis of Inherited von Willebrand Disease

Connie H. Miller, PhD

von Willebrand factor (VWF) is a large adhesive glycoprotein required for platelet adhesion to subendothelium at the site of vessel injury, platelet–platelet interaction to form the platelet plug, and stabilization of factor VIII (FVIII) in the circulation. Deficiency or defect of VWF leads to the common disorder of hemostasis, von Willebrand disease (VWD). Many cases of VWD, but not all, are due to defects in the structural gene for VWF, located on chromosome 12p. VWF forms a noncovalently bound complex with FVIII, protecting it from clearance. Because unbound FVIII has a shortened half-life, decrease in normal VWF may lead to a secondary decrease of FVIII coagulant activity (VIII:C), although the *F8* gene on the X chromosome is normal.

The *VWF* gene includes 52 exons, covering 178 kilobases (kb) of DNA and encoding mRNA of about 9 kb. In addition, there is a pseudogene for *VWF* located on chromosome 22q, which is a copy of exons 23–34. The *VWF* gene produces a 2813-amino acid (AA) polypeptide (See Chapter 109). During processing in the endoplasmic reticulum, a 22-AA signal peptide is cleaved, dimers are formed by disulphide bonds at the carboxyl termini, and N-linked glycosylation occurs. In the Golgi, O-linked glycosylation occurs, large polymers called multimers are formed by disulphide bonds at the amino termini, and a 741-AA propolypeptide (VWFpp) is cleaved. The protein is secreted or stored as multimers reaching >20 million Daltons. After release from cells, the ultralarge multimers are reduced in size by proteolytic cleavage by the metalloproteinase ADAMTS13. Failure of cleavage results in the ultralarge multimers seen in thrombotic thrombocytopenic purpura. Normal VWF function requires the presence of high-molecular-weight (HMW) multimers.

Over 250 mutations in the *VWF* gene have been shown to cause VWD. They are listed in the International Society on Thrombosis and Haemostasis VWF database at http://vwf.group.shef.ac.uk. Many mutations are unique to individual families, while a few recur.

VWD is diagnosed and classified based on quantitative and qualitative differences in VWF. Type 1 VWD is a partial quantitative deficiency of VWF. Type 2 VWD results from a qualitative defect in VWF. Type 3 is a complete or nearly complete deficiency of VWF.

An algorithm for VWD testing is shown in Fig. 134.1. Routine screening tests are not useful for most forms of VWD. The activated partial thromboplastin time is prolonged only in the minority of VWD patients who have significantly reduced VIII:C. Prothrombin time, thrombin time, and fibrinogen are normal. The platelet function analyzer will detect moderate to severe VWD, but it is not sufficiently sensitive to allow exclusion of the diagnosis. Accurate diagnosis and characterization of VWD subtype require a panel of tests measuring different aspects of the FVIII/VWF complex. Most

Transfusion Medicine and Hemostasis. https://doi.org/10.1016/B978-0-12-813726-0.00134-3

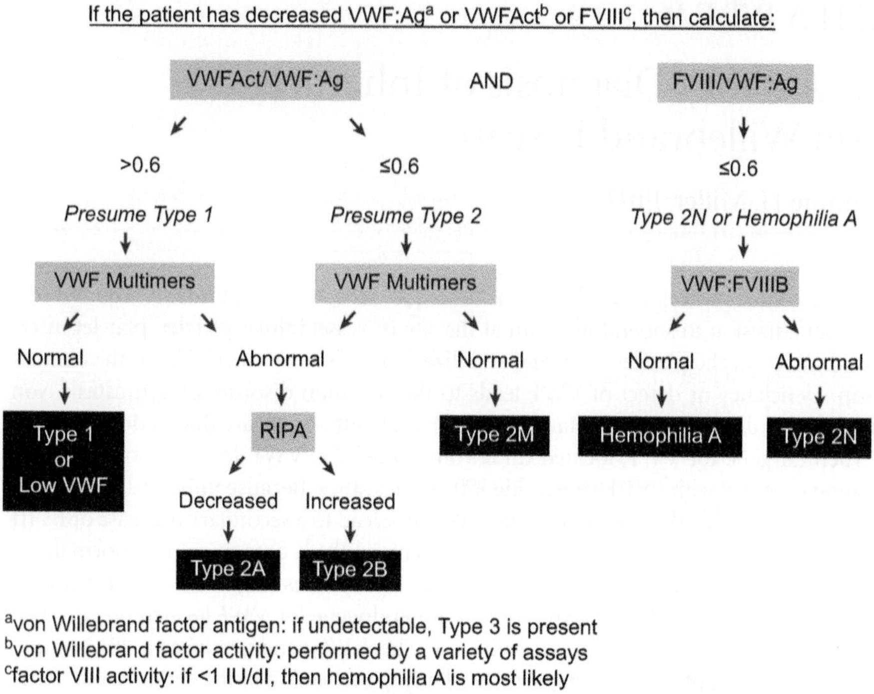

FIGURE 134.1 **Laboratory diagnosis of von Willebrand disease.** ***FVIII*, factor VIII; *RIPA*, ristocetin-induced platelet aggregation; *VWF*, von Willebrand factor; *FVIIIB*, FVIII binding to VWF.**

commonly, VWF antigen (VWF:Ag), VWF activity, and FVIII activity are used for the initial diagnosis. A decrease in any of these tests warrants further investigation. The ratio of results on these tests may be used to choose additional tests to be performed, which may be available only in reference laboratories. The more specialized tests are used to classify the type of VWD present.

Ristocetin cofactor (VWF:RCo) has been the most widely used method for measuring the activity of VWF. Platelet aggregation or agglutination by VWF occurs in vitro only when the antibiotic ristocetin induces a conformational change in VWF allowing binding to GPIb, a change that is induced by shear stress in vivo. Single-nucleotide polymorphisms (SNPs) have been identified in VWF near the ristocetin binding site, which inhibit binding of ristocetin and give a falsely low measure of functional activity in tests using ristocetin. These SNPs are more common in African Americans and may account for the reduced ratio of VWF:RCo/VWF:Ag often seen in that group. Use of GPIb binding assays without ristocetin overcomes this effect.

Methods

von Willebrand Factor Antigen: VWF:Ag is measured immunologically by its reaction with heterologous antibodies using enzyme-linked immunosorbent assay (ELISA) or latex immunoassay (LIA). VWF:Ag is expressed in units per milliliter (U/mL) or in units per deciliter (U/dL), which is equivalent to a percent of "normal" (%).

Ristocetin Cofactor: VWF:RCo is measured by agglutination of washed, lyophilized, or formalin-fixed normal platelets by patient plasma in the presence of ristocetin. Agglutination can be measured by aggregometry, automated coagulation analyzer, or visual agglutination. VWF:RCo is expressed in U/mL or U/dL, which is equivalent to %.

Glycoprotein Ib Binding Assays: GPIb binding assays have been developed to measure VWF binding to isolated GPIb rather than to whole platelets. Because they are based on ELISA or particle agglutination methods, these tests are more sensitive and reproducible than platelet-based assays and can more easily be automated to provide rapid results. Those using ristocetin (VWF:GPIbR) resemble VWF:RCo, but those utilizing a modified "gain of function" GPIb (VWF:GPIbM) can be conducted without ristocetin, eliminating the falsely low values seen in some individuals with ristocetin-based tests.

Monoclonal Antibody Assays: Tests using monoclonal antibodies directed against functional domains of VWF in an ELISA or LIA format (VWF:Ab) have been marketed as activity assays. Their ability to detect qualitative differences in VWF has been questioned, and they may actually be quantitative, more closely resembling VWF:Ag than VWF:RCo. They are not recommended as a replacement for VWF:RCo.

Factor VIII Coagulant Activity: VIII:C is secondarily reduced in VWD because of lack of sufficient VWF to stabilize it in the circulation or decreased ability of VWF to bind FVIII. Measurement of VIII:C is by clotting or chromogenic assays (see Chapter 131). VIII:C is expressed in U/mL or U/dL, which is equivalent to %.

Collagen Binding Assay: The collagen binding assay (VWF:CB) is determined by ELISA of VWF binding to collagen on microtiter plates. This test is highly dependent on the type of collagen used. It is sometimes performed as a substitute for VWF:RCo, but it measures a different function of VWF: its ability to bind to subendothelial collagen to allow platelet adhesion. VWF:CB is expressed in U/mL or U/dL, which is equivalent to %. This test is not widely available for clinical use.

Ristocetin-Induced Platelet Aggregation: Ristocetin-induced platelet aggregation (RIPA) is measured by aggregation of patient's platelets in platelet-rich plasma by addition of ristocetin in two concentrations. RIPA is also decreased in Bernard–Soulier syndrome (BSS). Decreased RIPA from VWD can be corrected by addition of normal plasma, whereas that from BSS cannot. RIPA is not sensitive to all forms of VWD and is frequently low in African Americans without VWD because of the ristocetin insensitivity discussed above. When RIPA is performed with low concentrations of ristocetin, at which platelets from normal individuals fail to aggregate, increased aggregation occurs in two disorders: type 2B VWD and platelet-type VWD (PT-VWD). A test measuring binding of VWF to fixed normal platelets can differentiate the two disorders.

von Willebrand Factor Multimers: VWF multimer analysis is used to reveal VWF structural abnormalities. VWF multimers are separated by SDS-agarose gel electrophoresis and reacted with polyclonal anti-VWF antibodies, usually by Western blot.

Type 1 has normal-sized multimers but may have decreased intensity of bands. Type 2A has decreased HMW multimers as a result of mutations that prevent multimerization or lead to increased clearance. Type 2B can also have decreased HMW multimers because they are removed from circulation as a result of their increased binding affinity for platelets. Normal multimers are seen in types 2M and 2N.

von Willebrand Factor Binding to Platelets: A monoclonal antibody to VWF is used to detect binding of patient's VWF to formalin-fixed normal platelets in the presence of low concentrations of ristocetin by ELISA or fluid-phase binding assay. This assay is used to distinguish between type 2B and PT-VWD. Plasma VWF from a type 2B patient will show increased binding to normal platelets. Plasma VWF from a PT-VWD patient will bind normally to platelets because the defect is in the patient's platelets. The test may be less sensitive if the patient's VWF:Ag is less than 10 U/dL.

Factor VIII Binding to von Willebrand Factor: Patient VWF is captured by a monoclonal antibody and reacted with normal FVIII. FVIII bound is measured by chromogenic methods, ELISA, or fluid-phase binding assay. This test is used to distinguish between type 2N VWD and mild hemophilia or hemophilia carrier. Results are expressed as the ratio between bound FVIII and bound VWF. Patients with type 2N VWD show a reduced ratio.

von Willebrand Factor Propeptide: VWFpp is measured by ELISA using specific monoclonal antibodies. Because VWFpp and VWF:Ag are initially released in a 1:1 ratio, the VWFpp/VWF:Ag ratio can be used to assess VWF survival. Individuals with increased VWFpp/VWF:Ag have increased clearance of their own VWF, as demonstrated by reduced half-life of VWF following desmopressin administration. Increased VWFpp/VWF:Ag may occur in type 1C, 2B, platelet-type, and some cases of 2A VWD, as well as in acquired von Willebrand syndrome (AVWS).

DNA Sequencing: Specific regions of involved genes are sequenced to identify mutations associated with subtypes of VWD by polymerase chain reaction and direct sequencing. These tests are used to confirm a diagnosis of type 2A, 2B, 2M, 2N, or PT-VWD. As sequencing of the entire *VWF* gene is not usually performed for clinical purposes because of its large size, mutations occurring outside of the targeted area may be missed.

von Willebrand Factor Inhibitor: Measurement of VWF inhibitor is used to detect antibodies interfering with VWF function in patients with type 3 VWD or AVWS (see Chapter 135). The test is analogous to the Bethesda method for FVIII inhibitors (see Chapter 132). Decrease in a measure of VWF activity in a mix of patient plasma and normal pool plasma (NPP) when compared with a mix of buffer and NPP suggests the presence of an inhibitor. Use of VWF activity methods other than VWF:RCo in this test has not been reported. Inhibition of VIII:C in the Bethesda assay may also occur in the presence of a VWF inhibitor.

Test Interpretation: Comparison of all VWF tests to a single standard, preferably one measured against an international standard, is important to allow use of ratios of the test results to reach a diagnosis. All VWF measures are influenced by ABO blood group, with group O individuals having 20%–25% lower VWF and VIII:C than those with non-O groups in the relationship group O < A < B < AB. African Americans show significantly higher levels of VWF:Ag, VWF:GPIbM, VWF:Ab, and VIII:C but may show decreases in ristocetin-based tests and the VWF:RCo/VWF:Ag ratio. VWF is an acute-phase reactant and is elevated with pregnancy, inflammation, and chronic disease. VWF is lowest in days 1–4 of the menstrual cycle, and testing women during menses has been advocated to minimize variability. If test results are inconclusive, repeat testing is warranted. Expected test results for different types of VWD are shown in Table 134.1.

Type 1 von Willebrand Disease: In type 1 VWD, the most common form, VWF:Ag, VWF:RCo, and VWF:CB are usually decreased proportionately to 10–50 IU/dL. VIII:C may be normal or reduced. US National Heart, Lung, and Blood Institute guidelines recommend that VWF:RCo and/or VWF:Ag < 30 IU/dL be considered diagnostic of VWD; individuals with levels of 30–50 IU/dL may also be given a VWD diagnosis if there is additional clinical or familial evidence of VWD. Most type 1 patients with VWF:Ag ≤ 30 IU/dL but only 50% of those with 30–50 IU/dL have VWF mutations. Those without VWF mutations may have defects in other genes; however, it has been suggested that a low level of normal VWF may not always represent a genetic defect. Lack of identification of a mutation does not preclude treatment for "low VWF." VWF multimers are normal with the usual methods; there may be subtle defects detectable only with more sensitive techniques. Most type 1 VWD is caused by decreased synthesis with normal VWFpp/VWF:Ag ratio, except type 1C, which is due to increased clearance and shows increased VWFpp/VWF:Ag.

TABLE 134.1 Diagnostic Test Results in Various Types of von Willebrand Disease

Feature	1	2A	2B	2M	2N	3
VWF[a] antigen (VWF:Ag)	D[b]	N/D	N/D	N/D	N	A
VWF activity[c]	D	D	D	D	N	A
FVIII[d] activity (VIII:C)	N/D	N/D	N/D	N/D	D	D
VWF activity/VWF:Ag	N	D	D	D	N	–
RIPA[e]	N/D	D	I	D	N	A
FVIII binding	N	N	N	N	D	–
VWF multimers	N	AB	AB	N	N	A
PFA-100	N/AB	AB	AB	AB	N	AB
Inheritance pattern	AD	AD/AR	AD/AR	AD/AR	AR	AR

FVIII, factor VIII; *PFA*, platelet function analyzer; *RIPA*, ristocetin-induced platelet aggregation; *VWF*, von Willebrand factor.

[a]von Willebrand factor.

[b]*A*, absent; *AB*, abnormal; *AD*, autosomal dominant; *AR*, autosomal recessive; *D*, decreased; *I*, increased; *N*, normal.

[c]Measured by various assays.

[d]Factor VIII.

[e]Ristocetin-induced platelet aggregation.

Type 2 von Willebrand Disease: Type 2 VWD forms have structurally abnormal VWF, which may be present in normal or reduced quantities. The VWF:RCo to VWF:Ag ratio is usually below 0.6–0.7, except in type 2N, which shows a reduced FVIII/VWF:Ag ratio.

Type 2A: HMW multimers are missing because of abnormal synthesis or increased proteolysis. Mutations for the former are found in multiple domains. Mutations causing increased proteolysis are often found in the A2 domain. RIPA is decreased.

Type 2B: Increased binding of VWF to platelets leads to loss of HMW multimers and mild thrombocytopenia, resulting from mutations in the GPIb binding region of the A1 domain. RIPA is increased with low concentrations of ristocetin.

Type 2M: Decreased VWF activity occurs without loss of HMW multimers because of a defect in the binding site for GPIb. The ratio of VWF:RCo to VWF:Ag is reduced, but multimers appear normal. Clinically, this is a subtle distinction from type 1 and may be missed if tests are not appropriately standardized. Mutations are located primarily in the A1 domain but can also be in the A3 domain.

Type 2N: This form mimics hemophilia A and may be misdiagnosed as mild hemophilia or hemophilia carrier. VWF is present at normal levels and functions normally in its interaction with platelets but fails to bind FVIII, leading to reduction in VIII:C to 5–20 IU/dL. It is most often caused by mutations in the VWF binding region of FVIII in the D′ and D3 domains. Affected individuals are usually homozygous or compound heterozygous for 2N mutations. Type 2N may also be seen in compound heterozygosity with other forms of VWD, in which case it is suspected because of the reduced VIII:C to VWF:Ag ratio. It may be confirmed by measurement of VIII:C binding to the patient's VWF or detection of a 2N mutation.

Type 3 von Willebrand Disease: In type 3 VWD, which is the most severe type, levels of VWF:Ag, VWF activity, and VWF:CB are usually undetectable, and VIII:C is <10 U/dL due to the presence of two abnormal alleles. Mutations are scattered throughout the gene, with nonsense and frameshift mutations being the most common. Large deletions have been linked to formation of alloantibodies, for which type 3 patients should be monitored. Some patients with this degree of VWF reduction have measurable or normal levels of VWF:pp, indicating that they produce VWF that is rapidly cleared from the circulation. These patients may be more properly referred to as "severe type 1" because they have two mutations that can cause type 1 VWD in heterozygous form, whereas true type 3 due to lack of VWF production is recessive, with unaffected heterozygotes. Severe type 1 mutations affecting clearance are most commonly found in the D3 and D4 domains.

Platelet-Type or Pseudo–von Willebrand Disease: Platelet-type or pseudo-VWD mimics type 2B VWD but is due to a defect in platelet GPIbα not VWF. The two can be distinguished by testing of VWF binding to platelets or by DNA sequencing.

Sources of Error: Errors in specimen processing and shipping may cause false positive results. The elevated VWF levels seen with stress, exercise, inflammation, chronic disease, pregnancy, and estrogen use may cause a diagnosis of VWD to be missed. Elevated rheumatoid factor (RF) may falsely elevate VWF:Ag using LIA but does not affect other measurement methods. Rarely, RF may result in a VWF:Ag level significantly below the VWF:RCo. Some forms of VWD are very similar to hemophilia A and its carrier state. Distinguishing them is important for 1) appropriate therapy, because they should be treated with products containing VWF, and 2) genetic counseling, because the inheritance patterns are different.

Further Reading

Baronciani, L., Goodeve, A., & Peyvandi, F. (2017). Molecular diagnosis of von Willebrand disease. *Haemophilia, 23*, 188–197.

Bodó, I., Eikenboom, J., Montgomery, R., Patzke, J., Schneppenheim, R., & Di Paola, J. (2015). Platelet-dependent von Willebrand factor activity. Nomenclature and methodology: Communication from the SSC of the ISTH. *J Thromb Haemost, 13*, 1345–1350.

Flood, V. H., Gill, J. C., Morateck, P. A., Christopherson, P. A., Friedman, K. D., Haberichter, S. L., et al. (2010). Common VWF exon 28 polymorphisms in African Americans affecting the VWF activity assay by ristocetin cofactor. *Blood, 116*, 280–286.

Haberichter, S. L. (2015). Von Willebrand factor propeptide: Biology and clinical utility. *Blood, 126*, 1753–1761.

ISTH VWF. Online Database. Available at: http://vwf.group.shef.ac.uk.

Nichols, W. L., Hultin, M., James, A. H., Manco-Johnson, M. J., Montgomery, R. R., Ortel, T. L., et al. (2008). von Willebrand disease (VWD): Evidence-based diagnosis and management guidelines, the National Heart, Lung, and Blood Institute (NHLBI) Expert Panel report (USA). *Haemophilia, 14*, 171–232.

Ng, C., Motto, D. G., & Di Paola, J. (2015). Diagnostic approach to von Willebrand disease. *Blood, 125*, 2029–2037.

Sanders, Y. V., Groeneveld, D., Meijer, K., Fijnvandraat, K., Cnossen, M. H., van der Bom, J. G., et al. (2015). Von Willebrand factor propeptide and the phenotypic classification of von Willebrand disease. *Blood, 125*, 3006–3013.

CHAPTER 135

Laboratory Diagnosis of Acquired von Willebrand's Syndrome

Connie H. Miller, PhD

Acquired von Willebrand syndrome (AVWS) is a rare disorder in which laboratory findings and clinical symptoms mimic various types of von Willebrand disease (VWD). It may be suspected in a patient with abnormal VWD tests and no previous history of excessive bleeding, who has one of the disorders and conditions listed in Table 135.1 or one with a similar mechanism. The International Registry on AVWS (http://www.intreavws.com/) lists over 100 cases, only two of which are idiopathic. More than 300 cases have been reported in the literature. AVWS may be due to the presence of an autoantibody, adsorption of von Willebrand factor (VWF) by cells or surfaces, increased VWF proteolysis or clearance, or decreased or aberrant VWF production. Autoantibodies specific to VWF may inhibit its function or form immune complexes that are cleared from the circulation. Nonspecific antibodies may also cause VWF clearance. VWF may be adsorbed to tumor cells, lymphocytes, plasma cells, or activated platelets. In congenital or acquired cardiac defects, AVWS may result from adsorption of high-molecular-weight (HMW) multimers to aberrant structures, or from increased proteolysis due to high shear stress, and often resolves with surgical correction. Hypothyroidism leads to decreased VWF production, which is restored by treatment. VWF propeptide (VWFpp), a marker for VWF synthesis, is decreased in hypothyroidism. It is usually normal in AVWS from other causes, often leading to an increased VWFpp/VWF:Ag ratio, indicating normal synthesis and subsequent loss.

Methods: The diagnosis of AVWS uses the assays described in Chapter 134 for inherited VWD: VWF antigen (VWF:Ag), VWF activity, collagen binding (VWF:CB), VWF multimers, VWFpp, and VWF inhibitor.

Test Interpretation: Measurement of ristocetin cofactor (VWF:RCo) is reported to be the most sensitive test, although other VWF activity assays may be used. Diagnostic algorithms for VWD may be followed. In contrast to inherited VWD, more than 80% of AVWS cases show loss of HMW VWF multimers. Inhibitory autoantibodies directed against VWF or the factor VIII/VWF complex are detected in fewer than 20% of AVWS cases by inhibition of VWF:RCo or VWF:CB in mixing studies. Nonneutralizing antibodies may lead to an increased clearance of immune complexes, yet be undetectable in vitro. Approximately 40% of cases have antibodies by enzyme-linked immunosorbent assay. The presence of immune complexes containing VWF has sometimes been demonstrable by gel techniques, such as crossed immunoelectrophoresis, or by their adsorption to staphylococcal protein A. Adsorption of VWF to cells and surfaces may be manifested only as loss of HMW multimers. VWFpp/VWF:Ag will be increased in

Transfusion Medicine and Hemostasis. https://doi.org/10.1016/B978-0-12-813726-0.00135-5

TABLE 135.1 Disorders and Conditions Associated With Acquired von Willebrand Syndrome

Lymphoproliferative disorders	Monoclonal gammopathy of undetermined significance, multiple myeloma, non-Hodgkin lymphoma, hairy cell leukemia, chronic lymphocytic leukemia, Waldenstrom macroglobulinemia, acute lymphocytic leukemia
Myeloproliferative disorders	Polycythemia vera, chronic myeloid leukemia, essential thrombocythemia, myelofibrosis, chronic granulocytic leukemia
Neoplastic disorders	Wilms tumor (nephroblastoma), peripheral neuroectodermal tumor, adrenocortical carcinoma, gastric carcinoma, acute lymphoblastic leukemia, lung cancer, acute myeloid leukemia
Autoimmune disorders	Systemic lupus erythematosus, scleroderma, mixed connective tissue disease, Ehlers Danlos syndrome, autoimmune hemolytic anemia, Felty syndrome
Endocrine disorders	Hypothyroidism, diabetes mellitus
Cardiovascular diseases	Cardiac defects (VSD, ASD), aortic stenosis, angiodysplasia, mitral valve prolapse, patent ductus arteriosus, hypertrophic obstructive cardiomyopathy, left ventricular assist device, primary pulmonary hypertension
Infectious diseases	Epstein–Barr virus, hydatid cyst
Drugs	Ciprofloxacin, valproic acid, griseofulvin, hydroxyethyl starch
Others	Uremia, hemoglobinopathies, reactive thrombocytosis, pesticide ingestion, glycogen storage disease, sarcoidosis, telangiectasia, ulcerative colitis, bone marrow transplant, graft-versus-host disease, transplacental transfer of maternal antibodies

most cases because of normal synthesis with subsequent loss, even when VWF:Ag is within normal limits and may be useful in monitoring success of treatment.

Sources of Error: AVWS is distinguished from inherited VWD primarily by the absence of a lifelong history of excessive bleeding. This may be difficult to discern in young children. A family history of excessive bleeding should be sought. Testing of symptomatic family members may confirm the diagnosis of an inherited defect, as would the identification of a VWD mutation, as discussed in Chapter 134.

Further Reading

Budde, U., Scheppenheim, S., & Ditmer, R. (2015). Treatment of the acquired von Willebrand syndrome. *Expert Rev Hematol*, *8*, 799–818.

Lee, A., Sinclair, G., Valentine, K., James, P., & Poon, M. C. (2014). Acquired von Willebrand syndrome: Von Willebrand factor propeptide to von Willebrand factor antigen ratio predicts remission status. *Blood*, *124*, e1–e3.

Nichols, W. L., Hultin, M., James, A. H., Manco-Johnson, M. J., Montgomery, R. R., Ortel, T. L., et al. (2008). von Willebrand disease (VWD): Evidence-based diagnosis and management guidelines, the National Heart, Lung, and Blood Institute (NHLBI) Expert Panel report (USA). *Haemophilia*, *14*, 171–232.

Tiede, A., Rand, J. H., Budde, U., Ganser, A., & Federici, A. B. (2011). How I treat the acquired von Willebrand syndrome. *Blood*, *117*, 6777–6785.

CHAPTER 136

Laboratory Assessment of Treatment of von Willebrand Disease

Connie H. Miller, PhD

von Willebrand disease (VWD) may be treated by 1) raising the patient's own von Willebrand factor (VWF), 2) replacement of VWF, or 3) use of adjuvant therapies with a global effect on hemostasis. A treatment plan must be based on an accurate diagnosis of the type and severity of VWD present in the patient, as described in Chapter 134, and the specific clinical situation (e.g., active bleeding, minor surgical or dental procedure, or major surgery).

Methods: Laboratory tests used to monitor treatment of VWD, including VWF activity assays, factor VIII activity (VIII:C), and platelet function analyzer (PFA), are described in Chapter 134 and 138. VIII:C level is thought to be the most important determinant for surgical, soft tissue, and joint bleeding, whereas VWF level is more important for mucous membrane bleeding. Most reports describe use of ristocetin cofactor (VWF:RCo) for measuring VWF activity rather than the newer VWF activity assays discussed in Chapter 134. Newer assays have advantages for monitoring treatment because they can be performed more rapidly. Although few comparisons in posttreatment specimens have been reported, results are expected to correlate with VWF:RCo. The PFA closure time (CT) has also been used to monitor therapy.

Treatment Modalities

Desmopressin: Desmopressin (1-deamino-8-D-arginine vasopressin, DDAVP) has been used for more than 30 years to treat VWD. It acts by causing release of stored VWF from endothelial cells; therefore, it is most effective in those patients with type 1 VWD, who have stores of normal VWF available and normal clearance. It is less effective in patients with type 1C, who have increased clearance, and in type 2 variants, in which the VWF stored is abnormal in structure. It is generally ineffective in type 3 patients. Desmopressin is contraindicated in type 2B VWD because release of more avidly binding VWF may cause or worsen thrombocytopenia. Because there are individual differences in response to desmopressin, a test dose is often given to gauge the patient's response before clinical use. A three- to fivefold increase over baseline level of VWF:RCo is expected. VWF:RCo should rise to >30 U/dL. Measurements of VWF:RCo and VIII:C should be made at baseline, within 1 hour and 2–4 hours later. The maximum levels of VIII:C and VWF occur 30–60 minutes after DDAVP administration. Patients with increased clearance of VWF may have shortened VWF and FVIII survival. The PFA CT has been shown to correlate with VWF:RCo levels following DDAVP in patients with type 1 and type 2 VWD and can be used to measure response rapidly; however, monitoring of VWF activity is preferred. Fluids should be restricted and serum electrolytes should be monitored for hyponatremia.

Transfusion Medicine and Hemostasis. https://doi.org/10.1016/B978-0-12-813726-0.00136-7

von Willebrand Factor Replacement: VWF replacement requires use of VWF concentrates or FVIII/VWF concentrates specifically labeled as containing VWF. Most recombinant and monoclonally purified FVIII concentrates do not contain VWF. Available FVIII/VWF concentrates differ in their ratios of VWF:RCo to VIII:C and the presence of high-molecular-weight multimers. Recombinant FVIII and recombinant VWF are sometimes given together. If VWF alone is given, FVIII rise will occur but may be delayed. Concentrate use should be monitored with VWF:RCo and VIII:C levels at least every 12 hours. PFA CT does not correct when VWF:RCo rises in type 3 patients, probably due to lack of platelet VWF. Peak levels of VWF:RCo and VIII:C over 200 U/dL should be avoided because of increased thrombotic risk. Those receiving concentrates for prophylaxis also should have factor recovery measured to determine that they do not have sustained high levels. Type 3 patients treated with VWF-containing products should be monitored for the appearance of inhibitors directed against VWF and/or FVIII.

Adjunctive Therapies: Adjunctive therapies, such as antifibrinolytic and topical agents, require no laboratory monitoring. Estrogens, usually given as oral contraceptives, may increase factor levels; however, they have additional effects promoting hemostasis that are not detectable in the laboratory. Some women with type 3 VWD respond to hormones with decreased bleeding, even if VWF levels do not rise. People with VWD should avoid aspirin, nonsteroidal antiinflammatory drugs, and other drugs known to interfere with platelet function (see Chapter 141).

Further Reading

Curnow, J., Pasalic, L., & Favaloro, E. J. (2016). Treatment of von Willebrand disease. *Semin Thromb Hemost*, *42*, 133–146.

Mannucci, P. M., & Franchini, M. (2017). Laboratory monitoring of replacement therapy for major surgery in von Willebrand disease. *Haemophilia*, *23*, 182–187.

Nichols, W. L., Hultin, M., James, A. H., Manco-Johnson, M. J., Montgomery, R. R., Ortel, T. L., et al. (2008). Von Willebrand disease (VWD): Evidence-based diagnosis and management guidelines, the National Heart, Lung, and Blood Institute (NHLBI) Expert Panel report (USA). *Haemophilia*, *14*, 171–232.

Tosetto, A., & Castaman, G. (2015). How I treat type 2 variant forms of von Willebrand disease. *Blood*, *125*, 907–914.

CHAPTER 137

Measurement of Platelet Count, Mean Platelet Volume, and Reticulated Platelets

Mikhail Roshal, MD, PhD and Morayma Reyes Gil, MD, PhD

Numerical platelet determination is a critical step in assessing bleeding and thrombosis risk. Additionally, platelet count serves as a biomarker for a number of pathologic states. Modern automated hematology analyzers produce fast, accurate, and precise platelet determination under most circumstances. Accuracy of automated platelet counts may be compromised by platelet clumping, platelets of unusual size, small red blood cells (RBC), or cell fragments. Depending on methodology, size and/or light scatter properties are commonly used. Very low platelet counts can also pose a challenge due to problems in distinguishing small platelets from background noise. Manual platelet counts can alleviate some of the problems inherent in the automated counts but may be imprecise and labor-intensive. The current gold standard method is based on a combination of light-scatter gating and staining with antibodies directed against platelet-specific antigens.

Platelets are central to proper hemostatic function. They participate both in primary hemostasis through formation of the temporary plug after injury and in secondary hemostasis by providing the phospholipid surface needed for clot formation and secreting factors that promote the clotting cascade.

Platelets, at 2–3 μm in diameter, are the smallest cellular elements in the plasma. They are produced in the bone marrow through fragmentation of megakaryocyte cytoplasmic projections that invade marrow vascular spaces. Under normal steady-state conditions, roughly 100–200 billion platelets are produced and destroyed per day. Platelet life span is 5–10 days under normal circumstances.

Newly produced platelets are larger in size and contain residual messenger RNA. They are considered to be the platelet analogue of red cell reticulocytes and are termed "reticulated" platelets. Because mRNA degrades within 24 hours in vivo, reticulated platelet counts are a good measure of megakaryocyte synthetic activity.

Increased average platelet size is also a sign of increased platelet turnover and activity and is associated with inflammation. It has also been linked with elevated risk of arterial thrombotic events. Decreased platelet volume may be seen in some congenital platelet disorders and depressed platelet production states. The platelet volume can be captured in the mean platelet volume (MPV) parameter.

Like most hemostatic factors in circulation, platelets are present in excess of what is minimally required to maintain hemostasis under normal circumstances. A healthy adult has roughly 150–450 thousand platelets per microliter of blood. In addition to the platelets present in circulation, the spleen holds approximately one-third of the total number of platelets in the body. Numerous pathologic conditions can reduce platelet numbers in the blood. However, reduction in circulating platelets does not always result in a substantially increased risk of hemorrhage. Recent studies have shown that as few as 5000/μL

Transfusion Medicine and Hemostasis. https://doi.org/10.1016/B978-0-12-813726-0.00137-9

platelets are adequate to maintain vascular integrity in the absence of significant hemostatic stress. Under minimal stress, such as a minor surgical procedure, 30–50,000/μL platelets appear sufficient. 100,000/μL platelets are desirable for major surgery.

On the other end of the spectrum, increased platelet counts may contribute to the pathogenesis of some hematologic disorders. Abnormally high platelet counts in myeloproliferative disorders correlate with a risk of thrombosis and may signal a more aggressive course of the disease. Moreover, abnormally elevated platelet counts may reduce concentration of functional von Willebrand factor through clearance, leading to acquired von Willebrand disease. This can paradoxically elevate the risk of hemorrhage.

As studies show the transfusion threshold should be as low as 10,000/μL platelets for most thrombocytopenic patients, it is imperative for laboratories to achieve adequate precision at these low platelet counts. This requirement still poses challenges in routine practice.

Methodology Overview

Manual Platelet Counts: Platelet counting poses unique challenges for automated analyzers because of their small size. Despite tremendous progress in instrumentation over the last 50 or so years, manual platelet determination is still employed to verify questionable platelet counts in many laboratories. The determination is done by placing a small volume of diluted whole blood that was treated with a red cell lysing reagent, such as ammonium oxalate, in a counting chamber (hemocytometer) and counting platelets using phase-contrast light microscopy. The count is then adjusted by the dilution factor. The method is reasonably accurate and does not suffer from most of the interferences that are relatively common in the instrument counts. As a tool for instrument calibration, calibration verification, or routine platelet determination, this method is not ideal. It suffers from substantial imprecision because of interobserver variation and the relatively small number of events counted.

Platelet count can also be estimated from microscopic examination of Wright–Giemsa or similar stained peripheral blood smear under high magnification. At least 10 representative fields should be counted with the average platelet number, and these are then multiplied by a "field factor" to account for the size of the microscopic field of the particular microscope. An alternative, more laborious, method uses a ratio of platelets to RBC in several high power fields with the platelet count (10,000/μL) being the number of platelets per 1000 RBC multiplied by the automated RBC count. Both of the methods tend to produce results that are within 20% of the automated platelet count when used properly.

Immunoreactive Platelet Count: Flow-cytometric determination based on light scatter and staining for platelet-specific glycoproteins CD41 and CD61 using fluorescently tagged antibodies is currently suggested as a gold standard method. This new gold standard method utilizes the platelet to RBC ratio to produce an accurate platelet count. The procedure can be performed on a flow cytometer but currently is not yet fully available on dedicated hematology analyzers. At this time, only one routine hematology analyzer (Cell-Dyn Sapphire, Abbott Diagnostics, Santa Clara, California, US) has the capability to perform platelet counts based on antibody staining, but employs only one antibody (CD61), and is thus not fully compliant with the reference method.

This methodology does appear to have best agreement with the CD61/CD41 staining method for low platelet counts.

Coulter Impedance Method: The first automated instruments to achieve reasonably accurate platelet estimates were based on the Wallace Coulter method of orifice impedance. The method is based on increased impedance of conductive solution when a nonconducting object is introduced into a conductive stream of fixed cross-sectional area. The increase in impedance is proportional to the volume of the object, as it is the same as the volume of conducting fluid it displaces. Using this principle, nonconducting particles—including RBCs and platelets—can be separated based on their volume. As a platelet volume (roughly 10 femtoliters [fL]) is substantially lower than that of a RBC (roughly 90 fL) and that of a white blood cell, impedance counting works well under most circumstances.

Optical Count: Optical counting methods have been devised to overcome some of the limitations of impedance counting. Unlike impedance-based methods, the particles are counted and separated based on their light scattering properties and in some cases refractive indices. In principle, this method should be more comparable to light microscopy. The counting is based on light scatter of polarized light (laser light) at various angles and refractive indices to separate platelets from other particles. This method is comparable in precision to impedance under most circumstances. Some instruments utilize both impedance and optical counting methods and have algorithms to choose between modes and to alert the operator about a possible interference.

Test Limitations

General Considerations: All platelet counts require single-cell suspensions that can flow in a single particle stream or be identified as individual particles if counted manually. Clotted samples, even those with microscopic clots, can lead to severe underestimation of platelet counts and must be rejected for analysis. Anticoagulant-induced platelet clumping and adherence of platelets to white blood cells is a well-known phenomenon that occurs in some patient samples. Either ethylenediaminetetraacetic (EDTA) or heparin may be the cause. If this feature is identified, the sample should be redrawn using an alternative anticoagulant. Sodium citrate is a good alternative but results in a 10% dilution effect because of the presence of liquid anticoagulant in the tube.

The lower limit of platelet count detection varies somewhat from instrument to instrument depending on the amount of noise intrinsic to the instrument, the methodology (electronic noise), and the presence of particles in the plasma with size similar to that of platelets, termed pseudoplatelets (protein aggregates, red and white cell fragments). At very low platelet levels (less than 20,000/μL or so) and under some pathological circumstances where the "pseudoplatelet" numbers are increased, the number of signals due to noise and platelets may become comparable, and precision becomes unacceptably low.

Laboratories should have a protocol to address possible inaccuracies in automated platelet count. Most modern instruments are capable of alerting the user that the platelet count may be compromised. When the instrument is unable to obtain an adequate separation between platelets and possible interfering material, or when platelet clumps are present, an alert (flag) is issued and the laboratory can then use an alternative method for platelet determination. The International Society of Laboratory Hematology (ISLH) also recommends manual review of the blood smear when first time significant thrombocytopenia or thrombocytosis (platelets less than 100,000/μL or greater than 1,000,000/μL) is suspected, when there is a significant change in the platelet count, and when platelet count falls outside the linear range of platelet count determination of the instrument. These rules are meant to address both the possibility of inaccurate count and to alert the laboratory to look for other pathologies that are associated with altered platelet counts.

Coulter Impedance: Analysis that is purely based on impedance (size) has several drawbacks when used under pathological circumstances. Firstly, such an approach can undercount platelets of unusual size. This effect can be particularly evident with reduced half-life of the platelets or macrothrombocytopenia because of other causes when larger platelet species comparable in volume to red cells predominate. These platelets can be counted as red cells or excluded from the count. Small platelets that occur in some congenital and acquired conditions can also be undercounted as they may merge with "noise" in the instrument.

Secondly, artifactually elevated platelet counts may be seen when pseudoplatelets (objects in the same size range as platelets) are present. This can occur in patients with very low mean corpuscular volume (MCV) such as iron deficiency or thalassemia. The same is true when large numbers of red cell or white cell fragments are present, as in disseminated intravascular coagulopathy (DIC) and chemotherapy. Large immune complexes can on occasion also be counted as pseudoplatelets. Different data analysis algorithms have been designed to deal with the limitations of impedance counting. These have improved the discrimination between red cells and platelets to some degree, but problems still arise for some samples.

Optical Counting: Although optical counting methodology has shown superiority under a limited set of circumstances (low RBC MCV, DIC), it has not been able to significantly improve platelet counting in the low range under most other circumstances.

Both optical and impedance methods produce elevated platelet counts compared with the CD41- and CD61-based reference method when low platelet counts (under 20,000/μL) are present.

Mean Platelet Volume: Nearly every modern instrument is capable of reporting MPV. Impedance counting is naturally suited for MPV determination, as it is inherently volumetric. Optical scatter–based measurements of volume are a bit more complicated. Various optical properties, including the shape of the object and its internal refractive index, influence light scatter in addition to the object's volume. These limitations may be overcome to some extent by first making the platelet spherical using mild detergent in an isotonic solution and measuring both scatter and refraction index to obtain a good estimate of platelet volume.

Test Limitations: Unfortunately, MPV determination has not been standardized between instruments. A recent study has shown that instrument-to-instrument variability of MPV can be as high as 25%. In addition, platelets undergo swelling when stored in EDTA and, to a lesser extent, in citrate-containing solutions. This makes determination of MPV time-dependent. Over time, platelet swelling can increase the measured volume by an additional 20%–25%. Suggestions for using either time-dependent reference ranges or alternative anticoagulants for MPV determination have been made to resolve this issue.

Reticulated Platelets: Reticulated platelet counts can be produced using cell-permeant nucleic acid dyes that stain residual RNA in both RBCs and platelets. In combination with platelet gating for large platelets, this measurement provides a good gauge of platelet synthesis. At this time, the methodology is available only on XE and XN analyzers from Sysmex, Kobe, Japan. Platelets are stained with two dyes, polymethine and oxazine, and then gated on light scatter or impedance and fluorescence. RNA-containing platelets that meet size criteria are classified as immature platelet fraction (IPF). Results can be reported as either IPF % or IPF absolute count. IPF is expected to be high in immune thrombocytopenia and hereditary macrothrombocytopenia but low in hypothrombocytopenia because of bone marrow failure or chemotherapy.

Test Limitations: Owing to poor availability of the testing in the clinical setting, the testing has not gained wide clinical use. In the research setting, testing has not been standardized. Research studies have used differing methodologies, including different dyes for RNA and different gating strategies for platelet identification to assess the IPFs. Results of these studies are predictably varied; making comparisons between them is difficult.

Further Reading

Cid, J., Nascimento, J. D., Vicent, A., et al. (2010). Evaluation of low platelet counts by optical, impedance, and CD61-immunoplatelet methods: Estimation of possible inappropriate platelet transfusion. *Transfusion, 50*, 795–800.

Ferreira, F. L. B., Colella, M. P., Medina, S. S., et al. (2017). Evaluation of the immature platelet fraction contribute to the differential diagnosis of hereditary, immune and other acquired thrombocytopenias. *Sci Rep, 7*, 3355–3362.

Harrison, P., Ault, K. A., Chapman, S., et al. (2005). An inter laboratory study of a candidate reference method for platelet counting. *Am J Clin Path, 115*, 448–459.

International Council for Standardization in Haematology Expert Panel on Cytometry and International Society of Laboratory Hematology Task Force on Platelet Counting. (2001). Platelet counting by the RBC/platelet ratio method: A reference method. *Am J Clin Pathol, 115*, 460–464.

Lancé, M. D., Sloep, M., Henskens, Y. M., & Marcus, M. A. (2012). Mean platelet volume as a diagnostic marker for cardiovascular disease: Drawbacks of preanalytical conditions and measuring techniques. *Clin Appl Thromb Hemost, 18*, 561–568.

Latger-Cannard, V., Hoarau, M., Salignac, S., et al. (2012). Mean platelet volume: Comparison of three analysers towards standardization of platelet morphological phenotype. *Int J Lab Hematol, 34*, 300–310.

Ruisi, M. M., Psaila, B., Ward, M. J., Villarica, G., & Bussel, J. B. (2010). Stability of measurement of the immature platelet fraction. *Am J Hematol, 85*, 622–624.

CHAPTER 138

Platelet Function Analyzer

Connie H. Miller, PhD

The platelet function analyzer (PFA)-100 is the most widely used global test of primary hemostasis. It performs an in vitro test of platelet plug formation, referred to as the PFA, by measuring the time to occlusion of a window in a coated membrane through which blood is forced at high shear rate. It was designed to replace the bleeding time (BT) test, which is no longer recommended.

Methods: In the PFA-100 instrument (Siemens, Tarrytown, NY), 0.8 mL of citrated whole blood is forced at high shear rate (5000–6000/second) through a window in a membrane coated with collagen and epinephrine or collagen and adenosine diphosphate. The time for platelets to adhere and aggregate to produce closure is referred to as closure time (CT). The PFA-200 retains the same measurement system.

Test Interpretation: CT appears to be inversely correlated with functional von Willebrand factor, in particular with ristocetin cofactor, and is equivalent to or better than BT as a screening tool for von Willebrand disease (VWD). Its sensitivity for VWD Types 2A, 2B, and 3 is greater than 98%, but it is less sensitive to Type 1 VWD. The PFA is sensitive to severe platelet function defects but lacks sensitivity for more common platelet function disorders. Like the BT, the CT lacks sufficient sensitivity and specificity to be used alone for screening of individuals for platelet disorders and should be used along with clinical history to determine which patients should undergo diagnostic studies. Few studies have as yet been undertaken to establish a clear role for the PFA in predicting clinical outcomes and monitoring therapy. The use of the PFA in assessing response to antiplatelet medications is discussed in Chapter 143.

Sources of Error

False Positive: CT may be prolonged in specimens older than 4 hours, those sent through pneumatic tubes, those with low platelet count (<100,000/μL), and those with low hematocrit (<30%). Drugs interfering with platelet function, such as aspirin, may also prolong the CT (see Chapter 141).

False Negative: PFA may not detect mild VWD and many platelet function defects. In vitro tests will fail to detect defects of the vessel wall in connective tissue disorders, such as Ehlers–Danlos syndrome, which may result in prolonged BT.

Quality Assurance: No quality control materials are provided. Reference ranges should be calculated by each laboratory.

Transfusion Medicine and Hemostasis. https://doi.org/10.1016/B978-0-12-813726-0.00138-0

Further Reading

Favaloro, E. J., Lippi, G., & Franchini, M. (2010). Contemporary platelet function testing. *Clin Chem Lab Med*, *48*, 579–598.

Harrison, P. (2005). The role of PFA-100® testing in the investigation and management of haemostatic defects in children and adults. *Br J Haematol*, *130*, 3–10.

Hayward, C. P. M., Harrison, P., Cattaneo, M., Ortel, T. L., & Rao, A. K. (2006). Platelet function analyzer (PFA)-100® closure time in the evaluation of platelet disorders and platelet function. *J Thromb Haemost*, *4*, 312–319.

Peterson, P., Hayes, T. E., Arkin, C. F., et al. (1998). The preoperative bleeding time lacks clinical benefit. *Arch Surg*, *133*, 134–139.

CHAPTER 139

Thromboelastography/Thromboelastometry

Mikhail Roshal, MD, PhD and Morayma Reyes Gil, MD, PhD

Recent years have seen a significant increase in the utilization of rotational visco-electric global hemostasis testing in point-of-care settings in support of complicated surgical procedures and trauma. The current instruments allow near real-time data gathering, with many results being available between 10 and 30 minutes post blood draw. The tests are sensitive to large alterations of clotting factors, inhibitors, fibrinogen, platelets, and fibrinolytic cascade. The short time to results and the sensitivity to multiple components of the hemostasis system make such testing particularly attractive for guiding transfusion protocols and pharmacologic therapy. Recent advances in digital instrumentation and options for testing have simplified the simultaneous assessment of multiple components of hemostasis and have somewhat reduced the interpretative challenges associated with the testing. The predictive power of the test results as applied to hemorrhage or thrombosis in nonacute settings has not yet been firmly established.

Recent years have seen a significant increase in the utilization of rotational visco-electric global hemostasis testing in point-of-care settings in support of complicated surgical procedures and trauma. The current instruments allow near real-time data gathering with many results available between 10 and 30 minutes post blood draw. The tests are sensitive to large alterations in clotting factors, inhibitors, fibrinogen, platelets, and fibrinolytic cascade. The short time to results and sensitivity to multiple components of the hemostasis system make such testing particularly attractive to guide transfusion protocols and pharmacologic therapy. The methodology itself is not new, having been pioneered by Hartert in 1948, but until recently it had limited applicability, mainly in the settings of cardiopulmonary bypass and liver transplantation. Recent advances in digital instrumentation and options for testing have simplified the simultaneous assessment of multiple components of hemostasis and somewhat reduced the interpretative challenges associated with the testing. These improvements have gained the technique wider acceptance in trauma resuscitation and additional surgical interventions. Less clear is the applicability of the test results in nonacute settings. The predictive power of the test results as applied to hemorrhage or thrombosis in nonacute settings has not yet been firmly established.

Principles of Testing: Two systems are currently on the market: TEG 5000/TEG 6S (Haemonetics, Braintree, MA, US) and ROTEM delta (TEM Systems, Durham, NC, US). The TEG system refers to the test modality as thromboelastography, whereas ROTEM terms it thromboelastometry. Conceptually the tests are quite similar, but do differ in implementation, nomenclature of the output, and options. Both tests measure torque applied to a pin by a whole-blood or plasma sample in a cup heated to 37°C during oscillating rotation through a 4 degrees 45′ angle. The torque is minimal in an unclotted sample but increases with the formation of the clot because of clot-mediated

Transfusion Medicine and Hemostasis. https://doi.org/10.1016/B978-0-12-813726-0.00139-2

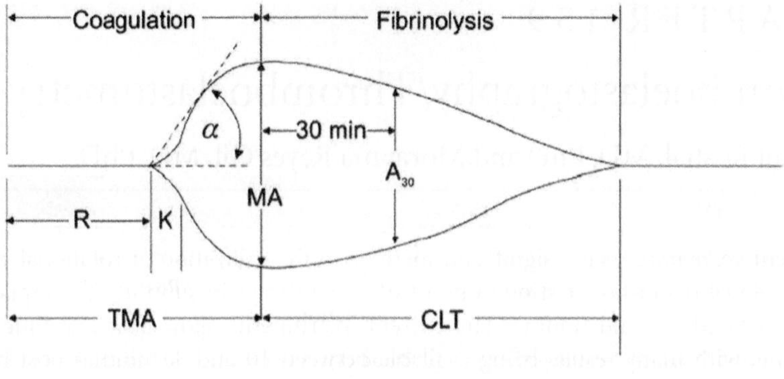

FIGURE 139.1 An illustration of graphical parameters captured in a TEG trace. Refer to Table 139.1 for equivalent parameters in ROTEM. (From Haemoscope Corporation. TEG®Haemostasis Analyzer tracing images are used by permission of Haemoscope Corporation, Niles, IL.)

bridge formation between the pin and the walls of the cup and increased sample viscosity. TEG utilizes a rotating cup and stationary pin connected to a torsion wire. As a clot forms, the torque due to rotation of the cup is increasingly transmitted to the pin. ROTEM has a rotating pin in a stationary cup. Clot formation impedes pin rotation. An optical detector is used to monitor the position of the pin. Output of both instruments is presented graphically and numerically in nearly real time. An illustration of the graphical output is provided in Fig. 139.1.

Generally TEG and ROTEM use citrated recalcified whole blood. Classic thromboelastography used whole blood and no activators of clotting. This option is still available on both the TEG 5000 and ROTEM systems but because of very large variability in procedures and lack of standardization this option is not recommended for clinical use. Combinations of activators and hemostasis component inhibitors have been developed to allow assessment of parts of the hemostatic system in isolation. A summary of testing options available on ROTEM and TEG systems is provided in Tables 139.1 and 139.2, respectively. TEG 6S is a new point-of-care device that integrates all combinations of activators into cartridges named "global hemostasis" and platelet mapping.

Result Interpretation: Commonly reported parameters that are measured in TEG and ROTEM are summarized in Table 139.3 and illustrated in Fig. 139.1. Additional calculated parameters are available on both the TEG and ROTEM instruments. The TEG coagulation index (CI) parameter, for instance, attempts to capture an overall status of the hemostatic system in a single number based on combination of r, k, MA, and α. The parameter is used in either native or celite-activated whole blood and aims to discriminate between hypocoagulable $CI < -3$ and hypercoagulable $CI > 3$ patients. The validity of this and other calculated parameters for predicting either hemorrhage or thrombosis is yet to be clearly established.

Often additional information can be gained from analyzing the shape of the output curves. The typical TEG findings in a range of disorders of hemostasis are illustrated in Fig. 139.2. Similar findings are expected with ROTEM traces.

TABLE 139.1 ROTEM Testing Options

Reagent	Activators and Inhibitors	Application Notes
Natem	No activator	Mainly used in research, very sensitive to hemostasis abnormalities, but slow and nonspecific. Varying contact and platelet activation during blood draw can influence results.
Intem	PTT-type contact activator composed of phospholipid made of rabbit brain, ellagic acid	Sensitive to low dose (0.2–1.0U) unfractionated heparin, rapid coagulation activation with CT reached in 100–240seconds.
Extem	PT-type activator recombinant tissue factor and phospholipids, heparin inhibitor	Not sensitive up to 5U of heparin. CT becomes prolonged at INR >3.5 in vitamin K deficiency.
Heptem	Intem reagent + heparinase	In comparison with Intem, measured the effect of heparin reversal able to detect coagulation disorders in the presence of heparin.
Fibtem	Extem reagent + cytochalasin D	Cytochalasin D inhibits platelets. Clot is composed mainly of fibrin. Allows assessment of functional fibrinogen and in platelet contribution in comparison with Extem.
Aptem	Extem reagent + aprotinin	Aprotinin inhibits plasmin allowing to confirm plasmin-dependent clot degradation in comparison to Extem. Can also be used to monitor antifibrinolytic therapy.
ROTEM platelet (effect of aspirin, clopidogrel or GP IIb/IIIa antagonists) on platelet aggregation	Adp-tem uses ADP to induce platelet aggregation and detection of ADP receptor inhibitors (clopidrogel) Ara-tem uses arachidonic acid to induced platelet aggregation and detection of cycloxygenase inhibitors (aspirin) TRAP-tem uses thrombin analogue to detect GP IIb/IIIa inhibitors (abciximab)	Platelet aggregation is measured by impedance for 6minutes and 3 parameters are reported: AUC (area under the curve, Ohm*min); A6 (amplitude at 6minutes in Ohm); MS (maximum slope of the aggregation graph in Ohm/minute).

Reference Ranges: Although reference ranges based on multiinstitutional clinical trials have been published for both ROTEM and TEG instruments, recent proficiency data indicate very large interinstitutional variation in results, with coefficients of variation of up to 80%. Thus generic reference ranges may not be applicable to individual centers and should be locally evaluated. Many parameters show age-related changes, likely due to increasing concentrations of fibrinogen with age.

Use of Results to Manage Transfusion Protocols and Pharmacologic Therapy: Many institutions have developed procedure-specific transfusion protocols that use the test results as guidance for specific product replacement and/or

TABLE 139.2 TEG Testing Options

Reagent	Activators and Inhibitors	Application Notes
TEG native	No activator Not available in TEG 6S	Mainly for research use, similar to Natem.
TEG kaolin	PTT-like activation Kaolin + phospholipid Included in "global hemostasis" for TEG 6S	Similar to item, but uses slower activation with kaolin rather then ellagic acid, may be sensitive to lupus anticoagulant.
TEG rapid	PT-like clot activation with tissue factor and phospholipid Included in "global hemostasis" for TEG 6S	Fast results, but less sensitive to factor deficiencies and platelet function disorders.
TEG heparin with heparinase	TEG kaolin + heparinase Included in "global hemostasis" for TEG 6S	In combination with TEG, kaolin allows monitoring of heparin reversal and allows detection of coagulation disorders in the presence of heparin.
TEG functional fibrinogen	Reptilase + FXIII + heparin (reagent K) Included in "global hemostasis" for TEG 6S	The reagent does not significantly activate platelets due to thrombin inhibition by heparin. Thus allows isolating fibrinogen contribution to clot strength.
TEG global hemostasis	This cartridge includes TEG kaolin, TEG rapid, TEG with heparinase cup, and TEG functional fibrinogen	Available in TEG 6S instruments. It does not provide % lysis (LY30, LY60).
TEG platelet mapping (effect of aspirin and/or clopidogrel) on platelet activation	For aspirin: reagent K (RK)+arachidonic acid For clopidogrel: reagent K + ADP	In comparison to reagent K activation alone and TEG kaolin allows calculation of percent platelet aggregation as follows $(MA_{ADP(AA)} - MA_{RK})/(MA_{Kaolin} - MA_{RK}) \times 100$

pharmacologic therapy. Thus prolonged r (CT) without evidence of heparin interference is used to trigger plasma transfusion, low fibrinogen value for cryoprecipitate, and low MA (MCF) for platelet transfusion. In addition, if overheparinization is noted (unexpected prolonged r [CT] value corrected on heparin removal) protamine may be used. Primary fibrinolysis as evidenced by prolonged or normal r (CT) with low clot strength and early clot lysis triggers consideration for antifibrinolytic therapy, whereas secondary (compensatory) fibrinolysis in a hypercoagulable patient (short r [CT] and high MA [MCF]) usually does not. Clearly, clinical considerations often take priority over reported test values in therapy considerations. Values for transfusion and intervention triggers vary significantly with specific clinical situation (e.g., trauma vs. liver transplant) and institution. Overall, TEG- and ROTEM-guided transfusion protocol usage in massive transfusion settings has resulted in a decrease in blood product utilization and blood loss but has not yet been proven to decrease donor exposure or improve long-term morbidity or mortality. The field is rapidly evolving.

TABLE 139.3 Commonly Reported Parameters

Instrumentation	TEG	ROTEM	
Clot time (period to 2 mm amplitude)	*r*	CT	Similar to PT/PTT, measures time between sample activation and clot initiation, prolonged in heparin treatment, clotting factor deficiencies and severe thrombocytopenia (lack of surface substrate)
Period from 2 to 20 mm amplitude	k	CFT	Time of clot propagation depends on thrombin activity, fibrinogen, platelet function, and number
Alpha angle	α (tangent between r and k)	α (angle at r)	Velocity of clot propagation depends on similar factors as k with large fibrinogen contribution
Maximal angle	–	CFR	Maximal velocity of clot propagation
Maximum strength of the clot	MA	MCF	Maximal amplitude achieved through formation of the clot, primarily due to platelet number and function with contribution from fibrinogen and thrombin activity
Amplitude at a set time	A (A30, A60)	same	Allows measure of residual clot strength at a given time and allows assessment of fibrinolysis
Percent clot remaining at a set time	CL (CL30, 60)	LI30, LI60	Calculated as (A(t)/MA)*100
Rate of clot lysis	LY (LY30, LY60)	–	Calculated as % reduction of the area under the curve in the amplitude trace, representation of total fibrinolytic activity
Shear-elastic modulus strength	G	MCE	Calculated by converting A to measure of shear. TEG calculation is 5*A/(100–A), reported in Kdynes/cm^2

CT, Clot time; *CFR*, Clot formation rate; *CFT*, Clot formation time; *LI30, LI60*, Lysis index at 30 or 60 min; *MCE*, Maximum clot elasticity; *MCF*, Maximum clot firmness.

Fibrinolysis Detection: See Chapter 146.

Hypercoagulability Assessment: TEG results have been used to assess prothrombotic tendency in intensive care units and surgical settings. MA, G, and CI parameters were generally used. Most studies were retrospective and of variable quality. Many have shown a variable positive association of thrombosis and hypercoagulability by TEG. A laboratory-based study in patients with a personal or family history of thrombosis has shown that TEG does not identify the same subset of patients as identified by routine thrombophilia screening tests. It has been suggested that TEG may prove a useful adjunct test in patients with suspected thrombophilia who have a negative conventional thrombophilia screen. Prospective studies are needed to further evaluate association of TEG and ROTEM findings with thrombosis risk.

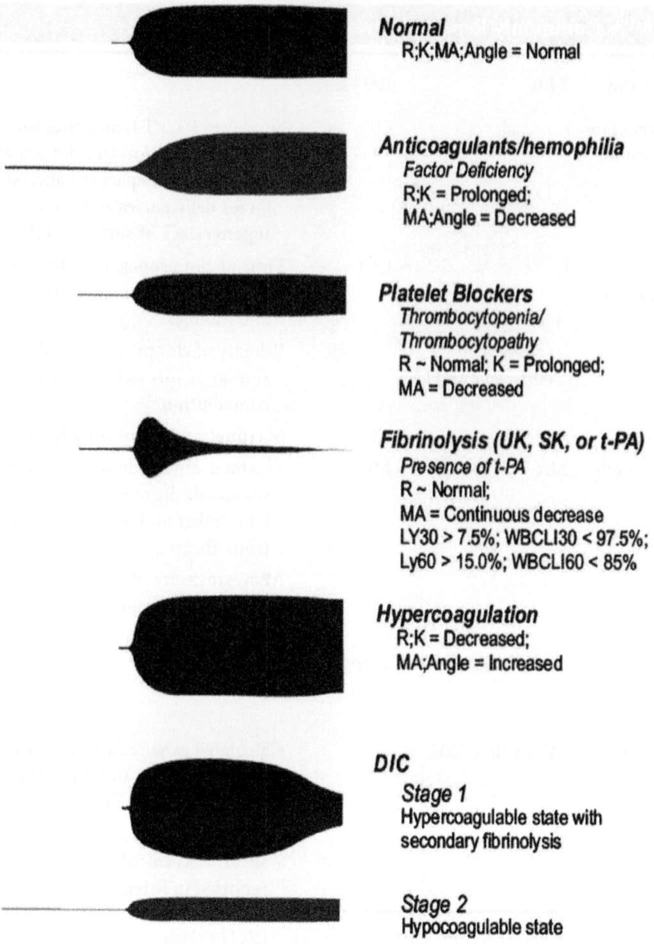

FIGURE 139.2 An illustration of expected pathological and normal TEG traces. Similar pattern of traces is expected as ROTEM output. (From Haemoscope Corporation. TEG®Haemostasis Analyzer tracing images are used by permission of Haemoscope Corporation, Niles, IL.)

Platelet Function Assessment

TEG Platelet Mapping and ROTEM Platelet: A limited number of studies have been published, which compare TEG platelet mapping with other assessment modalities of platelet function. TEG was shown to moderately correlate with light transmission aggregometry (LTA) and VerifyNow in assessment of aspirin and clopidogrel effect on platelets. Compared to LTA, TEG significantly overestimates resistance to aspirin.

ROTEM recently developed a new whole-blood aggregation module that can be integrated to the ROTEM delta system. It measures platelet aggregation by impedance (ohm/min) with ADP, arachidonic acid, and TRAP (thrombin receptor-activating peptide). It is used to detect the effects of clopidogrel, aspirin, and GP IIb/IIIa receptor antagonists. However, no correlation with other assays such as TEG platelet mapping, VerifyNow, or LTA has been published.

Test Limitations

Quality of Data: The usage of complex testing in the point-of-care setting raises unique questions about quality control (QC), quality assurance (QA), personnel training, and proficiency. The limited QC typical of point-of-care devices places a premium on assuring reliability of testing in day-to-day operations outside the usual laboratory setting. Performance of the test by clinicians rather than laboratory personnel has been shown to decrease the precision of the testing. Wide interinstitutional variability, including not infrequent misidentification of normal samples as abnormal, raises significant concern about the quality of results in some centers. Continued proficiency exercise participation may lead to improved reliability by highlighting problem areas.

Sample Handling: Standardization of sample handling is of paramount importance. Citrated recalcified whole-blood results differ significantly from nonanticoagulated whole blood. These samples should not be considered interchangeable. The very short (3 minutes) handing time for nonanticoagulated whole blood must be strictly observed. Use of citrated samples allows for longer storage times, and even transportation to local central testing sites, but lack of inhibition of the upstream factors of the contact system and perhaps platelet activation may significantly change results compared with nonanticoagulated samples with short handling times. The TEG system in particular is also susceptible to interference from vibration during test performance. This problem appears to be substantially reduced in ROTEM.

Correlation With Conventional Test Modalities: There is currently insufficient evidence to prefer TEG/ROTEM to standard coagulation tests in the massive transfusion setting. Moderate to poor correlation is observed between fibrinogen estimates by ROTEM and TEG and Clauss fibrinogen, and between r (CT) and PT/PTT. The findings are not entirely surprising given the integrative nature of TEG/ROTEM parameters. It appears that large variation of factor levels is reliably detected if an appropriate test modality is utilized. In case of fibrinogen, these findings raise concerns about accuracy of viscoelectric estimates in determining fibrinogen concentration. Correlations improve significantly when testing is performed in the absence of red blood cells. Hematocrit has been shown to affect fibrinogen results in whole-blood samples. This discrepancy, however, does not necessarily argue against the validity of the results in assessing fibrinogen function.

Further Reading

Afshari, A., Wikkelsø, A., Brok, J., Møller, A. M., & Wetterslev, J. (2011). Thrombelastography (TEG®) or thromboelastometry (ROTEM®) to monitor haemotherapy versus usual care in patients with massive transfusion. *Cochrane Database Syst Rev*, *16*, CD007871.

Blais, N., Pharand, C., Lordkipanidzé, M., et al. (2009). Response to aspirin in healthy individuals. Cross-comparison of light transmission aggregometry, VerifyNow system, platelet count drop, thromboelastography (TEG®) and urinary 11-dehydrothromboxane B(2). *Thromb Haemost*, *102*, 404–411.

Carroll, R. C., Craft, R. M., Chavez, J. J., et al. (2008). Measurement of functional fibrinogen levels using the thrombelastograph. *J Clin Anesth*, *20*, 186–190.

Cotton, B. A., Faz, G., Hatch, Q. M., et al. (2011). Rapid thrombelastography delivers real-time results that predict transfusion within 1 hour of admission. *J Trauma*, *71*, 407–414.

Dai, Y., Lee, A., Critchley, L. A., & White, P. F. (2009). Does thromboelastography predict postoperative thromboembolic events? A systematic review of the literature. *Anesth Analg*, *108*, 734–742.

Kashuk, J. L., Moore, E. E., Sabel, A., et al. (2009). Rapid thrombelastography (r-TEG®) identifies hypercoagulability and predicts thromboembolic events in surgical patients. *Surgery*, *146*, 764–772.

Kitchen, D. P., Kitchen, S., Jennings, I., Woods, T., & Walker, I. (2010). Quality assurance and quality control of thromboelastography and rotational thromboelastometry: The UK NEQAS for blood coagulation experience. *Semin Thromb Hemost*, *36*, 757–763.

Luddington, R. J. (2005). Thrombelastography/thromboelastometry. *Clin Lab Haematol*, *27*, 81–90.

O'Donnell, J., Riddell, A., Owens, D., et al. (2004). Role of the thrombelastograph as an adjunctive test in thrombophilia screening. *Blood Coagul Fibrinolysis*, *15*, 207–211.

Shen, L., Tabaie, S., & Ivascu, N. (2017). Viscoelastic testing inside and beyond the operating room. *J Thorac Dis*, *9*(Suppl. 4), S299–S308.

Whiting, P., Al, M., Westwood, M., et al. (2015). Viscoelastic point-of-care testing to assist with the diagnosis, management and monitoring of haemostasis: A systematic review and cost-effectiveness analysis. *Health Technol Assess*, *19*, 1–228.

CHAPTER 140

Platelet Aggregation Studies

Connie H. Miller, PhD

The platelet is a dynamic structure, covered externally with glycoprotein (GP) receptors. Internally, the platelet possesses alpha (α) granules, which contain a number of proteins, including von Willebrand factor (VWF), fibronectin, and fibrinogen; and dense (δ) granules, containing adenosine triphosphate (ATP), adenosine diphosphate (ADP), serotonin, pyrophosphate, magnesium, and calcium. Primary hemostasis is initiated by platelet adhesion to vessel wall components at the site of vessel injury. VWF is key for this process at a high shear rate. Platelet activation occurs through exposure to collagen and thrombin and leads to release of granule contents. Aggregation of additional platelets produces a physical barrier. Platelets participate in clot formation through release of stored factors and provision of a surface for assembly of coagulation factor complexes to complete the platelet plug.

In vitro platelet aggregation (PAGG) testing measures the response of platelets to added aggregating agents, or agonists, with measurements that primarily reflect activation, aggregation, and, with some instruments, secretion. Although PAGG testing is both time- and labor-intensive, requires use of fresh blood, and may be difficult to interpret, it remains the gold standard for assessing platelet function.

Methods

Light-Transmission Aggregometry: Traditionally, PAGG has been tested by light-transmission aggregometry (LTA), which measures change in optical density (OD) of stirred platelet-rich plasma (PRP) when an aggregating agent (agonist) is added (Fig. 140.1). As platelets aggregate, more light is transmitted because of decreased absorbance as a result of decreased platelet dispersion. Patient's PRP is set at 0% aggregation and autologous platelet-poor plasma (PPP) at 100%. OD versus time is recorded, which may reveal lag phase, shape change, and first and second waves of aggregation. Calculated parameters include the aggregation rate (slope), maximum amplitude, and percent aggregation.

Whole-Blood Aggregometry: In LTA, centrifugation of plasma to produce PRP and use of additionally centrifuged PPP or buffer to standardize platelet count are manipulations that may alter platelet response or result in loss of platelet subpopulations. Measurement of PAGG in whole blood eliminates these steps and maintains physiologic blood cell milieu. Also whole-blood aggregometry (WBA) requires only one-fourth the volume of blood required for LTA and thus is valuable in the pediatric setting.

WBA is measured by an impedance method. Platelets adhere in a monolayer to wires separated by a gap across which a voltage is applied. When an agonist is added, additional platelets aggregate to the layer, increasing the resistance across the gap, measured in ohms. Change in impedance is plotted as a function of time, giving similar

Transfusion Medicine and Hemostasis. https://doi.org/10.1016/B978-0-12-813726-0.00140-9

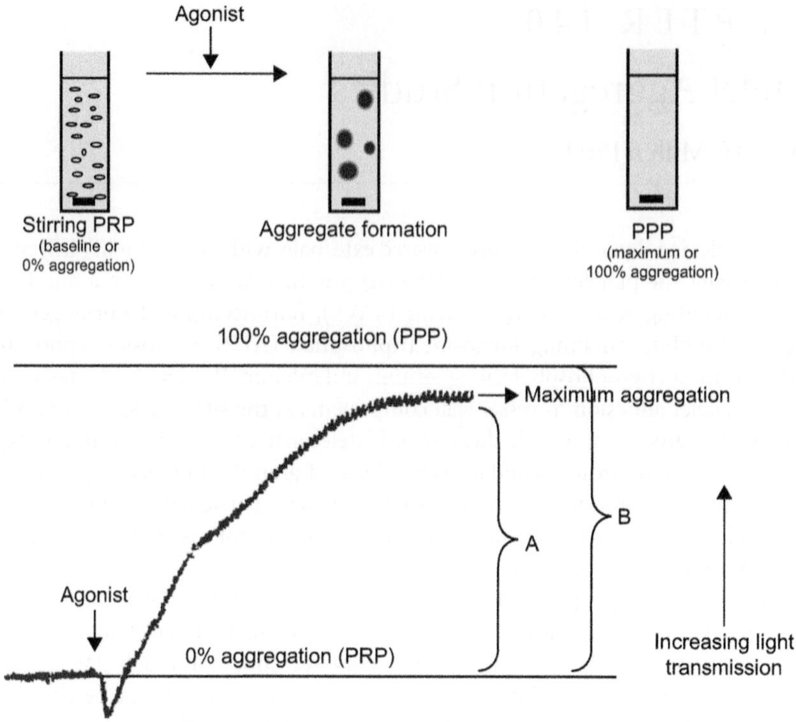

FIGURE 140.1 Light-transmission aggregometry. (A) Stirred platelet-rich plasma (PRP) is set as 0% aggregation. After an agonist is added, aggregates form and light transmission increases. The maximal amount of light transmission is seen in autologous platelet-poor plasma (PPP), which is set as 100% aggregation. (B) % platelet aggregation is calculated as the distance from 0 to maximum aggregation (A) divided by the distance from 0% to 100% aggregation (B) ×100. (Adapted from Jennings, L. K., White, M. M. (2007). Platelet aggregation. In A. D. Michelson (Ed.), *Platelets* (2nd ed.) (pp. 495–508). San Diego, CA: Academic Press.)

curves to those produced by OD changes in LTA. WBA is more sensitive to most agonists and requires lower concentrations for aggregation. Second wave of aggregation is not seen in WBA curves, but such visualization is not necessary if release is measured directly rather than inferred from the curves. Appearance of curves and numerical results differ somewhat between LTA and WBA, requiring method-specific interpretation. Expected results of WBA have not been described for many genetic platelet function disorders, and its substitution for LTA for diagnosis of bleeding disorders has been questioned.

Platelet Adenosine Triphosphate Release: Neither LTA nor WBA is sensitive to all cases of storage pool and release defects. PAGG and platelet nucleotide release may be measured simultaneously in specialized instruments. The secretion of adenosine triphosphate (ATP) is quantitated by its ability to cause the firefly enzyme luciferase to cleave its substrate luciferin and generate luminescence. The amount of luminescence

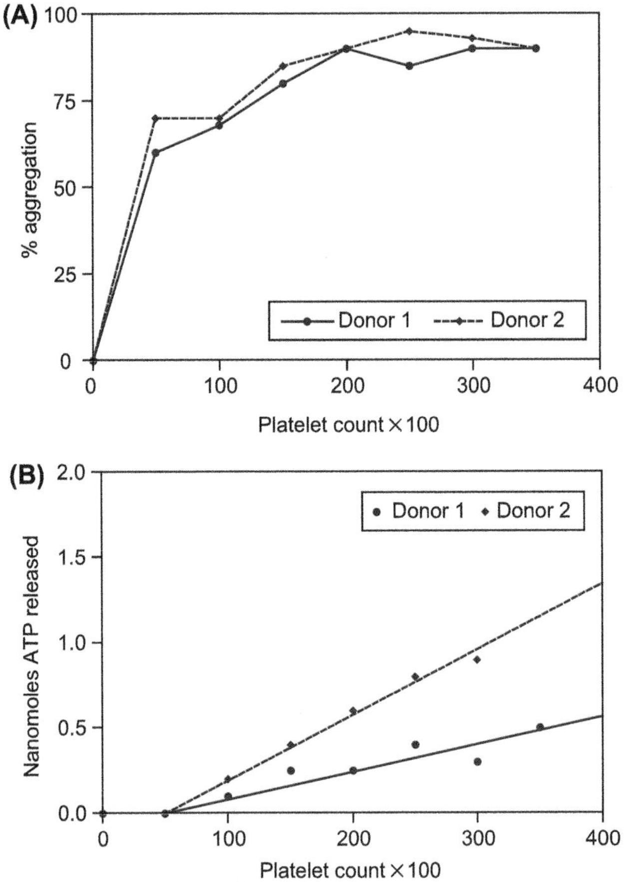

FIGURE 140.2 Platelet aggregation (A) and adenosine triphosphate (ATP) release (B) with 10 μM adenosine diphosphate as a function of platelet count. Aggregation occurs normally above a platelet count of 50,000/μL. ATP release is linearly related to platelet count. (From Jennings, L. K., White, M. M. (1999). *Platelet Protocols*. San Diego, CA: Academic Press.)

is compared with that generated by a standard amount of ATP added to patient blood or PRP and is reported in nanomoles (nM) (Fig. 140.2).

Test Interpretation: PAGG and ATP release are evaluated with a panel of agonists. For some agonists, several concentrations may be used to define the threshold concentration at which aggregation occurs, or standard concentrations may be set. The latter facilitates comparison with ATP release data. The curves should be inspected visually, but today's instruments provide quantitative measures, which can be compared with reference ranges. Reference ranges differ significantly among laboratories and should be established in each laboratory with the instrument and reagent system used.

Certain agonists are commonly used. Platelet function is evaluated by both response to a given agonist and general response patterns to multiple agonists. Examples of the results of WBA and ATP release with specific disorders and detailed description of the corresponding diagnoses are in Chapter 141.

ADP may be used at concentrations of 1–10 μM. In LTA, low concentrations produce a monophasic or biphasic curve with disaggregation. Higher concentrations show a single wave of aggregation, postrelease. Studies of anti-GPIIb/IIIa drugs have used 10–20 μM.

Epinephrine is a mild agonist used at 5–10 μM. In LTA, it produces a slight initial response and a second-wave response that is highly variable. The presence of decreased aggregation with epinephrine alone is common in normal individuals, and its continued use has been questioned because of its high false-positive rate, particularly in males. Epinephrine is not used in WBA.

Collagen is a strong agonist used at concentrations of 1–5 μg/mL. It produces adhesion to collagen fibrils, shape change, and release, followed by aggregation.

Arachidonic acid, used at 0.5 mM, reacts with cyclooxygenase to produce thromboxane A_2. This reaction is inhibited by aspirin and some other antiplatelet drugs. The effect of aspirin is not reversible and continues through the life of the platelet. Effects may still be seen at 2 weeks postingestion. WBA is more sensitive to aspirin than LTA.

Ristocetin is sometimes used to screen for von Willebrand disease (VWD), but it is not very efficient for that purpose. VWD type 1 patients often have normal PAGG with standard concentrations of ristocetin. Plasma tests for VWF antigen and activity are more sensitive and specific (see Chapter 134). Ristocetin is useful for diagnosis of Bernard–Soulier syndrome at concentrations ≥1.0 mg/mL. Low concentrations, ≤0.5 mg/mL, are used to detect the hyperaggregability of platelets in type 2B and platelet-type VWD. ATP release is not measured with ristocetin because its function is passive agglutination of platelets rather than aggregation. As in other tests using ristocetin, falsely low values may occur in patients with certain VWF sequence changes common in African Americans.

Thrombin at 1 U/mL produces maximum release of ATP from storage. As thrombin clots plasma or blood, aggregation cannot be measured unless a thrombin analogue is used. Failure to respond to strong thrombin indicates a platelet storage pool defect.

Thrombin receptor-activating peptide, or TRAP, used at 5–10 μM, is a synthetic peptide mimicking the sequence of the thrombin protease-activated receptor after thrombin hydrolysis. It is sometimes used in place of thrombin to generate activation without clotting, primarily in the measurement of the effects of antiplatelet drugs.

Sources of Error: Valid platelet function testing requires a clean venipuncture with minimal venous stasis, maintenance of the blood at room temperature before testing, and timely completion of tests. WBA but not LTA can be performed on icteric or lipemic specimens; however, hemolyzed specimens should not be used for either. Test results are significantly influenced by a variety of drugs and disorders (see Chapter 141) and may be difficult to interpret in hospitalized patients. Gender, race, and flavonoid-rich foods have been shown to affect findings in healthy drug-free subjects and may lead to false-positive results, which occur in about 20% of healthy subjects.

Further Reading

Favaloro, E. J., Lippi, G., & Franchini, M. (2010). Contemporary platelet function testing. *Clin Chem Lab Med*, *48*, 579–598.

Harrison, P., Mackie, I., Mumford, A., Briggs, C., Liesner, R., Winter, M., et al. (2011). Guidelines for the laboratory investigation of heritable disorders of platelet function. *Br J Haematol*, *155*, 30–44.

Jennings, L. K., & White, M. M. (1999). *Platelet Protocols*. San Diego, CA: Academic Press.

Jennings, L. K., & White, M. M. (2007). Platelet aggregation. In A. D. Michelson (Ed.), *Platelets* (2nd ed.) (pp. 495–508). San Diego, CA: Academic Press.

Miller, C. H., Rice, A. S., Garrett, K., & Stein, S. F. (2014). Gender, race and diet affect platelet function tests in normal subjects, contributing to a high rate of abnormal results. *Br J Haematol*, *165*, 842–853.

Pai, M., Wang, G., Moffat, K. A., Liu, Y., Seecharan, J., Webert, K., et al. (2011). Diagnostic usefulness of a lumi-aggregometer adenosine triphosphate release assay for the assessment of platelet function disorders. *Am J Clin Pathol*, *136*, 350–358.

CHAPTER 141

Laboratory Diagnosis of Platelet Functional Defects

Connie H. Miller, PhD

Platelet function defects (PFD) should be considered in patients with symptoms suggestive of a defect in primary hemostasis. A genetic disorder may be present, particularly if symptoms are lifelong or familial; however, most PFD encountered in clinical practice are acquired. An acquired PFD may be the cause of excessive bleeding, but it may also be an incidental finding or a purely in vitro phenomenon. The complex laboratory diagnosis of a specific PFD should be undertaken only when other more readily identifiable causes for the clinical observations have been ruled out using blood cell counts, peripheral smear, coagulation screening tests, von Willebrand disease (VWD) profile, and a careful history.

Normal platelet function involves over 100 gene products, defects in many of which could potentially cause PFD. Disorders of major clinical significance, such as Glanzmann thrombasthenia (GT) and Bernard–Soulier syndrome (BSS), or with striking clinical presentations, such as Hermansky–Pudlak syndrome, have been well characterized but are rare in most populations. Table 141.1 provides a detailed but not exhaustive list of genetic disorders categorized by mechanism. Many disorders listed have been described in only a few individuals. Examples of the platelet aggregation results expected in some genetic disorders are shown in Table 141.2.

Acquired PFD are most commonly due to drug effects but may result from the effects of an underlying disorder. Hospitalized patients are likely to have at least one drug or intervention affecting platelet function. Outpatients may take prescription or over-the-counter drugs. Herbal remedies have also been implicated in PFD. Sensitive platelet aggregation and release tests will be influenced by common drugs such as aspirin, which causes irreversible effects as long as 2 weeks post ingestion, or foods containing flavonoids. Ideally, abnormal studies should be repeated after 10–14 days with no medications and with flavonoid-rich foods avoided for 24 hours. Marked improvement in platelet function tests should cause the drug to be considered not only as an assay confounder but also as a potential cause of the patient's symptoms. Abnormal test results have occurred in about 20% of healthy individuals, including those documented to be drug-free. Gender, race, and flavonoid-rich foods have been shown to affect their results, and a high false-positive rate is inherent in some test systems. Table 141.3 gives a list of drugs influencing platelet function. It is not exhaustive, and because the impact of many drugs is unknown, exclusion from published lists does not indicate a lack of effect. PFD may also be secondary to a variety of disorders. Bleeding due to acquired PFD is usually mild to moderate; however, it can be life-threatening in the presence of other hemostatic defects, such as thrombocytopenia and coagulation factor deficiencies.

Transfusion Medicine and Hemostasis. https://doi.org/10.1016/B978-0-12-813726-0.00141-0

TABLE 141.1 Genetic Disorders of Platelet Function

Type of Defect	Disorder	Defect In
Adhesion defects	von Willebrand disease (vWD)	Plasma von Willebrand factor
	Bernard–Soulier syndrome	GPIbα, GPIbβ, GPIX
	Platelet-type vWD	GPIb
Aggregation defects	Glanzmann thrombasthenia	GPIIb/IIIa
	Afibrinogenemia	Plasma fibrinogen
Storage pool disorders		
Dense granules	Primary δ SPD	Heterogeneous
	Hermansky–Pudlak syndrome	HPS protein complex
	Chediak–Higashi syndrome	Membrane structure
	Wiskott–Aldrich syndrome[a]	WAS protein
α granules	Gray platelet syndrome	Granule packaging
	Quebec platelet disorder[a]	u-tPA
α/dense granules	αδ SPD	P-selectin
Receptor defects	Thromboxane A_2	Thromboxane A_2 receptor
	Collagen	Integrin α2β1, GPVI
	Adenosine diphosphate	$P2Y_{12}$
	Epinephrine	α-adrenergic receptor
	Platelet activating factor	PAF receptor
	Serotonin	Serotonin receptor
Release defects	Cyclooxygenase deficiency	Cyclooxygenase
	Thromboxane synthetase deficiency	Thromboxane synthetase
G protein defects	Gαδ deficiency	Gαq
Cytoskeletal defects	May–Hegglin anomaly	MYH9, non-muscle myosin
	Wiskott–Aldrich syndrome[a]	WAS protein
Procoagulant function	Scott syndrome	Phosphatidylserine transport
	Stormorken syndrome	Phosphatidylserine transport
	Quebec platelet disorder[a]	Multimerin, FV, u-tPA

[a]Entered in two categories.

Methods

Platelet Function Analyzer: Platelet function analyzer (PFA) is described in Chapter 138. The test has been reported not to be sensitive or specific enough to use for screening to determine who should have platelet function testing. Although it may exclude more severe PFD, such as GT and BSS, milder defects may be missed. There are insufficient data to support use of the PFA to predict clinical bleeding in most settings.

Platelet Aggregation Studies: Platelet aggregation studies are used to determine the cause of a defect in primary hemostasis (see Chapter 140). When sensitive platelet aggregation studies are performed in large numbers of individuals with mild bleeding symptoms, such as women with menorrhagia, a wide spectrum of functional defects is seen, most uncharacterized. Results suggesting a specific genetic diagnosis should be

TABLE 141.2 Examples of Whole-Blood Platelet Aggregation (AGG) and ATP Release (REL) Findings in Patients With Known Platelet Function Disorders

	ADP 20 mM	Collagen 2 μg/mL	Arachidonic Acid 0.5 mM	Thrombin 1 U/mL	Ristocetin 1.0 mg/mL	Ristocetin 0.25 mg/mL
Disorder	**AGG/REL**	**AGG/REL**	**AGG/REL**	**REL**	**AGG**	**AGG**
Glanzmann	A[a]/N	A/N	A/N	N	N[b]	D
Hermansky–Pudlak	N/D	N/D	N/D	D	N[b]	D
Bernard–Soulier	N/N	N/N	N/N	N	A	D
VWD[c] Type 2A	N/N	N/N	N/N	N	A	A
VWD Type 2B	N/N	N/N	N/N	N	N	I
Collagen receptor defect	N/N	D/D	N/N	N	N	D
Scott syndrome	N/N	N/N	N/N	N	N	D

[a]*A*, absent; *D*, decreased; *I*, increased; *N*, normal.
[b]Qualitatively abnormal pattern with disaggregation.
[c]von Willebrand disease.

confirmed by repetition, with drug effects carefully excluded. Most disorders cannot be diagnosed reliably without confirmatory tests such as visualization of granules, receptor studies, or gene sequencing.

Electron Microscopy: Electron microscopy allows for visualization of dense granules (whole mount) or α-granules (transmission) and is used for the diagnosis of storage pool disorders (SPD) and granule defects.

Flow Cytometry: Flow cytometry is used to assess expression of specific glycoprotein receptors on the platelet surface, to detect platelet activation, and to measure platelet procoagulant activity.

DNA Sequence Analysis: DNA sequence analysis identifies specific regions of involved genes that are associated with PFD through polymerase chain reaction and Sanger sequencing or next-generation sequencing. This method may be used to confirm a diagnosis of BSS, GT, platelet-type VWD, and others.

Test Interpretation—Genetic Platelet Function Disorders

Primary Aggregation Defects: GT exhibits a primary aggregation defect with all agonists, except ristocetin, because of a defect or deficiency of the receptor glycoprotein (GP) IIb/IIIa (see Chapter 97). Adenosine triphosphate (ATP) release usually occurs

TABLE 141.3 Drugs that Inhibit Platelet Function

Anesthetics and narcotics	Halothane, benoxinate, benzocaine, butacaine, cyclaine, dibucaine, lidocaine, metycaine, nupercaine, piperocaine, proparacaine, procaine, tetracaine, cocaine, heroin
Antibiotics	Penicillins, cephalosporins, hydroxychloroquine, miconazole, nitrofurantoin
Anticoagulants, fibrinolytics, and antifibrinolytics	UF and LMW heparin, streptokinase, tissue plasminogen activator, urokinase, ε-aminocaproic acid
Antihistamines	Diphenhydramine, chlorpheniramine, mepyramine
Cardiovascular drugs	
Nitrates	Nitroglycerin, nitroprusside, isosorbide dinitrate
Beta blockers	Propranolol, atenolol, nebivolol, pindolol
Calcium channel blockers	Verapamil, diltiazem, nifedipine
ACE inhibitors	Captopril, enalapril, lisinopril, ramipril
Angiotensin receptor	Valsartan, losartan, olmesartan
Antiarrhythmic	Quinidine
Diuretic	Furosemide
Chemotherapeutic agents	Bis-chloroethylnitrosourea, carmustine, daunorubicin, HD (cisplatin, cyclophosphamide, melphalan), mithramycin
Lipid-lowering drugs	Atorvastin, cerivastin, clofibrate, etofibrate, fluvastatin, lovastatin, pitavastatin, pravastatin, rosuvastatin, simvastatin
Nonsteroidal anti-inflammatory drugs	Aspirin, ibuprofen, naproxen, diflunisal, indomethacin, meclofenamic acid, mefenamic acid, phenylbutazone, piroxicam, sulindac, sulfinpyrazone, tolmetin, zomepirac
Platelet cAMP or cGMP modifiers	Prostacyclin, cilostazol, dipyridamole, glostazol, iloprost, sildenafil, aminophylline, theophylline, caffeine, nitric oxide
Psychotropic drugs	Amitriptyline, chlorpromazine, fluphenazine, fluoxetine, haloperidol, imipramine, nortriptyline, paroxetine, promethazine, trifluoperazine,
Volume expanders	Dextran, hydroxyethyl starch
Miscellaneous	Clofibrate, halofenate, hydroxychloroquine, radiographic contrast media, ethanol
Foods and food supplements	Vitamin E, omega-3 fatty acids, fish oil, ginkgo biloba, onions, garlic, ginger, cumin, tumeric, cloves, black tree fungus, chocolate, grape products, caffeine, alcohol

normally. Curves with ristocetin may be irregular and show disaggregation. The definitive diagnosis of GT is made by detection of defects in GPIIb and GPIII by flow cytometry or mutation analysis. Afibrinogenemia also results in a similar primary aggregation defect; however, unlike GT, afibrinogenemia will also result in prolongation in coagulation studies (i.e., prothrombin time, activated partial thromboplastin time, thrombin time).

Storage Pool Disorders: SPD may involve granules other than those in platelets, producing multisystem disease. Hermansky–Pudlak and Chediak–Higashi syndromes also produce oculocutaneous albinism. The latter also involves a severe immunodeficiency and lymphoproliferative syndrome. Wiskott–Aldrich syndrome (WAS) includes immune deficiency and eczema, but not albinism. WAS also is classified as a platelet cytoskeletal defect. SPD affecting dense granules without other features may be among the most common PFD; however, the diagnosis may be missed if ATP release is not measured. Use of thrombin at 1 U/mL causes maximum release of dense body contents, documenting their presence or absence, even when the release mechanism is faulty (see Chapter 98).

Alpha-granule disorders include the gray platelet syndrome, so named for its colorless platelets lacking granules; the similar white platelet syndrome; and Quebec platelet disorder, a complex disorder in which increased urokinase-type plasminogen activator in platelets degrades multiple alpha-granule proteins and sometimes produces fibrinolysis. Flow cytometry may be used to demonstrate surface P-selectin expression in these disorders.

Release Defects: Release defects or "aspirin-like" defects result from impairment of specific enzymes in the thromboxane A_2 pathway, such as cyclooxygenase or thromboxane synthetase, leading to faulty release of granule contents (see Chapter 99). ATP release is decreased. Electron microscopy demonstrates granules to be intact.

Adhesion Defects: Adhesion defects are demonstrated by selective decrease in response to ristocetin, which is most often due to VWD, a plasma deficiency or defect of von Willebrand factor (see Chapter 109). Typically, platelets will aggregate normally to other agonists. The same results are seen in BSS because of a defect in the GPIb-IX-V complex (see Chapter 96). The aggregation defect in VWD will correct with addition of normal plasma, whereas that of BSS will not. BSS can also be distinguished by the presence of large platelets and mild thrombocytopenia, as well as documentation of a characteristic mutation. Flow cytometry will show decreased GPIbα surface density.

Unlike most forms of VWD, type 2B VWD and platelet-type VWD (PT-VWD) show increased aggregation in response to ristocetin at low concentrations that fail to produce aggregation in normal subjects. Both disorders have gain-of-function mutations (VWF in the case of type 2B VWD and VWF ligand GPIb in the case of PT-VWD) and can be distinguished by sequencing of the relevant genes.

Receptor Defects: This heterogeneous group of rare PFD usually demonstrates absent or decreased aggregation with a single agonist with normal ATP release. Aggregation with other agonists may be minimally reduced. The group includes defects in the receptors for thromboxane A_2, collagen, ADP, and others.

Platelet Procoagulant Activity Defects: Scott syndrome is a disorder of phosphatidylserine transport that limits the construction of the prothrombinase complex on the platelet membrane, leading to insufficient thrombin generation. Platelet function studies are normal. Flow cytometric assessment of annexin V binding is diagnostic. Stormorken syndrome is due to an increase in platelet procoagulant activity.

Test Interpretation—Acquired Platelet Function Defects

Uremia: Uremia results in platelet aggregation and ATP release defects with a variety of agonists, including adenosine diphosphate (ADP), collagen, epinephrine, and arachidonic acid. Thrombin and ristocetin results are often normal. In vitro platelet function does not correlate well with clinical bleeding. Uremic plasma can inhibit platelet aggregation, perhaps by guanidinosuccinic acid mediated through nitric oxide. Findings are reported to return to normal after renal transplant (see Chapter 122).

Myeloproliferative Disorders: Myeloproliferative disorders, including essential thrombocythemia, polycythemia vera, chronic myelogenous leukemia, and agnogenic myeloid metaplasia, result in PFD that may lead to excessive bleeding. Some effects are asymptomatic, however, and thrombosis is often a greater risk. Abnormalities in aggregation, ATP release, and an acquired SPD have been described, as well as acquired VWD, due to loss of high-molecular-weight multimers.

Acute leukemias and myelodysplastic syndrome produce bleeding primarily through thrombocytopenia and disseminated intravascular coagulation (DIC); however, PFD also play a role. Decreased platelet aggregation with ADP, epinephrine, and collagen, as well as ATP release defects, have been reported. Acquired VWD and BSS also have been described.

Hepatic Disease: In liver disease, platelet dysfunction plays a secondary role to coagulation factor deficiencies and DIC as a cause of bleeding. Aggregation and release may be decreased with collagen, thrombin, ADP, epinephrine, and ristocetin (see Chapter 119).

Gammopathies: In paraproteinemias, PFD are thought to be due to nonspecific binding of immunoglobulins to the platelet membrane, producing primarily aggregation defects, which may occur as part of the bleeding diathesis in multiple myeloma, Waldenstrom macroglobulinemia, IgA myeloma, and monoclonal gammopathy of undetermined significance.

Autoimmunity to Platelet Proteins: Antiplatelet antibodies may interfere with normal platelet function by blocking glycoprotein receptors, inducing an acquired GT, BSS, or other receptor deficiency in patients with immune thrombocytopenia. Aggregation may be decreased with ristocetin, ADP, epinephrine, and collagen. Antibody-mediated platelet activation may also produce an acquired storage pool deficiency. Interpretation of platelet function tests may be difficult due to low platelet count (see Chapter 140).

Disseminated Intravascular Coagulopathy: DIC may result in activation of platelets by thrombin or other agonists, producing an acquired storage pool deficiency or exhausted platelet syndrome. This effect is less important clinically than the consumption of coagulation factors and platelets (see Chapter 124).

Extracorporeal Membrane Oxygenation: Cardiopulmonary bypass produces PFD in most patients because of platelet activation and physical damage. Platelet aggregation by most agonists is reduced, and alpha and dense granules are depleted. The defects usually resolve within 24 hours.

Further Reading

Bolton-Maggs, P. H. B., Chalmers, E. A., Collins, P. W., Harrison, P., Kitchen, S., Liesner, R. J., et al. (2006). A review of inherited platelet disorders with guidelines for their management on behalf of the UKHCDO. *Br J Haematol, 135*, 603–633.

Freeson, K., & Turro, E. (2017). High-throughput sequencing approaches for diagnosing hereditary bleeding and platelet disorders. *J Thromb Haemost*, *15*, 1262–1272.

Grisele, P., & Subcommittee on Platelet Physiology (2015). Diagnosis of inherited platelet function disorders: Guidance from the SSC of the ISTH. *J Thromb Haemost*, *13*, 314–322.

Israels, S. J., Kahr, W. H. A., Blanchette, V. S., Luban, N. L. C., Rivard, G. E., & Rand, M. L. (2011). Platelet disorders in children: A diagnostic approach. *Pediatr Blood Cancer*, *56*, 975–983.

Nurden, A., & Nurden, P. (2011). Advances in our understanding of the molecular basis of disorders of platelet function. *J Thromb Haemost*, *9*(Suppl. 1), 76–91.

Scharf, R. E. (2012). Drugs that affect platelet function. *Semin Thromb Hemost*, *38*, 865–883.

Shen, Y. M. P., & Frenkel, E. P. (2007). Acquired platelet dysfunction. *Hematol Oncol Clin N Am*, *21*, 647–661.

CHAPTER 142

Confirmatory Testing for Diagnosis of Platelet Disorders

Rong He, MD and Dong Chen, MD, PhD

Laboratory investigation of a suspected inherited platelet disorder (IPD) requires both standard and confirmatory platelet laboratory testing. The former includes platelet count, mean platelet volume and other indices by a hematology analyzer, peripheral blood smear review by light microscopy, and platelet functional analysis. The latter encompasses platelet flow cytometry, platelet transmission electron microscopy (PTEM), and molecular genetic testing. Although the standard laboratory testing is an important first step to investigate a suspected IPD, the confirmatory testing is usually required to render a definitive diagnosis. This chapter will focus on these three confirmatory tests.

Platelet Transmission Electron Microscopy: Platelets are the smallest enucleated cellular fragments in peripheral blood with a mean volume ranging from 7.2 to 11.5 fL. Their intracellular organelles are mostly invisible by light microscopy. The transmission electron microscopy (TEM) can yield magnifications up to 500,000× with a resolution of 0.1 nm, which is sufficient to visualize platelet subcellular structures. PTEM was first used to visualize fibrin-platelet clot formation in 1955, and it was later used to evaluate platelet ultrastructural abnormalities. PTEM generally employs two methods: thin section (TS) (Fig. 142.1A and B) and whole mount (WM) TEM (Fig. 142.1C and D). The former method can also be used to examine white blood cells.

The TS-PTEM involves platelet fixation, embedding, and sectioning. Platelets from platelet-rich plasma (PRP) or white blood cells from buffy coat are pelleted by centrifugation and then fixed with glutaraldehyde. The fixed pellets are stained in 1% osmic acid and embedded in plastic blocks. TSs can be cut via an ultramicrotome and stained with uranyl acetate and lead citrate to enhance contrast. TS-PTEM visualizes ultrastructures and abnormal inclusions in both platelets and white blood cells. Platelets are composed of the plasma membrane, cytoplasm, cytoplasmic organelles, and cytoskeleton. The plasma membrane has complex invagination throughout the entire platelet, to form a network called surface-connected canalicular system (CS). Platelets contain two main types of cytoplasmic granules: α-granules (AG) and dense granules (or δ-granules, DG). AG is the primary storage site of platelet secretory proteins (e.g., coagulation factor V, proteoglycans, platelet factor 4, and von Willebrand factor). Under TS-PTEM, AG usually present as uniform-sized round granules with mildly electron-dense content and a single layer membrane. On platelet activation, AG fuse with their nearby CS and release their content into the canalicular lumen, whereas the DG migrate to and fuse directly with the plasma membrane. Platelet cytoskeleton is essential for maintaining the discoid shape of platelets (Fig. 142.1A) by forming a circumferential microtubule scaffold underneath the plasma membrane (Fig. 142.1B).

Transfusion Medicine and Hemostasis. https://doi.org/10.1016/B978-0-12-813726-0.00142-2

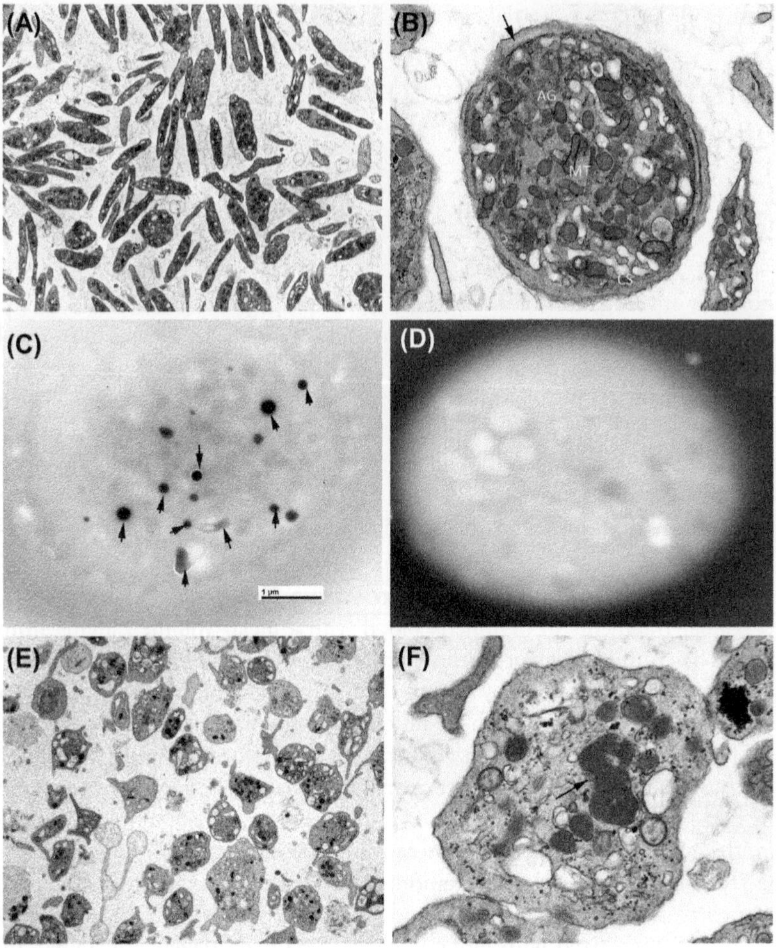

FIGURE 142.1 Examples of platelet and white blood cell electron microscopy. (A and B) Thin section electron microscopy of platelets from a healthy donor. (C) Whole mount electron microscopy of a platelet from a healthy donor shows dense granules (*arrows*). (D) Whole mount electron microscopy of a platelet from a patient with Hermansky–Pudlak syndrome. (E) Thin section electron microscopy of activated platelets due to suboptimal sample transportation and storage. (F) Large and irregular-shaped α-granule in a platelet (*arrow*) from a patient with Paris-Trousseau–Jacobsen syndrome.

The platelet submembrane cytoskeleton is composed of actin, which also anchors platelet transmembrane glycoproteins (GP). Actin molecules can polymerize and form actin filaments, and regulate platelet shape change and pseudopodia formation on activation. Nevertheless, actin filaments are invisible by standard TS-PTEM.

The WM-PTEM examines platelet electron-dense structures such as DG and abnormal opaque inclusions in rare IPD such as York platelet syndrome. WM-PTEM preparation is made by applying a drop of PRP directly on a carbon-stabilized formvar grid. The grid is air-dried and then directly examined under an electron microscope.

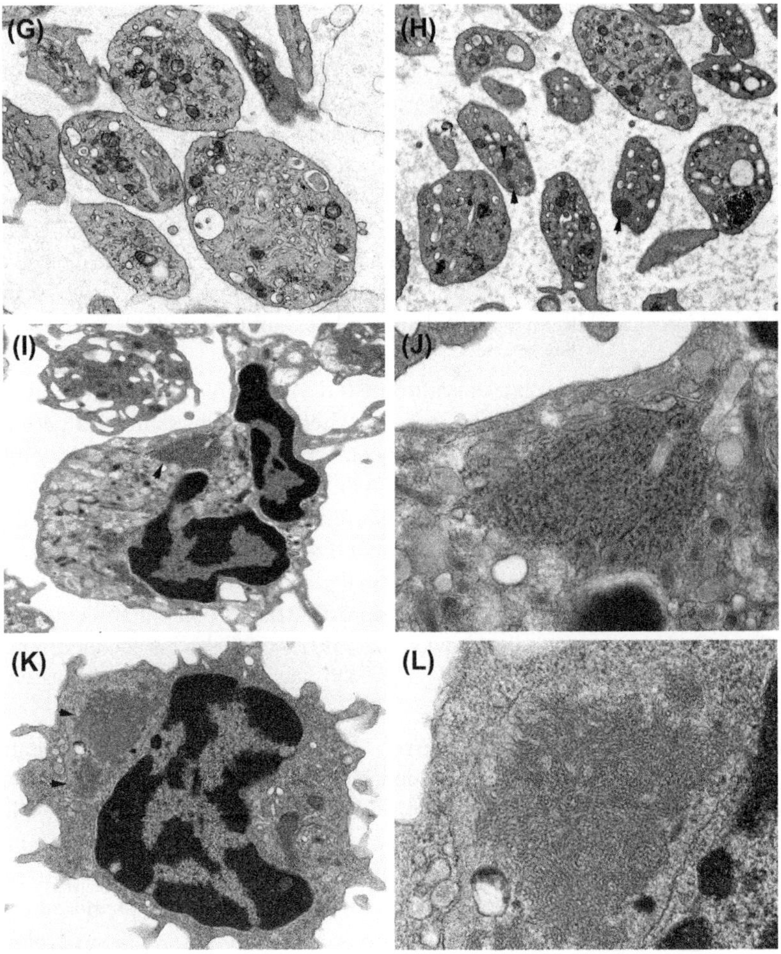

FIGURE 142.1, cont'd (G) Thin section electron microscopy of platelets from a patient with gray platelet syndrome due to *NBEAL2* mutation. Virtually no α-granule is present. (H) Thin section electron microscopy of platelets with partial α-granule deficiency due to *VPS33b* mutation with *arrows* pointing to scattered rare α-granules and a large α-granule (on the right). (I and J) Low and high magnification of the cytoplasmic inclusions (*arrow*) in a white blood cell from a patient with *MYH9* mutation–associated disorder. (K and L) Low and high magnification of the cytoplasmic inclusions (*arrows*) in a white blood cell from a patient with Chediak–Higashi syndrome.

Platelet cytoplasm is transparent to the electron beam, whereas DG and abnormal electron-dense objects are inherently opaque (Fig. 142.1C). Normal platelets contain about 1–8 dense bodies per platelet. WM-PTEM is considered the gold standard method for diagnosing platelet DG deficiencies, e.g., Hermansky–Pudlak syndrome (HPS), Paris-Trousseau–Jacobsen syndrome (PTJS), Wiskott–Aldrich syndrome (WAS), thrombocytopenia with absent radii (TAR) syndrome, Chediak–Higashi syndrome (CHS), and combined alpha–delta platelet storage pool deficiency.

Platelet Ultrastructural Defects

Shape and Size Abnormalities: The average size of platelets is about 2–4 μm in diameter. Resting platelets are discoid in shape with a smooth cell border (Fig. 142.1A and B). Activated platelets, such as suboptimal sample collection, transportation, and storage, show irregular shapes and pseudopodia formation (Fig. 142.1E). Giant platelets of Bernard–Soulier syndrome (BSS) and MYH9 mutation–related disorders (MYH9-RD) are often spheroid in shape. However, we need to be aware that round and enlarged platelets can be a laboratory artifact when ethylenediaminetetraacetic acid (EDTA) is used as the anticoagulant for whole-blood collection and storage. Conversely, platelets of WAS are microcytic with an average diameter less than 1.8 μm.

α-Granule Defects: Gray platelet syndrome (GPS) is a rare IPD caused by AG deficiency. The platelets are gray appearing by the Giemsa–Wright stain on a peripheral blood smear. GPS is a genetically heterogeneous entity with its pathogenicity attributable to mutations in at least three genes, *NBEAL2*, *GFI1b*, and *VPS33b*. Clinically, patients with GPS show heterogeneity in bleeding symptoms and platelet function assessed by platelet aggregation analysis. Under TS-PTEM, platelets contain virtually no AG in *NBEAL2* mutation–associated GPS (Fig. 142.1G), whereas partial AG deficiency and occasional giant AG are observed in *GFI1B* or *VSP33b* mutation–associated GPS (Fig. 142.1H). Recent studies showed that *GFI1B*-GPS also has a concurrent DG deficiency.

Abnormal α-Granule Ultrastructure: Giant irregular-shaped alpha granules (Fig. 142.1F) and combined AG and DG deficiencies are the hallmarks of PTJS, which is caused by chromosome 11q23 deletion where *FLI-1* gene is located. Other clinical features include congenital heart defects, trigonocephaly, mental retardation, and multiple organ malfunctions. Recent studies demonstrated that similar PTEM ultrastructural abnormalities are also present in *FLI-1* mutation–associated IPD and other rare IPDs. It should be noted that occasional large and fuse AG may be due to prolonged and suboptimal whole-blood storage.

Dense Granule Deficiency: Platelet dense granule defects are a heterogeneous group of diseases. Some diseases such as HPS, CHS, and PTJS have a virtually complete absence of DG. Other disorders may have mild to moderate deficiency. Platelet light transmission aggregation and PFA-100 tests have about 50% sensitivity of detecting a DG deficiency. ATP or serotonin release test may serve as a screening testing for DG deficiencies. However, because of the significant variability of ATP release in the healthy population, the ATP release test still has insufficient sensitivity and specificity for diagnosing DG deficiencies. Because of these limitations of the standard platelet functional testing, WM-PTEM remains the preferred method for assessing platelet DG deficiency.

Hermansky–Pudlak Syndrome: HPS was described in 1959 when Hermansky and Pudlak reported two patients with the mild bleeding tendency and oculocutaneous albinism. There are 10 distinct subtypes of HPS caused by mutations in nine genes. Under WM-PTEM, platelets from HPS patients have virtually zero DG (Fig. 142.1D).

Wiskott–Aldrich Syndrome and X-Linked Microthrombocytopenia: WAS is an X-linked thrombocytopenia characterized by abnormally small platelet size, associated problems with eczema, and an increased susceptibility to infections. A milder form, without the immune deficiency symptoms, is referred to as X-linked microthrombocytopenia. Both diseases are caused by mutations in the *WASP* gene on chromosome X that encodes the WAS protein, which plays a role in actin polymerization. Similar to HPS, WAS platelets exhibit severe δ-granule deficiency.

Abnormal Inclusions in Neutrophils

***MYH9* Mutation–Related Disorders:** This group of disorders includes May–Hegglin anomaly, Fechtner syndrome, Sebastian syndrome, and Epstein syndrome. Patients may present with mild bleeding, macrothrombocytopenia, and characteristic inclusions in neutrophils. All four syndromes are caused by mutations in the *MYH9* gene encoding the nonmuscle myosin heavy chain IIA (NMMHC-IIA). Mutated NMMHC-IIA is unstable and causes platelet macrocytosis and forms abnormal deposits in the cytoplasm of neutrophils resembling Döhle bodies. At the ultrastructural level, giant platelets from MYH9-RD are large and often spheroid. Microtubules in these giant platelets are disorganized. TEM evaluation of neutrophils can visualize the abnormal focal, linear, and parallel filaments oriented in the long axis of the inclusion. Some filaments are perpendicular to the long axis, forming small dotlike structures (Fig. 142.1I and J).

Chediak–Higashi Syndrome: CHS is a disorder that exhibits partial oculocutaneous albinism, frequent pyogenic infections with neutropenia, and a mild bleeding tendency. Besides severe DG deficiency, CHS exhibits abnormal giant inclusions in leukocytes. The abnormal cytoplasmic inclusions can be detected in buffy coat TS-TEM (Fig. 142.1K and L).

Flow Cytometric Studies: Flow cytometry can be used to evaluate the platelet surface antigen expression and activation state. Platelet surface antigens include the VWF binding receptor GP complex Ibα/β-IX-V (CD42a-d), fibrinogen receptor GPIIb/IIIa (CD41/CD61), collagen binding receptors GPIa (CD49b) and GPVI. GP expression levels are measured with the use of fluorescent-conjugated GP-specific antibodies.

Glanzmann Thrombasthenia: GT is an autosomal-recessive disorder characterized by a lifelong bleeding tendency and defective platelet aggregation responses to most agonists in platelet light transmission aggregation assays, due to quantitative or qualitative abnormalities of the platelet GPIIb/IIIa. GPIIb/IIIa is a calcium-dependent heterodimer that binds preferentially to fibrinogen and serves as its receptor. A genetic defect in either *GPIIb* or *GPIIIa* can prevent heterodimer assembly and impede the function of the receptor, which can result in decreased GPIIb/IIIa surface expression (Fig. 142.2) or/and defective fibrinogen binding after platelet activation. Hemorrhagic

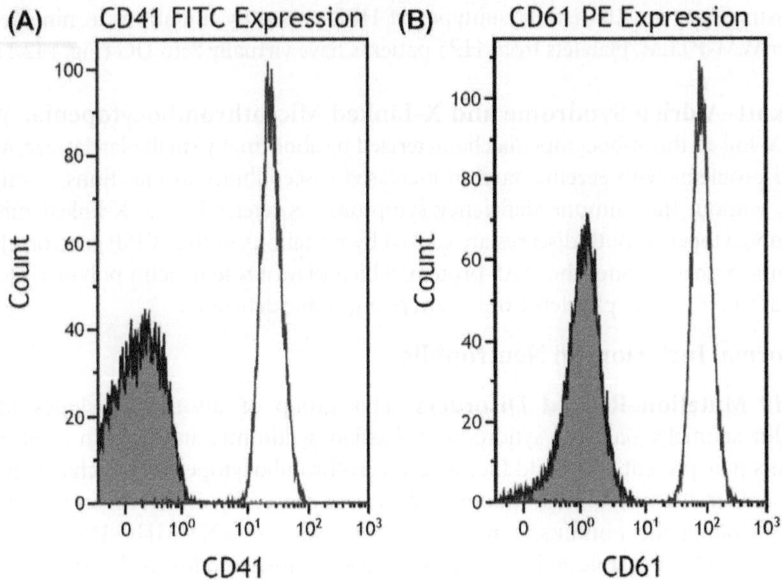

FIGURE 142.2 Flow cytometric study of the decreased glycoprotein IIB (CD41) and IIIA (CD61) in Glanzmann thrombasthenia. (A and B) Normal and decreased fluorescent intensities of CD41 (A) and CD61 (B) antibody staining on a healthy donor's (peak on the right) and patient's platelets (shaded peak on the left).

symptoms only occur in patients who are homozygous for GT mutations, whereas the heterozygous carriers are mostly asymptomatic.

Macrothrombocytopenia in Bernard–Soulier Syndrome, Mediterranean Macrothrombocytopenia, and DiGeorge or Velocardiofacial Syndrome: BSS, first described by Bernard and Soulier in 1948, is a severe bleeding disorder characterized by macrothrombocytopenia. BSS platelets show markedly decreased response to ristocetin-induced aggregation. BSS is caused by homozygous or compound heterozygous mutations described in the *GpIba*, *GPIbβ*, or *GPIX* genes. Most of these genetic defects cause either quantitatively decreased GPIbα/β, V, and IX expression or functional defects of the receptor complex. DiGeorge or Velocardiofacial syndrome macrothrombocytopenia is a condition characterized by cardiac anomalies, cleft palate, learning disabilities, typical facies, and thymic hypoplasia with immunodeficiency. The syndrome is associated with a hemizygous deletion of chromosome 22q11.2, which includes the *GPIbβ* gene.

Collagen Receptor Deficiencies: There are two sets of collagen receptors on the surface of platelets, GPIa and GPVI. Patients with severe collagen receptor deficiencies usually present with a lifelong bleeding diathesis and absence of platelet aggregation response to collagen. GPIa is the minor platelet collagen receptor, and its expression levels can vary with different genetic variants of the $\alpha 2$ subunit gene *ITGA2*. GPVI is the primary platelet collagen receptor, and its deficiency is frequently associated with decreased collagen-induced platelet aggregation.

Platelet Functional Test by Flow Cytometry: Flow cytometry can be used to evaluate platelet function by measuring activation markers on platelet activation. The most commonly used platelet activation markers are P-selectin (CD62P), activated GPIIb/IIIa complex (PAC-1), and lysosomal-associated membrane protein (CD63). CD62P is located on the AG membrane, whereas CD63 is on the membrane of DG and lysosome. Both CD62P and CD63 translocate to the surface of plasma membrane after platelet activation with ensued granule release. Decreased expression of CD62P and CD63 on platelet activation can be seen in AG or DG deficiencies, respectively. PAC-1 is a mouse monoclonal antibody that detects the neoepitope of active GPIIb/IIIa conformation on platelet activation. PAC-1 binding on the platelet surface is indicative of platelet activation. A lack of PAC-1 binding by activated platelets suggests GT or other platelet disorders.

Molecular Genetic Testing: Genotyping has a confirmatory role in the diagnosis of IPDs. Molecular genetic analysis by Sanger sequencing was first employed in the early 1990s to identify the causal mutations in the genes coding for GPIIb/IIIa and GPIbα/β-IX-V in GT and BSS, respectively. With the advent of next-generation sequencing (NGS), several novel platelet disorder–associated gene mutations have been identified, for example, *NBEAL2*, *GFI1B*, and *VPS33b* for GPS, *ACTN1* for autosomal dominant thrombocytopenia, *RBM8A* for thrombocytopenia absent radii, *STIM1* for York platelet syndrome, and *ANKRD26*, *ETV6*, *GATA-1*, *GATA-2*, and *RUNX1* for thrombocytopenia with predisposition to hematologic malignancy. Although NGS-based approaches may potentially revolutionize the diagnosis of IPDs, preliminary clinical and laboratory phenotypic characterization remain crucial and indispensable.

Confirmatory Testing Algorithm: It is crucial to utilize these confirmatory tests strategically. For some IPDs, a combination of clinical and phenotypic laboratory results is sufficient for a conclusive diagnosis, for example, BSS, GT, collagen receptor deficiency, GPS, etc. For others, a genotype/phenotype correlation is helpful, and genotyping can further substantiate and confirm the diagnosis and provide more accurate prognostication. The examples include MYH9-RD, WAS, CHS, HPS, etc. Genetic testing is advisable for diseases whose laboratory phenotypes are unremarkable or indistinct, as can be seen in various storage pool deficiencies. Finally, genetic analysis is recommended for IPD associated with a myeloid neoplasm.

Further Reading

Israels, S. J. (2015). Laboratory testing for platelet function disorders. *Int J Lab Hematol*, *37*(Suppl. 1), 18–24.

Markello, T., et al. (2015). York platelet syndrome is a CRAC channelopathy due to gain-of-function mutations in STIM1. *Mol Genet Metab*, *114*(3), 474–482.

Nurden, A. T. (2014). Platelet membrane glycoproteins: A historical review. *Semin Thromb Hemost*, *40*(5), 577–584.

Nurden, A. T., & Nurden, P. (2014). Congenital platelet disorders and understanding of platelet function. *Br J Haematol*, *165*(2), 165–178.

Perez Botero, J., et al. (July 1, 2017). Comprehensive platelet phenotypic laboratory testing and bleeding history scoring for diagnosis of suspected hereditary platelet disorders: A single-institution experience. *Am J Clin Pathol*, *148*(1), 23–32.

Simeoni, I., et al. (2016). A high-throughput sequencing test for diagnosing inherited bleeding, thrombotic, and platelet disorders. *Blood, 127*(23), 2791–2803.
Sivapalaratnam, S., Collins, J., & Gomez, K. (2017). Diagnosis of inherited bleeding disorders in the genomic era. *Br J Haematol, 179*, 363–376.
White, J. G. (1969). The dense bodies of human platelets: Inherent electron opacity of the serotonin storage particles. *Blood, 33*(4), 598–606.
White, J. G. (1972). Ultrastructural defects in congenital disorders of platelet function. *Ann NY Acad Sci, 201*, 205–233.
Witkop, C. J., et al. (1987). Reliability of absent platelet dense bodies as a diagnostic criterion for Hermansky-Pudlak syndrome. *Am J Hematol, 26*(4), 305–311.

CHAPTER 143

Antiplatelet Therapy Monitoring

Connie H. Miller, PhD and Mikhail Roshal, MD, PhD

Antiplatelet medications are widely used to treat or prevent thrombosis in a variety of cardiovascular disorders. Currently available antiplatelet drugs comprise four categories:

1. Acetylsalicylic acid (aspirin) inhibits cyclooxygenase (COX-1),
2. P2Y12 receptor antagonists (clopidogrel, ticlopidine, prasugrel, cangrelor, ticagrelor) block the platelet adenosine diphosphate (ADP) receptor,
3. Phosphodiesterase inhibitors (dipyridamole, anagrelide, cilostazol) inhibit the platelet phosphodiesterases, and
4. Glycoprotein (GP)IIb/IIIa inhibitors (abciximab, eptifibatide, tirofiban, lamifiban) block the platelet GPIIb/IIIa receptor.

The response to these medications varies among individuals, with some patients failing to achieve the expected decrease in platelet activation. This phenomenon, called high on-treatment residual platelet reactivity (HOPR), may be attributable to either the drug failing to inhibit the target (true resistance) or extrinsic factors that increase platelet reactivity despite the drug's presence. Detection of true drug resistance requires highly specific tests that directly measure the effect of the drug on the target. On the other hand, evaluation of HOPR for any reason can be performed with nonspecific tests that measure the end result of platelet activation. Unfortunately, major problems arise from the weak correlation among the currently available methods in detecting poor drug response. Nonspecific tests tend to identify significantly more patients with HOPR than the highly specific ones. This problem makes it difficult to compare results utilizing different assessment methods. Furthermore, while initial studies generally showed favorable outcomes in patients whose treatments with antiplatelet medications were adjusted based on a given test modality, larger prospective studies tended to produce less impressive results. At present, the indications for testing outside of monitoring compliance with therapy and assessment of the presence of medication before invasive procedures remain controversial. The field is rapidly evolving, and further clinical studies may well provide an impetus to expand platelet therapy monitoring. Because HOPR is mostly described in clopidogrel and aspirin therapy, the chapter mostly focuses on monitoring the effect of these medications.

Methods: Although light-transmission aggregometry (LTA) is considered by many to be the gold standard for evaluating primary hemostatic function, the test is time-consuming, difficult to standardize, and not available in many clinical settings. Whole-blood aggregometry (WBA), particularly in newer, multiple electrode instruments, is faster and has less user variability. WBA is better suited for use outside of a specialized laboratory because it does not require centrifugation. More

Transfusion Medicine and Hemostasis. https://doi.org/10.1016/B978-0-12-813726-0.00143-4

specific tests designed for particular drugs and for point-of-care (POC) use have been developed. The overview of techniques that are not covered in depth in other chapters is given below.

Light-Transmission Aggregometry: LTA methods are discussed in Chapter 140.

Whole-Blood Aggregometry: WBA methods are discussed in Chapter 140.

Platelet Function Analyzer-100: PFA is discussed in Chapter 138.

VerifyNow: This fully automated POC instrument measures whole-blood platelet aggregation via fibrinogen-coated beads in cartridges designed to measure response to specific drugs. The VerifyNow aspirin cartridge uses arachidonic acid (AA). When present, aspirin prevents COX-1-mediated conversion of AA into platelet activating thromboxane A2 (TxA2) and reduces AA-induced platelet aggregation. The IIb/IIIa cartridge uses modified thrombin receptor-activating peptide (iso-TRAP) to maximally activate platelets without inducing clot formation. Under these conditions, the binding of platelets to fibrinogen-coated beads becomes proportional to the number of unblocked platelet GPIIb/IIIa inhibitors. In its newest iteration, the P2Y12 cartridge uses high-dose ADP in the presence of prostaglandin E1 (PGE). PGE provides an additional layer of specificity because it blocks P2Y1, but not P2Y12 receptor activity, making the assay P2Y12 receptor dependent. Results for all cartridges are now reported in platelets response units (PRU) that are based on changes in light transmittance due to platelet aggregation.

Plateletworks: This kit compares platelet count before and after aggregation with various agonists, including ADP, AA, or collagen. Blood is drawn into special tubes containing a specific agonist and thrombin inhibitor, D-phenylalanyl-L-prolyl-L-arginine chloromethyl ketone (PPACK). In addition, a reference K3-EDTA tube is drawn for baseline platelet count. Following a short incubation with the agonist, the count of the remaining unaggregated platelets is compared to the baseline and the ratio (also called % aggregation) is calculated. The test is used infrequently for detecting HOPR.

Thromboelastography: The TEG 5000 instrument from Haemonetics (Braintree, MA, USA) has a platelet mapping modulus that allows evaluation of clopidogrel and aspirin response in POC setting. Details of the test are provided in Chapter 139.

Flow Cytometry: The flow cytometer can be used to measure numerous aspects of platelet function, including those blocked by platelet-inhibitory drugs. In these assays, platelets or platelet-derived fragments are first identified through a combination of light scattering parameters and antigen binding by platelet-specific, fluorescently tagged, monoclonal antibodies (usually anti-CD42b). Expression of activation-related antigens on the platelets can be measured using the antigen-specific reagents labeled with different fluorescent tags. Some assays, such as those measuring P-selectin expression, fibrinogen, and PAC-1 binding, reflect the end result of platelet activation through numerous pathways and thus can be adapted to measure inhibition of platelet activation by many drugs. Others, such as vasodilator-stimulated phosphoprotein content assessment, are more drug class–specific. Some of the assays are described below:

- Expression density of P-selectin (CD62P) and CD63 components of the platelet granule membrane expressed on the platelet surface on activated, but not on resting, platelets.
- PAC-1 binding uses a monoclonal antibody that recognizes the fibrinogen binding site exposed on the activated form of the GPIIb/IIIa receptor.
- Platelet fibrinogen binding assay assesses platelet-bound fibrinogen using anti-fibrinogen antibodies.
- Platelet vasodilator-stimulated phosphoprotein (VASP) is an intracellular signaling protein that is non-phosphorylated in the basal state and phosphorylated through the cyclic adenosine monophosphate (cAMP)–dependent cascade in PGE-inhibited platelets and dephosphorylated through ADP-induced signal. The test can be utilized to measure inhibition of PGE-induced VASP phosphorylation by phosphodiesterase inhibitors that block cAMP formation and VASP inhibition of ADP-induced VASP dephosphorylation in P2Y12 antagonist–treated platelets. Commercially produced VASP kits are available.

Flow cytometry is an extremely versatile technique. Many additional assays have been developed and are used, primarily in research settings. Unfortunately, platelet flow cytometry is rarely available in clinical laboratories. It requires expensive instrumentation, sophisticated dedicated operators, and special expertise for processing and interpretation of results.

Measurement of Thromboxane A2 Metabolites: Aspirin blocks COX-1-dependent TxA2 production in platelets. Thus, assessment of TxA2 production is an attractive therapy monitoring tool for patients on aspirin. It is now considered the gold standard in assessing true aspirin resistance. TxA2 itself is not stable enough for monitoring purposes. It is rapidly metabolized into a more stable metabolite, thromboxane B2 (TxB2) that can be detected in the serum. TxB2 is further converted into 11-dehydro TxB2 that is excreted in the urine. Enzyme-linked immunosorbent assay (ELISA)–based plasma TxB2 assessments have been described but unfortunately are not widely available outside of research studies. The urinary 11-dehydro TxB2 can be assessed by either ELISA using monoclonal antibodies or liquid chromatography–tandem mass spectrometry (LC-MS/MS) methods.

Bleeding Time: The bleeding time is not recommended because of poor standardization.

Test Interpretation and Sources of Error: All platelet function tests are limited by difficulties in standardization across laboratories. LTA and WBA are particularly variable. Establishment of local reference ranges for the types of patients being studied is crucial to their use. With most drugs, the benefit of altering care on the basis of the results of platelet function tests has not been definitively established. Most tests of platelet function are influenced by a number of characteristics of the patient, including the congenital and acquired defects reviewed in Chapter 141. In most cases, the sensitivity and specificity of the tests for the drugs monitored have not been established.

Aspirin: True aspirin resistance appears to be a very rare phenomenon. Most cases of poor aspirin response may be due to noncompliance with therapy or delayed absorption of enteric-coated aspirin. Because of the rarity of true resistance, most studies comparing methodologies did not have enough aspirin-resistant patients to come to definite conclusions. The effects of aspirin are detectable in numerous tests of platelet function. LTA and WBA using AA, low-dose ADP, and low-dose collagen usually show reduced aggregation in patients on aspirin therapy. Platelet aggregation tends to be reduced to less than 20% by LTA using AA as agonist and remains close to normal in aspirin-resistant patients. WBA results show similar patterns. ADP- and collagen-induced aggregation are nonspecific and tend to "overcall" HOPR. PFA collagen/epinephrine closure time (CT) is usually elevated in patients on aspirin therapy; however, the measure is nonspecific and is influenced by other variables, such as von Willebrand factor levels and hematocrit. The test also has a wide normal range that can mask real changes from baseline when pretreatment CT is on the shorter end of the range. Because of these factors, PFA-100 also tends to "overcall" HOPR.

The VerifyNow aspirin test using AA is more specific than PFA and shows better agreement with LTA AA and TxB2 levels for aspirin resistance in many studies. Values greater than 550 aspirin reaction units are considered indicative of aspirin HOPR. Urinary 11-dehydro TxB2 measurements have better precision than either LTA or WBA and provide time-integrated measure of TxA2 production. They are also not affected by in vitro platelet activation. Commonly, a drop in 11-dehydro TxB2 is expected within 5 days of initiation of platelet therapy. ELISA and LC-MS/MS measurements correlate well, but LC-MS/MS shows values that are on average 66% lower than ELISA due to increased specificity for detection of the metabolite. The assay is not absolutely specific for platelet TxA2 production. Monocytes and mast cells may contribute significantly to the measured metabolite level, with about 30% of the measured level being due to extra-platelet sources. The plasma TxB2 level is more specific, but may in part be due to in vitro platelet activation because of traumatic blood draw.

P2Y12 Antagonists: Clopidogrel resistance and clopidogrel HOPR are relatively common and can be mediated by a range of factors, including cytochrome isoform polymorphisms (CYP2C19*2 allele is associated with poor response due to poor conversion of the pro-drug to the active form), inadequate absorption, drug interactions, and disease states. When LTA and WBA are performed on samples from patients treated with P2Y12 antagonists, ADP-induced aggregation is expected to be reduced. Unfortunately, there is up to an 80% overlap between treated and untreated patients in these assays. This is likely due to the effect of ADP on P2Y1 and the relatively low precision of the assay. Thus, the cutoff values for the detection of HOPR by aggregation are not clear. PFA-100 is relatively insensitive to these agents and should not be used to detect resistance. A newer PFA-200 P2Y cartridge has been developed and shows better sensitivity, but its performance characteristics are unclear at present. The VerifyNow P2Y12 cartridge is relatively specific but may be influenced by simultaneous presence of GPIIb/IIIA inhibitors and platelet counts outside the validated range of

119,000–502,000/μL. The performance characteristics of the new iteration of the assay that is solely based on PRU values are not clear at present. VASP is considered a gold standard for detecting true resistance but is expensive and requires technical expertise.

Phosphodiesterase Inhibitors: Relatively little information is available regarding the test response to these agents. LTA shows reduced aggregation in response to collagen, ADP, AA, and adrenalin. VASP demonstrates inhibition of PGE-induced phosphorylation.

GPIIb/IIIa Inhibitors: LTA and WBA show inhibition of aggregation with all agonists, except ristocetin-induced platelet agglutination. These drugs prolong PFA CT with both cartridges. The VerifyNow IIb/IIIa cartridge is designed to be specific for these drugs and shows reduction in PRU.

Further Reading

Bonello, L., Tantry, U. S., Marcucci, R., Blindt, R., Angiolillo, D. J., Becker, R., et al. (2010). Consensus and future directions on the definition of high on-treatment platelet reactivity to adenosine diphosphate. *J Am Coll Cardiol, 56*, 919–933.

Collet, J. P., Cuisset, T., Rangé, G., Cayla, G., Elhadad, S., Pouillot, C., et al. (2012). Bedside monitoring to adjust antiplatelet therapy for coronary stenting. *N Engl J Med, 367*, 2100–2109.

Hankey, G. J., & Eikelboom, J. W. (2006). Aspirin resistance. *Lancet, 367*, 606–617.

Harrison, P., Frelinger, A. L., Furman, M. I., & Michelson, A. D. (2007). Measuring antiplatelet drug effects in the laboratory. *Thromb Res, 120*, 323–336.

Hayward, C. P. M., Harrison, P., Cattaneo, M., et al. (2006). Platelet function analyzer (PFA)-100® closure time in the evaluation of platelet disorders and platelet function. *J Thromb Haemost, 4*, 312–319.

Jovanovic, L., Antonijevic, N., Novakovic, T., Savic, N., Terzic, B., et al. (2017). Practical aspects of monitoring of antiplatelet therapy. *Semin Thromb Hemost, 43*, 14–23.

Linden, M. D., Tran, H., Woods, R., & Tonkin, A. (2012). High platelet reactivity and antiplatelet therapy resistance. *Semin Thromb Hemost, 38*, 200–212.

Lippi, G., Favaloro, E. J., Salvagno, G. L., & Franchini, M. (2009). Laboratory assessment and perioperative management of patients on antiplatelet therapy: From the bench to the bedside. *Clin Chim Acta, 405*, 8–16.

Lordkipanidze, M., Pharand, C., Schampaert, E., Turgeon, J., Palisaitis, D. A., & Diodati, J. G. (2007). A comparison of six major platelet function tests to determine the prevalence of aspirin resistance in patients with stable coronary artery disease. *Eur Heart J, 28*, 1702–1708.

Michelson, A. D., Cattaneo, M., Eikelboom, J. W., Gurbel, P., Kottke-Marchant, K., Kunicki, T. J., et al. (2005). Aspirin resistance: Position paper of the Working Group on Aspirin Resistance. *J Thromb Haemost, 3*, 1309–1311.

Muir, A. R., McMullin, M. F., Patterson, C., & McKeown, P. P. (2009). Assessment of aspirin resistance varies on a temporal basis in patients with ischaemic heart disease. *Heart, 95*, 1225–1229.

CHAPTER 144

Laboratory Evaluation of Factor XIII Deficiency

Mikhail Roshal, MD, PhD and Morayma Reyes Gil, MD, PhD

FXIII is a hemostatic protein that plays a critical role in stabilizing fibrin networks at the site of injury and thus preventing premature fibrinolysis. In the plasma, FXIII circulates as a tetrameric zymogen composed of two catalytic alpha chains (FXIII-A) and two carrier/inhibitory beta chains (FXIII-B) that serve to stabilize FXIII-A in the plasma pool. In addition to the plasma pool, platelet FXIII pool composed only of FXIII-A homodimers plays an important role in increasing local FXIII concentration at sites of injury. FXIII-A and FXIII-B are produced in different tissues. FXIII-A is produced by the cells in the bone marrow, including monocyte/macrophage lineage cells and megakaryocytes, whereas FXIII-B is produced in hepatocytes. The tetrameric A_2B_2 complex is subsequently assembled in the plasma. Under the usual circumstances, circulating FXIII-B chains are in 50% excess of the tetrameric complex. FXIII activation requires thrombin-induced proteolysis and is greatly accelerated in the presence of fibrin. Thrombin cleaves inhibitory peptides on the alpha chains causing dissociation of catalytically active FXIII-A_2 from FXIII-B_2. Once activated, FXIII acts as a transglutaminase, catalyzing the formation of covalent cross-links between specific peptide lysine and glutamine residues on fibrin and alpha-2 plasmin inhibitor. Ammonia is released as a byproduct of these enzymatic reactions. The enzyme has three principle clot stabilizing activities; first it cross-links fibrin γ chains, leading to formation of γ chain dimers thus preventing dissociation of fibrin monomers. Second it forms multiple cross-links between fibrin alpha chains, thus increasing the rigidity of the clot. Finally, it cross-links alpha-2 plasmin inhibitor to fibrin, leading to a decrease in fibrin digestion by plasmin. These tasks are critically important for proper hemostatic function, as evidenced by the severe bleeding seen in cases of FXIII deficiency.

Both activity-based and antigenic FXIII assays play a significant role in evaluating FXIII status in bleeding patients. Initial screening relies on determination of FXIII activity. If the activity is significantly decreased, further classification into quantitative FXIII-A deficiency, qualitative FXIII-A deficiency, and FXIII-B deficiency can be made by performing antigenic tests for FXIII A_2B_2, FXIII-A, FXIII-B antigens, and occasionally testing FXIII status in the patient's platelet lysate. Quantitative FXIII testing has also enabled FXIII replacement monitoring.

Factor XIII Activity Assays

Clot Solubility: Historically clot solubility assays were utilized to screen for FXIII deficiency. The test uses either calcium or thrombin to induce fibrin clot. The clot is incubated for 30 minutes at 37°C to allow FXIII-dependent cross-linking and then

Transfusion Medicine and Hemostasis. https://doi.org/10.1016/B978-0-12-813726-0.00144-6

placed in either 5 or 8 M urea, 1% acetic, or 1%–5% monochloroacetic acid. Clot dissolution within a 24 hour incubation period is suggestive of an absence of FXIII activity.

Quantitative Factor XIII Activity Assays: Until recently, quantitative determination of FXIII activity was laborious and difficult to automate. The assays relied on FXIII-mediated labeled amine (used in place of lysine residue) incorporation into glutamine residue as a readout of FXIII activity. Rare laboratories could perform the laborious assays in daily practice. Recent introduction of automated, quantitative measurements of the ammonia released as a byproduct of FXIII activity has dramatically improved the prospect for introduction of reliable FXIII tests into the clinical laboratory. Recent introduction of a WHO (World Health Organization) FXIII activity standard for use in calibration of the assay standard material has enabled further harmonization of the assay performance across different platforms and laboratories.

Amine Incorporation Assay: The older methodology measures the incorporation of radiolabeled, fluorescently labeled, or biotinylated low-molecular-weight amines into peptide glutamine residue. In the end point assay, the amount of incorporated label is proportional to the FXIII activity present in the test plasma. The assay demonstrates good analytical sensitivity on the low end of the measurement range and is able to measure FXIII activity down to ~1%. There are two major difficulties that arise during performance of the assay. First is the need to separate free from incorporated amines on completion of the amine incorporation reaction. This problem is solved by either acid precipitation of acceptor peptides and repeated washings to remove free amines or by attaching the acceptor peptides to solid support to remove the need for acid precipitation. Neither of these solutions lend themselves to easy automation. The second problem is interference of fibrin clot with the performance of the assay. If allowed to form, fibrin clot makes it very difficult to wash out unincorporated amines and interferes with photooptical detection of the fluorescent label through inner filter effect. Newer iterations of the assay prevent fibrin formation through either high plasma dilution or using peptides that interfere with fibrin clot assembly.

Ammonia Release Assay: Ammonia release from the glutamine side chain happens during the first step of the transglutaminase reaction and is proportionate to the FXIII enzymatic activity. Generation of ammonia can be quantified by using NADH or NADPH consumption in glutamate dehydrogenase–mediated conversion of 2-oxoglutarate to L-glutamate in the reaction shown below:

$$\text{2-oxoglutarate} + NH_3 + NAD(P)H + H^+ \rightleftharpoons \text{L-glutamate} + H_2O + NAD(P)^+$$

NADPH or NADH consumption is measured by the decrease in absorption at 340 nm. In this true kinetic assay, the rate of NAD(P)H consumption is proportionate to FXIII activity. As in amine incorporation reaction, fibrin clot can pose problems for readout. Thus, clot formation is prevented by using peptides designed to inhibit fibrin polymerization.

Antigenic Assays: Measuring FXIII A_2B_2, A, and B antigen levels in the setting of significant decreased FXIII activity can help determine the specific deficiency type. There are few commercial reagents available for FXIII antigenic evaluation. Laboratory developed enzyme-linked immunoassay (ELISA) using either monoclonal or polyclonal antibodies is used in many laboratories. An A_2B_2 antigen detection kit is marketed (Reanal-ker, Budapest, Hungary). The kit uses one step sandwich ELISA with a peroxidase-labeled monoclonal detection antibody against FXIII-A and a biotinylated monoclonal capture antibody mixed with diluted patient plasma in a streptavidin coated well of a microplate. Following multiple washes, the complex is detected using peroxidase substrate. The assay is reported to be sensitive to very low levels of FXIII A_2B_2 tetramer (0.1%) with good precision. A new commercial immunoturbinometric assay from Instrument Laboratories (Milano, Italy) uses latex-enhanced agglutination to measure FXIII-A antigen. The assay can be used on many automatic coagulation analyzers.

Test Limitations

Clot Solubility: The clot solubility test is poorly standardized for clot induction conditions (calcium vs. thrombin), chemical composition of incubation solution, and methods for final clot detection. Assays that utilize thrombin induction seem to be more accurate. The lack of standardization in the qualitative test leads to variable sensitivity cutoffs between 0.5% and 5% FXIII. Thus many severe and most milder FXIII deficiencies are missed if the test is used for screening purposes. The assay can produce false-positive results with dys- or hypofibrinogenemia. Historical use of this test is considered the primary reason why FXIII deficiency is the most underdiagnosed hemorrhagic condition. Recently, the International Society on Thrombosis and Hemostasis, Scientific, and Standardization Committee advised against using this methodology for screening for FXIII deficiencies.

Amine Incorporation: When performed correctly, the assay appears to be fairly robust. A high degree of technical sophistication and good technique is needed to avoid erroneous results because of insufficient removal of unincorporated amines. The choice of acyl group acceptor and donor pairs, methodology of avoiding fibrin clot formation during the assay performance, and approach to separation of incorporated from unincorporated amines may cause substantial difference in analytical performance of different assays. A major problem arises with FXIII-A Val34Leu polymorphism. Because of the mutation near the thrombin cleavage site, FXIII Val34 variant has slower activation kinetics but shows normal enzymatic activity after activation. Some versions of amine incorporation assays produce significantly decreased activity values for Val34 polymorphism homozygotes and heterozygotes. These values do not match FXIII antigenic levels and may be misleading.

Ammonia Release Assays: Predictably, the assay sensitivity may be compromised when other sources of ammonia production and consumption of NADH or NADPH are present in the plasma. This is particularly true when low concentrations of FXIII are being measured. One commercial iteration of the assay (Berichrom FXIII, Siemens, Marburg, Germany) has been shown to overestimate FXIII activity when it was present at low concentration. The effect was much less pronounced when the FXIII levels were near normal. To overcome this problem, a plasma "blank," which has FXIII activity inhibited by 1 mM iodoacetamide, has been suggested. This assay blank is integrated into two other commercial systems (REA-chrom FXIII, Reanal-ker, Budapest, Hungary and Technochrom FXIII, Technoclone, Vienna, Austria). Use of the plasma blank eliminates most concerns of external reactions influencing results of the assay. The assays are not influenced by FXIII Val34Leu polymorphism. The wide adoption of the ammonia release assay is somewhat hampered by the requirement of monitoring the reaction at 340 nm wavelength. This capability to monitor absorbance at this wavelength is absent on most coagulation analyzers.

Antigenic Assays: Limitations of antigenic assays primarily pertain to nonspecific reactivity of the antibodies used in the assay and the presence of nonspecific anti-antibody reactivity. In immunoturbinometric assays, plasma discoloration or turbidity may also interfere with the test. Results outside the usual physiologic range or those that are otherwise improbable should be investigated for evidence of interference.

Test Interpretation: Congenital severe homozygous FXIII deficiency (generally <1% FXIII activity) is an extremely rare hemorrhagic condition with approximate frequency of 1 in 2 million live births. Like many autosomal recessive disorders, the frequency of severe FXIII deficiency is elevated in populations with a high degree of consanguinity. This disorder usually presents in the neonatal period with abnormally prolonged bleeding from the umbilical stump. Later in life, intracranial, gastrointestinal, intramuscular, subcutaneous bleeding, and poor wound healing become major concerns. Women with severe FXIII deficiency have very high rates of spontaneous abortions.

The deficiencies can be further classified into quantitative FXIII-A deficiencies, qualitative FXIII-A deficiencies, and deficiencies of FXIII-B. Of these, FXIII-B deficiency has been described in less than 10 families and appears to have a milder clinical phenotype then FXIII-A deficiency. Of note, FXIII-B deficiencies appear to have no effect on platelet FXIII-A pool. Expected test results for these deficiencies are summarized in Table 144.1.

Recent data, which use newer FXIII assays that are sensitive to milder deficiencies (3%–5% FXIII activity), suggest the frequency of less severe deficiencies may be significantly underestimated. While data linking different FXIII levels to bleeding propensity are sparse, patients with FXIII levels above the older method sensitivity threshold of 1%–5% may demonstrate propensity for bleeding.

Acquired FXIII deficiencies are much more frequent than the congenital form. Cases of severe acquired deficiencies associated with bleeding have been reported because of

TABLE 144.1 Classification of Congenital Factor XIII (FXIII) Deficiencies

	FXIII Activity	FXIII A_2B_2 Antigen	FXIII-A Antigen	FXIII-B Antigen
FXIII-A quantitative	<3%	<3%	<3%	Normal
FXIII-A qualitative	<3%	Mildly reduced to normal	Mildly reduced to normal	Normal
FXIII-B	5%–10%	<3%	5%–10%	<3%

autologous anti-FXIII antibodies in association with autoimmune disease and spontaneously in the elderly. Both neutralizing and clearance-inducing anti-FXIII antibodies have been reported. Neutralizing antibodies can be demonstrated by performing a mix of the patient plasma with a normal plasma pool. If neutralizing antibodies are present, the mix FXIII activity is expected to be dramatically reduced. Nonneutralizing antibodies induce clearance of FXIII from plasma but do not interfere with the FXIII activity in the 1:1 mix. Because the antibodies do not have access to the platelet pool of FXIII, FXIII activity performed on the platelet lysate is expected to be normal in cases of autoantibody-induced deficiency (generally above 60%). Milder deficiencies due to factor consumption or reduced synthesis are not infrequently seen in patients with disseminated intravascular coagulation, inflammatory bowel disorders, Henoch–Schönlein purpura (due to consumption), myeloid stem cell disorders (due to reduced synthesis of FXIII-A), and hepatic failure (due to reduced synthesis of FXIII-B). Association of these milder deficiencies with bleeding risk is not entirely clear. Recently FDA approved the use of recombinant FXIII-A to treat FXIII deficiency.

Further Reading

Dorgalaleh, A., Tabibian, S., Hosseini, S., & Shamsizadeh, M. (2016). Guidelines for laboratory diagnosis of factor XIII deficiency. *Blood Coagul Fibrinolysis*, *27*, 361–364.

Hsu, P., Zantek, N. D., Meijer, P., et al. (2014). Factor XIII assays and associated problems for laboratory diagnosis of factor XIII deficiency: An analysis of International Proficiency testing results. *Semin Thromb Hemost*, *40*, 232–238.

Karimi, M., Bereczky, Z., Cohan, N., et al. (2009). Factor XIII deficiency. *Semin Thromb Hemost*, *35*, 426–438.

Katona, E., Pénzes, K., Molnár, E., et al. (2012). Measurement of factor XIII activity in plasma. *Clin Chem Lab Med*, *50*, 1191–1202.

Lawrie, A. S., Green, L., Mackie, I. J., et al. (2010). Factor XIII – an under diagnosed deficiency – are we using the right assays? *J Thromb Haemost*, *8*, 2478–2482.

Peyvandi, F., Palla, R., Menegatti, M., European Network of Rare Bleeding Disorders Group, et al. (2012). Coagulation factor activity and clinical bleeding severity in rare bleeding disorders: Results from the European Network of Rare Bleeding Disorders. *J Thromb Haemost*, *10*, 615–621.

CHAPTER 145

Fibrinolytic Testing

Wayne L. Chandler, MD

Physiology: In healthy subjects, there is a dynamic equilibrium between coagulation and platelets forming a hemostatic plug to close the wound and the fibrinolytic system removing fibrin to prevent vascular occlusion. Fibrinolysis is initiated by the formation of fibrin. Tissue plasminogen activator (tPA) and plasminogen bind to fibrin. Once on the fibrin surface, tPA converts plasminogen into the active enzyme plasmin, which in turn lyses the fibrin, forming fibrin degradation products, including D-dimer. The rate of fibrinolysis is controlled by the concentration of active tPA in blood, which in turn is regulated in three ways:

1. By the rate of endothelial tPA secretion,
2. By the concentration of the tPA inhibitor plasminogen activator inhibitor 1 (PAI-1), and
3. By the rate of hepatic clearance of tPA.

Changes in the relative balance of tPA secretion, inhibition and clearance can lead to alterations in the overall rate of fibrinolysis and clinical risk of bleeding or thrombosis.

Hereditary Fibrinolytic Bleeding

Plasminogen Activator Inhibitor 1 Deficiency: Patients with heterozygous PAI-1 deficiency usually do not have increased bleeding. Patients with homozygous PAI-1 deficiency, resulting in no PAI-1 activity, present with a mild bleeding disorder, including easy bruising, menorrhagia, and moderate to severe delayed bleeding after surgery, often with postoperative wound hematomas. To evaluate draw citrate anticoagulated blood in the morning (PAI-1 peaks in the morning due to a circadian rhythm) with minimal tourniquet time, tourniquets trap tPA secretion in the arm, falsely lowering PAI-1 activity. Patients with PAI-1 deficiency typically show undetectable PAI-1 activity, undetectable or low PAI-1 antigen, and low tPA antigen. It is important to measure tPA antigen and demonstrate a low level to correctly diagnose PAI-1 deficiency. Free tPA in the absence of PAI-1 is cleared faster, leading to lower levels of tPA antigen in patients with PAI-1 deficiency. Low PAI-1 activity with normal or increased tPA antigen is usually due to prolonged application of the tourniquet during blood draw, leading to increased tPA in the sample and falsely low PAI-1 activity.

Antiplasmin Deficiency: Antiplasmin is the primary inhibitor of plasmin. Homozygous antiplasmin deficiency is a rare disorder resulting in moderate to severe bleeding characterized by umbilical stump bleeding, hemarthrosis, easy bruising, gingival bleeding, and bleeding after minor dental or surgical procedures. Only about 20% of patients with heterozygous antiplasmin deficiency typically report bleeding problems,

Transfusion Medicine and Hemostasis. https://doi.org/10.1016/B978-0-12-813726-0.00145-8

most are asymptomatic. Symptomatic patients typically report mild bleeding similar to that found with PAI-1 deficiency. To evaluate, draw citrate anticoagulated blood and measure antiplasmin activity. Patients with homozygous antiplasmin deficiency have close to undetectable levels of activity, whereas patients with heterozygous deficiency typically have antiplasmin activity levels from 35% to 70% of normal.

Acquired Fibrinolytic Bleeding Disorders: Acquired fibrinolytic bleeding typically occurs during thrombolytic therapy, or during or after surgery. It is usually associated with increased release of tPA, reduced PAI-1 activity, and enhanced fibrinolysis on whole-blood viscoelastic assays.

Cardiopulmonary Bypass Surgery: Passage of blood over the artificial surface of the bypass pump oxygenator activates the contact system, leading to increased bradykinin formation, which stimulates endothelial release of tPA. Increased tPA leads to enhanced fibrinolysis and increased bleeding during surgery. tPA levels rapidly fall once cardiopulmonary bypass ends. Clinically significant fibrinolysis during cardiopulmonary bypass can be monitored using whole-blood viscoelastic assays and treated with antifibrinolytic agents.

Liver Transplantation: Patients undergoing liver transplantation often have cirrhosis, which can lead to increased tPA levels, due to reduced clearance, and reduced PAI-1 levels, due to decreased hepatic secretion. This is made worse during the anhepatic phase of transplantation when hepatic tPA clearance is transiently eliminated, causing a rapid rise in tPA, increased lysis, and potentially increased fibrinolytic bleeding. Increased fibrinolysis during liver transplantation can be monitored using whole-blood viscoelastic measurements and treated with antifibrinolytic agents.

Shock/Trauma/Disseminated Intravascular Coagulation: Severe trauma, shock from multiple causes, and disseminated intravascular coagulation (DIC) can all lead to increased levels of circulating tPA, enhanced fibrinolysis, and increased bleeding. Shock, trauma, and DIC are often overlapping problems that tend to feed one another. Symptomatic increases in fibrinolysis associated with bleeding can be detected with whole-blood viscoelastic clot lysis measurements, but these assays can take 1–2 hours to show lysis and so may not provide information fast enough in a rapidly evolving situation.

Fibrinolytic Thrombosis

Increased Plasminogen Activator Inhibitor 1: Elevated levels of PAI-1 activity in blood are associated with decreased tPA activity, decreased fibrinolysis, and an increased risk of arterial thrombosis. PAI-1 is an acute phase factor with elevated levels during infection, cancer, and other inflammatory disorders. Transient acquired elevations of PAI-1 may play a role in increased thrombosis associated with sepsis and cancer. Persistent or hereditary increases in PAI-1 most commonly occur in association with the metabolic syndrome/insulin resistance and type 2 diabetes. PAI-1 in this situation is not an independent predictor of arterial thrombosis. Because of the highly variable nature of PAI-1 levels and lack of independent predictive power, PAI-1 levels are not typically used clinically to assess risk of arterial thrombosis.

Dysfibrinogen: Most dysfibrinogens are either asymptomatic or associated with an increased risk of bleeding. Very rare dysfibrinogens have been described that are associated with an increased risk of thrombosis. Thrombotic dysfibrinogens have been reported that are less susceptible to lysis by the fibrinolytic system. There are no simple screening assays for this type of dysfibrinogen, it typically requires specialized research assays to detect.

Plasminogen Deficiency: Heterozygous plasminogen deficiency is usually asymptomatic. Homozygous plasminogen deficiency is a rare disorder most often associated with pseudomembrane formation on mucosal surfaces (example ligneous conjunctivitis), with subsequent damage to the affected organs. Heterozygous and homozygous plasminogen deficiencies are not associated with an increased risk of thrombosis.

Increased Thrombin-Activatable Fibrinolysis Inhibitor: Thrombin-activatable fibrinolysis inhibitor (TAFI) is a carboxypeptidase that removes C-terminal lysine binding sites for tPA and plasminogen from fibrin, decreasing the ability of fibrin to stimulate fibrinolysis. Elevated levels of TAFI have been associated with an increased risk of venous thrombosis. Measurement of TAFI is complex, as there are multiple forms of the molecule, including the proenzyme (TAFI), activated enzyme (TAFIa), and an inactive form (TAFIi). Currently standardization between methods is poor and the assay is not routinely used for clinical evaluation of patients with thrombophilia.

Further Reading

Alessi, M. C., & Juhan-Vague, I. (2008). Metabolic syndrome, haemostasis and thrombosis. *Thromb Haemost*, *99*, 995–1000.

Carpenter, S. L., & Mathew, P. (2008). Alpha2-antiplasmin and its deficiency: Fibrinolysis out of balance. *Haemophilia*, *14*, 1250–1254.

Haverkate, F., & Samama, M. (1995). Familial dysfibrinogenemia and thrombophilia. Report on a study of the SSC Subcommittee on Fibrinogen. *Thromb Haemost*, *73*, 151–161.

Lisman, T. (2017). Decreased plasma fibrinolytic potential as a risk for venous and arterial thrombosis. *Semin Thromb Hemost*, *43*, 178–184.

Mehta, R., & Shapiro, A. D. (2008). Plasminogen activator inhibitor type 1 deficiency. *Haemophilia*, *14*, 1255–1260.

Sniecinski, R. M., & Chandler, W. L. (2011). Activation of the hemostatic system during cardiopulmonary bypass. *Anesth Analg*, *113*, 1319–1333.

Theusinger, O. M., Wanner, G. A., Emmert, M. Y., et al. (2011). Hyperfibrinolysis diagnosed by rotational thromboelastometry (ROTEM) is associated with higher mortality in patients with severe trauma. *Anesth Analg*, *113*, 1003–1012.

CHAPTER 146

Laboratory Techniques in Fibrinolysis Testing

Wayne L. Chandler, MD

The purpose of this chapter is to provide an overview of laboratory techniques involved in fibrinolysis testing. Chapter 145 provides a general introduction to fibrinolytic testing and common fibrinolytic disorders (Fig. 146.1).

Screening Tests

Euglobulin Lysis Time: The euglobulin lysis time (ELT) is a historical screening test for hyperfibrinolysis. Newer, more specific tests for individual fibrinolytic factors and more rapid tests, such as thromboelastometry, have supplanted ELT. To measure ELT, the euglobulin fraction of the plasma-containing plasminogen, plasminogen activators, and fibrinogen is acid-precipitated then clotted with thrombin. Time to clot disappearance is determined. A positive test is premature clot disappearance in 60–120 minutes, suggesting increased fibrinolysis, although the specific cause of the excessive fibrinolysis is not determined by this methodology.

Thromboelastometry/Thromboelastography: Thromboelastometry (and related thromboelastography) has largely replaced ELT as a screening test for hyperfibrinolysis in acute settings. Whole blood or plasma is clotted. Fibrinolysis is measured as the difference between the maximum viscoelasticity achieved after the clot formation and the clot viscoelasticity at 30 and 60 minutes after the maximum. Either percent clot lysis or rate of clot lysis can be measured. Proof of fibrinolytic activity can be shown by adding a fibrinolytic inhibitor such as aprotinin. This methodology is sensitive to moderate to severe increases in fibrinolytic activity but not to mild abnormalities such as PAI-1 or antiplasmin deficiency or modest increases in tissue plasminogen activator (t-PA) activity. When fibrinolysis is obvious on thromboelastometry, clinical bleeding is often evident.

Specific Tests: Fibrinogen—see Chapter 123.

D-Dimer: D-dimer levels are specific for lysed cross-linked fibrin based on an epitope that is not present on fibrinogen. D-dimer levels are primarily related to intravascular fibrin levels. Most assays use liquid-phase latex bead agglutination technology on automated instruments. The assay can be reported in either fibrinogen equivalent units/mL (FEU/mL) or D-dimer units/mL, with 2 FEU/mL roughly equal to 1 D-dimer unit/mL. D-dimer assays vary in their sensitivity to potential interfering substances.

Elevated D-dimer levels occur in situations of increased intravascular fibrin such as disseminated intravascular coagulation, surgery, burns, trauma, infection, recent or current thromboembolic event, malignancy, pregnancy, and many others clinical situations. During fibrinolytic therapy, D-dimer can increase because of increased

Transfusion Medicine and Hemostasis. https://doi.org/10.1016/B978-0-12-813726-0.00146-X

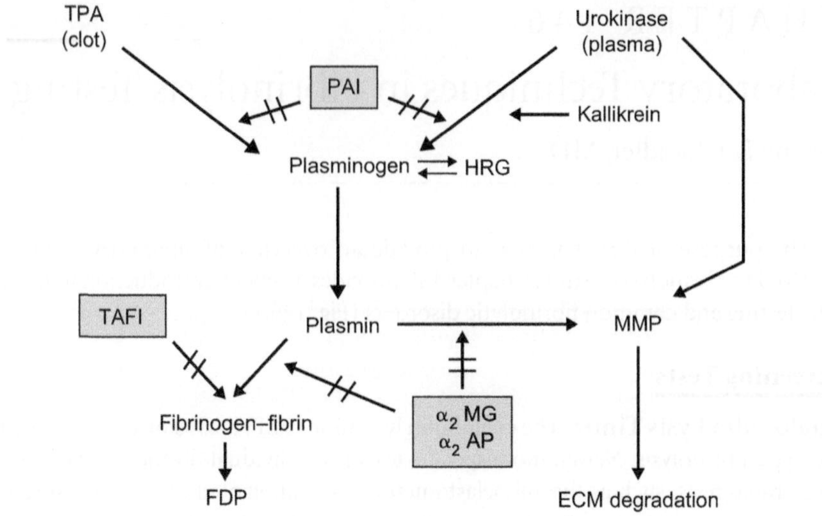

FIGURE 146.1 The fibrinolytic pathway in coagulation. *α_2AP*, α_2 antiplasmin; *α_2MG*, α_2 macroglobulin; *ECM*, extracellular matrix; *FDP*, fibrin degradation products; *HRG*, histadine-rich glycoprotein; *MMP*, matrix metalloproteinases; *PAI-1*, plasminogen activator inhibitor-1; *TAFI*, thrombin activatable thrombinolysis inhibitor; *TPA*, tissue plasminogen activator. (From Abshire, T. (2009). Laboratory techniques in fibrinolysis testing. In C. D. Hillyer, B. H. Shaz, J. C. Zimring, & T. C. Abshire (Eds.), *Transfusion medicine and hemostasis: clinical and laboratory aspects* (pp. 671–676). Elsevier.)

rates of degradation. Reduced liver clearance can also increase D-dimer levels. D-dimer levels increase after a thrombotic event, such as venous thrombosis, then fall if the thrombus begins resolving or rise if it extends further.

D-dimer is frequently used to rule out recent deep venous thrombosis and pulmonary embolus in patients with low pretest probability for the event. In this setting, a test result below the threshold level has a high negative predictive value. When the pretest probability is high or intermediate, however, D-dimer is insufficient to rule out a thrombotic event. Because of differences in antibodies, techniques, instruments, and reporting units, direct comparison of D-dimer values obtained by different methods is challenging. It is preferable that assay-specific thresholds for ruling out thrombotic events be established in the context of clinical trials.

Plasminogen: Plasminogen is typically measured with an activity assay based on plasminogen activation by streptokinase, followed by measurement of plasmin-like activity using a chromogenic substrate. Acquired plasminogen deficiency is seen in consumptive processes, surgery, liver failure, thrombolytic therapy, asparaginase treatment, and Argentine hemorrhagic fever caused by Junin virus.

Antiplasmin: Antiplasmin is typically measured with an activity assay based on addition of plasmin to plasma followed by measurement of residual plasmin with a chromogenic substrate. The amount of residual plasmin activity is inversely proportional to the concentration of antiplasmin in the sample.

Tissue Plasminogen Activator: t-PA exists in two forms in plasma: (1) free t-PA capable of binding to and activating plasminogen, and (2) t-PA/PAI-1 complex an inactive form of inhibited t-PA. t-PA may be measured by either functional or antigenic assays. t-PA activity measures only the free, functional form of t-PA. t-PA in complex with PAI-1 and other inhibitors is not measured. Free t-PA in blood must be stabilized by drawing the sample into an acidified citrate solution to prevent inactivation of free t-PA by active PAI-1 in blood. t-PA activity is measured by capturing t-PA on a microtiter plate using monoclonal antibodies, washing off PAI-1 and antiplasmin inhibitors, and then measuring activity by adding plasminogen, a fibrinlike activator that enhances plasminogen activation by t-PA and a plasmin sensitive chromogenic substrate. t-PA converts plasminogen to plasmin, and the plasmin cleaves the substrate. Total t-PA antigen representing both active and inactive fractions can be measured using an enzyme immunoassay.

Plasminogen Activator Inhibitor 1: PAI-1 exists in three forms in plasma: (1) free active PAI-1, (2) inactive PAI-1 complexed with t-PA and latent PAI-1 (an inactive PAI-1 conformation). PAI-1 may be measured using a functional assay by adding excess exogenous t-PA to the patient's plasma, allowing PAI-1 to bind to and inactivate t-PA, then measuring residual t-PA activity or by measuring the formation of t-PA/PAI-1 complex (immunofunctional assay). Total PAI-1 antigen assays measure the sum of active PAI-1, t-PA/PAI-1, and latent PAI-1 typically using an enzyme immunoassay. PAI-1 testing is complex. Evaluation of PAI-1 deficiency should ideally involve testing for PAI-1 antigen, activity, and t-PA antigen. PAI-1 activity alone is not sufficient because of poor test analytic sensitivity at low PAI-1 levels. Some assays demonstrate undetectable PAI-1 activity in patients with low normal PAI-1 levels. Thus, additional measures such as PAI-1 antigen and t-PA antigen are needed to confirm a suspicion of deficiency. Elevated PAI-1 levels can be readily detected by the activity assays, but their significance under most circumstances is unclear.

Thrombin Activatable Fibrinolysis Inhibitor: Thrombin activatable fibrinolysis inhibitor (TAFI) may be measured by functional assay or immunologically. The testing is complex owing to the presence of multiple forms of TAFI and a very short half-life of the active form of TAFI in the plasma. Standardization of the assays is needed before the utility of TAFI testing in the clinical setting may be confirmed.

Further Reading

Agren, A., Wiman, B., & Schulman, S. (2007). Laboratory evidence of hyperfibrinolysis in association with low plasminogen activator inhibitor type 1 activity. *Blood Coagul Fibrinolysis*, *18*, 657–660.

Bates, S. M., Jaeschke, R., Stevens, S. M., et al. (2012). Diagnosis of DVT: Antithrombotic therapy and prevention of thrombosis, 9th ed: American College of Chest Physicians Evidence-Based Clinical Practice Guidelines. *Chest*, *141*, e351S–418S.

Chandler, W. L., Jascur, M. L., & Henderson, P. J. (2000). Measurement of different forms of tissue plasminogen activator in plasma. *Clin Chem*, *46*, 38–46.

Levrat, A., Gros, A., Rugeri, L., et al. (2008). Evaluation of rotation thrombelastography for the diagnosis of hyperfibrinolysis in trauma patients. *Br J Anaesth*, *100*, 792–797.

Linkins, L. A., & Takach Lapner, S. (2017). Review of D-dimer testing: Good, bad, and ugly. *Int J Lab Hematol*, *39*(Suppl. 1), 98–103.

Vercauteren, E., Gils, A., & Declerck, P. J. (2013). Thrombin activatable fibrinolysis inhibitor: A putative target to enhance fibrinolysis. *Semin Thromb Hemost*, *39*, 365–372.

CHAPTER 147

Laboratory Evaluation of Long-Term Thrombophilic Disorders

Mikhail Roshal, MD, PhD and Morayma Reyes Gil, MD, PhD

Abnormal thrombosis is a result of a pathologic shift in the balance between antihemorrhagic (prothrombotic) and antithrombotic factors toward conditions that favor thrombus formation at the wrong time and/or in the wrong location. The Virchow triad of blood stasis, inflammation, endothelial injury, and alterations of blood components, can all contribute to the pathologic event. Frequently the cause of a venous clot is not difficult to discern and is clearly due to the physiologic state of the patient, possibly significant trauma, neoplasia, and prolonged immobility. At other times there may not be an entirely obvious explanation for a pathologic thrombus, and a laboratory evaluation for thrombophilic tendency is considered.

Long-term thrombophilia may occur secondary to acquired chronic disorders such as antiphospholipid antibody syndrome and paroxysmal nocturnal hemoglobinuria (PNH) due to inherited gene mutations that cause alterations of factors involved in the clotting cascade, or due to inherited conditions that influence clotting, or the fibrinolysis cascade indirectly. The strongly prothrombotic congenital defects such as most homozygous or compound heterozygous antithrombin (AT), protein C (PC), and protein S (PS) deficiencies are very rare and are either lethal in the embryonic stage or cause dramatic thrombotic complications in infants. Deficiency of another natural anticoagulant, the tissue factor pathway inhibitor, is not known to occur in humans, presumably due to early embryonic lethality. For more frequently encountered congenital defects, the excess risk of venous thromboembolism (VTE) is relatively modest (Table 147.1). While these defects can cause thrombosis without obvious provocation, frequently it is the interplay between the mild physiologic provocation and relatively modest congenital or acquired risk factors that tilts the hemostatic system to produce a pathological thromboembolic event. For instance, pregnant women who are carriers of factor V Leiden (FVL) mutation (a modest risk factor for venous thromboembolic event) are at substantially increased risk of serious thrombotic complications compared with either noncarrier pregnant women or nonpregnant carriers of FVL. The presence of multiple congenital thrombophilia risk factors greatly increases the risk of VTE. The estimates of risk of VTE for carriers of compound defects are as high as 50–80 times than those of unaffected subjects.

Testing Recommendations: Defining which patients to test for thrombophilia tendencies is challenging. Screening the general population is clearly not indicated, as the low prevalence of any of the risk factors (Table 147.1) would lead to an unacceptably high rate of false-positive results. Moreover, the significance of some risk factors for thrombophilia in otherwise asymptomatic subjects outside of the thrombophilic

Transfusion Medicine and Hemostasis. https://doi.org/10.1016/B978-0-12-813726-0.00147-1

TABLE 147.1 Congenital Venous Thrombosis Risk Factors in Caucasians

	Approximate Lifetime Relative Risk (Fold)	Approximate Population Prevalence
Factor V Leiden (heterozygous)	2–5	3%–7%
Prothrombin G20210A mutation	2–4	1%–3%
Protein C deficiency	3–15	0.2%–0.4%
Protein S deficiency	3–15	0.03%–0.17%
Antithrombin deficiency	3–15	0.02%–0.04%
Persistent elevation of FVIII (>150 IU/dL)	1.5–5.5	2%–10%
Congenital dysfibrinogenemia	Depends on specific mutation	Very rare

kindreds is unclear. Current testing guidelines are primarily focused on selecting appropriate groups of patients with a high prevalence of long-term thrombophilia. Broadly, this includes patients who have suffered a VTE or thrombotic pregnancy complication without a sufficient external provocation to fully explain the event. For this patient population with unprovoked thrombosis, thrombophilia testing should be considered only if results would change management and testing should be performed after routine anticoagulation treatment (e.g., 3 months) is completed. Alternatively, in patients in which longer anticoagulation treatment is being considered, a two-stage approach for thrombophilia testing may be indicated: performed tests that can provide reliable results while on anticoagulation (Table 147.2) and if these results are normal then remaining thrombophilia testing can be performed once anticoagulation treatment is completed. An additional group is the female relatives of patients with known congenital high-risk thrombophilia such as PC, PS, and AT deficiency who are considering estrogen use or pregnancy. Again, in these cases, thrombophilia testing should be considered only if the results would change management and testing should be limited to the known thrombophilia in the affected family members.

Approximately 60% of patients from carefully chosen populations with unexpected thrombosis will have one or more of the identifiable heritable thrombophilia risk factors. Conversely, although the prevalence of risk factors is by definition elevated in patients with unexpected thromboembolic events, the risk of VTE recurrence among carriers and noncarriers is either similar or only mildly elevated in most instances. The phenomenon is likely attributable to the presence of yet undiscovered or multifactorial risk determinants in a substantial subset of the remaining VTE patients. Moreover, neither thrombophilia testing lowers the risk of thrombosis recurrence nor does the presence of known thrombophilia has an impact on the long-term survival of VTE patients. Thus, clinical management of patients with identified thrombophilia is often driven by clinical rather than laboratory findings. In many situations, the greatest beneficiaries of testing may be the asymptomatic relatives of patients in whom a congenital risk factor for thrombosis is demonstrated. In that cohort, appropriate measures may be taken to prevent the first provoked thrombosis in those who test positive.

TABLE 147.2 Anticoagulant Interference in Thrombophilia Testing

Thrombophilia	Available Tests	Interference	No Interference
APC resistance	FVL DNA test	None	All
	APCR	DTI, Xa inhibitors	Heparin[a], coumadin
Prothrombin G20210A mutation	Mutational analysis	None	All
Protein C deficiency	Chromogenic protein C function	Coumadin[b]	Xa inhibitors, DTI, heparin
	Protein C clottable	All	None
Protein S deficiency	Protein S free Antigen	Coumadin[b]	Xa inhibitors, DTI, heparin
	Protein S clottable	All	None
Antithrombin deficiency	Chromogenic (residual Xa activity)	Xa inhibitors, heparin[c]	DTI, coumadin
	Chromogenic (residual IIa activity)	DTI, heparin[c]	Xa inhibitors, coumadin
Persistence elevation of FVIII	Chromogenic FVIII	Xa inhibitors	Heparin[a], DTI, coumadin
	Clottable FVIII	All	None
Lupus anticoagulant	DRVVT	Xa inhibitors, DTI	Heparin[a], coumadin
	PTT-based (STACLOT-LA)	Xa inhibitors, DTI	Heparin[a], coumadin
Dysfibrinogenemia	Fibrinogen activity	DTI	Heparin[a], coumadin, Xa inhibitors
	thrombin time	Heparin, DTI	Coumadin, Xa inhibitors

APCR, activated protein C resistance; *AT*, antithrombin; *FVL*, factor V Leiden.
[a]Therapeutic levels of heparin should not interfere.
[b]Physiologic decrease of vitamin K–dependent factors.
[c]Approximately 30% decrease in AT within 3–5 days of heparin treatment due to increased clearance of heparin–bound AT.

Testable Thrombophilia Risk Factors: The most commonly tested thrombophilia risk factors are deficiencies of PC, PS, AT, activated protein C resistance (APCR), prothrombin gene mutation *G20210A*, and persistent increase in FVIII and lupus anticoagulant. Under certain clinical circumstances, such as abdominal vein thrombosis, testing for PNH is indicated. Homocysteine and dysfibrinogenemia testing indications are controversial, and only dysfibrinogenemia testing may be warranted. The risk factors can be loosely divided into the four categories described below.

Deleterious Mutations in Genes Encoding Natural Anticoagulants: Natural anticoagulants dampen the coagulation cascade to prevent thrombosis at a wrong site and to reduce propagation of the clot beyond the area of injury. Details of specific tests for the natural anticoagulant AT, PC, and PS deficiencies are provided in subsequent chapters of this section. Some common general principles, which are important for understanding testing approaches for all three, are described below.

First, protein activity level can strongly affect the clinical manifestation of the lesion. Homozygous or compound heterozygous deleterious mutations that result in absence or near absence of one of the anticoagulants' activity can result in embryonic lethality or severe thrombosis, blindness, and disseminated intravascular coagulation in the early postnatal period. These types of presentations are uncommon. More common are milder heterozygous states. Patients carrying one deleterious allele in natural anticoagulants are at variably increased risk for venous thrombosis during their lifetime. The activity levels for the heterozygous-deficient patients usually average around 50% of the median for age and sex, but may have substantially lower or higher levels.

Second, congenital deficiencies result from mutations that decrease plasma anticoagulant activity through either decreasing the quantity of the protein (type I and sometimes type III) or through structural alterations that have a pronounced effect on the anticoagulant activity without significantly affecting the antigenic levels (type II). Thus, measuring total antigenic protein level is not enough to rule out functional deficiency due to the potential of missing type II deficiency. Recommended screening tests are based on evaluating the functional fraction of the natural anticoagulant in the plasma. Subsequently, antigenic tests for total protein level may be used to designate the deficiency type.

Third, timing of the tests relative to thrombosis, anticoagulation therapy, and other physiologic and pathologic states can influence the reliability of the results. In most laboratory practices, acquired deficiencies of natural anticoagulants are encountered far more frequently than heritable disease. For instance, patients who receive warfarin are expected to have low PC and PS activity, and patients with ongoing thrombosis have low levels of AT, PC, and PS activities. Laboratorians and clinicians should be mindful of these and other potential physiologic confounding factors when ordering, performing, and interpreting results. Even when potential confounding factors are ruled out, a second test—generally in 4–6 weeks—is advised to confirm persistence of the deficiency. On the other hand, repeating a normal result is generally not required unless suspicion for the deficiency is very high and results fall into borderline range.

Finally, clot-based activity tests for natural anticoagulants are fraught with numerous potential interferences from other factors involved in the clotting pathway, drugs, and lupus anticoagulant. Alternative tests that isolate measurements of the active form of the protein from clot formation are usually preferred. Such tests include chromogenic PC and AT activity and determination of PS free antigen concentration. These tests offer less analytically challenging alternatives compared with clot-based methodologies. The increased specificity of these tests often comes with a price of reduced sensitivity to mostly rare defects that would be detected with natural substrate-based activity measures.

Increased Concentration, Gain of Function, or Delayed Inactivation of Procoagulant Proteins: Increased activity of procoagulants may also shift the balance toward thrombosis. APCR, prothrombin *G20210A* mutation, and persistent elevation of FVIII are the most frequently encountered defects in this category. These defects cause increased thrombotic risk through diverse mechanisms but ultimately lead to increased thrombin generation at the site of clotting initiation.

APCR is usually, but not always, due to *G1691A* mutation in factor V (FVL). The mutation results in delayed inactivation of FVa and presumably extended time of thrombin production through the complex of FXa and FVa. The *G20210A* mutation in the 3′ untranslated region of the prothrombin gene results in increased concentrations of circulating prothrombin. Testing for APCR is covered in Chapter 152, and genetic testing for FVL, other FV mutations, and prothrombin *G20210A* is discussed in Chapter 153.

A persistently elevated level of FVIII levels has complex genetic and physiologic determinants. Unlike FVL and the *G20210A* prothrombin mutation, where risk is determined by either heterozygous or homozygous state, the level of FVIII acts as a continuous variable in increasing thrombotic risk, with levels above 150% considered pathologic. For every 10% increase in FVIII level, there is a ~10% increase in risk of thrombosis. General details for testing for clotting factors, including FVIII, are provided in Chapter 154. Determination of FVIII via chromogenic assay appears to offer the most robust way of measuring elevated FVIII levels compared with either clot-based or antigenic assays.

Hereditary dysfibrinogenemia is a very rare condition. Depending on the specific mutation, dysfibrinogenemia can either be asymptomatic, lead to increased hemorrhage or to increased risk of thrombosis. Mechanisms leading to a prothrombotic state vary with different mutations and are often not known. Resistance to clot degradation is observed with some thrombotic dysfibrinogens. It should be noted that specialized DNA-based testing in a research laboratory is frequently required for definitive diagnosis.

Hereditary Nonhemostatic Factors That Result in Prothrombotic State: This category is broad and lies mostly outside the range of regular coagulation laboratory testing. Among the most common, but underappreciated, risk factors for thrombosis in this category is blood type, with non-O blood group subjects having approximately twofold increased risk of thrombosis compared with those of blood type O. The congenital defects leading to prothrombotic states range from hemoglobin derangements, such as sickle cell anemia and beta thalassemia, to metabolic derangements such as inherited hyperlipidemias, hypercystenuria, and hyperhomocysteinemia. An elevated level of plasma homocysteine (hyperhomocysteinemia) has been linked to a modest increase in risk of venous and arterial thrombosis. However, elevated homocysteine is likely a consequence rather than a cause of thrombosis, due to metabolic disturbance such as increased oxidant stress. Treatment to lower homocysteine levels does not reduce persistent thrombotic risk despite normalization of homocysteine levels.

Acquired Testable Risk Factors: Acquired thrombosis risk factors are quite a diverse group. Many severe disease states are associated with thrombosis. However, for some illnesses venous thrombosis may be either the first sign or most severe manifestation. Acquired prothrombotic conditions for which testing is available are discussed in Chapter 158.

Further Reading

Ataga, K. I., Cappellini, M. D., & Rachmilewitz, E. A. (2007). Beta-thalassaemia and sickle cell anaemia as paradigms of hypercoagulability. *Br J Haematol*, *139*, 3–13.

Borowitz, M. J., Craig, F. E., Digiuseppe, J. A., Clinical Cytometry Society, et al. (2010). Guidelines for the diagnosis and monitoring of paroxysmal nocturnal hemoglobinuria and related disorders by flow cytometry. *Cytom Part B Clin Cytom*, *78*, 211–230.

Chandler, W. L., Ferrell, C., Lee, J., Tun, T., & Kha, H. (2003). Comparison of three methods for measuring factor VIII levels in plasma. *Am J Clin Pathol, 120*, 34–39.

Christiansen, S. C., Cannegieter, S. C., Koster, T., Vandenbroucke, J. P., & Rosendaal, F. R. (2005). Thrombophilia, clinical factors, and recurrent venous thrombotic events. *JAMA, 293*, 2352–2361.

Coppens, M., Reijnders, J. H., Middeldorp, S., Doggen, C. J., & Rosendaal, F. R. (2008). Testing for inherited thrombophilia does not reduce the recurrence of venous thrombosis. *J Thromb Haemost, 6*, 1474–1477.

Coppola, A., Tufano, A., Cerbone, A. M., & Di Minno, G. (2009). Inherited thrombophilia: Implications for prevention and treatment of venous thromboembolism. *Semin Thromb Hemost, 35*, 683–694.

den Heijer, M., Willems, H. P. J., Blom, H. J., et al. (2007). Homocysteine lowering by B vitamins and the secondary prevention of deep vein thrombosis and pulmonary embolism: A randomized, placebo-controlled, double blind trial. *Blood, 109*, 139–144.

Jenkins, P. V., Rawley, O., Smith, O. P., & O'Donnell, J. S. (2012). Elevated factor VIII levels and risk of venous thrombosis. *Br J Haematol, 157*, 653–663.

Kearon, C., Julian, J. A., Kovacs, M. J., et al. (2008). Influence of thrombophilia on risk of recurrent venous thromboembolism while on warfarin: Results from a randomized trial. *Blood, 112*, 4432–4436.

Reitter-Pfoertner, S., Waldhoer, T., Mayerhofer, M., et al. (2013). The influence of thrombophilia on the long-term survival of patients with a history of venous thromboembolism. *Thromb Haemost, 109*, 79–84.

Stevens, S. M., Woller, S. C., Bauer, K. A., et al. (2016). Guidance for the evaluation and treatment of hereditary and acquired thrombophilia. *J Thromb Thrombolysis, 41*, 154–164.

CHAPTER 148

Thrombophilia Testing in the Pediatric Population

Jennifer Davila, MD

Introduction: Thrombophilia refers to the propensity to develop thrombosis and can be a result of acquired and/or inherited conditions. Inherited conditions are usually identified by a blood test, where acquired conditions are characterized by a patient's clinical state or associated comorbidities. There has been a documented increase in the incidence of venous thromboembolism (VTE) in the pediatric population over the last two decades. The incidence follows a bimodal pattern with neonates having the greatest risk of thromboembolism (TE) and a second peak in incidence during puberty and adolescence. The rise in VTE has been attributed to increase in medical availability and expertise in the care of children with fatal conditions and complex chronic disease. The incidence of idiopathic VTE is only 5% in children and <1% in neonates, in contrast to 40% in the adult population. The majority of VTE in children is associated with predisposing factors such as a central venous catheter (CVC) and acute or chronic illness. Although there are some published recommendations on thrombophilia testing in children, there is no general consensus. Additionally, historical data show that there is no benefit to testing children.

Laboratory Thrombophilia: Inheritable thrombophilia (IT) is an inherited tendency predisposing to TE. The most common inherited thrombophilias are listed below:

- Factor V Leiden mutation
- Prothrombin 20210A mutation
- Antithrombin deficiency
- Protein C (PC) deficiency
- Protein S (PS) deficiency

Factor V Leiden (FVL) and the *20210A* variation in the prothrombin gene are the most common genetic mutations associated with thrombophilia, comprising an integral part of the thrombophilia workup. The American College of Medical Genetics recommends testing for these mutations in several situations, including a history of recurrent VTE, a family history of VTE before age 50, or VTE during pregnancy. The testing is not recommended for general population screening or as a routine test during pregnancy. FVL is present almost exclusively among Caucasians, and risk of VTE is 40- to 80-fold higher among homozygotes. The risk of spontaneous VTE risk in *20210A* is about two- to sevenfold higher in heterozygotes. For both mutations, the risk of VTE increases with age and often occurs after 45 years of age.

Transfusion Medicine and Hemostasis. https://doi.org/10.1016/B978-0-12-813726-0.00148-3

Antithrombin III (ATIII), PC, and PS deficiency occurs less frequently in the general population, compared with FVL and 20210A; however, it puts patients at a five- to eightfold higher risk for developing their first VTE. Familiarity with developmental hemostasis is essential for the interpretation of thrombophilia testing, specifically in children under 6 months of age, who account for the largest proportion of VTE in pediatrics. At birth, levels of ATIII, PC, and PS are at about 50% of adult levels, and normal levels are not achieved until 6 months of life or older. Therefore care must be taken to ensure that measured plasma levels are interpreted based on age-adjusted norms. Ideally, laboratories processing large numbers of neonatal samples should derive their own in-house reference ranges, as these are both instrument and reagent specific. Among children with VTE, ~1% have PC and PS deficiencies and 5%–15% have FVL mutation. This is consistent with what is seen in the general adult population. Young et al., reported in a meta-analysis of infants and children with a first episode of VTE a three- to ninefold increased likelihood of having an IT compared with controls. However, in this analysis, more than 70% of the patients had one or more risk factors for VTE (i.e., CVC, immobility, use of oral contraceptive, etc.). In heterozygous patients with PC, PS, or ATIII deficiency, approximately half of the thrombotic events occur in association with circumstantial risk factors (surgery, pregnancy, immobility) with the first VTE more frequently occurring before 40 years of age.

Other laboratory assays that are part of the "thrombophilia panel" but that are not necessarily genetically determined are

- Antiphospholipid antibodies (APLA)
- Factor VIII levels
- Fibrinogen
- Lipoprotein

APLA have been associated with both arterial and venous thrombosis. Assessment of a lupus anticoagulant (LAC) may be helpful before initiation of anticoagulation (AC) because AC may affect the interpretation of LAC screens, in particular warfarin. Elevated Factor VIII has been linked with TE in adults and is thought to contribute to a thrombophilic state as a direct result of increased expression and not due to acute-phase response. In children, elevated levels of plasma factor VIII, d-dimer, or both at diagnosis and a persistent elevation of at least one of these factors after standard duration anticoagulant therapy predict a poor outcome in children with thrombosis.

Lipoprotein A (Lp(a)) competes with plasminogen for its binding site to fibrin, leading to reduced fibrinolysis. Elevated Lp(a) may also promote atherosclerosis because of its high low-density lipoprotein component. The clinical features of congenital dysfibrinogenemias are heterogeneous, unpredictable, and can be hemorrhagic or thrombotic in nature. The majority do not have thrombosis, although the cumulative incidence is ~30% at 50 years of age. For those who do have a thrombosis, venous sites are more common than arterial, but both can occur. The prevalence of dysfibrinogenemia in patients with venous thrombosis is very low, and systematic testing for it in patients with thrombophilia is not recommended. However, there are thrombophilia-associated fibrinogen mutations, which increase the risk of arterial and venous thrombosis, often at a young age.

Impact of Thrombophilia in Childhood Thrombosis: The reported prevalence of thrombophilia in children with venous and arterial thrombotic events varies greatly, from as low as 13% to as high as 79%. This variation in the reported prevalence most likely reflects the differences in study design and laboratory assays, as well as the clinical variability among the patients studied. Nonetheless, this discrepancy elucidates the confusion regarding the role of testing in children. The majority of VTE occurs in hospitalized children with >1 underlying medical condition such as prematurity, cancer, sepsis, congenital heart disease, trauma, and short gut syndrome, among others. However, the single most common risk factor for thrombosis in children continues to be the presence of a CVC, with rates as high as 90% in the neonatal population and greater than 50% in all other age groups. The high frequency in those <1 year old is likely due to the increased ratio of catheter to vessel size and difficult venous access. There have been divergent results among large European and Canadian registries regarding the prevalence of thrombophilia in CVC-related VTE. At this time, there are no consistent data to make recommendations for testing in this population.

Screening Recommendations: In 2002 the Subcommittee for Perinatal and Pediatric Thrombosis of the Scientific and Standardization Committee (SSC) of the International Society of Thrombosis and Hemostasis (ISTH) recommended that all pediatric patients with venous or arterial thrombosis be tested for a full panel of genetic and acquired prothrombotic traits. The rationale for this recommendation was that pediatric patients often have several risk factors for thrombosis; therefore, detection of one thrombophilic factor does not exclude the existence of a second or a third. The committee, however, also recognized that future studies in children of all ages, races, and ethnicities were needed so that risk factors can be identified and applied appropriately. Over the last 20 years, however, our knowledge of thrombosis in pediatrics has grown. We understand that there is great heterogeneity in pediatric thrombosis patients and that there is a multifactorial pathogenesis of thrombosis in this population. It is therefore difficult to present a single, definitive guideline for laboratory thrombophilia testing in these patients. Table 148.1 includes suggestions for thrombophilia testing in children in certain clinical situations.

When thrombophilia testing is undertaken, testing for Factor V Leiden mutation, prothrombin 20210 mutation, deficiencies of ATIII, PC, and PS, homocysteine, factor VIII level, antiphospholipid antibodies, and fibrinogen are suggested. In terms of timing of testing, it is best to perform testing after recovery from the acute VTE, to avoid an incorrect diagnosis. Levels of ATIII, PC, and PS may transiently decrease during acute thrombosis and factor VIII, and fibrinogen can be elevated in inflammatory conditions. Therefore, any non-DNA-based test that is abnormal during the acute setting should be repeated later, ideally off AC.

TABLE 148.1 Thrombophilia Testing in the Pediatric Population

Clinical Scenario	Should You Test?	Comments/Rationale
Adolescents with unprovoked TE	Strongly consider	• these patients have the highest prevalence of inherited thrombophilia in pediatrics
Neonates/children with unprovoked TE/ATE	Consider/Strongly consider	• may help determine future risk of recurrence • may identify need for prophylaxis under high-risk situations • if positive, may allow for screening of family members
Children/adolescents with provoked TE	Consider especially if recurrent	
Neonates/children/adolescents with recurrent TE	Strongly consider	• may help determine future risk of recurrence • may identify need for prophylaxis under high-risk situations • if positive, may allow for screening of family members
Neonates/children/adolescents with CVC-related TE	Consider if recurrent	• may identify need for prophylaxis under high-risk situations
Children/Adolescent with family history of VTE and known or unknown thrombophilia	Consider in certain situations	• to guide prophylaxis in high-risk situations • for guidance with concurrent prothrombotic conditions (i.e., CVC placement, contraception, major surgery)

ATE, arterial thrombotic event; *CVC*, central venous catheter; *TE*, thromboembolism, *VTE*, venous thromboembolism.

Further Reading

Casini, A., Neerman-Arbez, M., Ariens, et al. (2015). Dysfibrinogenemia: From molecular anomalies to clinical manifestations and management. *J Thromb Haemost*, *13*, 909–919.

De Stefano, V., Leone, G., Mastrangelo, S., et al. (1994). Clinical manifestations and management of inherited thrombophilia: Retrospective analysis and follow-up after diagnosis of 238 patients with congenital deficiency of antithrombin III, protein C, protein S. *Thromb Haemost*, *72*, 352–358.

De Stefano, V., Rossi, E., Paciaroni, K., & Leone, G. (2002). Screening for inherited thrombophilia: Indications and therapeutic implications. *Haematologica*, *87*(10), 1095–1108.

Falcon, C. R., Cattaneo, M., Panzeri, D., et al. (1994). High prevalence of hyperhomocyst(e)inemia in patients with juvenile venous thrombosis. *Arterioscler Thromb*, *14*, 1080–1083.

Goldenberg, N. A., Knapp-Clevenger, R., & Manco-Johnson, M. J. (2004). Elevated plasma factor VIII and D-dimer levels as predictors of poor outcomes of thrombosis in children. *N Engl J Med*, *351*, 1081–1088.

Jaffray, J., & Young, G. (2013). Developmental hemostasis: Clinical implications from the fetus to the adolescent. *Pediatr Clin N Am, 60,* 1407–1417.

Kraaijenhagen, R. A., in't Anker, P. S., Koopman, M. M., et al. (2000). High plasma concentration of factor VIIIc is a major risk factor for venous thromboembolism. *Thromb Haemost, 83,* 5–9.

Mahajerin, A., Obasaju, P., Eckert, G., et al. (2014). Thrombophilia testing in children: A 7 year experience. *Pediatr Blood Cancer, 61,* 523–527.

Manco-Johnson, M. J., Grabowski, E. F., Hellgreen, M., et al. (2002). Laboratory testing for thrombophilia in pediatric patients. On behalf of the Subcommittee for Perinatal and Pediatric Thrombosis of the Scientific and Standardization Committee of the International Society of Thrombosis and Haemostasis (ISTH). *Thromb Haemost, 88,* 155–156.

Prentiss, A. S. (2012). Early recognition of pediatric venous thromboembolism: A risk-assessment tool. *Am J Crit Care, 21,* 178–183.

Raffini, L., Huang, Y. S., Witmer, C., et al. (2009). Dramatic increase in venous thromboembolism in children's hospitals in the United States from 2001 to 2007. *Pediatrics, 124,* 1001–1008.

Revel-Vilk, S., Chan, A., Bauman, M., et al. (2003). Prothrombotic conditions in an unselected cohort of children with venous thromboembolic disease. *J Thromb Haemost, 1,* 915–921.

Spentzouris, G., Scriven, R. J., Lee, T. K., et al. (2012). Pediatric venous thromboembolism in relation to adults. *J Vasc Surg, 55,* 1785–1793.

van Ommen, C. H., Heijboer, H., van den Dool, E. J., et al. (2003). Pediatric venous thromboembolic disease in one single center: Congenital prothrombotic disorders and the clinical outcome. *J Thromb Haemost, 1,* 2516–2522.

Young, G., Albisetti, M., Bonduel, M., et al. (2008). Impact of inherited thrombophilia on venous thromboembolism in children: A systematic review and meta-analysis of observational studies. *Circulation, 118,* 1373–1382.

CHAPTER 149

Antithrombin Testing

Mikhail Roshal, MD, PhD and Morayma Reyes Gil, MD, PhD

Congenital antithrombin (AT) deficiency occurs in 0.2%–0.02% of the general population. However, in patients who had suffered a venous thromboembolic event (VTE), the prevalence is approximately 1%–3%. Preanalytical variables must be taken into account when deciding on the timing of AT testing and pretest probability, as false-positive and false-negative results may occur if testing is done in the wrong patient or wrong time. Common causes of acquired decline in AT activity include recent thrombosis, trauma, burns, and surgical procedures. Nephrotic syndrome may result in significant renal AT loss. Warfarin may elevate AT levels. Functional testing for AT activity is the dominant testing modality. AT antigen levels may be used to distinguish type I (quantitative) from type II (qualitative) deficiency. Among qualitative deficiencies, heparin-binding site abnormalities may be further distinguished from reactive site (RS) defects by measuring AT activity both in the absence and presence of heparin.

AT (previously called antithrombin III) is a natural anticoagulant, which on activation by heparin, is a potent inactivator of coagulation. The predominant activities of AT are inhibition of the common pathway factors thrombin (factor IIa) and factor Xa. However, AT also has some activity against intrinsic factors (factors IXa, XIa), contact factors (XIIa), and in some cases factor VIIa.

AT functions normally in the absence of administered heparin, presumably by activation by heparan sulfate associated with the vascular endothelium. However, therapeutic administration of heparin greatly enhances this activity. Inherited AT deficiency can be quantitative (type I), in which AT protein is normal but present at decreased levels, or qualitative (type II) deficiencies, in which AT protein amount is normal but dysfunctional. AT deficiency can also be acquired through mechanisms resulting in decreased synthesis (e.g., hepatic disease) or increased loss (e.g., consumptive coagulopathies and nephrotic syndrome). We recommend to screen for AT deficiency by testing only AT activity, and in circumstances where differentiation of the two phenotypes is indicated, then AT antigen test can follow a low AT activity result.

Both thrombosis itself and treatment with certain anticoagulants can substantially alter measureable AT levels. Thus, when possible, AT testing should be performed after a thrombotic event has resolved. Understanding which AT activity assay is performed in your laboratory is important to determine if it is appropriate to measure AT activity while your patient is on anticoagulation therapy. AT testing should not be done when patients are therapeutic on heparin as heparin physiologically reduces AT levels because of increased clearance of AT bound to heparin. AT activity assays that measure residual thrombin activity are affected by direct thrombin inhibitors, whereas AT activity assays that measure residual Xa activity are affected by direct Xa inhibitors. The risk of decreasing anticoagulation to assess AT deficiency must be evaluated on a case-by-case basis based on clinical judgment. In addition, AT levels can be decreased in the neonatal

Transfusion Medicine and Hemostasis. https://doi.org/10.1016/B978-0-12-813726-0.00149-5

period; during pregnancy; in the presence of liver disease, burn injury, trauma, sepsis, disseminated intravascular coagulation, and nephrotic syndrome; and secondary to L-asparaginase and estrogens.

A recent AT resistance, analogue to Factor V Leiden, has been described, whereas mutations in Arg596 in the prothrombin protein render it resistant to AT inactivation and is associated with increased risk of venous thrombosis.

Physiologic Role and Mechanism of Action: AT is the most important and potent physiologic inhibitor of thrombin. AT is essential for postembryonic life, with complete knockout in mice resulting in embryonic lethality due to thrombosis. The protein belongs to the class of serine protease inhibitors (serpins). Serpins are suicide substrates that mimic the natural substrates of the target protease. On partial digestion, these inhibitors form covalent bonds with the target, irreversibly blocking further protease activity by steric hindrance. In addition to thrombin, AT inhibits numerous other proteases involved in the clotting cascade, including factors IXa, Xa, XIa, and tissue factor bound factor VIIa. AT activity is enhanced approximately 1000-fold by heparin-like glucosaminoglycans (GAG). This potentiation requirement provides a mechanism for localizing AT activity to the endothelial surface where heparan sulfate is expressed. A single pentasaccharide moiety of heparin-like GAG is sufficient for potentiation of AT activity toward most proteases other than thrombin. Inactivation of thrombin requires additional interaction of the GAG with thrombin. For this to occur, at least 18 pentasaccharide moieties are necessary.

Congenital Antithrombin Deficiency: Congenital AT deficiency occurs in 0.2%–0.02% of the general population. The deficiency is overrepresented in subjects who have suffered a VTE. The prevalence of AT deficiency in younger patients with unprovoked first VTE is approximately 1%–3%. Heterozygous AT deficiency is a moderately strong risk factor for venous thrombosis and pregnancy complications, with deficient subjects experiencing a 5- to 50-fold increased risk over their relatives without AT deficiency. The peak incidence of thrombotic events in AT deficiency may be earlier than for other congenital risk factors. VTE incidence peaks between ages of 15 and 35 years. Role of AT deficiency in arterial thrombosis is controversial and may be specific mutation-dependent.

Most AT deficiencies are a result of an inheritance of a single defective allele of the *SERPIN1* gene located on chromosome 1. Thus, an autosomal dominant inheritance pattern is the norm. Complete AT deficiency, resulting from inheritance of two defective AT alleles with no protein activity, appears to be incompatible with postembryonic life in humans. Cases of double heterozygosity for defective alleles with a milder phenotype for each have been reported. Deficiencies resulting from such mutations have variable presentation, with a subset of patients having severe thrombotic complications in the neonatal period. Most AT deficiencies (about 80%) are classified as type I (quantitative) defects with declines in AT protein levels and corresponding declines in AT inhibitory activity. The remaining deficiencies are classified as type II (qualitative) defects where critical aspects of the protein function are compromised. In type II deficiencies, a decline of protein activity is out of proportion to the protein levels.

Among type II deficiencies there are three recognized subtypes. Type II RS occurs due to mutations that directly affect AT interaction with the target proteases. Mutations of this type are expected to affect AT inhibitory activity toward its targets regardless of the presence of heparin. Type II heparin-binding site (HBS) mutations affect potentiation of the thrombin activity by heparin. These types of defective AT show normal inhibitory activity when heparin is not present but will show substantial defect when heparin is added to potentiate the reaction. Pleiotropic (PE) subtypes are due to mutations that affect the structure of the protein in the way that compromises both functions. Overall, some HBS mutations show milder clinical phenotype than the RS mutations.

Testing Indications: AT deficiency is quite rare in the general population. Assuming that low end of the reference range excludes ~2.5% of the population and the incidence of AT deficiency is at most 0.2%, false-positive results would far exceed true positives in a general population screening. AT deficiency testing should be reserved for patients with a high pretest probability of a true-positive result. Confirmation of AT deficiency in a first-degree relative of a suspected case greatly improves the confidence in diagnosis. Outside of a diagnosis of congenital AT deficiency, AT measurements may be warranted in patients undergoing very high-dose heparin therapy for extended periods, such as neonates receiving extracorporeal membrane oxygenation. Under those circumstances, the testing is performed to assess the need for AT replacement using AT concentrates.

Preanalytical Considerations: Preanalytical variables must be taken into consideration when deciding on the timing of AT testing, as both false-positive and false-negative results may occur if testing is mistimed. Like other factors involved in coagulation, AT is consumed in thrombosis, burns, trauma, disseminated intravascular coagulation, and postoperatively. AT is reduced by long-term full-dose unfractionated heparin therapy. It is advisable to delay testing after a recent thrombotic event or long-term heparin administration. Liver insufficiency and treatment with L-asparaginase can lead to substantial decreases in AT levels due to decreased liver synthetic function. In addition, renal loss in nephrotic syndrome can lower AT levels. Mild declines have been reported in diabetics and in women taking oral contraceptives or hormone replacement therapy. AT levels may be mildly increased by oral anticoagulant therapy with warfarin.

Testing Overview: Over 200 mutations spread over most of the length of the *SERPINC1* gene encoding for AT can cause AT deficiency. Thus DNA-based detection would require whole-gene sequencing if the mutation site is not known from a prior study. Because of this, DNA-based tests are not commonly performed for diagnosis of AT outside research testing. The situation may change with the rapidly decreasing costs of genetic testing. Functional testing for AT activity is the dominant testing modality at present. AT antigen quantitation may be used to distinguish type I from type II deficiency.

Among qualitative deficiencies, HBS abnormalities may be further distinguished from RS defects by measuring AT activity both in the absence and presence of heparin. Type II HBS deficiencies will show normal activity in a heparin-independent test.

Testing Methodologies

Activity Determination in the Presence of Heparin: Most common laboratory preparations for detection of AT activity measure residual activity of exogenously added thrombin or factor Xa, following a short incubation with the diluted test plasma in the presence of heparin. AT demonstrates concentration-dependent inhibition of the target protease activity. Most modern tests utilize chromogenic substrate readout of the target protease activity. The assay measures release of a chromophore, typically paranitroanilide, from a peptide that resembles either thrombin or factor Xa natural substrate. The intact substrate is colorless, but peptide bond cleavage by the protease leads to release of the chromophore. Readout is typically at a wavelength of 405 nm. Either end point or kinetic activity measurements can be made. Results are compared to calibrated pooled plasma dilution with known amount of AT activity. Calibrators traceable to the World Health Organization (WHO) standard are recommended.

Test Interpretation: In general, 75–85 international units (IU) per deciliter of activity is the low end of normal range for adults. AT activity is about one half of the normal adult range in full-term infants at birth. Normal adult levels are reached by 3–6 months of life. Each laboratory should verify the normal range for their method and institution. Use of age-specific normal ranges in infants is advised.

Sources of Error: Chromogenic tests generally suffer from few analytic interferences. New nonheparin factor Xa inhibitors are expected to interfere with factor Xa–based assays, whereas direct thrombin inhibitors interfere with thrombin-based assays. In those cases, AT activity readout may be falsely elevated. As a matter of clinical practice, the authors' laboratories routinely perform prothrombin time, partial thromboplastin time, and thrombin time testing on samples sent for AT evaluation to pick up potential preanalytic variables and analytic interferences. Nonspecific proteolytic activity in rare plasma samples may result in dramatically decreased values (usually negative or zero AT activity depending on whether the negative values are allowed). Discolored plasma samples may theoretically cause interference, although very few interferences are seen at 405 nm wavelength.

Selection of the test substrate and time of initial incubation has been shown to affect the diagnostic performance of the test, and specifically the ability to detect certain AT defects. Human thrombin–based tests lack complete specificity for AT as heparin cofactor II, a minor heparin-activated thrombin inhibitor may also contribute to heparin-dependent AT activity. Overestimation of AT activity in human thrombin–based assays has been reported. Bovine thrombin cannot be inhibited by heparin cofactor II and is now commonly used in thrombin-based assays. Factor Xa–based assays miss some type II deficiencies, including a relatively common Cambridge II (*A384S*) deficiency and AT Denver (*S426L*). Shorter initial incubation times were shown to improve sensitivity of the assay to some, but not all, AT defects. It appears that no single activity test is sensitive to all known AT defects.

Antigenic Tests: Antigenic tests are performed by a limited number of laboratories. The test is primarily used to distinguish quantitative from qualitative deficiencies. AT antigen can be measured by radial immunodiffusion assay, usually using precalibrated commercial plates or Laurell-type (rocket) assay. In the Laurell-type assay, plasma undergoes electrophoresis, whereas in the radial immune diffusions, it is allowed to diffuse through agarose gel. Both assays rely on monoclonal antibodies embedded in the agarose gel to precipitate the antigen (in this case AT) as the antigen moves through the gel. Under the condition of excess of antigen, the antibody/antigen precipitates are redissolved until such a point when antigen and antibody reach equivalence. AT concentration can then be read from the standard curve where the log of the antigen is plotted against either the length of the precipitin "rocket" in the Laurell-type assay or the diameter of the precipitin circle in radial immune diffusion assay. Other types of immunoassays for AT include latex agglutination and enzyme-linked immunosorbent assay.

Test Interpretation: Results can be reported in either IU/deciliter or milligrams/liter of plasma. Antigenic AT levels are best interpreted in the context of AT activity. In the presence of low AT activity, a decrease of AT antigen to activity ratio would signify a type II deficiency, whereas normal ratio would be suggestive of type I defects. Generally, ratios of less than 0.9 would be considered abnormal. However, each laboratory must establish its own cutoff. Occasionally, type II defects could present with low normal activity and low AT activity to antigen ratios. Screening for such defects is challenging, so some experts advocate performing both AT activity and antigen on all samples. The potential gain in sensitivity, however, should be balanced against the potential of false positives in a typical population with low prevalence of AT deficiency and the significantly increased cost of testing.

Sources of Error: As in most antigenic tests, nonspecific antibody interactions may cause either a falsely high or a falsely low AT reading. Values that are significantly higher or lower than the physiologically probable variation should be treated as suspect for interference. Demonstration of a nonlinear result on plasma dilution will confirm the presence of interference.

Determining the Subtype of Type II Antithrombin Deficiency: Once the diagnosis of type II AT deficiency is established, further classification of the deficiency can be performed. RS and HBS deficiencies can be distinguished by measuring AT activity in the absence of heparin. This test measures the so-called progressive activity of AT because very slow inhibition kinetics are observed. The assay requires much longer incubation times than the usual heparin-dependent assay. Pure type II HBS AT shows normal heparin-independent activity, whereas RS defect shows low activity. Pleiotropic defects show intermediate phenotype. This test is rarely performed outside of specialized reference laboratories. This assay is sensitive to the same potential interferences as the heparin-dependent AT activity determination. In addition, this assay requires the absence of heparin in the plasma.

Further Reading

Cooper, P. C., Coath, F., Daly, M. E., & Makris, M. (2011). The phenotypic and genetic assessment of antithrombin deficiency. *Int J Lab Hematol*, *33*, 227–237.

Girolami, A., Cosi, E., Ferrari, S., Lombardi, A. M., & Girolami, B. (2017). New clotting disorders that cast new light on blood coagulation and may play a role in clinical practice. *J Thromb Thrombolysis*, *44*, 71–75.

Halbmayer, W. M., Weigel, G., Quehenberger, P., et al. (2012). Interference of the new oral anticoagulant dabigatran with frequently used coagulation tests. *Clin Chem Lab Med*, *50*, 1601–1605.

Ignjatovic, V., Kenet, G., & Monagle, P. (2012). Developmental hemostasis: Recommendations for laboratories reporting pediatric samples. *J Thromb Haemost*, *10*, 298–300.

Khor, B., & Van Cott, E. M. (2010). Laboratory tests for antithrombin deficiency. *Am J Hematol*, *85*, 947–950.

Lijfering, W. M., Brouwer, J. L., Veeger, N. J., et al. (2009). Selective testing for thrombophilia in patients with first venous thrombosis: Results from a retrospective family cohort study on absolute thrombotic risk for currently known thrombophilic defects in 2479 relatives. *Blood*, *113*, 5314–5322.

Muszbek, L., Bereczky, Z., Kovács, B., & Komáromi, I. (2010). Antithrombin deficiency and its laboratory diagnosis. *Clin Chem Lab Med*, *48*, S67–S78.

CHAPTER 150

Protein C Deficiency Evaluation

Mikhail Roshal, MD, PhD and Morayma Reyes Gil, MD, PhD

Protein C (PC) is a vitamin K–dependent, heterodimeric, plasma glycoprotein that is synthesized as a zymogen in the liver. It is activated on the endothelial surface by the thrombin–thrombomodulin (TM) complex. Once produced, the activated protein C (APC) cleaves and inactivates factors Va and VIIIa. Protein S (PS) cofactor activity is required for optimal proteolytic activity of APC in vivo. PC deficiency is rare in healthy populations, with reported incidence of 0.2%–0.4% of unselected subjects. Most commonly it is transmitted as an autosomal dominant trait. It can be demonstrated in ~3% of younger patients with venous thrombosis. Diagnosis of congenital PC deficiency is usually made by demonstrating reduced PC activity on at least two separate occasions and exclusion of the numerous acquired causes that can also lead to reduced PC activity. At present, amydolic (chromogenic) PC activity test is usually performed first in the context of screening for PC deficiency. Antigenic PC test may be performed to distinguish quantitative (type I) from qualitative (type II) deficiencies.

PC is a vitamin K–dependent, heterodimeric, plasma glycoprotein that is synthesized as a zymogen in the liver. PC activation requires a complex of an endothelial thrombin receptor TM with thrombin. TM binding induces a confirmation change in thrombin. This change leads to suppression of thrombin procoagulant activity and expression of both anticoagulant and antifibrinolytic activity. PC is recruited to the endothelial surface by the endothelial protein C receptor (EPCR). Thrombin–TM complex cleaves the heavy chain of the EPCR-bound PC leading to formation of proteolytically active APC. Once produced, APC cleaves and inactivates factor (F) Va and FVIIIa, leading to a dampening of the clotting cascade. Inactivation of both factors by APC requires cofactor PS. PS serves an important function of directing APC to the phospholipid surface. Additionally, inactivation of FVIIIa requires the presence of intact FV in the multimolecular complex with APC, PS, and FVIIIa. The APC-containing complex cleaves phospholipid bound FVa sequentially at Arginine 506 (site of the mutation in FV Leiden), Arginine 306, and Arginine 679. Arginine 506 proteolysis is kinetically favored and leads to partial inactivation of FVa. The subsequent cleavage at Arginine 306 leads to the complete inactivation of the catalytic activity of FVa. The significance of the Arginine 679 proteolysis site is not entirely clear. APC complex also cleaves FVIIIa at three Arginine residues (336, 562, and 740) leading to the complete loss of FVIIIa cofactor activity toward FIXa.

Epidemiology and Clinical Presentation: PC deficiency is rare in healthy populations, with a reported incidence of 0.2%–0.4% in unselected subjects. PC deficiency is more frequent in thrombophilic patients. Most commonly it is transmitted as an autosomal dominant trait. The inheritance of a single defective allele of the gene encoding PC (*PROC*) is a moderate risk factor for thrombosis. It can be demonstrated in ~3% of patients with venous thrombosis. Epidemiologic studies suggest that subjects

Transfusion Medicine and Hemostasis. https://doi.org/10.1016/B978-0-12-813726-0.00150-1

with a defective allele have approximately a sevenfold increased risk of venous thrombosis compared with their nondeficient relatives in symptomatic families. Additional thrombophilia-associated mutations, such as FV Leiden and prothrombin *G20210A* mutation, appear to substantially increase thrombotic risk in PC-deficient subjects in a more than additive fashion. Overall, 30%–50% of PC-deficient subjects within symptomatic kindreds develop venous thromboembolism by the age of 50. Association with arterial thrombosis is controversial, and if present, appears weak. Coinheritance of two defective PC alleles results in a severe thrombotic disorder in the early neonatal period, usually within hours of birth. Infants with severe PC deficiency usually present with thrombotic purpura (purpura fulminants [PF]) and disseminated intravascular coagulation (DIC). Delayed presentations of PF and DIC have been reported in adolescent and young adult patients with moderately severe PC deficiency (PC levels between 1% and 20% per mL). PC has a relatively short plasma half-life and its levels drop early in the onset of oral anticoagulant therapy with warfarin. Heterozygous PC-deficient patients who are not otherwise anticoagulated may experience warfarin-induced skin necrosis because of rapid drop of PC levels below 20% per mL in settings when coagulation factors other than FVII are not yet substantially decreased.

Testing Recommendations: Routine screening testing for PC deficiency is not recommended. This is due to the low prevalence of the deficiency in the general population, which gives an increased probability of false-positive results. General thrombophilia testing guidelines are provided in the introductory chapter to this section and are applicable to investigation of PC deficiency. However, testing is appropriate for family members of a known PC deficiency case.

Diagnosis of PC deficiency requires two decreased values significantly in separate occasions. In addition, a finding of decreased PC levels in a first-degree relative is helpful for diagnosis of congenital deficiency. A single normal value of PC activity outside the borderline range is usually sufficient to rule out PC deficiency, and follow-up testing is usually not necessary for a normal result.

Protein C Deficiency Testing Overview: The diagnosis is usually made through determination of plasma PC activity. The type of PC deficiency can be determined through follow-up antigenic PC level testing. Approximately three quarters of deficiency-linked mutations are associated with quantitative (type I) deficiencies, resulting in proportionate decrease in both PC antigenic plasma levels and activity. The remainder of mutations results in qualitative (type II) deficiency resulting from disproportionate decline of PC activity compared with antigenic PC levels. Types of deficiency and the expected test results are summarized in Table 150.1.

TABLE 150.1 Protein C (PC) Deficiency Types

	Amydolic PC Activity	Clot-Based PC Activity	PC Antigen
Type I	Low	Low	Low
Type IIa	Low	Low	Normal
Type IIb (rare)	Normal	Low	Normal

Over 160 different mutations in the *PROC* gene, which encodes PC, have been linked to PC deficiency. In addition, genome-wide association studies have demonstrated a number of loci outside *PROC* that influence PC levels in human subjects. The association between lower PC levels due to loci outside *PROC* and thrombosis has not been established. Because of the large variety of mutations linked to PC deficiency, DNA-based testing is currently reserved for a low proportion of cases.

Diagnostic Guidelines

Timing of the Test: The timing of the PC test relative to acquired conditions may have a significant impact on its diagnostic performance and the need for repeat testing. In general, testing for PC deficiency should not be performed when the patient is vitamin K–deficient or has been taking warfarin within the past 10 and preferably 30 days. Numerous additional factors, including liver disease, active thrombosis, or L-asparaginase therapy, may lower PC levels. Testing in the immediate aftermath of a thrombotic event and before initiation of warfarin therapy can be used to rule out, but not to rule, in PC deficiency. A low value at a single time point should not be overinterpreted as an indication of PC deficiency, as multiple physiologic, pathologic, and iatrogenic conditions have a strong negative effect on PC activity levels in the plasma. Some, but not all, studies have reported mild increases in PC activity in nephrotic syndrome, ischemic heart disease, and in women who are pregnant, using oral contraceptives or hormone replacement therapy. It may be prudent to repeat borderline normal results in patients falling in this category to avoid false-negative results. For a summary of preanalytical considerations please refer to Table 150.2.

Test Selection: Determination of PC activity can be performed via either clot-based or chromogenic tests. Clot-based tests are subject to a large number of potential interferences and therefore are less accurate for diagnosis. Recent recommendation from the College of American Pathologists suggests PC activity via chromogenic test as the primary screen for PC deficiency.

Antigenic testing for PC is not sensitive to type II PC deficiency and is therefore inappropriate for initial screening. It may be performed to distinguish between type I and type II deficiency.

TABLE 150.2 Preanalytical Variables That Can Influence Protein C Activity Levels

Factors reported to decrease protein C (PC) activity	Warfarin*, vitamin K deficiency*, liver disease active thrombosis, disseminated intravascular coagulation, recent surgical procedure, L-asparaginase therapy, breast cancer therapy, acute respiratory distress syndrome, PC inhibitor (rare)
Factors reported to increase PC activity	Nephrotic syndrome†, ischemic heart disease, myocardial infarction, pregnancy†, oral contraceptives,† hormone replacement therapy†

*Early vitamin K deficiency and warfarin therapy have a greater effect on PC clottable assays then on chromogenic and antigenic assays.

†The effect of these factors has not been consistent in different studies and appears mild.

Protein C Testing Methodologies

Clottable Protein C: Clottable PC assay tests the ability of APC to prolong clotting time in either activated partial thromboplastin time (PTT) or Russell's viper venom (RVV) clotting time format. In a typical assay, PC in the patient's plasma is first converted to APC by the venom of the *Agkistrodon contortrix* snake (Protac). The prolongation of clotting time due to APC activity is then measured relative to PC-deficient plasma in the presence of phospholipid and calcium.

In PTT format the prolongation of the clotting time depends on the ability of APC to inactivate both FVa and FVIIIa. On the other hand, only APC activity against FVa is measured in RVV format. This is because RVV directly activates FX and bypasses the requirement for FVIIIa for clot formation. Some test manufacturers also supply excess bovine FV in their test preparations to minimize interference from relatively common FV Leiden. PC activity is reported based on the standard curve that uses dilutions of plasma with known PC activity as a calibrator.

Chromogenic Protein C Activity: In a chromogenic assay, PC is also usually activated by Protac. The assay measures the release of a chromophore, typically paranitroanilide, from a peptide that resembles APC's natural substrate. The intact substrate is colorless, but peptide bond cleavage by the protease leads to release of the chromophore. Readout is typically at a wavelength of 405 nm. Results are compared with calibrated pooled plasma dilution with known amount of APC activity. Because the synthetic substrate is not localized to the phospholipid surface and is structurally less complex than either FVa or FVIIIa, neither PS nor additional phospholipid plays a role in the assay.

Protein C Antigen: PC antigen is typically measured using monoclonal antibodies in enzyme-linked immunosorbent assay (ELISA) or radial immunodiffusion assays. ELISA assays are sensitive down to less than 5% PC concentrations, whereas radial immunodiffusion shows less sensitivity at the low end. Pooled plasma with known PC concentration is used to calibrate the assays.

DNA-Based Tests: *PCOS1* spans nine exons with a total length of over 11 kb. Because over 160 mutations have been identified, which span almost its entire length, sequencing of the entire gene coding region may be required for the initial genetic diagnosis. Currently, such testing is rarely performed in clinical settings. In the future, molecular testing may become increasingly available, as newer technology drives down the cost of DNA-based tests.

Limitations of Protein C Testing

Clot-Based Protein C Activity Determination: Clottable PC assays suffer from a variety of interferences. APC resistance due to FV Leiden variably affects the assays with some assays sensitive to homozygous, but not heterozygous FV Leiden. The interference may produce a falsely low result. Direct thrombin inhibitors and lupus inhibitor can cause false elevations of PC activity. Although no published data exist, new oral anti-Xa agents are likely to affect all versions of the clottable assay to some degree.

Many versions of the assay contain heparin inactivator and are usually not affected by therapeutic heparin concentrations (under 1–2 units depending on the assay) but will produce artifactually increased values at higher doses. Package inserts for individual assays can be consulted for details on sensitivity to heparin. It may be prudent to exclude heparin and other thrombin inhibitor interference by performing thrombin time before PC testing. In addition, PTT-based assays tend to suffer substantial interferences when very high levels of FVIII are present in the patient's plasma. The substantially elevated FVIII levels may lead to substantial underestimation of PC activity. RVV-based tests do not demonstrate interferences from high FVIII levels and may have reduced sensitivity to lupus inhibitors. When interference from lupus inhibitors or high FVIII is present, additional dilutions decrease the interference.

Chromogenic Protein C Activity Determination: Because the APC activity is measured directly using a synthetic PC substrate, the potential for analytic interference is greatly reduced in this format. In particular, lupus inhibitors, high FVIII levels, and FV Leiden have no effect. Chromogenic tests are limited to measuring the ability of PC to cleave a synthetic substrate that does not fully recapture all features of the natural substrate. This makes the assay insensitive to rare type II (termed type IIb) mutations that do not affect the active site but alter interactions with cofactors, phospholipids, or natural substrates. In practice, chromogenic tests are sensitive to all type I PC deficiencies and a vast majority of type II deficiencies. Clot-based tests may be performed to pick up an additional rare subset of type II mutations in special circumstances. Increased precision and reduced potential interferences in the chromogenic tests appear to justify the choice of this testing methodology over clot-based tests, despite some loss of sensitivity under typical laboratory circumstances. Chromogenic PC assays may be less sensitive to vitamin K deficiency and oral anticoagulants, allowing exclusion of PC deficiency in a subset of patients early in the course of oral anticoagulants. Chromogenic PC assays may theoretically give inaccurate results in either discolored plasma or with exogenous factors breaking down the chromogenic substrate.

Protein C Antigen: Nonspecific antibodies present in the plasma may interfere with the assay. Thus, values outside the usual range of PC variation should be interpreted with caution.

Test Interpretation: In general, patients with heterozygous PC deficiency show PC activity levels between 30% and 60%. Up to 15% of heterozygous-deficient patients may have PC levels within the normal range, generally >70%. Borderline values must be interpreted with caution, as ~15% of heterozygous-deficient samples may have a value within the reference interval. When testing within a general population, a normal PC value has a very high negative predictive value and no follow-up is needed. However, a borderline low normal value in a patient with thrombosis may need to be measured again because of the relatively high prevalence of PC deficiency within that group.

Because acquired PC deficiency due to vitamin K deficiency, warfarin therapy, and pathologic and iatrogenic causes is a lot more common then congenital deficiency, acquired PC deficiency must be ruled out before a diagnosis of congenital PC deficiency is made. Correlation with clinical history, medications, prothrombin time (PT), and/or FVII level may be needed. PC activity drops relatively early in the course of warfarin therapy and may be decreased out of proportion to PS and PT/international normalized ratio early in therapy. A low value must be confirmed at a separate time point if a congenital deficiency is suspected.

A case of acquired PC inhibitor leading to catastrophic thrombosis has been reported. If extremely low levels of PC activity are detected in a previously normal individual inhibitor workup, including PC, mixing study should be considered to rule out this rare, but apparently very serious complication.

PC levels are low in newborns (in the range of 17%–53%). They increase rapidly in the neonatal period but do not reach adult levels until 16–18 years of age. Age-specific ranges for infants and children should be used to avoid false diagnoses of deficiency.

Further Reading

Dahlbäck, B. (1997). Factor V and protein S as cofactors to activated protein C. *Haematologica, 82*, 91–95.

Danese, S., Vetrano, S., Zhang, L., Poplis, V. A., & Castellino, F. J. (2010). The protein C pathway in tissue inflammation and injury: Pathogenic role and therapeutic implications. *Blood, 115*, 1121.

Griffin, J. H., Zlokovic, B. V., & Mosnier, L. O. (2012). Protein C anticoagulant and cytoprotective pathways. *Int J Hematol, 95*, 333–455.

Khor, B., & Van Cott, E. M. (2010). Laboratory tests for protein C deficiency. *Am J Hematol, 85*, 440–442.

Kottke-Marchant, K., & Comp, P. (2001). Laboratory issues in diagnosing abnormalities of protein C, thrombomodulin, and endothelial cell protein C receptor. In *College of American Pathologists Consensus Conference XXXVI: Diagnostic Issues in Thrombophilia. Atlanta: Georgia Conference Synopsis.*

Mitchell, C. A., Rowell, J. A., Hau, L., et al. (1987). A fatal thrombotic disorder associated with an acquired inhibitor of protein C. *N Engl J Med, 317*, 1638–1642.

Reitsma, P. H., Bernardi, F., Doig, R. G., et al. (1995). Protein C deficiency: A database of mutations, 1995 update. On behalf of the subcommittee on plasma coagulation inhibitors of the scientific and standardization committee of the ISTH. *Thromb Haemost, 73*, 876–889.

Simioni, P., Tormene, D., Spiezia, L., et al. (2006). Inherited thrombophilia and venous thromboembolism. *Semin Thromb Hemost, 32*, 700–708.

CHAPTER 151

Protein S Deficiency Evaluation

Mikhail Roshal, MD, PhD and Morayma Reyes Gil, MD, PhD

Protein S (PS) is a vitamin K–dependent plasma glycoprotein that serves as a cofactor of activated protein C (APC). When bound to APC, it accelerates protein C–dependent degradation of factor (F) Va and FVIIIa. PS deficiency is inherited in an autosomal dominant fashion and has highly variable penetrance. It is a moderately strong risk factor for venous thrombosis, with a reported adjusted lifetime relative risk increase of 5- to 32-fold in PS-deficient subjects compared with relatives without the deficiency. Congenital PS deficiency is subdivided into three types. Type I is a quantitative deficiency with resulting low levels of total PS, PS activity, and PS free fraction. It is the most frequent of the subtypes and is a result of over 130 different *PROS1* (gene encoding PS) mutations. Type II qualitative deficiency is rare (<5% of all PS-deficient patients) and presents with a disproportionate decrease in APC-dependent PS activity relative to PS free fraction and total antigen. Type III deficiency demonstrates normal total PS, with low free PS and activity values. Screening for PS deficiency is done either by activity or by antigenic tests for PS free antigen.

Physiologic Role: PS is a vitamin K–dependent plasma glycoprotein that serves as a cofactor of APC. When bound to APC it accelerates protein C–dependent degradation of FVa and FVIIIa. In addition, PS may have a protein C–independent antithrombotic activity by serving as a cofactor for tissue factor pathway inhibitor that inactivates FVIIa. Like other vitamin K–dependent coagulation cofactors, it is synthesized primarily in the liver. Approximately 60%–70% of plasma PS is bound to complement factor 4b (C4b) binding protein (C4BP). The remaining PS is not bound and is thus referred to as the PS free fraction. Previously only the free fraction of PS was considered active in anticoagulation, but it has been recently reported that the C4BP bound fraction also has some anticoagulant activity. The significance of the APC-independent activity of PS and that of the C4b-bound PS fraction is a focus of investigation. Current clinical assays measure only APC-dependent PS activity and use the PS free fraction to estimate the concentration of active PS.

Nonhemostatic Function of Protein S: PS has innate immunity functions due to its interaction with C4BP or directly given its capacity to bind negatively charged phospholipids on the surface of apoptotic cells or activated cells. PS serves as a bridge for phagocytic macrophages and potentiates clearance of apoptotic cells. This pathway is so important that bacteria such as group A streptococci have developed toxins (streptococcal pyrogenic exotoxin B) that inhibit PS-mediated phagocytosis. PS also binds the nascent C5, C6, C7 complement complex and prevents inappropriate complement activation and decreases systemic inflammation.

Epidemiology: PS deficiency is inherited in an autosomal dominant fashion and has highly variable penetrance. It is a moderately strong risk factor for venous thrombosis,

Transfusion Medicine and Hemostasis. https://doi.org/10.1016/B978-0-12-813726-0.00151-3

with a reported adjusted lifetime relative risk increase of about sevenfold in PS-deficient subjects compared with relatives without the deficiency. The first episode of venous thromboembolism (VTE) in patients with PS deficiency generally occurs between the ages of 10 and 50. Heterozygous PS-deficient patients have ~50% cumulative risk of having a VTE episode before age 50. A role of PS deficiency in arterial thrombosis is not well established with some, but not all studies reporting a mildly increased risk. PS deficiency is rare in the normal Caucasian population, with recent estimates of 0.03%–0.17%. East Asians have a substantially higher prevalence of PS deficiency, with 1%–2% being reported in Japanese healthy donors. PS deficiency is reported in 1%–13% of Caucasian patients with VTE depending on selection criteria and test modality, and as high as 22%–36% of East Asian VTE patients.

Homozygous or compound heterozygous PS deficiency is very rare and often causes catastrophic thrombosis in newborns (called Purpura Fulminants), as well as disseminated intravascular coagulation. It requires lifelong replacement of PS and anticoagulation therapy.

Test Recommendations: Routine screening testing for PS deficiency is not recommended, due to its low prevalence in the general population, leading to the increased probability of false-positive results. General thrombophilia testing guidelines are provided elsewhere in the text (see Chapter 147). In addition, testing is appropriate in family members of a known PS deficiency patient. Testing to rule in diagnosis of congenital deficiency should not be performed when the patient is pregnant or within 3 months postpartum, taking oral contraceptives, or has been taking warfarin within the past 30 days. Testing in the immediate aftermath of a thrombotic event and before initiation of warfarin can be used to rule out, but not to rule in, PS deficiency. Because of the multiple physiologic, pathologic, and iatrogenic conditions having a significant effect on PS free and activity levels in the plasma, a low PS value at a single time point should not be overinterpreted as an indication of congenital PS deficiency. Thus, the diagnosis of PS deficiency requires two decreased functional test values, significantly separated in time. In addition, a finding of decreased PS levels in a first-degree relative is helpful for diagnosis of congenital deficiency. A single normal value of PS free antigen or activity test outside of borderline range is usually sufficient to rule out PS deficiency, and follow-up testing is not usually necessary. Testing to assess a risk of arterial thrombotic events is currently not recommended.

Types of Protein S Deficiency: Congenital PS deficiency is subdivided into three types (Table 151.1). Type I is a quantitative deficiency with resulting low levels of total PS, PS activity, and PS free fraction. It is the most frequent of the subtypes and is a result of over 130 different *PROS1* mutations. Type II qualitative deficiency is rare (<5% of all PS-deficient patients) and presents with a disproportionate decrease in APC-dependent PS activity relative to PS free fraction and total antigen. Type III deficiency demonstrates normal total PS with low free PS and activity values. Type III PS is a very heterogeneous entity. A large subset of type III deficiencies does not have a clear link to mutations in *PROS1* and may be the result of other genetic or epigenetic factors. Another subset cosegregates with type I deficiency within the affected families. Conversions between a type I and type III PS phenotype have been documented with

TABLE 151.1 Protein S (PS) Deficiency Types

	PS Total	PS Free	PS Activity
Type I	Low	Low	Low
Type II	Normal	Normal	Low
Type III	Normal	Low	Low

aging in women and are related to increased PS protein levels as women age. Still, a few mutations in *PROS1* have been shown to lead to a pure type III phenotype.

Diagnostic Guidelines: Screening for PS deficiency is usually performed via either determination of PS activity or free fraction of PS in the citrated plasma. A recent recommendation from the College of American Pathologists (CAP) recommends PS free antigenic test or PS activity tests as primary screening modalities. Although PS activity is sensitive to a small subset of additional deficiencies (type II), which would be missed by PS free test, it suffers from numerous interferences and thus should be used with caution as a primary screening modality. It is recommended that abnormal PS activity results should be confirmed with PS free antigen testing. In the authors' opinion, PS free antigen should be used as a primary screening tool to increase the overall diagnostic yield. PS free antigen should definitely be used as the primary screen in patients with lupus anticoagulant. Activity determination in addition to PS free antigen may be considered in a subset of cases with high pretest probability of PS deficiency. Antigenic measure of total PS is not recommended for routine screening but can be used to distinguish types of PS deficiency. Utilizing PS total antigen as a primary screening method would miss a clinically significant type III deficiency. It is rarely indicated to test C4bBP levels by immunoassay to determine the type of PS deficiency.

Testing Modalities

Protein S Activity: PS activity is measured by the ability of the patient's plasma to enhance APC anticlotting activity via inactivation of FVa, in some versions of the assay, and FVIII, in PS-depleted plasma. Multiple variants of this test are marketed for clinical use. Because APC is not present in sufficient amounts in plasma samples, it is either added exogenously or is generated in the sample with snake venom Protac. Under these circumstances, PS becomes a limiting factor for APC activity. Prolongation of the clotting time then depends on the amount of active PS from the patient plasma. The activity is reported based on the standard curve that uses dilutions of plasma with known PS activity as a calibrator. Clotting reaction may be initiated with either prothrombin time (PT) reagents, activated partial thromboplastin time (PTT) reagents, or via FXa that is either added exogenously or is generated using Russell viper venom. Some versions of the assay supply excess FV to overcome effects of FV Leiden.

Analytical Interferences of Protein S Activity Testing: Many analytical interferences have been reported that can produce inaccurate PS activity results. Because many versions of the assay are currently on the market, some interfering substances can have a greater effect in some versions of the assay than in others. Direct thrombin inhibitors are expected to cause falsely elevated results in most PS activity assays. Although the data are sparse, new oral anti-Xa agents are likely to affect all versions of the clottable assay to some degree. Presence of FV Leiden can cause underestimation of PS activity levels in many versions of the assay. Lupus anticoagulants are heterogeneous and mostly cause elevated PS activity readout. Many versions of the assay contain heparin inactivator and are usually not affected by therapeutic heparin concentrations (under 1–2 units depending on the assay) but will produce artifactually increased values at higher heparin doses. It may be prudent to exclude heparin by performing thrombin time before PS testing. In addition, PTT-based assays tend to suffer from substantial interferences from elevations of FVIII in the patients' plasma that lead to substantial underestimation of PS activity. Recombinant FVIIa therapy will substantially affect PT-based tests. In addition, low factor levels particularly those of FV, FX, and PC may cause spurious results when not added exogenously.

Protein S Free Antigen: Modern PS free antigen assays rely on monoclonal antibodies that can recognize only the unbound form of PS. Two major formats are currently commercially available. Enzyme-linked immunosorbent assay (ELISA) formats were first to be introduced. Automated turbinometric assays, also known as latex immunoassay (LIA), utilize latex beads coated with the antibody to capture free PS in solution. The presence of PS causes aggregation of the beads in amounts proportional to the amount of antigen that is present. Automated LIA tests provide substantial labor and time savings but suffer from additional interferences. In the recent CAP and ECAT (External quality Control of diagnostic Assays and Tests) proficiency testing and quality control surveys, both assay formats produced acceptable results, but LIA tests showed substantially increased imprecision compared with ELISA.

Protein S Total Antigen: LIA and ELISA tests using antibodies recognizing all forms of PS are available and suffer from similar limitations as PS free antigen. Overall PS total assay shows less interlaboratory variation compared with PS free determination.

Analytical Interferences of Antigen Testing: Antigenic tests are not subject to analytical interferences that affect the clotting assays. Falsely increased values may be seen in LIA test in association with rheumatoid factor and other nonspecific antibodies are used. As in most laboratory tests, unexpected results, or results falling outside the usual variation of expected analyte levels, should prompt additional investigation. Platelet contamination may cause spurious results because of increased plasma turbidity. Discoloration of the plasma for any reason may also affect the test depending on the wavelength used for readout.

C4b-Binding Protein: C4BP testing can be done by either radial immune diffusion or ELISA.

Molecular Testing: *PROS1* spans 15 exons with a total length of over 80 kb. In the recent years, over 200 mutations in *PROS1*, most of them point mutations, have been linked to PS deficiency. These span almost the entire length of the gene. In addition, large deletions and insertions in *PROS1* have been demonstrated. Such a variety of genetic hits would necessitate sequencing of the entire gene coding region and performing deletion analysis for maximum sensitivity if the genetic alteration was not known from a prior study. Moreover, many familial PS type III deficiencies are not linked to *PROS1* genetic mutations and may have complex genetic causes. Genetic testing of *PROS1* is therefore rarely performed in a clinical setting. This situation may change in the future as genetic testing evolves and becomes less costly.

Test Interpretation: PS levels are low in newborns and reach the adult range by 6 months to 1 year of age. The fraction of free PS is increased in newborns, which may partially compensate for decreased total PS fraction. There is a substantial difference in PS levels between healthy men and women of reproductive age. In women, PS levels are also influenced by hormone replacement therapy and oral contraceptives. Values during pregnancy and postpartum are decreased compared with nonpregnant women. In addition, while PS levels stay relatively constant with age in men, levels rise in women. The use of gender and age specific reference ranges is thus appropriate. Borderline values should not be overinterpreted because of the existence of overlap between normal and heterozygous-deficient PS levels, and because under the usual testing scenarios most decreased PS free and PS free antigen values are due to acquired rather than hereditary conditions. If PS activity is used as a primary screening test, potential analytical interferences should be ruled out. A low value on either PS activity test or PS free antigen test must be confirmed at a separate time point if a congenital deficiency is suspected.

Further Reading

Castoldi, E., & Hackeng, T. M. (2008). Regulation of coagulation by protein S. *Curr Opin Hematol, 15*, 529–536.

Cunningham, M. T., Olson, J. D., Chandler, W. L., et al. (2011). External quality assurance of antithrombin, protein C, and protein S assays: Results of the College of American Pathologists proficiency testing program in thrombophilia. *Arch Pathol Lab Med, 135*, 227–232.

Goodwin, A. J., Rosendaal, F. R., Kottke-Marchant, K., & Bovill, E. G. (2001). A review of the technical, diagnostic, and epidemiological considerations for protein S assays. In *College of American Pathologists Consensus Conference XXXVI: Diagnostic Issues in Thrombophilia. Atlanta, Georgia: Conference Synopsis.*

Marlar, R. A., & Gausman, J. N. (2017). Assessment of hereditary thrombophilia: Performance of protein S (PS) testing. *Methods Mol Biol, 1646*, 153–160.

Mulder, R., ten Kate, M. K., Kluin-Nelemans, H. C., & Mulder, A. B. (2010). Low cut-off values increase diagnostic performance of protein S assays. *Thromb Haemost, 104*, 618–625.

Nomura, T., Suehisa, E., Kawasaki, T., & Okada, A. (2000). Frequency of protein S deficiency in general Japanese population. *Thromb Res, 100*, 367–371.

Seligsohn, U., & Lubetsky, A. (2001). Genetic susceptibility to venous thrombosis. *N Engl J Med, 344*, 1222–1231.

ten Kate, M. K., & van der Meer, J. (2008). Protein S deficiency: A clinical perspective. *Haemophilia, 14*, 1222–1228.

CHAPTER 152

Testing for Activated Protein C Resistance

Anne M. Winkler, MD

Activated protein C (APC) resistance has a reported prevalence of 10%–15% and is a common cause of thrombophilia. The majority of APC resistance is directly attributable to heritable mutations in coagulation factor V (FV), primarily the FV Leiden (FVL) R506Q mutation, which results in decreased efficiency of FVa inactivation by APC due to alteration of one of the APC cleavage sites. However, other less common mutations in FV have also been associated with in vitro APC resistance and have variable associations with thrombosis. APC resistance that is independent of FVL has also been reported and can be attributable to increased levels of factor VIII or prothrombin, presence of a lupus anticoagulant, oral contraceptive use, or pregnancy.

Activated Protein C Resistance Assays: APC resistance assays are fluid-phase functional assays that measure the ability of protein C to inactivate FVa and FVIIIa. The primary indication for this assay is the clinical assessment of thrombophilia.

Methods: Most commercially available assays for the detection of APC resistance utilize activated partial thromboplastin time (APTT)–based or prothrombin time (PT)–based assays for detection. However, Russell viper venom–based assays with or without the addition of Noscarin, a snake venom that activates prothrombin in a FVa-dependent manner, assays utilizing the endogenous thrombin potential, and an assay utilizing a diluted plasma supplemented with APC and nonactivated clotting factors triggered with purified Xa, phospholipids, and calcium have also been employed for this purpose. Regardless of the activator, the principle of these assays involves measuring a clotting time; i.e., APTT before and after the addition of APC and calcium to a plasma sample. Theoretically, the addition of APC to a normal plasma sample should prolong the clotting time (e.g., APTT) because of inactivation of FVa and FVIIIa, whereas in an APC-resistant individual the degree of prolongation of the clotting time (e.g., APTT) should be decreased. As a result, a calculation known as the APC sensitivity ratio is performed:

$$\text{APC sensitivity ratio} = \frac{\text{APTT in the presence of APC}}{\text{APTT in the absence of APC}}$$

A resultant ratio less than or equal to the cutoff value is consistent with resistance to APC. For instance, it has been reported that most patients with ratios less than 0.71 using an APTT-based assay are heterozygous or homozygous for FVL.

Test Interpretation: A positive APC resistance assay may indicate FVL and should lead to direct molecular testing in most cases. It is essential to remember that there are mutations other than FVL that can result in plasma APC resistance (Table 152.1). FV Liverpool and *R485K* cause a positive APC resistance result and may contribute to

Transfusion Medicine and Hemostasis. https://doi.org/10.1016/B978-0-12-813726-0.00152-5

TABLE 152.1 Variant Factor V Mutations

Variant	Nucleotide Variation	Amino Acid Substitution	Activated Protein C Resistance	Increased Risk of Thrombosis
FV Cambridge	G1091C	R306T	Yes	No
FV Hong Kong	A1090G	R306G	No	No
FV Liverpool	T1250C	I359T	Yes	Yes
R485K mutation	G1628A	R485K	Yes	Yes
R2 haplotype	T1328C	M385T	Yes	Unknown
	A4070G	H1299R		
	A5380G	M1736V		
	A6755G	D2194G		
A/G allele	A2391G	S739S	Yes	Unknown
	A2663G	K830R		
	A2684G	H837R		
	A2863G	K897D		

thrombosis. In contrast, FV Cambridge is associated with APC resistance but does not appear to increase the risk of thrombosis. Additional mutations associated with APC resistance have unknown clinical effects. Thus, a positive APC resistance assay in the absence of the FVL mutation can still be an independent risk factor for thrombosis. In addition, there are other acquired causes of APC resistance, such as oral contraceptive use, hormone replacement therapy, pregnancy, cancer, and the antiphospholipid syndrome that are associated with an increased risk of a thrombotic event.

A negative APC resistance assay has been used to exclude the presence of FVL. False negatives are very rare, especially in second-generation assays.

Test Performance: In first-generation assays, an additional normalization step was required; however, the addition of FV-deficient plasma to normalize other factor levels in current assay systems (second-generation assays) has abrogated this requirement. Current-generation assays have a sensitivity approaching 100% and a variable clinical specificity depending on the method used.

Sources of Error

False Positives: Use of FV-deficient plasma has substantially reduced interference from non-FV factor deficiencies and use of oral anticoagulants such as warfarin that were previously the causes of false-positive results. However, false positives can and still do occur in the second-generation assays because of baseline alterations of the APTT resulting from the presence of a lupus anticoagulant, other factor inhibitors, FV deficiency, increased FVIII levels, or as a result of other anticoagulants such as direct thrombin and FXa inhibitors. Prolongation of the APTT due to heparin can be avoided by the use of heparin-neutralizing agents, which is incorporated into most commercial assays, and further dilution with normal plasma or addition of exogenous phospholipids can be utilized to neutralize lupus anticoagulants. However, any result consistent with APC resistance should be confirmed with a direct test for FVL. Despite their limitations, APC resistance assays are cost-effective when used in an algorithm with genetic testing and are also able to detect acquired and rare genetic causes of APC resistance other than FVL.

Further Reading

Amiral, J., Vissac, A. M., & Seghatchian, J. (2017). Laboratory assessment of activated protein C resistance/factor V-Leiden and performance characteristics of a new quantitative assay. *Transfus Apher Sci, 56*, 906–913.

Botero, J. P., Majerus, J. A., Strege, A. K., et al. (2017). Diagnostic testing approaches for activated protein C resistance and factor V Leiden. *Am J Clin Pathol, 147*, 604–610.

Castoldi, E., & Rosing, J. (2010). APC resistance: Biological basis and acquired influences. *J Thromb Haemost, 8*, 445–453.

Herskovits, A. Z., Lemire, S. J., Longtine, J., & Dorfman, D. M. (2008). Comparison of Russell viper venom-based and activated partial thromboplastin time-based screening assays for resistance to activated protein C. *Am J Clin Pathol, 130*, 796–804.

Kadauke, S., Khor, B., & Van Cott, E. M. (2014). Activated protein C resistance testing for factor V Leiden. *Am J Hematol, 89*, 1147–1150.

Mohammed, S., & Favaloro, E. J. (2017). Laboratory testing for activated protein C resistance (APCR). *Methods Mol Biol, 1646*, 37–143.

Saenz, A. J., Johnson, N. V., & Van Cott, E. M. (2011). Activated protein C resistance caused by lupus anticoagulants. *Am J Clin Pathol, 136*, 344–349.

Van Cott, E. M., Khor, B., & Zehnder, J. L. (2016). Factor V Leiden. *Am J Hematol, 91*, 46–49.

CHAPTER 153

Molecular Testing for Factor V Leiden and Prothrombin Gene Mutations in Inherited Thrombophilia

Hanna Rennert, PhD and Robert A. DeSimone, MD

Factor V Leiden: Factor V Leiden (FVL) refers to the specific G-to-A transition at nucleotide 1601 (c.1601G>A) in exon 10 of the factor V gene, which results in arginine to glutamine substitution at amino acid 534 of the protein (p.Arg534Gln), previously known as p.Arg506Gln. Activated factor V serves as a cofactor for the conversion of prothrombin to the active enzyme thrombin. Activated protein C (APC) is a natural anticoagulant, which cleaves and inactivates activated factor V. The FVL mutation abolishes one of the three APC cleavage sites, leading to APC resistance and increased thrombin generation. FVL thrombophilia is suspected in individuals with a history of venous thromboembolism (VTE) manifesting as deep vein thrombosis or pulmonary embolism, especially in women with a history of VTE during pregnancy or in association with oral contraceptive use and in individuals with a personal or family history of recurrent thrombosis.

FVL is the most common genetic cause of hereditary thrombophilia. In the United States, 3%–7% of the Caucasian population is heterozygous for this mutation. The prevalence is lower in other populations. Homozygous FVL mutations are found in about 1 in 5000 Caucasians. The FVL mutation is identified in ~20% with first VTE episode and in 50%–60% of patients with recurrent VTE. In addition to FVL, other rare variants have been observed, including factor V Cambridge (p.Arg306Thr), factor V Liverpool (p.Ile359Thr), and p.Glu666Asp, which have been associated with thrombosis risk. Other rare polymorphisms, including c.1692A>C and c.1696A>G, are silent, and their clinical significance is unknown.

Prothrombin Gene Mutation: Prothrombin (factor II) is a vitamin K–dependent coagulation factor. On activation, prothrombin is proteolytically cleaved to form thrombin, and in turn acts as a serine protease that converts fibrinogen to fibrin. In addition, thrombin catalyzes many other coagulation-related reactions. Hyperthrombinemia has been mainly associated with a c.*97G>A mutation (also known as G20210A) located in the 3′ untranslated region of the prothrombin gene. This mutation results in increased production through increased prothrombin mRNA expression and stabilization. High levels of prothrombin can lead to increased thrombin generation in the plasma, coagulation activation, and thrombosis. High prothrombin levels also inhibit APC-mediated inactivation of activated factor V and factor VIII.

Prothrombin mutations are the second most common inherited thrombophilia. In the United States, the heterozygous carrier frequency is about 1%–2%, accounting for approximately 6%–18% of VTE cases. This mutation is also more common

Transfusion Medicine and Hemostasis. https://doi.org/10.1016/B978-0-12-813726-0.00153-7

in the Caucasian population and is rare in other ethnic groups. Homozygosity for this mutation is found in about 1 in 10,000 individuals. Transheterozygosity for FVL and prothrombin c.*97G>A affects about 1 in 1000 individuals. Additional variations identified in the 3′-untranslated region of the prothrombin gene include changes at positions 20207, 20209, 20218, and 20221. The clinical significance of these polymorphisms is unclear.

Diagnostic Testing

Factor V Leiden: The APC resistance assay is the preferred initial test for FVL in most circumstances. About 90%–95% of individuals with APC resistance carry the FVL mutation. Depending on methodology, the assay may be uninterpretable when the baseline aPTT is prolonged due to inhibitors such as lupus anticoagulant. Therefore, a genetic assay has been recommended to confirm positive results and, most importantly, to distinguish heterozygotes from homozygotes.

Prothrombin Mutation: Increased plasma prothrombin levels are not specific for the mutation. Direct DNA testing is required.

Methods: A variety of DNA-based methods to detect FVL and prothrombin gene mutations are available. These methods can be divided into PCR-based and non-PCR-based methods. PCR-based methods employ primers to amplify the area with potential mutation. Different methods are then used to determine the genotype of the amplified alleles. Restriction fragment length polymorphism analysis was the first method used for detecting the FVL mutation, employing the *Mnl*I restriction enzyme to digest the PCR products. The enzyme recognition site is abolished by the FVL mutation, which is present in the normal allele. Subsequently, as DNA tests have become less labor intensive, more automated newer methods have been developed. This includes real-time PCR and DNA-based hybridization methodologies, using genotype-specific oligonucleotide probes coupled with fluorescence detection. Emerging mutation detection approaches also involve direct DNA sequencing.

The Invader assay (Hologic, Inc., Bedford, MA) utilizes a single-tube, two-phase reaction, in a microtiter plate format. In this assay, specific upstream invading oligonucleotide probes and primary downstream probes are designed to bind in tandem to either the normal or mutated alleles. The primary probes contain 5′ flaps, which can be cleaved off by a unique endonuclease (Cleavase VIII) when a partially overlapping complex is formed with the invading and the primary probes and the DNA template. The 5′ flaps can then invade secondary probes containing fluorescence resonance energy transfer (FRET) in a secondary reaction. This invasion results in a complex that is recognized by cleavase and subsequent release of the fluorophore from the quencher. The signal is then detected using a multiwell fluorometer. Different genotypes, normal, heterozygous, and homozygous, can be determined based on the normal to mutant signal ratio.

The eSensor Thrombophilia Risk Test (GenMark Dx, Carlsbad, CA) technology uses a solid-phase electrochemical method and allele-specific probes for genotyping. Following PCR amplification, single-stranded target DNA is generated by exonuclease digestion and mixed with allele-specific oligonucleotide signal probes labeled with a ferrocene derivative. The mixture is then added to a cartridge with electrodes containing capture probes complementary to the genomic sequence adjacent to the mutation site. The cartridge is inserted into the XT-8 instrument where the hybridization takes place, and the genotype of each variation is determined by voltammetry, which generates specific electrical signals from the allele-specific signal probes.

The Roche LightCycler assay (Roche Diagnostics, Indianapolis, IN) is a real-time PCR assay. Two probes containing fluorophores are used. The probes bind to sequence adjacent to the mutation site and the mutation site, respectively. A signal is detected only when the two probes come into close proximity resulting in FRET. Increased temperature during the hybridization step melts off the probes and decreases the fluorescence signal generated. The melting temperature difference between the normal and the mutant allele enables one to discriminate these two alleles.

Direct DNA sequencing can detect other polymorphisms or rare private mutations in these genes. The Illumina extension ligation-based microbead assay also received FDA clearance, which is more valuable for laboratories with large volumes. Overall, the different molecular diagnostic tests have a high rate of diagnostic concordance.

Test and Result Interpretation: The molecular tests demonstrate the presence or absence of FVL or prothrombin mutation, respectively, and the results are interpreted as normal, heterozygous, or homozygous for the respective mutation.

Heterozygosity for FVL is associated with a three- to eightfold increased relative risk of VTE, whereas the risk for homozygous individuals is increased up to 80- to 100-fold. Individuals who are compound heterozygous for FVL and a null mutation in factor V manifest a phenomenon called "pseudohomozygous" FVL. A molecular test demonstrates the presence of a heterozygous FVL mutation, whereas the APC resistance assay indicates the presence of homozygous mutations. Clinically, these individuals have an increased thrombotic risk similar to that of FVL homozygotes. Similarly, coinheritance of a non-FVL mutation and a FVL further increases the VTE risk when compared with harboring the FVL mutation alone, producing a more severe APC resistance phenotype.

The prothrombin c.*97G>A confers a two- to threefold increased risk for VTE. *Trans*-heterozygosity for FVL and c.*97G>A is associated with a higher risk of VTE compared with the risk in carriers of the individual mutations.

The risk for VTE is further increased by the presence of other genetic or acquired risk factors, including antithrombin, protein C or protein S deficiency, smoking, use of oral contraceptives, estrogen replacement therapy, and cancer.

Sources of Error: PCR-based methods are susceptible to contamination of previously amplified DNA material, which can result in false-positive results. It is important to employ procedures to minimize contamination. False-negative results can stem from poor sample quality or PCR inhibitors in the samples. It can also result from some rare conditions such as allogeneic hematopoietic stem cell transplantation because the peripheral blood is used as a source of DNA. Rare polymorphisms in FV and prothrombin genes that affect primer or probe binding sites can also interfere with the analysis and yield erroneous results. Finally, negative results do not rule out increased thrombotic risk from other genetic causes not detected with targeted molecular testing.

Further Reading

Baglin, T., Gray, E., Greaves, M., et al. (2010). Clinical guidelines for testing for heritable thrombophilia. *Br J of Haematol, 149*, 209–220.

Connors, J. M. (2017). Thrombophilia testing and venous thrombosis. *N Engl J Med, 377*, 1177–1187.

Cooper, P. C., & Rezende, S. M. (2007). An overview of methods for detection of factor V Leiden and the prothrombin G20210A mutations. *Int J Lab Hematol, 29*, 153–162.

Emaldi, A., Crim, M. T., Brotman, D. J., et al. (2010). Analytic validity of genetic tests to identify factor V Leiden and prothrombin G20210A. *Am J Hematol, 85*, 264–270.

Grody, W. W., Griffin, J. H., Taylor, A. K., Korf, B. R., Heit, J. A., & ACMG Factor V & Leiden Working Group (2007). *American College of Medical Genetics Consensus Statement on Factor V Leiden Mutation Testing. 2006 Edition*. Available at: http://www.acmg.net/Pages/ACMG_Activities/stds-2002/fv-pt.htm.

Johnson, N. V., Khor, B., & Van Cott, E. M. (2012). Advances in laboratory testing for thrombophilia. *Am J Hematol, 87*, S108–S112.

Kottke-Marchant, K. (2002). Genetic polymorphisms associated with venous and arterial thrombosis: An overview. *Arch Pathol Lab Med, 126*, 295–304.

Kujovich, J. L. (2011). Factor V Leiden thrombophilia. *Genet Med, 13*, 1–16.

Ledford, M., Friedman, K. D., Hessner, M. J., et al. (2000). A multi-site study for detection of the factor V (Leiden) mutation from genomic DNA using a homogeneous invader microtiter plate fluorescence resonance energy transfer (FRET) assay. *J Mol Diagn, 2*, 97–104.

Oh, H., & Smith, C. L. (2011). Evolving methods for single nucleotide polymorphism detection: Factor V Leiden mutation detection. *J Clin Lab Anal, 25*, 259–288.

Varga, E. A., & Kujovich, J. L. (2012). Management of inherited thrombophilia: Guide for genetics professionals. *Clin Genet, 81*, 7–17.

CHAPTER 154

Chronic Elevated Levels of Factor VIII and Other Coagulation Factors

Wayne L. Chandler, MD

Chronic elevation of some coagulation factor levels is associated with an increased risk of venous thrombosis (factors VIII, IX, XI, II). Some of these show level dependence; the risk increases linearly with the plasma factor level (factors VIII, IX, II). For other coagulation factors, elevated levels were either not confirmed as a venous thrombotic risk factor or the risk association was not independent; it disappeared when the risk was adjusted for other known risk factors. The utility of routine thrombophilia testing in patients with venous thromboembolism has been questioned in multiple studies, including the Choose Wisely campaign. Many studies are now showing that mild thrombophilic risk factors, particularly in the setting of provoked thrombosis, have no effect on therapy or risk of recurrence and should be abandoned.

Elevated Factor VIII: Factor VIII is an acute-phase factor that rises two- to fourfold during an inflammatory response to infection, cancer, surgery, trauma, and other stimuli. Factor VIII levels increase with age. Some individuals show a persistent elevation of factor VIII activity in the absence of an acute-phase response. Chronically elevated factor VIII is a level-related independent risk factor for venous thromboembolism with an odds ratio (OR) of two- to sixfold for factor VIII activities above 150–200 IU/dL after adjustment of other known risk factors. The risk of venous thrombosis associated with high factor VIII is present in children and the elderly. Children are also at greater risk of catheter-associated thrombosis if they have chronically elevated factor VIII activity. There is a fourfold increased risk of recurrent venous thrombosis in patients with chronic elevated factor VIII and unprovoked initial thrombosis, but no increase in recurrence in patients with provoked thrombosis. Risk associated with chronic factor VIII elevation is synergistic with acquired risk factors such as oral contraceptive use. It is recommended that evaluation for persistent factor VIII elevation be performed 3–6 months after the last episode of thrombosis, when the patient is otherwise well with no evidence of acute-phase response and off antithrombotic therapy. That being said, it is not clear that knowing the factor VIII status of a patient with venous thrombosis has any effect on either initial antithrombotic therapy or therapy during recurrence, and therefore routine evaluation of factor VIII in patients with venous thrombosis is not recommended.

Factor VIII levels are related to von Willebrand factor levels and blood group, but these are not independent predictors of thrombosis, their association appears to be mediated through elevated factor VIII levels. The genetic basis for persistent elevation of factor VIII is currently unknown; no polymorphisms have been detected in the factor VIII gene.

Elevated Factor IX: Factor IX is a vitamin K–dependent protein that does not show an acute-phase response. Elevated levels of factor IX above 130–150 IU/dL have been

Transfusion Medicine and Hemostasis. https://doi.org/10.1016/B978-0-12-813726-0.00154-9

associated with a twofold increased risk of venous thrombosis after adjustment for factor VIII, factor XI, and other known risk factors. There is no information on risk of recurrence associated with elevated factor XI. Nothing is known about the genetic cause of elevated factor IX levels. One family has been described (Factor IX Padua) with a modified factor IX that shows a gain of function mutation, leading to an eightfold increase in factor IX activity and an increased risk of venous thrombosis.

Elevated Factor XI: Elevated levels of factor XI above 110–120 IU/dL have been associated with a twofold increased risk of venous thrombosis after adjustment for factor VIII, factor IX, and other known risk factors. A genetic mechanism has not been confirmed for elevated factor XI levels.

Elevated Prothrombin: Elevated factor II (prothrombin) activity above 115% has been associated with a twofold increased risk of venous thrombosis. Some or most patients with elevated factor II have a mutation in the prothrombin gene, the most common of which is the 20210A allele. This polymorphism is associated with increased prothrombin levels and an increased risk of venous thrombosis. The 20210A mutation is not associated with an increased risk of recurrence.

Other Coagulation Factors: After correction for other hemostatic risk factors (FVIII, FIX, FXI, FV Leiden, PT20210A) and C-reactive protein, the OR was only 1.5 for venous thrombosis risk related to elevated fibrinogen in one study and no association in another study suggesting that the overall effect of fibrinogen on venous thrombotic risk is very small.

Factor V levels were not associated with venous thrombosis in a case–control study. The level of factor X was associated with an increased risk of venous thrombosis, but the risk disappeared when the results were adjusted for the levels of other vitamin K–dependent factors (factors II, VII, and IX). Factor VII levels above the 95th percentile, as compared with the lowest quartile, conveyed a higher risk of venous thrombosis in one study but not confirmed in other studies. Elevated levels of thrombin-activatable fibrinolysis inhibitor (TAFI) have been associated with an increased risk of venous thrombosis in some studies but not confirmed in a multivariable analysis of population-based cohort study. Measurement of TAFI is complex with multiple forms in plasma and multiple different assays that are not well calibrated. No association between TAFI gene polymorphisms, levels, and venous thrombosis has been confirmed.

Further Reading

Bertina, R. M. (2003). Elevated clotting factor levels and venous thrombosis. *Pathophysiol Haemost Thromb*, *33*, 395–400.

Franchini, M., Martinelli, I., & Mannucci, P. M. (2016). Uncertain thrombophilia markers. *Thromb Haemost*, *115*, 25–30.

Hicks, L. K., Bering, H., Carson, K. R., Kleinerman, J., Kukreti, V., Ma, A., et al. (2013). The ASH choosing Wisely® campaign: Five hematologic tests and treatments to question. *Blood*, *122*, 3879–3883.

Jenkins, P. V., Rawley, O., Smith, O. P., & O'Donnell, J. S. (2012). Elevated factor VIII levels and risk of venous thrombosis. *Br J Haematol*, *157*, 653–663.

Neshat-Vahid, S., Pierce, R., Hersey, D., Raffini, L. J., & Faustino, E. V. (2016). Association of thrombophilia and catheter-associated thrombosis in children: A systematic review and meta-analysis. *J Thromb Haemost*, *14*, 1749–1758.

CHAPTER 155

Acquired Prothrombotic Conditions

Sabrina Racine-Brzostek, MD, PhD and Morayma Reyes Gil, MD, PhD

Classic Acquired Prothrombotic Conditions: Although there is a growing number of known established thrombophilias with an underlying genetic basis, the majority of venous thromboembolism events are provoked in the context of an acquired hypercoagulability state, leading to venous thrombosis. These acquired cases can be associated with clinical events, such as surgery, malignancy, pregnancy, drugs, kidney disease, infection, and inflammatory states. The most classic acquired conditions include hemolytic uremic syndrome (HUS), thrombotic thrombocytopenic purpura, antiphospholipid syndrome (APS), disseminated intravascular coagulopathy (DIC), and heparin-induced thrombocytopenia and are discussed in further detail in other chapters (see Table 155.1). A full assessment may be indicated in patients with recurrent thrombotic events and may begin with platelet count, fibrinogen, prothrombin time, activated partial thromboplastin time, thrombin time, reptilase time, activated protein C, immunological and functional ATIII, protein C, and protein S.

Cancer, especially in metastasis is a prothrombotic state. Particularly certain hematologic malignancies are strongly prothrombotic such as paroxysmal nocturnal hemoglobinuria (PNH), polycythemia vera, and essential thrombocythemia and may present with thrombosis as a first clinical sign. Evaluation for these disorders is usually triggered by abnormal complete blood count results. PNH is an extremely rare acquired hematopoietic stem cell disorder. It arises because of mutations in phosphatidylinositol glycan A, resulting in deficiency of proteins anchored to the cell membrane via glycophosphatidylinositol (GPI). This defect in turn leads to complement-induced hemolysis and a prothrombotic state. Flow cytometry assessing for the presence of GPI-linked proteins on neutrophils, monocytes, and red blood cells is the test modality of choice for PNH investigation.

Role of Inflammation in Acquired Prothrombotic Conditions: Although it has been well accepted that inflammation is associated with a prothrombotic state, its exact pathogenesis continues to be complicated. By inducing endothelial damage, inflammation could increase procoagulant factors, yet at the same time inhibit anticoagulation pathways and fibrinolysis. Systemic inflammatory disease that could induce prothrombotic states includes rheumatoid arthritis, systemic lupus erythematosus, and antiphospholipid syndrome. Many questions remain in defining the complex relationship between proinflammatory states and hemostasis. As such, there are no clear guidelines of how to assess thrombosis in the setting of chronic and active/transient inflammation. For example, C-reactive protein (CRP) is an acute-phase reactant, which acts as a marker for underlying systemic inflammation. As with D-dimer, CRP may be a negative predictor of thrombosis, but its clinical utility has neither been clearly evaluated nor have any guidelines been set in its interpretation in the context of other laboratory testing.

Transfusion Medicine and Hemostasis. https://doi.org/10.1016/B978-0-12-813726-0.00155-0

TABLE 155.1 Classic Acquired Prothombotic Conditions and Recommended Testing

Condition	Recommended Testing[a]	Ref. Chapters
Hemolytic uremic syndrome	ADAMTS13, CBC, fibrinogen, VWF, FVIII, D-dimer, BUN, Cr, complement	106
Thrombotic thrombocytopenic purpura	ADAMTS13 activity, ADAMTS13 inhibitor, Bethesda inhibitor assay	107/157
Antiphospholipid syndrome	Lupus anticoagulant, anticardiolipin, anti-β2-glycoprotein 1, platelet count	108/158
Disseminated intravascular coagulopathy	Fibrinogen, TT, D-dimer, AT, protein C and S, factor	124
Heparin-induced thrombocytopenia	Platelet count, anti-PF4 heparin	156

[a]In addition to routine coagulation testing.

Role of Infection in Acquired Prothrombotic Conditions: Although the classic prothrombotic scenarios induced with infection include such states as HUS and DIC, an emerging area of study focuses on the role of neutrophil activation with resulting NETosis and its role in intravascular thrombosis. This innate immune response involves the nuclear extrusion of DNA and histones by neutrophils culminating in the formation of *neutrophil extracellular traps* (NETS). These NETS allow trapping of bacteria and the interconnection of neutrophils creating a concentrated bactericidal microenvironment of neutrophilic proteases and reactive oxygen species. The dysregulation of NETosis has become an emerging explanation for the underlying cause of thrombotic events in a variety of clinical scenarios, including myocardial infarctions, acute lung injury, and postchemotherapy complications. Although the laboratory evaluation of NETS in prothrombotic states is still in its investigatory state, the clinical testing for the evaluation and quantification of NETS may become available in the near future.

Recently Utilized (Nontraditional) Laboratory Assays (Table 155.2)

von Willebrand Factor: von Willebrand factor (VWF), a large multimeric protein involved with platelet adhesion, has been implemented as a major player in acquired prothrombotic conditions, such as in the thrombotic microangiopathies. The failure of ADAMTS13 (due to the autoantibodies against the enzyme) to cleave VWF leads to the accumulation of large VWF multimers and the resultant microvascular platelet thrombi with vascular occlusion (Chapter 109). The functional assays, which detect cleavage products on VWF degradation, include the VWF multimer, Western blot, immunoradiometric assays in microtiter plates, and assays evaluating VWF's ability to bind collagen and/or ristocetin. These assays take days to complete and are not readily available to most laboratories. Emerging assays reduce the turnaround time to a few hours. These assays focus on the direct detection of ADAMTS13 activity and utilize a recombinant ADAMST13 substrate, whose cleavage product could be detected by enzyme-linked immunosorbent assay, Western blot, or fluorescence resonance energy transfer (Chapter 161). Certain intrinsic factors, but in particular FVIII, have been recognized as independent risk factor of venous thrombosis (Chapter 154). Elevated VWF

TABLE 155.2 Recently Recognized Acquired Prothombotic Conditions and Recommended Testing

Prothrombotic Condition	Recommended Testing[a]
Infection	Institutional sepsis workup, neutrophil extracellular traps (NETS)[b]
Inflammation	C-reactive protein (CRP), microparticles[b]
Autoimmune disorders	Lupus anticoagulant, CRP, antinuclear antibody, Double-stranded DNA, Rheumatoid factor, Extractable nuclear antigen antibodies panel, NETS[b]
Solid and hematological malignancies	JAK-2 mutational analysis, flow cytometry, microparticles[b]
Surgical/vascular devices	von Willebrand factor, platelet functional assays[b]
Nephrotic syndrome	Creatinine, antithrombin, selective renal venography

[a]In addition to routine coagulation testing.
[b]Laboratory testing without set clinical guidelines.

levels have been associated with venous thrombosis by virtue of its association with FVIII. On the other hand, recent studies point to an association of elevated levels of VWF in arterial thrombosis, especially strokes. However, large randomized prospective clinical and epidemiologic studies are needed to determine if elevated VWF is an independent risk factor of arterial thrombosis. In addition, establishment of guidelines in assessing the risk of thrombosis during the evaluation of VWF levels will be especially critical, as VWF levels are heavily influenced during inflammatory states.

Thromboelastography and Thrombin Generation Testing: Whereas platelet qualitative or quantitative defects are associated with bleeding diatheses, platelet hyperactivity and/or decreased fibrinolysis could increase the risk of venous and arterial thrombosis (Chapter 145). The increased role of platelet function testing has shown promise in assessing global hemostasis. The global thrombosis test (GTT), thromboelastography (TEG), rotational TEG, and Sonoclot all assess global platelet function (Chapter 139). Although these tests traditionally have been utilized for the monitoring of antiplatelet therapy and management of perioperative hemostasis, it has vast potential in identifying patients at increased risk of thrombosis. Although platelet hyperactivity can be monitored by these platforms, guidelines on how to interpret these results have yet to be established.

Microparticles: Cellular membrane vesicles, known as microparticles, are released from the cell surface of apoptotic cells, promoting thrombin formation by either displaying tissue factor itself on its surface or, due to its high phosphatidylserine content, by promoting assembly of the coagulation factors. As these circulating microparticles have become known to promote coagulation, it is hypothesized that they could act as triggers to enhance prothrombotic conditions in acquired thrombotic conditions such as APS and cancer. Therefore, microparticles could theoretically act as a prognostic biomarker capable of estimating risk of thrombosis in primary and secondary prevention. Unfortunately, to date, there are no prospective studies on the use of microparticles in evaluating the risk of thrombosis.

Concluding Remarks: There is an urgent need to standardize these newer testing technologies. This includes the establishment of testing guidelines and proper reference ranges that are based on the local population. Furthermore, as with all laboratory testing, these emerging thrombophilia assays should be performed only if the results are likely to change medical management. Any new guidelines should also take into consideration the infectious and inflammatory states. Therefore, the clinical utility of information gathered from these test results needs to be further explored so that they may be properly implemented for clinical diagnosis.

Further Reading

Campello, E., Radu, C. M., Spiezia, L., et al. (2017). Modulating thrombotic diathesis in hereditary thrombophilia and antiphospholipid antibody syndrome: A role for circulating microparticles? *Clin Chem Lab Med*, *55*, 934–943.

Foley, J. H., & Conway, E. M. (2016). Cross talk pathways between coagulation and inflammation. *Circ Res*, *118*, 1392–1408.

Kimball, A. S., Obi, A. T., Diaz, J. A., et al. (2016). The emerging role of NETs in venous thrombosis and immunothrombosis. *Front Immunol*, *7*, 236–243.

Vazquez-Garza, E., Jerjes-Sanchez, C., Navarrete, A., et al. (2017). Venous thromboembolism: Thrombosis, inflammation, and immunothrombosis for clinicians. *J Thromb Thrombolysis*, *44*, 377–385.

CHAPTER 156

Laboratory Evaluation of Heparin-Induced Thrombocytopenia

Anne M. Winkler, MD

Heparin-induced thrombocytopenia (HIT) is a clinicopathologic syndrome characterized by a constellation of clinical findings and identification of antiplatelet factor 4 (PF4)/heparin antibodies. Classically in HIT, progressive thrombocytopenia is observed, reaching a nadir within 5–10 days following the immunizing heparin exposure. In addition, because the mechanism by which thrombocytopenia ensues involves platelet activation, thrombosis occurs in a significant proportion of cases. Thrombosis in HIT can be severe and life-threatening. As a result, a decrease in platelet count by 50% after heparin administration is sufficient evidence to replace heparin with an alternative anticoagulant pending laboratory evaluation of HIT unless a rapid assay is available. As heparin is usually a safe and often preferred drug for inpatient thromboprophylaxis, cardiac surgery, and other interventional procedures, laboratory assessment of HIT is valuable in distinguishing true cases of HIT from thrombocytopenia of other etiologies.

Understanding of the molecular mechanisms of HIT has led to a model in which antibodies are formed against a complex between PF4 and heparin. These pathogenic anti-PF4/heparin antibodies provoke platelet activation by interaction with the platelet FcγIIa receptor, resulting in platelet aggregation, thromboxane generation, granule release, formation of platelet-derived microparticles, and overall stimulation of a procoagulant response. Other mechanisms potentially contributing to hypercoagulability include activation of endothelial cells and monocytes with resultant exposure of tissue factor and neutralization of the anticoagulant effect of heparin by released PF4, which further propagates thrombogenesis.

Functional assays such as the serotonin release assay (SRA) or heparin-induced platelet activation assay (HIPA) are considered the gold standard for laboratory diagnosis of HIT because of high sensitivity and specificity. However, these assays are difficult to perform, requiring, in some cases, specialized equipment and use of radioactivity, as well as well-characterized platelet donors and experienced operators. Thus, SRA and HIPA are not feasible as general screening tests for anti-PF4/heparin antibodies. Accordingly, a number of different methodologies have been developed to allow routine screening for anti-PF4/heparin antibodies. These assays have high sensitivity, but in some cases can result in a substantial rate of false positives because of lower specificity. Currently, the best approach is considered to be identification of anti-PF4/heparin antibodies through a combination of an antigen assay and functional platelet activation assay, if required, interpreted in conjunction with the clinical pretest probability (e.g., 4Ts score).

Antigen Assays: In general, most clinical laboratories utilize commercially available immunoassays for detection of anti-PF4/heparin antibodies. These assays detect

Transfusion Medicine and Hemostasis. https://doi.org/10.1016/B978-0-12-813726-0.00156-2

anti-PF4/heparin antibodies using PF4 as a substrate. To simulate in vivo epitopes, PF4 is complexed with either heparin or other polyanion molecules that mimic heparin.

Solid-Phase Enzyme Immunoassays: In solid-phase enzyme immunoassays (EIA), PF4/polyanion complexes are bound to microtiter plates, and after incubation with patient sample and addition of an enzyme-linked antihuman globulin reagent, a colorimetric response can be quantified using an automated microplate reader. The resultant optical density (OD) correlates with a qualitative (positive or negative) result depending on a prevalidated cutoff. However, subtle differences in commercially available kits exist, regarding the source of PF4 (recombinant vs. platelet-derived) and the use of heparin or other polyanions such as polyvinyl sulfonate (PVS). One commercially available EIA incorporates the addition of interleukin-8 (IL-8) and neutrophil-activating peptide 2 (NAP-2), two chemokines that have been implicated in cases of HIT. Despite these differences, solid-phase EIAs have sensitivities approaching 94%–100%; however, there is a significant variability in clinical specificity. Two recent systematic reviews and meta-analyses have examined the diagnostic value of anti-PF4/heparin immunoassays.

In the past, immunoassays incorporated combinations of antihuman IgG, IgA, and IgM, and thus detected anti-PF4/heparin antibodies of IgG, IgM, and IgA isotypes. However, it is now understood that the pathogenesis of HIT relies on antibody binding to the platelet FcγIIa receptor, which is IgG dependent. Although isolated cases of HIT associated with IgA and IgM anti-PF4/heparin have been described, these antibody isotypes are not typically considered pathogenic and acceptance of positive results without confirmation by a functional assay may be partly responsible for the overdiagnosis of HIT. As a result, use of IgG-specific assays have increased the clinical specificity of antigen assays. Guidelines from the International Society on Thrombosis and Haemostasis and British Committee for Standards in Haematology/British Society for Haematology advocate for the use of IgG-specific EIAs. In addition, to increase specificity EIAs can also be performed following incubation with high concentrations of heparin (10–100 IU/mL). Reactivity of anti-PF4/heparin antibodies is inhibited at high heparin concentrations.

As an adjunct to guide further testing, a recent publication correlated OD measurements of solid-phase EIAs to SRA. In this study, the probability of pathogenic antibodies as defined by strong positivity (>50% release) in the SRA increased in relation to the magnitude of OD measurements of the EIA. It was determined that most cases of HIT were associated with an EIA OD 1.40 units, whereas weak positive results (0.40–1.00 OD units) typically excluded the diagnosis of HIT. Because of lower EIA OD cutoffs supplied by manufacturers, the author recommends reporting the qualitative and quantitative result.

Fluid-Phase Enzyme Immunoassays: As a result of the inherent problem of protein denaturation with solid-phase EIAs, a fluid-phase EIA in which anti-PF4/heparin IgG bind PF4/heparin antigens followed by capture with beads was developed. Although not commonly used clinically, this assay may have greater sensitivity than the aforementioned solid phase EIAs. In addition, this method avoids nonspecific binding by minimizing exposure of cryptic antigens affiliated with denatured PF4.

Rapid Immunoassays: For rapid detection of PF4/heparin antibodies, additional assays such as the particle gel immunoassay (PaGIA), particle immunofiltration assay (PIFA), a latex agglutination assay, and chemiluminescent immunoassays have been added to the armamentarium of HIT testing. With analytical turnaround times less than 30 minutes and on demand availability, these assays allow clinicians to make an informed decision before switching to alternative anticoagulation, thereby improving outcomes while decreasing costs.

Particle Gel Immunoassay: In the PaGIA, PF4/heparin complexes are bound to high-density polystyrene beads. If present, anti-PF4/heparin antibodies bind to the beads and are detected through agglutination that occurs as a result of the addition of a secondary antihuman globulin reagent. Similar to gel testing widely used in blood banks, agglutinated beads fail to migrate through the gel, producing a visible band that indicates a positive result. Because of limited data, the sensitivity and specificity of this assay is currently uncertain; however, it is postulated to be intermediate between solid-phase EIAs and functional platelet assays.

Particle Immunofiltration Assay: PIFA is an alternative but similar approach, which utilizes a reaction well that contains dyed particles coated with PF4 without heparin. The lack of heparin does not adversely impact detection of anti-PF4/heparin antibodies as a result of close approximation of PF4. If the patient specimen contains PF4/heparin antibodies, particles agglutinate and fail to migrate through the membrane filter, resulting in lack of detectable color and thus indicating a positive result. This assay has been cleared for use by the United States Food and Drug Administration; however, because of the lack of performance characteristics in large studies, it is not commonly used.

Latex Immunoassay: The latex immunoassay utilizes platelet-activating anti-PF4/heparin murine monoclonal antibodies, KKO, coated onto polystyrene latex nanoparticles. Addition of PF4/PVS complexes to the latex nanoparticles results in particle agglutination and increased absorbance. The amount of absorbance is inversely proportional to the concentration of anti-PF4/heparin antibodies in the sample. If anti-PF/heparin antibodies of any immunoglobulin class are present, there is competition with KKO on the particles for PF4/PVS and no or minimal change in absorbance. The concentration of anti-PF4/heparin antibodies is reported in units per milliliter (U/mL) with values equal to or greater than 1.0 U/mL considered positive. The cutoff of 1.0 U/mL was determined through multiple studies, including an expected value study with healthy donors, an expected value study with heparin-exposed patients, and a receiver operating characteristic curve analysis compared with SRA. The performance characteristics of the latex immunoassay were recently evaluated in 429 patients from a prospective cohort study of 4Ts scoring and consecutive HIT patients at a single institution using reference SRA. The authors demonstrated a high negative predictive value (NPV, 99.7%) and positive predictive value (PPV, 55.6%), and a diagnostic specificity and PPV higher than that of two EIAs and PaGIA. In addition, the magnitude of a positive latex immunoassay predicted greater probability of a positive SRA, and a probable diagnosis of HIT similar to what has been shown for EIA. Unlike other antigen assays, the latex immunoassay is fully automated and standardized with a monoclonal antibody calibrator, which allows for the possibility of comparable test results from different laboratories.

Chemiluminescent Immunoassays: Two chemiluminescent immunoassays are also available. In both assays, PF4-coated magnetic particles capture, if present, anti-PF4/heparin antibodies. After incubation, magnetic separation, and a wash step, a tracer consisting of an isoluminol-labeled secondary antibody is added and may bind with the captured PF4/heparin on the particles. The two assays differ in their secondary antibody; one has a secondary anti-IgG antibody and the other a mixture of antihuman IgG, IgA, and IgM. After a second incubation, magnetic separation, and a wash step, reagents that trigger the luminescent reaction are added, and the emitted light is measured. The emitted light intensity is directly proportional to the concentration of anti-PF4/heparin antibodies. These assays also have a very high NPV and PPV. Similar to the latex immunoassay, the chemiluminescent immunoassays are fully automated and standardized with a monoclonal antibody calibrator, which allows for the possibility of comparable test results from different laboratories.

Antigen Assay Interpretation: As a general rule, due to the sensitivity of antigen assays, a negative result essentially excludes a diagnosis of HIT in the majority of cases; however, results should always be interpreted in conjunction with the clinical pretest probability. If the clinical suspicion indicates intermediate or high probability, all positive immunoassays should be confirmed by a functional assay; however, in patients with a high pretest probability, a strongly positive immunoassay result may be sufficient.

Platelet Activation (Functional) Assays: Functional platelet assays are almost exclusively performed in specialized laboratories, which are capable of performing high-quality platelet activation assays with the SRA being more commonly performed in North America compared with the HIPA being performed in Europe.

Platelet Aggregation Test: Historically, conventional aggregometry using citrated platelet-rich plasma was used to assess platelet aggregation; however, in comparison to the functional assays (SRA and HIPA), which incorporate addition of washed platelets to eliminate nonspecific aggregation due to heparin, fibrinogen, or other acute-phase reactants, the platelet aggregation test (PAT) method demonstrated suboptimal sensitivity (50%–80%). As a result, the PAT is not advocated for initial detection of anti-PF4/heparin antibodies. However, new methods utilizing whole-blood impedance aggregometry are starting to be used and under evaluation in clinical studies. A protocol has recently been published using heparin-induced multielectrode aggregometry by the International Society on Thrombosis and Haemostasis Subcommittee on Platelet Immunology.

Serotonin Release Assay: Since first described in 1986, a radiolabeled ^{14}C-SRA has been accepted as the gold standard for detection of "pathogenic" anti-PF4/heparin antibodies. The ^{14}C-SRA is performed by incubating a patient's sample with radiolabeled washed donor platelets, which have been determined to be susceptible to platelet activation by HIT antibodies. After activation, radiolabeled serotonin is released and detected; a positive result requires at least 20% release of serotonin over background reactivity. However, experts promote that a higher threshold of at least 50% serotonin release should be used. Specificity can be increased through including incubation with high concentrations of heparin (10–100 IU/mL) and addition of a monoclonal antibody to block platelet FcγIIa receptors. Positive, negative, and "weak positive" controls to ensure platelet reactivity are necessary to further confirm results. The ^{14}C-SRA is a time-consuming and technically challenging assay that requires strict adherence to established procedures. In addition, special licensure is required in some geographies for use of radioisotopes, limiting the ability of many laboratories to perform this test. There can be substantial variability in this assay between laboratories because of the lack of standardization and intrinsic differences in platelet donors. However, when performed correctly, the ^{14}C-SRA has a sensitivity of 92%–100% with a specificity of 98% for detection of pathogenic PF4/heparin antibodies.

Other end points have been analyzed using an SRA, which includes quantification of serotonin release by high-performance liquid chromatography, EIA, and flow cytometry. These detection methods have resulted in similar sensitivity and specificity to the ^{14}C-SRA.

Heparin-Induced Platelet Activation Test: In the HIPA assay, patient sample and buffer are added to microtiter wells containing washed donor platelets. A magnetic stirrer agitates the mixture to maintain suspension of unaggregated platelets. At 5-minute time intervals, the wells are examined for platelet aggregation using an indirect light source, and change of the mixture from turbid to transparent indicates a positive result. Both pharmacologic and high doses of heparin can be used as previously described. This technique is sensitive and specific; however, visual interpretation can result in substantial interobserver variability. On the other hand, an advantage of the HIPA is possibility for repeat testing over time.

Interpretation of Functional Platelet Assays: Despite advances in laboratory testing for HIT, ability to detect pathogenic HIT antibodies relies on their detection using functional platelet assays. Immunoassays are useful for screening; however, it is important to be aware of the differences of antibody specificity used in coagulation laboratories.

Other Methodologies: Quantification of platelet microparticles and annexin V binding assays has also been published as effective methods of anti-PF4/heparin antibody detection; however, these are not in widespread clinical use.

Further Reading

Althaus, K., Hron, G., Strobel, U., Abbate, R., Rogolino, A., Davidson, S., et al. (2013). Evaluation of automated immunoassays in the diagnosis of heparin induced thrombocytopenia. *Thromb Res*, *131*, e85–90.

Husseinzadeh, H. D., Gimotty, P. A., Pishko, A. M., Buckley, M., Warkentin, T. E., & Cuker, A. (2017). Diagnostic accuracy of IgG-specific versus polyspecific enzyme-linked immunoassays in heparin-induced thrombocytopenia: A systematic review and meta-analysis. *J Thromb Haemost*, *15*, 1203–1212.

Morel-Kopp, M. C., Mullier, F., Gkalea, V., Bakchoul, T., Minet, V., Elalamy, I., et al. (2016). Heparin-induced multi-electrode aggregometry method for heparin-induced thrombocytopenia testing: Communication from the SSC of the ISTH. *J Thromb Haemost*, *14*, 2548–2552.

Nagler, M., Bachmann, L. M., ten Cate, H., & ten Cate-Hoek, A. (2016). Diagnostic value of immunoassays for heparin-induced thrombocytopenia; systematic review and meta-analysis. *Blood*, *127*, 546–557.

Nagler, M., & Bakchoul (2016). Clinical and laboratory tests for the diagnosis of heparin-induced thrombocytopenia. *Thromb Haemost*, *116*, 823–834.

Sun, L., Gimotty, P. A., Lakshamana, S., & Cuker, A. (2016). Diagnostic accuracy of rapid immunoassays for heparin induced thrombocytopenia: A systematic review and meta-analysis. *Thromb Haemost*, *115*, 1044–1055.

Warkentin, T. E. (2011). HIT paradigms and paradoxes. *J Thromb Haemost*, *9*(Suppl. 1), 105–117.

Warkentin, T. E., Greinacher, A., Gruel, Y., Aster, R. H., & Chong, B. H. (2011). Laboratory testing for heparin-induced thrombocytopenia: A conceptual framework and implications for diagnosis. *J Thromb Haemost*, *9*, 2498–5000.

Warkentin, T. E., & Sheppard, J. I. (2006). Testing for heparin-induced thrombocytopenia antibodies. *Transfus Med Rev*, *20*, 259–272.

Warkentin, T. E., Sheppard, J. I., Linkins, L. A., Arnold, D. M., & Nazy, I. (2017). Performance characteristics of an automated latex immunoturbidimetric assay [HemosIL® HIT-Ab(PF4-H)] for the diagnosis of immune heparin-induced thrombocytopenia. *Thromb Res*, *153*, 108–117.

Warkentin, T. E., Sheppard, J. I., Moore, J. C., et al. (2008). Quantitative interpretation of optical density measurements using PF4-dependent enzyme immunoassays. *J Thromb Haemost*, *6*, 1304–1312.

Watson, H., Davidson, S., & Keeling, D. (2012). Guidelines on the diagnosis and management of heparin-induced thrombocytopenia. *Br J Haematol*, *159*, 528–540.

CHAPTER 157

Laboratory Evaluation of Thrombotic Thrombocytopenic Purpura

Christine L. Kempton, MD, MSc and Ana G. Antun, MD, MSc

A rapid and accurate diagnosis of thrombotic thrombocytopenic purpura (TTP) is critical, as initiation of therapeutic plasma exchange (TPE) can be lifesaving. To date, the diagnosis of TTP is dependent on the demonstration of an otherwise unexplained microangiopathic hemolytic anemia and thrombocytopenia, which may or may not have concurrent end organ damage, such as neurologic symptoms or renal dysfunction. Laboratory findings consistent with a hemolytic anemia include an elevated lactate dehydrogenase, increased reticulocyte count, and elevated indirect bilirubin. A microangiopathic cause of the hemolytic anemia is supported by the presence of schistocytes on peripheral blood smear (see Chapter 107). This chapter will review both the assays for ADAMTS13 activity and the assessment of inhibitory antibodies against the ADAMTS13 enzyme.

Measurement of ADAMTS13 Activity: As early symptoms and laboratory findings can also be found in thrombotic microangiopathies, such as disseminated intravascular coagulopathy, a specific and sensitive test for the diagnosis of TTP is desirable. With the discovery of ADAMTS13 and its association with TTP, laboratory assessment of ADAMTS13 enzyme activity has been an attractive test to fill this role. Unfortunately, no assay is sensitive or specific enough to allow reliable diagnostic use; however, there is evidence that ADAMTS13 activity and inhibitor titer have prognostic value.

Different preanalytical variables are crucial for the quality of ADAMTS13 testing. First, as treatment alters ADAMTS13 levels, diagnostic samples should be collected before starting therapy. Second, EDTA plasma is discouraged as EDTA chelates the metal ions of the metalloproteinase that are necessary for ADAMTS13 function. Third, plasma samples are preferred because thrombin degrades ADAMTS13. Fourth, although there is some degree of plasma-free hemoglobin during the initial diagnosis, hemolyzed plasma induced in vitro (collection and preparation) with a hemoglobin concentration >2 g/L should be avoided because free hemoglobin inhibits ADAMTS13. Fifth, plasma should be frozen if it cannot be tested within 4 hours of collection and multiple freeze-thaw cycles should be avoided.

Whole Substrate Methods: A variety of tests of ADAMTS13 activity have been developed using the natural substrate, VWF. Such assays require two steps. The first step incubates test plasma obtained from the patient with purified VWF that has been treated to cause partial unfolding. The second step determines the ADAMTS13 activity in the patient's plasma by measuring proteolysis of VWF. This has been accomplished through direct measurement of the VWF protein itself (multimers or proteolytic fragments) or via indirect measurement of the VWF activity (glycoprotein Ib-binding,

Transfusion Medicine and Hemostasis. https://doi.org/10.1016/B978-0-12-813726-0.00157-4

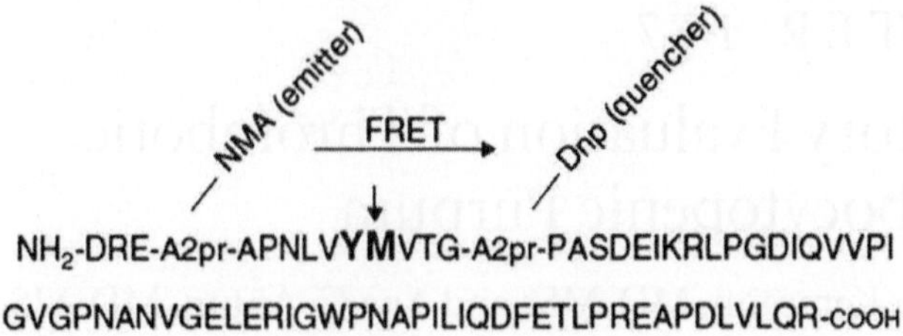

FIGURE 157.1 Structure of FRET-VWF73. 73 amino acids required for ADAMTS13 cleavage. P7 and P5′ are replaced with A2pr, and Nma and Dnp are attached. Nma emits energy that is transferred to Dnp when the sequence is intact. The arrowhead indicates the ADAMTS13 cleavage site. *FRET*, fluorescence resonance energy transfer.

ristocetin cofactor activity, or collagen binding). Many of these assays are time-consuming and cumbersome to perform. However, a semiautomated method for indirect measurement has been developed and shows promise, with an interassay coefficient of variation of 14% and an intraassay coefficient of variation of 9%.

Peptide Substrate Methods: The cloning of ADAMTS13 and detailed mapping of its recognition sites on VWF has facilitated the development of modified ADAMTS13 substrates, leading to rapid and reproducible assays.

FRET-vWF73 Assay: The most widely used assays are based on ADAMTS13 substrate, vWF73, which contains the 73 amino acid residues that are the minimum amino acid sequence required for ADAMTS13 cleavage of VWF between Y1605 and M1606 in the A2 domain (Fig. 157.1). Using the vWF73 substrate eliminates the need to treat purified plasma or recombinant VWF to cause unfolding. To facilitate detection of ADAMTS13 cleavage of vWF73, fluorescence resonance energy transfer (FRET) is utilized. One side of the peptide has a fluorescent molecule, whereas the other has a quenching molecule; thus, cleavage of the peptide dequenches the fluorochrome, resulting in a positive signal. The fluorescence signal is detected every 5 minutes for 1 hour and, as more substrate is cleaved, the fluorescence increases. The change in fluorescence over time (reaction rate) is calculated by linear regression analysis. Pooled normal plasma is used as a reference. ADAMTS13 activity is expressed as a percentage of the reaction rate found in the test plasma compared with the reaction rate found in pooled normal plasma.

The assay reproducibility is excellent (coefficient of variation 6%). The FRET-vWF73 assay is now considered superior to other assays because it is a one-step assay that can be completed in 1 hour with excellent precision. When directly compared with other assays, there is good correlation with reported correlation coefficients ranging from 0.898 to 0.971.

Other Peptide Methods: Recombinant VWF peptide (vWF73) can be fused with a lysine tag. The lysine tag is released after cleavage of VWF by ADAMTS13. After release, the lysine tag is captured by CM10 cationic protein chip and measured by surface-enhanced laser desorption/ionization time-of-flight mass spectrometer. The amount of lysine tag that is released after cleavage by ADAMTS13 corresponds to the amount of ADAMTS13 activity. The reported coefficient of variation for this test is <20%. Alternatively, VWF73 has been conjugated to horseradish peroxidase (HRP) at the N-terminus and labeled with biotin at the C-terminus. ADAMTS13 in the test plasma cleaves the peptide substrate. Uncleaved substrate is removed by adsorption with streptavidin agarose. The amount of ADAMTS13 activity is determined by measuring the unabsorbed HRP activity remaining in solution. The coefficient of variation was reported as 5.8%.

Recombinant VWF peptide (VWF71) has similar characteristics to FRETS-VWF73, which is highly sensitive rapid assay without interference with plasma proteins such as bilirubin or hemoglobin.

Test Limitations: Severe hyperbilirubinemia has been reported to interfere with the assay, leading to a false reduction in the ADAMTS13 activity in the FRET assay. However, the degree of reduction is mild, and would not lead to a normal level being measured as severely reduced, but may change a mild/moderate reduction to severe. Free plasma hemoglobin greater than 2 g/L also reduces ADAMTS13 activity by inhibition of the enzymatic activity, although not specific to this assay. To avoid interference, plasma is diluted ≥1:20, which limits assay sensitivity to ~3%–5% and prevents the detection of some inhibitors.

ADAMTS13 Activity Test Interpretation: Although newer-generation assays give rapid and accurate measurement, the utility of such measurements in acute or long-term medical management continues to be a matter of debate.

Use in Diagnosis: Although the presence of ADAMTS13 deficiency supports the diagnosis of TTP, a prospective cohort showed that only 16 of 48 patients with idiopathic TTP had severe ADAMTS13 deficiency. Thus, a normal ADAMTS13 activity should not delay initiation of TPE in a patient with a high degree of suspicion for TTP (otherwise unexplained thrombocytopenia and microangiopathic hemolytic anemia). In prospective studies, severe deficiency of ADAMTS13 (<5%) had sensitivity for the diagnosis of TTP ranging from 33% to 80%. Accordingly, a severely reduced ADAMTS13 activity is neither necessary nor sufficient for the diagnosis of TTP.

Use in Prognosis: Those with severe ADAMTS13 deficiency are more likely to achieve a remission in response to TPE than those with nonsevere ADAMTS13 deficiency (82%–88% vs. 20%–75%); however, despite superior remission rates, the likelihood of relapse is greater (43% vs. 8%).

The utility of testing patients in follow-up after remission has been investigated in several small studies. In a retrospective study, patients with an undetectable ADAMTS13 activity at first remission had a greater risk of relapse (odds ratio 2.9) than those with detectable ADAMTS13 activity. In a prospective study of 35 patients with a first episode of idiopathic TTP and undetectable ADAMTS13 activity at presentation, the presence of undetectable ADAMTS13 activity at the time of remission had a positive and negative predictive value of 28% and 95%, respectively. Among subjects with an undetectable ADAMTS13 activity at the time of remission 38% relapsed, whereas only 5% of those with a detectable level relapsed. Accordingly, a detectable ADAMTS13 activity suggests relapse is unlikely, but an undetectable level is not predictive of relapse over the subsequent 18 months.

Use in Guiding Treatment: There is little literature on using ADAMTS13 to manage patients and tailor therapy according to activity levels. In a cross-sectional analysis where 48 patients with a least one episode of acquired TTP followed by severe ADAMTS13 deficiency either after remission or following an initial, partial, or complete enzyme activity recovery were actively monitored with ADAMTS13 activity levels. 30 patients received preemptive rituximab promptly after the detection of severe acquired ADAMTS13 deficiency and 18 did not. These therapeutic maneuvers were followed with ADAMTS13 activity levels at 1 and 3 months post therapy and then every 3 months for at least 24 months. The relapse incidence 17 months following rituximab was 0 episode/year with an increase in ADAMTS13 activity to 46%, compared with a relapse incidence of 0.5 episodes per year in those that did not receive rituximab. Although further studies with larger number of patient are needed, the persistent deficiency of ADAMTS13 activity during follow-up should be carefully monitored, particularly in patients who have experienced recurrent acute events.

Measurement of Anti-ADAMTS13 Autoantibodies: Several tests for anti-ADAMTS13 inhibitory antibodies exist, including mixing studies, using the same methodology for measurement of ADAMTS13 activity as above (similar to testing for coagulation factor inhibitors—see Chapter 132). These tests measure the inhibitory activity of the patient's plasma on the ADAMTS13 activity in normal pooled plasma thereby detecting neutralizing antibodies. In contrast, enzyme-linked immunosorbent assay–based procedures detect both neutralizing and nonneutralizing antibodies.

Interpretation: Inhibitors are present in 67%–87% of patients with severe ADAMTS13 deficiency and, if detected, improve the specificity of the diagnosis of acquired TTP; however, the benefit of TPE is not limited to acquired TTP, therefore, the identification of an inhibitor has little impact on initial therapy. The risk of relapse in patients with an undetectable ADAMTS13 activity and a positive inhibitor titer was increased (odds ratio 3.6). Overall, the presence of an inhibitor to ADAMTS13 has been associated with poorer outcomes: lower remission rates (67%–84% vs. 100%), slower response to plasma exchange (23 days vs. 7 days), greater chances of relapse (43%–62% vs. 5%–25%), and higher mortality (16%–33% vs. 0%).

Further Reading

Hie, M., et al. (2014). Preemptive rituximab infusions after remission efficiently prevent relapses in acquired thrombotic thrombocytopenic purpura. *Blood*, *124*, 204–210.

Knovich, M. A., Farland, A., & Owen, J. (2012). Longterm management of acquired thrombotic thrombocytopenia purpura using serial ADAMTS13 measurements. *Eur J Haematol*, *88*, 518–525.

Kremer Hovinga, J. A., Mottini, M., Lammle, B., et al. (2006). Measurement of ADAMTS-13 activity in plasma by the FRETS-VWF73 assay: Comparison with other assay methods. *J Thromb Haemost*, *4*, 1146–1148.

Lammle, B., Kremer Hovinga, J. A., & George, J. N. (2008). Acquired thrombotic thrombocytopenic purpura: ADAMTS13 activity, anti-ADAMTS13 autoantibodies and risk of recurrent disease. *Haematologica*, *93*, 172–177.

Peyvandi, F., Lavoretano, S., Palla, R., et al. (2008). ADAMTS13 and anti-ADAMTS13 antibodies as markers for recurrence of acquired thrombotic thrombocytopenic purpura during remission. *Haematologica*, *93*, 232–239.

Tripodi, A., Chantarangkul, V., Bohm, M., et al. (2004). Measurement of von Willebrand factor cleaving protease (ADAMTS-13): Results of an international collaborative study involving 11 methods testing the same set of coded plasmas. *J Thromb Haemost*, *2*, 1601–1609.

CHAPTER 158

Laboratory Diagnosis of Lupus Anticoagulant and Antiphospholipid Antibodies

Lucia R. Wolgast, MD

Introduction: Antiphospholipid (aPL) antibodies are a heterogeneous family of autoantibodies that are primarily directed against plasma proteins complexed with anionic phospholipids that are found on damaged/activated cellular membranes, including endothelial cells, trophoblasts, platelets, and monocytes. The formation of autoantibodies against many antigenic targets, interfering with many of the biochemical and cellular functions of phospholipids, has led to various pathological clinical manifestations in patients with high levels of these autoantibodies. The association of thrombosis and/or fetal loss with aPL antibodies is known as the *antiphospholipid syndrome* (APS). The current investigational criteria for APS (Table 158.1) are defined by a constellation of clinical and laboratory manifestations. These criteria include histories of documented thrombosis and/or defined pregnancy complications together with positivity of specific laboratory tests (i.e., "APS criteria assays"), including positivity of coagulation tests identifying a lupus anticoagulant (LA), and medium to high titer positivity in immunoassays for anti-β2-glycoprotein I (anti-β2GPI) IgG or IgM and anticardiolipin (aCL) IgG or IgM. Rare patients with aPL antibodies may develop disseminated thrombosis in large and small vessels with resulting multiorgan failure known as catastrophic APS (Table 158.2).

It is important to note that some patients may have typical clinical manifestations of APS but are entirely negative for APS criteria assays, (aCL or anti-β2GPI IgG/IgM and/or LA tests), a situation referred to as seronegative APS (SNAPS). Furthermore, there are other aPL antibody tests that are positive in patients with APS clinical manifestations, but are not yet accepted by consensus criteria. These tests are currently being referred to as "noncriteria laboratory assays." In this chapter we will review the assays involved in identifying the presence of aPL antibodies.

Antiphospholipid Syndrome Criteria Assays: The current APS criteria assays developed from two laboratory anomalies. First, the biologic false-positive serologic test for syphilis, which was known to be associated with autoimmunity, and the test's antigenic target cardiolipin led to the quantitative immunoassays to measure antibodies against cardiolipin and its cofactor β_2-glycoprotein I. Subsequent clinical trials associated elevated aCL and anti-β2GPI with thrombosis, spontaneous abortion, and neurologic disease. Meanwhile, patients with systemic lupus erythematosus (SLE) were found to have a *lupus anticoagulant,* an antibody-mediated inhibition of phospholipid-dependent coagulation, which caused prolonged partial thromboplastin time tests. The identification of a lupus anticoagulant (LA) in coagulation-based testing was subsequently associated with thrombosis and pregnancy complications. Both immunoassay and coagulation-based tests became crucial to identifying the aPL antibodies of APS.

Transfusion Medicine and Hemostasis. https://doi.org/10.1016/B978-0-12-813726-0.00158-6

TABLE 158.1 Sydney Investigational Criteria for the Diagnosis of the Antiphospholipid Syndrome[a]

Clinical

- Vascular thrombosis (one or more episodes of arterial, venous, or small vessel thrombosis). For histopathologic diagnosis, there should be no evidence of inflammation in the vessel wall.
- Pregnancy morbidities attributable to placental insufficiency, including (1) three or more otherwise unexplained recurrent spontaneous miscarriages, before 10 weeks of gestation, (2) one or more fetal losses after the 10th week of gestation, (3) stillbirth, and (4) episode of preeclampsia, preterm labor, placental abruption, intrauterine growth restriction, or oligohydramnios that are otherwise unexplained.

Laboratory

- Medium or high titer aCL or anti-β2GPI IgG and/or IgM antibody present on two or more occasions, at least 12 weeks apart, measured by standard ELISAs.
- Lupus anticoagulant in plasma, on two or more occasions, at least 12 weeks apart, detected according to the guidelines of the ISTH SSC Subcommittee on Lupus Anticoagulants and Phospholipid-Dependent Antibodies.

aCL, anticardiolipin; ***aPL***, antiphospholipid; ***ELISA***, enzyme-linked immunosorbent assay; ***Ig***, immunoglobulin; ***β2GPI***, β2-glycoprotein I.

[a]"Definite APS" is considered to be present if at least one of the clinical criteria and one of the laboratory criteria are met.

Modified from Miyakis, S., Lockshin, M.D., Atsumi, T., et al. (2006). International consensus statement on an update of the classification criteria for definite antiphospholipid syndrome (APS). *Thromb Haemost, 4*, 295–306.

TABLE 158.2 Proposed Criteria for the Classification of Catastrophic Antiphospholipid Syndrome (APS)

1. Evidence of involvement of three or more organs, systems, and/or tissues[a]
2. Development of manifestations simultaneously or in less than a week
3. Confirmation by histopathology of small vessel occlusion in at least one organ or tissue[b]
4. Laboratory confirmation of the presence of antiphospholipid antibodies (lupus anticoagulant and/or anticardiolipin antibodies)[c]

Definite Catastrophic APS

- All four criteria

Probable Catastrophic APS

- All four criteria, except for only two organs, systems, and/or tissues involvement
- All four criteria, except for the absence of laboratory confirmation at least 6 weeks apart due to the early death of a patient never previously tested for aPL before the catastrophic APS event
- Criteria 1, 2, and 4
- Criteria 1, 3, and 4 and the development of a third event in more than a week but less than a month, despite anticoagulation

[a]Usually, clinical evidence of vessel occlusions, confirmed by imaging techniques when appropriate. Renal involvement is defined by a 50% rise in serum creatinine, severe systemic hypertension (N180/100 mm Hg), and/or proteinuria (N500 mg/24 hours).

[b]For histopathological confirmation, significant evidence of thrombosis must be present, although, in contrast to Sydney criteria, vasculitis may coexist occasionally.

[c]If the patient had not been previously diagnosed as having an APS, the laboratory confirmation requires that the presence of antiphospholipid antibodies must be detected on two or more occasions at least 6 weeks apart (not necessarily at the time of the event), according to the proposed preliminary criteria for the classification of definite APS.

Modified from Asherson, R.A., Cevera, R., de Groot, P.G., et al. (2003). Catastrophic antiphospholipid syndrome: International consensus statement on classification criteria and treatment guidelines. *Lupus, 12*, 530–534.

However, these current APS criteria assay are inherently limited because they were not designed to measure known disease mechanisms. The aCL IgG and IgM assays are the most sensitive, but least specific assays, anti-β2GPI IgG and IgM assays are more specific but less sensitive, and LA assays, of which the dilute Russell viper venom time (dRVVT) is the most common, are the least sensitive but the most specific. Nevertheless, these assays have proven to be useful surrogates for thrombotic risk. Strong positivity for more than one of the APS criteria assays is indicative of an increased risk for clinical events.

Prevalence of Antiphospholipid Antibodies and Appropriateness of Testing: The prevalence of aPL antibodies in the asymptomatic "normal" population has generally been estimated at 3%–10% with a prevalence of 1%–5% for LA, 1%–5% for aCL antibodies, and 3% for anti-β2GPI antibodies. In one group of healthy young women, who served as controls in a study, 18% had elevated aCL antibodies and 13% tested LA positive. The development of aPL antibodies can occur after viral infections, syphilis, Lyme disease, hepatitis C, alcoholic liver disease, HIV infection, multiple sclerosis, and in patients taking medications such as chlorpromazine or procainamide. These elevations in aPL antibodies are generally not associated with thrombosis.

A significant proportion of SLE patients are positive for aPL antibodies; estimates range between 12% and 30% for aCL antibodies and 15% and 34% for LA antibodies. aPL antibodies have been associated with virtually all other autoimmune conditions. In some patients, the presence of aPL antibodies may herald the development of SLE.

Because of the prevalence of aPL antibodies in normal and reactive conditions, clinicians should determine the appropriateness of testing in their patients to avoid misdiagnosis. Patients in the **high appropriateness group** include patients with unprovoked and unexplained venous thromboembolism, arterial thrombosis in young patients (<50 years of age), thrombosis at unusual sites, late pregnancy loss, and any thrombosis or pregnancy morbidity in patients with autoimmune diseases (SLE, rheumatoid arthritis, autoimmune thrombocytopenia, autoimmune hemolytic anemia). The **moderate appropriateness group** includes asymptomatic patients who are incidentally found prolonged aPTT—often during routine testing—and young patients with recurrent spontaneous early pregnancy loss and provoked VTE. Patients in the **low appropriateness group** include elderly patients with venous or arterial thromboembolism.

Lupus Anticoagulant Tests: The various LA tests use different coagulation systems to report the inhibition of phospholipid-dependent blood coagulation reactions. These include modifications of the activated partial thromboplastin time (aPTT) with LA-sensitive and LA-insensitive reagents, the dRVVT, the kaolin clotting time, the tissue thromboplastin inhibition time, the hexagonal phase array test, and the platelet neutralization procedure. The results of LA tests can be variable among laboratories; although most of laboratories agree on identification of plasmas containing strong positive LA activity. The Subcommittee on Antiphospholipid Antibodies of the International Society of Thrombosis and Hemostasis has specific criteria for standardizing the diagnosis of LAs. They recommend performing two different coagulation tests with different assay principles (aPTT and dRVVT) as screening assays, then using mixing studies on abnormal results and repeat testing using phospholipids to confirm the presence of a phospholipid-dependent antibody.

Activated Partial Thromboplastin Time Tests: Different commercial aPTT reagents vary widely with respect to their sensitivities to LA, so it is important to know the characteristics of the particular reagent(s) that is being utilized. In general, LA-sensitive aPTT reagents are used in routine coagulation testing, so the identification of a prolonged aPTT in the absence of anticoagulation, bleeding, or known factor deficiency is a good screening test for an LA. When the aPTT is prolonged and not "correctable" by mixing studies with normal plasma, an LA should be suspected. The effects of incubation with normal plasma may be helpful in differentiating LAs from coagulation factor inhibitors. A factor VIII inhibitor usually requires incubation with normal plasma for 1–2 hours at 37°C to show prolongation on aPTT, whereas LA-containing plasmas usually prolong the aPTT immediately after mixing and show no further prolongation with incubation.

LA is confirmed by reassaying the aPTT with higher amount of phospholipids (e.g., hexagonal phospholipids) or frozen washed platelets as the source of phospholipid (i.e., *platelet neutralization procedure*). It is important to note that both types of anticoagulants—i.e., LA and specific coagulation factor inhibitors—may sometimes coexist and yield a confusing laboratory picture. LAs may cause artifactual decreases in certain coagulation-based factor levels that are based on aPTT reagents; these patients are sometimes erroneously misdiagnosed as having multiple coagulation factor deficiencies. This problem can be handled by repeating the coagulation by using an LA-insensitive aPTT reagent or chromogenic assays for coagulation factor assays.

Dilute Russell Viper Venom Time: The dRVVT is considered to be one of the most sensitive of the LA tests. The assay uses Russell viper venom (RVV) in a system containing limiting quantities of diluted rabbit brain phospholipid. RVV directly activates coagulation factor X, leading to formation of fibrin clot. aPL antibodies can prolong the dRVVT by interfering with assembly of the prothrombinase complex; this prolongation is reversed by adding excess phospholipid to the reaction (sometimes referred to a "confirmatory test"). The ability to prolong and then shorten the clotting time using excess phospholipid is a positive DRVVT test. In addition, to ensure that prolongation of the clotting time is not a result of a factor deficiency, the dRVVT test can be repeated with a mixture of patient and control plasmas. Anticoagulant therapy with direct Xa inhibitors or direct thrombin inhibitors can yield falsely abnormal (i.e., positive) test results.

Antiphospholipid Immunoassays: Correlation of aPL antibody levels determined in reference sera by different laboratories has been somewhat problematic, particularly when aPL antibodies are measured using different commercial or homemade enzyme-linked immunosorbent assay (ELISA) kits. Automated chemiluminescent assays will hopefully provide better standardization of aPL antibody testing. In the meantime, it is important that a patient's aPL antibody testing is performed at the same laboratory for both initial and repeat testing.

Anticardiolipin Antibody Assay: The quantities of aCL IgG and aCL IgM bound are expressed in GPL or MPL units, respectively, one unit representing the cardiolipin binding activity of 1 μg/mL of affinity-purified aPL antibodies from reference sera.

Binding reflects both the titer and affinity/avidity of the antibody. Most patients with APS are identified by elevated levels of aCL antibodies. High levels of aCL antibodies are associated with increased risk of thrombosis, including deep vein thrombosis, myocardial infarction and stroke, and increased frequency of fetal losses. .aCL antibody testing has high sensitivity but poor specificity, especially in asymptomatic patients. There is an interesting seasonal variability, with more normal healthy people having increased aPL antibodies in the winter than in the summer months.

Anti-β2GPI Antibody Assay: β2GPI is the major protein cofactor recognized by aPL antibodies. ELISAs for anti-β2GPI antibodies are considered to be more specific but less sensitive for APS than aCL antibody assays. Although these antibodies are usually present in conjunction with abnormal aCL and antiphosphatidylserine (aPS) antibodies, patients with APS can present with only antibodies to β2GPI. Despite their higher specificity for APS (98%), β2GPI antibodies alone cannot be relied on for the diagnosis because of their low sensitivity (40%–50%) and triple testing for aCL and anti-β2GPI antibodies, along with LA is advised.

In a systematic literature review, 34 of 60 studies, of which none were prospective, showed significant associations between anti-β2GPI antibodies and thrombosis. Of 10 studies that included multivariate analysis, only 2 confirmed that IgG anti-β2GPI antibodies were independent risk factors for venous thrombosis. Anti-β2GPI antibodies were more often associated with venous than arterial events.

Multipositivity for Antiphospholipid Syndrome Criteria Assays and Clinical Risk: Strong positivity for more than one of the aPL antibody criteria assays has been correlated with increased risk for developing clinical events in several retrospective and prospective studies.

Overall, both symptomatic and asymptomatic patients with triple-positive laboratory results: LA positive, aCL (IgG or IgM >40 GPL), and anti-β2GPI (IgG or IgM >99th percentile) should be considered high risk for future manifestations of APS. Meanwhile, patients who are double positive for LA, aCL (IgG or IgM >40 GPL), and/or anti-β2GPI (IgG or IgM >99th percentile) or single positive for LA should be considered medium risk for APS. Finally, single positivity for aCL (IgG or IgM >40 GPL) or anti-β2GPI (IgG or IgM >99th percentile) should be considered low risk for APS; however, treatment may be warranted in the presence of thrombosis or pregnancy complications or other high-risk factors.

"Noncriteria" Laboratory Assays: "Noncriteria" laboratory tests include IgA isotypes to aCL and anti-β2GPI, antibodies to other plasma proteins (i.e., prothrombin (PT), vimentin, and annexin V), as well as to anionic phospholipids (i.e., phosphatidylserine, phosphatidylethanolamine, phosphatidylinositol), and to protein/phospholipid complexes (i.e., aPS/PT). These newer assays have been associated with clinical manifestations of APS and have identified patients who would otherwise be considered SNAPS.

Anticardiolipin and Anti-β_2-Glycoprotein I IgA Assays: The clinical utility of testing aPL antibodies of IgA isotype remains controversial. The prevalence of true positivity to aCL IgA antibodies has been reported to be very low. However, anti-β2GPI IgA antibodies were reported to be significantly associated with thrombosis.

aCL and anti-β2GPI IgA testing were not included in consensus criteria for APS classification, however, testing for the IgA isotype (particularly anti-β2GPI IgA) was recommended in cases where APS is suspected, but the IgG and IgM tests are negative.

Antiphosphotidylserine/Prothrombin Assays: The data for individual anti-prothrombin (aPT) assays and antiphosphatidylserine (aPS) assays have been mixed in identifying patients with clinical manifestations of APS and correlating with other APS criteria assays. However, antibodies against the phosphatidylserine–prothrombin complex (aPS/PT) were closely associated with APS and LA. The presence of aPS/PT antibodies strongly correlated with the presence of LA had a higher sensitivity and specificity than aPT antibodies alone for the diagnosis of APS and had comparable sensitivity and specificity to aCL for the diagnosis of APS. In a recent study of 168 APS patients, 86% of these patients were positive for aPS/PT antibodies and 50% of the SNAPS patients were positive. There was a significant association of aPS/PT IgG with thrombotic events, and aPS/PT IgG and IgM with pregnancy loss. The aPS/PT IgG/IgM assays highly correlated with LA testing. The high diagnostic performance of this test may make it a useful test for SNAPS patients, to assess risk, and when LA testing is not available.

Other Antiphospholipid Immunoassays: Antibodies against the zwitterionic phospholipid, phosphatidylethanolamine, have been associated with thrombosis and with activated protein C resistance. Some studies have suggested that antiphosphatidylethanolamine antibodies can occur in APS in the absence of antibodies against cardiolipin or other anionic phospholipids. Antibodies against phosphatidylinositol antibodies were reported in young patients with cerebral ischemia. In a recent study of nearly 3000 women, IgG antibodies against phosphatidylinositol, phosphatidic acid, phosphatidylserine, and IgM antibodies against phosphatidylcholine were reported to be significantly elevated in patients with recurrent pregnancy loss, implantation failure, and unexplained fertility compared with fertile controls. Although some investigators have advocated testing for such antibodies, the most recent consensus statement concluded that there was no benefit in testing patients for antibodies against a large panel of phospholipids.

Antidomain I of β2GPI Assay: This immunoassay identifies IgG antibodies against a specific amino acid sequence, G40-R30, within domain I of β_2GPI. A recent multi-center study of 442 patients who tested positive for anti-β2GPI antibodies reported that the detection of specific antidomain I IgG antibodies was more strongly associated with thrombosis and obstetric complications than anti-β2GPI antibodies detected using the standard anti-β2GPI antibody assays. In a recent study, patients tested for antidomain I IgG and aPS/PT IgG/IgM showed high positive predictive value for the diagnosis APS.

Annexin A5 Assays: Annexin A5 is potent anticoagulant protein that prevents binding and activation of coagulation factors by shielding exposed anionic phospholipid surfaces. aPL antibodies disrupt anionic phospholipid surfaces by displacing annexin

A5 and allowing acceleration of coagulation reactions. The annexin A5 resistance (A5R) assay is a coagulation-based assay with added annexin A5 that identifies the presence of aPL antibodies by the shorter clotting times. A recent study showed that aPL antibody–positive patients had significantly shorter coagulation times (i.e., lower A5R) than aPL antibody–negative patients using this assay. The lower A5R also correlated with higher titers and multipositivity to aPL antibodies. An annexin A5 flow cytometric assay also quantifies the degree to which aPL antibodies compete with annexin A5 and correlates with APS criteria assays and is sensitive to APS patients.

Genetic, Genomic, and Proteomic Studies in Antiphospholipid Syndrome: Proteomic studies have also aimed to provide insights into the mechanism and to identify proteins that might predict thrombotic risk in APS patients. The proteins that were reported to be differentially expressed in monocytes of APS patient with a history thrombosis included annexin A1, annexin A2, ubiquitin Nedd8, Rho A protein, protein disulfide isomerase, and Hsp60. Genetic studies of APS patients have identified upregulated genes encoding many proteins known to be involved in thrombosis, such as apolipoprotein E (ApoE), coagulation factor X, and thromboxane. Further studies are needed to elucidate potential gene targets.

Further Reading

Bertolaccini, M., Amengual, O., Andreoli, L., et al. (2014). 14th international congress on antiphospholipid antibodies task force. Report on antiphospholipid syndrome laboratory diagnostics and trends. *Autoimmun Rev*, *3*(9), 917–930.

Chaturvedi, S., & McCrae, K. (2017). Diagnosis and management of the antiphospholipid syndrome. *Blood Rev*, *31*, 406–417.

Lakos, G., Favaloro, E. J., Harris, E. N., Meroni, P. L., et al. (2011). International consensus guidelines on anticardiolipin and anti-beta2-glycoprotein I testing: A report from the APL task force at the 13th international congress on antiphospholipid antibodies. *Arthritis Rheum*, *64*, 1–10.

Miyakis, S., Lockshin, M. D., Atsumi, T., et al. (2006). International consensus statement on an update of the classification criteria for definite antiphospholipid syndrome (APS). *J Thromb Haemost*, *4*(2), 295–306.

Pengo, V., Bison, E., Denas, G., et al. (2017). Laboratory diagnostics of antiphospholipid syndrome. *Int J Hematol*, *106*, 206–211.

Pengo, V., Tripodi, A., Reber, G., Rand, J. H., Ortel, T. L., Galli, M., et al. (2009). Update of the guidelines for lupus anticoagulant detection. Subcommittee on lupus anticoagulant/antiphospholipid antibody of the scientific and standardisation committee of the international society on thrombosis and haemostasis. *J Thromb Haemost*, *7*(10), 1737–1740.

Sciasca, S., Radin, M., Bazzan, M., et al. (2017). Novel diagnostic and therapeutic frontiers in thrombotic antiphospholipid syndrome. *Intern Emerg Med*, *12*, 1–7.

CHAPTER 159

Laboratory Monitoring for Heparins, Fondaparinux, Direct Thrombin Inhibitors, and Oral Anti-Xa Medications

Scott T. Avecilla, MD, PhD, Mikhail Roshal, MD, PhD,
Morayma Reyes Gil, MD, PhD and Patrick A. Erdman, DO

Laboratory testing modalities and recommendations for monitoring heparin, heparin-like medications, direct thrombin inhibitors (DTIs), and direct anti-Xa inhibitors (DXIs), including indications, testing recommendations, testing modalities, and test limitations are discussed below.

There are several options available for laboratory monitoring of these drugs, all with varying levels of utility. New DTI's and DXI's are in clinical trials and the number of patients undergoing direct oral anticoagulant therapy will continue to expand.

Unfractionated Heparin

Testing Recommendations: Therapeutic heparin is usually administered as a continuous IV infusion. Response to heparin is highly variable between patients, and heparin activity requires monitoring to prevent both over and under coagulation. Therapeutic monitoring should be performed every 6 hours after initiating or changing IV infusion until therapeutic levels are reached and every 24 hours thereafter. Low-dose subcutaneous heparin administration is usually not monitored.

Testing Modalities: In the United States, activated partial thromboplastin time (PTT) is the predominant modality for monitoring unfractionated heparin (UFH) therapy. This is primarily due to low cost and wide availability considerations and not to the specific advantages of the method. It is recommended that PTT-based tests be correlated to anti-Xa activity in each laboratory that performs PTT-based UFH monitoring. PTT has a number of limitations due to the highly variable response to UFH that is dependent on PTT reagents, instrumentation, patient-to-patient variation, lupus anticoagulant interference, factor deficiencies, high levels of factor (FVIII), and the heparin preparation itself. Alternatively, direct measures of heparin activity using Factor (FXa)–based assays or protamine titration can be used.

Partial Thromboplastin Time Correlation/Calibration for Heparin Monitoring: Most laboratories performing PTT-based testing calibrate the assay using anti-Xa activity. To do so at least 40 samples from patients receiving IV heparin are assayed in parallel for PTT and anti-Xa activity. The therapeutic PTT range is then established on the basis of the lower cutoff of 0.3 units (U) and upper cutoff of 0.7 U of anti-FXa activity. Because heparin has a variable effect in vivo, the calibration should

Transfusion Medicine and Hemostasis. https://doi.org/10.1016/B978-0-12-813726-0.00159-8

be done on samples taken from patients receiving the specific preparation of UFH rather than samples spiked with heparin in vitro. A linear regression curve fitting is used to establish the cutoffs. Only about 50%–60% of variability of PTT is attributable directly to the concentration of heparin in ex vivo samples. Thus r^2 values of 0.5–0.6 are to be expected.

Anti-Xa Assays for Direct Measurement of Heparin Activity: Chromogenic antithrombin (AT)–dependent FXa inhibition offers an alternative to PTT and has substantially fewer interferences. There are slight variations in these assays among different manufacturers. In a typical assay, the following sequence is used: (1) add fixed amount of FXa to patient plasma sample, (2) short incubation period, and (3) add chromogenic substrate to assay residual activity. Reduction in the FXa activity due to test plasma heparin can then be plotted on the standard curve using standardized heparin activity as a calibrator. Either kinetic or endpoint measurements of color change due to FXa activity in cleaving the chromogenic substrate can be taken. Readout is performed at the wavelength of 405 nm. The relationship between color change or development and heparin level is indirectly proportional in that higher levels of heparin result in lower levels of color formation and vice versa. This assay can be performed either with or without addition of excess exogenous AT. The assays relying on endogenous AT only are affected by low AT concentration. This variant may be advantageous if there is a concern for heparin resistance due to low AT.

Preanalytical Considerations and Result Interpretation: The specimen should be drawn from a different extremity than the one used for infusion. Timely testing is critical to obtain a proper heparin activity result. Heparin can be bound and inactivated by platelet factor 4 released from platelets during prolonged storage. Thus, it is recommended that plasma be separated from cellular components and platelets within 1 hour of collection to minimize this effect. Testing should be completed within 4 hours of collection.

Result Interpretation: Lupus anticoagulant, factor deficiencies, factor inhibitors, and pharmacological coagulation inhibitors other than heparin affect the PTT. The presence of those variables can cause overestimation of PTT-based heparin activity, leading to underdosing. Elevated levels of FVIII activity cause underestimation of the effect and possible overdosing. Direct anti-Xa-based heparin activity tests suffer from fewer interferences. Nonetheless simultaneous presence of other anti-factor Xa active medications such as low-molecular-weight heparin (LMWH), fondaparinux, and DXIs can produce overestimation of UFH activity. Lack of standardization between different anti-Xa tests has been shown to cause up to 30% difference in UFH concentration reporting between different kits.

Low-Molecular-Weight Heparin and Fondaparinux

Testing Recommendations: LMWH and the single pentasaccharide drug fondaparinux have limited or no AT activity and either no or a modest effect on the PTT, depending on the drug concentration and reagent sensitivity. Because of their reduced binding to platelet and endothelial surfaces, these drugs show substantially less variability in bioavailability compared with UFH. Thus unlike UFH, they generally show a predictable dose response. Indications for activity monitoring of these medications are therefore limited. The anticoagulants show renal clearance, and renal impairment may change drug levels. In addition, drug monitoring in particular clinical situations such as administration to infants, children, obese or underweight patients, those on long-term treatment, pregnant, or having unexpected bleeding or thrombosis may be needed because of differences in pharmacokinetic response. LMWH peak steady-state levels (at least after the third dose) should be drawn 4 hours after subcutaneous injection. Steady-state fondaparinux levels should be drawn 3 hours after last dose.

Testing Modalities: Chromogenic anti-Xa assays can be used for monitoring these drugs. LMWH heparin or hybrid with UFH standard curve is acceptable to monitor LMWH. A standard curve for each type of LMWH is not required. Reporting is in U/mL. Fondaparinux is calibrated separately and reported in mcg/mL.

Test Interpretations: LMWH recommended therapeutic range is 1–2 U/mL for once-daily dosing and 0.5–1 U/mL for twice-daily treatment, measured 4 hours after subcutaneous administration. For fondaparinux observed mean peak steady-state plasma concentration is approximately 0.39–0.50 mcg/mL for patients receiving 2.5 mg once-daily dose and 1.20–1.26 mcg/mL in patients receiving weight-adjusted treatment doses of 5–10 mg/day. Peak levels are generally drawn 3–4 hours after drug administration. The recommended prophylactic range is 0.1–0.5 mcg/mL, and the recommended therapeutic range is 0.6–1.5 mcg/mL. Simultaneous presence of other anti-Xa active drugs and low AT levels in tests relying on endogenous AT activity will affect results of testing.

Direct Anti-Xa Inhibitors

Testing Indications: Apixaban, edoxaban, and rivaroxaban are direct FXa inhibitors that do not require AT for an inhibitory effect. They show highly predictable bioavailability and wide therapeutic range. Routine monitoring is not needed. The indications for monitoring are currently unclear, but drug concentration measurement may be requested in special situations, such as suspected overdose, uncontrolled bleeding, urgent surgery, and suspected noncompliance (see Chapter 54).

Test Modalities: No US Food and Drug Administration (FDA)–approved assays to measure these medications are currently on the market. High-pressure liquid chromatography followed by tandem mass spectroscopy (HPLC–MS/MS) produces highly accurate concentration results but is not widely available. Modified anti-Xa assay without added AT using drug-specific calibrators was shown to provide a good measure of plasma concentration of these medications. This modification is not currently available commercially but is relatively straightforward. The drug also affects prothrombin time PT and PTT. Although chromogenic anti-Xa assays appear preferable, PT-based tests using drug-specific calibrators can be used to measure DXI concentration, provided no other factors affecting the PT are present. PT reagents have variable sensitivity to these medications, thus local calibration appears to be necessary.

Test Interpretation: Dosing for rivaroxaban includes 15/20 mg qD for nonvalvular atrial fibrillation, 15 mg BID for the first 21 days, then 20 mg qD thereafter for deep vein thrombosis and pulmonary embolism (DVT/PE), and 10 mg qD for DVT prophylaxis. Doing for apixaban includes 5 mg BID for most patients and 2.5 mg BID with any two of the following: (1) age ≥80 years, (2) body weight ≤60 kg, and (3) serum creatine ≥1.5 mg/dL. Apixaban is only approved for nonvalvular atrial fibrillation. Dosing for edoxaban includes 60 mg qD for nonvalvular atrial fibrillation and 60 mg qD for DVT/PE. There are currently no published therapeutic ranges for the oral factor X inhibitors and no recommendations on timing of testing. Peak plasma levels are reported to occur 1.5–4 hours after tablet ingestion. The maximum steady-state plasma concentrations for rivaroxaban were 112–184 ng/mL for a 10 mg daily dose and 111–294 ng/mL for a 20 mg daily dose. Twice-daily dose of 2.5 mg apixaban resulted in steady state through concentration of 55.5 ng/mL and 5 mg dose of 107 ng/mL.

Direct Thrombin Inhibitors

Testing Indications: DTIs have predictable pharmacokinetics and monitoring indications are not clear. Patients with impaired drug clearance and those at high risk for bleeding should be monitored for overdose and uncontrolled bleeding (see Chapter 54).

Testing Modalities: HPLC–MS/MS produces highly accurate concentration results but is not widely available. Activity-based tests are more common.

Activated Partial Thromboplastin Time: DTIs variably prolong both PT and PTT. Because the effect on PTT is greater, the test is occasionally used to assess the anticoagulant of effect of the medications. This approach suffers from numerous interferences that can cause both over- and underestimation of the actual anticoagulant effect. Vitamin K–dependent factor deficiencies, other factor and fibrinogen deficiencies, heparin, lupus anticoagulants, and elevated fibrin split products—among others—can cause overestimation of the anticoagulant effect. Elevated levels of FVIII and high fibrinogen levels on the other hand can cause underestimation. The new oral DTI dabigatran has a mild effect on the PTT, producing a rather shallow dose-response curve.

Thrombin-Based Assays: Thrombin-based assays (thrombin time, anti-IIa, and Ecarin-activated tests) offer a superior quantitative approach to measuring the drug activity in the plasma and are free from most interferences affecting PTT. Several thrombin-based test strategies are now in use.

Thrombin time has a predictable dose response to DTIs. Because thrombin time is very exquisitely to DTIs, it can be greatly prolonged beyond the usual measurement range of the instrument at the expected therapeutic and mildly supertherapeutic DTI concentrations. The issue can be overcome by a 1:4 to 1:10 dilution of the patient's plasma into the normal plasma pool or buffer. Plasma dilution has the added benefit of lowering potential interferences from fibrinogen concentration and fibrin split products, which often affect thrombin time. Commercial assay that uses saline, rather than plasma, for sample dilution is available. Chromogenic anti-FIIa (thrombin)–based assays without AT offer another potential avenue for DTI measurement.

Endogenous thrombin–based assays also offer a quantitative way to assess DTI concentrations. Ecarin (venom of the saw-scaled viper *Echis carinatus*) converts thrombin to mesothrombin (a partially processed thrombin form with clotting activity). Ecarin activation of thrombin does not require cofactors and is not influenced by DTIs. Mesothrombin activity itself is inhibited by DTI. The effect can be measured either in the clot-based (Ecarin clotting time) or chromogenic format (Ecarin chromogenic assay). Calibration of the assays must be done using active drug forms of the medications being measured. Dabigatran is supplied as a prodrug and cannot be used as an assay calibrator in that form. Commercial calibrators are available for most DTIs.

Results of the test are reported in concentration units of the medication or fold change from baseline. It is critical to ascertain absence of heparin in the sample, as heparin will lead to overestimation of the DTI concentration. If heparin contamination is suspected, it may be possible to neutralize heparin using enzymatic digestion or polybrene.

Test Interpretation: Published therapeutic ranges are not available for most DTIs. For dabigatran, expected peak steady-state concentration in patients with atrial fibrillation receiving 150 mg tablets, twice daily was ~180 ng/mL (0.5–2 hours after the oral dose) with a trough of ~90 ng/mL (11.5 hours after last dose). In the Ecarin clotting time assay, a fourfold increase of clotting time from baseline represents approximately a 200–300 ng/mL concentration of dabigatran. For argatroban, a 1.5- to 3-fold increase of baseline aPTT is considered therapeutic in many institutions for anticoagulation in heparin-induced thrombocytopenia. Published concentrations corresponding to this range vary, with 0.6–1.8 μg/mL approximating the mean of these ranges. Similarly for bivalirudin and lepirudin 1.5–2.5 times baseline aPTT are considered therapeutic, they correspond to approximately 0.25–1.5 and 0.5–1.5 mcg/mL concentrations, respectively.

Reversal Agents: Dabigatran anticoagulation can be reversed by the recently approved antidote, idarucizumab, which is a humanized monoclonal antibody fragment (Fab) (see Chapter 54). The antidote is specifically indicated for (1) emergency surgery/urgent procedures and (2) life-threatening or uncontrolled bleeding. Because the drug is bound to antidote, prolonged coagulation test parameters are expected to return to the patient's baseline.

Andexanet, a recombinant catalytically inactive form of FXa, is a drug under clinical investigation as a potential reversal agent for DXIs and in DTIs. Treatment with andexanet is also expected to return prolonged coagulation test parameters to baseline.

Further Reading

Avecilla, S. T., Ferrell, C., Chandler, W. L., & Reyes, M. (2012). Plasma-diluted thrombin time to measure dabigatran concentrations during dabigatran etexilate therapy. *Am J Clin Pathol*, *137*, 572–574.

Barrett, Y. C., Wang, Z., Frost, C., & Shenker, A. (2012). Clinical laboratory measurement of direct factor Xa inhibitors: Anti-Xa assay is preferable to prothrombin time assay. *Thromb Haemost*, *104*, 1263–1271.

Funk, D. M. (2012). Coagulation assays and anticoagulant monitoring. *Hematol Am Soc Hematol Educ Program*, *2012*, 460–465.

Harenberg, J., Marx, S., Kramer, R., Giese, C., & Weiss, C. (2011). Determination of an international sensitivity index of thromboplastin reagents using a WHO thromboplastin as calibrator for plasma spiked with rivaroxaban. *Blood Coagul Fibrinolysis, 22,* 637–641.

Hirsh, J., Warkentin, T. E., Shaughnessy, S. G., et al. (2001). Heparin and low-molecular-weight heparin: Mechanisms of action, pharmacokinetics, dosing, monitoring, efficacy, and safety. *Chest, 119,* 64S–94S.

Love, J. E., Ferrell, C., & Chandler, W. L. (2007). Monitoring direct thrombin inhibitors with a plasma diluted thrombin time. *Thromb Haemost, 98,* 234–242.

Mani, H., Rohde, G., Stratmann, G., et al. (2012). Accurate determination of rivaroxaban levels requires different calibrator sets but not addition of antithrombin. *Thromb Haemost, 108,* 191–198.

Olson, J. D., Arkin, C. F., Brandt, J. T., et al. (1998). Laboratory monitoring of unfractionated heparin therapy. *Arch Pathol Lab Med, 122,* 782–798.

Samama, M. M., Contant, G., Spiro, T. E., et al. (2012). Evaluation of the prothrombin time for measuring rivaroxaban plasma concentrations using calibrators and controls: Results of a multicenter field trial. *Clin Appl Thromb Hemost, 18,* 150–158.

van Ryn, J., Stangier, J., Haertter, S., et al. (2010). Dabigatran etexilate – a novel, reversible, oral direct thrombin inhibitor: Interpretation of coagulation assays and reversal of anticoagulant activity. *Thromb Haemost, 103,* 1116–1127.

CHAPTER 160

Laboratory Support for Warfarin Monitoring

Anne M. Winkler, MD

Warfarin use has declined with increased adoption of oral direct thrombin and direct factor (F) Xa inhibitors, but warfarin remains an important oral anticoagulant for patients with contraindications to or indications not currently covered by direct oral anticoagulant therapy, and for those with financial constraints that limit access to these medications. Warfarin is a vitamin K antagonist, which exerts its anticoagulant effect by interfering with γ-carboxylation of glutamate residues of the N-terminus of FII, FVII, FIX, and FX through inhibition of enzymatic reduction of vitamin K, resulting in reduced coagulant activity. However, warfarin also inhibits carboxylation of the natural occurring anticoagulants, proteins C, S, and Z, producing a less dominant, paradoxical procoagulant effect.

Warfarin is available as a racemic mixture of R and S enantiomers with a half-life of 36–42 hours, and its pharmacodynamics are strongly influenced by both environmental and genetic factors. As a result, frequent monitoring is required to maintain therapeutic dosing while minimizing the risk of sub- or supratherapeutic levels. The prothrombin time (PT) calculated in the form of the international normalized ratio (INR) is traditionally used to monitor warfarin treatment and remains the mainstay of warfarin monitoring. Genetic testing, including analysis of mutations of the cytochrome P450 complex CYP2C9 and vitamin K epoxide reductase (VKOR), has led to an explanation of the extremes of physiologic warfarin responses and a potential for more individualized treatment; however, prospective pharmacogenomic testing has not widely been adopted because of discrepant results from randomized controlled trials but may be reconsidered given results from a recent study.

Laboratory Testing for Warfarin Monitoring

Prothrombin Time: The PT measures the function of coagulation factors of the "*extrinsic*" and "*common*" pathways, namely fibrinogen, FII (prothrombin), FV, FVII, and FX. The PT is performed by adding thromboplastin (a source of tissue factor extracted from tissues such as brain, lung, or placenta and phospholipid or recombinant tissue factor and synthetic phospholipids) to recalcified plasma and assessing clotting times.

Interpretation: Traditionally, the PT has been used to monitor warfarin therapy because of the sensitivity to the variability of the vitamin K–dependent coagulation factors, FII, FVII, and FX. Initially, the PT reflects the marked reduction of FVII by warfarin; however, with continued treatment, further prolongation results from reduction of FII and FX.

Special Considerations and Sources of Error: Although the PT is a routine test that is widely used for monitoring, reporting the PT alone for this purpose has several distinct problems.

Transfusion Medicine and Hemostasis. https://doi.org/10.1016/B978-0-12-813726-0.00160-4

Variability in Responsiveness to Thromboplastin Reagents: Thromboplastin reagents vary in responsiveness to reduction of the vitamin K–dependent coagulation factors, phospholipid content, and preparation. More specifically, responsive thromboplastin reagents produce greater prolongation of the PT compared with unresponsive reagents, and this variability can be measured by determining the international sensitivity index (ISI) for each thromboplastin reagent. This is accomplished by comparison of the PT of the working reagent from normal controls and patients who have received stable vitamin K antagonist therapy for at least 6 weeks to results obtained using a standard reagent calibrated against the World Health Organization (WHO) Standard. In general, thromboplastin reagents with lower ISI values are more responsive and may result in lower coefficients of variation as compared with thromboplastin reagents with a higher ISI. As a result, the College of American Pathologists (CAP) recommends use of thromboplastin reagents with an ISI <1.7 (moderately responsive) that have been validated for a specific reagent/instrument combination and have added additional recommendations to the CAP laboratory accreditation program hematology checklist. These CAP requirements have been valuable in providing the best practices for the laboratories and were implemented almost entirely in all of the clinical laboratories participating in proficiency testing from a recent CAP Q probes study.

Standardization of Reporting: Owing to the variability of the PT, adoption of the INR has improved vitamin K antagonist monitoring and can be calculated using the following formula:

$$\mathrm{INR} = \left(\frac{\text{patient PT}}{\text{mean normal PT}}\right)^{\mathrm{ISI}}$$

The mean normal PT should be determined locally from normal individuals, and the PT geometric mean should be calculated for each new lot of reagent.

On initiation of warfarin anticoagulation, INR monitoring should begin after 2–3 doses; once stable, monitoring should occur at intervals no greater than 4–6 weeks. Patients with consistently stable INRs may be extended to 12 weeks; however, testing intervals are highly dependent on the time in therapeutic range. The American College of Chest Physicians (ACCP) publishes evidence-based clinical practice guidelines for warfarin treatment over a range of clinical indications, and additional information concerning the management of supratherapeutic INR values is published within the same supplement as recommended below.

The use of the INR to monitor patients with antiphospholipid syndrome has been criticized because of the varying effects of lupus anticoagulants on thromboplastin reagents. There is potential for spurious results without adequate reduction of vitamin K–dependent coagulation factors, and therefore subtherapeutic dosing. In patients with lupus anticoagulants, monitoring can be accomplished using chromogenic FX assays to avoid this artifact and chromogenic FX testing should be considered in patients with a lupus anticoagulant and abnormal baseline INR values or those who have had a lupus anticoagulant detected on more than one occasion. In addition, use of the INR in initial monitoring in patients with liver disease may be less reliable because of the variation in reduction of factors measured in the PT, and its use has arguably been questioned in the calculation of prognostic scores such as the Model for End-Stage Liver Disease. As a result, a new INR calibrated with plasma from patients with liver disease has been proposed for these patients. Despite all these criticisms, the INR remains the gold standard laboratory test for warfarin monitoring.

Point-of-Care International Normalized Ratio Testing: With the growing numbers of anticoagulation clinics and anticoagulation management services, patient self-testing and patient self-management with point-of-care INR testing has increased. However, point-of-care devices, which determine a thromboplastin-initiated clotting time that is electronically converted to a PT and/or INR, have limitations in accuracy and precision when compared with laboratory-based methods. These limitations include incorrect calibration of the ISI to the WHO standard, extrapolated mean normal PT values, and nonlinearity at supratherapeutic levels. Although the evidence does not support widespread use of point-of-care INR testing in general practice, patient self-testing and patient self-management has been associated with improved anticoagulation control and reduced incidence of thromboembolic or major bleeding events.

Abnormal Warfarin Responsiveness

Genetic Factors: As previously stated, warfarin is composed of R and S enantiomers, which are metabolized by different cytochrome P450 enzymes. The more potent S enantiomer is primarily metabolized by the CYP2C9 enzyme, whereas the R enantiomer is metabolized primarily by the CYP1A2 and CYP3A4 enzymes. As a result, mutations in the genes encoding the cytochrome P450 2C9 enzymes produce impaired metabolism of the S enantiomer, resulting in increased half-life and therefore reduced dosing requirements. The *CYP2C9*2* and *CYP2C9*3* have been the most characterized variants, and the specific demographics are displayed in Table 160.1.

In addition, mutations in the gene encoding the VKOR protein and comprising the vitamin K oxide reductase complex 1 (VKORC1) have been identified as a cause of hereditary warfarin resistance, which results in increased warfarin-dosing requirements. More specifically, three haplotypes have been identified, which have a reported prevalence of 58% in Europeans, 49% in Africans, and 10% in Asians.

Lastly, an additional mutation affecting FIX propeptide has been shown to affect warfarin pharmacokinetics, causing reduction of FIX, which is not measured using the PT. This mutation increases the risk of bleeding associated with warfarin and has been identified in less than 1.5% of the population.

Commercially available assays for the detection of 2C9 and VKORC1 mutants exist; however, universal screening of genetic mutations in patients beginning warfarin

TABLE 160.1 CYP2C9 Allele Prevalence

CYP2C9 Allele	CYP2C9*1	CYP2C9*2	CYP2C9*3
Ethnic Group			
White (%)	79–86	8–19.1	6–10
African American (%)	98.5	3	6
Asian (%)	95–98.3	0	1.7–5

TABLE 160.2 Drug, Food, and Dietary Supplement Interaction With Warfarin by Level of Supporting Evidence and Direction of Interaction

Level of Causation	Antiinfectives	Cardiovascular Drugs	Analgesics, Antiinflammaries, and Immunologics	Central Nervous System Drugs	Gastrointestinal Drugs and Food	Herbal Supplements	Other Drugs
Potentiation							
Highly probable	Ciprofloxacin Cotrimoxazole Erythromycin Fluconazole Isoniazid (600 mg/day) Metronidazole Miconazole oral gel Miconazole vaginal suppositories Voriconazole	Amiodarone Clofibrate Diltiazem Fenofibrate Propafenone Propranolol Sulfinpyrazone (biphasic with later inhibition)	Phenylbutazone Piroxicam	Alcohol (if concomitant liver disease) Citalopram Entacapone Sertraline	Cimetidine Fish oil Mango Omeprazole	Boldo-fenugreek Quilinggao	Anabolic steroids Zileuton
Probable	Amoxicillin/clavulanate Azithromycin Clarithromycin Itraconazole Levofloxacin Ritonavir Tetracycline	Acetylsalicylic acid Fluvastatin Quinidine Ropinirole Simvastatin	Acetaminophen Acetylsalicylic acid Celecoxib Dextropropoxyphene Interferon Tramadol	Disulfiram Choral hydrate Fluvoxamine Phenytoin (biphasic with later inhibition)	Grapefruit juice	Danshen Dong quai Lycium barbarum L PC-SPES	Fluorouracil Gemcitabine Levamisole/fluorouracil Paclitaxel Tamoxifen Tolterodine

Possible	Amoxicillin Amoxicillin/tranexamic rinse Chloramphenicol	Amiodarone-induced toxicosis Disopyramide Gemfibrozil	Celecoxib Indomethacin Leflunomide Propoxyphene Rofecoxib	Felbamate	Cranberry juice Orlistat	Danshen/methyl salicylate	Acarbose CMF (cyclo-phosphamide/methotrexate/fluorouracil)
	Gatifloxacin Miconazole topical gel Nalidixic acid Norfloxacin Ofloxacin Saquinavir Terbinafine	Metolazone	Sulindac Tolmetin Topical salicylates				Curbicin Danazol Ifosfamide Trastuzumab
Highly improbable	Cefamandole Cefazolin Sulfisoxazole	Bezafibrate Heparin	Levamisole Methylprednisolone Nabumetone	Fluoxetine/diazepam Quetiapine			Etoposide/carboplatin Levonorgestrel
Inhibition							
Highly probable	Griseofulvin Nafcillin Ribavirin Rifampin	Cholestyramine	Mesalamine	Barbiturates Carbamazepine	High vitamin K content foods/enteral feeds Avocado (large amounts)		Mercaptopurine
Probable	Dicloxacillin Ritonavir	Bosentan	Azathioprine	Chlordiazepoxide	Soy milk Sucralfate	Ginseng	Chelation therapy Influenza vaccine Multivitamin supplement Raloxifene hydrochloride
Possible	Terbinafine	Telmisartan	Sulfasalazine		Sushi-containing seaweed		Cyclosporine Etretinate Ubidecarenone
Highly improbable	Cloxacillin Nafcillin/dicloxacillin Teicoplanin	Furosemide		Propofol		Green tea	

Modified from Holbrook, A. M., Pereira, J. A., Labiris, R., et al. (2005). Systematic overview of warfarin and its drug and food interactions. *Arch Intern Med, 165*, 1095–1106.

therapy is not routinely performed, even though customized dosing algorithms exist and information concerning pharmacogenetic testing has been added to the warfarin-prescribing information. At present, data are equivocal as to whether or not the improvement in dosing prediction by pharmacogenetics leads to improved clinical outcomes. Recently, a randomized controlled trial of elderly patients undergoing elective hip and knee arthroplasty being treated with perioperative warfarin demonstrated that use of genotype-guided warfarin dosing in comparison with clinically guided dosing reduced the composite outcome, including risk of major bleeding, INR≥4, death within 30 days, and venous thromboembolism within 60 days. As a result, there may be a call for reconsideration of genotype-guided warfarin dosing.

Environmental and Drug Interactions: The pharmacokinetics of warfarin are heavily influenced by environmental factors and drug interactions, including prescribed medications, over-the-counter drugs, nutritional supplements, diet, and herbal products (Table 160.2).

Further Reading

Ageno, W., Gallu, A. S., & Wittkowsky, A. (2012). Oral anticoagulant therapy: Antithrombotic therapy and prevention of thrombosis: American College of Chest Physicians evidence-based clinical practice guidelines (9th edition). *Chest*, *141*, e44S–e88S.

Crowl, A., Schullo-Feulner, A., & Moon, J. Y. (2014). Warfarin monitoring in antiphospholipid syndrome and lupus anticoagulant. *Ann Pharmacother*, *48*, 1479–1483.

Gage, B. F., Bass, A. R., Lin, H., et al. (2017). Effect of genotype-guided warfarin dosing on clinical events and anticoagulation control among patients undergoing hip or knee Arthroplasty: The GIFT randomized clinical trial. *JAMA*, *318*, 1115–1124.

Holbrook, A., Schulman, S., & Witt, D. M. (2012). Evidence based management of anticoagulant therapy: antithrombotic therapy and prevention of thrombosis: American College of Chest Physicians evidence-based clinical practice guidelines (9th edition). *Chest*, *141*, e152S–e184S.

Howanitz, P. J., Darcy, T. P., Meier, F. A., & Bashleben, C. P. (2015). Assessing clinical laboratory quality: A College of American Pathologists Q-probes study of prothrombin time INR structures, processes, and outcomes in 98 laboratories. *Arch Pathol Lab Med*, *139*, 1108–1114.

Kimmel, S. E. (2015). Warfarin pharmacogenomics: Current best evidence. *J Thromb Haemost*, *13*, S266–S271.

Olson, J. D., Brandt, J. T., Chandler, W. L., et al. (2007). Laboratory reporting of the international normalized ratio. *Arch Pathol Lab Med*, *131*, 1641–1647.

Poller, L. (2004). International normalized ratios (INR): The first 20 years. *J Thromb Haemost*, *2*, 849–860.

CHAPTER 161

Molecular Testing in Coagulation

Jack Jacob, DO, Yitz Goldstein, MD, and Morayma Reyes Gil, MDPhD

John Bartlett, Director of Diagnostic Development at the Ontario Institute for Cancer Research writes that "accurate and appropriate diagnosis is fundamental to the successful treatment of disease." With the introduction and evolution of new technologies, we can evaluate coagulopathy at a molecular level like never before. Nevertheless, we must avoid ordering expensive and sometimes expansive molecular tests for rare disorders when alternative methods of diagnosis are available and recommended. The Core Laboratory (in-house or commercial) provides most tools for making many of these diagnoses, and it is only under certain circumstances, for example, with equivocal values or when we require additional information that may be important for the patient, the patient's family, or family planning, which we should proceed with molecular testing.

Although testing guidelines for certain coagulopathies, such as factor V Leiden (FVL) and prothrombin 20210A mutations, have been well established, testing for other conditions, such as von Willebrand disease (VWD), are more complicated, and with respect to methylenetetrahydrofolate reductase C677T (MTHFR) are generally refuted. In this discussion, we present two tables (Tables 161.1 and 161.2) which, along with the gold standard of clinical correlation, should serve as a molecular reference guide for identifying the etiology of either thrombosis or bleeding.

Bleeding: A focused patient and family history of bleeding can aid in the determination of factor deficiency versus platelet/vessel interaction bleeding. In the appropriate clinical setting and after excluding acquired etiologies, we should consider inherited bleeding disorders. Evaluation of inherited bleeding disorders should include Hemophilia A and B, VWD, and then the more rare, inherited coagulation factor deficiencies, dysfibrinogenemia, antiplasmin deficiency, and Plasminogen Activator Inhibitor Type I.

As was alluded to above, phenotypic assays in the Core Laboratory are widely sufficient for assessing bleeding disorders; however, certain circumstances may necessitate molecular testing. The most common utilization of molecular testing in the setting of inherited bleeding disorders involves prenatal diagnosis of hemophilic pregnancies, and to a lesser extent, VWD Type III. Genetic analysis could also be used in determining a patient's carrier status when he or she has a known family history of a defined inherited bleeding disorder. Finally, molecular testing could also expound on a phenotypic diagnosis when determination of the genotype can influence clinical management, most commonly, when concerned about inhibitor development in patients with hemophilia and possibly VWD Type III.

Thrombosis: Inherited or acquired thrombophilia should be evaluated from a multifactorial perspective. Although an inherited or acquired defect can rarely cause thrombosis without provocation, thromboembolism is more commonly the result of

Transfusion Medicine and Hemostasis. https://doi.org/10.1016/B978-0-12-813726-0.00161-6

TABLE 161.1 Inherited Prothrombotic Disorders

	Gene, Location, Most Common Mutation (SNP Identification)	Heterozygous Genotype Frequency or Prevalence	Homozygous Genotype Frequency or Prevalence	Odds Ratio	Is Molecular Testing Available?	Notes
Prothrombin mutation G20210A, heterozygous	*F2;* 11p11.2; 20210G>A (rs1799963)	(A; G), 0.006	–	2.80	Yes; Molecular analysis is the test of choice.	Accounts for 40%–50% of inherited thrombophilia
Prothrombin mutation G20210A, homozygous		–	(A; A), 0.0004	6.74		
Factor V Leiden, heterozygous	*F5*; 1q24.2; R506Q (rs6025)	(C; T), 0.011	–	4.38	Yes, although activated protein C (APC) resistance is the coagulation test of choice.	Present almost exclusively in Caucasian population. Small risk of arterial thrombosis.
Factor V Leiden, homozygous		–	(T; T); 0.0004	11.45		
Dysfibrinogenemia (F1)	FGA (exon 2) and FGG (exon 8); 4q35				Core/Commercial Laboratory testing; identified with FGA, FGB, and FGG sequencing.	Multiple SNPs are reported in the literature, with odds ratios <1.5 and minor allele frequencies >0.3
Factor XI	*F11*				Not routinely evaluated.	Multiple SNPs are reported in the literature, with odds ratios <1.5 and minor allele frequencies >0.25

Protein C deficiency	*PROC*; 2q14.3;	0.14%–1.5% (5%–9% of patients with VTE)	Homozygous or compound heterozygotes—seen with severe hereditary protein C deficiency (1 in 4 million newborns); milder manifestations have been observed	2- to 11-fold increased risk; also increased risk of arterial thrombosis	Core/Commercial Laboratory testing; molecular testing is generally unnecessary.	Account for <10% of inherited thrombophilia; many variations of mutations have been identified throughout these genes and thus, only if necessary, sequencing of the entire gene would be recommended
Protein S deficiency	*PROS1*; 3q11.1	0.1% (2% of patients with VTE)	–	See above discussion	Core/Commercial Laboratory measurement of free protein S is generally recommended, and molecular analysis is considered unnecessary.	
Antithrombin deficiency	*SERPINC1*; 1q25.1; autosomal dominant pattern of inheritance	0.02%–0.17% (0.5%–5% of patients with VTE)	–	5- to 50-fold increased risk of VTE	Core/Commercial Laboratory testing; molecular testing is generally unnecessary.	
MTHFR/ Hyperhomocysteinemia	*MTHFR*; 1p36.22; 677C>T (rs1801133) and 1298A>C (rs1801131)	677C>T: approximately 20%–40% of Caucasian and Hispanic patients (1%–2% in African Americans)	677C>T: >25% Hispanics, 10%–15% North American Caucasians	–	Not recommended	Mutation identified in 5%–7% of general population with mildly elevated homocysteine levels

Prevalence and allele frequency data are provided using general population–based statistics. Many of these variants have different prevalence depending on patient ethnicity. Additionally, in the setting of thrombophilia, the table represents relative risk of a single risk factor, and that in combination with these relative risk scores could portend a far higher risk of thromboembolism. Finally, the "Most Common" mutation is the one, which is most prevalent, not necessarily the mutation, which confers the greatest risk. For information regarding additional mutations, which may or may not confer varying risk from the abovementioned variant, consultation with your local coagulation expert is recommended.

TABLE 161.2 Bleeding Disorders

	Worldwide Prevalence	Gene, Location	Number of Associated Mutations	Is Molecular Testing Available?	Notes
Hemophilia A	1:10,000	*F8*; Xq28; ~45% of severe disease due to recurring intron 22 inversion	>2100	Yes, although Core/Commercial Laboratory testing is usually sufficient. Molecular testing may be recommended in severe disease, but this should be individualized and discussed with a coagulation expert.	• Genetic analysis of *F8* and *F9* genes results in identification of causative mutations in >95% of patients • Identifying multidomain mutations are at higher risk of inhibitor development
Hemophilia B	1:30,000	*F9*; Xq27.1	>1100		• Identifying multidomain mutations in Hemophilia B also has been associated with the potential for anaphylactic reactions to FIX concentrate infusion. • Hemophilia B Leiden is caused by a group of single-nucleotide substitutions in the area around the transcription start site of FIX and results in severe hemophilia in childhood that progresses to the normal range after puberty.
von Willebrand disease	1:100,000	Chromosome 12	~400	Yes, although Core/Commercial Laboratory testing is usually sufficient for the vast majority of patients; molecular testing should be individualized and discussed with a coagulation expert.	• Mutations for Type 1 and Type 3 VWD are spread throughout the entire VWF gene, whereas Type 2 mutations are predominantly identified within Exon 28.

α2-Antiplasmin deficiency	~40 known cases; autosomal recessive inheritance	*SERPINF2*; 17p13.3	–	Yes, although Core/Commercial Laboratory testing is generally sufficient. Euglobulin lysis time and/or a specific assay for α2-antiplasmin deficiency can be performed. (Carpenter, 2008)	• Homozygous patients may exhibit severe bleeding symptoms or may exhibit delayed bleeding. • Heterozygous patients may have milder bleeding or may be asymptomatic.
Plasminogen Activator Inhibitor Type I (PAI-1)	<10 families with complete PAI-1 deficiency have been reported	*SERPINE1*; 7q21.3-22	–	Yes, although Core/Commercial Laboratory testing (PAI-1 antigen and PAI-1 activity) is the initial tests of choice.	Association of PAI-1 deficiency with cardiac fibrosis. • Elevated levels of PAI-1: The *SERPINE1* c.-820_-817G(4_5) (also known as 4G/5G variant) is a common insertion/deletion of four or five G-nucleotides in the promoter region of *SERPINE1*. Studies have shown that the 4G allele results in higher levels of PAI-1 activity. This leads to a state of decreased fibrinolysis and therefore increases one's risk of developing arterial and venous thrombosis.
Factor I deficiency; fibrinogen disorders	1:1 million (AR), unknown AD	*FGA, FGB, GG*	>250	See above. Associated with both bleeding and thrombosis.	These rare inherited coagulation factor deficiencies represent 3%–5% of coagulation factor deficiency–associated bleeding. • FVII and FXI deficiencies represent of 30% of these rare bleeding disorders. • Molecular testing may be useful when evaluating borderline factor deficiency states or with combined FV/FVIII deficiency disease.
Factor II deficiency	1:1–2 million	*F2;* 11p11.2	>50	Yes, although as above, standard functional clotting assays are the diagnostic method of choice. If necessary, molecular testing should be individualized and discussed with a coagulation expert.	
Factor V deficiency	1:1 million	*F5*; 1q24.2	>130		
Factor VII deficiency	1:0.5–1 million	*F7; 13q34*	>240		
Factor X deficiency	1:0.5–1 million	*F10; 13q34*	>100		
Factor XI deficiency	1:100,000–1,000,000	*F11;* 4q35	>220		
Factor XIII deficiency	1:2 million	*F13A* (6p24.2-p23), *F13B* (1q31-q32.1)	>120		
Combined FV and FVIII deficiency	~200 known cases; often associated with consanguinity; concentrated in the Mediterranean, Middle Eastern, and South Asian countries (Zheng, 2013)	*LMAN1* (18q.21) and *MCFD2* (2p21)	>20		

the interplay among multiple risk factors, including physiologic (i.e., pregnancy or obesity), environmental (i.e., smoking or oral contraceptive use), acquired (i.e., antiphospholipid syndrome), and inherited (i.e., FVL or prothrombin) disorders. Additionally, incomplete penetrance of many of the inherited thrombophilia disorders only generates more confusion when evaluating thrombophilia risk.

Evaluation of thrombophilias should consider hereditary and acquired conditions, with molecular testing pertaining mostly to hereditary coagulopathies. Hereditary conditions generally affect the quality or quantity of a coagulation protein with gain or loss of function mutations. The gain of function mutations includes the more common FVL and prothrombin-associated mutations, and the loss of function category includes antithrombin (AT), protein C, and protein S (PS) mutations.

Platelet Disorders: The most common platelet disorders to be considered in coagulopathy include Glanzmann thrombasthenia, Bernard–Soulier syndrome, platelet-type VWD, May–Hegglin syndrome, Hermansky–Pudlak syndrome, Gray platelet syndrome, Wiskott–Aldrich syndrome, and Quebec syndrome. Platelet disorders are generally evaluated in the Core/Commercial laboratory. Methods of platelet disorder testing include light transmission aggregometry, whole-blood aggregometry, secretion assays, and flow cytometry. When diagnostic confirmation is requested because phenotypic studies are equivocal, or when genetic information can be important for familial purposes, molecular analysis can be utilized (see Chapter 142). For more information on platelet-derived coagulopathic disorders, please see "Genetic Loci Associated with Platelet Traits and Platelet Disorders."

Controversial Coagulopathic Entities

Hyperhomocysteinemia and Methylenetetrahydrofolate Reductase: Homocysteine is a by-product of amino acid metabolism, which is either recycled or excreted in the urine. An important recycling enzyme, and one that has been the subject of much controversy, is MTHFR. An inherited defect in MTHFR or a deficiency in vitamins B12, B6, or folate may lead to a disruption of homocysteine catabolism and may result in homocysteinemia or elevated levels of homocysteine. Classically, elevated levels of homocysteine have been linked to an increased risk of thrombosis. Recent studies, however, have found no such association.

Although severe variants in the MTHFR gene may result in extremely high levels of homocysteine accumulation potentially causing thrombosis, this is more likely related to cystathionine beta-synthase deficiency, which in addition to thrombosis would also result in developmental delay, eye disorders, thrombosis, and osteoporosis. Thus, MTHFR polymorphism genotyping should not be routinely ordered for thrombophilia evaluation, recurrent pregnancy loss, or for at risk family members. Additionally, studies using homocysteine-lowering therapy, folate, B6, or B12 supplementation have demonstrated persistent thrombotic risk despite normalization of homocysteine levels. In summary, homocysteine and MTHFR testing in thrombophilia workup is not recommended.

von Willebrand Disease: Core Laboratory testing of VWD can characterize the phenotype in most settings, and thus molecular sequencing is generally confirmatory (see Chapters 134–136). Molecular techniques are thus recommended to fully characterize the subtype when certain assays are not available within the Core Laboratory, to aid in the differential diagnosis between hemophilia A versus Type 2N and Type 2B versus pseudo-VWD, and to predict the response to desmopressin or for family planning purposes. The complication is that the *VWF* gene, located on chromosome 12, is highly polymorphic and it has a partial pseudogene on chromosome 22 that replicates the sequence between exons 23 and 34 with 3% variance. As of 2013, sequencing analysis of the VWF gene has been successful in patients with VWD subtypes 2B, 2M, 2N, and 3, with detection rates as high as 85%. This can be particularly important for diagnosis of Type 3 VWD where management can be initiated as quickly as possible, or in the setting of prenatal diagnosis, even before presenting symptoms arise. For VWD Type 1, however, the rate of detection was closer to 65%. This discrepancy could be attributed to a defect in clearance of VWF instead of an actual quantitative abnormality or it could be the result of not evaluating intron and distal transcriptional regulatory sequences. Complicating things further, many patients with a clinical diagnosis of VWD Type 1 are genotypically identical to patients with Type 3, owing to the possibility that these individuals may be Type 3 carriers with a Type 1 phenotype. A 2017 primer on molecular characterization of VWD recommends utilizing multiplex ligation-dependent probe amplification and mRNA analysis for further evaluation of VWD but again reports that nearly 30% of Type 1 patients will remain without an identifiable disease-causing mutation.

In the absence of a phenotypic diagnosis, or when a phenotype warrants further characterization, consultation with a local coagulation specialist is recommended for an appropriate individualized diagnostic approach.

Fibrinogen: Quantitative and qualitative abnormalities in fibrinogen have been implicated in bleeding and thrombosis. For a more in-depth discussion of the biochemical profile and pathophysiology of fibrinogen, please see Chapter 115. Afibrinogenemia and hypofibrinogenemia have both been implicated, most commonly, with bleeding, whereas dysfibrinogenemia and hypodysfibrinogenemia have been associated with both bleeding and/or thrombosis. Although the Core Laboratory is generally the best evaluator of quantitative and qualitative fibrinogen abnormalities, molecular analysis may be desired to study structural aberrations of the Aα, Bβ, and γ chains. Many genome-wide association studies, with differing results, have attempted to correlate particular single-nucleotide polymorphisms of either *FGA*, *FGB*, or *FGG* with risk of thrombosis and/or bleeding. Perhaps most importantly, however, is that many of these studies have found limited genotype–phenotype correlation. The point of all this is to demonstrate that sequencing analysis of fibrinogen may not reflect the true risk of developing thrombosis and/or bleeding. Thus, if one is concerned about a congenital qualitative or quantitative abnormality, additional testing should be individualized, and a coagulation specialist should be consulted.

Protein S: Studies correlating PS deficiency with thrombophilia have involved thrombophilic families. Although these studies have shown elevated risks of first thrombosis and recurrent thrombotic events, it is unclear whether this increased risk was a result of PS deficiency in a vacuum or if other prothrombotic risk factors were at fault. Although some studies have shown increased risk in the general population as high as 2.5-fold, others have shown no association at all. The MEGA case–control study studied free and total PS levels in their patient population and when available compared these data to *PROS1* genetic analyses. They found that very low levels of free PS (<0.10th percentile), but not total PS, correlated with an increased risk of unprovoked thrombosis. Additionally, they found reduced PS levels in patients utilizing vitamin K antagonist and/or oral contraceptive therapy, even though no known genetic deficiency could be identified. Finally, when comparing their findings to *PROS1* status in patients with very low levels of PS, they found that various mutations were present in equal distribution in both the control (known nondetrimental mutations) and subject (complete deletion of *PROS1,* 967delTinsGG, and Thr144Asn) populations. Considering those findings, they recommend that while low free PS, more likely, or low total PS can rarely identify subjects at increased risk of venous thromboembolism, molecular analysis of the *PROS1* should not be routinely performed as a screening tool of thrombophilia. Therefore, as has been stated above, it is our recommendation that if one is considering genotypic characterization, additional testing should be individualized, and a coagulation specialist should be consulted.

Conclusions: At this early stage of molecular diagnostic development and testing, it is becoming more apparent that over time these assays will hold the key to diagnosing disorders of primary and secondary hemostasis. Nevertheless, we are not quite there. We need guidelines that establish when molecular testing should be utilized and what in particular should be tested. Most importantly, at least this time, we must consider disadvantages such as expense to the patient or hospital, spurious results due to incorrect timing of testing, misinterpretation of results by the patient or even the health-care provider leading to anxiety and emotional distress, and the possibility of genetic discrimination. Because of the confusion and sometimes uncertainty surrounding these molecular assays, we strongly recommend an individualized approach and consultation with a local coagulation expert.

In Dr. Bartlett's book he writes that with all of these molecular advancements, the challenge is not to forget the past but to build on it. It is our duty as health-care providers to develop an interdisciplinary approach that incorporates both molecular diagnostics and coagulation testing in the Core Laboratory so as to ensure the highest standard of care for our patients.

Further Reading

Baronciani, L., Goodeve, A., & Peyvandi, F. (2017). Molecular diagnosis of von Willebrand disease. *Haemophilia, 23*(2), 188–197.

Bartlett, J. M. S., Shaaban, A., & Schmitt, F. (2015). Molecular pathology: A practical guide for the surgical pathologist and cytopathologist. (Eds.), Cambridge University Press. Available from https://www.cambridge.org/core/books/molecular-pathology/6334936E2C6DD862461B2163D28C8B96.

Bunimov, N., Fuller, N., & Hayward, C. P. (2013). Genetic loci associated with platelet traits and platelet disorders. *Semin Thromb Hemost, 39*(3), 291–305.

Carpenter, S. L., & Mathew, P. (2008). A2-antiplasmin and its deficiency: Fibrinolysis out of balance. *Haemophilia, 14*(6), 1250–1254.

Dean, L. (2012). Methylenetetrahydrofolate reductase deficiency. In V. Pratt, H. McLeod, L. Dean, et al. (Eds.), *Medical genetics summaries*. Bethesda, MD: National Center for Biotechnology Information (US). Available from https://www.ncbi.nlm.nih.gov/books/NBK66131/.

Favaloro, E. J., & Lippi, G. (2017). *Hemostasis and thrombosis: methods and protocols*. New York, NY: Humana Press. Available from https://www.springer.com/la/book/9781493971947.

Gresele, P. (2015). Diagnosis of inherited platelet function disorders: Guidance from the SSC of the ISTH. *Semin Thromb Hemost, 13*(2), 314–322.

Heiman, M., Gupta, S., Khan, S. S., et al. (August 3, 2017). Complete plasminogen activator inhibitor 1 deficiency. In M. P. Adam, H. H. Ardinger, R. A. Pagon, et al. (Eds.), *GeneReviews®*. Seattle, WA: University of Washington, Seattle 1993–2017. Available from https://www.ncbi.nlm.nih.gov/books/NBK447152/.

Hickey, S. E., Curry, C. J., & Toriello, H. V. (2013). Acmg practice guideline: Lack of evidence for Mthfr polymorphism testing. *Genet Med, 15*(2), 153–156.

Kujovich, J. L., Curry, C. J., & Toriello, H. V. (2011). Factor V Leiden thrombophilia. *Genet Med, 13*(1), 1–16.

Lillicrap, D. (2013). Molecular testing for disorders of hemostasis. *Int J Lab Hematol, 35*(3), 290–296.

Macrae, F., Domingues, M., Casini, A., & Ariëns, R. (2016). The (patho)physiology of fibrinogen Gamma′. *Semin Thromb Hemost, 42*(4), 344–355.

Mumford, A. D., Domingues, M., Casini, A., & Ariëns, R., et al. (2014). Guideline for the diagnosis and management of the rare coagulation disorders′. *Br J Haematol, 167*(3), 304–326.

Olson, J. D. (2002). College of American Pathologists Consensus Conference Xxxvi: Diagnostic issues in thrombophilia. *Arch Pathol Lab Med, 126*(11), 1277–1433.

Peyvandi, F., Kunicki, T., & Lillicrap, D. (2013). Genetic sequence analysis of inherited bleeding diseases. *Blood, 122*(20), 3423–3431.

Shapiro, A. D., & Peyvandi, F. (2014). *Rare coagulation disorders*. Available from http://www.rarecoagulationdisorders.org/.

Stevens, S. M., & Woller, S. C., Bauer, K. A., Kasthuri, R., Cushman, M., Streiff, M., Lim, W., et al. (2016). Guidance for the evaluation and treatment of hereditary and acquired thrombophilia. *J Thromb Thrombolysis, 41*(1), 154–164.

Winter, W. E., Sherri, D. F., & Harris, N. S., Kasthuri, R., Cushman, M., Streiff, M., Lim, W. (2017). Coagulation testing in the core laboratory. *Lab Med, 48*(4), 295–313.

Zheng, C., Sherri, D. F., & Zhang, B. (2013). Combined deficiency of coagulation factors V and Viii: An update. *Semin Thromb Hemost, 39*(6), 613–620.

CHAPTER 162

Circulating Microparticles

Florencia G. Jalikis, MD and Morayma Reyes Gil, MD, PhD

Microparticle (MP) analysis has become popular in the last decade because of mounting evidence of their active role in multiple cellular processes. Alterations in total MP numbers or changes in the relative abundance of MP populations have been correlated with multiple diseases. It is still unknown whether these MP alterations are markers of disease or whether MPs play an active role in the pathology, progression, or modulation of these diseases. Because of a high variety of protocols and methods available to analyze MPs, there is an urgent need to standardize these techniques to facilitate large-scale multicenter studies to understand the role of MPs in disease and their potential use for clinical diagnosis and prognosis.

MPs are a heterogeneous group of bioactive small vesicles (100–1000 nm) that can be found in blood and body fluids following activation, necrosis, or apoptosis of virtually any eukaryotic cell. It is thought that they play an important role in intercellular communication and participate in the maintenance of homeostasis under physiological conditions.

Healthy individuals have small amounts of circulating MPs, the majority of which are derived from platelets and erythrocytes. Marked elevations of total number of MPs or relative increases of specific MP populations have been associated with a variety of disorders, including deep venous thrombosis and pulmonary embolism, congestive heart failure (CHF), sepsis, heparin-induced thrombocytopenia (HIT), thrombotic thrombocytopenic purpura (TTP), paroxysmal thrombocytopenic purpura, and preeclampsia.

Composition of Microparticles: MPs are composed of a phospholipid bilayer, which carries several bioactive molecules, including transmembrane proteins, receptors, counterreceptors, and adhesion molecules, as well as cytoplasmic products such as nucleic acids, chemokines, cytokines, enzymes, growth factors, and signaling proteins. The membrane shows externalization of phosphatidylserine (PS) that provides a platform for the assembly of coagulation factors. The precise content of lipids and proteins depends on their cellular origin and the type of stimulus involved in their formation.

Classification of Microparticles: Living cells are capable of releasing different types of membrane vesicles that can be classified based on their size, sedimentation, and secretion mechanisms into exosomes, MPs, and apoptotic vesicles. The term MP encompasses a heterogeneous group of poorly characterized vesicles ranging in size from 100 to 1000 nm that are released by budding of the plasma membrane (ectocytosis). Based on their size, these MPs have been divided into microvesicles (100–1000 nm), ectosomes (50–200 nm), membrane particles (50–80 nm), and exosome-like vesicles (20–50 nm) (Fig. 162.1; Table 162.1). These large vesicles should be distinguished from exosomes (<100 nm) that are membrane vesicles that are stored intracellularly

Transfusion Medicine and Hemostasis. https://doi.org/10.1016/B978-0-12-813726-0.00162-8

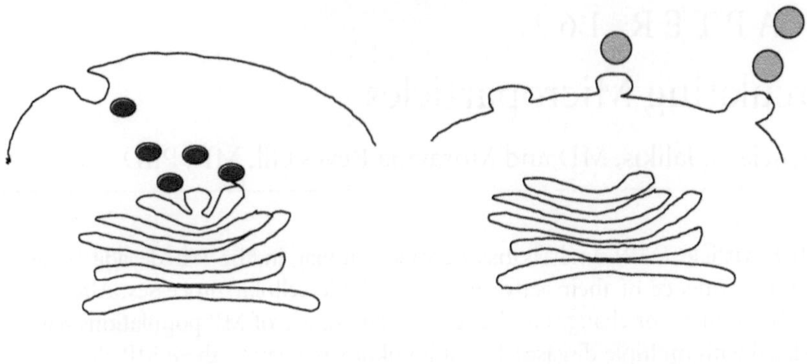

FIGURE 162.1 Microparticles should be differentiated from exosomes. Although microparticles are secreted by budding of the cell plasma membrane, exosomes are stored intracellularly in multivesicular compartments and are secreted following the fusion of the internal compartment with the cell plasma membrane.

TABLE 162.1 Characteristics of Different Types of Secreted Vesicles

Feature	Exosomes	Microparticles	Apoptotic Vesicles
Size	50–100 nm	100–1000 nm	50–500 nm
Density in sucrose	1.13–1.19 g/mL	1.04–1.23 g/mL	1.16–1.28 g/mL
Sedimentation	100,000 g	20 000 g or greater	1200, 10,000 or 100,000 g
Origin	Multivesicular, internal compartments	Plasma membrane	Cellular fragments
Release	Constitutive and/or cellular activation	Cellular activation and early apoptosis	Terminal apoptosis
Appearance on electron microscopy	Cup-shaped	Irregular shape, heterogeneous	Heterogeneous
Main protein markers	Tetraspanins (CD9, CD63), Alix, flotillin, TSG101	Integrins, selectins, other antigens of parental cell	Histones
Annexin V binding capacity	No or low	High	High

in multivesicular compartments and are secreted following the fusion of internal compartments with the cell plasma membrane (Fig. 162.1). Dying or apoptotic cells are also able to secrete membrane vesicles with different features to those from living cells.

Mechanisms of Microparticle Release: It is thought that MPs form when the asymmetric distribution of plasma membrane lipids is lost. As highlighted by Hugel et al., under resting conditions, phospholipids are asymmetrically distributed in the plasma membrane, with PS almost exclusively located in the inner membrane layer, and phosphatidylcholine and sphingomyelin located on the external one. Increased concentrations of cytosolic calcium, like those seen in response to cellular activation, may lead to lipid redistribution and surface exposure of PS, with membrane blebbing and subsequent MP shedding. Once exposed, PS promotes blood coagulation by serving as a scaffold for the assembly of the prothrombinase and thrombinase complexes. Of interest, some investigators have shown that platelet-derived MP surfaces have 50- to 100-fold higher procoagulant activity when compared with the normal platelet surface. The physiological importance of MP shedding is exemplified in a rare disorder known as Scott syndrome, which is characterized by moderate to severe bleeding episodes, usually provoked hemorrhages due to an enzymatic deficit that leads to decreased surface exposure of PS and decreased MP formation.

Circulating vesicles originating from apoptotic cells are thought to be produced by less-controlled mechanisms, possibly secondary to loss of membrane integrity or mechanical destruction. Although these apoptotic vesicles display PS on their surface, in contrast to the MPs, they show no or weak procoagulant activity. Surface exposure of PS also serves as a signal for the clearance of senescent cells by the reticuloendothelial system.

Role of Microparticles: MPs play an important role in homeostasis and intercellular communication through several mechanisms including

1. Transfer of surface receptors,
2. Transfer of mRNA,
3. Release of proteins or active lipids, and
4. Induction of adaptive immune response (Table 162.2).

Microparticles in Human Disease: Abnormal levels of circulating MPs have been associated with a variety of disorders (Table 162.3).

Venous Thromboembolism: Venous thromboembolism (VTE) is a multifactorial disease with a high incidence. Few studies have demonstrated increased levels of endothelial-derived MP (EMP) and platelet MP (PMP) in patients with VTE. A retrospective case–control study conducted by Bucciarelli et al. has demonstrated an association between high plasmatic levels of total MP and the risk of a first VTE. The study showed that compared with individuals with low levels of MP (<10th percentile), individuals with high levels (>90th percentile) had a fivefold increased risk of having had a previous VTE. The investigators concluded that elevated levels of MPs are associated with increased risk of VTE, which is independent of other known risk factors for VTE. A prospective study conducted by Rectenwald et al. suggested that combined detection of PMP, D-dimer, and P-selectin correlates with the diagnosis of deep venous thrombosis with a sensitivity of 73% and a specificity of 81%. Although the underlying mechanisms are still unknown, it is likely that VTE develops as a result of a complex interaction between MPs, endothelial cells, platelets, and inflammatory cells.

Cardiovascular Disease: Increased MPs have been described in association with cardiovascular risks factors (smoking, dyslipidemia, diabetes mellitus, and hypertension), coronary artery disease (CAD), and CHF. PMPs have been implicated in atherosclerotic plaque formation and development of arterial thrombosis.

TABLE 162.2 Biological Function of Microparticles

Biological Effect	Proposed Mechanisms
Procoagulant	Surface exposure of phosphatidylserine
	Expression of Tissue Factor (mostly monocyte-derived MPs)
	Transfer of GPIIb/IIIa
	Presence of membrane functional effectors (integrins, P-selectin, VWF)
Anticoagulant (in vitro)	Proteolytic inactivation of FVa by activated protein C
Fibrinolytic	Expression of u-PA and u-PAR
Proteolytic	Expression of matrix metalloproteinases
Vascular	Membrane expression of thromboxane A2
	Impairment of endothelium-dependent relaxation through eNOS downregulation
Proinflammatory	Release of proinflammatory endothelial cytokines (IL-6 or MCP-1)
	Induction of expression of ICAM-1, VCAM-1 and e-selectin
	Serve as a substrate for production of lysophosphatidic acid
Antiinflammatory	Secretion of TGFβ, potent inhibitor of macrophage activation
	Expression of Annexin A1, endogenous antiinflammatory protein
Immunity	Expression of major histocompatibility complex molecules
	Display of autoantigens, such as RNA and DNA, which may act as potent autoadjuvants, inducing B cell tolerance
	Expression of Fas-L leading to apoptosis, leading to immune evasion

Bernal-Mizrachi et al. reported that increased levels of EMP in patients with CAD appear to correlate directly with angiographic findings and severity of luminal stenosis. Bulut et al. found that EMP levels are significantly higher in patients with stable CAD (SCAD) and left ventricular dysfunction when compared with patients with SCAD and preserved left ventricular function, and hypothesized that EMP levels constitute an independent predictor of cardiovascular events in this population. Increased EMP has also been noted in healthy offspring of parents with CAD, suggesting an initial step in the development of endothelial dysfunction in genetically predisposed subjects. Although poorly understood, several mechanisms have been postulated, including binding of coagulation factors, platelet or endothelial cell activation, and increased macrophage adhesion. Circulating MPs may stimulate angiogenesis, increase vascular tone, impair vascular relaxation, activate free radical formation, and stimulate the production of proinflammatory mediators.

Sepsis: Increased levels of MPs have been seen in a variety of inflammatory disorders. In sepsis, MPs may serve as important bioeffectors of inflammation and thrombosis and contribute to tissue injury and organ dysfunction. MPs are thought to be responsible, at least in part, for the prothrombotic state associated with sepsis. Activation of coagulation during sepsis and septic shock serves two main purposes:

1. It limits the propagation of the infection and
2. It leads to intravascular fibrin deposition, which could progress to disseminated intravascular coagulation.

TABLE 162.3 Diseases Associated With Microparticle Alterations

Disease	Source of Microparticles
Thrombotic Disorders	
Venous thromboembolism	Platelet, endothelial cell
Thrombotic thrombocytopenic purpura	Platelet, endothelial cell
Antiphospholipid syndrome	Platelet, endothelial cell
Heparin-induced thrombocytopenia	Platelet
Sickle cell disease	Platelet, red blood cell, endothelial cell, monocyte
Paroxysmal nocturnal hemoglobinuria	Platelet
Bleeding Disorders	
Scott syndrome	Platelet
Castaman syndrome	Platelet
Cardio- and Cerebrovascular Diseases	
Acute coronary syndrome	Platelet, endothelial cell
Acute ischemic stroke	Endothelial cell
Arteriosclerosis obliterans	Platelet
Hypertension	Monocyte, platelet, endothelial cell
Hyperlipidemia	Endothelial cell
Atherosclerosis	Monocyte, platelet, endothelial cell
Congestive heart failure	Endothelial cell
Diabetes	Platelet, monocytes, endothelial cells
Infectious Diseases	
Escherichia coli hemolytic uremic syndrome	Platelet, leukocytes
HIV infection	Lymphocyte
Prion diseases	Platelet
Malaria	Platelet
Hepatitis C	Lymphocyte
Inflammatory Disorders	
Preeclampsia	Leukocytes
Sepsis	Platelet, endothelial cells, leukocytes
Sepsis-induced immunosuppression	Platelet
Other	
End-stage renal disease	Endothelial cell
Organ transplantation	Endothelial cell
Rheumatoid arthritis	Platelet
Immunosuppression	Platelet
Polycystic ovarian syndrome	Platelet

MPs represent a source of phospholipids, a substrate for phospholipase A2, which facilitates platelet aggregation. MPs may also lead to vascular inflammation and facilitate chemotactic migration of platelets and leukocytes. In addition, generation of monocyte-derived MPs bearing tissue factor may also contribute to the procoagulant state.

Thrombotic Thrombocytopenic Purpura: TTP is a rare but potentially life-threatening disease characterized by otherwise unexplained thrombocytopenia and microangiopathic hemolytic anemia. Many patients will also experience fever, as well as renal and neurological abnormalities. TTP is associated with systemic thrombi composed primarily of platelets. This disease has been associated with absence or decreased levels of ADAMTS13 activity, a von Willebrand factor–cleaving protease, which leads to accumulation of ultralarge von Willebrand factor multimers, which in turn leads to platelet adhesion, aggregation, and thrombi formation. Elevated levels of PMP have been seen in acute and chronic phases of TTP. In addition, elevated levels of EMP may be also implicated in the pathophysiology of this disease. Jimenez et al. demonstrated that EMP levels increase long before the onset of thrombocytopenia, postulating that these MPs may represent an early prognostic marker in chronic phases. It has also been postulated that EMPs facilitate cell adhesion and platelet aggregation, as suggested by a 13-fold increase in adhesion molecules ICAM-1, VCAM-1, and the presence of VWF-rich EMP.

Heparin-Induced Thrombocytopenia: HIT is one of the most common immune-mediated adverse drug reactions, with an estimated incidence of 1%–3%. It is characterized by thrombocytopenia usually occurring 5–10 days after the initiation of heparin treatment. Many patients will experience thrombotic complications. Heparin administration leads to the formation of antibodies to not only heparin but also to the complex formed by heparin and platelet factor 4 (PF4), a heparin-neutralizing protein present in the platelet alpha granules. The heparin–PF4 complex will bind to the platelet surface and will be recognized by the relevant antibodies. These newly formed complexes (IgG–heparin–PF4) will lead to platelet activation (with further release of PF4 creating a positive feedback), aggregation, and premature clearance from circulation leading to thrombocytopenia and MP formation. It is still unclear whether these MPs are primarily markers of platelet activation or if they play a role in the pathophysiology of this disease.

Preeclampsia: The term preeclampsia refers to a multisystemic disorder characterized by new onset of hypertension and proteinuria after 20 weeks of gestation in a previously normotensive woman. Normal pregnancy elicits a maternal inflammatory response, more significant toward the end of the pregnancy. In preeclampsia, this is exaggerated. As highlighted by Redman et al., the syncytial surface of the placenta sheds a variety of particles, including exosomes and syncytiotrophoblast MPs. These MPs circulate in the plasma of a normal pregnant woman and appear significantly increased in preeclampsia. These observations lead to the belief that MPs may contribute to the pathophysiology of preeclampsia. A recent study has speculated that in normal pregnancy, the placenta sheds predominantly small exosomes, which are immunoregulatory in nature, whereas in preeclampsia, oxidative and inflammatory stress activates the syncytiotrophoblast causing release of larger MPs, which have proinflammatory, antiangiogenic, and procoagulant activity. The proinflammatory effect may be due to the presence of Hsp70 and HMGB1, and the procoagulant effect may be related to the expression of tissue factor. In addition, the presence of Flt-1 and endoglin may contribute to endothelial dysfunction.

Processing of Microparticles: Samples for MP analysis should be collected in citrated tubes and processed for MP purification preferable within an hour, with minimal agitation, to minimize the release of additional MPs. Many protocols have been published for the centrifugation and purification of MP and unfortunately, a comprehensive comparison between these protocols has not been performed. Thus the lack of a standard (universal) protocol for preparation and analysis of MP has hampered its use in clinical applications. There is an urgent need to standardize and validate the MP techniques and procedures for clinical testing. The International Society on Thrombosis and Haemostasis has established some guidelines for the processing and analysis of MPs.

Methods for Detection of Microparticles: There are several methods for detecting MPs, including atomic force microscopy, electron microscopy, dynamic light scattering, reflection interference contrast microscopy, functional assays, immunoassays, and flow cytometry (FC). FC is the most widely used method and thus we will focus this discussion on challenges of MP analysis using FC.

Microparticle Size and Count: One of the major challenges in FC analysis is the accurate measure of size, especially of particles of less than 500 nm. Chandler et al. demonstrated that silica and polystyrene beads, which are commonly used to gate the size of MPs, scattered more light (using forward scatter by the 488 nm laser) than lipid vesicles and platelets. Thus a 0.4 μm polystyrene bead scatters similar light to a 1 μm lipid or cellular MP. These different light-scattering properties of beads and MPs should be taken into account when using calibration beads to measure the size of MP. There are new instruments, such as Gallios by Beckman Coulter, Influx by Becton Dickinson, and Apogee A50 by Apogee Flow System that can measure particles of small size more accurately. For instance, the Apogee A50 can distinguish 0.1 μm latex beads from noise.

Most of the MP alterations reported in association with several diseases involve increases in total MPs or relative increases in specific MP populations. Thus an accurate count of total MPs and their relative distribution is one of the most important parameters measured by FC. The original studies in MP analysis used a given amount of calibration beads added to the MP sample to quantify the relative number of MPs within a given amount of beads. Many new FCs use a volumetric fluidic system, which can directly measure the MP counts.

Cellular Derivation of Microparticles: MPs are derived from cells that are biologically active or undergoing apoptosis (Table 162.4). They are formed by blebbing or budding from the cell membrane, and they are rich in PS in their outer surface. Therefore, MPs should bind Annexin V in the presence of calcium. A good negative

TABLE 162.4 Microparticle Classification Based on Surface Antigens Specific for Cell of Origin

Cell of Origin	Markers
Platelet	CD31, CD41, CD41a, CD42a, CD42b, CD61, CD62P
Endothelial cell	CD31, CD34, CD54, CD62E, CD51, CD105, CD106, CD144, CD146
Red blood cells	CD235a
Leukocyte	CD45
Granulocyte	CD66b
Monocyte	CD14
Lymphocyte	CD4, CD8, CD20

control for Annexin V specific binding is Annexin V staining without calcium. Although the majority of MPs bind Annexin V, there is a significant proportion of MPs (defined by size) that do not bind Annexin V. This can be due to a different concentration of Annexin V, loss of PS in the MP surface, and, more importantly, generation of "false" MPs during the samples preparation. There are dyes that can be used to distinguish "real" MP (generated from biological processes) from "false" particles generated during the sample preparation. One of such dyes is phalloidin, a cyclic peptide that binds f-actin in platelets disrupted mechanically.

Platelet and Red Blood Cell Microparticles: The majority of MPs are platelet-derived or red blood cell–derived and can be identified as CD41+ or CD235a+, respectively.

Endothelial-Derived Microparticles: Although very scarce in the plasma, increases in EMP have been associated with multiple diseases, including sickle cell anemia, systemic lupus erythematosus, and cardiovascular diseases (Table 162.3). Unfortunately, EMPs are difficult to detect accurately because of lack of endothelial-specific markers. Furthermore, markers considered fairly specific for endothelial cells, such as CD144, are present at low density in the plasma membrane. The sensitivity of EMP detection can be increased by using a multicolored combination of endothelial markers, such as CD31, CD146, CD105, and CD144, but this is only applicable using FCs capable of detecting more than four colors, limiting the simultaneous detection of other MP populations in the same sample.

Monocyte-Derived Microparticles: Alterations in monocyte-derived microparticles (MMPs) have been associated with atherosclerosis, hypertension, and inflammatory conditions (Table 162.3). MMPs can be identified as CD14+.

Therapeutic Opportunities: Several strategies could be potentially used to attenuate microparticle-driven disease, including inhibiting the production or release of MPs, targeting MPs components, and/or inhibiting their uptake. The use of therapeutic extracellular vesicles is also been explored particularly in regenerative medicine.

Further Reading

Burnier, L., Fontana, P., Kwak, B. R., & Angelillo-Scherrer, A. (2009). Cell-derived microparticles in haemostasis and vascular medicine. *Thromb Haemost*, *101*, 439–451.

Chandler, W. L., Yeung, W., & Tait, J. F. (2011). A new microparticle size calibration standard for use in measuring smaller microparticles using a new flow cytometer. *J Thromb Haemost*, *9*, 1216–1224.

Hugel, B., Martínez, M. C., Kunzelmann, C., & Freyssinet, J. M. (2005). Membrane microparticles: Two sides of the coin. *Physiology*, *20*, 22–27.

Lacroix, R., Robert, S., Poncelet, P., & Dignat-George, F. (2010). Overcoming limitations of microparticle measurement by flow cytometry. *Semin Thromb Hemost*, *36*, 807–818.

Mause, S., & Weber, C. (2010). Microparticles: Protagonists of a novel communication network for intercellular information exchange. *Circ Res*, *107*, 1047–1057.

Montoro-Garcia, S., Orenes-Piñero, E., Marín, F., et al. (2011). Circulating microparticles: New insights into the biochemical basis of microparticle release and activity. *Basic Res Cardiol*, *106*, 911–923.

Morel, O., Jesel, L., Freyssinet, J. M., & Toti, F. (2011). Cellular mechanisms underlying the formation of circulating microparticles. *Arterioscler Thromb Vasc Biol*, *31*, 15–26.

Morel, O., Morel, N., Jesel, L., Freyssinet, J. M., & Toti, F. (2011). Microparticles: A critical component in the nexus between inflammation, immunity, and thrombosis. *Semin Immunopathol*, *33*, 469–486.

Piccin, A., Murphy, W. G., & Smith, O. P. (2007). Circulating microparticles: Pathophysiology and clinical implications. *Blood Rev*, *21*, 157–171.

Théry, C., Ostrowski, M., & Segura, E. (2009). Membrane vesicles as conveyors of immune responses. *Nat Rev Immunol*, *9*, 581–593.

CHAPTER 163

Thrombin Generation Assays

Mikhail Roshal, MD, PhD and Morayma Reyes Gil, MD, PhD

The TGA is a global test of coagulation systems, with the potential to contribute to the current arsenal of clinical prognostication and drug monitoring tools. By varying the conditions under which it is applied, the test can be made sensitive to virtually every known component of the clotting cascade. The current iterations of the test utilize a fluorogenic substrate to monitor thrombin generation beyond the time point when the fibrin clot is formed. Standardization of testing conditions and defining testing indications for TGA is likely needed before the test enters into widespread clinical use.

Thrombin is a multifunctional enzyme that catalyzes the conversion of fibrinogen to fibrin, promotes additional thrombin formation by activation of coagulation factors (F) XI, V, and VIII, stabilizes a clot via activation of antifibrinolytic factors, and activates platelets and endothelial surfaces. When combined with thrombomodulin (TM) it can activate the anticoagulant protein C pathway, leading to dampening of the coagulation cascade.

Fibrin clots form relatively early in the process of thrombin generation, when 95% of thrombin is yet to form. Thus, traditional coagulation tests that terminate with clot formation do not measure the full thrombin generation potential of the plasma sample. Because thrombin formation is controlled by numerous pro- and anticoagulant factors, antigenic prothrombin level is only weakly reflective of the amount of thrombin that can be generated in plasma samples.

The total amount of thrombin that plasma is capable of generating is potentially quite important. Numerous studies have shown that this quantity may serve as a predictor of a patient's bleeding or thrombosis risk. By varying test conditions, TGA measurements can be made sensitive to virtually all known risk factors for venous thrombosis, allowing assessment of their interaction in individual patients. The test can also be made sensitive to clotting factor abnormalities and factor replacement therapy. In the recent years it has been shown that elevated thrombin generation in vivo was predictive of recurrent venous thrombosis risk and of a first thrombotic event in high-risk subjects. It also could stratify severe FXI-, VIII-, and FIX-deficient patients into high- and low-risk groups for hemorrhage better than measuring individual factors in isolation.

TGAs have been performed in the research setting for many years but were cumbersome and suffered from numerous technical limitations. Recent advances in the development of fluorogenic substrates, software, and instrumentation have made the assays less laborious, potentially more reproducible outside of highly specialized settings and have allowed for automation of large portions of the test workflow. These developments portend well for adoption of TGA into general clinical practice.

Assay Methodology Overview: Currently, there are two commercially available fluorogenic thrombin generation assays; the calibrated automated thrombogram (thrombinoscope, originally from The Netherlands and recently acquired by STAGO) and the Technothrombin TGA (Technoclone, Vienna, Austria). Although the assays

Transfusion Medicine and Hemostasis. https://doi.org/10.1016/B978-0-12-813726-0.00163-X

differ in important technical details, they offer similar measurements. Both utilize a fluorogenic thrombin substrate that is added to recalcified plasma. Initiation of clotting can be performed with different amounts of tissue factor and phospholipid or can be carried out under "native" conditions where endogenous activators are allowed to initiate the clotting system.

Several important parameters can be measured. These include lag time, peak rate of thrombin generation, and endogenous thrombin potential (ETP). Lag time is the time between reaction initiation and the beginning of exponential phase of thrombin generation. As such, it is essentially a clotting time and is dependent on clotting factor concentrations and initiation conditions. Peak thrombin generation measures the maximal rate of formation of thrombin, as reflected by maximal acceleration of fluorogenic substrate cleavage. ETP is the total amount of thrombin that is generated throughout the reaction before thrombin inactivation by endogenous inhibitors. ETP can be measured by quantifying the amount of substrate that is cleaved in the whole reaction once it is adjusted for substrate exhaustion and other factors. Modifications of the assays that involve addition of TM or activated protein C to additionally assay the protein C system have also been described. Recently, a fluorogenic assay that uses whole blood rather than plasma has been introduced. The assay has the advantage of potentially accounting for the regulatory role of red blood cells, white blood cells, and intact platelets on thrombin generation in blood. Yet another version of the assay allows the simultaneous measurement of thrombin and plasmin activity. At present, however, the optimal clinical assay is not clear and will likely differ for different applications.

Test Interpretation: TGAs have been tested for applicability in numerous pathologic circumstances and have been found to be useful in the assessment of bleeding and clotting risk, and also in monitoring factor therapy in clinical trials. Conditions of reaction initiation can be varied to make the assay more or less sensitive to specific factors. For instance, adding a corn trypsin inhibitor to the plasma before reaction initiation makes it insensitive to factor XII, which activates coagulation, in vitro, but not in vivo. Varying amounts of tissue factor can alter the sensitivity of the reaction to concentration of the tissue factor pathway inhibitor. Addition of activated protein C makes the reaction sensitive to protein S concentration and factor V Leiden. Adding TM sensitizes TGA to protein C concentration but may alter sensitivity to procoagulant activity of thrombin.

The flexibility in reaction conditions makes TGA an attractive research tool but perhaps to an equal extent has prevented harmonization of result interpretation and widespread clinical adoption of the testing. Elevated peak thrombin generation and ETP have been associated with elevated risk of thrombosis under some, but not all experimental conditions. Decreased values have been linked to risk of bleeding. Slower inhibition of thrombin generation as reflected by decreased ratio of peak thrombin generation to ETP has been linked to coagulopathy of trauma and disseminated intravascular coagulation. TGAs have also been used to monitor coagulation factor replacement therapy and therapeutic drug monitoring.

Test Limitations: Ideally the rate of fluorescence increase would be linearly related to the amount of thrombin that is being generated. This is not the case through most of the time of the test. Several factors contribute to the breakdown of the relationship. First, because thrombin has relatively low affinity for the small molecule fluorogenic substrate, the substrate concentration becomes limiting in the later stages of the assay as it is consumed. Second, the inner filter effect (absorbance of either excitation or emission light by plasma elements) prevents equal excitation and capture of light from deeper layers of the assay liquid. Third, alpha-2 macroglobulin (A2M) is able to inhibit thrombin interaction with natural macromolecular substrates and inhibitors, but not with synthetic small molecule substrates. A2M is an abundant plasma protein that acts as a slow inhibitor of active thrombin in vivo. Thus, under the testing conditions, A2M does not inhibit thrombin-mediated proteolysis of the small molecule substrate but does prevent inactivation of thrombin by antithrombin resulting in residual activity of A2M-thrombin complexes that is not physiologically relevant. Endogenous thrombin activity is increased by about 20%–30% by this artifact. The A2M effect also results in a minor (about 5%–10%) increase in peak thrombin generation rate. In the calibrated automated thrombogram assays, these issues are dealt with by adjustment of the reading with a sample with a defined concentration of thrombin/ A2M complexes. This standard allows for real-time adjustments for substrate exhaustion, inner filter effect, and A2M effect. The Technoclone assay does not use an internal standard; however, barring significant inner filter effects due to unusual sample discoloration, a mathematical correction for other test limitations is possible.

In an attempt to standardize TGA, Dargaud et al. conducted a multicenter study using the same equipment, protocol, reagents, and most importantly a reference standard to normalize the patients' results and demonstrated significant reduction in variability of the results. Therefore, more studies like this, the development of a WHO reference standard and universal guidelines, are likely needed to make the TGA a clinical test for diagnosis and management of bleeding and prothrombotic conditions.

Further Reading

Al Dieri, R., de Laat, B., & Hemker, H. C. (2012). Thrombin generation: What have we learned? *Blood Rev*, *26*, 197–203.

Ay, C., Dunkler, D., Simanek, R., et al. (2011). Prediction of venous thromboembolism in patients with cancer by measuring thrombin generation: Results from the Vienna Cancer and Thrombosis Study. *J Clin Oncol*, *29*, 2099–2103.

Chandler, W. L., & Roshal, M. (2009). Optimization of plasma fluorogenic thrombin generation assays. *Am J Clin Pathol*, *132*, 169–179.

Dargaud, Y., Wolberg, A. S., Luddington, R. V., et al. (2012). Evaluation of a standardized protocol for thrombin generation measurement using the calibrated automated thrombogram: An international multicenter study. *Thromb Res*, *130*, 929–934.

Debaugnies, F., Azerad, M. A., Noubouossié, D., et al. (2010). Evaluation of the procoagulant activity in the plasma of cancer patients using a thrombin generation assay. *Thromb Res*, *126*, 531–535.

Dunbar, N. M., & Chandler, W. L. (2009). Thrombin generation in trauma patients. *Transfusion*, *49*, 2652–2660.

Hron, G., Kollars, M., Binder, B. R., Eichinger, S., & Kyrle, P. A. (2006). Identification of patients at low risk for recurrent venous thromboembolism by measuring thrombin generation. *JAMA*, *296*, 397–402.

Ninivaggi, M., Apitz-Castro, R., Dargaud, Y., et al. (2012). Whole-blood thrombin generation monitored with a calibrated automated thrombogram-based assay. *Clin Chem*, *58*, 1252–1259.

Tripodi, A. (2016). Thrombin generation assay and its application in the clinical laboratory. *Clin Chem*, *62*, 699–707.

Santagostino, E., Mancuso, M. E., Tripodi, A., et al. (2010). Severe hemophilia with mild bleeding phenotype: Molecular characterization and global coagulation profile. *J Thromb Haemost*, *8*, 737–743.

CHAPTER 164

Diagnostic Use of Venoms

Ivo M.B. Francischetti, MD, PhD and Morayma Reyes Gil, MD, PhD

The most common use of snake venom molecules as reagents in hemostasis is described below.

Fibrinogen Assessment: Fibrinogen promotes platelet aggregation and is converted to fibrin by thrombin, with formation of a hemostatic plug.

Reptilase Time: This is a modification of the thrombin time (TT) in which thrombin-like enzymes (serine proteases) replace thrombin. The most commonly used enzyme is Batroxobin, a 43 Kda protein from the venom of *Bothrops atrox*. It splits the Arg^{16}–Gly^{17} bond in the Aα-chain of fibrinogen with formation of fibrin monomer. Alternatively, Ancrod from *Calloselasma rhodostoma* is employed in the assay. Contrarily to thrombin, these enzymes cleave fibrinopeptide A, but not fibrinopeptide B, do not activate factors (F) V, FVIII, FXI, FXIII, and do not induce platelet aggregation. Importantly, they are not sensitive to heparin–antithrombin complex or hirudin. A second laboratory use for Batroxobin is to detect antithrombin activity. Using these enzymes, plasma can be prepared free of fibrinogen without addition of thrombin, which would otherwise react with antithrombin and interfere with the assay.

Principle of the Tests and Interpretation: Reptilase reagent is added to platelet poor plasma and clotting time (CT) is measured; phospholipid or Ca^{2+} is not required. The RT is usually performed together with TT, and results from both tests are complementary in the interpretation of a coagulation disorder. A prolonged RT and TT are seen in inherited and acquired dysfibrinogenemia. Both tests may be prolonged following thrombolytic therapy or in disseminated intravascular coagulation, due to the high levels of fibrin-degradation products, which inhibit the fibrin polymerization step, or due to hypofibrinogenemia. Other conditions that might be associated with prolonged RP and TT are hypoalbuminemia seen in nephrotic syndrome, in liver disease due to dysfibrinogenemia, and in multiple myeloma due to paraproteins affecting fibrinogen polymerization. The RT is not affected by low-molecular-weight heparin or unfractionated heparin, direct thrombin inhibitors, and warfarin, in contrast to the TT.

Meizothrombin Assessment: Meizothrombin is an intermediate that is produced during the conversion of prothrombin to thrombin in systems composed of FXa and FVa and phospholipid (prothrombinase complex).

Ecarin Clotting Time: Several prothrombin activators have been identified in snake venoms and classified in four different types, according to structure, function, and cofactor requirements. Ecarin is a 55 kDa metalloprotease from the venom of *Echis carinatus* and specific activator of prothrombin. It activates prothrombin

Transfusion Medicine and Hemostasis. https://doi.org/10.1016/B978-0-12-813726-0.00164-1

independently of Ca^{2+}, phospholipids, and FV (do not require a cofactor). Ecarin cleaves prothrombin at Arg^{323}–Ile^{324} producing meizothrombin, an intermediate which is finally converted into α-thrombin by autolysis. Thrombin is detectable in a clotting assay known as Ecarin clotting time (ECT) or with chromogenic substrates in an assay known as Ecarin chromogenic assay (ECA). Importantly, meizothrombin binds to and is inhibited by hirudin but is unaffected by heparin/antithrombin because of steric hindrance. Therefore, the ECT is a test to determine meizothrombin production and can be used to detect direct thrombin inhibitors (e.g., hirudin) in citrated and heparinized blood. ECT is also employed to determine lupus anticoagulant (LA).

Principles of the Test and Interpretation: Ecarin is added to plasma and ECT determined turbidimetrically. In the presence of hirudin, meizothrombin inhibition leads to prolongation of clot formation for hirudin concentrations of 0.05–5.0 μg/mL. However, plasma levels >2.5 μg/mL found in cardiac surgery require a modification of ECT in which the blood sample is diluted 1:1. ECT is particularly useful in clinical practice because the activated partial thromboplastin time (aPTT) shows a linear dose-response curve for hirudin concentrations up to 1 μg/mL, whereas TT and activated clotting time show a nonlinear correlation. The ECT is relatively insensitive to variations in clotting factors, including fibrinogen and prothrombin. In the ECA, generation of meizothrombin is measured using a specific chromogenic substrate.

Factor V Assessment: FVa is a cofactor for the prothrombinase complex. RVV-V is a 27 kDa serine proteinase from the Russell's viper venom *Daboia russelli* snake, which converts FV to FVa. RVV-V may be used to determine FV levels but is most commonly used in the determination of LA and resistance to activated protein C (APCR) (see below).

Factor X Assessment: FXa activates prothrombin in the presence FVa, phospholipid, and Ca^{2+} (prothrombinase complex).

RVV-X From Russell's Viper Venom: RVV-X is a 120 kDa metalloprotease from the *D. russelli* snake and consists of two peptide chains with a molecular weight of 60 kDa each. It activates factor X and is strictly dependent on Ca^{2+}, FV, prothrombin, and phospholipids. RVV-X is used to quantitatively convert FX to FXa, which can be determined by clotting-based assay or with chromogenic substrates. RVV-X activator is also used to test for LA.

Principles and Interpretation of the Test: The principle of the RVV-X is similar to that of prothrombin time (PT)–based FX determination. An aliquot of the FX-deficient substrate plasma is mixed with an aliquot of the control or test plasma, and RVV-X and phospholipid substitute are added. Clotting is initiated by the addition of Ca^{2+} and the time to clot formation is recorded. On a log–log graph, CT for the PT is plotted against each dilution and the results compared with a log–log graph plotted with FX standard and results interpolated. Low FX levels may be observed in patients with vitamin K deficiency, FX deficiency, acquired factor inhibitors, and amyloidosis due to absorption of FX in amyloid fibrils. The chromogenic assay utilizes a specific substrate, which is cleaved by FXa generated by RVV-X, producing color that is detected spectrophotometrically.

Lupus Anticoagulant Assessment: Antiphospholipid antibodies are a heterogeneous population of immunoglobulins, including anticardiolipin (aCL) antibody, anti-β2-glycoprotein (GP) I antibody (β2-GPI), and LA. Detection of LA relies on phospholipid-dependent coagulation assays and involves screening, mixing, and confirmatory tests using venom proteins. aCL and β2-GPI antibodies are measured by enzyme-linked immunosorbent assay (ELISA).

Dilute Russell's Viper Venom Time: LA in vitro diagnostic test utilizes the procoagulant activity of the venom of the Russell's viper. The venom contains two enzymes: RVV-V is a 27 kDa serine protease, which activates FV to FVa, and RVV-X, which activates FX to FXa. The time for clotting triggered by these enzymes is known as dRVVT (dilute Russell's viper venom time). The dRVVT is particularly useful in LA identification because it is quick, sensitive, and inexpensive. Addition of a reagent containing both enzymes to plasma converts prothrombin to thrombin in the presence of phospholipid and Ca^{2+}. Because LA interferes with the phospholipid-binding proteins required for prothrombinase assembly, their presence result in a prolonged dRVVT. A mixing study is needed to exclude that prolongation of the dRVVT is secondary to coagulation factor deficiencies: prolonged dRVVT that does not correct after mixing suggests the presence of an LA. Additionally, a dRVVT confirmatory test is performed to determine whether the inhibitory effect of LA can be overcome by adding an excess of phospholipid in the assay. The CT of both the screening dRVVT assay and confirmatory test (with phospholipids) is normalized, and a ratio ≥ 1.2 is considered a positive result. The dRVVT test has a higher specificity than the aPTT test to detect LA because it is not influenced by deficiencies of FVIII, FIX, and FXI. The test is affected by warfarin, direct FXa and thrombin inhibitors, and inhibitors or deficiency of FV. Although the dRVVT reagent contains heparin-neutralizing activity, very high concentration of heparin may interfere with the assay. The International Society of Thrombosis and Haemostasis (ISTH) recommends the dRVVT to determine LA.

Textarin/Ecarin Ratio: Textarin is a 50 kDa serine protease from the venom of the Australian Eastern brown snake (*Pseudonaja textilis*) that directly activates prothrombin and requires FV, Ca^{2+}, and phospholipid; time for clotting is the Textarin time (TTC). In contrast, Ecarin from *E. carinatus* activates prothrombin to form meizothrombin in the absence of cofactors and phospholipids; the time for clotting is the Ecarin time. Accordingly, Textarin/Ecarin ratio (T/E) ratio is useful to detect LA given the differential phospholipid dependence of these two prothrombin activators. In patients with LA, the TTC is prolonged because of interference of prothrombin activation by the antibodies targeting phospholipids; however, the ECT is not affected. The T/E ratio is both a sensitive and relatively specific test. T/E ratio time of greater than 1.3 is considered abnormal, and studies have supported its use in the diagnosis of LA. Prothrombin deficiency will prolong both TTC and ECT times, whereas FV deficiency/inhibitors will prolong the TTC only. These are the only factors deficiencies, which cause a false positive. Also, T/E ratio can help in LA determination for patients taking FXa anticoagulants (e.g., rivaroxaban). These anticoagulants confound LA testing because it prolongs the dRVVT and it also corrects with excess phospholipid. However, this test is no longer recommended by the ISTH because of limited availability and poorly standardized commercial kits.

Taipan Venom Time: Oscutarin (scutelarin) from coastal Taipan snake (*Oxyuranus scutellatus*) and pseutarin C from Australian brown snake (*P. textilis*) are large multi-subunit serine proteases consisting of both FXa-like and FVa-like subunits. These enzymes activate prothrombin in the presence of Ca^{2+} and phospholipids; therefore, they are sensitive to LA. In the Taipan venom time (TVT), diluted phospholipid is incubated with normal or test plasma in the presence of Taipan snake venom. Ca^{2+} is added to start reactions and TVT is recorded. To verify the dependency of phospholipids, lysed platelets containing membranes of phospholipids are added in the assay, a step known as platelet neutralization. The ratio of test and control TVT is calculated. A positive test for LA is defined as a ratio of ≥1.1 with correction by >10% (≤1.1) after platelet neutralization. The TVT requires only adequate concentration of prothrombin and fibrinogen but is unaffected by FIX and FX. However, studies have demonstrated that TVT is a specific but insensitive assay for LA because the test did not have a 100% detection rate. It has been used to detect LA in patients with warfarin or with direct FXa anticoagulants.

Protein C Assessment: Protein C (PC) is a vitamin K–dependent serine protease and physiological anticoagulant. It is activated by thrombin/thrombomodulin complex and together with cofactor protein S (PS) degrades FVa and FVIIIa.

Protac: It is a serine protease (36–40 kDa) and PC activator from the venom of the copperhead snake *Agkistrodon contortrix*. It rapidly converts PC into activated PC (APC), which is determined by prolongation of the aPTT or with chromogenic substrates. Protac is unaffected by the plasma PC inhibitor. It is also useful to determine PS levels.

Principles of the Test and Interpretation

Clot-Based Assay: Patient's plasma is incubated with aPTT reagent and Protac. Ca^{2+} is added to initiate clotting and the time for clotting recorded. A reference curve with APC is performed in parallel where results with test plasma can be interpolated. Conditions associated with low PC levels result in less APC generated by Protac. As a result, less FVa and FVIIIa are inactivated and the CT is shorter, when compared with control plasma. aPTT-based assay may show falsely low PC levels in patients with FV Leiden mutation or in conditions associated with elevated plasma FVIII and in hyperlipidemia (with the optical clot detection method). aPTT-based functional assay may also yield falsely normal results in the presence of LA and in patients on direct thrombin inhibitors.

Chromogenic Assay: Plasma is incubated with Protac, and after incubation a chromogenic substrate for APC is added. Results are determined spectrophotometrically, and PC levels are determined with a standard curve. Chromogenic assays are sensitive but do not detect variants of PC, which have *Gla* domain mutation resulting in impaired binding to phospholipids. It is also independent of and unaffected by PS deficiency. ELISA assay to determine APC antigen is available and does not require Protac. PC determinations are usually performed as part of thrombophilia testing, including PS, genetic testing for FV Leiden and the G20210A prothrombin gene mutations, and antithrombin activity.

Resistance to Activated Protein C: APCR is a hereditary or acquired condition characterized by the inability of APC to cleave cofactors FVa and/or FVIIIa. The most common example of APC resistance is FV Leiden, which occur in 95% of cases.

Pefakit APCR Factor V Leiden: At least two assays have been optimized to determine APCR. In the Pefakit APCR factor V Leiden assay, sample plasma is prediluted in plasma deficient in FV. Then, a reagent containing RVV-V activator (RVV-V) from *D. russelli* and APC is added to test and control plasma to trigger coagulation, in the absence of Ca^{2+}. RVV-V activates FV to FVa and in normal conditions, APC inactivates FVa. The same assay is performed in parallel in the absence of APC. In a second step, a reagent containing Noscarin, a 54 kDa FVa-dependent prothrombin activator from *Notechis scutatus scutatus*, is added and CT is recorded. In control plasma, cleavage of FVa by APC results in prolonged CT. However, in patients who are carriers for FVa Leiden, cleavage of FVa is impaired. Therefore, enough FVa cofactor activity remains to support prothrombin activation by Noscarin, resulting in a shortened CT. The ratio (CT with APC/CT without APC) is determined and allows identifying wild-type, heterozygous, and homozygous carriers. Neither heparin/pentasaccharide therapy nor LA interferes with this assay, which contains heparin-neutralizing activity. Deficiency of FV (<50%) and direct thrombin inhibitors interferes with the test. Another assay known as Acticlot APCR also uses RVV-V and Noscarin to detect resistance to APC.

dRVVT-Based APCR: Assay based on the activation of endogenous PC by incubation of plasma with Protac. A dRVVT is then performed on the plasma. The dRVVT is sensitive to prolongation in the presence of APC. In normal individuals, activation of protein C prolongs the result two- to three-fold. In individuals with FV Leiden, the venom activation of PC induces only marginal prolongation, usually less than 1.5-fold. The test is affected by PC levels below 50%.

von Willebrand Factor Assessment: VWF is a GP whose function is to promote platelet adhesion to collagen and to form a stabilizing complex with FVIIIa.

Botrocetin: C-type lectin (22 kDa) from *Bothrops* sp requires VWF to agglutinate/activate platelets. Its mechanism of action slightly differs from ristocetin, an antibiotic that promotes agglutination of platelets by a VWF-dependent mechanism and used for diagnostic workup of VWF disease. The distinct mechanism of both molecules helps to differentiate molecular variants of VWF. Accordingly, botrocetin partially aggregates platelets from patients with Bernard–Soulier disease, whereas ristocetin does not. Furthermore, the reactivity of botrocetin in platelet agglutination is essentially independent of the multimer size of VWF molecules in contrast to ristocetin.

Accordingly, plasma from patients with VWF, type IIA, which lacks the highest molecular weight multimers, does not agglutinate platelets with ristocetin but will respond to botrocetin.

Therapeutics Use of Exogenous Factors: Molecules from snake venoms or saliva from blood-sucking arthropods have been instrumental as tools in biochemistry and in the development of novel therapeutics. Although a complete description of these molecules is beyond the scope of this chapter, novel therapeutics derived from exogenous secretions and approved by the Food and Drug Administration (FDA) are listed below.

Capoten (Captopril): Peptides from *Bothrops jararaca* inhibit angiotensin-converting enzyme (ACE) and were used as prototypes to develop orally active ACE inhibitors. Captopril is currently used for the treatment of hypertension.

Aggrastat (Tirofiban): Snake venom disintegrins present in *Viperidae* sp. are reversible antagonists of platelet integrin $\alpha IIb\beta 3$ receptor and inhibit platelet aggregation. Aggrastat is a nonpeptide drug based on the RGD sequence found in *E. carinatus* (among other venoms) and is used for treatment of coronary artery disease.

Integrilin (Eptifibatide): Synthetic cyclic heptapeptide modeled on the biologically active KGD motif present in the disintegrin barborin from the snake *Sistrurus barbouri*. It is used in coronary angioplasty.

Angiomax (Bivalirudin): Synthetic congener containing 20 residues is produced based on hirudin peptide (65 residues) from the saliva of the leech *Hirudo medicinalis*. Hirudin binds thrombin and inhibits coagulation. Angiomax is indicated for patients with unstable angina undergoing coronary angioplasty and for heparin-induced thrombocytopenia (HIT). Other thrombin inhibitors based on hirudin sequence are lepirudin (approved for HIT, discontinued) and desirudin, which has been licensed for thromboprophylaxis status post hip arthroplasty.

Byetta (Exenatide): It is a 39-aa peptide and incretin mimetic from the saliva of the lizard *Heloderma suspectum*. It binds and activates GLP-1 receptor leading to reduced plasma glucose and is approved for treatment of *Diabetes mellitus*.

Prialt (Ziconotide): It is ω-comotoxin MVIIA from the cone snail *Conus magus*. It blocks N-type Ca^{2+} channels and is used for the treatment of severe chronic pain.

Further Reading

Arachchillage, D. R., Mackie, I. J., Efthymiou, M., Isenberg, D. A., Machin, S. J., & Cohen, H. (2015). Interactions between rivaroxaban and antiphospholipid antibodies in thrombotic antiphospholipid syndrome. *J Thromb Haemost*, *13*, 1264–1273.

Fox, J. W., & Serrano, S. M. (2007). Approaching the golden age of natural product pharmaceuticals from venom libraries: An overview of toxins and toxin-derivatives currently involved in therapeutic or diagnostic applications. *Curr Pharm Des*, *13*, 2927–2934.

Harvey, A. L. (2014). Toxins and drug discovery. *Toxicon, 92*, 193–200.

Kini, R. M., & Koh, C. Y. (2016). Metalloproteases affecting blood coagulation, fibrinolysis and platelet aggregation from snake venoms: Definition and nomenclature of interaction sites. *Toxins (Basel), 8*.

Mans, B. J., & Francischetti, I. M. B. (2010). Sialomic perspectives on the evolution of blood-feeding behavior in arthropods: Future therapeutics by natural design. In R. M. Kini, K. J. Clemetson, F. S. Markland, M. A. McLane, & T. Morita (Eds.), *Toxins and Hemostasis. From Bench to Bedside* (pp. 21–44). New York: Springer.

Marsh, N., & Williams, V. (2005). Practical applications of snake venom toxins in haemostasis. *Toxicon, 45*, 1171–1181.

Moore, G. W. (2015). Current controversies in Lupus anticoagulant detection. *Antibodies, 5*, 1–15.

Perchu, A. M., & Wilmer, M. (2010). Diagnostic use of snake venom component in the coagulation laboratory. In R. M. Kini, K. J. Clemetson, F. S. Markland, M. A. McLane, & T. Morita (Eds.), *Toxins and Hemostasis. From Bench to Bedside* (pp. 747–766). New York, NY: Springer.

Perry, D. J., & Todd, T. *Practical hemostasis, a practical guide to laboratory hemostasis*. Available at: http://www.practical-haemostasis.com.

Slagboom, J., Kool, J., Harrison, R. A., & Casewell, N. R. (2017). Haemotoxic snake venoms: Their functional activity, impact on snakebite victims and pharmaceutical promise. *Br J Haematol, 177*, 947–959.

Abbreviations

AAP	American Academy of Pediatrics
ABC	Assessment of Blood Consumption
ACE	Angiotensin-converting enzyme
ACT	Adoptive cell transfer
AD	Atopic dermatitis
aDHQ	Abbreviated Donor History Questionnaire
AIDS	Acquired immunodeficiency syndrome
AIHA	Autoimmune hemolytic anemias
ALL	Acute lymphoblastic leukemia
ALPS	Autoimmune lymphoproliferative syndrome
AMR	Antibody-mediated rejection
AMT	Applied muscle tension
ANH	Acute normovolemic hemodilution
APC	Activated protein C
aPCC	Activated prothrombin complex concentrate
APL	Acute promyelocytic leukemia
APS	Antiphospholipid syndrome
aPTT	Activated partial thromboplastin time
ARDP	American Rare Donor Program
ATRUS	Amegakaryocytic thrombocytopenia with radio/ulnar synostosis
AVWS	Acquired von Willebrand syndrome
β2GPI	β2-glycoprotein I
BB/TM	Blood banking and transfusion medicine
BD	Behcet's disease
BECS	Blood establishment computer software
BLA	Biologics License Application
BNP	Brain natriuretic peptide
BPAC	Blood Products Advisory Committee
BSS	Bernard–Soulier syndrome
BTIC	Babesia Testing Investigational Containment
CAD	Cold agglutinin disease
cAMP	Cyclic adenosine monophosphate
CAMT	Congenital amegakaryocytic thrombocytopenia
CAP	College of American Pathologists
CAPS	Catastrophic antiphospholipid syndrome
CB	Cord blood
CBER	Center for Biologics Evaluation and Research
CCI	Corrected count increment

CEO	Chief Executive Officer
CFR	Code of Federal Regulations
CGD	Chronic granulomatous disease
CHS	Chediak–Higashi syndrome
CIBMTR	Center for International Blood and Marrow Transplant Research
CIDP	Chronic inflammatory demyelinating polyradiculoneuropathy
CJD	Creutzfeldt–Jakob disease
CLIA	Clinical Laboratory Improvement Amendments
CLSI	Clinical and Laboratory Standards Institute
CMO	Chief Medical Officer
CMS	Center for Medicaid and Medicare Services
CMV	Cytomegalovirus
CPB	Cardiopulmonary bypass
CPD	Citrate phosphate dextrose
CPOE	Computerized physician order entry
CT	Clotting time
CWD	Chronic wasting disease
DAF	Decay accelerating factor
DAH	Diffuse alveolar hemorrhage
DAT	Direct antiglobulin test
DCM	Dilated cardiomyopathy
DIC	Disseminated intravascular coagulation
DIHA	Drug-induced immune hemolytic anemia
DIT	Drug-induced thrombocytopenia
DMSO	Dimethyl sulfoxide
DSTR	Delayed serologic transfusion reaction
DVT	Deep vein thrombosis
ECA	Ecarin chromogenic assay
ECMO	Extracorporeal membrane oxygenation
ECT	Ecarin clotting time
EIA	Enzyme immunoassay
ELISA	Enzyme-linked immunosorbent assay
ELT	Euglobulin lysis time
EMP	Endothelial-derived microparticles
EPCR	Endothelial protein C receptor
EPP	Erythropoietic protoporphyria
ETP	Endogenous thrombin potential
F2	Coagulation factor II gene
F5	Coagulation factor V gene
F10	Coagulation factor X gene
FACT	Foundation for the Accreditation of Cellular Therapy
FC	Flow cytometry
FDA	Food and Drug Administration
FFP	Fresh frozen plasma
FH	Familial hypercholesterolemia
FII	Coagulation factor II, prothrombin

FIX	Coagulation factor IX
FNAIT	Fetal and neonatal alloimmune thrombocytopenia
FRCT	Fixed ratio component therapy
FRET	Fluorescence resonance energy transfer
FV	Coagulation factor V
FVa	Activated coagulation factor V
FVIIa	Activated coagulation factor VII
FVIII	Coagulation factor VIII
FVIIIa	Activated coagulation factor VIII
FVL	Factor V Leiden
FX	Coagulation factor X
FXa	Activated coagulation factor X
G-CSF	Granulocyte-colony stimulating factor
GAG	Glucosaminoglycans
GGCX	G-glutamyl carboxylase
GLOB	Globoside
GPA	Glycophorin A
GPB	Glycophorin B
GPI	Glycophosphatidylinositol
GPS	Gray platelet syndrome
GT	Glanzmann thrombasthenia
GSA	Guanidinosuccinic acid
GVHD	Graft-versus-host disease
GVT	Graft-versus-tumor
HAV	Hepatitis A virus
HBV	Hepatitis B virus
Hct	Hematocrit
HCV	Hepatitis C virus
HDFN	Hemolytic disease of the fetus and newborn
HES	Hydroxyethyl starch
HEV	Hepatitis E virus
HGA	Human granulocytic anaplasmosis
Hgb	Hemoglobin
HH	Hereditary hemochromatosis
HHT	Hereditary hemorrhagic telangiectasia
HIPA	Heparin-induced platelet activation assay
HIT	Heparin-induced thrombocytopenia
HLA	Human leukocyte antigen
HMWK	High-molecular-weight kininogen
HNA	Human neutrophil antigen
HOPR	High on-treatment residual platelet reactivity
HPC	Hematopoietic progenitor cell
HPS	Hermansky–Pudlak syndrome
HRSA	Health Research and Services Administration
HSC	Hematopoietic stem cell
HSCT	Hematopoietic stem cell transplantation

HSP	Henoch–Schönlein purpura
HTLV	Human T-cell lymphotropic virus
HTR	Hemolytic transfusion reaction
HUS	Hemolytic uremic syndrome
IAT	Indirect antiglobulin test
IBD	Inflammatory bowel disease
ICH	Intracranial hemorrhage
ID	individual donor
IHD	Isovolemic hemodilution
IND	Investigational new drug
INR	International normalized ratio
IPD	Inherited platelet disorder
IPF	Immature platelet fraction
IRFR	Infusion-related febrile reactions
ISBT	International Society of Blood Transfusion
ISI	International sensitivity index
IT	Inheritable thrombophilia
ITP	Immune thrombocytopenic purpura
ITT	Immune tolerance therapy
IVIG	Intravenous immunoglobulin
KMS	Kasabach -Merritt syndrome
LA	Lupus anticoagulant
LDH	Lactate dehydrogenase
LEMS	Lambert–Eaton myasthenic syndrome
LIA	Latex immunoassay
LIBS	Ligand-induced binding sites
LMAN(1)	Lectin mannose-binding protein 1
LMWH	Low-molecular-weight heparin
LP	Liquid plasma
LTA	Light-transmission aggregometry
LVAD	Left ventricular assist device
MAHA	Microangiopathic hemolytic anemia
MG	Myasthenia gravis
MGUS	Monoclonal gammopathy of unknown significance
MP	Minipool
MPHA	Mixed passive hemagglutination assay
MPV	Mean platelet volume
MS	Multiple sclerosis
MTHFR	Methylenetetrahydrofolate reductase
NAIT	neonatal alloimmune thrombocytopenia
NAT	Nucleic acid testing
NCCLS	National Committee on Clinical and Laboratory Standards
NEC	Necrotizing enterocolitis
NETS	Neutrophil extracellular traps
NGS	Next-generation sequencing
NiSHOTs	Noninfectious serious hazards of transfusion

NMDP	National Marrow Donor Program
NMDAR	N-methyl d-aspartate receptor
NMMHC-IIA	Nonmuscle myosin heavy chain IIA
NSAIDs	Nonsteroidal antiinflammatory medications
NTBI	Nontransferrin-bound iron
PAD	Preoperative autologous donation
PAGGSM	Phosphate–adenine–glucose–guanosine–saline–mannitol
PAT	Platelet aggregation test
PBQ	Pediatric Bleeding Questionnaire
PCCs	Prothrombin complex concentrates
PCH	Paroxysmal cold hemoglobinuria
PCR	Polymerase chain reaction
pdFX	Plasma-derived factor X
PF	Purpura fulminants
PFA	Platelet function analyzer
PHS	Public Health Service
PIFA	Particle immunofiltration assay
PIVKA	Proteins induced by vitamin K absence
PK	Prekallikrein
PMP	Platelet microparticles
PNH	paroxysmal nocturnal hemoglobinuria
PPF	Plasma protein fraction
PPH	Postpartum hemorrhage
PR	Pathogen reduced
PRNT	Plaque reduction neutralization test
PS	Protein S
PT	Prothrombin time
PTT	Partial thromboplastin time
PTEM	Platelet transmission electron microscopy
PTLD	Posttransplantation lymphoproliferative disease
PTP	Posttransfusion purpura
PV	Pemphigus vulgaris
PVD	Presumptive viremic donation
QSE	Quality system essentials
RBC	Red blood cell
RIBA	Recombinant immunoblot assay
RIPA	Radioimmunoprecipitation assay
ROTEM	Rotational thromboelastometry
RT	Reptilase time
SBP	Spontaneous bacterial peritonitis
SBT	Sequencing-based typing
SCAD	Stable coronary artery disease
SCD	Sickle cell disease
SLE	Systemic lupus erythematosus
SMA	Spinal muscular atrophy
SNAPS	Seronegative antiphospholipid syndrome

SPA	Staphylococcal protein A
SQUID	Superconducting quantum interface device
SQUIPP	Safety, quality, identity, potency, and purity
SRA	Serotonin release assay
SSP	Sequence-specific primers
TA	Therapeutic apheresis
TACO	Transfusion-associated circulatory overload
TAD	Transfusion-associated dyspnea
TAFI	Thrombin activatable fibrinolytic inhibitor
TE	Thromboembolism
TFPI	Tissue factor pathway inhibitor
TGA	Thrombin generation assay
TA-GVHD	Transfusion-associated graft-versus-host disease
TM	Thrombomodulin
TMA	Thrombotic microangiopathy
TPE	Therapeutic plasma exchange
TPO	Thrombopoietin
TRALI	Transfusion-related acute lung injury
TRIM	Transfusion-related immunomodulation
TT	Thrombin time
TTB	Transfusion-transmitted babesiosis
TTP	Thrombotic thrombocytopenia purpura
TVT	Taipan venom time
TXA	Tranexamic acid
UC	Ulcerative colitis
UFH	Unfractionated heparin
uDHQ	Uniform Donor History Questionnaire
US	United States
VASP	Vasodilator-stimulated phosphoprotein
VEGF	Vascular endothelial growth factor
VK	Vitamin K
VKA	Vitamin K antagonist
VKDB	Vitamin K deficiency bleeding
VKOR	Vitamin K epoxide reductase
VNRD	Voluntary, nonremunerated donors
VTE	Venous thromboembolism
VWF	von Willebrand factor
WAIHA	Warm autoimmune hemolytic anemia
WAS	Wiskott–Aldrich syndrome
WB	Whole blood
WBA	Whole-blood aggregometry
WBDP	Whole blood-derived platelet
WM	Whole mount
WNV	West Nile virus
XLT	X-linked thrombocytopenia

Index

Note: Page numbers followed by "f" indicate figures and "t" indicate tables.

B

C

D

J

K

L

M

Q

R

S

T

W

X

Y

Z

Printed and bound by CPI Group (UK) Ltd, Croydon, CR0 4YY
13/10/2024
01773609-0001